Geometric Formulas

Conversion between Radians and Degrees: π radians $= 180°$

Triangle

$A = \frac{1}{2}bh$

$\quad = \frac{1}{2}ab\sin\theta$

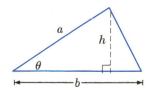

Circle

$A = \pi r^2$

$C = 2\pi r$

Sector of Circle

$A = \frac{1}{2}r^2\theta \quad$ (θ in radians)

$s = r\theta \quad$ (θ in radians)

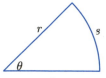

Sphere

$V = \frac{4}{3}\pi r^3 \quad A = 4\pi r^2$

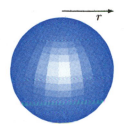

Cylinder

$V = \pi r^2 h$

Cone

$V = \frac{1}{3}\pi r^2 h$

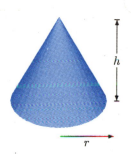

Trigonometric Functions

$\sin\theta = \dfrac{y}{r}$

$\cos\theta = \dfrac{x}{r}$

$\tan\theta = \dfrac{y}{x}$

$\tan\theta = \dfrac{\sin\theta}{\cos\theta}$

$\cos^2\theta + \sin^2\theta = 1$

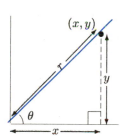

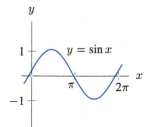

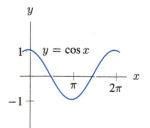

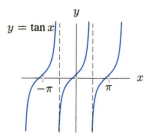

The Binomial Theorem

$$(x+y)^n = x^n + nx^{n-1}y + \frac{n(n-1)}{1\cdot 2}x^{n-2}y^2 + \frac{n(n-1)(n-2)}{1\cdot 2\cdot 3}x^{n-3}y^3 + \cdots + nxy^{n-1} + y^n$$

$$(x-y)^n = x^n - nx^{n-1}y + \frac{n(n-1)}{1\cdot 2}x^{n-2}y^2 - \frac{n(n-1)(n-2)}{1\cdot 2\cdot 3}x^{n-3}y^3 + \cdots \pm nxy^{n-1} \mp y^n$$

CALCULUS
Single and Multivariable

Produced by the Consortium based at Harvard and funded by a National Science Foundation Grant.

Deborah Hughes-Hallett
Harvard University

David Mumford
Brown University

Andrew M. Gleason
Harvard University

Brad G. Osgood
Stanford University

William G. McCallum
University of Arizona

Andrew Pasquale
Chelmsford High School

Daniel E. Flath
University of South Alabama

Douglas Quinney
University of Keele

Patti Frazer Lock
St. Lawrence University

Jeff Tecosky-Feldman
Haverford College

Sheldon P. Gordon
Suffolk County Community College

Joe B. Thrash
University of Southern Mississippi

David O. Lomen
University of Arizona

Karen R. Thrash
University of Southern Mississipi

David Lovelock
University of Arizona

Thomas W. Tucker
Colgate University

with the assistance of

Adrian Iovita
*Centre Interuniversitaire en
Calcul Mathématique
Algébraique*

Otto K. Bretscher
Harvard University

Brad Mann
Harvard University

WILEY

John Wiley & Sons, Inc.

New York   Chichester   Weinheim   Brisbane   Singapore   Toronto

Cover Photo: Dennis O'Clair/Tony Stone Images/New York, Inc.

Problems from *Calculus: The Analysis of Functions*, by Peter D. Taylor
(Toronto: Wall & Emerson, Inc., 1992). Reprinted with permission of the publisher.

EXECUTIVE EDITOR Ruth Baruth
DEVELOPMENTAL EDITOR Nancy Perry
MARKETING MANAGER Leslie Hines
SENIOR PRODUCTION MANAGER Lucille Buonocore
SENIOR PRODUCTION EDITOR Monique Calello
SENIOR DESIGNER Laura Nicholls Boucher
ILLUSTRATION DIRECTOR Ishaya Monokoff
MANUFACTURING MANAGER Monique Calello

This book was set in Times Roman by the Consortium based at Harvard using LATEX,
Mathematica, and the package AsTeX, which was written by Alex Kasman for
this project.

This book was printed and bound by Courier Westford, Inc. This cover was printed by
Phoenix Color Corporation.

This book is printed on acid-free paper. ∞

The paper in this book was manufactured by a mill whose forest management programs
include sustained yield harvesting of its timberlands. Sustained yield harvesting principles
ensure that the number of trees cut each year does not exceed the amount of new growth.

ISBN 0-471-19490-5

Printed in the United States of America.

10 9 8 7 6 5 4 3 2 1

This project was supported, in part,
by the
National Science Foundation
Opinions expressed are those of the authors
and not necessarily those of the Foundation

PREFACE

Calculus is one of the greatest achievements of the human intellect. Inspired by problems in astronomy, Newton and Leibniz developed the ideas of calculus 300 years ago. Since then, each century has demonstrated the power of calculus to illuminate questions in mathematics, the physical sciences, engineering, and the social and biological sciences.

Calculus has been so successful because of its extraordinary power to reduce complicated problems to simple rules and procedures. Therein lies the danger in teaching calculus: it is possible to teach the subject as nothing but the rules and procedures—thereby losing sight of both the mathematics and of its practical value. This combined edition of *Calculus: Single and Multivariable* continues our effort to refocus the teaching of calculus on concepts as well as procedures.

A Focused Vision: Conceptual Understanding

Our goal is to provide students with a clear understanding of the ideas of calculus as a solid foundation for subsequent courses in mathematics and other disciplines. When we designed this curriculum we started with a clean slate. We increased the emphasis on some topics, such as differential equations, and omitted some traditional topics whose inclusion we could not justify after discussions with mathematicians, engineers, physicists, chemists, biologists, and economists. We focused on a small number of key concepts, emphasizing depth of understanding rather than breadth of coverage.

The First Edition: A New Curriculum

The first edition of this book was the work of faculty at a consortium of eleven institutions, generously supported by the National Science Foundation. It represents the first consensus between such a diverse group of research mathematicians and instructors to have shaped a mainstream calculus text.

The Second Edition: Expanded Choices

The second edition has the same vision as the first edition and provides instructors with additional choices through the *Focus On* sections. Instructors can select a focus for their course which reflects their interests and the needs of their students. In particular:

- All *Focus On* sections are optional.
- Chapter 3 (the definite integral) may be covered immediately before Chapter 6 (the antiderivative).
- In Chapters 5 and 8, instructors may select the sections relevant to their students.
- Chapters 9 and 10 may be covered in either order.
- Multivariable calculus may be taught with or without the vector calculus in Chapters 18, 19, and 20.

Guiding Principles: Varied Problems and the Rule of Four

Since students usually learn most when they are active, we feel that the exercises in a text are of central importance. In addition, we have found that multiple representations encourage students to reflect on the meaning of the material. Consequently, we have been guided by the following principles.

- Our problems are varied and some are challenging. Most cannot be done by following a template in the text.
- The Rule of Four: Where appropriate, topics should be presented geometrically, numerically, analytically, and verbally.

The Development of Mathematical Thinking

The first stage in the development of mathematical thinking is the acquisition of a clear intuitive picture of the central ideas. In the next stage, the student learns to reason with the intuitive ideas and explain the reasoning clearly in plain English. After this foundation has been laid, there is a choice of direction. Some students, for example mathematics majors, may benefit from a more theoretical approach, while others, for example science and engineering majors, may benefit from a further exploration of modeling. The new supplementary sections, *Focus on Theory* and *Focus on Modeling*, provide material to support either choice.

Focus on Theory

Calculus as a logical structure of theorem and proof is a masterpiece of mathematics. Its beauty, if not glossed over in a broad but superficial treatment, attracts mathematically inclined students to their vocation. In the *Focus on Theory* sections we have chosen a few topics to cover in depth. We show how axioms, definitions, and theorems are formulated, and how proofs are constructed. Since we believe that to understand and appreciate this material students must work with it themselves, we have included challenging exercises that guide students to construct definitions and proofs on their own.

Focus on Modeling

Calculus is a powerful tool for analyzing the real world. Students gain an understanding of the power of calculus by focusing on its use in an extended problem. The *Focus on Modeling* sections take the time to explore selected applications of calculus in depth.

The Development of Mathematical Skills

To use calculus effectively, students need skill in both symbolic manipulation and the use of technology. The exact proportions of each may vary widely, depending on the preparation of the student and the wishes of the instructor. The book is adaptable to many different combinations.

Focus on Practice

These new sections increase students' skills in the mechanics of differentiation and integration.

Technology

The book does not require any specific software or technology. It has been used with graphing calculators, graphing software, and computer algebra systems. Any technology with the ability to graph functions and perform numerical integration will suffice. Students are expected to use their own judgement to determine where technology is useful. In multivariable calculus, even more so than in single variable calculus, computer technology can be put to great advantage to help students learn to think mathematically. For example, looking at surface graphs and contour diagrams is enormously helpful in understanding functions of many variables.

Content

This content represents our vision of how calculus can be taught. It is flexible enough to accommodate individual course needs and requirements. Topics can easily be added or deleted, or the order changed.

Chapter 1: A Library of Functions

This chapter introduces all the elementary functions to be used in the book. Although the functions are probably familiar, the graphical, numerical, verbal, and modeling approach to them is likely to be new. We introduce exponential functions at the earliest possible stage, since they are fundamental to the understanding of real-world processes.

Focus on Theory: These sections discuss the theoretical underpinnings of calculus, illustrate the process of formulating and proving a theorem using the example of the binomial theorem, and show how axioms are developed using the example of the completeness axiom.

Chapter 2: Key Concept---The Derivative

The purpose of this chapter is to give the student a practical understanding of the definition of the derivative and its interpretation as an instantaneous rate of change. The power rule is introduced; other rules are introduced in Chapter 4.

Focus on Theory: These sections present the definition of limit and continuity, and analyze the connection between differentiability and local linearity. Exercises lead students to a proof of the Intermediate Value Theorem.

Chapter 3: Key Concept---The Definite Integral

The purpose of this chapter is to give the student a practical understanding of the definite integral as a limit of Riemann sums and to bring out the connection between the derivative and the definite integral in the Fundamental Theorem of Calculus. Some instructors using the first edition of the book have delayed covering Chapter 3 until after Chapter 5 without any difficulty.

Focus on Theory: This section explores the formal definition of the definite integral using upper and lower sums.

Chapter 4: Short-Cuts to Differentiation

The derivatives of all the functions in Chapter 1 are introduced, as well as the rules for differentiating products, quotients, and composite functions.

Focus on Practice: This section provides a battery of differentiation problems for skill building.

Chapter 5: Using the Derivative

The aim of this chapter is to enable the student to use the derivative in solving problems, including optimization and graphing. It is not necessary to cover all the sections.

Focus on Theory: This section pulls together many of the strands of the earlier theory sections, showing how the completeness axiom and the definition of limit and continuity can be used to prove the Extreme Value Theorem, the Increasing Function Theorem, and the Mean Value Theorem.

Chapter 6: Constructing Antiderivatives

This chapter focuses on going backward from a derivative to the original function, first graphically and numerically, then analytically. It introduces the Second Fundamental Theorem of Calculus and the concept of a differential equation.

Focus on Modeling: This section discusses the equations of motion.

Chapter 7: Integration

This chapter includes several techniques of integration; others are included in the table of integrals. There is a discussion of numerical methods and of improper integrals.

Focus on Practice: This section provides a battery of integration problems for skill building.

Chapter 8: Using the Definite Integral

This chapter emphasizes the idea of subdividing a quantity to produce Riemann sums which, in the limit, yield a definite integral. It shows how the integral is used in geometry, physics, and economics; it is not necessary to cover all the sections.

Focus on Modeling: These sections study applications of integration to probability distributions.

Chapter 9: Approximations and Series

This chapter introduces Taylor Series and Fourier series via the idea of approximating functions with simpler functions. Geometric series, the ratio test, the harmonic series, and alternating series are discussed.

Focus on Theory: These sections present convergence tests in more detail, sketch a proof of the ratio test, and study error bounds.

Chapter 10: Differential Equations

This chapter introduces differential equations. The emphasis is on qualitative solutions, modeling, and interpretation.

Chapter 11: Functions of Many Variables

Functions of many variables are introduced using formulas, surface graphs, contour diagrams, and tables. Attention is paid to the idea of a cross-section of a function, obtained by varying one variable independently of the others, as this notion helps students understand partial derivatives.

Focus on Theory: This section investigates limits and continuity of functions of two variables.

Chapter 12: A Fundamental Tool: Vectors

Vectors are defined as geometric objects having direction and magnitude and then are represented in terms of coordinates. Equivalent geometric and algebraic definitions of the dot and cross products are given.

Chapter 13: Differentiating Functions of Many Variables

This chapter introduces the notions of partial derivative, directional derivative, gradient, and differential. Local linearity is used to introduce differentiability and the multivariable chain rule. Higher order partial derivatives and their application to quadratic Taylor approximations are discussed.

Focus on Theory: This section is on differentiability.

Chapter 14: Optimization

The ideas of the previous chapter are applied to optimization problems, both constrained and unconstrained. The section on constrained optimization discusses Lagrange multipliers, equality and inequality constraints.

Focus on Modeling: This section introduces the Lagrangian function and its interpretation.

Chapter 15: Integrating Functions of Many Variables

The multivariable definite integral is introduced graphically using a density function. Double and triple integrals in Cartesian, polar, spherical, and cylindrical coordinates are discussed.

Focus on Theory: This section introduces the change of variables formula for double integrals.

Chapter 16: Parametric Curves

This chapter starts by representing curves parametrically, and then uses parametric representations to describe motion and to analyze velocity and acceleration.

Chapter 17: Vector Fields

Vector fields are introduced in this chapter and the foundation is laid for the geometric approach in Chapters 18, 19, and 20 to line integrals, flux integrals, divergence, and curl.

Chapter 18: Line Integrals

A coordinate-free definition of line integrals is presented, and then line integrals are calculated using parameterizations. Path independent, or conservative fields, gradient fields, the Fundamental Theorem of Calculus for Line Integrals, and Green's Theorem are discussed.

Focus on Theory: This section presents a proof of Green's Theorem using the change of variables formula.

Chapter 19: Flux Integrals

The flux integral of a vector field through a surface is introduced in the same way as line integrals. First a coordinate-free definition is given, then flux integrals are calculated over surface graphs, portions of cylinders, and portions of spheres.

Chapter 20: Calculus of Vector Fields

The divergence and curl are introduced in a coordinate-free way; the divergence in terms of flux density, and curl in terms of circulation density. The formulas in Cartesian coordinates are then given. The Divergence Theorem and Stokes' Theorem are explained geometrically.

Focus on Theory: This section discusses the three fundamental theorems of multivariable calculus, and shows how they lead to the three-dimensional curl test for a path-independent vector field.

Appendices

There are appendices on polar coordinates, complex numbers, determinants, and some additional projects.

Our Experiences

In the process of developing the ideas incorporated in this book, we were conscious of the need to test the materials thoroughly in a wide variety of institutions serving many different types of students. Before the first editions were produced, consortium members and colleagues at over one hundred schools around the country class-tested preliminary versions of the book at large and small public universities, liberal arts colleges, two-year institutions, and high schools. The first edition was used by a very large and diverse group of schools in semester and quarter systems, in large lectures and small classes, in computer labs, small groups, and traditional settings, and with a number of different technologies. In preparing this edition, we solicited comments from a large number of mathematicians who had used the text. We continued to discuss with our colleagues in client disciplines the mathematical needs of their students. This included careful reviewing by a group of engineering faculty from highly regarded engineering programs. We were offered many valuable suggestions, which we have tried to incorporate, while maintaining our original commitment to a focused treatment of a limited number of topics.

Changes from the First Editions of Calculus and Multivariable Calculus

In this combined edition, we have streamlined some topics and added new sections on theory and on skill-building; we have moved some material into separate sections on modeling. The new arrangement divides the curriculum into a basic core and a set of supplementary sections from which the instructor can make choices to support a wide variety of courses.

General Changes
- There are more easy and medium level problems in each section, and a battery of drill problems in the *Focus on Practice* sections at the end of Chapters 4 (differentiation) and 7 (integration).
- There are short answers in the back of the book to odd-numbered problems where appropriate.

- Each chapter concludes with a review of the main points, and there is a list of useful formulas on the endpapers.

- Projects are included at the end of each chapter; a wider selection is in the appendix.

Summary of New Material

- New *Focus on Theory* sections discuss the concepts of axiom, definition, and mathematical proof, and give a more theoretical treatment of selected topics, such as limits, differentiability, and the definite integral. The first edition's emphasis on concepts and on geometric and numerical reasoning serves as a natural introduction to this thread. These sections are optional and may be selected by instructors who prefer a more theoretical course.

- Chapter 4 contains a new section on local linearity and limits, which includes l'Hopital's rule.

- Chapter 5 contains a new section on the hyperbolic functions.

- Chapter 7 contains an expanded treatment of partial fractions.

- Chapter 8 contains a new subsection on center of mass.

- Chapter 9 contains new material on using the ratio test to find the radius of convergence of a power series, as well as new material on the harmonic and alternating series.

- There is a new appendix containing additional projects.

Chapter-by-Chapter Description of Changes between Calculus (First Edition) and Multivariable Calculus (First Edition) and this Combined Edition

- Chapter 1. The introduction to continuity, formerly in an appendix, has been moved to this chapter. Notes on Compound Interest has been omitted. There are three new *Focus on Theory* sections: Underpinnings of Calculus, The Binomial Theorem, and Completeness of the Real Numbers.

- Chapter 2. The material on derivatives of constant and linear functions, and the statement of the power rule have been moved to Section 2.3. The sections on Approximations and Local Linearity, Notes on the Limit, and Notes on Differentiability have been replaced by two *Focus on Theory* sections: Limits and Continuity, and Differentiability and Linear Approximation.

- Chapter 3. The discussion in Section 3.2 of the relation between the value of n and the convergence of Riemann sums has been shortened and the material on visualizing the integral as area has been moved to this section. The applications of the definite integral formerly in Section 6.1 have been moved to Section 3.3. Section 3.4 has been expanded to include both the Fundamental Theorem and the properties of definite integrals formerly in Section 6.2. The section on Further Notes on the Limit has been replaced by a *Focus on Theory* section on the Definite Integral.

- Chapter 4. Notes on the Tangent Line Approximation has been incorporated into a new section on Local Linearity and Limits, which contains a treatment of l'Hopital's rule. More problems on the applications of rates have been included. A new *Focus on Practice* section provides drill problems in differentiation.

- Chapter 5. The material on local extrema has been shortened and put into Section 5.1. The material on families of functions has been simplified and is in Section 5.2. Section 5.3 is on global extrema and includes the first section on optimization (formerly Section 5.5). A new section on Hyperbolic Functions has been added, and the section on Newton's Method has been omitted. There is a new *Focus on Theory* section on Theorems about Continuous and Differentiable Functions.

- Chapter 6. This chapter has been refocused on antiderivatives; the material about definite integrals has been moved to Chapter 3. The chapter starts with a section on finding antiderivatives graphically and numerically. The material on finding antiderivatives analytically (formerly Section 7.1) has been moved to Section 6.2. The material on constructing antiderivatives using the definite integral has been moved here to form a new section on The Second Fundamental Theorem of Calculus. The section on equations of motion has become a *Focus on Modeling* section.

- Chapter 7 now starts with substitutions.

- Chapter 8. Applications to geometry have been put first, and new material on center of mass has been added to the material on density now in Section 8.2. The sections on distributions and probability have been moved to *Focus on Modeling* sections.

- Chapter 9. This chapter now covers Approximations and Series. There is new material on the ratio test for determining the radius of convergence of a power series, as well as on harmonic and alternating series. A new *Focus on Theory* section deals with the convergence of series. The Error in Taylor Approximations is now a *Focus on Theory* section.

- Chapter 10. This chapter now covers Differential Equations. The sections on Systems of Differential Equations and Analyzing the Phase Plane are *Focus on Modeling* sections in the Second Edition of Calculus, but are omitted in this Combined Edition.

- Chapter 11. Section 11.7 on Limits and Continuity has been moved to a *Focus on Theory* section.

- Chapter 13. Section 13.8 on Partial Differential Equations has been omitted; Section 13.10 on Differentiability has been moved to a *Focus on Theory* section.

- Chapter 14. The material in Section 14.3 on the Lagrangian and on the interpretation of λ has been moved to a *Focus on Modeling* section.

- Chapter 15. Section 15.4 on the Monte Carlo method and Section 15.7 on Probability have been omitted. Section 15.8, Notes on Change of Variables, has been moved to a *Focus on Theory* section.

- Chapter 16. This chapter now focuses on Parameterized Curves. Sections 16.3 and 16.4 on Parameterized Surfaces and the Implicit Function Theorem have been omitted. Some of the material on Kepler's Laws has been moved to a project.

- Chapter 18. The proof of Green's Theorem in Section 18.5 has been moved to a *Focus on Theory* section.

- Chapter 19. Section 19.3 on Flux integrals over parameterized surfaces has been omitted.

- Chapter 20. Section 20.5 on the Three Fundamental Theorems has been moved to a *Focus on Theory* section; Section 20.6, containing the proofs of the Divergence and Stokes' Theorems, has been omitted.

- Appendix: There are sections on Polar Coordinates, Complex Numbers, Determinants, and a new section containing additional projects.

Supplementary Materials

The following supplementary materials are available for the Combined Edition.
- **Instructor's Manual with Sample Exam Questions** containing teaching tips, calculator programs, some overhead transparency masters and test questions arranged according to section.
- **Instructor's Solution Manual** with complete solutions to all problems.
- **Student's Solution Manual** with complete solutions to half the odd-numbered problems.
- **Instructor's Resource CD-ROM** which contains the Instructor's Manual, Instructor's Solutions Manual, additional projects, as well as other valuable resources.

Acknowledgements

First and foremost, we want to express our appreciation to the National Science Foundation for their faith in our ability to produce a revitalized calculus curriculum and, in particular, to Louise Raphael, John Kenelly, John Bradley, and James Lightbourne. We also want to thank the members of our Advisory Board, Benita Albert, Lida Barrett, Simon Bernau, Robert Davis, M. Lavinia DeConge-Watson, John Dossey, Ron Douglas, Eli Fromm, William Haver, Seymour Parter and Stephen Rodi for their ongoing guidance and advice.

In addition, a host of other people around the country and abroad deserve our thanks for all that they did to help our project succeed. They include: Leonid Andreev, Huriye Arikan, Ralph Baierlein, Paul Balister, Ruth Baruth, Martin Batey, Melanie Bell, Thomas Bird, Paul Blanchard, Laura Boucher, Melkana Brakalova, John Bravman, David Bressoud, Stephen Boyd, Otto Bretscher, Mark Bridger, Morton Brown, Lucille Buonocore, Monique Calello, Ray Cannon, Marilyn Carlson, Phil Cheifetz, C.K. Cheung, Bob Condon, Jurie Conradie, Sterling G. Crossley, Caspar Curjel, Ehud de Shalit, Bob Decker, Tom Dick, Srdjan Divac, Wade Ellis, Hermann Flaschka, Leonid Fridman, Leonid Friedlander, Amanda Galtman, Lynn Garner, Tom Gearhart,

Howard Georgi, James F. Gibbons, Allan Gleason, Florence Gordon, Danny Goroff, Robin Gottlieb, Biddy Greene, Dudley Herschbach, JoEllen Hillyer, Leslie Hines, Luke Hunsberger, John Huth, Richard Iltis, Rob Indik, Adrian Iovita, Pete Janzow, Mary Johenk, Jerry Johnson, Mille Johnson, Matthias Kawski, David Kazhdan, Thomas Kerler, Mike Klucznik, Donna Krawczyk, Mark Kunka, Carl Leinbach, David Levermore, Don Lewis, Jim Lewis, John Lucas, Tom MacMahon, Dan Madden, Barry Mazur, Eric Mazur, Dan McGee, Ansie Meiring, Dave Meredith, Andrew Metrick, Ishaya Monokoff, Lang Moore, Don Myers, Bridget Neale, Alan Newell, Steve Olson, John Orr, Arnie Ostebee, Mike Pavloff, Howard Penn, Nancy Perry, Tony Phillips, Ago Pisztora, John Prados, Nancy Prinz, Ted Pyne, Amy Radunskaya, Wayne Raskind, Gabriella Ratay, Janet Ray, George Rublein, Anneliese Schauerte, Wilfried Schmid, Walter Seaman, Alida Setford, Pat Shure, Esther Silberstein, David Smith, Sharon Smith, Don Snow, Bob Speiser, Howard Stone, Steve Strogatz, "Suds" Sudholz, Cliff Taubes, Peter Taylor, Ralph Teixeira, Tom Timchek, Alan Tucker, Jerry Uhl, Bill Vélez, Charles Walter, Mary Jean Winter, Chris Yetman, Debbie Yoklic, Bruce Yoshiwara, Kathy Yoshiwara, Lee Zia, Paul Zorn, and all the people in the Harvard and Arizona mathematics departments who shared their computers and their space with us.

Reports from the following reviewers were most helpful in shaping the second edition. Mathematics faculty and students: Frank Beatrous, Dennis De Turck, G. Martin Forrest, John Hagood, Kenneth Hannsgen, Kerry Johnson, Lisa A. Mantini, Joe A. Marlin, Grace Orzech, Howard Penn, Anthony L. Peressini, David R. Pitts, Lynne Small, Michell Anne Stillman, Brian Tankersley.

Engineering faculty: Howie Choset, Michael R. Foster, Terry L. Kohutek, Dorian Liepmann, Howard Littman, Vijay Modi, J. W. Rogers, Jr., John Sanchez.

Most of all, to the remarkable team that worked day and night to get the text into the computer (and out again), to get the solutions written and the pictures labeled: we greatly appreciate your ingenuity, energy and dedication. Thanks to: Ebo Bentil, John Cho, Jie Cui, Mike Esposito, David Grenda, David Harris, Alex Kasman, Misha Kazhdan, Alex Mallozzi, Elliot Marks, Kyle Niedzwiecki, Dave Richards, Ann Ryu, Noah Syroid, and Xianbao Xu.

In particular, we would like to thank Paul Feehan for his remarkable contributions.

Deborah Hughes-Hallett	David O. Lomen	Douglas Quinney
Andrew M. Gleason	David Lovelock	Jeff Tecosky-Feldman
William G. McCallum	David Mumford	Joe B. Thrash
Daniel E. Flath	Brad G. Osgood	Karen R. Thrash
Patti Frazer Lock	Andrew Pasquale	Thomas W. Tucker
Sheldon P. Gordon		

To Students: How to Learn from this Book

- This book may be different from other math textbooks that you have used, so it may be helpful to know about some of the differences in advance. At every stage, this book emphasizes the *meaning* (in practical, graphical or numerical terms) of the symbols you are using. There is much less emphasis on "plug-and-chug" and using formulas, and much more emphasis on the interpretation of these formulas than you may expect. You will often be asked to explain your ideas in words or to explain an answer using graphs.

- The book contains the main ideas of calculus in plain English. Success in using this book will depend on reading, questioning, and thinking hard about the ideas presented. It will be helpful to read the text in detail, not just the worked examples.

- There are few examples in the text that are exactly like the homework problems, so homework problems can't be done by searching for similar–looking "worked out" examples. Success with the homework will come by grappling with the ideas of calculus.

- Many of the problems in the book are open-ended. This means that there is more than one correct approach and more than one correct solution. Sometimes, solving a problem relies on common sense ideas that are not stated in the problem explicitly but which you know from everyday life.

- This book assumes that you have access to a calculator or computer that can graph functions, find (approximate) roots of equations, and compute integrals numerically. There are many situations where you may not be able to find an exact solution to a problem, but can use a calculator or computer to get a reasonable approximation. An answer obtained this way is usually just as useful as an exact one. However, the problem does not always state that a calculator is required, so use your own judgement.

  If you mistrust technology, listen to this student, who started out the same way:

  > Using computers is strange, but surprisingly beneficial, and in my opinion is what leads to success in this class. I have difficulty visualizing graphs in my head, and this has always led to my downfall in calculus. With the assistance of the computers, that stress was no longer a factor, and I was able to concentrate on the concepts behind the shapes of the graphs, and since these became gradually more clear, I got increasingly better at picturing what the graphs should look like. It's the old story of not being able to get a job without previous experience, but not being able to get experience without a job. Relying on the computer to help me avoid graphing, I was tricked into focusing on what the graphs meant instead of how to make them look right, and what graphs symbolize is the fundamental basis of this class. By being able to see what I was trying to describe and learn from, I could understand a lot more about the concepts, because I could change the conditions and see the results. For the first time, I was able to see how everything works together . . . .

  That was a student at the University of Arizona who took calculus in Fall 1990, the first time we used the text. She was terrified of calculus, got a C on her first test, but finished with an A for the course.

- This book attempts to give equal weight to three methods for describing functions: graphical (a picture), numerical (a table of values) and algebraic (a formula). Sometimes it's easier to translate a problem given in one form into another. For example, you might replace the graph of a parabola with its equation, or plot a table of values to see its behavior. It is important to be flexible about your approach: if one way of looking at a problem doesn't work, try another.

- Students using this book have found discussing these problems in small groups helpful. There are a great many problems which are not cut-and-dried; it can help to attack them with the other perspectives your colleagues can provide. If group work is not feasible, see if your instructor can organize a discussion session in which additional problems can be worked on.

- You are probably wondering what you'll get from the book. The answer is, if you put in a solid effort, you will get a real understanding of one of the most important accomplishments of the millennium – calculus – as well as a real sense of how mathematics is used in the age of technology.

Table of Contents

6 CONSTRUCTING ANTIDERIVATIVES   293

7 INTEGRATION   321

8 USING THE DEFINITE INTEGRAL   375

12 A FUNDAMENTAL TOOL: VECTORS   617

13 DIFFERENTIATING FUNCTIONS OF MANY VARIABLES   653

14 OPTIMIZATION: LOCAL AND GLOBAL EXTREMA   717

19 FLUX INTEGRALS 875

20 CALCULUS OF VECTOR FIELDS 895

APPENDIX 931

ANSWERS TO ODD NUMBERED PROBLEMS 951

INDEX 973

CHAPTER ONE

A LIBRARY OF FUNCTIONS

Functions are truly fundamental to mathematics. In everyday language we say, "The price of a ticket is a function of where you sit," or "The fuel needed to launch a rocket is a function of its payload." In each case, the word *function* expresses the idea that knowledge of one fact tells us another. In mathematics, the most important functions are those in which knowledge of one number tells us another number. If we know the length of the side of a square, its area is determined. If the circumference of a circle is known, its radius is determined.

Calculus starts with the study of functions. This chapter will lay the foundation for calculus by surveying the behavior of the most common functions, including powers, exponentials, logarithms, and the trigonometric functions. We will also explore ways of handling the graphs, tables, and formulas that represent these functions.

1.1 WHAT'S A FUNCTION?

In mathematics, a *function* is used to represent the dependence of one quantity upon another.

Let's look at an example. In the summer of 1990, the temperatures in Arizona reached an all-time high (so high, in fact, that some airlines decided it might be unsafe to land their planes there). The daily high temperatures in Phoenix for June 19–29 are given in Table 1.1.

TABLE 1.1 *Daily High Temperature in Phoenix, Arizona, June 19–29, 1990*

Date (t, June 1990)	19	20	21	22	23	24	25	26	27	28	29
Temperature (H, °F)	109	113	114	113	113	113	120	122	118	118	108

Although you may not have thought of something so unpredictable as temperature as being a function, the temperature *is* a function of date, because each day gives rise to one and only one high temperature. There is no formula for temperature (otherwise we would not need the weather bureau), but nevertheless the temperature does satisfy the definition of a function: Each input date, t, has a unique output temperature, H, associated with it.

We define a function as follows:

> A **function** is a rule that takes certain numbers as inputs and assigns to each a definite output number. The set of all input numbers is called the **domain** of the function and the set of resulting output numbers is called the **range** of the function.

The input is called the *independent variable* and the output is called the *dependent variable*. In the temperature example the domain is the set of dates $t = \{19, 20, 21, 22, 23, 24, 25, 26, 27, 28, 29\}$ and the range is the set of temperatures $H = \{109, 113, 114, 120, 122, 118, 108\}$. Notice that a function may have identical outputs for different inputs (June 22, 23, and 24, for example).

Some quantities, such as date, are *discrete*, meaning they take only certain isolated values (dates must be integers). Other quantities, such as length, are *continuous* as they can be any number. For a continuous variable, domains and ranges are often written using interval notation:

$$a \leq t \leq b \ \text{ is written } \ [a, b]$$
$$a < t < b \ \text{ is written } \ (a, b).$$

Representation of Functions: Tables, Graphs, Formulas, and Words

Functions can be represented by tables, graphs, formulas, and descriptions in words. For example, the function giving the daily high temperatures in Phoenix, Arizona as a function of time can be represented by each of the graphs in Figure 1.1, as well as by Table 1.1.

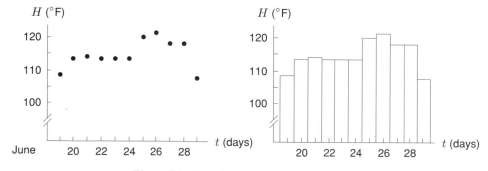

Figure 1.1: Phoenix temperatures, June 1990

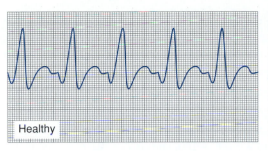
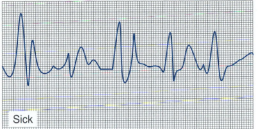

Figure 1.2: EKG readings on two patients

Other functions arise naturally as graphs. Figure 1.2 contains electrocardiogram (EKG) pictures showing the heartbeat patterns of two patients, one normal and one not. Although it is possible to construct a formula to approximate an EKG function, this is seldom done. The pattern of repetitions is what a doctor needs to know, and these are much more easily seen from a graph than from a formula. However, each EKG represents a function which gives electrical activity as a function of time.

As another example of a function, consider the snow tree cricket. Surprisingly enough, all such crickets chirp at essentially the same rate if they are at the same temperature. That means that the chirp rate is a function of temperature. In other words, if we know the temperature, we can determine the chirp rate. Even more surprisingly, the chirp rate, C, in chirps per minute, increases steadily with the temperature, T, in degrees Fahrenheit, and can be computed by the formula

$$C = 4T - 160$$

to a fair degree of accuracy. We write $C = f(T)$ to express the fact that we think of C as a function of T. The graph of this function is in Figure 1.3.

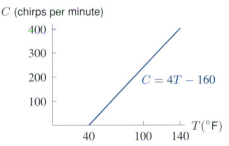

Figure 1.3: Cricket chirp rate versus temperature

Examples of Domain and Range

If the domain of a function is not specified, we usually take it to be the largest possible set of real numbers. For example, we usually think of the domain of the function $f(x) = x^2$ as all real numbers. However, the domain of the function $g(x) = 1/x$ is all real numbers except zero, since we cannot divide by zero.

Example 1 Find the domain and range of the function $y = f(x) = 1/(x-3)^2$.

Solution Since the formula makes sense for all values of x except 3, the domain is all x-values with $x \neq 3$. The range of f is all positive y-values.

Sometimes we specify the domain to be something smaller than the largest possible set of real numbers. This is called restricting the domain. For example, if the function $f(x) = x^2$ is used to represent the area of a square of side x, we consider only nonnegative values of x and restrict the domain to nonnegative numbers.

Example 2   Consider the function $C = f(T)$ giving chirp rate as a function of temperature. We restrict this function to temperatures for which the predicted chirp rate is positive, and up to the highest temperature ever recorded at a weather station, namely, $136°$F. What is the domain of this function f?

Solution   If we consider the equation

$$C = 4T - 160$$

simply as a mathematical relationship between two variables C and T, any T value is possible. However, if we think of it as a relationship between cricket chirps and temperature, then C cannot be less than 0. Since $C = 0$ leads to $0 = 4T - 160$, and so $T = 40°$F, we see that T cannot be less than $40°$F. (See Figure 1.3.) In addition, we are told that the function is not defined for temperatures above $136°$. Thus, for the function $C = f(T)$ we have

$$\text{Domain} = \text{All } T \text{ values between } 40°F \text{ and } 136°F$$
$$= \text{All } T \text{ values with } 40 \leq T \leq 136$$
$$= [40, 136].$$

Therefore, we say that the function $C = f(T)$ is represented by the formula

$$C = f(T) = 4T - 160 \quad \text{on the domain} \quad 40 \leq T \leq 136.$$

Example 3   Find the range of the function f, given the domain from Example 2. In other words, find all possible values of the chirp rate, C, in the equation $C = f(T)$.

Solution   Again, if we consider $C = 4T - 160$ simply as a mathematical relationship, its range is all real C values. However, when thinking of the meaning of $C = f(T)$ for crickets, we see that the function will predict cricket chirps per minute between 0 (when $T = 40°$F) and 384 (when $T = 136°$F). Hence,

$$\text{Range} = \text{All } C \text{ values from 0 to 384}$$
$$= \text{All } C \text{ values with } 0 \leq C \leq 384$$
$$= [0, 384].$$

So far we have used the temperature to predict the chirp rate and have thought of the temperature as the *independent variable* and the chirp rate as the *dependent variable*. However, we could do this backwards, and calculate the temperature from the chirp rate. From this point of view, the temperature is dependent on the chirp rate. Thus, which variable is dependent and which is independent may depend on your viewpoint.

Thinking of temperature as a function of chirp rate would enable us (in theory, at least) to use the chirp rate to measure temperature. The way we actually do measure temperature is based on another function: the relationship between the height of the liquid in a thermometer and temperature. Although the height of the mercury is certainly a function of temperature, we actually always use this relation the other way around and determine the temperature as a function of the height of the mercury.

Proportionality

A common functional relationship occurs when one quantity is *proportional* to another. For example, if apples are 60 cents a pound, we say the price you pay, p cents, is proportional to the weight you buy, w pounds, because

$$p = f(w) = 60w.$$

As another example, the area, A, of a circle is proportional to the square of the radius, r:

$$A = f(r) = \pi r^2.$$

> We say y is (directly) **proportional** to x if there is a constant k such that
>
> $$y = kx.$$
> This k is called the constant of proportionality.

We also say that one quantity is *inversely proportional* to another if one is proportional to the reciprocal of the other. For example, the speed, v, at which you make a 50-mile trip is inversely proportional to the time, t, taken, because v is proportional to $1/t$:

$$v = 50 \left(\frac{1}{t} \right) = \frac{50}{t}.$$

Notice that if y is directly proportional to x, then the magnitude of one variable increases (decreases) when the magnitude of the other increases (decreases). If, however, y is inversely proportional to x, then the magnitude of one variable increases when the value of the other decreases.

Problems for Section 1.1

1. Which of the graphs in Figure 1.4 best matches each of the following three stories?[1] Write a story for the remaining graph.

   (a) I had just left home when I realized I had forgotten my books, and so I went back to pick them up.
   (b) Things went fine until I had a flat tire.
   (c) I started out calmly but sped up when I realized I was going to be late.

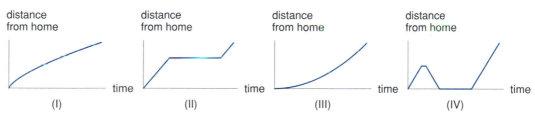

Figure 1.4

2. Match each story to the most appropriate graph. Write a story for the remaining graph.

   (a) During my journey, I had a flat tire. Then after fixing the flat tire, I had to speed up to avoid being late.
   (b) My car broke down and I left it parked at the road side.
   (c) As soon as I'd dropped off the package, I turned around and drove home.

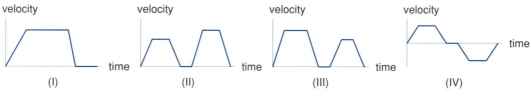

Figure 1.5

3. The population of a city, P (in millions), is a function of t, the number of years since 1950. We have $P = f(t)$. Explain the meaning of the statement $P(35) = 12$ in terms of the population of this city.

---

[1]Adapted from Jan Terwel, "Real Math in Cooperative Groups in Secondary Education." *Cooperative Learning in Mathematics*, ed. Neal Davidson, p. 234. (Reading: Addison Wesley, 1990).

In Problems 4–7, write a formula representing the function described.

4.   Kinetic energy, K, is proportional to the square of velocity, v.

5.   The gravitational force, F, between two bodies is inversely proportional to the square of the distance d between them.

6.   The average velocity, v, for a trip over a fixed distance is inversely proportional to the time of travel.

7.   The volume of a sphere is proportional to the cube of its radius, r.

8.   It warmed up throughout the morning, and then suddenly got much cooler around noon, when a storm came through. After the storm, it warmed up before cooling off at sunset. Sketch a possible graph of this day's temperature as a function of time.

9.   A doctor rides his bicycle from his home to the clinic, a trip of about ten miles. He goes along at a constant speed until he reaches the big hill which starts at mile 4. He slows down as he goes up the hill, which peaks at mile 5. He is barely moving when he reaches the top and stops to rest for a bit. He then goes down the hill (fairly quickly thanks to the steep grade), reaching the bottom at mile 7. The last three miles to his office are flat. Sketch a graph which shows the doctor's *speed* as a function of the *distance* from home.

10. When a nerve impulse reaches the end of a neuron, neurotransmitter is released, initiating a nerve impulse in the next neuron. The neurotransmitter is then enzymatically degraded. Sketch a graph showing the concentration of neurotransmitter in the space between the neurons as a function of time.

11. A flight from Dulles Airport in Washington, DC to LaGuardia Airport in New York City has to circle LaGuardia several times before being allowed to land. Plot a graph of the distance of the plane from Washington against time, from the moment of takeoff until landing.

12. Plot the distance of the plane in Problem 11 from LaGuardia as a function of time, from the moment of take-off until landing.

13. An object is put outside on a cold day and its temperature, H, in °C is a function of the time, t, in minutes since it was put outside. A graph of the function $H = f(t)$ is given in Figure 1.6.

  (a) What does the statement $f(30) = 10$ mean in terms of temperature? Include units for 30 and for 10 in your answer.

  (b) Explain what the vertical intercept, a, and the horizontal intercept, b, represent in terms of temperature of the object and time outside.

Problems 14–15 are about supply and demand curves. Economists are interested in how the quantity, q, of an item which is manufactured and sold depends on the price, p, per unit. They think of quantity as a function of price. However, for historical reasons[2] the economists put price (the independent variable) on the vertical axis and quantity (the dependent variable) on the horizontal axis. Since manufacturers and customers react differently to changes in price, there are two functions relating p and q. The *supply curve* represents how the quantity of an item that manufacturers are willing to supply depends on the price for which the item can be sold. The *demand curve* represents how the quantity of an item demanded by consumers depends on its price.

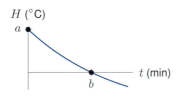

Figure 1.6

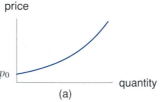

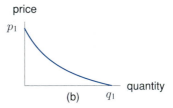

Figure 1.7: Supply and demand curves

14. One of the graphs in Figure 1.7 is a supply curve, and the other is a demand curve. Which is which? Why?

15. The price p_0 in Figure 1.7(a) represents the price below which the manufacturers are unwilling to produce any of the item. What do the price p_1 and the quantity q_1 in Figure 1.7(b) represent in practical economic terms?

---

[2]Originally, the economists thought of price as the dependent variable and put it on the vertical axis. Unfortunately, when their point of view changed, the axes did not.

For Problems 16–18, give the approximate domain and range of each function. Assume the entire graph is shown.

16.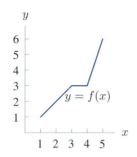

17.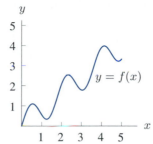

18.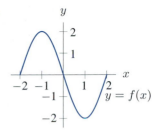

Find the domain and range of the functions in Problems 19–21.

19. $y = x^2 + 2$

20. $y = \dfrac{1}{x - 2}$

21. $y = \dfrac{1}{x^2 + 2}$

22. If $f(t) = \sqrt{t^2 - 16}$, find all values of t for which $f(t)$ is a real number. Solve $f(t) = 3$.

23. If $g(x) = (4 - x^2)/(x^2 + x)$, find the domain of $g(x)$. Solve $g(x) = 0$.

24. When Galileo was formulating the laws of motion, he considered the motion of a body starting from rest and falling under gravity. He originally thought that the velocity of such a falling body was proportional to the distance it had fallen. What do the experimental data in Table 1.2 tell you about Galileo's hypothesis? What alternative hypothesis is suggested by the two sets of data in Table 1.2 and Table 1.3?

TABLE 1.2

Distance (ft)	0	1	2	3	4
Velocity (ft/sec)	0	8	11.3	13.9	16

TABLE 1.3

Time (sec)	0	1	2	3	4
Velocity (ft/sec)	0	32	64	96	128

1.2 LINEAR FUNCTIONS

Probably the most commonly used functions are the *linear functions*. These are functions that have a constant rate of increase or decrease. A function is linear if its slope, or rate of change, is the same everywhere. For a function that is not linear, the rate of change will vary from point to point.

The Olympic Pole Vault

During the early years of the Olympics, the height of the winning pole vault increased approximately 8 inches every four years. Table 1.4 shows that the height started at 130 inches in 1900, and increased by the equivalent of 2 inches every year. So the height was a linear function of time from 1900 to 1912. If y is the winning height in inches and t is the number of years since 1900, we can write

$$y = f(t) = 130 + 2t.$$

Since $y = f(t)$ increases with t, we say that f is *an increasing function*. The coefficient 2 tells us the rate, in inches per year, at which the height increases. This rate is the *slope* of the line $f(t) = 130 + 2t$.

TABLE 1.4   *Olympic pole vault records (approximate)*

Year	1900	1904	1908	1912
Height (inches)	130	138	146	154

You can visualize the slope in Figure 1.8 as the ratio

$$\text{Slope} = \frac{\text{Rise}}{\text{Run}} = \frac{138 - 130}{1904 - 1900} = \frac{8}{4} = 2 \text{ inches/year.}$$

Calculating the slope (rise/run) using any other two points on the line gives the same value.

What about the constant 130? This represents the initial height in 1900, when $t = 0$. Geometrically, 130 is the *intercept* on the vertical axis.

You may wonder whether the linear trend continues beyond 1912. Not surprisingly, it doesn't exactly. The formula $y = 130 + 2t$ predicts that the height in the 1996 Olympics would be 322 inches or 26 feet 10 inches, which is considerably higher than the actual value of 19 feet 5.25 inches. There is clearly a danger in *extrapolating* too far from the given data. You should also observe that the data in Table 1.4 is *discrete*, because it is given only at specific points (every four years). However, we have treated the variable t as though it were *continuous*, because the function $y = 130 + 2t$ makes sense for all values of t. The graph in Figure 1.8 is of the continuous function because it is a solid line, rather than four separate points representing the years in which the Olympics were held.

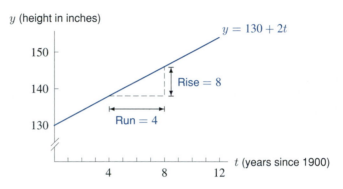

Figure 1.8: Olympic pole vault records

Linear Functions in General

> A **linear function** has the form
>
> $$y = f(x) = b + mx.$$
>
> Its graph is a line such that
> - m is the **slope**, or rate of change of y with respect to x.
> - b is the **vertical intercept**, or value of y when x is zero.

Notice that if the slope, m, is zero, we have $y = b$, a horizontal line.

> To recognize that a table of x and y values comes from a linear function, $y = b + mx$, look for differences in y-values that are constant for equal differences in x.

Difference Quotients and Delta Notation

We use the symbol Δ (the Greek letter capital delta) to mean "change in," so Δx means change in x and Δy means change in y.

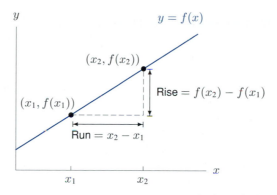

Figure 1.9: Difference quotient $= \dfrac{f(x_2) - f(x_1)}{x_2 - x_1}$

The slope of a linear function $y = f(x)$ can be calculated from values of the function at two points, given by x_1 and x_2, using the formula

$$m = \frac{\text{Rise}}{\text{Run}} = \frac{\Delta y}{\Delta x} = \frac{f(x_2) - f(x_1)}{x_2 - x_1}.$$

The quantity $(f(x_2) - f(x_1))/(x_2 - x_1)$ is called a *difference quotient* because it is the quotient of two differences. (See Figure 1.9). In Chapter 2, you will see that difference quotients play an important role in calculus.

The Success of Search and Rescue Teams

Consider the problem of the "search and rescue" teams working to find lost hikers in remote areas. To search for an individual, members of the search team separate and walk parallel to one another through the area to be searched. Experience has shown that the team's chance of finding a lost individual is related to the distance, d, by which team members are separated. For a particular type of terrain, the percent found[3] for various separations is recorded in Table 1.5.

TABLE 1.5 *Success rate versus separation of searchers*

Separation distance, d (ft)	Approximate percent found, P
20	90
40	80
60	70
80	60
100	50

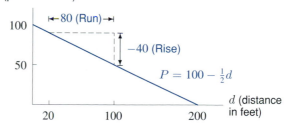

Figure 1.10: Success rate versus separation of searchers

As we would expect, the data in the table indicates that as the separation distance decreases, a larger percentage of the lost hikers is found. Since $P = f(d)$ decreases as d increases, we say that P is a *decreasing function* of d. In addition, the data shows that for each 20-foot increase in distance, the percentage found drops by 10. This indicates that the graph of P against d is a line. (See Figure 1.10.) Notice that the slope is $-40/80 = -1/2$. The negative sign shows that P decreases as d increases. The magnitude of the slope is the rate at which P is decreasing as d increases.

---

[3]From *An Experimental Analysis of Grid Sweep Searching*, by J. Wartes (Explorer Search and Rescue, Western Region, 1974).

What about the vertical intercept? If $d = 0$, the searchers are walking shoulder to shoulder and we'd expect everyone to be found, so $P = 100$. This is exactly what happens if the line is continued to the vertical axis (a decrease of 20 in d causes an increase of 10 in P). Therefore, the equation of the line is

$$P = f(d) = 100 - \frac{1}{2}d.$$

What about the horizontal intercept? When $P = 0$, or $0 = 100 - \frac{1}{2}d$, then $d = 200$. The value $d = 200$ represents the separation distance at which, according to the model, no one is found. This is unreasonable, because even when the searchers are far apart, the search will sometimes be successful. This suggests that at some point, the linear relationship ceases to hold. As in the pole vault example, extrapolating too far beyond the given data may not give accurate answers.

Increasing versus Decreasing Functions

The terms increasing and decreasing can be applied to other functions, not just linear ones. See Figure 1.11. In general,

> A function f is **increasing** if the values of $f(x)$ increase as x increases.
> A function f is **decreasing** if the values of $f(x)$ decrease as x increases.
>
> The graph of an *increasing* function *climbs* as we move from left to right.
> The graph of a *decreasing* function *descends* as we move from left to right.

Figure 1.11: Increasing and decreasing functions

Families of Linear Functions

Formulas such as $f(x) = b + mx$, in which the constants m and b can take on various values, are said to define a *family of functions*. All the functions in a family share certain properties—in this case, all the graphs are straight lines. Each of the functions in this section belongs to the linear family $f(x) = b + mx$. The constants m and b are called *parameters*; their meaning is shown in Figures 1.12 and 1.13. Notice the greater the magnitude of m is, the steeper the line is.

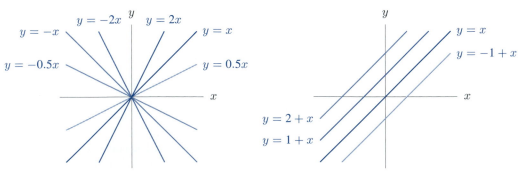

Figure 1.12: The family $y = mx$
(with $b = 0$)

Figure 1.13: The family $y = b + x$
(with $m = 1$)

Problems for Section 1.2

For Problems 1–3, determine the slope and the y-intercept of the line whose equation is given.

1. $7y + 12x - 2 = 0$ 2. $-4y + 2x + 8 = 0$ 3. $12x = 6y + 4$

For Problems 4–6, find the equation of the line that passes through the given points.

4. $(0, 0)$ and $(1, 1)$ 5. $(0, 2)$ and $(2, 3)$ 6. $(-2, 1)$ and $(2, 3)$

For Problems 7–9, use the facts that parallel lines have equal slopes and that the slopes of perpendicular lines are negative reciprocals of one another.

7. Find the equation of the line through the point $(2, 1)$ which is perpendicular to the line $y = 5x - 3$.

8. Find the equations of the lines through the point $(1, 5)$ that are parallel to and perpendicular to the line with equation $y + 4x = 7$.

9. Find the equations of the lines through the point (a, b) that are parallel and perpendicular to the line $y = mx + c$, assuming $m \neq 0$.

10. Match the graphs in Figure 1.14 with the equations below. (Note that the x and y scales may be unequal.)

    (a) $y = x - 5$ (c) $5 = y$ (e) $y = x + 6$
    (b) $-3x + 4 = y$ (d) $y = -4x - 5$ (f) $y = x/2$

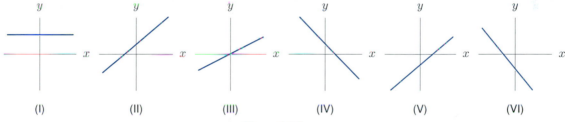

(I)   (II)   (III)   (IV)   (V)   (VI)

Figure 1.14

11. Match the graphs in Figure 1.15 with the equations below. (Note that the x and y scales may be unequal.)

    (a) $y = -2.72x$ (c) $y = 27.9 - 0.1x$ (e) $y = -5.7 - 200x$
    (b) $y = 0.01 + 0.001x$ (d) $y = 0.1x - 27.9$ (f) $y = x/3.14$

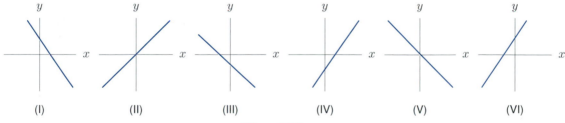

(I)   (II)   (III)   (IV)   (V)   (VI)

Figure 1.15

12. An equation of a line is $3x + 4y = -12$. Find the length of the portion of the line that lies between the x and y intercepts of the equation.

13. The table below gives values for a linear function. Find a formula for W in terms of R.

R	6	9	12	15	18
W	20	25	30	35	40

14. Suppose you are driving at a constant speed from Chicago to Detroit, a distance of 275 miles. About 120 miles from Chicago you pass through Kalamazoo, Michigan. Sketch a graph of your distance from Kalamazoo as a function of time.

15. Residents of the town of Maple Grove who are connected to the municipal water supply are billed a fixed amount yearly plus a charge for each cubic foot of water used. A household using 1000 cubic feet was billed $90, while one using 1600 cubic feet was billed $105.

    (a) What is the charge per cubic foot?
    (b) Write an equation for the total cost of a resident's water as a function of cubic feet of water used.
    (c) How many cubic feet of water used would lead to a bill of $130?

16. You are knitting a scarf of constant width. Let $s(y) = b + ay$ be a function that gives the length $s(y)$, in feet, of the scarf after y skeins of yarn have been used. Here a and b are constants.

    (a) What is the value of b? Why?
    (b) Is a positive, negative, or zero?
    (c) Your friend is also knitting a scarf using a pattern that alternates between stitches and empty spaces. Your own scarf has no such empty spaces, only solid stitches. Your friend's length formula is $p(y) = cy$, where c is a constant. Which constant is larger, a or c?

17. The graph of Fahrenheit temperature, °F, as a function of Celsius temperature, °C, is a line. You know that 212°F and 100°C both represent the temperature at which water boils. Similarly, 32°F and 0°C both represent water's freezing point.

    (a) What is the slope of the graph?
    (b) What is the equation of the line?
    (c) Use the equation to find what Fahrenheit temperature corresponds to 20°C.
    (d) What temperature is the same number of degrees in both Celsius and Fahrenheit?

18. Let p represent the selling price in dollars of a certain type of toy and q the number of toys (in thousands) which sell at that price. Market research indicates the following relationship between p and q:

p	1	2	3	4
q	950	900	850	800

    (a) Find a formula for q as a linear function of p.
    (b) Find a formula for p as a linear function of q.
    (c) According to this model, if toys are given away for free, how many will be taken?

19. When a solid is subjected to small stresses (forces of tension or compression), the resulting strain (the fractional change in length) is proportional to the stress. The elasticity of the material is represented by the constant E, called Young's modulus, where

$$E = \frac{\text{Stress}}{\text{Strain}}.$$

Figure 1.16 shows the stress-strain curve for bone. Estimate Young's modulus for bone.

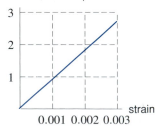

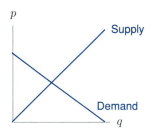

Figure 1.16   **Figure 1.17**

20. Linear supply and demand curves[4] are shown in Figure 1.17, with price on the vertical axis. The equilibrium point is defined to be where the supply and demand curves cross.

    (a) Label the equilibrium price p^* and the equilibrium quantity q^* on the axes.
    (b) Explain the effect on equilibrium price and quantity if the slope of the supply curve increases. Illustrate your answer graphically.

[4]For a discussion of supply and demand curves, see page 6.

    (c)   Explain the effect on equilibrium price and quantity if the slope of the demand curve becomes more negative, keeping the p-intercept constant. Illustrate your answer graphically.

21.   Since the opening up of the West, the US population has moved westward. To observe this, we look at the "population center" of the US, which is the point at which the country would balance if it were a flat plate with no weight, and every person had equal weight. In 1790 the population center was east of Baltimore, Maryland. It has been moving westward ever since, and in 1990 it crossed the Mississippi river to Steelville, Missouri (southwest of St. Louis). During the second half of this century, the population center has moved about 50 miles west every 10 years.

    (a)   Let us measure position westward from Steelville along the line running through Baltimore. Express the approximate position of the population center as a function of time, measured in years from 1990.

    (b)   The distance from Baltimore to Steelville is a bit over 700 miles. Could the population center have been moving at roughly the same rate for the last two centuries?

    (c)   Could the function in part (a) continue to apply for the next three centuries? Why or why not? [Hint: You may want to look at a map. Note that distances are in air miles and are not driving distances.]

22.   For small changes in temperature, the formula for the expansion of a metal rod under a change in temperature is:

$$l - l_0 = a l_0 (t - t_0),$$

where l is the length of the object at temperature t, and l_0 is the initial length at temperature t_0, and a is a constant which depends on the type of metal.

    (a)   Express l as a linear function of t. Find the slope and vertical intercept. [Hint: Treat the other quantities as constants.]

    (b)   Suppose you had a rod which was initially 100 cm long at $60°$F and made of a metal with a equal to 10^{-5}. Write an equation giving the length of this rod at temperature t.

    (c)   What does the sign of the slope of the graph tell you about the expansion of a metal under a change in temperature?

23.   When a cold yam is put into a hot oven to bake, the temperature of the yam rises. The rate, R (in degrees per minute), at which the temperature of the yam rises is governed by Newton's Law of Heating, which says that the rate is proportional to the temperature difference between the yam and the oven. If the oven is at $350°$F and the temperature of the yam is $H°$F,

    (a)   Write a formula giving R as a function of H.

    (b)   Sketch the graph of R against H.

24.   A body of mass m is falling downward with velocity v. Newton's Second Law of Motion, $F = ma$, says that the net downward force, F, on the body is proportional to its downward acceleration, a. The net force, F, consists of the force due to gravity, F_g, which acts downward, minus the air resistance, F_r, which acts upward. The force due to gravity is mg, where g is a constant. Assume the air resistance is proportional to the velocity of the body.

    (a)   Write an expression for the net force, F, as a function of the velocity, v.

    (b)   Write a formula giving a as a function of v.

    (c)   Sketch a against v.

1.3 EXPONENTIAL FUNCTIONS

Population Growth

Consider the data for the population of Mexico in the early 1980s given in Table 1.6. To see how the population is growing, we might look at the increase in population from one year to the next, as shown in the third column. If the population had been growing linearly, all the numbers in the third column would be the same. But populations usually grow faster as they get bigger, because there are more people to have babies. So we shouldn't be surprised to see the numbers in the third column increasing.

TABLE 1.6 *Population of Mexico (estimated), 1980–1986*

Year	Population (millions)	Change in population (millions)
1980	67.38	
		1.75
1981	69.13	
		1.80
1982	70.93	
		1.84
1983	72.77	
		1.89
1984	74.66	
		1.94
1985	76.60	
		1.99
1986	78.59	

Suppose we divide each year's population by the previous year's population. We get, approximately,

$$\frac{\text{Population in 1981}}{\text{Population in 1980}} = \frac{69.13 \text{ million}}{67.38 \text{ million}} = 1.026$$

$$\frac{\text{Population in 1982}}{\text{Population in 1981}} = \frac{70.93 \text{ million}}{69.13 \text{ million}} = 1.026.$$

The fact that both calculations give 1.026 shows the population grew by about 2.6% between 1980 and 1981 *and* between 1981 and 1982. If we do similar calculations for other years, we find that the population grew by a factor of about 1.026, or 2.6%, every year. Whenever we have a constant growth factor (here 1.026), we have *exponential growth*. If t is the number of years since 1980,

When $t = 0$, population $= 67.38 = 67.38(1.026)^0$.

When $t = 1$, population $= 69.13 = 67.38(1.026)^1$.

When $t = 2$, population $= 70.93 = 69.13(1.026) = 67.38(1.026)^2$.

When $t = 3$, population $= 72.77 = 70.93(1.026) = 67.38(1.026)^3$.

So in general, the population t years after 1980 is given by

$$P = 67.38(1.026)^t.$$

This is an *exponential function* with base 1.026. It is called exponential because the independent variable, t, is in the exponent. The base represents the factor by which the population grows each year. If we assume that the same formula will hold for the next 50 years, the population graph has the shape shown in Figure 1.18. Since the population is growing, the function is increasing. Notice also that the population grows faster and faster as time goes on. This behavior is typical of an exponential function. Compare this with the behavior of a linear function, which climbs at the same rate everywhere and so has a straight-line graph. Because the graph of an exponential function is bending upward, we say it is *concave up*. Even exponential functions which climb slowly at first, such as this one, climb extremely quickly eventually. That is why exponential population growth is considered by some to be such a threat to the world.

Even if it represents reliable data, the smooth graph in Figure 1.18 is actually only an approximation to the true graph of the population of Mexico. Since we can't have fractions of people, the graph should really be jagged, jumping up or down by one each time someone is born or dies. However, with a population in the millions, the jumps are so small as to be invisible at the scale we are using. Therefore, the smooth graph is an extremely good approximation.

Example 1   Predict the population of Mexico in the year

(a)   2007 (when $t = 27$).   (b)   2034 (when $t = 54$).   (c)   2061 (when $t = 81$).

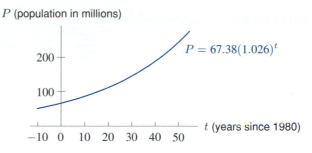

Figure 1.18: Population of Mexico (estimated): Exponential growth

Solution   Extrapolating far into the future can be risky because it assumes that the population continues to grow exponentially with the same constant growth factor. (There could, for example, be a medical breakthrough that would increase the growth factor, or an epidemic that would decrease it.) Writing "$\approx$" to represent approximately equal, the model we are using predicts
(a) $P = 67.38(1.026)^{27} \approx 67.38(2) = 134.76$ million.
(b) $P = 67.38(1.026)^{54} \approx 67.38(4) = 269.52$ million.
(c) $P = 67.38(1.026)^{81} \approx 67.38(8) = 539.04$ million.

If we look at the answers to Example 1, we see that after 27 years the population has doubled; after another 27 years (at $t = 54$), it has doubled again. In yet another 27 years (when $t = 81$), the population has doubled yet again. As a result, we say that the *doubling time* of the population of Mexico is 27 years. Every exponentially growing population has a fixed doubling time.

Concavity

We have used the term concave up to describe the graph in Figure 1.18. In Figure 1.22 on page 18, we will see a graph which is concave down [5]. In words:

> The graph of any function is **concave up** if it bends upward as we move left to right; it is **concave down** if it bends downward. (See Figure 1.19.) A line is neither concave up nor concave down.

Concave up   Concave down

Figure 1.19: Concavity of a graph

Musical Pitch

The pitch of a musical note is determined by the frequency of the vibration which causes it. Middle C on the piano, for example, corresponds to a vibration of 263 hertz (cycles per second). A note one octave above middle C vibrates at 526 hertz, and a note two octaves above middle C vibrates at 1052 hertz. (See Table 1.7.)

TABLE 1.7   *Pitch of notes above middle C*

Number, n, of octaves above middle C	Number of hertz, $V = f(n)$
0	263
1	526
2	1052
3	2104
4	4208

TABLE 1.8   *Pitch of notes below middle C*

n	$V = 263 \cdot 2^n$
-3	$263 \cdot 2^{-3} = 263(1/2^3) = 32.875$
-2	$263 \cdot 2^{-2} = 263(1/2^2) = 65.75$
-1	$263 \cdot 2^{-1} = 263(1/2) = 131.5$
0	$263 \cdot 2^0 = 263$

[5] In Chapter 5 we consider concavity in more depth.

Notice that the ratios of successive values of V are

$$\frac{526}{263} = 2 \quad \text{and} \quad \frac{1052}{526} = 2 \quad \text{and} \quad \frac{2104}{1052} = 2,$$

and so on. In other words, each value of V is twice the value before, so

$$f(1) = 526 = 263 \cdot 2 = 263 \cdot 2^1$$
$$f(2) = 1052 = 526 \cdot 2 = 263 \cdot 2^2$$
$$f(3) = 2104 = 1052 \cdot 2 = 263 \cdot 2^3.$$

In general

$$V = f(n) = 263 \cdot 2^n,$$

where n is the number of octaves above middle C. The base 2 represents the fact that as we go up an octave, the frequency of vibrations doubles. Indeed, our ears hear a note as one octave higher than another precisely because it vibrates twice as fast. For the negative values of n in Table 1.8, this function represents the octaves below middle C. The notes on a piano are represented by values of n between -3 and 4, and the human ear finds values of n between -4 and 7 audible.

Although $V = f(n) = 263 \cdot 2^n$ makes sense in musical terms only for certain values of n, values of the function $f(x) = 263 \cdot 2^x$ can be calculated for all real x. The graph of $f(x)$ has the typical exponential shape, as can be seen in Figure 1.20. It is increasing and concave up, climbing faster and faster as x increases.

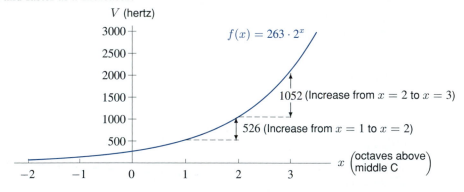

Figure 1.20: Pitch as a function of number of octaves above middle C

Removal of Pollutants from Jet Fuel

Now we look at an example in which a quantity is decreasing instead of increasing. Before kerosene can be used as jet fuel, federal regulations require that the pollutants in it be removed by passing the kerosene through clay. We will suppose the clay is in a pipe and that each foot of the pipe removes 20% of the pollutants that enter it. Therefore each foot leaves 80% of the pollution. If P_0 is the initial quantity of pollutant and $P = f(n)$ is the quantity left after n feet of pipe, then

$$f(0) = P_0$$
$$f(1) = (0.8)P_0$$
$$f(2) = (0.8)(0.8)P_0 = (0.8)^2 P_0$$
$$f(3) = (0.8)(0.8)^2 P_0 = (0.8)^3 P_0.$$

So, after n feet,

$$P = f(n) = P_0(0.8)^n.$$

In this example, n must be nonnegative. However, the *exponential decay function*

$$P = f(x) = P_0(0.8)^x$$

makes sense for any real x. We'll plot it with $P_0 = 1$ in Figure 1.21; some values of the function are shown in Table 1.9.

TABLE 1.9 *Values of decay function*

x	$P = (0.8)^x$
-2	1.56
-1	1.25
0	1
1	0.8
2	0.64
3	0.51
4	0.41

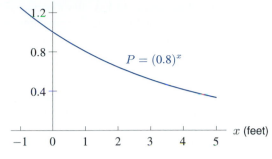

Figure 1.21: Pollutant removal: Exponential decay

Notice the way the function in Figure 1.21 is decreasing. Each additional foot of clay causes a smaller quantity of pollutant to be removed than the previous foot. This is because as the kerosene gets cleaner, there's less pollutant to remove, and so each foot of clay takes out less pollutant than the previous one. Compare this to the exponential growth in Figures 1.18 and 1.20 on pages 15 and 16, where each step upward is larger than the one before. Notice, however, that all three graphs are concave up.

The General Exponential Function

> We say P is an **exponential function** of t with base a if
>
> $$P = P_0 a^t,$$
>
> where P_0 is the initial quantity (when $t = 0$) and a is the factor by which P changes when t increases by 1. If $a > 1$, we have exponential growth; if $0 < a < 1$, we have exponential decay.

The largest possible domain for the exponential function is all real numbers, provided $a > 0$. The reason we do not want $a \leq 0$ is that, for example, we cannot define $a^{1/2}$ if $a \leq 0$. Also, we do not usually have $a = 1$, since $P = P_0 a^t = P_0 1^t = P_0$ is then a constant function.

> To recognize that a table of t and P values comes from an exponential function $P = P_0 a^t$, look for ratios of P values that are constant for equally spaced t values.

Radioactive Decay

Radioactive substances, such as uranium, decay by a certain percentage of their mass in a given unit of time. The most common way to express this rate of decay is to give the time period it takes for half the mass to decay. This period of time is called the *half-life* of the substance.

One of the most well-known radioactive substances is carbon-14, which is used to date organic objects. When an object, such as a piece of wood or bone, was part of a living organism, it accumulated small amounts of radioactive carbon-14. Once the organism dies, it no longer picks up carbon-14 through interaction with its environment (for example, through respiration). By measuring the proportion of carbon-14 in the object and comparing that to the proportion in living material, we can estimate how much of the original carbon-14 has decayed.

The half-life of carbon-14 is about 5730 years. We can write an exponential function for the amount of carbon-14 left after a period of t years. If $T = t/5730$ is the number of half-lives which have elapsed and C_0 is the original amount of carbon-14, then the amount, C, of carbon-14 left is given by

$$C = C_0 \left(\frac{1}{2} \right)^T = C_0 \left(\frac{1}{2} \right)^{(t/5730)}.$$

In general, suppose a substance has a half-life of h years (or minutes or seconds). Then, if Q_0 was the original quantity of the substance, the quantity, Q, of the substance left after t units of time is given by

$$Q = Q_0 \left(\frac{1}{2} \right)^{(t/h)}.$$

In summary, we use the following definitions:

> The **doubling time** of an exponentially increasing population is the time required for the population to double.
>
> The **half-life** of an exponentially decaying quantity is the time required for the quantity to be reduced by a factor of one half.

Drug Buildup

Suppose that we want to model the amount of a certain drug in the body. Imagine that initially there is none, but that the quantity slowly starts to increase via a continuous intravenous injection. As the quantity of the drug in the body increases, so does the rate at which the body excretes the drug, so that eventually the quantity levels off at a saturation level, S. The graph of quantity against time will look something like Figure 1.22.

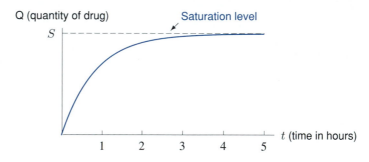

Figure 1.22: Buildup of drug in body

Suppose we want to construct a mathematical model of this situation; that is, suppose we want to find a formula giving the quantity, Q, in terms of time, t. Making a mathematical model often involves looking at a graph and deciding what kind of function has that shape. The graph in Figure 1.22 looks like an exponential decay function, but upside down. What actually decays is the difference between the saturation level, S, and the quantity, Q, in the blood. Suppose the difference between the saturation level and the quantity in the body is given by the formula

$$\text{Difference} = (\text{Initial difference}) \cdot (0.3)^t$$

with t in hours. Since the difference is $S - Q$, and the initial value of this difference is $S - 0 = S$, we have

$$S - Q = S \cdot (0.3)^t.$$

Solving for Q as a function of t gives

$$Q = S - S \cdot (0.3)^t$$
$$Q = f(t) = S \cdot \left(1 - (0.3)^t\right).$$

The graph of this function looks like exponential decay, but upside-down. Notice that the quantity, Q, starts at zero and increases toward S. Since the rate at which the quantity of the drug increases slows as the quantity approaches S, this graph is bending downward; hence it is increasing and *concave down*. It is an example of a function of the form

$$Q = S(1 - a^t) \quad \text{with} \quad 0 < a < 1.$$

Asymptotes

We say that the line representing the saturation level is a *horizontal asymptote*, because the graph gets arbitrarily close to it as time increases. As t gets larger, $(0.3)^t$ gets smaller, so Q gets closer to S. Using "$\rightarrow$" to mean "tends to," we can write $(0.3)^t \rightarrow 0$ as $t \rightarrow \infty$. Then

$$Q = S(1 - (0.3)^t) \rightarrow S(1 - 0) = S \quad \text{as} \quad t \rightarrow \infty,$$

so the graph of $Q = S(1 - (0.3)^t)$ has a horizontal asymptote at $Q = S$.

> If the graph of $y = f(x)$ approaches a horizontal line $y = L$ as $x \rightarrow \infty$ or $x \rightarrow -\infty$, then the line $y = L$ is called a **horizontal asymptote**.[6] This occurs when
>
> $$f(x) \rightarrow L \quad \text{as} \quad x \rightarrow \infty \qquad \text{or} \qquad f(x) \rightarrow L \quad \text{as} \quad x \rightarrow -\infty.$$
>
> If the graph of $y = f(x)$ approaches the vertical line $x = K$ as $x \rightarrow K$ from one side or the other, that is, if
>
> $$y \rightarrow \infty \quad \text{or} \quad y \rightarrow -\infty \quad \text{when} \quad x \rightarrow K,$$
>
> then the line $x = K$ is called a **vertical asymptote**.

As an example, the graph of $f(x) = 2 + 1/(x - 3)$ has a vertical asymptote at $x = 3$ (See Figure 1.23.)

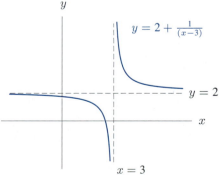

Figure 1.23: Graph with a horizontal and a vertical asymptote

---

[6]We are assuming that $f(x)$ gets arbitrarily close to L as $x \rightarrow \infty$.

The Family of Exponential Functions

The formula $P = P_0 a^t$ gives a family of exponential functions with parameters P_0 (the initial quantity) and a (the base, or growth/decay factor). The base tells us whether the function is increasing ($a > 1$) or decreasing ($0 < a < 1$). Since a is the factor by which P changes when t is increased by 1, large values of a mean fast growth; values of a near 0 mean fast decay. (See Figures 1.24 and 1.25.) All members of the family $P = P_0 a^t$ are concave up.

Alternative Formula for the Exponential Function

Exponential growth is often described in terms of growth rates in percent. For example, the population of Mexico is growing at 2.6% per year; in other words, the growth factor is $a = 1 + 0.026 = 1.026$. Similarly, each foot of clay removes 20% of the pollution from jet fuel, so the decay factor is $a = 1 - 0.20 = 0.8$. In general, the following formulas apply.

If r is the *growth* rate, then $a = 1 + r$, and

$$P = P_0 a^t = P_0(1 + r)^t.$$

If r is the *decay* rate, then $a = 1 - r$, and

$$P = P_0 a^t = P_0(1 - r)^t.$$

Note, for example, that $r = 0.05$ when the growth rate is 5%.

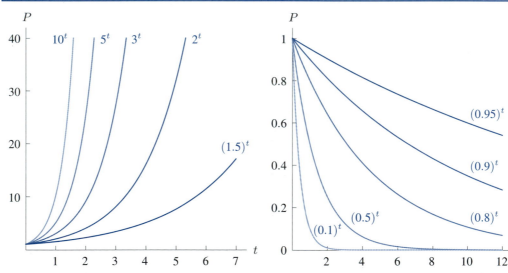

Figure 1.24: Exponential growth: $P = a^t$, for $a > 1$

Figure 1.25: Exponential decay: $P = a^t$, for $0 < a < 1$

Example 2   Suppose that $Q = f(t)$ is an exponential function of t. If $f(20) = 88.2$ and $f(23) = 91.4$:

    (a) Find the base.   (b) Find the growth rate.   (c) Evaluate $f(25)$.

Solution (a) Let
$$Q = Q_0 a^t.$$
Substituting $t = 20, Q = 88.2$ and $t = 23, Q = 91.4$ gives two equations for Q_0 and a:
$$88.2 = Q_0 a^{20} \quad \text{and} \quad 91.4 = Q_0 a^{23}.$$

Dividing the two equations enables us to eliminate Q_0:

$$\frac{91.4}{88.2} = \frac{Q_0 a^{23}}{Q_0 a^{20}} = a^3.$$

Solving for the base, a, gives

$$a = \left(\frac{91.4}{88.2}\right)^{1/3} \approx 1.012.$$

(b) Since $a \approx 1.012$, the growth rate is $r \approx 0.012 = 1.2\%$.

(c) We want to evaluate $f(25) = Q_0 a^{25} = Q_0(1.012)^{25}$. First we find Q_0 from the equation

$$88.2 = Q_0(1.012)^{20}.$$

Solving gives $Q_0 \approx 69.48$. Thus,

$$f(25) = 69.48(1.012)^{25} = 93.62.$$

Review: Definition and Properties of Exponents

Below we list the definitions and properties that are used to manipulate exponents.

Definition of Zero, Negative, and Fractional Exponents

$$a^0 = 1, \quad a^{-1} = \frac{1}{a}, \quad \text{and, in general, } a^{-x} = \frac{1}{a^x}$$

$$a^{1/2} = \sqrt{a}, \quad a^{1/3} = \sqrt[3]{a}, \quad \text{and, in general, } a^{1/n} = \sqrt[n]{a}.$$

Also, $a^{m/n} = \sqrt[n]{a^m} = (\sqrt[n]{a})^m$.

Properties of Exponents

1. $a^x \cdot a^t = a^{x+t}$ For example, $2^4 \cdot 2^3 = (2 \cdot 2 \cdot 2 \cdot 2) \cdot (2 \cdot 2 \cdot 2) = 2^7$.

2. $\dfrac{a^x}{a^t} = a^{x-t}$ For example, $\dfrac{2^4}{2^3} = \dfrac{2 \cdot 2 \cdot 2 \cdot 2}{2 \cdot 2 \cdot 2} = 2^1$.

3. $(a^x)^t = a^{xt}$ For example, $(2^3)^2 = 2^3 \cdot 2^3 = 2^6$.

Problems for Section 1.3

In Problems 1–4, decide whether each graph is concave up, concave down, or neither.

1.   2.   3.   4.

In Problems 5–6, explain whether the given function represents exponential growth or decay.

5. $p = 100(7)^t$

6. $p = 75(0.25)^t$

For Problems 7–9, say whether the data set appears to show exponential growth, exponential decay, or neither. If it does show exponential behavior, give a possible formula for the function.

7.   **TABLE 1.10**   8.   **TABLE 1.11**   9.   **TABLE 1.12**

t	$g(t)$
0	1
1	2
2	4
3	8
4	16

t	$f(t)$
2	9
3	18
4	81
5	243
6	486

t	$h(t)$
3	2096
4	1048
5	524
6	262
7	131

10.   If a yam is put in the oven, the temperature of the yam increases quickly at first, and then increases more and more slowly. Draw a graph of the temperature of the yam against time.

11.   Each year the world's annual consumption of electricity rises. In addition, each year the increase in annual consumption also rises. Sketch a possible graph of the annual world consumption of electricity as a function of time.

12.   A drug is injected into a patient's bloodstream over a five-minute interval. During this time, the quantity in the blood increases linearly. After five minutes the injection is discontinued, and the quantity then decays exponentially. Sketch a graph of the quantity versus time.

13.   The number of cancer cells in a tumor grows slowly at first but then grows with increasing rapidity. Draw a possible graph of the number of cancer cells against time.

Decide if each of the functions in Problems 14–17 has a horizontal asymptote. If there is an asymptote, give its equation.

14. $y = 5(0.8)^x$ 15. $P = 5(1 - (0.8)^t)$ 16. $Q = 2^t + 2^{-t}$ 17. $Z = 3 + (0.1)^t$

In Problems 18–19, assume that $f(t) = Q_0 a^t = Q_0(1 + r)^t$. Given two values of f:

(a)   Find the base, a.
(b)   Find the percentage growth rate, r.

18. $f(5) = 75.94$ and $f(7) = 170.86$
19. $f(0.02) = 25.02$ and $f(0.05) = 25.06$

Give a possible formula for each function graphed in Problems 20–23.

20.

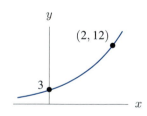

21.

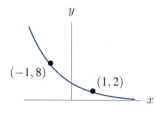

22.

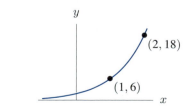

23.

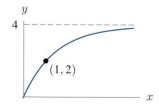

24. Each of the functions in Table 1.13 is increasing, but each increases in a different way. Which of the graphs in Figure 1.26 below best fits each function?

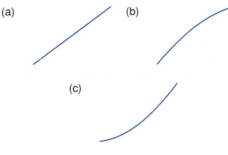

(a)

(b)

(c)

Figure 1.26

TABLE 1.13

t	$g(t)$	$h(t)$	$k(t)$
1	23	10	2.2
2	24	20	2.5
3	26	29	2.8
4	29	37	3.1
5	33	44	3.4
6	38	50	3.7

25. Match up the functions $h(s)$, $f(s)$, and $g(s)$, whose values are in Table 1.14, with the formulas

$$y = a(1.1)^s , \quad y = b(1.05)^s , \quad y = c(1.03)^s ,$$

assuming a, b, and c are constants. Note that the function values have been rounded to two decimal places.

TABLE 1.14

s	$h(s)$	s	$f(s)$	s	$g(s)$
2	1.06	1	2.20	3	3.47
3	1.09	2	2.42	4	3.65
4	1.13	3	2.66	5	3.83
5	1.16	4	2.93	6	4.02
6	1.19	5	3.22	7	4.22

26. A certain region has a population of 10,000,000 and an annual growth rate of 2%. Estimate the doubling time by guessing and checking.

27. (a) The half-life of radium-226 is 1620 years. Write a formula for the quantity, Q, of radium left after t years, if the initial quantity is Q_0.
     (b) What percentage of an original amount of radium is left after 500 years?

28. In the early 1960s, radioactive strontium-90 was released during atmospheric testing of nuclear weapons and got into the bones of people alive at the time. If the half-life of strontium-90 is 29 years, what fraction of the strontium-90 absorbed in 1960 remained in people's bones in 1990?

29. When the Olympic Games were held outside Mexico City in 1968, there was much discussion about the effect the high altitude (7340 feet) would have on the athletes. Assuming air pressure decays exponentially by 0.4% every 100 feet, by what percentage is air pressure reduced by moving from sea level to Mexico City?

30. (a) During 1996, the US population grew by 0.9% to 266.5 million. Assuming the population continues to grow at the same rate, what will it be at the start of the year 2000?
     (b) What was the annual percentage growth rate of the US population over the period from the start of 1990, when the population was 248.7 million, until the end of 1996?

31. During 1988, Nicaragua's inflation rate averaged 1.3% a day. This means that, on average, prices went up by 1.3% from one day to the next.

     (a) By what percentage did Nicaraguan prices increase in June of 1988?
     (b) What was Nicaragua's annual inflation rate during 1988?

32.   Assume that the median price, P, of a home rose from \$50,000 in 1970 to \$100,000 in 1990. Let t be the number of years since 1970.

(a)   Assume the increase in housing prices has been linear. Give an equation for the line representing price, P, in terms of t. Use this equation to complete column (a) of Table 1.15. Work with the price in units of \$1000.

(b)   If instead the housing prices have been rising exponentially, determine an equation of the form $P = P_0 a^t$ which would represent the change in housing prices from 1970–1990, and complete column (b) of Table 1.15.

(c)   On the same set of axes, sketch the functions represented in column (a) and column (b) of Table 1.15.

(d)   Which model for the price growth do you think is more realistic?

TABLE 1.15

t	(a) Linear growth price in \$1000 units	(b) Exponential growth price in \$1000 units
0	50	50
10		
20	100	100
30		
40		

1.4 POWER FUNCTIONS

A *power function* is a function in which the dependent variable is proportional to a power of the independent variable. For example, the area, A, of a square of side s is given by

$$A = f(s) = s^2.$$

The volume, V, of a sphere of radius r is given by

$$V = g(r) = \frac{4}{3}\pi r^3.$$

Both of these are *power functions*. Another example is the function which describes how the gravitational attraction of the earth varies with distance. If F is the force of gravitational attraction on a unit mass at a distance r from the earth, Newton's Inverse Square Law of Gravitation says that

$$F = \frac{k}{r^2} \quad \text{or} \quad F = kr^{-2}$$

where k is a positive constant.

> A **power function** has the form
>
> $$f(x) = kx^p$$
>
> where k and p are any constants.

In this section we compare various power functions with one another and with the exponential functions.

Positive Integral Powers: $y = x$, $y = x^2$, $y = x^3$, ...

First, we look at functions of the form $f(x) = x^n$, with n a positive integer. Figures 1.27 and 1.28 show that the graphs of these functions fall into two groups: the odd powers and the even powers. All the odd powers (x, x^3, x^5, and so on) are increasing everywhere and their graphs are symmetric about the origin. All odd powers above $n = 1$ have a "seat" at the origin. The even powers, on the other hand, are first decreasing and then increasing, making them $\bigcup$-shaped with symmetry about the y-axis. The even powers are concave up everywhere, whereas those with odd powers greater than 1 are concave down for negative x and concave up for positive x. For large x, the higher the power of x, the faster the function climbs. (See Figure 1.29.) For large values of x (in fact, for all $x > 1$), $y = x^5$ is above $y = x^4$, which is above $y = x^3$, and so on. Not only are the higher powers larger, but they are *much* larger. This is because if $x = 100$, for example, 100^5 is one hundred times as big as 100^4 which is one hundred times as big as 100^3. As x gets larger (written as $x \to \infty$), any positive power of x completely swamps all lower powers of x. We say that, as $x \to \infty$, higher powers of x *dominate* lower powers.

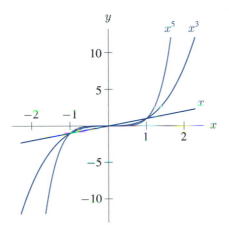

Figure 1.27: Odd powers of x: "Seat" shaped for $n > 1$

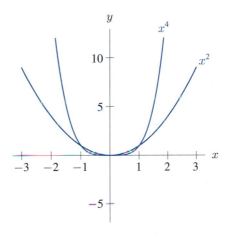

Figure 1.28: Even powers of x: $\bigcup$-shaped

As $x \to 0$, the story is entirely different. See Figure 1.30, which is a close-up view near the origin. For x between 0 and 1, x^3 is bigger than x^4, which is bigger than x^5. (Try $x = 0.1$ to confirm this.) For values of x near zero, smaller powers dominate.

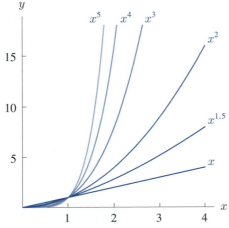

Figure 1.29: For large x: Large powers of x dominate

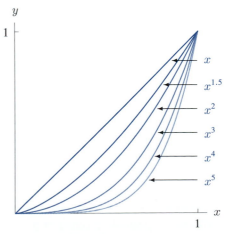

Figure 1.30: For $0 \le x \le 1$: Small powers of x dominate

Zero and Negative Integral Powers: $y = x^0, y = x^{-1}, y = x^{-2}, \ldots$

The function $y = x^0 = 1$ has a graph that is a horizontal line. For negative powers, rewriting

$$y = x^{-1} = \frac{1}{x} \quad \text{and} \quad y = x^{-2} = \frac{1}{x^2}$$

makes it clear that as $x > 0$ increases, the denominators increase and the functions decrease. The graphs of $y = x^{-1}$ and $y = x^{-2}$ have both the x- and y-axes as asymptotes. (See Figure 1.31.)

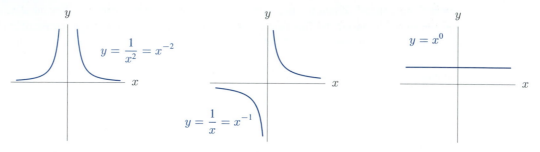

Figure 1.31: Comparison of zero and negative powers of x

Example 1   Plot a graph of pressure against volume for a fixed quantity of gas at a constant temperature. Use the fact that pressure is inversely proportional to volume.

Solution   Think of a fixed quantity of air—for example, inside a cylinder in a car engine. If the volume, V, of the air is decreased (by moving the pistons), the pressure, P, of the air is increased. Conversely, if the volume is increased, the pressure will decrease. For an ideal gas, Boyle's law gives the exact relationship between pressure and volume, provided the temperature is constant. It says P is inversely proportional to V, so

$$P = \frac{k}{V} = kV^{-1}, \qquad \text{where } k \text{ is a positive constant.}$$

Both axes are asymptotes to the graph in Figure 1.32, because as the volume tends to infinity, the pressure tends to zero, and vice versa. The shape is known as a *hyperbola*. The power function $P = k/V = kV^{-1}$ differs from exponential decay in that it is undefined for $V = 0$, so this graph does not cross the vertical axis. In addition, it approaches the horizontal axis more slowly than the exponential function.

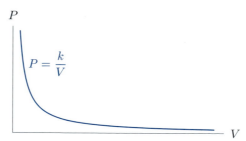

Figure 1.32: Graph of pressure, P, against volume, V, for Boyle's law

---

Example 2   Quantum mechanics predicts that the force between two gas molecules has two components: an attractive force which is proportional to r^{-7} (where r is the distance between the molecules) and a repulsive force which is proportional to r^{-13}. How does the net force vary with r?

Solution   A repulsive force is usually considered to be positive, whereas an attractive force is negative. So we have

$$F = -ar^{-7} + br^{-13} = -\frac{a}{r^7} + \frac{b}{r^{13}}, \quad \text{where } a, b \text{ are positive constants.}$$

If r is very small, $1/r^{13}$ is much larger than $1/r^7$, so

$$F \approx \frac{b}{r^{13}}$$

and the net force is repulsive. (This occurs when the molecules are so close together that the protons in the nuclei repel one another.) For very big r, $1/r^7$ is much larger than $1/r^{13}$, and so

$$F \approx -\frac{a}{r^7},$$

making the net force attractive. As $r \to \infty$, all the forces die away to zero.

Positive Fractional Powers: $y = x^{1/2}, y = x^{1/3}, y = x^{3/2}, \ldots$

The function giving the side of a square, s, in terms of its area, A, involves a root, or fractional power:

$$s = \sqrt{A} = A^{1/2}.$$

Similarly, the equation relating the average number of species on an island to its area, A, is approximately

$$N = k\sqrt[3]{A} = kA^{1/3},$$

where k is a constant depending on the region of the world in which the island is found.[7]

We now look at functions of the form $y = x^{m/n} = \sqrt[n]{x^m}$. Since some fractional powers such as $x^{1/2}$ involve roots and are defined only for positive x and 0, we restrict the domain of these functions to $x \geq 0$.

Figure 1.33 shows that for large x (in fact, all $x > 1$), the graph of $y = x^{1/2}$ is below the graph of $y = x$, and $y = x^{1/3}$ is below $y = x^{1/2}$. This is reasonable since, for $x > 1$, squaring x makes it bigger and cubing it makes it bigger still; thus taking the square root of x makes it smaller and taking its cube root makes it smaller still. Between $x = 0$ and $x = 1$, the situation is reversed and $y = x^{1/3}$ is on top. (Why?) Not surprisingly, $y = x^{3/2}$ is between $y = x$ and $y = x^2$ for all x.

The other important feature to notice about the graphs of $y = x^{1/2}$ and $y = x^{1/3}$ is that they bend in a direction opposite to that of the graphs of $y = x^2$ and x^3. For example, the graph of $y = x^2$ is concave up. On the other hand, the graphs of $y = x^{1/2}$ and $y = x^{1/3}$ are concave down. Despite this, all these functions do become infinitely large as x increases. Notice also that all these functions pass through the point $(1, 1)$.

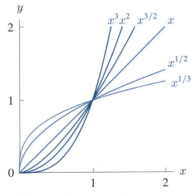

Figure 1.33: Comparison of some fractional powers of x

[7]*Scientific American*, September 1989, p. 112.

What Effect Do Coefficients Have?

We know that $x^2 < x^3$ for all $x > 1$. But which is larger, $50x^2$ or x^3? For moderate values of x, like $x = 10$, we have $50x^2 > x^3$. But eventually, $50x^2 < x^3$. In fact, $50x^2 < x^3$ for all $x > 50$. The graphs of $y = x^2$ and $y = x^3$ cross at $x = 1$, whereas the graphs of $y = 50x^2$ and $y = x^3$ cross at $x = 50$. (See Figure 1.34.) Thus, the effect of the factor of 50 is to change the point at which the graphs cross. However, x^3 ends up on top in both cases; provided the coefficients are positive, as $x \to \infty$, the higher power is always larger eventually.

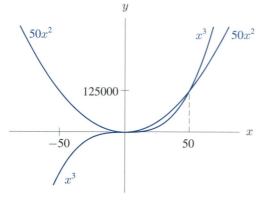

Figure 1.34: Graph of $y = x^3$ lies above graph of $y = 50x^2$ for large positive x

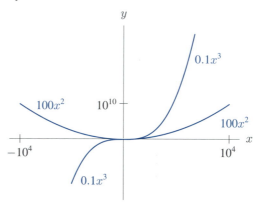

Figure 1.35: For large positive x, $y = 0.1x^3$ dominates $y = 100x^2$

Example 3   Which of $y = 100x^2$ and $y = 0.1x^3$ is larger as $x \to \infty$?

Solution   Since $x \to \infty$, we are looking at large positive values of x, where the higher power will eventually be larger, or dominate. This suggests that $y = 0.1x^3$ will be larger. In fact, $0.1x^3 > 100x^2$ when $x > 1000$. (See Figure 1.35.)

Exponentials and Power Functions: Which Dominate?

In everyday language, the word exponential is often used to imply very fast growth. But do exponential functions always grow faster than power functions? To determine what happens "in the long run," we often want to know which functions dominate as $x \to \infty$.

Let's consider $y = 2^x$ and $y = x^3$. The close-up view in Figure 1.36(a) shows that between $x = 2$ and $x = 4$, the graph of $y = 2^x$ lies below the graph of $y = x^3$. But the far-away view in Figure 1.36(b) shows that the exponential function $y = 2^x$ eventually overtakes $y = x^3$. And Figure 1.36(c), which gives a very far–away view, shows that, for large x, x^3 is insignificant compared to 2^x. Indeed, 2^x is growing so much faster than x^3 that the graph of 2^x appears almost vertical in comparison to the more leisurely climb of x^3.

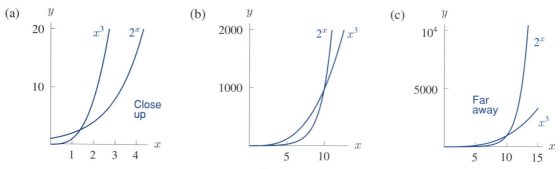

Figure 1.36: Comparison of $y = 2^x$ and $y = x^3$: Notice that $y = 2^x$ eventually dominates $y = x^3$

In fact, *every* exponential growth function eventually dominates *every* power function. Although an exponential function may be below a power function for some values of x, if we look at large enough x-values, a^x (with $a > 1$) will eventually dominate x^n, no matter what n is. Two more examples are presented in Figure 1.37 and Table 1.16.

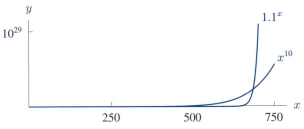

Figure 1.37: Exponential function eventually dominates power function

TABLE 1.16   *Comparison of* x^{100} *and* 1.01^x

x	x^{100}	1.01^x
10^4	10^{400}	$1.6 \cdot 10^{43}$
10^5	10^{500}	$1.4 \cdot 10^{432}$
10^6	10^{600}	$2.4 \cdot 10^{4321}$

Can you guess what happens in the case of negative powers and negative exponents? For example, consider $y = 2^{-x}$ and $y = x^{-2}$. Since $y = 2^{-x} = 1/2^x$ and $y = x^{-2} = 1/x^2$, knowing that 2^x is eventually larger than x^2 tells you that $y = 2^{-x}$ is eventually smaller than $y = x^{-2}$. Hence $y = 2^{-x}$ is eventually below $y = x^{-2}$. (See Figure 1.38.) This behavior is also typical: Every exponential decay function will eventually approach 0 faster than every power function with a negative exponent.

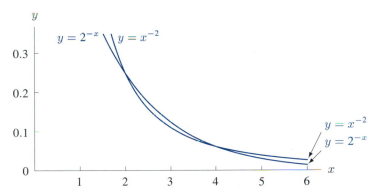

Figure 1.38: Comparison of $y = 2^{-x}$ and $y = x^{-2}$: Exponential dies away faster in the long run

Problems for Section 1.4

For Problems 1–8, simplify the given expression.

1. $8^{4/3}$
2. $27^{1/3}$
3. $10{,}000^{5/4}$
4. $16^{1/4}$
5. $27^{-1/3}$
6. $16^{-0.25}$
7. $8^{-2/3}$
8. $4^{-3/2}$

For Problems 9–11, determine which function dominates as $x \to \infty$.

9. $10 \cdot 2^x$ or $72{,}000x^{12}$
10. $0.1x^2$ or $10^{10}x$
11. $0.25\sqrt{x}$ or $25{,}000x^{-3}$

For Problems 12–14, determine what happens to the value of the function as $x \to \infty$ and as $x \to -\infty$.

12. $y = x^6$
13. $y = 0.25x^3 + 3$
14. $y = 2 \cdot 10^{4x}$

15. On a graphing calculator or computer, plot graphs of the following functions, first for $-5 \le x \le 5$, $-100 \le y \le 100$, and then for $-1.2 \le x \le 1.2$, $-2 \le y \le 2$.
    (a) $y = x, y = x^3, y = x^6, y = x^9$

(b) $y = x, y = x^4, y = x^7, y = x^{10}$

Observe the general shape of these functions: Do the odd powers have the same general shape? What about the even powers? Which function is largest in magnitude for big x? For x near 0? Is this what you expected?

16. (a) Use a graphing calculator or a computer to plot the graphs of $y = x^3$, $y = x^4$, and $y = x^5$ on the interval $-0.1 \leq x \leq 0.1$. Determine an appropriate range for y so that all powers will be distinguishable in the viewing rectangle.

  (b) Plot the same graphs for $-100 \leq x \leq 100$, and determine an appropriate range for y.

17. By hand, sketch global graphs of $f(x) = x^3$ and $g(x) = 20x^2$ on the same axes. Which function has larger values as $x \to \infty$?

18. By hand, sketch graphs of $f(x) = x^5$, $g(x) = -x^3$, and $h(x) = 5x^2$ on the same axes. Which has the largest positive values as $x \to \infty$? As $x \to -\infty$?

19. In physiology, the DuBois formula relates a person's surface area, s, in m², to weight w, in kg, and height h, in cm, by

$$s = 0.01w^{0.25}h^{0.75}.$$

  (a) What is the surface area of a person who weighs 65 kg and is 160 cm tall?
  (b) What is the weight of a person whose height is 180 cm and who has a surface area of 1.5 m²?
  (c) For people of fixed weight 70 kg, solve for h as a function of s. Simplify your answer.

20. Suppose a force between two atoms a distance r apart is given by

$$F = \frac{A}{r^3} - \frac{B}{r^2} \quad \text{where } A, B \text{ are positive constants.}$$

  (a) For small positive r, is F positive or negative?
  (b) For large positive r, is F positive or negative?
  (c) For what, if any, value(s) of r is F zero?
  (d) What are the vertical and horizontal asymptotes of F?

21. According to the April 1991 issue of *Car and Driver*, an Alfa Romeo going at 70 mph requires 177 feet to stop. Assuming that the stopping distance is proportional to the square of velocity, find the stopping distances required by an Alfa Romeo going at 35 mph and at 140 mph (its top speed).

22. Poiseuille's law gives the rate of flow, R, of a gas through a cylindrical pipe in terms of the radius of the pipe, r, for a fixed drop in pressure. Assume a constant drop in pressure throughout the remainder of this problem.

  (a) Determine a formula for Poiseuille's Law, given that the rate of flow is proportional to the fourth power of the radius.
  (b) If $R = 400$ cm³/sec in a pipe of radius 3 cm for a certain gas, determine an explicit formula for the rate of flow of that gas through a pipe of radius r cm.
  (c) What is the rate of flow of the same gas through a pipe with a 5 cm radius?

23. Values of three functions are given in Table 1.17, rounded to two decimal places. One function is of the form $y = ab^t$, one is of the form $y = ct^2$, and one is of the form $y = kt^3$. Which function is which?

TABLE 1.17

t	$f(t)$	t	$g(t)$	t	$h(t)$
2.0	4.40	1.0	3.00	0.0	2.04
2.2	5.32	1.2	5.18	1.0	3.06
2.4	6.34	1.4	8.23	2.0	4.59
2.6	7.44	1.6	12.29	3.0	6.89
2.8	8.62	1.8	17.50	4.0	10.33
3.0	9.90	2.0	24.00	5.0	15.49

TABLE 1.18

x	$f(x)$	x	$g(x)$	x	$k(x)$
8.4	5.93	5.0	3.12	0.6	3.24
9.0	7.29	5.5	3.74	1.0	9.01
9.6	8.85	6.0	4.49	1.4	17.66
10.2	10.61	6.5	5.39	1.8	29.19
10.8	12.60	7.0	6.47	2.2	43.61
11.4	14.82	7.5	7.76	2.6	60.91

24. Values of three functions are contained in Table 1.18. (The numbers have been rounded to two decimal places.) Two are power functions and one is an exponential. One of the power functions is a quadratic and one a cubic. Which one is exponential? Which one is quadratic? Which one is cubic?

25.   Use a graphing calculator or a computer to graph $y = x^4$ and $y = 3^x$. Determine approximate domains and ranges that give each of the graphs in Figure 1.39.

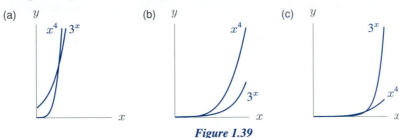

Figure 1.39

1.5 NOTES ON INVERSE FUNCTIONS

From Distance to Time and Back

On August 18, 1989, the runner Arturo Barrios of Mexico set a world record for 10,000-meters. His times, in seconds, at 2000-meter intervals are recorded in Table 1.19, where $f(d)$ is the number of seconds Barrios took to complete the first d meters of the race. For example, Barrios ran the first 4000 meters in 650.1 seconds, so $f(4000) = 650.1$. The function f is useful to athletes planning to compete with Barrios.

Let us now change our point of view and ask for distances rather than times. If we ask how far Barrios ran during the first 650.1 seconds of his race, the answer is clearly 4000 meters. Going backwards in this way from numbers of seconds to numbers of meters gives a function called the *inverse function* of f, denoted by f^{-1}. Thus, $f^{-1}(t)$ is the number of meters that Barrios ran during the first t seconds of his race. To find values of f^{-1}, we can either read Table 1.19 backwards, or use Table 1.20 which contains values of f^{-1}.

The two functions f and f^{-1} convey the same information, but they express it differently. For example, the fact that Barrios ran the first 6000 meters in 975.5 seconds can be written with either f or f^{-1}:

$$f(6000) = 975.5 \quad \text{or} \quad f^{-1}(975.5) = 6000.$$

The independent variable for f is the dependent variable for f^{-1}, and vice versa. The domains and ranges of f and f^{-1} are also interchanged. The domain of f is all distances d such that $0 \le d \le 10000$, which is the range of f^{-1}. The range of f is all times t, such that $0 \le t \le 1628.23$, which is the domain of f^{-1}.

TABLE 1.19 *Barrios's running time*

d (meters)	$f(d)$ (seconds)
0	0.00
2000	325.90
4000	650.10
6000	975.50
8000	1307.00
10000	1628.23

TABLE 1.20 *Distance run by Barrios*

t (seconds)	$f^{-1}(t)$ (meters)
0.00	0
325.90	2000
650.10	4000
975.50	6000
1307.00	8000
1628.23	10000

Which Functions Have Inverses?

If a function has an inverse, we say it is *invertible*. Let's first look at a function which is not invertible.

Consider the flight of the Mercury spacecraft *Freedom 7*, which carried Alan Shepard, Jr. into space in May 1961. Shepard was the first American to journey into space. After launch, his spacecraft rose to an altitude of 116 miles, and then came down into the sea. The function $f(t)$ giving the altitude in miles t minutes after lift-off does not have an inverse. To see why not, try to decide on a value for $f^{-1}(100)$. Clearly, $f^{-1}(100)$ should be the time when the altitude of the spacecraft was 100 miles.

However, there are two such times, one when the spacecraft was ascending and one when it was descending. (See Figure 1.40.)

The reason the altitude function does not have an inverse is that the altitude first increases and then decreases. The reason the Barrios time function did have an inverse is that it increased everywhere. Each running time, t, corresponds to some unique distance, d.

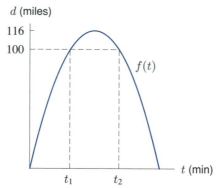

Figure 1.40: Two times, t_1 and t_2, at which altitude of spacecraft is 100 miles

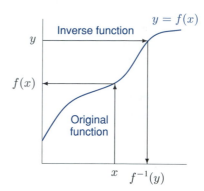

Figure 1.41: An increasing function has an inverse

A function does not have to be increasing everywhere to have an inverse. Figure 1.41 suggests when an inverse will exist. The original function, f, takes us from an x-value to a y-value, as shown in Figure 1.41. Since having an inverse means there is a function going from a y-value to an x-value, the crucial question is whether we can get back. In other words, does each y-value correspond to a unique x-value? If so, there's an inverse; if not, there isn't. This principle may be stated geometrically, as follows:

A function has an inverse if (and only if) its graph intersects any horizontal line at most once.

For example, the function $f(x) = x^2$ does not have an inverse because many horizontal lines intersect the parabola twice.

Definition of an Inverse Function

If the function f is invertible, its inverse is defined as follows:

$$f^{-1}(t) = d \quad \text{means} \quad f(d) = t.$$

The notation f^{-1} for an inverse function is perhaps misleading, as it is easy to confuse with a reciprocal, which the inverse function is not. However, there's no changing such well-established notation!

Example 1   The temperature at which water boils decreases as altitude increases, a fact important to cooks. Let $f(h)$ be the boiling point of water (in °C) at an altitude of h meters above sea level during standard atmospheric conditions. What is the meaning in practical terms of $f^{-1}(90)$ and of $f^{-1}(90) = 3000$? Evaluate $f^{-1}(100)$.

Solution   The function f goes from altitude to temperature, so f^{-1} goes from temperature to altitude. The expression $f^{-1}(90)$ represents the altitude in meters at which the boiling point of water is 90°C. The equation $f^{-1}(90) = 3000$ means that the boiling point of water is 90°C at an altitude of 3000 meters. The equation $f(3000) = 90$ has the same meaning. Since the boiling point of water is 100°C at sea level (where the altitude is 0 meters), we must have $f^{-1}(100) = 0$.

Formulas for Inverse Functions

If a function is defined by a formula, it is sometimes possible to find a formula for the inverse function as well. In Section 1.1, we looked at the snow tree cricket, whose chirp rate, C, in chirps per minute, is approximated by the formula

$$C = f(T) = 4T - 160,$$

where T is the temperature in degrees Fahrenheit. So far we have used this formula to predict the chirp rate from the temperature. But it is also possible to use this formula backwards to calculate the temperature from the chirp rate.

Example 2   Find the formula for the function giving temperature in terms of the number of cricket chirps per minute; that is, find the inverse function f^{-1} such that

$$T = f^{-1}(C).$$

Solution   Since C is an increasing function, f is invertible. We know $C = 4T - 160$. We solve for T, giving

$$T = \frac{C}{4} + 40,$$

so

$$f^{-1}(C) = \frac{C}{4} + 40.$$

Graphs of Inverse Functions

The function $f(x) = x^3$ is increasing everywhere and so has an inverse. To find the inverse, we solve

$$y = x^3$$

for x, giving

$$x = y^{1/3}.$$

The inverse function is

$$f^{-1}(y) = y^{1/3}$$

or, if we want to call the variable x,

$$f^{-1}(x) = x^{1/3}.$$

The graphs of $y = x^3$ and $y = x^{1/3}$ are shown in Figure 1.42. Notice that these graphs are the reflections of one another about the line $y = x$. For example, $(8, 2)$ is on the graph of $y = x^{1/3}$ because $2 = 8^{1/3}$, and $(2, 8)$ is on the graph of $y = x^3$ because $8 = 2^3$. The points $(8, 2)$ and $(2, 8)$ are reflections of one another about the line $y = x$. In general, if the x- and y-axes have the same scales:

> The graph of f^{-1} is the reflection of the graph of f about the line $y = x$.

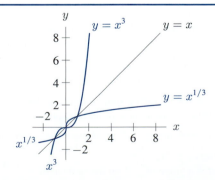

Figure 1.42: Graphs of inverse functions, $y = x^3$ and $y = x^{1/3}$, are reflections about the line $y = x$

Problems for Section 1.5

1. Let p be the price of an item and q be the number of items sold at that price. Assume $q = f(p)$. Explain what the following quantities mean in terms of prices and quantities sold.

    (a) $f(25)$ (b) $f^{-1}(30)$

2. Suppose $C = f(A)$, where C is the cost in dollars, of building a store of area A square feet. Explain in terms of cost and square feet what the following quantities represent.

    (a) $f(10{,}000)$ (b) $f^{-1}(20{,}000)$

3. Let $f(x)$ equal the temperature (°F) when the column of mercury in a particular thermometer is x inches long. What is the meaning of $f^{-1}(75)$ in practical terms?

For Problems 4–7, decide whether the function f is invertible.

4. $f(t)$ equals the number of customers in Macy's department store at t minutes past noon on December 18, 1993.

5. $f(n)$ is the number of students in your calculus class whose birthday is on the nth day of the year.

6. $f(x)$ is the volume in liters of x kilograms of water at 4°C.

7. $f(w)$ is the cost in cents of mailing a letter that weighs w grams.

For Problems 8–11, determine whether the functions $y = f(x)$ are invertible.

8.   9. 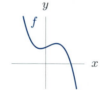   10.   11.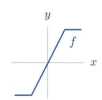

12. Graph each function below, and decide if it has an inverse. If so, approximate the inverse at $x = 20$.

    (a) $f(x) = x^2 + 3^x$ (b) $g(x) = x^3 + 3^x$

13. The cost of producing q articles is given by the function $C = f(q) = 100 + 2q$.

    (a) Find a formula for the inverse function.
    (b) Explain in practical terms what the inverse function tells you.

14. A kilogram weighs about 2.2 pounds.

    (a) Write a formula for the function, f, which gives an object's mass in kilograms, k, as a function of its weight in pounds, p.
    (b) Find a formula for the inverse function of f. What does this inverse function tell you, in practical terms?

15. A function $y = t(x)$ is always increasing, has a domain of all real numbers, and has a range of $0 < t(x) < 4$. Sketch a possible graph of $t^{-1}(x)$.

16. Determine if the following statement is true or false, and explain your answer: If f is an increasing function, then f^{-1} is an increasing function.

17. Consider the function $y = f(x)$ whose graph is given in Figure 1.43.

    (a) Estimate $f^{-1}(2)$.
    (b) Sketch a graph of f^{-1} on the same axes.

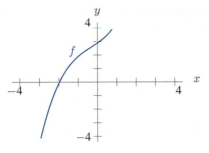

Figure 1.43

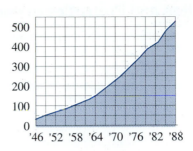

Figure 1.44

18. Figure 1.44 is the graph[8] of the function f, where $f(t)$ is the number (in millions) of motor vehicles registered in the world in the year t. (In 1988, one-third of the registered vehicles in the world were in the United States.)

(a) Is f invertible? Explain.
(b) What is the meaning of $f^{-1}(400)$ in practical terms? Evaluate $f^{-1}(400)$.
(c) Sketch the graph of f^{-1}.

1.6 LOGARITHMS

In Section 1.3, we set up a function approximating the population of Mexico (in millions) as

$$P = f(t) = 67.38(1.026)^t,$$

where t is the number of years since 1980. Writing the function this way shows that we are thinking of the population as a function of time, that the population was 67.38 million in 1980, and that it grows by 2.6% every year.

Now suppose that instead of calculating the population, we want to find when the population is expected to reach 200 million. This means we want to find the value of t for which

$$200 = f(t) = 67.38(1.026)^t.$$

Since the exponential function is always increasing and is eventually more than 200, there's exactly one value of t making $P = 200$. How should we find it? A reasonable way to start is trial and error. Taking $t = 40$ and $t = 50$ we get

$$P = f(40) = 67.38(1.026)^{40} = 188.115\ldots \quad \text{(so } t = 40 \text{ is too small),}$$
$$P = f(50) = 67.38(1.026)^{50} = 243.163\ldots \quad \text{(so } t = 50 \text{ is too large).}$$

Some more experimenting leads to

$$P = f(42) = 67.38(1.026)^{42} \approx 198.0$$
$$P = f(43) = 67.38(1.026)^{43} \approx 203.2$$

so t is between 42 and 43. In other words, the population is projected to reach 200 million sometime during the year 2022.

Although it is always possible to approximate t by trial and error like this, it would clearly be better to have a formula that gives t in terms of P. The *logarithm* function will enable us to do this.

---

[8]From D. Blerics and P. Walzer, "Energy for Motor Vehicles," *Scientific American*, September 1990.

Definition and Properties of Logs to Base 10

We define the *logarithm* function, $\log_{10} x$, to be the inverse of the exponential function, 10^x. We call 10 the *base*. So

> The **logarithm** to base 10 of x, written **log**$_{10}$ x, is the power of 10 we need to get x. In other words,
>
> $$\log_{10} x = c \quad \text{means} \quad 10^c = x,$$
>
> We often write $\log x$ in place of $\log_{10} x$.

For example, $\log 1000 = \log 10^3 = 3$ since 3 is the power of 10 needed to get 1000. Similarly, $\log(0.1) = -1$ because $0.1 = 1/10 = 10^{-1}$. However, if we try to find $\log(-7)$ on a calculator, we get an error message because 10 to any power is never negative (or 0). In general,

> $\log x$ is not defined if x is negative or 0.

In working with logarithms, we need the following properties:

> ### Properties of Logarithms
> 1. $\log(AB) = \log A + \log B$
> 2. $\log\left(\frac{A}{B}\right) = \log A - \log B$
> 3. $\log(A^p) = p \log A$
> 4. $\log(10^x) = x$
> 5. $10^{\log x} = x$
>
> In addition, $\log 1 = 0$ because $10^0 = 1$.

The fact that $\log x$ is the power of 10 giving x justifies rule 5.

Solving Equations Using Logarithms

Logs are frequently useful when we have to solve for unknown exponents, as in the next examples.

Example 1   Find t such that $2^t = 7$.

Solution   First, notice that we expect t to be between 2 and 3 (because $2^2 = 4$ and $2^3 = 8$). To calculate t, take logs to base 10:

$$\log(2^t) = \log 7.$$

Then use the third property of logs, which says $\log(2^t) = t \log 2$, and get:

$$t \log 2 = \log 7.$$

Using a calculator to find the logs gives

$$t = \frac{\log 7}{\log 2} \approx 2.81.$$

Example 2   We return to the question of when the population of Mexico reaches 200 million. To get an answer, solve $200 = 67.38(1.026)^t$ for t, using logs.

Solution   Dividing both sides of the equation by 67.38, we get

$$\frac{200}{67.38} = (1.026)^t.$$

Now take logs of both sides:

$$\log\left(\frac{200}{67.38}\right) = \log(1.026^t).$$

Using the fact that $\log(A^t) = t \log A$, we get

$$\log\left(\frac{200}{67.38}\right) = t \log(1.026).$$

Solving this equation using a calculator to find the logs, we get

$$t = \frac{\log(200/63.78)}{\log(1.026)} \approx 42.4 \text{ years}$$

which is between $t = 42$ and $t = 43$, as we found at the beginning of this section. This value of t corresponds to the year 2022.

Example 3   Give a formula for the inverse of the following function (that is, solve for t in terms of P):

$$P = f(t) = 67.38(1.026)^t.$$

Solution   We want a formula expressing t as a function of P. Take logs:

$$\log P = \log(67.38(1.026)^t).$$

Since $\log(AB) = \log A + \log B$, we have

$$\log P = \log 67.38 + \log((1.026)^t).$$

Now use $\log(A^t) = t \log A$:

$$\log P = \log 67.38 + t \log 1.026.$$

Solve for t in two steps, using a calculator at the final stage:

$$t \log 1.026 = \log P - \log 67.38$$

$$t = \frac{\log P}{\log 1.026} - \frac{\log 67.38}{\log 1.026} \approx 89.7 \log P - 164.0.$$

Thus,

$$f^{-1}(P) = 89.7 \log P - 164.0.$$

Note that

$$f^{-1}(200) = 89.7(\log 200) - 164.0 \approx (89.7)(2.301) - 164.0 \approx 42.4,$$

which agrees with the result of Example 2.

Example 4   Find the half-life of the decaying exponential $P = P_0(0.8)^x$ that we used to model the removal of pollutants in jet fuel on page 16. What does your answer mean in practical terms?

Solution   We seek the value of x such that

$$P = \frac{1}{2}P_0,$$

so we must solve the equation

$$P_0(0.8)^x = \frac{1}{2}P_0.$$

Dividing both sides by P_0 leaves

$$(0.8)^x = \frac{1}{2}.$$

Taking logs of both sides gives

$$\log(0.8)^x = \log\left(\frac{1}{2}\right)$$

$$x(\log 0.8) = \log\left(\frac{1}{2}\right)$$

$$x = \frac{\log(1/2)}{\log 0.8}$$

$$x \approx 3.1 \text{ feet.}$$

In practical terms, this tells us that forcing kerosene through 3.1 feet of clay pipe removes half of its impurities.

The Graph of $\log x$

Example 5   Graph $f(x) = \log x$, and compare it to the graph of $g(x) = 10^x$.

Solution   The values of $\log x$ in Table 1.21 were found using a calculator. Because no power of 10 gives 0, $\log 0$ is undefined. See Figure 1.45. The domain of $y = \log x$ is positive x-values; the range is all y values. In contrast, $y = 10^x$ has domain all x-values and range all positive y-values. The graph of $y = \log x$ has a vertical asymptote at $x = 0$, whereas $y = 10^x$ has a horizontal asymptote at $x = 0$.

One of the big differences between $g(x) = 10^x$ and $f(x) = \log x$ is that the exponential function grows extremely quickly whereas the log function grows extremely slowly. However, the log function does go to infinity, albeit slowly, as x increases.

TABLE 1.21   *Values for $\log x$ and 10^x*

x	$\log x$	x	10^x
0	undefined	0	1
1	0	1	10
2	0.3	2	100
3	0.5	3	10^3
4	0.6	4	10^4
$\vdots$	$\vdots$	$\vdots$	$\vdots$
10	1	10	10^{10}

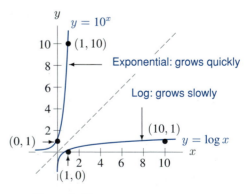

Figure 1.45: Graphs of $\log x$ and 10^x

Since $f(x) = \log x$ and $g(x) = 10^x$ are inverse functions, the graphs of the two functions are reflections of one another about the line $y = x$, provided the scales along the x- and y-axes are equal. (See Figure 1.45.)

Comparison between Logarithms and Power Functions: $A \log x$ and x^p

Although there is a superficial resemblance between the graphs of $y = x^{1/3}$ and $y = \log x$, shown in Figure 1.46, there are important differences. The graph of $x^{1/3}$ includes the origin $(0, 0)$, whereas the graph of $\log x$ never reaches the vertical axis $x = 0$, which is an asymptote. Another difference is the rate of growth of the two functions for large x. The fact that $\log(1{,}000{,}000) = 6$ and $\log(10{,}000{,}000) = 7$ shows how slowly the logarithm climbs when x is large. Indeed, the logarithm climbs more slowly than any positive power of x.

Figure 1.47 compares $x^{1/3}$ and $100 \log x$; Table 1.22 compares $x^{0.001}$ and $1000 \log x$. In both cases, the logarithmic function is eventually less than the power function. In fact, x^p dominates $A \log x$ for large x no matter what the values of $p > 0$ and $A > 0$. This follows from the fact that 10^x dominates Bx^q for large x for any $q > 0$ and $B > 0$.

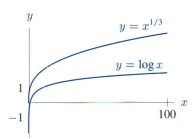

Figure 1.46: Comparison of $y = \log x$ and $y = x^{1/3}$

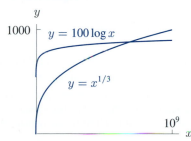

Figure 1.47: Comparison of $y = x^{1/3}$ and $y = 100 \log x$

TABLE 1.22 *Comparison of $x^{0.001}$ and $1000 \log x$*

x	$x^{0.001}$	$1000 \log x$
10^{5000}	10^5	$5 \cdot 10^6$
10^{6000}	10^6	$6 \cdot 10^6$
10^{7000}	10^7	$7 \cdot 10^6$

Problems for Section 1.6

1. Construct a table of values to compare the values of $f(x) = \log x$ and $g(x) = \sqrt{x}$ for $x = 1, 2, \ldots, 10$. Round to two decimal places. Use these values to graph both functions.

For Problems 2–8, solve for x using logs. (You can check your answer using a graphing calculator or computer.)

2. $3^x = 11$
3. $17^x = 2$
4. $10 = 4^x$
5. $20 = 50(1.04)^x$
6. $25 = 2(5)^x$
7. $2 \cdot 5^x = 11 \cdot 7^x$
8. $4 \cdot 3^x = 7 \cdot 5^x$

In Problems 9–12, solve for t.

9. $a = b^t$
10. $P = P_0 \, a^t$
11. $Q = Q_0 \, a^{nt}$
12. $P_0 \, a^t = Q_0 \, b^t$

For Problems 13–17, simplify the expression as much as possible.

13. $\log A^3 - \log B^{2/3} + \log A^{1/3} + \log B^{5/3}$

14. $\dfrac{\log(ABC)}{\log\left(\frac{1}{ABC}\right)}$

15. $\dfrac{\log A^2 - 2\log B}{\log A^2 + \log\left(\frac{1}{AB}\right)}$

16. $10^{\log(AB)}$

17. $100^{(\log A - \log B)}$

Plot a graph of the functions in Problems 18–19 on a calculator or computer. Explain what you see.

18. $y = \log(10^x)$

19. $y = 10^{\log x}$

For Problems 20–21, determine which function has larger values as $x \to \infty$.

20. $x^{1/3}$ or $75{,}000 \log x$

21. $\log x$ or $10^{-0.01x}$

22.   Find the equation of the line l in Figure 1.48.

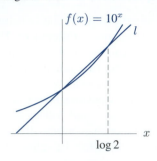

Figure 1.48

23.   What is the doubling time of prices which are increasing by 5% a year?

24.   The population of a region is growing exponentially. If there were 40,000,000 people in 1980 ($t = 0$) and 56,000,000 in 1990, find an expression for the population at any time t, in years. What would you predict for the year 2000? What is the doubling time?

25.   In 1980, there were about 170 million vehicles (cars and trucks) and about 227 million people in the United States. If the number of vehicles has been growing at 4% a year, while the population has been growing at 1% a year, in what year will there be, on average, one vehicle per person?

26.   Owing to an innovative rural public health program, infant mortality in Senegal, West Africa, is being reduced at a rate of 10% per year. How long will it take for infant mortality to be reduced by 50%?

27.   The half-life of a certain radioactive substance is 12 days. If there are 10.32 grams initially,

   (a)   Write an equation to determine the amount, A, of the substance as a function of time.
   (b)   When will the substance be reduced to 1 gram?

28.   (a)   Find the doubling time, D, for annual growth rates, $i\%$, of 2%, 3%, 4%, and 5%.
   (b)   Since D decreases as i increases, we might guess that D is inversely proportional to i, that is, $D = k/i$. Use your answers to part (a) to confirm that, approximately, $D = 70/i$. This is the "Rule of 70" used by bankers. To estimate the doubling time of an investment, the banker divides 70 by the yearly interest rate.

29.   Find the inverse function of $p(t) = (1.04)^t$.

1.7 THE NUMBER e AND NATURAL LOGARITHMS

The calculations in the previous section were made with *common logarithms*, which means logarithms with base 10. However, in science, the most frequently used base is the famous number $e = 2.71828\ldots$. In fact, this base is used so often that the logarithm with base e is called the *natural logarithm* and is denoted by "ln." You will find a ln button on most scientific calculators as well as an e^x button; that should be some indication of how important the base e is. At first glance, this is all somewhat mysterious. What can possibly be natural about using logarithms to the base $2.71828\ldots$? The full answer to that question must wait until Chapter 4, where we show that many calculus formulas come out neater when e is used as the base rather than any other base.

Definition and Properties of the Natural Logarithm

The natural logarithm of x, written $\ln x$, is defined to be the logarithm to base e of x. This means $\ln x$ is the inverse function of e^x. So

> The **natural logarithm** of x, written $\ln x$, is the power of e needed to get x. In other words,
>
> $$\ln x = c \quad \text{means} \quad e^c = x.$$

The rules for manipulating natural logs are the same as those for logs to base 10, and $\ln x$ is not defined when x is negative or 0.

Properties of Natural Logarithms

1. $\ln(AB) = \ln A + \ln B$

2. $\ln\left(\dfrac{A}{B}\right) = \ln A - \ln B$

3. $\ln(A^p) = p \ln A$

4. $\ln e^x = x$

5. $e^{\ln x} = x$

In addition, $\ln 1 = 0$ because $e^0 = 1$.

Using the $\boxed{\text{LN}}$ button on a calculator to plot a graph of $f(x) = \ln x$ for $0 < x \leq 10$, we get Figure 1.49. Observe that the graph of $y = \ln x$, like the graph of $y = \log x$, climbs very slowly as x increases. Also, note the x-axis intercept is $x = 1$, since $\ln 1 = 0$.

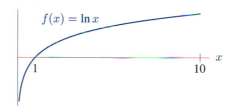

Figure 1.49: Graph of the natural logarithm

Using Exponential Functions with Base e

We can show that any exponential function can be written using e.

Any **exponential growth** function can be written in the form

$$P = P_0 e^{kt}$$

and any **exponential decay** function can be written as

$$Q = Q_0 e^{-kt},$$

where P_0 and Q_0 are the initial quantities and k is a positive constant.

We say that P and Q are growing or decaying at a *continuous rate* of k. (For example, $k = 0.02$ corresponds to a continuous rate of 2%.)

The reason that k is called the continuous rate is explored in detail in Chapter 10.

Example 1   Sketch the graphs of $P = e^{0.5t}$ and $Q = e^{-0.2t}$.

Solution

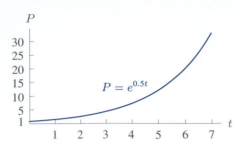

Figure 1.50: An exponential growth function   **Figure 1.51:** An exponential decay function

The graph of $P = e^{0.5t}$ is in Figure 1.50. Notice that the graph is the same shape as the previous exponential growth curves: increasing and concave up. Indeed, the families a^t, $a > 1$, and e^{kt}, $k > 0$, are one and the same. The graph of $Q = e^{-0.2t}$ is in Figure 1.51; it has the same shape as other exponential decay functions. The families a^t, $0 < a < 1$, and e^{kt}, $k < 0$, are one and the same.

Example 2   The release of chlorofluorocarbons used in air conditioners and, to a lesser extent, in household sprays (hair spray, shaving cream, etc.) destroys the ozone in the upper atmosphere. At the present time, the amount of ozone, Q, is decaying exponentially at a continuous rate of 0.25% per year. What is the half-life of ozone? In other words, at this rate, how long will it take for half the ozone to disappear?

Solution   If Q_0 is the initial quantity of ozone, then

$$Q = Q_0 e^{-0.0025t}.$$

We want to find T, the value of t making $Q = Q_0/2$, so

$$Q_0 e^{-0.0025T} = \frac{Q_0}{2}.$$

Dividing by Q_0 and then taking natural logs yields

$$\ln\left(e^{-0.0025T}\right) = -0.0025T = \ln\left(\frac{1}{2}\right) \approx -0.6931,$$

so

$$T \approx 277 \text{ years.}$$

In Example 2 the decay rate was given. However, in many situations where we expect to find exponential growth or decay, the rate is not given. To find it, we must know the quantity at two different times and then solve for the growth or decay rate, as in the next example.

Example 3   The population of Kenya was 19.5 million in 1984 and 21.2 million in 1986. Assuming it increases exponentially, find a formula for the population of Kenya as a function of time.

Solution   If we measure the population, P, in millions and time, t, in years since 1984, we can say
$$P = P_0 e^{kt} = 19.5 e^{kt},$$
where $P_0 = 19.5$ is the initial value of P. We find k by using the fact that $P = 21.2$ when $t = 2$, so
$$21.2 = 19.5 e^{k \cdot 2}.$$

To find k, we divide both sides by 19.5, giving

$$\frac{21.2}{19.5} = 1.087 = e^{2k}.$$

Now take natural logs of both sides:

$$\ln(1.087) = \ln(e^{2k}).$$

Using a calculator and the fact that $\ln(e^{2k}) = 2k$, this becomes

$$0.0834 = 2k.$$

So

$$k \approx 0.042,$$

and therefore

$$P = 19.5e^{0.042t}.$$

Since $k = 0.042 = 4.2\%$, we say the population of Kenya was growing at a continuous rate of 4.2% per year.

In Example 3 we chose to use e for the base of the exponential function representing Kenya's population, making clear that the continuous growth rate was 4.2%. If, however, we had wanted to emphasize the annual growth rate, we could have expressed the exponential function in the form

$$P = P_0 a^t.$$

We now show in Example 4 that the annual growth rate is about 4.3%, which is just slightly more than the continuous growth rate of 4.2% found in Example 3.

Example 4   Express the population of Kenya in the form

$$P = P_0 a^t.$$

Solution   Since the population grew from 19.5 to 21.2 million in 2 years, we know that

$$21.2 = 19.5a^2,$$

so

$$a = \left(\frac{21.2}{19.5} \right)^{1/2} \approx 1.043.$$

This gives

$$P = P_0 (1.043)^t.$$

Relationship between a^t and e^{kt}

In general, an exponential function of the form $P_0 e^{kt}$ can always be written in the form $P_0 a^t$. The reason is that

$$P_0 e^{kt} = P_0 (e^k)^t,$$

which suggests taking

$$a = e^k, \quad \text{so} \quad k = \ln a.$$

The two different formulas, $P = P_0 e^{kt}$ and $P = P_0 a^t$, have the same graph and represent the same function.

Exponential growth or decay can always be written in two ways,

$$P = P_0 a^t \quad \text{or} \quad P = P_0 e^{kt}.$$

Here a is the annual growth factor, $a - 1$ is the *annual growth rate*, and k is the *continuous growth rate*. Then

$$a = e^k, \quad \text{so} \quad k = \ln a.$$

Note $k > 0$ gives exponential growth; $k < 0$ gives decay.

If a comes from a percentage growth rate r, that is, $a = 1 + r$, then the continuous growth rate $k = \ln(1 + r)$ will be slightly less than, but very close to, r, provided r is small.

Time Constants and Half Lives

Engineers measure the speed at which an exponentially decaying quantity decreases using a *time constant*. This is the time taken for the quantity to decrease to $1/e$ times its initial value.

Example 5   Find the time constant of a circuit in which the charge, Q, is given by

$$Q = Q_0 e^{-t/RC}.$$

Here Q_0 is the initial charge, t is time, and R and C are constants which depend on the circuit.

Solution   If T is the time constant, then $Q = Q_0/e$ at time T, so

$$\frac{Q_0}{e} = Q_0 e^{-T/RC}.$$

Canceling Q_0 gives

$$e^{-T/RC} = \frac{1}{e} = e^{-1}.$$

Taking natural logs gives

$$-\frac{T}{RC} = -1,$$

so

$$T = RC.$$

Since $1/2 > 1/e$, there is greater decay during a time constant than a half-life, so the time constant is longer than the half-life. See Figure 1.52.

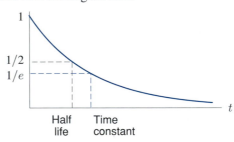

Figure 1.52: Comparing the half-life and the time constant

Problems for Section 1.7

1. Construct a table of values to compare $f(x) = \log x$ and $g(x) = \ln x$ for $x = 1, 2, \ldots, 10$. Round to two decimal places. Use these values to graph both functions.

Solve for x in Problems 2–7.

2. $2^x = e^{x+1}$

3. $2e^{3x} = 4e^{5x}$

4. $7^{x+2} = e^{17x}$

5. $10^{x+3} = 5e^{7-x}$

6. $2x - 1 = e^{\ln x^2}$

7. $9^x = 2e^{x^2}$

For Problems 8–10, solve for t. Assume a and b are positive constants and k is nonzero.

8. $a = be^t$

9. $P = P_0 e^{kt}$

10. $ae^{kt} = e^{bt}$, where $k \neq b$

Simplify the expressions in Problems 11–17 completely.

11. $e^{\ln(1/2)}$

12. $5e^{\ln(A^2)}$

13. $\ln(e^{2AB})$

14. $\ln e^{\sqrt{A}}$

15. $\ln\left(\dfrac{1}{e}\right) + \ln AB$ 16. $2\ln\left(e^A\right) + 3\ln B^e$ 17. $2\ln(AB) - \ln\left(\dfrac{B}{A}\right)$

Convert the functions in Problems 18–23 into the form $P = P_0 a^t$. Which represent exponential growth and which represent exponential decay?

18. $P = 15e^{0.25t}$ 19. $P = 2e^{-0.5t}$ 20. $P = 10e^{0.917t}$

21. $P = 79e^{-2.5t}$ 22. $P = P_0 e^{0.2t}$ 23. $P = 7e^{-\pi t}$

Convert the functions in Problems 24–27 into the form $P = P_0 e^{kt}$.

24. $P = 15(1.5)^t$ 25. $P = 10(1.7)^t$ 26. $P = 174(0.9)^t$ 27. $P = 4(0.55)^t$

28. (a) Evaluate the quantity $(\ln x)/(\log x)$ for several values of x (say $x = 0.5, 5, 10, 100$). What do you notice?

(b) Plot a graph of $y = (\ln x)/(\log x)$ on a computer or calculator. What do you see? How does it relate to your results in part (a)?

(c) Take the natural logs of both sides of the equation $x = 10^{\log x}$ to show that $\dfrac{\ln x}{\log x} = \ln 10$.

29. Find the inverse function of $f(t) = 50e^{0.1t}$.

30. Find the inverse function of $f(t) = 1 + \ln t$.

31. Define $g(x) = \dfrac{2}{e^x + 5}$.

(a) Is g increasing or decreasing?

(b) Explain why g is invertible and find a formula for g^{-1}.

32. Define $f(x) = \dfrac{1}{1 + e^{-x}}$.

(a) Is f increasing or decreasing?

(b) Explain why f is invertible, and find a formula for $f^{-1}(x)$.

(c) What is the domain of f^{-1}?

(d) Sketch the graphs of f and f^{-1} on the same axes, and explain their relationship.

33. (a) A population grows according to the equation $P = P_0 e^{kt}$ (with P_0, k constants). Find the population as a function of time, t, if it grows at a continuous rate of 2% a year and starts at 1 million.

(b) Plot a graph of the population you found in part (a) against time.

34. The air in a factory is being filtered so that the quantity of a pollutant, P (measured in mg/liter), is decreasing according to the equation $P = P_0 e^{-kt}$, where t represents time in hours . If 10% of the pollution is removed in the first five hours:

(a) What percentage of the pollution is left after 10 hours?

(b) How long will it take before the pollution is reduced by 50%?

(c) Plot a graph of pollution against time. Show the results of your calculations on the graph.

(d) Explain why the quantity of pollutant might decrease in this way.

35. Air pressure, P, decreases exponentially with the height above the surface of the earth, h:

$$P = P_0 e^{-0.00012h}$$

where P_0 is the air pressure at sea level and h is in meters.

(a) If you go to the top of Mount McKinley, height 6198 meters (about 20,320 feet), what is the air pressure, as a percent of the pressure at sea level?

(b) The maximum cruising altitude of an ordinary commercial jet is around 12,000 meters (about 40,000 feet). At that height, what is the air pressure, as a percent of the sea level value?

36. Under certain circumstances, the velocity, V, of a falling raindrop is given by $V = V_0(1 - e^{-t})$, where t is time and V_0 is a positive constant.

(a) Sketch a rough graph of V against t, for $t \geq 0$.

(b) What does V_0 represent?

37. The population of the United States was 226.5 million in 1980 and 248.7 million in 1990. In what year is the population expected to go over 300 million?

38. The half-life of radioactive strontium-90 is 29 years. In 1960, radioactive strontium-90 was released into the atmosphere during testing of nuclear weapons, and was absorbed into people's bones. How many years does it take until only 10% of the original amount absorbed remains?

39. If the size of a bacteria colony doubles in 5 hours, how long will it take for the number of bacteria to triple?

40. Predict the earth's population in the year 2000. An almanac lists the population in 1980 as 4.478 billion, and in 1994 as 5.642 billion. What is the doubling time of the world's population?

41. Suppose that a certain radioactive substance has a half-life of 5 years. An object starts with 20 kg of the radioactive material.

    (a) How much of the radioactive material is left after 10 years?
    (b) The object can be moved safely when the quantity of the radioactive material is 0.1 kg or less. How much time must pass before the object can be moved?

42. One of the main contaminants of a nuclear accident, such as that at Chernobyl, is strontium-90, which decays exponentially at a continuous rate of approximately 2.47% per year. Preliminary estimates after the Chernobyl disaster suggested that it would be about 100 years before the region would again be safe for human habitation. What percent of the original strontium-90 would still remain by this time?

43. The population, P, in millions, of Nicaragua was 3.6 million in 1990 and growing at an annual rate of 3.4%. Let t be time in years since 1990.

    (a) Express P as a function in the form $P = P_0 a^t$.
    (b) Express P as an exponential function using the base e.
    (c) Compare the annual and continuous growth rates.

44. Geological dating of rocks is done using potassium-40 rather than carbon-14 because potassium has a longer half-life. The potassium decays to argon, which remains trapped in the rocks and can be measured; thus the original quantity of potassium can be calculated. The half-life of potassium-40 is $1.28 \cdot 10^9$ years. Find a formula giving the quantity, P, of potassium-40 remaining as a function of time in years, assuming the initial quantity is P_0:

    (a) Using base 1/2.
    (b) Using base e.

45. A picture supposedly painted by Vermeer (1632–1675) contains 99.5% of its carbon-14 (half-life 5730 years). From this information decide whether the picture is a fake. Explain your reasoning.

46. Assume that the amount of electric charge on a capacitor is an exponentially decaying function of time:

$$Q = Q_0 e^{-t/k}.$$

    (a) What is the time constant?
    (b) Show that after a period of 3 time constants, over 95% of the charge has decayed.
    (c) Show that after 5 time constants, less than 1% of the charge remains.

1.8 NEW FUNCTIONS FROM OLD

Shifts and Stretches

The graph of a constant multiple of a given function is easy to visualize: each y-value is stretched or shrunk by that multiple. For example, consider the function $f(x)$ and its multiples $y = 3f(x)$ and $y = -2f(x)$. Their graphs are shown in Figure 1.53. The factor 3 in the function $y = 3f(x)$ stretches each $f(x)$ value by multiplying it by 3; the factor -2 in the function $y = -2f(x)$ stretches $f(x)$ by multiplying by 2 and reflects it about the x-axis. You can think of the multiples of a given function as a family of functions.

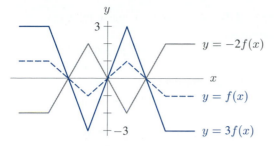

Figure 1.53: Multiples of the function $f(x)$

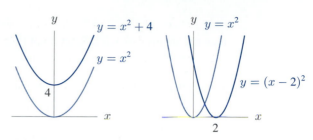

Figure 1.54: Graphs of $y = x^2$ with $y = x^2 + 4$ and $y = (x - 2)^2$

It is also easy to create families of functions by shifting graphs. For example, $y - 4 = x^2$ is the same as $y = x^2 + 4$, which is the graph of $y = x^2$ shifted up by 4. Similarly, $y = (x - 2)^2$ is the graph of $y = x^2$ shifted right by 2. (See Figure 1.54.)

- Multiplying a function by a constant, c, stretches the graph vertically (if $c > 1$) or shrinks the graph vertically (if $0 < c < 1$). A negative sign (if $c < 0$) reflects the graph about the x-axis, in addition to shrinking or stretching.
- Replacing y by $(y - k)$ moves a graph up by k (down if k is negative).
- Replacing x by $(x - h)$ moves a graph to the right by h (to the left if h is negative).

We can sketch the graph of the sum of two functions by imagining the y-values of the two graphs stacked on top of each other.

Example 1   Graph the function $y = 2x^2 + 1/x$ for $x > 0$.

Solution   We graph the functions $y = 2x^2$ and $y = 1/x$ separately first. Now imagine the corresponding y-values stacked on top of one another, and you will get the graph of $y = 2x^2 + 1/x$. (See Figure 1.55.)

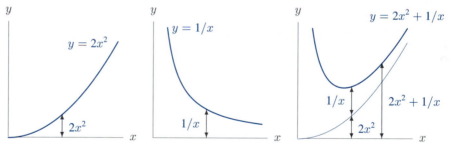

Figure 1.55: Summing two functions from their graphs

Notice that, for x near 0, the graph of $y = 2x^2 + 1/x$ looks much like the graph of $y = 1/x$ (because $1/x$ is much larger than $2x^2$ there). For large x, the graph of $y = 2x^2 + 1/x$ looks like the graph of $y = 2x^2$.

Composite Functions

If oil is spilled from a tanker, the area of the oil slick will grow with time. Suppose that the oil slick is always a perfect circle. (In practice, this does not happen because of winds and tides and the location of the coastline.) The area of the oil slick is a function of its radius. This function is

$$A = f(r) = \pi r^2.$$

The radius is also a function of time, because the radius increases as more oil spills. Thus, the area, being a function of the radius, is also a function of time. If, for example, the radius is given by

$$r = g(t) = 1 + t,$$

then the area is given as a function of time by substitution:

$$A = \pi r^2 = \pi(1 + t)^2.$$

We say that A is a *composite function* or a "function of a function," which is written

$$A = \underbrace{f(g(t))}_{} = \pi(g(t))^2 = \pi(1 + t)^2.$$

Composite function;
f is outside function,
g is inside function

Let's look at this example and think about how we would calculate A using the formula $\pi(1 + t)^2$. For any given t, the first step is to find $1 + t$, and the second step is to square and multiply by π. So the first step corresponds to the inside function $g(t) = 1 + t$, and the second step corresponds to the outside function $f(r) = \pi r^2$.

Example 2 If $f(x) = x^2$ and $g(x) = x + 1$, find each of the following:
(a) $f(g(2))$ (b) $g(f(2))$ (c) $f(g(x))$ (d) $g(f(x))$

Solution (a) Since $g(2) = 3$, we have $f(g(2)) = f(3) = 9$.
(b) Since $f(2) = 4$, we have $g(f(2)) = g(4) = 5$. Notice that $f(g(2)) \neq g(f(2))$.
(c) $f(g(x)) = f(x + 1) = (x + 1)^2$.
(d) $g(f(x)) = g(x^2) = x^2 + 1$. Again, notice that $f(g(x)) \neq g(f(x))$.

Example 3 Express each of the following functions as a composition:

(a) $h(t) = (1 + t^3)^{27}$ (b) $k(x) = \log(x^2)$ (c) $l(x) = (\log x)^2$ (d) $n(y) = e^{-y^2}$

Solution In each case think about how you would calculate a value of the function. The first stage of the calculation will give you the inside function, and the second stage will give you the outside one.
(a) For $(1 + t^3)^{27}$, the first stage is cubing and adding 1, so the inside function is $g(t) = 1 + t^3$. The second stage is taking the 27th power, so the outside function is $f(y) = y^{27}$. Then

$$f(g(t)) = f(1 + t^3) = (1 + t^3)^{27}.$$

In fact, there are lots of different answers: $g(t) = t^3$ and $f(y) = (1+y)^{27}$ is another possibility.
(b) Evaluating this function involves squaring x first and then taking the log. So if $g(x) = x^2$ is the inside function and $f(y) = \log y$ is the outside function, then

$$f(g(x)) = \log(x^2).$$

(c) This function involves taking the log first and then squaring. Using the same definitions of f and g as in part (b), namely $f(x) = \log x$ and $g(t) = t^2$, the composition is

$$g(f(x)) = (\log x)^2.$$

By evaluating the functions in parts (b) and (c) for $x = 2$, say, giving $\log(2^2) = 0.602$, and $\log^2(2) = 0.091$, you can see that the order in which two functions are composed can make a difference.
(d) To calculate e^{-y^2} you square y, take its negative, and then take e to that power. So if we set $g(y) = -y^2$ and $f(z) = e^z$, then we can write

$$f(g(y)) = e^{-y^2}.$$

Alternatively, you could take $g(y) = y^2$ and $f(z) = e^{-z}$.

Odd and Even Functions: Symmetry

There is a certain symmetry apparent in the graphs of $f(x) = x^2$ and $g(x) = x^3$ in Figure 1.56. For each point (x, x^2) on the graph of f, the point $(-x, x^2)$ is also on the graph; for each point (x, x^3) on the graph of g, the point $(-x, -x^3)$ is also on the graph. The graph of $f(x) = x^2$ is symmetric about the y-axis, whereas the graph of $g(x) = x^3$ is symmetric about the origin. The graph of any polynomial involving only even powers has symmetry about the y-axis, while polynomials with only odd powers are symmetric about the origin. Consequently, any functions with these symmetry properties are called *even* and *odd*.

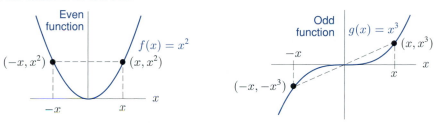

Figure 1.56: Symmetry of even and odd functions

For any function f,
f is an **even** function if $f(-x) = f(x)$ for all x.
f is an **odd** function if $f(-x) = -f(x)$ for all x.

For example, $g(x) = e^{x^2}$ is even and $h(x) = x^{1/3}$ is odd. However, you should be aware that many functions do not have any symmetry and are neither even nor odd.

Problems for Section 1.8

1. The graph of $y = f(x)$ is shown in Figure 1.57. Sketch graphs of each of the following. Label any intercepts or asymptotes that can be determined.

   (a) $y = f(x) + 3$ (b) $y = 2f(x)$ (c) $y = f(x + 4)$ (d) $y = 4 - f(x)$

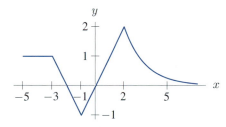

Figure 1.57

2. (a) Write an equation for a graph obtained by vertically stretching the graph of $y = x^2$ by a factor of 2, followed by a vertical upward shift of 1 unit. Sketch the graph.
   (b) What is the equation if the order of the transformations (stretching and shifting) in part (a) is interchanged?
   (c) Are the two graphs the same? Explain the effect of reversing the order of transformations.

For Problems 3–11, let $f(x) = 2x + 1$, $g(x) = \ln(x + 3)$, and $h(x) = e^{4x+7}$. Find the following:

3. $f(g(x))$ 4. $g(f(x))$ 5. $f(h(x))$
6. $h(f(x))$ 7. $g(h(x))$ 8. $h(g(x))$
9. $g(h(x) - 3)$ 10. $f(g(h(x)))$ 11. $h(g(f(x)))$

For Problems 12–15, does the graph suggest the function is even, odd or neither?

12. 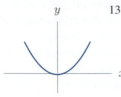   13. 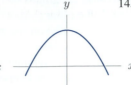   14. 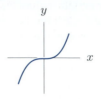   15.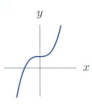

Are the functions in Problems 16–17 even, odd, or neither?

16. $f(x) = x^6 + x^3 + 1$ 17. $f(x) = x^3 + x^2 + x$

For Problems 18–21, determine functions f and g such that $h(x) = f(g(x))$. [Note: There is more than one correct answer. Do not choose $f(x) = x$ or $g(x) = x$.]

18. $h(x) = (x+1)^3$ 19. $h(x) = x^3 + 1$ 20. $h(x) = \ln(x^3)$ 21. $h(x) = (\ln x)^3$

For Problems 22–25, let $m(z) = z^2$. Find and simplify the following quantities:

22. $m(z+1) - m(z)$ 23. $m(z+h) - m(z)$

24. $m(z) - m(z-h)$ 25. $m(z+h) - m(z-h)$

26.   Find a calculator window in which the graph of $f(x) = 9x^2 + 5$ looks exactly like the graph of $g(x) = x^2$ in the window $-3 < x < 3$, $-1 < y < 9$.

27.   Find a calculator window in which the graph of $f(x) = 10x^2 + 1000x$ looks exactly like the graph of $g(x) = x^2$ in the window $-10 < x < 10$, $-10 < y < 100$. [Hint: Use completing the square.]

28.   Graph $f(x) = 3^x$ in the calculator window $0 < x < 3$, $0 < y < 27$. Then graph $g(x) = 2(3^{x/2})$ in the window $0 < x < 6$, $0 < y < 54$. Explain why the two graphs look exactly the same.

29.   (a)   You are given the following table of values for a function f. For what values of x can we evaluate $g(x) = f(x/2)$ exactly?

x	0	1	2	3	4	5	6
$f(x)$	0	2	6	12	20	30	42

(b)   Construct a table of values for $g(x) = f(x/2)$.
(c)   How will the graph of g differ from the graph of f?

For Problems 30–35, suppose that f and g are given by the graphs in Figure 1.58:

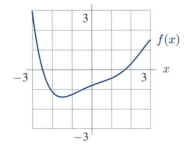   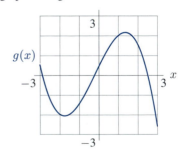

Figure 1.58

30.   Estimate $f(g(1))$.   31.   Estimate $g(f(2))$.
32.   Estimate $f(f(1))$.   33.   Sketch a graph of $f(g(x))$.
34.   Sketch a graph of $g(f(x))$.   35.   Sketch a graph of $f(f(x))$.

36. Let $h(x)$ and $j(x)$ be the functions graphed in Figures 1.59 and 1.60.

(a) Sketch graphs of $h(j(x))$ and $j(h(x))$. Label all intercepts.
(b) If possible, sketch a graph of a function $k(x)$ such that $j(k(x)) = k(j(x)) = x$. If this is not possible, explain why.

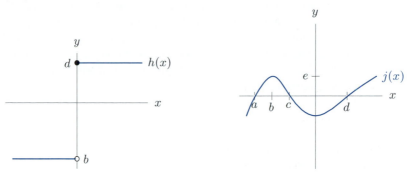

Figure 1.59   *Figure 1.60*

37. The Heaviside step function, H, is graphed in Figure 1.61. Sketch graphs of the following functions.

(a) $2H(x)$

(b) $H(x) + 1$

(c) $H(x + 1)$

(d) $-H(x)$

(e) $H(-x)$

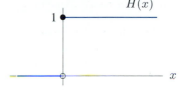

Figure 1.61

For Problems 38–40, use the graph of $y = f(x)$ in Figure 1.62 to graph the indicated functions:

38. $y = 2f(x)$ 39. $y = 2 - f(x)$ 40. $y = 1/f(x)$

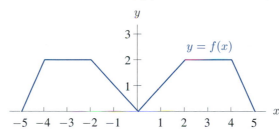

Figure 1.62

41. Complete Table 1.23 to show values for functions f, g, and h, given the following conditions:

(a) f is symmetric about the y-axis.
(b) g is symmetric about the origin.
(c) h is the composition of g with f [that is, $h(x) = g(f(x))$].

TABLE 1.23

x	$f(x)$	$g(x)$	$h(x)$
-3	0	0	
-2	2	2	
-1	2	2	
0	0	0	
1			
2			
3			

1.9 THE TRIGONOMETRIC FUNCTIONS

Trigonometry originated as part of the study of triangles. Indeed, the name *tri-gon-o-metry* means the measurement of three-cornered figures. Thus, the first definitions of the trigonometric functions were in terms of triangles. However, the trigonometric functions can also be defined using the unit circle, a definition that makes them periodic, or repeating. Many naturally occurring processes are also periodic. The water level in a tidal basin, the blood pressure in a heart, an alternating current, and the position of the air molecules transmitting a musical note all fluctuate regularly. Such phenomena can be represented by trigonometric functions.

We will use only the three trigonometric functions found on a calculator: the sine, the cosine, and the tangent.

Radians

There are two commonly used ways to represent the input of the trigonometric functions: radians and degrees. The formulas of calculus, as you will see, are neater in radians than in degrees.

> An angle of 1 **radian** is defined to be the angle at the center of a unit circle which cuts off an arc of length 1, measured counterclockwise. (See Figure 1.63(a).) A unit circle has radius 1.

An angle of 2 radians cuts off an arc of length 2 on a unit circle. A negative angle, such as $-1/2$ radians, cuts off an arc of length $1/2$, but measured clockwise. (See Figure 1.63(b).)

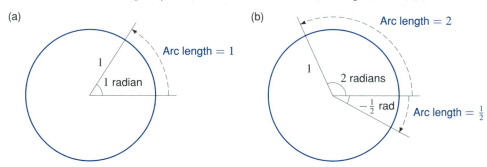

Figure 1.63: Radians defined using unit circle

It is useful to think of angles as rotations, since then we can make sense of angles of over $360°$. Since one full rotation of $360°$ cuts off an arc of length 2π, the circumference of the unit circle, it follows that

$$360° = 2\pi \text{ radians}, \quad \text{so} \quad 180° = \pi \text{ radians}.$$

In other words, 1 radian $= 180°/\pi$, so one radian is about $60°$. The word radians is often dropped, so if an angle or rotation is referred to without units, it is understood to be in radians.

Arc Length

Radians are also useful for computing the length of an arc of a circle of radius other than 1. If the circle has radius r and the arc cuts off an angle θ, as in Figure 1.64, then we have the following relation:

> Arc length $= s = r\theta.$

Figure 1.64: Arc length of a sector of a circle

The Sine and Cosine

The two basic trigonometric functions—the sine and cosine—are defined using a unit circle. In Figure 1.65, an angle of t radians is measured counterclockwise around the circle from the point $(1,0)$. If P has coordinates (x, y), we define

$$\cos t = x \quad \text{and} \quad \sin t = y.$$

Note that we will assume that the angles are *always* in radians unless specified otherwise.

Since the equation of the unit circle is $x^2 + y^2 = 1$, we have the following fundamental identity

$$\cos^2 t + \sin^2 t = 1.$$

As t increases and P moves around the circle, the values of $\sin t$ and $\cos t$ oscillate between 1 and -1, and eventually repeat as P moves through points where it has been before. If t is negative, the angle is measured clockwise around the circle.

Amplitude, Period, and Phase

The graphs of sine and cosine are shown in Figure 1.66. Notice that sine is an odd function, and cosine is even. The maximum and minimum values of sine and cosine are $+1$ and -1, because those are the maximum and minimum values of y and x on the unit circle. After the point P has moved around the complete circle once, the values of $\cos t$ and $\sin t$ start to repeat; we say the functions are *periodic*.

> For any periodic function of time:
> - The **amplitude** is half the distance between maximum and minimum values (if it exists).
> - The **period** is the time needed for the function to execute one complete cycle.

The amplitude of $\cos t$ and $\sin t$ is 1, and the period is 2π. Why 2π? Because that's the value of t when the point P has gone exactly once around the circle. (Remember that $360° = 2\pi$ radians.)

In looking at Figure 1.66 you can see that the sine and cosine graphs are exactly the same shape, only shifted horizontally. Since the cosine is the sine shifted $\pi/2$ to the left,

$$\cos t = \sin(t + \pi/2).$$

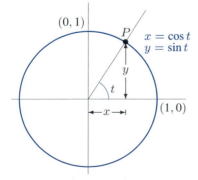

Figure 1.65: The definitions of $\sin t$ and $\cos t$

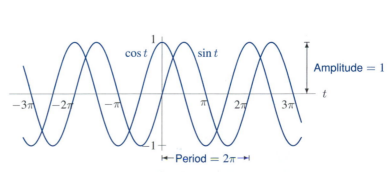

Figure 1.66: Graphs of $\cos t$ and $\sin t$

This says that the cosine of any number is the same as the sine of the number that is $\pi/2$ further to the right—which is exactly what the graphs show. Equivalently, the sine is the cosine shifted $\pi/2$ to the right:

$$\sin t = \cos(t - \pi/2).$$

We say that the *phase difference* between $\sin t$ and $\cos t$ is $\pi/2$.

Functions whose graphs are the shape of a sine or cosine curve are called *sinusoidal* functions.

> To describe arbitrary amplitudes and periods of sinusoidal functions, we use functions of the form
>
> $$f(t) = A \sin(Bt) \qquad \text{and} \qquad g(t) = A \cos(Bt),$$
>
> where $|A|$ is the amplitude and $2\pi/|B|$ is the period.
> To represent arbitrary phase differences, we shift a graph of the correct amplitude and period horizontally by replacing t with $t - h$ or $t + h$.

Example 1   Find and show on a graph the amplitude and period of the functions

(a) $y = 5 \sin(2t)$ (b) $y = -5 \sin \left(\dfrac{t}{2}\right)$ (c) $y = 1 + 2 \sin t$

Solution   (a)   From Figure 1.67, you can see that the amplitude of $y = 5 \sin(2t)$ is 5 because the factor of 5 stretches the oscillations up to 5 and down to -5. The period of $y = \sin(2t)$ is π, because when t changes from 0 to π, the quantity $2t$ changes from 0 to 2π, and so the sine function will have gone through one complete oscillation.

(b)   Figure 1.68 shows that the amplitude of $y = -5 \sin\left(t/2\right)$ is again 5, because the negative sign reflects the oscillations in the t-axis, but does not change how far up or down they go. The period of $y = -5 \sin\left(t/2\right)$ is 4π because when t changes from 0 to 4π, the quantity $t/2$ changes from 0 to 2π, and so the sine function goes through one complete oscillation.

(c)   The 1 shifts the graph $y = 2 \sin t$ up by 1. Since $y = 2 \sin t$ has an amplitude of 2 and a period of 2π, the graph of $y = 1 + 2 \sin t$ goes up to 3 and down to -1, and has a period of 2π. (See Figure 1.69.) Thus, $y = 1 + 2 \sin t$ also has amplitude 2.

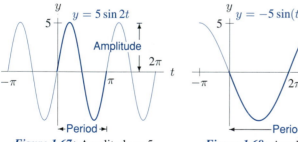

Figure 1.67: Amplitude = 5, period = π

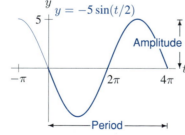

Figure 1.68: Amplitude = 5, period = 4π

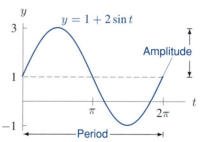

Figure 1.69: Amplitude = 2, period = 2π

Example 2   Find formulas for the functions describing the oscillations in Figures 1.70–1.72.

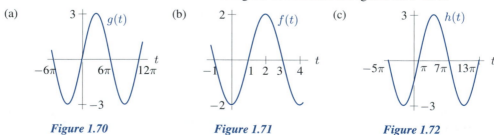

(a) $g(t)$

(b) $f(t)$

(c) $h(t)$

Figure 1.70   *Figure 1.71*   *Figure 1.72*

Solution (a) The function in Figure 1.70 looks like a sine function of amplitude 3, so $g(t) = 3\sin(Bt)$. Since the function executes one full oscillation between $t = 0$ and $t = 12\pi$, when t changes by 12π, the quantity Bt changes by 2π. This means $B \cdot 12\pi = 2\pi$, so $B = 1/6$. Therefore, $g(t) = 3\sin(t/6)$ has the graph shown.

(b) The function in Figure 1.71 looks like an upside down cosine function with amplitude 2, so $f(t) = -2\cos(Bt)$. The function completes one oscillation between $t = 0$ and $t = 4$. Thus, when t changes by 4, the quantity Bt changes by 2π, so $B \cdot 4 = 2\pi$, or $B = \pi/2$. Therefore, $f(t) = -2\cos(\pi t/2)$ has the graph shown.

(c) The function in Figure 1.72 looks like the function $g(t)$ in Figure 1.70, but shifted a distance of π to the right. Since $g(t) = 3\sin(t/6)$, we replace t by $(t - \pi)$ to obtain $h(t) = 3\sin[(t - \pi)/6]$.

There are many possible equations for these graphs; you may be able to find others.

Example 3 On February 10, 1990, high tide in Boston was at midnight. The water level at high tide was 9.9 feet; later, at low tide, it was 0.1 feet. Assuming the next high tide is at exactly 12 noon and that the height of the water is given by a sine or cosine curve, find a formula for the water level in Boston as a function of time.

Solution Let y be the water level in feet, and let t be the time measured in hours from midnight. Then the oscillations are to have amplitude 4.9 feet ($= (9.9 - 0.1)/2$) and period 12, so $12B = 2\pi$ and $B = \pi/6$. Since the water is highest at midnight, when $t = 0$, the oscillations are best represented by a cosine, because the cosine is at its maximum at the beginning of the cycle. (See Figure 1.73.) We can say

$$\text{Height above average} = 4.9\cos\left(\frac{\pi}{6}t\right).$$

Since the average level of the water was 5 feet ($= (9.9 + 0.1)/2$), we want a cosine curve shifted up by 5. We get this by adding 5:

$$y = 5 + 4.9\cos\left(\frac{\pi}{6}t\right).$$

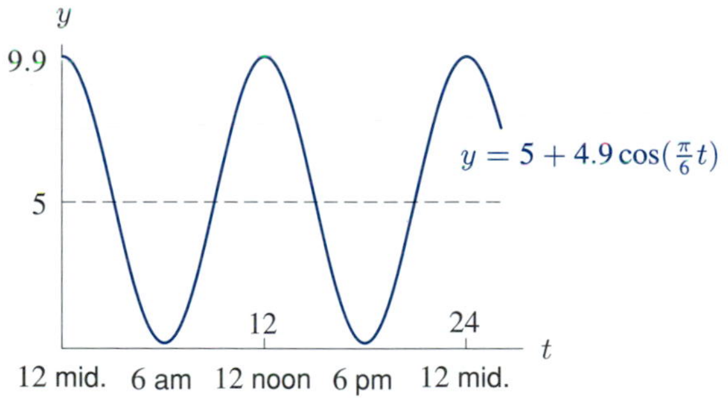

Figure 1.73: Function approximating the tide in Boston on February 10, 1990

Example 4 Of course, there's something wrong with the assumption in Example 3 that the next high tide will be at noon. If so, the high tide would always be at noon or midnight, instead of progressing slowly through the day, as in fact it does. The interval between successive high tides actually averages about 12 hours 24 minutes. Using this, give a more accurate formula for the height of the water as a function of time.

Solution The period is 12 hours 24 minutes $= 12.4$ hours, so $B = 2\pi/12.4$, giving

$$y = 5 + 4.9\cos\left(\frac{2\pi}{12.4}t\right) = 5 + 4.9\cos(0.507t).$$

Example 5 Use the information from Example 3 to write a formula for the water level in Boston on a day when the high tide is at 2 pm.

Solution   When the high tide is at midnight

$$y = 5 + 4.9 \cos(0.507t).$$

Since 2 pm is 14 hours after midnight, we replace t by $(t - 14)$. Therefore, on a day when the high tide is at 2 pm,

$$y = 5 + 4.9 \cos(0.507(t - 14)).$$

Example 6   Explain why the following function represents an oscillation which is dying out, called a *damped oscillation*:

$$f(x) = (0.9)^x \sin x.$$

Solution   Think of $f(x)$ as a sine function with an amplitude of $(0.9)^x$, which decays as x increases. See Figure 1.74. This is the kind of motion we'd expect from a weight on a spring, where the oscillations are dying out because of friction.

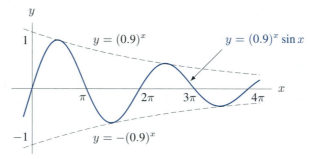

Figure 1.74: A damped oscillation

The Tangent Function

If t is any number with $\cos t \neq 0$, we define the tangent function by

$$\tan t = \frac{\sin t}{\cos t}.$$

Figure 1.65 on page 53 shows the geometrical meaning of the tangent function: $\tan t$ is the slope of the line through the origin $(0, 0)$ and the point $P = (\cos t, \sin t)$ on the unit circle.

The tangent function is undefined wherever $\cos t = 0$, namely, at $t = \pm\pi/2, \pm3\pi/2, \ldots$, and it has a vertical asymptote at each of these points. Now think about where $\tan t$ will be positive and where it will be negative. It will be positive where $\sin t$ and $\cos t$ have the same sign. The graph of the tangent is shown in Figure 1.75.

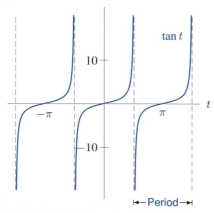

Figure 1.75: The tangent function

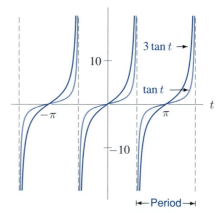

Figure 1.76: Multiple of tangent

Notice that the tangent function has period π, because it repeats every π units. Does it make sense to talk about the amplitude of the tangent function? Not if we're thinking of the amplitude as a measure of the size of the oscillation, because the tangent becomes infinitely large near each vertical asymptote. We can still multiply the tangent by a constant, but that constant no longer represents an amplitude. (See Figure 1.76.)

The Inverse Trigonometric Functions

On occasion, you may need to find a number with a given sine. For example, you might want to find x such that

$$\sin x = 0$$

or such that

$$\sin x = 0.3.$$

The first of these equations can be solved by inspection; the solutions are $x = 0, \pm\pi, \pm 2\pi, \ldots$. Solving the second equation requires a calculator, and it also has infinitely many solutions. They are

$$x \approx 0.305, 2.84, 0.305 \pm 2\pi, 2.84 \pm 2\pi, \ldots$$

For each equation, we pick out the only solution between $-\pi/2$ and $\pi/2$ as the preferred solution. For example, the preferred solution to $\sin x = 0$ is $x = 0$, and the preferred solution to $\sin x = 0.3$ is $x = 0.305$. We define the inverse sine, written "arcsin" or "$\sin^{-1}$," as the function which gives the preferred solution.

For $-1 \leq y \leq 1$,
$$\arcsin y = x$$
means
$$\sin x = y \quad \text{with} \quad -\frac{\pi}{2} \leq x \leq \frac{\pi}{2}.$$

Thus the arcsine is the inverse function to the piece of the sine function having domain $[-\pi/2, \pi/2]$. (See Table 1.24 and Figure 1.77.) On a calculator, the arcsine function[9] is usually denoted by $\boxed{\sin^{-1}}$.

TABLE 1.24   *Values of* $\sin x$ *and* $\sin^{-1} x$

x	$\sin x$	x	$\sin^{-1} x$
$-\frac{\pi}{2}$	-1.000	-1.000	$-\frac{\pi}{2}$
-1.0	-0.841	-0.841	-1.0
-0.5	-0.479	-0.479	-0.5
0.0	0.000	0.000	0.0
0.5	0.479	0.479	0.5
1.0	0.841	0.841	1.0
$\frac{\pi}{2}$	1.000	1.000	$\frac{\pi}{2}$

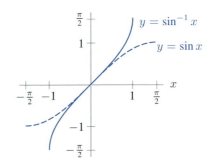

Figure 1.77: The arcsine function

The inverse tangent, written "arctan" or "$\tan^{-1}$," is the inverse function for the piece of the tangent function having the domain $-\pi/2 < x < \pi/2$. On a calculator, the inverse tangent is usually denoted by $\boxed{\tan^{-1}}$. The graph of the arctangent is shown in Figure 1.79.

---

[9]Note that $\sin^{-1} x = \arcsin x$ is not the same as $(\sin x)^{-1} = 1/\sin x$.

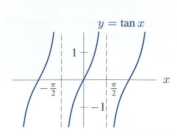

Figure 1.78: The tangent function   **Figure 1.79:** The arctangent function

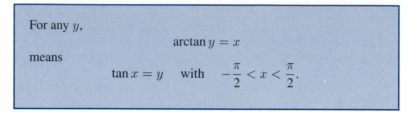

The inverse cosine function, written "arccos" or "$\cos^{-1}$," is discussed in Problem 34.

Problems for Section 1.9

For Problems 1–9, draw the angle using a ray through the origin, and determine whether the sine, cosine, and tangent of that angle are positive, negative, zero, or undefined.

1. $\frac{3\pi}{2}$ 2. 2π 3. $\frac{\pi}{4}$ 4. 3π 5. $\frac{\pi}{6}$

6. $\frac{4\pi}{3}$ 7. $\frac{-4\pi}{3}$ 8. 4 9. -1

Given that $\sin(\pi/12) = 0.259$ and $\cos(\pi/5) = 0.809$, compute (without using the trigonometric functions on your calculator) the quantities in Problems 10–13. You may want to draw a picture showing the angles involved, and then check your answers on a calculator.

10. $\cos\left(-\frac{\pi}{5}\right)$ 11. $\sin\frac{11\pi}{12}$ 12. $\sin\frac{\pi}{5}$ 13. $\cos\frac{\pi}{12}$

14. Use the solution to Example 3 on page 55 to estimate the water level in Boston Harbor at 3:00 am, 4:00 am, and 5:00 pm on February 10, 1990.

15. A compact disk spins at a rate of 200 to 500 revolutions per minute. What are the equivalent rates measured in radians per second?

16. When a car's engine makes less than about 200 revolutions per minute, it stalls. What is the period of the rotation of the engine when it is about to stall?

17. What is the difference between $\sin x^2$, $\sin^2 x$, and $\sin(\sin x)$? Express each of the three as a composition. (Note: $\sin^2 x = (\sin x)^2$.)

18. Consider the function $y = 5 + \cos(3x)$.

    (a) What is its amplitude?   (b) What is its period?   (c) Sketch its graph.

19. Match the functions below with the graphs in Figure 1.80.
    (a) $y = 2\cos(t - \pi/2)$ (b) $y = 2\cos t$ (c) $y = 2\cos(t + \pi/2)$

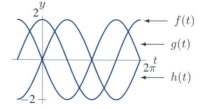

Figure 1.80

For Problems 20–28, find a possible formula for each graph.

20.

21.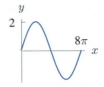

22.

23.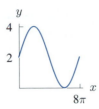

24.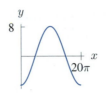

25.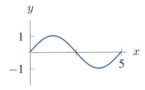

26.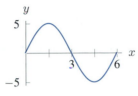

27.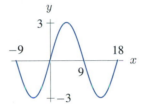

28.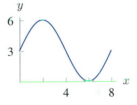

29. The voltage, V, of an electrical outlet in a home is given as a function of time, t (in seconds), by $V = V_0 \cos(120\pi t)$.

   (a) What is the period of the oscillation?
   (b) What does V_0 represent?
   (c) Sketch the graph of V against t. Label the axes.

30. A population of animals oscillates sinusoidally between a low of 700 on January 1 and a high of 900 on July 1.

   (a) Graph the population against time.
   (b) Find a formula for the population as a function of time, t, measured in months since the start of the year.

31. The visitors' guide to St. Petersburg, Florida, contains the chart shown in Figure 1.81 to advertise their good weather. Fit a trigonometric function approximately to the data. The independent variable should be time in months. In order to do this, you will need to estimate the amplitude and period of the data, and when the maximum occurs. (There are many possible answers to this problem, depending on how you read the graph.)

Temp(°F)	Jan	Feb	Mar	Apr	May	June	July	Aug	Sept	Oct	Nov	Dec
100°												
90°												
80°												
70°												
60°												
50°												

Figure 1.81: "St. Petersburg...where we're famous for our wonderful weather and year-round sunshine." (Reprinted with permission)

32.   The Bay of Fundy in Canada is reputed to have the largest tides in the world. The difference between low and high water levels is 15 meters (nearly 50 feet). Suppose at a particular point in the Bay of Fundy, the depth of the water, y meters, is given as a function of time, t, in hours since midnight on January 1, 1997, by

$$y = D + A \cos \left(B(t - C) \right).$$

(a)   What is the physical meaning of D?
(b)   What is the value of A?
(c)   What is the value of B? Assume the time between successive high tides is 12.4 hours.
(d)   What is the physical meaning of C?

33.   (a)   Using a calculator set in radians, make a table of values, to two decimal places, of $f(x) = \arcsin x$, for $x = -1, -0.8, -0.6, \ldots, 0, \ldots, 0.8, 1$. (Arcsine is denoted by $\boxed{\sin^{-1}}$ on most calculators).
      (b)   Sketch $f(x) = \arcsin x$. Mark on the graph the domain and range of f.

34.   This problem introduces the arccosine function, or inverse cosine, denoted by $\boxed{\cos^{-1}}$ on most calculators.

(a)   Using a calculator set in radians, make a table of values, to two decimal places, of $g(x) = \arccos x$, for $x = -1, -0.8, -0.6, \ldots, 0, \ldots, 0.8, 1$.
(b)   Sketch the graph of $g(x) = \arccos x$.
(c)   Based on your graph, what are the domain and range of the arccosine?
(d)   Why is the domain of the arccosine the same as the domain of the arcsine?
(e)   Why is the range of the arccosine *not* the same as the range of the arcsine? To answer this, look at how the domain of the original sine function was restricted to construct the arcsine. Why can't the domain of the cosine be restricted in exactly the same way to construct the arccosine? How should the domain of the cosine be restricted?

35.   Using a calculator or computer, estimate all points of intersection of the graphs of $f(x) = x + \sin x$ and $g(x) = x^3$. How do you know you have found all of them?

36.   (a)   Use a graphing calculator or computer to estimate the period of $2 \sin \theta + 3 \cos(2\theta)$.
      (b)   Explain your answer, given that the period of $\sin \theta$ is 2π and the period of $\cos(2\theta)$ is π.

37.   A baseball hit at an angle of θ to the horizontal with initial velocity v_0 has horizontal range, R, given by

$$R = \frac{v_0^2}{g} \sin(2\theta).$$

Here g is a constant, the acceleration due to gravity. Sketch R as a function of θ for $0 \leq \theta \leq \pi/2$. What angle gives the maximum range? What is the maximum range?

38.   Consider a mass oscillating on the end of a spring. The distance, y, of the mass from its equilibrium position is given by

$$y = y_0 \cos(2\pi\omega t).$$

Here y is in centimeters, t is time in seconds, and y_0 and ω are positive constants.

(a)   What is the meaning of y_0 in terms of the oscillation?
(b)   How many oscillations are completed in 1 second?

1.10 POLYNOMIALS AND RATIONAL FUNCTIONS

Polynomials

Some of the best-known functions for which there are formulas are the polynomials:

$$y = p(x) = a_n x^n + a_{n-1} x^{n-1} + \cdots + a_1 x + a_0.$$

Here n is a positive integer called the *degree* of the polynomial, and $a_n, a_{n-1}, \ldots, a_1, a_0$ are constants, with leading coefficient $a_n \neq 0$. An example of a polynomial of degree $n = 3$ is

$$y = p(x) = 2x^3 - x^2 - 5x - 7.$$

In this case $a_3 = 2, a_2 = -1, a_1 = -5$, and $a_0 = -7$. The shape of the graph of a polynomial

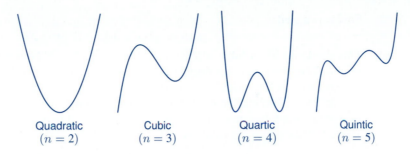

Quadratic
$(n = 2)$

Cubic
$(n = 3)$

Quartic
$(n = 4)$

Quintic
$(n = 5)$

Figure 1.82: Graphs of typical polynomials of degree n

depends on its degree; typical graphs are shown in Figure 1.82. These graphs correspond to a positive coefficient for x^n; a negative coefficient turns the graph upside down. Notice that the graph of the quadratic "turns around" once, the cubic "turns around" twice, and the quartic (fourth degree) "turns around" three times. An nth degree polynomial "turns around" at most $n - 1$ times (where n is a positive integer), but there may be fewer turns.

Example 1   Find possible formulas for the polynomials whose graphs are in Figure 1.83.

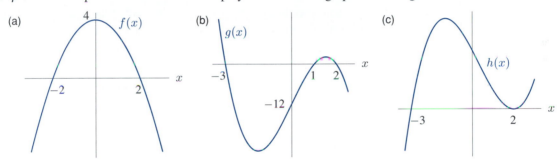

(a)   (b)   (c)

Figure 1.83: Graphs of polynomials

Solution   (a)   This graph appears to be a parabola, turned upside down, and moved up by 4, so

$$f(x) = -x^2 + 4.$$

The negative sign turns the parabola upside down and the $+4$ moves it up by 4. Notice that this formula does give the correct x-intercepts since $0 = -x^2 + 4$ has solutions $x = \pm 2$. These values of x are called *zeros* of f.

We can also solve this problem by looking at the x-intercepts first, which tell us that $f(x)$ has factors of $(x + 2)$ and $(x - 2)$. So

$$f(x) = k(x + 2)(x - 2).$$

To find k, use the fact that the graph has a y-intercept of 4, so $f(0) = 4$, giving

$$4 = k(0 + 2)(0 - 2),$$

or $k = -1$. Therefore, $f(x) = -(x + 2)(x - 2)$, which multiplies out to $-x^2 + 4$. Note that

$$f(x) = 4 - \frac{x^4}{4}$$

also fits the requirements, but its "shoulders" are sharper. There are many possible answers to these questions.

(b)   This looks like a cubic with factors $(x + 3)$, $(x - 1)$, and $(x - 2)$, one for each intercept:

$$g(x) = k(x + 3)(x - 1)(x - 2).$$

Since the y-intercept is -12,

$$-12 = k(0 + 3)(0 - 1)(0 - 2).$$

So $k = -2$, and

$$g(x) = -2(x+3)(x-1)(x-2).$$

(c) This also looks like a cubic with zeros at $x = 2$ and $x = -3$. Notice that at $x = 2$ the graph of $h(x)$ touches the x-axis but doesn't cross it, whereas at $x = -3$ the graph crosses the x-axis. We say that $x = 2$ is a *double zero*, but that $x = -3$ is a single zero.

To find a formula for $h(x)$, first imagine the graph of $h(x)$ to be slightly lower down, so that the graph has one x-intercept near $x = -3$ and two near $x = 2$, say at $x = 1.9$ and $x = 2.1$. Then a formula would be

$$h(x) \approx k(x+3)(x-1.9)(x-2.1).$$

Now move the graph back to its original position. The zeros at $x = 1.9$ and $x = 2.1$ move toward $x = 2$, giving

$$h(x) = k(x+3)(x-2)(x-2) = k(x+3)(x-2)^2.$$

The double zero leads to a repeated factor, $(x-2)^2$. Notice that when $x > 2$, the factor $(x-2)^2$ is positive, and when $x < 2$, the factor $(x-2)^2$ is still positive. This reflects the fact that $h(x)$ doesn't change sign near $x = 2$. Compare this with the behavior near the single zero at $x = -3$, where h does change sign.

We cannot find k, as no coordinates are given for points off of the x-axis. Inserting any positive value of k will vertically stretch the graph but not change the zeros, and so any positive k will work.

Example 2   Using a calculator or computer, sketch graphs of $y = x^4$ and $y = x^4 - 15x^2 - 15x$ for $-4 \leq x \leq 4$ and for $-20 \leq x \leq 20$. Set the y range to $-100 \leq y \leq 100$ for the first domain, and to $-100 \leq y \leq 200,000$ for the second. What do you observe?

Solution   From the graphs in Figure 1.84 we see that close up (for $-4 \leq x \leq 4$) the graphs look different; from far away, however, they are almost indistinguishable. The reason is that the leading terms (those with the highest power of x) are the same, namely x^4, and for large values of x, the leading term dominates the other terms.

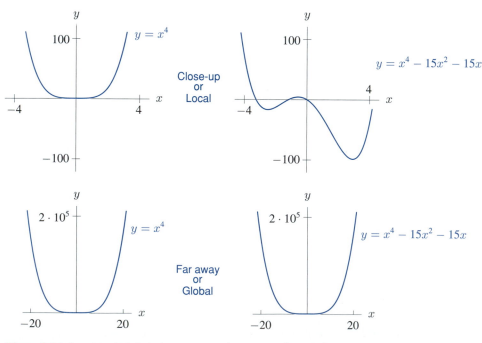

Figure 1.84: Local and global views of $y = x^4$ and $y = x^4 - 15x^2 - 15x$

When $x = \pm 20$, the differences in the values of the two functions, although large, are tiny compared with the vertical scale (-100 to $200{,}000$). See Table 1.25. So these differences can't be seen on the graph.

TABLE 1.25   *Numerical values of $y = x^4$ and $y = x^4 - 15x^2 - 15x$*

x	$y = x^4$	$y = x^4 - 15x^2 - 15x$	Difference
-20	160,000	154,300	5700
-15	50,625	47,475	3150
15	50,625	47,025	3600
20	160,000	153,700	6300

Rational Functions

Rational functions are those of the form

$$f(x) = \frac{p(x)}{q(x)},$$

where p and q are polynomials. Their graphs often have vertical asymptotes where the denominator is zero. If the denominator is nowhere zero, there are no vertical asymptotes. Rational functions may also have horizontal asymptotes; these occur if $f(x)$ approaches a finite number as $x \to \infty$ or $x \to -\infty$. We call the behavior of a function as $x \to \pm\infty$ its *end behavior*.

Example 3   Graph and describe the behavior of $y = \dfrac{1}{x^2 + 4}$.

Solution   This graph has no vertical asymptotes as the denominator is never zero. The graph is symmetric about the y-axis, and the x-axis is a horizontal asymptote since

$$y \to 0 \quad \text{as} \quad x \to \pm\infty.$$

See Figure 1.85.

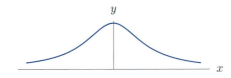

Figure 1.85: Graph of $y = \frac{1}{x^2+4}$

Example 4   Plot and describe the graph of $y = \dfrac{x^2 - 4}{x^2 - 1}$, including end behavior.

Solution   Factoring gives

$$y = \frac{x^2 - 4}{x^2 - 1} = \frac{(x + 2)(x - 2)}{(x + 1)(x - 1)}$$

so $x = \pm 1$ are vertical asymptotes. If $y = 0$, then $(x + 2)(x - 2) = 0$ or $x = \pm 2$; these are the x-intercepts. Note that the vertical asymptotes arise from zeros of the denominator, while zeros of the numerator give rise to x-intercepts. Substituting $x = 0$ gives $y = 4$; this is the y-intercept. Notice that, in this example, positive and negative x's give the same y value, so the graph is symmetric about the y-axis. This is because $(-x)^2 = x^2$.

TABLE 1.26   *Values*
of $y = \frac{x^2-4}{x^2-1}$

x	$y = \dfrac{x^2-4}{x^2-1}$
± 10	0.969697
± 100	0.999700
± 1000	0.999997

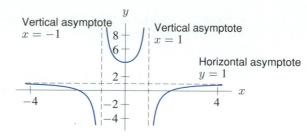

Figure 1.86: Graph of the function $y = \frac{x^2-4}{x^2-1}$

To see what happens as $x \to \pm\infty$, look at the y-values in Table 1.26. Clearly y is getting closer to 1 as x gets large positively or negatively. This can also be seen by realizing that as $x \to \pm\infty$, only the highest powers of x really matter. For large x, the 4 and the 1 are insignificant compared to x^2, so

$$y = \frac{x^2 - 4}{x^2 - 1} \approx \frac{x^2}{x^2} = 1 \quad \text{for large } x.$$

So $y \to 1$ as $x \to \pm\infty$, and therefore the horizontal asymptote is $y = 1$. Since, for $x > 1$, the denominator is positive and the numerator is less than the denominator, the graph lies *below* its asymptote. (Why doesn't the graph lie below $y = 1$ when $-1 < x < 1$?) See Figure 1.86.

Problems for Section 1.10

1. Determine the end behavior of each function below, as $x \to +\infty$ and as $x \to -\infty$.
   (a) $f(x) = x^7$
   (b) $f(x) = 5 + 24x + 78x^3 - 12x^4$
   (c) $f(x) = x^{-4}$
   (d) $f(x) = \left(6x^3 - 5x^2 + 12\right) / \left(x^3 - 8\right)$

2. Assume that each of the graphs in Figure 1.87 is of a polynomial. For each graph:
   (a) What is the minimum possible degree of the polynomial?
   (b) Is the *leading coefficient* of the polynomial positive or negative? (You may assume that the graphs are in windows large enough to show the global behavior.)

(I)    (II)    (III)    (IV) 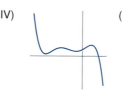   (V)

Figure 1.87

For Problems 3–6, sketch graphs of the polynomials by hand. Check your work using a calculator or computer.

3. $f(x) = (x - 3)(x - 4)(x - 5)$
4. $f(x) = (x + 3)(x + 4)(x + 1)(x + 2)$
5. $f(x) = 5(x^2 - 4)(x^2 - 25)$
6. $f(x) = -5(x^2 - 4)(25 - x^2)$

7. Graph each of the following polynomials on a graphing calculator or computer and use the graphs to determine which ones are odd functions, even functions, or neither.
   (a) $a(x) = x^2$
   (b) $b(x) = x^3$
   (c) $c(x) = x^4$
   (d) $d(x) = -10x^5$
   (e) $e(x) = x^3 + 3x^2$
   (f) $f(x) = x^4 - x^2$
   (g) $g(x) = x^5 - 2x^3$
   (h) $h(x) = 2x^4 + 5$
   (i) $i(x) = 7x + 5$

8. Based on your answers to Problem 7, how can you tell whether a polynomial is an odd or even function by just looking at its formula?

For each of the rational functions in Problems 9–11, find asymptotes, the behavior of the function as $x \to \pm\infty$ and the behavior of the function near any vertical asymptotes. Then use this information to sketch a graph by hand. Check your work using a computer or graphing calculator.

9. $y = \dfrac{1 - x^2}{x - 2}$

10. $y = \dfrac{1 - 4x}{2x + 2}$

11. $y = \dfrac{x^2 + 2x + 1}{x^2 - 4}$

12. Which of the functions I–III meet each of the following descriptions? There may be more than one function for each description, or none at all.

    (a) Horizontal asymptote of $y = 1$.
    (b) The x-axis is a horizontal asymptote.
    (c) Symmetric about the y-axis.
    (d) An odd function.
    (e) Vertical asymptotes at $x = \pm 1$.

    I. $y = \dfrac{x - 1}{x^2 + 1}$
    II. $y = \dfrac{x^2 - 1}{x^2 + 1}$
    III. $y = \dfrac{x^2 + 1}{x^2 - 1}$

13. The height of an object above the ground at time t is given by

$$s = v_0 t - \frac{g}{2}t^2,$$

where v_0 represents the initial velocity and g is a constant called the acceleration due to gravity.

    (a) At what height is the object initially?
    (b) How long is the object in the air before it hits the ground?
    (c) When will the object reach its maximum height?
    (d) What is that maximum height?

14. A pomegranate is thrown from ground level straight up into the air at time $t = 0$ with velocity 64 feet per second. Its height at time t is $f(t) = -16t^2 + 64t$. Find the time it hits the ground and the instant that it reaches its highest point. What is the maximum height?

15. If $f(x) = ax^2 + bx + c$, what do you know about the values of a, b, and c if:

    (a) $(1, 1)$ is on the graph of $f(x)$?
    (b) $(1, 1)$ is the vertex of the graph of $f(x)$? (You may want to use the fact that the equation for the axis of symmetry of the parabola $y = ax^2 + bx + c$ is $x = -b/2a$.)
    (c) The y intercept of the graph is $(0, 6)$?
    (d) Find a quadratic function that satisfies all three conditions.

Determine cubic polynomials that represent each of the graphs in Problems 16–17.

16.

17.

For Problems 18–21:
(a) Find a possible formula for the graph.
(b) For each graph, read off approximate intervals over which the function is increasing and over which it is decreasing.

18.

19.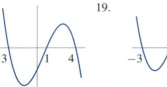

20.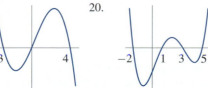

21.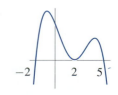

22. The graph of a rational function $y = f(x)$ is given in Figure 1.88. If $f(x) = g(x)/h(x)$ with $g(x)$ and $h(x)$ both quadratic functions, give possible formulas for $g(x)$ and $h(x)$. (There are many possible answers.)

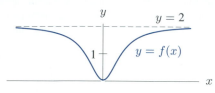

Figure 1.88

23. Match the following functions with the graphs in Figure 1.89. Assume $0 < b < a$.

(a) $y = \dfrac{a}{x} - x$

(b) $y = \dfrac{(x-a)(x+a)}{x}$

(c) $y = \dfrac{(x-a)(x^2+a)}{x^2}$

(d) $y = \dfrac{(x-a)(x+a)}{(x-b)(x+b)}$

(I) 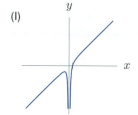   (II) 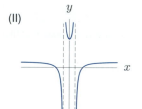   (III) 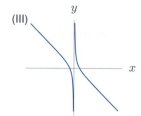   (IV)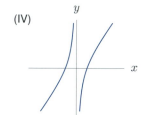

Figure 1.89

24. The rate, R, at which a population in a confined space increases is proportional to the product of the current population, P, and the difference between the *carrying capacity*, L, and the current population. (The carrying capacity is the maximum population the environment can sustain.)

    (a) Write R as a function of P.
    (b) Sketch R as a function of P.

25. Newton's Second Law of Motion, $F = ma$, tells us that the net force, F, on a train of mass m is proportional to its acceleration, a. Suppose that the only forces are those of the engine, which exerts a constant force, F_E, in the direction of motion, and the wind resistance, which exerts a force proportional to the square of the train's velocity, v, but in the opposite direction.

    (a) Write a formula giving a as a function of v.
    (b) Sketch a graph of a against v.

26. Consider a box-shaped package with square ends. The US postal service will accept such a package if the sum of its length plus girth is less than 108 inches. (The girth is the perimeter of the square cross-section perpendicular to the length.) Find a formula for the volume, V, of a package which just meets the post office criteria in terms of s, the length of the side of the square end. Sketch a graph of V against s.

27. A cylindrical can of fixed volume V has closed ends and radius r for $r > 0$.

    (a) Find the surface area, S, as a function of r.
    (b) What happens to the value of S as $r \to \infty$?
    (c) Sketch a graph of S against r, assuming $V = 10$ cm^3.

28. After running 3 miles at a speed of x mph, a man walked the next 6 miles at a speed that was 2 mph slower. Express the total time spent on the trip as a function of x. What horizontal and vertical asymptotes does the graph of this function have?

29. Consider the point P at the intersection of the circle $x^2 + y^2 = 2a^2$, and the parabola $y = x^2/a$ in Figure 1.90. If a is increased, the point P will trace out a curve. Find the equation of this curve.

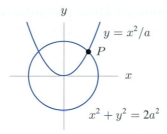

Figure 1.90

30. Suppose c is the velocity of light. An object of mass m moving with a velocity v, which is small compared to c, has energy given approximately by

$$E \approx \frac{1}{2}mv^2.$$

If v is comparable in size to c, then the energy must be computed by the exact formula

$$E = mc^2 \left(\frac{1}{\sqrt{1 - v^2/c^2}} - 1 \right).$$

(a) Plot a graph of both formulas for E against v for $0 \le v \le 5 \cdot 10^8$ and $0 \le E \le 5 \cdot 10^{17}$. Take $m = 1$ kg and $c = 3 \cdot 10^8$ m/sec. Explain how you can predict from the exact formula the position of the vertical asymptote.

(b) What do the graphs tell you about the approximation? For what values of v does the first formula give a good approximation to E?

1.11 INTRODUCTION TO CONTINUITY

This section introduces the concept of *continuity* from graphical and numerical points of view. The section on continuity and limits on page 127 investigates the same idea in more depth.

Continuity of a Function on an Interval

Roughly speaking, a function is said to be *continuous* on an interval if its graph has no breaks, jumps, or holes in that interval. Continuity is important because, as we shall see, functions with this property have many other desirable properties.

For example, to locate the zeros of a function, we often look for intervals where the function changes sign. In the case of the function $f(x) = 3x^3 - x^2 + 2x - 1$, for instance, we expect[10] to find a zero between 0 and 1 because $f(0) = -1$ and $f(1) = 3$. (See Figure 1.91.) To be sure that $f(x)$ has a zero there, we need to know that the graph of the function has no breaks or jumps in it. Otherwise the graph could jump across the x-axis, changing sign but not creating a zero. For example, $f(x) = 1/x$ has opposite signs at $x = -1$ and $x = 1$, but no zeros because of the break at $x = 0$. (See Figure 1.92.) In order to be certain that a function has a zero in an interval on which it changes sign, we need to know that the function is defined and continuous.

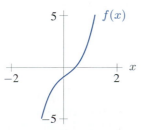

Figure 1.91: The graph of $f(x) = 3x^3 - x^2 + 2x - 1$

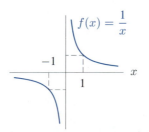

Figure 1.92: No zero although $f(-1)$ and $f(1)$ have opposite signs

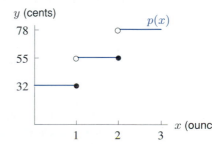

Figure 1.93: Cost of mailing a letter

---

[10]This is due to the Intermediate Value Theorem, which is discussed on page 84.

What Does Continuity Mean Graphically?

A continuous function has a graph which can be drawn without lifting the pencil from the paper.

Example: The function $f(x) = 3x^2 - x^2 + 2x + 1$ is continuous on any interval. (See Figure 1.91.)

Example: The function $f(x) = 1/x$ is not defined at $x = 0$. It is continuous on any interval not containing the origin. (See Figure 1.92.)

Example: Suppose $p(x)$ is the price of mailing a first-class letter weighing x ounces. It costs 32¢ for one ounce or less, 55¢ between the first and second ounces, and so on. So the graph (in Figure 1.93) is a series of steps. This function is not continuous on any interval containing a positive integer because the graph jumps at these points.

What Does Continuity Mean Numerically?

A function is continuous if nearby values of the independent variable give nearby values of the function. In practical work, continuity is important because it means that small errors in the independent variable lead to small errors in the value of the function.

Example: Suppose that $f(x) = x^2$ and that we want to compute $f(\pi)$. Knowing f is continuous tells us that taking $x = 3.14$ should give a good approximation to $f(\pi)$, and that we can get a better approximation to $f(\pi)$ by using more decimals of π.

Example: If $p(x)$ is the cost of mailing a letter weighing x ounces, then $p(0.99) = p(1) = 32¢$, whereas $p(1.01) = 55¢$, because as soon as we get over 1 ounce, the price jumps up to 55¢. So a small difference in the weight of a letter can lead to a significant difference in its mailing cost. Hence p is not continuous.

Which Functions are Continuous?

Requiring a function to be continuous on an interval is not asking very much, as any function whose graph is an unbroken curve over the interval is continuous. For example, exponential functions, polynomials, and sine and cosine are continuous on every interval. Rational functions are continuous on any interval in which their denominators are not zero. Functions created by adding, multiplying, or composing continuous functions are also continuous.

Example 1   What do the values in Table 1.27 tell you about the zeros of $f(x) = \cos x - 2x^2$?

TABLE 1.27

x	$f(x)$
0	1.00
0.2	0.90
0.4	0.60
0.6	0.11
0.8	-0.58
1.0	-1.46

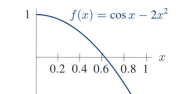

Figure 1.94: Zeros occur where the graph of a continuous function crosses the horizontal axis

Solution   Since $f(x)$ is the difference of the two continuous functions, it is continuous. We conclude that $f(x)$ has at least one zero in the interval $0.6 < x < 0.8$, since $f(x)$ changes from positive to negative on that interval. The graph of $f(x)$ in Figure 1.94 suggests that there is only one zero in the interval $0 \le x \le 1$, but we cannot be sure of this from the graph or the table of values.

Problems for Section 1.11

Are the functions in Problems 1–8 continuous on the given intervals?

1. $\dfrac{1}{x-2}$ on $[0,3]$

2. $\dfrac{1}{x-2}$ on $[-1,1]$

3. $\dfrac{x}{x^2+2}$ on $[-2,2]$

4. $\dfrac{1}{\sqrt{2x-5}}$ on $[3,4]$

5. $\dfrac{1}{\sin x}$ on $[-\frac{\pi}{2},\frac{\pi}{2}]$

6. $\dfrac{1}{\cos x}$ on $[0,\pi]$

7. $\dfrac{e^{\sin\theta}}{\cos\theta}$ on $[-\frac{\pi}{4},\frac{\pi}{4}]$

8. $\dfrac{e^x}{e^x-1}$ on $[-1,1]$

9. Discuss the continuity of the function g graphed in Figure 1.95 and defined as follows:

$$g(\theta) = \begin{cases} \dfrac{\sin\theta}{\theta} & \text{for } \theta \neq 0 \\[2mm] 1/2 & \text{for } \theta = 0. \end{cases}$$

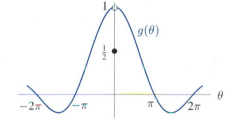

Figure 1.95

10. Sketch the graphs of three different functions that are continuous on $0 \le x \le 1$ and that have the values given in the table. The first function is to have exactly one zero in $[0,1]$, the second is to have at least two zeros in the interval $[0.6, 0.8]$, and the third is to have at least two zeros in the interval $[0, 0.6]$.

x	0	0.2	0.4	0.6	0.8	1.0
$f(x)$	1.00	0.90	0.60	0.11	-0.58	-1.46

11. Use a computer or calculator to sketch the functions $y(x) = \sin x$ and $z_k(x) = ke^{-x}$ for $k = 1, 2, 4, 6, 8, 10$. In each case find the smallest positive solution of the equation $y(x) = z_k(x)$. Now define a new function f by

$$f(k) = \{\text{Smallest positive solution of } y(x) = z_k(x)\}.$$

Explain why the function $f(k)$ is not continuous on the interval $0 \le k \le 10$.

CHAPTER SUMMARY

- **Function terminology**
  Domain/range, increasing/decreasing, concavity, zeros (roots), even/odd, end behavior, asymptotes.
- **Linear Functions**
  Slope, y-intercept. Grow by equal amounts in equal times.
- **Exponential Functions**
  Exponential growth and decay, growth rate, continuous growth rate, doubling time, half life. Grow by equal percentages in equal times.

- **Power Functions**
  Fractional powers, negative powers.

- **Logarithmic Functions**
  Log base 10, natural logarithm.

- **Trigonometric Functions**
  Sine, cosine, tangent, amplitude, period, arcsine, arctangent.

- **Polynomials and Rational Functions**

- **New Functions from Old**
  Inverse functions, composition of functions, shifting, stretching, shrinking, flipping.

- **Working with Functions**
  Find a formula for a linear, exponential, power, logarithmic, or trigonometric function, given graph, table of values, or verbal description. Fit functions to data. Find roots, vertical and horizontal asymptotes.

- **Comparisons Between Functions**
  Exponential functions dominate power and linear functions, higher powers dominate lower powers, power and linear functions dominate logarithmic functions.

- **Continuity**

REVIEW PROBLEMS FOR CHAPTER ONE

1. The entire graph of $y = f(x)$ is shown in Figure 1.96.

    (a) What is the domain of $f(x)$?
    (b) What is the range of $f(x)$?
    (c) List all roots of $f(x)$.
    (d) List all intervals on which $f(x)$ is decreasing.
    (e) Is $f(x)$ concave up or concave down at $x = 6$?
    (f) What is $f(4)$?
    (g) Is this function invertible? Explain.

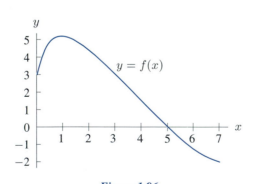

Figure 1.96

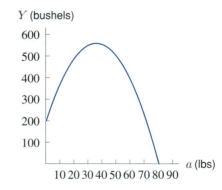

Figure 1.97

2. A graph of the yield, Y, of an apple orchard (in bushels) against the amount, a, of fertilizer (in pounds) used on the orchard is shown in Figure 1.97.

    (a) Describe the effect of the amount of fertilizer on the yield of the orchard.
    (b) What is the vertical intercept? Explain what it means in terms of apples and fertilizer.
    (c) What is the horizontal intercept? Explain what it means in terms of apples and fertilizer.
    (d) What is the range of this function for $0 \le a \le 80$?
    (e) Is the function increasing or decreasing at $a = 60$?
    (f) Is the function concave up or concave down at $a = 40$?

3. When a new product is advertised, more and more people try it. However, the rate at which new people try it slows as time goes on.

    (a) Sketch a graph of the total number of people who have tried such a product against time.
    (b) What do you know about the concavity of the graph?

4. A car starts out slowly and then goes faster and faster until a tire blows out. Sketch a possible graph of the distance the car has traveled as a function of time.

5. Sketch reasonable graphs for the following. Pay particular attention to the concavity of the graphs, and explain your reasoning.

    (a) The total revenue generated by a car rental business, plotted against the amount spent on advertising.
    (b) The temperature of a cup of hot coffee standing in a room, plotted as a function of time.

6. In each pair below, which function will eventually be larger as x goes to infinity?

    (a) $2x^5$ or $200x^4$
    (b) $10x^3$ or e^x
    (c) x^{-2} or x^{-5}
    (d) $x^{1/2}$ or $\ln x$

Find a possible equation involving an exponential for the graph given in Problems 7–8.

7.

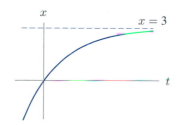

8.

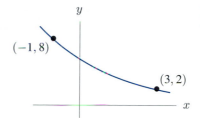

9. By shifting the graph of $y = x^3$, find a cubic polynomial with the graph in Figure 1.98.

10. A population is known to be growing exponentially. Estimate the doubling time of the population shown by the graph in Figure 1.99, and verify graphically that the doubling time is independent of where you start on the graph. Show algebraically that if $P = P_0 a^t$ doubles between time t and time $t + d$, then d is the same number no matter what t is.

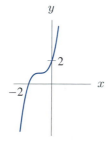

Figure 1.98

population

Figure 1.99

TABLE 1.28

x	$f(x)$	$g(x)$	$h(x)$
-2	12	16	37
-1	17	24	34
0	20	36	31
1	21	54	28
2	18	81	25

11. Table 1.28 contains values for three different functions.

    (a) Which (if any) of these functions are linear functions? For those functions which are linear, find the formula.
    (b) Which (if any) of these functions are exponential functions? For those functions which are exponential, find the formula.

12. Draw the graph of a function $f(x)$ satisfying the following conditions:
- As $x \to \infty$, $f(x) \to 0$.
- As $x \to -\infty$, $f(x) \to +\infty$.
- The roots of $f(x)$ are $-3, 1$, and 2.

13. An airplane uses a fixed amount of fuel for takeoff, a (different) fixed amount for landing, and a third fixed amount per mile when it is in the air. How does the total quantity of fuel required depend on the length of the trip? Write a formula for the function involved. Explain the meaning of the constants in your formula.

14. For tax purposes, you may have to report the value of your assets, such as cars or refrigerators. The value you report depreciates, or drops, with time. The idea is that a car you originally paid $10,000 for may be worth only $5000 a few years later. The simplest way to calculate the value of your asset is using "straight-line depreciation," which assumes that the value is a linear function of time. If a $950 refrigerator depreciates completely in seven years, find a formula for its value as a function of time.

15. For $g(x) = x^2 + 2x + 3$, find and simplify:
    (a) $g(2 + h)$ (b) $g(2)$ (c) $g(2 + h) - g(2)$

16. If $f(x) = x^2 + 1$, find and simplify:
    (a) $f(t + 1)$ (b) $f(t^2 + 1)$ (c) $f(2)$ (d) $2f(t)$ (e) $[f(t)]^2 + 1$

17. For $f(n) = 3n^2 - 2$ and $g(n) = n + 1$, find and simplify:
    (a) $f(n) + g(n)$ (d) $f(g(n))$
    (b) $f(n)g(n)$ (e) $g(f(n))$
    (c) The domain of $f(n)/g(n)$.

State whether the functions in Problems 18–21 are continuous on the interval $[-1, 1]$.

18. $f(x) = |x|$ **19.** $g(x) = \dfrac{|x|}{x}$ **20.** $h(\theta) = \theta \sin \theta$ **21.** $f(t) = \dfrac{\sin t}{t^2}$

Convert the functions in Problems 22–23 into the form $P = P_0 a^t$.

22. $P = 2.91 e^{0.55t}$ **23.** $P = (5 \cdot 10^{-3})e^{-1.9 \cdot 10^{-2}t}$

24. Different kinds of the same element (called different *isotopes*) can have very different half-lives. The decay of plutonium-240 is described by the formula

$$Q = Q_0 e^{-0.00011t},$$

whereas the decay of plutonium-242 is described by

$$Q = Q_0 e^{-0.0000018t}.$$

Find the half-lives of plutonium-240 and plutonium-242.

25. (a) Use the data from Table 1.29 to determine a formula of the form

$$Q = Q_0 e^{rt}$$

which would give the number of rabbits, Q, at time t (in months).
    (b) What is the approximate doubling time for this population of rabbits?
    (c) Use your equation to predict when the rabbit population will reach 1000.

TABLE 1.29

t	0	1	2	3	4	5
Q	25	43	75	130	226	391

26.   A culture of bacteria originally numbers 500. After 2 hours there are 1500 bacteria in the culture. Assuming exponential growth, find how many are present after 6 hours.

27.   A culture of 100 bacteria doubles after 2 hours. How long will it take for the number of bacteria to reach 3,200?

28.   One hundred kilograms of a certain radioactive substance decay to 40 kg after 10 years. Find how much remains after 20 years.

29.   Find the half-life of a radioactive substance that is reduced by 30% in 20 hours.

30.   A radioactive substance has a half-life of 8 years. If 200 gm are present initially, how much remains at the end of 12 years? How long will it be until 90% of the original amount has decayed?

31.   Suppose prices are increasing by 0.1% a day.

   (a)   By what percent do prices increase a year?
   (b)   Looking at your answer to part (a), guess the approximate doubling time of prices increasing at this rate. Check your guess.

32.   In the early 1920s, Germany had tremendously high inflation, called hyperinflation. Photographs of the time show people going to the store with wheelbarrows full of money. If a loaf of bread cost $1/4$ RM in 1919 and 2,400,000 RM in 1922, what was the average yearly inflation rate between 1919 and 1922?

33.   Each planet moves around the sun in an elliptical orbit. The orbital period, T, of a planet is the time it takes the planet to go once around the sun. The semimajor axis of each planet's orbit is the average of the largest and the smallest distances between the planet and the sun. Johannes Kepler (1571-1630) discovered that the period of a planet is proportional to the $\frac{3}{2}$ power of its semimajor axis. What is the orbiting period (in days) of Mercury, the closest planet to the sun, with a semimajor axis of 58 million km? What is the period (in years) of Pluto, the farthest planet, with a semimajor axis of 6000 million km? The semimajor axis of the earth is 150 million km. [Hint: What is the earth's period?]

34.   (a)   Consider the functions graphed in Figure 1.100(a). Find the coordinates of C.
   (b)   Consider the functions in Figure 1.100(b). Find the coordinates of C in terms of b.

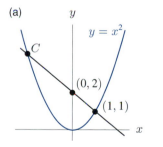 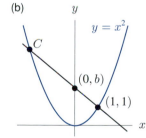

Figure 1.100

Find possible formulas for the graphs in Problems 35–37.

35. 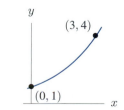   **36.** 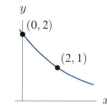   **37.**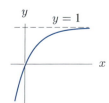

Find possible formulas for the graphs in Problems 38–43.

38.
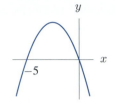

39.
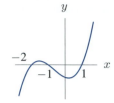

40.
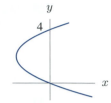

41.
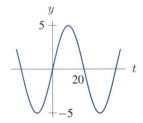

42.
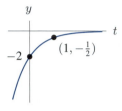

43.
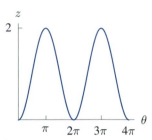

44. The depth of water in a tank oscillates sinusoidally once every 6 hours. If the smallest depth is 5.5 feet and the largest depth is 8.5 feet, find a formula for the depth in terms of time, measured in hours. (There are many possible answers.)

45. Given the graph of $y = h(x)$ in Figure 1.101:

(a) Sketch a graph of:

   (i) $y = h^{-1}(x)$

   (ii) $y = \dfrac{1}{h(x)}$

(b) What becomes of the asymptote when you sketch the inverse function?

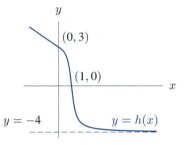

Figure 1.101

46. Each of the functions described by the data in Table 1.30 is increasing over its domain, but each increases in a different way. Which of the graphs in Figure 1.102 best fits each function?

TABLE 1.30

x	$f(x)$	x	$g(x)$	x	$h(x)$
1	1	3.0	1	10	1
2	2	3.2	2	20	2
4	3	3.4	3	28	3
7	4	3.6	4	34	4
11	5	3.8	5	39	5
16	6	4.0	6	43	6
22	7	4.2	7	46.5	7
29	8	4.4	8	49	8
37	9	4.6	9	51	9
47	10	4.8	10	52	10

(a)   (b)   (c)

Figure 1.102

47.

Figure 1.103 is a graph of the function $f(t)$. Here $f(t)$ is the depth in meters below the Atlantic Ocean floor where t million-year-old rock can be found.[11]

(a)   Evaluate $f(15)$, and say what it means in practical terms.

(b)   Is f invertible? Explain.

(c)   Evaluate $f^{-1}(120)$, and say what it means in practical terms.

(d)   Sketch a graph of f^{-1}.

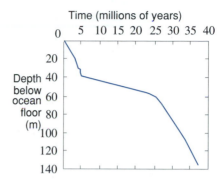

Figure 1.103

48.   A fish population is reproducing at an annual rate equal to 5% of the current population, P. Meanwhile, fish are being caught by fishermen at a constant rate, Y (measured in fish per year).

(a)   Write a formula for the rate, R, at which the fish population is increasing as a function of P.
(b)   Sketch a graph of R against P.

49.   Glucose is fed by intravenous injection at a constant rate, k, into a patient's bloodstream. Once there, the glucose is removed at a rate proportional to the amount of glucose present. If R is the net rate at which the quantity, G, of glucose in the blood is increasing:

(a)   Write a formula giving R as a function of G.
(b)   Sketch a graph of R against G.

50.   A *catalyst* in a chemical reaction is a substance which speeds up the reaction but which does not itself change. If the product of a reaction is itself a catalyst, the reaction is said to be *autocatalytic*. Suppose the rate, r, of a particular autocatalytic reaction is proportional to the quantity of the original material remaining times the quantity of product, p, produced. If the initial quantity of the original material is A and the amount remaining is $A - p$:

(a)   Express r as a function of p.
(b)   What is the value of p when the reaction is proceeding fastest?

51.   What approximate domains and ranges make the graphs of $y = x^2$ and $y = 0.01e^{0.01x}$ look like each of the graphs in Figure 1.104?

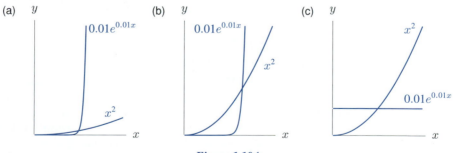

Figure 1.104

---

[11]Data of Dr. Murlene Clark based on core samples drilled by the research ship *Glomar Challenger*, taken from *Initial Reports of the Deep Sea Drilling Project*.

PROJECTS

1. **Matching Functions to Data**

   From the data in Table 1.31, determine a possible formula for each function.[12] Write an explanation of your reasoning.

   TABLE 1.31

x	$f(x)$	$g(x)$	$h(x)$	$F(x)$	$G(x)$	$H(x)$
−5	−10	20	25	0.958924	0.544021	2.958924
−4.5	−9	19	20.25	0.97753	−0.412118	2.97753
−4	−8	18	16	0.756802	−0.989358	2.756802
−3.5	−7	17	12.25	0.350783	−0.656987	2.350783
−3	−6	16	9	−0.14112	0.279415	1.85888
−2.5	−5	15	6.25	−0.598472	0.958924	1.401528
−2	−4	14	4	−0.909297	0.756802	1.090703
−1.5	−3	13	2.25	−0.997495	−0.14112	1.002505
−1	−2	12	1	−0.841471	−0.909297	1.158529
−0.5	−1	11	0.25	−0.479426	−0.841471	1.520574
0	0	10	0	0	0	2
0.5	1	9	0.25	0.479426	0.841471	2.479426
1	2	8	1	0.841471	0.909297	2.841471
1.5	3	7	2.25	0.997495	0.14112	2.997495
2	4	6	4	0.909297	−0.756802	2.909297
2.5	5	5	6.25	0.598472	−0.958924	2.598472
3	6	4	9	0.14112	−0.279415	2.14112
3.5	7	3	12.25	−0.350783	0.656987	1.649217
4	8	2	16	−0.756802	0.989358	1.243198
4.5	9	1	20.25	−0.97753	0.412118	1.02247
5	10	0	25	−0.958924	−0.544021	1.041076

2. **Compound Interest**

   The newspaper article below is from *The New York Times*, May 27, 1990. Fill in the three blanks. (For the first blank, assume that daily compounding is essentially the same as continuous compounding. For the last blank, assume the interest has been compounded yearly, and give your answer in dollars. Exclude the occurrence of leap years.)

   ## 213 Years After Loan, Uncle Sam Is Dunned

   ### By LISA BELKIN

   Special to The New York Times

   SAN ANTONIO, May 26 — More than 200 years ago, a wealthy Pennsylvania merchant named Jacob DeHaven lent $450,000 to the Continental Congress to rescue the troops at Valley Forge. That loan was apparently never repaid.

   So Mr. DeHaven's descendants are taking the United States Government to court to collect what they believe they are owed.

   The total: ____ in today's dollars if the interest is compounded daily at 6 percent, the going rate at the time. If compounded yearly, the bill is only ____.

   ### Family Is Flexible

   The descendants say that they are willing to be flexible about the amount of a settlement and that they might even accept a heartfelt thank you or perhaps a DeHaven statue. But they also note that interest is accumulating at ____ a second.

---

[12]Based on a problem by Lee Zia

FOCUS ON THEORY

UNDERPINNINGS OF CALCULUS

Mathematics in its applied form has existed from time immemorial. Commercial arithmetic and the geometry of land-surveying and building construction were well-developed by 1500 B.C. Gradually people realized that simple mathematical facts may be interrelated in non-obvious ways, and that the interrelationships themselves were worthy of study. Thales (640-546 B.C.) is said to have *proved* that the sum of the angles of a triangle is two right angles. This is the oldest indication we have of the idea of proof in geometry. In the next section we look at an example of a proof.

Over the next few centuries, people began to think of geometry as being not about points, lines, and circles drawn with chalk on a slate, but about abstract entities: points tinier than the smallest speck, lines that are perfectly straight, and circles that are perfectly round. In other words, what one draws on a slate or carves into stone is merely an imperfect model of the abstract reality. Plato (427-347 B.C.) extended this view. He believed that the entire world of experience was an imperfect shadow of the true reality. However, one cannot reason about abstract things without taking some of their properties for granted. In mathematics, these assumptions are called *axioms* or *postulates*.

About 300 B.C. Euclid wrote a textbook, *The Elements,* covering a good deal of geometry, some number theory, and some more advanced topics concerning irrational numbers. It was and is the most successful textbook ever written. (It is still in print.) Euclid begins his treatment of geometry by stating several axioms concerning lines and circles. For example:

If A and B are two points, there is a circle having center A that passes through B.

This is surely a reasonable property to ascribe to the abstract points and circles that we imagine. He goes on to prove many facts about figures in the plane and in space, including the celebrated Pythagorean theorem: The area of the square drawn on the hypotenuse of a right triangle is equal to the sum of the areas of the squares drawn on the two legs.

For many years *The Elements* was revered as the pinnacle of logical reasoning. It is quite rightly regarded as a masterpiece, but its reasoning is no longer thought to be airtight. In fact, there is an error (by modern standards) in the proof of the very first proposition. The argument goes as follows.

Starting from two points A and B, consider the circle with center A passing through B and the circle with center B passing through A. (See Figure 1.105.) Let these circles cut one another at C...

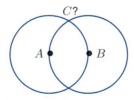

Figure 1.105

But why must there be a point at which these two circles intersect? It is clear from the figure that they do, but the figure is drawn in the real world, not the abstract world of pure geometry. Perhaps when the circles we have drawn are replaced by their abstract representations, it might turn out that there isn't any point where C ought to be. Here and in several other places Euclid seems to have relied on a figure. Perhaps he did not know how to describe clearly those properties of drawings that he believed carried over to abstract geometry and guaranteed the existence of C, and so he left it to his readers to decide whether they believed that C would exist in the ideal realm. Although these deficiencies were noted in classical times, *The Elements* retained its status as the ultimate example of mathematical rigor until well into the nineteenth century. Finally, after centuries of study by many mathematicians, Hilbert (1863-1943) gave what is regarded today as the definitive treatment of Euclidean geometry. It is important to realize that none of the theorems stated by Euclid have been found to be wrong in substance; the difficulties lie entirely in Euclid's incomplete statement of the axioms on which he was relying.

Calculus belongs to a different branch of mathematics than geometry. Instead of lines, circles, and angles, calculus studies the behavior of numerical functions: specifically, functions that represent a rate of change. Calculus also provides a language for expressing laws of nature that govern everything from the behavior of the atomic nucleus to the life cycles of stars.

Some anticipations of calculus can be seen in Euclid and other classical writers, but most of the ideas appear first in the seventeenth century. Newton (1642-1727) and Leibniz (1646-1716) are generally credited with shaping the subject into a coherent theory. Newton's most famous work, *Philosophiae Naturalis Principia Mathematica* (in three volumes) appeared in 1686-1687. Its best known result is that the Laws of Planetary Motion, which had been announced by Kepler (1571-1630) on purely empirical evidence, can be deduced from simpler universal laws, such as the Law of Gravity. In addition, Newton's theory explained other astronomical phenomena, such as the irregularities in the motion of the moon, and terrestrial phenomena, such as the tides. The real significance of *Principia* lies in its demonstration that very complicated physical systems can be successfully modeled by pure mathematics. Although *Principia* uses geometrical arguments, not calculus, the ideas in it were, by Newton's own statement, generated with the aid of calculus.

After its start in the seventeenth century, calculus went for over a century without a proper axiomatic foundation. Newton wrote that it could be rigorously founded on the idea of *limits*, but he never presented his ideas in detail. A limit is, roughly speaking, the value approached by a function near a given point. During the eighteenth century many mathematicians based their work on limits, but their definition of limit was not clear. In 1784, Lagrange (1736-1813) at the Berlin Academy proposed a prize for a successful axiomatic foundation for calculus. He and others were interested in being as certain of the internal consistency of calculus as they were about algebra and geometry. No one was able to successfully respond to the challenge. It remained for Cauchy (1789-1857) to show, around 1820, that limits can be defined rigorously by means of inequalities. The modern definition of the limit, given on page 128, is essentially due to Cauchy.[13]

This rigorous definition of the limit was the advance that was needed in order to begin the axiomatic foundations of calculus, where every result is carefully proved from axioms, or from theorems that have already been proved. Courses in *analysis* follow this chain of logical reasoning.

In this book, we concentrate on developing the solid intuitive understanding on which the rigorous approach depends. We emphasize plausibility arguments, not proofs, but give some glimpses of the theoretical underpinnings of calculus in the "Focus on Theory" sections. We hope that these brief excursions into a more theoretical world encourage you to investigate further.

THE BINOMIAL THEOREM

In everyday life we are often content to believe things simply by observing that they seem to be true. In mathematics, however, we decide what is true by means of logical arguments. Mathematicians attempt to eliminate all possible sources of disagreement by carefully stating *axioms* (assumptions) and *definitions*, formulating precise *theorems* (statements to be proved), and using strict rules of logic. In this section we illustrate how theorems are formulated and proved by studying the example

---

[13]Grabiner, Judith V. "Who Gave You the Epsilon? Cauchy and the Origins of Rigorous Calculus." *American Mathematical Monthly* 90 (1983) pp. 185-194.

of the binomial theorem. In the next section we see how, and why, an axiom is introduced, using as an example the completeness of the real numbers.

Recall the algebraic formulas for squaring and cubing $x + y$:

$$(x + y)^2 = x^2 + 2xy + y^2$$
$$(x + y)^3 = x^3 + 3x^2y + 3xy^2 + y^3.$$

We want to find a general formula for $(x + y)^n$ for any positive integer n. We do this in three stages:

- Find a pattern.
- Formulate a statement, called a *conjecture*, describing the pattern.
- Prove the conjecture.

Once we have proved the statement, it becomes a *theorem*.

Finding a pattern

First we look at some more examples. Multiplying out $(x + y)^n$ for $n = 4, 5, 6$ gives:

$$(x + y)^4 = x^4 + 4x^3y + 6x^2y^2 + 4xy^3 + y^4$$
$$(x + y)^5 = x^5 + 5x^4y + 10x^3y^2 + 10x^2y^3 + 5xy^4 + y^5$$
$$(x + y)^6 = x^6 + 6x^5y + 15x^4y^2 + 20x^3y^3 + 15x^2y^4 + 6xy^5 + y^6.$$

Notice that the exponents of x and y in each term on the right always add up to n. The reason for this is that in the expansion of

$$(x + y)^n = \underbrace{(x + y)(x + y) \cdots (x + y)}_{n \text{ times}},$$

each term comes from choosing x's from some of the factors and y's from the others. The total number of x's and y's chosen equals the total number of $(x + y)$'s, which is n. For example, in the expansion of $(x + y)^3$, choosing x from one of the factors and y from the other two yields a term xy^2. There are three different ways of doing this (depending on which factor x is chosen from), so there are three terms of this form, giving $3xy^2$.

We arrange the coefficients in the expansion of $(x + y)^n$ in a triangle called Pascal's triangle, after the French mathematician Blaise Pascal.

$$
\begin{array}{ccccccccccc}
 & & & & & 1 & & 1 & & & \\
 & & & & 1 & & 2 & & 1 & & \\
 & & & 1 & & 3 & & 3 & & 1 & \\
 & & 1 & & 4 & & 6 & & 4 & & 1 \\
 & 1 & & 5 & & 10 & & 10 & & 5 & & 1 \\
1 & & 6 & & 15 & & 20 & & 15 & & 6 & & 1
\end{array}
$$

The second row in this triangle gives the coefficients in the expansion of $(x+y)^2 = x^2+2xy+y^2$, namely 1, 2, and 1. The next row gives the coefficients for $(x + y)^3$, and so on. The top row gives the coefficients for the expansion $(x + y)^1 = x + y$.

There appears to be a pattern to the triangle: The outside entries are all 1s; each inside entry is equal to the sum of the entries immediately to its left and right in the row above. For example, each 10 in the fourth row has a 4 and a 6 immediately above it, and $10 = 4 + 6$.

Formulating the theorem

We want to prove that, for any n, the coefficients in the expansion of $(x + y)^n$ satisfy the pattern we have observed for $n = 1, \ldots, 6$. The general case is made easier by writing C_k^n for the coefficient of $x^{n-k}y^k$ in the expansion of $(x+y)^n$, so

$$(x+y)^n = C_0^n x^n + C_1^n x^{n-1}y + C_2^n x^{n-2}y^2 + \cdots + C_{n-1}^n xy^{n-1} + C_n^n y^n.$$

Thus, for example, $C_3^5 = 10$ because the $x^2 y^3$ term in the expansion of $(x+y)^5$ is $10x^2 y^3$.
Now

$$C_0^n \quad C_1^n \quad C_2^n \cdots \quad C_{n-1}^n \quad C_n^n$$

is the n-th row in Pascal's triangle. There are two rules that described the pattern we have observed: first, the outside entries are all 1s; second, each inside entry is the sum of the two above it. Thus, we must show, first, that $C_0^n = 1$ and $C_n^n = 1$ for all n, and, second, that

$$C_k^n = C_{k-1}^{n-1} + C_k^{n-1}, \quad 0 < k < n.$$

Notice, if $0 < k < n$, then C_k^n is an inside entry in the triangle, and C_{k-1}^{n-1} and C_k^{n-1} are the entries immediately above it. Now we can state the theorem we want to prove as follows:

The Binomial Theorem

If n is a positive integer and we write

$$(x+y)^n = C_0^n x^n + C_1^n x^{n-1}y + C_2^n x^{n-2}y^2 + \cdots + C_{n-1}^n xy^{n-1} + C_n^n y^n,$$

then

$$C_0^n = C_n^n = 1, \quad \text{for } n \geq 1,$$

and

$$C_k^n = C_{k-1}^{n-1} + C_k^{n-1}, \quad \text{for} \quad n \geq 2 \quad \text{and} \quad 0 < k < n.$$

Proof In the expansion of

$$(x+y)^n = \underbrace{(x+y)(x+y)\cdots(x+y)}_{n \text{ times}},$$

there is only one way of getting the term x^n, and that is by choosing an x from each factor. So, the coefficient of x^n is 1. By the same argument, the coefficient of y^n is also 1, so

$$C_0^n = C_n^n = 1.$$

To prove $C_k^n = C_{k-1}^{n-1} + C_k^{n-1}$, we write

$$(x+y)^n = C_0^n x^n + C_1^n x^{n-1}y + C_2^n x^{n-2}y^2 + \cdots + C_{n-1}^n xy^{n-1} + C_n^n y^n,$$

and we write

$$(x+y)^{n-1} = C_0^{n-1} x^{n-1} + C_1^{n-1} x^{n-2}y + C_2^{n-1} x^{n-3}y^2 + \cdots + C_{n-2}^{n-1} xy^{n-2} + C_{n-1}^{n-1} y^{n-1}.$$

Now, we will use the fact that

$$(x+y)^n = (x+y)(x+y)^{n-1}.$$

Substituting in the expressions for $(x+y)^{n-1}$ and $(x+y)^n$ gives

$$
\begin{aligned}
& C_0^n x^n + C_1^n x^{n-1}y + \quad \cdots \quad + C_{n-1}^n xy^{n-1} + C_n^n y^n \\
&= (x+y)\left(C_0^{n-1} x^{n-1} + C_1^{n-1} x^{n-2}y + \quad \cdots \quad + C_{n-2}^{n-1} xy^{n-2} + C_{n-1}^{n-1} y^{n-1}\right) \\
&= x\left(C_0^{n-1} x^{n-1} + C_1^{n-1} x^{n-2}y + \quad \cdots \quad + C_{n-2}^{n-1} xy^{n-2} + C_{n-1}^{n-1} y^{n-1}\right) \\
&\quad + y\left(C_0^{n-1} x^{n-1} + C_1^{n-1} x^{n-2}y + \quad \cdots \quad + C_{n-2}^{n-1} xy^{n-2} + C_{n-1}^{n-1} y^{n-1}\right) \\
&= C_0^{n-1} x^n + \left(C_1^{n-1} + C_0^{n-1}\right) x^{n-1}y + \cdots + \left(C_{n-1}^{n-1} + C_{n-2}^{n-1}\right) xy^{n-1} + C_{n-1}^{n-1} y^n.
\end{aligned}
$$

The inside terms in this expression have the form $(C_k^{n-1} + C_{k-1}^{n-1})x^{n-k}y^k$, for $k = 1, \ldots, n-1$. The corresponding term in the expansion of $(x+y)^n$ is $C_k^n x^{n-k}y^k$. Since the expressions are equal, the coefficients of like terms must be equal, so

$$C_k^n = C_k^{n-1} + C_{k-1}^{n-1} = C_{k-1}^{n-1} + C_k^{n-1},$$

which is what we wanted to show.

A Formula for the Binomial Coefficients

The numbers C_k^n are called *binomial coefficients*. They are usually computed using the following formula, rather than by writing out Pascal's triangle. (Note that $k! = k(k-1)\cdots 3 \cdot 2 \cdot 1$.)

$$C_k^n = \frac{n!}{k!(n-k)!} = \frac{n(n-1)\cdots(n-k+1)}{k!}$$

This formula holds for $k = 0$ and $k = n$ if we adopt the convention that $0! = 1$:

$$C_0^n = \frac{n!}{0!(n-0)!} = \frac{n!}{n!} = 1 \qquad \text{and} \qquad C_n^n = \frac{n!}{n!(n-n)!} = \frac{n!}{n!} = 1.$$

To prove the formula in general, we use an important technique called *mathematical induction*. There are two steps:

- Prove the formula in the case $n = 1$.

- Prove that if it holds for a specific positive integer n then it holds for $n + 1$.

The second step, called the induction step, enables us to deduce that the formula is true for all n, as follows. Since we know by the first step that it is true for $n = 1$, by the induction step it is true for $n = 2$. But then, by the induction step again, it is true for $n = 3$, and so on.

Proof We have already proved that the formula holds for $n = 1$ since the only binomial coefficients in that case are C_0^1 and C_1^1.

Now we prove the induction step. Suppose that our formula is true for n. That is, suppose that

$$C_k^n = \frac{n!}{k!(n-k)!}, \qquad 0 \le k \le n.$$

We want to deduce the formula for $n + 1$. That is, we want to show that

$$C_k^{n+1} = \frac{(n+1)!}{k!(n+1-k)!}, \qquad 0 \le k \le n+1.$$

We already know that this is true if $k = 0$ or $k = n + 1$. If $0 < k < n + 1$, then, using the Binomial Theorem,

$$
\begin{aligned}
C_k^{n+1} = C_k^n + C_{k-1}^n &= \frac{n!}{k!(n-k)!} + \frac{n!}{(k-1)!(n-k+1)!} \\
&= \frac{n!}{k(k-1)!(n-k)!} + \frac{n!}{(k-1)!(n-k+1)(n-k)!} \\
&= \frac{n!}{(k-1)!(n-k)!}\left(\frac{1}{k} + \frac{1}{n-k+1}\right) \\
&= \frac{n!}{(k-1)!(n-k)!}\left(\frac{n-k+1+k}{k(n-k+1)}\right) \\
&= \frac{n!}{(k-1)!(n-k)!}\frac{(n+1)}{k(n-k+1)} = \frac{(n+1)!}{k!(n-k+1)!},
\end{aligned}
$$

which is what we wanted to prove.

Problems on the Binomial Theorem ━━━━━━━━━━━━━━━━━━━━━━━━━━━━━

1. The formula for the binomial coefficients gives C_k^n as a ratio of integers. Is C_k^n necessarily an integer? Could C_k^n be a fraction? Justify your answer.

2. Look at the entries in the first few rows of Pascal's triangle on page 79. You should see a pattern of symmetry.

   (a) Describe the pattern in words.
   (b) Formulate a conjecture about the binomial coefficients C_k^n that describes the pattern mathematically.
   (c) Prove your conjecture.

3. Add the entries across the rows in Pascal's triangle for the first six rows. You should notice a pattern to the sequence of numbers that you obtain.

   (a) Formulate a general conjecture that describes the pattern.
   (b) Prove your conjecture.

COMPLETENESS OF THE REAL NUMBERS
━━━

If two people argue about something long enough, they may eventually reveal the hidden assumptions that are the root of the argument. In the same way, mathematicians arrive at axioms by a process which is somewhat like arguing with themselves. In attempting to understand something, they question every seemingly obvious statement, hoping to eventually arrive at fundamental axioms.

We apply this method to the process of finding a root of a polynomial by zooming in on its graph. This will lead us to a subtle property of the real numbers, called *completeness*. Many proofs involving limits depend on this property.

Case Study: Finding the Roots of a Polynomial

Consider the polynomial $f(x) = 3x^3 - x^2 + 2x - 1$ on the interval $[0, 1]$. Since $f(0) = -1$ and $f(1) = 3$, we expect that the graph of f crosses the x-axis at some point $x = r$ between $x = 0$ and $x = 1$. Since the coordinates of this point are $(r, 0)$, we have $f(r) = 0$. Suppose we estimate this root by graphing the polynomial on a calculator or computer and zooming in. We start by knowing that

$$0 \leq r \leq 1.$$

By zooming in, we find $f(0.4) < 0$ and $f(0.5) > 0$, so r is trapped in a smaller interval

$$0.4 \leq r \leq 0.5.$$

Successive zooming in shows that

$$0.45 \leq r \leq 0.46$$

$$0.459 \leq r \leq 0.460$$

$$0.4598 \leq r \leq 0.4599.$$

At each stage, we divide the interval into tenths and pick any one for which f is negative at the left end point and positive at the right. (If f is 0 at any of the endpoints, we have found r and can stop.) Continuing this way we obtain a sequence of intervals, each one one-tenth the length of the previous one and each one containing r. (See Figure 1.106.) Although a calculator will only give us a finite number of digits, we could in principle continue forever, generating an infinite sequence of intervals.

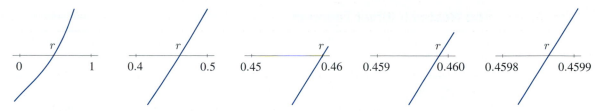

Figure 1.106: Zooming in on a zero of $f(x) = 3x^3 - x^2 + 2x - 1$

It seems that this process leads us to a number r such that $f(r) = 0$. However, there are two questions that can be raised:

- How do we know[14] that this process of zooming in really does close in on a number, r?
- How do we know that $f(r) = 0$?

The Completeness Axiom

Consider the first question above (the answer to the second question is worked out in Problem 26 on page 135). The left hand endpoints of the nested intervals form an ever increasing sequence of decimals; clearly the number r that we are looking for is the smallest number that is greater than all these decimals. The completeness axiom says that, given any nonempty set of numbers, if there is any number which is greater than or equal to all of the numbers in the set, then there is a smallest such number.[15] A number which is greater than or equal to all the numbers in a set is called an *upper bound* for the set. We have the following:

The Completeness Axiom

Any nonempty set of real numbers which has an upper bound has a *least* upper bound.

Example 1   For each of the following sets, say whether it has an upper bound. If so, give the least upper bound.

(a) The set of x such that $-2 < x < 3$.
(b) The set of x such that $-2 \leq x \leq 3$.
(c) The set of all integers.
(d) The sequence $0.9, \ 0.99, \ 0.999, \dots$.

Solution   (a) The numbers 3, 4, and π are all upper bounds; 3 is the least upper bound.
(b) Same as part (a); an upper bound for a set can be in the set, since it only has to be greater than *or equal to* each number in the set.
(c) There is no upper bound for this set; no matter how large a number we choose for the upper bound, there will always be some integer bigger than it.
(d) All the numbers are less than 1, and 1 is the smallest number with this property. Thus, this sequence has a least upper bound of 1.

In the previous example we could see directly what the least upper bounds were. In other situations it may not be obvious. The completeness axiom guarantees the existence of the least upper bound but offers no help in finding it.

---

[14]This is related to the question raised on page 77 about the existence of an intersection point C between two circles.

[15]It is possible to give a definition of the real numbers in which the completeness axiom becomes a theorem.

The Nested Interval Theorem

Now we see how the completeness axiom ensures that the process of zooming in closes in on a specific number.

> **Nested Interval Theorem**
>
> Given an infinite sequence of closed intervals, $[a_n, b_n]$, each one contained within the previous one, then there is at least one number in all the intervals.

Proof Because each interval $[a_n, b_n]$ is contained in the previous one, each a_n must be at least as large as the previous one, so

$$a_1 \leq a_2 \leq a_3 \leq \cdots \leq a_n \leq \cdots.$$

Similarly, each b_n is no larger than its predecessor, so

$$\cdots \leq b_n \leq \cdots \leq b_3 \leq b_2 \leq b_1.$$

All the a_n are bounded above by b_1, and in fact by every b_n. Therefore, by the completeness axiom, we know that the a_n have a least upper bound; call it r. Since r is an upper bound, $r \geq a_n$ for all n. Since r is the least upper bound, r must be less than or equal to each of the upper bounds b_n. Thus, r is in all the intervals.

Note that in our statement of the Nested Interval Property, we did not assume that the lengths of the intervals approached zero, so in general there may be more than one number r in all the intervals. However, when the lengths do approach zero, as in the case of finding a root by zooming in, there is a *unique* number r (see Problem 2).

The Intermediate Value Theorem

When we considered the polynomial $3x^3 - x^2 + 2x - 1$, we assumed it must have a root between $x = 0$ and $x = 1$ because it was negative at $x = 0$ and positive at $x = 1$. More generally, our intuitive notion of continuity tells us that, as we follow the graph of a continuous function f from some point $(a, f(a))$ to another point $(b, f(b))$, then f must take on all intermediate values between $f(a)$ and $f(b)$. (See Figure 1.107.) This is:

> **Intermediate Value Theorem**
>
> Suppose f is continuous on a closed interval $[a, b]$. If k is any number between $f(a)$ and $f(b)$, then there is at least one number c in $[a, b]$ such that $f(c) = k$.

Problems 26 and 27 on page 135 suggest a way of proving the Intermediate Value Theorem using the Nested Interval Theorem.

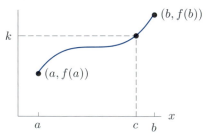

Figure 1.107: The Intermediate Value Theorem

Problems on the Completeness of Real Numbers

1. (a) Using the definitions in this section as a guide, define the following terms:

   (i) A lower bound of a set of numbers

   (ii) The greatest lower bound of a set of numbers

   (b) State the completeness axiom in terms of lower bounds.

2. Let r be a number contained in each of a sequence of nested intervals $[a_n, b_n]$. Suppose that the width of the intervals, $|b_n - a_n|$, goes to 0 as $n \to \infty$. Prove that r is unique. [Hint: Suppose there were two such numbers, and argue to a contradiction.]

3. In this problem we will use the completeness axiom to show that an infinite decimal expansion actually defines a real number, and that the first n digits of the expansion give the number accurate to n decimal places. Let x_n be the number defined by the first n digits of the expansion; we call x_n the n-th *truncation* of the expansion.

   (a) For any n, show that $x_n + (1/10^n)$ is an upper bound for the set of all the truncations.

   (b) Deduce that there is a real number, c, such that $x_n \leq c \leq x_n + (1/10)^n$ for all n. Thus x_n represents c accurate to n places, so it's reasonable to say that c is the number represented by the infinite decimal expansion.

CHAPTER TWO

KEY CONCEPT: THE DERIVATIVE

We begin this chapter by investigating the problem of speed: How can we measure the speed of a moving object at a given instant in time? Or, more fundamentally, what do we mean by the term *speed*? We'll come up with a definition of speed that has wide-ranging implications — not just for the speed problem, but for measuring the rate of change of any quantity. Our journey will lead us to the key concept of *derivative*, which forms the basis for our study of calculus.

The derivative can be interpreted geometrically as the slope of a curve, and physically as a rate of change. Because derivatives can be used to represent everything from fluctuations in interest rates to the rates at which fish populations vary and gas molecules move, they have applications throughout the sciences.

2.1 HOW DO WE MEASURE SPEED?

The speed of an object at an instant in time is surprisingly difficult to define precisely. Consider the statement "At the instant it crossed the finish line, the horse was traveling at 42 mph." How can such a claim be substantiated? A photograph taken at that instant will show the horse motionless—it is no help at all. There is some paradox in trying to study the horse's motion at a particular instant in time, since by focusing on a single instant we stop the motion!

Problems of motion were of central concern to Zeno and other philosophers as early as the fifth century B.C. The modern approach, made famous by Newton's calculus, is to stop looking for a simple notion of speed at an instant, and instead to look at speed over small intervals containing the instant. This method sidesteps the philosophical problems mentioned earlier but introduces new ones of its own.

We illustrate the ideas discussed above by an idealized example, called a thought experiment. It is idealized in the sense that we assume that we can make measurements of distance and time as accurately as we wish.

A Thought Experiment: Average and Instantaneous Velocity

We look at the speed of a small object (say, a grapefruit) that is thrown straight upward into the air at $t = 0$ seconds. The grapefruit leaves the thrower's hand at high speed, slows down until it reaches its maximum height, and then gradually speeds up in the downward direction and finally, "Splat!" (See Figure 2.1.)

But suppose that we would like to be more precise in our determination of the speed, say, at $t = 1$ second. We assume that we can measure the height of the grapefruit above the ground at any time t; we think of the height, y, as a function of time. (See Table 2.1.) "Splat!" comes sometime between 6 and 7 seconds. The numbers show the behavior noted above: During the first second the grapefruit travels $90 - 6 = 84$ feet, and during the second second it travels only $142 - 90 = 52$ feet. Hence the grapefruit traveled faster over the first interval, $0 \le t \le 1$, than the second interval, $1 \le t \le 2$.

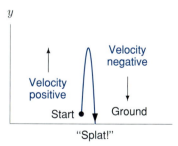

Figure 2.1: The grapefruit's path is straight up and down

TABLE 2.1   *Height of the grapefruit above the ground*

t (sec)	0	1	2	3	4	5	6
y (feet)	6	90	142	162	150	106	30

Velocity versus Speed

From now on, we will make a distinction between velocity and speed. Suppose an object moves along a line. We pick one direction to be positive and say that the *velocity* is positive if it is in the same direction, and negative if it is in the opposite direction. For the grapefruit, upward is positive and downward is negative. (See Figure 2.1.) *Speed* is the magnitude of the velocity and so is always positive or zero.

If $s(t)$ is the position of an object at time t, then the **average velocity** of the object over the interval $a \leq t \leq b$ is

$$\text{Average velocity} = \frac{\text{Change in position}}{\text{Change in time}} = \frac{s(b) - s(a)}{b - a}.$$

In words, the **average velocity** of an object over an interval is the net change in position during the interval divided by the change in time.

Example 1   Compute the average velocity of the grapefruit over the interval $4 \leq t \leq 5$. What is the significance of the sign of your answer?

Solution   During this interval, the grapefruit moves $(106 - 150) = -44$ feet. Therefore the average velocity is -44 ft/sec. The negative sign means the height is decreasing and the grapefruit is moving downward.

Example 2   Compute the average velocity of the grapefruit over the interval $1 \leq t \leq 3$.

Solution   Average velocity $= (162 - 90)/(3 - 1) = 72/2 = 36$ ft/sec.

The average velocity is a useful concept since it gives a rough idea of the behavior of the grapefruit: if two grapefruits are hurled into the air, and one has an average velocity of 10 ft/sec over the interval $0 \leq t \leq 1$ while the second has an average velocity of 100 ft/sec over the same interval, clearly the second one is moving faster.

But average velocity over an interval doesn't solve the problem of measuring the velocity of the grapefruit at *exactly* $t = 1$ second. To get closer to an answer to that question, we have to look at what happens near $t = 1$ in more detail. The data[1] in Figure 2.2 shows the average velocity over small intervals on either side of $t = 1$.

Notice that the average velocity before $t = 1$ is slightly more than the average velocity after $t = 1$. We expect to define the velocity *at* $t = 1$ to be something between these two average velocities. As the size of the interval shrinks, the values of the velocity before $t = 1$ and the velocity after $t = 1$ get closer together. In the smallest interval in Figure 2.2, both velocities are 68.0 ft/sec (to one decimal place), so we will define the velocity at $t = 1$ to be 68.0 ft/sec (to one decimal place).

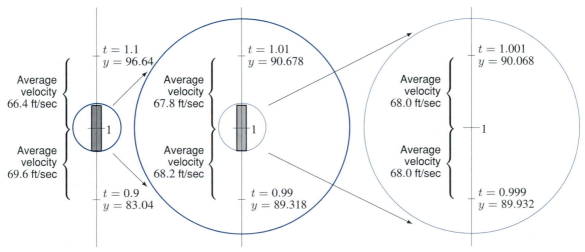

Figure 2.2: Average velocities over intervals on either side of $t = 1$: Showing successively smaller intervals

[1]The data is in fact calculated from the formula $y = 6 + 100t - 16t^2$.

Of course, if we calculate to more decimal places, the average velocities before and after $t = 1$ would no longer agree. To calculate the velocity at $t = 1$ to more decimal places of accuracy, we would take smaller and smaller intervals on either side of $t = 1$ until the average velocities agree to the number of decimal places we wanted. In this way, we could estimate the velocity at $t = 1$ to any accuracy.

Defining Instantaneous Velocity Using Limit Notation

When we take smaller and smaller intervals, it turns out that the average velocities are always just above or just below 68 ft/sec. It seems natural, then, to define velocity at the instant $t = 1$ to be 68 ft/sec. This is called the *instantaneous velocity* at this point. Its definition depends on our being convinced that smaller and smaller intervals will provide average speeds that come arbitrarily close to 68. This process is called *taking the limit*.

Notice how we have replaced the original difficulty of computing velocity at a point by a search for an argument to convince ourselves that the average velocities do approach a number as the time intervals shrink in size. In a sense, we have traded one hard question for another, since we don't yet have any idea how to be certain what number the average velocities are approaching. In the thought experiment, the number seems to be exactly 68, but what if it were 68.000001? How can we be sure that we have taken small enough intervals? Showing that the limit is exactly 68 requires more precise knowledge of the limiting process.

We now define instantaneous velocity at an arbitrary point $t = a$. We use the same method as for $t = 1$: we look at smaller and smaller intervals of size h around $t = a$. Then, over the interval $a \leq t \leq a + h$,

$$\text{Average velocity} = \frac{s(a + h) - s(a)}{h}.$$

The same formula holds when $h < 0$. The instantaneous velocity is the number that the average velocities approach as the intervals decrease in size, that is, as h becomes smaller. So we define

$$\text{Instantaneous velocity} = \text{Limit, as } h \text{ approaches } 0, \text{ of } \frac{s(a + h) - s(a)}{h}.$$

This is written more compactly using limit notation, as follows:

Let $s(t)$ give the position at time t. Then the **instantaneous velocity** at $t = a$ is

$$\begin{matrix} \text{Instantaneous velocity} \\ \text{at } t = a \end{matrix} = \lim_{h \to 0} \frac{s(a + h) - s(a)}{h}.$$

In words, the **instantaneous velocity** of an object at time $t = a$ is given by the limit of the average velocity over an interval, as the interval shrinks around a.

This expression forms the foundation of the rest of calculus. Be sure that you are not confused by the notation and recognize it for what it is: the number that the average velocities approach as the intervals shrink. To estimate that number, the limit, we look at intervals of smaller and smaller, but never zero, length.

Visualizing Velocity: Slope of Curve

Now we will see how to visualize velocity using a graph of height. Let's go back to the grapefruit. Suppose that Figure 2.3 shows the height of the grapefruit plotted against time. (Note that this is not a picture of the grapefruit's path, which is straight up and down.)

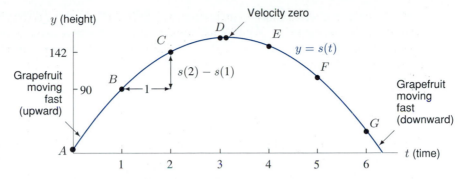

Figure 2.3: The height, y, of the grapefruit at time t

How can we visualize the average velocity on this graph? Suppose $y = s(t)$. Let's consider the interval $1 \leq t \leq 2$ and the expression

$$\text{Average velocity} = \frac{\text{Change in position}}{\text{Change in time}} = \frac{s(2) - s(1)}{2 - 1} = \frac{142 - 90}{1} = 52 \text{ ft/sec.}$$

Now $s(2) - s(1)$ is the change in position over the interval, and it is marked vertically in Figure 2.3. The 1 in the denominator is the time elapsed and is marked horizontally in Figure 2.3. Therefore,

$$\text{Average velocity} = \frac{\text{Change in position}}{\text{Change in time}} = \text{Slope of line joining } B \text{ and } C.$$

(See Figure 2.3.) A similar argument shows the following:

The **average velocity** over any time interval $a \leq t \leq b$ is the slope of the line joining the points on the graph of $s(t)$ corresponding to $t = a$ and $t = b$.

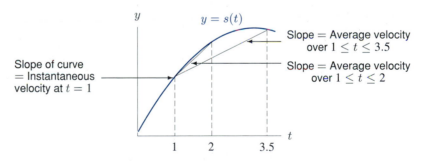

Figure 2.4: Average velocities over small intervals

The next question is how to visualize the instantaneous velocity. Let's think about how we found the instantaneous velocity. We took average velocities over smaller and smaller intervals beginning at the point $t = 1$. Two such velocities are represented by the slopes of the lines in Figure 2.4. As the length of the interval shrinks, the slope of the line gets closer to the slope of the curve at $t = 1$.

The cornerstone of the idea is the fact that, on a very small scale, most functions look almost like straight lines. Imagine taking the graph of a function near a point and "zooming in" to get a close-up view. (See Figure 2.5.) The more we zoom in, the more the curve will appear to be a straight line. In other words, if we repeatedly zoom in on a section of the curve centered at a point of interest, the section of curve will eventually look like a straight line. We call the slope of this line the *slope of the curve* at the point. Therefore, the slope of the magnified line is the instantaneous velocity. Thus, we can say:

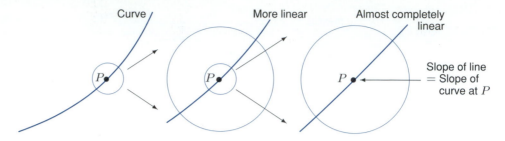

Figure 2.5: Estimating the slope of the curve at the point by "zooming in"

> The **instantaneous velocity** is the slope of the curve at a point.

Look back at the graph of the grapefruit's height as a function of time in Figure 2.3. If we think of the velocity at any point as the slope of the curve there, we can see how the grapefruit's velocity varies during its journey. At points A and B the curve has a large positive slope, indicating that the grapefruit is traveling up rapidly. Point D is almost at the top: the grapefruit is slowing down. At the peak, the slope of the curve is zero: the fruit has slowed to zero velocity for an instant in preparation for its return to earth. At point E the curve has a small negative slope, indicating a slow velocity of descent. Finally, the slope of the curve at point G is large and negative, indicating a large downward velocity that is responsible for the "Splat."

The Idea of a Limit

In the process of defining the instantaneous velocity, we oberved average velocities as time intervals shrank around a point and introduced the notation for a limit. Now we look a bit more at the idea of a limit, which is considered in more detail in the Focus on Theory section on page 127. We talk about the *limit* of the function at the point c.

> We write $\lim_{x \to c} f(x)$ to represent the number L approached by $f(x)$ as x approaches c.

Example 3   Investigate $\lim_{x \to 2} x^2$.

Solution   Notice that we can make x^2 as close to 4 as we like by taking x sufficiently close to 2. (Look at the values of 1.9^2, 1.99^2, 1.999^2, and 2.1^2, 2.01^2, 2.001^2; they seem to be approaching 4 in Table 2.2.) We write

$$\lim_{x \to 2} x^2 = 4,$$

which is read "the limit, as x approaches 2, of x^2 is 4." Notice that the limit doesn't ask what happens *at* $x = 2$, so it is not sufficient to substitute 2 to find the answer. The limit describes behavior of a function *near* a point, not *at* the point.

TABLE 2.2   *Values of x^2*

x	1.9	1.99	1.999	2.001	2.01	2.1
x^2	3.61	3.96	3.996	4.004	4.04	4.41

Example 4   Use a graph to estimate $\lim\limits_{\theta \to 0} \dfrac{\sin \theta}{\theta}$. (Be sure to use radians.)

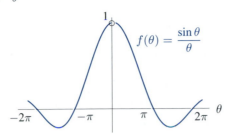

Figure 2.6: Find the limit as $\theta \to 0$

Solution   Figure 2.6 shows that as θ approaches 0 from either side, the value of $\dfrac{\sin \theta}{\theta}$ appears to approach 1, suggesting that $\lim\limits_{\theta \to 0} \dfrac{\sin \theta}{\theta} = 1$. Zooming in on the graph near $\theta = 0$ provides further support for this conclusion. Notice that $\dfrac{\sin \theta}{\theta}$ is undefined at $\theta = 0$.

Example 5   Estimate $\lim\limits_{h \to 0} \dfrac{(3+h)^2 - 9}{h}$ numerically.

Solution   The limit is the value approached by this expression as h approaches 0. The values in Table 2.3 seem to be converging to 6 as $h \to 0$. So it is a reasonable guess that

$$\lim_{h \to 0} \frac{(3+h)^2 - 9}{h} = 6.$$

However, we cannot be sure that the limit is *exactly* 6 by looking at the table. To calculate the limit exactly requires algebra.

TABLE 2.3   *Values of* $\left((3+h)^2 - 9\right)/h$

h	-0.1	-0.01	-0.001	0.001	0.01	0.1
$\left((3+h)^2 - 9\right)/h$	5.9	5.99	5.999	6.001	6.01	6.1

Example 6   Use algebra to find $\lim\limits_{h \to 0} \dfrac{(3+h)^2 - 9}{h}$.

Solution   Expanding the numerator gives

$$\frac{(3+h)^2 - 9}{h} = \frac{9 + 6h + h^2 - 9}{h} = \frac{6h + h^2}{h}.$$

Since taking the limit as $h \to 0$ means looking at values of h near, but not equal, to 0, we can cancel h, giving

$$\lim_{h \to 0} \frac{(3+h)^2 - 9}{h} = \lim_{h \to 0}(6 + h).$$

As h approaches 0, the values of $(6 + h)$ approach 6, so

$$\lim_{h \to 0} \frac{(3+h)^2 - 9}{h} = \lim_{h \to 0}(6 + h) = 6.$$

Problems for Section 2.1

1. A car is driven at a constant speed. Sketch a graph of the distance the car has traveled as a function of time.

2. A car is driven at an increasing speed. Sketch a graph of the distance the car has traveled as a function of time.

3. A car starts at a high speed, and its speed then decreases slowly. Sketch a graph of the distance the car has traveled as a function of time.

4. For the function shown in Figure 2.7, at what labeled points is the slope of the graph positive? Negative? At which labeled point does the graph have the greatest (i.e., most positive) slope? The least slope (i.e., negative and with the largest magnitude)?

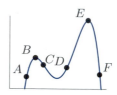

Figure 2.7

5. Match the points labeled on the curve in Figure 2.8 with the given slopes.

Slope	Point
−3	
−1	
0	
1/2	
1	
2	

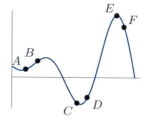

Figure 2.8

6. For the graph $y = f(x)$ shown in Figure 2.9, arrange the following numbers in ascending (i.e., smallest to largest) order:

    - The slope of the graph at A.
    - The slope of the graph at B.
    - The slope of the graph at C.
    - The slope of the line AB.
    - The number 0.
    - The number 1.

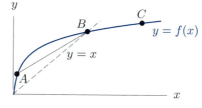

Figure 2.9

7. The graph of $f(t)$ in Figure 2.10 gives the position of a particle at time t. List the following quantities in order, smallest to largest.
    - A, the average velocity between $t = 1$ and $t = 3$,
    - B, the average velocity between $t = 5$ and $t = 6$,
    - C, the instantaneous velocity at $t = 1$,
    - D, the instantaneous velocity at $t = 3$,
    - E, the instantaneous velocity at $t = 5$,
    - F, the instantaneous velocity at $t = 6$.

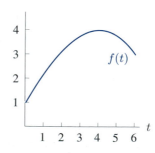

Figure 2.10

8. Suppose a particle is moving at varying velocity along a straight line and that $s = f(t)$ represents the distance of the particle from a point as a function of time, t. Sketch a possible graph for f if the average velocity of the particle between $t = 2$ and $t = 6$ is the same as the instantaneous velocity at $t = 5$.

Estimate the limits in Problems 9–12 by substituting smaller and smaller values of h. Give your answers to one decimal place.

9. $\lim\limits_{h \to 0} \dfrac{(3 + h)^3 - 27}{h}$ 10. $\lim\limits_{h \to 0} \dfrac{\cos h - 1}{h}$ 11. $\lim\limits_{h \to 0} \dfrac{7^h - 1}{h}$ 12. $\lim\limits_{h \to 0} \dfrac{e^{1+h} - e}{h}$

For each of the functions in Problems 13–22, do the following four things:

(a) Make a table of values of $f(x)$ for $x = 0.1, 0.01, 0.001, 0.0001, -0.1, -0.01, -0.001$, and -0.0001.

(b) Make a conjecture about the value of $\lim\limits_{x \to 0} f(x)$.

(c) Graph the function to see if it is consistent with your answers to parts (a) and (b).

(d) Find an interval for x near 0 such that the difference between your conjectured limit and the value of the function is less than 0.01. (In other words, find a window of height 0.02 such that the graph exits the sides of the window and not the top or bottom of the window.)

13. $f(x) = 3x + 1$ 14. $f(x) = x^2 - 1$

15. $f(x) = \sin 2x$ 16. $f(x) = \sin 3x$

17. $f(x) = \dfrac{\sin 2x}{x}$ 18. $f(x) = \dfrac{\sin 3x}{x}$

19. $f(x) = \dfrac{e^x - 1}{x}$ 20. $f(x) = \dfrac{e^{2x} - 1}{x}$

21. $f(x) = \dfrac{\cos 2x - 1 + 2x^2}{x^3}$ 22. $f(x) = \dfrac{\cos 3x - 1 + 4.5x^2}{x^3}$

2.2 THE DERIVATIVE AT A POINT

Average Rate of Change

Now we will apply the analysis of Section 2.1 to any function $y = f(x)$, not just to height as a function of time. In the case of height, we looked at the change in height divided by the change in time, which tells us

$$\text{Average rate of change of height with respect to time} = \frac{s(a + h) - s(a)}{h}.$$

This ratio is called the *difference quotient*. For any function f, we say:

$$\text{Average rate of change of } f \text{ over the interval from } a \text{ to } a + h = \frac{f(a + h) - f(a)}{h}.$$

The numerator, $f(a + h) - f(a)$, measures the change in the value of f over the interval from a to $a + h$. Therefore, the difference quotient is the change in f divided by the change in x. See Figure 2.11.

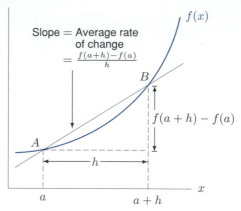

Figure 2.11: Visualizing the average rate of change of f

Although the interval is no longer necessarily a time interval, we still talk about the *average rate of change* of f over the interval. If we want to emphasize the independent variable, we talk about the average rate of change of f *with respect to x*.

Average Rate of Change versus Absolute Change

The average rate of change of a function over an interval is not the same as the absolute change. Absolute change is just the difference in the values of f at the ends of the interval, that is,

$$f(a+h) - f(a).$$

The average rate of change, however, is the absolute change divided by the size of the interval,

$$\frac{f(a+h) - f(a)}{h}.$$

The average rate of change tells how quickly (or slowly) the function changes from one end of the interval to the other, relative to the size of the interval. It is often more useful to know the rate of change than the absolute change. For example, if someone offers you a \$150 return on a \$100 investment, you will want to know how long it is going to take to make that money. Just knowing the absolute change in your money, \$50, is not enough, but knowing the rate of change (i.e., \$50 divided by the time it takes to make it) will help you decide whether or not to make the investment.

Blowing Up a Balloon

Consider the function which gives the radius of a sphere in terms of its volume. For example, think of blowing air into a balloon. You've probably noticed that a balloon seems to blow up faster at the start and then slows down as you blow more air into it. What you're seeing is variation in the rate of change of the radius with respect to volume.

Example 1   The volume, V, of a sphere is given by $V = 4\pi r^3/3$. Solving for r in terms of V gives

$$r = f(V) = \left(\frac{3V}{4\pi}\right)^{1/3}.$$

Calculate the average rate of change of r with respect to V over the intervals $0.5 \leq V \leq 1$ and $1 \leq V \leq 1.5$.

Solution Using the formula for the average rate of change gives

$$\begin{array}{l} \text{Average rate of change} \\ \text{of radius for } 0.5 \leq V \leq 1 \end{array} = \frac{f(1) - f(0.5)}{0.5} = 2\left(\left(\frac{3}{4\pi}\right)^{1/3} - \left(\frac{1.5}{4\pi}\right)^{1/3} \right) \approx 0.26$$

$$\begin{array}{l} \text{Average rate of change} \\ \text{of radius for } 1 \leq V \leq 1.5 \end{array} = \frac{f(1.5) - f(1)}{0.5} = 2\left(\left(\frac{4.5}{4\pi}\right)^{1/3} - \left(\frac{3}{4\pi}\right)^{1/3} \right) \approx 0.18.$$

So we see that the rate decreases as the volume increases.

Instantaneous Rate of Change: The Derivative

We can also define the *instantaneous rate of change* of a function at a point in the same way that we defined instantaneous velocity: we look at the average rate of change over smaller and smaller intervals. This instantaneous rate of change is so important that it is given its own name, the *derivative of f at a*, denoted by $f'(a)$. We define it as follows:

> The **derivative of f at a**, written $f'(a)$, is defined as
>
> $$\begin{array}{l} \text{Rate of change} \\ \text{of } f \text{ at } a \end{array} = f'(a) = \lim_{h \to 0} \frac{f(a+h) - f(a)}{h}.$$
>
> If the limit exists, then f is said to be **differentiable at a**.

To emphasize that $f'(a)$ is the rate of change of $f(x)$ as the variable x changes, we call $f'(a)$ the derivative of f *with respect to x* at $x = a$. When the function $y = s(t)$ represents the position of an object, the derivative $s'(t)$ is the velocity.

Example 2 By choosing small values for h, estimate the instantaneous rate of change of the radius of a sphere with respect to change in volume at $V = 1$.

Solution With $h = 0.01$ and $h = -0.01$, we have the difference quotients

$$\frac{f(1.01) - f(1)}{0.01} \approx 0.2061 \qquad \text{and} \qquad \frac{f(0.99) - f(1)}{-0.01} \approx 0.2075.$$

With $h = 0.001$ and $h = -0.001$,

$$\frac{f(1.001) - f(1)}{0.001} \approx 0.2067 \qquad \text{and} \qquad \frac{f(0.999) - f(1)}{-0.001} \approx 0.2069.$$

The values of these difference quotients suggest that the limit is between 0.2067 and 0.2069. We conclude that the value is about 0.207; taking smaller h values confirms this. So we say

$$f'(1) = \begin{array}{c} \text{Instantaneous rate of change} \\ \text{of radius with respect} \\ \text{to volume at } V = 1 \end{array} \approx 0.207.$$

In this example we found an approximation to the instantaneous rate of change, or derivative, by substituting in smaller and smaller values of h. Now we see how to visualize the derivative.

Visualizing the Derivative: Slope of Curve and Slope of Tangent

As with velocity, we can visualize the derivative $f'(a)$ as the slope of the graph of f at $x = a$. In addition, there is another way to think of $f'(a)$. Consider the difference quotient $(f(a+h)-f(a))/h$. The numerator, $f(a+h)-f(a)$, is the vertical distance marked in Figure 2.12 and h is the horizontal distance, so

$$\text{Average rate of change of } f = \frac{f(a+h) - f(a)}{h} = \text{Slope of line } AB.$$

As h becomes smaller, the line AB approaches the tangent line to the curve at A. (See Figure 2.13.) We say

$$\begin{array}{l}\text{Instantaneous} \\ \text{rate of change of } f \\ \text{at } a\end{array} = \lim_{h \to 0} \frac{f(a+h) - f(a)}{h} = \text{Slope of tangent at } A.$$

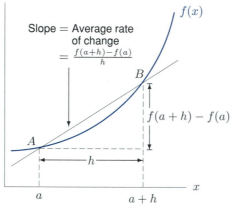

Figure 2.12: Visualizing the average rate of change of f

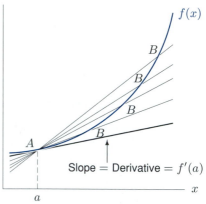

Figure 2.13: Visualizing the instantaneous rate of change of f

> The derivative at point A can be interpreted as:
>
> - The slope of the curve at A.
> - The slope of the tangent line to the curve at A.

The slope interpretation is often useful in gaining rough information about the derivative, as the following examples show.

Example 3   Is the derivative of $\sin x$ at $x = \pi$ positive or negative?

Solution   Looking at a graph of $\sin x$ in Figure 2.14 (remember, x is in radians), we see that a tangent line drawn at $x = \pi$ has negative slope. So the derivative at this point is negative.

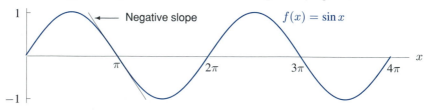

Figure 2.14: Tangent line to $\sin x$ at $x = \pi$

Recall that if we zoom in on the graph of a function $y = f(x)$ at the point where $x = a$, we usually find that the graph looks more and more like a straight line with slope $f'(a)$.

Example 4   By zooming in on the point $(0,0)$ on the graph of the sine function, estimate the value of the derivative of $\sin x$ at $x = 0$, with x in radians.

Solution   Figure 2.15 shows successive graphs of $\sin x$, with smaller and smaller scales. On the interval $-0.1 \leq x \leq 0.1$, the graph looks like a straight line of slope 1. Thus, the derivative of $\sin x$ at $x = 0$ is about 1.

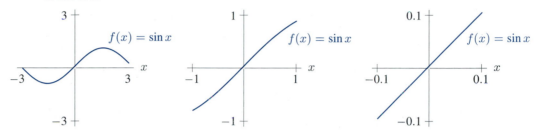

Figure 2.15: Zooming in on the graph of $\sin x$ near $x = 0$

Later we will show that the derivative of $\sin x$ at $x = 0$ is exactly 1. (See page 216 in Section 4.5.) From now on we will assume that this is so.

Example 5   Use the tangent line at $x = 0$ to estimate values of $\sin x$ near $x = 0$.

Solution   In the previous example we see that near $x = 0$, the graph of $y = \sin x$ looks like the graph of the straight line $y = x$; we can use this line to estimate values of $\sin x$ when x is close to 0. For example, the point on the straight line $y = x$ with x coordinate 0.32 is $(0.32, 0.32)$. Since the line is close to the graph of $y = \sin x$, we estimate that $\sin 0.32 \approx 0.32$. (See Figure 2.16.) Checking on a calculator, we find that $\sin 0.32 \approx 0.3146$, so our estimate is quite close. Notice that the graph suggests that the real value of $\sin 0.32$ is slightly less than 0.32.

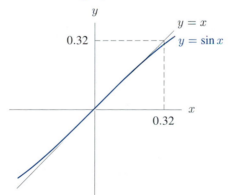

Figure 2.16: Approximating $y = \sin x$ by $y = x$

Why Do We Use Radians and Not Degrees?

After Example 4 we stated that the derivative of $\sin x$ at $x = 0$ is 1, when x is in radians. This is the reason we choose to use radians. If we had done Example 4 in degrees, the derivative of $\sin x$ would have turned out to be a much messier number. (See Problem 25, page 104.)

Estimating the Derivative of an Exponential Function

Example 6   Estimate the value of the derivative of $f(x) = 2^x$ at $x = 0$ graphically and numerically.

Solution   Graphically: Figure 2.17 indicates that the graph is concave up. Assuming this, the slope at A is between the slope of BA and the slope of AC. Since

$$\text{Slope of line } BA = \frac{(2^0 - 2^{-1})}{(0 - (-1))} = \frac{1}{2} \quad \text{and} \quad \text{Slope of line } AC = \frac{(2^1 - 2^0)}{(1 - 0)} = 1,$$

we know that the derivative is between $1/2$ and 1.

Numerically: To estimate the derivative at $x = 0$, we need to look at values of the difference quotient

$$\frac{f(0 + h) - f(0)}{h} = \frac{2^h - 2^0}{h} = \frac{2^h - 1}{h}$$

for small h. Table 2.4 shows some values of 2^h together with values of the difference quotients. (See Problem 33 on page 104 for what happens for very small values of h.)

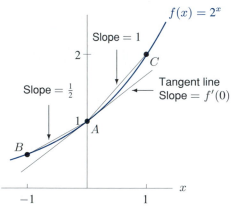

Figure 2.17: Graph of $y = 2^x$ showing the derivative at $x = 0$

TABLE 2.4   *Numerical values for difference quotient of 2^x at $x = 0$*

h	2^h	Difference quotient: $\frac{2^h - 1}{h}$
-0.0003	0.999792078	0.693075
-0.0002	0.999861380	0.693099
-0.0001	0.999930688	0.693123
0	1	
0.0001	1.00006932	0.693171
0.0002	1.00013864	0.693195
0.0003	1.00020797	0.693219

Our concavity assumption tells us that difference quotients calculated with negative h's are smaller than the derivative, and those calculated with positive h's are larger. From Table 2.4 we see that the derivative is between 0.693123 and 0.693171. To three decimal places, $f'(0) = 0.693$.

Example 7   Find an approximate equation for the tangent line to $f(x) = 2^x$ at $x = 0$.

Solution   From the previous example, we know the slope of the tangent line is about 0.693. Since we also know the line has y-intercept 1, its equation is

$$y = 0.693x + 1.$$

Computing the Derivative of x^2

In previous examples we got an approximation to the derivative by using smaller and smaller values of h. We now see how to find the derivative of x^2 exactly.

Example 8   Find the derivative of the function $f(x) = x^2$ at the point $x = 1$.

Solution   We need to look at

$$f'(1) = \lim_{h \to 0} \frac{f(1 + h) - f(1)}{h}.$$

This is the same as

$$\lim_{h \to 0} \frac{(1 + h)^2 - 1^2}{h} = \lim_{h \to 0} \frac{(1 + 2h + h^2) - 1}{h} = \lim_{h \to 0} \frac{2h + h^2}{h}.$$

Since the limit only examines values of h close to, but not equal to zero, we can divide by h in the expression $(2h + h^2)/h$. We get

$$\lim_{h \to 0} \frac{h(2 + h)}{h} = \lim_{h \to 0}(2 + h).$$

This limit is 2, so $f'(1) = 2$. At $x = 1$ the rate of change of x^2 is 2.

Since the derivative is the rate of change, $f'(1) = 2$ means that for small changes in x near $x = 1$, the change in $f(x) = x^2$ is about twice as big as the change in x. As an example, if x changes from 1 to 1.1, a net change of 0.1, then $f(x)$ changes by about 0.2. Figure 2.18 shows this geometrically.

Table 2.5 shows the derivative of $f(x) = x^2$ numerically. Notice that near $x = 1$, every time the value of x increases by 0.001, the value of x^2 increases by approximately 0.002. Near $x = 1$ the function is approximately linear with slope $0.002/0.001 = 2$.

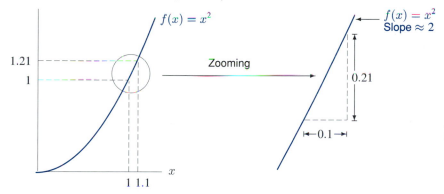

Figure 2.18: Graph of $f(x) = x^2$ near $x = 1$ has slope ≈ 2

TABLE 2.5   *Values of $f(x) = x^2$ near $x = 1$*

x	x^2	Difference in successive x^2 values
0.998	0.996004	
		0.001997
0.999	0.998001	
		0.001999
1.000	1.000000	
		0.002001
1.001	1.002001	
		0.002003
1.002	1.004004	
↑		↑
x increments of 0.001		All approximately 0.002

Problems for Section 2.2

1. Sketch a rough graph of $f(x) = \sin x$, and use the graph to decide whether the derivative of $f(x)$ at $x = 3\pi$ is positive or negative. Give reasons for your decision.

2. (a) Make a table of values rounded to two decimal places for the function $f(x) = e^x$ for $x = 1$, 1.5, 2, 2.5, and 3. Then use the table to answer parts (b) and (c).
    (b) Find the average rate of change of $f(x)$ between $x = 1$ and $x = 3$.
    (c) Use average rates of change to approximate the instantaneous rate of change of $f(x)$ at $x = 2$.

3. Label points A, B, C, D, E, and F on the graph of $y = f(x)$ in Figure 2.19.

    (a) Point A is a point on the curve where the derivative is negative.
    (b) Point B is a point on the curve where the value of the function is negative.
    (c) Point C is a point on the curve where the derivative is largest.
    (d) Point D is a point on the curve where the derivative is zero.
    (e) Points E and F are different points on the curve where the derivative is about the same.

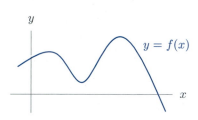

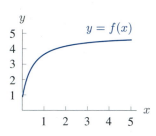

| Figure 2.19 | Figure 2.20 |

4. The graph of $y = f(x)$ is shown in Figure 2.20. Which is larger in each of the following pairs?

    (a) Average rate of change: Between $x = 1$ and $x = 3$? Or between $x = 3$ and $x = 5$?
    (b) $f(2)$ or $f(5)$?
    (c) $f'(1)$ or $f'(4)$?

5. On a copy of Figure 2.21, mark lengths that represent the quantities in parts (a) – (d). (Pick any convenient x, and assume $h > 0$.)
    (a) $f(x)$ (b) $f(x + h)$ (c) $f(x + h) - f(x)$ (d) h

    (e) Using your answers to parts (a)–(d), show how the quantity $\dfrac{f(x + h) - f(x)}{h}$ can be represented as the slope of a line on the graph.

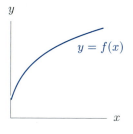

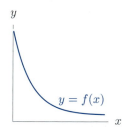

| Figure 2.21 | Figure 2.22 |

6. On a copy of Figure 2.22, mark lengths that represent the quantities in parts (a) – (d). (Pick any convenient x, and assume $h > 0$.)
    (a) $f(x)$ (b) $f(x + h)$ (c) $f(x + h) - f(x)$ (d) h

    (e) Using your answers to parts (a)–(d), show how the quantity $\dfrac{f(x + h) - f(x)}{h}$ can be represented as the slope of a line on the graph.

7. Show how you can represent the following on a copy of Figure 2.23.

    (a) $f(4)$ (b) $f(4) - f(2)$ (c) $\dfrac{f(5) - f(2)}{5 - 2}$ (d) $f'(3)$

8. Consider the function $y = f(x)$ shown in Figure 2.23. For each of the following pairs of numbers, decide which is larger. Explain your answer.

    (a) $f(3)$ or $f(4)$?   (b) $f(3) - f(2)$ or $f(2) - f(1)$?

    (c) $\dfrac{f(2) - f(1)}{2 - 1}$ or $\dfrac{f(3) - f(1)}{3 - 1}$?   (d) $f'(1)$ or $f'(4)$?

9. With the function f given by Figure 2.23, arrange the following quantities in ascending order:

    $$0, \quad 1, \quad f'(2), \quad f'(3), \quad f(3) - f(2)$$

10. Suppose $y = f(x)$ graphed in Figure 2.23 represents the cost of manufacturing x kilograms of a chemical. Then $f(x)/x$ represents the average cost of producing 1 kilogram when x kilograms are made. This problem asks you to visualize these averages graphically.

    (a) Show how to represent $f(4)/4$ as the slope of a line.
    (b) Which is larger, $f(3)/3$ or $f(4)/4$?

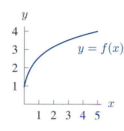

Figure 2.23

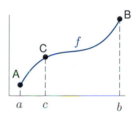

Figure 2.24

11. Consider the function shown in Figure 2.24.

    (a) Write an expression involving f for the slope of the line joining A and B.
    (b) Draw the tangent line at C. Compare its slope to the slope of the line in part (a).
    (c) Are there any other points on the curve at which the slope of the tangent line is the same as the slope of the tangent line at C? If so, mark them on the graph. If not, why not?

12. Table 2.6 shows values of $f(x) = x^3$ near $x = 2$ (to three decimal places). Use it to estimate $f'(2)$.

    TABLE 2.6

x	1.998	1.999	2.000	2.001	2.002
x^3	7.976	7.988	8.000	8.012	8.024

Find the derivatives in Problem 13–18 algebraically.

13. $g(t) = 3t^2 + 5t$ at $t = -1$ 14. $f(x) = 5x^2$ at $x = 10$ 15. $f(x) = x^3$ at $x = -2$
16. $f(x) = x^3 + 5$ at $x = 1$ 17. $g(x) = 1/x$ at $x = 2$ 18. $g(z) = z^{-2}$, find $g'(2)$

For Problems 19–22, find the equation of the line tangent to the function at the given point.

19. $f(x) = x^3$ at $x = -2$ 20. $f(x) = 5x^2$ at $x = 10$
21. $f(x) = x$ at $x = 20$ 22. $f(x) = 1/x^2$ at $(1, 1)$

23. Find the value of the derivative of $f(x) = x^2 + 1$ at $x = 3$ algebraically. Find the equation of the tangent line to f at $x = 3$.

24. Find the equation of the tangent line to $f(x) = x^2 + x$ at $x = 3$.
    Sketch a graph of the function and this tangent line.

25. (a) Estimate $f'(0)$ if $f(x) = \sin x$, with x in degrees.
    (b) In Example 4 on page 99, we found that the derivative of $\sin x$ at $x = 0$ was 1. Why do we get a different result here? (This problem shows why radians are almost always used in calculus.)

26. Estimate the derivative of $f(x) = x^x$ at $x = 2$.

27. For $y = f(x) = 3x^{3/2} - x$, use your calculator to construct a graph of $y = f(x)$, for $0 \leq x \leq 2$. From your graph, estimate $f'(0)$ and $f'(1)$.

28. Let $f(x) = \ln(\cos x)$. Use your calculator to approximate the instantaneous rate of change of f at the point $x = 1$. Do the same thing for $x = \pi/4$. (Note: Be sure that your calculator is set in radians.)

29. There is a function called the error function, $y = \text{erf}(x)$. Suppose your calculator has a button for $\text{erf}(x)$ that gives the following values:

    $$\text{erf}(0) = 0 \quad \text{erf}(1) = 0.84270079 \quad \text{erf}(0.1) = 0.11246292 \quad \text{erf}(0.01) = 0.01128342.$$

    (a) Use all this information to determine your best estimate for $\text{erf}'(0)$. (Give only those digits of which you feel reasonably certain.)
    (b) Suppose you find that $\text{erf}(0.001) = 0.00112838$. How does this extra information change your answer to part (a)?

30. (a) Use your calculator to approximate the derivative of the hyperbolic sine function (written $\sinh x$) at the points $0, 0.3, 0.7$, and 1.
    (b) Can you find a relation between the values of this derivative and the values of the hyperbolic cosine (written $\cosh x$)?

31. The population, P, of China, in billions, can be approximated by the function

    $$P = 1.15(1.014)^t,$$

    where t is the number of years since the start of 1993. According to this model, how fast is the population growing at the start of 1993 and at the start of 1995? Give your answers in millions of people per year.

32. (a) Sketch graphs of the functions $f(x) = \frac{1}{2}x^2$ and $g(x) = f(x) + 3$ on the same set of axes. What can you say about the slopes of the tangent lines to the two graphs at the point $x = 0$? $x = 2$? Any point $x = x_0$?
    (b) Explain why adding a constant value, C, to any function does not change the value of the slope of its graph at any point. [Hint: Let $g(x) = f(x) + C$, and calculate the difference quotients for f and g.]

33. Suppose Table 2.4 on page 100 is continued with smaller values of h. A particular calculator gives the results in Table 2.7. (Your calculator may give slightly different results.) Comment on the values of the difference quotient in Table 2.7. In particular, why is the last value of $(2^h - 1)/h$ zero? What do you expect the calculated value of $(2^h - 1)/h$ to be when $h = 10^{-20}$?

TABLE 2.7   *Questionable values of difference quotients of 2^x near $x = 0$*

h	Difference quotient: $(2^h - 1)/h$
10^{-4}	0.6931712
10^{-6}	0.693147
10^{-8}	0.6931
10^{-10}	0.69
10^{-12}	0

2.3 THE DERIVATIVE FUNCTION

In the last section we looked at the derivative of a function at a fixed point. Now we consider what happens at a variety of points. The derivative generally takes on different values at different points and is itself a function.

First, remember that the derivative of a function at a point tells us the rate at which the value of the function is changing at that point. Geometrically, if we "zoom in" on a point in the graph until it looks like a straight line, the slope of that line is the derivative at that point. Equivalently, we can think of the derivative as the slope of the tangent line at the point, because as we "zoom in," the curve and the tangent line become indistinguishable.

Example 1   Estimate the derivative of the function $f(x)$ graphed in Figure 2.25 at $x = -2, -1, 0, 1, 2, 3, 4, 5$.

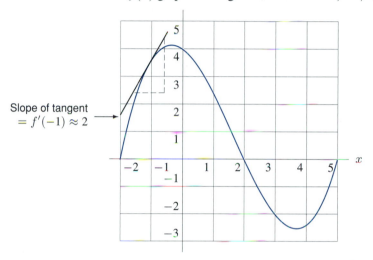

Figure 2.25: Estimating the derivative graphically as the slope of the tangent line

Solution   From the graph we estimate the derivative at any point by placing a straightedge so that it forms the tangent line at that point, and then using the grid squares to estimate the slope of the straightedge. For example, the tangent at $x = -1$ is drawn in Figure 2.25, and has a slope of about 2, so $f'(-1) \approx 2$. Notice that the slope at $x = -2$ is positive and fairly large; the slope at $x = -1$ is positive but smaller. At $x = 0$, the slope is negative, by $x = 1$ it has become more negative, and so on. Some estimates of the derivative are listed in Table 2.8. You should check these values. Are they reasonable? Is the derivative positive where you expect? Negative?

TABLE 2.8   *Estimated values of derivative of function in Figure 2.25*

x	-2	-1	0	1	2	3	4	5
$f'(x)$	6	2	-1	-2	-2	-1	1	4

The important point to notice is that for every x-value, there's a corresponding value of the derivative. Therefore, the derivative is itself a function of x.

For any function f, we define the **derivative function**, f', by

$$f'(x) = \text{ Rate of change of } f \text{ at } x = \lim_{h \to 0} \frac{f(x+h) - f(x)}{h}.$$

For every x-value for which this limit exists, we say f is *differentiable at* that x-value. If the limit exists for all x in the domain of f, we say f is *differentiable everywhere*. The functions we shall deal with will be differentiable at every point in their domain, except perhaps for a few isolated points.

The Derivative Function: Graphically

Example 2   Sketch the graph of the derivative of the function shown in Figure 2.25.

Solution   We plot the values of this derivative given in Table 2.8. We obtain Figure 2.26, which shows a graph of the derivative (the black curve), along with the original function (color).

You should check for yourself that this graph of f' makes sense. The values of f' are positive where f is increasing ($x < -0.3$ or $x > 3.8$) and negative where f is decreasing. Notice that at the points where f has large positive slope, such as $x = -2$, the graph of the derivative is far above the x-axis, as it should be, since the value of the derivative is large there. On the other hand, at points where the slope is gentler, such as $x = -1$, the graph of f' is closer to the x-axis, since the derivative is smaller.

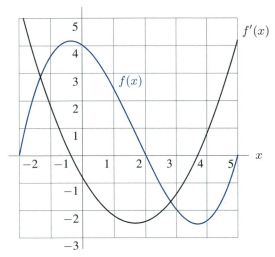

Figure 2.26: Function (colored) and derivative (black) from Example 1

What Does the Derivative Tell Us Graphically?

When f' is positive, the tangent line to f is sloping up; when f' is negative, the tangent line to f is sloping down. If $f' = 0$ everywhere, then the tangent line to f is horizontal everywhere, and so f is constant. We see that the sign of f' tells us whether f is increasing or decreasing.

> If $f' > 0$ on an interval, then f is *increasing* over that interval.
> If $f' < 0$ on an interval, then f is *decreasing* over that interval.

Moreover, the magnitude of the derivative gives us the magnitude of the rate of change; so if f' is large (positive or negative), then the graph of f is steep (up or down), whereas if f' is small the graph of f slopes gently. With this in mind, we can deduce a lot about the behavior of a function from the behavior of its derivative.

The Derivative Function: Numerically

If we are given a table of function values instead of a graph of the function, we can estimate values of the derivative.

Example 3   Table 2.9 gives values of $c(t)$, the concentration (μg/cm^3) of a drug in the bloodstream at time t (min). Construct a table of estimated values for $c'(t)$, the rate of change of $c(t)$ with respect to time.

TABLE 2.9   *Concentration as a function of time*

t (min)	0	0.1	0.2	0.3	0.4	0.5	0.6	0.7	0.8	0.9	1.0
$c(t)$ (μg/cm^3)	0.84	0.89	0.94	0.98	1.00	1.00	0.97	0.90	0.79	0.63	0.41

Solution   We want to estimate values of c' using the values in the table. To do this, we have to assume that the data points are close enough together that the concentration doesn't change wildly between them. From the table, we can see that the concentration is increasing between $t = 0$ and $t = 0.4$, so we expect a positive derivative there. However, the increase is quite slow, so we expect the derivative to be small. The concentration doesn't change between 0.4 and 0.5, so we expect the derivative to be roughly 0 there. From $t = 0.5$ to $t = 1.0$, the concentration starts to decrease, and the rate of decrease gets larger and larger, so we expect the derivative to be negative and of greater and greater magnitude.

Using the data in the table, we can estimate the derivative using the difference quotient:

$$c'(t) \approx \frac{c(t+h) - c(t)}{h}.$$

Since the data points are 0.1 apart, we use $h = 0.1$. We estimate that,

$$c'(0) \approx \frac{c(0.1) - c(0)}{0.1} = \frac{0.89 - 0.84}{0.1} = 0.5 \ \mu\text{g/cm}^3\text{/min}$$

$$c'(0.1) \approx \frac{c(0.2) - c(0.1)}{0.1} = \frac{0.94 - 0.89}{0.1} = 0.5 \ \mu\text{g/cm}^3\text{/min}$$

$$c'(0.2) \approx \frac{c(0.3) - c(0.2)}{0.1} = \frac{0.98 - 0.94}{0.1} = 0.4 \ \mu\text{g/cm}^3\text{/min}$$

$$c'(0.3) \approx \frac{c(0.4) - c(0.3)}{0.1} = \frac{1.00 - 0.98}{0.1} = 0.2 \ \mu\text{g/cm}^3\text{/min}$$

$$c'(0.4) \approx \frac{c(0.5) - c(0.4)}{0.1} = \frac{1.00 - 1.00}{0.1} = 0.0 \ \mu\text{g/cm}^3\text{/min}$$

and so on. These values are tabulated in Table 2.10. Notice that the derivative has small positive values up until $t = 0.4$, where it is roughly 0, and then it gets more and more negative, as we expected. The slopes are shown on the graph of $c(t)$ in Figure 2.27.

TABLE 2.10
Estimated derivative of concentration

t	$c'(t)$
0	0.5
0.1	0.5
0.2	0.4
0.3	0.2
0.4	0.0
0.5	-0.3
0.6	-0.7
0.7	-1.1
0.8	-1.6
0.9	-2.2

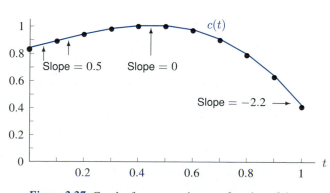

Figure 2.27: Graph of concentration as a function of time

Improving Numerical Estimates for the Derivative

In the previous example, the estimate for the derivative at 0.2 used the interval to the right; we found the average rate of change between $t = 0.2$ and $t = 0.3$. However, we could equally well have gone to the left and used the rate of change between $t = 0.1$ and $t = 0.2$ to approximate the derivative at 0.2. For a more accurate result, we could average these slopes and say

$$c'(0.2) \approx \frac{1}{2} \left(\begin{matrix} \text{Slope to left} \\ \text{of } 0.2 \end{matrix} + \begin{matrix} \text{Slope to right} \\ \text{of } 0.2 \end{matrix} \right) = \frac{0.5 + 0.4}{2} = 0.45.$$

In general, averaging the slopes leads to a more accurate answer.

Derivative Function: From a Formula

If we are given a formula for f, can we come up with a formula for f'? We often can, as shown in the next example. Indeed, much of the power of calculus depends on our ability to find formulas for the derivatives of all the functions we described earlier. This is done systematically in Chapter 4.

Derivative of a Constant Function

The graph of a constant function $f(x) = k$ is a horizontal line, with a slope of 0 everywhere. Therefore, its derivative is 0 everywhere. (See Figure 2.28.)

$$\boxed{\text{If } f(x) = k, \text{ then } f'(x) = 0.}$$

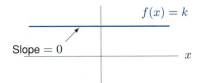

Figure 2.28: A constant function

Derivative of a Linear Function

We already know that the slope of a straight line is constant. This tells us that the derivative of a linear function is constant.

$$\boxed{\text{If } f(x) = b + mx, \text{ then } f'(x) = \text{Slope} = m.}$$

Derivative of a Power Function

Example 4   Find a formula for the derivative of $f(x) = x^2$.

Solution   Before computing the formula for $f'(x)$ algebraically, let's try to guess the formula by looking for a pattern in the values of $f'(x)$. Table 2.11 contains values of $f(x) = x^2$ (rounded to three decimals), which we can use to estimate the values of $f'(1)$, $f'(2)$, and $f'(3)$.

TABLE 2.11 *Values of $f(x) = x^2$ near $x = 1$, $x = 2$, $x = 3$ (rounded to three decimals)*

x	x^2	x	x^2	x	x^2
0.999	0.998	1.999	3.996	2.999	8.994
1.000	1.000	2.000	4.000	3.000	9.000
1.001	1.002	2.001	4.004	3.001	9.006
1.002	1.004	2.002	4.008	3.002	9.012

Near $x = 1$, the value of x^2 increases by about 0.002 each time x increases by 0.001, so

$$f'(1) \approx \frac{0.002}{0.001} = 2.$$

Similarly, near $x = 2$ and $x = 3$, the value of x^2 increases by about 0.004 and 0.006, respectively, when x increases by 0.001. So

$$f'(2) \approx \frac{0.004}{0.001} = 4 \quad \text{and} \quad f'(3) \approx \frac{0.006}{0.001} = 6.$$

Knowing the value of f' at specific points can never tell us the formula for f', but it certainly can be suggestive: Knowing $f'(1) \approx 2$, $f'(2) \approx 4$, $f'(3) \approx 6$ suggests that $f'(x) = 2x$.

The derivative is calculated by forming the difference quotient and taking the limit as h goes to zero. The difference quotient is

$$\frac{f(x+h) - f(x)}{h} = \frac{(x+h)^2 - x^2}{h} = \frac{x^2 + 2xh + h^2 - x^2}{h} = \frac{2xh + h^2}{h}.$$

Since h never actually reaches zero, we can divide it out in the last expression to get $2x + h$. The limit of this as h goes to zero is $2x$, so

$$f'(x) = \lim_{h \to 0}(2x + h) = 2x.$$

Example 5 Calculate $f'(x)$ if $f(x) = x^3$.

Solution We look at the difference quotient

$$\frac{f(x+h) - f(x)}{h} = \frac{(x+h)^3 - x^3}{h}.$$

Multiplying out gives $(x+h)^3 = x^3 + 3x^2h + 3xh^2 + h^3$, so

$$f'(x) = \lim_{h \to 0} \frac{x^3 + 3x^2h + 3xh^2 + h^3 - x^3}{h} = \lim_{h \to 0} \frac{3x^2h + 3xh^2 + h^3}{h}.$$

Since in taking the limit as $h \to 0$, we consider values of h near, but not equal to, zero, we can divide by h giving

$$f'(x) = \lim_{h \to 0} \frac{3x^2h + 3xh^2 + h^3}{h} = \lim_{h \to 0}(3x^2 + 3xh + h^2).$$

As $h \to 0$, the value of $(3xh + h^2) \to 0$ so

$$f'(x) = \lim_{h \to 0}(3x^2 + 3xh + h^2) = 3x^2.$$

The previous two examples show how to compute the derivatives of power functions of the form $f(x) = x^n$, when n is 2 or 3. We can use the Binomial Theorem (from the Focus on Theory section on page 78) to show that, for n a positive integer,

$$\boxed{\text{If } f(x) = x^n \text{ then } f'(x) = nx^{n-1}.}$$

This result is in fact valid for any real value of n.

Problems for Section 2.3

For Problems 1–9, sketch a graph of the derivative function of each of the given functions.

1.

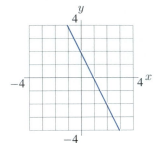

2.

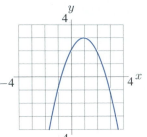

3.

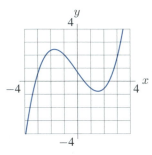

4.

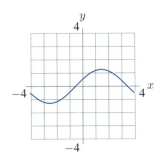

5.

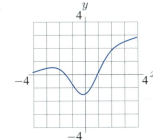

6.

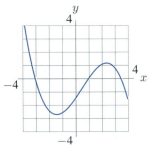

7.

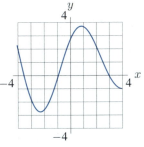

8.

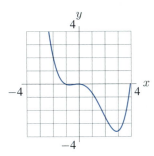

9.

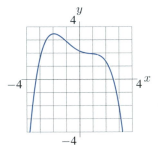

10. (a) Sketch a smooth curve whose slope is both everywhere positive and increasing gradually.
    (b) Sketch a smooth curve whose slope is both everywhere positive and decreasing gradually.
    (c) Sketch a smooth curve whose slope is both everywhere negative and increasing gradually (i.e., becoming less and less negative).
    (d) Sketch a smooth curve whose slope is both everywhere negative and decreasing gradually (i.e., becoming more and more negative).

11. Given the numerical values shown, find approximate values for the derivative of $f(x)$ at each of the x-values given. Where is the rate of change of $f(x)$ positive? Where is it negative? Where does the rate of change of $f(x)$ seem to be greatest?

x	0	1	2	3	4	5	6	7	8
$f(x)$	18	13	10	9	9	11	15	21	30

12. The values of x and the corresponding values of $g(x)$ are given in the table. For what value of x is $g'(x)$ closest to 3?

x	2.7	3.2	3.7	4.2	4.7	5.2	5.7	6.2
$g(x)$	3.4	4.4	5.0	5.4	6.0	7.4	9.0	11.0

Find a formula for the derivatives of the functions in Problems 13–16 algebraically.

13. $g(x) = 2x^2 - 3$ 14. $k(x) = 1/x$ 15. $l(x) = 1/x^2$ 16. $m(x) = 1/(x+1)$

17. Draw the graph of a continuous function $y = f(x)$ that satisfies the following three conditions.

   - $f'(x) > 0$ for $x < -2$,
   - $f'(x) < 0$ for $-2 < x < 2$,
   - $f'(x) = 0$ for $x > 2$.

18. Draw the graph of a continuous function $y = f(x)$ that satisfies the following three conditions.

   - $f'(x) > 0$ for $-\frac{\pi}{2} < x < \frac{\pi}{2}$,
   - $f'(x) < 0$ for $-\pi < x < -\frac{\pi}{2}$ and $\frac{\pi}{2} < x < \pi$,
   - $f'(x) = 0$ at $x = -\frac{\pi}{2}$ and $x = \frac{\pi}{2}$.

For Problems 19–22, sketch the graph of $f(x)$, and use this graph to sketch the graph of $f'(x)$.

19. $f(x) = x^2$ 20. $f(x) = x(x - 1)$ 21. $f(x) = \cos x$ 22. $f(x) = \log x$

For Problems 23–28, sketch the graph of $y = f'(x)$ for the function given.

23.

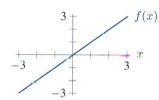

24.

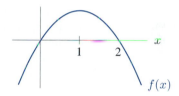

25.

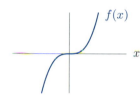

26.

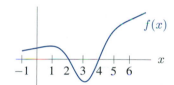

27.

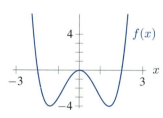

28.

29. In the graph of f in Figure 2.29, at which of the labeled x-values is

   (a) $f(x)$ greatest?   (b) $f(x)$ least?
   (c) $f'(x)$ greatest?   (d) $f'(x)$ least?

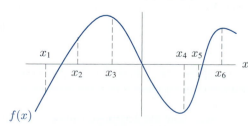

Figure 2.29

30.   Consider a vehicle moving along a straight road. Suppose $f(t)$ gives the vehicle's distance from its starting point at time t. Which of the graphs in Figure 2.30 could be $f'(t)$ for the following scenarios? (Assume the scales on the vertical axes are all the same.)

(a)   A bus on a popular route, with no traffic

(b)   A car with no traffic and all green lights

(c)   A car in heavy traffic conditions

Figure 2.30

31.   A child inflates a balloon, admires it for a while and then lets the air out at a constant rate. If $V(t)$ gives the volume of the balloon at time t, then Figure 2.31 shows $V'(t)$ as a function of t. At what time does the child:

(a)   Begin to inflate the balloon?

(b)   Finish inflating the balloon?

(c)   Begin to let the air out?

(d)   What would the graph of $V'(t)$ look like if the child had alternated between pinching and releasing the open end of the balloon, instead of letting the air out at a constant rate?

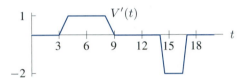

Figure 2.31

32.   The population of a herd of deer is modeled by

$$P(t) = 4000 + 500 \sin\left(2\pi t - \frac{\pi}{2}\right)$$

where t is measured in years from January 1.

(a)   How does this population vary with time? Sketch a graph of $P(t)$ for one year.

(b)   Use the graph to decide when in the year the population is a maximum. What is that maximum? Is there a minimum? If so, when?

(c)   Use the graph to decide when the population is growing fastest. When is it decreasing fastest?

(d)   Estimate roughly how fast the population is changing on the first of July.

33.   Looking at the graph, explain why if $f(x)$ is an even function, then $f'(x)$ is odd.

34.   Looking at the graph, explain why if $g(x)$ is an odd function, then $g'(x)$ is even.

2.4   INTERPRETATIONS OF THE DERIVATIVE

We have already seen how the derivative can be interpreted as a slope and as a rate of change. In this section, we will see examples of other interpretations. The purpose of these examples is not to make a catalog of interpretations but to illustrate the process of obtaining them.

An Alternative Notation for the Derivative

So far we have used the notation f' to stand for the derivative of the function f. An alternative notation for derivatives was introduced by the German mathematician Wilhelm Gottfried Leibniz (1646–1716) when calculus was first being developed. If the variable y depends on the variable x, that is, if

$$y = f(x),$$

then he wrote dy/dx for the derivative, so

$$\frac{dy}{dx} = f'(x).$$

Leibniz's notation is quite suggestive, especially if we think of the letter d in dy/dx as standing for "small difference in . . . ." The notation dy/dx reminds us that the derivative is a limit of ratios of the form

$$\frac{\text{Difference in } y\text{-values}}{\text{Difference in } x\text{-values}}.$$

The notation dy/dx is useful for determining the units for the derivative: The units for dy/dx are the units for y divided by the units for x. The separate entities dy and dx officially have no independent meaning: they are all part of one notation. In fact, a good formal way to view the notation dy/dx is to think of d/dx as a single symbol meaning "the derivative with respect to x of . . .". So dy/dx can be viewed as

$$\frac{d}{dx}(y), \quad \text{meaning "the derivative with respect to } x \text{ of } y."$$

On the other hand, many scientists and mathematicians really do think of dy and dx as separate entities representing "infinitesimally" small differences in y and x, even though it is difficult to say exactly how small "infinitesimal" is. Although not formally correct, it can be very helpful to think intuitively of dy/dx as a very small change in y divided by a very small change in x.

For example, recall that if $s = f(t)$ is the position of a moving object at time t, then $v = f'(t)$ is the velocity of the object at time t. Writing

$$v = \frac{ds}{dt}$$

reminds us that v is a velocity, since the notation suggests a distance, ds, over a time, dt, and we know that distance over time is velocity. Similarly, we recognize

$$\frac{dy}{dx} = f'(x)$$

as the slope of the graph of $y = f(x)$ by remembering that slope is vertical rise, dy, over horizontal run, dx.

The disadvantage of Leibniz's notation is that it is awkward to specify the x-value at which we are evaluating the derivative. To specify $f'(2)$, for example, we have to write

$$\left.\frac{dy}{dx}\right|_{x=2}.$$

Using Units to Interpret the Derivative

The following examples illustrate how useful units can be in suggesting interpretations of the derivative.

For example, suppose $s = f(t)$ gives the distance, in meters, of a body from a fixed point as a function of time, t, in seconds. Then knowing that

$$\left.\frac{ds}{dt}\right|_{t=2} = f'(2) = 10 \text{ meters/sec}$$

tells us that when $t = 2$ sec, the body is moving at an instantaneous velocity of 10 meters/sec. This means that if the body continued to move at this speed for a whole second, it would cover 10 meters. In practice, however, the velocity of the body may be changing and so may not remain 10 meters/sec for long. Notice that the units of instantaneous velocity and of average velocity are the same. The units of the instantaneous and the average rate of change are *always* the same.

Example 1   The cost C (in dollars) of building a house A square feet in area is given by the function $C = f(A)$. What is the practical interpretation of the function $f'(A)$?

Solution   In the alternative notation,

$$f'(A) = \frac{dC}{dA}.$$

This is a cost divided by an area, so it is measured in dollars per square foot. You can think of dC as the extra cost of building an extra dA square feet of house. Then you can think of dC/dA as the additional cost per square foot. So if you are planning to build a house roughly A square feet in area, $f'(A)$ is the cost per square foot of the *extra* area involved in building a slightly larger house, and is called the *marginal cost*. The marginal cost is probably smaller than the average cost per square foot for the entire house, since once you are already set up to build a large house, the cost of adding a few square feet is likely to be small.

Example 2   The cost of extracting T tons of ore from a copper mine is $C = f(T)$ dollars. What does it mean to say that $f'(2000) = 100$?

Solution   In the alternative notation,

$$f'(2000) = \left.\frac{dC}{dT}\right|_{T=2000} = 100.$$

Since C is measured in dollars and T is measured in tons, dC/dT must be measured in dollars per ton. So the statement $f'(2000) = 100$ says that when 2000 tons of ore have already been extracted from the mine, the cost of extracting the next ton is approximately \$100.

Example 3   If $q = f(p)$ gives the number of pounds of sugar produced when the price per unit is p dollars, then what are the units and the meaning of

$$f'(3) = 50?$$

Solution   Since $f'(3)$ is the limit as $h \to 0$ of

$$\frac{f(3 + h) - f(3)}{h}$$

and $f(3 + h) - f(3)$ is in pounds, whereas h is in dollars, the units of the difference quotient are pounds/dollar. Since $f'(3)$ is the limit of the difference quotient, its units are also pounds/dollar. The statement

$$f'(3) = 50 \text{ pounds/dollar}$$

tells us that the rate of change of q with respect to p is 50 when $p = 3$. Rephrasing, this means that when the price is \$3, the quantity produced is increasing at 50 pounds/dollar. This is an instantaneous rate of change, meaning that, if the rate were to remain 50 pounds/dollar, and if the price were to increase by a whole dollar, the quantity produced would increase by approximately 50 pounds. In fact, the rate probably doesn't remain constant and so the quantity produced would probably not be exactly 50 pounds. Notice that the units of the derivative and of the average rate of change are again the same.

Example 4   You are told that water is flowing through a pipe at a constant rate of 10 cubic feet per second. Interpret this rate as the derivative of some function.

Solution   You might think at first that the statement has something to do with the velocity of the water, but in fact a flow rate of 10 cubic feet per second could be achieved either with very slowly moving water through a large pipe, or with very rapidly moving water through a narrow pipe. If we look at the units—cubic feet per second—we realize that we are being given the rate of change of a quantity measured in cubic feet. But a cubic foot is a measure of volume, so we are being told the rate of change of a volume. One way to visualize this is to imagine all the water that is flowing through the pipe ending up in a tank somewhere. Let $V(t)$ be the volume of water in the tank at time t. Then we are being told that the rate of change of $V(t)$ is 10, or

$$V'(t) = \frac{dV}{dt} = 10.$$

Example 5   Suppose $P = f(t)$ is the population of Mexico in millions, where t is the number of years since 1980. Explain the meaning of the statements:

(a) $f'(6) = 2$ (b) $f^{-1}(95.5) = 16$ (c) $(f^{-1})'(95.5) = 0.46$

Solution   (a) The units of P are millions of people, the units of t are years, so the units of $f'(t)$ are millions of people per year. Therefore the statement $f'(6) = 2$ tells us that at $t = 6$ (that is, in 1986), the population of Mexico was increasing at 2 million people per year.

(b) The statement $f^{-1}(95.5) = 16$ tells us that the year when the population was 95.5 million was $t = 16$ (that is, in 1996).

(c) The units of the derivative $(f^{-1})'(P)$ are years per million of population. The statement $(f^{-1})'(95.5) = 0.46$ tells us that when the population was 95.5 million, it took about 0.46 years for the population to increase by 1 million.

Problems for Section 2.4

1. Let $f(x)$ be the elevation in feet of the Mississippi river x miles from its source. What are the units of $f'(x)$? What can you say about the sign of $f'(x)$?

2. The temperature, T, in degrees Fahrenheit, of a cold yam placed in a hot oven is given by $T = f(t)$, where t is the time in minutes since the yam was put in the oven.

    (a) What is the sign of $f'(t)$? Why?
    (b) What are the units of $f'(20)$? What is the practical meaning of the statement $f'(20) = 2$?

3. Suppose $P(t)$ is the monthly payment, in dollars, on a mortgage which will take t years to pay off. What are the units of $P'(t)$? What is the practical meaning of $P'(t)$? What is its sign?

4. Suppose $C(r)$ is the total cost of paying off a car loan borrowed at an annual interest rate of $r\%$. What are the units of $C'(r)$? What is the practical meaning of $C'(r)$? What is its sign?

5. After investing $1000 at an annual interest rate of 7% compounded continuously for t years, your balance is $\$B$, where $B = f(t)$. What are the units of dB/dt? What is the financial interpretation of dB/dt?

6. Investing $1000 at an annual interest rate of $r\%$, compounded continuously, for 10 years gives you a balance of $\$B$, where $B = g(r)$. What is a financial interpretation of the statements

    (a) $g(5) \approx 1649$?
    (b) $g'(5) \approx 165$? What are the units of $g'(5)$?

7. An ice cream company knows that the cost, C (in dollars), to produce g quarts of cookie dough ice cream is a function of g, so $C = f(g)$.

    (a) If $f(200) = 70$, what are the units of the 200? What are the units of the 70? Explain clearly what this equation is telling you.
    (b) If $f'(200) = 3$, what are the units of the 200? What are the units of the 3? Explain clearly what this equation is telling you.

8. An economist is interested in how the price of a certain item affects its sales. Suppose that at a price of $p, a quantity, q, of the item is sold. If $q = f(p)$, explain the meaning of each of the following statements:

   (a) $f(150) = 2000$ (b) $f'(150) = -25$

9. Let $p(h)$ be the pressure in dynes per cm^2 on a diver at a depth of h meters below the surface of the ocean. What do each of the following quantities mean to the diver? Give units for the quantities.

   (a) $p(100)$ (b) h such that $p(h) = 1.2 \cdot 10^6$
   (c) $p(h) + 20$ (d) $p(h + 20)$
   (e) $p'(100)$ (f) h such that $p'(h) = 20$

10. Let $f(t)$ be the number of centimeters of rainfall that has fallen since midnight, where t is the time in hours. Interpret the following in practical terms, giving units.

    (a) $f(10) = 3.1$ (b) $f^{-1}(10) = 16$ (c) $f'(8) = 0.4$ (d) $(f^{-1})'(5) = 2$

11. Let W be the amount of water, in gallons, in a bathtub at time t, in minutes.

    (a) What are the meaning and units of dW/dt?
    (b) Suppose the bathtub is full of water at time t_0, so that $W(t_0) > 0$. Subsequently, at time $t_p > t_0$, the plug is pulled. Is dW/dt positive, negative, or zero:
        (i) For $t_0 < t < t_p$?
        (ii) After the plug is pulled?
        (iii) When all the water has drained from the tub?

12. Suppose w is the weight of a stack of papers in an open container, and t represents time. A lighted match is thrown into the stack at time $t = 0$.

    (a) What is the sign of dw/dt for t in the period of time while the stack of paper is in the process of burning?
    (b) What behavior of dw/dt indicates that the fire is no longer burning?
    (c) If the fire starts small but increases in vigor over a certain time interval, then does dw/dt increase or decrease over that interval? What about $|dw/dt|$?

13. If $g(v)$ is the fuel efficiency, in miles per gallon, of a car going at v miles per hour, what are the units of $g'(55)$? What is the practical meaning of the statement $g'(55) = -0.54$?

14. (a) If you jump out of an airplane without a parachute, you fall faster and faster until wind resistance causes you to approach a steady velocity, called the *terminal* velocity. Sketch a graph of your velocity against time.
    (b) Explain the concavity of your graph.
    (c) Assuming wind resistance to be negligible at $t = 0$, what natural phenomenon is represented by the slope of the graph at $t = 0$?

15. A company's revenue from car sales, C (measured in thousands of dollars), is a function of advertising expenditure, a, also measured in thousands of dollars. Suppose $C = f(a)$.

    (a) What does the company hope is true about the sign of f'?
    (b) What does the statement $f'(100) = 2$ mean in practical terms? How about $f'(100) = 0.5$?
    (c) Suppose the company plans to spend about $100,000 on advertising. If $f'(100) = 2$, should the company spend more or less than $100,000 on advertising? What if $f'(100) = 0.5$?

16. Let $P(x) =$ the number of people in the US of height $\leq x$ inches. What is the meaning of $P'(66)$? What are its units? Estimate $P'(66)$ (using common sense). Is $P'(x)$ ever negative? [Hint: You may want to approximate $P'(66)$ by a difference quotient, using $h = 1$. Also, you may assume the US population is about 250 million, and that 66 inches $=$ 5 feet 6 inches.]

2.5 THE SECOND DERIVATIVE

Since the derivative is itself a function, we can consider its derivative. For a function f, the derivative of its derivative is called the *second derivative*, and written f'' (read "f double-prime"). If $y = f(x)$, the second derivative can also be written as $\dfrac{d^2y}{dx^2}$, which means $\dfrac{d}{dx}\left(\dfrac{dy}{dx}\right)$, the derivative of $\dfrac{dy}{dx}$.

What Do Derivatives Tell Us?

Recall that the derivative of a function tells you whether a function is increasing or decreasing:

- If $f' > 0$ on an interval, then f is *increasing* over that interval.

- If $f' < 0$ on an interval, then f is *decreasing* over that interval.

Since f'' is the derivative of f',

- If $f'' > 0$ on an interval, then f' is *increasing* over that interval.

- If $f'' < 0$ on an interval, then f' is *decreasing* over that interval.

What does it mean for f' to be increasing or decreasing? An example in which f' is increasing is shown in Figure 2.32, where the curve is bending upward, or is *concave up*. In the example shown in Figure 2.33, in which f' is decreasing, the graph is bending downward, or is *concave down*. These figures suggest

---

> If $f'' > 0$ on an interval, then f' is increasing, so the graph of f is concave up there.
>
> If $f'' < 0$ on an interval, then f' is decreasing, so the graph of f is concave down there.

Concave up
$f'' > 0$

$f' < 0$ $f' > 0$

Figure 2.32: Meaning of f'': The slope increases from left to right, so f'' is positive and f is concave up

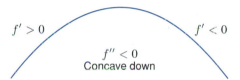

$f' > 0$ $f' < 0$

$f'' < 0$
Concave down

Figure 2.33: Meaning of f'': The slope decreases from left to right, so f'' is negative and f is concave down

---

Example 1   For the functions whose graphs are given in Figure 2.34, decide where their second derivatives are positive and where they are negative.

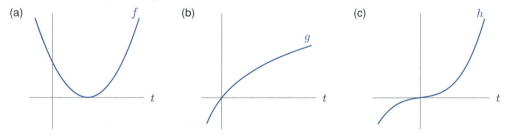

(a) f (b) g (c) h

Figure 2.34: What signs do the second derivatives have?

Solution   From the graphs it appears that:

(a) $f'' > 0$ everywhere, because the graph of f is concave up everywhere.

(b) $g'' < 0$ everywhere, because the graph is concave down everywhere.

(c) $h'' < 0$ for $t < 0$, because the graph of h is concave down there; $h'' > 0$ for $t > 0$, because the graph of h is concave up there.

Interpretation of the Second Derivative as a Rate of Change

If we think of the derivative as a rate of change, then the second derivative is a rate of change of a rate of change. If the second derivative is positive, the rate of change of f is increasing; if the second derivative is negative, the rate of change of f is decreasing.

The second derivative can be a matter of practical concern. In 1985 a newspaper headline reported the Secretary of Defense as saying that Congress had cut the defense budget. As his opponents pointed out, however, Congress had merely cut the rate at which the defense budget was increasing.[2] In other words, the derivative of the defense budget was still positive (the budget was increasing), but the second derivative was negative (the budget's rate of increase had slowed).

Example 2   A population, P, growing in a confined environment often follows a *logistic* growth curve, like that shown in Figure 2.35. Use d^2P/dt^2 to describe how the rate at which the population is increasing changes over time. What are practical interpretations of t_0 and L?

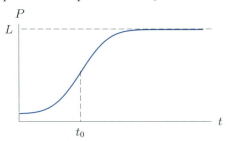

Figure 2.35: Logistic growth curve

Solution   Initially, the population is increasing, and at an increasing rate. So, initially dP/dt is increasing and so $d^2P/dt^2 > 0$. At t_0, the rate at which the population is increasing is a maximum. In other words, at time t_0 the population is growing fastest. Beyond t_0, the rate at which the population is growing is decreasing, and so $d^2P/dt^2 < 0$. At t_0, the curve changes from concave up to concave down, and $d^2P/dt^2 = 0$ there.

The quantity L represents the limiting value of the population as $t \to \infty$. Biologists call L the *carrying capacity* of the environment.

Example 3   Tests on the C5 Chevy Corvette sports car gave the results[3] in Table 2.12.

(a)   Estimate dv/dt for the time intervals shown.

(b)   What can you say about the sign of d^2v/dt^2 over the period shown?

TABLE 2.12   *Velocity of C5 Chevy Corvette*

Time, t (sec)	0	3	6	9	12
Velocity, v (meters/sec)	0	20	33	43	51

Solution   (a)   For each time interval we can calculate the average rate of change of velocity. For example, from $t = 0$ to $t = 3$ we have

$$\frac{dv}{dt} \approx \text{Average rate of change of } v = \frac{20 - 0}{3 - 0} = 6.67 \, \frac{\text{m/sec}}{\text{sec}}.$$

Estimated values of dv/dt are in Table 2.13.

(b)   Since the values of dv/dt are decreasing, $d^2v/dt^2 < 0$. The graph of v against t in Figure 2.36 confirms this; it is concave down. The fact that $dv/dt > 0$ tells us that the car is speeding up;

---

[2]In the *Boston Globe*, March 13, 1985, Representative William Gray (D–Pa.) was reported as saying: "It's confusing to the American people to imply that Congress threatens national security with reductions when you're really talking about a reduction in the increase."

[3]Adapted from the report in *Car and Driver*, February 1997.

the fact that $d^2v/dt^2 < 0$ tells us that the rate of increase decreased over this time period.

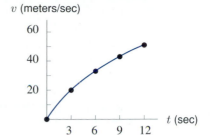

TABLE 2.13   *Estimates for dv/dt (meters/sec/sec)*

Time interval (sec)	$0-3$	$3-6$	$6-9$	$9-12$
Average rate of change (dv/dt)	6.67	4.33	3.33	2.67

Figure 2.36: Velocity of C5 Chevy Corvette

Velocity and Acceleration

The car in Example 3 is speeding up; we say that it is accelerating. We define *acceleration* as the rate of change of velocity with respect to time. If $v(t)$ is the velocity of an object at time t, we have

$$\text{Average acceleration from } t \text{ to } t+h = \frac{v(t+h) - v(t)}{h},$$

and

$$\text{Instantaneous acceleration} = v'(t) = \lim_{h \to 0} \frac{v(t+h) - v(t)}{h}.$$

If the term velocity or acceleration is used alone, it is assumed to be instantaneous. Since velocity is the derivative of position, acceleration is the second derivative of position. Summarizing:

> If $y = s(t)$ is the position of an object at time t, then
>
> - Velocity: $v(t) = \dfrac{dy}{dt} = s'(t)$.
>
> - Acceleration: $a(t) = \dfrac{d^2y}{dt^2} = s''(t) = v'(t)$.

Example 4   A particle is moving along a straight line. If its distance, s, to the right of a fixed point is given by Figure 2.37, estimate:

(a)   When the particle is moving to the right and when it is moving to the left.

(b)   When the particle has positive acceleration and when it has negative acceleration.

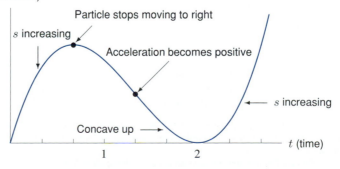

Figure 2.37: Distance of particle to right of a fixed point

Solution (a) The particle is moving to the right whenever s is increasing. From the graph, this appears to be for $0 < t < \frac{2}{3}$ and for $t > 2$. For $\frac{2}{3} < t < 2$, the value of s is decreasing, so the particle is moving to the left.

(b) The particle has positive acceleration whenever the curve is concave up, which appears to be for $t > \frac{4}{3}$. The particle has negative acceleration when the curve is concave down, for $t < \frac{4}{3}$.

Problems for Section 2.5

For Problems 1–6, give the signs of the first and second derivatives for each of the following functions.

1.
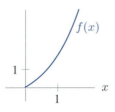

2.
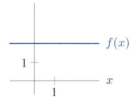

3.
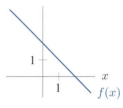

4.
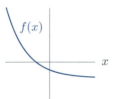

5.
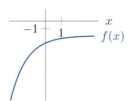

6.
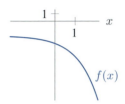

7. An accelerating sports car goes from 0 mph to 60 mph in five seconds. Its velocity is given in Table 2.14, converted from miles per hour to feet per second, so that all time measurements are in seconds. (Note: 1 mph is 22/15 ft/sec.) Find the average acceleration of the car over each of the first two seconds.

TABLE 2.14

Time, t (sec)	0	1	2	3	4	5
Velocity, $v(t)$ (ft/sec)	0	30	52	68	80	88

8. Let $P(t)$ represent the price of a share of stock of a corporation at time t. What does each of the following statements tell us about the signs of the first and second derivatives of $P(t)$?

(a) "The price of the stock is rising faster and faster."
(b) "The price of the stock is close to bottoming out."

9. "Winning the war on poverty" has been described cynically as slowing the rate at which people are slipping below the poverty line. Assuming that this is happening:

(a) Sketch a graph of the total number of people in poverty against time.
(b) If N is the number of people below the poverty line at time t, what are the signs of dN/dt and d^2N/dt^2?

10. In economics, *total utility* refers to the total satisfaction from consuming some commodity. According to the economist Samuelson[4]:

As you consume more of the same good, the total (psychological) utility increases. However, . . . with successive new units of the good, your total utility will grow at a slower and slower rate because of a fundamental tendency for your psychological ability to appreciate more of the good to become less keen.

(a) Sketch the total utility as a function of the number of units consumed.
(b) In terms of derivatives, what is Samuelson telling us?

[4]From Paul A. Samuelson, *Economics*, 11th edition (New York: McGraw-Hill, 1981).

11. In April 1991, the *Economist* carried an article[5] which said:

    > Suddenly, everywhere, it is not the rate of change of things that matters, it is the rate of change of rates of change. Nobody cares much about inflation; only whether it is going up or down. Or rather, whether it is going up fast or down fast. "Inflation drops by disappointing two points," cries the billboard. Which roughly translated means that prices are still rising, but less fast than they were, though not quite as much less fast as everybody had hoped.

    In the last sentence, there are three statements about prices. Rewrite these as statements about derivatives.

12. IBM-Peru uses second derivatives to assess the relative success of various advertising campaigns. They assume that all campaigns produce some increase in sales. If a graph of sales against time shows a positive second derivative during a new advertising campaign, what does this suggest to IBM management? Why? What does a negative second derivative during a new campaign suggest?

13. An industry is being charged by the Environmental Protection Agency (EPA) with dumping unacceptable levels of toxic pollutants in a lake. Over a period of several months, an engineering firm makes daily measurements of the rate at which pollutants are being discharged into the lake.

    Suppose the engineers produce a graph similar to either Figure 2.38(a) or Figure 2.38(b). For each case, give an idea of what argument the EPA might make in court against the industry and of the industry's defense.

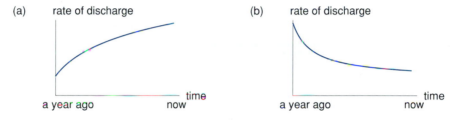

Figure 2.38: Discharge of pollutants

14. Given the following data:

x	0	0.2	0.4	0.6	0.8	1.0
$f(x)$	3.7	3.5	3.5	3.9	4.0	3.9

    (a) Estimate $f'(0.6)$ and $f'(0.5)$.   (b) Estimate $f''(0.6)$.

    (c) Where do you think the maximum and minimum values of f occur in the interval $0 \le x \le 1$?

15. The graph of a function $f(x)$ is shown in Figure 2.39. On a copy of the table indicate whether f, f', f'' at each marked point is positive, negative, or zero.

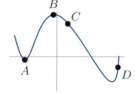

Figure 2.39

Point	f	f'	f''
A			
B			
C			
D			

---

[5]From "The Tyranny of Differential Calculus: $d^2P/dt^2 > 0 =$ misery." *The Economist* (London: April 6, 1991).

16. The graph of f' (not f) is given in Figure 2.40. At which of the marked values of x is
    (a) $f(x)$ greatest?   (b) $f(x)$ least?   (c) $f'(x)$ greatest?   (d) $f'(x)$ least?
    (e) $f''(x)$ greatest?   (f) $f''(x)$ least?

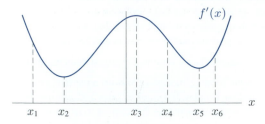

Figure 2.40: Note that this is a graph of f' **Figure 2.41**

17. Which of the points labeled by letters in the graph of f in Figure 2.41 have

    (a) f' and f'' nonzero and of the same sign?
    (b) At least two of f, f', and f'' equal to zero?

18. The distance of a car from its initial position t minutes after setting out is given by $s(t) = 5t^2 + 3$ kilometers. What are the car's velocity and acceleration at time t? Give units.

CHAPTER SUMMARY

- **Rate of change**
    Average, instantaneous
- **Definition of Derivative**
    Difference quotient, limit
- **Estimating and computing derivatives**
    Estimate derivatives from a graph, table of values, or formula. Use definition to find derivatives of simple functions algebraically. Know derivatives of constant, linear, and power functions.
- **Interpretation of Derivatives**
    Rate of change, instantaneous velocity, slope, using units
- **Second Derivative**
    Concavity, acceleration
- **Working with Derivatives**
    Understand relation between sign of f' and whether f is increasing or decreasing. Sketch graph of f' from graph of f.

REVIEW PROBLEMS FOR CHAPTER TWO

1. A bicyclist pedals at a fairly constant rate, with evenly spaced intervals of coasting. Sketch a graph of the distance she has traveled as a function of time.

2. As you drive away from home, your speed is fast, then slow, then fast again. Draw a graph of your distance from home as a function of time.

3. (a) Make a table of values, rounded to two decimal places, for $f(x) = \log x$ (that is, log base 10) with $x = 1$, 1.5, 2, 2.5, 3. Then use this table to answer parts (b) and (c).
    (b) Find the average rate of change of $f(x)$ between $x = 1$ and $x = 3$.
    (c) Use average rates of change to approximate the instantaneous rate of change of $f(x)$ at $x = 2$.

4. For the function $f(x) = \log x$, estimate $f'(1)$. From the graph of $f(x)$, would you expect your estimate to be greater than or less than $f'(1)$?

5. For $f(x) = \ln x$, construct tables, rounded to four decimals, near $x = 1$, $x = 2$, $x = 5$, and $x = 10$. Use the tables to estimate $f'(1)$, $f'(2)$, $f'(5)$, and $f'(10)$. Then guess a general formula for $f'(x)$.

6. Given the values for a Bessel function $J_0(x)$ in Table 2.15, what is your best estimate for the derivative of $J_0(x)$ at $x = 0.5$?

TABLE 2.15

x	0	0.1	0.2	0.3	0.4	0.5
$J_0(x)$	1.0	0.9975	0.9900	0.9776	0.9604	0.9385

x	0.6	0.7	0.8	0.9	1.0	
$J_0(x)$	0.9120	0.8812	0.8463	0.8075	0.7652	

Sketch the graphs of the derivatives of the functions shown in Problems 7–12. Be sure your sketches are consistent with the important features of the graphs of the original functions.

7.

8.
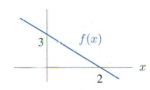

9.

10.
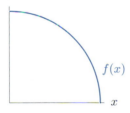

11.
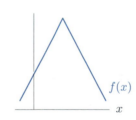

12.
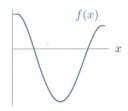

Find a formula for the derivatives of the functions in Problems 13–14 algebraically.

13. $f(x) = 5x^2 + x$

14. $n(x) = \dfrac{1}{x} + 1.$

15. (a) On the same set of axes, graph $f(x) = \sin x$ and $g(x) = \sin 2x$ from $x = 0$ to $x = 2\pi$.
    (b) On a second set of axes, sketch the graphs of $f'(x)$ and $g'(x)$ and compare them. (Be careful when comparing the slopes of $f(x)$ and $g(x)$ at each point.)

16. Do the values for the function $y = k(x)$ in Table 2.16 suggest that the graph of $k(x)$ is concave up or concave down for $1 \leq x \leq 3.3$? Write a sentence in support of your conclusion.

TABLE 2.16

x	1.0	1.2	1.5	1.9	2.5	3.3
$k(x)$	4.0	3.8	3.6	3.4	3.2	3.0

17. A yam has just been taken out of the oven and is cooling off before being eaten. The temperature, T, of the yam (measured in degrees Fahrenheit) is a function of how long it has been out of the oven, t (measured in minutes). Thus, we have $T = f(t)$.

    (a) Is $f'(t)$ positive or negative? Why?
    (b) What are the units for $f'(t)$?

18. An economist is interested in how the price of a certain commodity affects its sales. Suppose that at a price of $\$p$, a quantity q of the commodity is sold. If $q = f(p)$, explain in economic terms the meaning of the statements $f(10) = 240,000$ and $f'(10) = -29,000$.

19. Students were asked to evaluate $f'(4)$ from the following table which shows values of the function f:

x	1	2	3	4	5	6
$f(x)$	4.2	4.1	4.2	4.5	5.0	5.7

    • Student A estimated the derivative as $f'(4) \approx \dfrac{f(5) - f(4)}{5 - 4} = 0.5$.

    • Student B estimated the derivative as $f'(4) \approx \dfrac{f(4) - f(3)}{4 - 3} = 0.3$.

    • Student C suggested that they should split the difference and estimate the average of these two results, that is, $f'(4) \approx \frac{1}{2}(0.5 + 0.3) = 0.4$.

    (a) Sketch the graph of f, and indicate how the three estimates are represented on the graph.
    (b) Explain which answer is likely to be best.
    (c) Use Student C's method to find an algebraic formula which approximates $f'(x)$ using increments of size h.

20. Each of the graphs in Figure 2.42 shows the position of a particle moving along the x-axis as a function of time, $0 \le t \le 5$. The vertical scales of the graphs are the same. During this time interval, which particle has

    (a) Constant velocity?   (d) Zero average velocity?
    (b) The greatest initial velocity?   (e) Zero acceleration?
    (c) The greatest average velocity?   (f) Positive acceleration throughout?

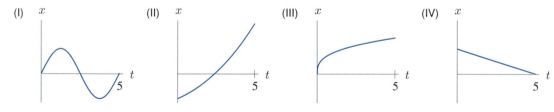

Figure 2.42

21. A person with a certain liver disease first exhibits larger and larger concentrations of certain enzymes (called SGOT and SGPT) in the blood. As the disease progresses, the concentration of these enzymes drops, first to the predisease level and eventually to zero (when almost all of the liver cells have died). Monitoring the levels of these enzymes allows doctors to track the progress of a patient with this disease. If $C = f(t)$ is the concentration of the enzymes in the blood as a function of time,

    (a) Sketch a possible graph of $C = f(t)$.
    (b) Mark on the graph the intervals where $f' > 0$ and where $f' < 0$.
    (c) What does $f'(t)$ represent, in practical terms?

22. The population of a herd of deer is modeled by

$$P(t) = 4000 + 400 \sin\left(\frac{\pi}{6}t\right) + 180 \sin\left(\frac{\pi}{3}t\right)$$

where t is measured in months from the first of April.

    (a) Use a calculator or computer to sketch a graph showing how this population varies with time.

Use the graph to answer the following questions.

    (b)   When is the herd largest? How many deer are in it at that time?

    (c)   When is the herd smallest? How many deer are in it then?

    (d)   When is the herd growing the fastest? When is it shrinking the fastest?

    (e)   How fast is the herd growing on April 1?

23.   Let $g(x) = \sqrt{x}$ and $f(x) = kx^2$, where k is a constant.

    (a)   Find the slope of the tangent line to the graph of g at the point $(4, 2)$.

    (b)   Find the equation of this tangent line.

    (c)   If the graph of f contains the point $(4, 2)$, find k.

    (d)   Where does the graph of f intersect the tangent line?

24.   A circle with center at the origin and radius of length $\sqrt{19}$ has equation $x^2 + y^2 = 19$. Graph the circle.

    (a)   Just from looking at the graph, what can you say about the slope of the line tangent to the circle at the point $(0, \sqrt{19})$? What about the slope of the tangent at $(\sqrt{19}, 0)$?

    (b)   Estimate the slope of the tangent to the circle at the point $(2, -\sqrt{15})$ by graphing the tangent carefully at that point.

    (c)   Use the result of part (b) and the symmetry of the circle to find slopes of the tangents drawn to the circle at $(-2, \sqrt{15})$, $(-2, -\sqrt{15})$, and $(2, \sqrt{15})$.

25.   A continuous function defined for all x has the following properties:

    • f is increasing.   • f is concave down.   • $f(5) = 2$.   • $f'(5) = \frac{1}{2}$.

    (a)   Sketch a possible graph for f.   (b)   How many zeros does f have?

    (c)   What can you say about the location of the zeros?   (d)   What is $\lim\limits_{x \to -\infty} f(x)$?

    (e)   Is it possible that $f'(1) = 1$?   (f)   Is it possible that $f'(1) = \frac{1}{4}$?

26.   The number of hours, H, of daylight in Madrid is a function of t, the number of days since the start of the year. Figure 2.43 shows a one-month portion of the graph of H.

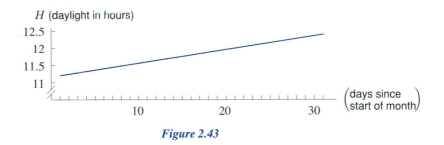

Figure 2.43

    (a)   Comment on the shape of the graph. Why does it look like a straight line?

    (b)   What month does this graph show? How do you know?

    (c)   What is the approximate slope of this line? What does the slope represent in practical terms?

27.   Suppose you put a yam in a hot oven, maintained at a constant temperature of $200°$C. As the yam picks up heat from the oven, its temperature rises.[6]

    (a)   Draw a possible graph of the temperature T of the yam against time t (minutes) since it is put into the oven. Explain any interesting features of the graph, and in particular explain its concavity.

    (b)   Suppose that, at $t = 30$, the temperature T of the yam is $120°$ and increasing at the (instantaneous) rate of $2°$/min. Using this information, plus what you know about the shape of the T graph, estimate the temperature at time $t = 40$.

    (c)   Suppose in addition you are told that at $t = 60$, the temperature of the yam is $165°$. Can you improve your estimate of the temperature at $t = 40$?

    (d)   Assuming all the data given so far, estimate the time at which the temperature of the yam is $150°$.

---

[6]From Peter D. Taylor, *Calculus: The Analysis of Functions* (Toronto: Wall & Emerson, Inc., 1992)

PROJECTS

1. **Hours of Daylight as a Function of Latitude**

    Let $S(x)$ be the number of sunlight hours on a cloudless June 21, as a function of latitude, x, measured in degrees.

    (a) What is $S(0)$?

    (b) Let x_0 be the latitude of the Arctic Circle ($x_0 \approx 66°30'$). In the northern hemisphere, $S(x)$ is given, for some constants a and b, by the formula:

    $$S(x) = \begin{cases} a + b\arcsin\left(\dfrac{\tan x}{\tan x_0}\right) & \text{for } 0 \le x < x_0 \\ 24 & \text{for } x_0 \le x \le 90. \end{cases}$$

    Find a and b so that $S(x)$ is continuous.

    (c) Calculate $S(x)$ for Tucson, Arizona ($x = 32°13'$) and Walla Walla, Washington ($46°4'$).

    (d) Graph $S(x)$, for $0 \le x \le 90$.

    (e) Does $S(x)$ appear to be differentiable?

2. **US Population**

    Census figures for the US population (in millions) are listed in Table 2.17. Let f be the function such that $P = f(t)$ is the population (in millions) in year t.

    TABLE 2.17   *US population (in millions), 1790–1990*

Year	Population	Year	Population	Year	Population	Year	Population
1790	3.9	1850	23.1	1910	92.0	1970	205.0
1800	5.3	1860	31.4	1920	105.7	1980	226.5
1810	7.2	1870	38.6	1930	122.8	1990	248.7
1820	9.6	1880	50.2	1940	131.7		
1830	12.9	1890	62.9	1950	150.7		
1840	17.1	1900	76.0	1960	179.0		

    (a) (i) Estimate the rate of change of the population for the years 1900, 1945, and 1990.

    (ii) When, approximately, was the rate of change of the population greatest?

    (iii) Estimate the US population in 1956.

    (iv) Based on the data from the table, what would you predict for the census in the year 2000?

    (b) Assume that f is increasing (as the values in the table suggest). Then f is invertible.

    (i) What is the meaning of $f^{-1}(100)$?

    (ii) What does the derivative of $f^{-1}(P)$ at $P = 100$ represent? What are its units?

    (iii) Estimate $f^{-1}(100)$.

    (iv) Estimate the derivative of $f^{-1}(P)$ at $P = 100$.

    (c) (i) Usually we think the US population $P = f(t)$ as a smooth function of time. To what extent is this justified? What happens if we zoom in at a point of the graph? What about events such as the Louisiana Purchase? Or the moment of your birth?

    (ii) What do we in fact mean by the rate of change of the population at a particular time t?

    (iii) Give another example of a real-world function which is not smooth but is usually treated as such.

FOCUS ON THEORY

LIMITS AND CONTINUITY

In this section we use the example of limits and continuity to illustrate how formal definitions are developed from intuitive ideas.

Definition of Limit

By the beginning of the 19th century, calculus had proved its worth, and there was no doubt about the correctness of its answers. However, it was not until the work of the French mathematician Augustin Cauchy (1789–1857) that a formal definition of the limit was given, similar to the following:

> Suppose a function f, is defined on an interval around c, except perhaps not at the point $x = c$. We define the **limit** of the function $f(x)$ as x approaches c, written $\lim_{x \to c} f(x)$, to be a number L (if one exists) such that $f(x)$ is as close to L as we please whenever x is sufficiently close to c (but $x \neq c$). If L exists, we write
> $$\lim_{x \to c} f(x) = L.$$

Shortly, we will see how "as close as we please" and "sufficiently close" can be given a precise meaning using inequalities. First, we look at $\lim_{\theta \to 0} (\sin \theta / \theta)$ more closely (see Example 4 on page 93).

Example 1   By graphing $y = (\sin \theta)/\theta$ in an appropriate window, find how close θ should be to 0 in order to make $(\sin \theta)/\theta$ within 0.01 of 1.

Solution   Since we want $(\sin \theta)/\theta$ to be within 0.01 of 1, we set the y-range on the graphing window to go from 0.99 to 1.01. Our first attempt with $-0.5 \leq \theta \leq 0.5$ yields the graph in Figure 2.44. Since we want the y-values to stay within the range $0.99 < y < 1.01$, we do not want the graph to leave the window through the top or bottom. By trial and error, we find that changing the θ-range to $-0.2 \leq \theta \leq 0.2$ gives the graph in Figure 2.45. Thus, the graph suggests that $(\sin \theta)/\theta$ will be within 0.01 of 1 whenever θ is within 0.2 of 0. Proving this requires an analytical argument, not just graphs from a calculator.

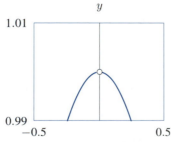

Figure 2.44: $(\sin \theta)/\theta$ with $-0.5 \leq \theta \leq 0.5$

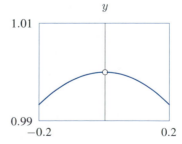

Figure 2.45: $(\sin \theta)/\theta$ with $-0.2 \leq \theta \leq 0.2$

When we say "$f(x)$ is as close to L as we please," we mean that we can specify a maximum distance between $f(x)$ and L. We express the distance using absolute values:

$$|f(x) - L| = \text{Distance between } f(x) \text{ and } L.$$

Using ϵ (the Greek letter epsilon) to stand for the distance we have specified, we write

$$|f(x) - L| < \epsilon$$

to indicate that the maximum distance between $f(x)$ and L is less than ϵ. In Example 1 we used $\epsilon = 0.01$. In a similar manner we interpret "x is sufficiently close to c" as specifying a maximum distance between x and c:

$$|x - c| < \delta,$$

where δ (the Greek letter delta) tells us how close x should be to c. In Example 1 we found $\delta = 0.2$.

If $\lim\limits_{x \to c} f(x) = L$, we know that no matter how narrow the band determined by ϵ in Figure 2.46, there's always a δ which makes the graph stay within the band for $c - \delta < x < c + \delta$.

Thus we restate the definition of a limit, using symbols:

Definition of Limit

We define $\lim\limits_{x \to c} f(x)$ to be the number L (if one exists) such that for any $\epsilon > 0$ (as small as we want), there is a $\delta > 0$ (sufficiently small) such that if $|x - c| < \delta$ and $x \neq c$, then $|f(x) - L| < \epsilon$.

Realize that the point of this definition is that for any ϵ we are given, we need to be able to determine a corresponding δ. One way to do this is to give an explicit expression for δ in terms of ϵ.

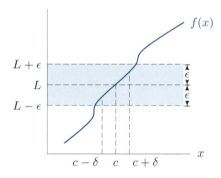

Figure 2.46: What the definition of the limit means practically

---

Example 2   Use algebra to find a maximum distance between x and 2 which ensures that x^2 is within 0.1 of 4. Use a similar argument to show that $\lim\limits_{x \to 2} x^2 = 4$.

Solution   We write $x = 2 + h$. We want to find the values of h making x^2 within 0.1 of 4. We know

$$x^2 = (2 + h)^2 = 4 + 4h + h^2,$$

so x^2 differs from 4 by $4h + h^2$. Since we want x^2 to be within 0.1 of 4, we need

$$|x^2 - 4| = |4h + h^2| = |h| \cdot |4 + h| < 0.1.$$

Assuming $0 < |h| < 1$, we know $|4 + h| < 5$, so we need

$$|x^2 - 4| < 5 |h| < 0.1.$$

Thus, if we choose h such that $0 < |h| < 0.1/5 = 0.02$, then x^2 is less than 0.1 from 4.

An analogous argument using any small ϵ instead of 0.1 shows that if we take $\delta = \epsilon/5$, then

$$\left| x^2 - 4 \right| < \epsilon \quad \text{for all} \quad |x - 2| < \epsilon/5.$$

Thus, we have used the definition to show that

$$\lim_{x \to 2} x^2 = 4.$$

It is important to understand that the ϵ, δ definition by itself does not make it easier to calculate limits. The advantage of the ϵ, δ definition is that it makes it possible to put calculus on a rigorous foundation. From this foundation, we can prove the following properties. See Problems 13–16.

Theorem: Properties of Limits

Assuming all the limits on the right hand side exist:

1. If b is a constant, then $\lim\limits_{x \to c} \left(bf(x) \right) = b \left(\lim\limits_{x \to c} f(x) \right)$.

2. $\lim\limits_{x \to c} \left(f(x) + g(x) \right) = \lim\limits_{x \to c} f(x) + \lim\limits_{x \to c} g(x)$.

3. $\lim\limits_{x \to c} \left(f(x)g(x) \right) = \left(\lim\limits_{x \to c} f(x) \right) \left(\lim\limits_{x \to c} g(x) \right)$.

4. $\lim\limits_{x \to c} \dfrac{f(x)}{g(x)} = \dfrac{\lim_{x \to c} f(x)}{\lim_{x \to c} g(x)}$, provided $\lim\limits_{x \to c} g(x) \neq 0$.

5. For any constant k, $\lim\limits_{x \to c} k = k$.

6. $\lim\limits_{x \to c} x = c$.

These properties underlie many limit calculations, though we seldom acknowledge them explicitly.

Example 3   Explain how the limit properties are used in the following calculation:

$$\lim_{x \to 3} \frac{x^2 + 5x}{x + 9} = \frac{3^2 + (5)(3)}{3 + 9} = 2.$$

Solution   We calculate this limit in stages, using the limit properties to justify each step:

$$\lim_{x \to 3} \frac{x^2 + 5x}{x + 9} = \frac{\lim\limits_{x \to 3} \left(x^2 + 5x \right)}{\lim\limits_{x \to 3} \left(x + 9 \right)} \qquad \text{Property 4 (since } \lim_{x \to 3} (x + 9) \neq 0\text{)}$$

$$= \frac{\lim\limits_{x \to 3} \left(x^2 \right) + \lim\limits_{x \to 3} \left(5x \right)}{\lim\limits_{x \to 3} x + \lim\limits_{x \to 3} 9} \qquad \text{Property 2}$$

$$= \frac{\left(\lim\limits_{x \to 3} x \right)^2 + 5 \left(\lim\limits_{x \to 3} x \right)}{\lim\limits_{x \to 3} x + \lim\limits_{x \to 3} 9} \qquad \text{Properties 1 and 3}$$

$$= \frac{3^2 + (5)(3)}{3 + 9} = 2. \qquad \text{Properties 5 and 6}$$

One- and Two-Sided Limits

When we write

$$\lim_{x \to 2} f(x),$$

we mean the number that $f(x)$ approaches as x approaches 2 *from both sides*. We examine values of $f(x)$ as x approaches 2 through values greater than 2 (such as 2.1, 2.01, 2.003) and values less than 2 (such as 1.9, 1.99, 1.994). If we want x to approach 2 only through values greater than 2, we write

$$\lim_{x \to 2^+} f(x)$$

for the number that $f(x)$ approaches (assuming such a number exists). Similarly,

$$\lim_{x \to 2^-} f(x)$$

denotes the number (if it exists) obtained by letting x approach 2 through values less than 2. We call $\lim_{x \to 2^+} f(x)$ a *right-hand limit* and $\lim_{x \to 2^-} f(x)$ a *left-hand limit*.

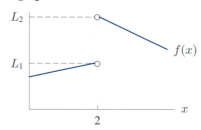

Figure 2.47: Left- and right-hand limits at $x = 2$

For the function graphed in Figure 2.47, we have

$$\lim_{x \to 2^-} f(x) = L_1 \qquad \lim_{x \to 2^+} f(x) = L_2.$$

If the left- and right-hand limits were equal, that is, if $L_1 = L_2$, then it could easily be proved that $\lim_{x \to 2} f(x)$ exists and $\lim_{x \to 2} f(x) = L_1 = L_2$. Since $L_1 \neq L_2$ in Figure 2.47, we see that $\lim_{x \to 2} f(x)$ does not exist in this case.

When Limits Do Not Exist

Whenever there is not a number L such that $\lim_{x \to c} f(x) = L$, we say $\lim_{x \to c} f(x)$ does not exist. In addition to cases in which the left- and right-hand limits are not equal, there are some other cases in which limits fail to exist. Here are three examples.

Example 4   Explain why $\lim_{x \to 2} \dfrac{|x - 2|}{x - 2}$ doesn't exist.

Solution   Figure 2.48 shows the problem: The right-hand limit and the left-hand limit are different. For $x > 2$, we have $|x - 2| = x - 2$, so as x approaches 2 from the right,

$$\lim_{x \to 2^+} \frac{|x - 2|}{x - 2} = \lim_{x \to 2^+} \frac{x - 2}{x - 2} = \lim_{x \to 2^+} 1 = 1.$$

Similarly, if $x < 2$, then $|x - 2| = 2 - x$ so

$$\lim_{x \to 2^-} \frac{|x - 2|}{x - 2} = \lim_{x \to 2^-} \frac{2 - x}{x - 2} = \lim_{x \to 2^-} (-1) = -1.$$

So if $\lim_{x \to 2} \dfrac{|x - 2|}{x - 2} = L$ then L would have to be both 1 and -1. Since L cannot have two different values, the limit does not exist.

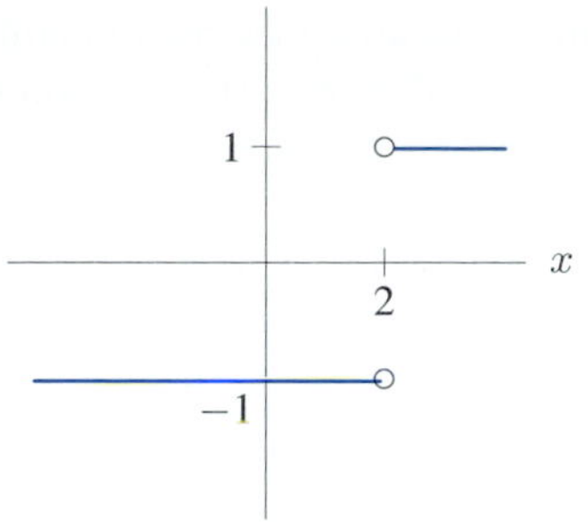

Figure 2.48: Graph of $\frac{|x-2|}{x-2}$ 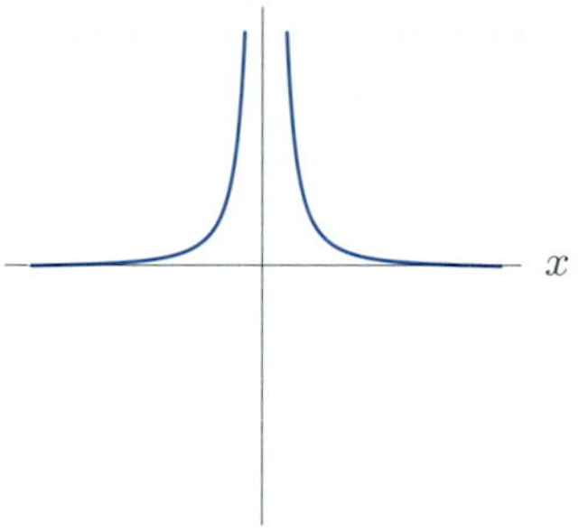   ***Figure 2.49:*** Graph of $\frac{1}{x^2}$ 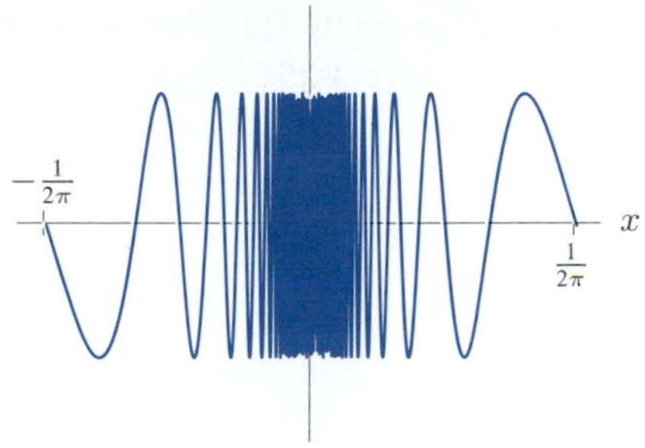   ***Figure 2.50:*** Graph of $\sin\left(\frac{1}{x}\right)$

Example 5   Explain why $\displaystyle\lim_{x\to 0}\frac{1}{x^2}$ doesn't exist.

Solution   As x approaches zero, $1/x^2$ becomes arbitrarily large, so it can't stay close to any finite number L. See Figure 2.49. Therefore we say $1/x^2$ has no limit as $x \to 0$.

Example 6   Explain why $\displaystyle\lim_{x\to 0}\sin\left(\frac{1}{x}\right)$ doesn't exist.

Solution   We know that the sine function has values between -1 and 1. The graph in Figure 2.50 oscillates more and more rapidly as $x \to 0$. There are x-values as close to 0 as we like where $\sin(1/x) = 0$. There are also x-values as close to 0 as we like where $\sin(1/x) = 1$. So if the limit existed, it would have to be both 0 and 1. Thus, the limit does not exist.

Limits at Infinity

Sometimes we want to know what happens to $f(x)$ as x gets large, that is, the end behavior of f.

> If $f(x)$ gets as close to a number L as we please when x gets sufficiently large, then we write
>
> $$\lim_{x\to\infty} f(x) = L.$$
>
> Similarly, if $f(x)$ approaches L as x gets more and more negative, then we write
>
> $$\lim_{x\to-\infty} f(x) = L.$$

The symbol ∞ does not represent a number. Writing $x \to \infty$ means that we consider arbitrarily large values of x. If the limit of $f(x)$ as $x \to \infty$ or $x \to -\infty$ is L, we say that the graph of f has a *horizontal asymptote* $y = L$.

Example 7   Investigate $\displaystyle\lim_{x\to\infty}\frac{1}{x}$ and $\displaystyle\lim_{x\to-\infty}\frac{1}{x}$.

Solution   A graph of $f(x) = \frac{1}{x}$ in a large window shows $1/x$ approaching zero as x increases in either the positive or the negative direction (See Figure 2.51). This is as we would expect, since dividing 1 by larger and larger numbers yields answers which are smaller and smaller. This suggests that

$$\lim_{x \to \infty} \frac{1}{x} = \lim_{x \to -\infty} \frac{1}{x} = 0,$$

and $f(x) = 1/x$ has $y = 0$ as a horizontal asymptote as $x \to \pm\infty$.

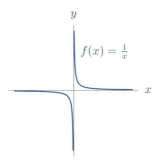

Figure 2.51: The end behavior of $f(x) = \frac{1}{x}$

Definition of Continuity

We can now define continuity. Recall that the idea of continuity rules out breaks, jumps, or holes by demanding that the behavior of a function *near* a point be consistent with its behavior *at* the point:

> The function f is **continuous** at $x = c$ if f is defined at $x = c$ and
>
> $$\lim_{x \to c} f(x) = f(c).$$
>
> In other words, $f(x)$ is as close as we please to $f(c)$ provided x is close enough to c. The function is **continuous on an interval** $[a, b]$ if it is continuous at every point in the interval.[7]

Constant functions and $f(x) = x$ are continuous. (See Problem 16.) Using the continuity of sums and products, we can show that any polynomial is continuous. Proving that $\sin x$, $\cos x$, and e^x are continuous is more difficult. The following theorem, based on the properties of limits on page 129, makes it easier to decide whether a given function is continuous.

> ### Theorem: Continuity of Sums, Products, and Quotients of Functions
>
> Suppose that f and g are continuous on an interval and that b is a constant. Then, on that same interval,
> 1. $bf(x)$ is continuous
> 2. $f(x) + g(x)$ is continuous
> 3. $f(x)g(x)$ is continuous
> 4. $f(x)/g(x)$ is continuous, provided $g(x) \neq 0$ on the interval.

We prove the first of these properties.

---

[7]If c is an endpoint of the interval, we define continuity at $x = c$ using one-sided limits at c.

Proof To prove that $bf(x)$ is continuous, pick any point c in the interval. We must show that $\lim_{x \to c} bf(x) = bf(c)$. Since $f(x)$ is continuous, we already know that $\lim_{x \to c} f(x) = f(c)$. So, by the first property of limits,

$$\lim_{x \to c} (bf(x)) = b \left(\lim_{x \to c} f(x) \right) = bf(c).$$

Since c was chosen arbitrarily, we have shown that $bf(x)$ is continuous at every point in the interval.

Theorem: Continuity of Composite Functions

If f and g are continuous, and if the composite function $f(g(x))$ is defined on an interval, then $f(g(x))$ is continuous on that interval.

Assuming the continuity of $\sin x$ and e^x, this result shows us, for example, that $\sin(e^x)$ and $e^{\sin x}$ are both continuous. A proof of the continuity of composite functions is outlined in Problem 17.

Problems on Limits and Continuity

1. Consider the function $(\sin \theta)/\theta$. Estimate how close θ should be to 0 to make $(\sin \theta)/\theta$ stay within 0.001 of 1.

2. The function $g(\theta) = (\sin \theta)/\theta$ is not defined at $\theta = 0$. Is it possible to define $g(0)$ in such a way that g is continuous at $\theta = 0$? Explain your answer.

Use a graph to estimate each of the limits in Problems 3–6.

3. $\lim_{\theta \to 0} \dfrac{\sin (2\theta)}{\theta}$ (use radians)

4. $\lim_{\theta \to 0} \dfrac{\cos \theta - 1}{\theta}$ (use radians)

5. $\lim_{\theta \to 0} \dfrac{\sin \theta}{\theta}$ (use degrees)

6. $\lim_{\theta \to 0} \dfrac{\theta}{\tan(3\theta)}$ (use radians)

7. Consider the limit

$$\lim_{x \to 0^+} x^x.$$

Estimate this limit either by evaluating x^x for smaller and smaller positive values of x (say $x = 0.1, 0.01, 0.001, \ldots$) or by zooming in on the graph of $y = x^x$ near $x = 0$.

8. (a) Give an example of a function such that $\lim_{x \to 2} f(x) = \infty$.
   (b) Give an example of a function such that $\lim_{x \to 2} f(x) = -\infty$.

9. Consider the function $f(x) = \sin(1/x)$.

   (a) Find a sequence of x-values that approach 0 such that $\sin(1/x) = 0$.
   [Hint: Use the fact that $\sin(\pi) = \sin(2\pi) = \sin(3\pi) = \ldots = \sin(n\pi) = 0$.]
   (b) Find a sequence of x-values that approach 0 such that $\sin(1/x) = 1$.
   [Hint: Use the fact that $\sin(n\pi/2) = 1$ if $n = 1, 5, 9, \ldots$.]
   (c) Find a sequence of x-values that approach 0 such that $\sin(1/x) = -1$.
   (d) Explain why your answers to any two of parts (a)–(c) show that $\lim_{x \to 0} \sin(1/x)$ does not exist.

10. Write the definition of the following statement both in words and in symbols:

$$\lim_{h \to a} g(h) = K.$$

11. For each of the following functions do the following:

    (i) Make a table of values of $f(x)$ for $x = a + 0.1$, $a + 0.01$, $a + 0.001$, $a + 0.0001$, $a - 0.1$, $a - 0.01$, $a - 0.001$, and $a - 0.0001$.

    (ii) Make a conjecture about the value of $\lim_{x \to a} f(x)$.

    (iii) Graph the function to see if it is consistent with your answers to parts (i) and (ii).

    (iv) Find an interval for x containing a such that the difference between your conjectured limit and the value of the function is less than 0.01 on that interval. (In other words, find a window of height 0.02 such that the graph exits the sides of the window and not the top or bottom of the window.)

    (a) $f(x) = \dfrac{x^2 - 4}{x - 2}$, $\quad a = 2$

    (b) $f(x) = \dfrac{x^2 - 9}{x - 3}$, $\quad a = 3$

    (c) $f(x) = \dfrac{\sin x - 1}{x - \pi/2}$, $\quad a = \dfrac{\pi}{2}$

    (d) $f(x) = \dfrac{\sin 5x - 1}{x - \pi/2}$, $\quad a = \dfrac{\pi}{2}$

    (e) $f(x) = \dfrac{e^{2x-2} - 1}{x - 1}$, $\quad a = 1$

    (f) $f(x) = \dfrac{e^{0.5x-1} - 1}{x - 2}$, $\quad a = 2$

12. Assuming that limits as $x \to \infty$ have the properties listed for limits as $x \to c$ on page 129, use algebraic manipulations to evaluate $\lim_{x \to \infty}$ for the following functions:

    (a) $f(x) = \dfrac{x + 3}{2 - x}$

    (b) $f(x) = \dfrac{x^2 + 2x - 1}{3 + 3x^2}$

    (c) $f(x) = \dfrac{x^2 + 4}{x + 3}$

    (d) $f(x) = \dfrac{2x^3 - 16x^2}{4x^2 + 3x^3}$

    (e) $f(x) = \dfrac{x^4 + 3x}{x^4 + 2x^5}$

    (f) $f(x) = \dfrac{3e^x + 2}{2e^x + 3}$

    (g) $f(x) = \dfrac{2e^{-x} + 3}{3e^{-x} + 2}$

13. This problem suggests a proof of the first property of limits on page 129: $\lim_{x \to c} bf(x) = b \lim_{x \to c} f(x)$.

    (a) First, prove the property in the case $b = 0$.

    (b) Now suppose that $b \neq 0$. Let $\epsilon > 0$. Show that if $|f(x) - L| < \epsilon/|b|$, then $|bf(x) - bL| < \epsilon$.

    (c) Finally, prove that if $\lim_{x \to c} f(x) = L$ then $\lim_{x \to c} bf(x) = bL$. [Hint: Choose δ so that if $|x - c| < \delta$, then $|f(x) - L| < \epsilon/|b|$.]

14. Prove the second property of limits: $\lim_{x \to c} \big(f(x) + g(x)\big) = \lim_{x \to c} f(x) + \lim_{x \to c} g(x)$. Assume that the limits on the right exist.

15. This problem suggests a proof of the third property of limits:

$$\lim_{x \to c} \big(f(x)g(x)\big) = \left(\lim_{x \to c} f(x)\right)\left(\lim_{x \to c} g(x)\right)$$

(assuming the limits on the right exist). Let $L_1 = \lim_{x \to c} f(x)$ and $L_2 = \lim_{x \to c} g(x)$.

    (a) First, show that if $\lim_{x \to c} f(x) = \lim_{x \to c} g(x) = 0$, then $\lim_{x \to c} \big(f(x)g(x)\big) = 0$.

    (b) Show algebraically that $f(x)g(x) = \big(f(x) - L_1\big)\big(g(x) - L_2\big) + L_1 g(x) + L_2 f(x) - L_1 L_2$.

    (c) Use the second limit property (see Problem 14) to explain why $\lim_{x \to c} \big(f(x) - L_1\big) = \lim_{x \to c} \big(g(x) - L_2\big) = 0$.

    (d) Use parts (a) and (c) to explain why $\lim_{x \to c} \big(f(x) - L_1\big)\big(g(x) - L_2\big) = 0$.

    (e) Finally, use parts (b) and (d) and the first and second limit properties to show that

$$\lim_{x \to c} \big(f(x)g(x)\big) = \left(\lim_{x \to c} f(x)\right)\left(\lim_{x \to c} g(x)\right).$$

16. Show that the following functions are both continuous everywhere.

    (a) $f(x) = k$ (a constant)   (b) $g(x) = x$

17. This problem suggests a proof of the theorem on continuity of composite functions on page 133: If f and g are continuous and the composite function $f\left(g(x)\right)$ is defined on an interval, then $f\left(g(x)\right)$ is continuous on that interval.

Let c be a point inside the interval where $f\left(g(x)\right)$ is defined. We must show that $\lim_{x \to c} f\left(g(x)\right) = f\left(g(c)\right)$. Let $d = g(c)$. Then the continuity of f at d means that $\lim_{y \to d} f(y) = f(d)$. Thus, for a given $\epsilon > 0$, we can choose $\delta > 0$ so that $|y - d| < \delta$ implies $|f(y) - f(d)| < \epsilon$.

Now, take $y = g(x)$ and show that the continuity of g means that we can find a $\delta_1 > 0$ such that, if $|x - c| < \delta_1$, then $|g(x) - d| < \delta$. Explain how this establishes the continuity of $f\left(g(x)\right)$ at $x = c$.

For each value of ϵ in Problems 18–19, find a positive value of δ such that the graph of the function leaves the window $a - \delta < x < a + \delta$, $b - \epsilon < y < b + \epsilon$ by the sides and not through the top or bottom.

18. $f(x) = -2x + 3$; $a = 0$; $b = 3$; $\epsilon = 0.2, 0.1, 0.02, 0.01, 0.002, 0.001$.

19. $g(x) = -x^3 + 2$; $a = 0$; $b = 2$; $\epsilon = 0.1, 0.01, 0.001$.

20. Show that $\lim_{x \to 0} (-2x + 3) = 3$. You may use the result of Problem 18.

21. Show that $\lim_{x \to 0} (-x^3 + 2) = 2$. [Hint: Try $\delta = \epsilon^{1/3}$.]

In Problems 22–24, modify the definition of limit on page 128 to give a definition of each of the following.

22. A right-hand limit   23. A left-hand limit   24. $\lim_{x \to \infty} f(x) = L$

25. Consider the function $f(x) = \begin{cases} x \sin\left(\dfrac{1}{x}\right) & x \neq 0 \\ 0 & x = 0. \end{cases}$

Show that f is continuous everywhere, but that it is neither always increasing nor always decreasing on the interval $[0, \epsilon]$ for any $\epsilon > 0$, no matter how small.

26. On page 82 we showed how to find a sequence of intervals $[a_n, b_n]$ that close in on a root r of $f(x) = 3x^3 - x^2 + 2x - 1$. In this problem we use the continuity of f to prove that r is in fact a root, that is, that $f(r) = 0$. You may assume that all the intervals have been chosen so that $f(a_n) < 0$ and $f(b_n) > 0$.

(a) Suppose $f(r) = L > 0$. Use the definition of continuity with any ϵ such that $\epsilon < L$ to choose a δ such that

$$|f(x) - L| < \epsilon \quad \text{for all} \quad |x - r| < \delta.$$

Find an a_n in the interval $[r - \delta, r + \delta]$ and arrive at a contradiction involving $f(a_n)$.

(b) Suppose $f(r) = L < 0$. Make a similar argument to arrive at a contradiction involving a b_n.

(c) Conclude that $f(r) = 0$.

27. Adapt the zooming argument on page 82 and the argument in Problem 26 to prove the Intermediate Value Theorem: If f is continuous on $[a, b]$ and k is between $f(a)$ and $f(b)$, there is a point c in $[a, b]$ with $f(c) = k$. [Hint: Consider $g(x) = f(x) - k$ and look for a zero of g.]

DIFFERENTIABILITY AND LINEAR APPROXIMATION

In this section we analyze the tangent line approximation and its error. This leads us to a different way of looking at differentiability. Recall that:

> A function f is said to be **differentiable at** $x = a$ if $f'(a)$ exists.

Most functions we deal with have a derivative at every point in their domain; they are said to be *differentiable everywhere*.

How Can We Recognize Whether a Function Is Differentiable?

If a function has a derivative at a point, its graph must have a tangent line there; the slope of the tangent line is the derivative. When we zoom in on the graph of the function, we see a nonvertical straight line.

Occasionally we meet a function which fails to have a derivative at a few points. For example, a discontinuous function whose graph has a break at some point cannot have a derivative at that point. Some of the ways in which a function can fail to be differentiable at a point are if:

- The function is not continuous at the point.
- The graph has a sharp corner at that point.
- The graph has a vertical tangent line.

Figure 2.52 shows a function which appears to be differentiable at all points except $x = a$ and $x = b$. There is no tangent at A because the graph has a corner there. As x approaches a from the left, the slope of the line joining P to A converges to some positive number. As x approaches a from the right, the slope of the line joining P to A converges to some negative number. Thus the slopes approach different numbers as we approach $x = a$ from different sides. Therefore the function is not differentiable at $x = a$. At B, there is no sharp corner, but as x approaches b, the slope of the line joining B to Q does not converge; it just keeps growing larger and larger. This reflects the fact that the graph has a vertical tangent at B. Since the slope of a vertical line is not defined, the function is not differentiable at $x = b$.

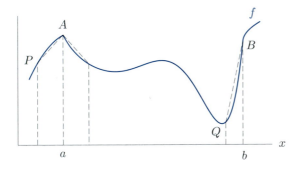

Figure 2.52: A function which is not differentiable at A or B

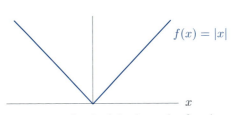

Figure 2.53: Graph of absolute value function, showing point of non-differentiability at $x = 0$

Examples of Nondifferentiable Functions

The best-known function with a corner is the *absolute value* function defined as follows:

$$f(x) = |x| = \begin{cases} x & \text{if } x \geq 0, \\ -x & \text{if } x < 0. \end{cases}$$

The graph of this function is in Figure 2.53. Near $x = 0$, even close-up views of the graph of $f(x)$ look the same, so this is a corner which can't be straightened out by zooming in.

Example 8   Try to compute the derivative of the function $f(x) = |x|$ at $x = 0$. Is f differentiable there?

Solution   To find the slope at $x = 0$, we want to look at

$$\lim_{h \to 0} \frac{f(h) - f(0)}{h} = \lim_{h \to 0} \frac{|h| - 0}{h} = \lim_{h \to 0} \frac{|h|}{h}.$$

As h approaches 0 from the right, h is always positive, so $|h| = h$, and the ratio is always 1. As h approaches 0 from the left, h is negative, so $|h| = -h$, and the ratio is -1. Since the limits are different from each side, the limit of the difference quotient does not exist. Thus, the absolute value function is not differentiable at $x = 0$. The limits of 1 and -1 correspond to the fact that the slope of the right-hand part of the graph is 1, and the slope of the left-hand part is -1.

Example 9   Investigate the differentiability of $f(x) = x^{1/3}$ at $x = 0$.

Solution   This function is smooth at $x = 0$ (no sharp corners) but appears to have a vertical tangent there. (See Figure 2.54.) Looking at the difference quotient at $x = 0$, we see

$$\lim_{h \to 0} \frac{(0 + h)^{1/3} - 0^{1/3}}{h} = \lim_{h \to 0} \frac{h^{1/3}}{h} = \lim_{h \to 0} \frac{1}{h^{2/3}}.$$

As $h \to 0$ the denominator becomes small, so the fraction grows without bound. Hence, the function fails to have a derivative at $x = 0$.

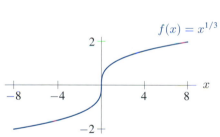

Figure 2.54: Continuous function not differentiable at $x = 0$: Vertical tangent

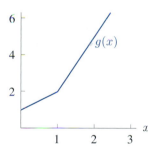

Figure 2.55: Continuous function not differentiable at $x = 1$

Example 10   Consider the function given by the formulas

$$g(x) = \begin{cases} x + 1 & \text{if} \quad x \le 1 \\ 3x - 1 & \text{if} \quad x > 1. \end{cases}$$

This kind of function is called *piecewise linear* because each part of it is linear. Draw the graph of g. Is g continuous? Is g differentiable at $x = 1$?

Solution   The graph in Figure 2.55 has no breaks in it, so the function is continuous. However, the graph has a corner at $x = 1$ which no amount of magnification will remove. To the left of $x = 1$, the slope is 1; to the right of $x = 1$, the slope is 3. Thus, the difference quotient at $x = 1$ will fail to have a limit, and so the function g is not differentiable at $x = 1$.

A great deal of interest has been sparked in the last few years in the study of curves which do not possess derivatives *anywhere*. These curves, known as *fractals*, arise in the modeling of natural processes that are random and chaotic, such as the path of a water molecule in a glass of water. As the molecule bounces off of its neighbors haphazardly, it traces out a path with many jagged, nondifferentiable corners. Although the path may be smooth between collisions, it can be modeled effectively by a curve that is not differentiable anywhere. The coastlines of Maine and Washington are also examples. They never straighten out, no matter how close we look.

Differentiability and Linear Approximation

When we zoom in on the graph of a differentiable function, it looks like a straight line. In fact, the graph is not exactly a straight line when we zoom in; however, its deviation from straightness is so small that it can't be detected by the naked eye. Let's examine what this means. The straight line that we think we see when we zoom in on the graph of $f(x)$ at $x = a$ has slope equal to the derivative, $f'(a)$, so the equation is

$$y = f(a) + f'(a)(x - a).$$

The fact that the graph looks like a line means that y is a good approximation to $f(x)$. (See Figure 2.56.) This suggests the following definition:

The Tangent Line Approximation

Suppose f is differentiable at a. Then, for values of x near a, the tangent line approximation to $f(x)$ is

$$f(x) \approx f(a) + f'(a)(x - a).$$

The expression $f(a) + f'(a)(x - a)$ is called the *local linearization* of f near $x = a$. We are thinking of a as fixed, so that $f(a)$ and $f'(a)$ are constant.
The **error**, $E(x)$, in the approximation is defined by

$$E(x) = f(x) - f(a) - f'(a)(x - a).$$

It can be shown that the tangent line approximation is the best linear approximation to f near a. See Problem 15.

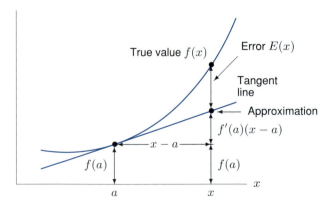

Figure 2.56: The tangent line approximation and its error

Example 11   What is the tangent line approximation for $f(x) = \sin x$ near $x = 0$? Assume that $f'(0) = 1$.

Solution   The tangent line approximation of f near $x = 0$ is

$$f(x) \approx f(0) + f'(0)(x - 0).$$

If $f(x) = \sin x$, then $f(0) = \sin 0 = 0$. Using the given fact that $f'(0) = 1$, the approximation is

$$\sin x \approx x.$$

This means that, near $x = 0$, the function $f(x) = \sin x$ is well approximated by the function $y = x$. If we zoom in on the graphs of the functions $\sin x$ and x near the origin, we won't be able to tell them apart. (See Figure 2.57.)

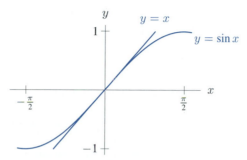

Figure 2.57: Tangent line approximation to $y = \sin x$

Estimating the Error in the Approximation

Let us look at the error, $E(x)$, which is the difference between $f(x)$ and the local linearization. (Look back at Figure 2.56.) The fact that the graph of f looks like a line as we zoom in means that not only is $E(x)$ small for x near a, but also that $E(x)$ is small relative to $(x - a)$. To demonstrate this, we prove the following theorem about the ratio $E(x)/(x - a)$.

Theorem: Differentiability and Local Linearity

Suppose f is differentiable at $x = a$ and $E(x)$ is the error in the tangent line approximation, that is:

$$E(x) = f(x) - f(a) - f'(a)(x - a).$$

Then

$$\lim_{x \to a} \frac{E(x)}{x - a} = 0.$$

Proof   Using the definition of $E(x)$, we have

$$\frac{E(x)}{x - a} = \frac{f(x) - f(a) - f'(a)(x - a)}{x - a} = \frac{f(x) - f(a)}{x - a} - f'(a).$$

Taking the limit as $x \to a$ and using the definition of the derivative, we see that

$$\lim_{x \to a} \frac{E(x)}{x - a} = \lim_{x \to a} \left(\frac{f(x) - f(a)}{x - a} - f'(a) \right) = f'(a) - f'(a) = 0.$$

Why Differentiability Makes A Graph Look Straight

We can use the error $E(x)$ to understand why differentiability makes a graph look straight when we zoom in.

Example 12   Consider the graph of $f(x) = \sin x$ near $x = 0$, and its linear approximation computed in Example 11. Show that there is an interval around 0 with the property that the distance from $f(x) = \sin x$ to the linear approximation is less than $0.1|x|$ for all x in the interval.

Solution   The linear approximation of $f(x) = \sin x$ near 0 is $y = x$, so we write

$$\sin x = x + E(x).$$

Since $\sin x$ is differentiable at $x = 0$, the theorem tells us that

$$\lim_{x \to 0} \frac{E(x)}{x} = 0.$$

If we take $\epsilon = 1/10$, then the definition of limit guarantees that there is a $\delta > 0$ such that

$$\left| \frac{E(x)}{x} \right| < 0.1 \quad \text{for all} \quad |x| < \delta.$$

In other words, for x in the interval $(-\delta, \delta)$, we have $|x| < \delta$, so

$$|E(x)| < 0.1|x|.$$

(See Figure 2.58.)

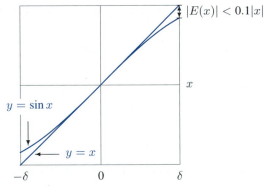

Figure 2.58: Graph of $y = \sin x$ and its linear approximation $y = x$, showing a window in which the magnitude of the error, $|E(x)|$, is less than $0.1|x|$ for all x in the window

We can generalize from this example to explain why differentiability makes the graph of f look straight when viewed over a small graphing window. Suppose f is differentiable at $x = a$. Then we know $\lim_{x \to a} \left| \frac{E(x)}{x - a} \right| = 0$. So, for any $\epsilon > 0$, we can find a δ small enough so that

$$\left| \frac{E(x)}{x - a} \right| < \epsilon, \qquad \text{for} \quad a - \delta < x < a + \delta.$$

So, for any x in the interval $(a - \delta, a + \delta)$, we have

$$|E(x)| < \epsilon|x - a|.$$

Thus, the error, $E(x)$, is less than ϵ times $|x - a|$, the distance between x and a. So, as we zoom in on the graph by choosing smaller ϵ, the deviation, $|E(x)|$, of f from its tangent line shrinks, even relative to the scale on the x-axis. So, zooming makes a differentiable function look straight.

Differentiability and Continuity

The fact that a function which is differentiable at a point has a tangent line suggests that the function is continuous there, as the next theorem shows.

> ### Theorem: A Differentiable Function Is Continuous
>
> If $f(x)$ is differentiable at a point $x = a$, then $f(x)$ is continuous at $x = a$.

Proof   We assume $f(x)$ is differentiable at $x = a$. Then we know

$$f'(a) = \lim_{x \to a} \frac{f(x) - f(a)}{x - a}.$$

So we must have

$$\lim_{x \to a} (f(x) - f(a)) = \lim_{x \to a} \left((x - a) \frac{f(x) - f(a)}{x - a} \right) = \left(\lim_{x \to a} (x - a) \right) \cdot \left(\lim_{x \to a} \frac{f(x) - f(a)}{x - a} \right)$$

$$= 0 \cdot f'(a) = 0.$$

Then,

$$\lim_{x \to a} f(x) = f(a),$$

which means $f(x)$ is continuous at $x = a$.

Problems on Differentiability and Linear Approximation

1. For each of the graphs in Figure 2.59, list the x-values for which the function appears to be
   (i) Not continuous and (ii) Not differentiable.

(a)
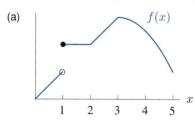

(b)
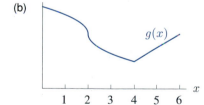

Figure 2.59

2. Look at the graph of $f(x) = (x^2 + 0.0001)^{1/2}$ shown in Figure 2.60. The graph of f appears to have a sharp corner at $x = 0$. Do you think f has a derivative at $x = 0$?

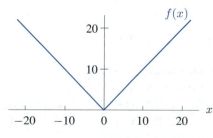

Figure 2.60

Decide if the functions in Problems 3–5 are differentiable at $x = 0$. Try zooming in on a graphing calculator, or calculating the derivative $f'(0)$ from the definition.

3. $f(x) = (x + |x|)^2 + 1$

4. $f(x) = \begin{cases} x\sin(1/x) + x & \text{for } x \neq 0 \\ 0 & \text{for } x = 0 \end{cases}$

5. $f(x) = \begin{cases} x^2\sin(1/x) & \text{for } x \neq 0 \\ 0 & \text{for } x = 0 \end{cases}$

6. An electric charge, Q, in a circuit is given as a function of time, t, by

$$Q = \begin{cases} C & \text{for } t \leq 0 \\ Ce^{-t/RC} & \text{for } t > 0, \end{cases}$$

where C and R are positive constants. The electric current, I, is the rate of change of charge, so

$$I = \frac{dQ}{dt}.$$

   (a) Is the charge, Q, a continuous function of time?
   (b) Do you think the current, I, is defined for all times, t? [Hint: To graph this function, take, for example, $C = 1$ and $R = 1$.]

7. A magnetic field, B, is given as a function of the distance, r, from the center of a wire as follows:

$$B = \begin{cases} \dfrac{r}{r_0}B_0 & \text{for } r \leq r_0 \\[2mm] \dfrac{r_0}{r}B_0 & \text{for } r > r_0. \end{cases}$$

   (a) Sketch a graph of B against r. What is the meaning of the constant B_0?
   (b) Is B continuous at $r = r_0$? Give reasons.
   (c) Is B differentiable at $r = r_0$? Give reasons.

8. A cable is made of an insulating material in the shape of a long, thin cylinder of radius r_0. It has electric charge distributed evenly throughout it. The electric field, E, at a distance r from the center of the cable is given by

$$E = \begin{cases} kr & \text{for } r \leq r_0 \\[2mm] k\dfrac{r_0^2}{r} & \text{for } r > r_0. \end{cases}$$

   (a) Is E continuous at r_0?
   (b) Is E differentiable at r_0?
   (c) Sketch a graph of E as a function of r.

9. Graph the function defined by

$$g(r) = \begin{cases} 1 + \cos(\pi r/2) & \text{for } -2 \leq r \leq 2 \\ 0 & \text{for } r < -2 \quad \text{or} \quad r > 2. \end{cases}$$

   (a) Is g continuous at $r = 2$? Explain your answer.
   (b) Do you think g is differentiable at $r = 2$? Explain your answer.

10. The potential, ϕ, of a charge distribution at a point on the y-axis is given by

$$\phi = \begin{cases} 2\pi\sigma\left(\sqrt{y^2 + a^2} - y\right) & \text{for } y \geq 0 \\[2mm] 2\pi\sigma\left(\sqrt{y^2 + a^2} + y\right) & \text{for } y < 0 \end{cases}$$

where σ and a are positive constants. [Hint: To graph this function, take, for example, $2\pi\sigma = 1$ and $a = 1$.]

   (a) Is ϕ continuous at $y = 0$?
   (b) Do you think ϕ is differentiable at $y = 0$?

11. Consider the function $f(x) = \sqrt{x}$. Assume $f'(4) = 1/4$.

    (a) Find and sketch $f(x)$ and the tangent line approximation to $f(x)$ near $x = 4$.

    (b) Compare the true value of $f(4.1)$ with the value obtained by using the tangent line approximation.

    (c) Compare the true and approximate values of $f(16)$.

    (d) Using a graph, explain why the tangent line approximation is a good one when $x = 4.1$ but not when $x = 16$.

12. Local linearization will give values too small for the function x^2 and too large for the function $\sqrt{x}$. Draw pictures to explain why.

13. Find the local linearization of $f(x) = x^2$ near $x = 1$.

14. Consider the graph of $f(x) = x^2$ near $x = 1$. Find an interval around $x = 1$ with the property that in any smaller interval, the graph of $f(x) = x^2$ never diverges from its local linearization by more than $0.1|x - 1|$ for all x in the interval.

15. Consider a function f and a point a. Suppose there is a number L such that the linear function g

$$g(x) = f(a) + L(x - a)$$

is a good approximation to f. By good approximation, we mean that

$$\lim_{x \to a} \frac{E_L(x)}{x - a} = 0,$$

where $E_L(x)$ is the approximation error defined by

$$f(x) = g(x) + E_L(x) = f(a) + L(x - a) + E_L(x).$$

Show that f is differentiable at $x = a$ and that $f'(a) = L$. Thus the tangent line approximation is the only good linear approximation.

CHAPTER THREE

KEY CONCEPT: THE DEFINITE INTEGRAL

We started Chapter 2 by calculating the velocity from the distance traveled. This led us to the notion of the derivative, or rate of change, of a function. Now we consider the reverse problem: given the velocity, how can we calculate the distance traveled? This will lead us to our second key concept, the *definite integral*, which computes the total change in a function from its rate of change. We will then discover that the definite integral can be used to calculate not only distance but also other quantities, such as the area under a curve and the average value of a function.

We end this chapter with the Fundamental Theorem of Calculus, which tells us that we can use a definite integral to get information about a function from its derivative. Calculating derivatives and calculating definite integrals are, in a sense, reverse processes.

3.1 HOW DO WE MEASURE DISTANCE TRAVELED?

If the velocity of a moving object is a constant, we can find the distance it travels using the formula

$$\text{Distance} = \text{Velocity} \times \text{Time}.$$

In this section we see how to estimate the distance when the velocity is not a constant.

A Thought Experiment: How Far Did the Car Go?

Velocity Data Every Two Seconds

Suppose a car is moving with increasing velocity. Suppose we measure the car's velocity every two seconds, and obtain the data in Table 3.1:

TABLE 3.1   *Velocity of car every two seconds*

Time (sec)	0	2	4	6	8	10
Velocity (ft/sec)	20	30	38	44	48	50

How far has the car traveled? Since we don't know how fast the car is moving at every moment, we can't calculate the distance exactly, but we can make an estimate. The velocity is increasing, so the car is going at least 20 ft/sec for the first two seconds. Since Distance = Velocity × Time, the car goes at least $(20)(2) = 40$ feet during the first two seconds. Likewise, it goes at least $(30)(2) = 60$ feet during the next two seconds, and so on. During the ten-second period it goes at least

$$(20)(2) + (30)(2) + (38)(2) + (44)(2) + (48)(2) = 360 \text{ feet}.$$

Thus, 360 feet is an underestimate of the total distance traveled during the ten seconds.

To get an overestimate, we can reason this way: During the first two seconds, the car's velocity is at most 30 ft/sec, so it moved at most $(30)(2) = 60$ feet. In the next two seconds it moved at most $(38)(2) = 76$ feet, and so on. Therefore, over the ten-second period it moved at most

$$(30)(2) + (38)(2) + (44)(2) + (48)(2) + (50)(2) = 420 \text{ feet}.$$

Therefore,

$$360 \text{ feet} \leq \text{Total distance traveled} \leq 420 \text{ feet}.$$

There is a difference of 60 feet between the upper and lower estimates.

Velocity Data Every One Second

What if we want a more accurate estimate? Then we should make more frequent velocity measurements, say every second. The data is in Table 3.2.

As before, we get a lower estimate for each second by using the velocity at the beginning of that second. During the first second the velocity is at least 20 ft/sec, so the car travels at least $(20)(1) = 20$ feet. During the next second the car moves at least 26 feet, and so on. So now we can say

$$\begin{aligned}
\text{New lower estimate} &= (20)(1) + (26)(1) + (30)(1) + (35)(1) + (38)(1) \\
&\quad + (42)(1) + (44)(1) + (46)(1) + (48)(1) + (49)(1) \\
&= 378 \text{ feet}.
\end{aligned}$$

TABLE 3.2   *Velocity of car every second*

Time (sec)	0	1	2	3	4	5	6	7	8	9	10
Velocity (ft/sec)	20	26	30	35	38	42	44	46	48	49	50

Notice that this is greater than the old lower estimate of 360 feet.

We get a new upper estimate by considering the velocity at the end of each second. During the first second the velocity is at most 26 ft/sec, and so the car moves at most $(26)(1) = 26$ feet; in the next second it moves at most 30 feet, and so on.

$$\text{New upper estimate} = (26)(1) + (30)(1) + (35)(1) + (38)(1) + (42)(1)$$
$$+ (44)(1) + (46)(1) + (48)(1) + (49)(1) + (50)(1)$$
$$= 408 \text{ feet.}$$

This is less than the old upper estimate of 420 feet. Now we know that

$$378 \text{ feet} \leq \text{Total distance traveled} \leq 408 \text{ feet.}$$

Notice that the difference between the new upper and lower estimates is now 30 feet, half of what it was before. By halving the interval of measurement, we have halved the difference between the upper and lower estimates.

Visualizing Distance on the Velocity Graph: Two Second Data

We can represent both upper and lower estimates on a graph of the velocity. On such a graph we can also see how changing the time interval between velocity measurements changes the accuracy of our estimates.

The velocity can be graphed by plotting the two second data in Table 3.1 and drawing a smooth curve through the data points. (See Figure 3.1.) The area of the first dark rectangle is $(20)(2) = 40$, the lower estimate of the distance moved during the first two seconds. The area of the second dark rectangle is $(30)(2) = 60$, the lower estimate for the distance moved in the next two seconds. The total area of the dark rectangles represents the lower estimate for the total distance moved during the ten seconds.

If the dark and light rectangles are considered together, the first area is $(30)(2) = 60$, the upper estimate for the distance moved in the first two seconds. The second area is $(38)(2) = 76$, the upper estimate for the next two seconds. Continuing this calculation suggests that the upper estimate for the total distance is represented by the sum of the areas of the dark and light rectangles. Therefore, the area of the light rectangles alone represents the difference between the two estimates.

To calculate the difference between the two estimates, look at Figure 3.1 and imagine the light rectangles all pushed to the right and stacked on top of each other. This gives a rectangle of width 2 and height 30. Notice that the height, 30, is just the difference between the initial and final values of the velocity: $30 = 50 - 20$. The width, 2, is the time interval between velocity measurements.

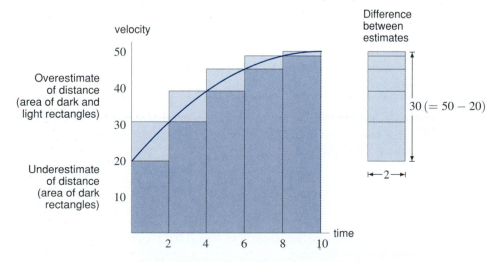

Figure 3.1: Velocity measured every 2 seconds

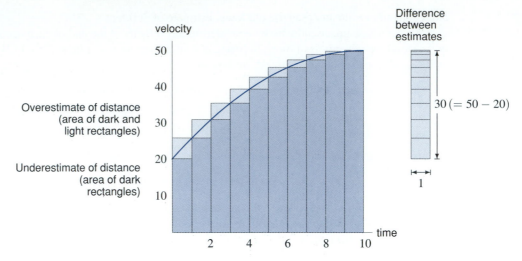

Figure 3.2: Velocity measured every second

Visualizing Distance on the Velocity Graph: One Second Data

The data for the velocities measured every second is graphed in Figure 3.2. The area of the dark rectangles again represents the lower estimate, and the dark and light rectangles together represent the upper estimate. As before, the difference between the two estimates is represented by the area of the light rectangles. This difference can be calculated by stacking the light rectangles vertically, giving a rectangle of the same height as before but of half the width. Its area is therefore half what it was before. Again, the height of this stack is $50 - 20 = 30$, but its width is the time interval between measurements, which is 1.

Example 1   What would be the difference between the upper and lower estimates if the velocity were given every tenth of a second? Every hundredth of a second? Every thousandth of a second?

Solution   Every tenth of a second: Difference between estimates $= (50 - 20)(1/10) = 3$ feet.
Every hundredth of a second: Difference between estimates $= (50 - 20)(1/100) = 0.3$ feet.
Every thousandth of a second: Difference between estimates $= (50 - 20)(1/1000) = 0.03$ feet.

Example 2   How frequently must the velocity be recorded in order to estimate the total distance traveled to within 0.1 feet?

Solution   The difference between the velocity at the beginning and end of the observation period is $50 - 20 = 30$. If the time between the measurements is h, then the difference between the upper and lower estimates is $(30)h$. We want
$$(30)h < 0.1,$$
or
$$h < \frac{0.1}{30} \approx 0.0033.$$
So if the measurements are made less than 0.0033 seconds apart, the distance estimate will be accurate to within 0.1 feet.

Making the Estimates for Distance Precise

We now obtain an exact expression for the total distance traveled. We express the exact total distance traveled as a limit of upper or lower estimates.

We want to know the distance traveled by a moving object over the time interval $a \leq t \leq b$. Let the velocity at time t be given by the function $v = f(t)$. We take measurements of $f(t)$ at equally spaced times $t_0, t_1, t_2, \ldots, t_n$, with time $t_0 = a$ and time $t_n = b$. The time interval between any two consecutive measurements is

$$\Delta t = \frac{b-a}{n},$$

where Δt means the change, or increment, in t.

During the first time interval, the velocity can be approximated by $f(t_0)$, so the distance traveled is approximately

$$f(t_0)\Delta t.$$

During the second time interval, the velocity is about $f(t_1)$, so the distance traveled is about

$$f(t_1)\Delta t.$$

Continuing in this way and adding up all the estimates, we get an estimate for the total distance. In the last interval, the velocity is approximately $f(t_{n-1})$, so the last term is $f(t_{n-1})\Delta t$:

$$\text{Total distance traveled between } t = a \text{ and } t = b \approx f(t_0)\Delta t + f(t_1)\Delta t + f(t_2)\Delta t + \cdots + f(t_{n-1})\Delta t.$$

This is called a *left-hand sum* because we used the value of velocity from the left end of each time interval. It is represented by the sum of the areas of the rectangles in Figure 3.3.

We can also calculate a *right-hand sum* by using the value of the velocity at the right end of each time interval. In that case the estimate for the first interval is $f(t_1)\Delta t$, for the second interval it is $f(t_2)\Delta t$, and so on. The estimate for the last interval is now $f(t_n)\Delta t$, so

$$\text{Total distance traveled between } t = a \text{ and } t = b \approx f(t_1)\Delta t + f(t_2)\Delta t + f(t_3)\Delta t + \cdots + f(t_n)\Delta t.$$

The right-hand sum is represented by the area of the rectangles in Figure 3.4.

If f is an increasing function, the left-hand sum is an underestimate of the total distance traveled. The reason is that a left-hand sum uses the velocity at the start of each interval to compute the distance

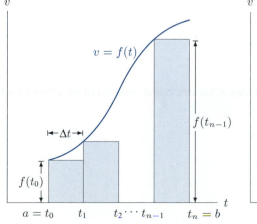

Figure 3.3: Left-hand sums

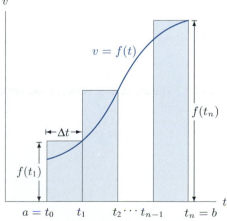

Figure 3.4: Right-hand sums

traveled, whereas in fact the velocity continues to increase after that measurement. Similarly, if f is increasing, the right-hand sum is an overestimate. If f is decreasing, as in Figure 3.5, then the roles of the two sums are reversed.

For either increasing or decreasing functions, the exact value of the distance traveled lies somewhere between the two estimates. Thus the accuracy of our estimate depends on how close these two sums are. For a function which is increasing throughout or decreasing throughout the interval $[a, b]$:

$$\left| \begin{array}{c} \text{Difference between} \\ \text{upper and lower estimates} \end{array} \right| = \left| \begin{array}{c} \text{Difference between} \\ f(a) \text{ and } f(b) \end{array} \right| \times \Delta t = |f(b) - f(a)| \cdot \Delta t.$$

(Absolute values are used to make the difference nonnegative.) In Figure 3.5, the area of the light rectangles is the difference between estimates. By making the time interval, Δt, between measurements small enough, we can make this difference between lower and upper estimates as small as we like.

In the car example, as n increased, the overestimates of the distance traveled decreased and the underestimates increased, trapping the exact distance between them. This suggests that to find exactly the total distance traveled between $t = a$ and $t = b$, we take the limit of the sums, as n goes to infinity.

$$\begin{array}{rl} \text{Total distance traveled} & = \lim_{n \to \infty} (\text{Left-hand sum}) \\ \text{between } t = a \text{ and } t = b & \\ & = \lim_{n \to \infty} \left[f(t_0)\Delta t + f(t_1)\Delta t + \cdots + f(t_{n-1})\Delta t \right] \\ & = \text{Area under curve } f(t) \text{ from } t = a \text{ to } t = b \end{array}$$

and

$$\begin{array}{rl} \text{Total distance traveled} & = \lim_{n \to \infty} (\text{Right-hand sum}) \\ \text{between } t = a \text{ and } t = b & \\ & = \lim_{n \to \infty} \left[f(t_1)\Delta t + f(t_2)\Delta t + \cdots + f(t_n)\Delta t \right] \\ & = \text{Area under curve } f(t) \text{ from } t = a \text{ to } t = b. \end{array}$$

Provided f is continuous, the limits of the left and the right sums are both equal to the total distance traveled. This method of calculating the distance by taking the limit of a sum works even if the velocity is not increasing throughout, or decreasing throughout, the interval.

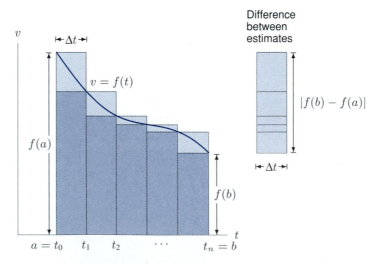

Figure 3.5: Left and right sums if f is decreasing

Problems for Section 3.1

1. A car comes to a stop six seconds after the driver applies the brakes. While the brakes are on, the following velocities are recorded:

Time since brakes applied (sec)	0	2	4	6
Velocity (ft/sec)	88	45	16	0

    (a) Give lower and upper estimates for the distance the car traveled after the brakes were applied.
    (b) On a sketch of velocity against time, show the lower and upper estimates of part (a).

2. A student is speeding down Route 11 in his fancy red Porche when his radar system warns him of an obstacle 400 feet ahead. He immediately applies the brakes, starts to slow down, and spots a skunk in the road directly ahead of him.

    Suppose that the "black box" in the Porche records the car's speed every two seconds, producing the following table. Assume that the speed decreases throughout the 10 seconds it takes to stop, although not necessarily at a uniform rate.

Time since brakes applied (sec)	0	2	4	6	8	10
Speed (ft/sec)	100	80	50	25	10	0

    (a) Using the information in this table, what is your best estimate of the total distance that the student's car traveled before coming to rest?
    (b) Which statement below can you justify from the information given in the story and data table? (Choose one and justify it.)
       (i)   The car stopped before getting to the skunk.
       (ii) The "black box" data is inconclusive. The skunk may or may not have been hit.
       (iii) The unfortunate skunk was hit by the car.

3. Roger decides to run a marathon. Roger's friend Jeff rides behind him on a bicycle and clocks his pace every 15 minutes. Roger starts out strong, but after an hour and a half he is so exhausted that he has to stop. The data Jeff collected are summarized below:

Time spent running (min)	0	15	30	45	60	75	90
Speed (mph)	12	11	10	10	8	7	0

    (a) Assuming that Roger's speed is never increasing, give upper and lower estimates for the distance Roger ran during the first half hour.
    (b) Give upper and lower estimates for the distance Roger ran in total during the entire hour and a half.
    (c) How often would Jeff have needed to measure Roger's pace in order to find lower and upper estimates within 0.1 mile of the actual distance that he ran?

4. Coal gas is produced at a gasworks. Pollutants in the gas are removed by scrubbers, which become less and less efficient as time goes on. The following measurements, made at the start of each month, show the rate at which pollutants are escaping in the gas:

Time (months)	0	1	2	3	4	5	6
Rate pollutants are escaping (tons/month)	5	7	8	10	13	16	20

    (a) Make an overestimate and an underestimate of the total quantity of pollutants that escaped during the first month.

(b)   Make an overestimate and an underestimate of the total quantity of pollutants that escaped during the six months.

(c)   How often would measurements have to be made in order to find overestimates and underestimates which differ by less than 1 ton from the exact quantity of pollutants that escaped during the first six months?

5.   Suppose time, t, is given in seconds and your velocity, v, in meters/second, is given by

$$v(t) = 1 + t^2 \quad \text{for} \quad 0 \le t \le 6.$$

Use $\Delta t = 2$ to estimate the distance traveled during this time. Find the left- and right-hand sums, and then average the two.

6.   For $0 \le t \le 1$, a bug is crawling at a velocity, v, determined by the formula

$$v = \frac{1}{1+t},$$

where t is in hours and v is in meters/hour. Use $\Delta t = 0.2$ to estimate the distance that the bug crawls during this hour. Find an overestimate and an underestimate. Then average the two to get a new estimate.

7.   In Figure 3.6, use the grid to estimate the area of the region bounded by the curve, the horizontal axis and the vertical lines $x = 3$ and $x = -3$. Get an upper and a lower estimate that are within 4 square units of one another. Explain your answer.

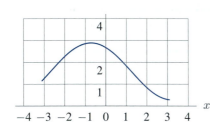

Figure 3.6

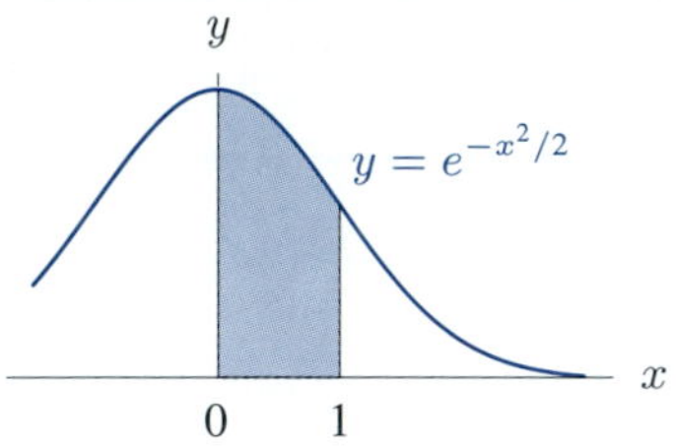

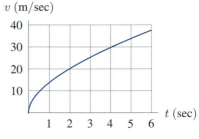

Figure 3.7

v (m/sec)

Figure 3.8

8.   (a)   In Figure 3.7 estimate the shaded area with an error of at most 0.1.
     (b)   How can you approximate this shaded area to any desired degree of accuracy?

9.   Figure 3.8 shows the graph of the velocity, v, of an object (in m/sec). Estimate the total distance the object traveled between $t = 0$ and $t = 6$.

10.   You jump out of an airplane. Before your parachute opens you fall faster and faster, but your acceleration decreases as you fall because of air resistance. The table below gives your acceleration, a (in m/sec^2), after t seconds.

t	0	1	2	3	4	5
a	9.81	8.03	6.53	5.38	4.41	3.61

(a)   Give upper and lower estimates of your speed at $t = 5$.
(b)   Get a new estimate by taking the average of your upper and lower estimates. What does the concavity of the graph of acceleration tell you about your new estimate?

11.   When an aircraft attempts to climb as rapidly as possible, its climb rate decreases with altitude. (This occurs because the air is less dense at higher altitudes.) Table 3.3 shows performance data for a certain single-engine aircraft.

TABLE 3.3

Altitude (1000 ft)	0	1	2	3	4	5	6	7	8	9	10
Climb rate (ft/min)	925	875	830	780	730	685	635	585	535	490	440

(a)   Calculate upper and lower estimates for the time required for this aircraft to climb from sea level to 10,000 ft.
(b)   If climb rate data were available in increments of 500 ft, what would be the difference between a lower and upper estimate of climb time based on 20 subdivisions?

3.2 THE DEFINITE INTEGRAL

In Section 3.1 we saw how distance traveled can be approximated by sums and expressed exactly as the limit of a sum. In this section we show how these sums can be defined for any function f, whether or not it represents a velocity. Suppose we have a function $f(t)$ which is continuous for $a \leq t \leq b$. We divide the interval from a to b into n equal subdivisions, and we call the width of an individual subdivision Δt, so

$$\Delta t = \frac{b - a}{n}.$$

We let $t_0, t_1, t_2, \ldots, t_n$ be endpoints of the subdivisions, as in Figures 3.9 and 3.10. As before, we construct two sums:

$$\text{Left-hand sum} = f(t_0)\Delta t + f(t_1)\Delta t + \cdots + f(t_{n-1})\Delta t$$

and

$$\text{Right-hand sum} = f(t_1)\Delta t + f(t_2)\Delta t + \cdots + f(t_n)\Delta t.$$

These sums represent the areas in Figures 3.9 and 3.10, provided $f(t) \geq 0$. In Figure 3.9, the first rectangle has width Δt and height $f(t_0)$, since the top of its left edge just touches the curve, and hence it has area $f(t_0)\Delta t$. The second rectangle has width Δt and height $f(t_1)$, and hence has area $f(t_1)\Delta t$, and so on. The sum of all these areas is the left-hand sum. The right-hand sum, shown in Figure 3.10, is constructed in the same way, except that each rectangle touches the curve on its right edge instead of its left.

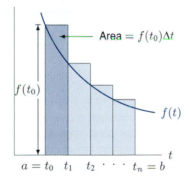

Figure 3.9: Left-hand sum

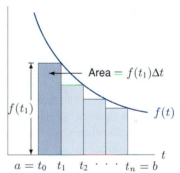

Figure 3.10: Right-hand sum

Writing Left and Right Sums Using Sigma Notation

Both the left-hand and right-hand sums can be written more compactly using *sigma*, or summation, notation. The symbol $\sum$ is a capital sigma, or Greek letter "S." We write

$$\text{Right-hand sum} = \sum_{i=1}^{n} f(t_i)\Delta t = f(t_1)\Delta t + f(t_2)\Delta t + \cdots + f(t_n)\Delta t.$$

The $\sum$ tells us to add terms of the form $f(t_i)\Delta t$. The "$i = 1$" at the base of the sigma sign tells us to start at $i = 1$, and the "n" at the top tells us to stop at $i = n$.

In the left-hand sum we start at $i = 0$ and stop at $i = n - 1$, so we write

$$\text{Left-hand sum} = \sum_{i=0}^{n-1} f(t_i)\Delta t = f(t_0)\Delta t + f(t_1)\Delta t + \cdots + f(t_{n-1})\Delta t.$$

Taking the Limit to Obtain the Definite Integral

In the previous section we took the limit of these sums as n went to infinity, and we do the same here. If f is continuous for $a \leq t \leq b$, the limits of the left- and right-hand sums exist and are equal. The *definite integral* is the limit of these sums. A formal definition of the definite integral is given in the Focus on Theory section on page 181.

Suppose f is continuous for $a \leq t \leq b$. The **definite integral** of f from a to b, written

$$\int_a^b f(t)\, dt,$$

is the limit of the left-hand or right-hand sums with n subdivisions of $[a, b]$ as n gets arbitrarily large. In other words,

$$\int_a^b f(t)\, dt = \lim_{n \to \infty} (\text{Left-hand sum}) = \lim_{n \to \infty} \left(\sum_{i=0}^{n-1} f(t_i) \Delta t \right)$$

and

$$\int_a^b f(t)\, dt = \lim_{n \to \infty} (\text{Right-hand sum}) = \lim_{n \to \infty} \left(\sum_{i=1}^{n} f(t_i) \Delta t \right).$$

Each of these sums is called a *Riemann sum*, f is called the *integrand*, and a and b are called the *limits of integration*.

The "$\int$" notation comes from an old-fashioned "S," which stands for "sum" in the same way that $\sum$ does. The "dt" in the integral comes from the factor Δt. Notice that the limits on the $\sum$ symbol are 0 and $n - 1$ for the left-hand sum, and 1 and n for the right-hand sum, whereas the limits on the $\int$ sign are a and b.

Computing a Definite Integral

In practice, we often approximate definite integrals numerically using a calculator or computer. They use programs which compute sums for larger and larger values of n, and eventually give a value for the integral. Different calculators and computers may give slightly different estimates, owing to round-off error and the fact that they may use different approximation methods. Some (but not all) definite integrals can be computed exactly. However, any definite integral can be approximated numerically.

In the next example, we see how numerical approximation works. For each value of n, we calculate an over- and an under-estimate for the integral. As we increase the value of n the over- and under-estimates get closer together, trapping the value of the integral between them. By increasing the value of n sufficiently, we can calculate the integral to any desired accuracy.

Example 1   Calculate the left-hand and right-hand sums with $n = 2$ and $n = 10$ for $\int_1^2 \frac{1}{t}\, dt$. How do the values of these sums compare with the exact value of the integral?

Solution   Here $a = 1$ and $b = 2$, so for $n = 2$, $\Delta t = (2 - 1)/2 = 0.5$. Therefore, $t_0 = 1$, $t_1 = 1.5$ and $t_2 = 2$. (See Figure 3.11.) We have

$$\text{Left-hand sum} = f(1)\Delta t + f(1.5)\Delta t$$

$$= 1(0.5) + \frac{1}{1.5}(0.5)$$

$$\approx 0.8333,$$

$$\text{Right-hand sum} = f(1.5)\Delta t + f(2)\Delta t$$

$$= \frac{1}{1.5}(0.5) + \frac{1}{2}(0.5)$$

$$\approx 0.5833.$$

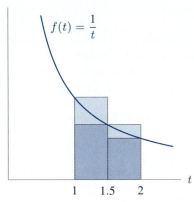

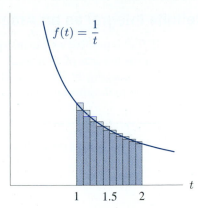

Figure 3.11: Approximating $\int_1^2 \frac{1}{t}\, dt$ with $n = 2$

Figure 3.12: Approximating $\int_1^2 \frac{1}{t}\, dt$ with $n = 10$

From Figure 3.11 we see that the left-hand sum is bigger than the area under the curve and the right-hand sum is smaller. So the area under the curve $f(t) = 1/t$ from $t = 1$ to $t = 2$ is between 0.5833 and 0.8333 :

$$0.5833 < \int_1^2 \frac{1}{t}\, dt < 0.8333.$$

When $n = 10$, we have $\Delta t = (2 - 1)/10 = 0.1$ (see Figure 3.12), so

$$\text{Left-hand sum} = f(1)\Delta t + f(1.1)\Delta t + \cdots + f(1.9)\Delta t$$

$$= \left(1 + \frac{1}{1.1} + \cdots + \frac{1}{1.9}\right) 0.1$$

$$\approx 0.7188,$$

$$\text{Right-hand sum} = f(1.1)\Delta t + f(1.2)\Delta t + \cdots + f(2)\Delta t$$

$$= \left(\frac{1}{1.1} + \frac{1}{1.2} + \cdots + \frac{1}{2}\right) 0.1$$

$$\approx 0.6688.$$

From Figure 3.12 you can see that the left-hand sum is larger than the area under the curve, and the right-hand sum smaller, so

$$0.6688 < \int_1^2 \frac{1}{t}\, dt < 0.7188.$$

Notice that the left- and right-hand sums trap the exact value of the integral between them. As the subdivisions become finer, the left- and right-hand sums get closer together.

Example 2   Use left and right sums with $n = 250$ for $\int_1^2 \frac{1}{t}\, dt$ to estimate the value of the integral.

Solution   Using a program on a calculator or computer, we see that

$$0.6921 < \int_1^2 \frac{1}{t}\, dt < 0.6941.$$

So we can say that

$$\int_1^2 \frac{1}{t}\, dt \approx 0.69.$$

to two decimal places. The exact value is known to be $\int_1^2 \frac{1}{t}\, dt = \ln 2 = 0.693147\ldots$.

The Definite Integral as an Area

If $f(x)$ is positive we can interpret each term $f(x_0)\Delta x, f(x_1)\Delta x, \ldots$ in a left- or right-hand Riemann sum as the area of a rectangle. See Figure 3.13. As the width Δx of the rectangles approaches zero, the rectangles fit the curve of the graph more exactly, and the sum of their areas gets closer and closer to the area under the curve shaded in Figure 3.14. This suggests that:

When $f(x)$ is positive and $a < b$:

$$\text{Area under graph of } f \text{ between } a \text{ and } b = \int_a^b f(x)\, dx.$$

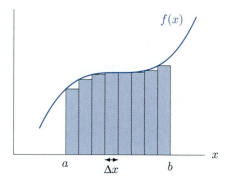

Figure 3.13: Area of rectangles approximating the area under the curve

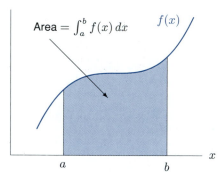

Figure 3.14: The definite integral $\int_a^b f(x)\, dx$

Example 3   Consider the integral $\displaystyle\int_{-1}^{1} \sqrt{1 - x^2}\, dx$.

(a)   Interpret the integral as an area, and find its exact value.

(b)   Estimate the integral using a calculator or computer. Compare your answer to the exact value.

Solution   (a)   The integral is the area under the graph of $y = \sqrt{1 - x^2}$ between -1 and 1. Rewriting this equation as $x^2 + y^2 = 1$, we see that the graph is a semicircle of radius 1 and area $\pi/2$ (see Figure 3.15).

(b)   A calculator estimates the integral as $1.5707966\ldots$. For comparison, $\pi/2 = 1.5707963\ldots$.

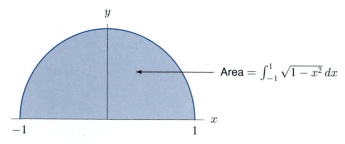

Figure 3.15: Area interpretation of $\int_{-1}^{1} \sqrt{1 - x^2}\, dx$

When $f(x)$ is not positive

We have assumed in drawing Figure 3.14 that the graph of $f(x)$ lies above the x-axis. If the graph lies below the x-axis, then each value of $f(x)$ is negative, so each $f(x)\Delta x$ is negative, and the area gets counted negatively. In that case, the definite integral is the negative of the area.

Example 4   How does the definite integral $\int_{-1}^{1} (x^2 - 1)\, dx$ relate to the area between the parabola $y = x^2 - 1$ and the x-axis?

Solution   Using a calculator, we find $\int_{-1}^{1} (x^2 - 1)\, dx \approx -1.33$. The parabola lies below the axis between $x = -1$ and $x = 1$. (See Figure 3.16.) So the area between the parabola and the x-axis is approximately 1.33.

When $f(x)$ is positive for some x values and negative for others, and $a < b$:

$$\int_{a}^{b} f(x)\, dx$$ is the sum of the areas above the x-axis, counted positively, and the areas below the x-axis, counted negatively.

Example 5   Interpret the definite integral $\int_{0}^{\sqrt{2\pi}} \sin(x^2)\, dx$ in terms of areas.

Solution   The integral is the area above the x-axis, A_1, minus the area below the x-axis, A_2. See Figure 3.17. Approximating the integral with a calculator gives

$$\int_{0}^{\sqrt{2\pi}} \sin(x^2)\, dx \approx 0.43.$$

The graph of $y = \sin(x^2)$ crosses the x-axis where $x^2 = \pi$, that is, at $x = \sqrt{\pi}$. The next crossing is at $x = \sqrt{2\pi}$. Breaking the integral into two parts and calculating each one separately gives

$$\int_{0}^{\sqrt{\pi}} \sin(x^2)\, dx \approx 0.89 \quad \text{and} \quad \int_{\sqrt{\pi}}^{\sqrt{2\pi}} \sin(x^2)\, dx \approx -0.46.$$

So $A_1 \approx 0.89$ and $A_2 \approx 0.46$. Then, as we would expect,

$$\int_{0}^{\sqrt{2\pi}} \sin(x^2)\, dx = A_1 - A_2 \approx 0.89 - 0.46 = 0.43.$$

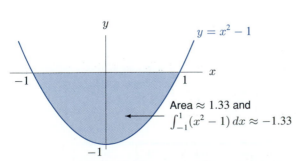

Figure 3.16: Integral $\int_{-1}^{1}(x^2 - 1)\, dx$ is negative of shaded area

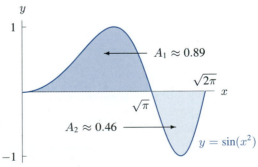

Figure 3.17: Integral $\int_{0}^{\sqrt{2\pi}} \sin(x^2)\, dx = A_1 - A_2$

Problems for Section 3.2

1. On a copy of Figure 3.18, draw rectangles representing each of the following Riemann sums for the function f on the interval $0 \le t \le 8$. Calculate the value of each sum.

    (a) Left-hand sum with $\Delta t = 4$
    (b) Right-hand sum with $\Delta t = 4$
    (c) Left-hand sum with $\Delta t = 2$
    (d) Right-hand sum with $\Delta t = 2$

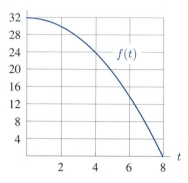

Figure 3.18

2. Write out the terms of the right-hand sum with $n = 5$ that could be used to approximate $\displaystyle\int_{3}^{7} \frac{1}{1+x}\, dx$.

    Do not evaluate the terms or the sum.

For Problems 3–8, construct a table of left- and right-hand sums with 2, 10, 50, and 250 subdivisions. Observe the limit to which your sums are tending as the number of subdivisions gets larger, and hence estimate the value of the definite integral.

3. $\displaystyle\int_{0}^{1} x^3\, dx$ 4. $\displaystyle\int_{0}^{\pi/2} \cos x\, dx$ 5. $\displaystyle\int_{2}^{3} \sin(t^2)\, dt$

6. $\displaystyle\int_{0}^{1} e^{t^2}\, dt$ 7. $\displaystyle\int_{0.2}^{3} \sin\left(1/x\right)\, dx$ 8. $\displaystyle\int_{1}^{2} x^x\, dx$

9. Estimate $\displaystyle\int_{1}^{2} x^2\, dx$ using left- and right-hand sums with four subdivisions. How far from the true value of the integral could your estimate be?

10. (a) Use a calculator or computer to find $\displaystyle\int_{0}^{6} (x^2 + 1)\, dx$. Represent this value as the area under a curve.

     (b) Estimate $\displaystyle\int_{0}^{6} (x^2 + 1)\, dx$ using a left-hand sum with $n = 3$. Represent this sum graphically on a sketch of $f(x) = x^2 + 1$. Is this sum an overestimate or underestimate of the true value found in part (a)?

     (c) Estimate $\displaystyle\int_{0}^{6} (x^2 + 1)\, dx$ using a right-hand sum with $n = 3$. Represent this sum on your sketch. Is this sum an overestimate or underestimate?

11. A table of values for $f(t)$ is given. Estimate $\displaystyle\int_{0}^{100} f(t)\, dt$.

t	0	20	40	60	80	100
$f(t)$	1.2	2.8	4.0	4.7	5.1	5.2

Estimate the area of the regions in Problems 12–17.

12. Under the curve $y = \cos t$ for $0 \leq t \leq \pi/2$.

13. Under the curve $y = 7 - x^2$ and above the x-axis.

14. Under the curve $y = \cos \sqrt{x}$ for $0 \leq x \leq 2$.

15. Under the curve $y = e^x$ and above the line $y = 1$ for $0 \leq x \leq 2$.

16. Between $y = x^2$ and $y = x^3$ for $0 \leq x \leq 1$.

17. Between $y = x^{1/2}$ and $y = x^{1/3}$ for $0 \leq x \leq 1$.

18. In Problems 16 and 17 we calculated the areas between $y = x^2$ and $y = x^3$ and between $y = x^{1/2}$ and $y = x^{1/3}$ on $0 \leq x \leq 1$. Explain why you would expect these two areas to be equal.

19. (a) Sketch a graph of $f(x) = x(x+2)(x-1)$.
    (b) Find the total area between the graph and the x-axis between $x = -2$ and $x = 1$.
    (c) Find $\displaystyle\int_{-2}^{1} f(x)\, dx$ and interpret it in terms of areas.

20. Compute the definite integral $\displaystyle\int_{0}^{4} \cos \sqrt{x}\, dx$ and interpret the result in terms of areas.

21. Without computing the integral, decide if

$$\int_{0}^{2\pi} e^{-x} \sin x\, dx$$

is positive or negative, and explain your decision. [Hint: Sketch $e^{-x} \sin x$.]

22. Suppose that we use $n = 500$ subintervals to approximate $\displaystyle\int_{-1}^{1} (2x^3 + 4)\, dx$. Without computing the Riemann sums, find the difference between the right- and left-hand Riemann sums.

23. The graph of a function $f(t)$ is given in Figure 3.19. Which of the following four numbers could be an estimate of $\displaystyle\int_{0}^{1} f(t)\,dt$ accurate to two decimal places? Explain how you chose your answer.

    (a) -98.35 (b) 71.84 (c) 100.12 (d) 93.47

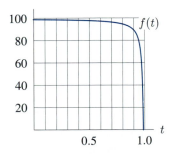

Figure 3.19

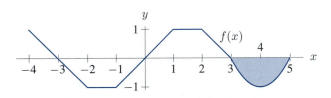

Figure 3.20

24. The graph of $y = f(x)$ is given in Figure 3.20.

    (a) What is $\displaystyle\int_{-3}^{0} f(x)\, dx$?

    (b) If the area of the shaded region is A, estimate $\displaystyle\int_{-3}^{4} f(x)\, dx$.

25.   Consider the definite integral $\int_0^1 x^4 \, dx$.

    (a)   Write an expression for a right-hand Riemann sum approximation for this integral using n subdivisions. Express each x_i, $i = 1,2,\ldots,n$, in terms of i.

    (b)   Use a computer algebra system to obtain a formula for the sum you wrote in part (a) in terms of n.

    (c)   Take the limit of this expression for the sum as $n \to \infty$, thereby finding the exact value of this integral.

26.   Repeat Problem 25, using the definite integral $\int_0^1 x^5 \, dx$.

27.   Three terms of a left-hand sum used to approximate a definite integral $\int_a^b f(x)\,dx$ are as follows.

$$\left(2 + 0 \cdot \frac{4}{3}\right)^2 \cdot \frac{4}{3} + \left(2 + 1 \cdot \frac{4}{3}\right)^2 \cdot \frac{4}{3} + \left(2 + 2 \cdot \frac{4}{3}\right)^2 \cdot \frac{4}{3}.$$

Find possible values for a and b and a possible formula for $f(x)$.

28.   Consider the integral $\int_1^2 \frac{1}{t}\, dt$. In Example 1, by dividing the interval $1 \le t \le 2$ into 10 equal parts, we showed that

$$0.1 \left[\frac{1}{1.1} + \frac{1}{1.2} + \ldots + \frac{1}{2.0} \right] \quad \le \quad \int_1^2 \frac{1}{t}\, dt \quad \le \quad 0.1 \left[\frac{1}{1} + \frac{1}{1.1} + \ldots + \frac{1}{1.9} \right].$$

    (a)   Now divide the interval $1 \le t \le 2$ into n equal parts to show that

$$\sum_{r=1}^{n} \frac{1}{n+r} \quad < \quad \int_1^2 \frac{1}{t}\, dt \quad < \quad \sum_{r=0}^{n-1} \frac{1}{n+r}.$$

    (b)   Show that the difference between the upper and lower sums in part (a) is $1/2n$.

    (c)   The exact value of $\int_1^2 (1/t)\, dt$ is $\ln(2)$. How large should n be to approximate $\ln(2)$ with an error of at most $5 \cdot 10^{-6}$, using one of the sums in part (a)?

3.3   INTERPRETATIONS OF THE DEFINITE INTEGRAL

The Notation and Units for the Definite Integral

Just as the Leibniz notation dy/dx for the derivative reminds us that the derivative is the limit of a ratio of differences, the notation for the definite integral helps us recall the meaning of the integral. The symbol

$$\int_a^b f(x)\, dx$$

reminds us that an integral is a limit of sums (the integral sign is an old-fashioned S) of terms of the form "$f(x)$ times a small difference of x." Officially, dx is not a separate entity, but a part of the whole integral symbol. Just as one thinks of d/dx as a single symbol meaning "the derivative with respect to x of $\ldots$," one can think of $\int_a^b \ldots dx$ as a single symbol meaning "the integral of $\ldots$ with respect to x."

However, many scientists and mathematicians informally think of dx as an "infinitesimally" small bit of x which in this context is multiplied by a function value $f(x)$. This viewpoint is often the key to interpreting the meaning of a definite integral. For example, if $f(t)$ is the velocity of a moving particle at time t, then $f(t)dt$ may by thought of informally as velocity $\times$ time, giving the distance traveled by the particle during a small bit of time dt. The integral $\int_a^b f(t)\, dt$ may then be thought of as the sum of all these small distances, giving us the net change in position of the particle between $t = a$ and $t = b$.

The notation for the integral also helps us determine what units should be used for the value of the integral. Since the terms being added up are products of the form "$f(x)$ times a difference in x," the unit of measurement for $\int_a^b f(x)\,dx$ is the product of the units for $f(x)$ and the units for x. For example, if $f(t)$ is velocity measured in meters/second and t is time measured in seconds, then

$$\int_a^b f(t)\,dt$$

has units of (meters/sec)×(sec) = meters. This is what we expect, since the value of this integral represents change in position.

As another example, graph $y = f(x)$ with the same units of measurement of length along the x- and y-axes, say cm. Then $f(x)$ and x are measured in the same units, so

$$\int_a^b f(x)\,dx$$

is measured in square units of cm × cm = cm². Again, this is what we would expect since in this context the integral represents an area.

The Definite Integral as an Average

We know how to find the average of n numbers: Add them and divide by n. But how do we find the average value of a continuously varying function? Let us consider an example. Suppose $f(t)$ is the temperature at time t, measured in hours since midnight, and that we want to calculate the average temperature over a 24-hour period. One way to start would be to average the temperatures at n equally spaced times, $t_1, t_2, \ldots, t_n$, during the day.

$$\text{Average temperature} \approx \frac{f(t_1) + f(t_2) + \cdots + f(t_n)}{n}.$$

The larger we make n, the better the approximation. We can rewrite this expression as a Riemann sum over the interval $0 \leq t \leq 24$ if we use the fact that $\Delta t = 24/n$, so $n = 24/\Delta t$:

$$\text{Average temperature} \approx \frac{f(t_1) + f(t_2) + \cdots + f(t_n)}{24/\Delta t}$$

$$= \frac{f(t_1)\Delta t + f(t_2)\Delta t + \cdots + f(t_n)\Delta t}{24}$$

$$= \frac{1}{24}\sum_{i=1}^{n} f(t_i)\Delta t.$$

As $n \to \infty$, the Riemann sum tends towards an integral, and $1/24$ of the sum also approximates the average temperature better. It makes sense, then, to write

$$\text{Average temperature} = \lim_{n\to\infty} \frac{1}{24}\sum_{i=1}^{n} f(t_i)\Delta t$$

$$= \frac{1}{24}\int_0^{24} f(t)\,dt.$$

We have found a way of expressing the average temperature over an interval in terms of an integral. Generalizing for any function f, if $a < b$, we define

$$\boxed{\text{Average value of } f \text{ from } a \text{ to } b = \frac{1}{b-a}\int_a^b f(x)\,dx.}$$

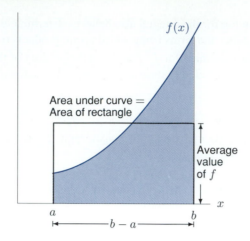

Figure 3.21: Area and average value

How to Visualize the Average on a Graph

The definition of average value tells us that

$$(\text{Average value of } f) \cdot (b - a) = \int_a^b f(x)\, dx.$$

Let's interpret the integral as the area under the graph of f. Then the average value of f is the height of a rectangle whose base is $(b - a)$ and whose area is the same as the area under the graph of f. (See Figure 3.21.)

Example 1 Suppose that $C(t)$ represents the daily cost of heating your house, measured in dollars per day, where t is time measured in days and $t = 0$ corresponds to January 1, 1997. Interpret $\int_0^{90} C(t)\, dt$ and $\dfrac{1}{90 - 0} \int_0^{90} C(t)\, dt$.

Solution The units for the integral $\int_0^{90} C(t)\, dt$ are (dollars/day) $\times$ (days) = dollars. The integral represents the total cost in dollars to heat your house for the first 90 days of 1997, namely the months of January, February, and March. The second expression is measured in (1/days)(dollars) or dollars per day, the same units as $C(t)$. It represents the average cost per day to heat your house during the first 90 days of 1997.

Example 2 On page 14, we saw that the population of Mexico could be modeled by the function

$$P = f(t) = 67.38(1.026)^t,$$

where P is in millions of people and t is in years since 1980. Use this function to predict the average population of Mexico between the years 2000 and 2020.

Solution We want the average value of $f(t)$ between $t = 20$ and $t = 40$. This is given by

$$\text{Average population} = \frac{1}{40 - 20} \int_{20}^{40} f(t)\, dt \approx \frac{1}{20}(2942.66) = 147.1.$$

We used a calculator to evaluate the integral. We see that the average population of Mexico between 2000 and 2020 is predicted to be about 147 million people.

Applications of the Definite Integral

The total distance traveled by a moving object in a given time interval may be represented by a definite integral of the velocity. The following examples show how representing a quantity as a definite integral, and thereby as an area, can be helpful even if we don't evaluate the integral.

Example 3   Two cars start from rest at a traffic light and accelerate for several minutes. Figure 3.22 shows their velocities as a function of time.   (a) Which car is ahead after one minute?   (b) Which car is ahead after two minutes?

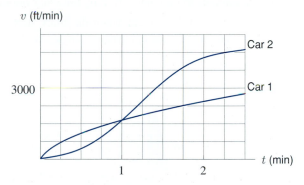

Figure 3.22: Velocities of two cars. Which is ahead when?

Solution   (a)   For the first minute car 1 goes faster than car 2, and therefore car 1 must be ahead at the end of one minute.

(b)   At the end of two minutes the situation is less clear, since car 1 was going faster for the first minute and car 2 for the second. However, if $v = f(t)$ is the velocity of a car after t minutes, then we know that

$$\text{Distance traveled in two minutes} = \int_0^2 f(t)\, dt,$$

since the integral of velocity is distance traveled. This definite integral may also be interpreted as the area under the graph of f between 0 and 2. Since the area representing the distance traveled by car 2 is clearly larger than the area for car 1 (see Figure 3.22), we know that car 2 has traveled farther than car 1.

Example 4   A car starts at noon and travels with the velocity shown in Figure 3.23. A truck starts at 1 pm from the same place and travels at a constant velocity of 50 mph.

(a)   How far away is the car when the truck starts?

(b)   During the period when the car is ahead of the truck, when is the distance between them greatest, and what is that greatest distance?

(c)   When does the truck overtake the car, and how far have both traveled then?

Figure 3.23: Velocity of car for Example 4

Solution   To find distances from the velocity graph, we use the fact that if t is the time measured from noon, and v is the velocity, then

$$\begin{array}{c} \text{Distance traveled} \\ \text{by car up to time } T \end{array} = \int_0^T v\,dt = \begin{array}{c} \text{Area under velocity} \\ \text{graph between 0 and } T. \end{array}$$

The truck's motion can be represented on the same graph by the horizontal line $v = 50$, starting at $t = 1$. The distance traveled by the truck is then the rectangular area under this line, and the distance between the two vehicles is the difference between these areas. Note each small rectangle on the graph corresponds to moving at 10 mph for a half hour (i.e., to a distance of 5 miles).

(a)   The distance traveled by the car when the truck starts is represented by the shaded area in Figure 3.24, which totals about seven rectangles or about 35 miles.

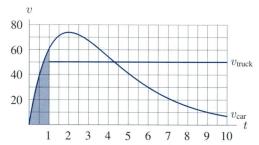

Figure 3.24: Shaded area represents distance
traveled by car from noon to 1pm

(b)   The car starts ahead of the truck, and the distance between them increases as long as the velocity of the car is greater than the velocity of the truck. Later, when the truck's velocity exceeds the car's, the truck starts to gain on the car. In other words, the distance between the car and the truck will increase as long as $v_{\text{car}} > v_{\text{truck}}$, and it will decrease when $v_{\text{car}} < v_{\text{truck}}$. Therefore, the maximum distance occurs when $v_{\text{car}} = v_{\text{truck}}$, that is, when $t \approx 4.3$ hours (at about 4:20 pm). (See Figure 3.25.) The distance traveled by the car is the area under the v_{car} graph between $t = 0$ and $t = 4.3$; the distance traveled by the truck is the area under the v_{truck} line between $t = 1$ (when it started) and $t = 4.3$. So the distance between the car and truck is represented by the shaded area in Figure 3.25, which is approximately

$$35 \text{ miles} + 50 \text{ miles} = 85 \text{ miles}.$$

(c)   The truck overtakes the car when both have traveled the same distance. This occurs when the area under the curve (v_{car}) up to that time equals the area under the line (v_{truck}) up to that time. Since the areas under v_{car} and v_{truck} overlap (see Figure 3.26), they are equal when the lightly shaded area equals the heavily shaded area (which we know is about 85 miles). This happens when $t \approx 8.3$ hours, or about 8:20 pm.

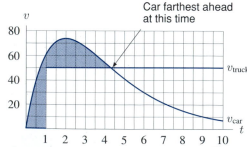

Figure 3.25: Shaded area = distance by which car is
ahead at about 4:20 pm

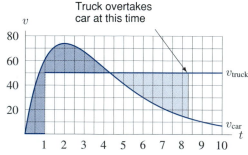

Figure 3.26: Truck overtakes car when dark and
light shaded areas are equal

Problems for Section 3.3

1. For the two cars in Example 3, page 163, estimate:

    (a) The distances moved by car 1 and car 2 during the first minute.
    (b) The time at which the two cars have gone the same distance.

2. Consider the car and the truck in Example 4, page 163.

    (a) How fast is the distance between the car and the truck increasing or decreasing at 3 pm?
    (b) What is the practical significance (in terms of the distance between the car and the truck) of the fact that the car's velocity is maximized at about 2 pm?

3. Consider the car and the truck in Example 4, page 163, but suppose the truck starts at noon. (Everything else remains the same.)

    (a) Sketch a new graph showing the velocities of both car and truck against time.
    (b) How many times do the two graphs intersect? What does each intersection mean in terms of the distance between the two?

4. If $f(t)$ is measured in meters/second^2 and t is measured in seconds, what are the units of $\int_a^b f(t)\,dt$?

5. If $f(t)$ is measured in dollars per year and t is measured in years, what are the units of $\int_a^b f(t)\,dt$?

6. If $f(x)$ is measured in pounds and x is measured in feet, what are the units of $\int_a^b f(x)\,dx$?

7. Oil is leaking out of a ruptured tanker at a rate of $r = f(t)$ gallons per minute, where t is in minutes. Write a definite integral expressing the total quantity of oil which leaks out of the tanker in the first hour.

In Problems 8–10, find the average value of the function over the given interval.

8. $g(t) = 1 + t$ over $[0, 2]$ 9. $f(x) = 4x + 7$ over $[1, 3]$ 10. $g(t) = e^t$ over $[0, 10]$

11. If the graph of f is in Figure 3.27:

    (a) What is $\int_1^6 f(x)\,dx$?   (b) What is the average value of f on $[1, 6]$?

12. A two-day environmental clean up operation started at 9 am on the first day. The number of workers fluctuated as shown in Figure 3.28. If the workers were paid $10 per hour, how much was the total personnel cost of the operation?

Figure 3.27

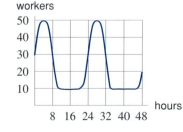

Figure 3.28

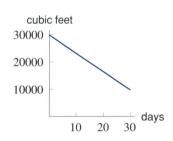

Figure 3.29

13. Suppose in Problem 12 that the workers were paid $10 per hour for work during the time period 9 am to 5 pm and were paid $15 per hour for work during the rest of the day. What would the total personnel costs of the operation have been under these conditions?

14. A warehouse charges it customers $5 per day for every 10 cubic feet of space used for storage. Figure 3.29 records the storage used by one company over a month. How much will the company have to pay?

15. If $f(x) = 2$, show that the average value of $f(x)$ over the interval $[a, b]$ is 2.

16. (a) Without computing any integrals, explain why the average value of $f(x) = \sin x$ on $[0, \pi]$ must be between 0.5 and 1.
    (b) Compute this average. Give 2 decimal places in your answer.

17. (a) What is the average value of $f(x) = \sqrt{1 - x^2}$ over the interval $0 \le x \le 1$?
    (b) How can you tell whether this average value is more or less than 0.5 without doing any calculations?

18. How do the units for the average value relate to the units for $f(x)$ and the units for x?

19. A bar of metal is cooling from $1000°C$ to room temperature, $20°C$. The temperature, H, of the bar t minutes after it starts cooling is given, in $°C$, by

$$H = 20 + 980e^{-0.1t}.$$

    (a) Find the temperature of the bar at the end of one hour.
    (b) Find the average value of the temperature over the first hour.
    (c) Is your answer to part (b) greater or smaller than the average of the temperatures at the beginning and the end of the hour? Explain this in terms of the concavity of the graph of H.

20. The value, V, of a Tiffany lamp, worth \$225 in 1965, increases at 15% per year. Its value in dollars t years after 1965 is given by
$$V = 225(1.15)^t.$$
    Find the average value of the lamp over the period 1965–2000.

21. The number of hours, H, of daylight in Madrid as a function of date is approximated by the formula

$$H = 12 + 2.4 \sin[0.0172(t - 80)],$$

    where t is the number of days since the start of the year. Find the average number of hours of daylight in Madrid:

    (a) in January   (b) in June   (c) over a whole year

    (d) Comment on the relative magnitudes of your answers to parts (a), (b), and (c). Why are they reasonable?

22. A bicyclist is pedaling along a straight road with velocity, v, given in Figure 3.30. Suppose the cyclist starts 5 miles from a lake, and that positive velocities take her away from the lake and negative velocities towards the lake. When is the cyclist farthest from the lake, and how far away is she then?

23. A force F parallel to the x-axis is given by the graph in Figure 3.31. Estimate the work, W, done by the force, where $W = \int_0^{16} F(x)\, dx$.

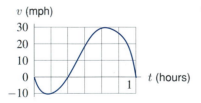

Figure 3.30

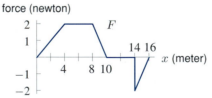

Figure 3.31

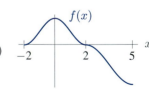

Figure 3.32

24. For the even function f in Figure 3.32, write an expression involving one or more definite integrals that denotes:
    (a) The average value of f for $0 \le x \le 5$.   (b) The average value of $|f|$ for $0 \le x \le 5$.

25. For the even function f in Figure 3.32, consider the average value of f over the following intervals:

    (I) $0 \le x \le 1$ (II) $0 \le x \le 2$ (III) $0 \le x \le 5$ (IV) $-2 \le x \le 2$

    (a) For which interval is the average value of f least?
    (b) For which interval is the average value of f greatest?
    (c) For which pair of intervals are the average values equal?

3.4 THEOREMS ABOUT DEFINITE INTEGRALS

The Definite Integral of a Rate Gives Total Change

We have seen how the definite integral of a velocity function can be interpreted as total distance traveled. If $v(t)$ is velocity and $s(t)$ is position, then $v(t) = s'(t)$ and we know that

$$\text{Total change in position} = s(b) - s(a) = \int_a^b s'(t)\,dt.$$

In this section we generalize this result to explain why the integral of the rate of change of any quantity gives the total change in that quantity.

Suppose $F'(t)$ is the rate of change of some quantity $F(t)$ with respect to time, and that we are interested in the total change in $F(t)$ between $t = a$ and $t = b$. We divide the interval $a \le t \le b$ into n subintervals, each of length Δt. For each small interval, we estimate the change in $F(t)$, written ΔF, and then add all these up. In each subinterval we assume the rate of change of $F(t)$ is approximately constant, so that we can say

$$\Delta F \approx \text{Rate of change of } F \times \text{Time elapsed}.$$

For the first subinterval, from t_0 to t_1, the rate of change of $F(t)$ is approximately $F'(t_0)$, so

$$\Delta F \approx F'(t_0)\,\Delta t.$$

Similarly, for the second interval

$$\Delta F \approx F'(t_1)\,\Delta t.$$

Summing over all the subintervals, we get

$$\text{Total change in } F = \sum_{i=0}^{n-1} \Delta F \approx \sum_{i=0}^{n-1} F'(t_i)\,\Delta t.$$

We have approximated the change in $F(t)$ as a left-hand sum.

However, the total change in $F(t)$ between the times $t = a$ and $t = b$ is simply $F(b) - F(a)$. Taking the limit as n goes to infinity suggests the following result:[1]

$$F(b) - F(a) = \begin{array}{c} \text{Total change in } F(t) \\ \text{between } t = a \text{ and } t = b \end{array} = \lim_{n \to \infty} \sum_{i=0}^{n-1} F(t_i)\Delta t = \int_a^b F'(t)\,dt.$$

This result is one of the most important in calculus because it makes the connection between the derivative and the definite integral. It is called the Fundamental Theorem of Calculus and is often stated as follows:

> ### The Fundamental Theorem of Calculus
>
> If f is continuous on the interval $[a, b]$ and $f(t) = F'(t)$, then
>
> $$\int_a^b f(t)\,dt = F(b) - F(a).$$
>
> In words:
>
> The definite integral of a rate of change gives total change.

---

[1] We could equally well have used a right-hand sum, since the definite integral is their common limit.

What Is Involved in Proving the Fundamental Theorem?

The argument we have given makes the Fundamental Theorem plausible. However, it is not a mathematical proof. Each of the approximations

$$\Delta F \approx F'(t_i)\,\Delta t, \quad i = 1, \ldots, n,$$

involves a small error. To prove that the total change in F is well-approximated by the Riemann sum $\sum_{i=0}^{n-1} F'(t_i)\,\Delta t$, we must show that the sum of all the errors is small (in fact, that it can be made as small as we like by choosing n large enough). We will not give the details of this argument. However, the key idea brings out an important fact about local linearization: the error in the local linearization $\Delta F \approx F'(t_i)\,\Delta t$ is not merely small, it is small relative to the size of Δt. This follows from the definition of the derivative: if we choose Δt small enough, we can ensure that the error in the approximation

$$F'(t_i) \approx \frac{\Delta F}{\Delta t}$$

is as small as we like. In other words,

$$F'(t_i) = \frac{\Delta F}{\Delta t} + \epsilon,$$

where ϵ can be made as small as we like. Multiplying through by Δt, we get an approximation

$$F'(t_i)\Delta t = \Delta F + \epsilon \Delta t.$$

Thus, in the approximation $F'(t_i)\Delta t \approx \Delta F$, the error is $\epsilon \Delta t$, which is small even relative to Δt. When we add these small errors, we still get a small total error. For a proof of the Fundamental Theorem from another point of view, see Problem 14 on page 291.

Using the Fundamental Theorem

The Fundamental Theorem provides a precise way of computing certain definite integrals.

Example 1 Compute $\displaystyle\int_{1}^{3} 2x\,dx$ by two different methods.

Solution Using left- and right-hand sums, we can approximate this integral as accurately as we want. With $n = 100$, for example, the left-sum is 7.96 and the right sum is 8.04. Using $n = 500$ we learn

$$7.992 < \int_{1}^{3} 2x\,dx < 8.008.$$

The Fundamental Theorem, on the other hand, allows us to compute the integral exactly. We take $f(x) = 2x$. By Example 4, on page 108, we know that if $F(x) = x^2$, then $F'(x) = 2x$. So we use $f(x) = 2x$ and $F(x) = x^2$ and obtain

$$\int_{1}^{3} 2x\,dx = F(3) - F(1) = 3^2 - 1^2 = 8.$$

The Fundamental Theorem can also be used when the rate, $F'(t)$, is known and we want to find the total change $F(b) - F(a)$. If we also know $F(a)$, the theorem enables us to reconstruct the function F from knowledge about its derivative $F' = f$.

Example 2 Let $F(t)$ represent a bacteria population which is 5 million at time $t = 0$. Suppose that after t hours, the population is growing at an instantaneous rate of 2^t million bacteria per hour. Estimate the total increase in the bacteria population during the first hour, and the population at $t = 1$.

Solution Since the rate at which the population is growing is $F'(t) = 2^t$, we have

$$\text{Change in population} = F(1) - F(0) = \int_{0}^{1} 2^t\,dt.$$

Using a calculator gives

$$\text{Change in population} = \int_0^1 2^t \, dt \approx 1.44 \text{ million bacteria.}$$

Since $F(0) = 5$, the population at $t = 1$ is given by

$$\text{Population} = F(1) = F(0) + \int_0^1 2^t \, dt \approx 5 + 1.44 = 6.44 \text{ million.}$$

Properties of the Definite Integral

For the definite integral $\int_a^b f(x) \, dx$, we have so far only considered the case $a < b$. We now allow $a \geq b$. We still set $x_0 = a$, $x_n = b$, and $\Delta x = (b-a)/n$. As before, we have $\int_a^b f(x)dx = \lim_{n\to\infty} \sum_{i=1}^n f(x_i)\Delta x$.

Theorem: Properties of Limits of Integration

If a, b, and c are any numbers and f is a continuous function, then

1. $\displaystyle \int_b^a f(x) \, dx = - \int_a^b f(x) \, dx.$

2. $\displaystyle \int_a^c f(x) \, dx + \int_c^b f(x) \, dx = \int_a^b f(x) \, dx.$

In words:
1. The integral from b to a is the negative of the integral from a to b.
2. The integral from a to c plus the integral from c to b is the integral from a to b.

By interpreting the integrals as areas, we can justify these results for $f \geq 0$. In fact, they are true for all functions for which the integrals make sense.

Why is $\displaystyle \int_b^a f(x) \, dx = - \int_a^b f(x) \, dx$?

By definition, both integrals are approximated by sums of the form $\sum f(x_i)\Delta x$. The x_is are the same in each case: the only difference in the sums for $\int_b^a f(x) \, dx$ and $\int_a^b f(x) \, dx$ is that in the first $\Delta x = (a-b)/n = -(b-a)/n$ and in the second $\Delta x = (b-a)/n$. Since everything else about the sums is the same, we must have $\int_b^a f(x) \, dx = - \int_a^b f(x) \, dx$.

Why is $\displaystyle \int_a^c f(x) \, dx + \int_c^b f(x) \, dx = \int_a^b f(x) \, dx$?

Suppose $a < c < b$. Figure 3.33 suggests that $\int_a^c f(x) \, dx + \int_c^b f(x) \, dx = \int_a^b f(x) \, dx$ since the area under f from a to c plus the area under f from c to b together make up the whole area under f from a to b.

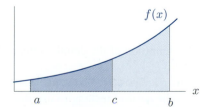

Figure 3.33: Additivity of the definite integral $(a < c < b)$

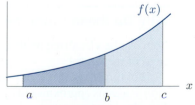

Figure 3.34: Additivity of the definite integral $(a < b < c)$

Actually, this property holds for all numbers a, b, and c, not just those satisfying $a < c < b$. (See Figure 3.34.) For example, the area under f from 3 to 6 is equal to the area from 3 to 8 *minus* the area from 6 to 8, so

$$\int_3^6 f(x)\,dx = \int_3^8 f(x)\,dx - \int_6^8 f(x)\,dx = \int_3^8 f(x)\,dx + \int_8^6 f(x)\,dx.$$

Example 3   Suppose you know that $\int_0^{1.25} \cos(x^2)\,dx = 0.98$ and $\int_0^1 \cos(x^2)\,dx = 0.90$. (See Figure 3.35.) What are the values of the following integrals?

(a) $\displaystyle\int_1^{1.25} \cos(x^2)\,dx$ (b) $\displaystyle\int_{-1}^1 \cos(x^2)\,dx$ (c) $\displaystyle\int_{1.25}^{-1} \cos(x^2)\,dx$

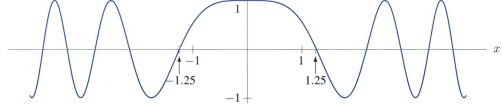

Figure 3.35: Graph of $f(x) = \cos(x^2)$

Solution   (a) Since $\int_0^{1.25} \cos(x^2)\,dx = \int_0^1 \cos(x^2)\,dx + \int_1^{1.25} \cos(x^2)\,dx$ by the additivity property, we get $0.98 = 0.90 + \int_1^{1.25} \cos(x^2)\,dx$, so $\int_1^{1.25} \cos(x^2)\,dx = 0.08$.

(b) $\int_{-1}^1 \cos(x^2)\,dx = \int_{-1}^0 \cos(x^2)\,dx + \int_0^1 \cos(x^2)\,dx$.
By the symmetry of $\cos(x^2)$ about the y-axis, $\int_{-1}^0 \cos(x^2)\,dx = \int_0^1 \cos(x^2)\,dx = 0.90$.
So $\int_{-1}^1 \cos(x^2)\,dx = 0.90 + 0.90 = 1.80$.

(c) $\int_{1.25}^{-1} \cos(x^2)\,dx = -\int_{-1}^{1.25} \cos(x^2)\,dx = -(\int_{-1}^0 \cos(x^2)\,dx + \int_0^{1.25} \cos(x^2)\,dx) = -(0.90 + 0.98) = -1.88$.

Theorem: Properties of Sums and Constant Multiples of the Integrand

Let f and g be continuous functions and let c be a constant.

1. $\displaystyle\int_a^b (f(x) \pm g(x))\,dx = \int_a^b f(x)\,dx \pm \int_a^b g(x)\,dx.$

2. $\displaystyle\int_a^b cf(x)\,dx = c\int_a^b f(x)\,dx.$

In words:

1. The integral of the sum (or difference) of two functions is the sum (or difference) of their integrals.

2. The integral of a constant times a function is that constant times the integral of the function.

Why Do these Properties Hold?

Both can be visualized by thinking of the definition of the definite integral as the limit of a sum of areas of rectangles.

For property 1, let's suppose that f and g are positive on the interval $[a, b]$ so that the area under $f(x) + g(x)$ is approximated by the sum of the areas of rectangles like the one shaded in Figure 3.36. The area of this rectangle is

$$[f(x_i) + g(x_i)]\Delta x = f(x_i)\Delta x + g(x_i)\Delta x.$$

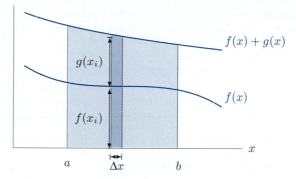

Figure 3.36: Area $= \int_a^b [f(x) + g(x)] \, dx = \int_a^b f(x) \, dx + \int_a^b g(x) \, dx$

Since $f(x_i)\Delta x$ is the area of a rectangle under the graph of f, and $g(x_i)\Delta x$ is the area of a rectangle under the graph of g, the area under $f(x) + g(x)$ is the sum of the areas under $f(x)$ and $g(x)$.

For property 2, notice that multiplying a function by c stretches or flattens the graph in the vertical direction by a factor of c. Thus, it stretches or flattens the height of each approximating rectangle by c, and hence multiplies the area by c.

Example 4   Evaluate the definite integral $\displaystyle\int_0^2 (1 + 3x) \, dx$ exactly.

Solution   We can break this integral up as follows:

$$\int_0^2 (1 + 3x) \, dx = \int_0^2 1 \, dx \; + \; \int_0^2 3x \, dx = \int_0^2 1 \, dx \; + \; 3 \int_0^2 x \, dx.$$

This expresses our original integral in terms of two simpler integrals. From the area interpretation of the integral, we see that

$$\int_0^2 1 \, dx = 2,$$

since it represents the area under the horizontal line $y = 1$ between $x = 0$ and $x = 2$ (see Figure 3.37). Similarly,

$$\int_0^2 x \, dx = \frac{1}{2} \cdot 2 \cdot 2 = 2$$

because it is the area of the triangle in Figure 3.38. Therefore,

$$\int_0^2 (1 + 3x) \, dx = \int_0^2 1 \, dx + 3 \int_0^2 x \, dx = 2 + 3(2) = 8.$$

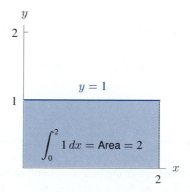

Figure 3.37: Area representing $\int_0^2 1 \, dx$

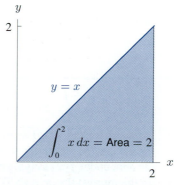

Figure 3.38: Area representing $\int_0^2 x \, dx$

Comparing Integrals

Suppose we have constants m and M such that $m \leq f(x) \leq M$ for $a \leq x \leq b$. We say f is *bounded above* by M and *bounded below* by m. Then the graph of f lies between the horizontal lines $y = m$ and $y = M$. So the definite integral lies between $m(b-a)$ and $M(b-a)$. See Figure 3.39.

Suppose $f(x) \leq g(x)$ for $a \leq x \leq b$, as in Figure 3.40. By a similar argument, the definite integral of f is less than or equal to the definite integral of g. This leads us to the following results:

Theorem: Results About Comparison of Definite Integrals

Let f and g be continuous functions.

1. If $m \leq f(x) \leq M$ for $a \leq x \leq b$, then $\quad m(b-a) \leq \displaystyle\int_a^b f(x)\,dx \leq M(b-a)$.

2. If $f(x) \leq g(x)$ for $a \leq x \leq b$, then $\quad \displaystyle\int_a^b f(x)\,dx \leq \int_a^b g(x)\,dx$.

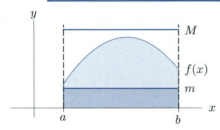

Figure 3.39: The area under the graph of f lies between the areas of the rectangles

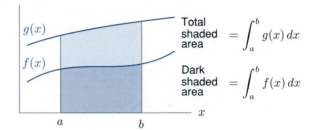

$$\text{Total shaded area} = \int_a^b g(x)\,dx$$

$$\text{Dark shaded area} = \int_a^b f(x)\,dx$$

Figure 3.40: If $f(x) \leq g(x)$ then $\int_a^b f(x)\,dx \leq \int_a^b g(x)\,dx$

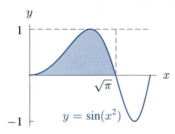

Figure 3.41: Graph showing that $\int_0^{\sqrt{\pi}} \sin(x^2)\,dx < \sqrt{\pi}$

Example 5 Explain why $\displaystyle\int_0^{\sqrt{\pi}} \sin(x^2)\,dx \leq \sqrt{\pi}$.

Solution Since $\sin(x^2) \leq 1$ for all x (see Figure 3.41), part 2 of the theorem gives

$$\int_0^{\sqrt{\pi}} \sin(x^2)\,dx \leq \int_0^{\sqrt{\pi}} 1\,dx = \sqrt{\pi}.$$

Example 6 Show that $2 \leq \displaystyle\int_0^2 \sqrt{1+x^3}\,dx \leq 6$.

Solution Notice that $f(x) = \sqrt{1+x^3}$ is increasing for $0 \leq x \leq 2$, since x^3 gets bigger as x increases. This means that $f(0) \leq f(x) \leq f(2)$. For this function, $f(0) = 1$ and $f(2) = 3$. Thus we can imagine the area under $f(x)$ as lying between the area under the line $y = 1$ and the area under the line $y = 3$ on the interval $0 \leq x \leq 2$. That is,

$$1(2-0) \;\leq\; \int_0^2 \sqrt{1+x^3}\,dx \;\leq\; 3(2-0).$$

Problems for Section 3.4

1. The graph of a derivative $f'(x)$ is shown in Figure 3.42. Fill in the table of values for $f(x)$ given that $f(0) = 2$.

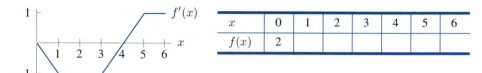

x	0	1	2	3	4	5	6
$f(x)$	2						

Figure 3.42: Graph of f', not f

2. Water is leaking out of a tank at a rate of $R(t)$ gallons/hour, where t is measured in hours.

    (a) Write a definite integral that expresses the total amount of water that leaks out in the first two hours.

    (b) Figure 3.43 is a graph of $R(t)$. On a sketch, shade in the region whose area represents the total amount of water that leaks out in the first two hours.

    (c) Give an upper and lower estimate of the total amount of water that leaks out in the first two hours.

3. Figure 3.44 shows the graph of f. If $F' = f$ and $F(0) = 0$, find $F(b)$ for $b = 1, 2, 3, 4, 5, 6$.

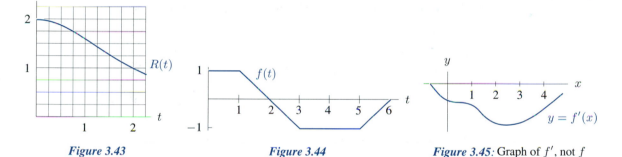

Figure 3.43   *Figure 3.44*   *Figure 3.45:* Graph of f', not f

Problems 4–5 concern the graph of f' in Figure 3.45.

4. Which is greater, $f(0)$ or $f(1)$?

5. List the following in increasing order: $\dfrac{f(4) - f(2)}{2}$, $\quad f(3) - f(2)$, $\quad f(4) - f(3)$.

6. A news broadcast in early 1993 said the average American's annual income is changing at a rate of $r(t) = 40(1.002)^t$ dollars per month, where t is in months from January 1, 1993. How much did the average American's income change during 1993?

7. A cup of coffee at $90°C$ is put into a $20°C$ room when $t = 0$. If the coffee's temperature is changing at a rate given in $°C$ per minute by

$$r(t) = -7e^{-0.1t}, \quad t \text{ in minutes,}$$

    estimate, to one decimal place, the coffee's temperature when $t = 10$.

8. The rate at which the world's oil is being consumed is continuously increasing. Suppose the rate (in billions of barrels per year) is given by the function $r = f(t)$, where t is measured in years and $t = 0$ is the start of 1990.

    (a) Write a definite integral which represents the total quantity of oil used between the start of 1990 and the start of 1995.

    (b) Suppose $r = 32e^{0.05t}$. Using a left-hand sum with five subdivisions, find an approximate value for the total quantity of oil used between the start of 1990 and the start of 1995.

    (c) Interpret each of the five terms in the sum from part (b) in terms of oil consumption.

9. The graph of a function $y = f(x)$ is given in Figure 3.46. Suppose $f(x)$ is the rate (in thousands of algae per hour) at which a population of algae is growing, where x is in hours.

   (a) Estimate the average value of the rate of growth of this population over the interval $x = -1$ to $x = 3$. Explain how you arrived at your answer.

   (b) Estimate the total change in the algae population over the interval $x = -3$ to $x = 3$.

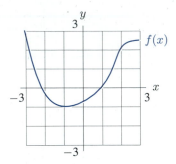

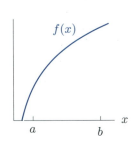

Figure 3.46   **Figure 3.47**

For Problems 10–13, mark the following quantities on a copy of the graph of f in Figure 3.47.

10. A length representing $f(b) - f(a)$.

11. A slope representing $\dfrac{f(b) - f(a)}{b - a}$.

12. An area representing $F(b) - F(a)$, where $F' = f$.

13. A length roughly approximating

$$\frac{F(b) - F(a)}{b - a}, \text{ where } F' = f.$$

Suppose $\int_a^b f(x)\,dx = 8$, $\int_a^b (f(x))^2\,dx = 12$, $\int_a^b g(t)\,dt = 2$, and $\int_a^b (g(t))^2\,dt = 3$. Find the integrals in Problems 14–19.

14. $\int_a^b (f(x) + g(x))\,dx$

15. $\int_a^b \left((f(x))^2 - (g(x))^2\right)\,dx$

16. $\int_a^b (f(x))^2\,dx - \left(\int_a^b f(x)\,dx\right)^2$

17. $\int_a^b cf(z)\,dz$

18. $\int_a^b \left(c_1 g(x) + (c_2 f(x))^2\right)\,dx$

19. $\int_{a+5}^{b+5} f(x - 5)\,dx$

20. The function for the *standard normal distribution*, which is often used in statistics, has the formula

$$\frac{1}{\sqrt{2\pi}} e^{-x^2/2}$$

and the graph in Figure 3.48. Statistics books usually contain tables such as the one below, showing only the area under the curve from 0 to b, for different values of b.

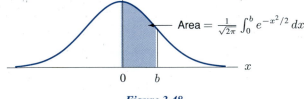

Area $= \dfrac{1}{\sqrt{2\pi}} \int_0^b e^{-x^2/2}\,dx$

b	$\frac{1}{\sqrt{2\pi}} \int_0^b e^{-x^2/2}\,dx$
1	0.3413
2	0.4772
3	0.4987
4	0.5000

Figure 3.48

Use the information given in the table and the symmetry of the standard normal curve about the y-axis to find:

(a) $\dfrac{1}{\sqrt{2\pi}} \int_1^3 e^{-x^2/2}\,dx$.

(b) $\dfrac{1}{\sqrt{2\pi}} \int_{-2}^3 e^{-x^2/2}\,dx$.

21. (a) Is $\int_{-1}^{1} e^{x^2}\, dx$ positive, negative, or zero? Explain.

    (b) Explain why $0 < \int_{0}^{1} e^{x^2}\, dx < 3$.

22. Without calculating the integral, explain why the following statements are false.

    (a) $\int_{-2}^{-1} e^{x^2}\, dx = -3$

    (b) $\int_{-1}^{1} \left| \dfrac{\cos(x+2)}{1+\tan^2 x} \right| dx = 0$

23. Without doing any computations, find the values of

    (a) $\int_{-2}^{2} \sin x\, dx$,

    (b) $\int_{-\pi}^{\pi} x^{113}\, dx$.

24. Use the property $\int_{b}^{a} f(x)\, dx = -\int_{a}^{b} f(x)\, dx$ to show that $\int_{a}^{a} f(x)\, dx = 0$.

25. The average value of $y = v(x)$ equals 4 for $1 \le x \le 6$, and equals 5 for $6 \le x \le 8$. What is the average value of $v(x)$ for $1 \le x \le 8$?

CHAPTER SUMMARY

- **Definite integral as limit of right-hand or left-hand sums**
- **Interpretations of the definite integral**
  Total change from rate of change, change in position given velocity, area, $(b-a)\cdot$ Average value
- **Properties of the definite integral**
  Properties involving integrand, properties involving limits, comparison between integrals.
- **Fundamental Theorem of Calculus**
- **Working with the definite integral**
  Estimate definite integral from graph, table of values, or formula.

REVIEW PROBLEMS FOR CHAPTER THREE

1. A car going 80 ft/sec (about 55 mph) brakes to a stop in 8 seconds. Its velocity is recorded every 2 seconds and is given in the following table.

   (a) Give your best estimate of the distance traveled by the car during the 8 seconds.
   (b) To estimate the distance traveled accurate to within 20 feet, how often should you record the velocity?

t (seconds)	0	2	4	6	8
$v(t)$ (ft/sec)	80	52	28	10	0

2. Fill in Table 3.4 with the appropriate left- and right-hand sums with n subdivisions for the integral $\int_{-1}^{2} 5e^{-x^2}\, dx$. Use the table to estimate the value of the definite integral.

   TABLE 3.4

n	Left	Right
5		
25		
500		

3. A graph of $y = f(x)$ is given in Figure 3.49. Estimate $\int_0^{20} f(x)\,dx$.

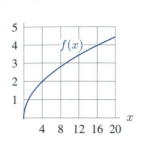

Figure 3.49

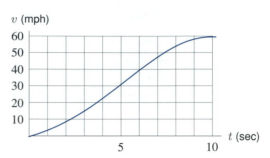

Figure 3.50

4. A car accelerates smoothly from 0 to 60 mph in 10 seconds. Suppose the car's velocity as a function of time is given in Figure 3.50. Estimate how far the car travels during the 10-second period.

5. Your velocity is given by
$$v(t) = \sin(t^2) \quad \text{for } 0 \le t \le 1.1.$$
Represent the distance traveled during this time by an integral, and use a calculator or computer to find the distance traveled.

Find the area of the regions in Problems 6–9.

6. Between $y = x^2 - 9$ and the x-axis.

7. Under one arch of the curve $y = \sin x$.

8. Between the parabola $y = 4 - x^2$ and the x-axis.

9. Between the line $y = 1$ and one arch of the curve $y = \sin\theta$.

10. If $\int_2^5 (2f(x) + 3)\,dx = 17$, find $\int_2^5 f(x)\,dx$.

11. The graph of a continuous function f is given in Figure 3.51. Rank the following integrals in ascending numerical order. Explain your reasons.

(i) $\int_0^2 f(x)\,dx$ (ii) $\int_0^1 f(x)\,dx$ (iii) $\int_0^2 (f(x))^{1/2}\,dx$ (iv) $\int_0^2 (f(x))^2\,dx$.

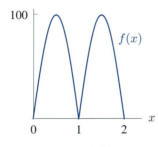

Figure 3.51

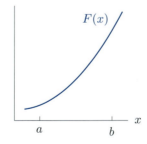

Figure 3.52

For Problems 12–14, assuming $F' = f$, mark the following quantities on a copy of Figure 3.52.

12. A slope representing $f(a)$.

13. A length representing $\int_a^b f(x)\,dx$.

14. A slope representing $\dfrac{1}{b-a}\int_a^b f(x)\,dx$.

15. (a) Sketch a graph of $f(x) = \sin(x^2)$ and mark on it the points $x = \sqrt{\pi}, \sqrt{2\pi}, \sqrt{3\pi}, \sqrt{4\pi}$.

(b) Use your graph to decide which of the four numbers

$$\int_0^{\sqrt{n\pi}} \sin(x^2)\,dx \quad n = 1, 2, 3, 4$$

is largest. Which is smallest? How many of the numbers are positive?

16. The Environmental Protection Agency was recently asked to investigate a spill of radioactive iodine. Measurements showed the ambient radiation levels at the site to be four times the maximum acceptable limit, so the EPA ordered an evacuation of the surrounding neighborhood.

It is known that the level of radiation from an iodine source decreases according to the formula

$$R(t) = R_0 e^{-0.004t}$$

where R is the radiation level (in millirems/hour) at time t, R_0 is the initial radiation level (at $t = 0$), and t is the time measured in hours.

(a) How long will it take for the site to reach an acceptable level of radiation?

(b) How much total radiation (in millirems) will have been emitted by that time, assuming the maximum acceptable limit is 0.6 millirems/hour?

17. Assume the population, P, of Mexico (in millions), is given by

$$P = 67.38(1.026)^t,$$

where t is the number of years since 1980.

(a) What was the average population of Mexico between 1980 and 1990?

(b) What is the average of the population of Mexico in 1980 and the population in 1990?

(c) Explain, in terms of the concavity of the graph of P (see Figure 1.18 on page 15), why your answer to part (b) is larger or smaller than your answer to part (a).

18. The graphs in Figure 3.53 represent the velocity, v, of a particle moving along the x-axis for time $0 \le t \le 5$. The vertical scales of all graphs are the same. Identify the graph showing which particle

(a) has a constant acceleration.

(b) ends up farthest to the left of where it started.

(c) ends up the farthest from its starting point.

(d) experiences the greatest initial acceleration.

(e) has the greatest average velocity.

(f) has the greatest average acceleration.

(I) 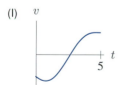   (II)   (III)   (IV)   (V)

Figure 3.53

19. A new sales agent finds that as she gains experience, she increases the number of large appliances she sells each month. In the first month she sells only seven, but each month she sells two more than the month before, so that the number she sells in month t is $2t + 5$.

(a) Find the average number of large appliances she sells per month over the first year arithmetically, by calculating the number of appliances sold each month and then taking the average over 12 months.

(b) Now find the average by integration as though the sales function applied for all values of t (instead of just for integers).

(c) How well do the two results compare?

(d) If the answer you found in part (a) is viewed as the true answer, and the integral answer as an approximation, why would anyone want to use the integral answer instead of the true answer?

(e) Draw a picture showing both answers as the area of some region. Mark on your picture a region representing the error in the integral answer.

20. Water is run into a large tank through a hose at a constant rate. After 5 minutes a hole is opened in the bottom of the tank, and water starts to flow out. Initially the flow rate through the hole is twice as great as the rate through the hose, but as the water level in the tank goes down, the flow rate through the hole decreases; after another 10 minutes the water level in the tank appears to be constant. Plot graphs of the flow rates through the hose and through the hole against time on the same pair of axes. Show how the volume of water in the tank at any time can be interpreted as an area (or the difference between two areas) on the graph. In particular, interpret the steady-state volume of water in the tank.[2]

21. The Glen Canyon Dam at the top of the Grand Canyon prevents natural flooding. In 1996, scientists decided an artificial flood was necessary to restore the environmental balance. Water was released through the dam at a controlled rate[3] shown in Figure 3.54. The figure also shows the rate of flow of the last natural flood in 1957.

(a) At what rate was water passing through the dam in 1996 before the artificial flood?
(b) At what rate was water passing down the river in the pre-flood season in 1957?
(c) Estimate the maximum rates of discharge for the 1996 and 1957 floods.
(d) Approximately how long did the 1996 flood last? How long did the 1957 flood last?
(e) Estimate how much additional water passed down the river in 1996 as a result of the artificial flood.
(f) Estimate how much additional water passed down the river in 1957 as a result of the flood.

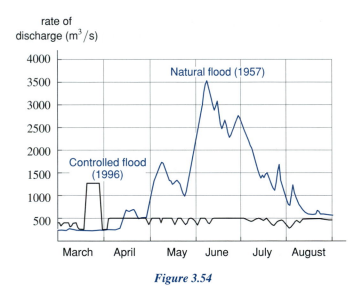

Figure 3.54

22. The Montgolfier brothers (Joseph and Etienne) were eighteenth-century pioneers in the field of hot-air ballooning. Had they had the appropriate instruments, they might have left us a record of one of their early experiments, like that shown in Figure 3.55. The graph shows their vertical velocity, v, with upward as positive.

(a) Over what intervals was the acceleration positive? Negative?
(b) What was the greatest altitude achieved, and at what time?
(c) At what time was the upward acceleration greatest?
(d) At what time was the deceleration greatest?
(e) What might have happened during this flight to explain the answer to part (d)?
(f) This particular flight ended on top of a hill. How do you know that it did, and what was the height of the hill above the starting point?

---

[2]From *Calculus: The Analysis of Functions*, by Peter D. Taylor (Toronto: Wall & Emerson, Inc., 1992)
[3]Adapted from M. Collier, R. Webb, E. Andrews "Experimental Flooding in Grand Canyon" in *Scientific American* (January 1997).

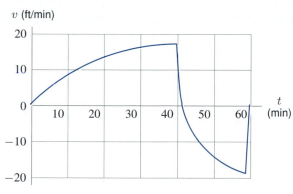

Figure 3.55

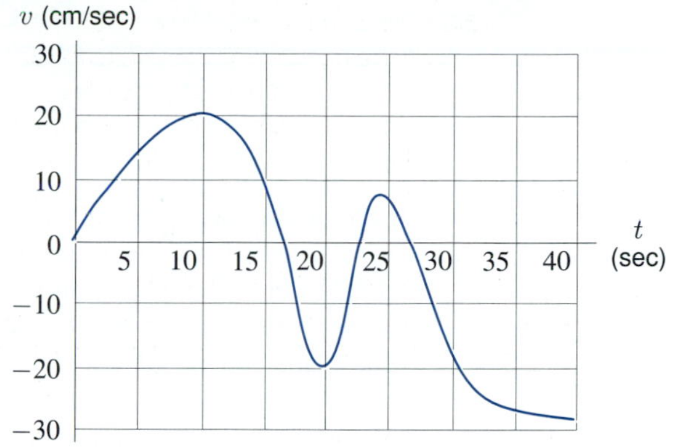

Figure 3.56

23. A mouse moves back and forth in a tunnel, attracted to bits of cheddar cheese alternately introduced to and removed from the ends (right and left) of the tunnel. The graph of the mouse's velocity, v, is given in Figure 3.56, with positive velocity corresponding to motion toward the right end. Assuming that the mouse starts $(t = 0)$ at the center of the tunnel, use the graph to estimate the time(s) at which

   (a) The mouse changes direction.
   (b) The mouse is moving most rapidly to the right; to the left.
   (c) The mouse is farthest to the right of center; farthest to the left.
   (d) The mouse's speed (i.e., the magnitude of its velocity) is decreasing.
   (e) The mouse is at the center of the tunnel.

24. For the even function f graphed in Figure 3.57:

   (a) Suppose you know $\int_0^2 f(x)\,dx$. What is $\int_{-2}^2 f(x)\,dx$?
   (b) Suppose you know $\int_0^5 f(x)\,dx$ and $\int_2^5 f(x)\,dx$. What is $\int_0^2 f(x)\,dx$?
   (c) Suppose you know $\int_{-2}^5 f(x)\,dx$ and $\int_{-2}^2 f(x)\,dx$. What is $\int_0^5 f(x)\,dx$?

25. For the even function f graphed in Figure 3.57:

   (a) Suppose you know $\int_{-2}^2 f(x)\,dx$ and $\int_0^5 f(x)\,dx$. What is $\int_2^5 f(x)\,dx$?
   (b) Suppose you know $\int_{-2}^5 f(x)\,dx$ and $\int_{-2}^0 f(x)\,dx$. What is $\int_2^5 f(x)\,dx$?
   (c) Suppose you know $\int_2^5 f(x)\,dx$ and $\int_{-2}^5 f(x)\,dx$. What is $\int_0^2 f(x)\,dx$?

26. The graph of some function f is given in Figure 3.58. List, from *least* to *greatest*,

   (a) $f'(1)$.
   (b) The average value of $f(x)$, $0 \le x \le a$.
   (c) The average value of the rate of change of $f(x)$, for $0 \le x \le a$.
   (d) $\displaystyle\int_0^a f(x)\,dx$.

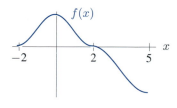

Figure 3.57

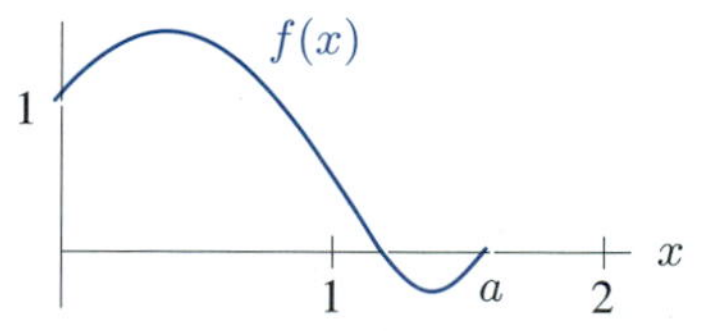

Figure 3.58

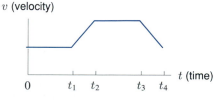

Figure 3.59

27. A force, F, acts on a particle of mass m, moving along a line. The velocity v of the particle is shown in Figure 3.59. Using $F = ma$, where a is the acceleration of the particle, decide if the work done by the force on the particle in each of the intervals $[0, t_1]$, $[t_1, t_2]$, $[t_2, t_3]$, $[t_3, t_4]$, is positive, negative, or zero. (Work done = Force · Distance, when force is constant.)

PROJECTS

1. **Carbon Dioxide in Pond Water**

    Biological activity in a pond is reflected in the rate at which carbon dioxide, CO_2, is added to or withdrawn from the water. Plants take CO_2 out of the water during the day for photosynthesis and put CO_2 into the water at night. Animals put CO_2 into the water all the time as they breathe. Biologists are interested in how the net rate at which CO_2 enters a pond varies during the day. Figure 3.60 shows this rate as a function of time of day.[4] The rate is measured in millimoles (mmol) of CO_2 per liter of water per hour; time is measured in hours past dawn. At dawn, there were 2.600 mmol of CO_2 per liter of water.

    (a) What can be concluded from the fact that the rate is negative during the day and positive at night?
    (b) Some scientists have suggested that plants respire (breathe) at a constant rate at night, and that they photosynthesize at a constant rate during the day. Does Figure 3.60 support this view?
    (c) When was the CO_2 content of the water at its lowest? How low did it go?
    (d) How much CO_2 was released into the water during the 12 hours of darkness? Compare this quantity with the amount of CO_2 withdrawn from the water during the 12 hours of daylight. How can you tell by looking at the graph whether the CO_2 in the pond is in equilibrium?
    (e) Estimate the CO_2 content of the water at three hour intervals throughout the day. Use your estimates to plot a graph of CO_2 content throughout the day.

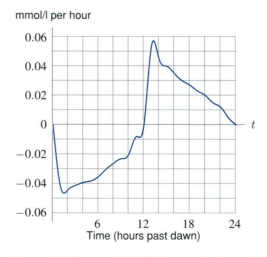

Figure 3.60: Rate at which CO_2 is entering the pond

2. **Average Velocity**

    In Section 2.1, page 89, we defined the average velocity of a particle over the interval $a \leq t \leq b$ as the change in position during this interval divided by $(b - a)$. If the velocity of the particle is $v = f(t)$, the average velocity is also given by $\frac{1}{b-a} \int_a^b f(t)dt$. Explain why these two ways of calculating the average velocity always lead to the same answer.

---

[4]Data from R. J. Beyers, *The Pattern of Photosynthesis and Respiration in Laboratory Microsystems* (Mem. 1st. Ital. Idrobiol., 1965).

FOCUS ON THEORY

THE DEFINITE INTEGRAL

Recall that if f is continuous on $[a, b]$ the definite integral is given by a limit of left or right sums:

$$\int_a^b f(x)\, dx = \lim_{n \to \infty} \sum_{i=0}^{n-1} f(x_i)\Delta x = \lim_{x \to \infty} \sum_{i=1}^{n} f(x_i)\Delta x.$$

This provides a method for approximating definite integrals numerically.[5] In this section we give a formal definition of the definite integral that makes use of more general sums.

A Special Case: Monotonic Functions

A function which is either increasing throughout an interval or decreasing throughout that interval is said to be *monotonic* on the interval. In Section 3.1 we saw that if f is monotonic, the left and right sums trap the exact value of the integral between them. Let us continue Example 1 on page 154, which looks at the value of

$$\int_1^2 \frac{1}{t}\, dt.$$

The left- and right-hand sums for $n = 2, 10, 50$, and 250 are listed in Table 3.5.

Because the function $f(t) = 1/t$ is decreasing, the left-hand sums converge to the integral from above, and the right-hand sums converge from below. From the last row of the table we can deduce that

$$0.6921 < \int_1^2 \frac{1}{t}\, dt < 0.6941,$$

so $\int_1^2 \frac{1}{t}\, dt \approx 0.69$ to two decimal places.

The Difference Between the Upper and Lower Estimates

To be sure that the left- and right-hand sums trap a unique number between them, we need to know that the difference between them approaches zero. On page 150 we saw that for a monotonic function f on the interval $[a, b]$:

$$\left| \begin{array}{c} \text{Difference between} \\ \text{upper and lower estimates} \end{array} \right| = |f(b) - f(a)| \cdot \Delta t,$$

where $\Delta t = (b - a)/n$. We can make this difference as small as we like by choosing Δt small enough.

TABLE 3.5 *Left- and right-hand sums for $\int_1^2 \frac{1}{t}\, dt$*

n	Left-hand sum	Right-hand sum
2	0.8333	0.5833
10	0.7188	0.6688
50	0.6982	0.6882
250	0.6941	0.6921

---

[5]In practice, we often approximate integrals using more sophisticated numerical methods.

TABLE 3.6   *Left- and right-hand sums for $\int_0^{2.5} \sin(t^2)\, dt$*

n	Left-hand sum	Right-hand sum
2	1.2500	1.2085
10	0.4614	0.4531
50	0.4324	0.4307
250	0.4307	0.4304
1000	0.4306	0.4305

When f Is Not Monotonic

If f is not monotonic, the definite integral is not always bracketed between the left- and right-hand sums. For example, Table 3.6 gives sums for the integral $\int_0^{2.5} \sin(t^2)\, dt$. Although $\sin(t^2)$ is certainly not monotonic on $[0, 2.5]$, by the time we get to $n = 250$, it is pretty clear that $\int_0^{2.5} \sin(t^2)\, dt \approx 0.43$ to two decimal places. Notice, however, that 0.43 does not lie between 1.2500 and 1.2085, the left- and right-hand sums for $n = 2$, or even between 0.4614 and 0.4531, the two sums for $n = 10$. If the integrand is not monotonic, the left- and right-hand sums may both be larger (or smaller) than the integral. (See Problems 2 and 3 for more examples of this behavior.)

Defining The Definite Integral by Upper and Lower Sums

When f is not monotonic, it is difficult to get upper and lower bounds for $\int_a^b f(x)\, dx$ from left and right sums. So instead we take the following approach for any function, f. If f is continuous, this new approach agrees with the previous approach. As before, we consider a subdivision of $[a, b]$ into n intervals; however, now we allow the subintervals to have different lengths. We let Δx_i be the length of the i-th interval, and make the following definition:

> Suppose that f is bounded above and below on $[a, b]$. A **lower sum** for f on the interval $[a, b]$ is a sum
> $$\sum_{i=1}^{n} m_i \Delta x_i,$$
> where m_i is the greatest lower bound for f on the i-th interval. An **upper sum** is
> $$\sum_{i=1}^{n} M_i \Delta x_i,$$
> where M_i is the least upper bound for f on the i-th interval.

See Figure 3.61. Now instead of taking a limit as $n \to \infty$, we consider the least upper and greatest lower bounds of these sums. We make the following definition.

> ### Definition of the Definite Integral
>
> Suppose that f is bounded above and below on $[a, b]$. Let L be the least upper bound for all the lower sums for f on $[a, b]$, and let U be the greatest lower bound for all the upper sums. If $L = U$, then we say that f is *integrable* and we define $\int_a^b f(x)\, dx$ to be equal to the common value of L and U.

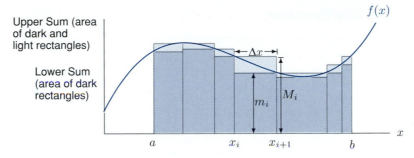

Figure 3.61: Lower and upper sums approximating $\int_a^b f(x)\,dx$

Using the Definition in a Proof

As an example, we will prove the theorem stated on page 172:

Theorem: The Mean Value Inequality for Integrals

If $m \le f(x) \le M$ for all x in $[a, b]$, and if f is integrable on $[a, b]$, then

$$m(b - a) \le \int_a^b f(x)\,dx \le M(b - a).$$

Geometrically, if f is positive, this theorem says that the area under the graph of f is less than the area of the rectangle of height M, and greater than the area of the rectangle of height m. See Figure 3.39 on page 172.

Proof The simplest subdivision of $[a, b]$ is the one that consists of one subinterval, namely, $[a, b]$ itself. Although it does not give a very good approximation for the definite integral, it still counts as a subdivision. The least upper bound for f on $[a, b]$ is less than or equal to M, and the length of the only subinterval in the subdivision is $b - a$. So the upper sum for this subdivision is less than or equal to $M(b - a)$. Since every upper sum is an upper estimate for $\int_a^b f(x)\,dx$, we have

$$\int_a^b f(x)\,dx \le \text{Upper sum} \le M(b - a).$$

The argument for the other inequality is similar, using lower sums. (See Problem 19.)

Problem 20 gives another example of a proof using the definition of the definite integral.

Continuous Functions Are Integrable

In this section we will prove the following:

Theorem: Continuous Functions are Integrable

If f is continuous on $[a, b]$, then $\int_a^b f(x)\,dx$ exists.

The Key Question

It can be shown (for example, using the Extreme Value Theorem in Chapter 5 on page 286) that a continuous function on a finite interval is bounded above and below. Since any lower sum is less than or equal to any upper sum (see Problems 13–17), it follows that $L \leq U$. (See Problem 18.) To show that f is integrable, we need only show that $L = U$, so the question is the following:

> Can we find a subdivision where the lower sum is as close as we like to the upper sum?

If we could, then $L < U$ would not be a possibility, since then $U - L$ would be a positive number, and we would be able to find lower and upper sums less than $U - L$ apart. In that case, either the lower sum would be bigger than L or the upper sum less than U. This can't happen, since L is an upper bound for the lower sums, and U is a lower bound for the upper sums.

We have already seen that if f is monotonic we can find upper and lower sums that are arbitrarily close; just take the left and right sums. This proves that monotonic functions are integrable. If f is not monotonic, we proceed differently.

Squeezing the Integral Between Lower and Upper Sums

For a subdivision of $[a, b]$ we have

$$\text{Difference between upper and lower sums} = \sum_{i=1}^{n} (M_i - m_i)\Delta x_i,$$

where M_i is the least upper bound for f on the i-th subinterval and m_i is the greatest lower bound. The number $M_i - m_i$ represents the amount by which f varies on the i-th subinterval; we call $M_i - m_i$ the variation[6] of f on this subinterval. Suppose that we could choose the subdivision so that the variation on each subinterval was less than some small positive number ϵ. Then

$$\text{Difference between upper and lower sums} < \sum_{i=1}^{n} \epsilon \Delta x_i = \epsilon \sum_{i=1}^{n} \Delta x_i = \epsilon(b - a).$$

By choosing ϵ small enough we would be able to make the difference as small as we liked. Thus the next question is:

> Can we find a subdivision where the maximum variation of f
> on each of the subintervals is as small as we like?

Making the Variation Small on Each Subinterval

Let ϵ be a positive number, as small as we like. We want to prove that there is a subdivision of $[a, b]$ such that the variation of f on each subinterval is less than ϵ. We will give an indirect proof: we assume that there is no such subdivision, and show that this leads to impossible consequences.

Divide the interval $[a, b]$ into two halves. If each half has a subdivision where the maximum variation on subintervals is less than ϵ, we can put the two subdivisions together to form a subdivision of $[a, b]$ with the same property.

So if $[a, b]$ fails to have such a subdivision, then so does one of the halves. Choose a half that does not have such a subdivision and divide it in half again. Once more, one of the halves must fail to have a subdivision where the variation of f on each subinterval is less than ϵ. Continuing in this way we find a nested sequence of intervals, each of which fails to have such a subdivision.

---

[6]This is not the same as the total variation used in more advanced texts.

By the Nested Interval Theorem on page 84, these intervals all contain some number c. Since f is continuous at c, we can find an interval around c on which the variation is less than ϵ. One of our nested intervals must be contained in this interval, since they get arbitrarily small. So the variation of f on one of the nested intervals is less than ϵ. This is impossible, given the way we chose each nested interval. So our supposition that $[a, b]$ fails to have the required subdivision is false; there must be a subdivision of $[a, b]$ such that the variation of f on each subinterval is less than ϵ.

Summary

We have shown by contradiction that for every positive number ϵ, no matter how small, there is a subdivision of $[a, b]$ such that the variation of f on each subinterval is less than ϵ. This means we can make the upper and lower sums as close as we like; hence $L = U$ and $\int_a^b f(x)\,dx = U = L$. That is, the continuous function f is integrable.

More General Riemann Sums

Left- and right-hand sums are special cases of Riemann sums. For a general Riemann sum, as with upper and lower sums, we allow subdivisions to have different lengths. Also, instead of evaluating f only at the left or right endpoint of each subdivision, we allow it to be evaluated anywhere in the subdivision. Thus, a general Riemann sum has the form

$$\sum_{i=1}^{n} (\text{Value of } f \text{ at some point in } i\text{-th subdivision}) \cdot (\text{Length of } i\text{-th subdivision}).$$

(See Figure 3.62.) As before, we let $x_0, x_1, \ldots, x_n$ be the endpoints of the subdivisions, so the length of the i-th subdivision is $\Delta x_i = x_i - x_{i-1}$. For each i we choose a point c_i in the i-th subinterval at which to evaluate f, leading to the following definition:

A general Riemann sum for f on the interval $[a, b]$ is a sum of the form

$$\sum_{i=1}^{n} f(c_i)\Delta x_i,$$

where $a = x_0 < x_1 < \cdots < x_n = b$, and, for $i = 1, \ldots, n$, $\Delta x_i = x_i - x_{i-1}$, and $x_{i-1} \leq c_i \leq x_i$.

We define the *error* in an approximation to be the magnitude of the difference between the approximate and the true values. (Notice that error doesn't mean mistake here.) Since the true value

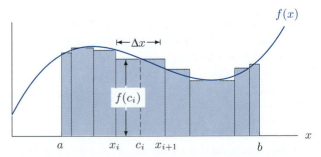

Figure 3.62: A general Riemann sum approximating
$$\int_a^b f(x)\,dx$$

of the integral is between any upper estimate and any lower estimate, the error in approximating a definite integral by a Riemann sum must be less than the difference between the upper and lower sums using the same subdivision. If $\int_a^b f(x)\,dx$ exists, there is a subdivision for which the upper and lower sums are as close as we like. So we can approximate the integral arbitrarily closely by Riemann sums.

Problems on the Definite Integral

1. Write a few sentences in support of or in opposition to the following statement:

   "If a left-hand sum underestimates a definite integral by a certain amount, then the corresponding right-hand sum will overestimate the integral by the same amount."

2. Using the graph of $2 + \cos x$, for $0 \le x \le 4\pi$, list the following quantities in increasing order: the value of the integral $\int_0^{4\pi} (2 + \cos x)\,dx$, the left-hand sum with $n = 2$ subdivisions, and the right-hand sum with $n = 2$ subdivisions.

3. Sketch the graph of a function f (you do not need to give a formula for f) on an interval $[a, b]$ with the property that with $n = 2$ subdivisions,

$$\int_a^b f(x)\,dx < \text{Left-hand sum} < \text{Right-hand sum}.$$

For Problems 4–12, find a subdivision using subintervals of equal length for which the lower and upper sums differ by less than 0.1. Give these sums and an estimate for the integral which is within 0.05 of the true value. Explain your reasoning. [Except for Problem 12, each function is monotonic over the given interval.]

4. $\displaystyle\int_0^5 x^2\,dx$

5. $\displaystyle\int_1^2 2^x\,dx$

6. $\displaystyle\int_1^4 \frac{1}{\sqrt{1 + x^2}}\,dx$

7. $\displaystyle\int_1^{1.5} \sin x\,dx$

8. $\displaystyle\int_0^{\pi/4} \frac{d\theta}{\cos\theta}$

9. $\displaystyle\int_{-2}^{-1} \cos^3 y\,dy$

10. $\displaystyle\int_1^5 (\ln x)^2\,dx$

11. $\displaystyle\int_{1.1}^{1.7} e^t \ln t\,dt$

12. $\displaystyle\int_{-3}^3 e^{-t^2}\,dt$

In Problems 13–17 you will show that every lower sum for a given bounded function f on an interval $[a, b]$ is less than every upper sum, using the idea of a *refinement* of a subdivision. Given a subdivision of the interval $[a, b]$, we can subdivide one or more of its subintervals to obtain a new subdivision. We say that the new subdivision is a *refinement* of the old one. Notice that if one subdivision's set of endpoints contains another's, then the first is a refinement of the second.

13. Show that the lower sum for f on $[a, b]$ using a given subdivision is less than or equal to the upper sum using the same subdivision.

14. In this problem we will show that refining a subdivision results in a lower sum which is larger than the original. Let f be a function defined and bounded from below on $[a, b]$, and choose a subdivision of $[a, b]$, with endpoints $a = x_0 < x_1 < \cdots < x_{n-1} < x_n = b$.

    (a) Suppose $x_{i-1} \le y \le x_i$. Let m_i be the greatest lower bound for f on $[x_{i-1}, x_i]$. Show that m_i is less than or equal to the greatest lower bound for f on $[x_{i-1}, y]$ and the greatest lower bound for f on $[y, x_i]$.

    (b) Show that the lower sum for f using the subdivision $a = x_0 < x_1 < \cdots < x_{n-1} < x_n = b$ is less than or equal to the lower sum using the same subdivision with y included.

    (c) Show that the lower sum for f using the subdivision $a = x_0 < x_1 < \cdots < x_{n-1} < x_n = b$ is less than or equal to the lower sum using any refinement of the subdivision.

15. In this problem we will show that refining a subdivision results in an upper sum which is smaller than the original. Let f be a function defined and bounded from above on $[a, b]$, and choose a subdivision of $[a, b]$, with endpoints $a = x_0 < x_1 < \cdots < x_{n-1} < x_n = b$.

    (a) Suppose $x_{i-1} \leq y \leq x_i$. Let M_i be the least upper bound for f on $[x_{i-1}, x_i]$. Show that M_i is greater than or equal to the least upper bound for f on $[x_{i-1}, y]$ and the least upper bound for f on $[y, x_i]$.

    (b) Show that the upper sum for f using the subdivision $a = x_0 < x_1 < \cdots < x_{n-1} < x_n = b$ is greater than or equal to the upper sum using the same subdivision with y included.

    (c) Show that the upper sum for f using the subdivision $a = x_0 < x_1 < \cdots < x_{n-1} < x_n = b$ is greater than or equal to the upper sum using any refinement of the subdivision.

16. Given two subdivisions of $[a, b]$, show that there is a third one which is a refinement of both.

17. Show that any lower sum for f on $[a, b]$ is less than or equal to any upper sum. [Hint: The lower sum uses one subdivision of $[a, b]$; the upper sum uses another. Use Problem 16 to choose a common refinement of the two subdivisions and then use Problems 13–15.]

18. Let f be a function defined and bounded on $[a, b]$, let L be the least upper bound for all the lower sums for f on $[a, b]$, and let U be the greatest lower bound for all the upper sums.

    (a) Show that if L were strictly greater than U, then there would be a lower sum that was strictly greater than an upper sum. [Hint: Let $\epsilon = L - U$, and find a lower sum within $\epsilon/3$ of L and an upper sum within $\epsilon/3$ of U.]

    (b) Deduce that $L \leq U$.

19. On page 183 we proved one half of the Mean Value Inequality for Integrals. Prove the other half: that is, if f is continuous on $[a, b]$ and $f(x) \geq m$ for x in $[a, b]$, then $m(b - a) \leq \int_a^b f(x)\, dx$.

20. In this problem we will prove that if f is continuous on $[a, b]$ and if c is in $[a, b]$, then

$$\int_a^b f(x)\, dx = \int_a^c f(x)\, dx + \int_c^b f(x)\, dx.$$

    (a) Show that if ℓ_1 is a lower sum for f on $[a, c]$, and if ℓ_2 is a lower sum for f on $[c, b]$, then $\ell_1 + \ell_2$ is a lower sum for f on $[a, b]$.

    (b) Show that if ℓ is a lower sum for f on $[a, b]$, then there is a lower sum ℓ_1 for f on $[a, c]$ and a lower sum ℓ_2 for f on $[c, b]$ such that $\ell \leq \ell_1 + \ell_2$.

    (c) Let L be the least upper bound of all the lower sums on $[a, b]$, let L_1 be the least upper bound of all the lower sums on $[a, c]$, and let L_2 be the least upper bound of all the lower sums on $[c, a]$. Use parts (a) and (b) to show that $L = L_1 + L_2$.

Since f is continuous on $[a, b]$, $L = \int_a^b f(x)\, dx$, $L_1 = \int_a^c f(x)\, dx$, and $L_2 = \int_c^b f(x)\, dx$. Thus you have proved the required statement.

CHAPTER FOUR

SHORT-CUTS TO DIFFERENTIATION

In Chapter 2, we defined the derivative function

$$f'(x) = \lim_{h \to 0} \frac{f(x+h) - f(x)}{h}$$

and saw how the derivative represents a slope and a rate of change. We learned how to approximate the derivative of a function given graphically (by estimating the slope of the tangent at each point) and numerically (by finding the average rate of change of the function between data values). We calculated the derivatives of x^2 and x^3 exactly using the definition.

In this chapter we make a systematic study of the derivatives of functions given by formulas. These functions include powers, polynomials, exponential, logarithmic, and trigonometric functions. The chapter also contains general rules, such as the product, quotient, and chain rules, which allow us to differentiate combinations of functions.

Useful Notation: We write $\frac{d}{dx}(x^3)$, for example, to mean the derivative of x^3 with respect to x. Similarly, $\frac{d}{d\theta}\left(\sin(\theta^2)\right)$ denotes the derivative of $\sin(\theta^2)$ with θ as the variable.

4.1 POWERS AND POLYNOMIALS

Derivative of a Constant Times a Function

Figure 4.1 shows the graph of $y = f(x)$ and of three multiples: $y = 3f(x)$, $y = \frac{1}{2}f(x)$, and $y = -2f(x)$. What is the relationship between the derivatives of these functions? In other words, for a particular x-value, how are the slopes of these graphs related?

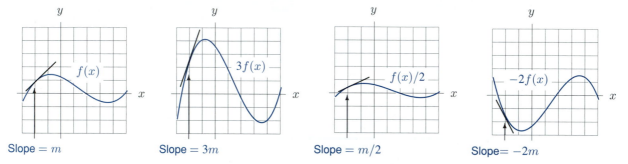

Figure 4.1: A function and its multiples: Derivative of multiple is multiple of derivative

Multiplying by a constant stretches or shrinks the graph (and reflects it in the x-axis if the constant is negative). This changes the slope of the curve at each point. If the graph has been stretched, the "rises" have all been increased by the same factor, whereas the "runs" remain the same. Thus, the slopes are all steeper by the same factor. If the graph has been shrunk, the slopes are all smaller by the same factor. If the graph has been reflected in the x-axis, the slopes will all have their signs reversed. In other words, if a function is multiplied by a constant, c, so is its derivative:

Derivative of a Constant Multiple

If f is differentiable and c is a constant, then

$$\frac{d}{dx}\left[cf(x)\right] = cf'(x).$$

This result can be obtained algebraically, too:

$$\frac{d}{dx}\left[cf(x)\right] = \lim_{h \to 0} \frac{cf(x+h) - cf(x)}{h} = \lim_{h \to 0} c\,\frac{f(x+h) - f(x)}{h}$$

$$= c \lim_{h \to 0} \frac{f(x+h) - f(x)}{h} = cf'(x).$$

You may wonder why c can be taken across the limit sign. The reason is that c is a constant. If a function's value gets close to a certain number, then c times that function gets close to c times that number. The function in this case is $\left[f(x+h) - f(x)\right]/h$.

Derivatives of Sums and Differences

Suppose we have two functions, $f(x)$ and $g(x)$, with the values listed in Table 4.1. Values of the sum $f(x) + g(x)$ are given in the same table.

TABLE 4.1   *Sum of Functions*

x	$f(x)$	$g(x)$	$f(x) + g(x)$
0	100	0	100
1	110	0.2	110.2
2	130	0.4	130.4
3	160	0.6	160.6
4	200	0.8	200.8

We see that adding the increments of $f(x)$ and the increments of $g(x)$ gives the increments of $f(x) + g(x)$. For example, as x increases from 0 to 1, $f(x)$ increases by 10 and $g(x)$ increases by 0.2, while $f(x) + g(x)$ increases by $110.2 - 100 = 10.2$. Similarly, as x increases from 3 to 4, $f(x)$ increases by 40 and $g(x)$ by 0.2, while $f(x) + g(x)$ increases by $200.8 - 160.6 = 40.2$.

From this example, we see that the rate at which $f(x) + g(x)$ is increasing is the sum of the rates at which $f(x)$ and $g(x)$ are increasing. Similar reasoning applies to the difference, $f(x) - g(x)$. In terms of derivatives:

Derivative of Sum and Difference

If f and g are differentiable, then

$$\frac{d}{dx}\left[f(x) + g(x)\right] = f'(x) + g'(x) \quad \text{and} \quad \frac{d}{dx}\left[f(x) - g(x)\right] = f'(x) - g'(x).$$

We can justify the sum rule using the definition of the derivative:

$$\frac{d}{dx}\left[f(x) + g(x)\right] = \lim_{h \to 0} \frac{\left[f(x+h) + g(x+h)\right] - \left[f(x) + g(x)\right]}{h}$$

$$= \lim_{h \to 0} \left[\underbrace{\frac{f(x+h) - f(x)}{h}}_{\text{Limit of this is } f'(x)} + \underbrace{\frac{g(x+h) - g(x)}{h}}_{\text{Limit of this is } g'(x)}\right]$$

$$= f'(x) + g'(x).$$

Powers of x

In Chapter 2 we showed that

$$f'(x) = \frac{d}{dx}(x^2) = 2x \quad \text{and} \quad g'(x) = \frac{d}{dx}(x^3) = 3x^2.$$

The graphs of $f(x) = x^2$ and $g(x) = x^3$ and their derivatives are shown in Figures 4.2 and 4.3. Notice $f'(x) = 2x$ has the behavior we expect. It is negative for $x < 0$ (when f is decreasing), zero for $x = 0$, and positive for $x > 0$ (when f is increasing). Similarly, $g'(x) = 3x^2$ is zero when $x = 0$, but positive everywhere else, as g is increasing everywhere else.

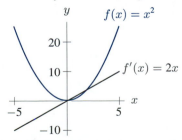

Figure 4.2: Graphs of $f(x) = x^2$ and its derivative $f'(x) = 2x$

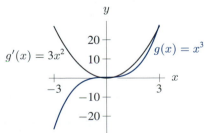

Figure 4.3: Graphs of $g(x) = x^3$ and its derivative $g'(x) = 3x^2$

These examples are special cases of the power rule which we justify for any positive n on page 193:

The Power Rule

For any constant real number n,

$$\frac{d}{dx}(x^n) = nx^{n-1}$$

Problem 52 asks you to show that this rule holds for negative integral powers. In Section 4.6 we indicate how to justify it for powers of the form $1/n$.

Example 1   Use the power rule to differentiate (a) $\dfrac{1}{x^3}$,   (b) $x^{1/2}$,   (c) $\dfrac{1}{\sqrt[3]{x}}$.

Solution   (a) For $n = -3$: $\dfrac{d}{dx}\left(\dfrac{1}{x^3}\right) = \dfrac{d}{dx}(x^{-3}) = -3x^{-3-1} = -3x^{-4} = -\dfrac{3}{x^4}$.

(b) For $n = 1/2$: $\dfrac{d}{dx}\left(x^{1/2}\right) = \dfrac{1}{2}x^{(1/2)-1} = \dfrac{1}{2}x^{-1/2} = \dfrac{1}{2\sqrt{x}}$.

(c) For $n = -1/3$: $\dfrac{d}{dx}\left(\dfrac{1}{\sqrt[3]{x}}\right) = \dfrac{d}{dx}\left(x^{-1/3}\right) = -\dfrac{1}{3}x^{(-1/3)-1} = -\dfrac{1}{3}x^{-4/3} = -\dfrac{1}{3x^{4/3}}$.

Example 2   Use the definition of the derivative to justify the power rule for $n = -2$: Show $\dfrac{d}{dx}(x^{-2}) = -2x^{-3}$.

Solution   Provided $x \neq 0$, we have

$$\frac{d}{dx}\left(x^{-2}\right) = \frac{d}{dx}\left(\frac{1}{x^2}\right) = \lim_{h \to 0}\left(\frac{\frac{1}{(x+h)^2} - \frac{1}{x^2}}{h}\right) = \lim_{h \to 0}\frac{1}{h}\left[\frac{x^2 - (x+h)^2}{(x+h)^2 x^2}\right] \quad \text{(Combining fractions over a common denominator)}$$

$$= \lim_{h \to 0}\frac{1}{h}\left[\frac{x^2 - (x^2 + 2xh + h^2)}{(x+h)^2 x^2}\right] \quad \text{(Multiplying out)}$$

$$= \lim_{h \to 0}\frac{-2xh - h^2}{h(x+h)^2 x^2} \quad \text{(Simplifying numerator)}$$

$$= \lim_{h \to 0}\frac{-2x - h}{(x+h)^2 x^2} \quad \text{(Dividing numerator and denominator by } h)$$

$$= \frac{-2x}{x^4} \quad \text{(Letting } h \to 0)$$

$$= -2x^{-3}.$$

The graphs of x^{-2} and its derivative, $-2x^{-3}$, are shown in Figure 4.4. Does the graph of the derivative have the features you expect?

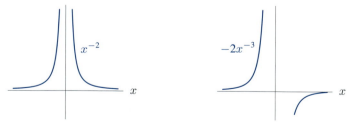

Figure 4.4: Graphs of x^{-2} and its derivative, $-2x^{-3}$

Justification of $\dfrac{d}{dx}(x^n) = nx^{n-1}$, for n a Positive Integer

To calculate the derivatives of x^2 and x^3, we had to expand $(x + h)^2$ and $(x + h)^3$. To calculate the derivative of x^n, we must expand $(x + h)^n$. Let's look back at the previous expansions:

$$(x + h)^2 = x^2 + 2xh + h^2, \qquad (x + h)^3 = x^3 + 3x^2h + 3xh^2 + h^3,$$

Using the Binomial Theorem on page 78 gives

$$(x + h)^n = x^n + nx^{n-1}h + \underbrace{\cdots\cdots + h^n}.$$

Terms involving h^2 and higher powers of h

Now to find the derivative,

$$\frac{d}{dx}(x^n) = \lim_{h \to 0} \frac{(x + h)^n - x^n}{h}$$

$$= \lim_{h \to 0} \frac{(x^n + nx^{n-1}h + \cdots + h^n) - x^n}{h}$$

Terms involving h^2 and higher powers of h

$$= \lim_{h \to 0} \frac{nx^{n-1}h + \overbrace{\cdots + h^n}}{h}.$$

When we factor out h from terms involving h^2 and higher powers of h, each term will still have an h in it. Factoring and dividing, we get:

Terms involving h and higher powers of h

$$\frac{d}{dx}(x^n) = \lim_{h \to 0} \frac{h(nx^{n-1} + \cdots + h^{n-1})}{h} = \lim_{h \to 0}(nx^{n-1} + \overbrace{\cdots + h^{n-1}}).$$

But as $h \to 0$, all terms involving an h will go to 0, so

$$\frac{d}{dx}(x^n) = \lim_{h \to 0}(nx^{n-1} + \underbrace{\cdots + h^{n-1}}) = nx^{n-1}.$$

These terms go to 0

Derivatives of Polynomials

Now that we know how to differentiate powers, constant multiples, and sums, we can differentiate any polynomial.

Example 3   Find the derivatives of (a) $5x^2 + 3x + 2$,   (b) $\sqrt{3}x^7 - \dfrac{x^5}{5} + \pi$.

Solution   (a)

$$\frac{d}{dx}(5x^2 + 3x + 2) = 5\frac{d}{dx}(x^2) + 3\frac{d}{dx}(x) + \frac{d}{dx}(2)$$

$$= 5 \cdot 2x + 3 \cdot 1 + 0 \qquad \text{(Since the derivative of a constant, $\frac{d}{dx}(2)$, is zero.)}$$

$$= 10x + 3.$$

(b)

$$\frac{d}{dx}\left(\sqrt{3}x^7 - \frac{x^5}{5} + \pi\right) = \sqrt{3}\frac{d}{dx}(x^7) - \frac{1}{5}\frac{d}{dx}(x^5) + \frac{d}{dx}(\pi)$$

$$= \sqrt{3} \cdot 7x^6 - \frac{1}{5} \cdot 5x^4 + 0 \qquad \text{(Since π is a constant, $d\pi/dx = 0$.)}$$

$$= 7\sqrt{3}x^6 - x^4.$$

We can also use the rules we have seen so far to differentiate expressions which are not polynomials.

Example 4   Differentiate (a) $5\sqrt{x} - \dfrac{10}{x^2} + \dfrac{1}{2\sqrt{x}}$ (b) $0.1x^3 + 2x^{\sqrt{2}}$

Solution   (a) $\dfrac{d}{dx}\left(5\sqrt{x} - \dfrac{10}{x^2} + \dfrac{1}{2\sqrt{x}}\right) = \dfrac{d}{dx}\left(5x^{1/2} - 10x^{-2} + \dfrac{1}{2}x^{-1/2}\right)$

$$= 5 \cdot \frac{1}{2}x^{-1/2} - 10(-2)x^{-3} + \frac{1}{2}\left(-\frac{1}{2}\right)x^{-3/2}$$

$$= \frac{5}{2\sqrt{x}} + \frac{20}{x^3} - \frac{1}{4x^{3/2}}.$$

(b) $\dfrac{d}{dx}(0.1x^3 + 2x^{\sqrt{2}}) = 0.1\dfrac{d}{dx}(x^3) + 2\dfrac{d}{dx}(x^{\sqrt{2}}) = 0.3x^2 + 2\sqrt{2}x^{\sqrt{2}-1}.$

Example 5   Find the second derivative and interpret its sign for
(a) $f(x) = x^2,$ (b) $g(x) = x^3,$ (c) $k(x) = x^{1/2}.$

Solution   (a)   If $f(x) = x^2$, then $f'(x) = 2x$, so $f''(x) = \dfrac{d}{dx}(2x) = 2$. Since f'' is always positive, f is concave up, as expected for a parabola opening upwards. (See Figure 4.5.)

(b)   If $g(x) = x^3$, then $g'(x) = 3x^2$, so $g''(x) = \dfrac{d}{dx}(3x^2) = 3\dfrac{d}{dx}(x^2) = 3 \cdot 2x = 6x$. This is positive for $x > 0$ and negative for $x < 0$, which means x^3 is concave up for $x > 0$ and concave down for $x < 0$. (See Figure 4.6.)

(c)   If $k(x) = x^{1/2}$, then $k'(x) = \frac{1}{2}x^{(1/2)-1} = \frac{1}{2}x^{-1/2}$, so

$$k''(x) = \frac{d}{dx}\left(\frac{1}{2}x^{-1/2}\right) = \frac{1}{2} \cdot \left(-\frac{1}{2}\right)x^{-(1/2)-1} = -\frac{1}{4}x^{-3/2}.$$

Now k' and k'' are only defined on the domain of k, that is, $x \geq 0$. When $x > 0$, we see that $k''(x)$ is negative, so k is concave down. (See Figure 4.7.)

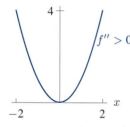

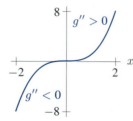

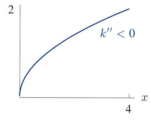

Figure 4.5: $f(x) = x^2$ and $f''(x) = 2$

Figure 4.6: $g(x) = x^3$ and $g''(x) = 6x$

Figure 4.7: $k(x) = x^{1/2}$ and $k''(x) = -\frac{1}{4}x^{-3/2}$

Example 6   If the position of a body, in meters, is given as a function of time t, in seconds, by

$$s = -4.9t^2 + 5t + 6,$$

find the velocity and acceleration of the body at time t.

Solution   The velocity, v, is the derivative of the position:

$$v = \frac{ds}{dt} = \frac{d}{dt}(-4.9t^2 + 5t + 6) = -9.8t + 5,$$

and the acceleration, a, is the derivative of the velocity:

$$a = \frac{dv}{dt} = \frac{d}{dt}(-9.8t + 5) = -9.8.$$

Notice that v is in meters/second and a is in meters/second^2.

Example 7   Figure 4.8 shows the graph of a cubic polynomial. Both graphically and algebraically, describe the behavior of the derivative of this cubic.

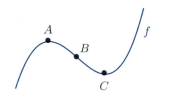

Figure 4.8: The cubic of Example 7   **Figure 4.9**: Derivative of the cubic of Example 7

Solution   Graphical approach: Suppose we move along the curve from left to right. To the left of A, the slope is positive; it starts very positive and decreases until the curve reaches A, where the slope is 0. Between A and C the slope is negative. Between A and B the slope is decreasing (getting more negative); it is most negative at B. Between B and C the slope is negative but increasing; at C the slope is zero. From C to the right, the slope is positive and increasing. The graph of the derivative function is shown in Figure 4.8.

Algebraic approach: f is a cubic that goes to $+\infty$ as $x \to +\infty$, so

$$f(x) = ax^3 + bx^2 + cx + d$$

with $a > 0$. Hence,

$$f'(x) = 3ax^2 + 2bx + c,$$

whose graph is a parabola opening upward, as in Figure 4.9.

Problems for Section 4.1

1. Let $f(x) = 7$. Using the definition of the derivative, show that $f'(x) = 0$ for all values of x.

2. Let $f(x) = -3x + 2$ and $g(x) = 2x + 1$.

    (a) If $k(x) = f(x) + g(x)$, find a formula for $k(x)$ and verify the sum rule by comparing $k'(x)$ with $f'(x) + g'(x)$.

    (b) If $j(x) = f(x) - g(x)$, find a formula for $j(x)$ and compare $j'(x)$ with $f'(x) - g'(x)$.

3. (a) If $f(x) = 5x - 3$, $g(x) = -2x + 1$, find the derivative of $f(g(x))$.

    (b) Use your answer to part (a) to make a conjecture about the derivative of the composition of two linear functions. Explain why what you say is true for any two linear functions.

For Problems 4–24, find the derivatives of the given functions.

4. $y = x^{11}$

5. $y = x^{12}$

6. $y = -x^{-11}$

7. $y = x^{3.2}$

8. $y = x^{-12}$

9. $y = x^{4/3}$

10. $y = x^{3/4}$

11. $y = x^{-3/4}$

12. $f(x) = \dfrac{1}{x^4}$

13. $f(x) = \sqrt[4]{x}$

14. $f(x) = x^e$

15. $y = 4x^{3/2} - 5x^{1/2}$

16. $y = 6x^3 + 4x^2 - 2x$

17. $y = -3x^4 - 4x^3 - 6x + 2$

18. $y = 3t^5 - 5\sqrt{t} + \dfrac{7}{t}$

19. $y = 3t^2 + \dfrac{12}{\sqrt{t}} - \dfrac{1}{t^2}$

20. $y = z^2 + \dfrac{1}{2z}$

21. $y = \dfrac{x^2 + 1}{x}$

22. $g(z) = \dfrac{z^7 + 5z^6 - z^3}{z^2}$

23. $f(t) = \dfrac{t^2 + t^3 - 1}{t^4}$

24. $y = \dfrac{\theta - 1}{\sqrt{\theta}}$

25.   Which functions in Problems 4–15 have derivatives that do not exist at $x = 0$?

For Problems 26–34, determine if the derivative rules from this section apply. If they do, find the derivative. If they don't apply, indicate why.

26. $y = \sqrt{x}$

27. $y = (x + 3)^{1/2}$

28. $y = 3x^2 + 4$

29. $y = \dfrac{1}{3z^2} + \dfrac{1}{4}$

30. $y = \dfrac{1}{3x^2 + 4}$

31. $y = 3^x$

32. $y = \dfrac{1}{3\sqrt{x}} + \dfrac{1}{4}$

33. $g(x) = 72\sqrt[6]{x} + \dfrac{27}{x^{2/3}}$

34. $g(x) = x^\pi - x^{-\pi}$

35.   If $f(t) = 2t^3 - 4t^2 + 3t - 1$, find $f'(t)$ and $f''(t)$.

36.   If $f(x) = 4x^3 + 6x^2 - 23x + 7$, find the intervals on which $f'(x) \geq 1$.

37.   On what intervals is the function $f(x) = x^4 - 4x^3$ both decreasing and concave up?

38.   For what values of x is the graph of $y = x^5 - 5x$ both increasing and concave up?

39.   If $f(x) = 13 - 8x + \sqrt{2}x^2$ and $f'(r) = 4$, find r.

40.   (a)   Find the *eighth* derivative of $f(x) = x^7 + 5x^5 - 4x^3 + 6x - 7$. Think ahead!
        (The nth derivative is the result of differentiating n times.)
       (b)   Find the seventh derivative of $f(x)$.

41.   Find the equation of the line tangent to the graph of f at $(1, 1)$, where f is given by $f(x) = 2x^3 - 2x^2 + 1$.

42.   Show that for any power function $f(x) = x^n$, we have $f'(1) = n$.

43.   Given a power function of the form $f(x) = ax^n$, with $f'(2) = 3$ and $f'(4) = 24$, find n and a.

44.   Is there a value of n which makes $y = x^n$ a solution to the equation $13x\dfrac{dy}{dx} = y$? If so, what value?

45.   Using a graph to help you, find the equations of all lines through the origin tangent to the parabola

$$y = x^2 - 2x + 4.$$

Sketch the lines on the graph.

46.   A ball is dropped from the top of the Empire State building to the ground below. The height, y, of the ball above the ground (in feet) is given as a function of time, t, (in seconds) by

$$y = 1250 - 16t^2.$$

(a)   Find the velocity of the ball at time t. What is the sign of the velocity? Why is this to be expected?
(b)   Show that the acceleration of the ball is a constant. What are the value and sign of this constant?
(c)   When does the ball hit the ground, and how fast is it going at that time? Give your answer in feet per second and in miles per hour (1 ft/sec = 15/22 mph).

47.   The gravitational attraction, F, between the earth and a satellite of mass m at a distance r from the center of the earth is given by

$$F = \frac{GMm}{r^2},$$

where M is the mass of the earth, and G is a constant. Find the rate of change of force with respect to distance.

48.   The period, T, of a pendulum is given in terms of its length, l, by

$$T = 2\pi\sqrt{\frac{l}{g}},$$

where g is the acceleration due to gravity (a constant).

(a)   Find $\dfrac{dT}{dl}$.

(b)   What is the sign of $\dfrac{dT}{dl}$? What does this tell you about the period of pendulums?

49. We know that the graph of a tangent line lies close to the graph of the function near the point of tangency. However, usually the further we go from the point of tangency, the greater the distance between the graph of the function and the tangent line. Let's investigate this for $f(x) = 1/x$. Find the value of f at $x = 2$. Find the tangent line to the curve at $x = 1$, and use it to approximate the value of f at $x = 2$. Now find the tangent to the curve at $x = 100$ and use it to approximate the value of f at $x = 2$. Which tangent line lies closer to the curve at $x = 2$? Is there something wrong with our basic idea? Explain.

50. (a) Use the formula for the area of a circle of radius r, $A = \pi r^2$, to find $\dfrac{dA}{dr}$.

     (b) The result from part (a) should look familiar. What does $\dfrac{dA}{dr}$ represent geometrically? Draw a picture.

     (c) Use the difference quotient to explain the observation you made in part (b).

51. What is the formula for $V(r)$, the volume of a sphere of radius r? Find $\dfrac{dV}{dr}$. What is the geometrical meaning of $\dfrac{dV}{dr}$?

52. Using the definition of derivative, justify the formula $\dfrac{d}{dx}(x^n) = nx^{n-1}$.

     (a) For $n = -1$; for $n = -3$.   (b) For any negative integer n.

4.2 THE EXPONENTIAL FUNCTION

What would we expect the graph of the derivative of the exponential function $f(x) = a^x$ to look like? The graph of the exponential function is shown in Figure 4.10. The function increases slowly for $x < 0$ and more rapidly for $x > 0$, so the values of f' are small for $x < 0$ and larger for $x > 0$. Since the function is increasing for all values of x, the graph of the derivative must lie above the x-axis. In fact, it appears that the graph of f' must resemble the graph of f itself. We will see how this observation holds for $f(x) = 2^x$ and $g(x) = 3^x$.

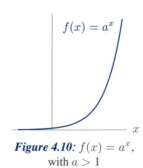

Figure 4.10: $f(x) = a^x$, with $a > 1$

The Derivatives of 2^x and 3^x

In Chapter 2, we saw that the derivative of $f(x) = 2^x$ at $x = 0$ is given by

$$f'(0) = \lim_{h \to 0} \frac{2^h - 2^0}{h} = \lim_{h \to 0} \frac{2^h - 1}{h} \approx 0.6931.$$

The estimate for this limit is obtained by evaluating $(2^h - 1)/h$ for small values of h. Similarly, we can estimate the derivative at $x = 1$ and $x = 2$:

$$f'(1) = \lim_{h \to 0} \frac{2^{1+h} - 2^1}{h} \approx 1.3863,$$

$$f'(2) = \lim_{h \to 0} \frac{2^{2+h} - 2^2}{h} \approx 2.7726.$$

Can you see the relationship between these values of the derivative? If we notice that $1.3863 \approx 2(0.6931)$ and $2.7726 \approx 4(0.6931)$, we see

$$f'(0) \approx 0.6931 = 0.6931 \cdot 2^0,$$
$$f'(1) \approx 1.3863 \approx 0.6931 \cdot 2^1,$$
$$f'(2) \approx 2.7726 \approx 0.6931 \cdot 2^2.$$

It looks as though we have $f'(x) \approx 0.6931 \cdot 2^x$, which is in fact true. Let's look at the derivative function to see why this happens:

$$f'(x) = \lim_{h \to 0} \left(\frac{2^{x+h} - 2^x}{h} \right) = \lim_{h \to 0} \left(\frac{2^x 2^h - 2^x}{h} \right) = \lim_{h \to 0} 2^x \left(\frac{2^h - 1}{h} \right)$$

$$= 2^x \lim_{h \to 0} \left(\frac{2^h - 1}{h} \right) \qquad \text{(Since } x \text{ and } 2^x \text{ are fixed during this calculation)}$$

$$= f'(0) 2^x.$$

We have already estimated $f'(0) \approx 0.6931$, so we get

$$\frac{d}{dx}(2^x) = f'(x) \approx (0.6931)2^x.$$

The graphs of $f(x) = 2^x$ and $f'(x) \approx (0.6931)2^x$ are shown in Figure 4.11. Notice that they do indeed resemble one another.

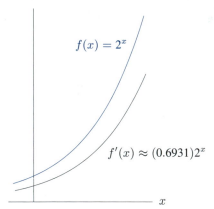

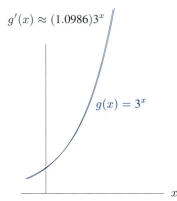

Figure 4.11: Graph of $f(x) = 2^x$ and its derivative

Figure 4.12: Graph of $g(x) = 3^x$ and its derivative

Example 1   Find the derivative of $g(x) = 3^x$ and plot g and g' on the same axes.

Solution   As before

$$g'(x) = \lim_{h \to 0} \frac{3^{x+h} - 3^x}{h} = \lim_{h \to 0} \frac{3^x 3^h - 3^x}{h} = 3^x \lim_{h \to 0} \left(\frac{3^h - 1}{h} \right).$$

Using a calculator gives

$$\lim_{h \to 0} \left(\frac{3^h - 1}{h} \right) \approx 1.0986,$$

so

$$g'(x) \approx (1.0986)3^x.$$

The graphs are shown in Figure 4.12.

Notice that for both $f(x) = 2^x$ and $g(x) = 3^x$, *the derivative is proportional to the original function*. For $f(x) = 2^x$, we have $f'(x) \approx (0.6931)2^x$, so the constant of proportionality is less than 1, and the graph of the derivative is below the graph of the original function. For $g(x) = 3^x$, we have $g'(x) \approx (1.0986)3^x$, so the constant is greater than 1 and the graph of the derivative is above that of the original function.

Note on Round-Off Error and Limits

If we try to evaluate $(2^h - 1)/h$ on a calculator by taking smaller and smaller values of h, the values of $(2^h - 1)/h$ will at first get closer to 0.6931. However, they will eventually move away from the correct value 0.6931... because of the *round-off error* (i.e., errors introduced by the fact that the calculator can only hold a certain number of digits).

As we try smaller and smaller values of h, how do we know when to stop? Unfortunately, there is no fixed rule. A calculator can only suggest the value of a limit, but can never confirm that this value is correct. In this case, it looks like the limit is about 0.6931 because the values of $(2^h - 1)/h$ hover around 0.6931 for a while. To be sure this is correct, we would have to find the limit by theoretical means.

The Derivative of a^x and the Definition of e

The calculation of the derivative of $f(x) = a^x$, for $a > 0$, is similar to that of 2^x and 3^x:

$$f'(x) = \lim_{h \to 0} \frac{a^{x+h} - a^x}{h} = a^x \lim_{h \to 0} \frac{a^h - 1}{h}.$$

The quantity $\lim_{h \to 0}(a^h - 1)/h$ doesn't depend on x, and so is a constant for any particular a. Therefore the derivative is again proportional to the original function, with constant of proportionality

$$\lim_{h \to 0} \frac{a^h - 1}{h}.$$

We can't use a calculator to evaluate this limit without knowing the value of a. However, when $a = 2$, we know that the limit (0.6931) is less than 1, and the derivative is smaller than the original function. When $a = 3$, the limit (1.0986) is more than 1, and the derivative is greater than the original function. Is there an in-between case, when derivative and function are exactly equal? In other words:

$$\text{Is there a value of } a \text{ that makes } \frac{d}{dx}(a^x) = a^x?$$

If so, we have found a function with the remarkable property that it is equal to its own derivative.

So let's look for such an a. This means we want to find a such that

$$\lim_{h \to 0} \frac{a^h - 1}{h} = 1, \qquad \text{or, for small } h, \qquad \frac{a^h - 1}{h} \approx 1.$$

Solving for a suggests that we can calculate a as follows:

$$a^h - 1 \approx h, \qquad \text{or} \qquad a^h \approx 1 + h, \qquad \text{so} \qquad a \approx (1 + h)^{1/h}.$$

Taking small values of h, as in Table 4.2, we can see $a \approx 2.718\ldots$, which looks like the number e introduced in Chapter 1.

TABLE 4.2

h	$(1 + h)^{1/h}$
0.001	2.7169239
0.0001	2.7181459
0.00001	2.7182682

In fact, it can be shown that

$$e = \lim_{h \to 0}(1 + h)^{1/h} = 2.718\ldots \quad \text{and} \quad \lim_{h \to 0}\frac{e^h - 1}{h} = 1.$$

This means that e^x is its own derivative:

$$\frac{d}{dx}(e^x) = e^x.$$

It turns out that the constants involved in the derivatives of 2^x and 3^x are natural logarithms. In fact, since $0.6931 \approx \ln 2$ and $1.0986 \approx \ln 3$, we (correctly) guess that

$$\frac{d}{dx}(2^x) = (\ln 2)2^x \quad \text{and} \quad \frac{d}{dx}(3^x) = (\ln 3)3^x.$$

In Section 4.6, we will show that, in general,

$$\frac{d}{dx}(a^x) = (\ln a)a^x.$$

Figure 4.13 shows the graph of the derivative of 2^x below the graph of the function, and the graph of the derivative of 3^x above the graph of the function. With $e \approx 2.718$, the function e^x and its derivative are identical.

Since $\ln a$ is a constant, the derivative of a^x is proportional to a^x. Many quantities have rates of change which are proportional to themselves; for example, the simplest model of population growth has this property. The fact that the constant of proportionality is 1 when $a = e$ makes e a particularly useful base for exponential functions.

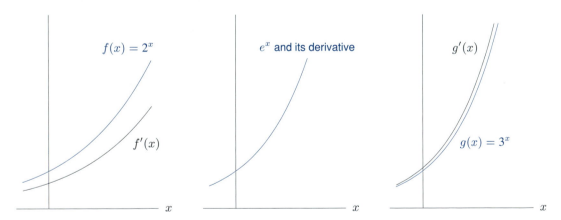

Figure 4.13: Graphs of the functions 2^x, e^x, and 3^x and their derivatives

Example 2   Differentiate $2 \cdot 3^x + 5e^x$.

Solution

$$\frac{d}{dx}(2 \cdot 3^x + 5e^x) = 2\frac{d}{dx}(3^x) + 5\frac{d}{dx}(e^x) = 2\ln 3 \cdot 3^x + 5e^x \approx (2.1972)3^x + 5e^x.$$

Problems for Section 4.2

Find the derivatives of the functions in Problems 1–21.

1. $f(x) = 2e^x + x^2$

2. $y = 5t^2 + 4e^t$

3. $y = 5^x + 2$

4. $f(x) = 2^x + 2 \cdot 3^x$

5. $y = 5x^2 + 2^x + 3$

6. $f(x) = 12e^x + 11^x$

7. $y = 4 \cdot 10^x - x^3$

8. $y = 3x - 2 \cdot 4^x$

9. $y = \dfrac{3^x}{3} + \dfrac{33}{\sqrt{x}}$

10. $f(x) = e^2 + x^e$

11. $f(x) = e^{1+x}$

12. $f(t) = e^{t+2}$

13. $y = e^{\theta - 1}$

14. $z = (\ln 4)e^x$

15. $z = (\ln 4)4^x$

16. $f(z) = (\ln 3)z^2 + (\ln 4)e^z$

17. $f(t) = (\ln 3)^t$

18. $f(x) = x^3 + 3^x$

19. $y = 5 \cdot 5^t + 6 \cdot 6^t$

20. $y = \pi^2 + \pi^x$

21. $f(x) = x^{\pi^2} + (\pi^2)^x$

Which of the functions in Problems 22–30 can be differentiated using the rules we have developed so far? Differentiate if you can; otherwise, indicate why the rules discussed so far do not apply.

22. $y = x^2 + 2^x$

23. $y = \sqrt{x} - \left(\tfrac{1}{2}\right)^x$

24. $y = x^2 \cdot 2^x$

25. $y = \dfrac{2^x}{x}$

26. $y = e^{x+5}$

27. $y = e^{5x}$

28. $y = 4^{(x^2)}$

29. $f(z) = (\sqrt{4})^z$

30. $f(\theta) = 4^{\sqrt{\theta}}$

31. Since January 1, 1960, the population of Slim Chance has been described by the formula

$$P = 35{,}000(0.98)^t,$$

where P is the population of the city t years after the start of 1960. At what rate was the population changing on January 1, 1983?

32. With a yearly inflation rate of 5%, prices are described by

$$P = P_0(1.05)^t,$$

where P_0 is the price in dollars when $t = 0$ and t is time in years. Suppose $P_0 = 1$. How fast (in cents/year) are prices rising when $t = 10$?

33. Certain pieces of antique furniture increased very rapidly in price in the 1970s and 1980s. For example, the value of a particular rocking chair is well approximated by

$$V = 75(1.35)^t,$$

where V is in dollars and t is the number of years since 1975. Find the rate, in dollars per year, at which the price is increasing.

34. The value of a certain automobile purchased in 1997 can be approximated by the function $V(t) = 25(0.85)^t$, where t is the time, in years, from the date of purchase, and V is the value, in thousands of dollars.

    (a) Evaluate and interpret $V(4)$.
    (b) Find an expression for $V'(t)$, including units.
    (c) Evaluate and interpret $V'(4)$.
    (d) Use $V(t)$, $V'(t)$, and any other considerations you think are relevant to write a paragraph in support of or in opposition to the following statement: "From a monetary point of view, it is best to keep this vehicle as long as possible."

35. (a) Find the slope of the graph of $f(x) = 1 - e^x$ at the point where it crosses the x-axis.
     (b) Find the equation of the tangent line to the curve at this point.
     (c) Find the equation of the line which is perpendicular to the tangent line at this point. (This line is known as the *normal* line.)

36. Find the value of c in Figure 4.14, where the line l tangent to the graph of $y = 2^x$ at $(0, 1)$ intersects the x-axis.

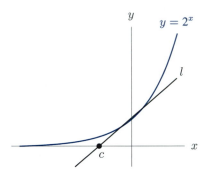

Figure 4.14

37. Find the quadratic polynomial $g(x) = ax^2 + bx + c$ which best fits the function $f(x) = e^x$ at $x = 0$, in the sense that

$$g(0) = f(0), \quad \text{and} \quad g'(0) = f'(0), \quad \text{and} \quad g''(0) = f''(0).$$

Using a computer or calculator, sketch graphs of f and g on the same axes. What do you notice?

38. Using the equation of the tangent line to the graph of e^x at $x = 0$, show that

$$e^x \geq 1 + x$$

for all values of x. A sketch may be helpful.

39. Find all solutions of the equation

$$2^x = 2x.$$

How do you know that you found all solutions?

4.3 THE PRODUCT AND QUOTIENT RULES

We now know how to find derivatives of powers and exponentials, and of sums and constant multiples of functions. This section will show how to find the derivatives of products and quotients.

Using Δ Notation

To express the difference quotients of general functions, some additional notation is helpful. We write Δf, read "delta f," for a small change in the value of f,

$$\Delta f = f(x + h) - f(x).$$

In this notation, the derivative is the limit of the ratio $\Delta f / h$:

$$f'(x) = \lim_{h \to 0} \frac{\Delta f}{h}.$$

The Product Rule

Suppose we know the derivatives of $f(x)$ and $g(x)$ and want to calculate the derivative of the product, $f(x)g(x)$. The derivative of the product is calculated by taking the limit, namely,

$$\frac{d[f(x)g(x)]}{dx} = \lim_{h \to 0} \frac{f(x+h)g(x+h) - f(x)g(x)}{h}.$$

To picture the quantity $f(x+h)g(x+h) - f(x)g(x)$, imagine the rectangle with sides $f(x+h)$ and $g(x+h)$ in Figure 4.15, where $\Delta f = f(x+h) - f(x)$ and $\Delta g = g(x+h) - g(x)$.

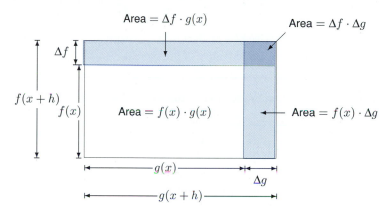

Figure 4.15: Illustration for the product rule (with Δf, Δg positive)

Then

$$f(x+h)g(x+h) - f(x)g(x) = (\text{Area of whole rectangle}) - (\text{Unshaded area})$$
$$= \text{Area of the three shaded rectangles}$$
$$= \Delta f \cdot g(x) + f(x) \cdot \Delta g + \Delta f \cdot \Delta g.$$

Now divide by h:

$$\frac{f(x+h)g(x+h) - f(x)g(x)}{h} = \frac{\Delta f}{h} \cdot g(x) + f(x) \cdot \frac{\Delta g}{h} + \frac{\Delta f \cdot \Delta g}{h}.$$

To evaluate the limit as $h \to 0$, we examine the three terms on the right separately. Notice that

$$\lim_{h \to 0} \frac{\Delta f}{h} \cdot g(x) = f'(x)g(x) \quad \text{and} \quad \lim_{h \to 0} f(x) \cdot \frac{\Delta g}{h} = f(x)g'(x).$$

In the third term we multiply the top and bottom by h to get $\dfrac{\Delta f}{h} \cdot \dfrac{\Delta g}{h} \cdot h$. Then,

$$\lim_{h \to 0} \frac{\Delta f \cdot \Delta g}{h} = \lim_{h \to 0} \frac{\Delta f}{h} \cdot \frac{\Delta g}{h} \cdot h = \lim_{h \to 0} \frac{\Delta f}{h} \cdot \lim_{h \to 0} \frac{\Delta g}{h} \cdot \lim_{h \to 0} h = f'(x) \cdot g'(x) \cdot 0 = 0.$$

Therefore, we conclude that

$$\lim_{h \to 0} \frac{f(x+h)g(x+h) - f(x)g(x)}{h} = \lim_{h \to 0} \left(\frac{\Delta f}{h} \cdot g(x) + f(x) \cdot \frac{\Delta g}{h} + \frac{\Delta f \cdot \Delta g}{h} \right)$$
$$= \lim_{h \to 0} \frac{\Delta f}{h} \cdot g(x) + \lim_{h \to 0} f(x) \cdot \frac{\Delta g}{h} + \lim_{h \to 0} \frac{\Delta f \cdot \Delta g}{h}$$
$$= f'(x)g(x) + f(x)g'(x).$$

Thus we have the following rule:

The Product Rule

If $u = f(x)$ and $v = g(x)$ are differentiable, then

$$(fg)' = f'g + fg'.$$

The product rule can also be written

$$\frac{d(uv)}{dx} = \frac{du}{dx} \cdot v + u \cdot \frac{dv}{dx}.$$

In words:

> The derivative of a product is the derivative of the first times the second, plus the first times the derivative of the second.

Another justification of the product rule is given in Problem 23 on page 233.

Example 1   Differentiate (a) $x^2 e^x$,   (b) $(3x^2 + 5x)e^x$,   (c) $\dfrac{e^x}{x^2}$.

Solution   (a)

$$\frac{d(x^2 e^x)}{dx} = \left(\frac{d(x^2)}{dx}\right)e^x + x^2\frac{d(e^x)}{dx} = 2xe^x + x^2 e^x = (2x + x^2)e^x.$$

(b)

$$\frac{d((3x^2 + 5x)e^x)}{dx} = \left(\frac{d(3x^2 + 5x)}{dx}\right)e^x + (3x^2 + 5x)\frac{d(e^x)}{dx}$$

$$= (6x + 5)e^x + (3x^2 + 5x)e^x = (3x^2 + 11x + 5)e^x.$$

(c)   First we must write $\dfrac{e^x}{x^2}$ as the product $x^{-2}e^x$:

$$\frac{d}{dx}\left(\frac{e^x}{x^2}\right) = \frac{d(x^{-2}e^x)}{dx} = \left(\frac{d(x^{-2})}{dx}\right)e^x + x^{-2}\frac{d(e^x)}{dx}$$

$$= -2x^{-3}e^x + x^{-2}e^x = (-2x^{-3} + x^{-2})e^x.$$

The Quotient Rule

Suppose we want to differentiate a function of the form $Q(x) = f(x)/g(x)$. (Of course, we have to avoid points where $g(x) = 0$.) We want a formula for Q' in terms of f' and g'.

Assuming that $Q(x)$ is differentiable,[1] we can use the product rule on $f(x) = Q(x)g(x)$:

$$f'(x) = Q'(x)g(x) + Q(x)g'(x)$$

$$= Q'(x)g(x) + \frac{f(x)}{g(x)}g'(x).$$

Solving for $Q'(x)$ gives

$$Q'(x) = \frac{f'(x) - \dfrac{f(x)}{g(x)}g'(x)}{g(x)}.$$

---

[1]The method in Example 6 on page 211 can be used to explain why $Q(x)$ must be differentiable.

Multiplying the top and bottom by $g(x)$ to simplify gives

$$\left(\frac{f(x)}{g(x)}\right)' = \frac{f'(x)g(x) - f(x)g'(x)}{(g(x))^2}.$$

So we have the following rule:

The Quotient Rule

If $u = f(x)$ and $v = g(x)$ are differentiable, then

$$\left(\frac{f}{g}\right)' = \frac{f'g - fg'}{g^2},$$

or equivalently,

$$\frac{d}{dx}\left(\frac{u}{v}\right) = \frac{\dfrac{du}{dx} \cdot v - u \cdot \dfrac{dv}{dx}}{v^2}.$$

In words:

  The derivative of a quotient is the derivative of the numerator times the denominator minus the numerator times the derivative of the denominator, all over the denominator squared.

Example 2 Differentiate (a) $\dfrac{5x^2}{x^3 + 1}$, (b) $\dfrac{1}{1 + e^x}$, (c) $\dfrac{e^x}{x^2}$.

Solution (a)

$$\frac{d}{dx}\left(\frac{5x^2}{x^3 + 1}\right) = \frac{\left(\dfrac{d}{dx}(5x^2)\right)(x^3 + 1) - 5x^2\dfrac{d}{dx}(x^3 + 1)}{(x^3 + 1)^2} = \frac{10x(x^3 + 1) - 5x^2(3x^2)}{(x^3 + 1)^2}$$

$$= \frac{-5x^4 + 10x}{(x^3 + 1)^2}.$$

(b)

$$\frac{d}{dx}\left(\frac{1}{1 + e^x}\right) = \frac{\left(\dfrac{d}{dx}(1)\right)(1 + e^x) - 1\dfrac{d}{dx}(1 + e^x)}{(1 + e^x)^2} = \frac{0(1 + e^x) - 1(0 + e^x)}{(1 + e^x)^2}$$

$$= \frac{-e^x}{(1 + e^x)^2}.$$

(c) This is the same as part (c) of Example 1, but this time we will do it by the quotient rule.

$$\frac{d}{dx}\left(\frac{e^x}{x^2}\right) = \frac{\left(\dfrac{d(e^x)}{dx}\right)x^2 - e^x\left(\dfrac{d(x^2)}{dx}\right)}{(x^2)^2} = \frac{e^x x^2 - e^x(2x)}{x^4}$$

$$= e^x\left(\frac{x^2 - 2x}{x^4}\right) = e^x\left(\frac{x - 2}{x^3}\right).$$

This is, in fact, the same answer as before, although it looks different. Can you show that it is the same?

Problems for Section 4.3

1. If $f(x) = x^2(x^3 + 5)$, find $f'(x)$ two ways: by using the product rule and by multiplying out before taking the derivative. Do you get the same result? Should you?

2. If $f(x) = 2^x \cdot 3^x$, find $f'(x)$ two ways: by using the product rule and by using the fact that $2^x \cdot 3^x = 6^x$. Do you get the same result?

For Problems 3–23, find the derivative. In some cases, it may be to your advantage to simplify first.

3. $f(x) = xe^x$

4. $y = x \cdot 2^x$

5. $y = \sqrt{x} \cdot 2^x$

6. $f(x) = (x^2 - \sqrt{x})3^x$

7. $z = (s^2 - \sqrt{s})(s^2 + \sqrt{s})$

8. $y = (t^2 + 3)e^t$

9. $w = (t^3 + 5t)(t^2 - 7t + 2)$

10. $y = (t^3 - 7t^2 + 1)e^t$

11. $f(x) = \dfrac{x}{e^x}$

12. $g(x) = \dfrac{25x^2}{e^x}$

13. $g(w) = \dfrac{w^{3.2}}{5^w}$

14. $h(t) = \dfrac{t + 4}{t - 4}$

15. $z = \dfrac{3t + 1}{5t + 2}$

16. $z = \dfrac{t^2 + 5t + 2}{t + 3}$

17. $f(x) = \dfrac{x^2 + 3}{x}$

18. $w = \dfrac{y^3 - 6y^2 + 7y}{y}$

19. $y = \dfrac{\sqrt{t}}{t^2 + 1}$

20. $f(z) = \dfrac{3z^2}{5z^2 + 7z}$

21. $w(x) = \dfrac{17e^x}{2^x}$

22. $h(p) = \dfrac{1 + p^2}{3 + 2p^2}$

23. $f(x) = \dfrac{1 + x}{2 + 3x + 4x^2}$

24. If $f(x) = (3x + 8)(2x - 5)$, find $f'(x)$ and $f''(x)$.

25. For what intervals is $f(x) = xe^{-x}$ concave down?

26. For what intervals is $g(x) = \dfrac{1}{x^2 + 1}$ concave down?

27. Differentiate $f(t) = e^{-t}$ by writing it as $f(t) = \dfrac{1}{e^t}$.

28. Differentiate $f(x) = e^{2x}$ by writing it as $f(x) = e^x \cdot e^x$.

29. Differentiate $f(x) = e^{3x}$ by writing it as $f(x) = e^x \cdot e^{2x}$ and using the result of Problem 28.

30. Based on your answers to Problems 28 and 29, guess the derivative of e^{4x}.

31. (a) Differentiate $y = \dfrac{e^x}{x}$, $y = \dfrac{e^x}{x^2}$, and $y = \dfrac{e^x}{x^3}$.

    (b) What do you anticipate the derivative of $y = \dfrac{e^x}{x^n}$ will be? Confirm your guess.

32. Using the product rule and the fact that $\dfrac{d(x)}{dx} = 1$, show that $\dfrac{d(x^2)}{dx} = 2x$ and $\dfrac{d(x^3)}{dx} = 3x^2$.

33. Use the product rule to show that $\dfrac{d}{dx}(x^{1/2}) = \dfrac{1}{2x^{1/2}}$. [Hint: Write $x = x^{1/2}x^{1/2}$.]

34. Suppose f and g are differentiable functions with the values shown in the following table. For each of the following functions h, find $h'(2)$.

    (a) $h(x) = f(x) + g(x)$
    (b) $h(x) = f(x)g(x)$
    (c) $h(x) = \dfrac{f(x)}{g(x)}$

x	$f(x)$	$g(x)$	$f'(x)$	$g'(x)$
2	3	4	5	-2

35. Given $\left\{\begin{array}{ll} H(3) = 1 & F(3) = 5 \\ H'(3) = 3 & F'(3) = 4 \end{array}\right\}$ find $\left\{\begin{array}{ll} \text{(a)} \ G'(3) & \text{if } G(z) = F(z) \cdot H(z) \\ \text{(b)} \ G'(3) & \text{if } G(w) = F(w)/H(w) \end{array}\right\}$.

36. Find a possible formula for a function $y = f(x)$ such that $f'(x) = 10x^9 e^x + x^{10} e^x$.

37. The quantity, q, of a certain skateboard sold depends on the selling price, p, in dollars, so we write $q = f(p)$. You are given that $f(140) = 15,000$ and $f'(140) = -100$.

    (a) What do $f(140) = 15,000$ and $f'(140) = -100$ tell you about the sales of skateboards?

    (b) The total revenue, R, earned by the sale of skateboards is given by $R = pq$. Find $\left.\dfrac{dR}{dp}\right|_{p=140}$.

    (c) What is the sign of $\left.\dfrac{dR}{dp}\right|_{p=140}$? If the skateboards are currently selling for $140, what happens to revenue if the price is increased to $141?

38. When an electric current passes through two resistors with resistance r_1 and r_2, connected in parallel, the combined resistance, R, can be calculated from the equation

$$\frac{1}{R} = \frac{1}{r_1} + \frac{1}{r_2}.$$

Find the rate at which the combined resistance changes with respect to changes in r_1. Assume that r_2 is constant.

39. A museum has decided to sell one of its paintings and to invest the proceeds. If the picture is sold between the years 2000 and 2020 and the money from the sale is invested in a bank account earning 5% annual interest compounded once a year, then $B(t)$, the balance in the year 2020, depends on the year, t, in which the painting is sold and the sale price $P(t)$. If t is measured from the year 2000 so that $0 < t < 20$ then

$$B(t) = P(t)(1.05)^{20-t}.$$

    (a) Explain why $B(t)$ is given by this formula.

    (b) Show that the formula for $B(t)$ is equivalent to

$$B(t) = (1.05)^{20} \frac{P(t)}{(1.05)^t}.$$

    (c) Find $B'(10)$, given that $P(10) = 150,000$ and $P'(10) = 5000$.

40. Let $f(v)$ be the gas consumption (in liters/km) of a car going at velocity v (in km/hr). In other words, $f(v)$ tells you how many liters of gas the car uses to go one kilometer, if it is going at velocity v. You are told that

$$f(80) = 0.05 \text{ and } f'(80) = 0.0005.$$

    (a) Let $g(v)$ be the distance the same car goes on one liter of gas at velocity v. What is the relationship between $f(v)$ and $g(v)$? Find $g(80)$ and $g'(80)$.

    (b) Let $h(v)$ be the gas consumption in liters per hour. In other words, $h(v)$ tells you how many liters of gas the car uses in one hour if it is going at velocity v. What is the relationship between $h(v)$ and $f(v)$? Find $h(80)$ and $h'(80)$.

    (c) How would you explain the practical meaning of the values of these functions and their derivatives to a driver who knows no calculus?

41. The function

$$f(x) = e^x$$

has the properties

$$f'(x) = f(x) \text{ and } f(0) = 1.$$

Explain why $f(x)$ is the only function with both these properties.
[Hint: Assume $g'(x) = g(x)$, and $g(0) = 1$, for some function $g(x)$. Define $h(x) = g(x)/e^x$, and compute $h'(x)$. Then use the fact that a function with a derivative of 0 must be a constant function.]

42. Find $f'(x)$ for the following functions with the product rule, rather than by multiplying out.

    (a) $f(x) = (x-1)(x-2)$.
    (b) $f(x) = (x-1)(x-2)(x-3)$.
    (c) $f(x) = (x-1)(x-2)(x-3)(x-4)$.

43.   Use the answer from Problem 42 to guess $f'(x)$ for the following function:

$$f(x) = (x - r_1)(x - r_2)(x - r_3) \cdots (x - r_n)$$

where $r_1, r_2, \ldots, r_n$ are any real numbers.

44.   (a)   Provide a three dimensional analogue for the geometrical demonstration of the formula for the derivative of a product, given in Figure 4.15 on page 203. In other words, find a formula for the derivative of $F(x) \cdot G(x) \cdot H(x)$ using Figure 4.16.

(b)   Verify your results by writing $F(x) \cdot G(x) \cdot H(x)$ as $[F(x) \cdot G(x)] \cdot H(x)$ and using the product rule twice.

(c)   Generalize your result to n functions: what is the derivative of

$$f_1(x) \cdot f_2(x) \cdot f_3(x) \cdots f_n(x)?$$

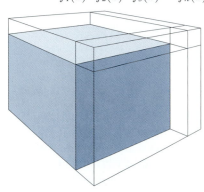

Figure 4.16: A graphical representation of the 3-dimensional product rule

4.4 THE CHAIN RULE

Composite functions such as $\sin(3t)$ or e^{-x^2} occur frequently in practice. In this section we will see how to differentiate such functions.

The Derivative of a Composition of Functions

Suppose $f(g(x))$ is a composite function, with f being the outside function and g being the inside. Let us write

$$z = g(x) \quad \text{and} \quad y = f(z), \quad \text{so} \quad y = f(g(x)).$$

Then a small change in x, called Δx, generates a small change in z, called Δz. In turn, Δz generates a small change in y called Δy. Provided Δx and Δz are not zero, we can say:

$$\frac{\Delta y}{\Delta x} = \frac{\Delta y}{\Delta z} \cdot \frac{\Delta z}{\Delta x}.$$

Since $\dfrac{dy}{dx} = \lim\limits_{\Delta x \to 0} \dfrac{\Delta y}{\Delta x}$, this suggests that in the limit as Δx, Δy, and Δz get smaller and smaller, we have:

The Chain Rule

$$\frac{dy}{dx} = \frac{dy}{dz} \cdot \frac{dz}{dx}.$$

Since $\dfrac{dy}{dz} = f'(z)$ and $\dfrac{dz}{dx} = g'(x)$, we can also write

$$\frac{d}{dx} f(g(x)) = f'(z) \cdot g'(x).$$

Substituting $z = g(x)$, we can rewrite this as follows:

The Chain Rule

$$\frac{d}{dx} f(g(x)) = f'(g(x)) \cdot g'(x).$$

In words:

The derivative of a composite function is the product of the derivatives of the outside and inside functions. The derivative of the outside function must be evaluated at the inside function.

A justification of the chain rule is given in Problem 24 on page 233.

Example 1   Suppose that the length, L, in cm, of a steel bar depends on the air temperature, $H°$C, and that H depends on time, t, measured in hours. If the length increases by 2 cm for every degree increase in temperature, and the temperature is increasing at $3°$C per hour, how fast is the length of the bar increasing? What are the units for your answer?

Solution   We expect the rate at which the length is increasing to be in cm/hr. We are told that

$$\text{Rate length increasing with respect to temperature} = \frac{dL}{dH} = 2 \text{ cm/°C}$$

$$\text{Rate temperature increasing with respect to time} = \frac{dH}{dt} = 3°\text{C/hr.}$$

We want to calculate the rate at which the length is increasing with respect to time, or dL/dt. We think of L as a function of H, and H as a function of t. By the chain rule we know that

$$\frac{dL}{dt} = \frac{dL}{dH} \cdot \frac{dH}{dt} = \left(2\frac{\text{cm}}{°\text{C}}\right) \cdot \left(3\frac{°\text{C}}{\text{hr}}\right) = 6 \text{ cm/hr.}$$

Thus, the length is increasing at 6 cm/hr.

Example 1 above shows us how to interpret the chain rule in practical terms. The next examples show how it is used to compute derivatives of functions given by formulas.

Example 2   Find the derivatives of the following functions: (a) $(4x^2 + 1)^7$ (b) e^{3x}

Solution   (a)   Here $z = g(x) = 4x^2 + 1$ is the inside function; $f(z) = z^7$ is the outside function. Since $g'(x) = 8x$ and $f'(z) = 7z^6$, we have

$$\frac{d}{dx}\left[(4x^2 + 1)^7\right] = 7z^6 \cdot 8x = 7(4x^2 + 1)^6 \cdot 8x = 56x(4x^2 + 1)^6.$$

(b)   Let $z = g(x) = 3x$ and $f(z) = e^z$. Then $g'(x) = 3$ and $f'(z) = e^z$, so

$$\frac{d}{dx}\left(e^{3x}\right) = e^z \cdot 3 = 3e^{3x}.$$

Example 3   Differentiate

(a) $(x^2+1)^{100}$ (b) $\sqrt{3x^2 + 5x - 2}$ (c) $\dfrac{1}{x^2 + x^4}$ (d) $\sqrt{e^x + 1}$ (e) e^{x^2}

Solution   (a)   Here $z = g(x) = x^2 + 1$ is the inside function; $f(z) = z^{100}$ is the outside function. Now $g'(x) = 2x$ and $f'(z) = 100z^{99}$, so

$$\frac{d[(x^2 + 1)^{100}]}{dx} = 100z^{99} \cdot 2x = 100(x^2 + 1)^{99} \cdot 2x = 200x(x^2 + 1)^{99}.$$

(b)   Here $z = g(x) = 3x^2 + 5x - 2$ and $f(z) = \sqrt{z}$, so $g'(x) = 6x + 5$ and $f'(z) = \dfrac{1}{2\sqrt{z}}$. Hence

$$\frac{d(\sqrt{3x^2 + 5x - 2})}{dx} = \frac{1}{2\sqrt{z}} \cdot (6x + 5) = \frac{1}{2\sqrt{3x^2 + 5x - 2}} \cdot (6x + 5).$$

(c)   Let $z = g(x) = x^2 + x^4$ and $f(z) = 1/z$, so $g'(x) = 2x + 4x^3$ and $f'(z) = -z^{-2} = -\dfrac{1}{z^2}$. Then

$$\frac{d}{dx}\left(\frac{1}{x^2 + x^4}\right) = -\frac{1}{z^2}(2x + 4x^3) = -\frac{2x + 4x^3}{(x^2 + x^4)^2}.$$

We could have done this problem using the quotient rule. Try it and see that you get the same answer!

(d)   Let $z = g(x) = e^x + 1$ and $f(z) = \sqrt{z}$. Therefore $g'(x) = e^x$ and $f'(z) = \dfrac{1}{2\sqrt{z}}$. We get

$$\frac{d(\sqrt{e^x + 1})}{dx} = \frac{1}{2\sqrt{z}}e^x = \frac{e^x}{2\sqrt{e^x + 1}}.$$

(e)   In order to figure out which is the inside function and which is the outside, notice that to evaluate e^{x^2} we first evaluate x^2 and then take e to that power. This tells us that the inside function is $z = g(x) = x^2$ and the outside function is $f(z) = e^z$. Therefore, $g'(x) = 2x$, and $f'(z) = e^z$, giving

$$\frac{d(e^{x^2})}{dx} = e^z \cdot 2x = e^{x^2} \cdot 2x = 2xe^{x^2}.$$

Example 4   Find the derivative of e^{2x} by the chain rule and by the product rule.

Solution   Chain rule: Let the inside function be $z = g(x) = 2x$ and the outside function be $f(z) = e^z$. Then

$$\frac{d(e^{2x})}{dx} = f'(g(x)) \cdot g'(x) = e^{2x} \cdot 2 = 2e^{2x}.$$

Product rule: Write $e^{2x} = e^x \cdot e^x$. Then

$$\frac{d(e^{2x})}{dx} = \frac{d(e^x e^x)}{dx} = \left(\frac{d(e^x)}{dx}\right)e^x + e^x\left(\frac{d(e^x)}{dx}\right) = e^x \cdot e^x + e^x \cdot e^x = 2e^{2x}.$$

The chain rule can often be used to find rates, as in the following example.

Example 5   A circular oil spill is spreading. If the radius is increasing at 0.2 km/hr when the radius is 3 km, find the rate at which the area is increasing at that time. Give units in your answer.

Solution   If A is the area of the spill in km^2 and r is the radius in km, then

$$A = \pi r^2.$$

Differentiating using the chain rule gives

$$\frac{dA}{dt} = \frac{dA}{dr} \cdot \frac{dr}{dt} = 2\pi r \cdot \frac{dr}{dt}.$$

We know that $dr/dt = 0.2$ km/hr when $r = 3$, so

$$\frac{dA}{dt} = (2\pi \cdot 3 \text{ km})(0.2 \text{ km/hr}) = 1.2\pi \approx 3.77 \text{ km}^2/\text{hr}.$$

Notice that the units of dA/dt are (km)(km/hr)=km^2/hr, which is area per unit time, as expected.

Using the Product and Chain Rules to Differentiate a Quotient

If you prefer, you can differentiate a quotient by the product and chain rules, instead of by the quotient rule. The resulting formulas may look different, but they will be equivalent.

Example 6   Find $k'(x)$ if $k(x) = \dfrac{x}{x^2 + 1}$.

Solution   One way is to use the quotient rule:

$$k'(x) = \frac{1 \cdot (x^2 + 1) - x \cdot (2x)}{(x^2 + 1)^2}$$

$$= \frac{1 - x^2}{(x^2 + 1)^2}.$$

Alternatively, we can write the original function as a product,

$$k(x) = x\frac{1}{x^2 + 1} = x \cdot (x^2 + 1)^{-1},$$

and use the product rule:

$$k'(x) = 1 \cdot (x^2 + 1)^{-1} + x \cdot \frac{d}{dx}\left[(x^2 + 1)^{-1}\right].$$

Now use the chain rule to differentiate $(x^2 + 1)^{-1}$. Let $z = x^2 + 1$, and $f(z) = z^{-1}$, so

$$\frac{d}{dx}\left[(x^2 + 1)^{-1}\right] = -z^{-2} \cdot 2x = -(x^2 + 1)^{-2} \cdot 2x = \frac{-2x}{(x^2 + 1)^2}.$$

Therefore,

$$k'(x) = \frac{1}{x^2 + 1} + x \cdot \frac{-2x}{(x^2 + 1)^2} = \frac{1}{x^2 + 1} - \frac{2x^2}{(x^2 + 1)^2}.$$

If we put these two fractions over a common denominator, we will get the same answer as from the quotient rule.

Problems for Section 4.4

Find the derivatives of the functions in Problems 1–30.

1. $f(x) = (x + 1)^{99}$

2. $f(x) = \sqrt{1 - x^2}$

3. $w = (t^2 + 1)^{100}$

4. $w = (t^3 + 1)^{100}$

5. $w = (\sqrt{t} + 1)^{100}$

6. $f(t) = e^{3t}$

7. $f(x) = 2^{(x+2)}$

8. $g(x) = 3^{(2x+7)}$

9. $k(x) = (x^3 + e^x)^4$

10. $z(x) = \sqrt[3]{2^x + 5}$

11. $y = \dfrac{\sqrt{z}}{2^z}$

12. $w = \sqrt{(x^2 \cdot 5^x)^3}$

13. $y = e^{3w/2}$

14. $y = e^{-4t}$

15. $y = \sqrt{s^3 + 1}$

16. $w = e^{\sqrt{s}}$

17. $y = te^{-t^2}$

18. $f(z) = \sqrt{z}e^{-z}$

19. $f(z) = \dfrac{\sqrt{z}}{e^z}$

20. $z = 2^{5t-3}$

21. $f(t) = te^{5-2t}$

22. $f(z) = \dfrac{1}{(e^z + 1)^2}$

23. $f(\theta) = \dfrac{1}{1 + e^{-\theta}}$

24. $f(x) = 6e^{5x} + e^{-x^2}$

25. $f(w) = (5w^2 + 3)e^{w^2}$ 26. $w = (t^2 + 3t)(1 - e^{-2t})$ 27. $f(y) = \sqrt{10^{(5-y)}}$

28. $f(x) = e^{-(x-1)^2}$ 29. $f(y) = e^{e^{(y^2)}}$ 30. $f(t) = 2 \cdot e^{-2e^{2t}}$

31. Find the equation of the line tangent to $y = f(x)$ at $x = 1$, where $f(x)$ is the function in Problem 24.

32. For what values of x is the graph of $y = e^{-x^2}$ concave down?

33. Suppose $f(x) = (2x + 1)^{10}(3x - 1)^7$. Find a formula for $f'(x)$. Then decide on a reasonable way to simplify your result, and find a formula for $f''(x)$.

34. Given
$$\left\{\begin{array}{ll} F(2) = 1 & G(4) = 2 \\ F(4) = 3 & G(3) = 4 \\ F'(2) = 5 & G'(4) = 6 \\ F'(4) = 7 & G'(3) = 8 \end{array}\right\}$$
find
$$\left\{\begin{array}{lll} \text{(a)} & H(4) & \text{if } H(x) = F(G(x)) \\ \text{(b)} & H'(4) & \text{if } H(x) = F(G(x)) \\ \text{(c)} & H(4) & \text{if } H(x) = G(F(x)) \\ \text{(d)} & H'(4) & \text{if } H(x) = G(F(x)) \\ \text{(e)} & H'(4) & \text{if } H(x) = F(x)/G(x) \end{array}\right\}$$

35. Suppose f and g are differentiable functions with the values given in the following table. For each of the following functions h, find $h'(2)$.

(a) $h(x) = f(g(x))$ (b) $h(x) = g(f(x))$ (c) $h(x) = f(f(x))$

x	$f(x)$	$g(x)$	$f'(x)$	$g'(x)$
2	5	5	e	$\sqrt{2}$
5	2	8	π	7

36. If the derivative of $y = k(x)$ equals 2 when $x = 1$, what is the derivative of

(a) $k(2x)$ when $x = \dfrac{1}{2}$?   (b) $k(x + 1)$ when $x = 0$?   (c) $k\left(\dfrac{1}{4}x\right)$ when $x = 4$?

37. Given $y = f(x)$ with $f(1) = 4$ and $f'(1) = 3$, find

(a) $g'(1)$ if $g(x) = \sqrt{f(x)}$.   (b) $h'(1)$ if $h(x) = f(\sqrt{x})$.

38. Is $x = \sqrt[3]{2t + 5}$ a solution to the equation $3x^2 \dfrac{dx}{dt} = 2$? Why or why not?

39. Find a possible formula for a function $m(x)$ such that $m'(x) = x^5 \cdot e^{(x^6)}$.

40. Suppose that the population of zebra mussels in a certain area of the St. Lawrence River is approximated by $P(t) = 10e^{0.6t}$, where t is measured in months since the zebra mussels first arrived in the area. Calculate the following quantities:

(a) $P(12)$.   (b) $P'(12)$.

Give units with your answers and explain what each quantity tells us in terms of zebra mussels.

41. One gram of radioactive carbon-14 decays according to the formula

$$Q = e^{-0.000121t}$$

where Q is the number of grams of carbon-14 remaining after t years.

(a) Find the rate at which carbon-14 is decaying (in grams/year).
(b) Sketch the rate you found in part (a) against time.

42. The temperature, H, in degrees Fahrenheit ($°$F), of a can of soda that is put into a refrigerator to cool is given as a function of time, t, in hours, by

$$H = 40 + 30e^{-2t}.$$

(a) Find the rate at which the temperature of the soda is changing (in $°$F/hour).
(b) What is the sign of $\dfrac{dH}{dt}$? Why does it have this sign?
(c) When, for $t \geq 0$, is the magnitude of $\dfrac{dH}{dt}$ largest? In terms of the can of soda, why is this?

43. If you invest P dollars in a bank account at an annual interest rate of $r\%$, then after t years you will have B dollars, where

$$B = P\left(1 + \frac{r}{100}\right)^t.$$

    (a)   Find dB/dt, assuming P and r are constant. In terms of money, what does dB/dt represent?

    (b)   Find dB/dr, assuming P and t are constant. In terms of money, what does dB/dr represent?

44. Suppose a pebble is dropped in still water, forming a circular ripple whose radius is expanding at a constant rate of 10 cm/sec. Find a formula giving the area enclosed by the ripple as a function of time. When the radius is 20 cm, how fast is the area enclosed by the ripple increasing?

45. The theory of relativity predicts that an object whose mass is m_0 when it is at rest will appear heavier when moving at speeds near the speed of light. When the object is moving at speed v, its mass m is given by

$$m = \frac{m_0}{\sqrt{1 - (v^2/c^2)}}, \qquad \text{where } c \text{ is the speed of light.}$$

    (a)   Find $\dfrac{dm}{dv}$.   (b)   In terms of physics, what does $\dfrac{dm}{dv}$ tell you?

46. The charge, Q, on a capacitor which starts discharging at time $t = 0$ is given by

$$Q = \begin{cases} Q_0 & \text{for } t \leq 0 \\ Q_0 e^{-t/RC} & \text{for } t > 0, \end{cases}$$

where R and C are positive constants depending on the circuit and Q_0 is the charge at $t = 0$, where $Q_0 \neq 0$. The current, I, flowing in the circuit is given by $I = dQ/dt$.

    (a)   Find the current I for $t < 0$ and for $t > 0$.

    (b)   Is it possible to define I at $t = 0$?

    (c)   Is the function Q differentiable at $t = 0$?

47. An electric charge is decaying exponentially according to the formula

$$Q = Q_0 e^{-t/RC}.$$

Show that the charge, Q, and the electric current, $I = dQ/dt$, have the same time constant. (The time constant is the time to decay to $1/e$ times the original value.)

48. A function f is said to have a *zero of multiplicity* m at $x = a$ if

$$f(x) = (x - a)^m h(x), \quad \text{with } h(a) \neq 0.$$

Explain why a function having a zero of multiplicity m at $x = a$ satisfies $f^{(p)}(a) = 0$, for $p = 1, 2, \ldots m - 1$.

4.5 THE TRIGONOMETRIC FUNCTIONS

Derivatives of the Sine and Cosine

Since the sine and cosine functions are periodic, their derivatives must be periodic also. (Why?) Let's look at the graph of $f(x) = \sin x$ in Figure 4.17 and estimate the derivative function graphically.

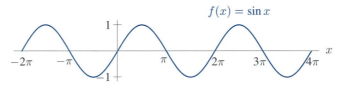

Figure 4.17: The sine function

First we might ask where the derivative is zero. (At $x = \pm\pi/2, \pm 3\pi/2, \pm 5\pi/2$, etc.) Then ask where the derivative is positive and where it is negative. (Positive for $-\pi/2 < x < \pi/2$; negative for $\pi/2 < x < 3\pi/2$, etc.) Since the largest positive slopes are at $x = 0, 2\pi$, and so on, and the largest negative slopes are at $x = \pi, 3\pi$, and so on, we get something like the graph in Figure 4.18.

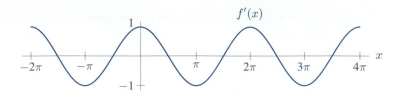

Figure 4.18: Derivative of $f(x) = \sin x$

The graph of the derivative in Figure 4.18 looks suspiciously like the graph of the cosine function. This might lead us to conjecture, quite correctly, that the derivative of the sine is the cosine.

Of course, we cannot be sure, just from the graphs, that the derivative of the sine really is the cosine. However, for now we'll assume that the derivative of the sine *is* the cosine and confirm the result at the end of the section.

One thing we can do now is to check that the derivative function in Figure 4.18 has amplitude 1 (as it ought to if it is the cosine). That means we have to convince ourselves that the derivative of $f(x) = \sin x$ is 1 when $x = 0$. The next example suggests that this is true when x is in radians.

Example 1   Using a calculator, estimate the derivative of $f(x) = \sin x$ at $x = 0$. Make sure your calculator is set in radians.

Solution   Since $f(x) = \sin x$,

$$f'(0) = \lim_{h \to 0} \frac{\sin(0 + h) - \sin 0}{h} = \lim_{h \to 0} \frac{\sin h}{h}.$$

Table 4.3 contains values of $(\sin h)/h$ which suggest that this limit is 1, so we estimate

$$f'(0) = \lim_{h \to 0} \frac{\sin h}{h} = 1.$$

TABLE 4.3

h (radians)	-0.1	-0.01	-0.001	-0.0001	0.0001	0.001	0.01	0.1
$(\sin h)/h$	0.99833	0.99998	1.0000	1.0000	1.0000	1.0000	0.99998	0.99833

Warning: It is important to notice that in the previous example h was in *radians*; any conclusions we have drawn about the derivative of $\sin x$ are valid *only* when x is in radians.

Example 2   Starting with the graph of the cosine function, sketch a graph of its derivative.

Solution   The graph of $g(x) = \cos x$ is in Figure 4.19(a). Its derivative is 0 at $x = 0, \pm\pi, \pm 2\pi$, and so on; it is positive for $-\pi < x < 0$, $\pi < x < 2\pi$, and so on; and it is negative for $0 < x < \pi$, $2\pi < x < 3\pi$, and so on. The derivative is in Figure 4.19(b).

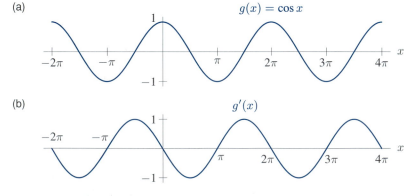

Figure 4.19: $g(x) = \cos x$ and its derivative, $g'(x)$

As we did with the sine, we use the graphs to make a conjecture. The derivative of the cosine in Figure 4.19(b) looks exactly like the graph of sine, except reflected about the x-axis. But how can we be sure that the derivative is $-\sin x$?

Example 3   Use the relation $\dfrac{d}{dx}(\sin x) = \cos x$ to show that $\dfrac{d}{dx}(\cos x) = -\sin x$.

Solution   Since the cosine function is the sine function shifted to the left by $\pi/2$, we would expect the derivative of the cosine to be the derivative of the sine, shifted to the left by $\pi/2$. Since

$$\cos x = \sin\left(x + \frac{\pi}{2}\right),$$

we can use the chain rule:

$$\frac{d}{dx}(\cos x) = \frac{d}{dx}\left(\sin\left(x + \frac{\pi}{2}\right)\right) = \cos\left(x + \frac{\pi}{2}\right).$$

But $\cos(x + \pi/2)$ is the cosine shifted to the left by $\pi/2$, which gives a sine curve reflected about the x-axis. So we have

$$\frac{d}{dx}(\cos x) = \cos\left(x + \frac{\pi}{2}\right) = -\sin x.$$

> For x in radians, $\dfrac{d}{dx}(\sin x) = \cos x$ and $\dfrac{d}{dx}(\cos x) = -\sin x.$

Example 4   Differentiate (a) $2\sin(3\theta)$, (b) $\cos^2 x$, (c) $\cos(x^2)$, (d) $e^{-\sin t}$.

Solution   Use the chain rule:

(a) $\dfrac{d}{d\theta}(2\sin(3\theta)) = 2\dfrac{d}{d\theta}(\sin(3\theta)) = 2(\cos(3\theta))\dfrac{d}{d\theta}(3\theta) = 2(\cos(3\theta))3 = 6\cos(3\theta).$

(b) $\dfrac{d}{dx}(\cos^2 x) = \dfrac{d}{dx}\left((\cos x)^2\right) = 2(\cos x)\cdot\dfrac{d}{dx}(\cos x) = 2(\cos x)(-\sin x) = -2\cos x\sin x.$

(c) $\dfrac{d}{dx}\left(\cos(x^2)\right) = -\sin(x^2)\cdot\dfrac{d}{dx}(x^2) = -2x\sin(x^2).$

(d) $\dfrac{d}{dt}(e^{-\sin t}) = e^{-\sin t}\dfrac{d}{dt}(-\sin t) = -(\cos t)e^{-\sin t}.$

Derivative of the Tangent Function

Since $\tan x = \sin x/\cos x$, we differentiate $\tan x$ using the quotient rule. Writing $(\sin x)'$ for $d(\sin x)/dx$, we have:

$$\frac{d}{dx}(\tan x) = \frac{d}{dx}\left(\frac{\sin x}{\cos x}\right) = \frac{(\sin x)'(\cos x) - (\sin x)(\cos x)'}{\cos^2 x} = \frac{\cos^2 x + \sin^2 x}{\cos^2 x} = \frac{1}{\cos^2 x}.$$

> For x in radians, $\dfrac{d}{dx}(\tan x) = \dfrac{1}{\cos^2 x}.$

The graphs of $f(x) = \tan x$ and $f'(x) = 1/\cos^2 x$ are in Figure 4.20. Is it reasonable that f' is always positive? Are the asymptotes of f' where we expect?

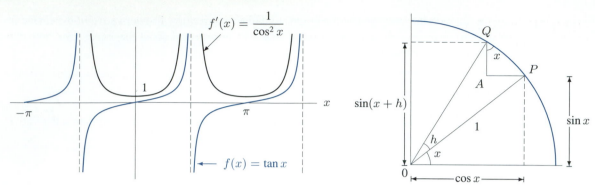

Figure 4.20: The function $\tan x$ and its derivative   **Figure 4.21:** Unit circle showing $\sin(x + h)$ and $\sin x$

Example 5   Differentiate (a) $2\tan(3t)$,   (b) $\tan(1 - \theta)$,   (c) $\dfrac{1 + \tan t}{1 - \tan t}$.

Solution   (a)   Use the chain rule:

$$\frac{d}{dt}(2\tan(3t)) = 2\frac{1}{\cos^2(3t)}\frac{d}{dt}(3t) = \frac{6}{\cos^2(3t)}.$$

(b)   Use the chain rule:

$$\frac{d}{d\theta}(\tan(1 - \theta)) = \frac{1}{\cos^2(1 - \theta)} \cdot \frac{d}{d\theta}(1 - \theta) = \frac{-1}{\cos^2(1 - \theta)}.$$

(c)   Use the quotient rule:

$$\frac{d}{dt}\left(\frac{1 + \tan t}{1 - \tan t}\right) = \frac{\left(\dfrac{d(1 + \tan t)}{dt}\right)(1 - \tan t) - (1 + \tan t)\dfrac{d(1 - \tan t)}{dt}}{(1 - \tan t)^2}$$

$$= \frac{\dfrac{1}{\cos^2 t}(1 - \tan t) - (1 + \tan t)\left(-\dfrac{1}{\cos^2 t}\right)}{(1 - \tan t)^2}$$

$$= \frac{2}{\cos^2 t \cdot (1 - \tan t)^2}.$$

Informal Justification of $\dfrac{d}{dx}(\sin x) = \cos x$

Consider the unit circle in Figure 4.21. To find the derivative of $\sin x$, we need to estimate

$$\frac{\sin(x + h) - \sin x}{h}.$$

In Figure 4.21, the quantity $\sin(x + h) - \sin x$ is represented by the length QA. The arc QP is of length h, so

$$\frac{\sin(x + h) - \sin x}{h} = \frac{QA}{\text{Arc } QP}.$$

Now, if h is small, QAP is approximately a right triangle because the arc QP is almost a straight line. Furthermore, using geometry, we can show that angle $AQP \approx x$. For small h, we have

$$\frac{\sin(x + h) - \sin x}{h} = \frac{QA}{\text{Arc } QP} \approx \cos x.$$

As $h \to 0$, the approximation gets better, so

$$\frac{d}{dx}(\sin x) = \lim_{h \to 0}\frac{\sin(x + h) - \sin x}{h} = \cos x.$$

Other derivations of this result are given in Problems 39 and 40 on page 218.

Problems for Section 4.5

1. Construct a table of values for $\cos x$, $x = 0, 0.1, 0.2, \ldots, 0.6$. Using the difference quotient, estimate the derivative at these points (use $h = 0.001$), and compare it with $(-\sin x)$.

Find the derivatives of the functions in Problems 2–29.

2. $r(\theta) = \sin \theta + \cos \theta$
3. $s(\theta) = \cos \theta \sin \theta$
4. $t(\theta) = \dfrac{\cos \theta}{\sin \theta}$

5. $z = \cos(4\theta)$
6. $f(x) = \sin(3x)$
7. $w = \sin(e^t)$

8. $f(x) = x^2 \cos x$
9. $f(x) = e^{\cos x}$
10. $f(y) = e^{\sin y}$

11. $f(x) = \sqrt{1 - \cos x}$
12. $f(x) = \cos(\sin x)$
13. $f(x) = \tan(\sin x)$

14. $k(x) = \sqrt{(\sin(2x))^3}$
15. $h(x) = 2^{\sin x}$
16. $w = 2^{2 \sin x + e^x}$

17. $z = \theta e^{\cos \theta}$
18. $f(x) = 2x \sin(3x)$
19. $f(x) = \sin(2x) \cdot \sin(3x)$

20. $y = e^{\theta} \sin(2\theta)$
21. $f(x) = e^{-2x} \cdot \sin x$
22. $z = \sqrt{\sin t}$

23. $y = \sin^5 \theta$
24. $g(z) = \tan(e^z)$
25. $z = \tan(e^{-3\theta})$

26. $w = e^{-\sin \theta}$
27. $h(t) = t \cos t + \tan t$
28. $f(\alpha) = \cos \alpha + 3 \sin \alpha$

29. $f(\theta) = \theta^2 \sin \theta + 2\theta \cos \theta - 2 \sin \theta$

30. Find the *fiftieth* derivative of $y = \cos x$.

31. Find a possible formula for the function $q(x)$ such that $q'(x) = \dfrac{e^x \cdot \sin x - e^x \cdot \cos x}{(\sin x)^2}$.

32. A boat at anchor is bobbing up and down in the sea. The vertical distance, y, in feet, between the sea floor and the boat is given as a function of time, t, in minutes, by

$$y = 15 + \sin(2\pi t).$$

   (a) Find the vertical velocity, v, of the boat at time t.
   (b) Make rough sketches of y and v against t.

33. On page 55 the depth, y, of water in Boston harbor was given by

$$y = 5 + 4.9 \cos\left(\frac{\pi}{6}t\right),$$

   where t is the number of hours since midnight.

   (a) Find $\dfrac{dy}{dt}$. What does $\dfrac{dy}{dt}$ represent, in terms of water level?
   (b) For $0 \leq t \leq 24$, when is $\dfrac{dy}{dt}$ zero? (Figure 1.73 on page 55 may be helpful.) Explain what it means (in terms of water level) for $\dfrac{dy}{dt}$ to be zero.

34. The voltage, V, in volts, in an electrical outlet is given as a function of time, t, in seconds, by the function $V = 156 \cos(120\pi t)$.

   (a) Give an expression for the rate of change of voltage with respect to time.
   (b) Is the rate of change ever zero? Explain.
   (c) What is the maximum value of the rate of change?

35. The function $y = A \sin\left(\sqrt{\frac{k}{m}}t\right)$ represents the oscillations of a mass m at the end of a spring. The constant k measures the stiffness of the spring.

   (a) Find a time at which the mass is farthest from its equilibrium position. Find a time at which the mass is moving fastest. Find a time at which the mass is accelerating fastest.
   (b) What is the period, T, of the oscillation?
   (c) Find dT/dm. What does the sign of dT/dm tell you?

36. Find the equations of the tangent lines to the graph of $f(x) = \sin x$ at $x = 0$ and at $x = \pi/3$. Use each tangent line to approximate $\sin(\pi/6)$. Would you expect these results to be equally accurate, since they are taken equally far away from $x = \pi/6$ but on opposite sides? If the accuracy is different, can you account for the difference?

37. A lighthouse is 2 km from the long, straight coastline shown in Figure 4.22. Find the rate of change of the distance of the spot of light from the point O with respect to the angle θ.

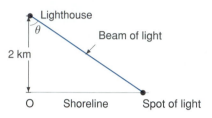

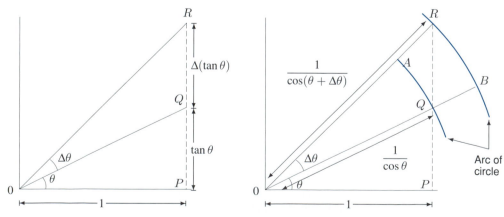

Figure 4.22   **Figure 4.23**

38. The metal bar of length l in Figure 4.23 has one end attached at the point P to a circle of radius a. Point Q at the other end can slide back and forth along the x-axis.

    (a) Find x as a function of θ.
    (b) Assume lengths are in centimeters and the angular speed ($d\theta/dt$) is 2 radians/second counter-clockwise. Find the speed at which the point Q is moving when

        (i) $\theta = \pi/2$,   (ii) $\theta = \pi/4$.

39. We will use the following identities to calculate the derivatives of $\sin x$ and $\cos x$:

$$\sin(a + b) = \sin a \cos b + \sin b \cos a$$
$$\cos(a + b) = \cos a \cos b - \sin a \sin b.$$

    (a) Use the definition of the derivative to show that if $f(x) = \sin x$,

$$f'(x) = \sin x \lim_{h \to 0} \frac{\cos h - 1}{h} + \cos x \lim_{h \to 0} \frac{\sin h}{h}.$$

    (b) Estimate the limits in part (a) with your calculator to explain why $f'(x) = \cos x$.
    (c) If $g(x) = \cos x$, use the definition of the derivative to show that $g'(x) = -\sin x$.

40. In this problem you will calculate the derivative of $\tan \theta$ rigorously (and without using the derivatives of $\sin \theta$ or $\cos \theta$). You will then use your result for $\tan \theta$ to calculate the derivatives of $\sin \theta$ and $\cos \theta$. Figure 4.24 shows $\tan \theta$ and $\Delta(\tan \theta)$, which is the change in $\tan \theta$, namely $\tan(\theta + \Delta\theta) - \tan \theta$.

Figure 4.24: $\tan \theta$ and $\Delta(\tan \theta)$

(a) By paying particular attention to how the two figures relate and using the fact that

Area of Sector OAQ $\leq$ Area of Triangle OQR $\leq$ Area of Sector OBR

explain why

$$\frac{\Delta\theta}{2\pi} \cdot \pi \left(\frac{1}{\cos\theta}\right)^2 \leq \frac{1}{2} \cdot 1 \cdot \Delta(\tan\theta) \leq \frac{\Delta\theta}{2\pi} \cdot \pi \left(\frac{1}{\cos(\theta+\Delta\theta)}\right)^2.$$

[Hint: A sector of a circle with angle α at the center has area $\alpha/(2\pi)$ times the area of the whole circle.]

(b) Use part (a) to show as $\Delta\theta \to 0$ that

$$\frac{\Delta\tan\theta}{\Delta\theta} \to \left(\frac{1}{\cos\theta}\right)^2,$$

and hence that $\dfrac{d(\tan\theta)}{d\theta} = \left(\dfrac{1}{\cos\theta}\right)^2.$

(c) Derive the identity $(\tan\theta)^2 + 1 = \left(\dfrac{1}{\cos\theta}\right)^2$. Then differentiate both sides of this identity with respect to θ, using the chain rule and the result of part (b) to show that $\dfrac{d}{d\theta}(\cos\theta) = -\sin\theta$.

(d) Differentiate both sides of the identity $(\sin\theta)^2 + (\cos\theta)^2 = 1$ and use the result of part (c) to show that $\dfrac{d}{d\theta}(\sin\theta) = \cos\theta$.

4.6 APPLICATIONS OF THE CHAIN RULE

In this section we will use the chain rule to calculate the derivatives of fractional powers, logarithms, exponentials, and the inverse trigonometric functions.[2]

Finding the Derivative of an Inverse Function: Derivative of $x^{1/2}$

Earlier we calculated the derivative of x^n with n an integer, but we have been using the result for non-integral values of n as well. We now confirm that the power rule holds for $n = 1/2$ by calculating the derivative of $f(x) = x^{1/2}$ using the chain rule. Since

$$[f(x)]^2 = x,$$

the derivative of $[f(x)]^2$ and the derivative of x must be equal, so

$$\frac{d}{dx}\left[f(x)\right]^2 = \frac{d}{dx}(x).$$

We can use the chain rule with $f(x)$ as the inside function to obtain:

$$\frac{d}{dx}[f(x)]^2 = 2f(x) \cdot f'(x) = 1.$$

Solving for $f'(x)$ gives

$$f'(x) = \frac{1}{2f(x)} = \frac{1}{2x^{1/2}},$$

or

$$\frac{d}{dx}(x^{1/2}) = \frac{1}{2x^{1/2}} = \frac{1}{2}x^{-1/2}.$$

A similar calculation can be used to obtain the derivative of $x^{1/n}$ where n is a positive integer.

---

[2]It requires a separate justification, not given here, that these functions are differentiable.

Derivative of $\ln x$

We use the chain rule to differentiate an identity involving $\ln x$. Since $e^{\ln x} = x$, we have

$$\frac{d}{dx}(e^{\ln x}) = \frac{d}{dx}(x),$$

$$e^{\ln x} \cdot \frac{d}{dx}(\ln x) = 1. \qquad \text{(Since } e^x \text{ is outside function and } \ln x \text{ is inside function)}$$

Solving for $d(\ln x)/dx$ gives

$$\frac{d}{dx}(\ln x) = \frac{1}{e^{\ln x}} = \frac{1}{x},$$

so

$$\boxed{\frac{d}{dx}(\ln x) = \frac{1}{x}.}$$

Example 1 Differentiate (a) $\ln(x^2 + 1)$, (b) $t^2 \ln t$, (c) $\sqrt{1 + \ln(1 - y)}$.

Solution (a) Using the chain rule:

$$\frac{d}{dx}\left(\ln(x^2 + 1)\right) = \frac{1}{x^2 + 1}\frac{d}{dx}(x^2 + 1) = \frac{2x}{x^2 + 1}.$$

(b) Using the product rule:

$$\frac{d}{dt}(t^2 \ln t) = \frac{d}{dt}(t^2) \cdot \ln t + t^2 \frac{d}{dt}(\ln t) = 2t \ln t + t^2 \cdot \frac{1}{t} = 2t \ln t + t.$$

(c) Using the chain rule:

$$\frac{d}{dy}\left(\sqrt{1 + \ln(1 - y)}\right) = \frac{d}{dy}\left(1 + \ln(1 - y)\right)^{1/2}$$

$$= \frac{1}{2}\left(1 + \ln(1 - y)\right)^{-1/2} \cdot \frac{d}{dy}\left(1 + \ln(1 - y)\right) \qquad \text{(Using the chain rule)}$$

$$= \frac{1}{2\sqrt{1 + \ln(1 - y)}} \cdot \frac{1}{1 - y} \cdot \frac{d}{dy}(1 - y) \qquad \text{(Using the chain rule again)}$$

$$= \frac{-1}{2(1 - y)\sqrt{1 + \ln(1 - y)}}.$$

Derivative of a^x

Earlier we showed that the derivative of a^x is proportional to a^x. Now we show that the constant of proportionality is $\ln a$. We use the identity

$$\ln(a^x) = x \ln a.$$

Differentiating, using $\frac{d}{dx}(\ln x) = \frac{1}{x}$ and the chain rule, and remembering that $\ln a$ is a constant, we obtain:

$$\frac{d}{dx}(\ln a^x) = \frac{1}{a^x} \cdot \frac{d}{dx}(a^x) = \ln a.$$

Solving gives the result we suggested in Section 4.2:

$$\boxed{\frac{d}{dx}(a^x) = (\ln a)a^x.}$$

Derivatives of Inverse Trigonometric Functions

In Section 1.9 we defined $\arcsin x$ as the angle between $-\pi/2$ and $\pi/2$ (inclusive) whose sine is x. Similarly, $\arctan x$ as the angle strictly between $-\pi/2$ and $\pi/2$ whose tangent is x. To find $\dfrac{d}{dx}(\arctan x)$ we use the identity $\tan(\arctan x) = x$. Differentiating using the chain rule gives

$$\frac{1}{\cos^2(\arctan x)} \cdot \frac{d}{dx}(\arctan x) = 1,$$

so

$$\frac{d}{dx}(\arctan x) = \cos^2(\arctan x).$$

Using the identity $1 + \tan^2\theta = \dfrac{1}{\cos^2\theta}$, and replacing θ by $\arctan x$, we have

$$\cos^2(\arctan x) = \frac{1}{1 + \tan^2(\arctan x)} = \frac{1}{1 + x^2}.$$

Thus we have

$$\frac{d}{dx}(\arctan x) = \frac{1}{1 + x^2}.$$

By a similar argument, we obtain the result:

$$\frac{d}{dx}(\arcsin x) = \frac{1}{\sqrt{1 - x^2}}.$$

Example 2   Differentiate (a) $\arctan(t^2)$,   (b) $\arcsin(\tan\theta)$.

Solution   Use the chain rule:

(a) $\quad \dfrac{d}{dt}\left(\arctan(t^2)\right) = \dfrac{1}{1 + (t^2)^2} \cdot \dfrac{d}{dt}(t^2) = \dfrac{2t}{1 + t^4}.$

(b) $\quad \dfrac{d}{dt}\left(\arcsin(\tan\theta)\right) = \dfrac{1}{\sqrt{1 - (\tan\theta)^2}} \cdot \dfrac{d}{d\theta}(\tan\theta) = \dfrac{1}{\sqrt{1 - \tan^2\theta}} \cdot \dfrac{1}{\cos^2\theta}.$

Example 3   An airplane, flying at 450 km/hr at a constant altitude of 5 km, is approaching a camera mounted on the ground. Let θ be the angle at which the camera is pointed. See Figure 4.25. When $\theta = \pi/3$, how fast does the camera have to rotate in order to keep the plane in view?

Solution   Suppose the plane is vertically above the point B. Let x be the distance between B and C. The fact that the plane is moving toward C at 450 km/hr means that x is decreasing and $dx/dt = -450$ km/hr. From Figure 4.25, we see that $\tan\theta = 5/x$.

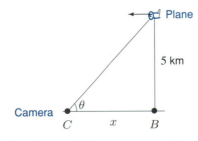

Figure 4.25: Plane approaching a camera at C

Differentiating $\tan\theta = 5/x$ with respect to t and using the chain rule gives

$$\frac{1}{\cos^2\theta}\frac{d\theta}{dt} = -5x^{-2}\frac{dx}{dt}.$$

We want to solve for $d\theta/dt$ when $\theta = \pi/3$. At that moment, $\cos\theta = 1/2$ and $\tan\theta = \sqrt{3}$; so $x = 5/\sqrt{3}$. Substituting gives

$$\frac{1}{(1/2)^2}\frac{d\theta}{dt} = -5\left(\frac{5}{\sqrt{3}}\right)^{-2}\cdot(-450)$$

$$\frac{d\theta}{dt} = 67.5 \text{ radians/hour.}$$

This answer tells us that the camera must turn at roughly 1 radian per minute if it is to remain pointed at the plane.

Problems for Section 4.6

For Problems 1–22, find the derivative of the given function. In some cases, it may be to your advantage to simplify before differentiating.

1. $f(t) = \ln(t^2 + 1)$

2. $f(x) = \ln(1 - x)$

3. $f(x) = \ln(e^{2x})$

4. $f(x) = e^{\ln(e^{2x^2+3})}$

5. $f(z) = \dfrac{1}{\ln z}$

6. $f(\theta) = \ln(\cos\theta)$

7. $f(x) = \ln(1 - e^{-x})$

8. $f(\alpha) = \ln(\sin\alpha)$

9. $f(x) = \ln(e^x + 1)$

10. $f(t) = \ln(\ln t) + \ln(\ln 2)$

11. $f(x) = \ln(e^{7x})$

12. $f(x) = e^{(\ln x)+1}$

13. $f(w) = \ln(\cos(w - 1))$

14. $f(t) = \ln(e^{\ln t})$

15. $f(y) = \arcsin(y^2)$

16. $g(t) = \arctan(3t - 4)$

17. $g(\alpha) = \sin(\arcsin\alpha)$

18. $g(t) = e^{\arctan(3t^2)}$

19. $g(t) = \cos(\ln t)$

20. $h(z) = z^{\ln 2}$

21. $h(w) = w \arcsin w$

22. $f(x) = \cos(\arcsin(x + 1))$

23. On what intervals is $\ln(x^2 + 1)$ concave up?

24. Using the chain rule, find $\dfrac{d}{dx}(\arcsin x)$.

25. Using the chain rule, find $\dfrac{d}{dx}(\log x)$. (Recall that $\log x = \log_{10} x$.)

26. To compare the acidity of different solutions, chemists use the pH (which is a single number, not the product of p and H). The pH is defined in terms of the concentration, x, of hydrogen ions in the solution as

$$\text{pH} = -\log x.$$

Find the rate of change of pH with respect to hydrogen ion concentration when the pH is 2. [Hint: Use the result of Problem 25.]

27. Hungary is one of the few countries in the world where the population is decreasing, currently at about 0.2% a year. Thus, if t is time in years since 1990, the population P, in millions, of Hungary can be approximated by

$$P = 10.8(0.998)^t.$$

(a) What does this model predict the population of Hungary will be in the year 2000?

(b) How fast (in people/year) does this model predict Hungary's population will be decreasing in the year 2000?

28. Imagine you are zooming in on the graph of each of the following functions near the origin:

$$y = x \qquad y = \sqrt{x} \qquad y = x^2 \qquad y = \sin x$$

$$y = x \sin x \qquad y = \tan x \qquad y = \sqrt{x/(x+1)} \qquad y = x^3$$

$$y = \ln(x+1) \qquad y = \tfrac{1}{2}\ln(x^2+1) \qquad y = 1 - \cos x \qquad y = \sqrt{2x - x^2}$$

Which of them look the same? Group together those functions which become indistinguishable, and give the equations of the lines they look like.

29. (a) For $x > 0$, find the derivative of $f(x) = \arctan x + \arctan(1/x)$ and simplify.
     (b) What does your result tell you about f?

30. (a) Find the equation of the tangent line to $y = \ln x$ at $x = 1$.
     (b) Use it to calculate approximate values for $\ln(1.1)$ and $\ln(2)$.
     (c) Using a graph, explain whether the approximate values you have calculated are smaller or larger than the true values. Would the same result have held if you had used the tangent line to estimate $\ln(0.9)$ and $\ln(0.5)$? Why?

31. (a) Find the equation of the best quadratic approximation to $y = \ln x$ at $x = 1$. The best quadratic approximation has the same first and second derivatives as $y = \ln x$ at $x = 1$.
     (b) Use a computer or calculator to graph the approximation and $y = \ln x$ on the same set of axes. What do you notice?
     (c) Use your quadratic approximation to calculate approximate values for $\ln(1.1)$ and $\ln(2)$.

32. The gravitational force, F, on a rocket at a distance, r, from the center of the earth is given by

$$F = \frac{k}{r^2},$$

where $k = 10^{13}$ newton $\cdot$ km^2. When the rocket is 10^4 km from the center of the earth, it is moving away at 0.2 km/sec. How fast is the gravitational force changing at that moment? Give units. (A newton is a unit of force.)

33. A train is traveling at 0.8 km/min along a long straight track, moving in the direction shown in Figure 4.26. A movie camera, 0.5 km away from the track, is focused on the train.

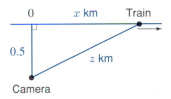

Figure 4.26

     (a) Express z, the distance between the camera and the train, as a function of x.
     (b) How fast is the distance from the camera to the train changing when the train is 1 km from the camera? Give units.
     (c) How fast is the camera rotating (in radians/min) at the moment when the train is 1 km from the camera?

34. The radius of a spherical balloon is increasing by 2 cm/sec. At what rate is air being blown into the balloon at the moment when the radius is 10 cm? Give units in your answer.

35. Coroners estimate time of death from body temperature using the simple rule of thumb that a body cools about 2°F during the first hour after death and about 1°F for each additional hour. (The temperature is measured using a small probe inserted into the liver which, being large and vascular, holds body heat the longest.)

Assuming an air temperature of 68°F and a living body temperature of 98.6°F, the temperature $T(t)$ in °F at time t in hours is given by:

$$T(t) = 68 + 30.6e^{-kt}$$

where $t = 0$ is the instant that death occurs.

     (a) For what value of k will the body cool by 2°F in the first hour?

(b) Using the value of k found in part (a), after how many hours will the temperature of the body be decreasing at a rate of $1°$F per hour?

(c) Using the value of k found in part (a), show that, 24 hours after death, the coroner's rule of thumb gives approximately the same temperature as the formula.

36. For the amusement of the guests, some hotels have elevators on the outside of the building. Suppose such a hotel is 300 feet high. You are standing by a window 100 feet above the ground and 150 feet away from the hotel, and the elevator descends at a constant speed of 30 ft/sec, starting at time $t = 0$, where t is time in seconds. Let θ be the angle between the line of your horizon and your line of sight to the elevator. (See Figure 4.27.)

(a) Find a formula for $h(t)$, the elevator's height above the ground as it descends from the top of the hotel.

(b) Using your answer to part (a), express θ as a function of time t and find the rate of change of θ with respect to t.

(c) If the rate of change of θ is a measure of how fast the elevator appears to you to be moving, at what height will the elevator be when it appears to be moving fastest?

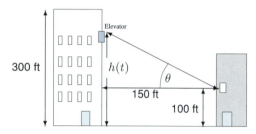

Figure 4.27: Descending elevator

4.7 IMPLICIT FUNCTIONS

In earlier chapters, most functions were written in the form $y = f(x)$; here y is said to be an *explicit* function of x. An equation such as

$$x^2 + y^2 = 4$$

is said to give y as an *implicit* function of x. Its graph is the circle in Figure 4.28. Since there are x-values which correspond to two y-values, y is not a function of x on the whole circle. Solving gives

$$y = \pm\sqrt{4 - x^2},$$

where $y = \sqrt{4 - x^2}$ represents the top half of the circle and $y = -\sqrt{4 - x^2}$ represents the bottom half. So y is a function of x on the top half, and y is a different function of x on the bottom half.

But let's consider the circle as a whole. The equation does represent a curve which has a tangent line at each point. The slope of this tangent can be found by differentiating the equation of the circle with respect to x:

$$\frac{d}{dx}(x^2) + \frac{d}{dx}(y^2) = \frac{d}{dx}(4).$$

If we think of y as a function of x and use the chain rule, we get

$$2x + 2y\frac{dy}{dx} = 0.$$

Solving gives

$$\frac{dy}{dx} = -\frac{x}{y}.$$

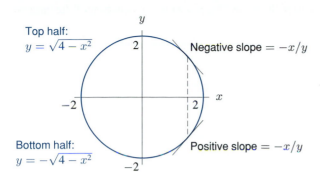

Figure 4.28: Graph of $x^2 + y^2 = 4$

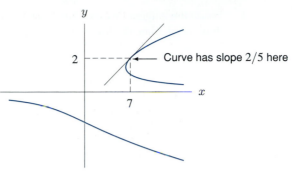

Figure 4.29: Graph of $y^3 - xy = -6$ and its tangent line at $(7, 2)$

The derivative here depends on both x and y (instead of just on x). This is because for each x-value (except for $x = \pm 2$) there are two y-values, and the curve has a different slope at each one. Figure 4.28 shows that for x and y both positive, we are on the top right quarter of the curve and the slope is negative (as the formula predicts). For x positive and y negative, we are on the bottom right quarter of the curve and the slope is positive (as the formula predicts). Differentiating the equation of the circle has given us the slope of the curve at all points except $(2, 0)$ and $(-2, 0)$, where the tangent is vertical. In general, this process of *implicit differentiation* leads to a derivative whenever the expression for the derivative does not have a zero in the denominator.

Example 1   Make a table of x and approximate y-values for the equation $y^3 - xy = -6$ near $x = 7, y = 2$. Your table should include the x-values $6.8, 6.9, 7.0, 7.1$, and 7.2.

Solution   We would like to solve for y in terms of x, but we cannot isolate y by factoring. There is a formula for solving cubics, somewhat like the quadratic formula, but it is too complicated to be useful here. Instead, first observe that $x = 7, y = 2$ does satisfy the equation. (Check this!) Now find dy/dx by implicit differentiation:

$$\frac{d}{dx}(y^3) - \frac{d}{dx}(xy) = \frac{d}{dx}(-6)$$

$$3y^2\frac{dy}{dx} - 1 \cdot y - x\frac{dy}{dx} = 0 \quad \text{(Differentiating with respect to x)}$$

$$3y^2\frac{dy}{dx} - x\frac{dy}{dx} = y$$

$$(3y^2 - x)\frac{dy}{dx} = y \quad \text{(Factoring out $\frac{dy}{dx}$)}$$

$$\frac{dy}{dx} = \frac{y}{3y^2 - x}.$$

When $x = 7$ and $y = 2$, we have

$$\frac{dy}{dx} = \frac{2}{12 - 7} = \frac{2}{5}.$$

(See Figure 4.29.) The equation of the tangent line at $(7, 2)$ is

$$y - 2 = \frac{2}{5}(x - 7)$$

or

$$y = 0.4x - 0.8.$$

Since the tangent lies very close to the curve near the point $(7, 2)$, we use the equation of the tangent line to calculate the following approximate y-values:

x	6.8	6.9	7.0	7.1	7.2
Approximate y	1.92	1.96	2.00	2.04	2.08

Notice that although the equation $y^3 - xy = -6$ leads to a curve which is difficult to deal with algebraically, it still looks like a straight line locally.

Example 2   Find all points where the tangent line to $y^3 - xy = -6$ is either horizontal or vertical.

Solution   From the previous example, $\dfrac{dy}{dx} = \dfrac{y}{3y^2 - x}$. The tangent is horizontal when the numerator of dy/dx equals 0, so $y = 0$. Since we also must satisfy $y^3 - xy = -6$, we get $0^3 - x \cdot 0 = -6$, which is impossible. We conclude that there are no points on the curve where the tangent line is horizontal.

The tangent is vertical when the denominator of dy/dx is 0, giving $3y^2 - x = 0$. Thus, $x = 3y^2$ at any point with a vertical tangent line. Again, we must also satisfy $y^3 - xy = -6$, so

$$y^3 - (3y^2)y = -6,$$
$$-2y^3 = -6,$$
$$y = \sqrt[3]{3} \approx 1.442.$$

We can then find x by substituting $y = \sqrt[3]{3}$ in $y^3 - xy = -6$. We get $3 - x(\sqrt[3]{3}) = -6$, so $x = 9/(\sqrt[3]{3}) \approx 6.240$. So the tangent line is vertical at $(6.240, 1.442)$.

Using implicit differentiation and the expression for dy/dx to locate the points where the tangent is vertical or horizontal, as in the previous example, is a first step in obtaining an overall picture of the curve $y^3 - xy = -6$. However, filling in the rest of the graph, even roughly, by using the sign of dy/dx to tell us where the curve is increasing or decreasing can be difficult.

Problems for Section 4.7

For Problems 1–12, find dy/dx.

1. $x^2 + y^2 = \sqrt{7}$

2. $x^2 + xy - y^3 = xy^2$

3. $\sqrt{x} = 5\sqrt{y}$

4. $\sqrt{x} + \sqrt{y} = 25$

5. $\ln x + \ln(y^2) = 3$

6. $e^{x^2} + \ln y = 0$

7. $\arctan(x^2 y) = xy^2$

8. $x \ln y + y^3 = \ln x$

9. $\sin(xy) = 2x + 5$

10. $x^{2/3} + y^{2/3} = a^{2/3}$ (a is a constant.)

11. $e^{\cos y} = x^3 \arctan y$

12. $\cos^2 y + \sin^2 y = y + 2$

For Problems 13–16, find the equations of the tangent lines to the following curves at the indicated points.

13. $xy^2 = 1$ at $(1, -1)$

14. $\ln(xy) = 2x$ at $(1, e^2)$

15. $y^2 = \dfrac{x^2}{xy - 4}$ at $(4, 2)$

16. $x^{2/3} + y^{2/3} = a^{2/3}$ at $(a, 0)$

17.   Show that the power rule for derivatives applies to rational powers of the form $y = x^{m/n}$ by raising both sides to the nth power and using implicit differentiation.

18. (a) Find the equations of the tangent lines to the circle $x^2 + y^2 = 25$ at the points where $x = 4$.
    (b) Find the equations of the normal lines to this circle at the same points. (The normal line to a curve at a point is perpendicular to the tangent line at that point.)
    (c) At what point do the two normal lines intersect?

19. (a) Find the slope of the tangent line to the ellipse $\dfrac{x^2}{25} + \dfrac{y^2}{9} = 1$ at the point (x, y).
    (b) Are there any points where the slope is not defined?

20. Consider the equation $x^3 + y^3 - xy^2 = 5$.

    (a) Find dy/dx by implicit differentiation.
    (b) Give a table of approximate y-values near $x = 1, y = 2$ for $x = 0.96, 0.98, 1, 1.02, 1.04$.
    (c) Find the y-value for $x = 0.96$ by substituting $x = 0.96$ in the equation and solving for y using a computer or calculator. Compare with your answer in part (b).
    (d) Find all points where the tangent line is horizontal or vertical.

21. Find the equation of the tangent line to the curve $y = x^2$ at $x = 1$. Show that this line is also a tangent to a circle centered at $(8, 0)$ and find the equation of this circle.

22. Sketch the circles $y^2 + x^2 = 1$ and $y^2 + (x - 3)^2 = 4$. There is a line with positive slope that is tangent to both circles. Determine the points at which this tangent line touches each circle.

4.8 LINEAR APPROXIMATIONS AND LIMITS

The Tangent Line Approximation

Since the graph of a function and its tangent line have the same slope at the point of tangency (namely, the derivative at that point), the tangent line stays close to the graph of the function near that point. This means that we can approximate values of the function by values on the tangent line. (See Figure 4.30.) The slope of the tangent to the graph of $y = f(x)$ at $x = a$ is $f'(a)$, so the equation of the tangent line is

$$y = f(a) + f'(a)(x - a).$$

Now we approximate the values of f by the y values from the tangent line, giving the following result.

> **The Tangent Line Approximation**
>
> For values of x near a,
> $$f(x) \approx f(a) + f'(a)(x - a)$$
> We are thinking of a as fixed, so $f(a)$ and $f'(a)$ are constant.

The expression $f(a) + f'(a)(x - a)$ is a linear function which approximates $f(x)$ well near a; it is called the *local linearization* of f near $x = a$.

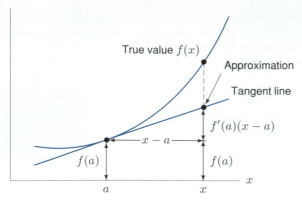

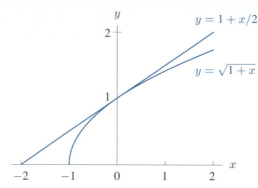

Figure 4.30: Local linearization: Approximation by the tangent line

Figure 4.31: Tangent line approximation to $\sqrt{1+x}$ near $x = 0$

Example 1   Find the tangent line approximation for $\sqrt{1+x}$ near $x = 0$.

Solution   With $f(x) = \sqrt{1+x}$, the chain rule gives $f'(x) = 1/(2\sqrt{1+x})$, so $f(0) = 1$ and $f'(0) = 1/2$. Therefore the tangent line approximation of f near $x = 0$,

$$f(x) \approx f(0) + f'(0)(x - 0),$$

becomes

$$\sqrt{1+x} \approx 1 + \frac{x}{2}.$$

This means that, near $x = 0$, the function $\sqrt{1+x}$ can be approximated by its tangent line $y = 1 + x/2$. (See Figure 4.31.)

Example 2   What is the local linearization of e^{kx} near $x = 0$?

Solution   If $f(x) = e^{kx}$, then $f(0) = 1$ and, by the chain rule, $f'(x) = ke^{kx}$, so $f'(0) = ke^{k \cdot 0} = k$. Thus

$$f(x) \approx f(0) + f'(0)(x - 0)$$

becomes

$$e^{kx} \approx 1 + kx.$$

This is the tangent line approximation to e^{kx} near $x = 0$. In other words, if we zoom in on the functions $f(x) = e^{kx}$ and $y = 1 + kx$ near the origin, we won't be able to tell them apart.

Using Local Linearity to Find Limits

Suppose we want to calculate the exact value of the limit

$$\lim_{x \to 0} \frac{e^{2x} - 1}{x}.$$

Substituting $x = 0$ gives us $0/0$, which is undefined:

$$\frac{e^{2(0)} - 1}{0} = \frac{1 - 1}{0} = \frac{0}{0}.$$

Substituting values of x near 0 gives us an approximate value for the limit.

However, the limit can be calculated exactly using local linearity. Suppose we let $f(x)$ be the numerator, so $f(x) = e^{2x} - 1$, and $g(x)$ be the denominator, so $g(x) = x$. Then $f(0) = 0$ and $f'(x) = 2e^{2x}$, so $f'(0) = 2$. When we zoom in on the graph of $f(x) = e^{2x} - 1$ near the origin, we see its tangent line $y = 2x$ shown in Figure 4.32. We are interested in the ratio $(e^{2x} - 1)/x = f(x)/g(x)$, which is approximately the ratio of the y-values in Figure 4.32. This ratio of y-values is just the ratio of the slopes of the lines. So

$$\frac{f(x)}{g(x)} = \frac{e^{2x} - 1}{x} \approx \frac{2}{1} = \frac{f'(0)}{g'(0)}.$$

As $x \to 0$, this approximation gets better, and we have

$$\lim_{x \to 0} \frac{e^{2x} - 1}{x} = 2.$$

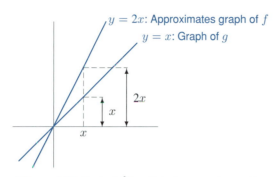

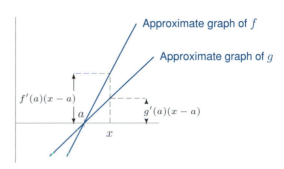

Figure 4.32: Ratio $(e^{2x} - 1)/x$ is approximated by ratio of slopes as we zoom in near the origin

Figure 4.33: Ratio $f(x)/g(x)$ is approximated by ratio of slopes, $f'(a)/g'(a)$, as we zoom in at a

L'Hopital's Rule

If $f(a) = g(a) = 0$, we can use the same method to investigate limits of the form

$$\lim_{x \to a} \frac{f(x)}{g(x)}.$$

As in the previous case, we zoom in on the graphs of $f(x)$ and $g(x)$. Figure 4.33 shows that both graphs cross the x-axis at $x = a$. This suggests that the limit of $f(x)/g(x)$ as $x \to a$ is the ratio of slopes, giving the following result:

L'Hopital's rule: If f and g are differentiable, $f(a) = g(a) = 0$, and $g'(a) \neq 0$, then

$$\lim_{x \to a} \frac{f(x)}{g(x)} = \frac{f'(a)}{g'(a)}.$$

To justify this result, let us assume $g'(a) \neq 0$ and consider the quantity $f'(a)/g'(a)$. Using the definition of the derivative and the fact that $f(a) = g(a) = 0$, we have

$$\frac{f'(a)}{g'(a)} = \frac{\displaystyle\lim_{h \to 0} \frac{f(a+h) - f(a)}{h}}{\displaystyle\lim_{h \to 0} \frac{g(a+h) - g(a)}{h}}$$

$$= \frac{\displaystyle\lim_{h \to 0} (f(a+h)/h)}{\displaystyle\lim_{h \to 0} (g(a+h)/h)}$$

$$= \lim_{h \to 0} \frac{f(a+h)}{g(a+h)} = \lim_{x \to a} \frac{f(x)}{g(x)}.$$

Note that if $f'(a) \neq 0$ and $g'(a) = 0$, the limit of $f(x)/g(x)$ does not exist.

Example 3   Use l'Hopital's rule to confirm that $\lim\limits_{x \to 0} \dfrac{\sin x}{x} = 1$.

Solution   Let $f(x) = \sin x$ and $g(x) = x$. Then $f(0) = g(0) = 0$ and $f'(x) = \cos x$ and $g'(x) = 1$. Thus,

$$\lim_{x \to 0} \frac{\sin x}{x} = \frac{\cos 0}{1} = 1.$$

If we also have $f'(a) = g'(a) = 0$, then we can use the following result:

> More general form of **l'Hopital's rule:** If f and g are differentiable and $f(a) = g(a) = 0$, then
> $$\lim_{x \to a} \frac{f(x)}{g(x)} = \lim_{x \to a} \frac{f'(x)}{g'(x)},$$
> provided the limit on the right exists.

Example 4   Calculate $\lim\limits_{t \to 0} \dfrac{e^t - 1 - t}{t^2}$.

Solution   Let $f(t) = e^t - 1 - t$ and $g(t) = t^2$. Then $f(0) = e^0 - 1 - 0 = 0$ and $g(0) = 0$, and $f'(t) = e^t - 1$ and $g'(t) = 2t$. So

$$\lim_{t \to 0} \frac{e^t - 1 - t}{t^2} = \lim_{t \to 0} \frac{e^t - 1}{2t}.$$

Since $f'(0) = g'(0) = 0$, the ratio $f'(0)/g'(0)$ is not defined. So we use l'Hopital's rule again:

$$\lim_{t \to 0} \frac{e^t - 1 - t}{t^2} = \lim_{t \to 0} \frac{e^t - 1}{2t} = \lim_{t \to 0} \frac{e^t}{2} = \frac{1}{2}.$$

The following forms of l'Hopital's rule apply to limits involving infinity.

> L'Hopital's rule can also be used
> - When $\lim_{x \to a} f(x) = \pm\infty$ and $\lim_{x \to a} g(x) = \pm\infty$,
>
> or
>
> - When $a = \pm\infty$.
> It can be shown that under these circumstances:
> $$\lim_{x \to a} \frac{f(x)}{g(x)} = \lim_{x \to a} \frac{f'(x)}{g'(x)}$$
> (where a may be $\pm\infty$), provided the limit on the right-hand side exists.

Notice that we cannot evaluate $f'(x)/g'(x)$ directly when $a = \pm\infty$. The next example shows how this version of l'Hopital's rule is used.

Example 5   Calculate $\lim\limits_{x \to \infty} \dfrac{5x + e^{-x}}{7x}$.

Solution   Let $f(x) = 5x + e^{-x}$ and $g(x) = 7x$. Then $\lim\limits_{x \to \infty} f(x) = \lim\limits_{x \to \infty} g(x) = \infty$, and $f'(x) = 5 - e^{-x}$ and $g'(x) = 7$, so

$$\lim_{x \to \infty} \frac{5x + e^{-x}}{7x} = \lim_{x \to \infty} \frac{(5 - e^{-x})}{7} = \frac{5}{7}.$$

We can also use l'Hopital's rule to calculate some limits of the form $\lim\limits_{x\to\infty} f(x)g(x)$, providing we rewrite them appropriately.

Example 6   Calculate $\lim\limits_{x\to\infty} xe^{-x}$.

Solution   Since $\lim\limits_{x\to\infty} x = \infty$ and $\lim\limits_{x\to\infty} e^{-x} = 0$, we see that

$$xe^{-x} \to 0 \cdot \infty \qquad \text{as } x \to \infty.$$

Since $0 \cdot \infty$ is undefined, we rewrite the function xe^{-x} as

$$xe^{-x} = \frac{x}{e^x}.$$

Now we use l'Hopital's rule, since

$$xe^{-x} = \frac{x}{e^x} \to \frac{\infty}{\infty} \qquad \text{as } x \to \infty.$$

Taking $f(x) = x$ and $g(x) = e^x$ gives $f'(x) = 1$ and $g'(x) = e^x$, so

$$\lim_{x\to\infty} xe^{-x} = \lim_{x\to\infty} \frac{x}{e^x} = \lim_{x\to\infty} \frac{1}{e^x} = 0.$$

Dominance: Powers, Polynomials, Exponentials, and Logarithms

In Chapter 1, we saw that some functions were much larger than others as $x \to \infty$. We say that g *dominates* f as $x \to \infty$ if $\lim\limits_{x\to\infty} \dfrac{f(x)}{g(x)} = 0$. L'Hopital's rule gives us an easy way of checking this.

Example 7   Show that $x^{1/2}$ dominates $\ln x$ as $x \to \infty$.

Solution   We apply l'Hopital's rule to $(\ln x)/x^{1/2}$:

$$\lim_{x\to\infty} \frac{\ln x}{x^{1/2}} = \lim_{x\to\infty} \frac{1/x}{\frac{1}{2}x^{-1/2}}.$$

To evaluate this limit, we simplify and get

$$\lim_{x\to\infty} \frac{1/x}{\frac{1}{2}x^{-1/2}} = \lim_{x\to\infty} \frac{2x^{1/2}}{x} = \lim_{x\to\infty} \frac{2}{x^{1/2}} = 0.$$

Therefore we have shown that

$$\lim_{x\to\infty} \frac{\ln x}{x^{1/2}} = 0,$$

which tells us that $x^{1/2}$ dominates $\ln x$ as $x \to \infty$.

Example 8   Show that any exponential function of the form e^{kx} (with $k > 0$) dominates any power function of the form Ax^p (with A and p positive) as $x \to \infty$.

Solution   We apply l'Hopital's rule repeatedly to Ax^p/e^{kx}:

$$\lim_{x\to\infty} \frac{Ax^p}{e^{kx}} = \lim_{x\to\infty} \frac{Apx^{p-1}}{ke^{kx}} = \lim_{x\to\infty} \frac{Ap(p-1)x^{p-2}}{k^2 e^{kx}} = \cdots$$

Keep applying l'Hopital's rule until the power of x is no longer positive. Then the limit of the numerator must be a finite number, while the limit of the denominator must be ∞. Therefore we have

$$\lim_{x\to\infty} \frac{Ax^p}{e^{kx}} = 0,$$

so e^{kx} dominates Ax^p.

Problems for Section 4.8

1. What is the tangent line approximation to $1/x$ near $x = 1$?

2. Show that $1 - x/2$ is the tangent line approximation to $1/\sqrt{1+x}$ near $x = 0$.

3. Show that $e^{-x} \approx 1 - x$ near $x = 0$.

4. Interpret the linear approximation $e^{rt} \approx 1 + rt$ in terms of interest in a bank account if r is the continuous annual interest rate and t is the time in years.

5. (a) Show that $1 + kx$ is the local linearization of $(1 + x)^k$ near $x = 0$.
   (b) Someone claims that the square root of 1.1 is about 1.05. Without using a calculator, do you think that this estimate is about right?
   (c) Is the actual number above or below 1.05?

For Problems 6–9, find the sign of $\lim\limits_{x \to a} \dfrac{f(x)}{g(x)}$ from the figure.

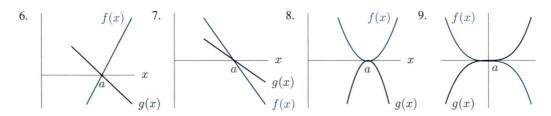

10. Evaluate $\lim\limits_{x \to 0^+} x \ln x$. [Hint: Write $x \ln x = \dfrac{\ln x}{1/x}$.]

11. Based on your knowledge of the behavior of the numerator and denominator, first predict the value of the following limits, and then find each limit using l'Hopital's rule.

   (a) $\lim\limits_{x \to 0} \dfrac{\sin x}{x^2}$
   (b) $\lim\limits_{x \to 0} \dfrac{\sin^2 x}{x}$
   (c) $\lim\limits_{x \to 0} \dfrac{\sin x}{x^{1/3}}$
   (d) $\lim\limits_{x \to 0} \dfrac{(\sin x)^{1/3}}{x}$

12. (a) What is the slope of $f(x) = \sin(3x)$ at $x = 0$?
   (b) What is the slope of $g(x) = 5x$ at $x = 0$?
   (c) Use the results of parts (a) and (b) to calculate $\lim\limits_{x \to 0} \dfrac{\sin(3x)}{5x}$.

In Problems 13–16, which function dominates as $x \to \infty$?

13. x^5 and $0.1x^7$
14. $0.01x^3$ and $50x^2$
15. $\ln(x + 3)$ and $x^{0.2}$
16. x^{10} and $e^{0.1x}$

17. The functions f and g and their tangent lines at $(4, 0)$ are shown in Figure 4.34. Find $\lim\limits_{x \to 4} \dfrac{f(x)}{g(x)}$.

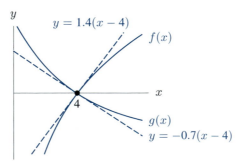

Figure 4.34

18. Explain why l'Hopital's rule cannot be used to calculate each of the following. Then evaluate the limit if it exists.

  (a) $\displaystyle\lim_{x \to 1} \frac{\sin(2x)}{x}$ (b) $\displaystyle\lim_{x \to 0} \frac{\cos x}{x}$ (c) $\displaystyle\lim_{x \to \infty} \frac{e^{-x}}{\sin x}$

19. Find the horizontal asymptote of $f(x) = \dfrac{2x^3 + 5x^2}{3x^3 - 1}$.

20. Multiply the local linearization of e^x near $x = 0$ by itself to obtain an approximation for e^{2x}. Compare this with the actual local linearization of e^{2x}. Explain why these two approximations are consistent, and discuss which one is more accurate.

21. (a) Show that $1 - x$ is the local linearization of $\dfrac{1}{1 + x}$ near $x = 0$.
  (b) From your answer to part (a), show that near $x = 0$,

$$\frac{1}{1 + x^2} \approx 1 - x^2.$$

  (c) Without differentiating, what do you think the derivative of $\dfrac{1}{1 + x^2}$ is at $x = 0$?

22. From the local linearizations of e^x and $\sin x$ near $x = 0$, write down the local linearization of the function $e^x \sin x$. From this result, write down the derivative of $e^x \sin x$ at $x = 0$. Using this technique, write down the derivative of $e^x \sin x / (1 + x)$ at $x = 0$.

23. Use local linearization to derive the product rule,

$$[f(x)g(x)]' = f'(x)g(x) + f(x)g'(x).$$

  [Hint: Use the definition of the derivative and the local linearizations $f(x + h) \approx f(x) + f'(x)h$ and $g(x + h) \approx g(x) + g'(x)h$.]

24. Derive the chain rule using local linearization. [Hint: In other words, differentiate $f(g(x))$, using $g(x + h) \approx g(x) + g'(x)h$ and $f(z + k) \approx f(z) + f'(z)k$.]

CHAPTER SUMMARY

- **Derivatives of elementary functions**
  Powers, polynomials, rational functions, exponential functions, logarithms, trigonometric functions, inverse trigonometric functions
- **Derivatives of sums, differences, and constant multiples**
- **Product and quotient rules**
- **Chain rule**
  Differentiation of implicitly defined functions, inverse functions.
- **Working with the derivative**
  Tangent line approximation, local linearity, l'Hopital's rule

REVIEW PROBLEMS FOR CHAPTER FOUR

1. Find the slope of the curve $x^2 + 3y^2 = 7$ at the point $(2, -1)$.

2. Assume y is a differentiable function of x and that $y + \sin y + x^2 = 9$. Find dy/dx at the point $x = 3, y = 0$.

3. Suppose we are given the data in the table about the functions f and g and their derivatives. Find the following values:
    (a) $h(4)$ if $h(x) = f(g(x))$
    (b) $h'(4)$ if $h(x) = f(g(x))$
    (c) $h(4)$ if $h(x) = g(f(x))$
    (d) $h'(4)$ if $h(x) = g(f(x))$
    (e) $h'(4)$ if $h(x) = g(x)/f(x)$
    (f) $h'(4)$ if $h(x) = f(x)g(x)$

x	1	2	3	4
$f(x)$	3	2	1	4
$f'(x)$	1	4	2	3
$g(x)$	2	1	4	3
$g'(x)$	4	2	3	1

4. Given: $r(2) = 4$, $s(2) = 1$, $s(4) = 2$, $r'(2) = -1$, $s'(2) = 3$, $s'(4) = 3$. Compute the following derivatives, or state what additional information you would need to be able to compute the derivative.
    (a) $H'(2)$ if $H(x) = r(x) \cdot s(x)$
    (b) $H'(2)$ if $H(x) = \sqrt{r(x)}$
    (c) $H'(2)$ if $H(x) = r(s(x))$
    (d) $H'(2)$ if $H(x) = s(r(x))$

5. Imagine you are zooming in on the graphs of the following functions near the origin:

$$y = \arcsin x \qquad y = \sin x - \tan x \qquad y = x - \sin x \qquad y = \arctan x$$

$$y = \frac{\sin x}{1 + \sin x} \qquad y = \frac{x^2}{x^2 + 1} \qquad y = \frac{1 - \cos x}{\cos x} \qquad y = \frac{x}{x^2 + 1}$$

$$y = \frac{\sin x}{x} - 1 \qquad y = -x \ln x \qquad y = e^x - 1 \qquad y = x^{10} + \sqrt[10]{x}$$

$$y = \frac{x}{x + 1}$$

Which of them look the same? Group together those functions which become indistinguishable, and give the equation of the line they look like. [Note: $(\sin x)/x - 1$ and $-x \ln x$ never quite make it to the origin.]

6. The graphs of $\sin x$ and $\cos x$ intersect once between 0 and $\pi/2$. What is the angle between the two curves at the point where they intersect? (You need to think about how the angle between two curves should be defined.)

7. The acceleration due to gravity, g, at a distance r from the center of the earth is given by

$$g = \frac{GM}{r^2},$$

where M is the mass of the earth and G is a constant.
    (a) Find $\frac{dg}{dr}$.
    (b) What is the practical interpretation (in terms of acceleration) of $\frac{dg}{dr}$? Why would you expect it to be negative?
    (c) You are told that $M = 6 \times 10^{24}$ and $G = 6.67 \times 10^{-20}$ where M is in kilograms and r in kilometers. What is the value of $\frac{dg}{dr}$ at the surface of the earth ($r = 6400$ km) ?
    (d) What does this tell you about whether or not it is reasonable to assume g is constant near the surface of the earth?

8. In 1990, the population of Mexico was about 84 million and growing at 2.6% annually, while the population of the US was about 250 million and growing at 0.7% annually. Which population was growing faster, if we measure growth rates in people/year? Explain your answer.

9. Suppose the distance, s, of a moving body from a fixed point is given as a function of time, t, by $s = 20e^{t/2}$.
    (a) Find the velocity, v, of the body as a function of t.

(b)   Find a relationship between v and s, then show that s satisfies the differential equation $s' = \frac{1}{2}s$.

10.   Air pressure at sea level is 30 inches of mercury. At an altitude of h feet above sea level, the air pressure, P, in inches of mercury, is given by

$$P = 30e^{-3.23 \times 10^{-5}h}$$

(a)   Sketch a rough graph of P against h.
(b)   Find the equation of the tangent line to the graph at $h = 0$.
(c)   A rough rule of thumb used by travelers is that air pressure drops about 1 inch for every 1000-foot increase in height above sea level. Write an approximate formula for the air pressure given by this rule of thumb.
(d)   What is the relation between your answers to parts (b) and (c)? Explain why the rule of thumb works.
(e)   Are the predictions made by the rule of thumb too large or too small? Why?

11.   Suppose the depth of the water, y, in meters, in the Bay of Fundy, Canada, is given as a function of time, t, in hours after midnight, by the function

$$y = 10 + 7.5\cos(0.507t).$$

How quickly is the tide rising or falling (in meters/hour) at each of the following times?
(a)   6:00 am   (b)   9:00 am   (c)   Noon   (d)   6:00 pm

12.   A yam is put in a hot oven, maintained at a constant temperature $200°$C. Suppose that at time $t = 30$ minutes, the temperature T of the yam is $120°$ and is increasing at an (instantaneous) rate of $2°$/min. Newton's law of cooling (or, in our case, warming) implies that the temperature at time t will be given by a formula of the form

$$T(t) = 200 - ae^{-bt}.$$

Find a and b.

13.   An object is oscillating at the end of a spring. Its position, in centimeters, relative to a fixed point, is given as a function of time, t, in seconds, by

$$y = y_0 \cos(2\pi\omega t), \quad \text{with } \omega \text{ a constant.}$$

(a)   Find an expression for the velocity and acceleration of the object.
(b)   How do the amplitudes of the position, velocity, and acceleration functions compare? How do the periods of these functions compare?
(c)   Show that the function y satisfies the differential equation

$$\frac{d^2y}{dt^2} + 4\pi^2\omega^2 y = 0.$$

14.   Suppose the total number of people, N, who have contracted a disease by a time t days after its outbreak is given by

$$N = \frac{1{,}000{,}000}{1 + 5{,}000e^{-0.1t}}.$$

(a)   In the long run, how many people will have had the disease?
(b)   Is there any day on which more than a million people fall sick? Half a million? Quarter of a million? (Note: You do not have to try to find out on what days these things happen.)

15.   A spherical cell is growing at a constant rate of 400 μm^3/day ($1\ \mu\text{m} = 10^{-6}$ m). At what rate is its radius increasing when the radius is 10 μm?

16.   When the growth of a spherical cell depends on the flow of nutrients through the surface, it is reasonable to assume that the growth rate, dV/dt, is proportional to the surface area, S. Assume that for a particular cell $dV/dt = \frac{1}{3} \cdot S$. At what rate is its radius r increasing?

17.   A radio navigation system used by aircraft gives a cockpit readout of the distance, s, in miles, between a fixed ground station and the aircraft. The system also gives a readout of the instantaneous rate of change, ds/dt, of this distance in miles/hour. An aircraft on a straight flight path at a constant altitude of 10,560 feet (2 miles) has passed directly over the ground station and is now flying away from it. What is the speed of this aircraft along its constant altitude flight path when the cockpit readouts are $s = 4.6$ miles and $ds/dt = 210$ miles/hour?

Problems 18–19 involve Boyle's Law which states that for a fixed quantity of gas at a constant temperature, the pressure, P, and the volume, V, are inversely related. Thus, for some constant k

$$PV = k.$$

18. Suppose a fixed quantity of gas is allowed to expand at constant temperature. Find the rate of change of pressure with respect to volume.

19. Suppose a certain quantity of gas occupies 10 cm³ at a pressure of 2 atmospheres. Suppose the pressure is increased, while keeping the temperature constant.

(a) Does the volume increase or decrease?

(b) If the pressure is increasing at a rate of 0.05 atmospheres/minute at a time when the pressure is 2 atmospheres, find the rate at which the volume is changing at that moment. What are the units of your answer?

20. Find the nth derivative of each of the following functions:

(a) $\ln x$ (b) xe^x (c) $e^x \cos x$

PROJECTS

1. Rule of 70

The "Rule of 70" is a rule of thumb to estimate how long it takes money in a bank to double. Suppose the money is in an account earning $i\%$ annual interest, compounded yearly. The Rule of 70 says that the time it takes the amount of money to double is approximately $70/i$ years, assuming i is small. Find the local linearization of $\ln(1 + x)$, and use it to explain why this rule works.

2. Newton's Method

Read about how to find roots using bisection and Newton's method.[3]

(a) What is the smallest positive zero of the function $f(x) = \sin x$? Apply Newton's method, with initial guess $x_0 = 3$, to see how fast it converges to $\pi = 3.1415926536\ldots$.

(i) Compute the first two approximations, x_1 and x_2; compare x_2 with π.

(ii) Newton's method works very well here. Explain why. To do this, you will have to outline the basic idea behind Newton's method.

(iii) Estimate the location of the zero using bisection, starting with the interval $[3, 4]$. How does bisection compare to Newton's method in terms of accuracy?

(b) Newton's method can be very sensitive to your initial estimate, x_0. For example, consider finding a zero of $f(x) = \sin x - \frac{2}{3}x$.

(i) Use Newton's method with the following initial estimates to find a zero:

$$x_0 = 0.904, \quad x_0 = 0.905, \quad x_0 = 0.906.$$

(ii) What happens?

---

[3] See, for example, D. Hughes-Hallett, A. M. Gleason, et al, *Calculus*, 2nd ed. (New York: John Wiley, 1998).

FOCUS ON PRACTICE

DIFFERENTIATION

Find derivatives for the functions in Problems 1–114. Assume $a, b, c,$ and k are constants.

1. $f(t) = 3t^2 - 4t + 1$

2. $y = 17x + 24x^{1/2}$

3. $g(x) = -\frac{1}{2}(x^5 + 2x - 9)$

4. $g(t) = \dfrac{t^3 + k}{t}$

5. $f(x) = 5x^4 + \dfrac{1}{x^2}$

6. $z = \dfrac{t^2 + 3t + 1}{t + 1}$

7. $y = \dfrac{e^{2x}}{x^2 + 1}$

8. $f(x) = \dfrac{x^2 + 3x + 2}{x + 1}$

9. $y = \left(\dfrac{x^2 + 2}{3}\right)^2$

10. $g(\theta) = \sin^2(2\theta) - \pi\theta$

11. $g(x) = \sin(2 - 3x).$

12. $R(x) = 10 - 3\cos(\pi x)$

13. $f(z) = \dfrac{z^2 + 1}{3z}$

14. $q(r) = \dfrac{3r}{5r + 2}$

15. $h(z) = \sqrt{\dfrac{\sin(2z)}{\cos(2z)}}$

16. $y = x \ln x - x + 2$

17. $j(x) = \ln(e^{ax} + b)$

18. $y = 2x(\ln x + \ln 2) - 2x + e$

19. $g(\theta) = \sin(\tan \theta)$

20. $w(x) = \tan(x^2)$

21. $f(x) = \sin(\sin x + \cos x)$

22. $j(x) = \cos\left(\sin^{-1} x\right)$

23. $k(\alpha) = \sin^5 \alpha \cos^3 \alpha$

24. $f(w) = \cos^2 w + \cos(w^2)$

25. $g(t) = \dfrac{4}{3 + \sqrt{t}}$

26. $g(t) = \dfrac{t - 4}{t + 4}$

27. $y = \dfrac{1}{e^{3x} + x^2}$

28. $h(w) = (w^4 - 2w)^5$

29. $q(\theta) = \sqrt{4\theta^2 - \sin^2(2\theta)}$

30. $g(t) = (t \cos t + \tan^3(t^5))^4$

31. $h(w) = w^3 \ln(10w)$

32. $f(x) = \ln(\sin x + \cos x)$

33. $g(x) = \arcsin(\sin \pi x)$

34. $r(t) = \arcsin(2t)$

35. $w(r) = \sqrt{r^4 + 1}$

36. $h(w) = -2w^{-3} + 3\sqrt{w}$

37. $h(x) = \sqrt{\dfrac{x^2 + 9}{x + 3}}$

38. $f(x) = \sqrt{\dfrac{1 - \sin x}{1 - \cos x}}$

39. $T(u) = \arctan\left(\dfrac{u}{1 + u}\right)$

40. $w = 2^{-4z} \sin(\pi z)$

41. $v(t) = t^2 e^{-ct}$

42. $f(x) = \pi^x + x^\pi$

43. $f(x) = \dfrac{x}{1 + \ln x}$

44. $G(x) = \dfrac{\sin^2 x + 1}{\cos^2 x + 1}$

45. $a(t) = \ln\left(\dfrac{1 - \cos t}{1 + \cos t}\right)^4$

46. $f(x) = e^{\ln(kx)}$

47. $R(\theta) = e^{\sin(3\theta)}$

48. $f(x) = e^\pi + \pi^x$

49. $y = \pi^{(x+2)}$

50. $g(x) = e^{\pi x}$

51. $g(\theta) = e^{\sin \theta}$

52. $f(\theta) = 2^{-\theta}$

53. $f(x) = e^{2x}\left(x^2 + 5^x\right)$

54. $h(x) = 2^{e^{3x}}$

55. $h(t) = \dfrac{4 - t}{4 + t}$

56. $r(y) = \dfrac{y}{\cos y + a}$

57. $h(z) = \left(\dfrac{b}{a + z^2}\right)^4$

58. $p(t) = e^{4t+2}$

59. $h(z) = (\ln 2)^z$

60. $j(x) = \dfrac{x^3}{a} + \dfrac{a}{b}x^2 - cx + k$

61. $f(x) = \cos(\arctan 3x)$

62. $f(x) = (3x^2 + \pi)(e^x - 4)$

63. $g(t) = e^{(1+3t)^2}$

64. $f(z) = \dfrac{z^2 + 1}{\sqrt{z}}$

65. $h(r) = \dfrac{r^2}{2r + 1}$

66. $g(x) = 2x - \dfrac{1}{\sqrt[3]{x}} + 3^x - e$

67. $f(t) = 2te^t - \dfrac{1}{\sqrt{t}}$

68. $w = \dfrac{5 - 3z}{5 + 3z}$

69. $g(w) = \dfrac{1}{2^w + e^w}$

70. $f(y) = \ln\left(\ln(2y^3)\right)$

71. $f(x) = \dfrac{x^3}{9}(3\ln x - 1)$

72. $g(x) = x^k + k^x$

73. $r(\theta) = \sin\left((3\theta - \pi)^2\right)$

74. $s(\theta) = \sin^2(3\theta - \pi)$

75. $h(t) = \ln\left(e^{-t} - t\right)$

76. $p(\theta) = \dfrac{\sin(5 - \theta)}{\theta^2}$

77. $w(\theta) = \dfrac{\theta}{\sin^2\theta}$

78. $g(x) = \dfrac{x^2 + \sqrt{x} + 1}{x^{3/2}}$

79. $s(x) = \arctan(2 - x)$

80. $r(\theta) = e^{(e^\theta + e^{-\theta})}$

81. $m(n) = \sin(e^n)$

82. $k(\alpha) = e^{\tan(\sin\alpha)}$

83. $g(t) = t\cos(\sqrt{t}e^t)$

84. $f(r) = (\tan 2 + \tan r)^e$

85. $y = e^{-\pi} + \pi^{-e}$

86. $y = \left(x^2 + 5\right)^3 \left(3x^3 - 2\right)^2$

87. $h(x) = xe^{\tan x}$

88. $y = e^{2x}\sin^2(3x)$

89. $g(x) = \tan^{-1}(3x^2 + 1)$

90. $y = 2^{\sin x}\cos x$

91. $h(x) = \ln e^{ax}$

92. $k(x) = \ln e^{ax} + \ln b$

93. $f(\theta) = e^{k\theta} - 1$

94. $N(\theta) = \tan(\arctan(k\theta))$

95. $f(t) = e^{-4kt}\sin t$

96. $f(x) = a^{5x}$

97. $f(x) = \dfrac{a^2 - x^2}{a^2 + x^2}$

98. $w(r) = \dfrac{ar^2}{b + r^3}$

99. $f(s) = \dfrac{a^2 - s^2}{\sqrt{a^2 + s^2}}$

100. $h(t) = e^{kt}(\sin at + \cos bt)$

101. $H(t) = (at^2 + b)e^{-ct}$

102. $g(\theta) = \sqrt{a^2 - \sin^2\theta}$

103. $y = \arctan\left(\dfrac{2}{x}\right)$

104. $r(t) = \ln\left(\sin\left(\dfrac{t}{k}\right)\right)$

105. $g(u) = \dfrac{e^{au}}{a^2 + b^2}$

106. $g(w) = \dfrac{5}{(a^2 - w^2)^2}$

107. $y = \dfrac{e^x - e^{-x}}{e^x + e^{-x}}$

108. $y = \dfrac{e^{ax} - e^{-ax}}{e^{ax} + e^{-ax}}$

109. $f(x) = (2 - 4x - 3x^2)(6x^e - 3\pi)$

110. $f(t) = (\sin(2t) - \cos(3t))^4$

111. $s(y) = \sqrt[3]{(\cos^2 y + 3 + \sin^2 y)}$

112. $f(x) = (4 - x^2 + 2x^3)(6 - 4x + x^7)$

113. $h(x) = \left(\dfrac{1}{x} - \dfrac{1}{x^2}\right)(2x^3 + 4)$

114. $f(z) = \sqrt{5z} + 5\sqrt{z} + \dfrac{5}{\sqrt{z}} - \sqrt{\dfrac{5}{z}} + \sqrt{5}$

115. If $g(2) = 3$ and $g'(2) = -4$, find $f'(2)$ for the following:

    (a) $f(x) = x^2 - 4g(x)$

    (b) $f(x) = \dfrac{x}{g(x)}$

    (c) $f(x) = x^2 g(x)$

    (d) $f(x) = (g(x))^2$

    (e) $f(x) = x\sin(g(x))$

    (f) $f(x) = x^2 \ln(g(x))$

116. For parts (a)–(f) of Problem 115, determine the equation of the line tangent to f at $x = 2$.

For Problems 117–122, assume that y is a differentiable function of x and find dy/dx.

117. $xy - x - 3y - 4 = 0$

118. $6x^2 + 4y^2 = 36$

119. $ax^2 - by^2 = c^2$

120. $x^2 y - 2y + 5 = 0$

121. $x^3 + y^3 - 4x^2 y = 0$

122. $\sin(ay) + \cos(bx) = xy$

CHAPTER FIVE

USING THE DERIVATIVE

In Chapter 2, we introduced the derivative and some of its interpretations. In Chapter 4, we saw how to differentiate all of the standard functions, including powers, exponentials, logarithms, and trigonometric functions. Now we will use first and second derivatives to analyze the behavior of families of functions and to solve optimization problems.

5.1 USING FIRST AND SECOND DERIVATIVES

What Derivatives Tell Us About a Function and its Graph

As we saw in Chapter 2, the connection between derivatives of a function and the function itself is given by the following:

- If $f' > 0$ on an interval, then f is increasing on that interval.
- If $f' < 0$ on an interval, then f is decreasing on that interval.
- If $f'' > 0$ on an interval, then the graph of f is concave up on that interval.
- If $f'' < 0$ on an interval, then the graph of f is concave down on that interval.

We can do more with these principles now than we could in Chapter 2 because we now have formulas for the derivatives of the elementary functions.

When we graph a function on a computer or calculator, we often see only part of the picture. Information given by the first and second derivatives can help identify regions with interesting behavior.

Example 1 Use a computer or calculator to sketch a useful graph of the function $f(x) = x^3 - 9x^2 - 48x + 52$.

Solution Since f is a cubic polynomial, we expect a graph that is roughly S-shaped. Graphing this function with $-10 \leq x \leq 10$, $-10 \leq y \leq 10$, gives the two nearly vertical lines in Figure 5.1. We know that there is more going on than this, but how do we know where to look?

Figure 5.1: Unhelpful graph of $f(x) = x^3 - 9x^2 - 48x + 52$

We use the derivative to determine where the function is increasing and where it is decreasing. The derivative of f is

$$f'(x) = 3x^2 - 18x - 48.$$

To find where $f' > 0$ or $f' < 0$, we first find where $f' = 0$, that is, where $3x^2 - 18x - 48 = 0$. Factoring, we get $3(x - 8)(x + 2) = 0$, so $x = -2$ or $x = 8$. Since $f' = 0$ *only* at $x = -2$ and $x = 8$, and since f' is continuous, f' cannot change sign on any of the three intervals $x < -2$, or $-2 < x < 8$, or $8 < x$. How can we tell the sign of f' on each of these intervals? The easiest way is to pick a point and substitute into f'. For example, since $f'(-3) = 33 > 0$, we know f' is positive for $x < -2$, so f is increasing for $x < -2$. Similarly, since $f'(0) = -48$ and $f'(10) = 72$, we know that f decreases between $x = -2$ and $x = 8$ and increases for $x > 8$. Summarizing:

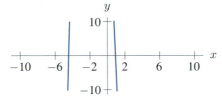

We find that $f(-2) = 104$ and $f(8) = -396$. Hence on the interval $-2 < x < 8$ the function decreases from a high of 104 to a low of -396. (Now we see why not much showed up in our first calculator graph.) One more point on the graph is easy to get: the y intercept, $f(0) = 52$. With just these three points we can get a much more helpful graph. By setting the plotting window to

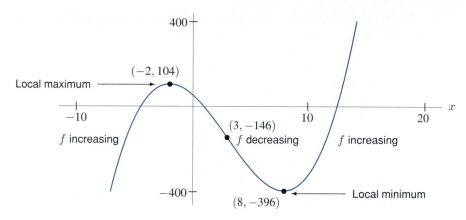

Figure 5.2: Useful graph of $f(x) = x^3 - 9x^2 - 48x + 52$

$-10 \le x \le 20$ and $-400 \le y \le 400$, we get Figure 5.2, which gives much more insight into the behavior of $f(x)$ than the graph in Figure 5.1.

In Figure 5.2, we see that part of the graph is concave up and part is concave down. We can use the second derivative to analyze concavity. We have

$$f''(x) = 6x - 18.$$

So, $f''(x) < 0$ when $x < 3$ and $f''(x) > 0$ when $x > 3$, and so the graph of f is concave down for $x < 3$ and concave up for $x > 3$. At $x = 3$, we have $f''(x) = 0$. Summarizing:

$$
\begin{array}{c|c}
 & x = 3 \\
f \text{ concave down } \cap & f \text{ concave up } \cup \\
\hline
f'' < 0 & f'' = 0 \quad f'' > 0
\end{array}
$$

Local Maxima and Minima

We are often interested in points such as those marked local maximum and local minimum in Figure 5.2. We have the following definition:

> Suppose p is a point in the domain of f:
> - f has a **local minimum** at p if $f(p)$ is less than or equal to the values of f for points near p.
> - f has a **local maximum** at p if $f(p)$ is greater than or equal to the values of f for points near p.

We use the adjective "local" because we are describing only what happens near p.

How Do We Detect a Local Maximum or Minimum?

In the preceding example, the points $x = -2$ and $x = 8$, where $f'(x) = 0$, played a key role in leading us to local maxima and minima. We give a name to such points:

> For any function f, a point p in the domain of f where $f'(p) = 0$ or $f'(p)$ is undefined is called a **critical point** of the function. In addition, the point $\big(p, f(p)\big)$ on the graph of f is also called a critical point. A **critical value** of f is the value, $f(p)$, of the function at a critical point, p.

Notice that "critical point of f" can refer either to points in the domain of f or to points on the graph of f. You will know which meaning is intended from the context.

Geometrically, at a critical point where $f'(p) = 0$, the line tangent to the graph of f at p is horizontal. At a critical point where $f'(p)$ is undefined, there is no horizontal tangent to the graph— there's either a vertical tangent or no tangent at all. (For example, $x = 0$ is a critical point for the absolute value function $f(x) = |x|$.) However, most of the functions we will work with will be differentiable everywhere, and therefore most of our critical points will be of the $f'(p) = 0$ variety.

The critical points divide the domain of f into intervals on which the sign of the derivative remains the same, either positive or negative. Therefore, if f is defined on the interval between two successive critical points, its graph cannot change direction on that interval; it is either increasing or decreasing. We have the following result, which is equivalent to the theorem proved on page 287.

Theorem: If a continuous function f has a local maximum or minimum at p, and if p is not an endpoint of the domain, then p is a critical point.

Warning! The sign of f' doesn't *have* to change at a critical point. Consider $f(x) = x^3$, which has a critical point at $x = 0$. (See Figure 5.3.) The derivative , $f'(x) = 3x^2$, is positive on both sides of $x = 0$, so f increases on both sides of $x = 0$, and there is neither a local maximum nor a local minimum at $x = 0$. In other words, not every critical point is a local maximum or local minimum.

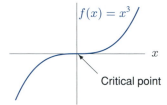

Figure 5.3: Critical point which is not a local maximum or minimum

Testing For Local Maxima and Minima

If f' has different signs on either side of a critical point p, with $f'(p) = 0$, then the graph changes direction at p and looks like one of those in Figure 5.4. So we have the following criterion:

The First-Derivative Test for Local Maxima and Minima

Suppose p is a critical point of a continuous function f.
- If f' changes from negative to positive at p, then f has a local minimum at p.
- If f' changes from positive to negative at p, then f has a local maximum at p.

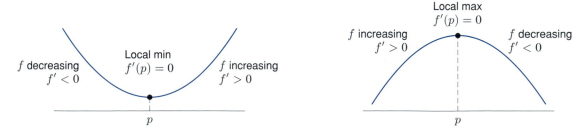

Figure 5.4: Changes in direction at a critical point, p: Local maxima and minima

---

Example 2   Use a graph of the function $f(x) = \dfrac{1}{x(x-1)}$ to observe its local maxima and minima. Confirm your observation analytically.

Solution The graph in Figure 5.5 suggests that this function has no local minima but that there is a local maximum at about $x = \frac{1}{2}$. Confirming this analytically means using the formula for the derivative to show that what we expect is true. Since $f(x) = (x^2 - x)^{-1}$, we have

$$f'(x) = -1(x^2 - x)^{-2}(2x - 1) = -\frac{2x - 1}{(x^2 - x)^2}.$$

So $f'(x) = 0$ where $2x - 1 = 0$. Thus, the only critical point in the domain of f is $x = \frac{1}{2}$.

Furthermore, $f'(x) > 0$ where $0 < x < 1/2$, and $f'(x) < 0$ where $1/2 < x < 1$. Thus, f increases for $0 < x < 1/2$ and decreases for $1/2 < x < 1$. According to the first derivative test, the critical point $x = 1/2$ is a local maximum.

For $-\infty < x < 0$ or $1 < x < \infty$, there are no critical points and no local maxima or minima. Although $1/(x(x - 1)) \to 0$ both as $x \to \infty$ and as $x \to -\infty$, we don't say 0 is a local minimum because $1/\left(x(x - 1)\right)$ never actually *equals* 0.

Notice that although $f' > 0$ everywhere that it is defined for $x < \frac{1}{2}$, the function f is not increasing throughout this interval. The problem is that f and f' are not defined at $x = 0$, so we cannot conclude that f is increasing when $x < 1/2$.

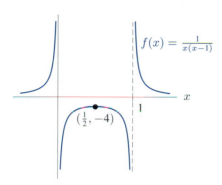

Figure 5.5: Find local maxima and minima

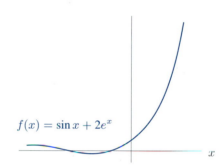

Figure 5.6: Explain the absence of local maxima and minima for $x \geq 0$

Example 3 The graph of $f(x) = \sin x + 2e^x$ is in Figure 5.6. Using the derivative, explain why there are no local maxima or minima for $x \geq 0$.

Solution Local maxima and minima can occur only at critical points. Now $f'(x) = \cos x + 2e^x$, which is defined for all x. We know $\cos x$ is always between -1 and 1, and $2e^x \geq 2$ for $x \geq 0$, so $f'(x)$ cannot be 0 for any $x \geq 0$. Therefore there are no local maxima or minima there.

The Second-Derivative Test for Local Maxima and Minima

Knowing the concavity of a function can also be useful in testing if a critical point is a local maximum or a local minimum. Suppose p is a critical point of f, with $f'(p) = 0$, so that the graph of f has a horizontal tangent line at p. If the graph is concave up at p, then f has a local minimum at p. Likewise, if the graph is concave down, f has a local maximum. (See Figure 5.7.) This suggests:

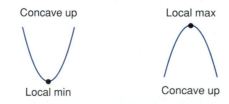

Figure 5.7: Local maxima and minima and concavity

> ### The Second-Derivative Test for Local Maxima and Minima
> - If $f'(p) = 0$ and $f''(p) > 0$ then f has a local minimum at p.
> - If $f'(p) = 0$ and $f''(p) < 0$ then f has a local maximum at p.
> - If $f'(p) = 0$ and $f''(p) = 0$ then   the test tells us nothing.

Example 4   Classify as local maxima or local minima the critical points of $f(x) = x^3 - 9x^2 - 48x + 52$.

Solution   As we saw in Example 1 on page 240,

$$f'(x) = 3x^2 - 18x - 48$$

and the critical points of f are $x = -2$ and $x = 8$. We have

$$f''(x) = 6x - 18.$$

Thus $f''(8) = 30 > 0$, so f has a local minimum at $x = 8$. Since $f''(-2) = -30 < 0$, f has a local maximum at $x = -2$.

Warning! The second-derivative test does not tell us anything if both $f'(p) = 0$ and $f''(p) = 0$. For example, if $f(x) = x^3$ and $g(x) = x^4$, both $f'(0) = g'(0) = 0$ and $f''(0) = g''(0) = 0$. The point $x = 0$ is a minimum for g but is neither a maximum nor a minimum for f. However, the first-derivative test is still useful. For example, g' changes sign from negative to positive at $x = 0$, so we know g has a local minimum there.

Concavity and Inflection Points

We have studied points where the slope changes sign, which led us to critical points. Now we look at points where the concavity changes.

> A point at which the graph of a function changes concavity is called an **inflection point** of f.

The words "inflection point of f" can refer either to a point in the domain of f or to a point on the graph of f. The context of the problem will tell you which is meant.

How Do We Locate an Inflection Point?

Since the concavity changes at an inflection point, the sign of f'' changes there. It is positive on one side of the inflection point, and negative on the other; so at the inflection point, f'' is zero or undefined. (See Figure 5.8.)

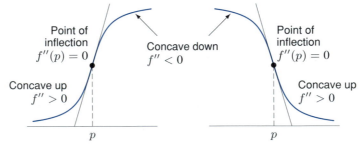

Figure 5.8: Change in concavity at p: Points of inflection are local maxima or minima of f'

Example 5   Find the critical and inflection points for $g(x) = xe^{-x}$ and sketch the graph of g for $x \geq 0$.

Solution   Taking derivatives and simplifying, we have

$$g'(x) = (1 - x)e^{-x} \quad \text{and} \quad g''(x) = (x - 2)e^{-x}.$$

So $x = 1$ is a critical point, and $g' > 0$ for $x < 1$ and $g' < 0$ for $x > 1$. Hence g increases to a local maximum at $x = 1$ and then decreases. Also, $g(x) \rightarrow 0$ as $x \rightarrow \infty$. There is an inflection point at $x = 2$ since $g'' < 0$ for $x < 2$ and $g'' > 0$ for $x > 2$. The graph is sketched in Figure 5.9.

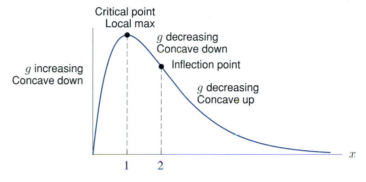

Figure 5.9: Graph of $g(x) = xe^{-x}$

Warning! Not every point x where $f''(x) = 0$ (or f'' is undefined) is an inflection point (just as not every point where $f' = 0$ is a local maximum or minimum). For instance $f(x) = x^4$ has $f''(x) = 12x^2$ so $f''(0) = 0$, but $f'' > 0$ when $x > 0$ and when $x < 0$, so there is *no* change in concavity at $x = 0$. See Figure 5.10.

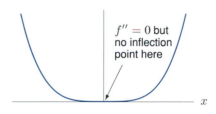

Figure 5.10: Graph of $f(x) = x^4$

Inflection Points and Local Maxima and Minima of the Derivative

Inflection points can also be interpreted in terms of first derivatives. If p is an inflection point, then $f''(p) = 0$ (or $f''(p)$ is undefined) and hence p is a critical point of the derivative function f'. If f' is continuous, this critical point is a local maximum or minimum of f', since f'' changes sign at p.

> A function f with a continuous derivative has an inflection point at p if either of the following conditions hold:
> - f' has a local minimum or a local maximum at p.
> - f'' changes sign at p.

Example 6   Graph $f(x) = x + \sin x$, and determine where f is increasing most rapidly, and least rapidly.

Solution   The graph of f is in Figure 5.11 and the graph of $f'(x) = 1 + \cos x$ is in Figure 5.12.

Where is f increasing most rapidly? At the points $x = \ldots, -2\pi, 0, 2\pi, 4\pi, 6\pi, \ldots$, because these points are local maxima for f', and f' has the same value at each of them. Likewise f is increasing least rapidly at the points $x = \ldots, -3\pi, -\pi, \pi, 3\pi, 5\pi, \ldots$, since these points are local minima for f'. Notice that the points where f is increasing most rapidly and the points where it is increasing least rapidly are inflection points of f.

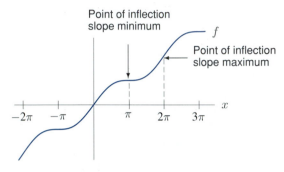

Figure 5.11: Graph of $f(x) = x + \sin x$

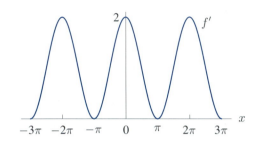

Figure 5.12: Graph of $f'(x) = 1 + \cos x$

Example 7   Suppose that water is being poured into the vase in Figure 5.13 at a constant rate, measured in volume per unit time. Graph $y = f(t)$, the depth of the water against time, t. Explain the concavity and indicate the inflection points.

Solution   At first the water level, y, rises quite slowly because the base of the vase is wide, and so it takes a lot of water to make the depth increase. However, as the vase narrows, the rate at which the water is rising increases. This means that initially y is increasing at an increasing rate and the graph is concave up. The rate of increase in the water level is at a maximum when the water reaches the middle of the vase, where the diameter is smallest; this is an inflection point. After that, the rate at which y increases starts to decrease again, and so the graph is concave down. (See Figure 5.14.)

Figure 5.13: A vase

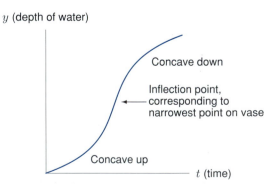

Figure 5.14: Graph of depth of water in the vase, y, against time, t

Problems for Section 5.1

1. Indicate all critical points on the graph of f in Figure 5.15 and determine which correspond to local maxima of f, which to local minima, and which to neither.

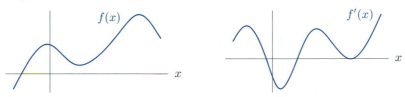

Figure 5.15   Figure 5.16

2. Indicate on the graph of the derivative function f' in Figure 5.16 the x-values that are critical points of the function f itself. At which critical points does f have local maxima, local minima, or neither?

3. Suppose f has a continuous derivative whose values are given in the following table.

    (a) Estimate the x-coordinates of critical points of f for $0 \le x \le 10$.
    (b) For each critical point, indicate if it is a local maximum of f, local minimum, or neither.

x	0	1	2	3	4	5	6	7	8	9	10
$f'(x)$	5	2	1	-2	-5	-3	-1	2	3	1	-1

4. Sketch graphs of two continuous functions f and g, each of which has exactly five critical points, the points A–E in Figure 5.17, and which satisfy the following conditions:

    (a) $\lim\limits_{x \to -\infty} f(x) = \infty$ and
         $\lim\limits_{x \to \infty} f(x) = \infty$

    (b) $\lim\limits_{x \to -\infty} g(x) = -\infty$ and
         $\lim\limits_{x \to \infty} g(x) = 0$

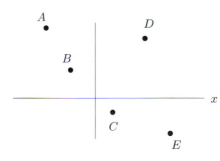

Figure 5.17

5. Indicate on the graph of the derivative f' in Figure 5.18 the x-values that are inflection points of the function f itself.

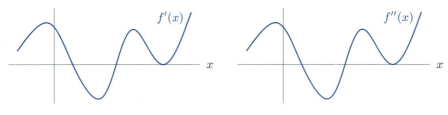

Figure 5.18   Figure 5.19

6. Indicate on the graph of the second derivative f'' in Figure 5.19 the x-values that are inflection points of the function f itself.

7. Find the inflection points of $f(x) = x^4 + x^3 - 3x^2 + 2$.

Using a calculator or computer, sketch the graph of each of the functions in Problems 8–15. Describe briefly in words the interesting features of the graph including the location of the critical points and where the function is increasing/decreasing. Then use the derivative and algebra to explain the shape of the graph.

8. $f(x) = x^3 - 6x + 1$ 9. $f(x) = x^3 + 6x + 1$ 10. $f(x) = 3x^5 - 5x^3$

11. $f(x) = x + 2\sin x$ 12. $f(x) = e^x - 10x$ 13. $f(x) = e^x + \sin x$

14. $f(x) = xe^{-x^2}$ 15. $f(x) = x\ln x, \quad x > 0$

16. Use the graphs you drew in Problems 8–15 to describe in words the concavity of each graph, including approximate x-coordinates for all points of inflection. Then use f'' and algebra to explain what you see.

17. You might think the graph of $f(x) = x^2 + \cos x$ should look like a parabola with some waves on it. Sketch the actual graph of $f(x)$ using a calculator or computer. Explain what you see using $f''(x)$.

18. Find values of a and b so that the function $f(x) = x^2 + ax + b$ has a local minimum at the point $(6, -5)$.

19. Find the value of a so that the function $f(x) = xe^{ax}$ has a critical point at $x = 3$.

20. Choose the constants a and b in the function

$$f(x) = axe^{bx}$$

such that $f(\frac{1}{3}) = 1$ and the function has a local maximum at $x = \frac{1}{3}$.

21. Assume the function f is differentiable everywhere and has just one critical point, at $x = 3$. In parts (a) through (d), you are given additional conditions. In each case decide whether $x = 3$ is a local maximum, a local minimum, or neither. Explain your reasoning. Also sketch possible graphs for all four cases.
    (a) $f'(1) = 3$ and $f'(5) = -1$ (c) $f(1) = 1, f(2) = 2, f(4) = 4, f(5) = 5$
    (b) $\lim_{x \to \infty} f(x) = \infty$ and   (d) $f'(2) = -1, f(3) = 1, \lim_{x \to \infty} f(x) = 3$
        $\lim_{x \to -\infty} f(x) = \infty$

22. Assume that the polynomial f has exactly two local maxima and one local minimum, and that these are the only critical points of f.

    (a) Sketch a possible graph of f.
    (b) What is the largest number of zeros f could have?
    (c) What is the least number of zeros f could have?
    (d) What is the least number of inflection points f could have?
    (e) What is the smallest degree f could have?
    (f) Find a possible formula for $f(x)$.

23. (a) Water is flowing at a constant rate into a cylindrical container standing vertically. Sketch a graph showing the depth of water against time.
    (b) Water is flowing at a constant rate into a cone-shaped container standing on its point. Sketch a graph showing the depth of the water against time.

24. If water is flowing at a constant rate (i.e., constant volume per unit time) into the Grecian urn in Figure 5.20, sketch a graph of the depth of the water against time. Mark on the graph the time at which the water reaches the widest point of the urn.

Figure 5.20

Figure 5.21

25. If water is flowing at a constant rate (i.e., constant volume per unit time) into the vase in Figure 5.21, sketch a graph of the depth of the water against time. Mark on the graph the time at which the water reaches the corner of the vase.

26. The rabbit population on a small Pacific island is approximated by

$$P(t) = \frac{2000}{1 + e^{(5.3 - 0.4t)}}$$

with t measured in years since 1774, when Captain James Cook left 10 rabbits on the island. Using a calculator or computer:

  (a) Graph P. Does the population level off?
  (b) Estimate when the rabbit population grew most rapidly. How large was the population at that time?
  (c) What natural causes could lead to the shape of the graph of P?

For Problems 27–32, sketch a possible graph of $y = f(x)$, using the given information about the derivatives $y' = f'(x)$ and $y'' = f''(x)$. Assume that the function is defined and continuous for all real x.

27.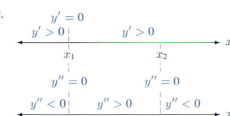

28.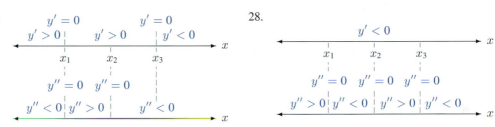

29.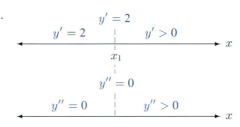

30.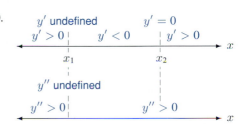

31.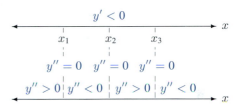

32.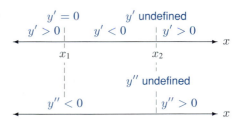

33. Let f be a function with $f(x) > 0$ for all x. Set $g = 1/f$.

  (a) If f is increasing in an interval around x_0, what about g?
  (b) If f has a local maximum at x_1, what about g?
  (c) If f is concave down at x_2, what about g?

34. What happens to concavity when functions are added?

  (a) If $f(x)$ and $g(x)$ are concave up for all x, is $f(x) + g(x)$ concave up for all x?
  (b) If $f(x)$ is concave up for all x and $g(x)$ is concave down for all x, what can you say about the concavity of $f(x) + g(x)$? For example, what happens if $f(x)$ and $g(x)$ are both polynomials of degree 2?
  (c) If $f(x)$ is concave up for all x and $g(x)$ is concave down for all x, is it possible for $f(x) + g(x)$ to change concavity infinitely often?

5.2 FAMILIES OF CURVES: A QUALITATIVE STUDY

We saw in Chapter 1 that knowledge of one function can provide knowledge of the graphs of many others. The shape of the graph of $y = x^2$ also tells us, indirectly, about the graphs of $y = x^2 + 2$, $y = (x + 2)^2$, $y = 2x^2$, and countless other functions. We say that all functions of the form $y = a(x + b)^2 + c$ form a *family of functions*; their graphs are like that of $y = x^2$, except for shifts and stretches determined by the values of $a, b,$ and c. The constants a, b, c are called *parameters*. Different values of the parameters give different members of the family.

One reason for studying families of functions is their use in mathematical modeling. Confronted with the problem of modeling some phenomenon, a crucial first step involves recognizing families of functions which might fit the available data.

Motion Under Gravity: $y = -4.9t^2 + v_0 t + y_0$

The position of an object moving vertically under the influence of gravity can be described by a function in the two-parameter family

$$y = -4.9t^2 + v_0 t + y_0$$

where t is time in seconds and y is the distance in meters above the ground. Why do we need the parameters v_0 and y_0 to describe all such motions? Notice that at time $t = 0$ we have $y = y_0$. Thus the parameter y_0 gives the height above ground of the object at time $t = 0$. Since $dy/dt = -9.8t + v_0$, the parameter v_0 gives the velocity of the object at time $t = 0$.

Curves of the Form $y = A \sin(Bx)$

This family is used to model a wave. We saw in Section 1.9 that $|A|$ is the amplitude of the wave and that $2\pi/|B|$ is its period. Figures 5.22 and 5.23 illustrate these facts.

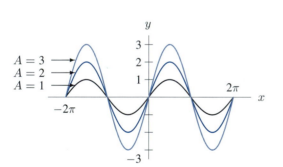

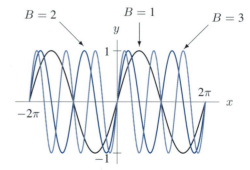

Figure 5.22: The family $y = A \sin x$
(with $B = 1$)

Figure 5.23: The family $y = \sin(Bx)$
(with $A = 1$)

Curves of the Form $y = e^{-(x-a)^2}$

The role of the parameter a in this family is to shift the graph of $y = e^{-x^2}$ to the right or left. Notice that the value of y is always positive. Since $y \to 0$ as $x \to \pm\infty$, the x-axis is a horizontal asymptote. Thus $y = e^{-(x-a)^2}$ is the family of horizontal shifts of the bell-shaped curve shown in Figure 5.24.

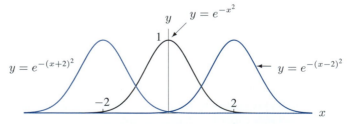

Figure 5.24: The family $y = e^{-(x-a)^2}$

Curves of the Form $y = e^{-(x-a)^2/b}$

This family is related to the *normal density* function, used in probability and statistics.[1] We assume that $b > 0$. The graph of a typical member of the family is a bell-shaped curve as in Figure 5.24. We know that the parameter a shifts the graph horizontally. We now consider the role of the parameter b by studying the family with $a = 0$:

$$y = e^{-x^2/b}.$$

To investigate the critical points and points of inflection, we calculate

$$\frac{dy}{dx} = -\frac{2x}{b}e^{-x^2/b}$$

and, using the product rule, we get

$$\frac{d^2y}{dx^2} = -\frac{2}{b}e^{-x^2/b} - \frac{2x}{b}\left(-\frac{2x}{b}e^{-x^2/b}\right) = \frac{2}{b}\left(\frac{2x^2}{b} - 1\right)e^{-x^2/b}.$$

There is a critical point where $dy/dx = 0$, that is, where

$$\frac{dy}{dx} = -\frac{2x}{b}\,e^{-x^2/b} = 0.$$

Since $e^{-x^2/b}$ is never zero, the only critical point is $x = 0$. At that point, $y = 1$ and $d^2y/dx^2 < 0$. Hence, by the second derivative test, there is a local maximum at $x = 0$.

Inflection points occur where the second derivative changes sign; thus, we start by finding values of x for which $d^2y/dx^2 = 0$. Since $e^{-x^2/b}$ is never zero, $d^2y/dx^2 = 0$ when

$$\frac{2x^2}{b} - 1 = 0.$$

Solving for x gives

$$x = \pm\sqrt{\frac{b}{2}}.$$

Looking at the expression for d^2y/dx^2, we see that d^2y/dx^2 is negative for $x = 0$, and positive as $x \to \pm\infty$. Therefore the concavity changes at $x = -\sqrt{b/2}$ and at $x = \sqrt{b/2}$, so we have inflection points there. Returning to the two-parameter family $y = e^{-(x-a)^2/b}$, we conclude that there is a maximum at $x = a$, obtained by horizontally shifting the maximum at $x = 0$ of $y = e^{-x^2/b}$ by a units. There are inflection points at $x = a \pm \sqrt{b/2}$ obtained by shifting the inflection points $x = \pm\sqrt{b/2}$ of $y = e^{-x^2/b}$ by a units. (See Figure 5.25.) At these points $y = e^{-1/2} \approx 0.6$.

With this information we can see the effect of the parameters. The parameter a determines the location of the center of the bell and the parameter b determines how narrow or wide the bell is. (See Figure 5.26.) If b is small, then the inflection points are close to a and the bell is sharply peaked near a; if b is large, the inflection points are further away from a and the bell is more spread out.

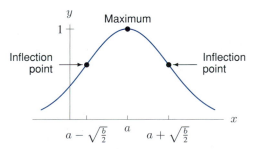

Figure 5.25: Graph of $y = e^{-(x-a)^2/b}$:
bell-shaped curve with peak at $x = a$

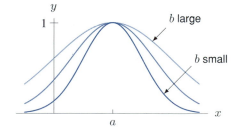

Figure 5.26: Graph of $y = e^{-(x-a)^2/b}$ for
fixed a and various b

[1]Probabilists divide our function by a constant, $\sqrt{\pi b}$, to get the normal density.

Curves of the Form $y = a\,(1 - e^{-bx})$

This is a two-parameter family. We will only consider $a, b > 0$. The graph of one member, with $a = 2$ and $b = 1$, is in Figure 5.27. Such a graph represents a quantity which is increasing but leveling off. For example, a body dropped in a thick fluid speeds up initially, but its velocity levels off as it approaches a terminal velocity. Similarly, if a pollutant pouring into a lake builds up toward a saturation level, its concentration may be described in this way. The graph might also represent the temperature of an object in an oven.

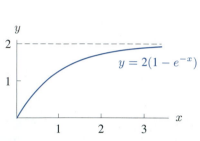

Figure 5.27: One member of the family $y = a(1 - e^{-bx})$, with $a = 2$

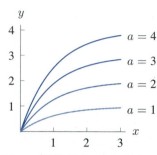

Figure 5.28: Fixing $b = 1$ gives $y = a(1 - e^{-x})$, graphed for various a

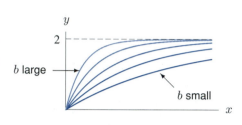

Figure 5.29: Fixing $a = 2$ gives $y = 2(1 - e^{-bx})$, graphed for various b

We examine the effect on the graph of varying a. Fix b at some positive number, say $b = 1$. Substitute different values for a and look at the graphs in Figure 5.28. We see that as x gets larger, y approaches a from below. The reason is $e^{-bx} \to 0$ as $x \to \infty$. Physically, the value of a represents the terminal velocity of a falling body or the saturation level of the pollutant in the lake.

Now examine the effect of varying b on the graph. Fix a at some positive number, say $a = 2$. Substitute different values for b and look at the graphs in Figure 5.29. The parameter b determines how sharply the curve rises and how quickly it gets close to the line $y = a$.

Let's confirm this last observation analytically. For $y = a(1 - e^{-bx})$, we have $dy/dx = abe^{-bx}$, so the slope of the tangent to the curve at $x = 0$ is ab. For larger b, the curve rises more rapidly at $x = 0$. How long does it take the curve to climb halfway up from $y = 0$ to $y = a$? When $y = a/2$, we have

$$a(1 - e^{-bx}) = \frac{a}{2}, \qquad \text{which leads to} \qquad x = \frac{\ln 2}{b}.$$

If b is large then $(\ln 2)/b$ is small, so in a short distance the curve is already half way up to a. If b is small, then $(\ln 2)/b$ is large and we have to go a long way out to get up to $a/2$. See Figure 5.30.

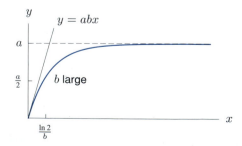

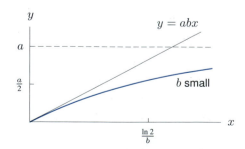

Figure 5.30: Tangent at $x = 0$ to $y = a(1 - e^{-bx})$, with fixed a, and large and small b

Problems for Section 5.2

1. Consider the function $p(x) = x^3 - ax$, where a is constant.

    (a) If $a < 0$, show that $p(x)$ is always increasing.
    (b) If $a > 0$, show that $p(x)$ has a local maximum and a local minimum.
    (c) Sketch and label typical graphs for the cases when $a < 0$ and when $a > 0$.

2. Consider the function $p(x) = x^3 - ax$, where a is constant and $a > 0$.

    (a) Find the local maxima and minima of p.
    (b) What effect does increasing the value of a have on the positions of the maxima and minima?
    (c) On the same axes, sketch and label the graphs of p for three positive values of a.

3. What effect does increasing the value of a have on the graph of $f(x) = x^2 + 2ax$? Consider roots, maxima and minima, and both positive and negative values of a.

4. The number, N, of people who have heard a rumor spread by mass media is modeled by the following function of time, t:
$$N(t) = a(1 - e^{-kt}).$$
    Suppose that there are 200,000 people in the population who hear the rumor eventually. If 10% of them heard it the first day, find a and k, assuming t is measured in days.

5. Suppose the temperature, T, of a yam put into a hot oven maintained at $200°$C is given as a function of time, t, by
$$T = a(1 - e^{-kt}) + b$$
    where T is in degrees Celsius and t is in minutes.

    (a) If the yam starts at $20°$C, find a and b.
    (b) If the temperature of the yam is initially increasing at $2°$C per minute, find k.

6. Consider the family of functions $y = f(x) = x - k\sqrt{x}$, with k a positive constant and $x \geq 0$. Show that the graph of $f(x)$ has a local minimum at a point whose x-coordinate is $1/4$ of the way between its x-intercepts.

7. (a) Find all critical points of $f(x) = x^4 + ax^2 + b$.
    (b) Under what conditions on a and b will this function have exactly one critical point? What is the one critical point, and is it a local maximum, a local minimum, or neither?
    (c) Under what conditions on a and b will this function have exactly three critical points? What are they and which are local maxima and which are local minima?
    (d) Is it ever possible for this function to have two critical points? No critical points? More than three critical points? Give an explanation in each case.

8. Sketch several members of the family $y = e^{-ax} \sin bx$ for $b = 1$, and describe the graphical significance of the parameter a.

9. Sketch several members of the family $e^{-ax} \sin bx$ for $a = 1$, and describe the graphical significance of the parameter b.

10. Sketch graphs of $y = xe^{-bx}$ for $b = 1, 2, 3, 4$. Describe the graphical significance of b.

11. Find the coordinates of the critical point of $y = xe^{-bx}$ and use it to confirm your answer to Problem 10.

12. If $a > 0$, $b > 0$, confirm analytically that $f(x) = a(1 - e^{-bx})$ is everywhere increasing and everywhere concave down.

13. Consider the family of curves
$$y = Ae^{-Bx^2}, \qquad \text{for positive } A, B.$$
    Analyze the effect of varying A and B on the shape of the curve. Illustrate your answer with sketches.

14. Consider the *surge function*
$$y = axe^{-bx}, \qquad \text{for positive } a, b.$$
    (a) Find the local maxima, local minima, and points of inflection.
    (b) How does varying a and b affect the shape of the graph?
    (c) On one set of axes, graph this function for several values of a and b.

15.   Consider the family of functions $f(x) = x^2 + \cos(kx)$, $k > 0$.

   (a)   Using a calculator or computer, sketch the graph of f for $k = 0.5, 1, 3, 5$. Try to find the smallest number k at which you see points of inflection in the graph of f.
   (b)   Explain why the graph of f has no points of inflection if $k \leq \sqrt{2}$, and infinitely many points of inflection if $k > \sqrt{2}$.
   (c)   Explain why f has only a finite number of critical points, no matter what the value of k.

16.   Consider the family of functions of the form $f(x) = e^x - kx$, for $k > 0$.

   (a)   Using a calculator or computer, sketch the graph of f for $k = 1/4, 1/2, 1, 2, 4$. Describe what happens as k changes.
   (b)   Show that f has a local minimum at $x = \ln k$.
   (c)   Find the value of k for which the local minimum is the largest.

17.   Consider the family

$$y = \frac{A}{x + B}.$$

   (a)   If $B = 0$, what is the effect of varying A on the graph?
   (b)   If $A = 1$, what is the effect of varying B?
   (c)   On one set of axes, graph the function for several values of A and B.

18.   The potential energy, U, of a particle moving along the x-axis is given by

$$U = b \left(\frac{a^2}{x^2} - \frac{a}{x} \right)$$

   where a and b are positive constants. Consider the graph of U against x for $x > 0$.

   (a)   Find the intercepts and asymptotes.
   (b)   Compute the local maxima and minima.
   (c)   Sketch the graph.

19.   The force, F, on a particle with potential energy U is given by

$$F = -\frac{dU}{dx}.$$

   Using the expression for U in Problem 18, graph F and U on the same axes, labeling intercepts and local maxima and minima.

20.   If the force between two atoms in a molecule is given as a function of the distance, r, between the molecules by

$$f(r) = -\frac{A}{r^2} + \frac{B}{r^3} \qquad r > 0,$$

   where A, B are positive constants:

   (a)   What are the zeros and asymptotes of f?
   (b)   Find the coordinates of the critical points and inflection points of f.
   (c)   Sketch a graph of f.
   (d)   Illustrating your answers with a sketch, describe the effect on the graph of f of:

       (i)   Increasing B, holding A fixed   (ii)   Increasing A, holding B fixed

5.3   OPTIMIZATION

It is often important to find the largest or smallest value of some quantity. For example, automobile engineers want to construct a car that uses the least amount of fuel, scientists want to calculate which wavelength carries the maximum radiation at a given temperature, and urban planners want to design traffic patterns to minimize delays. The techniques for solving such problems belong to the field of mathematics called *optimization*. The next three sections show how the derivative provides an efficient way of solving many optimization problems.

Global Maxima and Minima

The single greatest (or least) value of a function f over a specified domain is called the *global maximum* (or *minimum*) of f. Recall that the local maxima and minima tell us where a function is locally largest or smallest. Now we are interested in where the function is absolutely largest or smallest in a given domain. We define

> - f has a **global minimum** at p if $f(p)$ is less than or equal to all values of f.
> - f has a **global maximum** at p if $f(p)$ is greater than or equal to all values of f.

How Do We Find Global Maxima and Minima?

If f is a continuous function defined on a closed interval $a \leq x \leq b$ (that is, an interval containing its endpoints), Figure 5.31 illustrates that the global maximum or minimum of f occurs either at a local maximum or a local minimum, or at one of the endpoints of the interval, $x = a$ or $x = b$.

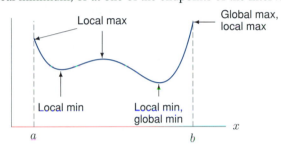

Figure 5.31: Global maximum and minimum on a closed interval $a \leq x \leq b$

> **To find the global maximum and minimum of a continuous function on a closed interval:**
> Compare values of the function at all the critical points in the interval and at the endpoints.

What if the function is defined on an open interval $a < x < b$ (that is, an interval not including its endpoints), or on all real numbers? Then there may or may not be a global maximum or minimum. For example, there is no global maximum in Figure 5.32 because the function has no actual largest value. The global minimum in Figure 5.32 coincides with the local minimum and is marked. There is a global minimum but no global maximum in Figure 5.33.

> **To find the global maximum and minimum of a continuous function on an open interval or on all real numbers:** Find the value of the function at all the critical points and sketch a graph. Look at the function values when x approaches the endpoints of the interval, or approaches $\pm\infty$, as appropriate.

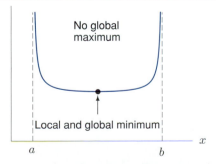

Figure 5.32: Global minimum on $a < x < b$

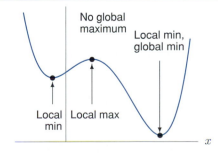

Figure 5.33: Global minimum when the domain is all real numbers

Example 1   Find the global maxima and minima of $f(x) = x^3 - 9x^2 - 48x + 52$ on the following intervals:

(a) $-5 \le x \le 12$ (b) $-5 \le x \le 14$ (c) $-5 \le x < \infty$.

Solution   (a)   We have previously obtained the critical points $x = -2$ and $x = 8$ using

$$f'(x) = 3x^2 - 18x - 48 = 3(x + 2)(x - 8).$$

We evaluate f at the critical points and the endpoints of the interval:

$$f(-5) = (-5)^3 - 9(-5)^2 - 48(-5) + 52 = -58$$
$$f(-2) = 104$$
$$f(8) = -396$$
$$f(12) = -92.$$

Comparing these function values, we see that the global maximum on $[-5, 12]$ is 104 and occurs at $x = -2$, and the global minimum on $[-5, 12]$ is -396 and occurs at $x = 8$.

(b)   For the interval $[-5, 14]$, we compare

$$f(-5) = -58, \quad f(-2) = 104, \quad f(8) = -396, \quad f(14) = 360.$$

The global maximum is now 360 and occurs at $x = 14$, and the global minimum is still -396 and occurs at $x = 8$. Notice that since the function is increasing for $x > 8$, changing the right-hand end of the interval from $x = 12$ to $x = 14$ alters the global maximum but not the global minimum. See Figure 5.34.

(c)   Figure 5.34 shows that for $-5 \le x < \infty$ there is no global maximum, because we can make $f(x)$ as large as we please by choosing x large enough. The global minimum remains -396 at $x = 8$.

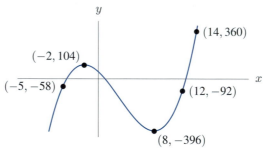

Figure 5.34: Graph of $f(x) = x^3 - 9x^2 - 48x + 52$

Example 2   When an arrow is shot into the air, its range, R, is defined as the horizontal distance from the archer to the point where the arrow hits the ground. If the ground is horizontal and we neglect air resistance, it can be shown that

$$R = \frac{v_0^2 \sin(2\theta)}{g},$$

where v_0 is the initial velocity of the arrow, g is the (constant) acceleration due to gravity, and θ is the angle above horizontal, so $0 \le \theta \le \pi/2$. (See Figure 5.35.) What initial angle maximizes R?

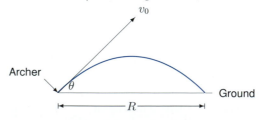

Figure 5.35: Arrow's path

Solution   We can find the maximum of this function without using calculus. The maximum value of R occurs when $\sin(2\theta) = 1$, so $\theta = \arcsin(1)/2 = \pi/4$, giving $R = v_0^2/g$.

Let's see how we can do the same problem with calculus. We want to find the global maximum of R for $0 \le \theta \le \pi/2$. First we look for critical points:

$$\frac{dR}{d\theta} = 2\frac{v_0^2 \cos(2\theta)}{g}.$$

Setting $dR/d\theta$ equal to 0, we get

$$0 = \cos(2\theta), \quad \text{or} \quad 2\theta = \pm\frac{\pi}{2}, \pm\frac{3\pi}{2}, \pm\frac{5\pi}{2}, \ldots$$

so $\pi/4$ is the only critical point in the interval $0 \le \theta \le \pi/2$. The range at $\theta = \pi/4$ is $R = v_0^2/g$.

Now we must check the value of R at the endpoints $\theta = 0$ and $\theta = \pi/2$. Since $R = 0$ at each endpoint, the critical point $\theta = \pi/4$ gives both a local and a global maximum on $0 \le \theta \le \pi/2$. Therefore, the arrow goes farthest if shot at an angle of $\pi/4$, or $45°$.

A Graphical Example: Minimizing Gas Consumption

Next we look at an example in which a function is given graphically and the optimum values are read from a graph. You already know how to estimate the optimum values of $f(x)$ from a graph of $f(x)$—read off the highest and lowest values. In this example, we see how to estimate the optimum value of the quantity $f(x)/x$ from a graph of $f(x)$ against x.

The question we investigate is how to set driving speeds to maximize fuel efficiency.[2] We assume that gas consumption, g (in gallons/hour), as a function of velocity, v (in mph) is as shown in Figure 5.36. We want to minimize the gas consumption per *mile*, not the gas consumption per hour. Let $G = g/v$ represent the average gas consumption per mile. (The units of G are gallons/mile.)

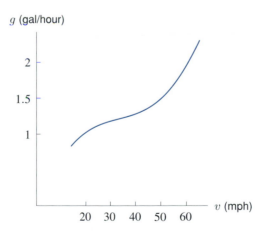

Figure 5.36: Gas consumption versus velocity

Example 3   Using Figure 5.36, estimate the velocity which minimizes G.

Solution   We want to find the minimum value of $G = g/v$ when g and v are related by the graph in Figure 5.36. We could use Figure 5.36 to sketch a graph of G against v and estimate a critical point. But there is an easier way.

---

[2]Adapted from Peter D. Taylor, *Calculus: The Analysis of Functions* (Toronto: Wall & Emerson, 1992).

Figure 5.37 shows that g/v is the slope of the line from the origin to the point P. Where on the curve should P be to make the slope a minimum? From the possible positions of the line shown in Figure 5.37, we see that the slope of the line is both a local and global minimum when the line is tangent to the curve. From Figure 5.38, we can see that the velocity at this point is about 50 mph. Thus to minimize gas consumption per mile, we should drive about 50 mph.

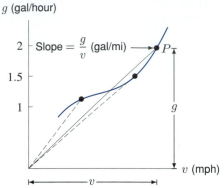

Figure 5.37: Graphical representation of gas consumption per mile, $G = \dfrac{g}{v}$

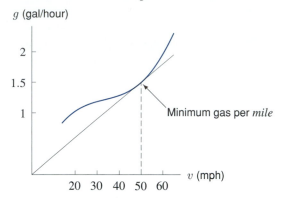

Figure 5.38: Velocity for maximum fuel efficiency

Finding Upper and Lower Bounds

A problem which is closely related to finding maxima and minima is finding the *bounds* of a function. In Example 1, the value of $f(x)$ on the interval $[-5, 12]$ ranges from -396 to 104. Thus

$$-396 \le f(x) \le 104$$

and we say that the function f is *bounded below* by -396 and *bounded above* by 104 on $[-5, 12]$. Of course, we could also say that

$$-400 \le f(x) \le 150,$$

so that f is also bounded below by -400 and above by 150 on $[-5, 12]$. However, we consider the -396 and 104 to be the *best possible bounds* because they describe more accurately how the function $f(x)$ behaves on $[-5, 12]$.

Example 4   Suppose an object on a spring oscillates about its equilibrium position at $y = 0$. Its distance from equilibrium is given as a function of time, t, by

$$y = e^{-t} \cos t.$$

Find the greatest distance the object goes above and below the equilibrium for $t \ge 0$.

Solution   We are looking for the bounds of the function. What does the graph of the function look like? We can think of it as a cosine curve with a decreasing amplitude of e^{-t}; in other words, it is a cosine curve squashed between the graphs of $y = e^{-t}$ and $y = -e^{-t}$, forming a wave with lower and lower crests and shallower and shallower troughs. (See Figure 5.39.)

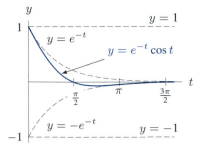

Figure 5.39: $f(t) = e^{-t} \cos t$ for $t \ge 0$

From the graph we can see that for $t \geq 0$, the graph lies between the horizontal lines $y = -1$ and $y = 1$. This means that -1 and 1 are bounds:

$$-1 \leq e^{-t} \cos t \leq 1.$$

The line $y = 1$ is the best possible upper bound because the graph does come up that high (at $t = 0$). However, we can find a better lower bound if we find the global minimum value of f for $t \geq 0$; this minimum occurs in the first trough between $t = \pi/2$ and $t = 3\pi/2$ because later troughs are squashed closer to the t-axis. At the minimum, $dy/dt = 0$. The product rule gives

$$\frac{dy}{dt} = (-e^{-t}) \cos t + e^{-t}(-\sin t) = -e^{-t}(\cos t + \sin t) = 0.$$

Since e^{-t} is never 0, we must have

$$\cos t + \sin t = 0, \quad \text{so} \quad \frac{\sin t}{\cos t} = -1.$$

Hence

$$\tan t = -1, \quad \text{giving} \quad t = \frac{3\pi}{4}.$$

Thus, the global minimum we see on the graph occurs at $t = 3\pi/4$. The value of y at that minimum is

$$y = e^{-3\pi/4} \cos\left(\frac{3\pi}{4}\right) \approx -0.067.$$

Rounding down so that the inequalities still hold for all $t \geq 0$ gives

$$-0.07 < e^{-t} \cos t \leq 1.$$

Notice how much smaller in magnitude the lower bound is than the upper. This is a reflection of how quickly the factor e^{-t} causes the oscillation to die out.

Problems for Section 5.3

For Problems 1–2, indicate all critical points on the given graphs. Determine which correspond to local minima, local maxima, global minima, global maxima, or none of these. (Note that the graphs are on closed intervals.)

1.

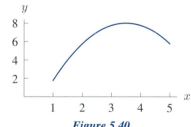

Figure 5.40

2.

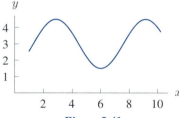

Figure 5.41

3. For $y = f(x) = x^{10} - 10x$, and $0 \leq x \leq 2$, find the value(s) of x for which:

   (a) $f(x)$ has a local maximum or local minimum. Indicate which ones are maxima and which are minima.

   (b) $f(x)$ has a global maximum or global minimum.

4. For $f(x) = x - \ln x$, and $0.1 \leq x \leq 2$, find the value(s) of x for which:

   (a) $f(x)$ has a local maximum or local minimum. Indicate which ones are maxima and which are minima.

   (b) $f(x)$ has a global maximum or global minimum.

5. For $f(x) = \sin^2 x - \cos x$, and $0 \leq x \leq \pi$, find, to two decimal places, the value(s) of x for which:

   (a) $f(x)$ has a local maximum or local minimum. Indicate which ones are maxima and which are minima.

   (b) $f(x)$ has a global maximum or global minimum.

6. The function $y = t(x)$ is positive and continuous with a global maximum at the point $(3, 3)$. Sketch a possible graph of $t(x)$ if $t'(x)$ and $t''(x)$ have the same sign for $x < 3$, but opposite signs for $x > 3$.

7. A function $y = g(x)$ has a derivative that is given in Figure 5.42 for $-2 \le x \le 2$.

    (a) Write a few sentences describing the behavior of $g(x)$ on this interval.
    (b) Does the graph of $g(x)$ have any inflection points? If so, give the approximate x-coordinates of their locations. Explain your reasoning.
    (c) What are the global maxima and minima of g on $[-2, 2]$?
    (d) If $g(-2) = 5$, what do you know about $g(0)$ and $g(2)$? Explain.

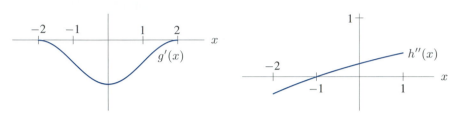

Figure 5.42   **Figure 5.43**

8. A function $y = h(x)$ has a second derivative shown in Figure 5.43 for $-2 \le x \le 1$. If $h'(-1) = 0$ and $h(-1) = 2$,

    (a) Explain why $h'(x)$ is never negative on this interval.
    (b) Explain why $h(x)$ must have a global maximum at $x = 1$.
    (c) Sketch a possible graph of $h(x)$ for $-2 \le x \le 1$.

9. A grapefruit is tossed straight up with an initial velocity of 50 ft/sec. The grapefruit is 5 feet above the ground when it is released. Its height at time t is given by

$$y = -16t^2 + 50t + 5.$$

How high does it go before returning to the ground?

10. For some positive constant C, the temperature change, T, in a patient generated by a dose, D, of a drug is given by

$$T = \left(\frac{C}{2} - \frac{D}{3}\right) D^2.$$

    (a) What dosage maximizes the temperature change?
    (b) The sensitivity of the body, at dosage D, to the drug is defined as dT/dD. What dosage maximizes sensitivity?

11. When you cough, your windpipe contracts. The speed, v, with which air comes out depends on the radius, r, of your windpipe. If R is the normal (rest) radius of your windpipe, then for $r \le R$, the speed is given by:

$$v = a(R - r)r^2$$

where a is a positive constant. What value of r maximizes the speed?

12. The bending moment M of a beam, supported at one end, at a distance x from the support is given by

$$M = \tfrac{1}{2}wLx - \tfrac{1}{2}wx^2,$$

where L is the length of the beam, and w is the uniform load per unit length. Find the point on the beam where the moment is greatest.

13. The efficiency of a screw, E, is given by

$$E = \frac{(\theta - \mu\theta^2)}{\mu + \theta}, \quad \theta > 0,$$

where θ is the angle of pitch of the thread and μ is the coefficient of friction of the material, a (positive) constant. What value of θ maximizes E?

14. A woman pulls a sled which, together with its load, has a mass of m kg. If her arm makes an angle of θ with her body (assumed vertical) and the coefficient of friction (a positive constant) is μ, the least force, F, she must exert to move the sled is given by

$$F = \frac{mg\mu}{\sin\theta + \mu\cos\theta}.$$

If $\mu = 0.15$, find the maximum and minimum values of F for $0 \le \theta \le \pi/2$. Give your answers as multiples of mg.

15. A circular ring of wire of radius r_0 lies in a plane perpendicular to the x-axis and is centered at the origin. The ring has a positive electric charge spread uniformly over it. The electric field in the x-direction, E, at the point x on the axis is given by

$$E = \frac{kx}{\left(x^2 + r_0^2\right)^{3/2}} \quad \text{for} \quad k > 0.$$

At what point on the x-axis is the field greatest? Least?

16. An electric current, I, in amps, is given by

$$I = \cos(wt) + \sqrt{3}\sin(wt),$$

where $w \ne 0$ is a constant. What are the maximum and minimum values of I?

17. (a) Show that $x > 2\ln x$ for all $x > 0$. [Hint: Find the minimum of $f(x) = x - 2\ln x$.]
     (b) Use the above result to show that $e^x > x^2$ for all positive x.
     (c) Is $x > 3\ln x$ for all positive x?

Find the best possible bounds for each of the functions in Problems 18–22.

18. e^{-x^2}, for $|x| \le 0.3$
19. $\ln(1+x)$, for $x \ge 0$
20. $\ln(1+x^2)$, for $-1 \le x \le 2$
21. $x^3 - 4x^2 + 4x$, for $0 \le x \le 4$
22. $x + \sin x$, for $0 \le x \le 2\pi$

23. When birds lay eggs, they do so in clutches of several at a time. When the eggs hatch, each clutch gives rise to a brood of baby birds. We want to determine the clutch size which maximizes the number of birds surviving to adulthood per brood. If the clutch is small, there are few baby birds in the brood; if the clutch is large, there are so many baby birds to feed that most die of starvation. The number of surviving birds per brood as a function of clutch size is shown by the benefit curve in Figure 5.44.[3]

     (a) Estimate the clutch size which maximizes the number of survivors per brood.
     (b) Suppose also that there is a biological cost to having a larger clutch: the female survival rate is reduced by large clutches. This cost is represented by the dotted line in Figure 5.44. If we take cost into account by assuming that the optimal clutch size in fact maximizes the vertical distance between the curves, what is the new optimal clutch size?

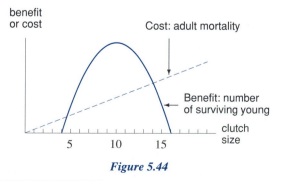

Figure 5.44

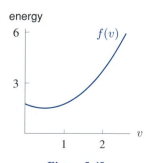

Figure 5.45

[3] Data from C. M. Perrins and D. Lack reported by J. R. Krebs and N. B. Davies in *An Introduction to Behavioural Ecology* (Oxford: Blackwell, 1987).

24. Let $f(v)$ be the amount of energy consumed by a flying bird, measured in joules per second (a joule is a unit of energy), as a function of its speed v (in meters/sec). See Figure 5.45.

   (a) Suggest a reason for the shape of this graph (in terms of the way birds fly).

   Now let $a(v)$ be the amount of energy consumed by the same bird, measured in joules *per meter*.

   (b) What is the relationship between $f(v)$ and $a(v)$?
   (c) Where is $a(v)$ a minimum?
   (d) Should the bird try to minimize $f(v)$ or $a(v)$ when it is flying? Why?

25. The forward motion of an aircraft in level flight is reduced by two kinds of forces, known as *induced drag* and *parasite drag*. Induced drag is a consequence of the downward deflection of air as the wings produce lift. Parasite drag results from friction between the air and the entire surface of the aircraft. Induced drag is inversely proportional to the square of speed and parasite drag is directly proportional to the square of speed. The sum of induced drag and parasite drag is called total drag. The graph in Figure 5.46 shows a certain aircraft's induced drag and parasite drag functions.

   (a) Sketch a graph of the total drag as a function of airspeed.
   (b) Estimate two different airspeeds which each result in a total drag of 1000 pounds. Does the total drag function have an inverse? What about the induced and parasite drag functions?
   (c) Fuel consumption (in gallons per hour) is roughly proportional to total drag. Suppose you are low on fuel and the control tower has just instructed you to enter a circular holding pattern of indefinite duration to await the passage of a storm at your landing field. At what airspeed should you fly the holding pattern? Why?

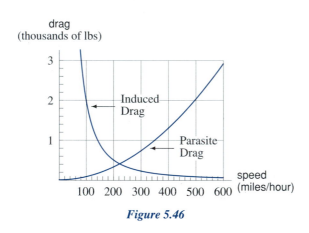

Figure 5.46

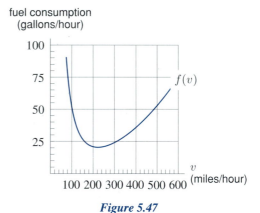

Figure 5.47

26. Let $f(v)$ be the fuel consumption, measured in gallons per hour, of a certain aircraft as a function of its airspeed, v, given in miles per hour. A graph of $f(v)$ is given in Figure 5.47.

   (a) Let $g(v)$ be the fuel consumption of the same aircraft, but measured in gallons per mile instead of gallons per hour. What is the relationship between $f(v)$ and $g(v)$?
   (b) For what value of v is $f(v)$ minimized?
   (c) For what value of v is $g(v)$ minimized?
   (d) Should a pilot try to minimize $f(v)$ or $g(v)$?

5.4 APPLICATIONS TO MARGINALITY

Management decisions within a particular business usually aim at maximizing profit for the company. In this section we will see how the derivative can be used to maximize profit. Profit depends on both production cost and revenue (or income) from sales. We begin by looking at the cost and revenue functions.

> The **cost function**, $C(q)$, gives the total cost of producing a quantity q of some good.

What sort of function do we expect C to be? The more goods that are made, the higher the total cost, so C is an increasing function. In fact, cost functions usually have the general shape shown in Figure 5.48. The intercept on the C-axis represents the *fixed costs*, which are incurred even if nothing is produced. (This includes, for instance, the machinery needed to begin production.) The cost function increases quickly at first and then more slowly because producing larger quantities of a good is usually more efficient than producing smaller quantities—this is called *economy of scale*. At still higher production levels, the cost function starts to increase faster again as resources become scarce, and sharp increases may occur when new factories have to be built. Thus, the graph of $C(q)$ may start out concave down and become concave up later on.

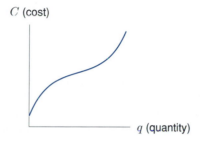

Figure 5.48: Cost as a function of quantity

> The **revenue function**, $R(q)$, gives the total revenue received by a firm from selling a quantity q of some good.

Revenue is income obtained from sales. If the price per item is p, and the quantity sold is q, then

$$\text{Revenue} = \text{Price} \times \text{Quantity}, \quad \text{so} \quad R = pq.$$

If the price per item does not depend on the quantity sold, then the graph of $R(q)$ is a straight line through the origin with slope equal to the price p. See Figure 5.49. In practice, for large values of q, the market may become glutted, causing the price to drop, giving $R(q)$ the shape in Figure 5.50.

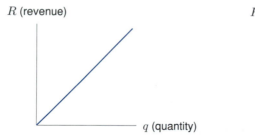

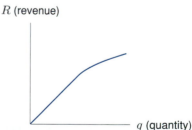

Figure 5.49: Revenue: Constant price   *Figure 5.50:* Revenue: Decreasing price

The profit is usually written as π (to distinguish it from the price, p; this π has nothing to do with the area of a circle, and merely stands for the Greek equivalent of the letter "p"). The profit resulting from producing and selling q items is defined by

> $$\text{Profit} = \text{Revenue} - \text{Cost}, \quad \text{so} \quad \pi(q) = R(q) - C(q).$$

Example 1   If cost, C, and revenue, R, are given by the graph in Figure 5.51, for what production quantities, q, does the firm make a profit? Approximately what production level maximizes profit?

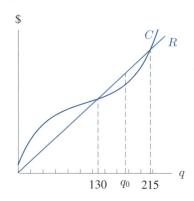

Figure 5.51: Costs and revenues for Example 1

Solution   The firm makes a profit whenever revenues are greater than costs, that is, when $R > C$. The graph of R is above the graph of C approximately when $130 < q < 215$. Production between $q = 130$ units and $q = 215$ units will generate a profit. The vertical distance between the cost and revenue curves is largest at q_0, so q_0 units gives maximum profit.

Marginal Analysis

Many economic decisions are based on an analysis of the costs and revenues "at the margin." Let's look at this idea through an example.

Suppose we are running an airline and we are trying to decide whether to offer an additional flight. How should we decide? We'll assume that the decision is to be made purely on financial grounds: if the flight will make money for the company, it should be added. Obviously we need to consider the costs and revenues involved. Since the choice is between adding this flight and leaving things the way they are, the crucial question is whether the *additional costs* incurred are greater or smaller than the *additional revenues* generated by the flight. These additional costs and revenues are called the *marginal costs* and *marginal revenues*.

Suppose $C(q)$ is the function giving the total cost of running q flights. If the airline had originally planned to run 100 flights, its costs would be $C(100)$. With the additional flight, its costs would be $C(101)$. Therefore,

$$\text{Marginal cost} = C(101) - C(100).$$

Now

$$C(101) - C(100) = \frac{C(101) - C(100)}{101 - 100},$$

and this quantity is the average rate of change of cost between 100 and 101 flights. In Figure 5.52 the average rate of change is the slope of the line joining the $C(100)$ and $C(101)$ points on the graph. If the graph of the cost function is not curving too fast near the point, the slope of this line is close to the slope of the tangent line there. Therefore, the average rate of change is close to the instantaneous rate of change. Since these rates of change are not very different, many economists choose to define marginal cost, MC, as the instantaneous rate of change of cost with respect to quantity:

$$\boxed{\text{Marginal cost} = MC = C'(q).}$$

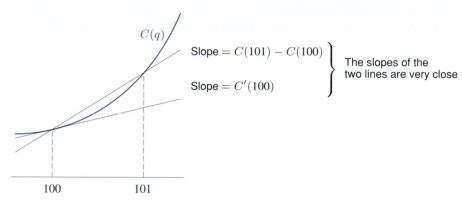

Figure 5.52: Marginal cost: Slope of one of these lines

Similarly if the revenue generated by q flights is $R(q)$, the additional revenue generated by increasing the number of flights from 100 to 101 is

$$\text{Marginal revenue} = R(101) - R(100).$$

Now $R(101) - R(100)$ is the average rate of change of revenue between 100 and 101 flights. As before, the average rate of change is usually almost equal to the instantaneous rate of change, so economists often define:

$$\boxed{\text{Marginal revenue} = MR = R'(q).}$$

We often refer to total cost and total revenue to distinguish them from marginal cost and marginal revenue. If the words cost and revenue are used alone, they are understood to mean total cost and total revenue.

Example 2   If $C(q)$ and $R(q)$ for the airline are given in Figure 5.53, should the company add the 101st flight?

Solution   The marginal revenue is the slope of the revenue curve, and the marginal cost is the slope of the cost curve at the point 100. From Figure 5.53, you can see that the slope at the point A is smaller than the slope at B, so $MC < MR$. This means that the airline will make more in extra revenue than it will spend in extra costs if it runs another flight, so it should go ahead and run the 101st flight.

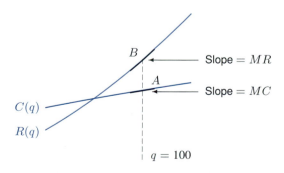

Figure 5.53: Cost and revenue for Example 2

Since MC and MR are derivative functions, they can be estimated from the graphs of total cost and total revenue.

Example 3   If R and C are given by the graphs in Figure 5.54, sketch graphs of $MR = R'(q)$ and $MC = C'(q)$.

Figure 5.54: Total revenue and total cost for Example 3

Solution   The revenue graph is a line through the origin, with equation

$$R = pq$$

where p is the price, which is a constant. The slope is p and

$$MR = R'(q) = p.$$

The total cost is increasing, so the marginal cost is always positive (above the q-axis). For small q values, the total cost curve is concave down, so the marginal cost is decreasing. For larger q, say $q > 100$, the total cost curve is concave up and the marginal cost is increasing. Thus the marginal cost has a minimum at about $q = 100$. (See Figure 5.55.)

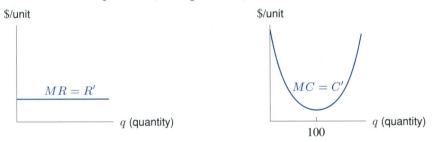

Figure 5.55: Marginal revenue and costs for Example 3

Maximizing Profit

Now let's look at how to maximize total profit, given functions for total revenue and total cost.

Example 4   Find the maximum profit if the total revenue and total cost are given, for $0 \leq q \leq 200$, by the curves R and C in Figure 5.56.

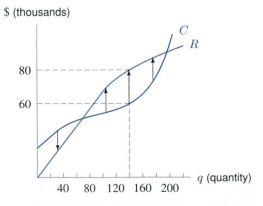

Figure 5.56: Maximum profit at $q = 140$

Solution   The profit is represented by the vertical difference between the curves and is marked by the vertical arrows on the graph. When revenue is below cost, the company is taking a loss; when revenue is above cost, the company is making a profit. We can see that the profit is largest at about $q = 140$, so this is the production level we're looking for. To be sure that the local maximum is a global maximum, we need to check the endpoints. At $q = 0$ and $q = 200$, the profit is negative, so the global maximum is indeed at $q = 140$.

To find the actual maximum profit, we estimate the vertical distance between the curves at $q = 140$. This gives a maximum profit of $\$80,000 - \$60,000 = \$20,000$.

Suppose we wanted to find the minimum profit. In this example, we must look at the endpoints, when $q = 0$ or $q = 200$. We see the minimum profit is negative (a loss), and it occurs at $q = 0$.

Maximum Profit Occurs Where *MR = MC*

In Example 4, observe that at $q = 140$ the slopes of the two curves in Figure 5.56 are equal. To the left of $q = 140$, the revenue curve has a larger slope than the cost curve, and the profit increases as q increases. The company will make more money by producing more units, so production should increase toward $q = 140$. To the right of $q = 140$, the slope of the revenue curve is less than the slope of the cost curve, and the profit is decreasing. The company will make more money by producing fewer units so production should decrease toward $q = 140$. At the point where the slopes are equal, the profit has a local maximum; otherwise the profit could be increased by moving toward that point. Since the slopes are equal at $q = 140$, we have $MR = MC$ there.

Now let's look at the general situation. To maximize or minimize profit over an interval, we optimize the profit, π, where

$$\pi(q) = R(q) - C(q).$$

We know that global maxima and minima can only occur at critical points or at endpoints of an interval. To find critical points of π, look for zeros of the derivative:

$$\pi'(q) = R'(q) - C'(q) = 0.$$

So

$$R'(q) = C'(q),$$

that is, the slopes of the revenue and cost curves are equal. This is the same observation that we made in the previous example. In economic language,

---

The maximum (or minimum) profit can occur where

Marginal cost = Marginal revenue.

Of course, maximal or minimal profit does not *have* to occur where $MR = MC$; there are also the endpoints to consider.

Example 5   Find the quantity q which maximizes profit if the total revenue and total cost are given by

$$R(q) = 5q - 0.003q^2$$
$$C(q) = 300 + 1.1q,$$

where $0 \leq q \leq 800$ units and $R(q)$ and $C(q)$ are in dollars. What production level gives the minimum profit?

Solution   We look for production levels that give marginal revenue = marginal cost:

$$MR = R'(q) = 5 - 0.006q$$
$$MC = C'(q) = 1.1.$$

So $5 - 0.006q = 1.1$, or

$$q = 3.9/0.006 = 650 \text{ units.}$$

Does this value of q represent a local maximum or minimum of π? We can tell by looking at production levels of 649 units and 651 units. When $q = 649$ we have $MR = \$1.106$, which is greater than the (constant) marginal cost of $\$1.1$. This means that producing one more unit will bring in more revenue than its cost, so profit will increase. When $q = 651$, $MR = \$1.094$, which is *less* than MC, so it is not profitable to produce the 651st unit. We conclude that $q = 650$ is a local maximum for the profit function π. The profit earned by producing and selling this quantity is $\pi(650) = R(650) - C(650) = \$967.50$.

To check for global maxima we need to look at the endpoints. If $q = 0$, the only cost is $\$300$ (the fixed costs) and there is no revenue, so $\pi(0) = -300$. At the upper limit of $q = 800$, we have $\pi(800) = \$900$. Therefore, the maximum profit is at the production level of 650 units, where $MR = MC$. The minimum profit (a loss) occurs when $q = 0$ and there is no production at all.

Problems for Section 5.4

1. A manufacturer reports the cost and revenue functions shown in Figure 5.57. Sketch graphs, as a function of quantity, of
    (a) Total profit   (b) Marginal cost   (c) Marginal revenue
    Label the points q_1 and q_2 on your graphs.

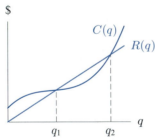

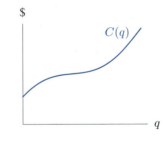

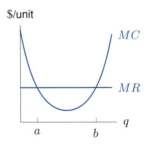

Figure 5.57   *Figure 5.58*   *Figure 5.59*

2. Assume a manufacturer's cost of producing a good is given by the graph of $C(q)$ in Figure 5.57. Assume also that the manufacturer can sell the product for a price p each (regardless of the quantity sold), so that the total revenue from selling a quantity q is $R(q) = pq$.

    (a) The difference $\pi(q) = R(q) - C(q)$ is the total profit. For which quantity q_0 is the profit a maximum? Mark your answer on a sketch of the graph.
    (b) What is the relationship between p and $C'(q_0)$? Explain your result both graphically and analytically. What does this mean in terms of economics? (Note that p is the slope of the line $R(q) = pq$. Note also that $\pi(q)$ has a maximum at $q = q_0$, so $\pi'(q_0) = 0$.)
    (c) Graph $C'(q)$ and p (as a horizontal line) on the same axes. Mark q_0 on the q-axis.

3. Let $C(q)$ be the total cost of producing a quantity q of a certain good. (See Figure 5.58.)

    (a) What is the meaning of $C(0)$?
    (b) Describe in words how the marginal cost changes as the quantity produced increases.
    (c) Explain the concavity of the graph (in terms of economics).
    (d) Explain the economic significance (in terms of marginal cost) of the point at which the concavity changes.
    (e) Do you expect the graph of $C(q)$ to look like this for all types of goods?

4. The marginal revenue and marginal cost for a certain item are graphed in Figure 5.59. Do the following quantities maximize revenue for the company? Explain your answer.

    (a) $q = a$ (b) $q = b$

5. Suppose the total cost $C(q)$ of producing q goods is given by:

    $$C(q) = 0.01q^3 - 0.6q^2 + 13q.$$

    (a) What is the fixed cost?
    (b) What is the maximum profit if each item is sold for $7? (Assume you sell everything you produce.)
    (c) Suppose exactly 34 goods are produced. They all sell when the price is $7 each, but for each $1 increase in price, 2 fewer goods are sold. Should the price be raised, and if so by how much?

6. Suppose a company manufactures only one product. The quantity, q, of this product produced per month depends on the amount of capital, K, invested (i.e., the number of machines the company owns, the size of its building, and so on) and the amount of labor, L, available each month. It is often assumed that q can be expressed as a function of K and L by a *Cobb-Douglas production function*:

    $$q = cK^\alpha L^\beta$$

    where c, α, β are positive constants, with $0 < \alpha < 1$ and $0 < \beta < 1$.

    In this problem we will see how the Russian government could use a Cobb-Douglas function to estimate how many people a newly privatized industry might employ. A company in such an industry will have only a small amount of capital available to it, and will need to use all of it; K is therefore fixed. Suppose L is measured in man-hours per month, and that each man-hour costs the company w rubles (a ruble is the unit of Russian currency). Suppose that the company has no other costs besides labor, and that each unit of the good can be sold for a fixed price of p rubles. How many man-hours of labor per month should the company use in order to maximize its profit?

7. An agricultural worker in Uganda is interested in planting clover to increase the number of bees making their home in the region. There are 100 bees in the region naturally, and for every acre put under clover, 20 more bees are found in the region.

    (a) Draw a graph of the total number, $N(x)$, of bees as a function of x, the number of acres devoted to clover.
    (b) Explain, both geometrically and algebraically, the shape of the graph of:

        (i)   The marginal rate of increase of the number of bees with acres of clover, $N'(x)$.

        (ii) The average number of bees per acre of clover, $N(x)/x$.

8. If you invest x dollars in a certain project, your return is $R(x)$. Suppose you want to choose x to maximize your return per dollar invested,[4] which is

    $$r(x) = \frac{R(x)}{x}.$$

    (a) Suppose the graph of $R(x)$ has the form in Figure 5.60 on page 270, with $R(0) = 0$. Illustrate on a copy of this graph that the maximum value of $r(x)$ is obtained at a point on the graph of $R(x)$ at which the line from the origin to the point is tangent to the graph.
    (b) It is also true that the maximum of $r(x)$ should occur at a point at which the slope of the graph of $r(x)$ is zero. On the same set of axes as part (a), draw a rough version of the graph of $r(x)$, corresponding to your R graph, and illustrate that the maximum occurs where the slope is 0.
    (c) Show, by taking the derivative of the preceding formula for $r(x)$, that the conditions in part (a) and (b) are equivalent: the point where the line from the origin is tangent to the graph of R is the same point where the graph of r has zero slope.

    ---
    [4]From Peter D. Taylor, *Calculus: The Analysis of Functions* (Toronto: Wall & Emerson, 1992).

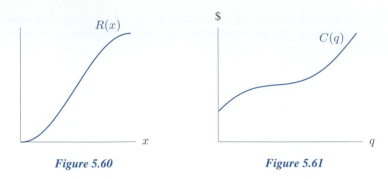

Figure 5.60   Figure 5.61

Problems 9–11 involve the *average cost* of manufacturing a quantity q of a good, which is defined to be

$$a(q) = \frac{C(q)}{q}.$$

9.   Figure 5.61 shows the cost of production, $C(q)$, as a function of quantity produced, q.

   (a)   For some q_0, sketch a line whose slope is the marginal cost, MC, at that point.
   (b)   For the same q_0, explain why the average cost $a(q_0)$ can be represented by the slope of the line from that point on the curve to the origin.
   (c)   Use the method of Example 3 on page 257 to explain why at the value of q which minimizes $a(q)$, the average and marginal costs are equal.

10.   Suppose a firm produces a quantity q of some good and that the average cost per item is given by:

$$a(q) = 0.01q^2 - 0.6q + 13, \quad \text{for} \quad q > 0.$$

   (a)   What is the total cost, $C(q)$, of producing q goods?
   (b)   What is the minimum marginal cost? What is the practical interpretation of this result?
   (c)   At what production level is the average cost a minimum? What is the lowest average cost?
   (d)   Compute the marginal cost at $q = 30$. How does this relate to your answer to part (c)? Explain this relationship both analytically and in words.

11.   A reasonably realistic model of a firm's costs is given by the *short-run Cobb-Douglas cost curve*

$$C(q) = Kq^{1/a} + F,$$

   where a is a positive constant, F is the fixed costs, and K measures the technology available to the firm.

   (a)   Show that C is concave down if $a > 1$.
   (b)   Assuming that average cost is minimized when average cost equals marginal cost, find what value of q minimizes the average cost.

5.5 MORE OPTIMIZATION: INTRODUCTION TO MODELING

Finding global maxima and minima is made much easier by having a formula for the function to be maximized or minimized. The process of translating a problem into a function whose formula we know is called *mathematical modeling*. By working through the examples that follow, you will get the flavor of some kinds of modeling.

Example 1   What are the dimensions of an aluminum can that holds 40 in³ of juice and that uses the least material (i.e., aluminum)? Assume that the can is cylindrical, and is capped on both ends.

Solution   It is often a good idea to think about a problem in general terms before trying to solve it. Since we're trying to use as little material as possible, why not make the can very small, say, the size of a peanut? We can't, since the can must hold 40 in^3. If we make the can short, to try to use less material in the sides, we'll have to make it fat as well, so that it can hold 40 in^3. Saving aluminum in the sides might actually use more aluminum in the top and bottom in a short, fat can. See Figure 5.62(a).

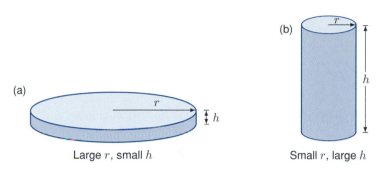

Figure 5.62: Various cylindrical-shaped cans

TABLE 5.1   *Amount of material, M, used in the can for various choices of radius, r, and height, h*

r (in)	h (in)	M (in^2)
0.2	318.31	400.25
1.0	12.73	86.27
2.0	3.18	65.09
3.0	1.41	83.13
4.0	0.80	120.64
10.0	0.13	636.49

If we try to save material by making the top and bottom small, the can has to be tall to accommodate the 40 in^3 of juice. So any savings we get by using a small top and bottom might be outweighed by the height of the sides. See Figure 5.62(b). This is, in fact, true, as Table 5.1 shows.

The table gives the amount of material used in the can for some choices of the radius, r, and the height, h. You can see that r and h change in opposite directions, and that more material is used at the extremes (very large or very small r and h) than in the middle. From the table it appears that the optimal radius for the can lies somewhere in $1.0 \leq r \leq 3.0$. If we consider the material used, M, as a function of the radius, r, a graph of this function looks like Figure 5.63. The graph shows that the global minimum we want is at a critical point.

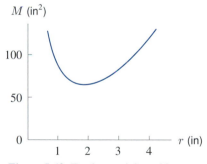

Figure 5.63: Total material used in can, M, as a function of radius, r

Both the table and the graph were obtained from a mathematical model, which in this case is a formula for the material used in making the can. Finding such a formula depends on knowing the geometry of a cylinder, in particular its area and volume. We have

$$M = \text{Material used in the can} = \text{Material in ends} + \text{Material in the side}$$

where

$$\text{Material in ends} = 2 \cdot \text{Area of a circle with radius } r = 2 \cdot \pi r^2,$$

$$\text{Material in the side} = \text{Area of cylinder with height } h \text{ and radius } r = 2\pi rh.$$

However, h is not independent of r: if r grows, h shrinks, and vice-versa. To find the relationship, we use the fact that the volume of the cylinder, $\pi r^2 h$, is equal to the constant 40 in^3:

$$\pi r^2 h = 40, \quad \text{giving} \quad h = \frac{40}{\pi r^2}.$$

This means

$$\text{Material in the side} = 2\pi rh = 2\pi r \frac{40}{\pi r^2} = \frac{80}{r}.$$

Thus we obtain the formula for the total material used in a can of radius r:

$$M(r) = 2\pi r^2 + \frac{80}{r}.$$

The domain of this function is all $r > 0$.

Now we use calculus to find the minimum of M. We look for critical points:

$$\frac{dM}{dr} = 4\pi r - \frac{80}{r^2} = 0 \quad \text{at a critical point,} \quad \text{so} \quad 4\pi r = \frac{80}{r^2}.$$

Therefore,

$$\pi r^3 = 20, \quad \text{so} \quad r = \left(\frac{20}{\pi}\right)^{1/3} \approx 1.85 \text{ inches,}$$

which agrees with our graph. We also have

$$h = \frac{40}{\pi r^2} \approx \frac{40}{\pi (1.85)^2} \approx 3.7 \text{ inches.}$$

Thus, the material used, $M(1.85)$, is about 64.7 in^2.

Practical Tips for Modeling Optimization Problems

1. Make sure that you know what quantity or function is to be optimized.

2. If possible, make several sketches showing how the elements that vary are related. Label your sketches clearly by assigning variables to quantities which change.

3. Try to obtain a formula for the function to be optimized in terms of the variables that you identified in the previous step. If necessary, eliminate from this formula all but one variable. Identify the domain over which this variable varies.

4. Find the critical points and evaluate the function at these points and the endpoints to find the global maxima and minima.

The next example, another problem in geometry, illustrates this approach.

Example 2   Alaina wants to get to the bus stop as quickly as possible. The bus stop is across a grassy park, 2000 feet west and 600 feet north of her starting position. Alaina can walk west along the edge of the park on the sidewalk at a speed of 6 ft/sec. She can also travel through the grass in the park, but only at a rate of 4 ft/sec (the park is a favorite place to walk dogs, so she must move with care). What path will get her to the bus stop the fastest?

Solution

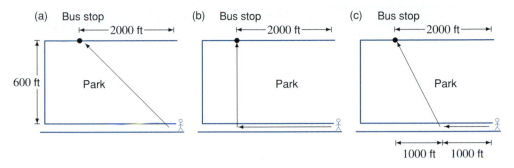

Figure 5.64: Three possible paths to the bus stop

We might first think that she should take a path that is the shortest distance. Unfortunately, the path that follows the shortest distance to the bus stop is entirely in the park, where her speed is slow. (See Figure 5.64(a).) That distance is $\sqrt{2000^2 + 600^2} \approx 2100$ feet, which will take her about 525 seconds to traverse. She could instead walk quickly the entire 2000 feet along the sidewalk, which would leave her just the 600-foot northward journey through the park. (See Figure 5.64(b).) This method would take $2000/6 + 600/4 \approx 483$ seconds total walking time.

But can she do even better? Perhaps another combination of sidewalk and park will give a shorter travel time. For example, what is the travel time if she walks 1000 feet west along the sidewalk and the rest of the way through the park? (See Figure 5.64(c).) The answer is about 458 seconds.

We make a model for this problem. We label the distance that Alaina walks west along the sidewalk x and the distance she walks through the park y, as in Figure 5.65. Then the total time, t, will be

$$t = t_{\text{sidewalk}} + t_{\text{park}}.$$

Since

$$\text{Time} = \text{Distance/Speed},$$

and she can walk 6 ft/sec on the sidewalk and 4 ft/sec in the park, we have that

$$t = \frac{x}{6} + \frac{y}{4}.$$

Now, by the Pythagorean Theorem, $y = \sqrt{(2000 - x)^2 + 600^2}$. Therefore

$$t = \frac{x}{6} + \frac{\sqrt{(2000 - x)^2 + 600^2}}{4}.$$

We can find the critical points of this function algebraically. (See Problem 1 on page 275.) Alternatively, we can graph the function on a calculator and estimate the critical point, which is $x \approx 1463$ feet. This gives a minimum total time of about 445 seconds.

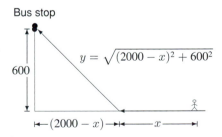

Figure 5.65: Modeling time to bus stop

Example 3   One hallway which is 4 feet wide meets another hallway which is 8 feet wide in a right angle. (See Figure 5.66.) What is the length of the longest ladder which can be carried horizontally around the corner?

Solution   We imagine the ladder being carried on its side and ignore its width. To allow the longest ladder possible we carry the ladder around the corner so that it just touches both walls (at A and C) and just touches the corner at B. Let's draw some lines that do this. (See Figure 5.66.) The length of the line $\overline{ABC}$ decreases as the corner is turned, then increases again. The minimum such length would be the length of the longest ladder that could make it around the corner. Certainly a smaller ladder would work (it wouldn't touch A, B, and C simultaneously), but a larger one would not fit.

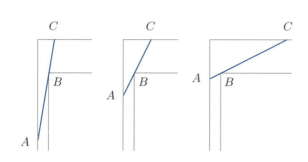

Figure 5.66: Various ladders that touch both walls and corner   *Figure 5.67:* Ladder and hallway

So we want the *smallest* length of the line $\overline{ABC}$. We will express the length, l, as a function of θ, the angle between the line and the wall of the narrow hall. (See Figure 5.67.)

We have
$$l = \overline{AB} + \overline{BC}.$$

Now $\overline{AB} = 4/\sin\theta$ and $\overline{BC} = 8/\cos\theta$, so
$$l = \frac{4}{\sin\theta} + \frac{8}{\cos\theta}.$$

Here the domain of θ is $0 < \theta < \pi/2$.

Now we differentiate:
$$\frac{dl}{d\theta} = -\frac{4}{(\sin\theta)^2}(\cos\theta) - \frac{8}{(\cos\theta)^2}(-\sin\theta).$$

To minimize l, we solve $dl/d\theta = 0$:
$$-4\frac{\cos\theta}{(\sin\theta)^2} + 8\frac{\sin\theta}{(\cos\theta)^2} = 0,$$

so
$$2(\sin\theta)^3 = (\cos\theta)^3, \quad \text{or} \quad \frac{(\sin\theta)^3}{(\cos\theta)^3} = \frac{1}{2}.$$

This means
$$\tan\theta = \sqrt[3]{0.5} \approx 0.79, \quad \text{so} \quad \theta \approx 0.67 \text{ radians.}$$

Thus $\theta \approx 0.67$ is a critical point, and an investigation of what happens for θ near 0 and θ near $\pi/2$ shows that l has a global minimum at $\theta = 0.67$. The minimum value of l is therefore
$$l = \frac{4}{\sin(0.67)} + \frac{8}{\cos(0.67)} \approx 16.65 \text{ feet,}$$

so the ladder can be at most 16.65 feet long and still make it around the corner.

Problems for Section 5.5

1. Find analytically the exact critical point of the function which represents the time, t, to walk to the bus stop in Example 2. Recall that t is given by

$$t = \frac{x}{6} + \frac{\sqrt{(2000 - x)^2 + 600^2}}{4}.$$

2. A smokestack deposits soot on the ground with a concentration inversely proportional to the square of the distance from the stack. With two smokestacks 20 miles apart, the concentration of the combined deposits on the line joining them, at a distance x from one stack, is given by

$$S = \frac{k_1}{x^2} + \frac{k_2}{(20 - x)^2}$$

where k_1 and k_2 are positive constants which depend on the quantity of smoke each stack is emitting. If $k_1 = 7k_2$, find the point on the line joining the stacks where the concentration of the deposit is a minimum.

3. A wave of wavelength λ traveling in deep water has speed, v, given by

$$v = k\sqrt{\frac{\lambda}{c} + \frac{c}{\lambda}},$$

where c and k are positive constants. As λ varies, does such a wave have a maximum or minimum velocity? If so, what is it? Explain.

4. If you have 100 feet of fencing and want to enclose a rectangular area up against a long, straight wall, what is the largest area you can enclose?

5. A closed box has a fixed surface area A and a square base with side x.

    (a) Find a formula for its volume, V, as a function of x.
    (b) Sketch a graph of V against x.
    (c) Find the maximum value of V.

6. A landscape architect plans to enclose a 3000 square foot rectangular region in a botanical garden. She will use shrubs costing \$25 per foot along three sides and fencing costing \$10 per foot along the fourth side. Find the minimum total cost.

7. A rectangular swimming pool is to be built with an area of 1800 square feet. The owner wants 5-foot wide decks along either side and 10-foot wide decks at the two ends. Find the dimensions of the smallest piece of property on which the pool can be built satisfying these conditions.

8. Suppose a rectangular beam is cut from a cylindrical log of radius 30 cm. The strength of a beam of width w and height h is proportional to wh^2. (See Figure 5.68.) Find the width and height of the beam of maximum strength.

9. A light is suspended at a height h above the floor. (See Figure 5.69.) The illumination at the point P is inversely proportional to the square of the distance from the point P to the light. In addition, the illumination is proportional to the cosine of the angle θ. How far from the floor should the light be to maximize the illumination at the point P?

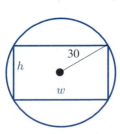

Figure 5.68

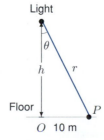

Figure 5.69

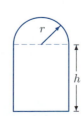

Figure 5.70

10. The cross-section of a tunnel is a rectangle of height h surmounted by a semicircular roof section of radius r (See Figure 5.70 on page 275). If the cross-sectional area is A, determine the dimensions of the cross section which minimize the perimeter.

11. Which point on the parabola $y = x^2$ is nearest to $(1,0)$? Find the coordinates to two decimals. [Hint: Minimize the square of the distance—this avoids square roots.]

12. Find the coordinates of the point on the parabola $y = x^2$ which is closest to the point (3,0).

13. Of all rectangles with given area, A, which has the shortest diagonals?

14. Suppose you run a small independent furniture business. Your assistant signs a deal with a customer to deliver up to 400 chairs, the exact number to be determined by the customer later. The price will be $90 per chair up to 300 chairs, and above 300, the price will be reduced by $0.25 per chair (on the whole order) for every additional chair over 300 ordered. What are the largest and smallest revenues your company can make under this deal?

15. The cost of fuel to propel a boat through the water (in dollars per hour) is proportional to the cube of the speed. A certain ferry boat uses $100 worth of fuel per hour when cruising at 10 miles per hour. Apart from fuel, the cost of running this ferry (labor, maintenance, and so on) is $675 per hour. At what speed should it travel so as to minimize the cost *per mile* traveled?

16. (a) For which positive number x is $x^{1/x}$ largest? Justify your answer.
    [Hint: You may want to write $x^{1/x} = e^{\ln(x^{1/x})}$.]
    (b) For which positive integer n is $n^{1/n}$ largest? Justify your answer.
    (c) Use your answer to parts (a) and (b) to decide which is larger: $3^{1/3}$ or $\pi^{1/\pi}$.

17. The *arithmetic mean* of two numbers a and b is defined as $(a+b)/2$; the *geometric mean* of two positive numbers a and b is defined as $\sqrt{ab}$.

    (a) For two positive numbers, which of the two means is larger? Justify your answer.
    [Hint: Define $f(x) = (a+x)/2 - \sqrt{ax}$ for fixed a.]
    (b) For three positive numbers a, b, c, the arithmetic and geometric mean are $(a+b+c)/3$ and $\sqrt[3]{abc}$, respectively. Which of the two means of three numbers is larger?
    [Hint: Redefine $f(x)$ for fixed a *and* b.]

18. On the same side of a straight river are two towns, and the townspeople want to build a pumping station, S, that supplies water to them. The pumping station is to be at the river's edge with pipes extending straight to the two towns. The distances are shown in Figure 5.71. Where should the pumping station be located to minimize the total length of pipe?

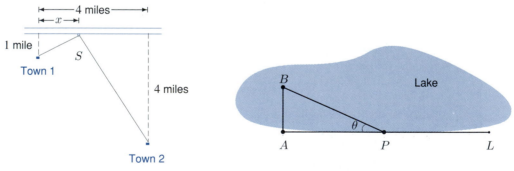

Figure 5.71   Figure 5.72

19. Assume that a pigeon is released from a boat (point B in Figure 5.72) floating on a lake. Because of falling air over the cool water, the energy required to fly one meter over the lake is twice the corresponding energy e required for flying over the bank ($e = 3$ joule/meter). To minimize the energy required to fly from B to the loft, L, the pigeon heads to a point P on the bank and then flies along the bank to L. The distance $\overline{AL}$ is 2000 m, and $\overline{AB}$ is 500 m.

    (a) Express the energy required to fly from B to L via P as a function of the angle θ (the angle BPA).
    (b) What is the optimal angle θ?
    (c) Does your answer change if $\overline{AL}$, $\overline{AB}$, and e have different numerical values?

20. To get the best view of the Statue of Liberty in Figure 5.73, you should be at the position where θ is a maximum. If the statue stands 92 meters high, including the pedestal, which is 46 meters high, how far from the base should you be? [Hint: Find a formula for θ in terms of your distance from the base. Use this function to maximize θ, noting that $0 \le \theta \le \pi/2$.]

Figure 5.73

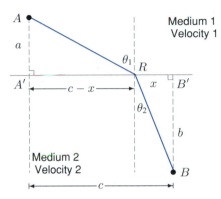

Figure 5.74

21. When a ray of light travels from one medium to another (for example, from air to water), it changes direction. This phenomenon is known as *refraction*. In Figure 5.74, light is traveling from A to B. The amount of refraction depends on the velocities, v_1 and v_2, of light in the two media and on *Fermat's Principle* which states that the light's time of travel, T, from A to B, is a minimum.

    (a) Find an expression for T in terms of x and the constants a, b, v_1, v_2, and c.
    (b) Show that if R is chosen so that the time of travel is minimized, then

    $$\frac{\sin \theta_1}{\sin \theta_2} = \frac{v_1}{v_2}.$$

    This result is known as *Snell's Law* and the ratio v_1/v_2 is called the *index of refraction* of the second medium with respect to the first medium.

22. Show that when the value of x in Figure 5.74 is chosen according to Snell's Law, the time taken by the light ray is a minimum.

5.6 HYPERBOLIC FUNCTIONS

There are two combinations of e^x and e^{-x} which are used so often in engineering that they are given their own name. They are the *hyperbolic sine*, abbreviated sinh, and the *hyperbolic cosine*, abbreviated cosh. They are defined as follows:

Hyperbolic Functions

$$\cosh x = \frac{e^x + e^{-x}}{2} \qquad \sinh x = \frac{e^x - e^{-x}}{2}$$

Properties of Hyperbolic Functions

The graphs of the $\cosh x$ and $\sinh x$ are given in Figures 5.75 and 5.76 together with the graphs of multiples of e^x and e^{-x}. The graph of $\cosh x$ is called a *catenary*; it is the shape of a hanging cable.

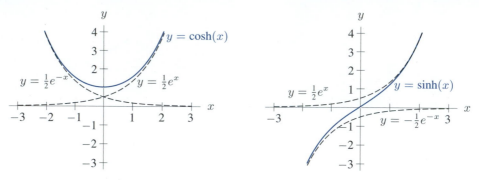

Figure 5.75: Graph of $y = \cosh x$ **Figure 5.76:** Graph of $y = \sinh x$

The graphs suggest that the following results hold:

$$\cosh 0 = 1 \qquad \sinh 0 = 0$$

$$\cosh(-x) = \cosh x \qquad \sinh(-x) = -\sinh x$$

To show that the hyperbolic functions really do have these properties, we use their formulas.

Example 1   Show that (a) $\cosh(0) = 1$ (b) $\cosh(-x) = \cosh x$

Solution   (a)   Substituting $x = 0$ into the formula for $\cosh x$ gives

$$\cosh 0 = \frac{e^0 + e^{-0}}{2} = \frac{1 + 1}{2} = 1.$$

(b)   Substituting $-x$ for x gives

$$\cosh(-x) = \frac{e^{-x} + e^{-(-x)}}{2} = \frac{e^{-x} + e^{x}}{2} = \cosh x.$$

Thus, we know that $\cosh x$ is an even function.

Example 2   Describe and explain the behavior of $\cosh x$ as $x \to \infty$ and $x \to -\infty$.

Solution   From Figure 5.75, it appears that as $x \to \infty$, the graph of $\cosh x$ resembles the graph of $\frac{1}{2}e^x$. Similarly, as $x \to -\infty$, the graph of $\cosh x$ resembles the graph of $\frac{1}{2}e^{-x}$. This behavior is explained by using the formula for $\cosh x$ and the facts that $e^{-x} \to 0$ as $x \to \infty$ and $e^x \to 0$ as $x \to -\infty$:

$$\text{As } x \to \infty, \qquad \cosh x = \frac{e^x + e^{-x}}{2} \to \frac{1}{2}e^x.$$

$$\text{As } x \to -\infty, \qquad \cosh x = \frac{e^x + e^{-x}}{2} \to \frac{1}{2}e^{-x}.$$

Identities Involving cosh *x* and sinh *x*

The reason the hyperbolic functions have names that remind us of the trigonometric functions is that they share some similar properties. A familiar identity for trigonometric functions is

$$(\cos x)^2 + (\sin x)^2 = 1.$$

To discover an analogous identity relating $(\cosh x)^2$ and $(\sinh x)^2$, we first calculate

$$(\cosh x)^2 = \left(\frac{e^x + e^{-x}}{2}\right)^2 = \frac{e^{2x} + 2e^x e^{-x} + e^{-2x}}{4} = \frac{e^{2x} + 2 + e^{-2x}}{4}$$

$$(\sinh x)^2 = \left(\frac{e^x - e^{-x}}{2}\right)^2 = \frac{e^{2x} - 2e^x e^{-x} + e^{-2x}}{4} = \frac{e^{2x} - 2 + e^{-2x}}{4}.$$

If we add these expressions, the resulting right-hand side contains terms involving both e^{2x} and e^{-2x}. If, however, we subtract the expressions for $(\cosh x)^2$ and $(\sinh x)^2$, we obtain a simple result:

$$(\cosh x)^2 - (\sinh x)^2 = \frac{e^{2x} + 2 + e^{-2x}}{4} - \frac{e^{2x} - 2 + e^{-2x}}{4} = \frac{4}{4} = 1.$$

Thus, writing $\cosh^2 x$ for $(\cosh x)^2$ and $\sinh^2 x$ for $(\sinh x)^2$, we have the identity

$$\boxed{\cosh^2 x - \sinh^2 x = 1}$$

The Hyperbolic Tangent

Extending the analogy to the trigonometric functions, we define

$$\boxed{\tanh x = \frac{\sinh x}{\cosh x}}$$

Derivatives of Hyperbolic Functions

We calculate the derivatives using the fact that $\frac{d}{dx}(e^x) = e^x$. The results are again reminiscent of the trigonometric functions. For example,

$$\frac{d}{dx}(\cosh x) = \frac{d}{dx}\left(\frac{e^x + e^{-x}}{2}\right) = \frac{e^x - e^{-x}}{2} = \sinh x.$$

We find $\frac{d}{dx}(\sinh x)$ similarly, giving the following results:

$$\boxed{\frac{d}{dx}(\cosh x) = \sinh x \qquad \frac{d}{dx}(\sinh x) = \cosh x}$$

Example 3   Compute the derivative of $\tanh x$.

Solution   Using the quotient rule gives

$$\frac{d}{dx}(\tanh x) = \frac{d}{dx}\left(\frac{\sinh x}{\cosh x}\right) = \frac{(\cosh x)^2 - (\sinh x)^2}{(\cosh x)^2} = \frac{1}{\cosh^2 x}.$$

Problems for Section 5.6

1. Show that $\sinh 0 = 0$.

2. Show that $\sinh(-x) = -\sinh(x)$.

3. Describe and explain the behavior of $\sinh x$ as $x \to \infty$ and as $x \to -\infty$.

4. Is there an identity analogous to $\sin(2x) = 2\sin x \cos x$ for the hyperbolic functions? Explain.

5. Is there an identity analogous to $\cos(2x) = \cos^2 x - \sin^2 x$ for the hyperbolic functions? Explain.

6. Show that $\dfrac{d}{dx}(\sinh x) = \cosh x$.

Find the derivatives of the functions in Problems 7–11.

7. $y = \cosh(2x)$
8. $y = \sinh(3z + 5)$
9. $f(t) = \cosh(e^{t^2})$

10. $f(y) = \sinh\left(\sinh(3y)\right)$
11. $g(\theta) = \ln\left(\cosh(1 + \theta)\right)$

12. Consider the family of functions $y = a\cosh(x/a)$ for $a > 0$. Sketch graphs for $a = 1, 2, 3$. Describe in words the effect of increasing a.

13. (a) Using a calculator or computer, sketch the graph of $y = 2e^x + 5e^{-x}$ for $-3 \le x \le 3, 0 \le y \le 20$. Observe that it looks like the graph of $y = \cosh x$. Approximately where is its minimum?
    (b) Show algebraically that $y = 2e^x + 5e^{-x}$ can be written in the form $y = A\cosh(x - c)$. Calculate the values of A and c. Explain what this tells you about the graph in part (a).

14. The following problem is a generalization of Problem 13. Show that any function of the form

$$y = Ae^x + Be^{-x}, \quad A > 0, \ B > 0,$$

can be written, for some K and c, in the form

$$y = K\cosh(x - c).$$

What does this tell you about the graph of $y = Ae^x + Be^{-x}$?

15. Consider the family of functions of the form $y = Ae^x + Be^{-x}$, for any constants A, B.

    (a) Sketch the function for

    (i) $A = 1, \ B = 1$ (ii) $A = 1, \ B = -1$ (iii) $A = 2, \ B = 1$
    (iv) $A = 2, \ B = -1$ (v) $A = -2, \ B = -1$ (vi) $A = -2, \ B = 1$

    (b) Describe in words the general shape of the graph if A and B have the same sign. What effect does the sign of A have on the graph?
    (c) Describe in words the general shape of the graph if A and B have different signs. What effect does the sign of A have on the graph?
    (d) For what values of A and B does the function have a local maximum? A local minimum? Justify your answer using derivatives.

16. The cable between the two towers of a suspension bridge hangs in the shape of the curve

$$y = \frac{T}{w}\cosh\left(\frac{wx}{T}\right),$$

where T is the tension in the cable at its lowest point and w is the weight of the cable per unit length. This curve is called a *catenary*.

    (a) Suppose the cable stretches between the points $x = -T/w$ and $x = T/w$. Find an expression for the "sag" in the cable. (That is, find the difference between the height of the cable at the highest and lowest points.)
    (b) Show that the shape of the cable satisfies the differential equation

$$\frac{d^2y}{dx^2} = \frac{w}{T}\sqrt{1 + \left(\frac{dy}{dx}\right)^2}.$$

17.   The Saint Louis arch can be approximated by using a function of the form $y = b - a\cosh(x/a)$. Putting the origin on the ground in the center of the arch and the y-axis upward, find an approximate equation for the arch given the dimensions shown in Figure 5.77. (In other words, find a and b.)

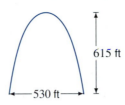

Figure 5.77

CHAPTER SUMMARY

- **Local extrema**
  Maximum, minimum, critical point, tests for local maxima/minima
- **Using the second derivative**
  Concavity, inflection point
- **Families of curves**
- **Optimization**
  Global extremum, modeling problems, graphical optimization, upper and lower bounds
- **Marginality**
  Cost/revenue functions, marginal cost/marginal revenue functions
- **Hyperbolic functions**

REVIEW PROBLEMS FOR CHAPTER FIVE

For Problems 1–2, indicate all critical points on the given graphs. Determine which correspond to local minima, local maxima, global maxima, global minima, or none of these. (Note that the graphs are on closed intervals.)

1.
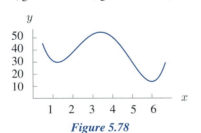

Figure 5.78

2.
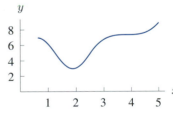

Figure 5.79

For the graphs of f' in Problems 3–6 decide:
(a)   Over what intervals is f increasing? Decreasing?
(b)   Does f have maxima or minima? If so, which, and where?

3.
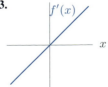

4.
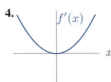

5.
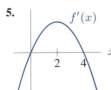

6.
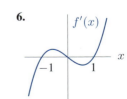

For each of the functions in Problems 7–10, do the following:

(a) Find f' and f''.

(b) Find the critical points of f.

(c) Find any inflection points.

(d) Evaluate f at the critical points and the endpoints. Identify the global maxima and minima of f.

(e) Sketch f. Indicate clearly where f is increasing or decreasing, and its concavity.

7. $f(x) = x^3 - 3x^2$ $(-1 \le x \le 3)$

8. $f(x) = x + \sin x$ $(0 \le x \le 2\pi)$

9. $f(x) = e^{-x} \sin x$ $(0 \le x \le 2\pi)$

10. $f(x) = x^{-2/3} + x^{1/3}$ $(1.2 \le x \le 3.5)$

For each of the functions in Problems 11–13, find the limits as x tends to $+\infty$ and $-\infty$, and then proceed as in Problems 7–10 (That is, find f', etc.).

11. $f(x) = 2x^3 - 9x^2 + 12x + 1$

12. $f(x) = \dfrac{4x^2}{x^2 + 1}$

13. $f(x) = xe^{-x}$

14. Determine local maxima and minima and points of inflection of $e^{-x^2/2}$. Sketch a graph.

For each of the functions in Problems 15–20, use derivatives to identify local maxima and minima and points of inflection. Sketch a graph of the function. Confirm your answers using a calculator or computer.

15. $f(x) = x^3 + 3x^2 - 9x - 15$ **16.** $f(x) = x^5 - 15x^3 + 10$ **17.** $f(x) = x - 2 \ln x$ for $x > 0$

18. $f(x) = x^2 e^{5x}$

19. $f(x) = e^{-x^2}$

20. $f(x) = \dfrac{x^2}{x^2 + 1}$

21. On the graph of the derivative function f' in Figure 5.80, indicate the x-values that are critical points of the function f itself. Are they local maxima, local minima, or neither?

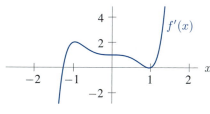

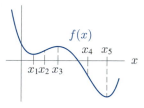

Figure 5.80: Graph of f', not f *Figure 5.81*

22. For the function, f, given in the graph in Figure 5.81:

(a) Sketch $f'(x)$.

(b) Where does $f'(x)$ change its sign?

(c) Where does $f'(x)$ have local maxima or minima?

23. Using your answer to Problem 22 as a guide, write a short paragraph (using complete sentences) which describes the relationships between the following features of a function f:

(a) The local maxima and minima of f.

(b) The points at which the graph of f changes concavity.

(c) The sign changes of f'.

(d) The local maxima and minima of f'.

24.   A square-bottomed box with a top has a fixed volume, V. What dimensions minimize the surface area?

25.   A square-bottomed box with no top has a fixed volume, V. What dimensions minimize the surface area?

Find the best possible bounds for the functions in Problems 26–27.

26. $e^{-x} \sin x$,   for $x \geq 0$

27. $x \sin x$,   for $0 \leq x \leq 2\pi$

28.   Sketch several members of the family $y = x^3 - ax^2$ on the same axes. Show that the critical points lie on the curve $y = -\frac{1}{2}x^3$.

29.   Suppose $g(t) = (\ln t)/t$ for $t > 0$.

   (a)   Does g have either a global maximum or a global minimum on $0 < t < \infty$? If so, where, and what are their values?

   (b)   What does your answer to part (a) tell you about the number of solutions to the equation

$$\frac{\ln x}{x} = \frac{\ln 5}{5}?$$

   (Note: There are many ways to investigate the number of solutions to this equation. We are asking you to draw a conclusion from your answer to part (a).)

   (c)   Estimate the solution(s).

30.   Populations with a growth limit have been modeled with the logistic family

$$y = \frac{A}{1 + Be^{-Cx}} \quad \text{for} \quad -\infty < x < \infty \quad \text{and} \quad A, B, C > 0.$$

   (a)   Sketch a graph of $g(x) = A/\left(1 + e^{-Cx}\right)$. What is the significance of the parameter A?
   (b)   Show that $g(-x) + g(x) = A$. What does this result mean about the graphs of $g(x)$ and $g(-x)$?
   (c)   What happens to the graph of g if A is kept constant and C is increased?
   (d)   Show that the curve $y = A/\left(1 + Be^{-Cx}\right)$ is a horizontal shift of the graph of g.

31.   For $a > 0$, the line

$$a(a^2 + 1)y = a - x$$

   forms a triangle in the first quadrant with the x- and y-axes.

   (a)   Find, in terms of a, the x- and y-intercepts of the line.
   (b)   Find the area of the triangle, as a function of a.
   (c)   Find the value of a making the area a maximum.
   (d)   What is this greatest area?
   (e)   If you want the triangle to have area $1/5$, what choices do you have for a?

32.   A ship is steaming due north at 12 knots (1 knot = 1.85 kilometers/hour) and sights a large tanker 3 kilometers away northwest steaming at 15 knots due east. For reasons of safety, the ships want to maintain a distance of at least 100 meters between them. Use a calculator or computer to determine the shortest distance between them if they remain on their current headings, and hence decide if they need to change course.

33.   Consider the vase in Figure 5.82 on page 284. Assume the vase is filled with water at a constant rate (i.e., constant volume per unit time).

   (a)   Graph $y = f(t)$, the depth of the water, against time, t. Show on your graph the points at which the concavity changes.
   (b)   At what depth is $y = f(t)$ growing most quickly? Most slowly? Estimate the ratio between the growth rates at these two depths.

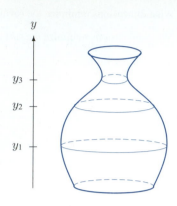

intensity of radiation
$r(\lambda)$ (MW/m^2/μm)

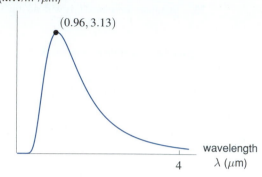

Figure 5.82   **Figure 5.83**

34.   Any body radiates energy at various wavelengths. The power of the radiation (per meter^2 of surface) and the distribution of the radiation among the wavelengths varies with temperature.

The function graphed in Figure 5.83 represents the intensity of the radiation of a black body at a temperature $T = 3000°$ Kelvin, as a function of the wavelength. The intensity of the radiation is highest in the infrared range, that is, at wavelengths longer than that of visible light (0.4–0.7μm). Max Planck's radiation law, announced to the Berlin Physical Society on October 19, 1900, states that

$$r(\lambda) = \frac{a}{\lambda^5(e^{b/\lambda} - 1)},$$

where a and b are empirical constants chosen to best fit the experimental data. Find a and b so that the formula fits the graph.

(Later in 1900 Planck was able to derive his radiation law entirely from theory. He found that $a = 2\pi c^2 h$ and $b = \frac{hc}{Tk}$ where c = speed of light, h = Planck's constant, and k = Boltzmann's constant.)

PROJECTS

1.   **Resonance**

The study of resonance leads to the family

$$y = \frac{1}{(1 - x^2)^2 + 2ax^2}, \qquad x \geq 0, \quad a > 0.$$

(a)   Find and classify the critical points, and then explain why the family is most interesting for $0 < a < 1$.

(b)   Show that for a very near 0, there is a critical point at approximately $(1, 1/2a)$.

(c)   Show that $y < 1/(1 - x^2)^2$ for $x > 0$.

(d)   Sketch the curves for $a = 0.05, 0.10, 1.0$, and 3.0 on one set of axes together with $y = 1/(1 - x^2)^2$, which is the curve with $a = 0$.

(e)   What is the geometric significance of the parameter a?

2. **Building a Greenhouse**[5]

Your parents are going to knock out the bottom of the entire length of the south wall of their house and turn it into a greenhouse by replacing the bottom portion of the wall with a huge sloped piece of glass (which is expensive). They have already decided they are going to spend a certain fixed amount. The triangular ends of the greenhouse will be made of various materials they already have lying around.

The floor space in the greenhouse is only considered usable if they can both stand up in it, so part of it will be unusable. They want to choose the dimensions of the greenhouse to get the most usable floor space. What should the dimensions of the greenhouse be and how much usable space will your parents get?

3. **Average and Marginal Costs**

Suppose $C(q)$ is the total cost of producing a quantity q. The average cost $a(q)$ is given in Figure 5.84. The following rule is used by economists to determine the marginal cost $C'(q_0)$, for any q_0:

- Construct the tangent t_1 to $a(q)$ at q_0.
- Let t_2 be the line with the same vertical intercept as t_1 but with twice the slope of t_1.

Then $C'(q_0)$ is as shown in Figure 5.84. Explain why this rule works.

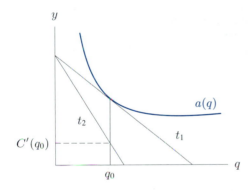

Figure 5.84

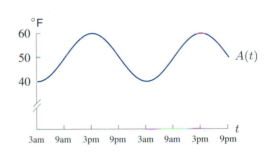

Figure 5.85

4. **Air and Water Temperature**

Consider a large tank of water, with temperature $W(t)$. The ambient temperature $A(t)$ (that is, the temperature of the surrounding air), is given by the graph in Figure 5.85. The temperature of the water is affected by the temperature of the surrounding air.

(a) How does the temperature of the water change if the water is colder than the surrounding air? What if the water is warmer?

(b) Assume that the temperature of the water is 50°F at 3 am. Using your answer in part (a), sketch a possible graph for $W(t)$ on the same axes as $A(t)$.

(c) Explain the relationship between the maxima and minima of $W(t)$ and the points where the two graphs intersect.

(d) What is the relationship between the rate at which the temperature of the water changes and the difference $A(t) - W(t)$?

(e) What is the relationship between the inflection points of $W(t)$ and the points where $A(t) - W(t)$ has a maxima or minima?

(f) Assume the tank is filled with cold water (35°F) at 3 am. Sketch a possible graph for $W(t)$. Pay attention to the concavity.

---

[5]Adapted from M. Cohen, E. Gaughan, A. Knoebel, D. Kurtz, D. Pengelley, *Student Research Projects in Calculus* (Washington DC: Mathematical Association of America, 1992).

FOCUS ON THEORY

THEOREMS ABOUT CONTINUOUS AND DIFFERENTIABLE FUNCTIONS

In this chapter we have used some basic facts without proof: for example, that a continuous function has a maximum on a bounded, closed interval, or that a function whose derivative is positive on an interval is increasing on that interval.

From a geometric point of view, these facts seem obvious. If we draw the graph of a continuous function, starting at one end of a bounded, closed interval and going to the other, it seems obvious that we must pass a highest point on the way. And if the derivative of a function is positive, then its graph must be sloping up, so the function has to be increasing.

However, this sort of graphical reasoning is not a rigorous proof, for two reasons. First, no matter how many pictures we imagine, we can't be sure we have covered all possibilities. Second, our pictures often depend on the theorems we are trying to prove.

A Continuous Function on a Closed Interval Has a Maximum

> **The Extreme Value Theorem**
> If f is continuous on the interval $[a, b]$, then f has a global maximum and a global minimum on that interval.

Our proof has two parts: The first is to show that f has an upper bound on $[a, b]$, the second is to show that if f has an upper bound then it has a global maximum on the interval. Here we prove the second part; the first part is proved in Problems 16 and 17. Then in Problem 5 we extend the result from maxima to minima.

Proof   We assume that f is continuous and has an upper bound on the interval $[a, b]$. This means f has a least upper bound M on $[a, b]$. Divide $[a, b]$ into two halves. Then, on one of the halves, the least upper bound for f is M, for if it were less than M on both halves, it would be less than M on the whole. Choose a half on which the least upper bound is equal to M. Continue bisecting and at each stage choose the half-interval where the least upper bound for f is M. See Figure 5.86. This results in a sequence of nested intervals. By the Nested Interval Theorem on page 84, there is a number c in $[a, b]$ which is contained in all these intervals.

Since M is the least upper bound for f, we have $f(c) \leq M$. It is not possible that $f(c) < M$. For if $f(c) < M$, then $f(c) < M_0$ for some number $M_0 < M$. (For example, we could take M_0 to be half-way between M and $f(c)$.) But then, since f is continuous, there would be a $\delta > 0$ such that $f(x) < M_0$ for all x in $[a, b]$ with $c - \delta < x < c + \delta$. (See Problem 15.) Since the nested

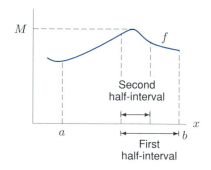

Figure 5.86: Successively choosing the half-interval where the least upper bound of f is M

intervals we constructed above have width tending to zero, one of them would be contained in the interval $c - \delta < x < c + \delta$. Therefore, f would be bounded above by M_0 on one of the nested intervals. However, we chose each nested interval so that the least upper bound for f is M. This is a contradiction of $M_0 < M$.

So it is not possible that $f(c) < M$; we must have $f(c) = M$. Thus, M is the global maximum of f on $[a, b]$, which is what we wanted to show.

The Extreme Value Theorem guarantees the existence of global maxima (and minima) on an interval. To actually find the global maxima, we look at all the local maxima. The following theorem tells us that inside an interval, local maxima only occur at critical points, where the derivative is either zero or undefined.

> **Theorem: Local Extrema and Critical Points**
> Suppose f is defined on an interval and has a local maximum or minimum at the point $x = a$, which is not an endpoint of the interval. If f is differentiable at $x = a$, then $f'(a) = 0$.

Proof We start with the definition of the derivative:

$$f'(a) = \lim_{h \to 0} \frac{f(a + h) - f(a)}{h}.$$

Remember that this is a two-sided limit:

$$f'(a) = \lim_{h \to 0^-} \frac{f(a + h) - f(a)}{h} = \lim_{h \to 0^+} \frac{f(a + h) - f(a)}{h}.$$

Suppose that f has a local maximum at $x = a$. By the definition of local maximum, $f(a+h) \leq f(a)$ for all sufficiently small h. Thus $f(a + h) - f(a) \leq 0$ for sufficiently small h. The denominator, h, is positive when we take the limit from the right and negative when we take the limit from the left. Thus

$$\lim_{h \to 0^-} \frac{f(a + h) - f(a)}{h} \geq 0 \quad \text{and} \quad \lim_{h \to 0^+} \frac{f(a + h) - f(a)}{h} \leq 0.$$

Since both these limits are equal to $f'(a)$, we have $f'(a) \geq 0$ and $f'(a) \leq 0$, so we must have $f'(a) = 0$.

A Relationship Between Local and Global: The Mean Value Theorem

We often want to infer a global conclusion (for example, f is increasing on an interval) from local information (f' is positive.) The following theorem relates the average rate of change of a function on an interval (global information) to the instantaneous rate of change at a point in the interval (local information).

> **The Mean Value Theorem**
> If f is continuous on $[a, b]$ and differentiable on (a, b), then there exists a number c, with $a < c < b$, such that
> $$f'(c) = \frac{f(b) - f(a)}{b - a}.$$
> In other words, $f(b) - f(a) = f'(c)(b - a)$.

To understand what this theorem is saying geometrically, consider the graph in Figure 5.87. Join the points on the curve where $x = a$ and $x = b$ with a line and observe that the slope of this secant line AB is given by

$$m = \frac{f(b) - f(a)}{b - a}.$$

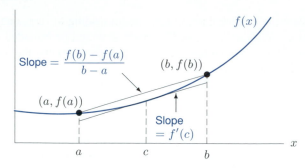

Figure 5.87: The point c with $f'(c) = \frac{f(b)-f(a)}{b-a}$

Now consider the tangent line drawn to the curve at each point between $x = a$ and $x = b$. In general, these lines will have different slopes. For the curve shown in Figure 5.87, the tangent line at $x = a$ is flatter than the secant line from A to B. Similarly, the tangent line at $x = b$ is steeper than the secant line. However, there is at least one point between a and b where the slope of the tangent line to the curve is precisely the same as the slope of the secant line. Suppose this occurs at $x = c$. Then

$$f'(c) = m = \frac{f(b) - f(a)}{b - a}.$$

The Mean Value Theorem tells us that the point $x = c$ exists, but it does not tell us how to find c.

Problems 18 and 19 show how the Mean Value Theorem can be deduced from the Extreme Value Theorem.

The Increasing Function Theorem

We say that a function f is *increasing* on an interval if, for any two numbers x_1 and x_2 in the interval such that $x_1 < x_2$, we have $f(x_1) < f(x_2)$. If instead we have $f(x_1) \leq f(x_2)$, we say f is *nondecreasing*.

> **The Increasing Function Theorem**
> Suppose that f is continuous on $[a, b]$ and differentiable on (a, b).
> - If $f'(x) > 0$ on (a, b), then f is increasing on $[a, b]$.
> - If $f'(x) \geq 0$ on (a, b), then f is nondecreasing on $[a, b]$.

Proof   Suppose $a \leq x_1 < x_2 \leq b$. By the Mean Value Theorem, there is a number c, with $x_1 < c < x_2$, such that

$$f(x_2) - f(x_1) = f'(c)(x_2 - x_1).$$

If $f'(c) > 0$, this says $f(x_2) - f(x_1) > 0$, which means f is increasing. If $f'(c) \geq 0$, this says $f(x_2) - f(x_1) \geq 0$, which means f is nondecreasing.

It may seem that something as simple as the Increasing Function Theorem should follow immediately from the definition of the derivative, and that the use of the Mean Value Theorem (which in turn depends on the Extreme Value Theorem) is surprising. A more direct, self-contained proof is given in Problem 11, but that proof also has some subtleties.

The Constant Function Theorem

If f is constant on an interval, then we know that $f'(x) = 0$ on the interval. The following theorem is the converse.

> **The Constant Function Theorem**
> Suppose that f is continuous on $[a, b]$ and differentiable on (a, b). If $f'(x) = 0$ on (a, b), then f is constant on $[a, b]$.

Proof The proof is the same as for the Increasing Function Theorem, only in this case $f'(c) = 0$ so $f(x_2) - f(x_1) = 0$. Thus $f(x_2) = f(x_1)$ for $a \leq x_1 < x_2 \leq b$, so f is constant.

A proof of the Constant Function Theorem using the Increasing Function Theorem is given in Problems 6 and 8.

The Racetrack Principle

> **The Racetrack Principle[6]**
> Suppose that g and h are continuous on $[a, b]$ and differentiable on (a, b), and that $g'(x) \leq h'(x)$ for $a < x < b$.
> - If $g(a) = h(a)$, then $g(x) \leq h(x)$ for $a \leq x \leq b$.
> - If $g(b) = h(b)$, then $g(x) \geq h(x)$ for $a \leq x \leq b$.

The Racetrack Principle has the following interpretation. We can think of $g(x)$ and $h(x)$ as the positions of two racehorses at time x, with horse h always moving faster than horse g. If they start together, horse h is ahead during the whole race. If they finish together, horse g was ahead during the whole race.

Proof Consider the function $f(x) = h(x) - g(x)$. Since $f'(x) = h'(x) - g'(x) \geq 0$, we know that f is nondecreasing by the Increasing Function Theorem. So $f(x) \geq f(a) = h(a) - g(a) = 0$. Thus $g(x) \leq h(x)$ for $a \leq x \leq b$. This proves the first part of the Racetrack Principle. Problem 7 asks for a proof of the second part.

Example 1 Explain graphically why $e^x \geq 1 + x$ for all values of x. Then use the Racetrack Principle to prove the inequality.

Solution The graph of the function $f(x) = e^x$ is concave up everywhere and the equation of its tangent line at the point $(0, 1)$ is $y = x + 1$. (See Figure 5.88.) Since the graph always lies above its tangent, we have the inequality

$$e^x \geq 1 + x.$$

Now we prove the inequality using the Racetrack Principle. Let $g(x) = 1 + x$ and $h(x) = e^x$. Then $g(0) = h(0) = 1$. Furthermore, $g'(x) = 1$ and $h'(x) = e^x$. Hence $g'(x) \leq h'(x)$ for $x \geq 0$. So by the Racetrack Principle, with $a = 0$, we have $g(x) \leq h(x)$, that is, $1 + x \leq e^x$.

For $x \leq 0$ we have $h'(x) \leq g'(x)$. So by the Racetrack Principle, with $b = 0$, we have $g(x) \leq h(x)$, that is, $1 + x \leq e^x$.

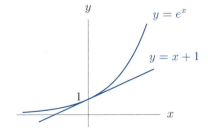

Figure 5.88: Graph showing that $e^x \geq 1 + x$

[6]Based on the Racetrack Principle in *Calculus&Mathematica*, by William Davis, Horacio Porta, Jerry Uhl (Reading: Addison Wesley, 1994).

Problems on the Theorems about Continuous and Differentiable Functions ━━━━━

1. Use the Racetrack Principle and the fact that $\sin 0 = 0$ to show that $\sin x \le x$ for all $x \ge 0$.

2. Use the Racetrack Principle to show that $\ln x \le x - 1$.

3. Use the fact that $\ln x$ and e^x are inverse functions to show that the inequalities $e^x \ge 1 + x$ and $\ln x \le x - 1$ are equivalent for $x > 0$.

4. Suppose that the position of a particle moving along the x-axis is given by $s = f(t)$, and that the initial position and velocity of the particle are $f(0) = 3$ and $f'(0) = 4$. Suppose that the acceleration is bounded by $5 \le f''(t) \le 7$ for $0 \le t \le 2$. What can we say about the position $f(2)$ of the particle at $t = 2$?

5. Show that if every continuous function on an interval $[a, b]$ has a global maximum, then every continuous function has a global minimum as well. [Hint: Consider $-f$.]

6. State a Decreasing Function Theorem, analogous to the Increasing Function Theorem. Deduce your theorem from the Increasing Function Theorem. [Hint: Apply the Increasing Function Theorem to $-f$.]

7. Suppose that g and h are continuous on $[a, b]$ and differentiable on (a, b). Prove that if $g'(x) \le h'(x)$ for $a < x < b$ and $g(b) = h(b)$, then $h(x) \le g(x)$ for $a \le x \le b$.

8. Deduce the Constant Function Theorem from the Increasing Function Theorem and the Decreasing Function Theorem (see problem 6).

9. Prove that if $f'(x) = g'(x)$ for all x in (a, b), then there is a constant C such that $f(x) = g(x) + C$ on (a, b). [Hint: Apply the Constant Function Theorem to $h(x) = f(x) - g(x)$.]

10. Suppose that $f'(x) = f(x)$ for all x. Prove that $f(x) = Ce^x$ for some constant C. [Hint: Consider the function $f(x)/e^x$.]

11. In this problem we give a proof of the Increasing Function Theorem that does not use the Mean Value Theorem. Suppose that f is continuous on $[a, b]$ and differentiable on (a, b), and that $f'(x) \ge 0$ for all x in (a, b).

    (a) We write $\text{Slope}(c, d)$ to denote the slope between $(c, f(c))$ and $(d, f(d))$, so

    $$\text{Slope}(c, d) = \frac{f(d) - f(c)}{d - c}.$$

    Show that if $c < e < d$, then

    $$\text{Slope}(c, d) = \left(\frac{e - c}{d - c}\right) \cdot \text{Slope}(c, e) + \left(\frac{d - e}{d - c}\right) \cdot \text{Slope}(e, d).$$

    Use this to show that $\text{Slope}(c, d)$ lies between $\text{Slope}(c, e)$ and $\text{Slope}(e, d)$.

    (b) First we prove f is nondecreasing on (a, b), that is, if $a < a_1 < b_1 < b$, then $f(a_1) \le f(b_1)$. Suppose, contrary to what we want, that $f(a_1) > f(b_1)$, so that $\text{Slope}(a_1, b_1)$ is negative. Show that there is a sequence of nested intervals $[a_n, b_n]$, each of which is one half of the previous one, such that $\text{Slope}(a_n, b_n) \le \text{Slope}(a_1, b_1)$ for all n.

    (c) By the Nested Interval Theorem, there is a number c in (a, b) which is contained in all the intervals $[a_n, b_n]$. Show that, for all n, either $\text{Slope}(a_n, c)$ or $\text{Slope}(c, b_n)$ is less than or equal to $\text{Slope}(a_1, b_1)$.

    (d) Using $\lim_{x \to c} \text{Slope}(x, c) = \lim_{x \to c} \text{Slope}(c, x) = f'(c) \ge 0$, show that for large enough n, both $\text{Slope}(a_n, c)$ and $\text{Slope}(c, b_n)$ are greater than $\text{Slope}(a_1, b_1)$, contradicting part (c).

    (e) The contradiction in part (d) proves that our assumption $f(a_1) > f(b_1)$ in part (b) must be false. Therefore, $f(a_1) \le f(b_1)$. Since this is true for all a_1 and b_1 such that $a < a_1 < b_1 < b$, we have shown that f is nondecreasing on (a, b). Use the continuity of f to deduce that f is nondecreasing on $[a, b]$.

    (f) Now suppose that $f'(x) > 0$ on (a, b). Show that if $a \le a_1 < b_1 \le b$, and $f(a_1) = f(b_1)$, then f is constant on $[a_1, b_1]$ so $f'(x) = 0$ on $[a_1, b_1]$. This contradicts $f'(x) > 0$, so it is not possible that $f(a_1) = f(b_1)$, and we must have $f(a_1) < f(b_1)$. Therefore, f is increasing on $[a, b]$.

12. Suppose that f is continuous on $[a,b]$ and differentiable on (a,b) and that $m \le f'(x) \le M$ on (a,b). Use the Racetrack Principle to prove that $f(x) - f(a) \le M(x-a)$ for all x in $[a,b]$, and that $m(x-a) \le f(x) - f(a)$ for all x in $[a,b]$. Conclude that $m \le (f(b) - f(a))/(b-a) \le M$. This is called the Mean Value Inequality. In words: If the instantaneous rate of change of f is between m and M on an interval, so is the average rate of change of f over the interval.

13. Suppose that $f''(x) \ge 0$ for all x in (a,b). We will show the graph of f lies above the tangent line at $(c, f(c))$ for any c with $a < c < b$.

    (a) Use the Increasing Function Theorem to prove that $f'(c) \le f'(x)$ for $c \le x < b$ and that $f'(x) \le f'(c)$ for $a < x \le c$.

    (b) Use (a) and the Racetrack Principle to conclude that $f(c) + f'(c)(x-c) \le f(x)$, for $a < x < b$.

14. In this problem we use the Mean Value Theorem to give a proof of the Fundamental Theorem of Calculus. Let f be continuous, with antiderivative F.

    (a) Let $[a,b]$ be an interval contained in the domain of f, and let
    $$a = x_0 < x_1 < \cdots < x_{n-1} < x_n = b$$
    be a subdivision of $[a,b]$. Show that there is a Riemann sum for f using this subdivision which is equal to $F(b) - F(a)$. [Hint: Apply the Mean Value Theorem to
    $$F(b) - F(a) = ((F(b) - F(x_{n-1})) + (F(x_{n-1}) - F(x_{n-2})) + \cdots + (F(x_1) - F(a))).]$$

    (b) Deduce that $F(b) - F(a) = \int_a^b f(x)\,dx$.

15. Suppose that f is continuous on $[a,b]$, and let c be in $[a,b]$. Show that if $f(c) < M$, then there is a δ such that $f(x) < M$ for all x in $[a,b]$ such that $c - \delta < x < c + \delta$. [Hint: Let $\epsilon = M - f(c)$, and choose δ such that $|f(x) - f(c)| < \epsilon$ if $|x - c| < \delta$.]

On page 286 we proved that a continuous function f has a global maximum on the interval $[a,b]$ under the assumption that f has an upper bound on $[a,b]$. In Problems 16–17 we prove this claim.

16. (a) Suppose that f has no upper bound on $[a,b]$. Bisect $[a,b]$ into two halves. Deduce that f has no upper bound on at least one of the halves. Call that half $[a_1, b_1]$.

    (b) Continue bisecting so that at the nth stage you obtain an interval $[a_n, b_n]$ on which f has no upper bound. By the Nested Interval Theorem on page 84, there is a point c in all the intervals $[a_n, b_n]$.

    (c) Use continuity of f at c to deduce that f has an upper bound on $[a_n, b_n]$ for n sufficiently large. This contradicts the original supposition, so f must have an upper bound on $[a,b]$.

17. (a) Show that if $y \ge 0$, then $y/(1+y) < 1$.

    (b) Suppose that f is continuous on $[a,b]$ and that $f(x) \ge 0$ on $[a,b]$. Define a function g by $g(x) = f(x)/(1 + f(x))$. Show that g is continuous and bounded on $[a,b]$. It follows from the partial proof of the Extreme Value Theorem on page 286 that g has a global maximum on $[a,b]$ at some point $x = c$.

    (c) Suppose that $y_1 \ge 0$ and $y_2 \ge 0$, and that $y_1/(1+y_1) \le y_2/(1+y_2)$. Show that $y_1 \le y_2$.

    (d) Use parts (c) and (d) to show that f has a global maximum at $x = c$.

    (e) We have shown that if f is continuous and non-negative on $[a,b]$, then it is bounded above on $[a,b]$. Now suppose that f is continuous, but not necessarily non-negative. By applying the argument to $|f|$, deduce that f is also bounded above.

18. In this problem we prove a special case of the Mean Value Theorem where $f(a) = f(b) = 0$. This special case is called Rolle's Theorem: If f is continuous on $[a,b]$ and differentiable on (a,b), and if $f(a) = f(b) = 0$, then there is a number c, with $a < c < b$, such that
$$f'(c) = 0.$$

By the Extreme Value Theorem, f has a global maximum and a global minimum on $[a,b]$.

    (a) Prove Rolle's theorem in the case that both the global maximum and the global minimum are at endpoints of $[a,b]$. [Hint: $f(x)$ must be a very simple function in this case.]

    (b) Prove Rolle's theorem in the case that either the global maximum or the global minimum is not at an endpoint. [Hint: Think about local maxima and minima.]

19. Use Rolle's Theorem to prove the Mean Value Theorem. Suppose that $f(x)$ is continuous on $[a, b]$ and differentiable on (a, b).

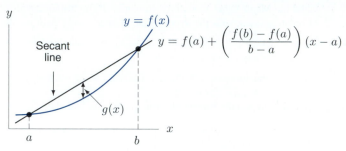

Figure 5.89: $g(x)$ is the difference between the secant line and the
graph of $f(x)$

(a) Let $g(x)$ be the difference between $f(x)$ and the y-value on the secant line joining $(a, f(a))$ to $(b, f(b))$. See Figure 5.89. Show that

$$g(x) = f(x) - f(a) - \frac{f(b) - f(a)}{b - a}(x - a).$$

(b) Use Rolle's Theorem to show that there must be a point c in (a, b) such that $g'(c) = 0$.
(c) Show that if c is the point in part (b), then

$$f'(c) = \frac{f(b) - f(a)}{b - a}.$$

CHAPTER SIX

CONSTRUCTING ANTIDERIVATIVES

In Chapter 2, we saw how to calculate velocity given position, and in Chapter 3, we saw how to reconstruct distance from velocity. In this chapter we will look in more detail at how to reconstruct a function from its derivative.

We already know how to get a rough idea of the graph of f from the graph of f' and how to use f' to estimate values of f numerically. We extend these ideas and we start to see how to go from f' to f analytically.

6.1 ANTIDERIVATIVES GRAPHICALLY AND NUMERICALLY

The Family of Antiderivatives

If the derivative of F is f, we call F an *antiderivative* of f. For example, since the derivative of x^2 is $2x$, we say that

$$x^2 \text{ is an antiderivative of } 2x.$$

Notice that $2x$ has many antiderivatives, since $x^2 + 1$, $x^2 + 2$, and $x^2 + 3$, all have derivative $2x$. In fact, if C is any constant, we have

$$\frac{d}{dx}(x^2 + C) = 2x + 0 = 2x,$$

and so any function of the form $x^2 + C$ is an antiderivative of $2x$. The function $f(x) = 2x$ has a *family of antiderivatives*.

Let us look at another example. If v is the velocity of a car and s is its position, then $v = ds/dt$ and s is an antiderivative of v. As before, $s + C$ is an antiderivative of v for any constant C. In terms of the car, adding C to s is equivalent to adding C to the odometer reading. Adding a fixed distance to the odometer reading simply means measuring distance from a different point, which doesn't alter the car's velocity.

Visualizing Antiderivatives Using Slopes

Suppose we have the graph of f', and we want to sketch an approximate graph of f. We are looking for the graph of f whose slope at any point is equal to the value of f' there. Where f' is above the x-axis, f is increasing; where f' is below the x-axis, f is decreasing. If f' is increasing, f is concave up; if f' is decreasing, f is concave down.

Example 1   The graph of f' is given in Figure 6.1. Sketch a graph of f in the cases when $f(0) = 0$ and $f(0) = 1$.

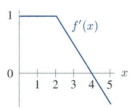

Figure 6.1: Graph of f'

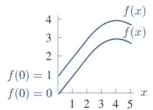

Figure 6.2: Two different f's which have the same derivative f'

Solution   For $0 \le x \le 2$, f has a constant slope of 1 and so the graph of f is a straight line. For $2 \le x \le 4$, f is increasing but more and more slowly; it has a maximum at $x = 4$ and decreases thereafter. (See Figure 6.2.)

Notice that the solutions with $f(0) = 0$ and $f(0) = 1$ start at different points on the vertical axis but have the same shape.

Example 2   Sketch a graph of the antiderivative F of $f(x) = e^{-x^2}$ satisfying $F(0) = 0$.

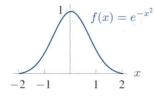

Figure 6.3: Graph of $f(x) = e^{-x^2}$

Figure 6.4: An antiderivative $F(x)$ of $f(x) = e^{-x^2}$

Solution   The graph of $f(x) = e^{-x^2}$ is shown in Figure 6.3. The slope of the antiderivative $F(x)$ is given by $f(x)$. Since $f(x)$ is always positive, the antiderivative $F(x)$ is always increasing. Since $f(x)$ is increasing for negative x, we know that $F(x)$ is concave up for negative x. Since $f(x)$ is decreasing for positive x, we know that $F(x)$ is concave down for positive x. Since $f(x) \to 0$ as $x \to \pm\infty$, $f(x) \to 0$ and so the graph of $F(x)$ levels off at both ends. A graph of $F(x)$ with $F(0) = 0$ is shown in Figure 6.4.

Example 3   For the function f' given in Figure 6.5, sketch a graph of three antiderivative functions f, where

(a) $f(0) = 0$,   (b) $f(0) = 1$,   (c) $f(0) = 2$.

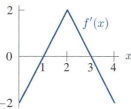

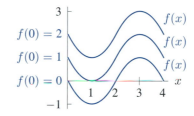

Figure 6.5: Slope function, f' *Figure 6.6:* Antiderivatives f

Solution   To draw a graph of f, start at the point on the vertical axis specified by the initial condition and move with slope given by the value of f'. (See Figure 6.6.) Different initial conditions lead to different graphs for f, but for a given x-value they all have the same slope (because the value of f' is the same for each). Thus, the different f curves are obtained from one another by a vertical shift.

- Where f' is positive ($1 < x < 3$), we see f is increasing; where f' is negative ($0 < x < 1$ or $3 < x < 4$), we see f is decreasing.
- Where f' is increasing ($0 < x < 2$), we see f is concave up; where f' is decreasing ($2 < x < 4$), we see f is concave down.
- Where $f' = 0$, we see f has a local maximum ($x = 3$) or local minimum ($x = 1$).
- Where f' has a maximum ($x = 2$), we see f has a point of inflection.

Computing Values of an Antiderivative Using the Fundamental Theorem

A graph of f' shows where f is increasing and where f is decreasing. We can find the actual value of the function f by using the Fundamental Theorem:

> **Fundamental Theorem of Calculus**
>
> If F is an antiderivative of a continuous function, f, then
>
> $$\int_a^b f(x)\,dx = F(b) - F(a).$$

Example 4   Figure 6.7 is the graph of the derivative $f'(x)$ of a function $f(x)$. It is given that $f(0) = 100$. Sketch the graph of $f(x)$, showing all critical points and inflection points of f and giving their coordinates.

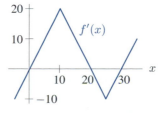

Figure 6.7: Graph of derivative

Solution   The critical points of f occur at $x = 0$, $x = 20$, and $x = 30$, where $f'(x) = 0$. The inflection points of f occur at $x = 10$ and $x = 25$, where $f'(x)$ has a maximum or minimum. To find the coordinates of the critical points and inflection points of f, we evaluate $f(x)$ for $x = 0, 10, 20, 25, 30$. We use the Fundamental Theorem to express the values of $f(x)$ in terms of definite integrals. We evaluate the definite integrals using the areas of triangular regions under the graph of $f'(x)$, remembering that areas below the x-axis are subtracted. (See Figure 6.8.)

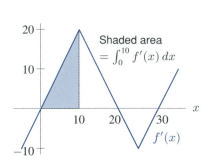

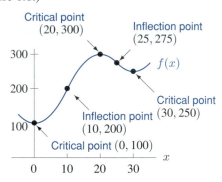

Figure 6.8: Finding $f(10) = f(0) + \int_0^{10} f'(x)\,dx$ **Figure 6.9:** Graph of $f(x)$

Since $f(0) = 100$, the Fundamental Theorem says

$$f(10) = f(0) + \int_0^{10} f'(x)\,dx = 100 + (\text{Shaded area in Figure 6.8}) = 100 + \frac{1}{2}(10)(20) = 200,$$

$$f(20) = f(10) + \int_{10}^{20} f'(x)\,dx = 200 + \frac{1}{2}(10)(20) = 300,$$

$$f(25) = f(20) + \int_{20}^{25} f'(x)\,dx = 300 - \frac{1}{2}(5)(10) = 275,$$

$$f(30) = f(25) + \int_{25}^{30} f'(x)\,dx = 275 - \frac{1}{2}(5)(10) = 250.$$

The graph of f is sketched in Figure 6.9.

Example 5   Suppose you are given that $F'(t) = t \cos t$ and $F(0) = 2$. Find the values of $F(b)$ at the points $b = 0, 0.1, 0.2, \ldots, 1.0$.

Solution   We apply the Fundamental Theorem with $f(t) = t \cos t$ and $a = 0$ to get values for $F(b)$. Since

$$F(b) - F(0) = \int_0^b F'(t)\,dt = \int_0^b t \cos t\,dt$$

and $F(0) = 2$, we have

$$F(b) = 2 + \int_0^b t \cos t\,dt.$$

Using numerical methods to estimate the definite integral $\int_0^b t \cos t\,dt$ for each of the values $b = 0$, $0.1, 0.2, \ldots, 1.0$ gives the approximate values for F in Table 6.1:

TABLE 6.1   *Approximate values for F*

b	0	0.1	0.2	0.3	0.4	0.5	0.6	0.7	0.8	0.9	1.0
$F(b)$	2.000	2.005	2.020	2.044	2.077	2.117	2.164	2.216	2.271	2.327	2.382

Notice from the table that the function $F(b)$ appears to be increasing between $b = 0$ and $b = 1$. This is in fact the case, and could have been predicted without the use of the Fundamental Theorem from the fact that $t \cos t$, the derivative of $F(t)$, is positive for t between 0 and 1.

Problems for Section 6.1

For each function in Problems 1–4, sketch two functions F such that $F' = f$. In one case let $F(0) = 0$ and in the other, let $F(0) = 1$.

1.

2.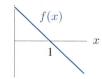

3.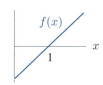

4.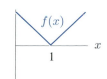

For each function $f(x)$ in Problems 5–7, assume $F(x)$ is such that $F'(x) = f(x)$.

(a) What are the critical points of $F(x)$?

(b) Which critical points are local maxima, local minima, or neither?

(c) Sketch a possible graph of $F(x)$.

5.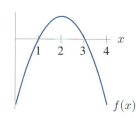

6.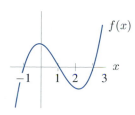

7.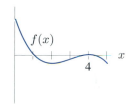

For each function in Problems 8–10, sketch two functions F where $F'(x) = f(x)$. In one case, let $F(0) = 0$, and in the other, let $F(0) = 1$. In each case, mark x_1, x_2, and x_3 on the x-axis of your graph. Identify local maxima, minima, and inflection points of $F(x)$.

8.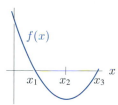

9.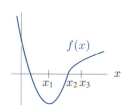

10.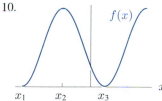

11. Assume f' is given by the graph in Figure 6.10. Suppose f is continuous and that $f(0) = 0$.

(a) Find $f(3)$ and $f(7)$.

(b) Find all x with $f(x) = 0$.

(c) Sketch a graph of f over the interval $0 \le x \le 7$.

12. Assume f' is given by the graph in Figure 6.10. Suppose f is continuous and that $f(3) = 0$.

(a) Sketch a graph of f.

(b) Find $f(0)$ and $f(7)$.

(c) Find $\int_0^7 f'(x)\, dx$ in two different ways.

13. Figure 6.11 shows the graph of the derivative $g'(x)$ of a function $g(x)$. It is given that $g(0) = 50$. Sketch the graph of $g(x)$, showing all critical points and inflection points of g and giving their coordinates.

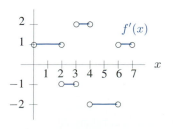

Figure 6.10

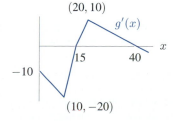

Figure 6.11

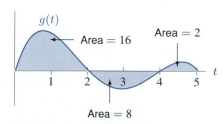

Figure 6.12

14.   A graph of $g(t)$ is given in Figure 6.12 on page 297. Sketch a graph of an antiderivative $G(t)$ of $g(t)$ satisfying $G(0) = 5$. Label each critical point of $G(t)$ with its coordinates.

15.   The vertical velocity of a cork bobbing up and down on the waves in the sea is given by Figure 6.13. Upward is considered positive. Describe the motion of the cork at each of the labeled points. At which point(s), if any, is the acceleration zero? Sketch a graph of the height of the cork above the sea floor as a function of time.

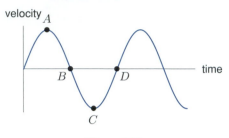

Figure 6.13

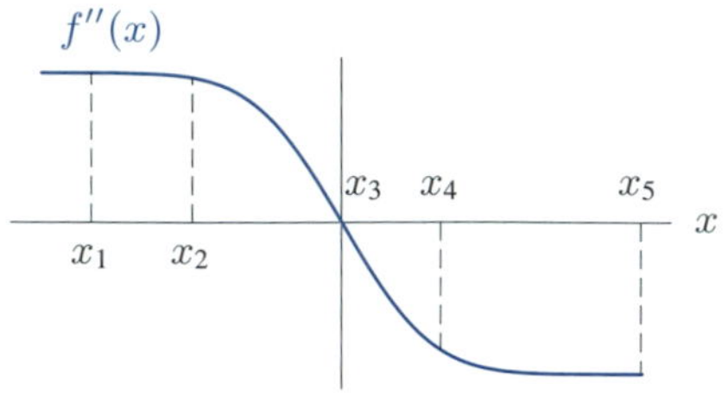

Figure 6.14

16.   The graph of f'' is given in Figure 6.14. Draw graphs of f and f', assuming both go through the origin, and use them to decide at which of the labeled x-values:

(a) $f(x)$ is greatest.   (c) $f'(x)$ is greatest.   (e) $f''(x)$ is greatest.
(b) $f(x)$ is least.   (d) $f'(x)$ is least.   (f) $f''(x)$ is least.

17.   Two functions, $f(x)$ and $g(x)$, are shown in Figure 6.15. Let $F(x)$ and $G(x)$ be antiderivatives, respectively, of these two functions. On the same axes, sketch graphs of the antiderivatives $F(x)$ and $G(x)$ satisfying $F(0) = 0$ and $G(0) = 0$. Compare the two antiderivatives, including a discussion of zeros and x- and y-coordinates of critical points.

18.   The Quabbin Reservoir in the western part of Massachusetts provides most of Boston's water. The graph in Figure 6.16 represents the flow of water in and out of the Quabbin Reservoir throughout 1993.

(a)   Sketch a possible graph for the quantity of water in the reservoir, as a function of time.
(b)   When, in the course of 1993, was the quantity of water in the reservoir largest? Smallest? Mark and label these points on the graph you drew in part (a).
(c)   When was the quantity of water increasing most rapidly? Decreasing most rapidly? Mark and label these times on both graphs.
(d)   By July 1994 the quantity of water in the reservoir was about the same as in January 1993. Draw plausible graphs for the flow into and the flow out of the reservoir for the first half of 1994. Explain your graphs.

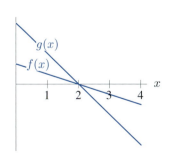

Figure 6.15

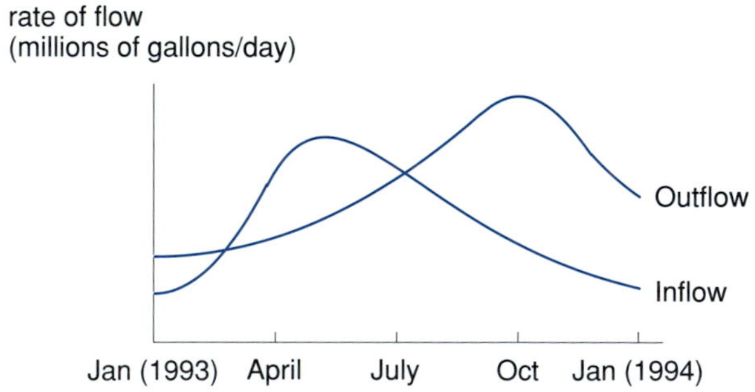

Figure 6.16

6.2 CONSTRUCTING ANTIDERIVATIVES ANALYTICALLY

What Is an Antiderivative of $f(x) = 0$?

A function whose derivative is zero everywhere on an interval must have a horizontal tangent line at every point of its graph, and the only way this can happen is if the function is constant. Alternatively, if we think of the derivative as a velocity, and if the velocity is always zero, then the object is standing still; the position function is constant. A rigorous proof of this result using the definition of the derivative is surprisingly subtle. (See the Constant Function Theorem on page 288.)

> If the derivative of a function is zero everywhere on an interval, then the function is constant on the interval:
>
> $$\text{If } F'(x) = 0 \text{ on an interval, then } F(x) = C \text{ on this interval.}$$

What Is the Most General Antiderivative of f?

We know that if a function f has an antiderivative F, then it has a family of antiderivatives of the form $F(x) + C$, where C is any constant. You might wonder if there are any others. To decide, suppose that we have two functions F and G with $F' = f$ and $G' = f$: that is, F and G are both antiderivatives of the same function f. Since $F' = G'$ we have $(F - G)' = 0$. But this means that we must have $F - G = C$, so $F(x) = G(x) + C$, where C is a constant. Thus, any two antiderivatives of the same function may only differ by a constant.

> If F and G are both antiderivatives of f, then $F(x) = G(x) + C$ for some constant C.

The Indefinite Integral

All antiderivatives of $f(x)$ are of the form $F(x) + C$. We introduce a notation for the general antiderivative that looks like the definite integral without the limits and is called the *indefinite integral*:

$$\int f(x)\, dx = F(x) + C.$$

It is important to understand the difference between

$$\int_a^b f(x)\, dx \qquad \text{and} \qquad \int f(x)\, dx.$$

The first is a number and the second is a family of *functions*. Because the notation is similar, the word "integration" is frequently used for the process of finding the antiderivative as well as of finding the definite integral. The context usually makes clear which is intended.

What is an Antiderivative of $f(x) = k$?

If k is a constant, the derivative of kx is k, so we have

$$\text{An antiderivative of } k \text{ is } kx.$$

Using the indefinite integral notation, we have

> If k is constant
>
> $$\int k\, dx = kx + C.$$

Finding Antiderivatives

Finding antiderivatives of functions is like taking square roots of numbers: if we pick a number at random, such as 7 or 493, we may have trouble saying what its square root is without a calculator. But if we happen to pick a number such as 25 or 64, which we know is a perfect square, then we can find its square root exactly. Similarly, if we pick a function which we recognize as a derivative, then we can find its antiderivative easily.

For example, to find an antiderivative of $f(x) = x$, notice that $2x$ is the derivative of x^2; this tells us that x^2 is an antiderivative of $2x$. If we divide by 2, then we guess that

$$\text{An antiderivative of } x \text{ is } \frac{x^2}{2}.$$

To check this statement, take the derivative of $x^2/2$:

$$\frac{d}{dx}\left(\frac{x^2}{2}\right) = \frac{1}{2} \cdot \frac{d}{dx} x^2 = \frac{1}{2} \cdot 2x = x.$$

What about an antiderivative of x^2? The derivative of x^3 is $3x^2$, so the derivative of $x^3/3$ is $3x^2/3 = x^2$. Thus,

$$\text{An antiderivative of } x^2 \text{ is } \frac{x^3}{3}.$$

Can you see the pattern? It looks like

$$\text{An antiderivative of } x^n \text{ is } \frac{x^{n+1}}{n+1}.$$

(We assume $n \neq -1$, or we would have $x^0/0$, which doesn't make sense.) It is easy to check this formula by differentiation:

$$\frac{d}{dx}\left(\frac{x^{n+1}}{n+1}\right) = \frac{(n+1)x^n}{n+1} = x^n.$$

Recall that the indefinite integral of a function is the family of all its antiderivatives. Thus, in indefinite integral notation, we have shown that

$$\boxed{\int x^n \, dx = \frac{x^{n+1}}{n+1} + C, \quad n \neq -1.}$$

So what about when $n = -1$? In other words, what is an antiderivative of $1/x$? Fortunately, we know a function whose derivative is $1/x$, namely, the natural logarithm. Thus, since

$$\frac{d}{dx}(\ln x) = \frac{1}{x},$$

we know that

$$\int \frac{1}{x} \, dx = \ln x + C, \quad \text{for } x > 0.$$

If $x < 0$, then $\ln x$ is not defined, so it can't be an antiderivative of $1/x$. In this case, we can try $\ln(-x)$:

$$\frac{d}{dx} \ln(-x) = (-1)\frac{1}{-x} = \frac{1}{x}$$

so

$$\int \frac{1}{x} \, dx = \ln(-x) + C, \quad \text{for } x < 0.$$

This means $\ln x$ is an antiderivative of $1/x$ if $x > 0$, and $\ln(-x)$ is an antiderivative of $1/x$ if $x < 0$. Since $|x| = x$ when $x > 0$ and $|x| = -x$ when $x < 0$ we can collapse these two formulas into:

$$\text{An antiderivative of } \frac{1}{x} \text{ is } \ln|x|.$$

Therefore

$$\boxed{\int \frac{1}{x} \, dx = \ln|x| + C.}$$

Since the exponential function is its own derivative, it is also its own antiderivative; thus

$$\int e^x \, dx = e^x + C.$$

Also, antiderivatives of the sine and cosine are easy to guess. Since

$$\frac{d}{dx} \sin x = \cos x \quad \text{and} \quad \frac{d}{dx} \cos x = -\sin x,$$

we get

$$\int \cos x \, dx = \sin x + C \quad \text{and} \quad \int \sin x \, dx = -\cos x + C.$$

Example 1 Find $\int (3x + x^2) \, dx$.

Solution We know that $x^2/2$ is an antiderivative of x and that $x^3/3$ is an antiderivative of x^2, so we expect

$$\int (3x + x^2) \, dx = 3 \left(\frac{x^2}{2} \right) + \frac{x^3}{3} + C.$$

You should always check your antiderivatives by differentiation—it's easy to do. Here

$$\frac{d}{dx} \left(\frac{3}{2} x^2 + \frac{x^3}{3} + C \right) = \frac{3}{2} \cdot 2x + \frac{3x^2}{3} = 3x + x^2.$$

The preceding example illustrates that the sum and constant multiplication rules of differentiation work in reverse:

Properties of Antiderivatives: Sums and Constant Multiples

In indefinite integral notation,

1. $\int \left[f(x) \pm g(x) \right] \, dx = \int f(x) \, dx \pm \int g(x) \, dx$

2. $\int cf(x) \, dx = c \int f(x) \, dx.$

In words,

1. An antiderivative of the sum (or difference) of two functions is the sum (or difference) of their antiderivatives.

2. An antiderivative of a constant times a function is the constant times an antiderivative of the function.

Example 2 Find $\int (\sin x + 3 \cos x) \, dx$.

Solution We break the antiderivative into two terms:

$$\int (\sin x + 3 \cos x) \, dx = \int \sin x \, dx + 3 \int \cos x \, dx = -\cos x + 3 \sin x + C.$$

Check by differentiating:

$$\frac{d}{dx} (-\cos x + 3 \sin x + C) = \sin x + 3 \cos x.$$

Using Antiderivatives to Compute Definite Integrals

The Fundamental Theorem of Calculus, which tells us that if $F' = f$, then

$$\int_a^b f(x)\, dx = F(b) - F(a),$$

gives us a new way of calculating definite integrals. To find $\int_a^b f(x)\, dx$, we first try to find F, and then calculate $F(b) - F(a)$. This method of computing definite integrals has an important advantage over using left- and right-hand sums: it gives an exact answer quickly. However, the method only works in situations where you can find the antiderivative $F(x)$. This is not always easy; for example, none of the functions we have encountered so far is an antiderivative of $\sin(x^2)$.

Example 3   Compute $\displaystyle\int_1^3 2x\, dx$ using the Fundamental Theorem.

Solution   Since $F(x) = x^2$ is an antiderivative of $f(x) = 2x$, we have

$$\int_1^3 2x\, dx = F(3) - F(1) = 3^2 - 1^2 = 8.$$

Notice in this example we used the antiderivative x^2, but $x^2 + C$ works just as well for any constant C. It makes no difference which antiderivative we use to compute a definite integral because the constant cancels out when we subtract $F(a)$ from $F(b)$. For example, if we had used $x^2 + C$, we would have arrived at the same answer:

$$\int_1^3 2x\, dx = F(3) - F(1) = (3^2 + C) - (1^2 + C) = 8.$$

It is helpful to introduce a shorthand notation for $F(b) - F(a)$: we write it as

$$F(x)\Big|_a^b$$

For example:

$$\int_1^3 2x\, dx = x^2\Big|_1^3 = 3^2 - 1^2 = 8.$$

Problems for Section 6.2

For each of the functions in Problems 1–21, find an antiderivative.

1. $f(x) = 5$

2. $f(x) = 5x$

3. $f(x) = x^2$

4. $g(t) = t^2 + t$

5. $h(t) = \cos t$

6. $g(z) = \sqrt{z}$

7. $h(z) = \dfrac{1}{z}$

8. $r(t) = \dfrac{1}{t^2}$

9. $g(z) = \dfrac{1}{z^3}$

10. $f(z) = e^z$

11. $g(t) = \sin t$

12. $f(t) = 2t^2 + 3t^3 + 4t^4$

13. $p(t) = t^3 - \dfrac{t^2}{2} - t$

14. $q(y) = y^4 + \dfrac{1}{y}$

15. $f(x) = 5x - \sqrt{x}$

16. $f(t) = \dfrac{t^2 + 1}{t}$

17. $p(\theta) = 2\sin(2\theta)$

18. $r(t) = e^t + 5e^{5t}$

19. $q(t) = (t + 1)^2$

20. $f(x) = 5^x$

21. $p(t) = \cos t + \dfrac{1}{\cos^2 t}$

For Problems 22–29, find an antiderivative $F(x)$ with $F'(x) = f(x)$ and $F(0) = 0$. Is there only one possible solution in each case?

22. $f(x) = 3$ 23. $f(x) = 2x$ 24. $f(x) = -7x$

25. $f(x) = \dfrac{1}{4}x$ 26. $f(x) = x^2$ 27. $f(x) = \sqrt{x}$

28. $f(x) = 2 + 4x + 5x^2$ 29. $f(x) = \sin x$

Find the indefinite integrals in Problems 30–45.

30. $\displaystyle\int 3x \, dx$

31. $\displaystyle\int (4t + 7) \, dt$

32. $\displaystyle\int \cos\theta \, d\theta$

33. $\displaystyle\int 5e^z \, dz$

34. $\displaystyle\int \left(x + \dfrac{1}{\sqrt{x}}\right) dx$

35. $\displaystyle\int \sin t \, dt$

36. $\displaystyle\int (\pi + x^{11}) \, dx$

37. $\displaystyle\int \left(t\sqrt{t} + \dfrac{1}{t\sqrt{t}}\right) dt$

38. $\displaystyle\int \cos(x + 1) \, dx$

39. $\displaystyle\int e^{2r} \, dr$

40. $\displaystyle\int \dfrac{1}{e^z} \, dz$

41. $\displaystyle\int \left(\dfrac{y^2 - 1}{y}\right)^2 dy$

42. $\displaystyle\int \dfrac{1}{x + 1} \, dx$ [Hint: $\frac{d}{dx}\left(\ln(x+1)\right) = ?$]

43. $\displaystyle\int \dfrac{1}{2x - 1} \, dx$ [Hint: $\frac{d}{dx}\left(\ln(2x-1)\right) = ?$]

44. $\displaystyle\int (e^{5+x} + e^{5x}) \, dx$ [Hint: $\frac{d}{dx}(e^{5x}) = ?$]

45. $\displaystyle\int (\cos 2x - 2\sin x) \, dx$ [Hint: $\frac{d}{dx}(\sin 2x) = ?$]

Using the Fundamental Theorem, evaluate the definite integrals in Problems 46–55 exactly [as in $\ln(3\pi)$] and numerically [$\ln(3\pi) \approx 2.243$]:

46. $\displaystyle\int_2^5 (x^3 - \pi x^2) \, dx$

47. $\displaystyle\int_0^1 \sin\theta \, d\theta$

48. $\displaystyle\int_1^2 \dfrac{1 + y^2}{y} \, dy$

49. $\displaystyle\int_0^2 \left(\dfrac{x^3}{3} + 2x\right) dx$

50. $\displaystyle\int_0^{\pi/4} (\sin t + \cos t) \, dt$

51. $\displaystyle\int_{-3}^{-1} \dfrac{2}{r^3} \, dr$

52. $\displaystyle\int_0^1 2e^x \, dx$

53. $\displaystyle\int_0^{\pi/4} \dfrac{1}{\cos^2 x} \, dx$

54. $\displaystyle\int_{-1}^1 2^x \, dx$

55. $\displaystyle\int_0^{\pi/6} (\sin x + \cos 2x) \, dx$ [Hint: What is $\frac{d}{dx}(\sin 2x)$?]

56. Calculate the exact area between the graph of $y = 7 - 8x + x^2$ and the x-axis.

57. Calculate the exact area above the graph of $y = \sin\theta$ and below the graph of $y = \cos\theta$ for $0 \leq \theta \leq \pi/4$.

58. Find the exact value of the area between the graphs of $y = \cos x$ and $y = e^x$ for $0 \leq x \leq 1$.

59. Find the exact positive value of c making the area between the graph of $y = x^2 - c^2$ and the x-axis equal to 36.

60. The average value of the function $v(x) = 6/x^2$ on the interval $[1, c]$ is equal to 1. Find the value of c.

61. (a) What is the average value of $f(t) = \sin t$ over $0 \leq t \leq 2\pi$? Why is this a reasonable answer?
     (b) Find the average value of $f(t) = \sin t$ over $0 \leq t \leq \pi$.

62. Consider the costs of drilling an oil well. There are two types of costs: *fixed costs* (independent of the depth of the well) and *marginal cost* (the increment in cost that comes from drilling one more meter). By using both pieces of information, you can determine the total cost, C. The marginal costs depend on the depth at which you are drilling; drilling becomes more expensive, per meter, as you dig deeper into the earth. Suppose the fixed costs are 1,000,000 riyals (the riyal is the unit of currency of Saudi Arabia), and the marginal costs are

$$C'(x) = 4000 + 10x$$

in riyals/meter, where x is the depth in meters. Find the total cost of drilling a well x meters deep.

63.   One of the earliest pollution problems brought to the attention of the Environmental Protection Agency (EPA) was the case of the Sioux Lake in eastern South Dakota. For years a small paper plant located nearby had been discharging waste containing carbon tetrachloride (CCl_4) into the waters of the lake. At the time the EPA learned of the situation, the chemical was entering at a rate of 16 cubic yards/year.

The agency ordered the installation of filters designed to slow (and eventually stop) the flow of CCl_4 from the mill. Implementation of this program took exactly three years, during which the flow of pollutant was steady at 16 cubic yards/year. Once the filters were installed, the flow declined. If t is time measured in years since the EPA learned of the situation, the rate of flow was well approximated by

$$\text{Rate (in cubic yards/year)} = t^2 - 14t + 49,$$

between the time the filters were installed and the time the flow stopped.

(a)   Draw a graph showing the rate of CCl_4 flow into the lake as a function of time, beginning at the time the EPA first learned of the situation.

(b)   How many years elapsed between the time the EPA learned of the situation and the time the pollution flow stopped entirely?

(c)   How much CCl_4 entered the waters during the time shown in the graph in part (a)?

6.3 DIFFERENTIAL EQUATIONS

In Chapter 2 we saw that velocity is the derivative of distance and that acceleration is the derivative of velocity. In this section we analyze the motion of an object falling freely under the influence of gravity. This involves going "backward" from acceleration to velocity to position.

Motion With Constant Velocity

Let's briefly consider a familiar problem: An object moving in a straight line with constant velocity. Imagine a car moving at 50 mph. How far does it go in a given time? The answer is given by

$$\text{Distance} = \text{Rate} \times \text{Time}$$

or

$$s = 50t$$

where s is the distance of the car (in miles) from a fixed reference point and t is the time in hours. Alternatively, we can describe the motion by writing the equation

$$\frac{ds}{dt} = 50.$$

This is called a *differential equation* for the *function* s. The solution to this equation is the general antiderivative

$$s = 50t + C.$$

The equation $s = 50t + C$ tells us that $s = C$ when $t = 0$. Thus, the constant C represents the initial distance, s_0, of the car from the reference point.

Uniformly Accelerated Motion

Now we consider an object moving with *constant acceleration* along a straight line, or *uniformly accelerated motion*. It has been known since Galileo's time that this is the motion of an object moving under the influence of gravity (ignoring air resistance). Thus, if v is the upward velocity and t is the time,

$$\frac{dv}{dt} = -g$$

where g is the constant *acceleration due to gravity*, whose value is determined by experiment. In the most frequently used units, its value is approximately

$$g = 9.81 \text{ m/sec}^2, \quad \text{or} \quad g = 32 \text{ ft/sec}^2.$$

The negative sign comes from the fact that velocity is measured upward, whereas gravity acts downward.

Example 1   A stone is dropped from a 100-foot-high building. Find, as a function of time, its position and velocity. When does it hit the ground, and how fast is it going at that time?

Solution   Suppose t is measured in seconds from when the stone was dropped. If we measure distance, s, in feet above the ground, then the velocity, v, is in ft/sec upward, and the acceleration due to gravity is 32 ft/sec^2 downward, so

$$\frac{dv}{dt} = -32.$$

From what we know about antiderivatives, we must have

$$v = -32t + C$$

where C is some constant. Since $v = C$ when $t = 0$, the constant C represents the initial velocity, v_0. The fact that the stone is dropped rather than thrown off the top of the building tells us that the initial velocity is zero, so $v_0 = 0$. Substituting gives

$$0 = -32(0) + C \quad \text{so} \quad C = 0.$$

Thus,

$$v = -32t.$$

But now we can write

$$v = \frac{ds}{dt} = -32t.$$

The general antiderivative of $-32t$ is

$$s = -16t^2 + K,$$

where K is another constant.

Since the stone starts at the top of the building, $s = 100$ when $t = 0$. Substituting into $s = -16t^2 + K$ gives

$$100 = -16(0^2) + K,$$

so

$$K = 100,$$

and therefore

$$s = -16t^2 + 100.$$

Thus, we have found both v and s as functions of time.

The stone hits the ground when $s = 0$, so we must solve

$$0 = -16t^2 + 100$$

giving $t^2 = 100/16$ or $t = \pm 10/4 = \pm 2.5$ sec. Since t must be positive, $t = 2.5$ sec. At that time, $v = -32(2.5) = -80$ ft/sec. (Note that the velocity is negative because we are considering up as the positive direction and down as the negative direction.)

Example 2   An object is thrown vertically upward with a speed of 10 m/sec from a height of 2 meters. Find the highest point it reaches and when it hits the ground.

Solution   We must find the position as a function of time. In this example, the velocity is in m/sec, so we use $g = 9.8$ m/sec^2. Measuring distance in meters upward from the ground, we have

$$\frac{dv}{dt} = -9.8.$$

As before, v is a function whose derivative is constant, so

$$v = -9.8t + C.$$

Since the initial velocity is 10 m/sec upward, we know that $v = 10$ when $t = 0$. Substituting gives

$$10 = -9.8(0) + C \quad \text{so} \quad C = 10.$$

Thus,

$$v = -9.8t + 10.$$

To find s, we use

$$v = \frac{ds}{dt} = -9.8t + 10$$

and look for a function that has $-9.8t + 10$ as its derivative. The general antiderivative of $-9.8t + 10$ is

$$s = -4.9t^2 + 10t + K,$$

where K is any constant. To find K, we use the fact that the object starts at a height of 2 meters, so $s = 2$ when $t = 0$. Substituting gives

$$2 = -4.9(0)^2 + 10(0) + K, \quad \text{so} \quad K = 2,$$

and therefore

$$s = -4.9t^2 + 10t + 2.$$

The object reaches its highest point when the velocity is 0, so at that time

$$v = -9.8t + 10 = 0.$$

This occurs when

$$t = \frac{10}{9.8} \approx 1.02 \text{ sec.}$$

When $t = 1.02$ seconds,

$$s = -4.9(1.02)^2 + 10(1.02) + 2 \approx 7.10 \text{ m.}$$

So the maximum height reached is 7.10 meters. The object reaches the ground when $s = 0$:

$$0 = -4.9t^2 + 10t + 2.$$

Solving this using the quadratic formula gives

$$t \approx -0.18 \quad \text{and} \quad t \approx 2.22 \text{ sec.}$$

Since the time at which the object hits the ground must be positive, $t \approx 2.22$ seconds.

Antiderivatives and Differential Equations

We solved the problem of uniformly accelerated motion by working backward from the derivative of a function to the function itself. If f is a known function, finding the *general solution* to the differential equation

$$\frac{dy}{dx} = f(x)$$

means finding the general antiderivative $y = F(x) + C$ with $F'(x) = f(x)$.

Example 3   Find and graph the general solution of the differential equation

$$\frac{dy}{dx} = \sin x + 2.$$

Solution   We are asking for a function whose derivative is $\sin x + 2$. An antiderivative of $\sin x + 2$ is

$$y = -\cos x + 2x.$$

The general solution is therefore

$$y = -\cos x + 2x + C,$$

where C is any constant. Figure 6.17 shows several curves in this family.

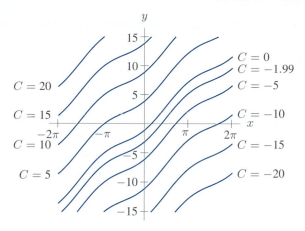

Figure 6.17: Solution curves of $\frac{dy}{dx} = \sin x + 2$

How Can We Pick One Solution to the Differential Equation $\dfrac{dy}{dx} = f(x)$?

Picking one antiderivative is equivalent to selecting a value of C. This requires knowing an extra piece of information, such as the initial velocity or initial position. In general, we pick a particular curve in the family of solutions by specifying that the curve should pass through a given point (x_0, y_0). The differential equation plus an extra condition

$$\frac{dy}{dx} = f(x), \qquad y(x_0) = y_0$$

is called an *initial value problem*. An initial value problem usually has a unique solution. Note that $y(x_0) = y_0$ is shorthand for $y = y_0$ when $x = x_0$.

Example 4   Find the solution of the initial value problem

$$\frac{dy}{dx} = \sin x + 2, \quad y(3) = 5.$$

Solution   We have already seen that the general solution to the differential equation is $y = -\cos x + 2x + C$. The initial condition allows us to determine the constant C. Substituting $y(3) = 5$ gives

$$5 = y(3) = -\cos 3 + 2 \cdot 3 + C,$$

so C is given by

$$C = 5 + \cos 3 - 6 \approx -1.99.$$

Thus, the (unique) solution is

$$y = -\cos x + 2x - 1.99.$$

Figure 6.17 shows this particular solution (marked with $C = -1.99$).

Problems for Section 6.3

Find the general solution of the differential equations in Problems 1–4.

1. $\dfrac{dy}{dx} = x^3 + 5$ 2. $\dfrac{dy}{dx} = 8x + \dfrac{1}{x}$ 3. $\dfrac{dW}{dt} = 4\sqrt{t}$ 4. $\dfrac{dr}{dp} = 3\sin p$

In Problems 5–7, find the solution of the initial value problem.

5. $\dfrac{dy}{dx} = 6x^2 + 4x,\quad y(2) = 10$ 6. $\dfrac{dP}{dt} = 10e^t,\quad P(0) = 25$

7. $\dfrac{ds}{dt} = -32t + 100,\quad s = 50$ when $t = 0$

8. Show that $y = x + \sin x - \pi$ satisfies the initial value problem

    $$\frac{dy}{dx} = 1 + \cos x, \quad y(\pi) = 0.$$

9. Show that $y = xe^{-x} + 2$ is a solution of the initial value problem

    $$\frac{dy}{dx} = (1 - x)e^{-x}, \quad y(0) = 2.$$

10. (a) Find the general solution of the differential equation $\dfrac{dy}{dx} = 2x + 1$.
    (b) Sketch a graph of at least three solutions.
    (c) Find the solution satisfying $y(1) = 5$. Graph this solution with the others from part (b).

11. Ice is forming on a pond at a rate given by

    $$\frac{dy}{dt} = k\sqrt{t},$$

    where y is the thickness of the ice in inches at time t measured in hours since the ice started forming, and k is a positive constant. Find y as a function of t.

12. If a car goes from 0 to 80 mph in six seconds with constant acceleration, what is that acceleration?

13. A car going 80 ft/sec (about 55 mph) brakes to a stop in five seconds. Assume the deceleration is constant.

    (a) Graph the velocity against time, t, for $0 \le t \le 5$ seconds.
    (b) Represent, as an area on the graph, the total distance traveled from the time the brakes are applied until the car comes to a stop.
    (c) Find this area and hence the distance traveled.
    (d) Now find the total distance traveled using antidifferentiation.

14. A 727 jet needs to be flying 200 mph to take off. If it can accelerate from 0 to 200 mph in 30 seconds, how long must the runway be? (Assume constant acceleration.)

15. Suppose a car going at 30 ft/sec decelerates at a constant rate of 5 ft/sec^2.

    (a) Draw up a table showing the velocity of the car every half second. When does the car come to rest?
    (b) Using your table, find left and right sums which estimate the total distance traveled before the car comes to rest. Which is an overestimate, and which is an underestimate?
    (c) Sketch a graph of velocity against time. On the graph, show an area representing the distance traveled before the car comes to rest. Use the graph to calculate this distance.
    (d) Now find a formula for the velocity of the car as a function of time, and then find the total distance traveled by antidifferentiation. What is the relationship between your answer to parts (c) and (d) and your estimates in part (b)?

16. A water balloon launcher has been installed on the roof of a building. The vertical velocity (in feet/sec) of the water balloon at time t seconds after launch is given by

    $$v(t) = -32t + 40,$$

    with $v > 0$ corresponding to upward motion.

    (a) If the roof of the building is 30 feet above the ground, find an expression for the height of the water balloon above the ground at any time t.
    (b) What is the average velocity of the balloon between $t = 1.5$ and $t = 3$ seconds?
    (c) A 6-foot person is standing on the ground. How fast is the water balloon falling when it strikes the person on the top of the head?

17. An object is shot vertically upward from the ground with a velocity of 160 ft/sec.

    (a) Sketch a graph of the velocity of the object (with upward as positive) against time.
    (b) Mark on the graph the points when the object reaches its highest point and when it lands.
    (c) Find the maximum height reached by the object by considering an area on the graph.
    (d) Now express velocity as a function of time, and find the greatest height by antidifferentiation.

18.   A stone thrown upwards from the top of a 320-foot cliff at 128 ft/sec eventually falls to the beach below.

    (a)   How long does the stone take to reach its highest point?
    (b)   What is its maximum height?
    (c)   How long before the stone hits the beach?
    (d)   What is the velocity of the stone on impact?

19.   On the moon, the acceleration due to gravity is about 1.6 m/sec^2 (compared to $g \approx 9.8$ m/sec^2 on earth). If you drop a rock on the moon (with initial velocity 0), find formulas for:

    (a)   Its velocity, $v(t)$, at time t.
    (b)   The distance, $s(t)$, it falls in time t.

20.   (a)   Imagine throwing a rock straight up in the air. What should its initial velocity be if the rock reaches a maximum height of 100 feet above its starting point?
    (b)   Now imagine being transplanted to the moon and throwing a moon rock vertically upward with the same velocity as in part (a). How high will it go? (On the moon, $g = 5$ ft/sec^2.)

21.   A cat, walking along the window ledge of a New York apartment, knocks off a flower pot, which falls to the street 200 feet below. How fast is the flower pot traveling when it hits the street? (Give your answer in ft/sec and in mph, given that 1 ft/sec $= 15/22$ mph.)

22.   Of all the cars surveyed in the April 1991 issue of *Car and Driver*, the Acura NSX, going at 70 mph, stops in the shortest distance: 157 feet. Find the acceleration, assuming it is constant.

6.4 SECOND FUNDAMENTAL THEOREM OF CALCULUS

Suppose f is an elementary function, that is, a combination of constants, powers of x, $\sin x$, $\cos x$, e^x, and $\ln x$. Then we have to be lucky to find an antiderivative F which is also an elementary function. But if we can't find F as an elementary function, how can we be sure that F exists at all? In this section we learn to use the definite integral to construct antiderivatives.

Construction of Antiderivatives Using the Definite Integral

Consider the function $f(x) = e^{-x^2}$. We would like to find a way of calculating values of its antiderivative. If the antiderivative, F, were an elementary function, we could evaluate it using a calculator, but it isn't. However, we know from the Fundamental Theorem of Calculus that

$$F(b) - F(a) = \int_a^b e^{-t^2}\, dt$$

so

$$F(x) - F(0) = \int_0^x e^{-t^2}\, dt.$$

All we have done here is to set $a = 0$ and replace the usual b in the upper limit with an x. Suppose we want the antiderivative that satisfies $F(0) = 0$. Then we get

$$F(x) = \int_0^x e^{-t^2}\, dt.$$

This is a formula for F. For any value of x, there is a unique value for $F(x)$, so F is a function. For any fixed x, we can calculate $F(x)$ numerically to any accuracy we like. For example, we can calculate

$$F(2) = \int_0^2 e^{-t^2}\, dt = 0.88208\ldots.$$

Notice that our expression for F is not an elementary function; we have *created* a new function using the definite integral. The next theorem says that this method of constructing antiderivatives works in general. This means that if we define F by

$$F(x) = \int_a^x f(t)\, dt$$

then F must be an antiderivative of f.

> **Construction Theorem for Antiderivatives (Second Fundamental Theorem of Calculus)**
> If f is a continuous function on an interval, and if a is any number in that interval, then the function F defined by
> $$F(x) = \int_a^x f(t)\, dt$$
> is an antiderivative of f.

Notice that if we already know an antiderivative of f, which we may call G, then the Fundamental Theorem of Calculus tells us that

$$G(x) - G(a) = \int_a^x f(t)\, dt.$$

By the definition of F, we have

$$F(x) = \int_a^x f(t)\, dt.$$

Therefore,

$$G(x) - G(a) = F(x),$$

so F and G differ by a constant (namely $G(a)$), and so F and G are *both* antiderivatives of f.

Justification of the Construction Theorem

Our task is to show that F, defined by this integral, is an antiderivative of f. We want to show that $F'(x) = f(x)$. By the definition of the derivative,

$$F'(x) = \lim_{h \to 0} \frac{F(x+h) - F(x)}{h}.$$

To gain some geometric insight, let's suppose f is positive and h is positive. Then we can visualize

$$F(x) = \int_a^x f(t)\, dt$$

and

$$F(x+h) = \int_a^{x+h} f(t)\, dt,$$

as areas, which leads to representing

$$F(x+h) - F(x) = \int_x^{x+h} f(t)\, dt$$

as a difference of two areas. From Figure 6.18, we see that $F(x+h) - F(x)$ is roughly the area of a rectangle of height $f(x)$ and width h (shaded darker in Figure 6.18), so we have

$$F(x+h) - F(x) \approx f(x)h,$$

hence

$$\frac{F(x+h) - F(x)}{h} \approx f(x).$$

More precisely, we can use the theorem on comparing integrals on page 172 to conclude that

$$mh \le \int_x^{x+h} f(t)\, dt \le Mh,$$

where m is the greatest lower bound for f on the interval from x to $x+h$ and M is the least upper

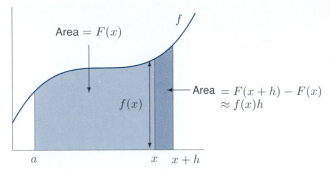

Figure 6.18: $F(x+h) - F(x)$ is area of roughly rectangular region

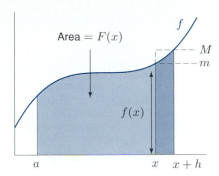

Figure 6.19: Upper and lower bounds for $F(x+h) - F(x)$

bound on that interval. (See Figure 6.19.) Hence

$$mh \leq F(x+h) - F(x) \leq Mh,$$

so

$$m \leq \frac{F(x+h) - F(x)}{h} \leq M.$$

Since f is continuous, both m and M approach $f(x)$ as h approaches zero. Thus

$$f(x) \leq \lim_{h \to 0} \frac{F(x+h) - F(x)}{h} \leq f(x).$$

Thus both inequalities must actually be equalities, so we have the result we want:

$$f(x) = \lim_{h \to 0} \frac{F(x+h) - F(x)}{h} = F'(x).$$

Using the Construction Theorem for Antiderivatives

The construction theorem enables us to write down antiderivatives of functions that do not have elementary antiderivatives. For example, an antiderivative of $(\sin x)/x$ is

$$F(x) = \int_0^x \frac{\sin t}{t}\, dt.$$

Notice that F is a function; we can calculate its values to any degree of accuracy. This function already has a name: it is called the *sine-integral*, and it is denoted $\text{Si}(x)$.

Example 1   Construct a table of values of $\text{Si}(x)$ for $x = 0, 1, 2, 3$.

Solution   Using numerical methods, we calculate the values of $\text{Si}(x) = \int_0^x \frac{\sin t}{t}\, dt$ given in Table 6.2. Since the integrand is undefined at $t = 0$, we took the lower limit as 0.00001 instead of 0.

TABLE 6.2   *A table of values of $\text{Si}(x)$*

x	0	1	2	3
$\text{Si}(x)$	0	0.95	1.61	1.85

The reason the sine-integral has a name is that scientists and engineers have found a use for it (for example, in optics). For people who use it all the time, it is just another common function like sine or cosine. Its derivative,

$$\frac{d}{dx}\,\mathrm{Si}(x) = \frac{\sin x}{x},$$

can be used with the other differentiation rules, as in the next example.

Example 2   Find the derivative of $x\,\mathrm{Si}(x)$.

Solution   Using the product rule,

$$\frac{d}{dx}\,(x\,\mathrm{Si}(x)) = \left(\frac{d}{dx}\,x\right)\mathrm{Si}(x) + x\left(\frac{d}{dx}\,\mathrm{Si}(x)\right)$$

$$= 1 \cdot \mathrm{Si}(x) + x\frac{\sin x}{x}$$

$$= \mathrm{Si}(x) + \sin x.$$

Problems for Section 6.4

1. The graph of the derivative F' of some function F is given in Figure 6.20. If $F(20) = 150$, estimate the maximum value attained by F.

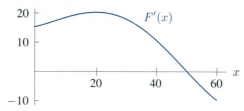

Figure 6.20

2. Assume that $F'(t) = \sin t \cos t$ and $F(0) = 1$. Find $F(b)$ for $b = 0,\ 0.5,\ 1,\ 1.5,\ 2,\ 2.5,$ and 3.

For the functions in Problems 3–5, let $F(x) = \int_0^x f(t)\,dt$. Graph $F(x)$ as a function of x.

3.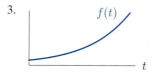

4.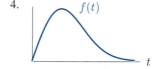

5.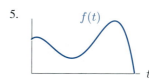

6. Make a table of values for the following function for $x = 0,\ 0.5,\ 1.0,\ 1.5,$ and 2.0.

$$I(x) = \int_0^x \sqrt{t^4 + 1}\,dt$$

7. (a) Continue the table of values for $\mathrm{Si}(x) = \int_0^x \dfrac{\sin t}{t}\,dt$, in Example 1, page 311 for $x = 4$ and $x = 5$.

    (b) Why is $\mathrm{Si}(x)$ decreasing between $x = 4$ and $x = 5$?

Find the derivatives in Problems 8–13.

8. $\dfrac{d}{dx}\displaystyle\int_0^x \sqrt{3 + \cos(t^2)}\,dt$

9. $\dfrac{d}{dx}\displaystyle\int_1^x (1 + t)^{200}\,dt$

10. $\dfrac{d}{dx}\displaystyle\int_{0.5}^x \arctan(t^2)\,dt$

11. $\dfrac{d}{dt}\displaystyle\int_t^{\pi} \cos(z^3)\,dz$

12. $\dfrac{d}{dx}\displaystyle\int_x^1 \ln t\,dt$

13. $\dfrac{d}{dx}\left[\mathrm{Si}(x^2)\right]$

14. Let $F(x) = \int_0^x \sin(2t)\,dt$.
    (a) Evaluate $F(\pi)$.
    (b) Draw a sketch to explain geometrically why the answer to part (a) is correct.
    (c) For what values of x is $F(x)$ positive? negative?

15. Let $F(x) = \int_2^x \frac{1}{\ln t}\,dt$ for $x \geq 2$.
    (a) Find $F'(x)$.
    (b) Is F increasing or decreasing? What can you say about the concavity of its graph?
    (c) Sketch a graph of $F(x)$.

Problems 16–19 concern the *error function*, erf(x), defined by

$$\text{erf}(x) = \frac{2}{\sqrt{\pi}} \int_0^x e^{-t^2}\,dt.$$

Find an expression in terms of erf(x) for each of the following quantities.

16. $\dfrac{d}{dx}(x\,\text{erf}(x))$ 17. $\dfrac{d}{dx}(\text{erf}(\sqrt{x}))$ 18. $\sqrt{\dfrac{2}{\pi}} \int_0^x e^{-t^2/2}\,dt$ 19. $\sqrt{\dfrac{2}{\pi}} \int_{x_1}^{x_2} e^{-t^2/2}\,dt$

CHAPTER SUMMARY

- **Constructing antiderivatives**
  Graphically, numerically, analytically
- **The family of antiderivatives**
  The indefinite integral
- **Differential equations**
  Initial value problems, uniform motion
- **Construction theorem (Second Fundamental Theorem of Calculus)**
  Constructing antiderivatives using definite integrals

REVIEW PROBLEMS FOR CHAPTER SIX

Find the indefinite integrals in Problems 1–15.

1. $\int (5x + 7)\,dx$ 2. $\int \left(\frac{3}{t} - \frac{2}{t^2}\right)dt$ 3. $\int (e^x + 5)\,dx$

4. $\int \left(x^{3/2} + \frac{\sqrt{x}}{5} - \frac{2}{x}\right)dx$ 5. $\int \frac{1}{\cos^2 x}\,dx$ 6. $\int 2^x\,dx$

7. $\int (x+1)^2\,dx$ 8. $\int (x+1)^3\,dx$ 9. $\int (x+1)^9\,dx$

10. $\int \left(\frac{x+1}{x}\right)dx$ 11. $\int \left(\frac{x^2+x+1}{x}\right)dx$ 12. $\int \left(3\cos\psi + 3\sqrt{\psi}\right)d\psi$

13. $\int (3\cos x - 7\sin x)\,dx$ 14. $\int \left(x + \frac{2}{x} + \pi\sin x\right)dx$ 15. $\int (2e^x - 8\cos x)\,dx$

Find antiderivatives for the functions in Problems 16–38. Check by differentiation.

16. $f(x) = x + x^5 + x^{-5}$

17. $g(x) = \dfrac{1}{x} + \dfrac{1}{x^2} + \dfrac{1}{x^3}$

18. $f(x) = x^6 - \dfrac{1}{7x^6}$

19. $g(t) = 5 + \cos t$

20. $g(\alpha) = 3\sin(3\alpha)$

21. $h(r) = 2\sqrt{r} + \dfrac{1}{2\sqrt{r}}$

22. $g(x) = (x+1)^3$

23. $h(t) = \dfrac{(t-1)^2}{t^2}$

24. $p(y) = \dfrac{1}{y} + y + 1$

25. $f(z) = e^z + 3$

26. $f(x) = 2xe^{x^2}$

27. $g(x) = e^x + e^{1+x}$

28. $g(\theta) = \sin\theta - 2\cos\theta$

29. $p(r) = 2\pi r$

30. $f(x) = 2^x + \dfrac{1}{2^x}$

31. $h(t) = \cos t(1 + \sin t)^{29}$

32. $h(t) = 2t\cos(t^2)$

33. $f(t) = t\cos(t^2)$

34. $r(x) = 3x^2\cos(x^3 + 7)$

35. $g(y) = e^2 + 2^y$

36. $g(\theta) = \sin\theta + \dfrac{1}{1+\theta^2}$

37. $f(x) = xe^{x^2}$

38. $f(x) = 6\sqrt{x} - \dfrac{1}{x^2} + \dfrac{10}{x}$

39. Calculate the exact area above the graph of $y = \frac{1}{2}\left(\frac{3}{\pi}x\right)^2$ and below the graph of $y = \cos x$. The curves intersect at $x = \pm\pi/3$.

40. Find the exact area of the region bounded by the x-axis and the graph of $y = x^3 - x$.

41. Find the exact positive value of c which makes the area under the graph of $y = c(1 - x^2)$ and above the x-axis equal to 1.

42. Sketch the parabola $y = x(x - \pi)$ and the curve $y = \sin x$, showing their points of intersection. Find the exact area between the two graphs.

For Problems 43–44 the graph of $f'(x)$ is given. Sketch a possible graph for $f(x)$. Mark the points $x_1 \ldots x_4$ on your graph and label local maxima, local minima and points of inflection.

43.

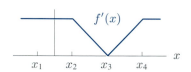

44.

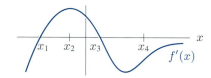

45. If the graph of f is as shown in Figure 6.21, sketch a graph of the function $F(x) = \displaystyle\int_0^x f(t)\,dt$.

46. An object is attached to a coiled spring which is suspended from the ceiling of a room. The function $h(t)$ gives the height of the object above the floor of the room at time, t. The graph of $h'(t)$ is given in Figure 6.22. Sketch a possible graph of $h(t)$.

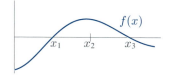

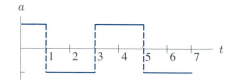

Figure 6.21   *Figure 6.22*   *Figure 6.23*

47. The acceleration, a, of a particle as a function of time, t, is shown in Figure 6.23. Sketch graphs of velocity and position against time. Assume the particle starts at rest at the origin.

48. An object thrown in the air on a planet in a distant galaxy is at height $s = -25t^2 + 72t + 40$ feet at time t seconds after it is thrown. What is the acceleration due to gravity on this planet? With what velocity was the object thrown? From what height?

49. The angular speed of a car engine increases from 1100 revs/min to 2500 revs/min in 6 sec. Assuming that the angular acceleration was constant,

   (a) Find the angular acceleration in revs/min^2.
   (b) How many revolutions does the engine make in this time?

50. A helicopter rotor slows down at a constant rate from 350 revs/min to 260 revs/min in 1.5 minutes.

   (a) Find the angular acceleration during this time interval. What are the units of this acceleration?
   (b) Assuming the angular acceleration remains constant, how long does it take for the rotor to stop? (Measure time from the moment when speed was 350 revs/min.)
   (c) How many revolutions does the rotor make between the time the angular speed was 350 revs/min and stopping?

51. An object is thrown vertically upward with a velocity of 80 ft/sec.

   (a) Make a table showing its velocity every second.
   (b) When does it reach its highest point? When does it hit the ground?
   (c) Using your table, write left and right sums which under- and overestimate the height the object attains.
   (d) Use antidifferentiation to find the greatest height it reaches.

52. A car, initially moving at 60 mph, has a constant deceleration and stops in a distance of 200 feet. What is its deceleration? (Give your answer in ft/sec^2. Note that 1 mph = 22/15 ft/sec.)

53. Consider a bacteria population whose birth rate, B, is given by the curve in Figure 6.24 as a function of time in hours. The birth rate is in births per hour. The curve marked D gives the death rate (in deaths per hour) of the same population.

   (a) Explain what the shape of each of these graphs tells you about the population.
   (b) Use the graphs to find the time at which the net rate of increase of the population is at a maximum.
   (c) Suppose at time $t = 0$ the population has size N. Sketch the graph of the total number born by time t. Also sketch the graph of the number alive at time t. Use the given graphs to find the time at which the population size is a maximum.

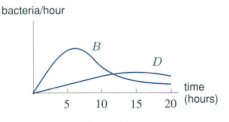

Figure 6.24

Figure 6.25

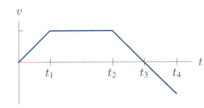

Figure 6.26

54. Water flows at a constant rate into the left side of the W-shaped container shown in Figure 6.25. Sketch a graph of the height, H, of the water in the left side of the container as a function of time, t. Suppose the container starts out empty.

55. Suppose a particle of mass, m, acted on by a force, F, moves in a straight line. Its acceleration, a, is given by Newton's Law:

$$F = ma.$$

The work, W, done by a *constant* force when the particle moves through a displacement, d, is

$$W = Fd.$$

Suppose the velocity, v, of the particle as a function of time, t, is given in Figure 6.26. What is the sign of the work done during each of the time intervals: $[0, t_1]$, $[t_1, t_2]$, $[t_2, t_3]$, $[t_3, t_4]$,$[t_2, t_4]$?

56. Let $F(x) = \int_{\pi/2}^{x} \frac{\sin t}{t}\, dt$. Find the value(s) of x between $\pi/2$ and $3\pi/2$ for which $F(x)$ has a global maximum and global minimum. Show your reasoning.

PROJECTS

1. **Distribution of Resources**

   Whether a resource is distributed evenly among members of a population is often an important political or economic question. How can we measure this? How can we decide if the distribution of wealth in this country is becoming more or less equitable over time? How can we measure which country has the most equitable income distribution? This problem describes a way of making such measurements.

   Suppose the resource is distributed evenly. Then any 20% of the population will have 20% of the resource. Similarly, any 30% will have 30% of the resource and so on. If, however, the resource is not distributed evenly, the poorest $p\%$ of the population (in terms of this resource) will not have $p\%$ of the goods. Suppose $F(x)$ represents the fraction of the resources owned by the poorest fraction x of the population. Thus $F(0.4) = 0.1$ means that the poorest 40% of the population owns 10% of the resource.

   (a) What would F be if the resource were distributed evenly?
   (b) What must be true of any such F? What must $F(0)$ and $F(1)$ equal? Is F increasing or decreasing? Is the graph of F concave up or concave down?
   (c) Gini's index of inequality, G, is one way to measure how evenly the resource is distributed. It is defined by

   $$G = 2 \int_0^1 [x - F(x)] \, dx.$$

   Show graphically what G represents.

2. **Yield from an Apple Orchard**

   Figure 6.27 is a graph of the annual yield, $y(t)$ (in bushels per year), from an orchard t years after planting. The trees take about 10 years to get established, but for the next 20 years they give a substantial yield. After about 30 years, however, age and disease start to take their toll, and the annual yield falls off.[1]

   (a) Represent on a sketch of Figure 6.27 the total yield, $Y(T)$, until time T. Write an expression for $Y(T)$ in terms of $y(t)$.
   (b) Sketch a graph of $Y(T)$ against T.
   (c) Write an expression for the average annual yield, $a(T)$, up until time T.
   (d) The important question is: When should the orchard be cut down and replanted? Assume that you want to maximize your average revenue per year, and that fruit prices remain constant, so that this is achieved by maximizing your average annual yield.
       (i) What condition on $Y(T)$ maximizes the average annual yield?
       (ii) What condition on $y(T)$ maximizes the average annual yield? Estimate the T value which yields the maximum average annual yield.

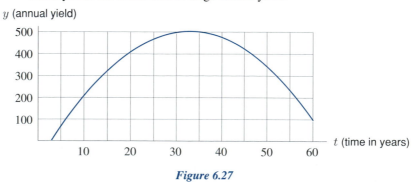

Figure 6.27

---

[1] From Peter D. Taylor, *Calculus: The Analysis of Functions*, (Toronto: Wall & Emerson, Inc., 1992).

FOCUS ON MODELING

THE EQUATIONS OF MOTION

The problem of a body moving freely under the influence of gravity near the surface of the earth intrigued mathematicians and philosophers from Greek times onward and was finally solved by Galileo and Newton. The question to be answered was: How do the velocity and the position of the body vary with time? We define s to be the position, or height, of the body above a fixed point (often ground level); v then is the velocity of the body measured upward. We assume that the acceleration of the body is a constant, $-g$ (the negative sign means that the acceleration is downward), so

$$\text{Acceleration} = \frac{dv}{dt} = -g.$$

Thus, velocity is the antiderivative of $-g$:

$$v = -gt + C.$$

If the initial velocity is v_0, then $C = v_0$, so

$$v = -gt + v_0.$$

How about the position? We know that

$$\frac{ds}{dt} = v = -gt + v_0.$$

Therefore, we can find s by antidifferentiating again, giving:

$$s = -\frac{gt^2}{2} + v_0 t + C.$$

If the initial position is s_0, then we must have

$$s = -\frac{gt^2}{2} + v_0 t + s_0.$$

Our derivation of the formulas for the velocity and the position of the body took little effort. It hides an almost 2000-year struggle to understand the mechanics of falling bodies, from Aristotle's *Physics* to Galileo's *Dialogues Concerning Two New Sciences*. Before sketching Galileo's solution of the problem, we make a few more observations on uniform and uniformly accelerated motion.

Though it is an oversimplification of his ideas, we can say that Aristotle's conception of motion was primarily in terms of *change of position*. This seems entirely reasonable; it is what we commonly observe, and this view dominated discussions of motion for centuries. But it misses a subtlety that was brought to light by Descartes, Galileo, and, with a different emphasis, by Newton. That subtlety is now usually referred to as the *principle of inertia*.

This principle holds that a body traveling undisturbed at constant velocity in a straight line will continue in this motion indefinitely. Stated another way, it says that one cannot distinguish in any absolute sense (that is, by performing an experiment), between being at rest and moving with constant velocity in a straight line. If you are reading this book in a closed room and have no external reference points, there is no experiment that will tell you, one way or the other, whether you are at rest or whether you, the room, and everything in it are moving with constant velocity in a straight line. Therefore, as Newton saw, an understanding of motion should be based on *change of velocity* rather than change of position. Since acceleration is the rate of change of velocity, it is acceleration that must play a central role in the description of motion.

How does acceleration come about? How does the velocity change? Through the action of *forces*. Newton placed a new emphasis on the importance of forces. Newton's laws of motion do not say what a force *is*, they say how it *acts*. His first law is the principle of inertia, which says what happens in the *absence* of a force—there is no change in velocity. His second law says that a force acts to produce a change in velocity, that is, an acceleration. It states that $F = ma$, where m is the mass of the object, F is the net force, and a is the acceleration produced by this force.

Let's return to Galileo for a moment. He demonstrated that a body falling under the influence of gravity does so with constant acceleration. Furthermore, assuming we can neglect air resistance, this constant acceleration is independent of the mass of the body. This last fact was the outcome of Galileo's famous observation that a heavy ball and a light ball dropped off the Leaning Tower of Pisa hit the ground at the same time. Whether or not he actually performed this experiment, Galileo presented a very clear thought experiment in the *Dialogues* to prove the same point. (This point was counter to Aristotle's more common sense notion that the heavier ball would reach the ground first.) Galileo showed that the mass of the object need not appear as a variable in the equation of motion. Thus, the same constant acceleration equation applies to all bodies falling under the influence of gravity.

Nearly a hundred years after Galileo's experiment, Newton formulated his laws of motion and gravity, which led to a differential equation describing the motion of a falling body. According to Newton, acceleration is caused by force, and in the case of falling bodies, that force is the force of gravity. Newton's law of gravity says that the gravitational force between two bodies is attractive and given by

$$F = \frac{GMm}{r^2},$$

where G is the gravitational constant, m and M are the masses of the two bodies, and r is the distance between them. This is the famous *inverse square law*. For a falling body, we take M to be the mass of the earth and r to be the distance from the body to the center of the earth. So, actually, r changes as the body falls, but for anything we can easily observe (say, a ball dropped from the Tower of Pisa), it won't change significantly over the course of the motion. Hence, as an approximation, it is reasonable to assume that the force is constant. According to Newton's second law,

$$\text{Force} = \text{Mass} \times \text{Acceleration}.$$

Since the gravitational force is acting downwards

$$-\frac{GMm}{r^2} = m\frac{d^2 s}{dt^2}.$$

Hence,

$$\frac{d^2 s}{dt^2} = -\frac{GM}{r^2} = \text{constant}.$$

If we define $g = GM/r^2$, then

$$\frac{d^2 s}{dt^2} = -g.$$

The fact that the mass cancels out of Newton's equations of motion reflects Galileo's experimental observation that the acceleration due to gravity is independent of the mass of the body.

Problems on the Equations of Motion

1. An object is dropped from a 400-foot tower. When does it hit the ground and how fast is it going at the time of the impact?

2. Consider the object in Problem 1, and suppose it falls off the same 400-foot tower. What would the acceleration due to gravity have to be to make it reach the ground in half the time?

3. A ball that is dropped from a window hits the ground in five seconds. How high is the window? (Give your answer in feet.)

4. On the moon the acceleration due to gravity is 5 ft/sec^2. An astronaut jumps into the air with an initial upward velocity of 10 ft/sec. How high does he go? How long is the astronaut off the ground?

5. Galileo was the first person to show that the distance traveled by a body falling from rest is proportional to the square of the time it has traveled, and independent of the mass of the body. Derive this result from the fact that the acceleration due to gravity is a constant.

6. While attempting to understand the motion of bodies under gravity, Galileo stated that:

    > The time in which any space is traversed by a body starting at rest and uniformly accelerated is equal to the time in which that same space would be traversed by the same body moving at a uniform speed whose value is the mean of the highest velocity and the velocity just before acceleration began.

    (a) Write Galileo's statement in symbols, defining all the symbols you use.
    (b) Verify Galileo's statement for a body dropped off a 100-foot building accelerating from rest under gravity until it hits the ground.
    (c) Show why Galileo's statement is true in general.

7. In his *Dialogues Concerning Two New Sciences*, Galileo wrote:

    > The distances traversed during equal intervals of time by a body falling from rest stand to one another in the same ratio as the odd numbers beginning with unity.

    Assume, as is now believed, that $s = -(gt^2)/2$, where s is the total distance traveled in time t, and g is the acceleration due to gravity.

    (a) How far does a falling body travel in the first second (between $t = 0$ and $t = 1$)? During the second second (between $t = 1$ and $t = 2$)? The third second? The fourth second?
    (b) What do your answers tell you about the truth of Galileo's statement?

8. The acceleration due to gravity 2 meters from the ground is 9.8 m/sec^2. What is the acceleration due to gravity 100 meters from the ground? At 100,000 meters? (The radius of the earth is 6.4×10^6 meters.)

CHAPTER SEVEN

INTEGRATION

In Chapter 3, we introduced the definite integral, and in Chapter 6, we introduced antiderivatives. In this chapter, we focus on methods of finding antiderivatives and definite integrals. We first learn additional ways to evaluate antiderivatives and definite integrals exactly, and we then develop more efficient numerical methods for approximating definite integrals. Finally, we introduce improper integrals and discuss methods of evaluating them.

7.1 INTEGRATION BY SUBSTITUTION: PART I

In Chapter 4, we learned rules which enabled us to differentiate any function obtained by combining constants, powers of x, $\sin x$, $\cos x$, e^x, and $\ln x$, using addition, multiplication, division, or composition of functions. These are called *elementary* functions.

Are there similar rules for antidifferentiation? For example, is there a product rule for antiderivatives? Or a chain rule? In the next few sections, we introduce two methods of antidifferentiation: substitution and integration by parts, which reverse the chain and product rules, respectively.

There is a great difference between looking for derivatives and looking for antiderivatives. Every elementary function has elementary derivatives. However, many elementary functions do not have elementary antiderivatives. Some examples are $\sqrt{x^3 + 1}$, $(\sin x)/x$, and e^{-x^2}. These are not exotic functions, but ordinary functions that arise naturally.

The Guess-and-Check Method

A good strategy for finding simple antiderivatives is to *guess* an answer (using knowledge of differentiation rules) and then *check* the answer by differentiating it. If we get the expected result, then we're done; otherwise, we revise the guess and check again.

The method of guess-and-check is most useful in reversing the chain rule. According to the chain rule,

$$\frac{d}{dx}(f(g(x))) = \underbrace{f'}_{\text{Derivative of outside}} \overbrace{(g(x))}^{\text{Inside}} \cdot \underbrace{g'(x)}_{\text{Derivative of inside}}.$$

Thus, any function which is the result of applying the chain rule will be the product of two factors: the "derivative of the outside" and the "derivative of the inside." If a function has this form, its antiderivative is $f(g(x))$.

Example 1 Find $\displaystyle\int 3x^2 \cos(x^3)\, dx$.

Solution The function $3x^2 \cos(x^3)$ looks like the result of applying the chain rule: there is an "inside" function x^3 and its derivative $3x^2$ appearing as a factor. We guess $\sin(x^3)$ for the antiderivative. Differentiating to check gives

$$\frac{d}{dx}(\sin(x^3)) = \cos(x^3) \cdot (3x^2).$$

Since this is what we began with, we have

$$\int 3x^2 \cos(x^3)\, dx = \sin(x^3) + C.$$

The basic idea of this method is to try to find an inside function whose derivative appears as a factor. This works even when the derivative is missing a constant factor, as in the next example.

Example 2 Find $\displaystyle\int t e^{(t^2+1)}\, dt$.

Solution It looks like $t^2 + 1$ is an inside function. So we guess $e^{(t^2+1)}$ for the antiderivative, since taking the derivative of an exponential results in the reappearance of the exponential together with other terms from the chain rule. Now we check:

$$\frac{d}{dt}\left(e^{(t^2+1)}\right) = \left(e^{(t^2+1)}\right) \cdot 2t.$$

The original guess was too large by a factor of 2. We change the guess to $\frac{1}{2}e^{(t^2+1)}$ and check again:

$$\frac{d}{dt}\left(\frac{1}{2}e^{(t^2+1)}\right) = \frac{1}{2}e^{(t^2+1)} \cdot 2t = e^{(t^2+1)} \cdot t.$$

Thus,

$$\int te^{(t^2+1)}\, dt = \frac{1}{2}e^{(t^2+1)} + C.$$

Example 3   Find $\displaystyle\int x^3\sqrt{x^4+5}\, dx.$

Solution   Here the inside function is x^4+5, and its derivative appears as a factor, with the exception of a missing 4. Thus, the integrand we have is more or less of the form

$$g'(x)\sqrt{g(x)},$$

with $g(x) = x^4 + 5$. We might guess that an antiderivative is

$$\frac{(g(x))^{3/2}}{3/2} = \frac{(x^4+5)^{3/2}}{3/2}.$$

Let's check and see:

$$\frac{d}{dx}\left(\frac{(x^4+5)^{3/2}}{3/2}\right) = \frac{3}{2}\frac{(x^4+5)^{1/2}}{3/2} \cdot 4x^3 = 4x^3(x^4+5)^{1/2},$$

so $\dfrac{(x^4+5)^{3/2}}{3/2}$ is too big by a factor of 4. The correct antiderivative is

$$\frac{1}{4}\frac{(x^4+5)^{3/2}}{3/2} = \frac{1}{6}(x^4+5)^{3/2}.$$

Thus

$$\int x^3\sqrt{x^4+5}\, dx = \frac{1}{6}(x^4+5)^{3/2} + C.$$

As a final check:

$$\frac{d}{dx}\left(\frac{1}{6}(x^4+5)^{3/2}\right) = \frac{1}{6}\cdot\frac{3}{2}(x^4+5)^{1/2}\cdot 4x^3 = x^3(x^4+5)^{1/2}.$$

As we have seen in the preceding examples, antidifferentiating a function often involves "correcting for" constant factors: if differentiation produces an extra factor of 4, antidifferentiation will require a factor of $\frac{1}{4}$.

The Method of Substitution

When the integrand is complicated, it helps to formalize this guess-and-check method as follows:

> **To Make a Substitution**
>
> Let w be the "inside function" and $dw = w'(x)\, dx = \dfrac{dw}{dx}dx.$

Let's redo the first example that we just did by guess-and-check.

Example 4   Find $\displaystyle\int 3x^2\cos(x^3)\, dx.$

Solution   As before, we look for an inside function whose derivative appears—in this case x^3. We let $w = x^3$. Then $dw = w'(x)\,dx = 3x^2\,dx$. The original integrand can now be completely rewritten in terms of the new variable w:

$$\int 3x^2 \cos(x^3)\,dx = \int \cos\underbrace{(x^3)}_{w} \cdot \underbrace{3x^2\,dx}_{dw} = \int \cos w\,dw = \sin w + C = \sin(x^3) + C.$$

By changing the variable to w, we can simplify the integrand. We now have $\cos w$, which can be antidifferentiated more easily. The final step, after antidifferentiating, is to convert back to the original variable, x.

Why Does Substitution Work?

The substitution method makes it look as if we can treat dw and dx as separate entities, even canceling them in the equation $dw = (dw/dx)dx$. Let's see why this works. Suppose we have an integral of the form $\int f(g(x))g'(x)\,dx$, where $g(x)$ is the inside function and $f(x)$ is the outside function. If F is an antiderivative of f, then $F' = f$, and by the chain rule $\frac{d}{dx}(F(g(x))) = f(g(x))g'(x)$. Therefore,

$$\int f(g(x))g'(x)\,dx = F(g(x)) + C.$$

Now write $w = g(x)$ and $dw/dx = g'(x)$ on both sides of this equation:

$$\int f(w)\frac{dw}{dx}\,dx = F(w) + C.$$

On the other hand, knowing that $F' = f$ tells us that

$$\int f(w)\,dw = F(w) + C.$$

Thus, the following two integrals are equal:

$$\int f(w)\frac{dw}{dx}\,dx = \int f(w)\,dw.$$

Substituting w for the inside function and writing $dw = w'(x)dx$ leaves the indefinite integral unchanged.

Let's revisit the second example that we did by guess-and-check.

Example 5   Find $\int te^{(t^2+1)}\,dt$.

Solution   Here the inside function is $t^2 + 1$, with derivative $2t$. Since there is a factor of t in the integrand, we try

$$w = t^2 + 1.$$

Then

$$dw = w'(t)\,dt = 2t\,dt.$$

Notice, however, that the original integrand has only $t\,dt$, not $2t\,dt$. We therefore write

$$\frac{1}{2}\,dw = t\,dt$$

and then substitute:

$$\int te^{(t^2+1)}\,dt = \int e^{\overbrace{(t^2+1)}^{w}} \cdot \underbrace{t\,dt}_{\frac{1}{2}dw} = \int e^{w}\frac{1}{2}\,dw = \frac{1}{2}\int e^{w}\,dw = \frac{1}{2}e^{w} + C = \frac{1}{2}e^{(t^2+1)} + C.$$

This gives the same answer as we found using guess-and-check.

Now let's redo the third example that we solved previously by guess-and-check.

Example 6   Find $\int x^3 \sqrt{x^4 + 5}\, dx$.

Solution   The inside function is $x^4 + 5$, with derivative $4x^3$. The integrand has a factor of x^3, and since the only thing missing is a constant factor, we try

$$w = x^4 + 5.$$

Then

$$dw = w'(x)\, dx = 4x^3\, dx,$$

giving

$$\frac{1}{4} dw = x^3\, dx.$$

Thus,

$$\int x^3 \sqrt{x^4 + 5}\, dx = \int \sqrt{w}\, \frac{1}{4} dw = \frac{1}{4} \int w^{1/2}\, dw = \frac{1}{4} \cdot \frac{w^{3/2}}{\frac{3}{2}} + C = \frac{1}{6}(x^4 + 5)^{3/2} + C.$$

Once again, we get the same result as with guess-and-check.

Warning

We saw in the preceding examples that we can apply the substitution method when a *constant* factor is missing from the derivative of the inside function. However, we may not be able to use substitution if anything other than a constant factor is missing. For example, setting $w = x^4 + 5$ to find

$$\int x^2 \sqrt{x^4 + 5}\, dx$$

does us no good because we can't easily get $x^2\, dx$ in terms of w. In order to use substitution, the integrand must contain the derivative of the inside function, *to within a constant factor*.

Example 7   Find $\int e^{\cos \theta} \sin \theta\, d\theta$.

Solution   We let $w = \cos \theta$ since its derivative is $-\sin \theta$ and there is a factor of $\sin \theta$ in the integrand. This gives

$$dw = w'(\theta)\, d\theta = -\sin \theta\, d\theta,$$

and so

$$-dw = \sin \theta\, d\theta.$$

Thus

$$\int e^{\cos \theta} \sin \theta\, d\theta = \int e^w (-dw) = (-1) \int e^w\, dw = -e^w + C = -e^{\cos \theta} + C.$$

Why didn't we put $(-1) \int e^w\, dw = -e^w - C$ in the preceding example? Since the constant C is arbitrary, it doesn't really matter whether we add or subtract it. However, the convention is always to add it on to whatever antiderivative we have calculated. If we think of $-e^{\cos \theta}$ as one of the antiderivatives of $e^{\cos \theta} \sin \theta$, then we write

$$-e^{\cos \theta} + C$$

to remind ourselves that there are lots of other antiderivatives.

Example 8 Find $\displaystyle\int \frac{e^t}{1+e^t}\,dt.$

Solution Observing that the derivative of $1 + e^t$ is e^t, we see $w = 1 + e^t$ is a good choice. Then $dw = e^t\,dt$, so that

$$\int \frac{e^t}{1+e^t}\,dt = \int \frac{1}{1+e^t}\,e^t\,dt = \int \frac{1}{w}\,dw = \ln|w| + C$$
$$= \ln|1+e^t| + C$$
$$= \ln(1+e^t) + C \qquad \text{(Since } (1 + e^t) \text{ is always positive.)}$$

Since the numerator is $e^t\,dt$, we might also have tried $w = e^t$. This substitution leads to the integral $\int (1/(1+w))\,dw$, which is better than the original integral but requires another substitution, $u = 1 + w$, to finish. There are often several different ways of doing an integral by substitution.

Notice the pattern in the previous example: having a function in the denominator and its derivative in the numerator leads to a natural logarithm. The next example follows the same pattern.

Example 9 Find $\displaystyle\int \tan\theta\,d\theta.$

Solution Recall that $\tan\theta = (\sin\theta)/(\cos\theta)$. If $w = \cos\theta$, then $dw = -\sin\theta\,d\theta$, so

$$\int \tan\theta\,d\theta = \int \frac{\sin\theta}{\cos\theta}\,d\theta = \int \frac{-dw}{w} = -\ln|w| + C = -\ln|\cos\theta| + C.$$

Some people prefer the substitution method over guess-and-check since it is more systematic, but both methods achieve the same result. For simple problems, guess-and-check can be faster.

Problems for Section 7.1

1.   (a) Find the derivatives of $\sin(x^2 + 1)$ and $\sin(x^3 + 1)$.
      (b) Use your answer to part (a) to find antiderivatives of:
          (i) $x\cos(x^2 + 1)$ (ii) $x^2\cos(x^3 + 1)$
      (c) Find the antiderivatives of:
          (i) $x\sin(x^2 + 1)$ (ii) $x^2\sin(x^3 + 1)$

For each of the functions in Problems 2–10, find the general antiderivative.

2. $p(t) = \pi t^3 + 4t$ 3. $f(x) = \sin 3x$ 4. $f(x) = 2x\cos(x^2)$

5. $r(t) = 12t^2\cos(t^3)$ 6. $f(x) = \sin(2 - 5x)$ 7. $f(x) = e^{\sin x}\cos x$

8. $g(t) = te^{t^2}$ 9. $f(x) = \dfrac{x}{x^2 + 1}$ 10. $f(x) = \dfrac{1}{3\cos^2(2x)}$

Find the general antiderivatives in Problems 11–41. Remember, you can check your answers.

11. $\displaystyle\int y(y^2 + 5)^8\,dy$ 12. $\displaystyle\int t^2(t^3 - 3)^{10}\,dt$ 13. $\displaystyle\int x(x^2 - 4)^{7/2}\,dx$

14. $\displaystyle\int \frac{1}{\sqrt{4 - x}}\,dx$ 15. $\displaystyle\int x(x^2 + 3)^2\,dx$ 16. $\displaystyle\int \frac{dy}{y + 5}$

17. $\displaystyle\int (2t - 7)^{73}\,dt$ 18. $\displaystyle\int (x^2 + 3)^2\,dx$ 19. $\displaystyle\int \sin\theta(\cos\theta + 5)^7\,d\theta$

20. $\int \sqrt{\cos 3t}\,\sin 3t\,dt$

21. $\int xe^{-x^2}\,dx$

22. $\int \sin^6\theta\cos\theta\,d\theta$

23. $\int \sin^6(5\theta)\cos(5\theta)\,d\theta$

24. $\int x^2 e^{x^3+1}\,dx$

25. $\int \sin^3\alpha\cos\alpha\,d\alpha$

26. $\int \frac{(\ln z)^2}{z}\,dz$

27. $\int \frac{e^t+1}{e^t+t}\,dt$

28. $\int \frac{y}{y^2+4}\,dy$

29. $\int \tan(2x)\,dx$

30. $\int \frac{\cos\sqrt{x}}{\sqrt{x}}\,dx$

31. $\int \frac{e^{\sqrt{y}}}{\sqrt{y}}\,dy$

32. $\int \frac{1+e^x}{\sqrt{x+e^x}}\,dx$

33. $\int \frac{e^x}{2+e^x}\,dx$

34. $\int \frac{x+1}{x^2+2x+19}\,dx$

35. $\int y^2(1+y)^2\,dy$

36. $\int x^2(1+2x^3)^2\,dx$

37. $\int \frac{t}{1+3t^2}\,dt$

38. $\int \frac{e^x-e^{-x}}{e^x+e^{-x}}\,dx$

39. $\int \frac{(t+1)^2}{t^2}\,dt$

40. $\int \frac{x\cos(x^2)}{\sqrt{\sin(x^2)}}\,dx$

41. $\int (2x+1)e^{x^2}e^x\,dx$ [Hint: Rewrite $e^{x^2}e^x = e^?$.]

42. Use the identity $\cos^2 x + \sin^2 x = 1$ to find $\int \sin^3 x\,dx$.

43. Find $\int 4x(x^2+1)\,dx$ using two methods:

   (a) Do the multiplication first, and then antidifferentiate.
   (b) Now do the integral using the substitution $w = x^2 + 1$.
   (c) Explain why the two expressions formed in parts (a) and (b) are different. Are they both correct?

44. Throughout much of this century, the yearly consumption of electricity in the US has been increasing exponentially at a continuous rate of 7% per year. Assuming this trend continues and that the electrical energy consumed in 1900 was 1.4 million megawatt-hours (a megawatt-hour is a measure of electrical energy),

   (a) Write an expression for yearly electricity consumption as a function of time, t, measured in years since 1900.
   (b) Find the average yearly electrical consumption throughout this century.
   (c) During what year was electrical consumption closest to the average for the century?
   (d) Without doing the calculation for part (c), how could you have predicted which half of the century the answer would be in?

45. If we assume that wind resistance is proportional to velocity, then the velocity, v, of a body of mass m falling vertically is given by

$$v = \frac{mg}{k}\left(1 - e^{-kt/m}\right),$$

   where g is the acceleration due to gravity and k is a constant. Find the height, h, above the surface of the earth as a function of time. Assume the body starts at height h_0.

46. (a) During the 1970s, ACME Widgets sold at a continuous rate of $R = R_0 e^{0.15t}$ widgets per year, where t is time in years since January 1, 1970. Suppose they were selling widgets at a rate of 1000 per year on the first day of the decade. How many widgets did they sell during the decade? How many did they sell if the rate on January 1, 1970 was 150,000,000 widgets per year?
   (b) In the first case above (1000 widgets per year on January 1, 1970), how long did it take for half the widgets in the 1970s to be sold? In the second case (150,000,000 widgets per year on January 1, 1970), when had half the widgets in the 1970s been sold?
   (c) In 1980 ACME began an advertising campaign claiming that half the widgets it had sold in the previous ten years were still in use. Based on your answer to part (b), about how long must a widget last in order to justify this claim?

7.2 INTEGRATION BY SUBSTITUTION: PART II

In this section we use substitution to evaluate *definite* integrals. We then look at some examples where it is harder to see which substitution to use, but where the method works nonetheless.

Definite Integrals by Substitution

Example 1   Compute $\displaystyle\int_0^2 xe^{x^2}\, dx.$

Solution   To evaluate this definite integral using the Fundamental Theorem of Calculus, we first need to find an antiderivative of $f(x) = xe^{x^2}$. The inside function is x^2, so let $w = x^2$. Then $dw = 2x\, dx$, so $\frac{1}{2}\, dw = x\, dx$. Thus,

$$\int xe^{x^2}\, dx = \int e^w \frac{1}{2}\, dw = \frac{1}{2}e^w + C = \frac{1}{2}e^{x^2} + C.$$

Now we find the definite integral

$$\int_0^2 xe^{x^2}\, dx = \frac{1}{2}e^{x^2}\Big|_0^2 = \frac{1}{2}(e^4 - e^0) = \frac{1}{2}(e^4 - 1).$$

There is another way to look at the same problem. After we established that

$$\int xe^{x^2}\, dx = \frac{1}{2}e^w + C,$$

our next two steps were to replace w by x^2, and then x by 2 and 0. We could have directly replaced the original limits of integration, $x = 0$ and $x = 2$, by the corresponding w limits. Since $w = x^2$, the w limits are $w = 0^2 = 0$ (when $x = 0$) and $w = 2^2 = 4$ (when $x = 2$), so we get

$$\int_{x=0}^{x=2} xe^{x^2}\, dx = \frac{1}{2}\int_{w=0}^{w=4} e^w\, dw = \frac{1}{2}e^w\Big|_0^4 = \frac{1}{2}\left(e^4 - e^0\right) = \frac{1}{2}(e^4 - 1).$$

As we would expect, both methods give the same answer.

To Use Substitution to Find Definite Integrals

Either
- Compute the indefinite integral, expressing an antiderivative in terms of the original variable, and then evaluate the result at the original limits,

or
- Convert the original limits to new limits in terms of the new variable and do not convert the antiderivative back to the original variable.

Example 2   Evaluate $\displaystyle\int_0^{\pi/4} \frac{\tan^3 \theta}{\cos^2 \theta}\, d\theta.$

Solution   To use substitution, we must decide what w should be. There are two possible inside functions, $\tan\theta$ and $\cos\theta$. Now

$$\frac{d}{d\theta}(\tan\theta) = \frac{1}{\cos^2\theta} \quad \text{and} \quad \frac{d}{d\theta}(\cos\theta) = -\sin\theta,$$

and since the integral contains a factor of $1/\cos^2\theta$ but not of $\sin\theta$, we let $w = \tan\theta$. Then $dw = (1/\cos^2\theta)d\theta$. When $\theta = 0$, $w = \tan 0 = 0$, and when $\theta = \pi/4$, $w = \tan(\pi/4) = 1$, so

$$\int_0^{\pi/4} \frac{\tan^3\theta}{\cos^2\theta}\,d\theta = \int_0^{\pi/4} (\tan\theta)^3 \cdot \frac{1}{\cos^2\theta}\,d\theta = \int_0^1 w^3\,dw = \frac{1}{4}w^4\Big|_0^1 = \frac{1}{4}.$$

Example 3   Evaluate $\displaystyle\int_1^3 \frac{dx}{5 - x}$.

Solution   Let $w = 5 - x$, so $dw = -dx$. When $x = 1$, $w = 4$, and when $x = 3$, $w = 2$, so

$$\int_1^3 \frac{dx}{5 - x} = \int_4^2 \frac{-dw}{w} = -\ln|w|\Big|_4^2 = -(\ln 2 - \ln 4) = \ln\left(\frac{4}{2}\right) = \ln 2 \approx 0.69.$$

Notice that we write the limit $w = 4$ at the bottom, even though it is larger than $w = 2$, because $w = 4$ corresponds to the lower limit $x = 1$.

More Complex Substitutions

In the examples of substitution presented so far, we had to guess an expression for w and then we had to be lucky enough to have dw (or some constant multiple of it) lying around in the integrand. What if we are not so lucky? It turns out that we can often get somewhere by letting w be some messy expression contained inside, say, a cosine or under a root, even if we cannot see immediately how such a substitution helps.

Example 4   Find $\displaystyle\int (x + 7)\sqrt[3]{3 - 2x}\,dx$.

Solution   Here, instead of the derivative of the inside function (which is -2), we have the factor $(x + 7)$. However, substituting $w = 3 - 2x$ turns out to help. Then $dw = -2\,dx$, so $(-1/2)\,dw = dx$. Now we must convert everything to w, including $x + 7$. If $w = 3 - 2x$, then $2x = 3 - w$, so $x = 3/2 - w/2$, and therefore we can write $x + 7$ in terms of w. Thus

$$\int (x + 7)\sqrt[3]{3 - 2x}\,dx = \int \left(\frac{3}{2} - \frac{w}{2} + 7\right)\sqrt[3]{w}\left(-\frac{1}{2}\right)dw$$

$$= -\frac{1}{2}\int \left(\frac{17}{2} - \frac{w}{2}\right)w^{1/3}\,dw$$

$$= -\frac{1}{4}\int (17 - w)w^{1/3}\,dw$$

$$= -\frac{1}{4}\int (17w^{1/3} - w^{4/3})\,dw$$

$$= -\frac{1}{4}\left(17\frac{w^{4/3}}{4/3} - \frac{w^{7/3}}{7/3}\right) + C$$

$$= -\frac{1}{4}\left(\frac{51}{4}(3 - 2x)^{4/3} - \frac{3}{7}(3 - 2x)^{7/3}\right) + C.$$

Looking back over the solution, the reason this substitution works is that it converts $\sqrt[3]{3 - 2x}$, the messiest part of the integrand, to $\sqrt[3]{w}$, which can be combined with the other term and then integrated.

Identities can sometimes help us simplify an integral, as the next example illustrates.

Example 5   Find $\int \cos^3 \theta \, d\theta$.

Solution   This integral can be simplified by using the Pythagorean identity $\cos^2 \theta + \sin^2 \theta = 1$ to rewrite the integrand:

$$\int \cos^3 \theta = \int \cos^2 \theta \cdot \cos \theta \, d\theta = \int (1 - \sin^2 \theta) \cos \theta \, d\theta$$

$$= \int \cos \theta \, d\theta - \int \sin^2 \theta \cos \theta \, d\theta.$$

The first integral is straightforward; for the second, we use substitution, letting $w = \sin \theta$ and $dw = \cos \theta \, d\theta$. We have

$$\int \sin^2 \theta \cos \theta \, d\theta = \int w^2 \, dw = \frac{1}{3} w^3 + C = \frac{1}{3} \sin^3 \theta + C.$$

Combining the constants from both integrals into C, we have

$$\int \cos^3 \theta \, d\theta = \int \cos \theta \, d\theta - \int \sin^2 \theta \cos \theta \, d\theta = \sin \theta - \frac{1}{3} \sin^3 \theta + C.$$

Example 6   Find $\int \sqrt{1 + \sqrt{x}} \, dx$.

Solution   This time, the derivative of the inside function is nowhere to be seen. Nevertheless, we try $w = 1 + \sqrt{x}$. Then $w - 1 = \sqrt{x}$, so $(w - 1)^2 = x$. Therefore $2(w - 1) \, dw = dx$. We then have

$$\int \sqrt{1 + \sqrt{x}} \, dx = \int \sqrt{w} \, 2(w - 1) \, dw = 2 \int w^{1/2}(w - 1) \, dw$$

$$= 2 \int (w^{3/2} - w^{1/2}) \, dw = 2 \left(\frac{2}{5} w^{5/2} - \frac{2}{3} w^{3/2} \right) + C$$

$$= 2 \left(\frac{2}{5} (1 + \sqrt{x})^{5/2} - \frac{2}{3} (1 + \sqrt{x})^{3/2} \right) + C.$$

Notice that the substitution in the preceding example again converts the inside of the messiest function into something simple. In addition, since the derivative of the inside function is not waiting for us, we have to solve for x so that we can get dx entirely in terms of w and dw.

Substitution can be used to express one definite integral in terms of another, even in situations where we don't know what the indefinite integral is.

Example 7   Derive a formula for the area of the ellipse

$$\frac{x^2}{a^2} + \frac{y^2}{b^2} = 1$$

assuming $a, b > 0$.

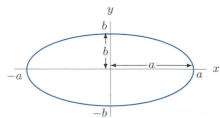

Figure 7.1: The ellipse: $\frac{x^2}{a^2} + \frac{y^2}{b^2} = 1$

Solution   Solving for y gives

$$y = \pm b\sqrt{1 - \frac{x^2}{a^2}},$$

where the positive square root gives the upper half of the ellipse, and the negative square root gives the lower. (See Figure 7.1.) Thus, the area of the ellipse is given by

$$\text{Area} = 2\int_{-a}^{a} b\sqrt{1 - \frac{x^2}{a^2}}\, dx$$

Substitution allows us to convert this integral to one whose value is known. Let $w = x/a$, so $dx = a\, dw$. When $x = -a$, $w = -1$, and when $x = a$, $w = 1$. Thus

$$2\int_{-a}^{a} b\sqrt{1 - \frac{x^2}{a^2}}\, dx = 2\int_{-1}^{1} b\sqrt{1 - w^2}\, a\, dw = 2ab\int_{-1}^{1}\sqrt{1 - w^2}\, dw.$$

Now the integral $\int_{-1}^{1}\sqrt{1 - w^2}\, dw$ is the area of half a circle of radius 1 (because $y = \sqrt{1 - w^2}$ is the top half of the circle $w^2 + y^2 = 1$). Thus,

$$\int_{-1}^{1}\sqrt{1 - w^2}\, dw = \frac{\pi}{2}$$

so

$$\text{Area of ellipse} = 2ab\int_{-1}^{1}\sqrt{1 - w^2}\, dw = \pi ab.$$

Problems for Section 7.2

1. Use substitution to express each of the following integrals as a multiple of $\int_{a}^{b}\frac{1}{w}\, dw$ for some a and b. Then evaluate the integrals.

   (a) $\displaystyle\int_{0}^{1}\frac{x}{1 + x^2}\, dx$ (b) $\displaystyle\int_{0}^{\pi/4}\frac{\sin x}{\cos x}\, dx$

For Problems 2–10, use the Fundamental Theorem to calculate the definite integrals. Show your work.

2. $\displaystyle\int_{0}^{\pi}\cos(x + \pi)\, dx$ 3. $\displaystyle\int_{0}^{1/2}\cos(\pi x)\, dx$ 4. $\displaystyle\int_{0}^{\pi/2} e^{-\cos\theta}\sin\theta\, d\theta$

5. $\displaystyle\int_{1}^{2} 2xe^{x^2}\, dx$ 6. $\displaystyle\int_{1}^{8}\frac{e^{\sqrt[3]{x}}}{\sqrt[3]{x^2}}\, dx$ 7. $\displaystyle\int_{-1}^{e-2}\frac{1}{t + 2}\, dt$

8. $\displaystyle\int_{1}^{4}\frac{\cos\sqrt{x}}{\sqrt{x}}\, dx$ 9. $\displaystyle\int_{0}^{2}\frac{x}{(1 + x^2)^2}\, dx$ 10. $\displaystyle\int_{-\pi/4}^{\pi/4}\cos^2\theta\sin^5\theta\, d\theta$

   [Hint: $\cos^2\theta + \sin^2\theta = 1$.]

For Problems 11–16, evaluate the definite integrals. Whenever possible, use the Fundamental Theorem of Calculus, perhaps after a substitution. Otherwise, use numerical methods.

11. $\displaystyle\int_{-1}^{3}(x^3 + 5x)\, dx$ 12. $\displaystyle\int_{-1}^{1}\frac{1}{1 + y^2}\, dy$ 13. $\displaystyle\int_{1}^{3}\frac{1}{x}\, dx$

14. $\displaystyle\int_{1}^{3}\frac{dt}{(t + 7)^2}$ 15. $\displaystyle\int_{-1}^{2}\sqrt{x + 2}\, dx$ 16. $\displaystyle\int_{1}^{2}\frac{\sin t}{t}\, dt$

For Problems 17–27, evaluate the definite integrals. Whenever possible, use the Fundamental Theorem of Calculus, perhaps after a substitution. Otherwise, use numerical methods.

17. $\displaystyle\int_0^1 x(1+x^2)^{20}\,dx$

18. $\displaystyle\int_0^\pi \sin\theta(\cos\theta+5)^7\,d\theta$

19. $\displaystyle\int_0^{\pi/12} \sin(3\alpha)\,d\alpha$

20. $\displaystyle\int_0^1 \frac{x}{1+5x^2}\,dx$

21. $\displaystyle\int_1^2 \frac{x^2+1}{x}\,dx$

22. $\displaystyle\int_0^1 \frac{1}{x^2+2x+1}\,dx$

23. $\displaystyle\int_0^1 \frac{(x+2)}{(x+2)^2+1}\,dx$

24. $\displaystyle\int_4^1 x\sqrt{x^2+4}\,dx$

25. $\displaystyle\int_0^{1/\sqrt{2}} \frac{x}{\sqrt{1-x^4}}\,dx$

26. $\displaystyle\int_{1/4}^1 \sin\frac{1}{t}\,dt$

27. $\displaystyle\int_{-2}^0 \frac{2x+4}{x^2+4x+5}\,dx$

In Problems 28–32, explain why the two antiderivatives are really, despite their apparent dissimilarity, different expressions of the same problem. You do not need to evaluate the integrals.

28. $\displaystyle\int \frac{e^x\,dx}{1+e^{2x}}$ and $\displaystyle\int \frac{\cos x\,dx}{1+\sin^2 x}$

29. $\displaystyle\int \frac{\ln x}{x}\,dx$ and $\displaystyle\int x\,dx$

30. $\displaystyle\int e^{\sin x}\cos x\,dx$ and $\displaystyle\int \frac{e^{\arcsin x}}{\sqrt{1-x^2}}\,dx$

31. $\displaystyle\int \sqrt{x+1}\,dx$ and $\displaystyle\int \frac{\sqrt{1+\sqrt{x}}}{\sqrt{x}}\,dx$

32. $\displaystyle\int (\sin x)^3\cos x\,dx$ and $\displaystyle\int (x^3+1)^3 x^2\,dx$

33. Find $\displaystyle\int \frac{dx}{x^2+4x+5}$ by completing the square and using the substitution $x+2=\tan\theta$.

34. Integrate:

   (a) $\displaystyle\int \frac{1}{\sqrt{x}}\,dx$ (b) $\displaystyle\int \frac{1}{\sqrt{x+1}}\,dx$ (c) $\displaystyle\int \frac{1}{\sqrt{x}+1}\,dx$

35. Find the exact area under the graph of $f(x)=xe^{x^2}$ between $x=0$ and $x=2$.

36. Find the exact value of the area under one arch of the curve $V(t)=V_0\sin(\omega t)$.

37.   (a) Calculate exactly: $\displaystyle\int_{-\pi}^\pi \cos^2\theta\sin\theta\,d\theta$.

   (b) Calculate the exact area under the curve $y=\cos^2\theta\sin\theta$ between $\theta=0$ and $\theta=\pi$.

38. Suppose $\displaystyle\int_0^1 f(t)\,dt=3$. Calculate each of the following integrals:

   (a) $\displaystyle\int_0^{0.5} f(2t)\,dt$ (b) $\displaystyle\int_0^1 f(1-t)\,dt$ (c) $\displaystyle\int_1^{1.5} f(3-2t)\,dt$

39. Find the average value of $f(x)=\dfrac{1}{x+1}$ on the interval $x=0$ to $x=2$. Sketch a graph showing the function and the average value.

40.   (a) Find $\displaystyle\int \sin\theta\cos\theta\,d\theta$.

   (b) You probably solved part (a) by making the substitution $w=\sin\theta$ or $w=\cos\theta$. (If not, go back and do it that way.) Now find $\int \sin\theta\cos\theta\,d\theta$ by making the *other* substitution.

   (c) There is yet another way of finding this integral which involves the trigonometric identities

$$\sin(2\theta)=2\sin\theta\cos\theta \quad\text{and}\quad \cos(2\theta)=\cos^2\theta-\sin^2\theta.$$

Find $\int \sin\theta\cos\theta\,d\theta$ using one of these identities and then the substitution $w=2\theta$.

   (d) You should now have three different expressions for the indefinite integral $\int \sin\theta\cos\theta\,d\theta$. Are they really different? Are they all correct? Explain the reasons for your answer.

41.   It is said that Charles Dickens was paid a penny per word for his novels. We can express the payment for a novel containing L words as $\int_0^L 1\,dw$. Instead of the constant 1 penny per word, suppose that for each new manuscript he produced he would be paid $\int_0^L f(w)dw$, where L is the total length of the novel (in words), and f is a function depending on w, the number of words, which represents the pay rate, given by:

$$f(w) = \begin{cases} (1/2) + w/2000 & 0 \le w \le 2000 \\ 3/2 & 2000 \le w \le 20{,}000 \\ (3/2)e^{20-w/1000} & w \ge 20{,}000. \end{cases}$$

   (a)   How much would Dickens be paid for a 40,000-word novel?
   (b)   Would he be paid more for writing one 30,000-word novel or for writing two 15,000-word novels? Try to answer part (b) without computing any integrals. [Hint: Does the exponentially decaying part of the definition of $f(w)$ encourage or discourage long novels?]

42.   The rate at which water is flowing into a tank is $r(t)$ gallons/minute, with t in minutes.

   (a)   Write an expression approximating the amount of water entering the tank during the interval from time t to time $t + \Delta t$, where Δt is small.
   (b)   Write a Riemann sum approximating the total amount of water entering the tank between $t = 0$ and $t = 5$. Write an exact expression for this amount.
   (c)   By how much has the amount of water in the tank changed between $t = 0$ and $t = 5$ if $r(t) = 20e^{0.02t}$?
   (d)   If $r(t)$ is as given in part (c), and if the tank contains 3000 gallons initially, find a formula for $Q(t)$, the amount of water in the tank at time t.

43.   If we assume that wind resistance is proportional to the square of velocity, then the downward velocity, v, of a falling body is given by

$$v = \sqrt{\frac{g}{k}} \left(\frac{e^{t\sqrt{gk}} - e^{-t\sqrt{gk}}}{e^{t\sqrt{gk}} + e^{-t\sqrt{gk}}} \right).$$

   Use the substitution $w = e^{t\sqrt{gk}} + e^{-t\sqrt{gk}}$ to find the height, h, of the body above the surface of the earth as a function of time. Assume the body starts at a height h_0.

7.3   INTEGRATION BY PARTS

We have seen how the method of substitution reverses the chain rule. Now we will introduce a method called *integration by parts* that is based on the product rule.

Example 1   Find $\int xe^x\,dx$.

Solution   We are looking for a function whose derivative is xe^x. If we think about how the product rule works, we might be led to guess xe^x, because we know that when we take the derivative we will get two terms, one of which will be xe^x:

$$\frac{d}{dx}(xe^x) = \frac{d}{dx}(x)e^x + x\frac{d}{dx}(e^x) = e^x + xe^x.$$

Of course, our guess is wrong because of the extra e^x. Maybe we can adjust our guess by subtracting from xe^x something to cancel out the extra e^x. This leads us to try $xe^x - e^x$. Let's check it:

$$\frac{d}{dx}(xe^x - e^x) = \frac{d}{dx}(xe^x) - \frac{d}{dx}(e^x) = e^x + xe^x - e^x = xe^x.$$

It works, so $\int xe^x\,dx = xe^x - e^x + C$.

Example 2   Find $\int \theta \cos \theta\,d\theta$.

Solution We guess the antiderivative is $\theta \sin \theta$ and use the product rule to check:

$$\frac{d}{d\theta}(\theta \sin \theta) = \frac{d(\theta)}{d\theta} \sin \theta + \theta \frac{d}{d\theta}(\sin \theta) = \sin \theta + \theta \cos \theta.$$

To correct for the extra $\sin \theta$ term, we must subtract from our original guess something whose derivative is $\sin \theta$. Since $\frac{d}{d\theta}(\cos \theta) = -\sin \theta$, we'll try:

$$\frac{d}{d\theta}(\theta \sin \theta + \cos \theta) = \frac{d}{d\theta}(\theta \sin \theta) + \frac{d}{d\theta}(\cos \theta) = \sin \theta + \theta \cos \theta - \sin \theta = \theta \cos \theta.$$

Thus, $\displaystyle\int \theta \cos \theta \, d\theta = \theta \sin \theta + \cos \theta + C.$

The General Formula for Integration by Parts

We can formalize the process illustrated in the last two examples in the following way. We begin with the product rule:

$$\frac{d}{dx}(uv) = u'v + uv'$$

where u and v are functions of x with derivatives u' and v', respectively. We rewrite this as:

$$uv' = \frac{d}{dx}(uv) - u'v$$

and then integrate both sides:

$$\int uv' \, dx = \int \frac{d}{dx}(uv) \, dx - \int u'v \, dx.$$

Since an antiderivative of $\dfrac{d}{dx}(uv)$ is just uv, we get the following formula:

Integration by Parts

$$\int uv' \, dx = uv - \int u'v \, dx.$$

This formula is useful when the integrand can be viewed as a product and when the integral on the right-hand side is simpler than that on the left. In effect, we were using integration by parts in the previous two examples. In Example 1, we let $xe^x = (x) \cdot (e^x) = uv'$, and choose $u = x$ and $v' = e^x$. Thus, $u' = 1$ and $v = e^x$, so

$$\int \underbrace{(x)}_{u} \underbrace{(e^x)}_{v'} \, dx = \underbrace{(x)}_{u} \underbrace{(e^x)}_{v} - \int \underbrace{(1)}_{u'} \underbrace{(e^x)}_{v} \, dx = xe^x - e^x + C.$$

So uv represents our first guess, and $\int u'v \, dx$ represents the correction to our guess.

Notice what would have happened in Example 1 if we took $u = x$ and $v = e^x + C_1$. Then

$$\int xe^x \, dx = x(e^x + C_1) - \int (e^x + C_1) \, dx$$
$$= xe^x + C_1 x - e^x - C_1 x + C$$
$$= xe^x - e^x + C,$$

as before. It is not necessary to include an arbitrary constant in the antiderivative for v; any antiderivative will do.

What would have happened in Example 1 if we had picked u and v' the other way around? If $u = e^x$ and $v' = x$, then $u' = e^x$ and $v = x^2/2$. The formula for integration by parts then gives

$$\int xe^x \, dx = \frac{x^2}{2} e^x - \int \frac{x^2}{2} \cdot e^x \, dx,$$

which is true but not helpful since the integral on the right is worse than the one on the left. To use this method, we must choose u and v' to make the integral on the right easier to find than the integral on the left.

Advice on How to Choose u and v'

- Whatever you let v' be, you need to be able to find v.
- It helps if u' is simpler than u (or at least no more complicated than u).
- It helps if v is simpler than v' (or at least no more complicated than v').

If we pick $v' = x$ in Example 1, then $v = x^2/2$, which is certainly "worse" than v'.

There are some examples which don't look like good candidates for integration by parts because they don't appear to involve products, but for which the process works well. Such examples often involve $\ln x$ or the inverse trigonometric functions. Here is one:

Example 3   Find $\displaystyle\int_2^3 \ln x \, dx$.

Solution   This doesn't look like a product unless we write $\ln x = (1)(\ln x)$. Then we might say $u = 1$ so $u' = 0$, which certainly makes things simpler. But if $v' = \ln x$, what is v? If we knew, we wouldn't need integration by parts. Let's try the other way: if $u = \ln x$, $u' = 1/x$ and if $v' = 1$, $v = x$, so

$$\int_2^3 \underbrace{(\ln x)}_{u}\,\underbrace{(1)}_{v'}\,dx = \underbrace{(\ln x)}_{u}\,\underbrace{(x)}_{v}\,\bigg|_2^3 - \int_2^3 \underbrace{\left(\frac{1}{x}\right)}_{u'} \cdot \underbrace{(x)}_{v}\,dx$$

$$= x \ln x \,\bigg|_2^3 - \int_2^3 1\,dx = (x \ln x - x)\,\bigg|_2^3$$

$$= 3\ln 3 - 3 - 2\ln 2 + 2 = 3\ln 3 - 2\ln 2 - 1.$$

Notice that when doing a definite integral by parts, we must remember to put the limits of integration (here 2 and 3) on the uv term (in this case $x \ln x$) as well as on the integral $\int u'v\,dx$.

Example 4   Find $\displaystyle\int x^6 \ln x \, dx$.

Solution   View $x^6 \ln x$ as uv' where $u = \ln x$ and $v' = x^6$. Then $v = \frac{1}{7}x^7$ and $u' = 1/x$, so integration by parts gives us:

$$\int x^6 \ln x\,dx = \int (\ln x)x^6\,dx = (\ln x)\left(\frac{1}{7}x^7\right) - \int \frac{1}{7}x^7 \cdot \frac{1}{x}\,dx$$

$$= \frac{1}{7}x^7 \ln x - \frac{1}{7}\int x^6\,dx$$

$$= \frac{1}{7}x^7 \ln x - \frac{1}{49}x^7 + C.$$

In Example 4 we did not choose $v' = \ln x$, because it is not immediately clear what v would be. In fact, we used integration by parts in Example 3 to find out what the antiderivative of $\ln x$ was. Also, using $u = \ln x$, as we have done, gives $u' = 1/x$, which most people would consider simpler than $u = \ln x$. This shows that u does not have to be the first factor in the integrand (here x^6).

Example 5 Find $\int x^2 \sin 4x \, dx$.

Solution If we let $v' = \sin 4x$, then $v = -\frac{1}{4} \cos 4x$, which is no worse than v'. Also letting $u = x^2$, we get $u' = 2x$, which is simpler than $u = x^2$. Using integration by parts:

$$\int x^2 \sin 4x \, dx = x^2 \left(-\frac{1}{4} \cos 4x \right) - \int 2x \left(-\frac{1}{4} \cos 4x \right) dx$$

$$= -\frac{1}{4} x^2 \cos 4x + \frac{1}{2} \int x \cos 4x \, dx.$$

The trouble is we still have to grapple with $\int x \cos 4x \, dx$. This can be done by using integration by parts again with a new u and v, namely $u = x$ and $v' = \cos 4x$:

$$\int x \cos 4x \, dx = x \left(\frac{1}{4} \sin 4x \right) - \int 1 \cdot \frac{1}{4} \sin 4x \, dx$$

$$= \frac{1}{4} x \sin 4x - \frac{1}{4} \cdot \left(-\frac{1}{4} \cos 4x \right) + C$$

$$= \frac{1}{4} x \sin 4x + \frac{1}{16} \cos 4x + C.$$

Thus,

$$\int x^2 \sin 4x \, dx = -\frac{1}{4} x^2 \cos 4x + \frac{1}{2} \int x \cos 4x \, dx$$

$$= -\frac{1}{4} x^2 \cos 4x + \frac{1}{2} \left(\frac{1}{4} x \sin 4x + \frac{1}{16} \cos 4x + C \right)$$

$$= -\frac{1}{4} x^2 \cos 4x + \frac{1}{8} x \sin 4x + \frac{1}{32} \cos 4x + C_1$$

where C_1 is the arbitrary constant $\frac{1}{2} C$. Notice that, in this example, each time we used integration by parts, the exponent of x went down by 1.

Example 6 Find $\int \cos^2 \theta \, d\theta$.

Solution Using integration by parts with $u = \cos \theta$, $v' = \cos \theta$ gives $u' = -\sin \theta$, $v = \sin \theta$, so we get

$$\int \cos^2 \theta \, d\theta = \cos \theta \sin \theta + \int \sin^2 \theta \, d\theta$$

Substituting $\sin^2 \theta = 1 - \cos^2 \theta$ leads to

$$\int \cos^2 \theta \, d\theta = \cos \theta \sin \theta + \int (1 - \cos^2 \theta) \, d\theta$$

$$= \cos \theta \sin \theta + \int 1 \, d\theta - \int \cos^2 \theta \, d\theta.$$

Looking at the right side, we see that the original integral has reappeared. If we move it to the left, we get

$$2 \int \cos^2 \theta \, d\theta = \cos \theta \sin \theta + \int 1 \, d\theta = \cos \theta \sin \theta + \theta + C.$$

Dividing by 2 gives

$$\int \cos^2 \theta \, d\theta = \frac{1}{2} \cos \theta \sin \theta + \frac{1}{2} \theta + C_1$$

where C_1 is the arbitrary constant $\frac{1}{2} C$. Problem 38 does this integral by another method.

You can check these results by differentiation.

Problems for Section 7.3

1. By writing $\arctan x = (1) \cdot (\arctan x)$, find $\int \arctan x \, dx$.

Find the indefinite integrals in Problems 2–27.

2. $\int t e^{5t} \, dt$

3. $\int t^2 e^{5t} \, dt$

4. $\int p e^{-0.1p} \, dp$

5. $\int t \sin t \, dt$

6. $\int y \ln y \, dy$

7. $\int x^3 \ln x \, dx$

8. $\int (z+1) e^{2z} \, dz$

9. $\int \frac{z}{e^z} \, dz$

10. $\int t^2 \sin t \, dt$

11. $\int \theta^2 \cos 3\theta \, d\theta$

12. $\int \sin^2 \theta \, d\theta$

13. $\int (\theta+1) \sin(\theta+1) \, d\theta$

14. $\int \cos^2(3\alpha + 1) \, d\alpha$

15. $\int \frac{\ln x}{x^2} \, dx$

16. $\int q^5 \ln 5q \, dq$

17. $\int y \sqrt{y+3} \, dy$

18. $\int (t+2) \sqrt{2+3t} \, dt$

19. $\int \frac{y}{\sqrt{5-y}} \, dy$

20. $\int \frac{t+7}{\sqrt{5-t}} \, dt$

21. $\int (\ln t)^2 \, dt$

22. $\int x (\ln x)^4 \, dx$

23. $\int \arcsin w \, dw$

24. $\int \arctan 7z \, dz$

25. $\int x \arctan x^2 \, dx$

26. $\int x^3 e^{x^2} \, dx$

27. $\int x^5 \cos x^3 \, dx$

Evaluate the integrals in Problems 28–36 both exactly [e.g., $\ln(3\pi)$] and numerically [e.g. $\ln(3\pi) \approx 2.243$]:

28. $\int_1^5 \ln t \, dt$

29. $\int_3^5 x \cos x \, dx$

30. $\int_0^{10} z e^{-z} \, dz$

31. $\int_1^3 t \ln t \, dt$

32. $\int_0^1 \arctan y \, dy$

33. $\int_0^5 \ln(1+t) \, dt$

34. $\int_0^1 x \arctan x^2 \, dx$

35. $\int_0^1 \arcsin z \, dz$

36. $\int_0^1 u \arcsin u^2 \, du$

37. In Problem 12, you evaluated $\int \sin^2 \theta \, d\theta$ using integration by parts. (If you didn't do it by parts, do so now!) Redo this integral by changing the form of the integrand using the identity $\sin^2 \theta = (1 - \cos 2\theta)/2$. Explain any differences in the form of the answer obtained by the two methods.

38. Compute $\int \cos^2 \theta \, d\theta$ in two different ways and explain any differences in the form of your answers. (The identity $\cos^2 \theta = (1 + \cos 2\theta)/2$ may be useful.)

39. Use integration by parts twice to find $\int e^x \sin x \, dx$.

40. Use integration by parts twice to find $\int e^\theta \cos \theta \, d\theta$.

41. Use the results from Problems 39 and 40 and integration by parts to find $\int x e^x \sin x \, dx$.

42. Use the results from Problems 39 and 40 and integration by parts to find $\int \theta e^\theta \cos \theta \, d\theta$.

43. Show that $\int x^n e^x \, dx = x^n e^x - n \int x^{n-1} e^x \, dx$.

44. Show that $\int x^n \cos ax \, dx = \frac{1}{a} x^n \sin ax - \frac{n}{a} \int x^{n-1} \sin ax \, dx$.

45. Show that $\int x^n \sin ax \, dx = -\frac{1}{a} x^n \cos ax + \frac{n}{a} \int x^{n-1} \cos ax \, dx$.

46. Show that $\int \cos^n x \, dx = \frac{1}{n} \cos^{n-1} x \sin x + \frac{n-1}{n} \int \cos^{n-2} x \, dx$.

47. Find the exact value of the area under the first arch of $f(x) = x \sin x$.

48. Let f be a twice differentiable function such that $f(0) = 6$, $f(1) = 5$, and $f'(1) = 2$. Evaluate the integral $\int_0^1 x f''(x) dx$.

49. Suppose $F(a)$ represents the area under the graph of $y = x^2 e^{-x}$ between $x = 0$ and $x = a$. (Assume $a > 0$.)

     (a) Find a formula for $F(a)$.
     (b) Is F an increasing or decreasing function?
     (c) Is F concave up or concave down for $0 < a < 2$?

50. The voltage, V, in an electric circuit is given as a function of time, t, by

$$V = V_0 \cos(\omega t + \phi).$$

     Suppose each of the positive constants, V_0, ω, ϕ is increased (while the other two are held constant). What is the effect of each increase on each of the following quantities:

     (a) The maximum value of V?
     (b) The maximum value of dV/dt?
     (c) The average value of V^2 over one period?

51. Integrating $e^{ax} \sin bx$ by parts twice yields a result of the form

$$\int e^{ax} \sin bx \, dx = e^{ax}(A \sin bx + B \cos bx) + C.$$

     (a) Find the constants A and B in terms of a and b. [Hint: Don't actually perform the integration by parts.]
     (b) Evaluate $\int e^{ax} \cos bx \, dx$ by modifying the result in part (a). [Again, it is not necessary to perform integration by parts, as the result is of the same form as that in part (a).]

52. During a surge in the demand for electricity, the rate, r, at which energy is used can be approximated by

$$r = te^{-at},$$

     where t is the time in hours and a is a positive constant.

     (a) Find the total energy, E, used in the first T hours. Give your answer as a function of a.
     (b) What happens to E as $T \to \infty$?

53. In describing the behavior of an electron, we use wave functions $\Psi_1, \Psi_2, \Psi_3, \ldots$ of the form

$$\Psi_n(x) = C_n \sin(n\pi x) \qquad \text{for } n = 1, 2, 3, \ldots$$

     where x is the distance from a fixed point and C_n is a positive constant.

     (a) Find C_1 so that Ψ_1 satisfies

$$\int_0^1 \left(\Psi_1(x) \right)^2 dx = 1.$$

     This process is called normalizing the wave function.
     (b) For any integer n, find C_n so that Ψ_n is normalized.

7.4 TABLES OF INTEGRALS

Since so few functions have elementary antiderivatives, they have been compiled in a list called a table of integrals. Such integral tables are available[1] so that if you need an antiderivative, you can simply look it up. A short table of indefinite integrals is given inside the back cover. The key to using these tables is being able to recognize the general class of function that you are trying to integrate, so you can know in what section of the table to look.

Warning: This section involves long division of polynomials and completing the square. You may want to review these topics!

Using the Table of Integrals

Part I of the table inside the back cover gives the antiderivatives of the basic functions x^n, a^x, $\ln x$, $\sin x$, $\cos x$, and $\tan x$. (The antiderivative for $\ln x$ is found using integration by parts and is a special case of the more general formula III-13.) Most of these are already familiar.

Part II of the table contains antiderivatives of functions involving products of e^x, $\sin x$, and $\cos x$. All of these antiderivatives were obtained using integration by parts.

Example 1   Find $\displaystyle\int \sin 7z \sin 3z \, dz$.

Solution   Since the integrand is the product of two sines, we should use II-10 in the table,

$$\int \sin 7z \sin 3z \, dz = -\frac{1}{40}(7\cos 7z \sin 3z - 3\cos 3z \sin 7z) + C.$$

Part III of the table contains antiderivatives for products of a polynomial and e^x, $\sin x$, or $\cos x$. It also has an antiderivative for $x^n \ln x$, which can easily be used to find the antiderivatives of the product of a general polynomial and $\ln x$. Each *reduction formula* is used repeatedly to reduce the degree of the polynomial until a zero degree polynomial is obtained.

Example 2   Find $\displaystyle\int (x^5 + 2x^3 - 8)e^{3x}\, dx$.

Solution   Since $p(x) = x^5 + 2x^3 - 8$ is a polynomial multiplied by e^{3x}, this is of the form in III-14. Now $p'(x) = 5x^4 + 6x^2$ and $p''(x) = 20x^3 + 12x$, and so on, giving

$$\int (x^5 + 2x^3 - 8)e^{3x}\, dx = e^{3x}\left[\frac{1}{3}(x^5 + 2x^3 - 8) - \frac{1}{9}(5x^4 + 6x^2) + \frac{1}{27}(20x^3 + 12x)\right.$$

$$\left. - \frac{1}{81}(60x^2 + 12) + \frac{1}{243}(120x) - \frac{1}{729}\cdot 120\right] + C.$$

Here we have the successive derivatives of the original polynomial $x^5 + 2x^3 - 8$, occurring with alternating signs and multiplied by successive powers of 1/3.

Part IV of the table contains reduction formulas for the antiderivatives of $\cos^n x$ and $\sin^n x$, which can be obtained by integration by parts. When n is a positive integer, formulas IV-17 and IV-18 can be used repeatedly to reduce the power n until it is 0 or 1.

---

[1] See, for example, *CRC Standard Mathematical Tables* (Boca Raton, Fl: CRC Press). There are now computer programs and calculators that can compute antiderivatives as well.

Example 3   Find $\int \sin^6 \theta \, d\theta$.

Solution   Use IV-17 repeatedly:

$$\int \sin^6 \theta \, d\theta = -\frac{1}{6} \sin^5 \theta \cos \theta + \frac{5}{6} \int \sin^4 \theta \, d\theta$$

$$\int \sin^4 \theta \, d\theta = -\frac{1}{4} \sin^3 \theta \cos \theta + \frac{3}{4} \int \sin^2 \theta \, d\theta$$

$$\int \sin^2 \theta \, d\theta = -\frac{1}{2} \sin \theta \cos \theta + \frac{1}{2} \int 1 \, d\theta.$$

Calculate $\int \sin^2 \theta \, d\theta$ first, and use this to find $\int \sin^4 \theta \, d\theta$; then calculate $\int \sin^6 \theta \, d\theta$. Putting this all together, we get

$$\int \sin^6 \theta \, d\theta = -\frac{1}{6} \sin^5 \theta \cos \theta - \frac{5}{24} \sin^3 \theta \cos \theta - \frac{15}{48} \sin \theta \cos \theta + \frac{15}{48}\theta + C.$$

The last item in **Part IV** of the table is not a formula at all: it is, instead, advice on how to antidifferentiate products of integer powers of $\sin x$ and $\cos x$. There are various techniques to choose from, depending on the nature (odd or even, positive or negative) of the exponents.

Example 4   Find $\int \cos^3 t \sin^4 t \, dt$.

Solution   Here the exponent of $\cos t$ is odd, so IV-23 recommends making the substitution $w = \sin t$. Then $dw = \cos t \, dt$. To make this work, we'll have to separate off one of the cosines to be part of dw. Also, the remaining even power of $\cos t$ can be rewritten in terms of $\sin t$ by using $\cos^2 t = 1 - \sin^2 t = 1 - w^2$, so that

$$\int \cos^3 t \sin^4 t \, dt = \int \cos^2 t \sin^4 t \cos t \, dt$$

$$= \int (1 - w^2)w^4 \, dw = \int (w^4 - w^6) \, dw$$

$$= \frac{1}{5}w^5 - \frac{1}{7}w^7 + C = \frac{1}{5}\sin^5 t - \frac{1}{7}\sin^7 t + C.$$

Example 5   Find $\int \cos^2 x \sin^4 x \, dx$.

Solution   In this example, both exponents are even. The advice given in IV-23 is to convert to all sines or all cosines. We'll convert to all sines by substituting $\cos^2 x = 1 - \sin^2 x$, and then we'll multiply out the integrand:

$$\int \cos^2 x \sin^4 x \, dx = \int (1 - \sin^2 x) \sin^4 x \, dx = \int \sin^4 x \, dx - \int \sin^6 x \, dx.$$

In Example 3 we found $\int \sin^4 x \, dx$ and $\int \sin^6 x \, dx$. Put them together to get

$$\int \cos^2 x \sin^4 x \, dx = -\frac{1}{4} \sin^3 x \cos x - \frac{3}{8} \sin x \cos x + \frac{3}{8}x$$

$$- \left(-\frac{1}{6} \sin^5 x \cos x - \frac{5}{24} \sin^3 x \cos x - \frac{15}{48} \sin x \cos x + \frac{15}{48}x \right) + C$$

$$= \frac{1}{6} \sin^5 x \cos x - \frac{1}{24} \sin^3 x \cos x - \frac{3}{48} \sin x \cos x + \frac{3}{48}x + C.$$

The last two parts of the table are concerned with quadratic functions: **Part V** has expressions with quadratic denominators; **Part VI** contains square roots of quadratics. The quadratics that appear in these formulas are of the form $x^2 \pm a^2$ or $a^2 - x^2$, or in factored form $(x-a)(x-b)$, where a and b are constants. Many integrands with quadratics in them are either in one of these forms or can be transformed into one of them by completing the square or factoring.

Using Factoring

Example 6 Find $\displaystyle\int \frac{3x+7}{x^2+6x+8}\,dx$.

Solution In this case the denominator factors

$$x^2 + 6x + 8 = (x+2)(x+4).$$

Now in V-27 we let $a = -2$, $b = -4$, $c = 3$, and $d = 7$, to obtain

$$\int \frac{3x+7}{x^2+6x+8}\,dx = \frac{1}{2}\left(\ln|x+2| - (-5)\ln|x+4|\right) + C.$$

Completing the Square to Rewrite the Quadratic in the Form $w^2 + a^2$

Example 7 Find $\displaystyle\int \frac{1}{x^2+6x+14}\,dx$.

Solution By completing the square, we get

$$x^2 + 6x + 14 = (x^2 + 6x + 9) - 9 + 14$$
$$= (x+3)^2 + 5.$$

Let $w = x + 3$. Then $dw = dx$ and so the substitution gives

$$\int \frac{1}{x^2+6x+14}\,dx = \int \frac{1}{w^2+5}\,dw = \frac{1}{\sqrt{5}}\arctan\frac{w}{\sqrt{5}} + C = \frac{1}{\sqrt{5}}\arctan\frac{x+3}{\sqrt{5}} + C,$$

where the antidifferentiation uses V-24 with $a^2 = 5$.

Preparing to Use the Table: Transforming the Integrand

Some integration problems will arise in forms that look nearly the same as the formulas in our table, but many will not. In order to use the table, we often need to manipulate or reshape integrands to fit entries in the table. The kinds of manipulation that tend to be useful are expansion, factoring, long division, completing the square, and substitution.

Example 8 Find $\displaystyle\int \frac{x^2}{x^2+4}\,dx$.

Solution A good rule of thumb when integrating a rational function whose numerator has a degree greater than or equal to that of the denominator is to start by doing *long division*. This results in a polynomial plus a simpler rational function as a remainder. Performing long division here, we obtain:

$$\frac{x^2}{x^2+4} = 1 - \frac{4}{x^2+4}.$$

Then, by V-24 with $a = 2$, we obtain:

$$\int \frac{x^2}{x^2+4}\,dx = \int 1\,dx - 4\int \frac{1}{x^2+4}\,dx = x - 4\cdot\frac{1}{2}\arctan\frac{x}{2} + C = x - 2\arctan\frac{x}{2} + C.$$

Example 9   Find $\int e^t \sin(5t + 7)\, dt$.

Solution   This looks pretty similar to II-8. To make the correspondence more complete, let's try the substitution $w = 5t + 7$. Then $dw = 5\, dt$, so $dt = \frac{1}{5}\, dw$. Also, $t = (w - 7)/5$. Then the integral becomes

$$\int e^t \sin(5t + 7)\, dt = \int e^{(w-7)/5} \sin w\, \frac{dw}{5}$$

$$= \frac{e^{-7/5}}{5} \int e^{w/5} \sin w\, dw \qquad \text{(Since } e^{(w-7)/5} = e^{w/5}e^{-7/5} \text{ and } e^{-7/5} \text{ is a constant)}$$

Now we can use II-8 with $a = \frac{1}{5}$ and $b = 1$ to write

$$\int e^{w/5} \sin w\, dw = \frac{1}{(\frac{1}{5})^2 + 1^2} e^{w/5} \left(\frac{\sin w}{5} - \cos w \right) + C,$$

so

$$\int e^t \sin(5t + 7)\, dt = \frac{e^{-7/5}}{5} \left(\frac{25}{26} e^{(5t+7)/5} \left(\frac{\sin(5t + 7)}{5} - \cos(5t + 7) \right) \right) + C$$

$$= \frac{5e^t}{26} \left(\frac{\sin(5t + 7)}{5} - \cos(5t + 7) \right) + C.$$

Method of Partial Fractions

The formulas in the table involving rational functions of the type in V-26 and V-27 were obtained by splitting the integrand into *partial fractions*. For example, to find

$$\int \frac{1}{(x - 2)(x - 5)}\, dx,$$

the integrand is split into partial fractions with denominators $(x - 2)$ and $(x - 5)$. We write

$$\frac{1}{(x - 2)(x - 5)} = \frac{A}{x - 2} + \frac{B}{x - 5}.$$

Multiplying by $(x - 2)(x - 5)$ gives

$$1 = A(x - 5) + B(x - 2)$$

so,

$$1 = (A + B)x - 5A - 2B.$$

Since this equation holds for all x, the constant terms on both sides must be equal. Similarly, the coefficient of x on both sides must be equal. So

$$-5A - 2B = 1 \quad \text{and} \quad A + B = 0.$$

Solving these equations gives $A = -1/3$, $B = 1/3$.

Example 10   Use partial fractions to rewrite the integrand in such a way that $\int \frac{1}{(x - 2)(x - 5)}\, dx$ can be integrated.

Solution   The calculation we have just done gives

$$\int \frac{1}{(x - 2)(x - 5)}\, dx = \int \left(\frac{-1/3}{x - 2} + \frac{1/3}{x - 5} \right)\, dx = -\frac{1}{3} \ln|x - 2| + \frac{1}{3} \ln|x - 5| + C.$$

You can check that using V-26 gives the same result.

A generalization of this method can be used to find an antiderivative of any rational function.

Example 11   Find $\displaystyle\int \frac{x+2}{x^2+x}\,dx$.

Solution   We factor the denominator and split the integrand into partial fractions:

$$\frac{x+2}{x^2+x} = \frac{x+2}{x(x+1)} = \frac{A}{x} + \frac{B}{x+1}.$$

Multiplying by $x(x+1)$ gives

$$x+2 = A(x+1) + Bx$$
$$= (A+B)x + A.$$

Equating constant terms and coefficients of x, we must have $A=2$ and $A+B=1$, so $B=-1$. Then

$$\int \frac{x+2}{x^2+x}\,dx = \int \left(\frac{2}{x} - \frac{1}{x+1}\right)dx = 2\ln|x| - \ln|x+1| + C.$$

Problems for Section 7.4

For Problems 1–30, antidifferentiate, using the table of integrals and a substitution if necessary.

1. $\displaystyle\int e^{-3\theta}\cos\theta\,d\theta$

2. $\displaystyle\int x^5 \ln x\,dx$

3. $\displaystyle\int \sin w \cos^4 w\,dw$

4. $\displaystyle\int \sin^4 x\,dx$

5. $\displaystyle\int \frac{1}{3+y^2}\,dy$

6. $\displaystyle\int \frac{1}{\cos^3 x}\,dx$

7. $\displaystyle\int x^3 e^{2x}\,dx$

8. $\displaystyle\int \sin 3\theta \cos 5\theta\,d\theta$

9. $\displaystyle\int \sin 3\theta \sin 5\theta\,d\theta$

10. $\displaystyle\int x^2 e^{3x}\,dx$

11. $\displaystyle\int x^2 e^{x^3}\,dx$

12. $\displaystyle\int x^4 e^{3x}\,dx$

13. $\displaystyle\int u^5 \ln(5u)\,du$

14. $\displaystyle\int \frac{t^2+1}{t^2-1}\,dt$

15. $\displaystyle\int x^3 \sin x^2\,dx$

16. $\displaystyle\int \cos 2y \cos 7y\,dy$

17. $\displaystyle\int y^2 \sin 2y\,dy$

18. $\displaystyle\int e^{5x} \sin 3x\,dx$

19. $\displaystyle\int \frac{1}{\sin^2 2\theta}\,d\theta$

20. $\displaystyle\int \frac{1}{\sin^3 3\theta}\,d\theta$

21. $\displaystyle\int \frac{1}{\cos^4 7x}\,dx$

22. $\displaystyle\int \frac{1}{x^2+4x+3}\,dx$

23. $\displaystyle\int \tan^4 x\,dx$

24. $\displaystyle\int \frac{dz}{z(z-3)}$

25. $\displaystyle\int \frac{dy}{4-y^2}$

26. $\displaystyle\int \frac{1}{1+(z+2)^2}\,dz$

27. $\displaystyle\int \frac{1}{y^2+4y+5}\,dy$

28. $\displaystyle\int \frac{1}{x^2+4x+4}\,dx$

29. $\displaystyle\int \sin^3 3\theta \cos^2 3\theta\,d\theta$

30. $\displaystyle\int ze^{2z^2}\cos(2z^2)\,dz$

31.   Show that for all integers m and n, with $m \neq \pm n$, $\displaystyle\int_{-\pi}^{\pi} \sin m\theta \sin n\theta\,d\theta = 0$.

32.   Show that for all integers m and n, with $m \neq \pm n$, $\displaystyle\int_{-\pi}^{\pi} \cos m\theta \cos n\theta\,d\theta = 0$.

33.   (a)   Show how combining terms in the expression $\displaystyle\frac{2}{x} + \frac{1}{x+3}$ yields $\displaystyle\frac{3x+6}{x^2+3x}$, and use this to evaluate
$$\int \frac{3x+6}{x^2+3x}\,dx.$$
      (b)   Show that your answer to part (a) agrees with the answer you get by using the integral tables.

34. Rewrite $\dfrac{1}{x^2-1}$ using partial fractions and find $\displaystyle\int \dfrac{1}{x^2-1}\,dx$.

35. (a) Use partial fractions to find $\displaystyle\int \dfrac{1}{x^2-x}\,dx$.
    (b) Show that your answer to part (a) agrees with the answer you get by using the integral tables.

36. Use partial fractions to find $\displaystyle\int \dfrac{1}{x(L-x)}\,dx$, where L is constant.

37. Use the method of partial fractions to evaluate $\displaystyle\int \dfrac{dP}{3P-3P^2}$.

38. Use the method of partial fractions to find $\displaystyle\int \dfrac{3x+1}{x^2-3x+2}\,dx$.

39. The voltage, V, in an electrical outlet is given as a function of time, t, by the function $V = V_0 \cos(120\pi t)$, where V is in volts and t is in seconds, and V_0 is a constant representing the maximum voltage.

    (a) What is the average value of the voltage over 1 second?
    (b) Engineers do not use the average voltage. They use the *root mean square* voltage defined by $\overline{V} = \sqrt{\text{average of }(V^2)}$. Find $\overline{V}$ in terms of V_0. (Take the average over 1 second.)
    (c) The standard voltage in an American house is 110 volts, meaning that $\overline{V} = 110$. What is V_0?

40. An economist studying the rate of production, $R(t)$, of oil in a new oil well has proposed the following model:
$$R(t) = A + Be^{-t}\sin(2\pi t)$$
where t is the time in years, A is the equilibrium rate (a constant), and B is the "variability" coefficient (a constant).

    (a) Find the total amount of oil produced in the first N years of operation. (Take N to be an integer.)
    (b) Find the average amount of oil produced per year over the first N years (where N is an integer).
    (c) From your answer to part (b), find the average amount of oil produced per year as $N \to \infty$.
    (d) Looking at the function $R(t)$, explain how you might have predicted your answer to part (c) without doing any calculations.
    (e) Do you think it is reasonable to expect this model to hold over a very long period? Why or why not?

41. Suppose n is a positive integer and $\Psi_n = C_n \sin(n\pi x)$ is the wave function used in describing the behavior of an electron. If n and m are different integers, find
$$\int_0^1 \Psi_n(x) \cdot \Psi_m(x)\,dx.$$

7.5 APPROXIMATING DEFINITE INTEGRALS

The methods of the last few sections allow us to get exact answers for definite integrals in a variety of special cases. However, many functions do not have elementary antiderivatives. To evaluate the definite integrals of such functions, we cannot use the Fundamental Theorem and so we must use numerical methods. In addition, many real-world applications of calculus do not require exact answers.

We already know how to approximate a definite integral numerically using left- and right-hand Riemann sums. In the next two sections we introduce better methods for approximating definite integrals—better in the sense that they give more accurate results with less work than that required to find the left- and right-hand sums.

The Midpoint Rule

In the left- and right-hand Riemann sums, the heights of the rectangles are found using the left-hand or right-hand endpoints, respectively, of the subintervals. For the *midpoint rule*, we use the midpoint of each of the subintervals.

For example, in approximating $\int_1^2 f(x)\,dx$ by a Riemann sum with two subdivisions, we first divide the interval $1 \leq x \leq 2$ into two pieces. The midpoint of the first subinterval is 1.25 and the midpoint of the second is 1.75. The heights of the two rectangles are $f(1.25)$ and $f(1.75)$, respectively. (See Figure 7.2.) The Riemann sum is

$$f(1.25)0.5 + f(1.75)0.5.$$

Figure 7.2 shows that evaluating f at the midpoint of each subdivision usually gives a better approximation to the area under the curve than evaluating f at either end. For this particular f, it appears that each rectangle is partly above and partly below the graph on each subinterval. Furthermore, the area under the curve which is not under the rectangle appears to be nearly equal to the area under the rectangle which is above the curve. In fact, this new midpoint Riemann sum is generally a better approximation to the definite integral than the left- or right-hand sum with the same number of subdivisions, n.

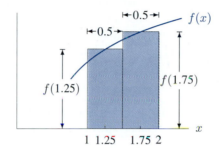

Figure 7.2: Midpoint rule with two subdivisions

So far, we have three ways of estimating an integral using a Riemann sum:

1. The **left rule** uses the left endpoint of each subinterval.
2. The **right rule** uses the right endpoint of each subinterval.
3. The **midpoint rule** use the midpoint of each subinterval.

We write LEFT(n), RIGHT(n), and MID(n) to denote the results obtained by using these rules with n subdivisions.

Example 1   For $\displaystyle\int_1^2 \frac{1}{x}\,dx$, compute LEFT(2), RIGHT(2) and MID(2), and compare your answers with the exact value of the integral.

Solution   For $n = 2$ subdivisions of the interval $[1, 2]$, we use $\Delta x = 0.5$. Then

$$\text{LEFT}(2) = f(1)(0.5) + f(1.5)(0.5) = \frac{1}{1}(0.5) + \frac{1}{1.5}(0.5) = 0.8333\ldots$$

$$\text{RIGHT}(2) = f(1.5)(0.5) + f(2)(0.5) = \frac{1}{1.5}(0.5) + \frac{1}{2}(0.5) = 0.5833\ldots$$

$$\text{MID}(2) = f(1.25)(0.5) + f(1.75)(0.5) = \frac{1}{1.25}(0.5) + \frac{1}{1.75}(0.5) = 0.6857\ldots$$

All three Riemann sums in this example are approximating

$$\int_1^2 \frac{1}{x}\,dx = \ln x \Big|_1^2 = \ln 2 - \ln 1 = \ln 2 = 0.6931\ldots.$$

With only two subdivisions, the left and right rules give quite poor approximations but the midpoint rule is already fairly close to the exact answer.

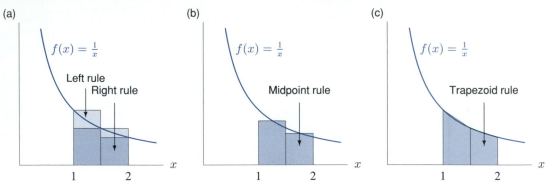

Figure 7.3: Left, right, midpoint, and trapezoid approximations to $\int_1^2 \frac{1}{x}\,dx$

Figure 7.3(a) illustrates why the left and right rules are so inaccurate. Since $f(x) = 1/x$ is decreasing from 1 to 2, the left rule overestimates on each subdivision while the right rule underestimates. However, the midpoint rule approximates with rectangles on each subdivision that are each partly above and partly below the graph, so the errors tend to balance out. (See Figure 7.3(b).)

The Trapezoid Rule

We have just seen how the midpoint rule can have the effect of balancing out the errors of the left and right rules. There is another way of balancing these errors: we average the results from the left and right rules. This approximation is called the *trapezoid rule*:

$$\text{TRAP}(n) = \frac{\text{LEFT}(n) + \text{RIGHT}(n)}{2}.$$

The trapezoid rule averages the values of f at the left and right endpoints of each subinterval and multiplies by Δx. This is the same as approximating the area under the graph of f in each subinterval by a trapezoid (see Figure 7.4).

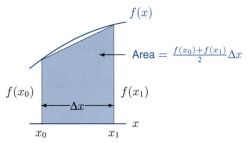

Figure 7.4: Area used in the trapezoid rule

Example 2   For $\displaystyle\int_1^2 \frac{1}{x}\,dx$, compare the trapezoid rule with two subdivisions with the left, right, and midpoint rules.

Solution   In the previous example we got $\text{LEFT}(2) = 0.8333\ldots$ and $\text{RIGHT}(2) = 0.5833\ldots$. The trapezoid rule is the average of these, so $\text{TRAP}(2) = 0.7083\ldots$. (See Figure 7.3(c).) The exact value of the integral is $0.6931\ldots$, so the trapezoid rule is better than the left or right rules. The midpoint rule is still the best, however, since $\text{MID}(2) = 0.6857\ldots$.

Is the Approximation an Over- or Underestimate?

It is useful to know when a rule is producing an overestimate and when it is producing an underestimate. In Chapter 3 we saw that the following relationship holds.

If f is increasing on $[a, b]$, then

$$\text{LEFT}(n) \leq \int_a^b f(x)\, dx \leq \text{RIGHT}(n).$$

If f is decreasing on $[a, b]$, then

$$\text{RIGHT}(n) \leq \int_a^b f(x)\, dx \leq \text{LEFT}(n).$$

The Trapezoid Rule

If the graph of the function is concave down on $[a, b]$, then each trapezoid lies below the graph and the trapezoid rule underestimates. If the graph is concave up on $[a, b]$, the trapezoid rule overestimates. (See Figure 7.5.)

f concave down:
Trapezoid underestimates

f concave up:
Trapezoid overestimates

Figure 7.5: Error in the trapezoid rule

The Midpoint Rule

To understand the relationship between the midpoint rule and concavity, take a rectangle whose top intersects the curve at the midpoint of a subinterval. Draw a tangent to the curve at the midpoint; this gives a trapezoid. See Figure 7.6. (This is *not* the same trapezoid as in the trapezoid rule.) The midpoint rectangle and the new trapezoid have the same area, because the shaded triangles in Figure 7.6 are congruent. Hence, if the graph of the function is concave down, the midpoint rule overestimates; if the graph is concave up, the midpoint rule underestimates. (See Figure 7.7.)

If the graph of f is concave down on $[a, b]$, then

$$\text{TRAP}(n) \leq \int_a^b f(x)\, dx \leq \text{MID}(n).$$

If the graph of f is concave up on $[a, b]$, then

$$\text{MID}(n) \leq \int_a^b f(x)\, dx \leq \text{TRAP}(n).$$

f concave down:
Midpoint overestimates

f concave up:
Midpoint underestimates

Figure 7.6: Midpoint rectangle and trapezoid with same area

Figure 7.7: Error in the midpoint rule

Problems for Section 7.5

1. (a) By hand, find LEFT(2) and RIGHT(2) for $\int_0^4 (x^2 + 1)\, dx$. Show your work.
    (b) Illustrate your answers to part (a) graphically. Is each estimate an underestimate or overestimate?

2. (a) By hand, find MID(2) and TRAP(2) for $\int_0^4 (x^2 + 1)\, dx$. Show your work.
    (b) Illustrate your answers to part (a) graphically. Is each estimate an underestimate or overestimate?

3. The following table contains approximations to $\int_0^4 \sqrt{100 + x^3}\, dx$. Fill in the rest of the table with values rounded to four decimal places.

	$n = 1$	$n = 2$	$n = 4$
LEFT			
RIGHT	51.2250		
TRAP		43.5909	
MID			

4. (a) Estimate $\int_0^1 \dfrac{1}{1 + x^2}\, dx$ by subdividing the interval into eight parts using:

    (i) The left-hand Riemann sum.   (ii) The right-hand Riemann sum.
    (iii) The trapezoidal rule.

    (b) Since the exact value of the integral is $\pi/4$, you can estimate the value of π. Explain why your first estimate is too large and your second estimate too small.

5. Consider the following integrals:

    $$(i)\quad \int_1^{10} \ln x\, dx \qquad (ii)\quad \int_0^4 e^x\, dx$$

    (a) For each integral, find LEFT(32), RIGHT(32), and TRAP(32). Also find the exact value of each integral.
    (b) For each integral, arrange LEFT(32), RIGHT(32), TRAP(32), and the true value in ascending order. Explain, using diagrams, how you could predict this ordering without doing the calculations in part (a).

6. Given the velocity-time data in the table, estimate the total distance traveled from time $t = 0$ to time $t = 6$ using LEFT, RIGHT, and TRAP.

t	0	1	2	3	4	5	6
v	3	4	5	4	7	8	11

7. The graph of g is shown in Figure 7.8. The results from the left, right, trapezoid, and midpoint rules used to approximate $\int_0^1 g(t)\, dt$, with the same number of subdivisions for each rule, are as follows: 0.601, 0.632, 0.633, 0.664.

    (a) Match each rule with its approximation.
    (b) Between which two consecutive approximations does the true value of the integral lie?

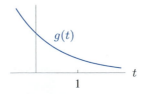

Figure 7.8

Figure 7.9

8. We approximate the integrals of the functions in Figure 7.9 on the interval shown, using a fixed number of subdivisions.

   (a) For which function, f or g, is LEFT more accurate? RIGHT? Explain.
   (b) For which function, f or g, is TRAP more accurate? MID? Explain.

9. Given the values for $y = f(x)$ in the table:

   (a) Which of the four approximation methods in this section is most likely to give the best estimate of $\int_0^{12} f(x)\, dx$? Estimate the integral using this method.
   (b) Assume $f(x)$ is continuous with no critical points or points of inflection on the interval $0 \leq x \leq 12$. Is the estimate found in part (a) an over- or underestimate? Explain.

x	0	3	6	9	12
$f(x)$	100	97	90	78	55

Suppose you want to approximate $\int_0^5 f(x)\, dx$ for the functions graphed in Problems 10–13. In each case, pick which of the approximations — left, right, trapezoid, or midpoint — is guaranteed to give an overestimate for the integral, and which is guaranteed to give an underestimate. (There may be more than one answer.)

10.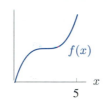

11.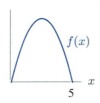

12.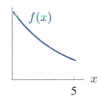

13.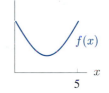

14. (a) Find the exact value of $\displaystyle\int_0^{2\pi} \sin\theta\, d\theta$.

    (b) Explain, using pictures, why the MID(1) and MID(2) approximations to this integral give the exact value.

    (c) Does MID(3) give the exact value of this integral? How about MID(n)? Explain.

15. (a) Show that $\displaystyle\int_0^1 \sqrt{2 - x^2}\, dx = \frac{\pi}{4} + \frac{1}{2}$. [Hint: Break up the area under $y = \sqrt{2 - x^2}$ from $x = 0$ to $x = 1$ into two pieces: a sector of a circle and a right triangle.]

    (b) Approximate $\int_0^1 \sqrt{2 - x^2}\, dx$ for $n = 5$ using the left, right, trapezoid, and midpoint rules. Compute the error in each case using the answer to part (a), and compare the errors.

16. Sketch the graph of a function for which LEFT(2) is more accurate than MID(2). (You don't need to give a formula for the function, just a graph.)

17. The width, in feet, at various points along the fairway of a hole on a golf course is given in Figure 7.10. If one pound of fertilizer covers 200 square feet, estimate the amount of fertilizer needed to fertilize the fairway.

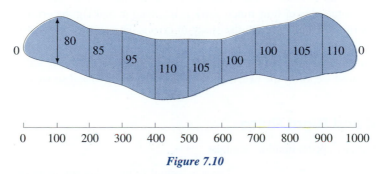

Figure 7.10

18. (a) Explain why the area of the trapezoid in Figure 7.11 is $h \cdot (l_1 + l_2)/2$.
     (b) On a graph similar to Figure 7.12, sketch areas representing each of the following quantities:

$$E = h \cdot f(0), \quad F = h \cdot f(h), \quad R = h \cdot f\left(\frac{h}{2}\right), \quad C = h \cdot \frac{f(0) + f(h)}{2} = \frac{E + F}{2}$$

$$N = \frac{h}{2} \cdot \frac{f(0) + f(\frac{h}{2})}{2} + \frac{h}{2} \cdot \frac{f(\frac{h}{2}) + f(h)}{2} = \frac{R + C}{2}.$$

     (c) Let A be the area under the function shown in Figure 7.12. Write the values A, E, F, R, C, and N in increasing order.
     (d) Which is the better approximation to A: E or F?
     (e) Which is the better approximation to A: R or C?

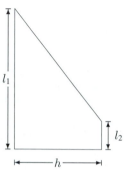

Figure 7.11

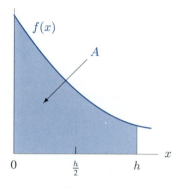

Figure 7.12

Problems 19–23 involve approximating $\int_a^b f(x)\, dx$.

19. Show that $\text{RIGHT}(n) = \text{LEFT}(n) + f(b)\Delta x - f(a)\Delta x$.

20. Show that $\text{TRAP}(n) = \text{LEFT}(n) + \frac{1}{2}\left(f(b) - f(a)\right)\Delta x$.

21. Show that $\text{LEFT}(2n) = \frac{1}{2}\left(\text{LEFT}(n) + \text{MID}(n)\right)$.

22. Using a computer or calculator, verify that the equations given in Problems 19 and 20 hold for $\int_1^2 (1/x)\, dx$, when $n = 10$.

23. Suppose that $a = 2$, $b = 5$, $f(2) = 13$, $f(5) = 21$ and that $\text{LEFT}(10) = 3.156$ and $\text{MID}(10) = 3.242$. Use the equations given in Problems 19–21 to compute $\text{RIGHT}(10)$, $\text{TRAP}(10)$, $\text{LEFT}(20)$, $\text{RIGHT}(20)$, and $\text{TRAP}(20)$.

7.6 APPROXIMATION ERRORS AND SIMPSON'S RULE

When we compute an approximation, we are always concerned about the error, namely the difference between the exact answer and the approximation. We usually do not know the exact error; if we did, we would also know the exact answer. Often the best we can get is an upper bound on the error and some idea of how much work is involved in making the error smaller. The study of numerical approximations is really the study of errors. The errors for some methods are much smaller than those for others. The errors for the midpoint and trapezoid rules are related to each other in a way that suggests an even better method, called Simpson's rule. We will work with the example $\int_1^2 \frac{1}{x}\, dx$ because we know the exact value of this integral ($\ln 2$) and so we can investigate the behavior of the errors.

Error in Left and Right Rules

For any approximation, we take

$$\text{Error} = \text{Actual value} - \text{Approximate value}.$$

TABLE 7.1   *Errors for the left and right rule approximation to* $\int_1^2 \frac{1}{x}\,dx = \ln 2 \approx 0.6931471806$

n	Error in left rule	Error in right rule
2	−0.1402	0.1098
10	−0.0256	0.0244
50	−0.0050	0.0050
250	−0.0010	0.0010

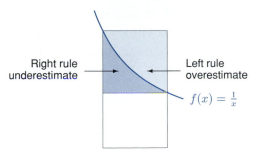

Figure 7.13: Errors in left and right sums

Let us see what happens to the error in the left and right rules as we increase n. We increase n each time by a factor of 5 starting at $n = 2$. The results are in Table 7.1. A positive error indicates that the Riemann sum is less than the exact value, ln 2. Notice that the errors for the left and right rules have opposite signs but are approximately equal in magnitude. (See Figure 7.13.) The best way to try to get the errors to cancel is to average the left and right rules; this average is the trapezoid rule. If we had not already thought of the trapezoid rule, we might have been led to invent it by this observation.

There is another pattern to the errors in Table 7.1. If we compute the *ratio* of the errors in Table 7.2, we see that the error[2] in both the left and right rules decreases by a factor of about 5 as n increases by a factor of 5.

There is nothing special about the number 5; the same holds for any factor. To get one extra digit of accuracy in any calculation, we must make the error $1/10$ as big, so we must increase n by a factor of 10. In fact, *for the left or right rules, each extra digit of accuracy requires about 10 times the work*. The calculator used to produce these tables took about half a second to compute the left rule approximation for $n = 50$, and this yields ln 2 to two digits. To get three correct digits, n would need to be around 500 and the time would be about 5 seconds. Four digits requires $n = 5000$ and 50 seconds. Ten digits requires $n = 5 \times 10^9$ and 5×10^7 seconds, which is more than a year! Clearly, the errors for the left and right rules do not decrease fast enough as n increases for practical use.

TABLE 7.2   *Ratio of the errors as n increases for* $\int_1^2 \frac{1}{x}\,dx$

	Ratio of errors in left rule	Ratio of errors in right rule
Error(2)/Error(10)	5.47	4.51
Error(10)/Error(50)	5.10	4.90
Error(50)/Error(250)	5.02	4.98

Error in Trapezoid and Midpoint Rules

Table 7.3 shows that the trapezoid and midpoint rules produce much better approximations to $\int_1^2 \frac{1}{x}\,dx$ than the left and right rules.

Again there is a pattern to the errors. For each n, the midpoint rule is noticeably better than the trapezoid rule; the error for the midpoint rule, in absolute value, seems to be about half the error of the trapezoid rule. To see why, compare the shaded areas in Figure 7.14. Also, notice in Table 7.3 that the errors for the two rules have opposite signs; this is due to concavity.

We are interested in how the errors behave as n increases. Table 7.4 gives the ratios of the errors for each rule. For each rule, we see that as n increases by a factor of 5, the error decreases by a factor of about $25 = 5^2$. In fact, it can be shown that this squaring relationship holds for any factor, so increasing n by a factor of 10 will decrease the error by a factor of about $100 = 10^2$. Reducing the error by a factor of 100 is equivalent to adding two more decimal places of accuracy to the result.

---

[2]The values in Table 7.1 are rounded to 4 decimal places; those in Table 7.2 were computed using more decimal places and then rounded.

TABLE 7.3   *The errors for the trapezoid and midpoint rules for $\int_1^2 \frac{1}{x}\,dx$*

n	Error in trapezoid rule	Error in midpoint rule
2	-0.0152	0.0074
10	-0.00062	0.00031
50	-0.0000250	0.0000125
250	-0.0000010	0.0000005

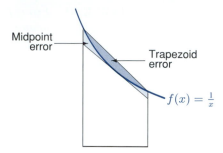

Figure 7.14: Errors in the midpoint and trapezoid rules

In other words: *In the trapezoid or midpoint rules, each extra 2 digits of accuracy requires about 10 times the work.*

This result shows the advantage of the midpoint and trapezoid rules over the left and right rules: less additional work needs to be done to get another decimal place of accuracy. The calculator used to produce these tables again took about half a second to compute the midpoint rule for $\int_1^2 \frac{1}{x}\,dx$ with $n = 50$, and this gets 4 digits correct. Thus to get 6 digits would take $n = 500$ and 5 seconds, to get 8 digits would take 50 seconds, and to get 10 digits would take 500 seconds, or about 10 minutes. That is still not great, but it is certainly better than the 1 year required by the left or right rule.

TABLE 7.4   *Ratios of the errors as n increases for $\int_1^2 \frac{1}{x}\,dx$*

	Ratio of errors in trapezoid rule	Ratio of errors in midpoint rule
Error(2)/Error(10)	24.33	23.84
Error(10)/Error(50)	24.97	24.95
Error(50)/Error(250)	25.00	25.00

Simpson's Rule

Still more improvement is possible. Observing that the trapezoid error has the opposite sign and about twice the magnitude of the midpoint error, we may guess that a weighted average of the two rules, with the midpoint rule weighted twice the trapezoid rule, will have a much smaller error. This approximation is called *Simpson's rule*[3]:

$$\text{SIMP}(n) = \frac{2 \cdot \text{MID}(n) + \text{TRAP}(n)}{3}.$$

Table 7.5 gives the errors for Simpson's rule. Notice how much smaller the errors are than the previous errors. Of course, it is a little unfair to compare Simpson's rule at $n = 50$, say, with the previous rules, because Simpson's rule must compute the value of f at both the midpoint and the endpoints of each subinterval and hence involves evaluating the function at twice as many points. We know by our previous analysis, however, that even if we did compute the other rules at $n = 100$ to compare with Simpson's rule at $n = 50$, the other errors would only decrease by a factor of 2 for the left and right rules and by a factor of 4 for the trapezoid and midpoint rules.

We see in Table 7.5 that as n increases by a factor of 5, the errors decrease by a factor of about 600, or about 5^4. Again this behavior holds for any factor, so increasing n by a factor of 10 decreases the error by a factor of about 10^4. In other words: *In Simpson's rule, each extra 4 digits of accuracy requires about 10 times the work.*

---

[3]Some books and computer programs use slightly different terminology for Simpson's rule; what we call $n = 50$, they call $n = 100$.

TABLE 7.5   *The errors for Simpson's rule and the ratios of the errors*

n	Error	Ratio
2	−0.0001067877	
		550.15
10	−0.0000001940	
		632.27
50	−0.0000000003	

This is a great improvement over either the midpoint or trapezoid rules, which only give two extra digits of accuracy when n is increased by a factor of 10. Simpson's rule is so efficient that we get 9 digits correct with $n = 50$ in about 1 second on our calculator. Doubling n will decrease the error by a factor of about $2^4 = 16$ and hence will give the tenth digit. The total time is 2 seconds, which is pretty good.

In general, Simpson's rule achieves a reasonable degree of accuracy when using relatively small values of n, and is a good choice for an all-purpose method for estimating definite integrals.

Analytical View of the Trapezoid and Simpson's Rules

Our approach to approximating $\int_a^b f(x)\, dx$ numerically has been empirical: try a method, see how the error behaves, and then try to improve it. We can also develop the various rules for numerical integration by making better and better approximations to the integrand, f. The left, right, and midpoint rules are all examples of approximating f by a constant (flat) function on each subinterval. The trapezoid rule is obtained by approximating f by a linear function on each subinterval. Simpson's rule can, in the same spirit, be obtained by approximating f by quadratic functions. The details are given in Problems 20 and 21 on page 356.

How the Error Depends on the Integrand

Other factors besides the size of n affect the size of the error in each of the rules. Instead of looking at how the error behaves as we increase n, let's leave n fixed and imagine trying our approximation methods on different functions. We observe that the error in the left or right rule depends on how steeply the graph of f rises or falls. A steep curve makes the triangular regions missed by the left or right rectangles tall and hence large in area. This observation suggests that the error in the left or right rules depends on the size of the derivative of f (see Figure 7.15).

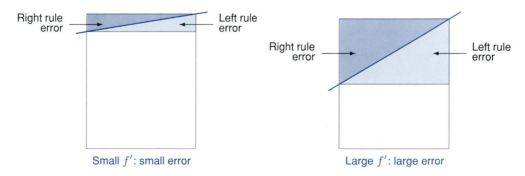

Small f': small error   Large f': large error

Figure 7.15: The error in the left and right rules depends on the steepness of the curve

From Figure 7.16 it appears that the errors in the trapezoid and midpoint rules depend on how much the curve is bent up or down. In other words, the concavity, and hence the size of the second derivative of f, has an effect on the errors of these two rules. Finally, it can be shown[4] that the error in Simpson's rule depends on the size of the *fourth* derivative of f, written $f^{(4)}$.

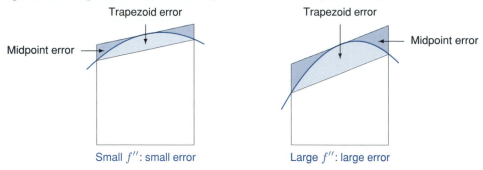

Figure 7.16: The error in the trapezoid and midpoint rules depends on how bent the curve is

Problems for Section 7.6

1. (a) Using the result of Problem 2 on page 348, compute SIMP(2) for $\int_0^4 (x^2 + 1)\, dx$.
    (b) Use the Fundamental Theorem of Calculus to find $\int_0^4 (x^2 + 1)\, dx$ exactly.
    (c) What is the error in SIMP(2) for this integral?

For Problems 2–7 use Simpson's rule with various values of n to evaluate the definite integrals with error less than 0.001. Explain why you think you have reached a large enough value of n.

2. $\displaystyle\int_0^1 \frac{1}{4 + x^2}\, dx$

3. $\displaystyle\int_0^2 e^{\sin t}\, dt$

4. $\displaystyle\int_0^3 \cos^2 \theta\, d\theta$

5. $\displaystyle\int_0^1 \sqrt{1 + x^4}\, dx$

6. $\displaystyle\int_0^{10} \ln(z^2 + 1)\, dz$

7. $\displaystyle\int_0^1 \cos(t^2)\, dt$

8. In this problem you will investigate the behavior of the errors in the approximation of the integral

$$\int_1^2 \frac{1}{x}\, dx \approx 0.6931471806\ldots .$$

    (a) For $n = 2, 4, 8, 16, 32, 64, 128$ subdivisions, compute the left and right approximations and the errors in each.
    (b) What are the signs of the errors in the left and right approximations? How do the errors change if n is doubled?
    (c) For the values of n in part (a), compute the midpoint and trapezoid approximations and the errors in each.
    (d) What are the signs of the errors in the midpoint and trapezoid approximations? How do the errors change if n is doubled?
    (e) For $n = 2, 4, 8, 16, 32$, compute Simpson's rule approximation and the error in each. How do these errors change as n doubles?

9. (a) What is the exact value of $\int_0^4 e^x\, dx$?
    (b) Find LEFT(2), RIGHT(2), TRAP(2), MID(2), and SIMP(2). Compute the error for each.
    (c) Repeat part (b) with $n = 4$ (instead of $n = 2$).
    (d) For each rule in part (b), as n goes from $n = 2$ to $n = 4$, does the error go down approximately as you would expect? Explain.

10. Consider Simpson's rule approximations to $\int_0^2 (x^3 + 3x^2)\, dx$.
    (a) What is the exact value of this integral?
    (b) Find SIMP(n) for $n = 2, 4$, and 100. What do you notice?

---

[4]See Kendall E. Atkinson, *An Introduction to Numerical Analysis* (New York: John Wiley and Sons, 1978).

11. Suppose that the approximation to a certain definite integral using $n = 10$ is 2.346 and that the exact value is 4.0. If the approximation was found using each of the following rules, use the same rule to estimate the integral with $n = 30$.

    (a) LEFT (b) TRAP (c) SIMP

12. (a) Suppose a certain computer takes two seconds to compute a certain definite integral accurate to 4 digits to the right of the decimal point, using the left rectangle rule. How long (in years) will it take to get 8 digits correct using the left rectangle rule? How about 12 digits? 20 digits?

    (b) Repeat part(a) but this time assume that the trapezoidal rule is being used throughout.

13. Suppose a certain computer takes 3 seconds to compute a certain definite integral accurate to 2 decimal places. How long will it take the computer to get 10 decimal places of accuracy if it is using the rule shown below? Give your answer in seconds and in appropriate time units (minutes, hours, days, or years).

    (a) LEFT (b) MID (c) SIMP

14. Suppose that for a certain definite integral, $\text{LEFT}(10) = 0.38745$ and $\text{LEFT}(20) = 0.36517$. Estimate the actual error for $\text{LEFT}(10)$ (and thus the actual value of the integral) by assuming that the error is reduced by a factor of 2 in going from $\text{LEFT}(10)$ to $\text{LEFT}(20)$.

15. Suppose for a certain definite integral that $\text{MID}(10) = 35.619$ and $\text{MID}(20) = 35.415$. Estimate the actual error for $\text{MID}(10)$ assuming that the error for $\text{MID}(10)$ is reduced by a factor of 4 in going to $\text{MID}(20)$.

16. Suppose that for a certain definite integral $\text{TRAP}(10) = 12.676$ and $\text{TRAP}(30) = 10.420$. Use this information to estimate the actual value of the integral. Justify your answer.

17. Suppose that for a certain definite integral, $\text{MID}(10) = 5.364$ and $\text{MID}(20) = 4.926$. Estimate as accurately as possible:

    (a) Actual value of the integral. (b) Value of $\text{MID}(60)$.

18. Some approximations for a certain definite integral are given in the following table. Use the fact that the exact value of the integral is 0.69315 to fill in the errors for $n = 2$. Then use what you know about errors to estimate the errors when $n = 20$.

	Approximation $n = 2$	Error $n = 2$	Error $n = 20$
LEFT	0.83333		
RIGHT	0.58333		
TRAP	0.70833		
MID	0.68571		
SIMP	0.69325		

19. Some approximations to an integral, using $n = 3$, are given in Table 7.6. The true value of the integral is 7.621372.

    (a) Does the integrand function appear to be increasing or decreasing? Concave up or concave down? Explain.

    (b) Find the error for each approximation, and fill in column (b) in Table 7.6.

    (c) Use what you know about errors and the errors for $n = 3$ to estimate errors for $n = 30$. Fill in column (c) in Table 7.6.

    TABLE 7.6

	Approximation $n = 3$	Error $n = 3$	Error $n = 30$
LEFT	5.416101		
RIGHT	9.307921		
TRAP	7.362011		
MID	7.742402		
SIMP	7.615605		

Problems 20–21 show how Simpson's rule can be obtained by approximating the integrand, f, by quadratic functions.

20. Suppose that $a < b$ and that m is the midpoint $m = (a + b)/2$. Let $h = b - a$. The purpose of this problem is to show that if f is a quadratic function, then

$$\int_a^b f(x)\, dx = \frac{h}{3} \left(\frac{f(a)}{2} + 2f(m) + \frac{f(b)}{2} \right).$$

(a) Show that this equation holds for the functions $f(x) = 1$, $f(x) = x$, and $f(x) = x^2$.
(b) Use part (a) and the "Properties about Sums and Constant Multiples of the Integrand" from Section 3.4, page 170, to show that the equation holds for any quadratic function, $f(x) = Ax^2 + Bx + C$.

21. Consider the following method for approximating $\int_a^b f(x)dx$. Divide the interval $[a, b]$ into n equal subintervals. On each subinterval approximate f by a quadratic function that agrees with f at both endpoints and at the midpoint of the subinterval.

(a) Explain why the integral of f on the subinterval $[x_i, x_{i+1}]$ is approximately equal to the expression

$$\frac{h}{3} \left(\frac{f(x_i)}{2} + 2f(m_i) + \frac{f(x_{i+1})}{2} \right),$$

where m_i is the midpoint of the subinterval, $m_i = (x_i + x_{i+1})/2$. (See Problem 20.)
(b) Show that if we add up these approximations for each subinterval, we get Simpson's rule:

$$\int_a^b f(x)dx \approx \frac{2 \cdot \text{MID}(n) + \text{TRAP}(n)}{3}.$$

7.7 IMPROPER INTEGRALS

Our original discussion of the definite integral $\int_a^b f(x)\, dx$ assumed that the interval $a \leq x \leq b$ was of finite length and that f was continuous. Integrals that arise in applications don't necessarily have these nice properties. In this section we investigate a class of integrals, called *improper* integrals, in which one limit of integration is infinite or the integrand is unbounded. As an example, to estimate the mass of the earth's atmosphere, we might calculate an integral which sums the mass of the air up to different heights. In order to represent the fact that the atmosphere doesn't end at a specific height, we let the upper limit of integration get larger and larger, or tend to infinity.

We will usually consider only improper integrals with positive integrands since they are the most common.

One Type of Improper Integral: When the Limit of Integration Is Infinite

Here is an example of an improper integral:

$$\int_1^\infty \frac{1}{x^2}\, dx.$$

To evaluate this integral, we first compute the definite integral $\int_1^b \frac{1}{x^2}\, dx$:

$$\int_1^b \frac{1}{x^2}\, dx = -x^{-1} \Big|_1^b = -\frac{1}{b} + \frac{1}{1}.$$

Now take the limit as $b \to \infty$. Since

$$\lim_{b \to \infty} \int_1^b \frac{1}{x^2}\, dx = \lim_{b \to \infty}\left(-\frac{1}{b} + 1\right) = 1,$$

we say that the improper integral $\int_1^\infty \frac{1}{x^2}\, dx$ *converges* to 1.

If we think of this in terms of areas, it may seem strange that the region whose area is computed by $\int_1^\infty \frac{1}{x^2}\, dx$ extends from $x = 1$ infinitely far to the right. How could it have finite area? (See Figure 7.17(a).) What our limit computations are saying is that

$$\text{When } b = 10: \qquad \int_1^{10} \frac{1}{x^2}\, dx = -\frac{1}{x}\bigg|_1^{10} = -\frac{1}{10} + 1 = 0.9$$

$$\text{When } b = 100: \qquad \int_1^{100} \frac{1}{x^2}\, dx = -\frac{1}{100} + 1 = 0.99$$

$$\text{When } b = 1000: \qquad \int_1^{1000} \frac{1}{x^2}\, dx = -\frac{1}{1000} + 1 = 0.999$$

and so on. In other words, as b gets larger and larger, the area between $x = 1$ and $x = b$ tends to 1. See Figure 7.17(b). Thus, it does make sense to declare that $\int_1^\infty \frac{1}{x^2}\, dx = 1$.

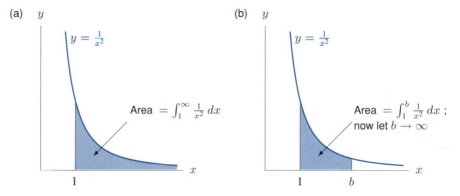

Figure 7.17: Area representation of improper integral

Of course, in another example, we might not get a finite limit as b gets larger and larger. In that case we say the improper integral *diverges*.

Suppose $f(x)$ is positive for $x \geq a$.

If $\displaystyle\lim_{b \to \infty} \int_a^b f(x)\, dx$ is a finite number, we say that $\displaystyle\int_a^\infty f(x)\, dx$ **converges** and define

$$\int_a^\infty f(x)\, dx = \lim_{b \to \infty} \int_a^b f(x)\, dx.$$

Otherwise, we say that $\displaystyle\int_a^\infty f(x)\, dx$ **diverges**. We define $\displaystyle\int_{-\infty}^b f(x)\, dx$ similarly.

Example 1   Does the improper integral $\int_1^\infty \frac{1}{\sqrt{x}}\,dx$ converge or diverge?

Solution   We consider

$$\int_1^b \frac{1}{\sqrt{x}}\,dx = \int_1^b x^{-1/2}\,dx = 2x^{1/2}\Big|_1^b = 2b^{1/2} - 2.$$

We see that $\int_1^b \frac{1}{\sqrt{x}}\,dx$ grows without bound as $b \to \infty$. Thus we say the integral $\int_1^\infty \frac{1}{\sqrt{x}}\,dx$ *diverges*. We have shown that the area under the curve in Figure 7.18 is not finite.

Notice that $f(x) \to 0$ as $x \to \infty$ does not guarantee convergence.

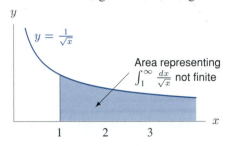

Figure 7.18: $\int_1^\infty \frac{1}{\sqrt{x}}\,dx$ diverges

What is the difference between the functions $1/x^2$ and $1/\sqrt{x}$ that makes the area under the graph of $1/x^2$ approach 1 as $x \to \infty$, whereas the area under $1/\sqrt{x}$ grows very large? Both functions approach 0 as x grows, so as b grows larger, smaller bits of area are being added to the definite integral. The difference between the functions is subtle: the values of the function $1/\sqrt{x}$ *don't shrink fast enough* for the integral to have a finite value. Of the two functions, $1/x^2$ drops to 0 much faster than $1/\sqrt{x}$, and this feature keeps the area under $1/x^2$ from growing beyond 1.

Example 2   Find $\int_0^\infty e^{-5x}\,dx$.

Solution   First we consider $\int_0^b e^{-5x}\,dx$:

$$\int_0^b e^{-5x}\,dx = -\frac{1}{5}e^{-5x}\Big|_0^b = -\frac{1}{5}e^{-5b} + \frac{1}{5}.$$

Since $e^{-5b} = \frac{1}{e^{5b}}$, this term tends to 0 as b approaches infinity, so $\int_0^\infty e^{-5x}\,dx$ converges. Its value is

$$\int_0^\infty e^{-5x}\,dx = \lim_{b\to\infty}\int_0^b e^{-5x}\,dx = \lim_{b\to\infty}\left(-\frac{1}{5}e^{-5b} + \frac{1}{5}\right) = 0 + \frac{1}{5} = \frac{1}{5}.$$

Since e^{5x} grows very rapidly, we expect that e^{-5x} will approach 0 rapidly. The fact that the area approaches $\frac{1}{5}$ instead of growing without bound is a consequence of the speed with which the integrand e^{-5x} approaches 0.

Example 3   Determine for which values of the exponent, p, the improper integral $\int_1^\infty \frac{1}{x^p}\,dx$ diverges.

Solution   For $p \neq 1$,

$$\int_1^b x^{-p}\,dx = \frac{1}{-p+1}x^{-p+1}\Big|_1^b = \left(\frac{1}{-p+1}b^{-p+1} - \frac{1}{-p+1}\right).$$

The important question is whether the exponent of b is positive or negative. If it is negative, then as b approaches infinity, b^{-p+1} approaches 0. If the exponent is positive, then b^{-p+1} grows without bound as b approaches infinity.

What happens if $p = 1$? In this case we get

$$\int_1^\infty \frac{1}{x} \, dx = \lim_{b \to \infty} \ln x \Big|_1^b = \lim_{b \to \infty} \ln b - \ln 1.$$

Since $\ln b$ becomes arbitrarily large as b approaches infinity, the integral grows without bound. We conclude that $\int_1^\infty \frac{1}{x^p} \, dx$ diverges precisely when $p \leq 1$. For $p > 1$ the integral has the value

$$\int_1^\infty \frac{1}{x^p} dx = \lim_{b \to \infty} \int_1^b \frac{1}{x^p} dx = \lim_{b \to \infty} \left(\frac{1}{-p+1} b^{-p+1} - \frac{1}{-p+1} \right) = -\left(\frac{1}{-p+1} \right) = \frac{1}{p-1}.$$

Application of Improper Integrals to Energy

The energy, E, required to separate two charged particles, originally a distance a apart, to a distance b, is given by the integral

$$E = \int_a^b \frac{k q_1 q_2}{r^2} \, dr$$

where q_1 and q_2 are the magnitudes of the charges and k is a constant. If q_1 and q_2 are in coulombs, a and b are in meters, and E is in joules, the value of the constant k is 9×10^9.

Example 4   A hydrogen atom consists of a proton and an electron, with opposite charges of magnitude 1.6×10^{-19} coulombs. Find the energy required to take a hydrogen atom apart (that is, to move the electron from its orbit to an infinite distance from the proton). Assume that the initial distance between the electron and the proton is the Bohr radius, $R_B = 5.3 \times 10^{-11}$ meter.

Solution   Since we are moving from an initial distance of R_B to a final distance of ∞, the energy is represented by the improper integral

$$E = \int_{R_B}^\infty k \frac{q_1 q_2}{r^2} \, dr = k q_1 q_2 \lim_{b \to \infty} \int_{R_B}^b \frac{1}{r^2} \, dr$$

$$= k q_1 q_2 \lim_{b \to \infty} -\frac{1}{r} \Big|_{R_B}^b = k q_1 q_2 \lim_{b \to \infty} \left(-\frac{1}{b} + \frac{1}{R_B} \right) = \frac{k q_1 q_2}{R_B}.$$

Substituting numerical values, we get

$$E = \frac{(9 \times 10^9)(1.6 \times 10^{-19})^2}{5.3 \times 10^{-11}} \approx 4.35 \times 10^{-18} \text{ joules}.$$

This is about the amount of energy needed to lift a speck of dust 0.000000025 inch off the ground. (In other words, not much!)

What happens if the limits of integration are $-\infty$ and ∞? In this case, we break the integral at any point and write the original integral as a sum of two new improper integrals.

We can use any (finite) number c to define

$$\int_{-\infty}^\infty f(x) \, dx = \int_{-\infty}^c f(x) \, dx + \int_c^\infty f(x) \, dx.$$

If *either* of the two new improper integrals diverges, we say the original integral diverges. Only if both of the new integrals have a finite value do we add the values to get a finite value for the original integral.

It is not hard to show that the preceding definition does not depend on the choice for c.

Another Type of Improper Integral: When the Integrand Becomes Infinite

There is another way for an integral to be improper. The interval may be finite but the function may be unbounded near some points in the interval. For example, consider $\int_0^1 \frac{1}{\sqrt{x}} \, dx$. Since the graph of $y = 1/\sqrt{x}$ has a vertical asymptote at $x = 0$, the region between the graph, the x-axis, and the lines $x = 0$ and $x = 1$ is unbounded. Instead of extending to infinity in the horizontal direction as in the previous improper integrals, this region extends to infinity in the vertical direction. See Figure 7.19(a). We handle this improper integral in a similar way as before: we compute $\int_a^1 \frac{1}{\sqrt{x}} \, dx$ for values of a slightly larger than 0 and look at what happens as a approaches 0 from the positive side. (This is written as $a \to 0^+$.)

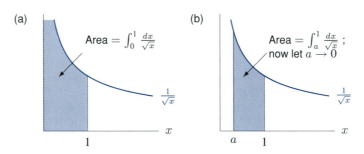

Figure 7.19: Area representation of improper integral

First we compute the integral:

$$\int_a^1 \frac{1}{\sqrt{x}} \, dx = 2x^{1/2} \Big|_a^1 = 2 - 2a^{1/2}.$$

Now we take the limit:

$$\lim_{a \to 0^+} \int_a^1 \frac{1}{\sqrt{x}} \, dx = \lim_{a \to 0^+} (2 - 2a^{1/2}) = 2.$$

Since the limit is finite, we say the improper integral converges, and that

$$\int_0^1 \frac{1}{\sqrt{x}} \, dx = 2.$$

Geometrically, what we have done is to calculate the finite area between $x = a$ and $x = 1$ and take the limit as a tends to 0 from the right. See Figure 7.19(b). Since the limit exists, the integral converges to 2. If the limit did not exist, we would say the improper integral diverges.

Example 5   Investigate the convergence of $\int_0^2 \frac{1}{(x-2)^2} \, dx$.

Solution   This is an improper integral since the integrand tends to infinity as x approaches 2, and is undefined at $x = 2$. Since the trouble is at the right endpoint, we replace the upper limit by b, and let b tend to 2 from the left. This is written $b \to 2^-$, with the "$-$" signifying that 2 is approached from below. See Figure 7.20.

$$\int_0^2 \frac{1}{(x-2)^2} \, dx = \lim_{b \to 2^-} \int_0^b \frac{1}{(x-2)^2} \, dx = \lim_{b \to 2^-} (-1)(x-2)^{-1} \Big|_0^b = \lim_{b \to 2^-} \left(-\frac{1}{(b-2)} - \frac{1}{2} \right).$$

Therefore, since $\lim_{b \to 2^-} \left(-\frac{1}{b-2} \right)$ does not exist, the integral diverges.

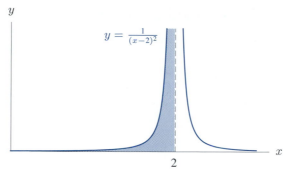

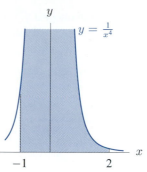

Figure 7.20: Shaded area represents $\int_0^2 \frac{1}{(x-2)^2}\, dx$

Figure 7.21: Shaded area represents $\int_{-1}^2 \frac{1}{x^4}\, dx$

Suppose $f(x)$ is positive and continuous on $a \leq x < b$ and tends to infinity as $x \to b$.
If $\displaystyle\lim_{c\to b^-} \int_a^c f(x)\, dx$ is a finite number, we say that $\displaystyle\int_a^b f(x)\, dx$ **converges** and define

$$\int_a^b f(x)\, dx = \lim_{c\to b^-} \int_a^c f(x)\, dx.$$

Otherwise, we say that $\displaystyle\int_a^b f(x)\, dx$ **diverges**.

When $f(x)$ tends to infinity as x approaches a, we define convergence in a similar way. In addition, an integral can be improper because the integrand tends to infinity *inside* the interval of integration rather than at an endpoint. In this case, we break the given integral into two (or more) improper integrals so that the integrand tends to infinity only at endpoints.

Suppose that $f(x)$ is positive and continuous on $[a, b]$ except at the point c. If $f(x)$ tends to infinity as $x \to c$, then we define

$$\int_a^b f(x)\, dx = \int_a^c f(x)\, dx + \int_c^b f(x)\, dx.$$

If *either* of the two new improper integrals diverges, we say the original integral diverges. Only if *both* of the new integrals have a finite value do we add the values to get a finite value for the original integral.

Example 6   Investigate the convergence of $\displaystyle\int_{-1}^2 \frac{1}{x^4}\, dx$.

Solution   See the graph in Figure 7.21. The trouble spot is $x = 0$, rather than $x = -1$ or $x = 2$. To handle this situation, we break the given improper integral into two other improper integrals each of which have $x = 0$ as one of the endpoints:

$$\int_{-1}^2 \frac{1}{x^4}\, dx = \int_{-1}^0 \frac{1}{x^4}\, dx + \int_0^2 \frac{1}{x^4}\, dx.$$

We can now use the previous technique to evaluate the new integrals, if they converge. Since

$$\int_0^2 \frac{1}{x^4}\, dx = \lim_{a\to 0^+} -\frac{1}{3}x^{-3}\Big|_a^2 = \lim_{a\to 0^+} \left(-\frac{1}{3}\right)\left(\frac{1}{8} - \frac{1}{a^3}\right)$$

the integral $\int_0^2 \frac{1}{x^4}\,dx$ does not exist. A similar computation shows that $\int_{-1}^0 \frac{1}{x^4}\,dx$ also diverges. Thus, the original integral diverges.

It is easy to miss an improper integral when the integrand tends to infinity inside the interval. For example, it is fundamentally incorrect to say that $\int_{-1}^2 \frac{1}{x^4}\,dx = -\frac{1}{3}x^{-3}\Big|_{-1}^2 = -\frac{1}{24} - \frac{1}{3} = -\frac{3}{8}$.

Example 7   Find $\displaystyle\int_0^6 \frac{1}{(x-4)^{2/3}}\,dx$.

Solution   Figure 7.22 shows that the trouble spot is at $x = 4$, so we break the integral at $x = 4$ and consider the separate parts.

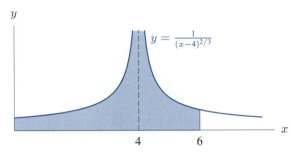

Figure 7.22: Shaded area represents $\int_0^6 \frac{1}{(x-4)^{2/3}}\,dx$

We have

$$\int_0^4 \frac{1}{(x-4)^{2/3}}\,dx = \lim_{b \to 4^-} 3(x-4)^{1/3}\Big|_0^b = \lim_{b \to 4^-}\left(3(b-4)^{1/3} - 3(-4)^{1/3}\right) = 3(4)^{1/3}.$$

Similarly,

$$\int_4^6 \frac{1}{(x-4)^{2/3}}\,dx = \lim_{a \to 4^+} 3(x-4)^{1/3}\Big|_a^6 = \lim_{a \to 4^+}\left(3 \cdot 2^{1/3} - 3(a-4)^{1/3}\right) = 3(2)^{1/3}.$$

Since both of these integrals converge, the original integral converges:

$$\int_0^6 \frac{1}{(x-4)^{2/3}}\,dx = 3(4)^{1/3} + 3(2)^{1/3} \approx 8.54.$$

Finally, there is a question of what to do when an integral is improper at both endpoints. In this case, we just break the integral at any interior point of the interval. The original integral diverges if either or both of the new integrals diverge.

Example 8   Investigate the convergence of $\displaystyle\int_0^\infty \frac{1}{x^2}\,dx$.

Solution   This integral is improper both because the upper limit is ∞ and because the function is undefined at $x = 0$. We break the integral into two parts at, say, $x = 1$. We know by Example 3 that $\int_1^\infty \frac{1}{x^2}\,dx$ has a finite value. However, the other part, $\int_0^1 \frac{1}{x^2}\,dx$, diverges since:

$$\int_0^1 \frac{1}{x^2}\,dx = \lim_{a \to 0^+} -x^{-1}\Big|_a^1 = \lim_{a \to 0^+}\left(\frac{1}{a} - 1\right).$$

Therefore $\displaystyle\int_0^\infty \frac{1}{x^2}\,dx$ diverges as well.

Problems for Section 7.7

Calculate the values of the integrals in Problems 1–24, if they converge.

1. $\displaystyle\int_1^\infty e^{-2x}\,dx$

2. $\displaystyle\int_0^\infty \frac{x}{e^x}\,dx$

3. $\displaystyle\int_1^\infty \frac{x}{4+x^2}\,dx$

4. $\displaystyle\int_{-\infty}^0 \frac{e^x}{1+e^x}\,dx$

5. $\displaystyle\int_{-\infty}^\infty \frac{dz}{z^2+25}$

6. $\displaystyle\int_0^4 \frac{dx}{\sqrt{16-x^2}}$

7. $\displaystyle\int_{\pi/4}^{\pi/2} \frac{\sin x}{\sqrt{\cos x}}\,dx$

8. $\displaystyle\int_{-1}^1 \frac{1}{v}\,dv$

9. $\displaystyle\int_0^1 \frac{x^4+1}{x}\,dx$

10. $\displaystyle\int_1^\infty \frac{1}{x^2+1}\,dx$

11. $\displaystyle\int_1^\infty \frac{1}{\sqrt{x^2+1}}\,dx$

12. $\displaystyle\int_0^4 \frac{1}{u^2-16}\,du$

13. $\displaystyle\int_1^\infty \frac{y}{y^4+1}\,dy$

14. $\displaystyle\int_2^\infty \frac{dx}{x\ln x}$

15. $\displaystyle\int_0^1 \frac{\ln x}{x}\,dx$

16. $\displaystyle\int_{16}^{20} \frac{1}{y^2-16}\,dy$

17. $\displaystyle\int_1^2 \frac{dx}{x\ln x}$

18. $\displaystyle\int_0^\pi \frac{1}{\sqrt{x}}e^{-\sqrt{x}}\,dx$

19. $\displaystyle\int_3^\infty \frac{dx}{x(\ln x)^2}$

20. $\displaystyle\int_0^2 \frac{1}{\sqrt{4-x^2}}\,dx$

21. $\displaystyle\int_4^\infty \frac{dx}{(x-1)^2}$

22. $\displaystyle\int_4^\infty \frac{dx}{x^2-1}$

23. $\displaystyle\int_7^\infty \frac{dy}{\sqrt{y-5}}$

24. $\displaystyle\int_\pi^\infty \sin y\,dy$

25. Find the area under the curve $y=xe^{-x}$ for $x\ge 0$.

26. Find the area under the curve $y=1/\cos^2 t$ between $t=0$ and $t=\pi/2$.

27. Suppose a function h is defined by $h(x)=\dfrac{1}{x\sqrt{x}}-\dfrac{1}{16}, 0<x\le 4$, and $h(x)=\dfrac{1}{x^2}, x>4$;

 (a) Evaluate $\int_0^\infty h(x)\,dx$.
 (b) Is h differentiable at $x=4$? If not, why not? If so, find $h'(4)$.

28. Given that $\displaystyle\int_{-\infty}^\infty e^{-x^2}\,dx=\sqrt{\pi}$, calculate the exact value of

$$\int_{-\infty}^\infty e^{-(x-a)^2/b}\,dx.$$

29. The gamma function is defined for all $x>0$ by the rule

$$\Gamma(x)=\int_0^\infty t^{x-1}e^{-t}\,dt.$$

 (a) Find $\Gamma(1)$ and $\Gamma(2)$.
 (b) Integrate by parts with respect to t to show that, for positive n,

$$\Gamma(n+1)=n\Gamma(n).$$

 (c) Find a simple expression for $\Gamma(n)$ for positive integers n.

30.   The rate, r, at which people get sick during an epidemic of the flu can be approximated by

$$r = 1000te^{-0.5t}$$

where r is measured in people/day and t is measured in days since the start of the epidemic.

(a)   Sketch a graph of r as a function of t.   (b)   When are people getting sick fastest?

(c)   How many people get sick altogether?

31.   Find the energy required to separate opposite electric charges of magnitude 1 coulomb. Assume the charges are initially 1 meter apart and that one is moved infinitely far away from the other. (The relevant definition of energy is on page 359.)

32.   For what values of p does the following integral converge?

$$\int_e^\infty x^p \ln x \, dx$$

What is the value of the integral when it converges?

33.   For what values of p does the following integral converge?

$$\int_0^e x^p \ln x \, dx$$

What is the value of the integral when it converges?

7.8 MORE ON IMPROPER INTEGRALS

Making Comparisons

Sometimes it is difficult to find the exact value of an improper integral by antidifferentiation, but it may be possible to determine whether an integral converges or diverges. The key is to *compare* the given integral to one whose behavior we already know. Let's look at an example.

Example 1   Determine whether $\int_1^\infty \frac{1}{\sqrt{x^3 + 5}} \, dx$ converges.

Solution   First, let's see what this integrand does as $x \to \infty$. For large x, the 5 becomes insignificant compared with the x^3, so

$$\frac{1}{\sqrt{x^3 + 5}} \approx \frac{1}{\sqrt{x^3}} = \frac{1}{x^{3/2}}.$$

Since

$$\int_1^\infty \frac{1}{\sqrt{x^3}} \, dx = \int_1^\infty \frac{1}{x^{3/2}} \, dx = \lim_{b \to \infty} \int_1^b \frac{1}{x^{3/2}} \, dx = \lim_{b \to \infty} \left. -2x^{-1/2} \right|_1^b = \lim_{b \to \infty} \left(2 - 2b^{-1/2} \right) = 2.$$

the integral $\int_1^\infty \frac{1}{x^{3/2}} \, dx$ converges. So we expect our integral to converge as well.

In order to confirm this, we observe that for $0 \le x^3 \le x^3 + 5$, we have

$$\frac{1}{\sqrt{x^3 + 5}} \le \frac{1}{\sqrt{x^3}}.$$

and so for $b \ge 1$,

$$\int_1^b \frac{1}{\sqrt{x^3 + 5}} \, dx \le \int_1^b \frac{1}{\sqrt{x^3}} \, dx.$$

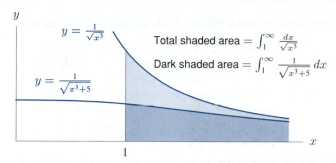

Figure 7.23: Graph showing $\int_1^\infty \frac{1}{\sqrt{x^3+5}}\,dx \le \int_1^\infty \frac{dx}{\sqrt{x^3}}$

(See Figure 7.23.) Since $\int_1^b \frac{1}{\sqrt{x^3+5}}\,dx$ increases as b approaches infinity but is always smaller than $\int_1^b \frac{1}{x^{3/2}}\,dx < \int_1^\infty \frac{1}{x^{3/2}}\,dx = 2$, we know $\int_1^\infty \frac{1}{\sqrt{x^3+5}}\,dx$ must have a finite value less than 2. Thus,

$$\int_1^\infty \frac{dx}{\sqrt{x^3+5}} \quad \text{converges to a value less than 2.}$$

Notice that we first looked at the behavior of the integrand as $x \to \infty$. This is useful because the convergence or divergence of the integral is determined by what happens as $x \to \infty$.

The Comparison Test for $\displaystyle\int_a^\infty f(x)\,dx$

Assume $f(x)$ is positive. Making a comparison involves two stages:

1. Guess, by looking at the behavior of the integrand for large x, whether the integral converges or not. (This is the "behaves like" principle.)

2. Confirm the guess by comparison:
    -   If $0 \le f(x) \le g(x)$ and $\int_a^\infty g(x)\,dx$ converges, then $\int_a^\infty f(x)\,dx$ converges
    -   If $0 \le g(x) \le f(x)$ and $\int_a^\infty g(x)\,dx$ diverges, then $\int_a^\infty f(x)\,dx$ diverges.

Example 2   Decide whether $\displaystyle\int_4^\infty \frac{dt}{(\ln t) - 1}$ converges or diverges.

Solution   Since $\ln t$ grows without bound as $t \to \infty$, the -1 is eventually going to be insignificant in comparison to $\ln t$. Thus, as far as convergence is concerned,

$$\int_4^\infty \frac{1}{(\ln t) - 1}\,dt \quad \text{behaves like} \quad \int_4^\infty \frac{1}{\ln t}\,dt.$$

Does $\int_4^\infty \frac{1}{\ln t}\,dt$ converge or diverge? Since $\ln t$ grows very slowly, $1/\ln t$ goes to zero very slowly, and so the integral probably doesn't converge. We know that $(\ln t) - 1 < \ln t < t$ for all positive t. So, provided $t > e$, we take reciprocals:

$$\frac{1}{(\ln t) - 1} > \frac{1}{\ln t} > \frac{1}{t}.$$

Since $\int_4^\infty \frac{1}{t}\,dt$ diverges, we conclude that

$$\int_4^\infty \frac{1}{(\ln t) - 1}\,dt \quad \text{diverges.}$$

How Do We Know What To Compare With?

In Examples 1 and 2, we investigated the convergence of an integral by comparing it with an easier integral. How did we pick the easier integral? This is a matter of trial and error, guided by any information we get by looking at the original integrand as $x \to \infty$. We want the comparison integrand to be easy and, in particular, to have a simple antiderivative.

Useful Integrals for Comparison

- $\displaystyle\int_1^\infty \frac{1}{x^p}\,dx$ converges for $p > 1$ and diverges for $p \leq 1$.

- $\displaystyle\int_0^1 \frac{1}{x^p}\,dx$ converges for $p < 1$ and diverges for $p \geq 1$.

- $\displaystyle\int_0^\infty e^{-ax}\,dx$ converges for $a > 0$.

Of course, we can use any function for comparison, provided we can determine its behavior.

Example 3   Investigate the convergence of $\displaystyle\int_1^\infty \frac{(\sin x) + 3}{\sqrt{x}}\,dx$.

Solution   Since it looks difficult to find an antiderivative of this function, we try comparison. What happens to this integrand as $x \to \infty$? Since $\sin x$ oscillates between -1 and 1,

$$\frac{2}{\sqrt{x}} = \frac{-1 + 3}{\sqrt{x}} \leq \frac{(\sin x) + 3}{\sqrt{x}} \leq \frac{1 + 3}{\sqrt{x}} = \frac{4}{\sqrt{x}},$$

the integrand oscillates between $2/\sqrt{x}$ and $4/\sqrt{x}$. (See Figure 7.24.)

   What do $\int_1^\infty \frac{2}{\sqrt{x}}\,dx$ and $\int_1^\infty \frac{4}{\sqrt{x}}\,dx$ do? As far as convergence is concerned, they certainly do the same thing, and whatever that is, the original integral does it too. It is important to notice that $\sqrt{x}$ grows very slowly. This means that $1/\sqrt{x}$ gets small slowly, which means that convergence is unlikely. Since $\sqrt{x} = x^{1/2}$, the result in the preceding box (with $p = \frac{1}{2}$) tells us that $\int_1^\infty \frac{dx}{\sqrt{x}}$ diverges. So the comparison test tells us that the original integral diverges.

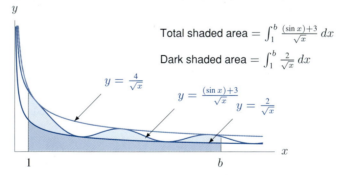

Figure 7.24: Graph showing $\int_1^b \frac{2}{\sqrt{x}}\,dx \leq \int_1^b \frac{(\sin x)+3}{\sqrt{x}}\,dx$, for $b \geq 1$

Notice that there are two possible comparisons we could have made in Example 3:

$$\frac{2}{\sqrt{x}} \leq \frac{(\sin x) + 3}{\sqrt{x}} \qquad \text{or} \qquad \frac{(\sin x) + 3}{\sqrt{x}} \leq \frac{4}{\sqrt{x}}.$$

Since both $\int_1^\infty \frac{2}{\sqrt{x}}\,dx$ and $\int_1^\infty \frac{4}{\sqrt{x}}\,dx$ diverge, only the first comparison is useful. Knowing that an integral is *smaller* than a divergent integral is of no help whatsoever!

   The next example shows what to do if the comparison does not hold throughout the interval of integration.

Example 4   Show $\int_{1}^{\infty} e^{-x^2/2}\, dx$ converges to a finite value.

Solution   We know that $e^{-x^2/2}$ goes very rapidly to zero as $x \to \infty$, so we expect this integral to converge. Hence we look for some larger integrand which has a convergent integral. One possibility is $\int_{1}^{\infty} e^{-x}\, dx$, because e^{-x} has an elementary antiderivative and $\int_{1}^{\infty} e^{-x}\, dx$ converges. What is the relationship between $e^{-x^2/2}$ and e^{-x}? We know that for $x \geq 2$,

$$x \leq \frac{x^2}{2} \quad \text{so} \quad -\frac{x^2}{2} \leq -x,$$

and so, for $x \geq 2$

$$e^{-x^2/2} \leq e^{-x}.$$

Since this inequality holds only for $x \geq 2$, we split the interval of integration into two pieces:

$$\int_{1}^{\infty} e^{-x^2/2}\, dx = \int_{1}^{2} e^{-x^2/2}\, dx + \int_{2}^{\infty} e^{-x^2/2}\, dx.$$

Now $\int_{1}^{2} e^{-x^2/2}\, dx$ is finite (it is not improper) and $\int_{2}^{\infty} e^{-x^2/2}\, dx$ is finite by comparison with $\int_{2}^{\infty} e^{-x}\, dx$. Therefore, $\int_{1}^{\infty} e^{-x^2/2}\, dx$ is the sum of two finite pieces and therefore must be finite.

Problems for Section 7.8

1. The graphs of $y = 1/x, y = 1/x^2$ and the functions $f(x)$, $g(x)$, $h(x)$, and $k(x)$ are shown in Figure 7.25.

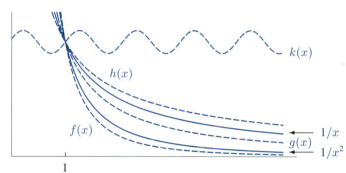

Figure 7.25

(a)   What is the area between $y = 1/x$ and $y = 1/x^2$ from $x = 1$ to $x = \infty$? Explain your answer.

(b)   Using the graph, decide whether the integral of each of the functions $f(x)$, $g(x)$, $h(x)$ and $k(x)$ from $x = 1$ to $x = \infty$ converges, diverges, or whether it is impossible to tell.

In Problems 2–16 decide if the improper integral converges or diverges. Explain your reasoning.

2. $\displaystyle\int_{50}^{\infty} \frac{dz}{z^3}$

3. $\displaystyle\int_{0.5}^{1} \frac{1}{x^{(19/20)}}\, dx$

4. $\displaystyle\int_{1}^{\infty} \frac{dx}{1 + x}$

5. $\displaystyle\int_{1}^{\infty} \frac{dx}{x^3 + 1}$

6. $\displaystyle\int_{2}^{\infty} \frac{d\theta}{\sqrt{\theta^3 + 1}}$

7. $\displaystyle\int_{-1}^{5} \frac{dt}{(t + 1)^2}$

8. $\displaystyle\int_{0}^{\infty} \frac{dy}{1 + e^y}$

9. $\displaystyle\int_{1}^{\infty} \frac{2 + \cos\phi}{\phi^2}\, d\phi$

10. $\displaystyle\int_{-\infty}^{\infty} \frac{du}{1 + u^2}$

11. $\displaystyle\int_{1}^{\infty} \frac{du}{u + u^2}$

12. $\displaystyle\int_{1}^{\infty} \frac{d\theta}{\sqrt{\theta^2 + 1}}$

13. $\displaystyle\int_{0}^{1} \frac{d\theta}{\sqrt{\theta^3 + \theta}}$

14. $\displaystyle\int_{0}^{\infty} \frac{dz}{e^z + 2^z}$

15. $\displaystyle\int_{0}^{\pi} \frac{2 - \sin\phi}{\phi^2}\, d\phi$

16. $\displaystyle\int_{4}^{\infty} \frac{3 + \sin\alpha}{\alpha}\, d\alpha$

Estimate the values of the integrals in Problems 17–18 correct to two decimal places by integrating the functions on your calculator or computer for large values of the upper bound of integration.

17. $\displaystyle\int_1^\infty e^{-x^2}\,dx$

18. $\displaystyle\int_0^\infty e^{-x^2}\cos^2 x\,dx$

19. The bell-shaped curve of statistics has the equation

$$f(x) = ae^{-x^2/2}.$$

Find the value of a (to three decimal places) that makes

$$\int_{-\infty}^\infty f(x)\,dx = 1.$$

20. Statisticians often use the function
$$g(x) = ae^{-(x-k)^2/2}.$$

(a) To three decimal places, what value of a should be chosen to ensure that

$$\int_{-\infty}^\infty g(x)\,dx = 1?$$

(b) Is your answer the same as or different from your answer to Problem 19? Why?

21. (a) Find an upper bound for

$$\int_3^\infty e^{-x^2}\,dx.$$

[Hint: $e^{-x^2} \le e^{-3x}$ for $x \ge 3$.]

(b) For any positive n, generalize the result of part (a) to find an upper bound for

$$\int_n^\infty e^{-x^2}\,dx$$

by noting that $nx \le x^2$ for $x \ge n$.

22. For what values of p does the integral $\displaystyle\int_2^\infty \frac{dx}{x(\ln x)^p}$ converge?

23. For what values of p does the integral $\displaystyle\int_1^2 \frac{dx}{x(\ln x)^p}$ converge?

24. Do the following integrals converge? If so, give an upper bound for the value of the integral.

(a) $\displaystyle\int_1^\infty \frac{2x^2+1}{4x^4+4x^2-2}\,dx$

(b) $\displaystyle\int_1^\infty \left(\frac{2x^2+1}{4x^4+4x^2-2}\right)^{1/4}\,dx$

25. In Planck's Radiation Law, we encounter the integral

$$\int_1^\infty \frac{dx}{x^5(e^{1/x}-1)}.$$

(a) Explain why a graph of the tangent line to e^t at $t = 0$ tells us that for all t

$$1 + t \le e^t.$$

(b) Substituting $t = 1/x$, show that for all x

$$e^{1/x} - 1 > \frac{1}{x}.$$

(c) Use the comparison test to show that the original integral converges.

CHAPTER SUMMARY

- **Integration techniques**
  Substitution, parts, partial fractions, using tables
- **Improper integrals**
  Convergence/divergence, comparison test for integrals
- **Numerical approximations**
  Riemann sums (left, right, midpoint), trapezoid rule, Simpson's rule, approximation errors

REVIEW PROBLEMS FOR CHAPTER SEVEN

For each region in Problems 1–3, write a definite integral which represents its area. Evaluate the integral to derive a formula for the area.

1. A rectangle with base b and height h:

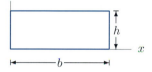

2. A right triangle of base b and height h:

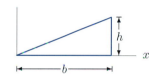

3. A circle of radius r:

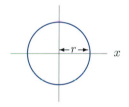

4. Suppose the function f is defined by $f(t) = t^2$ for $0 \le t \le 1$ and $f(t) = 2 - t$ for $1 < t \le 2$. Compute
$$\int_0^2 f(t)\,dt.$$

5. (a) Find the average value of the following functions over one cycle:

    (i) $f(t) = \cos t$ (ii) $g(t) = |\cos t|$ (iii) $k(t) = (\cos t)^2$

  (b) Write the averages you have just found in ascending order. Explain clearly, using words and graphs, why the averages should be expected to come out in the order they did.

6. Integrate:

  (a) $\displaystyle\int \frac{e^x}{1 + e^x}\,dx$ (b) $\displaystyle\int \frac{e^x}{1 + e^{2x}}\,dx$ (c) $\displaystyle\int \frac{1}{1 + e^x}\,dx$

7. (a) Using a substitution, find $\displaystyle\int \frac{dx}{\sqrt{1 - 4x^2}}$ from the table of integrals.

  (b) Use your answer to part (a) to find $\displaystyle\int_0^{\pi/8} \frac{dx}{\sqrt{1 - 4x^2}}$.

  (c) Check your answer to part (b) by numerical integration.

8. (a) Explain why you can rewrite x^x as $x^x = e^{x \ln x}$ for $x > 0$.

  (b) Use your answer to part (a) to find $\dfrac{d}{dx}(x^x)$.

  (c) Find $\displaystyle\int x^x (1 + \ln x)\,dx$.

  (d) Find $\displaystyle\int_1^2 x^x (1 + \ln x)\,dx$ using part (c) and check your answer using numerical methods.

In Problems 9–12 explain why the following pairs of antiderivatives are really, despite their apparent dissimilarity, different expressions of the same problem. You do not need to evaluate the integrals.

9. $\displaystyle\int \frac{dx}{x^2 + 4x + 4}$ and $\displaystyle\int \frac{x}{(x^2 + 1)^2}\,dx$ **10.** $\displaystyle\int \frac{1}{\sqrt{1 - x^2}}\,dx$ and $\displaystyle\int \frac{x\,dx}{\sqrt{1 - x^4}}$

11. $\displaystyle\int \frac{x}{1 - x^2}\,dx$ and $\displaystyle\int \frac{1}{x\ln x}\,dx$ **12.** $\displaystyle\int \frac{x}{x + 1}\,dx$ and $\displaystyle\int \frac{1}{x + 1}\,dx$

For Problems 13–21 decide if the integral converges or diverges. If the integral converges, find its value.

13. $\displaystyle\int_4^\infty \frac{dt}{t^{3/2}}$ **14.** $\displaystyle\int_{10}^\infty \frac{dx}{x\ln x}$ **15.** $\displaystyle\int_0^\infty we^{-w}\,dw$

16. $\displaystyle\int_{-1}^1 \frac{1}{x^4}\,dx$ **17.** $\displaystyle\int_{-\pi/4}^{\pi/4} \tan\theta\,d\theta$ **18.** $\displaystyle\int_2^\infty \frac{1}{4 + z^2}\,dz$

19. $\displaystyle\int_{10}^\infty \frac{1}{z^2 - 4}\,dz$ **20.** $\displaystyle\int_{-5}^{10} \frac{dt}{\sqrt{t + 5}}$ **21.** $\displaystyle\int_0^{\pi/2} \frac{1}{\sin\phi}\,d\phi$

For Problems 22–26, decide if the integral converges or diverges. If the integral converges, find its value, or give a bound for the value.

22. $\displaystyle\int_0^{\pi/4} \tan 2\theta\,d\theta$ **23.** $\displaystyle\int_1^\infty \frac{x}{x + 1}\,dx$ **24.** $\displaystyle\int_0^\infty \frac{\sin^2\theta}{\theta^2 + 1}\,d\theta$

25. $\displaystyle\int_0^\pi \tan^2\theta\,d\theta$ **26.** $\displaystyle\int_0^1 (\sin x)^{-3/2}\,dx$

27. Find the average (vertical) height of the shaded area in Figure 7.26.
28. Find the average (horizontal) width of the shaded area in Figure 7.26.

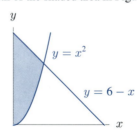

Figure 7.26

29. The curves $y = \sin x$ and $y = \cos x$ cross each other infinitely often. What is the area of the region bounded by these two curves between two consecutive crossings?

Are the statements in Problems 30–33 true or false? Why?

30. The midpoint rule approximation to $\displaystyle\int_0^1 (y^2 - 1)\,dy$ is always smaller than the exact value of the integral.

31. The trapezoid rule approximation is never exact.

32. If $\text{LEFT}(2) < \displaystyle\int_a^b f(x)\,dx$, then $\text{LEFT}(4) < \displaystyle\int_a^b f(x)\,dx$.

33. If $0 < f' < g'$ everywhere, then the error in approximating $\displaystyle\int_a^b f(x)\,dx$ by $\text{LEFT}(n)$ is less than the error in approximating $\displaystyle\int_a^b g(x)\,dx$ by $\text{LEFT}(n)$.

34. Does the improper integral $\int_0^1 \frac{e^x}{x}\,dx$ converge or diverge?

35. Suppose you estimate $\int_0^{0.5} f(x)dx$ by the trapezoid and midpoint rules with 100 steps. Explain which of the two estimates overestimates, and which underestimates, the true value of the integral if

(a) $f(x) = 1 + e^{-x}$ (b) $f(x) = e^{-x^2}$ (c) $f(x)$ is a straight line.

36. Suppose for a certain definite integral that TRAP(10) = 4.6891 and TRAP(50) = 4.6966. Estimate the actual error for TRAP(10) (and thus the actual value of the integral) by assuming that the error is reduced by a factor of roughly 25 in going from TRAP(10) to TRAP(50).

37. Suppose for a definite integral that SIMP(5) = 7.41562 and SIMP(10) = 7.41738. Estimate the actual value of the integral by using the fact that the error is reduced by a factor of roughly 16 in going from SIMP(5) to SIMP(10).

38. In 1987 the average per capita income in the US was $26,000. Suppose that average per capita income is increasing continuously at a rate in dollars per year given by

$$r(t) = 480(1.024)^t,$$

where t is the number of years since 1987.

(a) Estimate the average per capita income in 1995.
(b) Find a formula for the average per capita income as a function of time after 1987.

39. In 1990 humans generated 1.4×10^{20} joules of energy through the combustion of petroleum. It is estimated that all of the earth's petroleum will generate approximately 10^{22} joules. Assuming the use of energy generated by petroleum combustion will increase by 2% each year, how long will it be before all of our petroleum resources are used up?

PROJECTS

1. **Taylor Polynomial Inequalities**

(a) Use the fact that $e^x \geq 1 + x$ for all values of x and the formula

$$e^x = 1 + \int_0^x e^t\,dt$$

to show that

$$e^x \geq 1 + x + \frac{x^2}{2}$$

for all positive values of x. Generalize this idea to get inequalities involving higher-degree polynomials.

(b) Use the fact that $\cos x \leq 1$ for all x and repeated integration to show that

$$\cos x \leq 1 - \frac{x^2}{2!} + \frac{x^4}{4!}.$$

2. **Slope Fields**

Suppose we want to sketch the antiderivative, F, of the function f. To get an accurate graph of F, we must be careful about making F have the right slope at every point. The slope of F at any point (x, y) on its graph should be $f(x)$, since $F'(x) = f(x)$. We arrange this as follows: at the point (x, y) in the plane, draw a small line segment with slope $f(x)$. Do this at many points. We call such a diagram a *slope field*. If $f(x) = x$, we get the slope field in Figure 7.27.

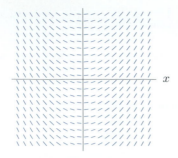
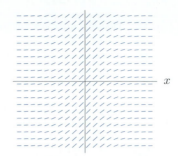

Figure 7.27: Slope field of
$f(x) = x$

Figure 7.28: Slope field of
$f(x) = e^{-x^2}$

Notice how the lines in Figure 7.27 seem to be arranged in a parabolic pattern. This is because the general antiderivative of x is $x^2/2 + C$, so the lines are all the tangent lines to the family of parabolas $y = x^2/2 + C$. This suggests a way of finding antiderivatives graphically even if we can't write down a formula for them: plot the slopes, and see if they suggest the graph of an antiderivative. For example, if you do this with $f(x) = e^{-x^2}$, which is one of the functions that does not have an elementary antiderivative, you get Figure 7.28.

You can see the ghost of the graph of a function lurking behind the slopes in Figure 7.28; in fact there is a whole stack of them. If you move across the plane in the direction suggested by the slope field at every point, you will trace out a curve. The slope field is tangent to the curve everywhere, so this is the graph of an antiderivative of e^{-x^2}.

(a) (i) Sketch a graph of $f(t) = \dfrac{\sin t}{t}$.

(ii) What does your graph tell you about the behavior of

$$\text{Si}(x) = \int_0^x \frac{\sin(t)}{t}\, dt$$

for $x > 0$? Is $\text{Si}(x)$ always increasing or always decreasing? Does $\text{Si}(x)$ cross the x-axis for $x > 0$?

(iii) By drawing the slope field for $f(t) = \dfrac{\sin t}{t}$, decide whether $\lim\limits_{x \to \infty} \text{Si}(x)$ exists.

(b) (i) Use your calculator or computer to sketch a graph of $y = x^{\sin x}$ for $0 < x \le 20$.

(ii) Using your answer to part (a), sketch by hand a graph of the function F, where

$$F(x) = \int_0^x t^{\sin t}\, dt.$$

(iii) Use a slope field program to check your answer to part (b).

(c) Let $F(x)$ be the antiderivative of $\sin(x^2)$ satisfying $F(0) = 0$.

(i) Describe any general features of the graph of F that you can deduce by looking at the graph of $\sin(x^2)$ in Figure 7.29.

(ii) By drawing a slope field (using a calculator or computer), sketch a graph of F. Does F ever cross the x-axis in the region $x > 0$? Does $\lim\limits_{x \to \infty} F(x)$ exist?

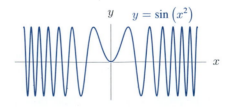

Figure 7.29

FOCUS ON PRACTICE

INTEGRATION

For Problems 1–113, evaluate the following integrals. Assume $a, b, c,$ and k are constants. Evaluate the definite integrals using the Fundamental Theorem of Calculus, if possible, and check your answers numerically. Problems 1–75 can be done without an integral table, as can some of the later problems.

1. $\displaystyle\int \sin t \, dt$

2. $\displaystyle\int \cos 2t \, dt$

3. $\displaystyle\int e^{5z} \, dz$

4. $\displaystyle\int (3w + 7) \, dw$

5. $\displaystyle\int \sin 2\theta \, d\theta$

6. $\displaystyle\int (x^3 - 1)^4 x^2 \, dx$

7. $\displaystyle\int \left(x^{3/2} + x^{2/3} \right) \, dx$

8. $\displaystyle\int \left(e^x + 3^x \right) \, dx$

9. $\displaystyle\int (r + 1)^3 \, dr$

10. $\displaystyle\int \left(\frac{4}{x^2} - \frac{3}{x^3} \right) \, dx$

11. $\displaystyle\int \left(\frac{x^3 + x + 1}{x^2} \right) \, dx$

12. $\displaystyle\int \frac{(1 + \ln x)^2}{x} \, dx$

13. $\displaystyle\int t e^{t^2} \, dt$

14. $\displaystyle\int x \cos x \, dx$

15. $\displaystyle\int \sin x (\sqrt{2 + 3\cos x}) dx$

16. $\displaystyle\int x^2 e^{2x} \, dx$

17. $\displaystyle\int x\sqrt{1 - x} \, dx$

18. $\displaystyle\int x \ln x \, dx$

19. $\displaystyle\int y \sin y \, dy$

20. $\displaystyle\int (\ln x)^2 \, dx$

21. $\displaystyle\int \ln(x^2) \, dx$

22. $\displaystyle\int e^{0.5 - 0.3t} \, dt$

23. $\displaystyle\int \sin^2 \theta \cos \theta \, d\theta$

24. $\displaystyle\int x\sqrt{4 - x^2} \, dx$

25. $\displaystyle\int \frac{(u + 1)^3}{u^2} \, du$

26. $\displaystyle\int \frac{\cos \sqrt{y}}{\sqrt{y}} \, dy$

27. $\displaystyle\int \frac{1}{\cos^2 z} \, dz$

28. $\displaystyle\int \cos^2 \theta \, d\theta$

29. $\displaystyle\int t^{10}(t - 10) \, dt$

30. $\displaystyle\int \tan(2x - 6) \, dx$

31. $\displaystyle\int_1^3 \ln(x^3) \, dx$

32. $\displaystyle\int_1^e (\ln x)^2 \, dx$

33. $\displaystyle\int_{-\pi}^{\pi} e^{2x} \sin 2x \, dx$

34. $\displaystyle\int_0^{10} z e^{-z} \, dz$

35. $\displaystyle\int_{-\pi/3}^{\pi/4} \sin^3 \theta \cos \theta \, d\theta$

36. $\displaystyle\int_{-\pi/4}^{\pi/4} x^3 \cos x^2 \, dx$

37. $\displaystyle\int_1^4 \frac{e^{\sqrt{x}}}{\sqrt{x}} \, dx$

38. $\displaystyle\int_0^1 \frac{dx}{x^2 + 1}$

39. $\displaystyle\int \frac{(\ln x)^2}{x} \, dx$

40. $\displaystyle\int \frac{(t + 2)^2}{t^3} \, dt$

41. $\displaystyle\int \left(x^2 + 2x + \frac{1}{x} \right) \, dx$

42. $\displaystyle\int \frac{t + 1}{t^2} \, dt$

43. $\displaystyle\int t e^{t^2 + 1} \, dt$

44. $\displaystyle\int \tan \theta \, d\theta$

45. $\displaystyle\int \sin(5\theta) \cos(5\theta) \, d\theta$

46. $\displaystyle\int \frac{x}{x^2 + 1} \, dx$

47. $\displaystyle\int \frac{dz}{1 + z^2}$

48. $\displaystyle\int \frac{dz}{1 + 4z^2}$

49. $\displaystyle\int \cos^3 2\theta \sin 2\theta \, d\theta$

50. $\displaystyle\int \sin 5\theta \cos^3 5\theta \, d\theta$

51. $\displaystyle\int \sin^3 z \cos^3 z \, dz$

52. $\displaystyle\int t(t-10)^{10}\,dt$

53. $\displaystyle\int \cos\theta\sqrt{1+\sin\theta}\,d\theta$

54. $\displaystyle\int xe^x\,dx$

55. $\displaystyle\int t^3 e^t\,dt$

56. $\displaystyle\int_1^3 x(x^2+1)^{70}\,dx$

57. $\displaystyle\int (3z+5)^3\,dz$

58. $\displaystyle\int \frac{du}{9+u^2}$

59. $\displaystyle\int \frac{\cos w}{1+\sin^2 w}\,dw$

60. $\displaystyle\int \frac{1}{x}\tan(\ln x)\,dx$

61. $\displaystyle\int \frac{1}{x}\sin(\ln x)\,dx$

62. $\displaystyle\int \frac{dx}{\sqrt{1-4x^2}}$

63. $\displaystyle\int \frac{w\,dw}{\sqrt{16-w^2}}$

64. $\displaystyle\int \frac{e^{2y}+1}{e^{2y}}\,dy$

65. $\displaystyle\int \frac{\sin w\,dw}{\sqrt{1-\cos w}}$

66. $\displaystyle\int \frac{dx}{x\ln x}$

67. $\displaystyle\int \frac{du}{3u+8}$

68. $\displaystyle\int \frac{x\cos\sqrt{x^2+1}}{\sqrt{x^2+1}}\,dx$

69. $\displaystyle\int \frac{t^3}{\sqrt{1+t^2}}\,dt$

70. $\displaystyle\int ue^{ku}\,du$

71. $\displaystyle\int (w+5)^4 w\,dw$

72. $\displaystyle\int e^{\sqrt{2x+3}}\,dx$

73. $\displaystyle\int r(\ln r)^2\,dr$

74. $\displaystyle\int (e^x+x)^2\,dx$

75. $\displaystyle\int u^2\ln u\,du$

76. $\displaystyle\int \frac{5x+6}{x^2+4}\,dx$

77. $\displaystyle\int \frac{1}{\sin^3(2x)}\,dx$

78. $\displaystyle\int \frac{dr}{r^2-100}$

79. $\displaystyle\int y^2\sin(cy)\,dy$

80. $\displaystyle\int e^{-ct}\sin kt\,dt$

81. $\displaystyle\int e^{5x}\cos(3x)\,dx$

82. $\displaystyle\int \left(x^{\sqrt{k}}+\sqrt{k}^x\right)dx$

83. $\displaystyle\int \sqrt{3+12x^2}\,dx.$

84. $\displaystyle\int (x^2-3x+2)e^{-4x}\,dx$

85. $\displaystyle\int \frac{dx}{x^2+5x+4}$

86. $\displaystyle\int \frac{1}{\sqrt{x^2-3x+2}}\,dx$

87. $\displaystyle\int \frac{x^3}{x^2+3x+2}\,dx$

88. $\displaystyle\int \frac{x^2+1}{x^2-3x+2}\,dx$

89. $\displaystyle\int \frac{dx}{ax^2+bx}$

90. $\displaystyle\int \frac{ax+b}{ax^2+2bx+c}\,dx$

91. $\displaystyle\int \frac{dz}{z^2+z}$

92. $\displaystyle\int \left(\frac{x}{3}+\frac{3}{x}\right)^2 dx$

93. $\displaystyle\int \frac{2^t}{2^t+1}\,dt$

94. $\displaystyle\int 10^{1-x}\,dx$

95. $\displaystyle\int (x^2+5)^3\,dx$

96. $\displaystyle\int v\arcsin v\,dv$

97. $\displaystyle\int \sin^2(2\theta)\cos^3(2\theta)\,d\theta$

98. $\displaystyle\int \cos(2\sin x)\cos x\,dx$

99. $\displaystyle\int \sqrt{4-x^2}\,dx$

100. $\displaystyle\int \frac{z^3}{z-5}\,dz$

101. $\displaystyle\int \frac{\sin w\cos w}{1+\cos^2 w}\,dw$

102. $\displaystyle\int \frac{1}{\tan(3\theta)}\,d\theta$

103. $\displaystyle\int \frac{x}{\cos^2 x}\,dx$

104. $\displaystyle\int \frac{x+1}{\sqrt{x}}\,dx$

105. $\displaystyle\int \frac{x}{\sqrt{x+1}}\,dx$

106. $\displaystyle\int \frac{\sqrt{\sqrt{x}+1}}{\sqrt{x}}\,dx$

107. $\displaystyle\int \frac{e^{2y}}{e^{2y}+1}\,dy$

108. $\displaystyle\int \frac{z}{(z^2-5)^3}\,dz$

109. $\displaystyle\int \frac{z}{(z-5)^3}\,dz$

110. $\displaystyle\int \frac{(1+\tan x)^3}{\cos^2 x}\,dx$

111. $\displaystyle\int \frac{(2x-1)e^{x^2}}{e^x}\,dx$

112. $\displaystyle\int (x+\sin x)^3(1+\cos x)\,dx$

113. $\displaystyle\int \left(2x^3+3x+4\right)\cos(2x)\,dx$

CHAPTER EIGHT

USING THE DEFINITE INTEGRAL

In Chapter 3, we saw how a definite integral can be used to compute an area, an average value, or a total change. In Chapter 7, we focused on different ways of evaluating integrals. In this chapter we will use definite integrals to solve problems in geometry, physics, economics, and probability. In each section, we use the same method to represent a quantity as a definite integral: we chop up the quantity and approximate it by a Riemann sum.

8.1 APPLICATIONS TO GEOMETRY

Setting Up Riemann Sums

There are many situations where a quantity can be expressed as a definite integral. Typically this will happen when the quantity can be approximated by dividing it into little pieces. We then solve the problem approximately for each piece and add the resulting contributions.

For example, in Section 3.1, the way we learned that distance traveled is the definite integral of velocity was by adding up the distances traveled over small time intervals. For a particle moving with velocity $v = f(t)$ at time t, we find the net distance traveled between $t = a$ and $t = b$ by cutting the time interval into n small subintervals each of length Δt. We approximate the distance traveled during each subinterval Δt by assuming that the velocity is constant over the subinterval. Let's suppose the velocity throughout the ith interval is $v = f(t_i)$, the velocity at the start of the interval. The distance moved during that time is approximately

$$\text{Velocity} \times \text{Time} = f(t_i) \, \Delta t.$$

Therefore, the net change in position is approximated by the sum of all of these small distances, so

$$\text{Change in position} \approx \sum_{i=0}^{n-1} f(t_i) \, \Delta t.$$

In the limit as Δt tends to zero, this sum becomes an integral:

$$\text{Change in position} = \lim_{\Delta t \to 0} \sum_{i=0}^{n-1} f(t_i) \, \Delta t = \int_a^b f(t) \, dt.$$

Although the dt in the definite integral is not a number, it behaves as if it were a very small version of Δt. Informally, $f(t) \, dt$ may be thought of as the distance traveled by the particle over a very small time interval, since it is the velocity, $f(t)$, multiplied by the time, dt. The integral may be thought of as the sum of all these small distances.

In this chapter we will see how the same strategy of subdividing and summing can be used to calculate many other quantities. In this section we show how the definite integral can be used to compute the volume of a solid and the length of a curve .

To Compute a Volume or Length Using an Integral

- Divide the solid (or curve) into small pieces whose volume (or length) we can easily approximate;

- Add the contributions of all the pieces, obtaining a Riemann sum that approximates the total volume (or length);

- Take the limit as the number of terms in the sum tends to infinity, giving a definite integral for the total volume (or total length).

Finding Volumes by Slicing

When calculating the volume of a solid using Riemann sums, we chop the solid into smaller pieces whose volumes we can estimate.

Figure 8.1: Cone cut into vertical slices

Figure 8.2: Cone cut into horizontal slices

How to Slice a Cone

Let's see how we might slice a cone standing with the vertex uppermost. We could divide the cone into parallel vertical slices. These slices will be arch-shaped (they are actually hyperbolas); see Figure 8.1. We could also divide the cone horizontally, giving coin-shaped slices; see Figure 8.2.

To construct a Riemann sum, we prefer to cut the cone into the horizontal (circular) slices because it would be difficult to estimate the volumes of the arch-shaped slices.

Example 1   Compute the volume, in cubic feet, of the Great Pyramid of Egypt, whose base is a square 755 feet by 755 feet and whose height is 410 feet.

Solution   You may already know the formula $V = \frac{1}{3}b^2 \cdot h$ for the volume V of a pyramid with height h and base length b. We will not use the formula, but our approach can be used to prove that $V = \frac{1}{3}b^2 h$.

We view the pyramid as built up in layers starting from the base. Each layer is a square with thickness Δh. The bottom layer is a square slab 755 feet by 755 feet and volume about $(755)^2 \Delta h$ ft^3. As we move up the pyramid, the layers have shorter side lengths. We pick points, $h_0, h_1, \ldots, h_n$, which divide the height into n subintervals of length Δh, with $h_0 = 0$ and $h_n = 410$. If we let s_i denote the side length of the ith layer, then the volume of this layer is approximately $s_i^2 \Delta h$ ft^3. (The volume is only approximately $s_i^2 \Delta h$ because the sides of the slab are not vertical.) See Figure 8.3.

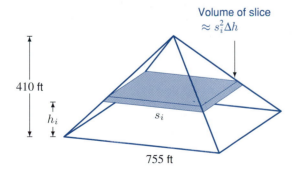

Figure 8.3: The Great Pyramid

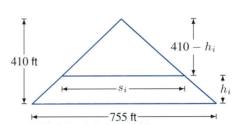

Figure 8.4: Cross-section relating s_i and h_i

The total volume of the pyramid is the sum of the volumes, $s_i^2 \Delta h$, of each layer. We express s_i as a function of h_i using the triangular cross-section in Figure 8.4. By similar triangles, we get $s_i/755 = (410 - h_i)/410$. Thus, $s_i = (755/410)(410 - h_i)$, and the total volume, V, is approximated by adding the volumes of the n layers:

$$V \approx \sum_{i=0}^{n-1} s_i^2 \, \Delta h = \sum_{i=0}^{n-1} \left[\left(\frac{755}{410}\right)(410 - h_i) \right]^2 \Delta h \text{ ft}^3.$$

As the thickness of each slice tends to zero, the sum becomes a definite integral. Finally, since h varies from 0 to 410, the height of the pyramid, we have

$$V = \int_{h=0}^{h=410} \left[\left(\frac{755}{410} \right) (410 - h) \right]^2 dh = \left(\frac{755}{410} \right)^2 \int_0^{410} (410 - h)^2 \, dh$$

$$= \left(\frac{755}{410} \right)^2 \left[-\frac{(410-h)^3}{3} \right] \Bigg|_0^{410} = \left(\frac{755}{410} \right)^2 \frac{(410)^3}{3} = \frac{1}{3} (755)^2 (410) \approx 78 \text{ million ft}^3.$$

Note that $V = \frac{1}{3}(755)^2(410) = \frac{1}{3}b^2 \cdot h$, as expected.

Notice that the limits in the definite integral are the limits for the variable h. Once we decided to slice the pyramid horizontally, we knew that a typical slice has thickness Δh, so h is the variable in our definite integral, and the limits must be values of h.

Example 2 Find the volume of a hemisphere of radius 7 inches.

Solution Again, you probably know the formula $\frac{4}{3}\pi r^3$ for the volume of a sphere; we will not use it here.

Imagine the hemisphere sitting flat on a plane and divide it into horizontal layers of thickness Δh inches. (See Figure 8.5.) Each layer is a circular slab with radius r, where r varies from 7 inches for the bottom layer to 0 inches for the top layer. The volume of each layer is approximately $\pi r^2 \Delta h$ in^3.

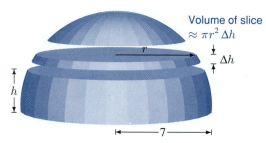

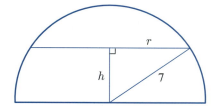

Volume of slice
$\approx \pi r^2 \Delta h$

Figure 8.5: Slicing to find the volume of a hemisphere

Figure 8.6: Vertical slice through center of hemisphere showing relation between r and h

Our problem is to express r in terms of h so that we can write a Riemann sum (whose terms depend only on h) representing the sum of the volumes of the layers. From Figure 8.6, showing a vertical slice through the center of the hemisphere, we find that $r^2 = 7^2 - h^2$. So at height h,

$$\text{Volume of slice} \approx \pi r^2 \Delta h = \pi (7^2 - h^2) \Delta h \text{ in}^3.$$

Summing the volumes of all slices, we have

$$\text{Volume} = V \approx \sum \pi r^2 \Delta h = \sum \pi (7^2 - h^2) \Delta h \text{ in}^3.$$

As the thickness of each slab tends to zero, the sum becomes a definite integral. Since h varies from 0 to 7, the radius of the hemisphere, we have

$$V = \int_{h=0}^{h=7} \pi r^2 \, dh = \pi \int_0^7 (7^2 - h^2) \, dh = \pi \left(7^2 h - \frac{1}{3} h^3 \right) \Bigg|_0^7 = \pi \left(7^3 - \frac{1}{3} \cdot 7^3 \right) = \frac{2}{3} \pi 7^3 \text{ in}^3.$$

Notice that V is half of $\frac{4}{3}\pi 7^3$ in^3, as we expected.

Note that the sum represented by the $\sum$ sign is over all the layers. In this example we did not bother to write down the points $h_0, h_1 \ldots,$ that are needed to write the Riemann sum precisely. This is unnecessary if all we want is the final expression for the definite integral.

In Examples 1 and 2 all cross-sections of the solid perpendicular to a certain direction were the same geometric figure—a square in the case of the pyramid and a circle in the case of the hemisphere. The volumes of other solids with known cross-sections can be calculated by a similar method.

Example 3   Find the volume of the solid whose base is the region in the xy-plane bounded by the curves $y = x^2$ and $y = 8 - x^2$ and whose cross-sections perpendicular to the x-axis are squares with one side in the xy-plane. (See Figure 8.7.)

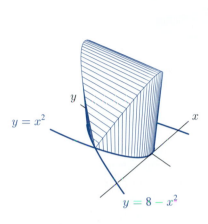

Figure 8.7: The solid for Example 3   **Figure 8.8**: A slice of the solid for Example 3

Solution   We view the solid as a loaf of bread sitting on the xy-plane and made up of square slices. These slices are analogous to the square layers making up the pyramid in our first example. A typical slice is shown in Figure 8.8. The thickness of each slice is Δx. The side length, s, of the square face is the distance (in the y direction) between the two curves. Therefore, in terms of x we have $s = (8 - x^2) - x^2 = 8 - 2x^2$.

Notice that this side length grows from $s = 0$ at $x = -2$ to $s = 8$ at $x = 0$, and then it shrinks back to $s = 0$ at $x = 2$. The volume of the solid is obtained by summing up the volumes of all the slices. Each slice has volume $s^2 \Delta x = (8 - 2x^2)^2 \Delta x$, so the total volume is

$$V \approx \sum s^2 \, \Delta x = \sum (8 - 2x^2)^2 \, \Delta x.$$

As the thickness Δx of each slice tends to zero, the sum becomes a definite integral. Since x varies between -2 and 2, we have

$$V = \int_{-2}^{2} (8 - 2x^2)^2 \, dx$$

$$= \int_{-2}^{2} (64 - 32x^2 + 4x^4) \, dx$$

$$= 64x - \frac{32}{3}x^3 + 4\frac{x^5}{5} \Big|_{-2}^{2}$$

$$= \frac{2048}{15} \approx 136.5.$$

We can check the plausibility of this answer by observing that the given solid is somewhat larger than the solid obtained by putting together two pyramids with 8×8 bases at $x = 0$ and apexes (highest points) at $x = 2$ and $x = -2$. The total volume of these two pyramids is $2(\frac{1}{3})b^2 h = 2(\frac{1}{3})8^2 \cdot 2 \approx 85.3$, which is, as expected, less than our answer, 136.5.

Volumes of Revolution

Another way to create a solid having known cross-sections is to revolve a region in the plane around some line in space, giving a *solid of revolution,* as in the following examples.

Example 4   The region bounded by the curve $y = e^{-x}$ and the x-axis between $x = 0$ and $x = 1$ is revolved around the x-axis. Find the volume of this solid of revolution.

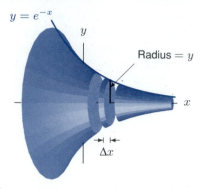

Figure 8.9: A thin strip rotated around the x-axis to form a circular slice

Solution   Imagine the plane region divided into thin strips perpendicular to the x-axis. Each strip has width Δx. As we rotate the whole region about the x-axis, each strip sweeps out a circular disk of thickness Δx. (See Figure 8.9.) The radius of the slice, as can be seen in Figure 8.9, is the distance from the x-axis to the curve, which is just the value of $y = e^{-x}$ on the slice. Thus the volume of the slice is $\pi y^2 \, \Delta x = \pi (e^{-x})^2 \, \Delta x$, and the total volume, V, of the solid is given by

$$V \approx \sum \pi y^2 \, \Delta x = \sum \pi \left(e^{-x}\right)^2 \Delta x.$$

As the thickness of each slice tends to zero, we get

$$V = \int_0^1 \pi (e^{-x})^2 \, dx = \pi \int_0^1 e^{-2x} \, dx = \pi \left(-\frac{1}{2}\right) e^{-2x} \Big|_0^1$$

$$= \pi \left(-\frac{1}{2}\right) (e^{-2} - e^0) = \frac{\pi}{2}(1 - e^{-2}) \approx 1.36.$$

Example 5   The region bounded by the curves $y = x$ and $y = x^2$ is rotated about the line $y = 3$. Compute the volume of the resulting solid.

Solution   Again imagine the region divided into thin vertical strips of thickness Δx, as in Figure 8.10.

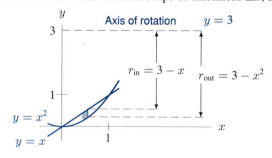

Figure 8.10: The region for Example 5

As each strip is rotated around the line $y = 3$, it sweeps out a slice of the total solid shaped like a disk with a hole in it (see Figures 8.11 and 8.12). This disk-with-a-hole has two radii: an inner radius, r_{in}, which is the distance from the line $y = 3$ to the nearer curve $y = x$, and an outer radius, r_{out}, which is the distance from the line $y = 3$ to the more distant curve, $y = x^2$. So the slice at x has inner radius $r_{\text{in}} = 3 - x$ and outer radius $r_{\text{out}} = 3 - x^2$. Think of the slice as a circular disk of radius r_{out} from which has been removed a smaller circular disk of radius r_{in}. Thus, the volume of the slice is

$$\pi r_{\text{out}}^2 \, \Delta x - \pi r_{\text{in}}^2 \, \Delta x = \pi (3 - x^2)^2 \, \Delta x - \pi (3 - x)^2 \, \Delta x.$$

Adding up the volumes of all the slices, we have

$$V \approx \sum \left(\pi r_{\text{out}}^2 - \pi r_{\text{in}}^2 \right) \Delta x = \sum \left[\pi (3 - x^2)^2 - \pi (3 - x)^2 \right] \Delta x.$$

We let Δx, the thickness of each slice, tend to zero to obtain the definite integral. Since the curves $y = x$ and $y = x^2$ intersect at $x = 0$ and $x = 1$, we have

$$V = \int_0^1 \left[\pi (3 - x^2)^2 - \pi (3 - x)^2 \right] dx = \pi \int_0^1 \left[(9 - 6x^2 + x^4) - (9 - 6x + x^2) \right] dx$$

$$= \pi \int_0^1 (6x - 7x^2 + x^4) \, dx = \pi \left(3x^2 - \frac{7x^3}{3} + \frac{x^5}{5} \right) \Big|_0^1 \approx 2.72.$$

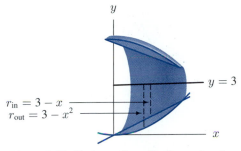

Figure 8.11: Cutaway view of volume showing inner and outer radii

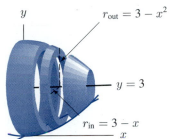

Figure 8.12: One slice (a disk-with-a-hole)

Arc Length

The definite integral can also be used to compute the lengths of various curves. Suppose we wish to compute the *arc length*, as it is usually called, of a curve $y = f(x)$ from $x = a$ to $x = b$, where $a < b$. As usual, we divide the curve into many little pieces; each one is approximately straight.

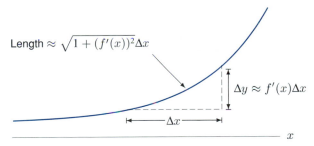

Figure 8.13: Length of a small piece of curve approximated using Pythagoras' theorem

Figure 8.13 shows that a small change Δx in the x coordinate produces a corresponding small change in the y coordinate of approximately $\Delta y \approx f'(x) \Delta x$. The length of the little piece of the curve is then approximately

$$\Delta L \approx \sqrt{(\Delta x)^2 + (\Delta y)^2} \approx \sqrt{(\Delta x)^2 + \left(f'(x) \, \Delta x \right)^2} = \sqrt{1 + (f'(x))^2} \, \Delta x.$$

Thus, for $a < b$, the arc length of the curve $y = f(x)$ from $x = a$ to $x = b$ is approximated by a Riemann sum:

$$\text{Arc length} = L \approx \sum \sqrt{1 + \left(f'(x) \right)^2} \, \Delta x.$$

As we let Δx tend to zero, the sum becomes a definite integral. Since x varies between a and b, we have the following expression for the arc length:

$$\boxed{\text{Arc length} = L = \int_a^b \sqrt{1 + (f'(x))^2} \, dx.}$$

Example 6   Set up and evaluate an integral to compute the length of the curve $y = x^3$ from $x = 0$ to $x = 5$.

Solution   If $f(x) = x^3$, then $f'(x) = 3x^2$, so

$$L = \int_0^5 \sqrt{1 + (3x^2)^2}\, dx.$$

Although the formula for the arc length of a curve is easy to apply, the integrands it generates often do not have elementary antiderivatives. Using numerical methods, we get $L = 125.68$. The curve starts at $(0, 0)$ and goes to $(5, 125)$, so its length must be at least the length of a straight line between these points, or $\sqrt{5^2 + 125^2} = 125.10$. (See Figure 8.14.)

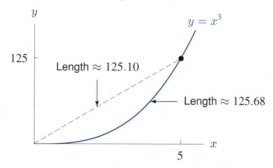

Figure 8.14: Arc length of $y = x^3$ (Note: the picture is distorted because the scales on the two axes are quite different)

Problems for Section 8.1

1. Imagine a hard-boiled egg lying on its side cut into thin slices. First think about vertical slices and then horizontal ones. What would these slices look like? Sketch them.

2. Find, by slicing, the volume of a cone whose height is 3 cm and whose base radius is 1 cm. Slice the cone as shown in Figure 8.2 on page 377.

3. Find, by slicing, a formula for the volume of a cone of height h and base radius r.

4. Find the volume of a sphere of radius r by slicing up the sphere, setting up a Riemann sum that approximates the volume, and then obtaining a definite integral by letting the thickness of each slice tend to zero.

5. When the ellipse $x^2/a^2 + y^2/b^2 = 1$ is rotated about the x-axis, it generates an *ellipsoid*. Compute its volume.

For Problems 6–8, sketch the solid obtained by rotating each region around the indicated axis. Using the sketch, show how to approximate the volume of the solid by a Riemann sum, and hence find the volume.

6. Region bounded by $y = x^3$, $x = 1$, $y = -1$. Axis: $y = -1$.
7. Region bounded by $y = \sqrt{x}$, $x = 1$, $y = 0$. Axis: $x = 1$.
8. Region bounded by the first arch of $y = \sin x$, $y = 0$. Axis: x axis.

For Problems 9–13 consider the region bounded by $y = e^x$, the x-axis, and the lines $x = 0$ and $x = 1$. Find the volume of the following solids.

9. The solid obtained by rotating the region about the x-axis.
10. The solid obtained by rotating the region about the horizontal line $y = -3$.
11. The solid obtained by rotating the region about the horizontal line $y = 7$.
12. The solid whose base is the given region and whose cross-sections perpendicular to the x-axis are squares.
13. The solid whose base is the given region and whose cross-sections perpendicular to the x-axis are semicircles.

14. The design of boats is based on Archimedes' Principle, which states that the buoyant force on an object in water is equal to the weight of the water displaced. Suppose you want to build a sailboat whose hull is parabolic with cross section $y = ax^2$, where a is a constant. Your boat will have length L and its maximum draft (the maximum vertical depth of any point of the boat beneath the water line) will be H. See Figure 8.15. Every cubic meter of water weighs 10,000 newtons. What is the maximum possible weight for your boat and cargo?

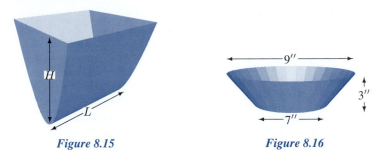

Figure 8.15   *Figure 8.16*

15. (a) A pie dish is 9 inches across the top, 7 inches across the bottom, and 3 inches deep. See Figure 8.16. Compute the volume of this dish.
     (b) Make a rough estimate of the volume in cubic inches of a single cut-up apple, and estimate the number of apples that is needed to make an apple pie that fills this dish.

16. Many communities expect to run out of space for their garbage soon. On Staten Island, solid garbage is packed into a pyramid-shaped dump with a square base of length 100 yards. One yard vertically above the base, the length of the side parallel to the base is 99 yards; the dump can be built up to a vertical height of 20 yards. (The top of the pyramid is never reached.) If 65 cubic yards of garbage arrive at the dump every day, how long will it be before the dump is full?

17. The circumference of a tree at different heights above the ground is given in the table below. Assume that all horizontal cross-sections of the tree are circles. Estimate the volume of the tree.

Height (inches)	0	20	40	60	80	100	120
Circumference (inches)	31	28	21	17	12	8	2

18. The hull of a certain boat has widths given by the following table. Reading across a row of the table gives widths at points $0, 10, \ldots, 60$ feet from the front to the back at a certain level below waterline. Reading down a column of the table gives widths at levels $0, 2, 4, 6, 8$ feet below waterline at a certain distance from the front. Use the trapezoidal rule to estimate the volume of the hull below waterline.

		Front of boat $\longrightarrow$ Back of boat						
		0	10	20	30	40	50	60
	0	2	8	13	16	17	16	10
Depth	2	1	4	8	10	11	10	8
below	4	0	3	4	6	7	6	4
waterline	6	0	1	2	3	4	3	2
(in feet)	8	0	0	1	1	1	1	1

For Problems 19–20, find the arc length of the given function from $x = 0$ to $x = 2$.

19. $f(x) = \sqrt{4 - x^2}$ 20. $f(x) = \sqrt{x^3}$

21. (a) Write an integral which represents the circumference of a circle of radius r.
     (b) Evaluate the integral, and show that you get the answer you expect.

22. Compute, with error at most 0.001, the perimeter of the region used for the base of the solids in Problems 9–13.

23. The graph of the function $y = \cosh x = \frac{1}{2}(e^x + e^{-x})$ is called a catenary and represents the shape of a hanging cable. Find the length of this catenary between $x = -1$ and $x = 1$.

24. Set up an integral for the circumference of an ellipse with semi-major axis $a = 2$ and semi-minor axis $b = 1$. Evaluating this integral numerically (i.e., using a calculator or computer) poses a problem. Describe this problem. Formulate a way to solve it.

25. There are very few elementary functions $y = f(x)$ for which arc length can be computed in elementary terms using the formula

$$\int_a^b \sqrt{1 + \left(\frac{dy}{dx}\right)^2} \, dx.$$

You have seen some such functions f in Problems 19, 20, and 23, namely, $f(x) = \sqrt{4 - x^2}$, $f(x) = \sqrt{x^3}$, and $f(x) = \frac{1}{2}(e^x + e^{-x})$. Try to find some other function that "works," that is, a function whose arc length you can find using this formula and antidifferentiation.

26. After doing Problem 25, you may wonder what sort of functions can represent arc length. If $g(0) = 0$ and g is differentiable and increasing, then can $g(x)$, $x \geq 0$, represent arc length? That is, can we find a function $f(t)$ such that

$$\int_0^x \sqrt{1 + (f'(t))^2} \, dt = g(x)?$$

(a) Show that $f(x) = \int_0^x \sqrt{(g'(t))^2 - 1} \, dt$ works as long as $g'(x) \geq 1$. In other words, show that the arc length of the graph of f from 0 to x is $g(x)$.

(b) Show that if $g'(x) < 1$ for some x, then $g(x)$ cannot represent the arc length of the graph of any function.

(c) Find a function f whose arc length from 0 to x is $2x$.

8.2 DENSITY AND CENTER OF MASS

Density and How to Slice a Region

The examples in this section involve the idea of *density*. For example,

- A population density is measured in, say, people per mile (along the edge of a road), or people per unit area (in a city), or bacteria per cubic centimeter (in a test tube).

- The density of a substance (e.g. air, wood, or metal), is the mass of a unit volume of the substance and is measured in, say, grams per cubic centimeter.

Suppose we want to calculate the total mass or total population, but the density is not constant over a region. Then to find the total quantity:

> Divide the region into small pieces in such a way that the density is approximately constant on each piece, and add up the contributions of all the pieces.

Example 1 The Massachusetts Turnpike ("the Pike") starts in the middle of Boston and heads west. The number of people living next to it varies as it gets further and further from the city. Suppose that, x miles out of town, the population density adjacent to the Pike is $P = f(x)$ people/mile. Express the total population living next to the Pike within 5 miles of Boston as a definite integral.

Solution Think about how we could estimate the population. Divide the Pike up into segments of length Δx. The population density at the center of Boston is $f(0)$; let's use that density for the first segment. This gives an estimate of

People living in first segment $\approx f(0)$ people/ mile $\cdot \Delta x$ mile $= f(0)\Delta x$ people.

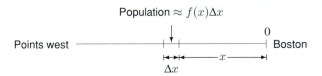

Figure 8.17: Population along the Massachusetts Turnpike

Similarly, the population in a typical segment is the population density times the length of the interval, or roughly $f(x)\,\Delta x$. (See Figure 8.17.) The sum of all these estimates gives the estimate

$$\text{Total population} \approx \sum f(x)\,\Delta x.$$

Letting $\Delta x \to 0$ shows that

$$\text{Total population} = \lim_{\Delta x \to \infty} \sum f(x)\,\Delta x = \int_0^5 f(x)\,dx.$$

The 5 and 0 in the limits of the integral are the upper and lower limits of the interval over which we are integrating.

Example 2   The air density (in kg/m^3) h meters above the earth's surface is $P = f(h)$. Find the mass of a cylindrical column of air 2 meters in diameter and 25 kilometers high.

Solution   The column of air is a circular cylinder 2 meters in diameter and 25 kilometers, or 25,000 meters, high. First we must decide how we are going to slice this column. Since the air density varies with altitude but remains constant horizontally, we take horizontal slices of air. That way, the density will be more or less constant over the whole slice, being close to its value at the bottom of the slice. (See Figure 8.18.)

Figure 8.18: Slicing a column of air horizontally

A slice is a cylinder of height Δh and diameter 2 m, so its radius will be 1 m. We find the approximate mass of the slice by multiplying its volume and its density together. If the thickness of the slice is Δh, then its volume is $\pi r^2 \cdot \Delta h = \pi 1^2 \cdot \Delta h = \pi\,\Delta h$ m^3 (where m^3 = cubic meter). The density of the slice is roughly $f(h)$. Thus,

$$\text{Mass of slice} \approx \text{Volume} \cdot \text{ Approximate density} = (\pi\Delta h \text{ m}^3)(f(h) \text{ kg/m}^3) = \pi\,\Delta h \cdot f(h) \text{ kg}.$$

Adding these slices up yields a Riemann sum:

$$\text{Total mass} \approx \sum \pi f(h)\,\Delta h \text{ kg}.$$

As $\Delta h \to 0$, this sum approximates the definite integral:

$$\text{Total mass} = \int_0^{25,000} \pi f(h)\,dh \text{ kg}.$$

In order to get a numerical value for the mass of air, we need an explicit formula for the density as a function of height, as in the next example.

Example 3   Find the mass of the column of air in Example 2 if the density of air at height h is given by

$$P = f(h) = 1.28e^{-0.000124h} \text{ kg/m}^3.$$

Solution   Using the result of the previous example, we have

$$\text{Mass} = \int_0^{25,000} \pi 1.28 e^{-0.000124h} \, dh = \frac{-1.28\pi}{0.000124} \left(e^{-0.000124h} \Big|_0^{25,000} \right)$$

$$\approx 32,429(e^0 - e^{-0.000124(25,000)}) \approx 31,000 \text{ kg}.$$

Sometimes it requires some thought to figure out how to slice a quantity up. The key point to remember is that you want the density to be nearly constant within each piece.

Example 4   The population density in Ringsburg is a function of the distance from the city center. At r miles from the center, the density is $P = f(r)$ people per square mile. Ringsburg has a radius of 5 miles. Write a definite integral that expresses the total population of Ringsburg.

Solution   We want to slice Ringsburg up and estimate the population on each slice. If we were to take straight-line slices, the population density would vary on each slice, since it depends on the distance from the city center. We want the population density to be pretty close to constant on each slice. We therefore take slices that are thin rings around the center. (See Figure 8.19.) Since the ring is very thin, we can approximate its area by straightening it into a thin rectangle. (See Figure 8.20.) The width of the rectangle is Δr miles, and its length is approximately equal to the circumference of the ring, $2\pi r$ miles, so its area is about $2\pi r \, \Delta r$ mi^2. Since

$$\text{Population on ring} \approx \text{Density} \cdot \text{Area},$$

we get

$$\text{Population on ring} \approx (f(r) \text{ people/mi}^2)(2\pi r \Delta r \text{ mi}^2) = f(r) \cdot 2\pi r \, \Delta r \text{ people}.$$

Adding the contributions from each ring, we get

$$\text{Total population} \approx \sum 2\pi r f(r) \, \Delta r \text{ people}.$$

So

$$\text{Total population} = \int_0^5 2\pi r f(r) \, dr \text{ people}.$$

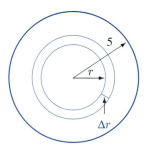

Figure 8.19: Ringsburg   *Figure 8.20:* Ring from Ringsburg (straightened out)

Width $= \Delta r$

$2\pi r$

Note: You may wonder what happens if we calculate the area of the ring by subtracting the area of the inner circle (πr^2) from the area of the outer circle [$\pi(r + \Delta r)^2$], giving

$$\text{Area} = \pi(r + \Delta r)^2 - \pi r^2.$$

Multiplying out and subtracting, we get

$$\text{Area} = \pi[r^2 + 2r\,\Delta r + (\Delta r)^2] - \pi r^2$$
$$= 2\pi r\,\Delta r + \pi(\Delta r)^2.$$

This expression differs from the one we used before by the $\pi(\Delta r)^2$ term. However, as Δr becomes very small, $\pi(\Delta r)^2$ becomes much, much smaller. We say it is *second order* of smallness, since the power of the small factor, Δr, is 2. In the limit as $\Delta r \to 0$, we can ignore $\pi(\Delta r)^2$.

Center of Mass

The *center of mass* of a mechanical system can be thought of as the balancing point. For a system consisting of n discrete masses m_i along the x-axis, each located at coordinate x_i, the center of mass is defined by the point whose x-coordinate is $\bar{x}$ where

$$\bar{x} = \frac{\sum_{i=1}^{n} x_i m_i}{\sum_{i=1}^{n} m_i}.$$

(See Figure 8.21.) The terms $x_i m_i$ in the numerator are called the *moments* of the masses m_i; the denominator is the total mass of the system. If the mechanical system is not composed of discrete masses, but is instead an object lying on the x-axis between $x = a$ and $x = b$, with mass density $\delta(x)$, then the center of mass is given by

$$\bar{x} = \frac{\int_a^b x\delta(x)\,dx}{\int_a^b \delta(x)\,dx}.$$

Again, the denominator is the total mass of the system.

We can see how this last formula relates to the formula for discrete masses using Riemann sums. If we divide the object into n small pieces, each of length Δx, then the ith piece has mass $m_i \approx \delta(x_i)\Delta x$ (where x_i is a point in the ith piece). Then the numerator of the formula for $\bar{x}$ is $\sum_{i=1}^{n} x_i \delta(x_i)\Delta x$, which approaches $\int_a^b x\delta(x)\,dx$ as $n \to \infty$. Similarly, the denominator $\sum_{i=1}^{n} m_i \approx \sum_{i=1}^{n} \delta(x_i)\Delta x$ approaches $\int_a^b \delta(x)\,dx$ as $n \to \infty$.

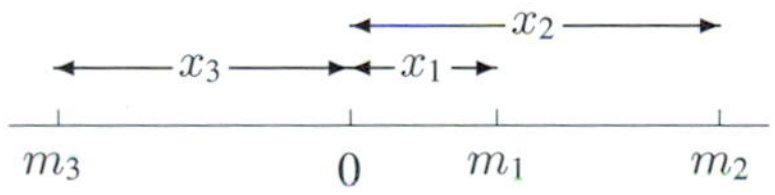

Figure 8.21: Center of mass of discrete masses, m_i

Example 5   Find the center of mass of a 2-meter rod lying on the x-axis with its left end at the origin if:
(a)   The density is constant and the total mass is 5 kg.   (b)   The density is $\delta(x) = 15x^2$ kg/m.

Solution   (a)   We expect that the balancing point will be in the middle, that is, $\bar{x} = 1$. To check this, we compute the density, which is the total mass divided by the length, or $\delta(x) = 5/2$ kg/m. Then

$$\bar{x} = \frac{\int_0^2 x \cdot \frac{5}{2}\,dx}{5} = \frac{1}{5} \cdot \frac{5}{2} \cdot \frac{x^2}{2}\Big|_0^2 = 1 \text{ meter.}$$

(b)   Since the rod is heavier in the interval $[1, 2]$ than in the interval $[0, 1]$, we expect the center of mass to be between $x = 1$ and $x = 2$. Now we have

$$\text{Total mass} = \int_0^2 15x^2\,dx = 5x^3\Big|_0^2 = 40 \text{ kg.}$$

Thus,

$$\bar{x} = \frac{\int_0^2 x \cdot 15x^2\,dx}{40} = \frac{15}{40} \cdot \frac{x^4}{4}\Big|_0^2 = \frac{3}{2} \text{ meter.}$$

The definition of $\bar{x}$ can be used to find the x-coordinate of the center of mass for objects that don't lie entirely along the x-axis. The key in these cases is to find an expression for the mass of a small slice of the object located at position x_i.

Example 6   An isosceles triangle of altitude 1 unit and base 1 unit has its base along the y-axis and the positive x-axis passing through its vertex. See Figure 8.22. If the mass of the triangle is m and the triangle has constant density, find the x-coordinate of its center of mass.

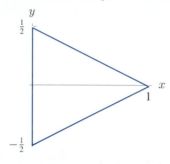

Figure 8.22: Find center of mass of this triangle   **Figure 8.23:** Sliced triangle

Solution   Because the mass of the triangle is symmetrically distributed with respect to the x-axis, the center of mass lies on the x-axis. We expect the center of mass to be closer to the origin than to the vertex at $x = 1$, since the triangle is wider near the origin.

   The area of the triangle is $\frac{1}{2} \cdot 1 \cdot 1 = \frac{1}{2}$. Thus Density = Mass/Area = $2m$. If we slice the triangle into strips of width Δx, then the piece at position x_i has height $2 \cdot \frac{1}{2}(1 - x_i) = (1 - x_i)$. (See Figure 8.23.) So

$$\text{Area of strip} \approx (1 - x_i)\Delta x,$$

giving

$$\text{Mass} = \text{Area} \times \text{Density} \approx 2m \cdot (1 - x_i)\Delta x.$$

So the center of mass is

$$\bar{x} \approx \frac{\sum x_i 2m \cdot (1 - x_i)\Delta x}{m}.$$

As $n \to \infty$, we have

$$\bar{x} = \frac{\int_0^1 2mx(1 - x)\,dx}{m} = 2\left(\frac{x^2}{2} - \frac{x^3}{3}\right)\Big|_0^1 = \frac{1}{3}.$$

   Similar ideas are used to find the coordinates of the center of mass for objects in the xy-plane.

Problems for Section 8.2

1.   A rod has length 2 meters. At a distance x meters from its left end, the density of the rod is given by

$$\rho(x) = 2 + 6x \text{ g/m}.$$

   (a)   Write a Riemann sum approximating the total mass of the rod.
   (b)   Find the exact mass by converting the sum into an integral.

2.   The density of cars (in cars per mile) down a 20-mile stretch of the Pennsylvania Turnpike can be approximated by

$$\rho(x) = 300\left(2 + \sin\left(4\sqrt{x + 0.15}\right)\right),$$

   where x is the distance in miles from the Breezewood toll plaza.

   (a)   Sketch a graph of this function for $0 \le x \le 20$.
   (b)   Write a sum that approximates the total number of cars on this 20-mile stretch.
   (c)   Find the total number of cars on the 20-mile stretch.

3. Suppose you want to find the total mass of a 3×5 rectangular sheet, whose density per unit area at a distance x from one of the sides of length 5 is $1/(1 + x^4)$.

    (a) Find a Riemann sum which approximates the total mass.
    (b) Find the mass to one decimal place. How do you know your answer is accurate enough?

4. The density of oil in a circular oil slick on the surface of the ocean at a distance r meters from the center of the slick is given by $\rho(r) = 50/(1 + r)$ kg/m^2.

    (a) If the slick extends from $r = 0$ to $r = 10{,}000$ m, find a Riemann sum approximating the total mass of oil in the slick.
    (b) Find the exact value of the mass of oil in the slick by turning your sum into an integral and evaluating it.
    (c) Within what distance r is half the oil of the slick contained?

5. The soot produced by a garbage incinerator spreads out in a circular pattern. The depth, $H(r)$, in millimeters, of the soot deposited each month at a distance r kilometers from the incinerator is given by $H(r) = 0.115e^{-2r}$.

    (a) Write a definite integral giving the total volume of soot deposited within 5 kilometers of the incinerator each month.
    (b) Evaluate the integral you found in part (a), giving your answer in cubic meters.

6. If a rod lies along the x-axis between a and b, the moment of the rod is $\int_a^b x\rho(x)\,dx$, where $\rho(x)$ is its density at a position x. Find the moment of the rod in Problem 1.

7. A rod of length 3 with density $\delta(x) = 1 + x^2$ is positioned along the positive x-axis, with its left end at the origin. Find the total mass and the center of mass of the rod.

8. Consider a rod of length 1, with density $\delta(x) = 1 + kx^2$, where k is a positive constant. Suppose the rod is lying on the positive x-axis with one end at the origin.

    (a) Find the center of mass as a function of k.
    (b) Show that the center of mass of the rod satisfies $0.5 < \bar{x} < 0.75$.

9. A rod of length 2 meters and density $\delta(x) = 3 - e^{-x}$ kilograms per meter is placed on the x-axis with its ends at $x = \pm 1$.

    (a) Will the center of mass of the rod be on the left or right of the origin? Explain.
    (b) Find the coordinate of the center of mass.

10. One half of a uniform circular disk of radius 1 lies with its diameter along the y-axis and center at the origin. The mass of the half-disk is 1. Find the location of the center of mass $\bar{x}$.

11. Suppose an isosceles triangle with uniform density, altitude a, and base b is placed in the xy-plane as in Figure 8.24. Show that the center of mass is at $\bar{x} = a/3$. Hence show that the center of mass is independent of the triangle's base.

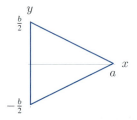

Figure 8.24

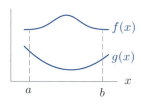

Figure 8.25

Figure 8.26

12. A cardboard figure has the shape shown in Figure 8.25. The region is bounded on the left by the line $x = a$, on the right by the line $x = b$, above by $f(x)$, and below by $g(x)$. If the density $\rho(x)$ gm/cm^2 varies only with x, find an expression for the total mass of the figure, in terms of $f(x)$, $g(x)$, and $\rho(x)$.

13. The storage shed in Figure 8.26 is the shape of a half-cylinder of radius r and length l.

    (a) What is the volume of the shed?
    (b) Suppose the shed is filled with sawdust whose density at any point is proportional to the distance of that point from the floor. The constant of proportionality is k. Calculate the total mass of sawdust in the shed.

14. The following table gives the density D (in gm/cm^3) of the earth at a depth x km below the earth's surface. The radius of the earth is about 6370 km. Find an upper and a lower bound for the earth's mass such that the upper bound is less than twice the lower bound. Explain your reasoning; in particular, what assumptions have you made about the density?

x	0	1000	2000	2900	3000	4000	5000	6000	6370
D	3.3	4.5	5.1	5.6	10.1	11.4	12.6	13.0	13.0

15. An exponential model for the density of the earth's atmosphere says that if the temperature of the atmosphere were constant, then the density of the atmosphere as a function of height, h (in meters), above the surface of the earth would be given by

$$\rho(h) = 1.28e^{-0.000124h} \text{ kg/m}^3.$$

(a) Write (but do not evaluate) a sum that approximates the mass of the portion of the atmosphere from $h = 0$ to $h = 100$ m (i.e., the first 100 meters above sea level). Assume the radius of the earth is 6370 km.

(b) Find the exact answer by turning your sum in part (a) into an integral. Evaluate the integral.

16. Water is flowing in a cylindrical pipe of radius 1 inch. Because water is viscous and sticks to the pipe, the rate of flow varies with the distance from the center. The speed of the water at a distance r inches from the center is $10(1 - r^2)$ inches per second. What is the rate (in cubic inches per second) at which water is flowing through the pipe?

8.3 APPLICATIONS TO PHYSICS

Although geometric problems of area, length, and volume were a driving force for the development of the calculus in the seventeenth century, it was Newton's spectacularly successful applications of the calculus to physics that most clearly demonstrated the power of this new mathematics.

Work

Our first application involves the concept of work. In physics the word "work" has a technical meaning which is different from its everyday meaning. Physicists say that if a constant force, F, is applied to some object to move it a distance, d, then the force has done work on the object. The force must be in the same direction as the motion. We make the following definition:

Work done = Force × Distance
$$W = F \cdot d$$

Notice that if we walk across a room holding a book, we do not accomplish any work on the book, since the only force we exert on the book is vertical, but the motion of the book is horizontal. On the other hand, if we lift the book from the floor to a height of 3 feet above the floor, we accomplish work.

Note on Units: Mass versus Weight

When we talk about how much something weighs, we're talking about the force that gravity exerts on that object. A lump of iron, for example, might weigh 10 pounds on the surface of the earth. This is because the earth's gravitational field is exerting 10 pounds of force on that lump of iron. If the same lump, however, is adrift in interstellar space, it is weightless because there is no gravitational field in its vicinity to exert a force on it. Of course, the quantity of iron present hasn't changed; there's no less iron there than there was before. So we make a distinction between the *mass* of the lump (the quantity of matter it comprises) and the *weight* of that mass (the force exerted on it due to gravity).

In the British system of units, a *pound* is a unit of weight. In the metric system, however, a *kilogram* is a unit of mass, not weight. Thus knowing we have 1 kilogram of iron means that we know the mass of iron present. To figure out how much our iron weighs, we have to figure out how much force is being exerted on it by gravity. Since Newton's second law tells us that Force = Mass × Acceleration, we see that we have to multiply the iron's mass by the acceleration due to gravity, which (on the surface of the earth) is $9.8 \text{ m}/\sec^2$. Thus the kilogram of iron weighs $1 \text{ kg} \cdot 9.8 \text{ m}/\sec^2 = 9.8 \text{ kg m}/\sec^2$, which is almost always written 9.8 newtons. (A newton = $1 \text{ kg} \cdot \text{ m}/\sec^2$ is a unit of force or weight. It is abbreviated nt.)

Example 1   How much work is done in lifting

    (a)   A 5-pound book 3 feet off the floor?   (b)   A 1.5-kilogram book 2 meters off the floor?

Solution   (a) $W = F \cdot d = (5 \text{ lb})(3 \text{ ft}) = 15$ foot-pounds.

    (b)   In this case we have not done $(1.5)(2) = 3$ units of work. We need to use the book's weight (not its mass), namely $(1.5 \text{ kg})(g \text{ m/sec}^2) = (1.5)(9.8)$ newtons. Thus, the work done is

$$W = F \cdot d = [(1.5 \text{ kg})(9.8 \text{ m/sec}^2)] \cdot (2 \text{ m}) = 29.4 \text{ newton-meters.}$$

A newton-meter is usually called a *joule*, so we did 29.4 joules of work in the metric system of units. In the British system the unit of work is the foot-pound, the work done when a 1-pound force moves an object a distance of 1 foot.

Calculating the Work Done

In the following examples, we calculate the work done where both the force and distance moved vary. To find the work done on a complicated object, *we slice the object up in such a way that we can find the work done on each piece.* We calculate the work for each piece using $W = Fd$, and we sum these pieces to approximate the total work as a Riemann sum. Letting the size of each piece tend to zero, we obtain a definite integral that represents the total work.

Example 2   A 28-meter uniform chain with a mass of 20 kilograms is dangling from the roof of a building. How much work is needed to pull the chain up to the top of the building?

Solution   Since a 20-kilogram chain weighs $(20 \text{ kg})(9.8 \text{ m}/\sec^2) = 196$ newtons, it might seem that the answer should be $(196 \text{ nt})(28 \text{ m}) = 5488$ joules. But remember that not all of the chain has to move 28 meters—links near the top move less.

Let's divide the chain into small sections of length Δy, each weighing $7\,\Delta y$ newtons. (A length of 28 meters weighs 196 newtons, so 1 meter weighs 7 newtons.) See Figure 8.27 on page 392. If Δy is small, all of this piece is hauled up approximately the same distance, namely y, against the force due to gravity of $7\,\Delta y$ newtons. Thus,

$$\text{Work done on the small piece} \approx (7\,\Delta y \text{ newtons})(y \text{ meters}) = 7y\,\Delta y \text{ joules.}$$

The work done on the entire chain is given by the total of the work done on each piece:

$$W \approx \sum 7y\,\Delta y \text{ joules.}$$

As $\Delta y \to 0$, we obtain a definite integral. Since y varies from 0 to 28 meters, the total work is

$$W = \int_0^{28} (7y)\, dy = \left. \frac{7}{2}y^2 \right|_0^{28} = 2744 \text{ joules.}$$

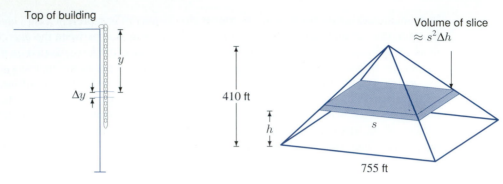

Figure 8.27: Chain for Example 2   **Figure 8.28:** Pyramid for Example 3

Example 3   It is reported that the Great Pyramid of Egypt was built in 20 years. If the stone making up the pyramid has density 200 pounds per cubic foot, find the total amount of work done in building the pyramid. Assume that the pyramid is 410 feet high and has a square base 755 feet by 755 feet. Estimate how many workers were needed to build the pyramid.

Solution   We assume that the stones were originally located at the approximate height of the construction site. Imagine the pyramid built in layers as we did in Example 1 on page 377.

We found by similar triangles that the layer at height h has a side length $s = \frac{755}{410}(410 - h)$ ft. The layer at height h has a volume of $s^2 \, \Delta h$ ft^3 and hence a weight of $200s^2 \, \Delta h$ lb. Constructing this layer involved lifting a weight of $200s^2 \, \Delta h$ through a height of h, so the work required is $(200s^2 \, \Delta h$ lb$))(h$ ft$) = 200s^2 h \, \Delta h$ ft-lb. (See Figure 8.28.) The total work, W, of constructing all the layers is then

$$W \approx \sum 200s^2 \, h\Delta h = \sum 200 \left(\frac{755}{410}\right)^2 (410 - h)^2 h \, \Delta h \text{ ft-lb.}$$

As the thickness of each layer, Δh, tends to zero, we obtain a definite integral. Since h varies from 0 to 410, we have

$$W = \int_0^{410} 200s^2 h \, dh = 200 \int_0^{410} \left(\frac{755}{410}\right)^2 (410 - h)^2 h \, dh = 200 \left(\frac{755}{410}\right)^2 \int_0^{410} (410 - h)^2 h \, dh$$

$$= 200 \left(\frac{755}{410}\right)^2 \left(410^2 \frac{h^2}{2} - 2(410)\frac{h^3}{3} + \frac{h^4}{4}\right)\Bigg|_0^{410} = 200 \left(\frac{755}{410}\right)^2 (410)^4 \left(\frac{1}{12}\right)$$

$$\approx 1{,}597{,}020{,}000{,}000 \approx 1.6 \times 10^{12} \text{ foot-pounds.}$$

We have calculated the total work done in building the pyramid; now we want to estimate the total number of workers needed. Let's assume every laborer worked 10 hours a day, 300 days a year, for 20 years. Assume that a typical worker lifted ten 50 pound blocks a distance of 4 feet every hour, thus performing 2000 foot-pounds of work per hour (this is a very rough estimate). Then each laborer performed $(10)(300)(20)(2000) = 1.2 \times 10^8$ foot-pounds of work over a twenty year period. Thus, the number of workers needed was about $(1.6 \times 10^{12})/(12 \times 10^7)$, or about 13,000.

Escape from the Earth's Gravitational Field

If you throw an object into the air, you expect gravity to bring it back to earth. However, the harder you throw it, the longer you expect to wait before it returns. You may wonder if it is possible to throw something hard enough so that it never returns to earth. We will calculate the total amount of work done to move an object infinitely far away from the earth. Amazingly enough, this total amount of work is finite, and it *is* possible to launch something hard enough that it never returns to earth.

Previously, when we analyzed the motion of a freely falling body, we assumed that the force of gravity is constant. Now, since we are thinking of a body moving into outer space, we have to take into account the fact that, as we move away from the earth, its gravitational force gets weaker. Newton's Law of Gravitation says that the force, F, exerted on a mass m at a distance, r, from the center of the earth is given by

$$F = \frac{GMm}{r^2}.$$

Here M is the mass of the earth, and G is called the gravitational constant, whose value is about 6.67×10^{-11} if mass is measured in kilograms, distance in meters, and force in newtons. We take $r > R$, the earth's radius.

Example 4   Find the total work done to move an object of mass m infinitely far from the earth.

Solution   The force pulling the object back to the earth is GMm/r^2. Thus,

$$\text{Work needed to move an object a distance } \Delta r \text{ further away } \approx \left(\frac{GMm}{r^2}\right)\Delta r.$$

Therefore the total work required to move the object from the surface of the earth $r = R$ to a point "infinitely far away" is approximated by

$$W \approx \sum \frac{GMm}{r^2}\Delta r$$

where the value of r runs from $r = R$ (at the earth's surface) to $r = \infty$ (very far away). As Δr tends to zero, we get the definite integral

$$W = \int_R^\infty \frac{GMm}{r^2}\,dr = \lim_{b \to \infty} \left. -\frac{GMm}{r}\right|_R^b = \frac{GMm}{R}.$$

Imagine an object shot upward with initial velocity v. The energy of the body by virtue of its motion is called its *kinetic energy* and is equal to $\frac{1}{2}mv^2$. When the object moves upward, it slows down, losing kinetic energy as work is done against gravity. The body moves upward until all the kinetic energy is expended; it will stop at the height at which

$$\text{Initial kinetic energy} = \text{Work done against gravity}.$$

Since the work done to move an object infinitely far from the earth is finite, a finite starting velocity can give the object enough kinetic energy to move infinitely far away. The *escape velocity* from the earth is defined to be the minimum vertical velocity that must be given to an object so that it is never drawn back to earth by gravity.

Example 5   Find the escape velocity of an object from the earth.

Solution   If v is the initial velocity, then setting the initial kinetic energy equal to the work done to escape infinitely far from the earth, we get

$$\frac{1}{2}mv^2 = \frac{GMm}{R}.$$

Solving for v gives

$$v = \sqrt{\frac{2GM}{R}}.$$

Since the mass of the earth is $M \approx 6 \times 10^{24}$ kg and its radius is $R \approx 6.4 \times 10^6$ meters, the escape velocity is about 11,000 meters/sec $\approx$ 25,000 mph, or about 30 times the speed of sound. Thus, the escape velocity is finite, though it is much higher than you can achieve by throwing or by shooting a gun. (A bullet's speed is slightly faster than sound.)

Force and Pressure

We can use the definite integral to compute the force exerted by a liquid on a surface, for example, the force of water on a dam. The idea is to get the force from the *pressure*. The pressure in a liquid is the force per unit area exerted by the liquid. Two things you need to know about pressure are:

- At any point, pressure is exerted equally in all directions—up, down, sideways.

- Pressure increases with depth. (That is one of the reasons why deep sea divers have to take much greater precautions than scuba divers.)

At a depth of h feet, the pressure, p, exerted by the liquid, measured in pounds per square foot, is given by computing the total weight of a column of liquid h feet high with a base of 1 square foot. The volume of such a column of liquid is just h cubic feet.

If the liquid has density ρ (mass per unit volume), then its weight per unit volume is ρg (where g is the acceleration due to gravity). The weight of the column of liquid is $\rho g h$, so

$$\text{Pressure} = \text{Density} \times g \times \text{Depth} \quad \text{or} \quad p = \rho g h$$

Sometimes we may be given the density of the liquid as a weight per unit volume, rather than a mass per unit volume. In that case, we do not need to multiply by g because it has already been done. Since pounds are units of weight, and not mass, knowing that water weighs 62.4 lb/ft^3 tells us that for water, $\rho g = 62.4$ lb/ft^3, and pressure at depth h is $62.4h$ lb/ft^2. See Figure 8.29.

Next we need to know the relation between force and pressure. Provided the pressure is constant over a given area, we have the following relation:

$$\text{Force} = \text{Pressure} \times \text{Area}$$

When the pressure isn't constant over a surface, we divide the surface into small pieces *in such a way that the pressure is nearly constant on each one*. Then we can use this formula on each piece and obtain a definite integral to get the force over an entire surface. Since the pressure varies with depth, we should divide the surface into horizontal strips, each of which is at an approximately constant depth.

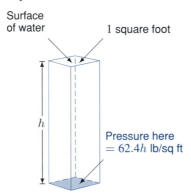

Figure 8.29: Pressure exerted by a column of water

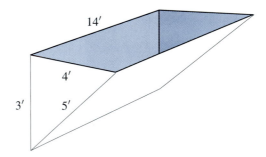

Figure 8.30: The trough for Example 6

Example 6   A trough 14 feet long has one rectangular vertical side 14 feet by 3 feet, one rectangular inclined side 14 feet by 5 feet, and two triangular ends with sides 3 feet, 4 feet, and 5 feet (see Figure 8.30). If the trough is filled to the top with water, compute the force on each end and each side.

Solution   We will do the long vertical side first. Divide the side into thin horizontal strips of width Δh. Since pressure depends only on the depth h and the strips are thin and horizontal, we can assume the pressure is the same at every point in the strip. The area of the strip is $14 \Delta h$ ft^2, and the force on the

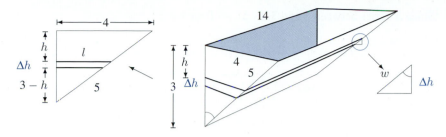

Figure 8.31: Calculating the widths of strips on the end and the inclined side

strip at depth h ft is the pressure at that depth times the area of the strip:

$$\text{Force on strip} = \text{Pressure} \times \text{Area} \approx (62.4h \text{ lb/ft}^2)(14 \,\Delta h \text{ ft}^2) = (62.4h)14 \,\Delta h \text{ lb}.$$

Summing the forces on these strips, we get an expression for the total force on the vertical side:

$$F_{\substack{\text{vertical} \\ \text{side}}} \approx \sum (62.4h)14 \,\Delta h \text{ lb}.$$

Letting $\Delta h \to 0$, we get a definite integral. Since h varies from 0 to 3 feet, we have

$$F_{\substack{\text{vertical} \\ \text{side}}} = \int_0^3 (62.4h)14 \,dh = (62.4)(14)\frac{h^2}{2}\bigg|_0^3 = (62.4)(14)\left(\frac{9}{2}\right) \approx 3930 \text{ pounds}.$$

To compute the force on the inclined side, we use an inclined horizontal strip. Each strip has length 14, but its width is not Δh. By similar triangles (see Figure 8.31) the width w of the strip satisfies

$$\frac{\Delta h}{w} = \frac{3}{5}, \quad \text{so} \quad w = \frac{5}{3}\,\Delta h.$$

Therefore,

$$\text{Area of strip} = 14w = 14\left(\frac{5}{3}\right)\Delta h \text{ ft}^2,$$

so

$$\text{Force on strip} = (62.4h)(14)\left(\frac{5}{3}\right)\Delta h \text{ lb}.$$

The total force on the inclined side is obtained by adding the forces on the strips, giving

$$F_{\substack{\text{inclined} \\ \text{side}}} \approx \sum (62.4h)(14)\left(\frac{5}{3}\right)\Delta h \text{ lb}.$$

We let $\Delta h \to 0$ to arrive at a definite integral. Again, h varies between 0 and 3 feet, so

$$F_{\substack{\text{inclined} \\ \text{side}}} = \int_0^3 (62.4h)(14)\left(\frac{5}{3}\right) dh = \left(\frac{5}{3}\right)(62.4)(14)\left(\frac{9}{2}\right) \approx 6550 \text{ pounds}.$$

Notice that this is just $\frac{5}{3}$ the force on the vertical side.

Finally, to compute the force on each triangular end, we again divide the end into horizontal strips. The width of the strip is Δh as with the vertical side, but this time the length of the strip varies from 4 feet at the top of the trough to 0 feet at the bottom. By similar triangles (see Figure 8.31) the length l of a strip is related to the depth h by

$$\frac{l}{4} = \frac{3-h}{3}, \quad \text{so} \quad l = \frac{4}{3}(3-h).$$

Thus, the area of the strip is $l\,\Delta h = \frac{4}{3}(3-h)\Delta h \text{ ft}^2$. As before, this means the total force is

$$F_{\text{end}} \approx \sum (62.4h)\left(\frac{4}{3}(3-h)\right)\Delta h \text{ lb}.$$

Since h varies between 0 and 3, as $\Delta h \to 0$ we have

$$F_{\text{end}} = \int_0^3 (62.4h)\left(\frac{4}{3}(3-h)\right) dh = (62.4)\left(\frac{4}{3}\right)\left(3\frac{h^2}{2} - \frac{h^3}{3}\right)\Big|_0^3$$

$$= (62.4)\left(\frac{4}{3}\right)\left(\frac{27}{6}\right) \approx 374 \text{ pounds.}$$

Problems for Section 8.3

1. Figure 8.32 shows thrust-time curves for two model rockets. The thrust or force, F, of the engine (in newtons) is plotted against time, t, (in seconds). The *total impulse* of the rocket's engine is defined as the definite integral of F with respect to t. The total impulse is a measure of the strength of the engine.

    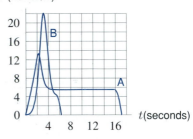

    (a)   For approximately how many seconds is the thrust of rocket B greater than 10 newtons?

    (b)   Estimate the total impulse for model rocket A.

    (c)   What are the units for the total impulse calculated in part (b)?

    (d)   Which rocket has the largest total impulse?

    (e)   Which rocket has the largest maximum thrust?

    Figure 8.32

2. A worker on a scaffolding 75 ft above the ground needs to lift a 500 lb bucket of cement from the ground to a point 30 ft above the ground by pulling on a rope weighing 5 lb/ft. How much work is required?

3. A 1000-lb weight is being lifted to a height 10 feet off the ground. It is lifted using a rope which weighs 4 lb per foot and which is being pulled up by construction workers standing on a roof 30 feet off the ground. Find the work done to lift the weight.

4. A rectangular water tank has length 20 ft, width 10 ft, and depth 15 ft. If the tank is full, how much work does it take to pump all the water out? (Note that 1 cubic foot of water weighs 62.4 lb.)

5. A water tank is in the form of a right circular cylinder with height 20 ft and radius 6 ft. If the tank is half full of water, find the work required to pump all of it over the top rim. (Note that 1 cubic foot of water weighs 62.4 lb.)

6. Suppose the tank in Problem 5 is full of water. Find the work required to pump all of it to a point 10 ft above the top of the tank.

7. A fuel oil tank is an upright cylinder, buried so that its circular top is 10 feet beneath ground level. The tank has a radius of 5 feet and is 15 feet high, although the current oil level is only 6 feet deep. Calculate the work required to pump all of the oil to the surface. Assume that the oil weighs 50 lb per cubic foot.

8. A gas station stores its gasoline in a tank under the ground. The tank is a cylinder lying horizontally on its side. (In other words, the tank is not standing vertically on one of its flat ends.) If the radius of the cylinder is 4 feet, its length is 12 feet, and its top is 10 feet under the ground, find the total amount of work needed to pump the gasoline out of the tank. (Gasoline weighs 42 lb/ft^3.)

9. How much work is required to lift a 1000-kg satellite to an altitude of 2×10^6 m above the surface of the earth? (The radius of the earth is 6.4×10^6 m, its mass is 6×10^{24} kg, and in these units the gravitational constant, G, is 6.67×10^{-11}.)

10. Show that the escape velocity needed by an object to escape the gravitational influence of a spherical planet of density ρ and radius R is proportional to R and to $\sqrt{\rho}$.

11. Calculate the escape velocity of an object from the moon. The acceleration due to gravity on the moon is 1.6 m/sec^2, whereas on earth it is 9.8 m/sec^2; the radius of the moon is approximately 1740 km. [Hint: If g is the acceleration due to gravity, the force on an object is $F = mg = GMm/R^2$. See page 393.]

12. A reservoir has a dam at one end. The dam is a rectangular wall, 1000 feet long and 50 feet high. The problem is to find the total force of the water acting on the dam.

(a) Show how to approximate this force by a Riemann sum, explaining your reasoning clearly.
(b) Write an integral which represents the force, and evaluate it.

13. Before any water was pumped out, what was the total force on the bottom and each side of the tank in Problem 4?

14. A lobster tank in a restaurant is 4 ft long by 3 ft wide by 2 ft deep. Find the water force on the bottom and on each of the four sides. (The density of water is 62.4 lb/ft^3.)

For Problems 15–16, find the kinetic energy of a rotating body, using the fact that the kinetic energy of a particle of mass m moving at a speed v is $\frac{1}{2}mv^2$. Slice the object into pieces in such a way that the velocity is approximately constant across each piece.

15. Find the kinetic energy of a rod of mass 10 kg and length 6 m rotating about an axis perpendicular to the rod at its midpoint, with an angular velocity of 2 radians per second. (Imagine a helicopter blade of uniform thickness.)

16. Find the kinetic energy of a phonograph record of uniform density, mass 50 gm and radius 10 cm rotating at $33\frac{1}{3}$ revolutions per minute.

17. We define the electric potential at a distance r from an electric charge q by q/r. The electric potential of a charge distribution is obtained by adding up the potential from each point. Suppose electric charge is sprayed (with constant density σ in units of charge/unit area) on to a circular disk of radius a. Consider the axis perpendicular to the disk and through its center. Find the electric potential at the point P on this axis at a distance R from the center. (See Figure 8.33.)

18. On a hot afternoon in Pétionville, Haiti, Lubonga is trying to cool off by sipping the glass of palm wine shown in Figure 8.34. How much work does she have to do to empty the glass? (The density of the palm wine is 1.2 gm/cm^3. Initially, the glass is filled to a vertical depth of 8 cm. Note that the force due to gravity acting on 1 gram is 980 dynes, and that in these units work is measured in ergs. The work done in lifting a small piece of fluid depends on the difference in height between the initial and final positions.)

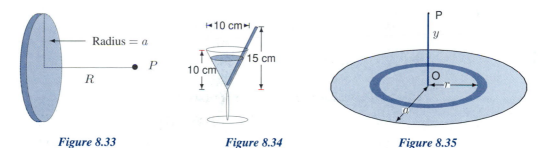

Figure 8.33   **Figure 8.34**   **Figure 8.35**

19. A uniform, thin, circular disk of radius a and mass M lies on a horizontal plane. The point P lies a distance y directly above O, the center of the disk. Calculate the gravitational force on a mass m at the point P. (See Figure 8.35.) Use the fact that the gravitational force exerted on the mass m by a thin horizontal ring of radius r, mass μ, and center O is toward O and given by

$$F = \frac{G\mu my}{(r^2 + y^2)^{3/2}}, \qquad \text{where } G \text{ is the gravitational constant.}$$

For Problems 20–21, find the gravitational force between two objects. Use the fact that the gravitational attraction between particles of mass m_1 and m_2 at a distance r apart is Gm_1m_2/r^2. For each problem, slice the objects into pieces, use this formula for the pieces, and sum using a definite integral.

20. What is the force of gravitational attraction between a thin uniform rod of mass M and length l and a particle of mass m lying in the same line as the rod at a distance a from one end?

21. Two long, thin, uniform rods of lengths l_1 and l_2 lie on a straight line with a gap between them of length a. Suppose their masses are M_1 and M_2, respectively, and the constant of the gravitation is G. What is the force of attraction between the rods? (Use the result of Problem 20.)

8.4 APPLICATIONS TO ECONOMICS

Present and Future Value

Many business deals involve payments in the future. If you buy a car or furniture, for example, you may buy it on credit and pay over a period of time. If you are going to accept payment in the future under such a deal, you obviously need to know how much you should be paid. Being paid $100 in the future is clearly worse than being paid $100 today for many reasons. If you are given the money today, you can do something else with it—for example, put it in the bank, invest it somewhere, or spend it. Thus, even without considering inflation, if you are to accept payment in the future, you would expect to be paid more to compensate for this loss of potential earnings. The question we will consider now is, how much more?

To simplify matters, we consider only what we would lose by not earning interest; we will not consider the effect of inflation. Let's look at some specific numbers. Suppose you deposit $100 in an account which earns 7% interest compounded annually, so that in a year's time you will have $107. Thus, $100 today will be worth $107 a year from now. We say that the $107 is the *future value* of the $100, and that the $100 is the *present value* of the $107. In general, we say the following:

- The **future value**, B, of a payment, P, is the amount to which the P would have grown if deposited in an interest bearing bank account.
- The **present value**, P, of a future payment, B, is the amount which would have to be deposited in a bank account today to produce exactly B in the account at the relevant time in the future.

Observe that the present value is always smaller than the future value. With an interest rate of r, compounded annually, and a time period of t years, a deposit of P grows to a future balance of B, where

$$B = P(1+r)^t, \quad \text{or equivalently,} \quad P = \frac{B}{(1+r)^t}.$$

If the interest is compounded n times a year for t years at rate r, and if B is the *future value* of P after t years and P is the *present value* of B, then

$$B = P\left(1 + \frac{r}{n}\right)^{nt}, \quad \text{or equivalently,} \quad P = \frac{B}{\left(1 + r/n\right)^{nt}}.$$

Note that for a 7% interest rate, $r = 0.07$. It can be shown that when n tends towards infinity, we say that the compounding becomes continuous, and we get the following result:

$$B = Pe^{rt}, \quad \text{or equivalently,} \quad P = \frac{B}{e^{rt}} = Be^{-rt}.$$

Example 1 You win the lottery and are offered the choice between $1 million in four yearly installments of $250,000 each, starting now, and a lump-sum payment of $920,000 now. Assuming a 6% interest rate, compounded continuously, and ignoring taxes, which should you choose?

Solution We will do the problem in two ways. First, we assume that you pick the option with the largest present value. The first of the four $250,000 payments is made now, so

Present value of first payment = $250,000.

The second payment is made one year from now and so

$$\text{Present value of second payment} = \$250,000e^{-0.06(1)}.$$

Calculating the present value of the third and fourth payments similarly, we find:

$$
\begin{aligned}
\text{Total present value} &= \$250,000 + \$250,000e^{-0.06(1)} + \$250,000e^{-0.06(2)} + \$250,000e^{-0.06(3)} \\
&\approx \$250,000 + \$235,441 + \$221,730 + \$208,818 \\
&= \$915,989.
\end{aligned}
$$

Since the present value of the four payments is less than \$920,000, you are better off taking the \$920,000 right now.

Alternatively, we can do the problem by comparing the future values of the two pay schemes. The scheme with the highest future value is the best from a purely financial point of view. We calculate the future value of both schemes three years from now, on the date of the last \$250,000 payment. At that time,

$$\text{Future value of the lump sum payment} = \$920,000e^{0.06(3)} \approx \$1,101,440.$$

Now we calculate the future value of the first \$250,000 payment:

$$\text{Future value of the first payment} = \$250,000e^{0.06(3)}.$$

Calculating the future value of the other payments similarly, we find:

$$
\begin{aligned}
\text{Total future value} &= \$250,000e^{0.06(3)} + \$250,000e^{0.06(2)} + \$250,000e^{0.06(1)} + \$250,000 \\
&\approx \$299,304 + \$281,874 + \$265,459 + \$250,000 \\
&= \$1,096,637.
\end{aligned}
$$

The future value of the \$920,000 payment is greater, so you are better off taking the \$920,000 right now. Of course, since the present value of the \$920,000 payment is greater than the present value of the four separate payments, you would expect the future value of the \$920,000 payment to be greater than the future value of the four separate payments.

(Note: If you read the fine print, you will find that many lotteries do not make their payments right away, but often spread them out, sometimes far into the future. This is to reduce the present value of the payments made, so that the value of the prizes is much less than it might first appear!)

Income Stream

When we consider payments made to or by an individual, we usually think of *discrete* payments, that is, payments made at specific moments in time. However, we may think of payments made by a company as being *continuous*. The revenues earned by a huge corporation, for example, come in essentially all the time, and therefore they can be represented by a continuous *income stream*. Since the rate at which revenue is earned may vary from time to time, the income stream is described by

$$P(t) \text{ dollars/year.}$$

Notice that $P(t)$ is a *rate* at which deposits are made (its units are dollars per year, for example) and that the rate depends on the time, t, usually measured in years from the present.

Present and Future Values of an Income Stream

Just as we can find the present and future values of a single payment, so we can find the present and future values of a stream of payments. As before, the future value represents the total amount of money obtained by depositing the income stream into a bank account and letting it gather interest. The present value represents the amount of money you would have to deposit today (in an interest-bearing bank account) in order to match what you would get from the income stream.

When we are working with a continuous income stream, we will assume that interest is compounded continuously. The reason for this is that the approximations we are going to make (of sums by integrals) are much simpler if both payments and interest are continuous.

Suppose that we want to calculate the present value of the income stream described by a rate of $P(t)$ dollars per year, and that we are interested in the period from now until T years in the future. In order to use what we know about single deposits to calculate the present values of an income stream, we must first divide the stream into many small deposits, each of which is made at approximately one instant. We divide the interval $0 \leq t \leq T$ into subintervals, each of length Δt:

Assuming Δt is small, the rate, $P(t)$, at which deposits are being made will not vary much within one subinterval. Thus, between t and $t + \Delta t$:

$$\text{Amount paid} \approx \text{Rate of deposits} \times \text{Time}$$
$$\approx (P(t) \text{ dollars/year})(\Delta t \text{ years})$$
$$= P(t)\Delta t \text{ dollars.}$$

Measured from the present, the deposit of $P(t)\Delta t$ is made t years in the future. Thus,

$$\begin{matrix} \text{Present value of money} \\ \text{deposited in interval } t \text{ to } t + \Delta t \end{matrix} \approx P(t)\Delta t e^{-rt}.$$

Summing over all subintervals gives

$$\text{Total present value} \approx \sum P(t)e^{-rt}\Delta t \text{ dollars.}$$

In the limit as $\Delta t \to 0$, we get the following integral:

$$\boxed{\text{Present value} = \int_0^T P(t)e^{-rt}\,dt.}$$

In computing future value, the deposit of $P(t)\Delta t$ has a period of $(T - t)$ years to earn interest, and therefore

$$\text{Future value of money deposited in interval } t \text{ to } t + \Delta t \approx \left[P(t)\Delta t\right]e^{r(T-t)}.$$

Summing over all subintervals, we get:

$$\text{Total future value} \approx \sum P(t)\Delta t e^{r(T-t)} \text{ dollars.}$$

As the length of the subdivisions tends toward zero, the sum becomes an integral:

$$\boxed{\text{Future value} = \int_0^T P(t)e^{r(T-t)}\,dt.}$$

Example 2   Find the present and future values of a constant income stream of \$100 per year over a period of 20 years, assuming an interest rate of 10% compounded continuously.

Solution   Using $P(t) = 100$ and $r = 0.1$, we have

$$\text{Present value} = \int_0^{20} 100e^{-0.1t}\,dt = 100\left(-\frac{e^{-0.1t}}{0.1}\right)\Big|_0^{20} = 1000(1 - e^{-2}) \approx \$864.66.$$

$$\text{Future value} = \int_0^{20} 100e^{0.1(20-t)}\,dt = \int_0^{20} 100e^2 e^{-0.1t}\,dt$$

$$= 100e^2\left(-\frac{e^{-0.1t}}{0.1}\right)\Big|_0^{20} = 1000e^2(1 - e^{-2}) \approx \$6389.06.$$

Example 3   What is the relationship between the present and future values in the previous example? Explain this relationship.

Solution   Since

$$\text{Present value} = 1000(1 - e^{-2}) \approx \$864.66$$
$$\text{Future value} = 1000e^2(1 - e^{-2}) \approx \$6389.06$$

you can see that

$$\text{Future value} = (\text{Present value})e^2.$$

The reason for this is that the stream of payments is equivalent to a single payment of $864.66 at time $t = 0$. With an interest rate of 10%, in 20 years that single payment will have grown to a future value of

$$864.66e^{0.1(20)} = 864.66e^2 \approx \$6389.02.$$

To within rounding error, this is the future value calculated previously.

Supply and Demand Curves

In a free market, the quantity of a certain item produced and sold can be described by the supply and demand curves of the item. The *supply curve* shows what quantity of the item the producers will supply at different price levels. It is usually assumed that as the price increases, the quantity supplied will increase. The consumers' behavior is reflected in the *demand curve*, which shows what quantity of goods are bought at various prices. An increase in price is usually assumed to cause a decrease in the quantity purchased. See Figure 8.36.

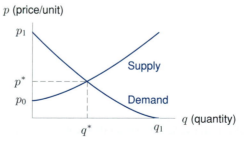

Figure 8.36: Supply and demand curves

It is assumed that the market will settle to the *equilibrium price* and *quantity*, p^* and q^*, where the graphs cross. This means that at the equilibrium point, a quantity q^* of an item will be produced and sold for a price of p^* each.

Consumer and Producer Surplus

Notice that at equilibrium, a number of consumers have bought the item at a lower price than they would have been willing to pay. (For example, there are some consumers who would have been willing to pay prices up to p_1.) Similarly, there are some suppliers who would have been willing to produce the item at a lower price (down to p_0, in fact). We define the following terms:

> - The **consumer surplus** measures the consumers' gain from trade. It is the total amount gained by consumers by buying the item at the current price rather than at the price they would have been willing to pay.
> - The **producer surplus** measures the suppliers' gain from trade. It is the total amount gained by producers by selling at the current price, rather than at the price they would have been willing to accept.
> In the absence of price controls, the current price is assumed to be the equilibrium price.

Both consumers and producers are richer for having traded. The consumer and producer surplus measure how much richer they are.

Suppose that all consumers buy the good at the maximum price they are willing to pay. Divide the interval from 0 to q^* into subintervals of length Δq. Figure 8.37 shows that a quantity Δq of items are sold at a price of about p_1, another Δq are sold for a slightly lower price of about p_2, the next Δq for a price of about p_3, and so on. Thus, the consumers' total expenditure is about

$$p_1 \Delta q + p_2 \Delta q + p_3 \Delta q + \cdots = \sum p_i \Delta q.$$

If D is the demand function given by $p = D(q)$, and if all consumers who were willing to pay more than p^* paid as much as they were willing, then as $\Delta q \to 0$, we would have

$$\begin{array}{c} \text{Consumer} \\ \text{expenditure} \end{array} = \int_0^{q^*} D(q)dq = \begin{array}{c} \text{Area under demand} \\ \text{curve from 0 to } q^*. \end{array}$$

Now if all goods are sold at the equilibrium price, the consumers' actual expenditure is p^*q^*, the area of the rectangle between the axes and the lines $q = q^*$ and $p = p^*$. Thus the consumer surplus may be calculated as follows:

$$\boxed{\begin{array}{c} \text{Consumer} \\ \text{surplus} \end{array} = \left(\int_0^{q^*} D(q)dq \right) - p^*q^* = \begin{array}{c} \text{Area under demand} \\ \text{curve above } p = p^*. \end{array}}$$

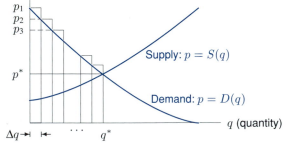

Figure 8.37: Calculation of consumer surplus   **Figure 8.38:** Consumer and producer surplus

See Figure 8.38. Similarly, if the supply curve is given by the function $p = S(q)$, we define the producer surplus as follows:

$$\boxed{\begin{array}{c} \text{Producer} \\ \text{Surplus} \end{array} = p^*q^* - \left(\int_0^{q^*} S(q)dq \right) = \begin{array}{c} \text{Area between supply} \\ \text{curve and line } p = p^*. \end{array}}$$

Problems for Section 8.4

1. Draw a graph, with time in years on the horizontal axis, of what an income stream might look like for a company that sells sunscreen in the northeast United States.

2. A business associate who owes you $3000 offers to pay you $2800 now, or else pay you three yearly installments of $1000 each, with the first installment paid now. If you use only financial reasons to make your decision, which option should you choose? Justify your answer, assuming a 6% market interest rate, compounded continuously.

3. Big Tree McGee is negotiating his rookie contract with a professional basketball team. They have agreed to a three-year deal which will pay Big Tree a fixed amount at the end of each of the three years, plus a signing bonus at the beginning of his first year. They are still haggling about the amounts and Big Tree must decide between a big signing bonus and fixed payments per year, or a smaller bonus with payments increasing each year. The two options are summarized in the table. All values are payments in millions of dollars.

	Signing bonus	Year 1	Year 2	Year 3
Option #1	6.0	2.0	2.0	2.0
Option #2	1.0	2.0	4.0	6.0

   (a) Big Tree decides to invest all income in stock funds which he expects to grow at a rate of 10% per year, compounded continuously. He would like to choose the contract option which gives him the greater future value at the end of the three years when the last payment is made. Which option should he choose? Justify your answer.

   (b) The local newspaper is running a story about Big Tree's contract. They wish to report the present value of each contract offer. Calculate these present values.

4. Find the present and future values of a constant income stream of $3000 per year over a period of 15 years, assuming a 6% annual interest rate compounded continuously.

5. (a) A bank account earns 10% interest compounded continuously. At what (constant, continuous) rate must a parent deposit money into such an account in order to save $100,000 in 10 years for a child's college expenses?

   (b) If the parent decides instead to deposit a lump sum now in order to attain the goal of $100,000 in 10 years, how much must be deposited now?

6. (a) If you deposit money continuously at a constant rate of $1000 per year into a bank account that earns 5% interest, how many years will it take for the balance to reach $10,000?

   (b) How many years would it take if the account had $2000 in it initially?

7. Suppose sales of Version 6.0 of a computer software package start out high and decrease exponentially. At time t, in years, the sales are $s(t) = 50e^{-t}$ thousands of dollars per year. After two years, Version 7.0 of the software is released and replaces Version 6.0. Assume that all income from software sales is immediately invested in government bonds which pay interest at a 6% rate compounded continuously, calculate the total value of sales of Version 6.0 over the two year period.

8. The value of good wine increases with age. Thus, if you are a wine dealer, you have the problem of deciding whether to sell your wine now, at a price of P a bottle, or to sell it later at a higher price. Suppose you know that the amount a wine-drinker is willing to pay for a bottle of this wine t years from now is $P(1+20\sqrt{t})$. Assuming continuous compounding and a prevailing interest rate of 5% per year, when is the best time to sell your wine?

9. An oil company discovered an oil reserve of 100 million barrels. Suppose for time $t > 0$, in years, the company's extraction plan is a linear declining function of time as follows:

$$q(t) = a - bt,$$

   where $q(t)$ is the rate of extraction of oil in millions of barrels per year at time t and $b = 0.1$ and $a = 10$.

   (a) How long does it take to exhaust the entire reserve?

   (b) If the oil price is a constant $20 per barrel, the extraction cost per barrel is a constant $10, and the market interest rate is 10%, what is the present value of the company's profit?

10. In 1980 West Germany made a loan of 20 billion Deutsche Marks to the Soviet Union, to be used for the construction of a natural gas pipeline connecting Siberia to Western Russia, and continuing to West Germany (Urengoi–Uschgorod–Berlin). Assume that the deal was as follows: In 1985, upon completion of the pipeline, the Soviet Union would deliver natural gas to West Germany, at a constant rate, for all future times. Assuming a constant price of natural gas of 0.10 Deutsche Mark per cubic meter, and assuming West Germany expects 10% annual interest on its investment (compounded continuously), at what rate does the Soviet Union have to deliver the gas, in billions of cubic meters per year? Keep in mind that delivery of gas could not begin until the pipeline was completed. Thus, West Germany received no return on its investment until after five years had passed. (Note: A more complex deal of this type was actually made between the two countries.)

11. In May 1991 *Car and Driver* described a Jaguar that sells for $980,000. At that price only 50 have been sold. It is estimated that 350 could have been sold if the price had been $560,000. Assuming that the demand curve is a straight line, and that $560,000 and 350 are the equilibrium price and quantity, find the consumer surplus.

12. Using Riemann sums, explain the economic significance of $\int_0^{q^*} S(q)\,dq$ to the producers.

13. Using Riemann sums, give an interpretation of producer surplus,

$$\int_0^{q^*} (p^* - S(q))\,dq$$

analogous to the interpretation of consumer surplus.

14. In Figure 8.38, page 402, find the regions with the following areas:

(a) $p^* q^*$ (b) $\int_0^{q^*} D(q)\,dq$ (c) $\int_0^{q^*} S(q)\,dq$

(d) $\left(\int_0^{q^*} D(q)\,dq\right) - p^* q^*$ (e) $p^* q^* - \int_0^{q^*} S(q)\,dq$ (f) $\int_0^{q^*} (D(q) - S(q))\,dq$

15. The dairy industry is an example of cartel pricing: the government has set milk prices artificially high. On a supply and demand graph, label p^+, a price above the equilibrium price. Using the graph, describe the effect of forcing the price up to p^+ on:

(a) The consumer surplus.   (b) The producer surplus.

(c) The total gains from trade (Consumer surplus + Producer surplus).

16. Rent controls on apartments are an example of price controls on a commodity. They keep the price of the good artificially low (below the equilibrium price). Sketch a graph of supply and demand curves, and label on it a price p^- below the equilibrium price. What effect does forcing the price down to p^- have on:

(a) The producer surplus?   (b) The consumer surplus?

(c) The total gains from trade (Consumer surplus + Producer surplus)?

CHAPTER SUMMARY

- **Geometry**
  Volume, arc length
- **Density**
  Finding total quantity from density, center of mass
- **Physics**
  Work, escape velocity, force and pressure
- **Economics**
  Present and future value of income stream, consumer and producer surplus

REVIEW PROBLEMS FOR CHAPTER EIGHT

1. (a) Sketch the solid obtained by rotating the region bounded by $y = \sqrt{x}$, $x = 1$, and $y = 0$ around the line $y = 0$.
    (b) Approximate its volume by Riemann sums, showing the volume represented by each term in your sum on the sketch.
    (c) Now find the volume of this solid using an integral.

2. Using the region of Problem 1, find the volume when it is rotated around
    (a) The line $y = 1$.   (b) The y-axis.

3. The reflector behind a car headlight is made in the shape of the parabola, $x = \frac{4}{9}y^2$, with a circular cross-section, as shown in Figure 8.39.

    (a) Find a Riemann sum approximating the volume contained by this headlight.
    (b) Find the volume exactly.

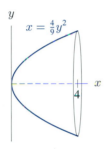

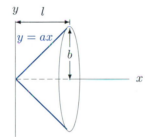

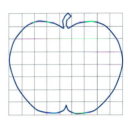

| Figure 8.39 | Figure 8.40 | Figure 8.41 |

4. In this problem, you will derive the formula for the volume of a right circular cone with height l and base radius b by rotating the line $y = ax$ from $x = 0$ to $x = l$ around the x-axis. See Figure 8.40.

    (a) What value should you choose for a such that the cone will have height l and base radius b?
    (b) Given this value of a, find the volume of the cone.

5. Figure 8.41 shows a cross section through an apple. (Scale: One division = 1/2 inch.)

    (a) Give a rough estimate for the volume of this apple (in cubic inches).
    (b) The density of these apples is about 0.03 lb/in³ (a little less than the density of water—as you might expect, since apples float). Estimate how much this apple would cost. (They go for 80 cents a pound.)

6. If the circle $x^2 + y^2 = 1$ is rotated about the line $y = 3$, it forms a torus (a doughnut-shaped figure). Find the volume of this torus.

For the curves described in Problems 7–8, write the integral that gives the exact length of the curve; you need not evaluate the integral.

7. One arch of the sine curve, from $x = 0$ to $x = \pi$.

8. The ellipse with equation $(x^2/a^2) + (y^2/b^2) = 1$.

9. A nuclear power plant produces strontium-90 at a rate of 1 kg/yr. How much of the strontium produced since 1971 (when the plant opened) was still around in 1992? (The half–life of strontium-90 is 28 years.)

10. A 200-lb weight is to be hoisted by a rope from the ground to a point on a flat roof 20 ft above the ground. If the rope weighs 2 lb/ft, find the work done in lifting the weight.

11. Water is to be raised from a well 40 ft deep by means of a bucket attached to a rope. When the bucket is full of water it weighs 30 lb, but the bucket has a leak that causes it to lose water at the constant rate of 1/4 lb for each foot that the bucket is raised. Neglecting the weight of the rope, find the work done in raising the bucket to the top.

12. A water tank is in the shape of a right circular cone with height 18 ft and radius 12 ft at the top. If it is filled with water to a depth of 15 ft, find the work done in pumping all of the water over the top of the tank. (The density of water is $\rho = 62.4$ lb/ft^3.)

13. The dam in Hannawa Falls, NY, on the Raquette River is approximately 60 feet across and 25 feet high. Find the water force on the dam. (The density of water is $\rho = 62.4$ lb/ft^3.)

14. (a) Find the present and future values of a constant income stream of $100 per year over a period of 20 years, assuming a 10% annual interest rate compounded continuously.
   (b) How many years will it take for the balance to reach $5000?

15. Suppose you are manufacturing a particular item, and that the rate at which you earn a profit on the item is decreasing with time according to the formula

$$\text{Rate profit earned} = (2 - 0.1t) \text{ thousand dollars per year}$$

after t years. (A negative profit represents a loss.) If interest is 10%, compounded continuously:

   (a) Write a Riemann sum approximating the present value of the total profit earned up to a time T years in the future.
   (b) Write an integral representing the present value in part (a). (You need not evaluate this integral.)
   (c) For what T is the present value of the stream of profits on this item maximized? What is the present value of the total profit earned up to that time?

16. A car moving at a speed of v mph achieves $25 + 0.1v$ mpg (miles per gallon) for v between 20 and 60 mph. Suppose your speed as a function of time, t, in hours, is given by

$$v = 50\frac{t}{t+1} \text{ mph.}$$

How many gallons do you consume between $t = 2$ and $t = 3$?

17. Mt. Shasta is a cone-like volcano whose radius at an elevation of h feet above sea level is approximately $(3.5 \cdot 10^5)/\sqrt{h + 600}$ feet. Its bottom is 400 feet above sea level, and its top is 14,400 feet above sea level. See Figure 8.42.

   (a) Give a Riemann sum approximating the volume of Mt. Shasta.
   (b) Find the volume in cubic feet. (Note: This data about Mt. Shasta is more or less accurate. Mt. Shasta lies in northern California, and for some time was thought to be the highest point in the US outside Alaska.)

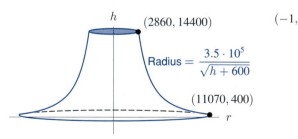

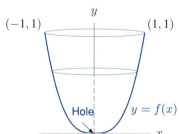

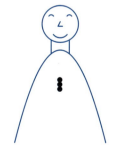

Figure 8.42   **Figure 8.43**   **Figure 8.44**

18. Figure 8.43 shows an ancient Greek water clock called a clepsydra, which is designed so that the depth of the water decreases at a constant rate as the water runs out a hole in the bottom. This design allows the hours to be marked by a uniform scale. The tank of the clepsydra is a volume of revolution about a vertical axis. According to Torricelli's law, the exit speed of the water flowing through the hole is proportional to the square root of the depth of the water. Use this to find the formula $y = f(x)$ for this profile, assuming that $f(1) = 1$.

19. Find the volume of the snowman in Figure 8.44. If x and y are in meters, and the origin is on the ground, the x-axis is horizontal and the y-axis is vertical, then the body is approximated by rotating the curve $y = 1 - 4x^2$ about the y-axis. The neck is a cylinder of radius 0.1 meter and length 0.15 meter; the head is spherical with radius 0.2 meter.

20. A blood vessel is cylindrical with radius R and length l. The blood near the boundary moves slowly; blood at the center moves the fastest. Assume that the velocity, v, of the blood at a distance r from the center of the artery is given by

$$v = \frac{P}{4\eta l}(R^2 - r^2)$$

where P is the pressure difference between the ends of the blood vessel and η is the viscosity of blood.

    (a)   Find the rate at which the blood is flowing down the blood vessel. (Give your answer as a volume per unit time.)

    (b)   Show that your result agrees with Poiseuille's Law which says that the rate at which blood is flowing down the blood vessel is proportional to the radius of the blood vessel to the fourth power.

21. Consider a bowl made by rotating the curve $y = ax^2$ around the y-axis (a is a constant).

    (a)   Assume the bowl is filled with water to depth h. What is the volume of water in the bowl? (Your answer will contain a and h.)

    (b)   What is the area of the surface of the water if the bowl is filled to depth h? (Your answer will contain a and h.)

    (c)   Suppose water is evaporating from the surface of the bowl at a rate proportional to the surface area, with proportionality constant k. Find a differential equation satisfied by h as a function of time, t. (That is, find an equation for dh/dt.)

    (d)   If the water starts at depth h_0, find the time taken for all the water to evaporate.

22. A cylindrical centrifuge of radius 1 m and height 2 m is filled with water to a depth of 1 meter (see Figure 8.45(I)). As the centrifuge accelerates, the water level rises along the wall and drops in the center; the cross-section will be a parabola. (See Figure 8.45(II).)

    (a)   Find the equation of the parabola in Figure 8.45(II) in terms of h, the depth of the water at its lowest point.

    (b)   As the centrifuge rotates faster and faster, either water will be spilled out the top, as in Figure 8.45(III), or the bottom of the centrifuge will be exposed, as in Figure 8.45(IV). Which happens first?

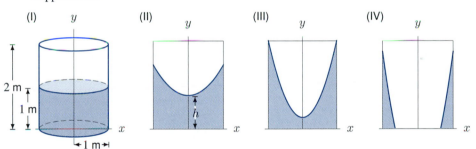

Figure 8.45

Problems 23–24 concern Figures 8.46 and 8.47. In each problem, you are given two objects which have the same mass M, the same radius R, and the same angular velocity about the indicated axes (say, one revolution per minute). For each problem, determine which object has the greater kinetic energy. (The kinetic energy of a particle of mass m with speed v is $\frac{1}{2}mv^2$.) Don't attempt to compute the kinetic energy of the objects to do this; just use reasoning.

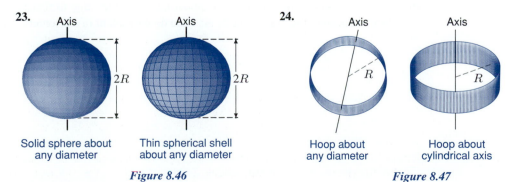

23. Solid sphere about any diameter · Thin spherical shell about any diameter

Figure 8.46

24. Hoop about any diameter · Hoop about cylindrical axis

Figure 8.47

PROJECTS

1. **Volume Enclosed by Two Cylinders**

    Two cylinders are inscribed in a cube of side length 2, as shown in Figure 8.48. What is the volume of the solid that the two cylinders enclose? [Hint: Use horizontal slices.] Note: The solution was known to Archimedes. The Chinese mathematician Liu Hui (third century A.D.) tried to find this volume, but he failed; he wrote a poem about his efforts calling the enclosed volume a "box-lid":

    > Look inside the cube
    > And outside the box-lid;
    > Though the dimension increases,
    > It doesn't quite fit.
    > The marriage preparations are complete;
    > But square and circle wrangle,
    > Thick and thin are treacherous plots,
    > They are incompatible.
    > I wish to give my humble reflections,
    > But fear that I will miss the correct principle;
    > I dare to let the doubtful points stand,
    > Waiting
    > For one who can expound them.

    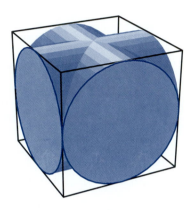

    Figure 8.48

2. **Length of a Hanging Cable**

    The distance between the towers of the main span of the Golden Gate Bridge is about 1280 m; the sag of the cable halfway between the towers on a cold winter day is about 143 m. See Figure 8.49.

    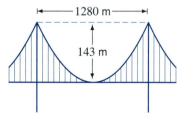

    Figure 8.49

    (a) How long is the cable, assuming it has an approximately parabolic shape? (Represent the cable as a parabola of the form $y = kx^2$ and determine k to at least 1 decimal place.)
    (b) On a hot summer day the cable is about 0.05% longer, due to thermal expansion. By how much does the sag increase? Assume no movement of the towers.

FOCUS ON MODELING

DISTRIBUTION FUNCTIONS

Understanding the distribution of various quantities through the population can be important to decision makers. For example, the income distribution gives useful information about the economic structure of a society. In this section we will look at the distribution of ages in the US. To allocate funding for education, health care, and social security, the government needs to know how many people are in each age group. We will see how to represent such information by a density function.

US Age Distribution

TABLE 8.1 *Distribution of ages in the US in 1995*

Age group	Percentage of total population
0–20	29%
20–40	31%
40–60	24%
60–80	13%
80–100	3%

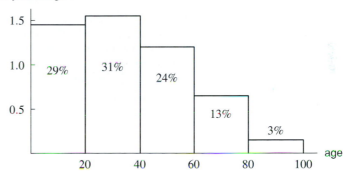

Figure 8.50: How ages were distributed in the US in 1995

Suppose we have the data in Table 8.1 showing how the ages of the US population were distributed in 1995. To represent this information graphically we use a type[1] of *histogram*, putting a vertical bar above each age group in such a way that the *area* of each bar represents the percentage in that age group. The total area of all the rectangles is 100% = 1. We only consider people who are less than 100 years old.[2] For the 0–20 age group, the base of the rectangle is 20, and we want the area to be 29%, so the height must be 29%/20 = 1.45%. We treat ages as though they were continuously distributed. The category 0–20, for example, contains people who are just one day short of their twentieth birthday. Notice that the vertical axis is measured in percent/year. (See Figure 8.50.)

Example 1   In 1995, estimate what percentage of the US population was:

    (a)   Between 20 and 60 years old?   (b)   Less than 10 years old?

    (c)   Between 75 and 80 years old?   (d)   Between 80 and 85 years old?

Solution   (a)   We add the percentages, so 31% + 24% = 55%.

        (b)   To find the percentage less than 10 years old, we could assume, for example, that the population was distributed evenly over the 0–20 group. (This means we are assuming that babies were born at a fairly constant rate over the last 20 years, which is probably reasonable.) If we make this assumption, then we can say that the population less than 10 years old was about half that in the 0–20 group, that is, 14.5%. Notice that we get the same result by computing the area of the rectangle from 0 to 10. (See Figure 8.51.)

---

[1]There are other types of histogram which have frequency on the vertical axis.

[2]In fact, 0.02% of the population is over 100, but this is too small to be visible on the histogram.

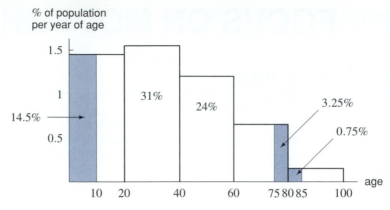

Figure 8.51: Ages in the US in 1995 — various subgroups (for Example 1)

(c)   To find the population between 75 and 80 years old, since 13% of Americans in 1990 were in the 60-80 group, we might apply the same reasoning and say that $\frac{1}{4}(13\%) = 3.25\%$ of the population was in this age group. This result is represented as an area in Figure 8.51. The assumption that the population was evenly distributed is not a good one here; certainly there were more people between the ages of 60 and 65 than between 75 and 80. Thus, the estimate of 3.25% is certainly too high.

(d)   Again using the (faulty) assumption that ages in each group were distributed uniformly, we would find that the percentage between 80 and 85 was $\frac{1}{4}(3\%) = 0.75\%$. (See Figure 8.51.) This estimate is also poor—there were certainly more people in the 80–85 group than, say, the 95–100 group, and so the 0.75% estimate is too low.

Smoothing Out the Histogram

We could get better estimates if we had smaller age groups (each age group in Figure 8.50 is 20 years, which is quite large) or if the histogram were smoother. Suppose we have the more detailed data in Table 8.2, which leads to the new histogram in Figure 8.52. As we get more detailed information, the upper silhouette of the histogram becomes smoother, but the area of any of the bars still represents the percentage of the population in that age group. Imagine, in the limit, replacing the upper silhouette of the histogram by a smooth curve in such a way that area under the curve above one age group is the same as the area in the corresponding rectangle. The total area under the whole curve is again 100% = 1. (See Figure 8.52.)

The Age Density Function

If t is age in years, we define $p(t)$, the age *density function*, to be a function which "smooths out" the age histogram. This function has the property that

$$\begin{matrix} \text{Fraction of population} \\ \text{between ages } a \text{ and } b \end{matrix} = \begin{matrix} \text{Area under} \\ \text{graph of } p \\ \text{between } a \text{ and } b \end{matrix} = \int_a^b p(t)dt.$$

If a and b are the smallest and largest possible ages (say, $a = 0$ and $b = 100$), so that the ages of all of the population are between a and b, then

$$\int_a^b p(t)dt = \int_0^{100} p(t)dt = 1.$$

What does the age density function p tell us? Notice that we have not talked about the meaning of $p(t)$ itself, but *only* of the integral $\int_a^b p(t)\,dt$. Let's look at this in a bit more detail. Suppose, for

TABLE 8.2 *Ages in the US in 1995 (more detailed)*

Age group	Percentage of total population
0–10	15%
10–20	14%
20–30	14%
30–40	17%
40–50	14%
50–60	10%
60–70	8%
70–80	5%
80–90	2%
90–100	1%

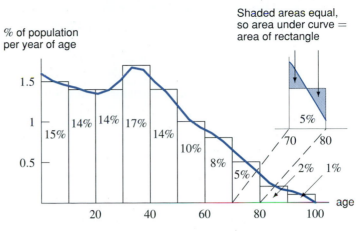

Figure 8.52: Smoothing out the age histogram

example, that $p(10) = 0.015 = 1.5\%$ per year. This is *not* telling us that 1.5% of the population is precisely 10 years old (where 10 years old means exactly 10, not $10\frac{1}{2}$, not $10\frac{1}{4}$, not 10.1). However, $p(10) = 0.015$ does tell us that for some small interval Δt around 10, the fraction of the population with ages in this interval is approximately $p(10)\,\Delta t = 0.015\,\Delta t$. Notice also that the units of $p(t)$ are *% per year*, so $p(t)$ must be multiplied by years to give a percentage of the population.

The Density Function

Suppose we are interested in how a certain characteristic, x, is distributed through a population. For example, x might be height or age if the population is people, or might be wattage for a population of light bulbs. Then we define a general density function with the following properties:

> The function, $p(x)$, is a **density function** if
>
> $$\begin{array}{ccc}
> \text{Fraction of population} & & \text{Area under} \\
> \text{for which } x \text{ is} & = & \text{graph of } p \quad = \int_a^b p(x)\,dx. \\
> \text{between } a \text{ and } b & & \text{between } a \text{ and } b
> \end{array}$$
>
> $$\int_{-\infty}^{\infty} p(x)\,dx = 1 \quad \text{and} \quad p(x) \geq 0 \quad \text{for all } x.$$

The density function must be nonnegative if its integral always gives a fraction of the population. Also, the fraction of the population with x between $-\infty$ and ∞ is 1 because the entire population has the characteristic x between $-\infty$ and ∞. The function p that was used to smooth out the age histogram satisfies this definition of a density function. We do not assign a meaning to the value $p(x)$ directly, but rather interpret $p(x)\,\Delta x$ as the fraction of the population with the characteristic in a short interval of length Δx around x.

The density function is often approximated by formulas, as in the next example.

Example 2   Find formulas to approximate the density function, p, for the US age distribution. Assume that p is continuous, that it is constant at 1.5% up to age 40 and then drops linearly.

Solution   We have $p(t) = 0.015$ for $0 \leq t < 40$. For $t \geq 40$, we need a linear function sloping downward. Because p is continuous, we have $p(40) = 1.5\% = 0.015$. Because p is a density function we have $\int_0^{100} p(t)dt = 1$. Suppose b is as in Figure 8.53 then

$$\int_0^{100} p(t)dt = \int_0^{40} p(t)dt + \int_{40}^{100} p(t)dt = 40(0.015) + \frac{1}{2}(0.015)b = 1,$$

where $\int_{40}^{100} p(t)dt$ is given by the area of the triangle. This gives

$$\frac{0.015}{2}b = 0.4, \quad \text{and so} \quad b \approx 53.3.$$

Thus the slope of the line is $-0.015/53.3 \approx -0.00028$, so for $40 \leq t \leq 40 + 53.3 = 93.3$, so

$$p(t) - 0.015 = -0.00028(t - 40),$$
$$p(t) = 0.0262 - 0.00028t.$$

According to this way of smoothing the data, there is no one over 93.3 years old, so $p(t) = 0$ for $t > 93.3$.

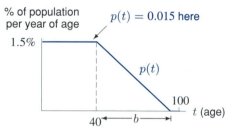

Figure 8.53: Age density function

Cumulative Distribution Function for Ages

Another way of showing how ages are distributed in the US is by using the *cumulative distribution function* $P(t)$, defined by

$$P(t) = \frac{\text{Fraction of population}}{\text{of age less than } t} = \int_0^t p(x)dx.$$

Thus, P is the antiderivative of p with $P(0) = 0$, and $P(t)$ gives the area under the density curve between 0 and t.

Notice that the cumulative distribution function is nonnegative and increasing (or at least non-decreasing), since the number of people younger than age t increases as t increases. Another way of seeing this is to notice that $P' = p$, and p is positive (or nonnegative). Thus the cumulative age distribution is a function which starts with $P(0) = 0$ and increases as t increases. $P(t) = 0$ for $t < 0$ because, when $t < 0$, there is no one whose age is less than t. The limiting value of P, as $t \to \infty$, is 1 since as t becomes very large (100 say), everyone will be younger than age t, so the fraction of people with age less than t tends towards 1. (See Figure 8.54.) Notice that for t less than 40, the graph of P is a straight line, because p is constant there. For $t > 40$, the graph of P levels off as p tends to 0.

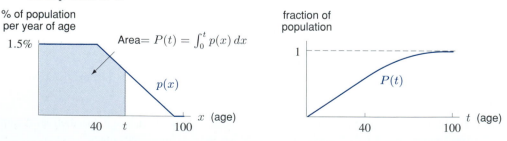

Figure 8.54: $P(t)$, the cumulative age distribution function, and its relation to $p(x)$, the age density function

Cumulative Distribution Function

A **cumulative distribution function**, $P(t)$, of a density function p, is defined by

$$P(t) = \int_{-\infty}^{t} p(x)\,dx = \begin{array}{l}\text{Fraction of population having}\\ \text{values of } x \text{ below } t.\end{array}$$

Thus, P is an antiderivative of p, that is, $P' = p$.
Any cumulative distribution has the following properties:
- P is increasing (or nondecreasing).
- $\displaystyle\lim_{t\to\infty} P(t) = 1$ and $\displaystyle\lim_{t\to-\infty} P(t) = 0.$

- $\begin{array}{l}\text{Fraction of population}\\ \text{having values of } x\\ \text{between } a \text{ and } b\end{array} = \int_{a}^{b} p(x)\,dx = P(b) - P(a).$

Problems on Distribution Functions

In Problems 1–3, sketch graphs of a density function and a cumulative distribution function which could represent the distribution of income through a population with the given characteristics.

1. A large middle class.

2. Small middle and upper classes and many poor people.

3. Small middle class, many poor and many rich people.

4. A large number of people take a standardized test, receiving scores described by the density function p graphed in Figure 8.55. Does the density function imply that most people receive a score near 50? Explain why or why not.

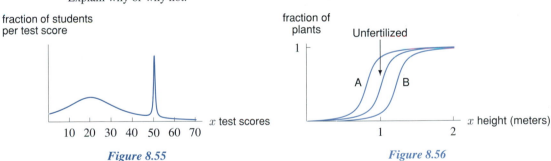

Figure 8.55   Figure 8.56

5. An experiment is done to determine the effect of two new fertilizers A and B on the growth of a species of peas. The cumulative distribution functions of the heights of the mature peas without treatment and treated with each of A and B are graphed in Figure 8.56.

    (a) About what height are most of the unfertilized plants?
    (b) Explain in words the effect of the fertilizers A and B on the mature height of the plants.

6. Suppose $F(x)$ is the cumulative distribution function for heights (in meters) of trees in a forest.

    (a) Explain in terms of trees the meaning of the statement $F(7) = 0.6$.
    (b) Which is greater, $F(6)$ or $F(7)$? Justify your answer in terms of trees.

7. Suppose that $p(x)$ is the density function for heights of American men in inches. What is the meaning of the statement $p(68) = 0.2$?

8. Suppose $P(t)$ is the fraction of the US population of age less than t. Using Table 8.2 on page 411, make a table of values for $P(t)$.

9. Figure 8.57 shows a density function and the corresponding cumulative distribution function.[3]

    (a) Which curve represents the density distribution functions and which represents the cumulative distribution? Give a reason for your choice.
    (b) Put reasonable values on the tick marks on each of the axes.

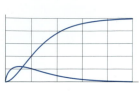

Figure 8.57

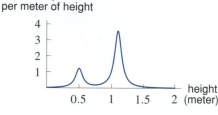

fraction of plants per meter of height

Figure 8.58

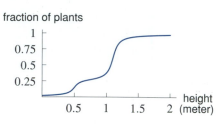

fraction of plants

Figure 8.59

10. The heights of grass plants in a meadow were measured and the density function and cumulative distribution function are graphed in Figure 8.58 and Figure 8.59, respectively.

    (a) You know that there are two species of grass in the meadow, a short grass and a tall grass. Explain how the graph of the density function reflects this fact.
    (b) Explain how the graph of the cumulative distribution functions reflects the fact that there are two species of grass in the meadow.
    (c) About what percentage of the grasses in the meadow belong to the short grass species?

11. After measuring the duration of many telephone calls, the telephone company found their data was well-approximated by the density function $p(x) = 0.4e^{-0.4x}$, where x is the duration of a call, in minutes.

    (a) What percentage of calls last between 1 and 2 minutes?
    (b) What percentage of calls last 1 minute or less?
    (c) What percentage of calls last 3 minutes or more?
    (d) Find the cumulative distribution function.

12. Consider a population of individuals with a disease. Suppose that t is the number of years since the onset of the disease. The death density function, $f(t) = cte^{-kt}$, approximates the fraction of the sick individuals who die in the time interval $[t, t + \Delta t]$ as follows:

$$\text{Fraction who die} \approx f(t)\Delta t = cte^{-kt}\Delta t$$

    where c and k are positive constants whose values depend on the particular disease.

    (a) Find the value of c in terms of k.
    (b) If 40% of the population dies within 5 years, find c and k.
    (c) Find the cumulative death distribution function, $C(t)$. Give your answer in terms of k.

13. Suppose that the time to conduct a routine maintenance check has a cumulative distribution function, $P(t)$, where $P(t)$ gives the fraction of maintenance checks completed in less than or equal to t minutes. Values of $P(t)$ are in Table 8.3.

    TABLE 8.3

t (minutes)	0	5	10	15	20	25	30
$P(t)$ (fraction completed)	0	0.03	0.08	0.21	0.38	0.80	0.98

    (a) What fraction of maintenance checks is completed in 15 minutes or less?
    (b) What fraction of maintenance checks takes longer than 30 minutes?
    (c) What fraction takes between 10 and 15 minutes?
    (d) Draw a histogram showing how times for routine maintenance checks are distributed.
    (e) In which of the given 5-minute intervals is the length of a maintenance check most likely to fall?
    (f) Give a rough sketch of the density function.
    (g) Sketch a graph of the cumulative distribution function.

---

[3] Adapted from *Calculus*, by David A. Smith and Lawerence C. Moore (Lexington: D.C. Heath, 1994).

14. Students at the University of California were surveyed and asked their grade point average . (The GPA ranges from 0 to 4, where 2 is just passing.) The distribution of GPAs is shown in Figure 8.60.[4]

 (a) Roughly what fraction of students are passing?
 (b) Roughly what fraction of the students have honor grades (GPAs above 3)?
 (c) Why do you think there is a peak around 2?
 (d) Sketch a graph of the cumulative distribution function.

fraction of students
per GPA

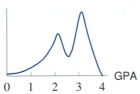

fraction of earth's surface
per mile of elevation

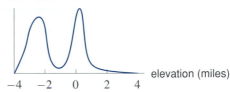

Figure 8.60   **Figure 8.61**

15. Figure 8.61[4] shows the distribution of elevation, in miles, across the earth's surface. Positive elevation denotes land above sea level; negative elevation shows land below sea level (i.e., the ocean floor).

 (a) Describe in words the elevation of most of the earth's surface.
 (b) Approximately what fraction of the earth's surface is below sea level?

16. A person who travels regularly on the 9:00 am bus from Oakland to San Francisco reports that the bus is almost always a few minutes late but rarely more than five minutes late. We are told further that the bus is never more than two minutes early, although it is on very rare occasions a little early.

 (a) Sketch a possible density function, $p(t)$, where t is the number of minutes that the bus is late. Shade the region under the graph between $t = 2$ minutes and $t = 4$ minutes. Explain what this region represents.
 (b) Now sketch the cumulative distribution function $P(t)$ for this situation. What measurement(s) on this graph correspond to the area shaded? What do the inflection point(s) on your graph of P correspond to on the graph of p? How can you interpret the inflection points on the graph of P without referring to the graph of p?

17. Suppose the density function for radii r (mm) of spherical raindrops during a storm is constant over the range $0 < r < 5$ and zero elsewhere.

 (a) Find the density function $f(r)$ for the radii.
 (b) Find the cumulative distribution function $F(r)$ for the radii.
 (c) Find the cumulative distribution function $G(v)$ for the volumes v (mm^3) of the raindrops.
 (d) Find the density function $g(v)$ for the volumes.

18. Consider a pendulum swinging through a small angle. The x-coordinate of the bob moves between $-a$ and a, as shown in Figure 8.62.

 (a) Draw the density function for the location of the x-coordinate of the pendulum bob (i.e., neglect up-and-down motion). In order to do this, imagine a camera taking pictures of the pendulum at random instants. Where is the bob most likely to be found? Least likely? [Hint: Consider the speed of the pendulum at different points in its path. Is the camera more likely to take a photograph of the bob at a point on its path where the bob is moving quickly, or where it is moving slowly?]

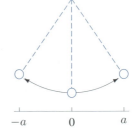

Figure 8.62

 (b) Now sketch the function

$$f(x) = \begin{cases} \dfrac{1}{\pi\sqrt{a^2 - x^2}} & -a < x < a; \\ 0 & |x| \geq a. \end{cases}$$

How does this graph compare with the one you drew in part (a)?

[4]Adapted from *Statistics*, by Freedman, Pisani, Purves, and Adikhari (New York: Norton, 1991).

(c)   Assuming that the function given in part (b) is the density function for the pendulum, what do you expect

$$\int_{-a}^{a} \frac{1}{\pi\sqrt{a^2 - x^2}} \, dx$$

to be? Check this by computing the integral.

(d)   Does it seem reasonable, physically speaking, that $f(x)$ "blows up" at a and $-a$? Explain.

PROBABILITY AND MORE ON DISTRIBUTIONS

Probability

Suppose we pick a member of the US population at random and ask what is the probability that the person is between, say, the ages of 70 and 80. We saw in Table 8.2 on page 411 that 5% of the population is in this age group. We say that the probability, or chance, that the person is between 70 and 80 is 0.05. Using any age density function $p(t)$, we can define probabilities as follows:

$$
\begin{array}{ccc}
\text{Probability that} \\
\text{a person is between} & = & \text{Fraction of population} \\
& & \text{between ages } a \text{ and } b
\end{array}
\ = \ \int_{a}^{b} p(t) \, dt.
$$
$$\text{ages } a \text{ and } b$$

Since the cumulative distribution function gives the fraction of the population younger than age t, the cumulative distribution can also be used to calculate the probability that a randomly selected person is in a given age group.

$$
\begin{array}{ccc}
\text{Probability that} \\
\text{a person is younger} & = & \text{Fraction of population} \\
& & \text{younger than age } t
\end{array}
\ = P(t) = \int_{0}^{t} p(x) \, dx.
$$
$$\text{than age } t$$

In the next example, both a density function and a cumulative distribution function are used to describe the same situation.

Example 3   Suppose you want to analyze the fishing industry in a small town. Each day, the boats bring back at least 2 tons of fish, and never more than 8 tons.

(a)   Using the density function describing the daily catch in Figure 8.63, find and graph the corresponding cumulative distribution function and explain its meaning.

(b)   What is the probability that the catch will be between 5 and 7 tons?

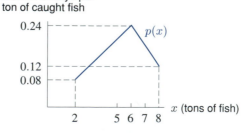

Figure 8.63: Density function of daily catch

Solution   (a)   The cumulative distribution function $P(t)$ is equal to the fraction of days on which the catch is less than t tons of fish. Since the catch is never less than 2 tons, we have $P(t) = 0$ for $t \leq 2$. Since the catch is always less than 8 tons, we have $P(t) = 1$ for $t \geq 8$. For t in the range $2 < t < 8$, we must evaluate the integral

$$P(t) = \int_{-\infty}^{t} p(x)dx = \int_{2}^{t} p(x)dx.$$

This integral equals the area under the graph of $p(x)$ between $x = 2$ and $x = t$. It can be calculated by noting that $p(x)$ is given by the formula

$$p(x) = \begin{cases} 0.04x & \text{for } 2 \leq x \leq 6 \\ -0.06x + 0.6 & \text{for } 6 < x \leq 8 \end{cases}$$

and $p(x) = 0$ for $x < 2$ or $x > 8$. Thus, for $2 \leq t \leq 6$,

$$P(t) = \int_{2}^{t} 0.04x \, dx = 0.04 \frac{x^2}{2} \bigg|_{2}^{t} = 0.02t^2 - 0.08.$$

And for $6 \leq t \leq 8$,

$$P(t) = \int_{2}^{t} p(x) \, dx = \int_{2}^{6} p(x) \, dx + \int_{6}^{t} p(x) \, dx$$

$$= 0.64 + \int_{6}^{t} (-0.06x + 0.6) \, dx = 0.64 + \left(-0.06 \frac{x^2}{2} + 0.6x \right) \bigg|_{6}^{t}$$

$$= -0.03t^2 + 0.6t - 1.88.$$

Thus

$$P(t) = \begin{cases} 0.02t^2 - 0.08 & \text{for } 2 \leq t \leq 6 \\ -0.03t^2 + 0.6t - 1.88 & \text{for } 6 < t \leq 8. \end{cases}$$

In addition $P(t) = 0$ for $t < 2$ and $P(t) = 1$ for $8 < t$. (See Figure 8.64.)

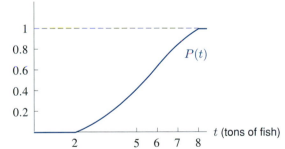

Figure 8.64: Cumulative distribution of daily catch

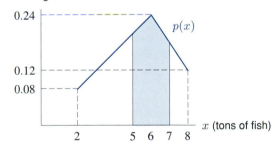

Figure 8.65: Shaded area represents the probability the catch is between 5 and 7 tons

(b)   The probability that the catch is between 5 and 7 tons can be found using either the density function, p, or the cumulative distribution function, P. If we use the density function, this probability can be represented by the shaded area in Figure 8.65, which is about 0.43.

$$\text{Probability catch is between 5 and 7 tons} = \int_{5}^{7} p(x) \, dx = 0.43.$$

The probability can be found from the cumulative distribution as follows:

$$\text{Probability catch is between 5 and 7 tons} = P(7) - P(5) = 0.85 - 0.42 = 0.43.$$

The Median and Mean

It is often useful to be able to give an "average" value for a distribution. Two measures that are in common use are the *median* and the *mean*.

The Median

> A **median** of a quantity x distributed through a population is a value T such that half the population has values of x less than (or equal to) T, and half the population has values of x greater than (or equal to) T. Thus, a median T satisfies
>
> $$\int_{-\infty}^{T} p(x)\,dx = 0.5,$$
>
> where p is the density function. In other words, half the area under the graph of p lies to the left of T.

Example 4   Find the median age in the US in 1995, using the age density function given by

$$p(t) = \begin{cases} 0.015 & \text{for } 0 \le t \le 40 \\ 0.0262 - 0.00028t & \text{for } 40 < t \le 93.3. \end{cases}$$

Solution   We want to find the value of T such that

$$\int_{-\infty}^{T} p(t)\,dt = \int_{0}^{T} p(t)\,dt = 0.5.$$

Since $p(t) = 1.5\%$ up to age 40, we have

$$\text{Median} = T = \frac{50\%}{1.5\%} \approx 33 \text{ years.}$$

(See Figure 8.66.)

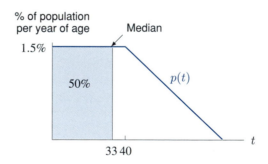

Figure 8.66: Median of age distribution

The Mean

Another commonly used average value is the *mean*. To find the mean of N numbers, you add the numbers and divide the sum by N. For example, the mean of the numbers 1, 2, 7, and 10 is $(1 + 2 + 7 + 10)/4 = 5$. The mean age of the entire US population is therefore defined as

$$\text{Mean age} = \frac{\sum \text{Ages of all people in the US}}{\text{Total number of people in the US}}.$$

Calculating the sum of all the ages directly would be an enormous task; we will approximate the sum by an integral. The idea is to "slice up" the age axis and consider the people whose age is between t and $t + \Delta t$. How many are there?

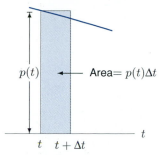

Figure 8.67: Shaded area is
percentage of population with age
between t and $t + \Delta t$

The fraction of the population between t and $t + \Delta t$ is the area under the graph of p between these points, which is well approximated by the area of the rectangle, $p(t)\Delta t$. (See Figure 8.67.) If the total number of people in the population is N, then

$$\text{Number of people with age between } t \text{ and } t + \Delta t \approx p(t)\Delta t N.$$

The age of all of these people is approximately t:

$$\text{Sum of ages of people between age } t \text{ and } t + \Delta t \approx tp(t)\Delta t N.$$

Therefore, adding and factoring out an N gives us

$$\text{Sum of ages of all people} \approx \left(\sum tp(t)\Delta t\right) N.$$

In the limit, as we allow Δt to shrink to 0, the sum becomes an integral, so

$$\text{Sum of ages of all people} = \left(\int_0^{100} tp(t)dt\right) N.$$

Therefore, with N equal to the total number of people in the US, and assuming no person is over 100 years old,

$$\text{Mean age} = \frac{\text{Sum of ages of all people in US}}{N} = \int_0^{100} tp(t)dt.$$

We can give the same argument for any[5] density function $p(x)$.

If a quantity has density function $p(x)$,

$$\textbf{Mean value} \text{ of the quantity} = \int_{-\infty}^{\infty} xp(x)\,dx.$$

It can be shown that the mean is the point on the horizontal axis where the region under the graph of the density function, if it were made out of cardboard, would balance.

[5]Provided all the relevant improper integrals converge.

Example 5   Find the mean age of the US population, using the density function of Example 4.

Solution   The formula for p is

$$p(t) = \begin{cases} 0.015 & \text{for } 0 \le t \le 40 \\ 0.0262 - 0.00028t & \text{for } 40 < t \le 93.3. \end{cases}$$

Using these formulas, we compute

$$\text{Mean age} = \int_0^{100} tp(t)dt = \int_0^{40} t(0.015)dt + \int_{40}^{93.3} t(0.0262 - 0.00028t)dt$$

$$= 0.015\frac{t^2}{2}\Big|_0^{40} + 0.0262\frac{t^2}{2}\Big|_{40}^{93.3} - 0.00028\frac{t^3}{3}\Big|_{40}^{93.3} \approx 35 \text{ years.}$$

The mean is shown is Figure 8.68.

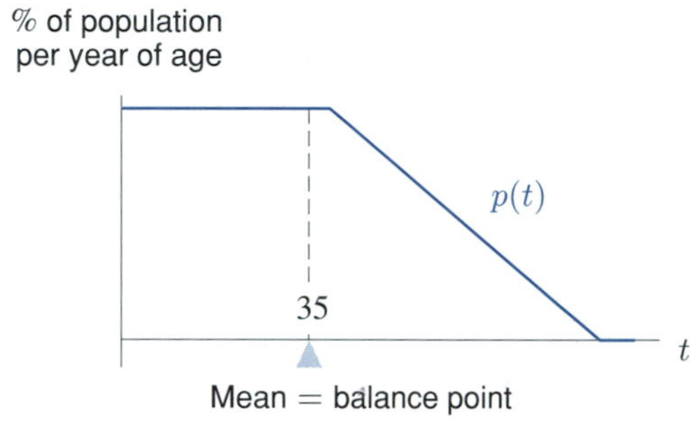

Figure 8.68: Mean of age distribution

Normal Distributions

How much rain do you expect will fall in your home town this year? If you live in Anchorage, Alaska, the answer is something close to 15 inches (including the snow). Of course, you don't expect exactly 15 inches. Some years there will be more than 15 inches, and some years there will be less. Most years, however, the amount of rainfall will be close to 15 inches; only rarely will it be well above or well below 15 inches. What does the density function for the rainfall look like? To answer this question, we look at rainfall data over many years. Records show that the distribution of rainfall is well-approximated by a *normal distribution*.Its graph is a bell-shaped curve which peaks at 15 inches and slopes downward approximately symmetrically on either side.

Normal distributions are frequently used to model real phenomena, from grades on an exam to the number of airline passengers on a particular flight. A normal distribution is characterized by its *mean*, μ, and its *standard deviation*, σ. The mean tells us the location of the central peak. The standard deviation tells us how closely the data is clustered around the mean. A small value of σ tells us that the data is close to the mean; a large σ tells us the data is spread out. The formula for a normal distribution is as follows. The factor of $1/(\sigma\sqrt{2\pi})$ makes the area under the graph equal to 1.

A **normal distribution** has a density function of the form

$$p(x) = \frac{1}{\sigma\sqrt{2\pi}}e^{-(x-\mu)^2/(2\sigma^2)},$$

where μ is the mean of the distribution and σ is the standard deviation, with $\sigma > 0$.

To model the rainfall in Anchorage, we use a normal distribution with $\mu = 15$. The standard deviation can be estimated by looking at the data; we will take it to be 1. (See Figure 8.69.)

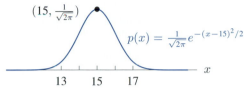

$(15, \frac{1}{\sqrt{2\pi}})$

$p(x) = \frac{1}{\sqrt{2\pi}}e^{-(x-15)^2/2}$

Figure 8.69: Normal distribution with $\mu = 15$ and $\sigma = 1$

Example 6   For Anchorage's rainfall, use the normal distribution with the density function with $\mu = 15$ and $\sigma = 1$ to compute the fraction of the years with rainfall between
(a)   14 and 16 inches,   (b)   13 and 17 inches,   (c)   12 and 18 inches.

Solution   (a)   The fraction of the years with annual rainfall between 14 and 16 inches is $\int_{14}^{16} \frac{1}{\sqrt{2\pi}}e^{-(x-15)^2/2}\,dx$.

Since there is no elementary antiderivative for $e^{-(x-15)^2/2}$, we find the integral numerically. Its value is about 0.68.

$$\begin{array}{c} \text{Fraction of years with rainfall} \\ \text{between 14 and 16 inches} \end{array} = \int_{14}^{16} \frac{1}{\sqrt{2\pi}}e^{-(x-15)^2/2}\,dx \approx 0.68.$$

(b)   Finding the integral numerically again:

$$\begin{array}{c} \text{Fraction of years with rainfall} \\ \text{between 13 and 17 inches} \end{array} = \int_{13}^{17} \frac{1}{\sqrt{2\pi}}e^{-(x-15)^2/2}\,dx \approx 0.95.$$

(c)

$$\begin{array}{c} \text{Fraction of years with rainfall} \\ \text{between 12 and 18 inches} \end{array} = \int_{12}^{18} \frac{1}{\sqrt{2\pi}}e^{-(x-15)^2/2}\,dx \approx 0.997.$$

Since 0.95 is so close to 1, we expect that most of the time the rainfall will be between 13 and 17 inches a year.

Among the normal distributions, the one having $\mu = 0$, $\sigma = 1$ is called the *standard normal distribution*. Values of the corresponding cumulative distribution function are published in tables.

Problems on Probability and Distributions

1.   Consider the fishing data given in Example 3 on page 416. Show that the area under the density function in Figure 8.63 is 1. Why is this to be expected?

2.   Find the mean daily catch for the fishing data in Figure 8.63, page 416.

3.   The probability of a transistor failing between $t = a$ months and $t = b$ months is given by $c\int_{a}^{b} e^{-ct}\,dt$ for some constant c.

(a)   If the probability of failure within the first six months is 10%, what is c?
(b)   Given the value of c in part (a), what is the probability the transistor will fail within the second six months?

4. Suppose that x measures the time (in hours) it takes for a student to complete an exam. Assume that all students are done within two hours and the density function for x is given by

$$p(x) = \begin{cases} x^3/4 & \text{if } 0 < x < 2 \\ 0 & \text{otherwise.} \end{cases}$$

    (a) What proportion of students take between 1.5 and 2.0 hours to finish the exam?
    (b) What is the mean time for students to complete the exam?
    (c) Compute the median of this distribution.

5. In 1950 an experiment was done observing the time gaps between successive cars on the Arroyo Seco Freeway.[6] The data show that the density function of these time gaps was given approximately by

$$p(x) = ae^{-0.122x}$$

    where x is the time in seconds and a is a constant.

    (a) Find a.
    (b) Find P, the cumulative distribution function.
    (c) Find the median and mean time gap.
    (d) Sketch rough graphs of p and P.

6. The distribution of IQ scores is often modeled by the normal distribution with mean 100 and standard deviation 15.

    (a) Write a formula for the density distribution of IQ scores.
    (b) Estimate the fraction of the population with IQ between 115 and 120.

7. For a normal population of mean 0, show that the fraction of the population within one standard deviation of the mean does not depend on the standard deviation.
    [Hint: Use the substitution $w = x/\sigma$.]

8. (a) Using a calculator or computer, sketch graphs of the density function of the normal distribution

$$p(x) = \frac{1}{\sigma\sqrt{2\pi}}e^{-(x-\mu)^2/(2\sigma^2)}.$$

    (i) For fixed μ (say, $\mu = 5$) and varying σ (say, $\sigma = 1, 2, 3$).
    (ii) For varying μ (say, $\mu = 4, 5, 6$) and fixed σ (say, $\sigma = 1$).

    (b) Explain how the graphs confirm that μ is the mean of the distribution and that σ is a measure of how closely the data is clustered around the mean.

9. Consider the normal distribution, $p(x)$.

    (a) Show that $p(x)$ is a maximum when $x = \mu$. What is that maximum value?
    (b) Show that $p(x)$ has points of inflection where $x = \mu + \sigma$ and $x = \mu - \sigma$.
    (c) Describe in your own words what μ and σ tell you about the distribution.

10. Consider a group of people who have received treatment for a disease such as cancer. Let t be the *survival time*, the number of years a person lives after receiving treatment. The density function giving the distribution of t is $p(t) = Ce^{-Ct}$ for some positive constant C.

    (a) What is the practical meaning for the cumulative distribution function $P(t) = \int_0^t p(x)\,dx$?
    (b) The survival function, $S(t)$, is the probability that a randomly selected person survives for at least t years. Find $S(t)$.
    (c) Suppose a patient has a 70% probability of surviving at least two years. Find C.

11. While taking a walk along the road where you live, you accidentally drop your glove. You don't know where you dropped it. Suppose the probability density $p(x)$ for having dropped the glove x kilometers from home (along the road) is

$$p(x) = 2e^{-2x} \quad \text{for } x \geq 0.$$

    (a) What is the probability that you dropped it within 1 kilometer of home?
    (b) At what distance y from home is the probability that you dropped it within y km of home equal to 0.95?

---

[6]Reported by Daniel Furlough and Frank Barnes.

12. Which of the following functions makes the most sense as a model for the probability density representing the time (in minutes, starting from $t = 0$) that the next customer walks into a store?

(a) $p(t) = \begin{cases} \cos t & 0 \le t \le 2\pi \\ e^{t-2\pi} & t \ge 2\pi \end{cases}$

(b) $p(t) = 3e^{-3t}$ for $t \ge 0$

(c) $p(t) = e^{-3t}$ for $t \ge 0$

(d) $p(t) = 1/4$ for $0 \le t \le 4$

13. Let $P(x)$ be the cumulative distribution function for the income distribution in the US in 1973 (income is measured in thousands of dollars). Some values of $P(x)$ are in the following table:

Income x (thousands)	1	4.4	7.8	12.6	20	50
$P(x)$ (%)	1	10	25	50	75	99

(a) What fraction of the population made between \$20,000 and \$50,000?

(b) What was the median income?

(c) Sketch a density function for this distribution. Where, approximately, does your density function have a maximum? What is the significance of this point, in terms of income distribution? How can you recognize this point on the graph of the density function and on the graph of the cumulative distribution?

14. Let v be the speed, in meters/second, of an oxygen molecule, and let $p(v)$ be the density function of the speed distribution of oxygen molecules at room temperature. Maxwell showed that

$$p(v) = av^2 e^{-mv^2/(2kT)},$$

where $k = 1.4 \times 10^{-23}$ is the Boltzmann constant, T is the temperature in Kelvin (at room temperature, $T = 293$), and $m = 5 \times 10^{-26}$ is the mass of the oxygen molecule in kilograms.

(a) Find the value of a.

(b) Estimate the median and the mean speed. Find the maximum of $p(v)$.

(c) How do your answers in part (b) for the mean and the maximum of $p(v)$ change as T changes?

15. If we think of an electron as a particle, the function

$$P(r) = 1 - (2r^2 + 2r + 1)e^{-2r}$$

is the cumulative distribution function of the distance, r, of the electron in a hydrogen atom from the center of the atom. The distance is measured in Bohr radii. (1 Bohr radius $= 5.29 \times 10^{-11}$ m. Niels Bohr (1885–1962) was a Danish physicist.)

For example, $P(1) = 1 - 5e^{-2} \approx 0.32$ means that the electron is within 1 Bohr radius from the center of the atom 32% of the time.

(a) Find a formula for the density function of this distribution. Sketch the density function and the cumulative distribution function.

(b) Find the median distance and the mean distance. Near what value of r is an electron most likely to be found?

(c) The Bohr radius is sometimes called the "radius of the hydrogen atom." Why?

CHAPTER NINE

APPROXIMATIONS
AND SERIES

In this chapter we will see how to approximate functions by simpler functions. Taylor approximations use polynomials, which may be considered the simplest functions. They are easy to use because they can be evaluated by hand, unlike the transcendental functions, such as e^x and $\ln x$.

Fourier approximations use sines and cosines, the simplest periodic functions, instead of polynomials. Taylor approximations are generally good approximations to the function locally (that is, near a specific point), whereas Fourier approximations are generally good approximations over an interval.

9.1 TAYLOR POLYNOMIALS AND SERIES

Taylor Polynomial Approximations

We approximate a function $f(x)$ near the point $x = a$ by polynomials. To get more accuracy, we take higher-degree polynomials. The approximation is usually good near the point $x = a$, but often not so good farther away from that point.

Linear Approximations

We have already seen how to approximate a function using a degree 1 polynomial, namely the tangent line approximation:

$$f(x) \approx f(a) + f'(a)(x - a).$$

The tangent line is the best linear approximation to the function near $x = a$. The tangent line and the curve share the same slope at $x = a$. (See Figure 9.1.) If $a = 0$, this approximation is called a *Taylor polynomial of degree 1*.

Taylor Polynomial of Degree 1 Approximating $f(x)$ for x near 0

$$f(x) \approx P_1(x) = f(0) + f'(0)x$$

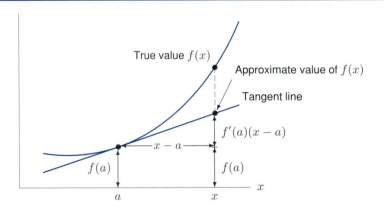

Figure 9.1: Tangent line approximation of $f(x)$ for x near a

---

Example 1 Approximate $f(x) = \cos x$, with x in radians, by its tangent line at $x = 0$.

Solution The tangent line at $x = 0$ is just the horizontal line $y = 1$, as shown in Figure 9.2, so

$$f(x) = \cos x \approx 1, \quad \text{for } x \text{ near } 0.$$

If we take $x = 0.05$, then

$$f(0.05) = \cos(0.05) = 0.999\ldots,$$

which is quite close to the approximation $\cos x \approx 1$. Similarly, if $x = -0.1$, then

$$f(-0.1) = \cos(-0.1) = 0.995\ldots$$

is close to the approximation $\cos x \approx 1$. However, if $x = 0.4$, then

$$f(0.4) = \cos(0.4) = 0.921\ldots,$$

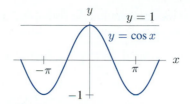

Figure 9.2: Graph of $\cos x$ and its tangent line at $x = 0$

so the approximation $\cos x \approx 1$ is less accurate. The graph suggests that the farther a point x is away from 0, the worse the approximation is likely to be.

Quadratic Approximations

Suppose we want a more accurate approximation to $f(x) = \cos x$ for x near 0. Instead of using a line, we use a quadratic function which not only has the same slope, but bends in the same way as the original curve. We require that at $x = 0$ the graphs of the original function f and the quadratic function have the same slope, $f'(0)$, and that they bend at the same rate—in other words, that they have the same second derivative, $f''(0)$.

Example 2   Find the quadratic approximation to $f(x) = \cos x$ for x near 0.

Solution   We take a quadratic polynomial

$$P_2(x) = C_0 + C_1 x + C_2 x^2,$$

and determine the values for C_0, C_1, and C_2. At $x = 0$, we want the two functions and their first and second derivatives to agree; that is, we want $P_2(0) = f(0)$, $P_2'(0) = f'(0)$, and $P_2''(0) = f''(0)$. Since

$$P_2(x) = C_0 + C_1 x + C_2 x^2 \qquad \text{and} \qquad f(x) = \cos x$$
$$P_2'(x) = C_1 + 2C_2 x \qquad\qquad\qquad\qquad f'(x) = -\sin x$$
$$P_2''(x) = 2C_2 \qquad\qquad\qquad\qquad\qquad f''(x) = -\cos x,$$

we have

$$C_0 = P_2(0) = f(0) = \cos 0 = 1 \qquad \text{so} \qquad C_0 = 1$$
$$C_1 = P_2'(0) = f'(0) = -\sin 0 = 0 \qquad\qquad\quad C_1 = 0$$
$$2C_2 = P_2''(0) = f''(0) = -\cos 0 = -1, \qquad\quad C_2 = -\tfrac{1}{2}.$$

Consequently, the quadratic approximation is

$$\cos x \approx P_2(x) = 1 + 0 \cdot x - \frac{1}{2}x^2 = 1 - \frac{x^2}{2}, \qquad \text{for } x \text{ near } 0.$$

From Figure 9.3, we see that the quadratic approximation $\cos x \approx P_2(x)$ is better than the linear approximation $\cos x \approx P_1(x)$ for x near 0. Let's compare the accuracy of the two approximations numerically. At $x = 0.4$, $\cos(0.4) = 0.921\ldots$ and $P_2(0.4) = 0.920$, so the quadratic approximation is a significant improvement over the linear approximation. (The magnitude of the error is about 0.001 instead of 0.08.) In addition, the quadratic approximation is much better than the linear approximation for x-values near 0: $\cos(0.1) = 0.995004\ldots$ and $P_2(0.1) = 0.995$.

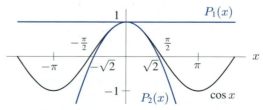

Figure 9.3: Graph of $\cos x$ and its linear, $P_1(x)$, and quadratic, $P_2(x)$, approximations for x near 0

We generalize the computations in Example 2 and define:

> ### Taylor Polynomial of Degree 2 Approximating $f(x)$ for x near 0
>
> $$f(x) \approx P_2(x) = f(0) + f'(0)x + \frac{f''(0)}{2}x^2$$

Higher-Degree Polynomials

As a rule, over a small interval around $x = 0$, the quadratic approximation to a function is a better approximation than the linear (tangent line) approximation. However, Figure 9.3 shows that even though we have matched up the function and the quadratic in terms of their values, their slopes, and their concavity at the point $x = 0$, the quadratic can still bend away for large x. We can fix this somewhat by using an approximating polynomial of higher degree. Suppose that we want to approximate a function $f(x)$ for x near 0 by a polynomial of degree n:

$$f(x) \approx P_n(x) = C_0 + C_1 x + C_2 x^2 + \cdots + C_{n-1}x^{n-1} + C_n x^n.$$

We need to find the values of the constants: C_0, C_1, C_2, ..., C_n. To do this, we require that the function $f(x)$ and each of its first n derivatives agree with those of the polynomial $P_n(x)$ at the point $x = 0$. Notice that the higher-order derivatives of a function contribute more subtle information about its graph than the first two derivatives do. (For instance, the third derivative measures how fast the concavity changes.) The more derivatives there are that agree at $x = 0$, the longer the function and the polynomial are likely to remain close to each other.

To see how to find the constants, let's take as an example

$$f(x) \approx P_3(x) = C_0 + C_1 x + C_2 x^2 + C_3 x^3.$$

Substituting $x = 0$ gives

$$f(0) = P_3(0) = C_0.$$

Differentiating $P_3(x)$ yields

$$P_3'(x) = C_1 + 2C_2 x + 3C_3 x^2,$$

so substituting $x = 0$ shows that

$$f'(0) = P_3'(0) = C_1.$$

Differentiating and substituting again, we get

$$P_3''(x) = 2 \cdot 1 C_2 + 3 \cdot 2 \cdot 1 C_3 x,$$

which gives

$$f''(0) = P_3''(0) = 2C_2,$$

so that

$$C_2 = \frac{f''(0)}{2}.$$

The third derivative, denoted by P_3''', is

$$P_3'''(x) = 3 \cdot 2 \cdot 1 C_3,$$

so

$$f'''(0) = P_3'''(0) = 3 \cdot 2 \cdot 1 C_3,$$

and then

$$C_3 = \frac{f'''(0)}{3 \cdot 2 \cdot 1}.$$

You can imagine a similar calculation starting with $P_4(x)$, using the fourth derivative $f^{(4)}$, which would give

$$C_4 = \frac{f^{(4)}(0)}{4 \cdot 3 \cdot 2 \cdot 1},$$

and so on. Using factorial notation,[1] we write these expressions as

$$C_3 = \frac{f'''(0)}{3!}, \quad C_4 = \frac{f^{(4)}(0)}{4!}.$$

In general, for any positive integer n,

$$C_n = \frac{f^{(n)}(0)}{n!},$$

where $f^{(n)}$ means the nth derivative of f. So we define:

Taylor Polynomial of Degree n Approximating $f(x)$ for x near 0

$$f(x) \approx P_n(x)$$

$$= f(0) + f'(0)x + \frac{f''(0)}{2!}x^2 + \frac{f'''(0)}{3!}x^3 + \frac{f^{(4)}(0)}{4!}x^4 + \cdots + \frac{f^{(n)}(0)}{n!}x^n$$

We call this a Taylor polynomial centered at $x = 0$, or a Taylor polynomial about $x = 0$.

Example 3 Construct the Taylor polynomial of degree 7 approximating the function $f(x) = \sin x$ for x near 0. Compare the value of the Taylor approximation with the true value of f at $x = \pi/3$.

Solution We have

$$
\begin{aligned}
f(x) &= \sin x & \text{giving} \quad f(0) &= 0 \\
f'(x) &= \cos x & f'(0) &= 1 \\
f''(x) &= -\sin x & f''(0) &= 0 \\
f'''(x) &= -\cos x & f'''(0) &= -1 \\
f^{(4)}(x) &= \sin x & f^{(4)}(0) &= 0 \\
f^{(5)}(x) &= \cos x & f^{(5)}(0) &= 1 \\
f^{(6)}(x) &= -\sin x & f^{(6)}(0) &= 0 \\
f^{(7)}(x) &= -\cos x, & f^{(7)}(0) &= -1.
\end{aligned}
$$

Using these values, we see that the Taylor polynomial approximation of degree 7 is

$$\sin x \approx P_7(x) = 0 + x + 0 \cdot \frac{x^2}{2!} - \frac{x^3}{3!} + 0 \cdot \frac{x^4}{4!} + \frac{x^5}{5!} + 0 \cdot \frac{x^6}{6!} - \frac{x^7}{7!}$$

$$= x - \frac{x^3}{3!} + \frac{x^5}{5!} - \frac{x^7}{7!}, \quad \text{for } x \text{ near } 0.$$

In Figure 9.4 we show the graphs of the sine function and the approximating polynomial of degree 7 for x near 0. They are indistinguishable where x is close to 0. However, as we look at values of x further away from 0 in either direction, the two graphs diverge. To check the accuracy of this approximation numerically, we see how well it approximates $\sin(\pi/3) = \sqrt{3}/2 = 0.8660254\ldots$.

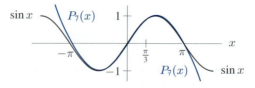

Figure 9.4: Graph of $\sin x$ and its seventh degree Taylor polynomial, $P_7(x)$, for x near 0

---

[1] We define 3! (read *three factorial*) to be equal to $3 \cdot 2 \cdot 1 = 6$. Similarly, $4! = 4 \cdot 3 \cdot 2 \cdot 1 = 24$ and $k! = k(k-1)\cdots 2 \cdot 1$. In addition, $0! = 1$.

When we substitute $\pi/3 = 1.0471976\ldots$ into the polynomial approximation, we obtain $P_7(\pi/3) = 0.8660213\ldots$, which is extremely accurate—to about four parts in a million.

Example 4   Graph the Taylor polynomial of degree 8 approximating $f(x) = \cos x$ for x near 0.

Solution   We can find the coefficients of the Taylor polynomial by the method of the preceding example. We have

$$\cos x \approx P_8(x) = 1 - \frac{x^2}{2!} + \frac{x^4}{4!} - \frac{x^6}{6!} + \frac{x^8}{8!}.$$

Figure 9.5 shows that $P_8(x)$ is close to the cosine function for a larger interval of x-values than the quadratic approximation $P_2(x) = 1 - x^2/2$ that we found in Example 2 on page 427.

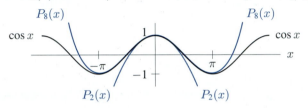

Figure 9.5: $P_8(x)$ approximates $\cos x$ better than $P_2(x)$ for x near 0

Example 5   Construct the Taylor polynomial of degree 10 about $x = 0$ for the function $f(x) = e^x$.

Solution   We have $f(0) = 1$. Since the derivative of e^x is equal to e^x, all the higher-order derivatives will be equal to e^x. Consequently, for any $k = 1, 2, \ldots, 10$, $f^{(k)}(x) = e^x$ and $f^{(k)}(0) = e^0 = 1$. Therefore, the Taylor polynomial approximation of degree 10 is given by

$$e^x \approx P_{10}(x) = 1 + x + \frac{x^2}{2!} + \frac{x^3}{3!} + \frac{x^4}{4!} + \cdots + \frac{x^{10}}{10!}, \quad \text{for } x \text{ near } 0.$$

To check the accuracy of this approximation, suppose we use it to approximate $e = e^1$ which is 2.718281828 correct to nine decimal places. If we substitute $x = 1$ into the approximating polynomial, we obtain $P_{10}(1) = 2.718281801$. Thus, this tenth-degree polynomial yields the first seven decimal places for e. For large values of x, however, the accuracy must diminish because e^x grows much faster than any polynomial as $x \to \infty$. Figure 9.6 shows graphs of $f(x) = e^x$ and the Taylor polynomials of degree $n = 0, 1, 2, 3, 4$. Notice that each successive approximation remains close to the exponential curve for a larger interval of x-values.

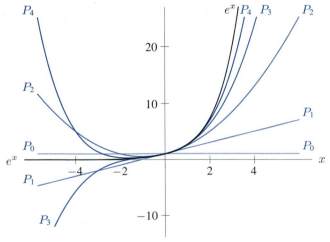

Figure 9.6: For x near 0, e^x is more closely approximated by higher-degree Taylor polynomials

Example 6   Construct the Taylor polynomial of degree n approximating $f(x) = \dfrac{1}{1-x}$ for x near 0.

Solution   You can check, by differentiating, that $f(0) = 1$, $f'(0) = 1$, $f''(0) = 2$, $f'''(0) = 3!$, $f^{(4)}(0) = 4!$, and so on. This means

$$\frac{1}{1-x} \approx P_n(x) = 1 + x + x^2 + x^3 + x^4 + x^5 + \cdots + x^n, \quad \text{for } x \text{ near } 0.$$

This is an example of a geometric series which we will study in Section 9.4.

Taylor Polynomials Around $x = a$

Suppose we want to approximate $f(x) = \ln x$ by a Taylor polynomial. This function has no Taylor polynomial about $x = 0$ because the function is not defined for $x = 0$ or for $x < 0$. Can we find another Taylor polynomial for $f(x) = \ln x$?

Rather than constructing a polynomial approximation for $f(x)$ about $x = 0$, let's construct a polynomial centered about some other point, $x = a$. First, let's look at the equation of the tangent line at $x = a$. Recall that since the tangent line goes through the point $(a, f(a))$ and has slope $f'(a)$, its equation is

$$y = f(a) + f'(a)(x - a).$$

This gives the approximation

$$f(x) \approx f(a) + f'(a)(x - a) \quad \text{for } x \text{ near } a.$$

The $f'(a)(x - a)$ term is a correction term which approximates how much $f(x)$ moves away from $f(a)$ as x moves away from a.

The approximating polynomial $P_n(x)$ centered at $x = a$ will be set up as $f(a)$ plus correction terms which depend on the derivatives of $f(x)$ and which are zero for $x = a$. This is achieved by writing the polynomial in powers of $(x - a)$ instead of powers of x:

$$f(x) \approx P_n(x) = C_0 + C_1(x - a) + C_2(x - a)^2 + \cdots + C_n(x - a)^n.$$

If we require the derivatives of the approximating polynomial $P_n(x)$ and the original function $f(x)$ to agree at $x = a$, we get the following result.

> ### Taylor Polynomial of Degree n Approximating $f(x)$ for x near a
>
> $$f(x) \approx P_n(x)$$
>
> $$= f(a) + f'(a)(x - a) + \frac{f''(a)}{2!}(x - a)^2 + \cdots\cdots + \frac{f^{(n)}(a)}{n!}(x - a)^n$$
>
> We call this a Taylor polynomial centered at $x = a$, or a Taylor polynomial about $x = a$.

You can derive the formula for these coefficients in the same way that we did for $a = 0$. (See Problem 33, page 435.)

Example 7   Construct the Taylor polynomial of degree 4 approximating the function $f(x) = \ln x$ for x near 1.

Solution   We have

$$
\begin{aligned}
f(x) &= \ln x & \text{so} \quad && f(1) &= \ln(1) = 0 \\
f'(x) &= 1/x & && f'(1) &= 1 \\
f''(x) &= -1/x^2 & && f''(1) &= -1 \\
f'''(x) &= 2/x^3 & && f'''(1) &= 2 \\
f^{(4)}(x) &= -6/x^4, & && f^{(4)}(1) &= -6.
\end{aligned}
$$

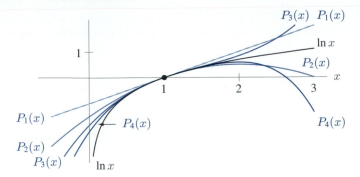

Figure 9.7: Taylor polynomials approximate $\ln x$ closely for x near 1, but not necessarily farther away

The Taylor polynomial is then

$$\ln x \approx P_4(x) = 0 + (x - 1) - \frac{(x - 1)^2}{2!} + 2\frac{(x - 1)^3}{3!} - 6\frac{(x - 1)^4}{4!}$$

$$= (x - 1) - \frac{(x - 1)^2}{2} + \frac{(x - 1)^3}{3} - \frac{(x - 1)^4}{4}, \quad \text{for } x \text{ near } 1.$$

We show $\ln x$ and several Taylor polynomials in Figure 9.7. Notice that $P_4(x)$ stays reasonably close to $\ln x$ for x near 1, but bends away as x gets farther from 1. Also, note that the Taylor polynomials are defined for $x \leq 0$, but $\ln x$ is not.

The examples in this section suggest that the following results are usually true:
- Taylor polynomials are a good approximation for x near a. Farther away, they may or may not be good.
- The higher the degree of the Taylor polynomial, the larger the interval over which it fits the function closely.

Taylor Series Approximations

We have just seen how to approximate a function near a point by Taylor polynomials. Now we define a Taylor series, which can be thought of as a Taylor polynomial that goes on forever. In the next section we will see at which points such a series serves as a good approximation to the function.

We have the following Taylor polynomials centered at $x = 0$ for $\cos x$:

$$\cos x \approx P_0(x) = 1$$

$$\cos x \approx P_2(x) = 1 - \frac{x^2}{2!}$$

$$\cos x \approx P_4(x) = 1 - \frac{x^2}{2!} + \frac{x^4}{4!}$$

$$\cos x \approx P_6(x) = 1 - \frac{x^2}{2!} + \frac{x^4}{4!} - \frac{x^6}{6!}$$

$$\cos x \approx P_8(x) = 1 - \frac{x^2}{2!} + \frac{x^4}{4!} - \frac{x^6}{6!} + \frac{x^8}{8!}.$$

Here we have a sequence of polynomials, $P_0(x)$, $P_2(x)$, $P_4(x)$, $P_6(x)$, $P_8(x)$, ..., each of which is a better approximation to $\cos x$ than the last, at least for x near 0. Notice that when we go to a higher-degree polynomial (say from P_6 to P_8), we add more terms ($x^8/8!$, for example), but the terms of lower degree don't change. Thus each polynomial includes the information from all the previous ones. We will write the *Taylor series* for $\cos x$:

$$T(x) = 1 - \frac{x^2}{2!} + \frac{x^4}{4!} - \frac{x^6}{6!} + \frac{x^8}{8!} - \cdots$$

to represent the whole sequence of Taylor polynomials.

What Does $\cdots$ **Mean?** The three dots indicate that the terms in the series continue further, possibly forever. This is the same idea as in $\pi = 3.1415\ldots$ where the three dots mean that the digits go on forever after the 5. For example,

$$1 - \frac{x^2}{2!} + \frac{x^4}{4!} \qquad \text{is a finite polynomial of degree 4,}$$

whereas

$$1 - \frac{x^2}{2!} + \frac{x^4}{4!} - \cdots \qquad \text{is an infinite series, and continues forever.}$$

Taylor Series for $\sin x$, $\cos x$, e^x

We can define the Taylor series for $\sin x$ and e^x in a similar way to the Taylor series for $\cos x$. In the next section we discuss what it means for a series to converge. It turns out that, for all x, we can write the following:

$$\sin x = x - \frac{x^3}{3!} + \frac{x^5}{5!} - \frac{x^7}{7!} + \frac{x^9}{9!} - \cdots$$

$$\cos x = 1 - \frac{x^2}{2!} + \frac{x^4}{4!} - \frac{x^6}{6!} + \frac{x^8}{8!} - \cdots$$

$$e^x = 1 + x + \frac{x^2}{2!} + \frac{x^3}{3!} + \frac{x^4}{4!} + \cdots$$

These series are also called *Taylor expansions* of the functions $\sin x$, $\cos x$, and e^x about $x = 0$.

Taylor Series in General

Any function f, all of whose derivatives exist at 0, has a Taylor series. However, for many functions, f, the series does not converge to $f(x)$ for all values of x. Assuming that the series does converge to $f(x)$, we have the following general formula.

Taylor Series for $f(x)$ about $x = 0$

$$f(x) = f(0) + f'(0)x + \frac{f''(0)}{2!}x^2 + \frac{f'''(0)}{3!}x^3 + \cdots + \frac{f^{(n)}(0)}{n!}x^n + \cdots$$

In addition, just as we have Taylor polynomials centered at points other than 0, we can also have a Taylor series centered at $x = a$ (provided all the derivatives exist at $x = a$). Assuming the series converges to $f(x)$, we have the following result.

Taylor Series for $f(x)$ about $x = a$

$$f(x) = f(a) + f'(a)(x - a) + \frac{f''(a)}{2!}(x - a)^2 + \frac{f'''(a)}{3!}(x - a)^3 + \cdots + \frac{f^{(n)}(a)}{n!}(x - a)^n + \cdots$$

For many functions f, the Taylor series converges to $f(x)$ only for x near a.

Problems for Section 9.1

For Problems 1–10, find the Taylor polynomials of degree n approximating the given functions for x near 0.

1. $\dfrac{1}{1+x}$, $n = 4, 6, 8$ 2. $\dfrac{1}{1-x}$, $n = 3, 5, 7$

3. $\sqrt{1+x}$, $n = 2, 3, 4$ 4. $\cos x$, $n = 2, 4, 6$

5. $\arctan x$, $n = 3, 4$ 6. $\tan x$, $n = 3, 4$

7. $\sqrt[3]{1-x}$, $n = 2, 3, 4$ 8. $\ln(1+x)$, $n = 5, 7, 9$

9. $\dfrac{1}{\sqrt{1+x}}$, $n = 2, 3, 4$ 10. $(1+x)^p$, $n = 2, 3, 4$ (p is a constant.)

11. Suppose the function $f(x)$ is approximated near $x = 0$ by a sixth degree Taylor polynomial
$$P_6(x) = 3x - 4x^3 + 5x^6.$$
Give the value of

(a) $f(0)$ (b) $f'(0)$ (c) $f'''(0)$ (d) $f^{(5)}(0)$ (e) $f^{(6)}(0)$

12. Suppose g is a function which has continuous derivatives, and that $g(5) = 3, g'(5) = -2, g''(5) = 1,$ $g'''(5) = -3$.

(a) What is the Taylor polynomial of degree 2 for g near 5? What is the Taylor polynomial of degree 3 for g near 5?

(b) Use the two polynomials that you found in part (a) to approximate $g(4.9)$.

For Problems 13–16, find the Taylor polynomial of degree n for x near the given point a.

13. $\sin x$, $a = \pi/2$, $n = 4$ 14. $\cos x$, $a = \pi/4$, $n = 3$

15. e^x, $a = 1$, $n = 4$ 16. $\sqrt{1+x}$, $a = 1$, $n = 3$

For Problems 17–20, suppose $P_2(x) = a + bx + cx^2$ is the second degree Taylor polynomial for the function f about $x = 0$. What can you say about the signs of a, b, c if f has the graph given below?

17. 18. 19. 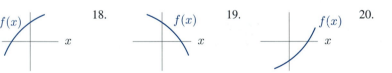 20.

For Problems 21–24, find the first four terms of the Taylor series for the function about the point a.

21. $\sin x$, $a = \pi/4$ 22. $\cos\theta$, $a = \pi/4$ 23. $\sin\theta$, $a = -\pi/4$

24. $\tan x$, $a = \pi/4$

25. Suppose that you are told that the Taylor series of $f(x) = x^2 e^{x^2}$ about $x = 0$ is
$$x^2 + x^4 + \frac{x^6}{2!} + \frac{x^8}{3!} + \frac{x^{10}}{4!} + \cdots.$$
Find $\dfrac{d}{dx}\left(x^2 e^{x^2}\right)\Big|_{x=0}$ and $\dfrac{d^6}{dx^6}\left(x^2 e^{x^2}\right)\Big|_{x=0}$.

26. Suppose you know that all the derivatives of some function f exist at 0, and that Taylor series for f about $x = 0$ is
$$x + \frac{x^2}{2} + \frac{x^3}{3} + \frac{x^4}{4} + \cdots + \frac{x^n}{n} + \cdots.$$
Find $f'(0), f''(0), f'''(0),$ and $f^{(10)}(0)$.

27. Find the second-degree Taylor polynomial for $f(x) = 4x^2 - 7x + 2$ about $x = 0$. What do you notice?

28. Find the third-degree Taylor polynomial for $f(x) = x^3 + 7x^2 - 5x + 1$ about $x = 0$. What do you notice?

29. (a) Based on your observations in Problems 27–28, make a conjecture about Taylor approximations in the case when f is itself a polynomial.
    (b) Show that your conjecture is true.

30. Show how you can use the Taylor approximation $\sin x \approx x - \dfrac{x^3}{3!}$, for x near 0, to explain why $\displaystyle\lim_{x \to 0} \frac{\sin x}{x} = 1$.

31. Use the fourth-degree Taylor approximation $\cos x \approx 1 - \dfrac{x^2}{2!} + \dfrac{x^4}{4!}$ for x near 0 to explain why $\displaystyle\lim_{x \to 0} \frac{1 - \cos x}{x^2} = \frac{1}{2}$.

32. Use a fourth degree Taylor approximation for e^h, for h near 0, to evaluate the following limits. Would your answer be different if you used a Taylor polynomial of higher degree?

    (a) $\displaystyle\lim_{h \to 0} \frac{e^h - 1 - h}{h^2}$

    (b) $\displaystyle\lim_{h \to 0} \frac{e^h - 1 - h - \frac{h^2}{2}}{h^3}$

33. Derive the formulas given in the box on page 431 for the coefficients of the Taylor polynomial approximating a function f for x near a.

34. (a) Find the Taylor polynomial approximation of degree 4 about $x = 0$ for the function $f(x) = e^{x^2}$.
    (b) Compare this result to the Taylor polynomial approximation of degree 2 for the function $f(x) = e^x$ about $x = 0$. What do you notice?
    (c) Use your observation in part (b) to write out the Taylor polynomial approximation of degree 20 to the function in part (a).
    (d) What is the Taylor polynomial approximation of degree 5 for the function $f(x) = e^{-2x}$?

35. (a) Find the terms up to degree 6 of the Taylor series for $f(x) = \sin(x^2)$ about $x = 0$ by taking derivatives.
    (b) Compare your result in part (a) to the series for $\sin x$. How could you have obtained your answer to part (a) from the series for $\sin x$?

36. The integral $\int_0^1 (\sin t / t)\, dt$ is difficult to approximate using, for example, left Riemann sums or the trapezoid rule because the integrand $(\sin t)/t$ is not defined at $t = 0$. However, this integral converges; its value is $0.94608\ldots$. Estimate the integral using Taylor polynomials for $\sin t$ about $t = 0$ of
    (a) Degree 3   (b) Degree 5

9.2   CONVERGENCE OF SERIES

Power Series

In this section we investigate the convergence of Taylor series. We will see the Taylor expansions for more functions, including $\ln(1 + x)$ and $(1 + x)^p$, and determine for which x-values we can safely use these series to make approximations.

We first introduce *power series*. Taylor series are a particular type of power series, just as Taylor polynomials are a particular type of polynomial.

> A **power series** is a sum of constants times powers of x:
>
> $$P(x) = C_0 + C_1 x + C_2 x^2 + \cdots + C_n x^n + \cdots = \sum_{n=0}^{\infty} C_n x^n.$$

We introduce the polynomials $P_n(x) = C_0 + C_1 x + C_2 x^2 + \ldots + C_n x^n$. For any particular values of x and n, the polynomial $P_n(x)$ is the sum of $n + 1$ numbers; it is a number and it always exists. Let us fix x and consider the sequence

$$P_0(x), \quad P_1(x), \quad P_2(x), \quad P_3(x), \quad \ldots \quad , P_n(x), \ldots$$

If this sequence converges to a limit L, that is, if $\lim\limits_{n \to \infty} P_n(x) = L$, then we say that the power series, $P(x)$, **converges** to L for this value of x.

In this section we see that some power series converge for all x, some converge only for $x = 0$, and some converge only on an interval. For a given function f and point x, the Taylor polynomials may or may not converge to a limit as $n \to \infty$, and if they do converge, it might not be to the original function $f(x)$, though it usually is. Fortunately, the Taylor polynomials for e^x, $\cos x$, and $\sin x$ converge to the original function for all x.

Convergence of Taylor Series for cos x

We have already seen that the Taylor polynomials centered at $x = 0$ for $\cos x$ are good approximations for x near 0. (See Figure 9.8.) Amazingly enough, for any value of x, if we take a Taylor polynomial centered at $x = 0$ of high enough degree, its graph is nearly indistinguishable from the graph of the cosine near that point.

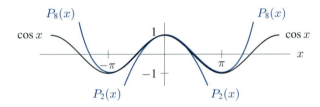

Figure 9.8: Graph of $\cos x$ and two Taylor polynomials for x near 0

Let's see what happens numerically. Suppose we take $x = \pi/2$. The successive Taylor polynomial approximations to $\cos(\pi/2) = 0$ are

$$P_2(\pi/2) = \qquad 1 - (\pi/2)^2/2! \qquad = -0.23370\ldots$$
$$P_4(\pi/2) = 1 - (\pi/2)^2/2! + (\pi/2)^4/4! = \quad 0.01997\ldots$$
$$P_6(\pi/2) = \qquad\qquad \cdots \qquad\qquad = -0.00089\ldots$$
$$P_8(\pi/2) = \qquad\qquad \cdots \qquad\qquad = \quad 0.00002\ldots,$$

and it appears that the approximations converge to the true value very rapidly. If we take a value of x somewhat further away from 0, say $x = \pi$, then $\cos \pi = -1$ and

$$P_2(\pi) = 1 - (\pi)^2/2! = \quad -3.93480\ldots$$
$$P_4(\pi) = \qquad \cdots \qquad = \quad 0.12391\ldots$$
$$P_6(\pi) = \qquad \cdots \qquad = \quad -1.21135\ldots$$
$$P_8(\pi) = \qquad \cdots \qquad = \quad -0.97602\ldots$$
$$P_{10}(\pi) = \qquad \cdots \qquad = \quad -1.00183\ldots$$
$$P_{12}(\pi) = \qquad \cdots \qquad = \quad -0.99990\ldots$$
$$P_{14}(\pi) = \qquad \cdots \qquad = -1.000004\ldots$$

We see that the rate of convergence is somewhat slower; it takes a 14th degree polynomial to approximate $\cos(\pi)$ as well as an 8th degree polynomial approximates $\cos(\pi/2)$. If x were taken still further away from 0, then we would need still more terms to obtain as accurate an approximation of $\cos x$. However it turns out that no matter how large x is, either positive or negative, the approximations $P_n(x)$ will converge to $\cos x$ as $n \to \infty$.

In this case, we are justified in writing an equality:

$$\cos x = 1 - \frac{x^2}{2!} + \frac{x^4}{4!} - \frac{x^6}{6!} + \frac{x^8}{8!} - \cdots \qquad \text{for all } x.$$

We say that the Taylor series converges to $\cos x$ for all x, because for all x, the numbers

$$P_1(x), \ P_2(x), \ P_3(x), \ldots, \ P_n(x), \ldots$$

have a limit of $\cos x$ as $n \to \infty$.

Intervals of Convergence

Let us look again at the Taylor polynomials for $\ln x$ about $x = 1$ that we derived on page 431.

Graphical Viewpoint

What happens if we attempt to improve the approximation to $\ln x$ by using higher-degree Taylor polynomials? Figure 9.9 suggests that for $0 < x < 2$, the polynomials fit the curve well, though outside that interval they are not close at all. In fact for $0 < x < 2$, the higher the degree of the polynomial, the better it fits the curve. It can be shown that, for $0 < x < 2$, the polynomials

$$P_5(x), P_6(x), P_7(x), \ldots, P_n(x), \ldots$$

converge to $\ln x$ as $n \to \infty$. For such x-values, a higher-degree polynomial will, in general, give a better approximation. (See Figure 9.9.)

However, when $x > 2$, the polynomials move away from the curve and the approximations get worse as the degree of the polynomial increases. For such x-values, we say the polynomials *diverge* from the original function.

The Taylor polynomial approximations about $x = 1$ are effective only as approximations to $\ln x$ for values of x between 0 and 2; beyond that, they definitely should not be used. We say the *interval of convergence* (without endpoints) of the Taylor series is $0 < x < 2$. As x gets near 0 or 2, the polynomials converge very slowly. This means we might have to take a polynomial of very high degree to get an accurate value for $\ln x$. At the endpoints of the interval of convergence, in this case $x = 0$ and $x = 2$, the series may or may not converge; we consider this question at the end of this section.

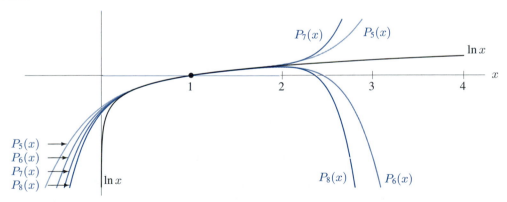

Figure 9.9: Taylor polynomials $P_5(x), P_6(x), P_7(x), P_8(x), \ldots$ converge to $\ln x$ for $0 < x < 2$ and diverge outside that interval

Numerical Viewpoint

To illustrate the interval of convergence numerically, consider the successive Taylor polynomial approximations:

$$\ln x \approx (x - 1) - \frac{(x-1)^2}{2} + \frac{(x-1)^3}{3} - \frac{(x-1)^4}{4} + \cdots + (-1)^{n-1}\frac{(x-1)^n}{n} \qquad \text{for } x \text{ near } 1.$$

If $x = 1.4$, we have the function value $\ln 1.4 = 0.33647\ldots$ and the approximations

$$P_2(1.4) = 0.32 \qquad\qquad P_6(1.4) = 0.33630\ldots$$
$$P_4(1.4) = 0.33493\ldots \qquad P_8(1.4) = 0.33645\ldots$$

So we see that the convergence is quite rapid.

In Table 9.1 we show the results of using $x = 1.9$ and $x = 2.3$ in the Taylor series for $\ln x$. Notice that for $x = 1.9$, which is inside the interval of convergence but close to an endpoint, the approximations converge, though rather slowly. For $x = 2.3$, which is outside the interval of convergence of the Taylor series, the "approximations" diverge: the larger the value for n, the less accurate the "approximations" become. In fact, the contribution of the twenty-fifth term is about 28; the contribution of the hundredth term is about $-2,500,000,000$. Thus, outside the interval of convergence, the "approximations" will oscillate ever more wildly away from the desired value. They are not approximations at all!

TABLE 9.1 *Approximations for*
$\ln 1.9 = 0.64185$ and $\ln 2.3 = 0.83291$

n	$P_n(1.9)$	n	$P_n(2.3)$
2	0.495	2	0.455
5	0.69021	5	1.21589
8	0.61802	8	0.28817
11	0.65473	11	1.71710
14	0.63440	14	-0.70701

Since the Talyor series about $x = 1$ for $\ln x$ converges to $\ln x$ for $0 < x < 2$, we write

$$\ln x = (x-1) - \frac{(x-1)^2}{2} + \frac{(x-1)^3}{3} - \frac{(x-1)^4}{4} + \cdots \qquad \text{for } 0 < x < 2.$$

For $x < 0$ or $x > 2$, the series does not converge. Notice that the interval of convergence is centered at $x = 1$. For most functions f that we will meet, a Taylor series about $x = a$ either converges for all x, or has an interval of convergence centered at $x = a$.

Example 1   Find the Taylor series for $\ln(1+x)$ about $x = 0$, and estimate its interval of convergence.

Solution   Taking derivatives of $\ln(1+x)$ and substituting $x = 0$ leads to the Taylor series

$$\ln(1+x) = x - \frac{x^2}{2} + \frac{x^3}{3} - \frac{x^4}{4} + \cdots.$$

Notice that this is the same series that we get by substituting $(1+x)$ for x in the series for $\ln x$:

$$\ln x = (x-1) - \frac{(x-1)^2}{2} + \frac{(x-1)^3}{3} - \frac{(x-1)^4}{4} + \cdots \qquad \text{for } 0 < x < 2.$$

Since the series for $\ln x$ about $x = 1$ converges for $0 < x < 2$, we shouldn't be surprised to see in Figure 9.10 that the interval of convergence for the Taylor series for $\ln(1+x)$ about $x = 0$ is $-1 < x < 1$. Thus we write

$$\ln(1+x) = x - \frac{x^2}{2} + \frac{x^3}{3} - \frac{x^4}{4} + \cdots \qquad \text{for } -1 < x < 1.$$

By the way, it should be clear that the series cannot possibly converge to $\ln(1+x)$ for $x \leq -1$ since $\ln(1+x)$ is not defined there.

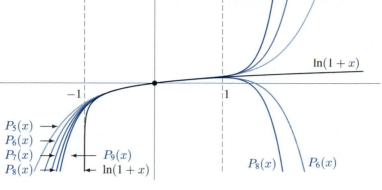

Figure 9.10: Interval of convergence for the Taylor series for $\ln(1+x)$ is $-1 < x < 1$

The Binomial Series Expansion

We now find the Taylor series about $x = 0$ for the function $f(x) = (1+x)^p$, with p a constant, not necessarily an integer. Taking derivatives:

$$f(x) = (1+x)^p \qquad\qquad \text{so} \qquad f(0) = 1$$
$$f'(x) = p(1+x)^{p-1} \qquad\qquad\qquad f'(0) = p$$
$$f''(x) = p(p-1)(1+x)^{p-2} \qquad\qquad f''(0) = p(p-1)$$
$$f'''(x) = p(p-1)(p-2)(1+x)^{p-3}, \qquad f'''(0) = p(p-1)(p-2).$$

Thus, the third-degree Taylor polynomial for x near 0 is

$$(1+x)^p \approx P_3(x) = 1 + px + \frac{p(p-1)}{2!}x^2 + \frac{p(p-1)(p-2)}{3!}x^3.$$

Graphing $P_3(x), P_4(x), \ldots$ for various specific values of p suggests that the Taylor polynomials converge to $f(x)$ for $-1 < x < 1$. (See Problems 13–15, page 443.) The Taylor series for $f(x) = (1+x)^p$ is as follows.

The Binomial Series

$$(1+x)^p = 1 + px + \frac{p(p-1)}{2!}x^2 + \frac{p(p-1)(p-2)}{3!}x^3 + \cdots \qquad \text{for } -1 < x < 1.$$

As we see in the next example, the binomial series gives the same result as multiplying $(1+x)^p$ out when p is a positive integer. For more details, see the section on the Binomial Theorem in the Focus on Theory for Chapter 1. (Newton discovered that the binomial series can be used for noninteger exponents.)

Example 2   Use the binomial series with $p = 3$ to expand $(1+x)^3$.

Solution   The series is

$$(1+x)^3 = 1 + 3x + \frac{3 \cdot 2}{2!}x^2 + \frac{3 \cdot 2 \cdot 1}{3!}x^3 + \frac{3 \cdot 2 \cdot 1 \cdot 0}{4!}x^4 + \cdots.$$

The x^4 term and all terms beyond it turn out to be zero, because each coefficient contains a factor of 0. Simplifying gives

$$(1+x)^3 = 1 + 3x + 3x^2 + x^3,$$

which is the usual expansion obtained by multiplying $(1+x)(1+x)(1+x)$ out.

Example 3   Find the Taylor series about $x = 0$ for $\dfrac{1}{1+x}$.

Solution   Since $\dfrac{1}{1+x} = (1+x)^{-1}$, let $p = -1$. Then

$$\frac{1}{1+x} = (1+x)^{-1} = 1 + (-1)x + \frac{(-1)(-2)}{2!}x^2 + \frac{(-1)(-2)(-3)}{3!}x^3 + \cdots$$

$$= 1 - x + x^2 - x^3 + \cdots \quad \text{for } -1 < x < 1.$$

This series is a special case of the binomial series and so converges for $-1 < x < 1$. It is also an example of a *geometric series,* which we will study in more detail in Section 9.4.

How Do We Compute Intervals of Convergence Analytically?

A graph can suggest an interval of convergence; the following result shows how the interval can be computed exactly.

Theorem: The Ratio Test

For the power series

$$P(x) = C_0 + C_1 x + C_2 x^2 + \cdots + C_n x^n + \cdots,$$

suppose that

$$\lim_{n \to \infty} \frac{|C_n|}{|C_{n+1}|} = R.$$

Then

- If[2] $R = \infty$, then the series converges for all x.
- If $0 < R < \infty$, then the series converges for $|x| < R$.
- If $R = 0$, then the series converges only for $x = 0$.

We call R the **radius of convergence**.

Note that the ratio test does not tell us anything if $\lim_{n \to \infty} |C_n|/|C_{n+1}|$ fails to exist.

Example 4   Show that the Taylor series for e^x converges for all x.

Solution   The Taylor series for e^x is

$$e^x = 1 + x + \frac{x^2}{2!} + \frac{x^3}{3!} + \cdots + \frac{x^n}{n!} + \cdots.$$

We have

$$\lim_{n \to \infty} \frac{|C_n|}{|C_{n+1}|} = \lim_{n \to \infty} \frac{1/n!}{1/(n+1)!} = \lim_{n \to \infty} \frac{(n+1)!}{n!} = \lim_{n \to \infty} (n+1) = \infty.$$

Since $(n+1)$ increases without bound, the series for e^x converges for all x.

Showing that the series for $\sin x$ and $\cos x$ converge for all x requires an extension of the ratio test because each series contains only odd or only even terms. See Example 4 on page 479.

---

[2]That is, if the sequence $|C_n|/|C_{n+1}|$ grows without bound.

Example 5   Determine the exact radius of convergence of the Taylor series for $\ln(1 + x)$ about $x = 0$. What does this tell us about the convergence of this series?

Solution   The general term of the series for $\ln(1 + x)$ is x^n/n if n is odd and $-x^n/n$ if n is even, so we can write

$$\ln(1 + x) = x - \frac{x^2}{2} + \frac{x^3}{3} - \frac{x^4}{4} + \cdots + (-1)^{n-1}\frac{x^n}{n} + \cdots.$$

We have

$$\lim_{n \to \infty} \frac{|C_n|}{|C_{n+1}|} = \lim_{n \to \infty} \frac{|(-1)^n/(n+1)|}{|(-1)^{n-1}/n|} = \lim_{n \to \infty} \frac{n}{n+1} = 1.$$

Therefore the radius of convergence is $R = 1$. This tells us that the Taylor series for $\ln(1 + x)$ converges for $|x| < 1$ and does not converge for $|x| > 1$. Notice that the radius of convergence does not tell us what happens at the endpoints, $x = \pm 1$.

For a Taylor series centered at $x = a$, we compute the radius of convergence using the ratio test. The interval of convergence with this radius is then centered at $x = a$.

Example 6   What does the ratio test tell us about the convergence of the Taylor series for $\ln x$ about $x = 1$?

Solution   The series is

$$\ln x = (x - 1) - \frac{(x-1)^2}{2} + \frac{(x-1)^3}{3} - \frac{(x-1)^4}{4} + \cdots + (-1)^{n-1}\frac{(x-1)^n}{n} + \cdots.$$

The coefficients, C_n, are the same as the coefficients in the series for $\ln(1 + x)$ in the previous example, so by the same computation, $R = 1$. This tells us that the series for $\ln x$

Converges for $|x - 1| < 1$;   that is $0 < x < 2$,

Does not converge for $|x - 1| > 1$;   that is $x < 0, x > 2$.

Note that the ratio test does not tell us what happens at the endpoints $x = 0$ and $x = 2$.

What Happens at the Endpoints of the Interval of Convergence?

The ratio test does not tell us whether the series converges at the endpoints of the interval of convergence, $x = \pm R$. There is no single theorem that will answer this question. Since substituting $x = \pm R$ converts the power series to a series of numbers, the following results are often useful.

Divergence of the Harmonic Series

The *harmonic series* is defined to be the infinite series of constants

$$1 + \frac{1}{2} + \frac{1}{3} + \frac{1}{4} + \frac{1}{5} + \cdots + \frac{1}{n} + \cdots.$$

Convergence of this sum would mean that the sequence of *partial sums*

$$S_1 = 1, \quad S_2 = 1 + \frac{1}{2}, \quad S_3 = 1 + \frac{1}{2} + \frac{1}{3}, \quad \cdots, \quad S_n = 1 + \frac{1}{2} + \cdots + \frac{1}{n}, \cdots$$

tends to a limit as $n \to \infty$. Let's look at some values:

$$S_1 = 1, \quad S_{10} \approx 2.93, \quad S_{100} \approx 5.19, \quad S_{1000} \approx 7.49, \quad S_{10000} \approx 9.79.$$

The growth of these partial sums is slow, but they do in fact grow without bound, and so the harmonic series does not converge. See the next example and Problem 21.

Example 7   Show that the series $1 + 1/2 + 1/3 + 1/4 + \ldots$ does not converge.

Solution   The idea is to think of the terms $1, 1/2, 1/3, \ldots$ as the heights of upper rectangles of base 1 being used to approximate $\int_1^\infty (1/x)\,dx$. (See Figure 9.11.)

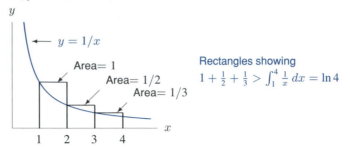

Figure 9.11: Comparing the harmonic series to $\int_1^\infty 1/x\,dx$

Since the sum of the areas of the rectangles is larger than the area under the curve, we have

$$S_n = 1 + \frac{1}{2} + \frac{1}{3} + \ldots + \frac{1}{n} > \int_1^{n+1} \frac{1}{x}\,dx = \ln(n+1)$$

for the nth partial sum. Since $\ln(n+1)$ gets arbitrarily large as $n \to \infty$, so must the partial sums. Thus, the partial sums have no limit, so the series does not converge.

Convergence of an Alternating Series

When we substitute for x in a power series, we often obtain an *alternating series* such as

$$1 - \frac{1}{2} + \frac{1}{3} - \frac{1}{4} + \ldots + \frac{(-1)^n}{n} + \ldots$$

$$2 - \frac{2}{3} + \frac{2}{9} - \frac{2}{27} + \ldots + (-1)^n \frac{2}{3^n} + \ldots$$

It is easy to check for the convergence of alternating series using the following test:

Alternating Series Test

A series of the form

$$a_0 - a_1 + a_2 - a_3 + \cdots + (-1)^n a_n + \cdots \qquad \text{with } a_n \geq 0 \text{ for all } n,$$

converges if $a_{n+1} < a_n$ for all n and

$$\lim_{n \to \infty} a_n = 0.$$

Although we do not prove this result, we can see why it is reasonable. The first partial sum, $S_0 = a_0$, is positive. The second $S_1 = a_0 - a_1$ is still positive, since $a_1 < a_0$. The next sum, $S_2 = a_0 - a_1 + a_2$, is to the right of S_1 but smaller than S_0. (See Figure 9.12.) The partial sums continue to oscillate back and forth, and since the distance between them tends to 0, they eventually converge.

Figure 9.12: Partial sums, S_0, S_1, S_2, S_3 of an alternating series

Example 8   Investigate convergence at the end points of the interval of convergence of the Taylor series for $\ln(1 + x)$ about $x = 0$.

Solution   In Example 5, we showed that the radius of convergence of the series

$$x - \frac{x^2}{2} + \frac{x^3}{3} - \frac{x^4}{4} + \cdots + (-1)^{n-1}\frac{x^n}{n} + \cdots$$

is $R = 1$. The endpoints of the interval are $x = \pm 1$. At $x = 1$, we have the series

$$1 - \frac{1}{2} + \frac{1}{3} - \frac{1}{4} + \cdots + \frac{(-1)^{n-1}}{n} + \cdots$$

This is an alternating series with $a_n = 1/n$, so by the alternating series test, it converges. At $x = -1$, we have the series

$$-1 - \frac{1}{2} - \frac{1}{3} - \frac{1}{4} - \cdots - \frac{1}{n} - \cdots$$

This is the negative of the harmonic series, so it does not converge. Therefore the right endpoint is included, and the left endpoint is not, in the interval of convergence, which is $-1 < x \leq 1$.

Problems for Section 9.2

Which of the series in Problems 1–4 are power series?

1. $x - x^3 + x^6 - x^{10} + x^{15} - \cdots$

2. $\frac{1}{x} + \frac{1}{x^2} + \frac{1}{x^3} + \frac{1}{x^4} + \cdots$

3. $1 + x + (x - 1)^2 + (x - 2)^3 + (x - 3)^4 + \cdots$

4. $x^7 + x + 2$

For Problems 5–8, find the first four terms of the Taylor series for the given function about 0.

5. $\dfrac{1}{1 - x}$

6. $\sqrt{1 + x}$

7. $\dfrac{1}{\sqrt{1 + x}}$

8. $\sqrt[3]{1 - y}$

For Problems 9–11, find the first four terms of the Taylor series for the function about the point a.

9. $1/x$, $\quad a = 1$

10. $1/x$, $\quad a = 2$

11. $1/x$, $\quad a = -1$

12. (a) Find the Taylor series for $f(x) = \ln(1 + 2x)$ about $x = 0$ by taking derivatives.
    (b) Compare your result in part (a) to the series for $\ln(1 + x)$. How could you have obtained your answer to part (a) from the series for $\ln(1 + x)$?
    (c) What do you expect the interval of convergence for the series for $\ln(1 + 2x)$ to be?

13. By graphing the function $f(x) = \dfrac{1}{1 - x}$ and several of its Taylor polynomials, estimate the interval of convergence of the series you found in Problem 5.

14. By graphing the function $f(x) = \sqrt{1 + x}$ and several of its Taylor polynomials, estimate the interval of convergence of the series you found in Problem 6.

15. By graphing the function $f(x) = \dfrac{1}{\sqrt{1 + x}}$ and several of its Taylor polynomials, estimate the interval of convergence of the series you found in Problem 7.

Use the ratio test to find the radius of convergence of the power series in Problems 16–20.

16. $x + 4x^2 + 9x^3 + 16x^4 + 25x^5 + \cdots$

17. $x - \dfrac{x^2}{4} + \dfrac{x^3}{9} - \dfrac{x^4}{16} + \dfrac{x^5}{25} - \cdots$

18. $1 + 2x + \dfrac{4x^2}{2!} + \dfrac{8x^3}{3!} + \dfrac{16x^4}{4!} + \dfrac{32x^5}{5!} + \cdots$

19. $\dfrac{x}{3} + \dfrac{2x^2}{5} + \dfrac{3x^3}{7} + \dfrac{4x^4}{9} + \dfrac{5x^5}{11} + \cdots$

20. $1 + 2x + \dfrac{4!x^2}{(2!)^2} + \dfrac{6!x^3}{(3!)^2} + \dfrac{8!x^4}{(4!)^2} + \dfrac{10!x^5}{(5!)^2} + \cdots$

21. Consider the following grouping of terms in the harmonic series:

$$1 + \left(\frac{1}{2}\right) + \left(\frac{1}{3} + \frac{1}{4}\right) + \left(\frac{1}{5} + \frac{1}{6} + \frac{1}{7} + \frac{1}{8}\right) + \left(\frac{1}{9} + \frac{1}{10} + \cdots + \frac{1}{16}\right) + \cdots$$

   (a) Show that the sum of each group of fractions is more than 1/2.
   (b) Explain why this shows that the harmonic series does not converge.

22. Estimate the sum of the first 100,000 terms of the harmonic series,

$$\sum_{k=1}^{100{,}000} \frac{1}{k},$$

   to the closest integer. [Hint: Use left- and right-hand sums of the function $f(x) = 1/x$ on the interval from 1 to 100,000 with $\Delta x = 1$.]

23. Although the harmonic series does not converge, the partial sums grow very, very slowly. Take a right-hand sum of $f(x) = 1/x$ with $\Delta x = 1$ on the interval $[1, n]$ to show that

$$\frac{1}{2} + \frac{1}{3} + \frac{1}{4} + \cdots + \frac{1}{n} < \ln n.$$

   If a computer could add a million terms of the harmonic series each second, estimate what the sum would be after one year.

Use a computer or calculator to investigate the behavior of the partial sums of the alternating series in Problems 24–26. Which ones converge? If a series converges, estimate its sum.

24. $1 - 0.1 + 0.01 - 0.001 + \cdots + (-1)^n 10^{-n} + \cdots$

25. $1 - 2 + 3 - 4 + 5 + \cdots + (-1)^n (n+1) + \cdots$

26. $1 - \dfrac{1}{1!} + \dfrac{1}{2!} - \dfrac{1}{3!} + \cdots + (-1)^n \dfrac{1}{n!} + \cdots$

By recognizing each series in Problems 27–31 as a Taylor series evaluated at a particular value of x, find the sum of each of the following convergent series.

27. $1 + \dfrac{2}{1!} + \dfrac{4}{2!} + \dfrac{8}{3!} + \cdots + \dfrac{2^n}{n!} + \cdots$

28. $1 - \dfrac{1}{3!} + \dfrac{1}{5!} - \dfrac{1}{7!} + \cdots + \dfrac{(-1)^n}{(2n+1)!} + \cdots$

29. $1 + \dfrac{1}{4} + \left(\dfrac{1}{4}\right)^2 + \left(\dfrac{1}{4}\right)^3 + \cdots + \left(\dfrac{1}{4}\right)^n + \cdots$

30. $1 - \dfrac{100}{2!} + \dfrac{10000}{4!} + \cdots + \dfrac{(-1)^n \cdot 10^{2n}}{(2n)!} + \cdots$

31. $\dfrac{1}{2} - \dfrac{(\frac{1}{2})^2}{2} + \dfrac{(\frac{1}{2})^3}{3} - \dfrac{(\frac{1}{2})^4}{4} + \cdots + \dfrac{(-1)^n \cdot (\frac{1}{2})^{n+1}}{(n+1)} + \cdots$

9.3 FINDING AND USING TAYLOR SERIES

Finding a Taylor series for a function means finding the coefficients. Assuming the function has all its derivatives defined, finding the coefficients can always be done, in theory at least, by differentiation. That is how we derived the four most important Taylor series, those for the functions e^x, $\sin x$, $\cos x$, and $(1+x)^p$.

For many functions, however, computing Taylor series coefficients by differentiation can be a very laborious business. The purpose of this section is to look for easier ways of finding Taylor series, at least if the series we want is closely related to a series that we already know.

New Series by Substitution

Suppose we want to find the Taylor series for e^{-x^2} about $x = 0$. We could find the coefficients by differentiation. Differentiating e^{-x^2} by the chain rule gives $-2xe^{-x^2}$, and differentiating again gives $-2e^{-x^2} + 4x^2e^{-x^2}$. Now, every time we differentiate we will need the product rule, and the number of terms will grow. Finding the tenth or twentieth derivative of e^{-x^2}, and thus the series for e^{-x^2} up to the x^{10} or x^{20} terms, by this method will be tiresome (at least without a computer or calculator that can differentiate).

Fortunately, there's a quicker way. Recall that

$$e^y = 1 + y + \frac{y^2}{2!} + \frac{y^3}{3!} + \frac{y^4}{4!} + \cdots, \quad \text{for all } y.$$

Substituting $y = -x^2$ tells us that

$$e^{-x^2} = 1 + (-x^2) + \frac{(-x^2)^2}{2!} + \frac{(-x^2)^3}{3!} + \frac{(-x^2)^4}{4!} + \cdots, \quad \text{for all } x.$$

Simplifying shows that the Taylor series for e^{-x^2} is

$$e^{-x^2} = 1 - x^2 + \frac{x^4}{2!} - \frac{x^6}{3!} + \frac{x^8}{4!} + \cdots, \quad \text{for all } x.$$

Using this method, it is easy to find the series up to the x^{10} or x^{20} terms.

Example 1   Find the Taylor series about $x = 0$ for $f(x) = \dfrac{1}{1 + x^2}$.

Solution   The binomial series tells us that

$$\frac{1}{1+y} = (1+y)^{-1} = 1 - y + y^2 - y^3 + y^4 + \cdots, \quad \text{for } -1 < y < 1.$$

Substituting $y = x^2$ gives

$$\frac{1}{1+x^2} = 1 - x^2 + x^4 - x^6 + x^8 + \cdots \quad \text{for } -1 < x < 1.$$

which is the Taylor series for $\dfrac{1}{1+x^2}$.

These examples show that we can get new series from old ones by substitution. Proving that the new series we get is the same one we would have obtained by direct calculation of the derivatives is more difficult, and is done in more advanced texts.

In Example 1, we made the substitution $y = x^2$. We can also substitute an entire series into another one, as in the next example.

Example 2   Find the Taylor series about $\theta = 0$ for $g(\theta) = e^{\sin \theta}$.

Solution   For all y and θ, we know that

$$e^y = 1 + y + \frac{y^2}{2!} + \frac{y^3}{3!} + \frac{y^4}{4!} + \cdots$$

and

$$\sin \theta = \theta - \frac{\theta^3}{3!} + \frac{\theta^5}{5!} - \cdots.$$

Let's substitute the series for $\sin\theta$ for y:

$$e^{\sin\theta} = 1 + \left(\theta - \frac{\theta^3}{3!} + \frac{\theta^5}{5!} - \cdots\right) + \frac{1}{2!}\left(\theta - \frac{\theta^3}{3!} + \frac{\theta^5}{5!} - \cdots\right)^2 + \frac{1}{3!}\left(\theta - \frac{\theta^3}{3!} + \frac{\theta^5}{5!} - \cdots\right)^3 + \cdots.$$

To simplify, we multiply out and collect terms. The only constant term is the 1, and there's only one θ term. The only θ^2 term is the first term you get by multiplying out the square, and it is $\theta^2/2!$. There are two contributors to the θ^3 term: the $-\theta^3/3!$ from within the first parentheses, and the first term you get from multiplying out the cube, which is $\theta^3/3!$. Thus the series starts

$$e^{\sin\theta} = 1 + \theta + \frac{\theta^2}{2!} + \left(-\frac{\theta^3}{3!} + \frac{\theta^3}{3!}\right) + \cdots$$

$$= 1 + \theta + \frac{\theta^2}{2!} + 0 \cdot \theta^3 + \cdots \quad \text{for all } \theta.$$

New Series by Integration

Just as we can get new series by substitution, we can also get new series by integration. Here again, proof that the new series so obtained is actually the Taylor series we want, and that it has the same interval of convergence as the original series, can be found in more advanced texts.

Example 3   Find the Taylor series about $x = 0$ for $\arctan x$ from the series for $\dfrac{1}{1 + x^2}$.

Solution   We know that $\dfrac{d(\arctan x)}{dx} = \dfrac{1}{1 + x^2}$, so we start with the series from Example 1:

$$\frac{d(\arctan x)}{dx} = \frac{1}{1 + x^2} = 1 - x^2 + x^4 - x^6 + x^8 - \cdots \quad \text{for } -1 < x < 1.$$

Antidifferentiating term by term (which turns out to be legal) gives

$$\arctan x = C + x - \frac{x^3}{3} + \frac{x^5}{5} - \frac{x^7}{7} + \frac{x^9}{9} - \cdots \quad \text{for } -1 < x < 1,$$

where C is the constant of integration. The fact that $\arctan 0 = 0$ tells us that we must have $C = 0$, so

$$\arctan x = x - \frac{x^3}{3} + \frac{x^5}{5} - \frac{x^7}{7} + \frac{x^9}{9} - \cdots \quad \text{for } -1 < x < 1.$$

Applications of Taylor Series

Example 4   Use a series to estimate the numerical value of π.

Solution   Since $\arctan 1 = \pi/4$, we use the series for $\arctan x$ that we worked out in Example 3. It can be shown that the series does converge to $\pi/4$ at $x = 1$; we will assume this fact. We therefore substitute $x = 1$ into the series for $\arctan x$, getting

$$\pi = 4\arctan 1 = 4\left(1 - \frac{1}{3} + \frac{1}{5} - \frac{1}{7} + \frac{1}{9} - \cdots\right).$$

Table 9.2 shows the value of the sum, S_n, obtained by summing the terms from 1 through n. The values of S_n do seem to converge to $\pi = 3.141\ldots$. Unfortunately, though, this series converges

TABLE 9.2   *Approximating π using the series for* $\arctan x$

n	7	9	25	100	500	1000	10,000
S_n	2.895	3.340	3.218	3.122	3.138	3.140	3.141

very slowly, meaning that we have to take a large number of terms to get an accurate estimate for π. So, this way of calculating π is not particularly practical (a better one is given in Problem 30, page 451). However, the expression for π given by this series is surprising and elegant.

One of the most basic questions we can ask about two functions is which one is larger. The first few terms of the Taylor series for the two functions can often be used to answer this question over a small interval. If the constant terms of the two series are the same, compare the linear terms; if the linear terms are the same, compare the quadratic terms, and so on.

Example 5   By looking at their Taylor series, decide which of the following functions is largest, and which is smallest, for a small positive θ.

(a) $1 + \sin\theta$ (b) e^θ (c) $\dfrac{1}{\sqrt{1-2\theta}}$

Solution   The Taylor expansion about $\theta = 0$ for $\sin\theta$ is

$$\sin\theta = \theta - \frac{\theta^3}{3!} + \frac{\theta^5}{5!} - \frac{\theta^7}{7!} + \cdots.$$

So

$$1 + \sin\theta = 1 + \theta - \frac{\theta^3}{3!} + \frac{\theta^5}{5!} - \frac{\theta^7}{7!} + \cdots.$$

The Taylor expansion about $\theta = 0$ for e^θ is

$$e^\theta = 1 + \theta + \frac{\theta^2}{2!} + \frac{\theta^3}{3!} + \frac{\theta^4}{4!} + \cdots.$$

The Taylor expansion about $\theta = 0$ for $1/\sqrt{1+\theta}$ is

$$\frac{1}{\sqrt{1+\theta}} = (1+\theta)^{-1/2} = 1 - \frac{1}{2}\theta + \frac{(-\frac{1}{2})(-\frac{3}{2})}{2!}\theta^2 + \frac{(-\frac{1}{2})(-\frac{3}{2})(-\frac{5}{2})}{3!}\theta^3 + \cdots$$

$$= 1 - \frac{1}{2}\theta + \frac{3}{8}\theta^2 - \frac{5}{16}\theta^3 + \cdots.$$

So, substituting -2θ for θ:

$$\frac{1}{\sqrt{1-2\theta}} = 1 - \frac{1}{2}(-2\theta) + \frac{3}{8}(-2\theta)^2 - \frac{5}{16}(-2\theta)^3 + \cdots$$

$$= 1 + \theta + \frac{3}{2}\theta^2 + \frac{5}{2}\theta^3 + \cdots.$$

For θ near 0, we can neglect the higher-order terms in these expansions. Keeping the constant, linear, and second-degree terms, we are left with three approximations, valid for θ near 0:

$$1 + \sin\theta \approx 1 + \theta$$

$$e^\theta \approx 1 + \theta + \frac{\theta^2}{2}$$

$$\frac{1}{\sqrt{1-2\theta}} \approx 1 + \theta + \frac{3}{2}\theta^2.$$

Since

$$1 + \theta < 1 + \theta + \frac{1}{2}\theta^2 < 1 + \theta + \frac{3}{2}\theta^2,$$

we conclude that, for small positive θ,

$$1 + \sin\theta < e^\theta < \frac{1}{\sqrt{1-2\theta}}.$$

Example 6   Two electrical charges of equal magnitude and opposite signs located near one another are called an electrical dipole. Suppose the charges Q and $-Q$ are a distance r apart. (See Figure 9.13.) The electric field, E, at the point P is given by

$$E = \frac{Q}{R^2} - \frac{Q}{(R+r)^2}.$$

Use series to investigate the behavior of the electric field far away from the dipole. Show that when R is large in comparison to r, the electric field is approximately proportional to $1/R^3$.

Figure 9.13: A dipole

Solution   In order to use a series approximation, we need to choose a variable whose value we know to be small. Although we know that r is much smaller than R, we do not know that r is itself small. The quantity r/R is, however, very small—much smaller than 1. Hence we expand $1/(R+r)^2$ in terms of r/R so that we can safely use only the first few terms:

$$\frac{1}{(R+r)^2} = \frac{1}{R^2(1+r/R)^2} = \frac{1}{R^2}\left(1+\frac{r}{R}\right)^{-2}$$

$$= \frac{1}{R^2}\left(1 + (-2)\left(\frac{r}{R}\right) + \frac{(-2)(-3)}{2!}\left(\frac{r}{R}\right)^2 + \frac{(-2)(-3)(-4)}{3!}\left(\frac{r}{R}\right)^3 + \cdots\right)$$

$$= \frac{1}{R^2}\left(1 - 2\frac{r}{R} + 3\frac{r^2}{R^2} - 4\frac{r^3}{R^3} + \cdots\right).$$

So

$$E = \frac{Q}{R^2} - \frac{Q}{(R+r)^2} = Q\left[\frac{1}{R^2} - \frac{1}{R^2}\left(1 - 2\frac{r}{R} + 3\frac{r^2}{R^2} - 4\frac{r^3}{R^3} + \cdots\right)\right]$$

$$= \frac{Q}{R^2}\left(2\frac{r}{R} - 3\frac{r^2}{R^2} + 4\frac{r^3}{R^3} - \cdots\right).$$

Since r/R is smaller than 1, the binomial expansion for $(1+r/R)^{-2}$ will converge. We are interested in the electric field far away from the dipole. The quantity r/R is small there, and $(r/R)^2$ and higher powers are much smaller still. Thus, we approximate by disregarding all terms except the first, giving

$$E \approx \frac{Q}{R^2}\left(\frac{2r}{R}\right), \quad \text{so} \quad E \approx \frac{2Qr}{R^3}.$$

Since Q and r are constants, this means that E is nearly proportional to $1/R^3$.

In the previous example, we say that E is *expanded in terms of r/R*, meaning that the independent variable in the expansion is r/R.

Problems for Section 9.3

Find the first four nonzero terms of the Taylor series about 0 for the functions in Problems 1–12.

1. $\sqrt{1-2x}$

2. $\cos(\theta^2)$

3. e^{-x}

4. $\dfrac{t}{1+t}$

5. $\ln(1-2y)$

6. $\arcsin x$

7. $\dfrac{1}{\sqrt{1-z^2}}$

8. $\phi^3 \cos(\phi^2)$

9. $\dfrac{z}{e^{z^2}}$

10. $\sqrt{(1+t)}\sin t$

11. $e^t \cos t$

12. $\sqrt{1+\sin\theta}$

13. By looking at the Taylor series, decide which of the following functions is largest, and which is smallest, for small positive θ.

    (a) $1 + \sin\theta$ (b) $\cos\theta$ (c) $\dfrac{1}{1 - \theta^2}$

14. For values of y near 0, put the following functions in increasing order, using their Taylor expansions.

    (a) $\ln(1 + y^2)$ (b) $\sin(y^2)$ (c) $1 - \cos y$

15. Figure 9.14 shows the graphs of the four functions below for values of x near 0. Use Taylor series to match graphs and formulas.

    (a) $\dfrac{1}{1 - x^2}$ (b) $(1 + x)^{1/4}$ (c) $\sqrt{1 + \dfrac{x}{2}}$ (d) $\dfrac{1}{\sqrt{1 - x}}$

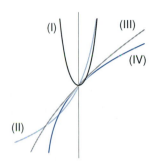

Figure 9.14

16. Consider the two functions $y = e^{-x^2}$ and $y = 1/(1 + x^2)$.

    (a) Write the Taylor expansions for the two functions about $x = 0$. What is similar about the two series? What is different?

    (b) Looking at the series, which function do you predict will be greater over the interval $(-1, 1)$? Graph both and see.

    (c) Are these functions even or odd? How might you see this by looking at the series expansions?

    (d) By looking at the coefficients, explain why it is reasonable that the series for $y = e^{-x^2}$ converges for all values of x, but the series for $y = 1/(1 + x^2)$ converges only on $(-1, 1)$.

For Problems 17–18, expand the quantity about 0 in terms of the variable given. Give four nonzero terms.

17. $\dfrac{1}{2 + x}$ in terms of $\dfrac{x}{2}$

18. $\dfrac{a}{\sqrt{a^2 + x^2}}$ in terms of $\dfrac{x}{a}$, where $a > 0$

19. Padé approximants are rational functions used to approximate more complicated functions. In this problem, you will derive the Padé approximant to the exponential function.

    (a) Let $f(x) = (1 + ax)/(1 + bx)$, where a and b are constants. Write down the first three terms of the Taylor series for $f(x)$ about $x = 0$.

    (b) By equating the first three terms of the Taylor series about $x = 0$ for $f(x)$ and for e^x, find a and b so that $f(x)$ approximates e^x as closely as possible near $x = 0$.

20. An electric dipole on the x-axis consists of a charge Q at $x = 1$ and a charge $-Q$ at $x = -1$. The electric field, E, at the point $x = R$ on the x-axis is given (for $R > 1$) by

$$E = \frac{kQ}{(R - 1)^2} - \frac{kQ}{(R + 1)^2}$$

where k is a positive constant whose value depends on the units. Expand E as a series in $1/R$, giving the first two nonzero terms.

21. Assume a is a positive constant. Suppose z is given by the expression

$$z = \sqrt{a^2 + x^2} - \sqrt{a^2 - x^2}.$$

Expand z as a series in x as far as the second nonzero term.

22. The electric potential, V, at a distance R along the axis perpendicular to the center of a charged disc with radius a and constant charge density σ, is given by

$$V = 2\pi\sigma(\sqrt{R^2 + a^2} - R).$$

Show that, for large R,

$$V \approx \frac{\pi a^2 \sigma}{R}.$$

23. One of Einstein's most amazing predictions was that light traveling from distant stars would bend around the sun on the way to earth. His calculations involved solving for ϕ in the equation

$$\sin\phi + b(1 + \cos^2\phi + \cos\phi) = 0$$

where b is a very small positive constant.

(a) Explain why the equation could have a solution for ϕ which is near 0.
(b) Expand the left-hand side of the equation in Taylor series about $\phi = 0$, disregarding terms of order ϕ^2 and higher. Solve for ϕ. (Your answer will involve b.)

24. The Michelson-Morley experiment, which contributed to the formulation of the Theory of Relativity, involved the difference between the two times t_1 and t_2 that light took to travel between two points. If v is the velocity of light; l_1, l_2, and c are constants; and $v < c$, then the times t_1 and t_2 are given by

$$t_1 = \frac{2l_2}{c(1 - v^2/c^2)} - \frac{2l_1}{c\sqrt{1 - v^2/c^2}} \qquad t_2 = \frac{2l_2}{c\sqrt{1 - v^2/c^2}} - \frac{2l_1}{c(1 - v^2/c^2)}.$$

(a) Find an expression for $\Delta t = t_1 - t_2$, and give its Taylor expansion in terms of v^2/c^2 up to the second nonzero term.
(b) For small v, to what power of v is Δt proportional? What is the constant of proportionality?

25. A hydrogen atom consists of an electron, of mass m, orbiting a proton, of mass M, where m is much smaller than M. The *reduced mass*, μ, of the hydrogen atom is defined by

$$\mu = \frac{mM}{m + M}.$$

(a) Show that $\mu \approx m$.
(b) To get a more accurate approximation for μ, express μ as m times a series in m/M.
(c) The approximation $\mu \approx m$ is obtained by disregarding all but the constant term in the series. The first-order correction is obtained by including the linear term but no higher terms. If $m \approx M/1836$, by what percentage does including the first-order correction change the estimate $\mu \approx m$?

26. A thin disk of radius a and mass M lies horizontally; a particle of mass m is at a height h directly above the center of the disk. The gravitational force, F, exerted by the disk on the mass m is given by

$$F = \frac{2GMmh}{a^2}\left(\frac{1}{h} - \frac{1}{(a^2 + h^2)^{1/2}}\right).$$

Assume $a < h$ and think of F as a function of a, with the other quantities constant.

(a) Expand F as a series in a/h. Give the first two nonzero terms.
(b) Show that the approximation for F obtained by using only the first nonzero term in the series is independent of the radius, a.
(c) If $a = 0.02h$, by what percentage does the approximation in part (a) differ from the approximation in part (b)?

27. When a body is near the surface of the earth, we usually assume that the force due to gravity on it is a constant mg, where m is the mass of the body and g is the acceleration due to gravity at sea level. For a body at a distance h above the surface of the earth, a more accurate expression for the force F is

$$F = \frac{mgR^2}{(R + h)^2}$$

where R is the radius of the earth. We will consider the situation in which the body is close to the surface of the earth so that h is much smaller than R.

(a) Show that $F \approx mg$.

(b) Express F as mg multiplied by a series in h/R.

(c) The first-order correction to the approximation $F \approx mg$ is obtained by taking the linear term in the series but no higher terms. How far above the surface of the earth can you go before the first-order correction changes the estimate $F \approx mg$ by more than 10%? (Assume $R = 6400$ km.)

28. (a) Estimate the value of $\int_0^1 e^{-x^2}\,dx$ using Riemann sums for both left-hand and right-hand sums with $n = 5$ subdivisions.

    (b) Approximate the function $f(x) = e^{-x^2}$ with a Taylor polynomial of degree 6.

    (c) Estimate the integral in part (a) by integrating the Taylor polynomial from part (b).

    (d) Indicate briefly how you could improve the results in each case.

29. Use Taylor series to explain how the following patterns arise:

    (a) $\dfrac{1}{0.98} = 1.020408163264\ldots$ (b) $\left(\dfrac{1}{0.99}\right)^2 = 1.020304050607\ldots$

30. Machin's formula says $\pi/4 = 4\arctan(1/5) - \arctan(1/239)$.

    (a) Verify that $\pi/4$ and the quantity $4\arctan(1/5) - \arctan(1/239)$ agree to as many decimal places as your calculator shows.

    (b) Use the Taylor polynomial approximation of degree 5 to the arctangent function to approximate the value of π. (Note: In 1873 William Shanks used this approach to calculate π to 707 decimal places. Unfortunately, in 1946 it was found that he made an error in the 528th place.)

    (c) Can you explain why the two series for arctangent converge so rapidly here while the series used in Example 4 on page 446 converges so slowly?

9.4 GEOMETRIC SERIES

In the last two sections we expressed a given function by a Taylor series. We started with the function and then derived the series. In this section we will work in the opposite direction, starting with a series and finding its sum, the function. This can't be done for most series, but for some particular types of series it can.

Repeated Drug Dosage

A person with an ear infection is told to take antibiotic tablets regularly for several days. Since the drug is being excreted by the body in between doses, how can we calculate the quantity of the drug remaining in the body at any particular time?

To be specific, let's suppose the drug is ampicillin (a common antibiotic) taken in 250 mg doses four times a day (that is, every six hours). It is known that at the end of six hours, about 4% of the drug is still in the body. What quantity of the drug is in the body right after the tenth tablet? The fortieth?

Let Q_n represent the quantity, in milligrams, of ampicillin in the blood right after the nth tablet. Then

$$Q_1 = 250 \qquad\qquad = 250\text{ mg}$$

$$Q_2 = \underbrace{250(0.04)}_{\text{Remnants of first tablet}} + \underbrace{250}_{\text{New tablet}} \qquad = 260\text{ mg}$$

$$Q_3 = Q_2(0.04) + 250 = \left(250(0.04) + 250\right)(0.04) + 250$$
$$= \underbrace{250(0.04)^2 + 250(0.04)}_{\text{Remnants of first and second tablets}} + \underbrace{250}_{\text{New tablet}} \qquad = 260.4\text{ mg}$$

$$Q_4 = Q_3(0.04) + 250 = \left(250(0.04)^2 + 250(0.04) + 250\right)(0.04) + 250$$
$$= \underbrace{250(0.04)^3 + 250(0.04)^2 + 250(0.04)}_{\text{Remnants of first, second, and third tablets}} + \underbrace{250}_{\text{New tablet}} \qquad = 260.416\text{ mg}$$

Looking at the pattern that is emerging, we can probably guess that

$$Q_5 = 250(0.04)^4 + 250(0.04)^3 + 250(0.04)^2 + 250(0.04) + 250$$
$$Q_{10} = 250(0.04)^9 + 250(0.04)^8 + \cdots + 250(0.04) + 250.$$

Notice that there are 10 terms in this sum—one for every tablet—but that the highest power of 0.04 is the ninth, because no tablet has been in the body for more than 9 six-hour time periods. (Check this: do you see why?) Now suppose we actually want to find the numerical value of Q_{10}. It seems that we have to add 10 terms—and if we want the value of Q_{40}, we would be faced with adding 40 terms:

$$Q_{40} = 250(0.04)^{39} + 250(0.04)^{38} + \cdots + 250(0.04) + 250.$$

Fortunately, there's a better way. Let's start with Q_{10}.

$$Q_{10} = 250(0.04)^9 + 250(0.04)^8 + 250(0.04)^7 + \cdots + 250(0.04)^2 + 250(0.04) + 250.$$

Notice the remarkable fact that if you subtract $(0.04)Q_{10}$ from Q_{10}, a great many terms (all but two, in fact) drop out. First multiplying by 0.04, we get

$$(0.04)Q_{10} = 250(0.04)^{10} + 250(0.04)^9 + 250(0.04)^8 + \cdots + 250(0.04)^3 + 250(0.04)^2 + 250(0.04).$$

Subtracting gives

$$Q_{10} - (0.04)Q_{10} = 250 - 250(0.04)^{10}.$$

Factoring Q_{10} on the left and solving for Q_{10} gives

$$Q_{10}(1 - 0.04) = 250 \left(1 - (0.04)^{10}\right)$$
$$Q_{10} = \frac{250 \left(1 - (0.04)^{10}\right)}{1 - 0.04}.$$

This is called the *closed-form* expression for Q_{10}. It is easy to evaluate on a calculator, giving $Q_{10} = 260.42$ (to two decimal places). Similarly, Q_{40} is given in closed-form by

$$Q_{40} = \frac{250 \left(1 - (0.04)^{40}\right)}{1 - 0.04}.$$

Evaluating this on a calculator shows $Q_{40} = 260.42$, which is the same (to two decimal places) as Q_{10}. Thus after ten tablets, the value of Q_n appears to have stabilized at just over 260 mg.

Looking at the closed-forms for Q_{10} and Q_{40}, we can see that, in general, Q_n must be given by

$$Q_n = \frac{250 \left(1 - (0.04)^n\right)}{1 - 0.04}.$$

What Happens as $n \to \infty$?

What does this closed-form for Q_n predict about the long-run level of ampicillin in the body right after a tablet is taken, assuming that 250 mg continue to be taken every six hours? As $n \to \infty$, the quantity $(0.04)^n \to 0$, so in the long run,

$$Q_n = \frac{250 \left(1 - (0.04)^n\right)}{1 - 0.04} \to \frac{250(1 - 0)}{1 - 0.04} \approx 260.42.$$

The Geometric Series in General

In the previous example we encountered sums of the form $a + ax + ax^2 + \cdots + ax^8 + ax^9$ (with $a = 250$ and $x = 0.04$). Such a sum is called a finite *geometric series*. In general, a geometric series is defined as one in which each term is a constant multiple of the one before. (In our example, the constant multiple was 0.04.) We assume $a \neq 0$ and $x \neq 0$.

> A **finite geometric series** has the form
>
> $$a + ax + ax^2 + \cdots + ax^{n-2} + ax^{n-1}.$$
>
> An **infinite geometric series** has the form
>
> $$a + ax + ax^2 + \cdots + ax^{n-2} + ax^{n-1} + ax^n + \cdots.$$

The "$\cdots$" at the end of the second series tells us that the series is going on forever—in other words, that it is infinite.

Sum of a Finite Geometric Series

The same remarkable trick that enabled us to find the closed-form for Q_{10} can be used to find the sum of any finite geometric series. Suppose we define S_n to be the sum of the first n terms, which means up to the term in x^{n-1}:

$$S_n = a + ax + ax^2 + \cdots + ax^{n-2} + ax^{n-1}.$$

Multiply S_n by x:

$$xS_n = ax + ax^2 + ax^3 + \cdots + ax^{n-1} + ax^n.$$

Now subtract xS_n from S_n, which cancels out all terms except for two, giving

$$S_n - xS_n = a - ax^n$$
$$(1-x)S_n = a(1-x^n).$$

Thus, provided $x \neq 1$, we can solve for S_n as follows.

> The **sum of a finite geometric series** is given by
>
> $$S_n = a + ax + ax^2 + \cdots + ax^{n-1} = \frac{a(1-x^n)}{1-x}, \qquad x \neq 1.$$

You may find it helpful to remember that the value of n which appears in the closed-form for S_n is the number of terms in the sum S_n.

Sum of an Infinite Geometric Series

We define convergence for an infinite geometric series in the same way as we defined convergence for a power series. In Example 9.4, we found the sum Q_n and then let $n \to \infty$. We do the same here. Suppose we want to find the sum S of the infinite series $a + ax + ax^2 + \cdots + ax^{n-1} + \cdots$. We consider the *partial sum, S_n,*

$$S_n = a + ax + ax^2 + \cdots + ax^{n-1} = \frac{a(1-x^n)}{1-x}$$

and we look at what happens to S_n as $n \to \infty$. What happens? It depends on the value of x. If $|x| < 1$, then $x^n \to 0$ as $n \to \infty$, and so

$$\lim_{n \to \infty} S_n = \lim_{n \to \infty} \frac{a(1-x^n)}{1-x} = \frac{a(1-0)}{1-x} = \frac{a}{1-x}.$$

Thus if $|x| < 1$, the partial sums approach a limit as $n \to \infty$ and we say the infinite geometric series *converges.*

> For $|x| < 1$, the **sum of the infinite geometric series** is given by
>
> $$S = a + ax + ax^2 + \cdots + ax^n + \cdots = \frac{a}{1-x}.$$

If, on the other hand, $|x| > 1$, then x^n and the partial sums have no limit as $n \to \infty$, and we say that the series does not converge. This corresponds to the fact that when $x > 1$, the terms in the series get larger and larger, so adding up more and more of them couldn't possibly give a finite sum. When $x < -1$, the partial sums oscillate more and more wildly and so do not converge.

What happens when $x = 1$? The series is

$$a + a + a + a \ldots,$$

and since $a \neq 0$, the partial sums grow without bound, and the series doesn't converge. When $x = -1$, the series is

$$a - a + a - a + a \ldots,$$

and, since $a \neq 0$, the partial sums oscillate between a and 0, and the series does not converge.

Example 1   For each of the following infinite series, find several partial sums and the sum (if the sum exists).

(a) $1 + \dfrac{1}{2} + \dfrac{1}{4} + \dfrac{1}{8} + \cdots$ (b) $1 + 2 + 4 + 8 + \cdots$ (c) $6 - 2 + \dfrac{2}{3} - \dfrac{2}{9} + \dfrac{2}{27} - \ldots$

Solution   (a)   This series may be written

$$1 + \frac{1}{2} + \left(\frac{1}{2}\right)^2 + \left(\frac{1}{2}\right)^3 + \cdots$$

which we can identify as a geometric series with $a = 1$ and $x = \frac{1}{2}$, so $S = \dfrac{1}{1 - (1/2)} = 2$.

Let's check this by finding the partial sums:

$$S_1 = 1$$
$$S_2 = 1 + \frac{1}{2} = \frac{3}{2} = 2 - \frac{1}{2}$$
$$S_3 = 1 + \frac{1}{2} + \frac{1}{4} = \frac{7}{4} = 2 - \frac{1}{4}$$
$$S_4 = 1 + \frac{1}{2} + \frac{1}{4} + \frac{1}{8} = \frac{15}{8} = 2 - \frac{1}{8}$$
$$S_5 = 1 + \frac{1}{2} + \frac{1}{4} + \frac{1}{8} + \frac{1}{16} = \frac{31}{16} = 2 - \frac{1}{16}.$$

Clearly the partial sums are creeping up on the sum $S = 2$, so $S_n \to 2$ as $n \to \infty$.

(b)   The partial sums of this geometric series (with $a = 1$ and $x = 2$) grow uncontrollably, so the series has no sum:

$$S_1 = 1$$
$$S_2 = 1 + 2 = 3$$
$$S_3 = 1 + 2 + 4 = 7$$
$$S_4 = 1 + 2 + 4 + 8 = 15$$
$$S_5 = 1 + 2 + 4 + 8 + 16 = 31.$$

(c)   This is an infinite geometric series with $a = 6$ and $x = -\frac{1}{3}$. The partial sums,

$$S_1 = 6.00, \quad S_2 = 4.00, \quad S_3 \approx 4.67, \quad S_4 \approx 4.44, \quad S_5 \approx 4.52, \quad S_6 \approx 4.49,$$

appear to be converging to 4.5 from alternatively above and below. The sum of the series is the limit of the partial sums,

$$S = \frac{6}{1 - (-1/3)} = 4.5.$$

Relationship between Geometric and Taylor Series

The geometric series *is* a Taylor series. For example, the fact that

$$1 + x + x^2 + x^3 + \cdots = \frac{1}{1-x}$$

suggests that this geometric series is the Taylor expansion of $f(x) = 1/(1-x)$ for x near 0. We can check that the geometric series really is the Taylor series by taking derivatives. Alternatively, this geometric series can also be obtained from the binomial series:

$$(1+x)^p = 1 + px + \frac{p(p-1)}{2!}x^2 + \frac{p(p-1)(p-2)}{3!}x^3 + \cdots.$$

First we substitute $p = -1$:

$$\frac{1}{1+x} = (1+x)^{-1} = 1 + (-1)x + \frac{(-1)(-2)}{2!}x^2 + \frac{(-1)(-2)(-3)}{3!}x^3 + \cdots$$

$$= 1 - x + x^2 - x^3 + \cdots.$$

Now replace x by $-x$:

$$\frac{1}{1-x} = (1-x)^{-1} = 1 - (-x) + (-x)^2 - (-x)^3 + \cdots$$

$$= 1 + x + x^2 + x^3 + \cdots.$$

Regular Deposits into a Savings Account

People who save money often do so by putting some fixed amount aside regularly. To be specific, suppose \$1000 is deposited every year in a savings account earning 5% a year, compounded annually. What is the balance, B_n, in dollars, in the savings account right after the nth deposit?

As before, let's start by looking at the first few years:

$$B_1 = 1000$$

$$B_2 = B_1(1.05) + 1000 = \underbrace{1000(1.05)}_{\text{Original deposit}} + \underbrace{1000}_{\text{New deposit}}$$

$$B_3 = B_2(1.05) + 1000 = \underbrace{1000(1.05)^2 + 1000(1.05)}_{\text{First two deposits}} + \underbrace{1000}_{\text{New deposit}}$$

$$B_4 = B_3(1.05) + 1000 = \underbrace{1000(1.05)^3 + 1000(1.05)^2 + 1000(1.05)}_{\text{First three deposits}} + \underbrace{1000}_{\text{New deposit}}$$

Observing the pattern, we see

$$B_n = 1000(1.05)^{n-1} + 1000(1.05)^{n-2} + \cdots + 1000(1.05) + 1000.$$

So B_n is a finite geometric series with $a = 1000$ and $x = 1.05$. Thus we have

$$B_n = \frac{1000\left(1 - (1.05)^n\right)}{1 - 1.05}.$$

Rewriting this so both the numerator and denominator of the fraction are positive gives

$$B_n = \frac{1000\left((1.05)^n - 1\right)}{1.05 - 1}.$$

What Happens as $n \to \infty$?

Common sense tells you that if you keep depositing \$1000 in an account and it keeps earning interest, your balance will grow without bound. This is what the formula for B_n shows also: $(1.05)^n \to \infty$ as $n \to \infty$, so B_n has no limit. (Alternatively, observe that the infinite geometric series of which B_n is a partial sum has $x > 1$ and so does not converge.)

Problems for Section 9.4 ━━━━━━━━━━━━━━━━━━━━━━━━━━━━━━━━━━━━━━

In Problems 1–10, decide which of the following are geometric series. For those which are, give the first term and the ratio between successive terms. For those which are not, explain why not.

1. $1 - \dfrac{1}{2} + \dfrac{1}{4} - \dfrac{1}{8} + \dfrac{1}{16} + \cdots$

2. $1 + \dfrac{1}{2} + \dfrac{1}{3} + \dfrac{1}{4} + \dfrac{1}{5} + \cdots$

3. $5 - 10 + 20 - 40 + 80 - \cdots$

4. $2 + 1 + \dfrac{1}{2} + \dfrac{1}{4} + \dfrac{1}{8} + \cdots$

5. $1 + x + 2x^2 + 3x^3 + 4x^4 + \cdots$

6. $y^2 + y^3 + y^4 + y^5 + \cdots$

7. $1 - x + x^2 - x^3 + x^4 - \cdots$

8. $1 - y^2 + y^4 - y^6 + \cdots$

9. $3 + 3z + 6z^2 + 9z^3 + 12z^4 + \cdots$

10. $1 + 2z + (2z)^2 + (2z)^3 + \cdots$

11. Find the sum of the series in Problem 6.

12. Find the sum of the series in Problem 7.

13. Find the sum of the series in Problem 8.

14. Find the sum of the series in Problem 10.

15. Find an exact value for each of the following sums.

(a) $7(1.02)^3 + 7(1.02)^2 + 7(1.02) + 7 + \dfrac{7}{(1.02)} + \dfrac{7}{(1.02)^2} + \cdots + \dfrac{7}{(1.02)^{100}}.$

(b) $7 + 7(0.1)^2 + \dfrac{7(0.1)^4}{2!} + \dfrac{7(0.1)^6}{3!} + \cdots .$

Find the sum of the series in Problems 16–19.

16. $-2 + 1 - \dfrac{1}{2} + \dfrac{1}{4} - \dfrac{1}{8} + \dfrac{1}{16} - \cdots$

17. $3 + \dfrac{3}{2} + \dfrac{3}{4} + \dfrac{3}{8} + \cdots + \dfrac{3}{2^{10}}$

18. $\displaystyle\sum_{n=4}^{\infty} \left(\dfrac{1}{3}\right)^n$

19. $\displaystyle\sum_{n=0}^{\infty} \dfrac{3^n + 5}{4^n}$

20. Figure 9.15 shows the quantity of the drug atenolol in the blood as a function of time, with the first dose at time $t = 0$. Atenolol is taken in 50 mg doses once a day to lower blood pressure.

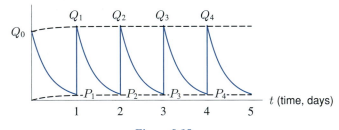

Figure 9.15

(a) If the half-life of atenolol in the blood is 6.3 hours, what percentage of the atenolol present at the start of a 24-hour period is still there at the end?

(b) Find expressions for the quantities $Q_0, Q_1, Q_2, Q_3, \ldots$, and Q_n shown in Figure 9.15. Write the expression for Q_n in closed-form.

(c) Find expressions for the quantities $P_1, P_2, P_3, \ldots$, and P_n shown in Figure 9.15. Write the expression for P_n in closed-form.

21. On page 451, you saw how to compute the quantity Q_n mg of ampicillin in the body right after the nth tablet of 250 mg, taken once every six hours.

(a) Do a similar calculation for P_n, the quantity of ampicillin (in mg) in the body right *before* the nth tablet is taken.

(b) Express P_n in closed form.

(c) What is $\displaystyle\lim_{n \to \infty} P_n$? Is this limit the same as $\displaystyle\lim_{n \to \infty} Q_n$? Explain in practical terms why your answer makes sense.

22. Draw a graph like that in Figure 9.15 for 250 mg of ampicillin taken every 6 hours, starting at time $t = 0$. Put on the graph the values of $Q_1, Q_2, Q_3, \ldots$ introduced in the text on page 451 and the values of $P_1, P_2, P_3, \ldots$ calculated in Problem 21.

23. A ball is dropped from a height of 10 feet and bounces. Each bounce is $\frac{3}{4}$ of the height of the bounce before. Thus after the ball hits the floor for the first time, the ball rises to a height of $10(\frac{3}{4}) = 7.5$ feet, and after it hits the floor for the second time, it rises to a height of $7.5(\frac{3}{4}) = 10(\frac{3}{4})^2 = 5.625$ feet.

    (a) Find an expression for the height to which the ball rises after it hits the floor for the nth time.
    (b) Find an expression for the total vertical distance the ball has traveled when it hits the floor for the first, second, third, and fourth times.
    (c) Find an expression for the total vertical distance the ball has traveled when it hits the floor for the nth time. Express your answer in closed-form.

24. You might think that the ball in Problem 23 keeps bouncing forever since it takes infinitely many bounces. This is not true!

    (a) Show that a ball dropped from a height of h feet reaches the ground in $\frac{1}{4}\sqrt{h}$ seconds.
    (b) Show that the ball in Problem 23 stops bouncing after

    $$\frac{1}{4}\sqrt{10} + \frac{1}{2}\sqrt{10}\sqrt{\frac{3}{4}}\left(\frac{1}{1 - \sqrt{3/4}}\right) \quad \text{seconds,}$$

    or approximately 11 seconds.

25. This problem illustrates how banks create credit and can thereby lend out more money than has been deposited. Suppose that initially \$100 is deposited in a bank. Experience has shown bankers that on average only 8% of the money deposited is withdrawn by the owner at any time. Consequently, bankers feel free to lend out 92% of their deposits. Thus \$92 of the original \$100 is loaned out to other customers (to start a business, for example). This \$92 will become someone else's income and, sooner or later, will be redeposited in the bank. Then 92% of \$92, or $\$92(0.92) = \$84.64$, is loaned out again and eventually redeposited. Of the \$84.64, the bank again loans out 92%, and so on.

    (a) Find the total amount of money deposited in the bank as a result of these transactions.
    (b) The total amount of money deposited divided by the original deposit is called the *credit multiplier*. Calculate the credit multiplier for this example and explain what this number tells us.

26. This problem deals with the question of estimating the cumulative effect of a tax cut on a country's economy. Suppose the government proposes a tax cut totaling \$100 million. We assume that all the people who have extra money to spend would spend 80% of it and save 20%. Thus, of the extra income generated by the tax cut, $\$100(0.8)$ million $= \$80$ million would be spent and so become extra income to someone else. Assume that these people also spend 80% of their additional income, or $\$80(0.8)$ million, and so on. Calculate the total additional spending created by such a tax cut.

27. Suppose the government proposes a tax cut of \$100 million as in Problem 26, but that economists now predict that people will spend 90% of their extra income and save only 10%. How much additional spending would be generated by the tax cut under these assumptions?

28. In an old puzzle, there are two trains, each moving at 10 km/hr toward one another. Initially the trains are 30 kilometers apart. At the same moment a fly, whose velocity is 20 km/hr, starts at one train and flies till it meets the other, then turns around and flies back till it meets the first train, and so on.

    (a) How far has the fly traveled the first time it turns around? The second time? The third time? The fourth time?
    (b) How far has the fly traveled by the nth time it turns around? Write your answer in closed-form.
    (c) Use you answer to part (b) to decide how far the fly has traveled by the time the trains meet in the middle and squash it.
    (d) How long does it take the trains to meet in the middle? Use this to answer part (c) without summing a series.

9.5 FOURIER SERIES

We have seen how to approximate a function by a Taylor polynomial of fixed degree. Such a polynomial is usually very close to the true value of the function near one point (the point at which the Taylor polynomial is centered), but not necessarily at all close anywhere else. In other words, Taylor polynomials are good approximations of a function *locally*, but not necessarily *globally*. In this section, we take another approach: we approximate the function by trigonometric functions, called *Fourier approximations*. The resulting approximation may not be as close to the original function at some points as the Taylor polynomial. However, the Fourier approximation is, in general, close over a larger interval. In other words, a Fourier approximation can be a better approximation globally. In addition, unlike Taylor approximations, Fourier approximations are periodic, so they are useful for approximating periodic functions.

Many processes in nature are periodic or repeating, so it makes sense to approximate them by periodic functions. For example, sound waves are made up of periodic oscillations of air molecules. Heartbeats, the movement of the lungs, and the electrical current that powers our homes are all periodic phenomena. Two of the simplest periodic functions are the square wave in Figure 9.16 and the triangular wave in Figure 9.17. Electrical engineers use the square wave as the model for the flow of electricity when a switch is repeatedly flicked on and off.

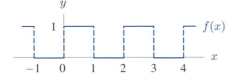

Figure 9.16: Square wave 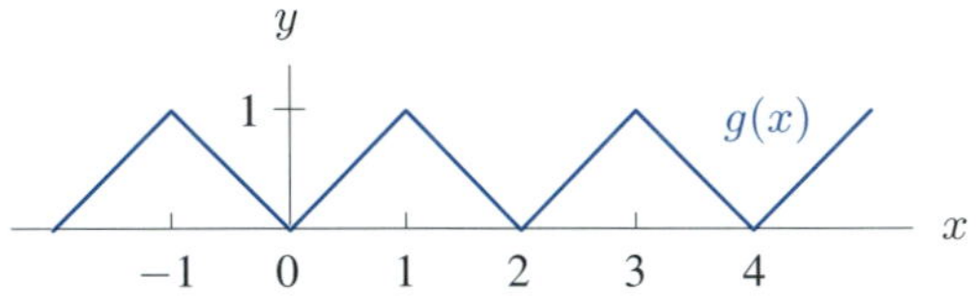   *Figure 9.17:* Triangular wave

Fourier Polynomials

We can express the square wave and the triangular wave by the formulas

$$f(x) = \begin{cases} \vdots & \vdots \\ 0 & -1 \leq x < 0 \\ 1 & 0 \leq x < 1 \\ 0 & 1 \leq x < 2 \\ 1 & 2 \leq x < 3 \\ 0 & 3 \leq x < 4 \\ \vdots & \vdots \end{cases} \qquad g(x) = \begin{cases} \vdots & \vdots \\ -x & -1 \leq x < 0 \\ x & 0 \leq x < 1 \\ 2-x & 1 \leq x < 2 \\ x-2 & 2 \leq x < 3 \\ 4-x & 3 \leq x < 4 \\ \vdots & \vdots \end{cases}$$

However, these formulas are not particularly easy to work with. Worse, the functions are not differentiable at various points. Here we show how to approximate such functions by differentiable, periodic functions.

Since the sine and cosine are the simplest periodic functions, they are the building blocks we use. Because they repeat every 2π, we assume that the function f we want to approximate repeats every 2π. (Later, we deal with the case where f has some other period.) We start by considering the square wave in Figure 9.18. Because of the periodicity of all the functions concerned, we only have to consider what happens in the course of a single period; the same behavior repeats in any other period.

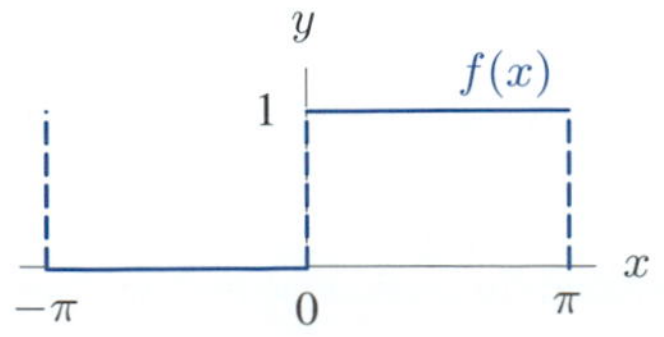

$$f(x) = \begin{cases} 0 & -\pi \leq x < 0 \\ 1 & 0 \leq x < \pi \end{cases}$$

Figure 9.18: Square wave on $[-\pi, \pi]$

We will attempt to approximate f with a sum of trigonometric functions of the form

$$f(x) \approx F_n(x)$$

$$= a_0 + a_1 \cos x + a_2 \cos(2x) + a_3 \cos(3x) + \cdots + a_n \cos(nx)$$

$$+ b_1 \sin x + b_2 \sin(2x) + b_3 \sin(3x) + \cdots + b_n \sin(nx)$$

$$= a_0 + \sum_{k=1}^{n} a_k \cos(kx) + \sum_{k=1}^{n} b_k \sin(kx).$$

$F_n(x)$ is known as a *Fourier polynomial of degree n*, named after the French mathematician Joseph Fourier (1768-1830), who was one of the first to investigate it.[3] The coefficients a_k and b_k are called *Fourier coefficients*. Since each of the component functions $\cos(kx)$ and $\sin(kx)$, $k = 1, 2, \ldots, n$, repeats every 2π, $F_n(x)$ must repeat every 2π and so is a potentially good match for $f(x)$, which also repeats every 2π. The problem is to determine values for the Fourier coefficients that achieve a close match between $f(x)$ and $F_n(x)$. We choose the following values:

The Fourier Coefficients for a Periodic Function f of Period 2π

$$a_0 = \frac{1}{2\pi} \int_{-\pi}^{\pi} f(x)\, dx,$$

$$a_k = \frac{1}{\pi} \int_{-\pi}^{\pi} f(x) \cos(kx)\, dx \quad \text{for } k > 0,$$

$$b_k = \frac{1}{\pi} \int_{-\pi}^{\pi} f(x) \sin(kx)\, dx \quad \text{for } k > 0.$$

Notice that a_0 is just the average value of f over the interval $[-\pi, \pi]$.

For an informal justification for the use of these values, see page 466.

Example 1 Construct successive Fourier polynomials for the square wave function f, with period 2π, given by

$$f(x) = \begin{cases} 0 & -\pi \le x < 0 \\ 1 & 0 \le x < \pi. \end{cases}$$

Solution Since a_0 is the average value of f on $[-\pi, \pi]$, we suspect from the graph of f that $a_0 = \frac{1}{2}$. We can verify this analytically:

$$a_0 = \frac{1}{2\pi} \int_{-\pi}^{\pi} f(x)\, dx = \frac{1}{2\pi} \int_{-\pi}^{0} 0\, dx + \frac{1}{2\pi} \int_{0}^{\pi} 1\, dx = 0 + \frac{1}{2\pi}(\pi) = \frac{1}{2}.$$

Furthermore,

$$a_1 = \frac{1}{\pi} \int_{-\pi}^{\pi} f(x) \cos x\, dx = \frac{1}{\pi} \int_{0}^{\pi} 1 \cos x\, dx = 0$$

and

$$b_1 = \frac{1}{\pi} \int_{-\pi}^{\pi} f(x) \sin x\, dx = \frac{1}{\pi} \int_{0}^{\pi} 1 \sin x\, dx = \frac{2}{\pi}.$$

Therefore, the Fourier polynomial of degree 1 is given by

$$f(x) \approx F_1(x) = \frac{1}{2} + \frac{2}{\pi} \sin x$$

---

[3] The Fourier polynomials are not polynomials in the usual sense of the word.

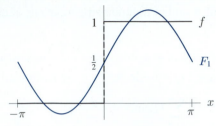

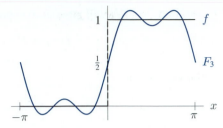

Figure 9.19: First Fourier approximation to the square wave

Figure 9.20: Third Fourier approximation to the square wave

and the graphs of the function and the first Fourier approximation are shown in Figure 9.19.

We next construct the Fourier polynomial of degree 2. The coefficients a_0, a_1, b_1 are the same as before. In addition,

$$a_2 = \frac{1}{\pi} \int_{-\pi}^{\pi} f(x) \cos(2x)\, dx = \frac{1}{\pi} \int_{0}^{\pi} 1 \cos(2x)\, dx = 0$$

and

$$b_2 = \frac{1}{\pi} \int_{-\pi}^{\pi} f(x) \sin(2x)\, dx = \frac{1}{\pi} \int_{0}^{\pi} 1 \sin(2x)\, dx = 0.$$

Since $a_2 = b_2 = 0$, the Fourier polynomial of degree 2 is identical to the Fourier polynomial of degree 1. Let's look at the Fourier polynomial of degree 3:

$$a_3 = \frac{1}{\pi} \int_{-\pi}^{\pi} f(x) \cos(3x)\, dx = \frac{1}{\pi} \int_{0}^{\pi} 1 \cos(3x)\, dx = 0$$

and

$$b_3 = \frac{1}{\pi} \int_{-\pi}^{\pi} f(x) \sin(3x)\, dx = \frac{1}{\pi} \int_{0}^{\pi} 1 \sin(3x)\, dx = \frac{2}{3\pi}.$$

So the approximation is given by

$$f(x) \approx F_3(x) = \frac{1}{2} + \frac{2}{\pi} \sin x + \frac{2}{3\pi} \sin(3x).$$

The graph of F_3 is shown in Figure 9.20. This approximation is better than $F_1(x) = \frac{1}{2} + \frac{2}{\pi} \sin x$, as comparing Figure 9.20 to Figure 9.19 shows.

Without going through the details, we calculate the coefficients for higher-degree Fourier approximations:

$$F_5(x) = \frac{1}{2} + \frac{2}{\pi} \sin x + \frac{2}{3\pi} \sin(3x) + \frac{2}{5\pi} \sin(5x)$$

$$F_7(x) = \frac{1}{2} + \frac{2}{\pi} \sin x + \frac{2}{3\pi} \sin(3x) + \frac{2}{5\pi} \sin(5x) + \frac{2}{7\pi} \sin(7x).$$

Figure 9.21 shows that higher-degree approximations match the step-like nature of the square wave function more and more closely.

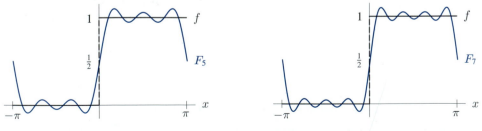

Figure 9.21: Fifth and seventh Fourier approximations to the square wave

We could have used a Taylor series to approximate the square wave, provided we did not center the series at a point of discontinuity. Since the square wave is a constant function on each interval, all its derivatives are zero, and so its Taylor series approximations are the constant functions: 0 or 1, depending on where the Taylor series is centered. They approximate the square wave perfectly on each piece, but they do not do a good job over the whole interval of length 2π. That is what Fourier polynomials succeed in doing: they approximate a curve fairly well everywhere, rather than just near a particular point. The Fourier approximations above look a lot like square waves, so they approximate well *globally*. However, they may not give good values near points of discontinuity. (For example, near $x = 0$, they all give values near $1/2$, which are incorrect.) Thus Fourier polynomials may not be good *local* approximations.

> Taylor polynomials give good *local* approximations to a function;
> Fourier polynomials give good *global* approximations to a function.

Fourier Series

As with Taylor polynomials, the higher the degree of the Fourier approximation, the more accurate it is. Therefore, we carry this procedure on indefinitely by letting $n \to \infty$, and we call the resulting infinite series a *Fourier series*.

> ### The Fourier Series for f on $[-\pi, \pi]$
>
> $$f(x) = a_0 + a_1 \cos x + a_2 \cos 2x + a_3 \cos 3x + \cdots$$
> $$+ b_1 \sin x + b_2 \sin 2x + b_3 \sin 3x + \cdots$$
>
> where a_k and b_k are the Fourier coefficients.

Thus, the Fourier series for the square wave is

$$f(x) = \frac{1}{2} + \frac{2}{\pi} \sin x + \frac{2}{3\pi} \sin 3x + \frac{2}{5\pi} \sin 5x + \frac{2}{7\pi} \sin 7x + \cdots .$$

Harmonics

Let us start with a function $f(x)$ that is periodic with period 2π, expanded in a Fourier series:

$$f(x) = a_0 + a_1 \cos x + a_2 \cos 2x + a_3 \cos 3x + \cdots$$
$$+ b_1 \sin x + b_2 \sin 2x + b_3 \sin 3x + \cdots$$

The function

$$a_k \cos kx + b_k \sin kx$$

is referred to as the kth *harmonic* of f, and it is customary to say that the Fourier series expresses f in terms of its harmonics. The first harmonic, $a_1 \cos x + b_1 \sin x$, is sometimes called the *fundamental harmonic* of f.

Example 2   Find a_0 and the first four harmonics of a *pulse train* function f of period 2π shown in Figure 9.22:

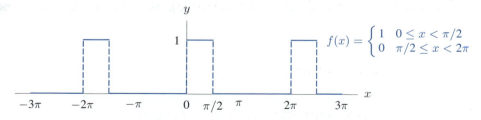

Figure 9.22: A train of pulses with period 2π

Solution   First, a_0 is the average value of the function, so

$$a_0 = \frac{1}{2\pi} \int_{-\pi}^{\pi} f(x)\, dx = \frac{1}{2\pi} \int_0^{\pi/2} 1\, dx = \frac{1}{4}.$$

Next, we compute a_k and b_k, $k = 1, 2, 3$, and 4. The formulas

$$a_k = \frac{1}{\pi} \int_{-\pi}^{\pi} f(x) \cos(kx)\, dx = \frac{1}{\pi} \int_0^{\pi/2} \cos(kx)\, dx$$

$$b_k = \frac{1}{\pi} \int_{-\pi}^{\pi} f(x) \sin(kx)\, dx = \frac{1}{\pi} \int_0^{\pi/2} \sin(kx)\, dx$$

lead to the harmonics

$$a_1 \cos x + b_1 \sin x = \frac{1}{\pi} \cos x + \frac{1}{\pi} \sin x$$

$$a_2 \cos(2x) + b_2 \sin(2x) = \frac{1}{\pi} \sin(2x)$$

$$a_3 \cos(3x) + b_3 \sin(3x) = -\frac{1}{3\pi} \cos(3x) + \frac{1}{3\pi} \sin(3x)$$

$$a_4 \cos(4x) + b_4 \sin(4x) = 0.$$

Figure 9.23 shows the graph of the sum of a_0 and these harmonics, which is the fourth Fourier approximation of f.

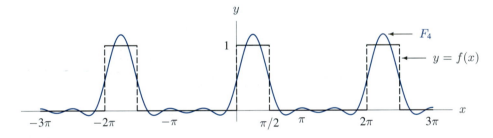

Figure 9.23: Fourth Fourier approximation to pulse train f equals the sum of a_0 and the first four harmonics

Energy and the Energy Theorem

The quantity $A_k = \sqrt{a_k^2 + b_k^2}$ is called the amplitude of the kth harmonic. The square of the amplitude has a useful interpretation. Adopting terminology from the study of periodic waves, we define the *energy* E of a periodic function f of period 2π to be the number

$$E = \frac{1}{\pi} \int_{-\pi}^{\pi} [f(x)]^2\, dx.$$

Problem 16 on page 469 asks you to verify that for all positive integers k,

$$\frac{1}{\pi} \int_{-\pi}^{\pi} (a_k \cos(kx) + b_k \sin(kx))^2\, dx = a_k^2 + b_k^2 = A_k^2.$$

This shows that the kth harmonic of f has energy A_k^2. The energy of the constant term a_0 of the Fourier series is $\frac{1}{\pi} \int_{-\pi}^{\pi} a_0^2\, dx = 2a_0^2$, so we make the definition

$$A_0 = \sqrt{2} a_0.$$

It turns out that for all reasonable periodic functions f, the energy of f equals the sum of the energy of its harmonics:

The Energy Theorem for a Periodic Function f of Period 2π

$$E = \frac{1}{\pi} \int_{-\pi}^{\pi} [f(x)]^2 \, dx = A_0^2 + A_1^2 + A_2^2 + \cdots$$

where $A_0 = \sqrt{2}a_0$ and $A_k = \sqrt{a_k^2 + b_k^2}$ (for all integers $k \geq 1$).

The graph of A_k^2 against k is called the *energy spectrum* of f. It shows how the energy of f is distributed among its harmonics.

Example 3　(a)　Graph the energy spectrum of the square wave of Example 1.

(b)　What fraction of the energy of the square wave is contained in the constant term and first three harmonics of its Fourier series?

Solution　(a)　We know from Example 1 that $a_0 = 1/2$, $a_k = 0$ for $k \geq 1$, $b_k = 0$ for k even, and $b_k = 2/(k\pi)$ for k odd. Thus

$$A_0^2 = 2a_0^2 = \frac{1}{2}$$

$$A_k^2 = 0 \quad \text{if } k \text{ is even}, \quad k \geq 1,$$

$$A_k^2 = \left(\frac{2}{k\pi}\right)^2 = \frac{4}{k^2\pi^2} \quad \text{if } k \text{ is odd}, \quad k \geq 1.$$

The energy spectrum is graphed in Figure 9.24. Notice that it is customary to represent the energy A_k^2 of the kth harmonic by a vertical line of length A_k^2. The graph shows that the constant term and first harmonic carry most of the energy of f.

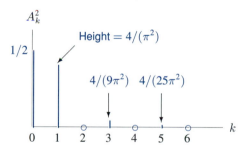

Figure 9.24: The energy spectrum of a square wave

(b)　The energy of the square wave $f(x)$ is

$$E = \frac{1}{\pi} \int_{-\pi}^{\pi} [f(x)]^2 \, dx = \frac{1}{\pi} \int_0^{\pi} 1 \, dx = 1.$$

The energy in the constant term and the first three harmonics of the Fourier series is

$$A_0^2 + A_1^2 + A_2^2 + A_3^2 = \frac{1}{2} + \frac{4}{\pi^2} + 0 + \frac{4}{9\pi^2} = 0.950.$$

The fraction of energy carried by the constant term and the first three harmonics is

$$0.95/1 = 0.95, \text{ or } 95\%.$$

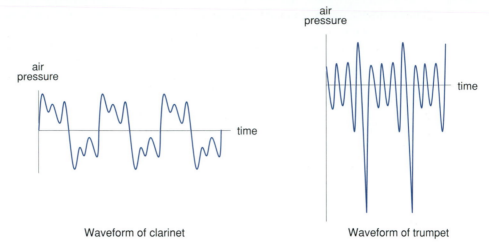

Figure 9.25: Sound waves of a clarinet and trumpet

Musical Instruments

You may have wondered why different musical instruments sound different, even when playing the same note. A first step might be to graph the periodic variations in air pressure that form the sound waves they produce. This has been done for clarinet and trumpet in Figure 9.25.[4] However, it is more revealing to graph the energy spectra of these functions, as in Figure 9.26. The most striking difference is the relative weakness of the second, fourth, and sixth harmonics for the clarinet, with the second harmonic completely absent. The trumpet sounds the second harmonic with as much energy as it does the fundamental.

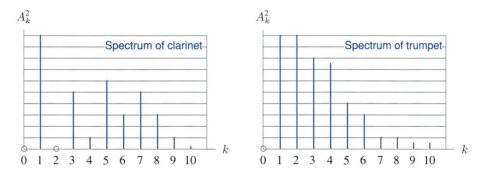

Figure 9.26: Energy spectra of a clarinet and trumpet

What Do We Do if Our Function Does Not Have Period 2π?

We can easily adapt what we have already done by changing variables. Suppose we have a function $f(x)$ that is periodic with period b. If we define $x = bt/2\pi$, then x varies over the interval $[-b/2, b/2]$ when t varies over the interval $[-\pi, \pi]$. Thus, if we substitute $x = bt/2\pi$ into f and define a new function g by

$$g(t) = f\left(\frac{bt}{2\pi}\right) = f(x),$$

then g has period 2π. We can find the Fourier series for g as before, giving

$$g(t) = a_0 + a_1 \cos t + a_2 \cos(2t) + a_3 \cos(3t) + \cdots$$
$$+ b_1 \sin t + b_2 \sin(2t) + b_3 \sin(3t) + \cdots.$$

---

[4] Adapted from C.A. Culver, *Musical Acoustics* (New York: McGraw-Hill, 1956) pp. 204, 220.

Substituting $t = 2\pi x/b$ allows us to convert back:

$$f(x) = g\left(\frac{2\pi x}{b}\right) = a_0 + a_1 \cos\left(\frac{2\pi x}{b}\right) + a_2 \cos\left(\frac{4\pi x}{b}\right) + a_3 \cos\left(\frac{6\pi x}{b}\right) + \cdots$$

$$+ b_1 \sin\left(\frac{2\pi x}{b}\right) + b_2 \sin\left(\frac{4\pi x}{b}\right) + b_3 \sin\left(\frac{6\pi x}{b}\right) + \cdots$$

Note that the terms in the Fourier series of a periodic function of period b are not $\cos(kx)$ and $\sin(kx)$, but instead are $\cos(2\pi kx/b)$ and $\sin(2\pi kx/b)$.

In addition, for a periodic function of period 2π, integrals over $[-\pi, \pi]$ can be replaced by integrals over any interval of length 2π. Thus the integrals for a_k and b_k can be evaluated over any interval of length 2π, not just $[-\pi, \pi]$.

Example 4   Find the fifth-degree Fourier polynomial of the square wave $f(x)$ graphed in Figure 9.27.

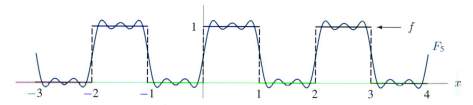

Figure 9.27: Square wave f and its fifth Fourier approximation F_5

Solution   Since f has period $b = 2$, we let $x = bt/2\pi = t/\pi$ and consider the function

$$g(t) = f\left(\frac{t}{\pi}\right) = f(x)$$

which has period 2π. Then

$$g(t) = \begin{cases} 0 & -\pi \leq t < 0 \\ 1 & 0 \leq t < \pi, \end{cases}$$

so $g(t)$ is the square wave of Example 1. We have, therefore,

$$g(t) \approx \frac{1}{2} + \frac{2}{\pi} \sin t + \frac{2}{3\pi} \sin(3t) + \frac{2}{5\pi} \sin(5t).$$

Hence, substituting $t = \pi x$,

$$f(x) = g(t) = g(\pi x) \approx \frac{1}{2} + \frac{2}{\pi} \sin(\pi x) + \frac{2}{3\pi} \sin(3\pi x) + \frac{2}{5\pi} \sin(5\pi x).$$

Seasonal Variation in the Incidence of Measles

Example 5   Fourier approximations have been used to analyze the seasonal variation in the incidence of diseases. One study[5] done in Baltimore, Maryland, for the years 1901–1931, studied $I(t)$, the average number of cases of measles per 10,000 susceptible children in the tth month of the year. The data points in Figure 9.28 show $f(t) = \log I(t)$. The curve in Figure 9.28 shows the second Fourier approximation of $f(t)$. Figure 9.29 contains the graphs of the first and second harmonics of $f(t)$, plotted separately as deviations about a_0, the average logarithmic incidence rate. Describe what these two harmonics tell you about incidence of measles.

[5]From C. I. Bliss and D. L. Blevins, *The Analysis of Seasonal Variation in Measles* (Am. J. Hyg. 70, 1959), reported by Edward Batschelet, *Introduction to Mathematics for the Life Sciences* (Springer-Verlag, Berlin, 1979).

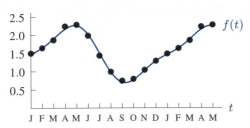

Figure 9.28: Logarithm of incidence of measles
per month (dots) and second Fourier
approximation (curve)

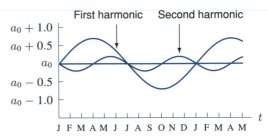

Figure 9.29: First and second harmonics of $f(t)$
plotted as deviations from average
log incidence rate, a_0

Solution   Taking the log of $I(t)$ has the effect of reducing the amplitude of the oscillations. However, since the log of a function increases when the function increases, and decreases when it decreases, oscillations in $f(t)$ correspond to oscillations in $I(t)$.

Figure 9.29 shows that the first harmonic in the Fourier series has a period of one year (the same period as the original function); the second harmonic has a period of six months. The graph in Figure 9.29 shows that the first harmonic is approximately a sine with amplitude about 0.7; the second harmonic is approximately the negative of a sine with amplitude about 0.2. Thus, for t in months ($t = 0$ in January),

$$\log I(t) = f(t) \approx a_0 + 0.7 \sin\left(\frac{\pi}{6}t\right) - 0.2 \sin\left(\frac{\pi}{3}t\right),$$

where $\pi/6$ and $\pi/3$ are introduced to make the periods 12 and 6 months, respectively. We can estimate a_0 from the original graph of f: it is the average value of approximately 1.5. Thus

$$f(t) \approx 1.5 + 0.7 \sin\left(\frac{\pi}{6}t\right) - 0.2 \sin\left(\frac{\pi}{3}t\right).$$

Figure 9.28 shows that the second Fourier approximation of $f(t)$ is quite good. The harmonics of $f(t)$ beyond the second must be rather insignificant. This suggests that the variation in incidence in measles comes from two sources, one with a yearly cycle that is reflected in the first harmonic and one with a half-yearly cycle reflected in the second harmonic. At this point the mathematics can tell us no more; we must turn to the epidemiologists for further explanation.

Informal Justification of the Formulas for the Fourier Coefficients

Recall that the coefficients in a Taylor series (which is a good approximation locally) are found by differentiation. In contrast, the coefficients in a Fourier series (which is a good approximation globally) are found by integration.

We will use a $\sum$ sign with ∞ at the top to denote an infinite series:

$$f(x) = a_0 + \sum_{k=1}^{\infty} a_k \cos(kx) + \sum_{k=1}^{\infty} b_k \sin(kx).$$

Consider the integral

$$\int_{-\pi}^{\pi} f(x)\, dx = \int_{-\pi}^{\pi} \left[a_0 + \sum_{k=1}^{\infty} a_k \cos(kx) + \sum_{k=1}^{\infty} b_k \sin(kx) \right] dx.$$

Splitting the integral into separate terms, and assuming we can interchange integration and summation, we get

$$\int_{-\pi}^{\pi} f(x)\,dx = \int_{-\pi}^{\pi} a_0\,dx + \int_{-\pi}^{\pi} \sum_{k=1}^{\infty} a_k \cos(kx)\,dx + \int_{-\pi}^{\pi} \sum_{k=1}^{\infty} b_k \sin(kx)\,dx$$

$$= \int_{-\pi}^{\pi} a_0\,dx + \sum_{k=1}^{\infty} \int_{-\pi}^{\pi} a_k \cos(kx)\,dx + \sum_{k=1}^{\infty} \int_{-\pi}^{\pi} b_k \sin(kx)\,dx.$$

But for $k \geq 1$, thinking of the integral as an area shows that

$$\int_{-\pi}^{\pi} \sin(kx)\,dx = 0 \ \text{ and } \ \int_{-\pi}^{\pi} \cos(kx)\,dx = 0,$$

so all terms drop out except the first, giving

$$\int_{-\pi}^{\pi} f(x)\,dx = \int_{-\pi}^{\pi} a_0\,dx = a_0 x \Big|_{-\pi}^{\pi} = 2\pi a_0$$

and so we get the following result:

$$\boxed{a_0 = \frac{1}{2\pi} \int_{-\pi}^{\pi} f(x)\,dx.}$$

Thus a_0 is the average value of f on the interval $[-\pi, \pi]$.

To determine the values of any of the other a_k (for positive k), we use a rather clever method which depends on the following facts. For all integers k and m,

$$\int_{-\pi}^{\pi} \sin(kx) \cos(mx)\,dx = 0,$$

and, provided $k \neq m$,

$$\int_{-\pi}^{\pi} \cos(kx) \cos(mx)\,dx = 0.$$

(See Problems 23–27 on page 470.) In addition, provided $m \neq 0$, we have

$$\int_{-\pi}^{\pi} \cos^2(mx)\,dx = \pi.$$

To use these facts, we multiply the Fourier series by $\cos(mx)$, where m is any positive integer:

$$f(x)\cos(mx) = a_0 \cos(mx) + \sum_{k=1}^{\infty} a_k \cos(kx)\cos(mx) + \sum_{k=1}^{\infty} b_k \sin(kx)\cos(mx).$$

We integrate this between $-\pi$ and π, term by term:

$$\int_{-\pi}^{\pi} f(x)\cos(mx)\,dx = \int_{-\pi}^{\pi} \left(a_0 \cos(mx) + \sum_{k=1}^{\infty} a_k \cos(kx)\cos(mx) + \sum_{k=1}^{\infty} b_k \sin(kx)\cos(mx) \right) dx$$

$$= a_0 \int_{-\pi}^{\pi} \cos(mx)\,dx + \sum_{k=1}^{\infty} \left(a_k \int_{-\pi}^{\pi} \cos(kx)\cos(mx)\,dx \right)$$

$$+ \sum_{k=1}^{\infty} \left(b_k \int_{-\pi}^{\pi} \sin(kx)\cos(mx)\,dx \right).$$

Provided $m \neq 0$, we have $\int_{-\pi}^{\pi} \cos(mx)\, dx = 0$. Since the integral $\int_{-\pi}^{\pi} \sin(kx)\cos(mx)\, dx = 0$, all the terms in the second sum are zero. Since $\int_{-\pi}^{\pi} \cos kx \cos mx\, dx = 0$ provided $k \neq m$, all the terms in the first sum are zero except where $k = m$. Thus the right-hand side reduces to one term:

$$\int_{-\pi}^{\pi} f(x)\cos(mx)\, dx = a_m \int_{-\pi}^{\pi} \cos(mx)\cos(mx)\, dx = \pi a_m.$$

This leads, for each value of $m = 1, 2, 3 \ldots$, to the following formula:

$$a_m = \frac{1}{\pi} \int_{-\pi}^{\pi} f(x)\cos(mx)\, dx.$$

Using a similar argument, we multiply through by $\sin(mx)$ instead of $\cos(mx)$ and eventually obtain, for each value of $m = 1, 2, 3 \ldots$, the following result:

$$b_m = \frac{1}{\pi} \int_{-\pi}^{\pi} f(x)\sin(mx) dx\, dx.$$

Problems for Section 9.5

Which of the series in Problems 1–4 are Fourier series?

1. $1 + \cos x + \cos^2 x + \cos^3 x + \cos^4 x + \cdots$

2. $\sin x + \sin(x + 1) + \sin(x + 2) + \cdots$

3. $\cos x + \sin x - \cos(2x) - \dfrac{\sin(2x)}{2} + \cos(3x) + \dfrac{\sin(3x)}{3} - \cdots$

4. $\dfrac{1}{2} - \sin x + \sin(2x) - \sin(3x) + \cdots$

5. Construct the first three Fourier approximations to the square wave function

    $$f(x) = \begin{cases} -1 & -\pi \leq x < 0 \\ 1 & 0 \leq x < \pi. \end{cases}$$

    Use a calculator or computer to draw the graph of each approximation.

6. Repeat Problem 5 with the function

    $$f(x) = \begin{cases} -x & -\pi \leq x < 0 \\ x & 0 \leq x < \pi. \end{cases}$$

7. What fraction of the energy of the function in Problem 6 is contained in the constant term and first three harmonics of its Fourier series?

For Problems 8–10, find the nth Fourier polynomial for the given functions, assuming them to be periodic with period 2π. Graph the first three approximations with the original function.

8. $f(x) = x^2, \quad -\pi < x \leq \pi.$ 9. $h(x) = \begin{cases} 0 & -\pi < x \leq 0 \\ x & 0 < x \leq \pi. \end{cases}$ 10. $g(x) = x, \quad -\pi < x \leq \pi.$

11. (a) Find and graph the third Fourier approximation of the square wave $g(x)$ of period 2π:

    $$g(x) = \begin{cases} 0 & -\pi \leq x < \pi/2 \\ 1 & -\pi/2 \leq x < \pi/2 \\ 0 & \pi/2 \leq x < \pi. \end{cases}$$

    (b) How does the result of part (a) differ from that of the square wave in Example 1?

12. Suppose we have a periodic function f with period 1 defined by $f(x) = x$ for $0 \leq x < 1$. Find the fourth degree Fourier polynomial for f and graph it on the interval $0 \leq x < 1$. [Hint: Remember that since the period is not 2π, you will have to start by doing a substitution. Notice that the terms in the sum are not $\sin(nx)$ and $\cos(nx)$, but instead turn out to be $\sin(2\pi nx)$ and $\cos(2\pi nx)$.]

13. Suppose f has period 2 and $f(x) = x$ for $0 \leq x < 2$. Find the fourth-degree Fourier polynomial and graph it on $0 \leq x < 2$. [Hint: See Problem 12.]

14. Suppose that a spacecraft near Neptune has measured a quantity A and sent it to earth in the form of a periodic signal $A \cos t$ of amplitude A. On its way to earth, the signal picks up periodic noise, containing only second and higher harmonics. Suppose that the signal $h(t)$ actually received on earth is graphed below in Figure 9.30. Determine the signal that the spacecraft originally sent and hence the value A of the measurement.

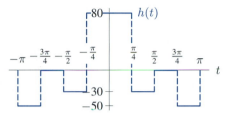

Figure 9.30

15. Figures 9.31 and 9.32 show the waveforms and energy spectra for notes produced by flute and bassoon.[6] Describe the principal differences between the two spectra.

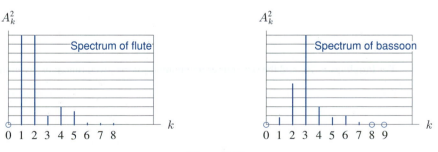

Figure 9.31

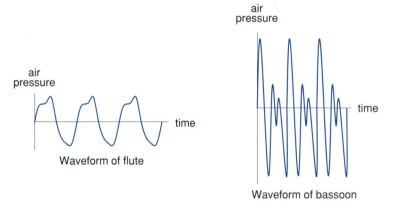

Figure 9.32

16. Show that for positive integers k, the periodic function $f(x) = a_k \cos kx + b_k \sin kx$ of period 2π has energy $a_k^2 + b_k^2$.

---

[6] Adapted from C.A. Culver, *Musical Acoustics* (New York: McGraw-Hill, 1956), pp. 200, 213.

17.   Given the graph of f in Figure 9.33, find the first two Fourier approximations numerically.

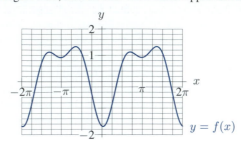

Figure 9.33

18.   Justify the formula $b_k = \frac{1}{\pi} \int_{-\pi}^{\pi} f(x) \sin(kx)\, dx$ for the Fourier coefficients, b_k, of a periodic function of period 2π. The argument is similar to that in the text for a_k. In addition to some of the formulas used there, you will need the formulas: $\int_{-\pi}^{\pi} \sin^2(kx)\, dx = \pi$ and $\int_{-\pi}^{\pi} \sin(kx) \sin(mx)\, dx = 0$ for $k \neq m$.

In Problems 19–22, the pulse train of width c is the periodic function f of period 2π given by

$$f(x) = \begin{cases} 0 & -\pi \leq x < -c/2 \\ 1 & -c/2 \leq x < c/2 \\ 0 & c/2 \leq x < \pi. \end{cases}$$

19.   Suppose that f is the pulse train of width 1.

   (a)   What fraction of the energy of f is contained in the constant term of its Fourier series? In the constant term and the first harmonic together?
   (b)   Find a formula for the energy of the kth harmonic of f. Use it to sketch the energy spectrum of f.
   (c)   How many terms of the Fourier series of f are needed to capture 90% of the energy of f?
   (d)   Graph f and its fifth Fourier approximation on the interval $[-3\pi, 3\pi]$.

20.   Suppose that f is the pulse train of width 0.4.

   (a)   What fraction of the energy of f is contained in the constant term of its Fourier series? In the constant term and the first harmonic together?
   (b)   Find a formula for the energy of the kth harmonic of f. Use it to sketch the energy spectrum of f.
   (c)   What fraction of the energy of f is contained in the constant term and the first five harmonics of f? (The constant term and the first thirteen harmonics are needed to capture 90% of the energy of f.)
   (d)   Graph f and its fifth Fourier approximation on the interval $[-3\pi, 3\pi]$.

21.   Suppose that f is the pulse train of width 2.

   (a)   What fraction of the energy of f is contained in the constant term of its Fourier series? In the constant term and the first harmonic together?
   (b)   How many terms of the Fourier series of f are needed to capture 90% of the energy of f?
   (c)   Graph f and its third Fourier approximation on the interval $[-3\pi, 3\pi]$.

22.   After working Problems 19–21, write a paragraph about the approximation of pulse trains by Fourier polynomials. Explain how the energy spectrum of a pulse train of width c will change as c gets closer and closer to 0 and how this affects the number of terms required for an accurate approximation.

For Problems 23–27, use the table of integrals inside the back cover to show that the following statements are true for positive integers k and m.

23. $\displaystyle\int_{-\pi}^{\pi} \cos(kx) \cos(mx)\, dx = 0, \quad \text{if } k \neq m.$ 24. $\displaystyle\int_{-\pi}^{\pi} \cos^2(mx)\, dx = \pi.$

25. $\displaystyle\int_{-\pi}^{\pi} \sin^2(mx)\, dx = \pi.$ 26. $\displaystyle\int_{-\pi}^{\pi} \sin(kx) \cos(mx)\, dx = 0.$

27. $\displaystyle\int_{-\pi}^{\pi} \sin(kx) \sin(mx)\, dx = 0, \quad \text{if } k \neq m.$

CHAPTER SUMMARY

- **Taylor series and polynomials**
  General expansion about $x = 0$ or $x = a$; specific series for $e^x, \sin x, \cos x, (1 + x)^p$; using known Taylor series to find others by substitution, integration, and differentiation; interval of convergence

- **Geometric series**
  Finite sum, infinite sum

- **Power series**
  Ratio test for radius of convergence

- **Series of constants**
  Harmonic series, alternating series

- **Fourier series**

REVIEW PROBLEMS FOR CHAPTER NINE

For Problems 1–3, find the second-degree Taylor polynomial about the given point.

1. $e^x,$ $x = 1$ **2.** $\ln x,$ $x = 2$ **3.** $\sin x,$ $x = -\pi/4$

4. Find the third-degree Taylor polynomial for $f(x) = x^3 + 7x^2 - 5x + 1$ at $x = 1$.

In Problems 5–8, find the first four nonzero terms of the Taylor series about the origin of the given functions.

5. $\theta^2 \cos \theta^2$ **6.** $\sin t^2$ **7.** $\dfrac{1}{\sqrt{4 - x}}$ **8.** $\dfrac{1}{1 - 4z^2}$

For Problems 9–10, expand the quantity in a Taylor series around the origin in terms of the variable given. Give the first four nonzero terms.

9. $\dfrac{a}{a + b}$ in terms of $\dfrac{b}{a}$ **10.** $\sqrt{R - r}$ in terms of $\dfrac{r}{R}$

Find the exact value of the sums of the series in Problems 11–14.

11. $1 - \dfrac{1}{3} + \dfrac{1}{9} - \dfrac{1}{27} + \dfrac{1}{81} - \cdots$ **12.** $8 + 4 + 2 + 1 + \dfrac{1}{2} + \dfrac{1}{4} + \cdots + \dfrac{1}{2^{10}}$

13. $3 + 3 + \dfrac{3}{2!} + \dfrac{3}{3!} + \dfrac{3}{4!} + \dfrac{3}{5!} + \cdots$ **14.** $(0.1)^2 - \dfrac{(0.1)^4}{3!} + \dfrac{(0.1)^6}{5!} - \dfrac{(0.1)^8}{7!} + \cdots$

15. Suppose x is positive but very small. Arrange the following expressions in increasing order:

$$x, \quad \sin x, \quad \ln(1 + x), \quad 1 - \cos x, \quad e^x - 1, \quad \arctan x, \quad x\sqrt{1 - x}.$$

16. Consider the Taylor expansion

$$f(x) = \frac{1}{1 + x} = 1 - x + x^2 - x^3 + x^4 - \cdots$$

By plotting several Taylor polynomials and the function $f(x) = 1/(1 + x)$, confirm that the interval of convergence of this series is $-1 < x < 1$.

17. Use Taylor series to evaluate $\displaystyle\lim_{x \to 0} \frac{\ln(1 + x + x^2) - x}{\sin^2 x}$.

18. (a) Find $\lim\limits_{\theta \to 0} \dfrac{\sin(2\theta)}{\theta}$. Explain your reasoning.

    (b) Use series to explain why $f(\theta) = \dfrac{\sin(2\theta)}{\theta}$ looks like a parabola near $\theta = 0$. What is the equation of the parabola?

19. (a) Find the Taylor series for $f(t) = te^t$ about $t = 0$.

    (b) Using your answer to part (a), find a Taylor series expansion about $x = 0$ for

    $$\int_0^x te^t \, dt.$$

    (c) Using your answer to part (b), show that

    $$\frac{1}{2} + \frac{1}{3} + \frac{1}{4(2!)} + \frac{1}{5(3!)} + \frac{1}{6(4!)} + \cdots = 1.$$

20. The theory of relativity predicts that when an object moves at speeds close to the speed of light, the object appears heavier. The apparent, or relativistic, mass, m, of the object when it is moving at speed v is given by the formula

    $$m = \frac{m_0}{\sqrt{1 - v^2/c^2}}$$

    where c is the speed of light and m_0 is the mass of the object when it is at rest.

    (a) Use the formula for m to decide what values of v are possible.
    (b) Sketch a rough graph of m against v, labeling intercepts and asymptotes.
    (c) Write the first three nonzero terms of the Taylor series for m in terms of v.
    (d) For what values of v do you expect the series to converge?

21. The potential energy, V, of two gas molecules separated by a distance r is given by

    $$V = -V_0 \left(2 \left(\frac{r_0}{r} \right)^6 - \left(\frac{r_0}{r} \right)^{12} \right),$$

    where V_0 and r_0 are positive constants.

    (a) Show that if $r = r_0$, then V takes on its minimum value, $-V_0$.
    (b) Write V as a series in $(r - r_0)$ up through the quadratic term.
    (c) For r near r_0, show that the difference between V and its minimum value is approximately proportional to $(r - r_0)^2$. In other words, show that $V - (-V_0) = V + V_0$ is approximately proportional to $(r - r_0)^2$.
    (d) The force, F, between the molecules is given by $F = -dV/dr$. What is F when $r = r_0$? For r near r_0, show that F is approximately proportional to $(r - r_0)$.

22. The *gravitational field* at a point in space is the gravitational force that would be exerted on a unit mass placed there. We will assume that the gravitational field strength at a distance d away from a mass M is

    $$\frac{GM}{d^2}$$

    where G is constant. In this problem you will investigate the gravitational field strength, F, exerted by a system consisting of a large mass M and a small mass m, with a distance r between them. (See Figure 9.34.)

Figure 9.34

    (a) Write an expression for the gravitational field strength, F, at the point P.
    (b) Assuming r is small in comparison to R, expand F in a series in r/R.
    (c) By discarding terms in $(r/R)^2$ and higher powers, explain why you can view the field as resulting from a single particle of mass $M + m$, plus a correction term. What is the position of the particle of mass $M + m$? Explain the sign of the correction term.

23. A repeating decimal can always be expressed as a fraction. This problem shows how writing a repeating decimal as a geometric series enables you to find the fraction. Consider the decimal $0.232323\ldots$.

    (a) Use the fact that $0.232323\ldots = 0.23 + 0.0023 + 0.000023 + \cdots$ to write $0.232323\ldots$ as a geometric series.
    (b) Use the formula for the sum of a geometric series to show that $0.232323\ldots = 23/99$.

24. Cephalexin is an antibiotic with a half-life in the body of 0.9 hours, taken in tablets of 250 mg every six hours.

    (a) What percentage of the cephalexin in the body at the start of a six-hour period is still there at the end (assuming no tablets are taken during that time)?
    (b) Write an expression for Q_1, Q_2, Q_3, Q_4, where Q_n mg, is the amount of cephalexin in the body right after the nth tablet is taken.
    (c) Express Q_3, Q_4 in closed-form and evaluate them.
    (d) Write an expression for Q_n and put it in closed-form.
    (e) If the patient keeps taking the tablets, use your answer to part (d) to find the quantity of cephalexin in the body in the long run, right after taking a pill.

25. Expand $f(x + h)$ and $g(x + h)$ in Taylor series and take a limit to confirm the product rule:

    $$\frac{d}{dx}(f(x)g(x)) = f'(x)g(x) + f(x)g'(x).$$

26. Use Taylor expansions for $f(y + k)$ and $g(x + h)$ to confirm the chain rule:

    $$\frac{d}{dx}(f(g(x))) = f'(g(x)) \cdot g'(x).$$

27. Suppose all the derivatives of g exist at $x = 0$ and that g has a critical point at $x = 0$.

    (a) Write the nth Taylor polynomial for g at $x = 0$.
    (b) What does the Second Derivative test for local maxima and minima say?
    (c) Use the Taylor polynomial to explain why the Second Derivative test works.

28. (Continuation of Problem 27) You may remember that the Second Derivative test tells us nothing when the second derivative is zero at the critical point. In this problem you will investigate that special case.

    Assume g has the same properties as in Problem 27, and that, in addition, $g''(0) = 0$. What does the Taylor polynomial tell you about whether g has a local maximum or minimum at $x = 0$?

29. Use the Fourier polynomials for the square wave

    $$f(x) = \begin{cases} -1 & -\pi < x \le 0 \\ 1 & 0 < x \le \pi \end{cases}$$

    to explain why

    $$1 - \frac{1}{3} + \frac{1}{5} - \frac{1}{7} + \cdots + (-1)^{2n+1}\frac{1}{2n+1}$$

    must approach $\frac{\pi}{4}$ as $n \to \infty$.

30. Find a Fourier polynomial of degree three for $f(x) = e^{2\pi x}$, for $0 \le x < 1$.

31. Suppose that $f(x)$ is a differentiable periodic function of period 2π. Assume the Fourier series of f is differentiable term by term.

    (a) If the Fourier coefficients of f are a_k and b_k, show that the Fourier coefficients of its derivative f' are kb_k and $-ka_k$.
    (b) How are the amplitudes of the harmonics of f and f' related?
    (c) How are the energy spectra of f and f' related?

32. If the Fourier coefficients of f are a_k and b_k, and the Fourier coefficients of g are c_k and d_k, and if A and B are real, show that the Fourier coefficients of $Af + Bg$ are $Aa_k + Bc_k$ and $Ab_k + Bd_k$.

33. Suppose that f is a periodic function of period 2π and that g is a horizontal shift of f, say $g(x) = f(x+c)$. Show that f and g have the same energy.

PROJECTS

1. **Approximating the Derivative**[7]

    In applications, the values of a function $f(x)$ are frequently known only at discrete values x_0, $x_0 \pm h$, $x_0 \pm 2h$,.... Suppose we are interested in approximating the derivative $f'(x_0)$. The definition

    $$f'(x_0) = \lim_{h \to 0} \frac{f(x_0 + h) - f(x_0)}{h}$$

    suggests that for small h we can approximate $f'(x)$ as follows:

    $$f'(x_0) \approx \frac{f(x_0 + h) - f(x_0)}{h}.$$

    Such *finite-difference approximations* are used frequently in programming a computer to solve differential equations.

    Taylor series can be used to analyze the error in this approximation. Substituting

    $$f(x_0 + h) = f(x_0) + f'(x_0)h + \frac{f''(x_0)}{2}h^2 + \cdots$$

    into the approximation for $f'(x_0)$, we find

    $$\frac{f(x_0 + h) - f(x_0)}{h} = f'(x_0) + \frac{f''(x_0)}{2}h + \cdots$$

    This suggests (and it can be proved) that the error in the approximation is bounded as follows:

    $$\left| \frac{f(x_0 + h) - f(x_0)}{h} - f'(x_0) \right| \leq \frac{Mh}{2},$$

    where

    $$|f''(x)| \leq M \qquad \text{for} \qquad |x - x_0| \leq |h|.$$

    Notice that as $h \to 0$, the error also goes to zero, provided M is bounded.

    As an example, we take $f(x) = e^x$ and $x_0 = 0$, so $f'(x_0) = 1$. The error for various values of h are given in Table 9.3. We see that decreasing h by a factor of 10 decreases the error by a factor of about 10, as predicted by the error bound $Mh/2$.

    TABLE 9.3

h	$(f(x_0 + h) - f(x_0))/h$	Error
10^{-1}	1.05171	5.171×10^{-2}
10^{-2}	1.00502	5.02×10^{-3}
10^{-3}	1.00050	5.0×10^{-4}
10^{-4}	1.00005	5.0×10^{-5}

    (a) Using Taylor series, suggest an error bound for each of the following finite-difference approximations.

    (i) $f'(x_0) \approx \dfrac{f(x_0) - f(x_0 - h)}{h}$

    (ii) $f'(x_0) \approx \dfrac{f(x_0 + h) - f(x_0 - h)}{2h}$

    (iii) $f'(x_0) \approx \dfrac{-f(x_0 + 2h) + 8f(x_0 + h) - 8f(x_0 - h) + f(x_0 - 2h)}{12h}$

---

[7]From Mark Kunka

(b) Use each of the formulas in part (a) to approximate the first derivative of e^x at $x = 0$ for $h = 10^{-1}, 10^{-2}, 10^{-3}, 10^{-4}$. As h is decreased by a factor of 10, how does the error decrease? Does this agree with the error bounds found in part (a)? Which is the most accurate formula?

(c) Repeat part (b) using $f(x) = 1/x$ and $x_0 = 10^{-5}$. Why are these formulas not good approximations anymore? Continue to decrease h by factors of 10. How small does h have to be before formula (iii) is the best approximation? At these smaller values of h, what changed to make the formulas accurate again?

2. **Probability of Winning in Sports**

In certain sports, winning a game requires a lead of two points. That is, if the score is tied you have to score two points in a row to win.

(a) For some sports (e.g. tennis), a point is scored every play. Suppose your probability of scoring the next point is always p. Then, your opponent's probability of scoring the next point is always $1 - p$.

   (i) What is the probability that you win the next two points?

   (ii) What is the probability that you and your opponent split the next two points, that is, that neither of you wins both points?

   (iii) What is the probability that you split the next two points but you win the two after that?

   (iv) What is the probability that you either win the next two points or split the next two and then win the next two after that?

   (v) Give a formula for your probability w of winning a tied game.

   (vi) Compute your probability of winning a tied game when $p = 0.5$; when $p = 0.6$; when $p = 0.7$; when $p = 0.4$. Comment on your answers.

(b) In other sports (e.g. volleyball), you can score a point only if it is your turn, with turns alternating until a point is scored. Suppose your probability of scoring a point when it is your turn is p, and your opponent's probability of scoring a point when it is her turn is q.

   (i) Find a formula for the probability S that you are the first to score the next point, assuming it is currently your turn.

   (ii) Suppose that if you score a point, the next turn is yours. Using your answers to part (a) and your formula for S, compute the probability of winning a tied game (if you need two points in a row to win).

   - Assume $p = 0.5$ and $q = 0.5$ and it is your turn.

   - Assume $p = 0.6$ and $q = 0.5$ and it is your turn.

FOCUS ON THEORY

CONVERGENCE THEOREMS

In this section[8] we study some methods for analyzing the convergence of the power series

$$\sum_{n=0}^{\infty} C_n x^n = C_0 + C_1 x + C_2 x^2 + \cdots + C_n x^n + \cdots.$$

For any particular value of x, the power series is a series of constants of the form

$$\sum_{n=1}^{\infty} a_n = a_1 + a_2 + a_3 + \cdots + a_n + \cdots.$$

So we first consider the convergence of series of constants. We can plot the terms of a series of constants as in Figure 9.35. If $a_n \geq 0$ for all n, then each rectangle has area a_n.

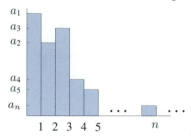

Figure 9.35: Height and area of the nth rectangle is a_n

Convergence of Series of Constants

We define the nth partial sum, S_n, as

$$S_n = a_1 + a_2 + \cdots + a_n,$$

and we say that

> If $S = \lim_{n \to \infty} S_n$ exists, then the series $\sum a_n$ **converges**, and its sum is S. If a series does not converge, we say that it **diverges**.

If $a_n \geq 0$ for all n, the series converges when the total area of the rectanges in Figure 9.35 is finite. In that case, the sum of the series is the total area of all the rectangles. This is similar to an improper integral $\int_0^{\infty} f(x)\,dx$, where we saw that it was possible for the area under the graph of f to be finite, even on an infinite interval.

Here are some properties that are useful in determining whether or not a series converges.

> ### Theorem: Convergence Properties of Series
>
> 1. If $\sum a_n$ and $\sum b_n$ converge and if k is a constant, then
>   - $\sum (a_n + b_n)$ converges to $\sum a_n + \sum b_n$.
>   - $\sum k a_n$ converges to $k \sum a_n$.
> 2. Changing a finite number of terms in a series does not change whether or not it converges, although it may change its sum if it does converge.
> 3. If $\sum a_n$ converges, then $\lim_{n \to \infty} a_n = 0$.

---

[8]Based on notes by Lynne Small.

For proofs of these properties, see Problems 1–3. Notice that Property 3 can often be used to tell us that a series does *not* converge. It says that if the terms do not go to 0, the series has no chance of converging. However, knowing that $\lim\limits_{n\to\infty} a_n = 0$ is *not* enough to ensure convergence. For example, in Example 7 on page 442 we showed that the harmonic series

$$1 + \frac{1}{2} + \frac{1}{3} + \frac{1}{4} + \cdots + \frac{1}{n} + \cdots$$

does not converge, even though

$$\lim_{n\to\infty} a_n = \lim_{n\to\infty} \frac{1}{n} = 0.$$

Example 1 Show that the geometric series

$$1 + x + x^2 + \cdots$$

does not converge if $x = \pm 1$.

Solution We can use Property 3 to see that the series does not converge when $x = \pm 1$. When $x = 1$, the nth term is $a_n = 1$, so $\lim_{n\to\infty} a_n = 1 \neq 0$. When $x = -1$, then $a_n = (-1)^{n-1}$, so $\lim_{n\to\infty} a_n$ does not exist. So, in either case, $\sum a_n$ does not converge.

Calculuating the partial sums confirms this. If $x = 1$, the series is $1 + 1 + \cdots$. Thus the nth partial sum is $S_n = n$, so the partial sums have no limit. Therefore the series diverges. If $x = -1$, the series is $1 - 1 + 1 - \cdots$. The nth partial sum is 1 if n is odd and 0 if n is even. So, again the partial sums have no limit and the series diverges.

Comparing Series

When we studied improper integrals in Chapter 7, we saw that it was sometimes useful to compare one integral with another in deciding whether the integral converges. In the same way, we can decide whether a series converges by comparing it with one that we already know converges or diverges.

Theorem: Comparison Test

Suppose $0 \leq a_n \leq b_n$ for all n.
- If $\sum b_n$ converges, then $\sum a_n$ converges.
- If $\sum a_n$ does not converge, then $\sum b_n$ does not converge.

Since $a_n \leq b_n$, the plot of the a_n lies under the plot of the b_n. (See Figure 9.36.) The comparison test says that if the total area for $\sum b_n$ is finite, then the total area for $\sum a_n$ is finite also; and that if the total area for $\sum a_n$ is not finite, then neither is the total area for $\sum b_n$.

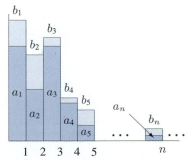

Figure 9.36: Each a_n is represented by the area of a dark rectangle, and each b_n by a dark plus a light rectangle

Example 2   Show that if $0 < p < 1$, then $\sum_{n=1}^{\infty} 1/n^p$ diverges.

Solution   Since $p < 1$, we have $n^p \leq n$, so $1/n^p \geq 1/n$. Since the harmonic series $\sum 1/n$ diverges, $\sum 1/n^p$ diverges also.

Infinite Series and Improper Integrals

Just as we can compare a series with a series, so we can compare a series with an improper integral as we did in Example 7 on page 442 to show that the harmonic series diverges. Here is an example where this type of comparison is used to prove convergence.

Example 3   Show that the series

$$\sum_{n=1}^{\infty} \frac{1}{n^2} = 1 + \frac{1}{4} + \frac{1}{9} + \cdots$$

converges by comparing it to the improper integral $\int_1^{\infty} (1/x^2)\, dx$.

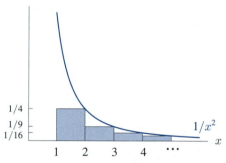

Figure 9.37: $\int_1^{\infty} (1/x^2)\, dx$ is greater than $\sum_{n=2}^{\infty} 1/n^2$

Solution   Figure 9.37 shows that the area under the graph of $1/x^2$ for $x \geq 1$ is greater than $1/4 + 1/9 + \cdots$. Since

$$\int_1^{\infty} \frac{1}{x^2}\, dx = \lim_{b \to \infty} \int_1^b \frac{1}{x^2}\, dx = \lim_{b \to \infty} \left(\frac{-1}{b} + 1 \right) = 1,$$

the area under the graph is finite, hence $1/4 + 1/9 + \cdots$ converges. So $1 + 1/4 + 1/9 + \cdots$ also converges.

Series of Both Positive and Negative Terms

If $\sum_n a_n$ has both positive and negative terms, then its plot has rectangles lying both above and below the x-axis. The total area of the rectangles is no longer equal to $\sum a_n$. However, it is still true that if the total area is finite, then the series converges. The area of the nth rectangle is $|a_n|$, so we have:

> If $\sum |a_n|$ converges, then so does $\sum a_n$.

Problem 10 shows how to prove this result.

Example 4   Explain why the series for $\cos x$ converges.

Solution   The series for $\cos x$ can be written

$$1 + 0x - \frac{x^2}{2!} + 0x^3 + \frac{x^4}{4!} + 0x^5 - \frac{x^6}{6!} + \cdots,$$

where the zero terms have been included to enable us to compare with the series for e^x. We consider the series

$$1 + 0|x| + \frac{|x^2|}{2!} + 0|x^3| + \frac{|x^4|}{4!} + 0|x^5| + \frac{|x^5|}{5!} + \cdots.$$

This series converges by comparison with the series for $e^{|x|}$. Hence the series for $\cos x$ converges.

The Ratio Test

In a geometric series, the ratio between the $(n+1)^{\text{st}}$ term and the nth term is a constant, r. If $|r| < 1$, the series converges, and if $|r| \geq 1$, the series diverges. We now state a similar test for a general series of constants, involving the limit of the ratio between the $(n+1)^{\text{st}}$ and n^{th} terms.

The Ratio Test for Series of Constants

For a series $\sum a_n$, suppose

$$\lim_{n \to \infty} \frac{|a_{n+1}|}{|a_n|} = L.$$

- If $L < 1$, then $\sum a_n$ converges.
- If $L > 1$, then $\sum a_n$ does not converge.
- If $L = 1$, the test does not tell us anything.

Proof   The basic idea is to compare the series with a geometric series.

Suppose $L < 1$. Let r be a number between L and 1, so that $L < r < 1$. Since

$$\lim_{n \to \infty} \frac{|a_{n+1}|}{|a_n|} = L,$$

for sufficiently large n, we have

$$\frac{|a_{n+1}|}{|a_n|} < r.$$

Then,

$$|a_{n+1}| < |a_n|r,$$
$$|a_{n+2}| < |a_{n+1}|r < |a_n|r^2,$$
$$|a_{n+3}| < |a_{n+2}|r < |a_n|r^3,$$

and so on. Thus, for a sufficiently large n, we have

$$|a_n| + |a_{n+1}| + |a_{n+2}| + \cdots < |a_n| + |a_n|r + |a_n|r^2 + |a_n|r^3 + \cdots$$
$$= |a_n|(1 + r + r^2 + r^3 + \cdots).$$

Thus, $|a_n| + |a_{n+1}| + |a_{n+2}| + \cdots$ converges by comparison with the convergent geometric series $|a_n|(1 + r + r^2 + r^3 + \cdots)$. Hence, $\sum |a_n|$ converges and so $\sum a_n$ converges.

If $L > 1$, then we choose r so that $1 < r < L$. Then, by a similar argument, the terms in $\sum a_n$ eventually have greater magnitude than the terms in a geometric series with common ratio r. Since $r > 1$, the geometric series does not converge, so $\sum a_n$ does not converge.

Power Series

We are now ready to prove the results about convergence of power series that we used in Section 9.2.

Interval of Convergence

For a power series $\sum C_n x^n$:

- If the series converges for $x = b$, then it converges for all x with $|x| < |b|$.
- If the series does not converge for $x = b$, then it does not converge for all x with $|x| > |b|$.

Proof   We assume[9] $b \neq 0$. If $\sum C_n b^n$ converges, then

$$\lim_{n \to \infty} C_n b^n = 0.$$

In particular, for sufficiently large n, we have

$$|C_n b^n| < 1.$$

For values of x with $|x| < |b|$, we have $|x|/|b| < 1$. Let $r = |x|/|b|$. Then, for sufficiently large n,

$$|C_n x^n| = |C_n b^n| \left| \frac{x^n}{b^n} \right| < \left| \frac{x}{b} \right|^n = r^n.$$

Therefore for x with $|x| < |b|$, the series $\sum |C_n x^n|$ converges by comparison with the convergent geometric series $\sum r^n$. Hence $\sum C_n x^n$ converges too.

For the second part of the theorem, we are told that $\sum C_n b^n$ does not converge. Let us assume that there is a value of x, with $|x| > |b|$, for which the series $\sum C_n x^n$ converges. Then, the first part of the theorem, with the roles of x and b reversed, says that the series $\sum C_n b^n$ would have to converge. But this contradicts the assumption and so completes the proof.

Ratio Test for Power Series

For the power series $\sum C_n x^n$, suppose that

$$\lim_{n \to \infty} \frac{|C_n|}{|C_{n+1}|} = R \quad \text{and } 0 < R < \infty.$$

Then the power series converges for x with $|x| < R$ and does not converge for x with $|x| > R$.

Proof   For $x = 0$, the power series has only one nonzero term and so is convergent. For any fixed x with $|x| \neq 0$, the series $\sum C_n x^n$ is a series of constants. We can apply the ratio test:

$$\lim_{n \to \infty} \frac{|a_{n+1}|}{|a_n|} = \lim_{n \to \infty} \frac{|C_{n+1} x^{n+1}|}{|C_n x^n|} = \lim_{n \to \infty} \frac{|C_{n+1}||x|}{|C_n|} = |x| \lim_{n \to \infty} \frac{|C_{n+1}|}{|C_n|} = \frac{|x|}{R}.$$

By the ratio test,

If $|x|/R < 1$, that is $|x| < R$, then the series converges.

If $|x|/R > 1$, that is, $|x| > R$, then the series does not converge.

---

[9]If $b = 0$, there are no x's where $|x| < |b|$, and $\sum C_n b^n$ must converge.

Problems on Convergence Theorems

1. Show that if $\sum a_n$ and $\sum b_n$ converge and if k is a constant, then $\sum (a_n + b_n)$, $\sum (a_n - b_n)$, and $\sum k a_n$ converge.

2. Let N be a positive integer. Show that if $a_n = b_n$ for $n \geq N$, then $\sum a_n$ and $\sum b_n$ either both converge, or both diverge. In other words, changing a finite number of terms in a series does not change whether or not it converges.

3. Show that if $\sum a_n$ converges, then $\lim_{n \to \infty} a_n = 0$. [Hint: Consider $\lim_{n \to \infty} (S_n - S_{n-1})$, where S_n is the nth partial sum.]

Use the Comparison Test to determine whether the series in Problems 4–5 converge.

4. $\displaystyle\sum_{n=1}^{\infty} \frac{1}{n^3}$

5. $\displaystyle\sum_{n=2}^{\infty} \frac{1}{\ln n}$

By comparing to an improper integral, decide whether the series in Problems 6–9 converge.

6. $1 + \dfrac{1}{5} + \dfrac{1}{9} + \dfrac{1}{13} + \dfrac{1}{17} + \cdots + \dfrac{1}{4n-3} + \cdots$

7. $\dfrac{1}{2} + \dfrac{2}{5} + \dfrac{3}{10} + \dfrac{4}{17} + \dfrac{5}{26} + \cdots + \dfrac{n}{n^2+1} + \cdots$

8. $1 + \dfrac{1}{2^{3/2}} + \dfrac{1}{3^{3/2}} + \dfrac{1}{4^{3/2}} + \dfrac{1}{5^{3/2}} + \cdots + \dfrac{1}{n^{3/2}} + \cdots$

9. $1 + \dfrac{1}{2^p} + \dfrac{1}{3^p} + \cdots + \dfrac{1}{n^p} + \cdots$, where $p > 1$.

10. For a series $\sum a_n$:
    (a) Show that $0 \leq a_n + |a_n| \leq 2|a_n|$.
    (b) Show that if $\sum |a_n|$ converges, then $\sum a_n$ converges.

11. Show that if $C_0 + C_1 x + C_2 x^2 + C_3 x^3 + \cdots$ converges by the ratio test with radius of convergence R, then so does $C_1 + 2C_2 x + 3C_3 x^2 + \cdots$.

THE ERROR IN TAYLOR APPROXIMATIONS

In order to use an approximation intelligently, we need to be able to estimate the size of the error, which is the difference between the exact answer (which we usually do not know) and the approximate value. In general, we cannot find the exact value of the error; if we could, we would not need the approximation. However, we can often find a maximum possible value for the error. This gives us a feel for how good the approximation is.

We now consider how to assess the size of the error involved in using $P_n(x)$, the nth degree Taylor polynomial, to approximate $f(x)$ for values of x near a. The error is the difference

$$E_n(x) = f(x) - P_n(x).$$

If E_n is positive, the approximation is smaller than the true value. If E_n is negative, the approximation is too large. Often we are only interested in the magnitude of the error, $|E_n|$.

The Error in $P_1(x)$

Let's try to bound the error in the first-degree Taylor polynomial approximation to a function $f(x)$. For simplicity, we will assume the Taylor polynomial is centered at $x = 0$. We have

$$f(x) \approx P_1(x) = f(0) + f'(0)x.$$

The error in using $P_1(x)$ depends on how straight or curved the graph of f is for values of x near 0. The more curved f is, the worse the approximation. Thus we expect that the size of the second derivative $f''(x)$ will be important. If we know how big $f''(x)$ is, we should be able to tell how good or bad the approximation can be.

Suppose we know that $f''(x)$ is bounded above by M and bounded below by L for values of x near 0, say for $0 \leq x \leq d$. Since $f''(x) < M$, we have

$$\int_0^x f''(x)\,dx \leq \int_0^x M\,dx, \quad \text{for} \quad 0 \leq x \leq d.$$

Since integrating the second derivative gives the first derivative, the Fundamental Theorem of Calculus tells us that $\int_0^x f''(x)\,dx = f'(x) - f'(0)$. The previous inequality gives us

$$f'(x) - f'(0) \leq M(x - 0),$$

so

$$f'(x) \leq f'(0) + Mx.$$

Integrating again gives

$$\int_0^x f'(x)\,dx \leq \int_0^x (f'(0) + Mx)\,dx.$$

On the left side, use the Fundamental Theorem. On the right side, observe that $f'(0)$ is just a constant, so $\int_0^x f'(0)\,dx = f'(0)(x - 0)$. We now have

$$f(x) - f(0) \leq f'(0)x + \frac{M}{2}x^2.$$

Thus we have shown that if $f''(x) \leq M$ for $0 \leq x \leq d$, then

$$f(x) \leq f(0) + f'(0)x + \frac{M}{2}x^2.$$

Now we consider the lower bound. By reversing the inequalities, it can be shown that if $L \leq f''(x)$ for $0 \leq x \leq d$, then

$$f(0) + f'(0)x + \frac{L}{2}x^2 \leq f(x).$$

If we subtract $P_1(x) = f(0) + f'(0)x$ from both sides of these last two inequalities, we get a bound on the error $E_1 = f(x) - P_1(x)$:

Bounding the Error in $P_1(x)$

If $L \leq f''(x) \leq M$ for $0 \leq x \leq d$,   then

$$\frac{L}{2}x^2 \leq f(x) - P_1(x) \leq \frac{M}{2}x^2 \quad \text{for} \quad 0 \leq x \leq d.$$

Error in $P_2(x)$

Suppose we now want to bound the error in the second-degree Taylor polynomial approximation. By analogy with the error estimate for $P_1(x)$, we expect that the size of $f'''(x)$ will control the size of the error in $P_2(x)$. Suppose we know that

$$f'''(x) \leq M \quad \text{for} \quad 0 \leq x \leq d$$

and try to find a bound on $E = f(x) - P_2(x)$. Integrating as before, we get

$$\int_0^x f'''(x)\,dx \leq \int_0^x M\,dx$$
$$f''(x) - f''(0) \leq M(x - 0)$$
$$f''(x) \leq f''(0) + Mx.$$

Integrate again:

$$\int_0^x f''(x)\,dx \leq \int_0^x (f''(0) + Mx)\,dx$$
$$f'(x) - f'(0) \leq f''(0)x + \frac{M}{2}x^2$$
$$f'(x) \leq f'(0) + f''(0)x + \frac{M}{2}x^2,$$

and again:

$$\int_0^x f'(x)\,dx \leq \int_0^x \left(f'(0) + f''(0)x + \frac{M}{2}x^2 \right) dx$$
$$f(x) - f(0) \leq f'(0)x + \frac{f''(0)}{2}x^2 + \frac{M}{3 \cdot 2}x^3.$$

We conclude that if $f'''(x) \leq M$ for $0 \leq x \leq d$, then

$$f(x) \leq f(0) + f'(0)x + \frac{f''(0)}{2}x^2 + \frac{M}{3!}x^3.$$

Similarly, if $L \leq f'''(x)$ for $0 \leq x \leq d$, then

$$f(0) + f'(0)x + \frac{f''(0)}{2!}x^2 + \frac{L}{3!}x^3 \leq f(x).$$

Thus, the error in the approximation by $P_2(x)$ is bounded as follows:

Bounding the Error in $P_2(x)$

If $\quad L \leq f'''(x) \leq M \quad$ for $\quad 0 \leq x \leq d, \quad$ then

$$\frac{L}{3!}x^3 \leq f(x) - P_2(x) \leq \frac{M}{3!}x^3 \quad \text{for} \quad 0 \leq x \leq d.$$

Error in $P_n(x)$

The error in higher-degree Taylor polynomials follows the same pattern. If $L \leq f^{(n+1)}(x) \leq M$ for $0 \leq x \leq d$, then

$$\frac{L}{(n+1)!}x^{n+1} \leq f(x) - P_n(x) \leq \frac{M}{(n+1)!}x^{n+1}.$$

General Error Formula

When x is to the left of 0, so $-d \leq x \leq 0$, and when the Taylor series is centered at $a \neq 0$, similar calculations give us the following statement:

> **Error Bound for Taylor Polynomial Approximations**
>
> If $|f^{(n+1)}(x)| \leq M$ for $|x - a| \leq d$,   then
>
> $$|E_n(x)| = |f(x) - P_n(x)| \leq \frac{M}{(n+1)!}|x - a|^{n+1} \quad \text{for} \quad |x - a| \leq d.$$

Example 5   Give a bound on the error, E_4, when e^x is approximated by its fourth-degree Taylor polynomial for $-0.5 \leq x \leq 0.5$.

Solution   Let $f(x) = e^x$. Then the fifth derivative is $f^{(5)}(x) = e^x$. Since e^x is increasing,

$$|f^{(5)}(x)| \leq e^{0.5} = \sqrt{e} < 2 \quad \text{for} \quad -0.5 \leq x \leq 0.5.$$

Thus

$$|E_4| = |f(x) - P_4(x)| \leq \frac{2}{5!}|x|^5.$$

This means, for example, that on $-0.5 \leq x \leq 0.5$, the approximation

$$e^x \approx 1 + x + \frac{x^2}{2!} + \frac{x^3}{3!} + \frac{x^4}{4!}$$

has an error of at most $\frac{2}{120}(0.5)^5 < 0.0006$.

Importance of the Error Bound for Taylor Polynomials

The error formula for Taylor polynomials can be used to bound the error in a particular numerical approximation, or to see how the accuracy of the approximation depends on the value of x or the value of n. Observe that the error for a Taylor polynomial of degree n depends on the $(n+1)^{\text{st}}$ power of x. That means, for example, with a Taylor polynomial of degree n centered at 0, if we decrease x by a factor of 2, the error bound decreases by a factor of 2^{n+1}.

Example 6   Compare the errors in the approximations

$$e^{0.1} \approx 1 + 0.1 + \frac{1}{2!}(0.1)^2 \quad \text{and} \quad e^{0.05} \approx 1 + (0.05) + \frac{1}{2!}(0.05)^2$$

Solution We are approximating e^x by its second-degree Taylor polynomial, first at $x = 0.1$, and then at $x = 0.05$. Since we have decreased x by a factor of 2, we expect that the error has decreased by a factor of $2^3 = 8$. Let's see what actually happens. Here are the approximate values of the two errors:

$$e^{0.1} - \left(1 + 0.1 + \frac{1}{2!}(0.1)^2\right) = 1.105171 - 1.105000 = 0.000171$$

$$e^{0.05} - \left(1 + 0.05 + \frac{1}{2!}(0.05)^2\right) = 1.051271 - 1.051250 = 0.000021$$

Thus the error has decreased by a factor of $(0.000171)/(0.000021) = 8.1$, which is about what we expected.

Convergence of Taylor Series

The error bound allows us to show that in most cases the Taylor series for a function converges to *that* function. Since the error represents the difference between the Taylor polynomial and the function, showing that the series converges to the function for a particular value of x means showing that the error in the nth degree Taylor approximation goes to 0 as $n \to \infty$ for that value of x. We illustrate this idea by showing that the Taylor series for e^x converges to e^x for all x.

Showing the Taylor Series for e^x Converges to e^x

When we write the equation

$$e^x = 1 + x + \frac{x^2}{2!} + \frac{x^3}{3!} + \cdots,$$

the equal sign and the "$\cdots$" mean that as we look at more and more terms of the series on the right, we eventually get arbitrarily close to e^x. Since

$$E_n(x) = e^x - P_n(x) = e^x - \left(1 + x + \frac{x^2}{2!} + \cdots + \frac{x^n}{n!}\right),$$

we can write

$$e^x = 1 + x + \frac{x^2}{2!} + \cdots + \frac{x^n}{n!} + E_n(x).$$

Thus, for the Taylor series to converge to e^x, we must have $E_n(x) \to 0$ as $n \to \infty$. We assumed this to be true when we defined Taylor series in Section 9.2; we will now prove it for any fixed x.

Showing $E_n(x) \to 0$ as $n \to \infty$

Proof Since $f(x) = e^x$, the $(n+1)^{\text{st}}$ derivative $f^{(n+1)}(x)$ is also e^x, no matter what n is. So if we let $M = e^x$, then $|f^{(n+1)}| \le M$ on the interval between 0 and x. (This works for $x \ge 0$; if $x < 0$ then we actually need to take $M = 1$.) The important observation is that for any x the *same* number M bounds all the higher derivatives $f^{(n+1)}(x)$.

By the error bound formula, we now have

$$|E_n(x)| = |e^x - P_n(x)| \le \frac{M|x|^{n+1}}{(n+1)!} \quad \text{for every } n.$$

To show that the errors go to zero, we must show that for a fixed x and a fixed number M,

$$\frac{M}{(n+1)!}|x|^{n+1} \to 0 \quad \text{as} \quad n \to \infty.$$

Since M is fixed, we need only show that

$$\frac{1}{(n+1)!}|x|^{n+1} \to 0 \quad \text{as} \quad n \to \infty.$$

To see why this is true, think of what happens when n is much larger than x. Suppose, for example, that $x = 17.3$. Let's look at the value of

$$\frac{1}{(n+1)!}(17.3)^{n+1}$$

for n at least twice as big as 17.3, that is $n = 35$, or $n = 36$, or $n = 37, \ldots$. We get the values

$$\frac{1}{35!}(17.3)^{35} \ ,$$

$$\frac{1}{36!}(17.3)^{36} = \frac{17.3}{36} \cdot \frac{1}{35!}(17.3)^{35},$$

$$\frac{1}{37!}(17.3)^{37} = \frac{17.3}{37} \cdot \frac{17.3}{36} \cdot \frac{1}{35!}(17.3)^{35}, \quad \ldots$$

The $n = 38$ term will be $17.3/38$ times the $n = 37$ term and so on. Each time we increase n by 1, the value of $\frac{1}{(n+1)!}(17.3)^{n+1}$ is multiplied by a number less than $\frac{1}{2}$. No matter what the value of $\frac{1}{35!}(17.3)^{35}$ is, if you keep on halving it, the result will get closer and closer to zero. Thus $\frac{1}{(n+1)!}(17.3)^{n+1}$ goes to 0 as n goes to infinity.

We can generalize this by replacing 17.3 with an arbitrary $|x|$. For $n > 2|x|$, as we increase n, we will keep multiplying $\frac{1}{(n+1)!}|x|^{n+1}$ by factors of $\frac{|x|}{n+1} < \frac{1}{2}$. So

$$\frac{M}{(n+1)!}|x|^{n+1} \to 0 \quad \text{as} \quad n \to \infty.$$

Therefore, the Taylor series $1 + x + x^2/2! + \cdots$ does converge to e^x.

Most functions have Taylor series which converge to the original function for x within the interval of convergence. Functions for which this is true are called *analytic functions*. The Taylor series for such a function can be interpreted when x is replaced by a complex number. This extends the domain of the function. See Problem 16.

Problems on the Error in Taylor Approximations

1. Suppose you approximate $f(t) = e^t$ by a Taylor polynomial of degree 0 about $t = 0$ on the interval $[0, 0.5]$.

    (a) Is the approximation an overestimate or an underestimate?
    (b) Estimate the magnitude of the largest possible error. Check your answer graphically on a computer or calculator.

2. Repeat Problem 1 using the second-degree Taylor approximation, $P_2(t)$, to e^t.

3. Consider the error in using the approximation $\sin \theta \approx \theta$ on the interval $[-1, 1]$.

    (a) Where is the approximation an overestimate, and where is it an underestimate?
    (b) Estimate the magnitude of the largest possible error. Check your answer graphically on a computer or calculator.

4. Repeat Problem 3 for the approximation $\sin\theta \approx \theta - \theta^3/3!$.

Use the methods of this section to show how you can estimate the magnitude of the error in approximating the quantities in Problems 5–8 using a third-degree Taylor polynomial about $x = 0$.

5. $\tan 1$ 6. $0.5^{1/3}$ 7. $\ln(1.5)$ 8. $1/\sqrt{3}$

9. Give a bound for the maximum possible error for the nth degree Taylor polynomial about $x = 0$ approximating $\cos x$ on the interval $[0, 1]$. What is the bound for $\sin x$?

10. What degree Taylor polynomial about $x = 0$ do you need to calculate $\cos 1$ to four decimal places? To six decimal places? Justify your answer using the results of Problem 9.

11. Show that the Taylor series about 0 for $\sin x$ converges to $\sin x$ for every x.

12. Show that the Taylor series about 0 for $\cos x$ converges to $\cos x$ for every x.

13. (a) Using a calculator, make a table of the values to four decimal places of $\sin x$ for

$$x = -0.5, -0.4, \ldots, -0.1, 0, 0.1, \ldots, 0.4, 0.5.$$

    (b) Add to your table the values of the error $E_1 = \sin x - x$ for these x-values.
    (c) Using a calculator or computer, draw a graph of the quantity $E_1 = \sin x - x$ showing that

$$|E_1| < 0.03 \quad \text{for} \quad -0.5 \le x \le 0.5.$$

14. In this problem, you will investigate the error in the nth degree Taylor approximation to e^x for various values of n.

    (a) Let $E_1 = e^x - P_1(x) = e^x - (1+x)$. Using a calculator or computer, graph E_1 for $-0.1 \le x \le 0.1$. What shape is the graph of E_1? Use the graph to confirm that

$$|E_1| \le x^2 \quad \text{for} \quad -0.1 \le x \le 0.1.$$

    (b) Let $E_2 = e^x - P_2(x) = e^x - (1 + x + x^2/2)$. Choose a suitable range and graph E_2 for $-0.1 \le x \le 0.1$. What shape is the graph of E_2? Use the graph to confirm that

$$|E_2| \le x^3 \quad \text{for} \quad -0.1 \le x \le 0.1.$$

    (c) Explain why the graphs of E_1 and E_2 have the shapes they do.

15. Graph the error
$$E_0 = \cos x - P_0(x) = \cos x - 1$$
for $|x| \le 0.1$. Explain the shape of the graph, using information given by the Taylor expansion of $\cos x$, and find a bound for $|E_0|$ for $|x| \le 0.1$.

16. Let $i = \sqrt{-1}$. We define $e^{i\theta}$ by substituting $i\theta$ in the Taylor series for e^x. Use this definition[10] to explain Euler's formula
$$e^{i\theta} = \cos\theta + i\sin\theta.$$

---

[10]Complex numbers are discussed in Appendix D.

CHAPTER TEN

DIFFERENTIAL EQUATIONS

The derivative was introduced as the rate of change of a function in Chapter 2. In Chapter 3, we saw how to use the definite integral to compute changes in the original function from the derivative. Chapter 6 introduced antiderivatives, differential equations, and going backward from the derivative to the original function.

Every time we antidifferentiate to find $\int f(x)\,dx$, we are actually solving the differential equation

$$\frac{dy}{dx} = f(x).$$

The new feature of this chapter is that the right-hand side of the differential equation may now be a function of y, or of both x and y, instead of just x.

We will see how to solve differential equations graphically first, then numerically, and, finally, in some of the special cases when it can be done, analytically. In the course of this chapter, we will treat first-order differential equations (involving first derivatives only) and some second-order differential equations (involving first and second derivatives).

10.1 WHAT IS A DIFFERENTIAL EQUATION?

How Fast Does a Person Learn?

Suppose we are interested in how fast an employee learns a new task. One theory claims that the more the employee already knows of the task, the slower he or she learns. In other words, if $y\%$ is the percentage of the task that has already been mastered, and dy/dt the rate at which the employee is learning, then dy/dt decreases as y increases.

What can we say about y as a function of time, t? Figure 10.1 shows three graphs whose slope, dy/dt, decreases as y increases. Figure 10.1(a) represents an employee who starts learning at $t = 0$ and who eventually masters 100% of the task. Figure 10.1(b) represents an employee who starts later but eventually masters 100% of the task. Figure 10.1(c), on the other hand, represents an employee who starts learning at $t = 0$, but who does not master the whole task (since y levels off at some percentage below 100%).

Notice the fact that y levels off does not follow from the fact that dy/dt decreases as y increases. However, a person's knowledge of the task cannot exceed 100%, and all the graphs in Figure 10.1 have been drawn to reflect this fact.

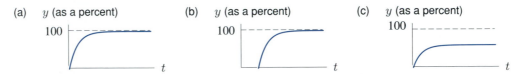

Figure 10.1: Possible graphs showing percentage of task learned, y, as a function of time, t

Setting up a Differential Equation to Model How a Person Learns

To describe more precisely how a person learns, we need more exact information about how dy/dt depends on y. Suppose we are told that if time, t, is measured in weeks, then

$$\frac{\text{Rate a person}}{\text{learns}} = \frac{\text{Percentage of task}}{\text{not yet learned.}}$$

Since y is the percentage learned by time t (in weeks), the percentage not yet learned by that time is $100 - y$. Thus we have

$$\frac{dy}{dt} = 100 - y.$$

Such an equation, which gives information about the rate of change of an unknown function, is called a *differential equation*.

Solving the Differential Equation Numerically

Suppose that the person starts learning at time zero, so $y = 0$ when $t = 0$. Then initially the person is learning at a rate

$$\frac{dy}{dt} = 100 - 0 = 100\% \text{ per week.}$$

In other words, if the person were to continue learning at this rate, the task would be mastered in a week. In fact, however, the rate at which the person learns decreases, so it takes more than a week to get close to mastering the task. Let's assume a five-day work week and that the 100% per week learning rate holds for the whole first day. (It doesn't, but we will assume this for now.) One day is 1/5 of a week, so during the first day the person learns $100(1/5) = 20\%$ of the task. By the end of the first day the rate at which the person learns has therefore been reduced to

$$\frac{dy}{dt} = 100 - 20 = 80\% \text{ per week.}$$

Thus, during the second day the person learns $80(1/5) = 16\%$, so by the end of the second day the person knows $20 + 16 = 36\%$ of the task. Continuing in this fashion, we compute the approximate values[1] of y in Table 10.1. These values for y represent an approximate numerical solution to the differential equation.

TABLE 10.1   *Approximate percentage of task learned as a function of time*

Time (working days)	0	1	2	3	4	5	10	20
Percentage learned	0	20	36	48.8	59.0	67.2	89.3	98.8

A Formula for the Solution to the Differential Equation

A function $y = f(t)$ which satisfies the differential equation

$$\frac{dy}{dt} = 100 - y$$

is called a *solution*. Figure 10.1 contains graphs of possible solutions, and Table 10.1 shows approximate numerical values of a solution. It is also sometimes (but not always) possible to find a formula for the solution. In this particular case there is a formula, which we will derive in Section 10.4. For now, we will simply check that the formula is correct, and look at its general form. The formula for the solution is

$$y = 100 + Ce^{-t},$$

where C is an *arbitrary constant*. Substituting this into the differential equation $dy/dt = 100 - y$ gives:

$$\text{Left side} = \frac{dy}{dt} = -Ce^{-t}$$

$$\text{Right side} = 100 - y = 100 - (100 + Ce^{-t}) = -Ce^{-t}.$$

Since we get the same thing on both sides, we say $y = 100 + Ce^{-t}$ is a solution to this differential equation.

Finding the Arbitrary Constant: Initial Conditions

To find a value for the arbitrary constant C, we need an additional piece of information—usually the initial value of y. If, for example, we are told that $y = 0$ when $t = 0$, then substituting into

$$y = 100 + Ce^{-t}$$

shows us that

$$0 = 100 + Ce^0, \quad \text{so} \quad C = -100.$$

So the function $y = 100 - 100e^{-t}$ satisfies the differential equation *and* the condition that $y = 0$ when $t = 0$.

The Family of Solutions

Later in this chapter we will show that any solution to this differential equation is of the form $y = 100 + Ce^{-t}$ for some constant C. Since any value of C gives a solution, we say that the *general solution* to the differential equation $dy/dt = 100 - y$ is the family of functions $y = 100 + Ce^{-t}$. The solution $y = 100 - 100e^{-t}$ that satisfies the differential equation together with the *initial condition* that $y = 0$ when $t = 0$ is called a *particular solution*. The differential equation and the initial condition together are called an *initial-value problem*. Several members of the family of solutions are graphed in Figure 10.2. Notice that, like a family of antiderivatives, this family contains one arbitrary constant, C.

---

[1]The values of y after 6, 7, ..., 9, 11, ..., 19 days were computed by the same method, but omitted from the table.

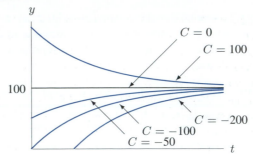

y

$C = 0$

$C = 100$

100

$C = -200$

$C = -100$

$C = -50$

t

Figure 10.2: Solution curves for $dy/dt = 100 - y$:
Members of the family $y = 100 + Ce^{-t}$

First- and Second-Order Differential Equations

First, some more definitions. The differential equation

$$\frac{dy}{dt} = 100 - y$$

is called *first-order* because it involves the first derivative, but no higher derivatives. By contrast, if *s* is the height (in feet) of a body moving under the force of gravity and *t* is time (in seconds), then

$$\frac{d^2s}{dt^2} = -32.$$

This is a *second-order* differential equation because it involves the second derivative of the unknown function, $s = f(t)$, but no higher derivatives.

Example 1   Show that $y = e^{2t}$ is not a solution to the second-order differential equation

$$\frac{d^2y}{dt^2} + 4y = 0.$$

Solution   To decide whether the function $y = e^{2t}$ is a solution, substitute it into the differential equation:

$$\frac{d^2y}{dt^2} + 4y = 2(2e^{2t}) + 4e^{2t} = 8e^{2t}.$$

Since $8e^{2t}$ is not identically zero, $y = e^{2t}$ is not a solution.

How Many Arbitrary Constants Should We Expect in the Family of Solutions?

Since a differential equation involves the derivative of an unknown function, solving it usually involves antidifferentiation, which introduces arbitrary constants. The solution to a first-order differential equation usually involves one antidifferentiation and one arbitrary constant (for example, the C in $y = 100 + Ce^{-t}$). Picking out one particular solution involves knowing an additional piece of information, such as an initial condition. Solving a second-order differential equation generally involves two antidifferentiations and so two arbitrary constants. Consequently, finding a particular solution usually involves two initial conditions.

For example, if *s* is the height (in feet) of a body above the surface of the earth at time *t* (in seconds), then

$$\frac{d^2s}{dt^2} = -32.$$

Integrating gives

$$\frac{ds}{dt} = -32t + C_1,$$

and integrating again gives

$$s = -16t^2 + C_1 t + C_2.$$

Thus the general solution for s involves the two arbitrary constants C_1 and C_2. We can find C_1 and C_2 if we are told, for example, that the initial velocity is 100 feet per second upward and that the initial position is 15 feet above the ground. In this case, $C_1 = 100$ and $C_2 = 15$, so

$$s = -16t^2 + 100t + 15.$$

Problems for Section 10.1

1. Match the graphs in Figure 10.3 with the following descriptions.

   (a) The population of a new species introduced onto a tropical island
   (b) The temperature of a metal ingot placed in a furnace and then removed
   (c) The speed of a car traveling at uniform speed and then braking uniformly
   (d) The mass of carbon-14 in a historical specimen
   (e) The concentration of tree pollen in the air over the course of a year.

(I)    (II)   (III)   (IV)   (V)

Figure 10.3

2. Suppose you know that $Q = Ce^{kt}$ satisfies the differential equation

$$\frac{dQ}{dt} = -0.03Q.$$

What (if anything) does this tell you about the values of C and k?

3. Show that, for any constant P_0, the function $P = P_0 e^t$ satisfies the differential equation

$$\frac{dP}{dt} = P.$$

4. Find the value(s) of ω for which $y = \cos \omega t$ satisfies

$$\frac{d^2y}{dt^2} + 9y = 0.$$

5. Show that $y = \sin 2t$ satisfies

$$\frac{d^2y}{dt^2} + 4y = 0.$$

6. (a) Show that $P = \dfrac{1}{1 + e^{-t}}$ satisfies the *logistic equation*

$$\frac{dP}{dt} = P(1 - P).$$

   (b) What is the limiting value of P as time, t, tends to infinity?

7. For what values of λ is $y = Ax^\lambda$ a solution of the equation

$$x^2 y'' + 2xy' - 6y = 0?$$

8. Pick out which functions are solutions to which differential equations. (Note: Functions may be solutions to more than one equation or to none; an equation may have more than one solution.)

   (a) $\dfrac{dy}{dx} = -2y$ (I) $y = 2\sin x$

   (b) $\dfrac{dy}{dx} = 2y$ (II) $y = \sin 2x$

   (c) $\dfrac{d^2y}{dx^2} = 4y$ (III) $y = e^{2x}$

   (d) $\dfrac{d^2y}{dx^2} = -4y$ (IV) $y = e^{-2x}$

9.   Which functions are solutions to which differential equations? (Note: a differential equation may have zero, one, or multiple solutions.)

(A) $\dfrac{dy}{dx} = \dfrac{y}{x}$

(I) $y = x^3$

(B) $\dfrac{dy}{dx} = 3\dfrac{y}{x}$

(II) $y = 3x$

(C) $\dfrac{dy}{dx} = 3x$

(III) $y = e^{3x}$

(D) $\dfrac{dy}{dx} = y$

(IV) $y = 3e^x$

(E) $\dfrac{dy}{dx} = 3y$

(V) $y = x$

10. Match the following differential equations and possible solutions. (Note the given functions may satisfy more than one equation or none, and some equations may have more than one solution.)

(a) $y'' = y$ (I) $y = \cos x$
(b) $y' = -y$ (II) $y = \cos(-x)$
(c) $y' = 1/y$ (III) $y = x^2$
(d) $y'' = -y$ (IV) $y = e^x + e^{-x}$
(e) $x^2 y'' - 2y = 0$ (V) $y = \sqrt{2x}$

11. Suppose y is the height of a hanging cable above a fixed horizontal line, as in Figure 10.4. It can be shown that

$$\frac{d^2 y}{dx^2} = k\sqrt{1 + \left(\frac{dy}{dx}\right)^2}.$$

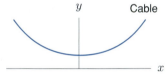

Figure 10.4

(a)   Show that $y = \dfrac{e^x + e^{-x}}{2}$ satisfies this differential equation if $k = 1$.
(b)   For general k, one solution to this differential equation is of the form

$$y = \frac{e^{Ax} + e^{-Ax}}{2A}.$$

Substitute this expression for y into the differential equation to find A in terms of k.

10.2 SLOPE FIELDS

In this section, we will visualize a first-order differential equation. Let's start with the equation

$$\frac{dy}{dx} = y.$$

We can check that each curve in the family of exponentials, $y = Ce^x$, is a solution to this differential equation.

Any solution to this differential equation has the property that the slope at any point is equal to the y coordinate at that point. (That's what the equation $dy/dx = y$ is telling us!) This means that if the solution goes through the point $(0, 0.5)$, its slope there is 0.5; where it goes through a point with $y = 1$ its slope is 1. A solution which goes through $(0, 1.5)$ has slope 1.5 there. (See Figure 10.5.)

In Figure 10.5 a small line segment is drawn at each of the marked points showing the slope of the curve there. Imagine drawing many of these line segments, but leaving out the curves; this gives

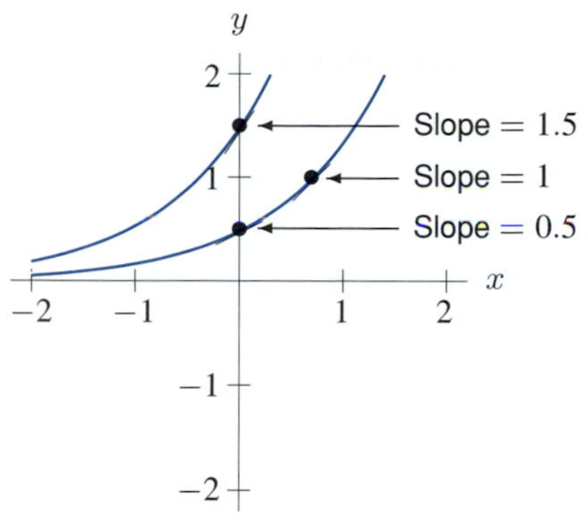

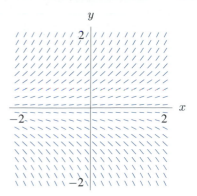

Figure 10.5: Solutions to $\frac{dy}{dx} = y$

Figure 10.6: Slope field for $\frac{dy}{dx} = y$

the *slope field* for the equation $dy/dx = y$ shown in Figure 10.6. From this picture, we can see that above the x-axis, the slopes are all positive (because y is positive there), and they increase as we move upward (as y increases). Below the x-axis, the slopes are all negative, and get more so as we move downward. Notice that on any horizontal line (where y is constant) the slopes are the same.

In the slope field we can see the ghost of the solution curve lurking. Start anywhere on the plane and move so that the slope lines are tangent to your path; you will trace out one of the solution curves. It may be helpful to think of the slope field as a set of signposts pointing in the direction you should go at each point.

Example 1   Figure 10.7 shows the slope field of the differential equation $\dfrac{dy}{dx} = 2x$.

     (a)   How does the slope field vary as you move around the xy-plane?

     (b)   Compare the solution curves sketched on the slope field with the formula for the solutions to this differential equation.

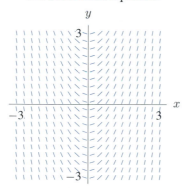

Figure 10.7: Slope field for $\frac{dy}{dx} = 2x$

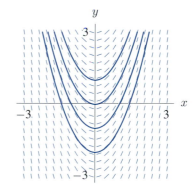

Figure 10.8: Some solutions to $\frac{dy}{dx} = 2x$

Solution   (a)   In Figure 10.7 you will notice that on a vertical line (where x is constant) the slopes are all the same. This is because in this differential equation dy/dx depends on x only. (In the previous example, $dy/dx = y$, the slopes depended on y only.)

     (b)   The solution curves in Figure 10.8 look like parabolas. Since the solution to the differential equation

$$\frac{dy}{dx} = 2x$$

is, by antidifferentiation,

$$y = \int 2x\,dx = x^2 + C,$$

we see that the solution curves really are parabolas.

Example 2   Using the slope field, guess the form of the solution curves of the differential equation

$$\frac{dy}{dx} = -\frac{x}{y}.$$

Solution   The slope field is shown in Figure 10.9. Notice that on the y-axis, where x is 0, the slope is 0. On the x-axis, where y is 0, the line segments are vertical and the slope is infinite. At the origin the slope is undefined, and there is no line segment.

What do the solution curves of this differential equation look like? The slope field suggests that they look like circles centered at the origin. Later we'll see how to obtain the solution analytically, but even without this, we can still check that the circle is a solution. Let's take the circle of radius r,

$$x^2 + y^2 = r^2,$$

and differentiate implicitly, thinking of y as a function of x. Using the chain rule, we get

$$2x + 2y \cdot \frac{dy}{dx} = 0.$$

Solving for dy/dx gives our differential equation,

$$\frac{dy}{dx} = -\frac{x}{y}.$$

This tells us that $x^2 + y^2 = r^2$ is a solution to the differential equation.

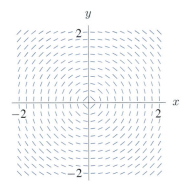

Figure 10.9: Slope field for $\frac{dy}{dx} = -\frac{x}{y}$

The previous example shows that the solution to a differential equation may often be an implicit function. (Recall that implicit functions are ones which have not been "solved" for y; in other words, the dependent variable is not expressed as an explicit function of x.)

Example 3   The slope fields for $\frac{dy}{dt} = 2 - y$ and $\frac{dy}{dt} = \frac{t}{y}$ are shown in Figures 10.10 and 10.11. For each slope field:
   (a)   Sketch solution curves with initial conditions
             (i) $y = 1$ when $t = 0$ (ii) $y = 0$ when $t = 1$ (iii) $y = 3$ when $t = 0$
   (b)   For each solution curve, can you say anything about the long-run behavior of y? For example, does $\lim_{t \to \infty} y$ exist? If so, what is its value?

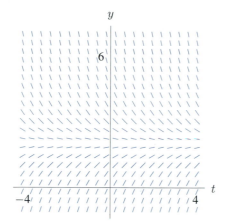

Figure 10.10: Slope field for $\frac{dy}{dt} = 2 - y$

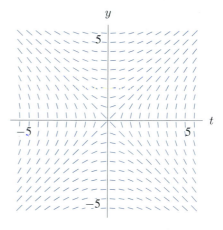

Figure 10.11: Slope field for $\frac{dy}{dt} = \frac{t}{y}$

Solution (a) See Figures 10.12 and 10.13.

(b) For $dy/dt = 2 - y$, all solution curves have $y = 2$ as a horizontal asymptote, so $\lim\limits_{t \to \infty} y = 2$.

For $dy/dt = t/y$, as $t \to \infty$, it appears that either $y \to t$ or $y \to -t$.

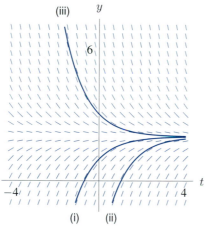

Figure 10.12: Solution curves for $\frac{dy}{dt} = 2 - y$

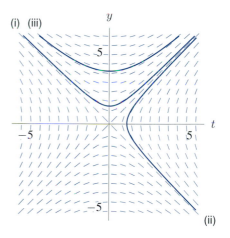

Figure 10.13: Solution curves for $\frac{dy}{dt} = \frac{t}{y}$

Existence and Uniqueness of Solutions

Since differential equations are used to model many real situations, the question of whether a solution is unique can have great practical importance. If we know how the velocity of a satellite is changing, can we know its velocity at any future time? If we know the initial population of a city, and we know how the population is changing, can we predict the population in the future? Common sense says yes: if we know the initial value of some quantity and we know exactly how it is changing, we should be able to figure out the future value of the quantity.

In the language of differential equations, an initial-value problem (that is, a differential equation and an initial condition) almost always has a unique solution. One way to see this is by looking at the slope field. Imagine starting at the point representing the initial condition. Through that point there will usually be a line segment pointing in the direction the solution curve must go. By following

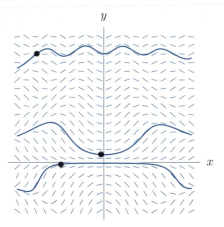

Figure 10.14: There's one and only one solution curve through each
point in the plane for this slope field (Dots represent initial conditions)

these line segments, we trace out the solution curve. See Figure 10.14. In general, at each point there
is one line segment and therefore only one direction for the solution curve to go. The solution curve
exists and is *unique* provided we are given an initial point. Notice that even though we can draw the
solution curves, we may have no simple formula for them.

It can be shown that if the slope field is continuous as we move from point to point in the
plane, we can be sure that a solution curve exists everywhere. Ensuring that each point has only one
solution curve through it requires a slightly stronger condition.

Problems for Section 10.2

1. The slope field for the equation $y' = x + y$ is shown in Figure 10.15.

    (a) Carefully sketch the solutions that pass through the points
         (i) $(0, 0)$ (ii) $(-3, 1)$ (iii) $(-1, 0)$
    (b) From your sketch, write the equation of the solution passing through $(-1, 0)$.
    (c) Verify your solution to part (b) by substituting it into the differential equation.

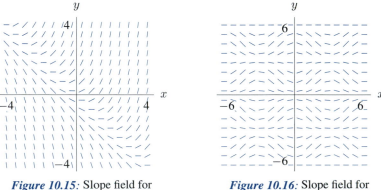

Figure 10.15: Slope field for   *Figure 10.16:* Slope field for
$y' = x + y$ $y' = (\sin x)(\sin y)$

2. Figure 10.16 shows the slope field for the equation $y' = (\sin x)(\sin y)$.

    (a) Carefully sketch the solutions that pass through the points: (i) $(0, -2)$ (ii) $(0, \pi)$.
    (b) What is the equation of the solution that passes through $(0, n\pi)$, where n is any integer?

3. Match the slope fields in Figure 10.17 with their differential equations:

    (a) $y' = -y$ (b) $y' = y$ (c) $y' = x$

    (d) $y' = 1/y$ (e) $y' = y^2$

(I)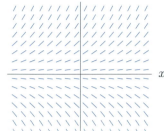
(II)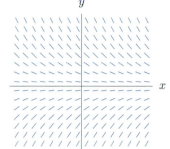
(III)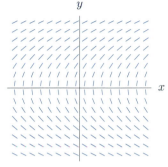

(IV)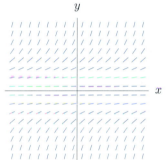
(V)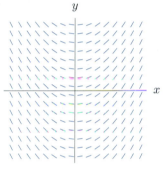

Figure 10.17

4. Match the slope fields in Figure 10.18 with their differential equations:

    (a) $y' = 1 + y^2$ (b) $y' = x$ (c) $y' = \sin x$

    (d) $y' = y$ (e) $y' = x - y$ (f) $y' = 4 - y$

(I)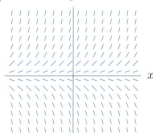
(II)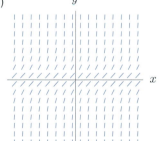
(III)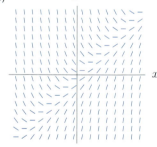

(IV)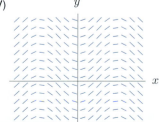
(V)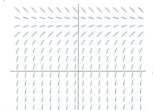
(VI)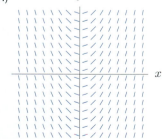

Figure 10.18: Each slope field is graphed for $-5 \leq x \leq 5$, $-5 \leq y \leq 5$

5. One of the slope fields in Figure 10.19 has the equation $y' = (x + y)/(x - y)$. Which one?

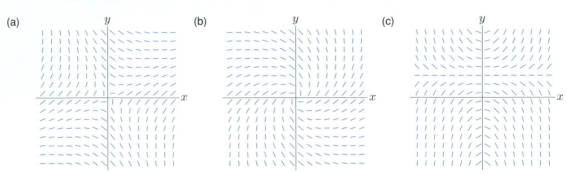

Figure 10.19

6. Match the slope fields in Figure 10.20 to the corresponding differential equations:

(a) $y' = xe^{-x}$ (b) $y' = \sin x$ (c) $y' = \cos x$

(d) $y' = x^2 e^{-x}$ (e) $y' = e^{-x^2}$ (f) $y' = e^{-x}$

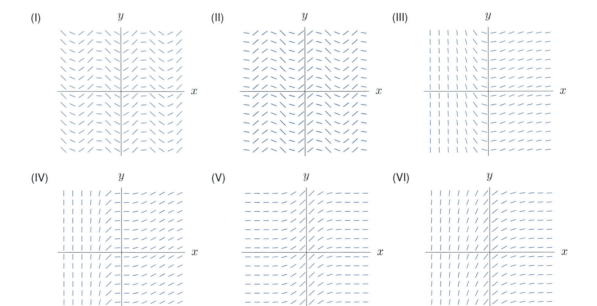

Figure 10.20

7. The slope field for $y' = 0.5(1 + y)(2 - y)$ is shown in Figure 10.21.

(a) Plot the following points on the slope field:

(i) the origin   (ii) $(0, 1)$ (iii) $(1, 0)$

(iv) $(0, -1)$ (v) $(0, -5/2)$ (vi) $(0, 5/2)$

(b) Plot solution curves through the points in part (a).

(c) For which regions are all solution curves increasing? For which regions are all solution curves decreasing? When can the solution curves have horizontal tangents? Explain why, using both the slope field and the differential equation.

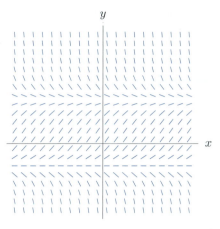

y

Figure 10.21: Note: x and y scales are equal

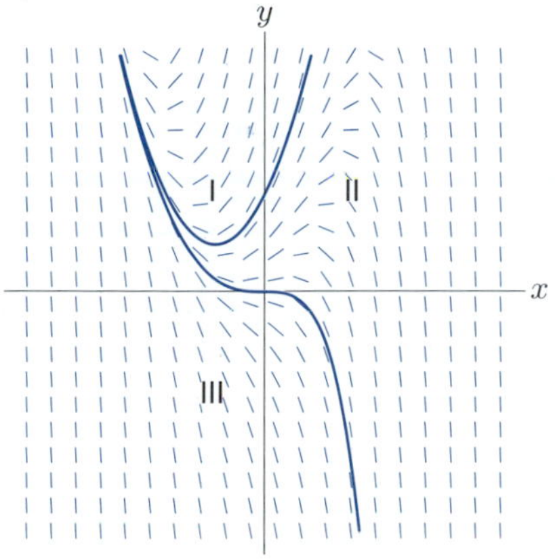

y

x

Figure 10.22

8. The slope field for the differential equation $y' = y - x^2$ together with two special solutions are shown in Figure 10.22.

    (a) The two curves shown in the figure are $y = x^2 + 2x + 2$ and $y = x^2 + 2x + 2 - 2e^x$. Verify that these formulas give solutions, and indicate which formula goes with which curve.
    (b) Sketch a few solutions to the differential equation in each of the three regions.
    (c) Describe the solution curves you have sketched in region I.
    (d) Describe the solution curves in region II.
    (e) Describe the solution curves in region III.

10.3 EULER'S METHOD

In the preceding section we saw how to sketch a solution curve to a differential equation using its slope field. In this section we compute points on a solution curve numerically using *Euler's method*. (Leonhard Euler was an eighteenth-century Swiss mathematician.) In some cases—but not many—we can find an equation for the solution curve. That will be the subject of Section 10.4.

Here's the concept behind Euler's method. Think of the slope field as a set of signposts directing you across the plane. Pick a starting point (corresponding to the initial value), and calculate the slope at that point using the differential equation. This slope is a signpost telling you the direction to take. Head off a small distance in that direction. Stop and look at the new signpost. Recalculate the slope from the differential equation, using the coordinates of the new point. Change direction to correspond to the new slope, and move another small distance, and so on.

Example 1 Use Euler's method to construct a solution for $\dfrac{dy}{dx} = y$. Start at the point $P_0 = (0, 1)$ and take $\Delta x = 0.1$.

Solution The slope at the point $P_0 = (0, 1)$ is $dy/dx = 1$. (See Figure 10.23.) As we move from P_0 to P_1, y will increase by Δy, where

$$\Delta y = (\text{slope at } P_0)\Delta x = 1(0.1) = 0.1.$$

So we have

$$y\text{-value at } P_1 = (y \text{ value at } P_0) + \Delta y = 1 + 0.1 = 1.1.$$

TABLE 10.2   *Euler's method for $dy/dx = y$, starting at $(0,1)$*

	x	y	$\Delta y = (\text{Slope})\Delta x$
P_0	0	1	$0.1 = (1)(0.1)$
P_1	0.1	1.1	$0.11 = (1.1)(0.1)$
P_2	0.2	1.21	$0.121 = (1.21)(0.1)$
P_3	0.3	1.331	$0.1331 = (1.331)(0.1)$
P_4	0.4	1.4641	$0.14641 = (1.4641)(0.1)$
P_5	0.5	1.61051	$0.161051 = (1.61051)(0.1)$

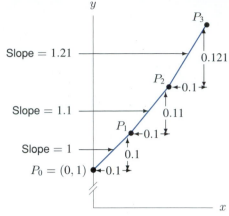

Figure 10.23: Euler's approximate solution to
$\frac{dy}{dx} = y$

Thus the point P_1 is $(0.1, 1.1)$. Now, using the differential equation again, we see that

$$\text{slope at } P_1 = 1.1,$$

so if we move to P_2, then y will change by

$$\Delta y = (\text{slope at } P_1)\Delta x = (1.1)(0.1) = 0.11.$$

This means

$$y\text{-value at } P_2 = (y \text{ value at } P_1) + \Delta y = 1.1 + 0.11 = 1.21.$$

Thus P_2 is $(0.2, 1.21)$. Continuing gives the results in Table 10.2.

Since the solution curves of $dy/dx = y$ are exponentials, they are concave up, and they curve upward away from the line segments of the slope field. Therefore, in this case, Euler's method produces y-values which are too small.

Notice that Euler's method calculates approximate values of y for points on a solution curve; it does *not* give a general formula for y in terms of x. The graph in Figure 10.23 approximates the solution curve.

Example 2   Show that Euler's method for $\dfrac{dy}{dx} = y$ starting at $(0, 1)$ and using two steps with $\Delta x = 0.05$ gives $y \approx 1.1025$ when $x = 0.1$.

Solution   At $(0, 1)$, the slope is 1 and $\Delta y = (1)(0.05) = 0.05$, so new $y = 1 + 0.05 = 1.05$. At $(0.05, 1.05)$, the slope is 1.05 and $\Delta y = (1.05)(0.05) = 0.0525$, so new $y = 1.05 + 0.0525 = 1.1025$ at $x = 0.1$.

In general, dy/dx may be a function of both x and y. Euler's method still works then, as the next example shows.

Example 3   Approximate four points on the solution curve to $\dfrac{dy}{dx} = -\dfrac{x}{y}$, starting at $(0, 1)$ and using $\Delta x = 0.1$.

Solution   The results from Euler's method are in Table 10.3, along with the exact y-values (to two decimals) calculated from the equation of the circle $x^2 + y^2 = 1$, which is the solution curve through $(0, 1)$. Since the curve is concave down, the approximate y-values are above the exact ones. (See Figure 10.24.)

TABLE 10.3 *Euler's method for $dy/dx = -x/y$, starting at $(0, 1)$*

x	Approximate y-value	$\Delta y = (\text{Slope})\Delta x$	True y-value
0	1	$0 = (0)(0.1)$	1
0.1	1	$-0.01 = (-0.1/1)(0.1)$	0.99
0.2	0.99	$-0.02 = (-0.2/0.99)(0.1)$	0.98
0.3	0.97		0.95

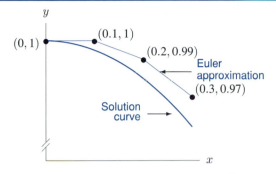

Figure 10.24: Euler's approximate solution to $\frac{dy}{dx} = -\frac{x}{y}$

The Accuracy of Euler's Method

To improve the accuracy of Euler's method, we choose Δx smaller. Let's go back to the differential equation $dy/dx = y$ and compare the exact and approximate values for different Δx's. The exact solution going through the point $(0, 1)$ is $y = e^x$, so the exact values are calculated using this function. (See Figure 10.25.) Where $x = 0.1$,

$$\text{Exact } y\text{-value} = e^{0.1} \approx 1.1051709.$$

In Example 1 we had $\Delta x = 0.1$, and where $x = 0.1$,

$$\text{Approximate } y\text{-value} = 1.1, \quad \text{so the error} \approx 0.005.$$

In Example 2 we decreased Δx to 0.05. After two steps, $x = 0.1$, and we had

$$\text{Approximate } y\text{-value} = 1.1025, \quad \text{so error} \approx 0.00267.$$

Thus, it appears that halving the step size has approximately halved the error.

> The *error* in using Euler's method is the difference between the approximate value and the exact value. If the number of steps used is n, the error is approximately proportional to $1/n$.

Just as there are more accurate numerical integration methods than left and right Riemann sums, there are more accurate methods than Euler's for approximating solution curves. However, Euler's method will be all we need.

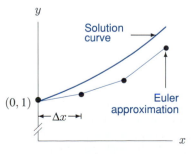

Figure 10.25: Euler's approximate solution to $\frac{dy}{dx} = y$

Problems for Section 10.3

1. Consider the solution of the differential equation $y' = y$ passing through $y(0) = 1$.

    (a) Sketch the slope field for this differential equation, and sketch the solution passing through the point (0,1).
    (b) Use Euler's method with step size $\Delta x = 0.1$ to estimate the solution at $x = 0.1, 0.2, \ldots, 1$.
    (c) Plot the estimated solution on the slope field; compare the solution and the slope field.
    (d) Check that $y = e^x$ is the solution of $y' = y$ with $y(0) = 1$.

2. Consider the differential equation $y' = (\sin x)(\sin y)$.

    (a) Calculate approximate y-values using Euler's method with three steps and $\Delta x = 0.1$, starting at each of the following points: (i) $(0, 2)$ (ii) $(0, \pi)$.
    (b) Use the slope field in Figure 10.16 on page 498 to explain your solution to part (a)(ii).

3. Consider the differential equation $y' = x + y$ whose slope field is in Figure 10.15 on page 498. Use Euler's method with $\Delta x = 0.1$ to estimate y when $x = 0.4$ for the solution curves satisfying

    (a) $y(0) = 1$ (b) $y(-1) = 0$.

4. (a) Use ten steps of Euler's method to determine an approximate solution for the differential equation $y' = x^3$, $y(0) = 0$, using a step size $\Delta x = 0.1$.
    (b) Guess the exact solution and compare it to the computed approximation.
    (c) Use a sketch of the slope field for this equation to explain the results of part (b).

5. (a) Use ten steps of Euler's method to approximate y-values for $dy/dt = 1/t$, starting at $(1, 0)$ and using $\Delta t = 0.1$.
    (b) Using integration, solve the differential equation to find the exact value of y at the end of these ten steps.
    (c) Is your approximate value of y at the end of ten steps bigger or smaller than the exact value? Use a slope field to explain your answer.

6. (a) Use Euler's method to approximate the value of y at $x = 1$ on the solution curve to the differential equation

    $$\frac{dy}{dx} = x^3 - y^3$$

    that passes through $(0, 0)$. Use $\Delta x = 1/5$ (i.e., 5 steps).

    (b) Figure 10.26 shows the slope field for the differential equation. Sketch the solution that passes through $(0, 0)$. Show on the graph the approximation you made in part (a).
    (c) From the graph of the slope field, can you say whether your answer to part (a) is an overestimate or an underestimate?

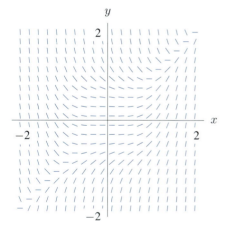

Figure 10.26: Slope field for $\frac{dy}{dx} = x^3 - y^3$

7. Consider the differential equation

    $$\frac{dy}{dx} = 2x, \quad \text{with initial condition } y(0) = 1.$$

    (a) Use Euler's method with two steps to estimate y when $x = 1$.
    (b) Use Euler's method with four steps to estimate y when $x = 1$.
    (c) What is the formula for the exact value of y?
    (d) Does the error in Euler's approximation behave as predicted in the box on page 503?

8. Consider the differential equation

$$\frac{dy}{dx} = \sin(xy), \qquad \text{with initial condition } y(1) = 1.$$

   Estimate $y(2)$, using Euler's method with step sizes $\Delta x = 0.2, 0.1, 0.05$.

   Plot the computed approximations for $y(2)$ against Δx. What do you conclude? Use your observations to estimate the exact value of $y(2)$.

9. Use Euler's method to solve

$$\frac{dB}{dt} = 0.05B$$

   with initial value $B = 1000$ when $t = 0$. Take:

   (a) $\Delta t = 1$ and 1 step.   (b) $\Delta t = 0.5$ and 2 steps.   (c) $\Delta t = 0.25$ and 4 steps.

   (d) Suppose B is the balance in a bank account earning interest. Explain why the result of your calculation in part (a) is equivalent to compounding the interest once a year instead of continuously.
   (e) Interpret the result of your calculations in parts (b) and (c) in terms of compound interest.

10. Consider the differential equation $dy/dx = f(x)$ with initial value $y(0) = 0$. Explain why using Euler's method to approximate the solution curve gives the same results as using left Riemann sums to approximate $\int_0^x f(t)\, dt$.

10.4 SEPARATION OF VARIABLES

We have seen how to sketch solution curves of a differential equation using a slope field and how to calculate approximate numerical solutions. Now we see how to find the equation of a solution curve for certain differential equations.

First, we'll look at a familiar example, the differential equation

$$\frac{dy}{dx} = -\frac{x}{y},$$

whose solution curves are the circles

$$x^2 + y^2 = C.$$

We can check that these circles are solutions by differentiation; the question now is how they were obtained. The method of *separation of variables* works by putting all the x's on one side of the equation and all the y's on the other, giving

$$y\, dy = -x\, dx.$$

We then integrate each side separately:

$$\int y\, dy = -\int x\, dx,$$

$$\frac{y^2}{2} = \frac{-x^2}{2} + k.$$

This gives the circles we were expecting:

$$x^2 + y^2 = C \qquad \text{where } C = 2k.$$

You might worry about whether it is legitimate to separate the dx and the dy. The reason it can be done is explained at the end of this section.

The Exponential Growth and Decay Equations

Let's use separation of variables on an equation that occurs frequently in practice:

$$\frac{dy}{dx} = ky.$$

Separating variables,

$$\frac{1}{y}dy = k\,dx,$$

and integrating,

$$\int \frac{1}{y}\,dy = \int k\,dx,$$

gives

$$\ln|y| = kx + C \quad \text{for some constant } C.$$

Solving for $|y|$ leads to

$$|y| = e^{kx+C} = e^{kx}e^C = Ae^{kx}$$

where $A = e^C$, so A is positive. Thus

$$y = (\pm A)e^{kx} = Be^{kx}$$

where $B = \pm A$, so B is any nonzero constant. Even though there's no C leading to $B = 0$, B can be 0 because $y = 0$ is a solution to the differential equation. We lost this solution when we divided through by y at the first step.

> The general solution to $\dfrac{dy}{dx} = ky$ is $y = Be^{kx}$ for any B.

The differential equation $dy/dx = ky$ always leads to exponential growth (if $k > 0$) or exponential decay (if $k < 0$). Graphs of solution curves for some fixed $k > 0$ are in Figure 10.27. For $k < 0$, the graphs are reflected in the y-axis.

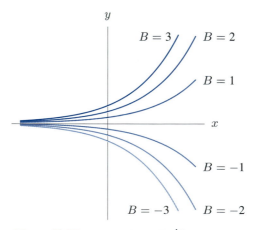

Figure 10.27: Graphs of $y = Be^{kx}$, which are solutions to $\frac{dy}{dx} = ky$, for some fixed $k > 0$

Example 1   For $k > 0$, find and graph solutions of

$$\frac{dH}{dt} = -k(H - 20).$$

Solution   The slope field in Figure 10.28 shows the qualitative behavior of the solutions. To find the equation of the solution curves, we separate variables and integrate:

$$\int \frac{1}{H-20}\, dH = -\int k\, dt.$$

This gives

$$\ln|H-20| = -kt + C.$$

Solving for H leads to:

$$|H-20| = e^{-kt+C} = e^{-kt}e^{C} = Ae^{-kt}$$

or

$$H - 20 = (\pm A)e^{-kt} = Be^{-kt}$$

$$H = 20 + Be^{-kt}.$$

Graphs for $k = 1$ and $B = -10, 0, 10$, with $t \geq 0$, are in Figure 10.28.

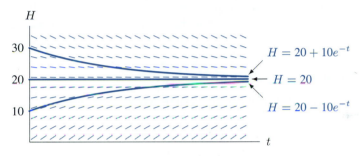

Figure 10.28: Slope field and some solution curves for $\frac{dH}{dt} = -k(H-20)$, with $k = 1$

Example 2   Find and sketch the solution to

$$\frac{dP}{dt} = 2P - 2Pt$$

satisfying $P = 5$ when $t = 0$.

Solution   Factoring the right-hand side gives

$$\frac{dP}{dt} = P(2 - 2t).$$

Separating variables, we get

$$\int \frac{dP}{P} = \int (2 - 2t)\, dt,$$

so

$$\ln|P| = 2t - t^2 + C.$$

Solving for P leads to

$$|P| = e^{2t - t^2 + C} = e^{C}e^{2t - t^2} = Ae^{2t - t^2}$$

with $A = e^{C}$, so $A > 0$. Thus the general solution to the differential equation is

$$P = Be^{2t - t^2} \quad \text{for any } B.$$

To find the value of B for this example, substitute $P = 5$ and $t = 0$ into the general solution, giving

$$5 = Be^{2(0) - 0^2} = B$$

so

$$P = 5e^{2t - t^2}.$$

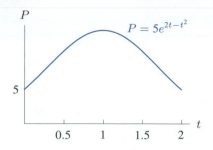

Figure 10.29: Bell-shaped solution curve

The graph of this function is in Figure 10.29.

Since the solution can be rewritten as

$$P = 5e^{1-1+2t-t^2} = 5e^1 e^{-1+2t-t^2} = (5e)e^{(t-1)^2},$$

the graph has the same shape as the graph of $y = e^{-t^2}$, the bell-shaped curve of statistics. Here the maximum, normally at $t = 0$, is shifted one unit to the right and occurs at $t = 1$.

Justification For Separation of Variables

Suppose a differential equation can be written in the form

$$\frac{dy}{dx} = g(x)f(y).$$

Then, provided $f(y) \neq 0$, we write $f(y) = 1/h(y)$ so that the right-hand side can be thought of as a fraction,

$$\frac{dy}{dx} = \frac{g(x)}{h(y)}.$$

If we multiply through by $h(y)$ we get

$$h(y)\frac{dy}{dx} = g(x).$$

Thinking of y as a function of x, so $y = y(x)$, and $dy/dx = y'(x)$, we can write the line above as

$$h(y(x)) \cdot y'(x) = g(x).$$

Now integrate both sides with respect to x:

$$\int h(y(x)) \cdot y'(x)\, dx = \int g(x)\, dx.$$

The form of the integral on the left suggests that we use the substitution $y = y(x)$. Since $dy = y'(x)\, dx$, we get

$$\int h(y)\, dy = \int g(x)\, dx.$$

If antiderivatives of h and g can be found, then this will give the equation of the solution curve.

Note that transforming the original differential equation,

$$\frac{dy}{dx} = \frac{g(x)}{h(y)},$$

into

$$\int h(y)\, dy = \int g(x)\, dx$$

looks as though we have treated dy/dx as a fraction, cross-multiplied and then integrated. Although that's not exactly what we've done, you may find this a helpful way of remembering the method. In fact, the dy/dx notation was introduced by Leibniz to allow shortcuts like this (more specifically, to make the chain rule look like cancelation).

Problems for Section 10.4

Find the solutions to the differential equations in Problems 1–22, subject to the given initial condition.

1. $\dfrac{dP}{dt} = 0.02P, \quad P(0) = 20$

2. $\dfrac{dQ}{dt} = \dfrac{Q}{5}, \quad Q = 50$ when $t = 0$

3. $\dfrac{dm}{dt} = 3m, \quad m = 5$ when $t = 1$

4. $\dfrac{dI}{dx} = 0.2I, \quad I = 6$ where $x = -1$

5. $\dfrac{dy}{dx} + \dfrac{y}{3} = 0, \quad y(0) = 10$

6. $\dfrac{1}{z}\dfrac{dz}{dt} = 5, \quad z(1) = 5$

7. $\dfrac{dP}{dt} = P + 4, \quad P = 100$ when $t = 0$

8. $\dfrac{dy}{dx} = 2y - 4, \quad$ through $(2, 5)$

9. $\dfrac{dm}{dt} = 0.1m + 200, \quad m(0) = 1000$

10. $\dfrac{dB}{dt} + 2B = 50, \quad B(1) = 100$

11. $\dfrac{dy}{dt} = 0.5(y - 200), \quad y = 50$ when $t = 0$

12. $\dfrac{dQ}{dt} = 0.3Q - 120, \quad Q = 50$ when $t = 0$

13. $\dfrac{dR}{dr} + R = 1, \quad R(1) = 0.1$

14. $\dfrac{dy}{dt} = \dfrac{y}{3 + t}, \quad y(0) = 1$

15. $\dfrac{dz}{dt} = te^z, \quad$ through the origin

16. $\dfrac{dy}{dx} = \dfrac{5y}{x}, \quad y = 3$ where $x = 1$

17. $\dfrac{dy}{dt} = y^2(1 + t), \quad y = 2$ when $t = 1$

18. $\dfrac{dz}{dt} = z + zt^2, \quad z = 5$ when $t = 0$

19. $\dfrac{dw}{d\theta} = \theta w^2 \sin \theta^2, \quad w(0) = 1$

20. $x(x + 1)\dfrac{du}{dx} = u^2, \quad u(1) = 1$

21. $\dfrac{dz}{dr} = z + zr^2, \quad z(0) = 1$

22. $\dfrac{dw}{d\psi} = \psi w^2 \cos(\psi^2), \quad w(0) = 1$

Solve the differential equations in Problems 23–28. Assume a, b, and k are constants.

23. $\dfrac{dR}{dt} = kR$

24. $\dfrac{dQ}{dt} - \dfrac{Q}{k} = 0$

25. $\dfrac{dP}{dt} = P - a$

26. $\dfrac{dQ}{dt} = b - Q$

27. $\dfrac{dP}{dt} = k(P - a)$

28. $\dfrac{dR}{dt} = aR + b$

Solve the differential equations in Problems 29–32. Assume $x \geq 0$, $y \geq 0$.

29. $\dfrac{dy}{dt} = y(2 - y), \quad y(0) = 1.$

30. $t\dfrac{dx}{dt} = (1 + 2\ln t)\tan x.$

31. $\dfrac{dx}{dt} = \dfrac{x \ln x}{t}.$

32. $\dfrac{dy}{dt} = -y \ln\left(\dfrac{y}{2}\right), \quad y(0) = 1.$

33. (a) Sketch the slope field for $y' = x/y$.
    (b) Sketch several solution curves.
    (c) Solve the differential equation analytically.

34. (a) Sketch the slope field for $y' = -y/x$.
    (b) Sketch several solution curves.
    (c) Solve the differential equation analytically.

35. Compare the slope field for $y' = x/y$, Problem 33, with that for $y' = -y/x$, Problem 34. Show that the solution curves of Problem 33 intersect the solution curves of Problem 34 at right angles.

36. The differential equation $\dfrac{dy}{dx} = \dfrac{x + y}{x - y}$ is not separable. Show that if we put $y(x) = xv(x)$, then the resulting differential equation is separable.

37. The differential equation $\dfrac{dy}{dx} = \dfrac{x+y}{x-y}$ can be rewritten as $\dfrac{dy}{dx} = \dfrac{1+(y/x)}{1-(y/x)} = f\left(\dfrac{y}{x}\right)$. In general, any differential equation that can be rewritten as $\dfrac{dy}{dx} = f\left(\dfrac{y}{x}\right)$ is called *homogeneous*. Show that every homogeneous differential equation can be reduced to a separable equation (in terms of x and v) by putting $y(x) = xv(x)$.

38. The differential equation $\dfrac{dt}{ds} = \dfrac{s+t+5}{s-t-1}$ is neither separable nor homogeneous (as defined in Problem 37). Show that making a change of variable of the form $s = x + a$, $t = y + b$, and choosing a and b appropriately, yields a homogeneous equation. The homogeneous equation can then be solved with the method of Problem 37.

10.5 GROWTH AND DECAY

In this section we look at exponential growth and decay equations. Consider the population of a region. If there is no immigration or emigration, the rate at which the population is changing is often proportional to the population. In other words, the larger the population, the faster it is growing, because there are more people to have babies. If the population at time t is P, and its continuous growth rate is 2% per unit time, then we know

$$\text{Rate of growth of population} = 2\%(\text{Current population})$$

and we can write this as

$$\frac{dP}{dt} = 0.02P.$$

The 2% growth rate is called the *relative growth rate* to distinguish it from the absolute growth rate, or rate of change of the population, dP/dt. Notice they are in different units. Since

$$\text{Relative growth rate} = 2\% = \frac{1}{P}\frac{dP}{dt}$$

the relative growth rate is in percent per unit time, while

$$\text{Absolute growth rate} = \text{Rate of change of population} = \frac{dP}{dt}$$

is in people per unit time.

The equation $dP/dt = 0.02P$ is of the form $dP/dt = kP$ for $k = 0.02$ and therefore has the solution

$$P = P_0 e^{0.02t}.$$

Other processes are described by differential equations similar to that for population growth, but with negative values for k. In summary, we have the following result from the preceding section:

Every solution to the equation

$$\frac{dP}{dt} = kP$$

can be written in the form

$$P = P_0 e^{kt},$$

where P_0 is the initial value of P, and $k > 0$ represents growth, while $k < 0$ represents decay.

Recall that the *doubling time* of an exponentially growing quantity is the time required for it to double. The *half-life* of an exponentially decaying quantity is the time for half of it to decay.

Continuously Compounded Interest

Continuous compounding means that, at any time, interest is being accrued at a rate that is a fixed percentage of the balance at that moment. Thus, the larger the balance the faster interest is being earned and the faster the balance grows.

Example 1 A bank account earns interest continuously at a rate of 5% of the current balance per year. Assume that the initial deposit is $1000, and that no other deposits or withdrawals are made.

    (a)   Write the differential equation satisfied by the balance in the account.

    (b)   Solve the differential equation and graph the solution.

Solution (a)   In this problem we are looking for B, the balance in the account in dollars, as a function of t, time in years. The fact that interest is being earned continuously means that interest is continuously being added to the account at a rate of 5% of the balance at that moment. Thus, at any instant,

$$\text{Rate at which interest is earned} = 5\%(\text{Current balance}).$$

If B is the balance at time t, and if no deposits or withdrawals are made, then the rate of change of the balance is exactly the rate the interest is earned, so

$$\text{Rate at which balance is increasing} = 5\%(\text{Current balance}),$$

which we write as

$$\frac{dB}{dt} = 0.05B.$$

This is the differential equation that describes the process. Notice that it does not involve the $1000, because the initial deposit does not affect the process by which interest is earned. The $1000 is the initial condition.

    (b)   Solving the differential equation by separation of variables gives

$$B = B_0 e^{0.05t},$$

where B_0 is the initial value of B, so $B_0 = 1000$. Thus

$$B = 1000 e^{0.05t}$$

and this function is graphed in Figure 10.30.

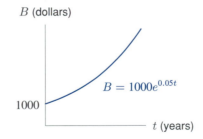

B (dollars)

$B = 1000e^{0.05t}$

1000

t (years)

Figure 10.30: Bank balance against time

You may wonder how we can represent an amount of money by a differential equation, since money can only take on discrete values (we can't have fractions of a cent). In fact, the differential equation is only an approximation, but for large amounts of money, it is a pretty good approximation.

The Difference Between Continuous and Annual Percentage Growth Rates

If $P = P_0(1 + r)^t$ with t in years, we say that r is the *annual* growth rate, while if $P = P_0 e^{kt}$, we say that k is the *continuous* or *instantaneous* growth rate.

It is important to note that the constant k in the differential equation $dP/dt = kP$ is not the annual growth rate, but rather the continuous growth rate. In Example 1, with a continuous interest rate of 5%, we obtain a balance at time t of $B = B_0 e^{0.05t}$, where t is measured in years. At the end of one year the balance is $B_0 e^{0.05}$. In that one year, our balance has changed from B_0 to $B_0 e^{0.05}$; that is, it has changed by the factor $e^{0.05} = 1.0513$. Thus the annual growth rate is 5.13%. This is what the bank means when it says "5% compounded continuously for an effective annual yield of 5.13%." Since $P_0 e^{0.05t} = P_0(1.0513)^t$, we are talking about two different ways to represent the same function of t.

Since most growth is measured over discrete time intervals, a continuous growth rate is an idealized concept. A demographer who says the population of some country is growing at the rate of 2% per year usually means just what you think: After one year the population will have increased by a factor of 1.02; and after t years the population will be given by $P = P_0(1.02)^t$. To find the continuous growth rate, k, for the population we proceed as follows. Suppose the population is expressed as $P = P_0 e^{kt}$. At the end of one year $P = P_0 e^k$. To get 2% growth in that year we need $e^k = 1.02$. Thus $k = \ln 1.02 \approx 0.0198$. The continuous growth rate $k = 1.98\%$ is close to the annual growth rate of 2%, but it is not the same. Again, we have two different representations of the same function, since $P_0(1.02)^t = P_0 e^{0.0198t}$.

Pollution in the Great Lakes

In the 1960s pollution in the Great Lakes became an issue of public concern. We set up a model for how long it would take for the lakes to flush themselves clean, assuming no further pollutants are being dumped in the lake.

Suppose Q is the total quantity of pollutant in a lake of volume V at time t. Suppose that clean water is flowing into the lake at a constant rate r and that water flows out at the same rate. Assume that the pollutant is evenly spread throughout the lake, and that the clean water coming into the lake immediately mixes with the rest of the water.

We investigate how Q varies with time. Since pollutants are being taken out of the lake but not added, Q decreases, and the water leaving the lake becomes less polluted, so the rate at which the pollutants leave decreases. This tells us that Q is decreasing and concave up. In addition, the pollutants are never completely removed from the lake though the quantity remaining becomes arbitrarily small: in other words, Q is asymptotic to the t-axis. (See Figure 10.31.)

Setting Up a Differential Equation for the Pollution

To model exactly how Q changes with time, we write an equation for the rate at which Q changes. We know that

$$\begin{pmatrix} \text{Rate } Q \\ \text{changes} \end{pmatrix} = -\begin{pmatrix} \text{Rate pollutants} \\ \text{leave in outflow} \end{pmatrix}.$$

where the negative sign represents the fact that Q is decreasing. At time t, the concentration of pollutants is Q/V and water containing this concentration is leaving at rate r. Thus

$$\begin{pmatrix} \text{Rate pollutants} \\ \text{leave in outflow} \end{pmatrix} = \begin{pmatrix} \text{Rate of} \\ \text{outflow} \end{pmatrix} \times \text{Concentration} = r \cdot \frac{Q}{V}.$$

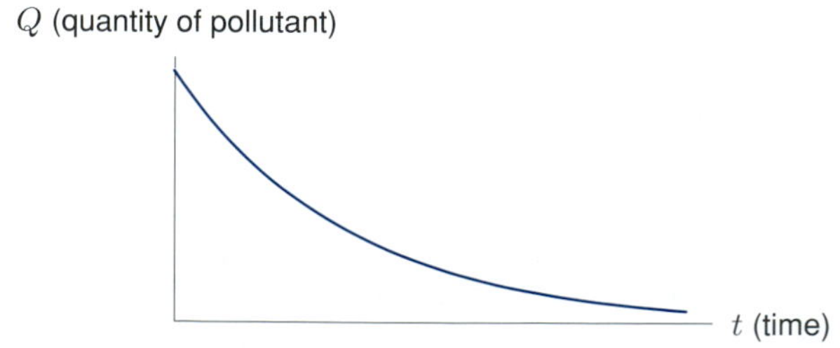

Figure 10.31: Pollutant in lake versus time

So the differential equation is

$$\frac{dQ}{dt} = -\frac{r}{V}Q$$

and its solution is

$$Q = Q_0 e^{-rt/V}.$$

Table 10.4 contains values of r and V for four of the Great Lakes.[3] We use this data to calculate how long it would take for certain fractions of the pollution to be removed.

TABLE 10.4 *Volume and outflow in Great Lakes*

	V (thousands of km^3)	r (km^3/year)
Superior	12.2	65.2
Michigan	4.9	158
Erie	0.46	175
Ontario	1.6	209

Example 2   According to this model, how long will it take for 90% of the pollution to be removed from Lake Erie? For 99% to be removed?

Solution   Substituting r and V for Lake Erie into the differential equation for Q gives

$$\frac{dQ}{dt} = -\frac{r}{V}Q = \frac{-175}{0.46 \times 10^3}Q = -0.38Q$$

where t is measured in years. Thus Q is given by

$$Q = Q_0 e^{-0.38t}.$$

When 90% of the pollution has been removed, 10% remains, so $Q = 0.1Q_0$. Substituting gives

$$0.1Q_0 = Q_0 e^{-0.38t}.$$

Canceling Q_0 and solving for t, we get

$$t = \frac{-\ln(0.1)}{0.38} \approx 6 \text{ years}.$$

When 99% of the pollution has been removed, $Q = 0.01Q_0$, so t satisfies

$$0.01Q_0 = Q_0 e^{-0.38t}.$$

Solving for t gives

$$t = \frac{-\ln(0.01)}{0.38} \approx 12 \text{ years}.$$

Newton's Law of Heating and Cooling

Newton proposed that the temperature of a hot object decreases at a rate proportional to the difference between its temperature and that of its surroundings. Similarly, a cold object heats up at a rate proportional to the temperature difference between the object and its surroundings.

For example, a hot cup of coffee standing on the kitchen table cools at a rate proportional to the temperature difference between the coffee and the surrounding air. As the coffee cools down, the rate at which it cools decreases, because the temperature difference between the coffee and the air decreases. In the long run, the rate of cooling tends to zero, and the temperature of the coffee

[3]Data from William E. Boyce and Richard C. DiPrima, *Elementary Differential Equations* (New York: Wiley, 1977).

approaches room temperature—as we would expect. Figure 10.32 shows the temperature of two cups of coffee against time, one starting at a higher temperature than the other, but both tending toward room temperature in the long run.

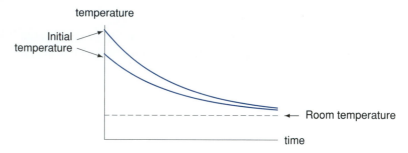

Figure 10.32: Temperature of coffee versus time

Example 3   When a murder is committed, the body, originally at 37°C, cools according to Newton's Law of Cooling. Suppose that after two hours the temperature is 35°C, and that the temperature of the surrounding air is a constant 20°C.

(a)   Find the temperature, H, of the body as a function of t, the time in hours since the murder was committed.

(b)   Sketch a graph of temperature against time.

(c)   What happens to the temperature in the long run? Show this on the graph and algebraically.

(d)   If the body is found at 4 pm at a temperature of 30°C, when was the murder committed?

Solution   (a)   We first find a differential equation for the temperature of the body as a function of time. Newton's Law of Cooling says that for some constant α,

Rate of change of temperature $= \alpha$(Temperature difference).

If H is the temperature of the body, then

Temperature difference $= H - 20$

so

$$\frac{dH}{dt} = \alpha(H - 20).$$

What about the sign of α? If the temperature difference is positive (i.e., $H > 20$), then H is falling, so the rate of change must be negative. Thus α should be negative, so we write:

$$\frac{dH}{dt} = -k(H - 20), \qquad \text{for some } k > 0.$$

Separating variables and solving, as in Example 1 on page 506, gives:

$$H - 20 = Be^{-kt}.$$

To find B, substitute the initial condition that $H = 37$ when $t = 0$:

$$37 - 20 = Be^{-k(0)} = B,$$

so $B = 17$. Thus,

$$H - 20 = 17e^{-kt}.$$

To find k, we use the fact that after 2 hours, the temperature is 35°C, so

$$35 - 20 = 17e^{-k(2)}.$$

Dividing by 17 and taking natural logs, we get:

$$\ln\left(\frac{15}{17}\right) = \ln(e^{-2k})$$
$$-0.125 = -2k$$
$$k \approx 0.063.$$

Therefore, the temperature is given by

$$H - 20 = 17e^{-0.063t}$$

or

$$H = 20 + 17e^{-0.063t}.$$

(b)   The graph of $H = 20 + 17e^{-0.063t}$ has a vertical intercept of $H = 37$, because the temperature of the body starts at 37°C. The temperature decays exponentially with $H = 20$ as the horizontal asymptote. (See Figure 10.33.)

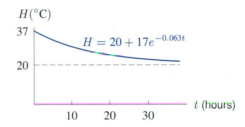

Figure 10.33: Temperature of dead body

(c)   "In the long run" means as $t \to \infty$. The graph shows that as $t \to \infty$, $H \to 20$. Algebraically, since $e^{-0.063t} \to 0$ as $t \to \infty$,

$$H = 20 + \underbrace{17e^{-0.063t}}_{\text{goes to 0 as } t \to \infty} \longrightarrow 20 \quad \text{as } t \to \infty.$$

(d)   We want to know when the temperature reaches 30°C. Substitute $H = 30$ and solve for t:

$$30 = 20 + 17e^{-0.063t}$$
$$\frac{10}{17} = e^{-0.063t}.$$

Taking natural logs:

$$-0.531 = -0.063t,$$

which gives

$$t \approx 8.4 \text{ hours.}$$

Thus the murder must have been committed about 8.4 hours before 4 pm. Since 8.4 hours = 8 hours 24 minutes, the murder was committed at about 7:30 am.

Equilibrium Solutions

Figure 10.34 shows the temperature of several objects in a 20°C room. One is initially hotter than 20°C and cools down toward 20°C; another is initially cooler and warms up toward 20°C. All these curves are solutions to the differential equation

$$\frac{dH}{dt} = -k(H - 20)$$

for some fixed $k > 0$, and all the solutions have the form

$$H = 20 + Ae^{-kt}$$

for some A. Notice that $H \to 20$ as $t \to \infty$ because $e^{-kt} \to 0$ as $t \to \infty$. In other words, in the long run, the temperature of the object always tends toward 20°C, the temperature of the room, no matter what the initial temperature. This means that what happens in the long run is independent of the initial condition.

In the special case when $A = 0$, we have the *equilibrium solution*

$$H = 20$$

for all t. This means that if the object starts at 20°C, it remains at 20°C for all time. Notice that such a solution can be found directly from the differential equation by solving $dH/dt = 0$:

$$\frac{dH}{dt} = -k(H - 20) = 0$$

giving

$$H = 20.$$

Regardless of the initial temperature, H always gets closer and closer to 20 as $t \to \infty$. As a result, $H = 20$ is called a *stable* equilibrium[4] for H.

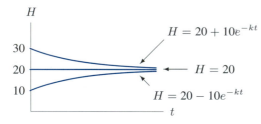

Figure 10.34: $H = 20$ is stable equilibrium $(k > 0)$

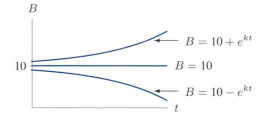

Figure 10.35: $B = 10$ is unstable equilibrium $(k > 0)$

A different situation is displayed in Figure 10.35, which shows solutions to the differential equation

$$\frac{dB}{dt} = k(B - 10)$$

for some fixed $k > 0$. Solving $dB/dt = 0$ gives the equilibrium $B = 10$, which is *unstable* because if B starts near 10, it moves away as $t \to \infty$.

In general, we have the following definitions.

- An **equilibrium solution** is constant for all values of the independent variable. The graph is a horizontal line. Equilibrium solutions can be identified by setting the derivative of the function to zero.
- An equilibrium is **stable** if a small change in the initial conditions gives a solution which tends toward the equilibrium as the independent variable tends to positive infinity.
- An equilibrium is **unstable** if a small change in the initial conditions gives a solution curve which veers away from the equilibrium as the independent variable tends to positive infinity.

---

[4]In more advanced work, this behavior is described as asymptotic stability.

Problems for Section 10.5

1. The amount of arable land (land that can be used for growing crops) in use increases as the world's population increases. Suppose $A(t)$ represents the total number of hectares of arable land in use in year t. (A hectare is about $2\frac{1}{2}$ acres.)

    (a) Explain why it is plausible that $A(t)$ satisfies the equation $A'(t) = kA(t)$. What assumptions are you making about the world's population and its relation to the amount of arable land used?

    (b) In 1950 about $1 \cdot 10^9$ hectares of arable land were in use; in 1980 the figure was $2 \cdot 10^9$. If the total amount of arable land available is thought to be $3.2 \cdot 10^9$ hectares, when does this model predict it is exhausted? (Let $t = 0$ in 1950.)

2. A yam is put in a 200°C oven and heats up according to the differential equation

    $$\frac{dH}{dt} = -k(H - 200),$$

    where k is positive.

    (a) If the yam is at 20°C when it is put in the oven, solve the differential equation.
    (b) Find k using the fact that after 30 minutes the temperature of the yam is 120°C.

3. Each of the curves in Figure 10.36 represents the balance in a bank account into which a single deposit was made at time zero. Assuming interest is compounded continuously, find

    (a) The curve representing the largest initial deposit.

    (b) The curve representing the largest interest rate.

    (c) Two curves representing the same initial deposit.

    (d) Two curves representing the same interest rate.

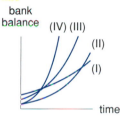

Figure 10.36

4. The graphs in Figure 10.37 represent the temperature, H°C, of four eggs as a function of time, t, measured in minutes. Match three of the graphs with the descriptions below. Write a similar description for the fourth graph, including an interpretation of any intercepts and asymptotes.

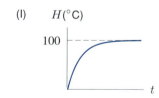 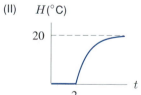 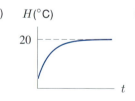 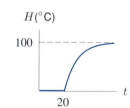

Figure 10.37

    (a) An egg is taken out of the refrigerator (just above 0°C) and put into boiling water.
    (b) Twenty minutes after the egg in part (a) is taken out of the fridge and put into boiling water, the same thing is done with another egg.
    (c) An egg is taken out of the refrigerator at the same time as the egg in part (a) and left to sit on the kitchen table.

5. (a) What are the equilibrium solutions for the differential equation

    $$\frac{dy}{dt} = 0.2(y - 3)(y + 2)?$$

    (b) Use a slope field to determine whether each equilibrium solution is stable or unstable.

6. The slope field for a differential equation is given in Figure 10.38. Estimate all equilibrium solutions for this differential equation, and indicate whether each is stable or unstable.

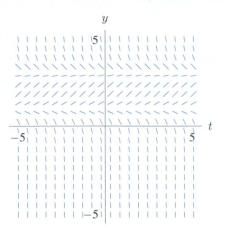

Figure 10.38

7. (a) Find the equilibrium solution to the differential equation

$$\frac{dy}{dt} = 0.5y - 250.$$

   (b) Find the general solution to this differential equation.
   (c) Sketch the graphs of several solutions to this differential equation, using different initial values for y.
   (d) Is the equilibrium solution stable or unstable?

8. Money in a bank account grows continuously at an annual rate of r (when the interest rate is 5%, $r = 0.05$, and so on). Suppose $1000 is put into the account in 1970.

   (a) Write a differential equation satisfied by M, the amount of money in the account at time t, measured in years since 1970.
   (b) Solve the equation.
   (c) Sketch the solution until the year 2000 for interest rates of 5% and 10%.

9. (a) If $B = f(t)$ is the balance of a bank account at time t that earns interest at a rate of $r\%$, compounded continuously, what is the differential equation that describes the rate at which the balance changes? What is the constant of proportionality, in terms of r?
   (b) What is the solution to this differential equation?
   (c) Sketch the graph of $B = f(t)$ for an account that starts with $1000 and earns interest at the following rates:

   (i) 4%   (ii) 10%   (iii) 15%

10. In Example 2 on page 513 we saw that it would take about 6 years for 90% of the pollution in Lake Erie to be removed, and about 12 years for 99% to be removed. Explain why one time is double the other.

11. Using the model in the text and data in Table 10.4 on page 513, find how long it would take for 90% of the pollution to be removed from Lake Michigan and from Lake Ontario, assuming no new pollutants are added. Explain how you can tell which lake will take longer to be purified just by looking at the data.

12. Use the model in the text and the data in Table 10.4 on page 513 to determine which of the Great Lakes would require the longest time and which would require the shortest time for 80% of the pollution to be removed, assuming no new pollutants are being added. Find the ratio of these two times.

13. Warfarin is a drug used as an anticoagulant. After stopping administration of the drug, the quantity remaining in a patient's body decreases at a rate proportional to the quantity remaining. The half-life of Warfarin in the body is 37 hours.

    (a)   Sketch a rough graph of the quantity, Q, of warfarin in a patient's body as function of the time, t, since stopping administration of the drug. Mark the 37 hours on your graph.

    (b)   Write a differential equation satisfied by Q.

    (c)   How many days does it take for the drug level in the body to be reduced to 25% of the original level?

14. Morphine is often used as a pain-relieving drug. The half-life of morphine in the body is 2 hours. Suppose morphine is administered to a patient intravenously at a rate of 2.5 mg per hour, and the rate at which the morphine is eliminated is proportional to the amount present.

    (a)   Show that the constant of proportionality for the rate at which morphine leaves the body (in mg/hour) is $k = -0.347$.

    (b)   Write a differential equation for the quantity, Q, of morphine in the blood after t hours.

    (c)   Use the differential equation to find the equilibrium solution. (This is the long-term amount of morphine in the body, once the system has stabilized.)

15. When the electromotive force (emf) is removed from a circuit containing inductance and resistance but no capacitors, the rate of decrease of current is proportional to the current. If the initial current is 30 amps but decays to 11 amps after 0.01 seconds, find an expression for the current as a function of time.

16. In some chemical reactions, the rate at which the amount of a substance changes with time is proportional to the amount present. For example, this is the case as δ-glucono-lactone changes into gluconic acid.

    (a)   Write a differential equation satisfied by y, the quantity of δ-glucono-lactone present at time t.

    (b)   If 100 grams of δ-glucono-lactone is reduced to 54.9 grams in one hour, how many grams will remain after 10 hours?

17. The rate at which barometric pressure decreases with altitude is proportional to the barometric pressure at that altitude. If the barometric pressure is measured in inches of mercury, and the altitude in feet, then the constant of proportionality is 3.7×10^{-5}. Suppose the barometric pressure at sea level is 29.92 inches of mercury.

    (a)   Calculate the barometric pressure at the top of Mount Whitney, 14,500 feet (the highest mountain in the US outside Alaska), and at the top of Mount Everest, 29,000 feet (the highest mountain in the world).

    (b)   People cannot easily survive at a pressure below 15 inches of mercury. What is the highest altitude to which people can safely go?

18. The rate (per foot) at which light is absorbed as it passes through water is proportional to the intensity, or brightness, at that point.

    (a)   Find the intensity as a function of the distance the light has traveled through the water.

    (b)   If 50% of the light is absorbed in 10 feet, how much is absorbed in 20 feet? 25 feet?

19. A detective finds a murder victim at 9 am. The temperature of the body is measured at $90.3°$F. One hour later, the temperature of the body is $89.0°$F. The temperature of the room has been maintained at a constant $68°$F.

    (a)   Assuming the temperature, T, of the body obeys Newton's Law of Cooling, write a differential equation for T.

    (b)   Solve the differential equation to estimate the time the murder occurred.

20. Suppose that at 1:00 pm one winter afternoon, there is a power failure at your house in Wisconsin, and your heat does not work without electricity. When the power goes out, it is $68°$F in your house. At 10:00 pm, it is $57°$F in the house, and you notice that it is $10°$F outside.

    (a)   Assuming that the temperature, T, in your home obeys Newton's Law of Cooling, write the differential equation satisfied by T.

    (b)   Solve the differential equation to estimate the temperature in the house when you get up at 7:00 am the next morning. Should you worry about your water pipes freezing?

    (c)   What assumption did you have to make in part (a) about the temperature outside? Given this (probably incorrect) assumption, would you revise your estimate up or down? Why?

21.   The amount of radioactive carbon-14 in a sample is measured using a Geiger counter, which records each disintegration of an atom. Living tissue disintegrates at a rate of about 13.5 atoms per minute per gram of carbon. In 1977 a charcoal fragment found at Stonehenge, England, recorded 8.2 disintegrations per minute per gram of carbon. Assuming that the half-life of carbon-14 is 5730 years and that the charcoal was formed during the building of the site, estimate the date at which Stonehenge was built.

22.   The radioactive isotope carbon-14 is present in small quantities in all life forms, and it is constantly replenished until the organism dies, after which it decays to stable carbon-12 at a rate proportional to the amount of carbon-14 present, with a half-life of 5730 years. Suppose $C(t)$ is the amount of carbon-14 present at time t.

(a)   Find the value of the constant k in the differential equation $C' = -kC$.

(b)   In 1988 three teams of scientists found that the Shroud of Turin, which was reputed to be the burial cloth of Jesus, contained 91% of the amount of carbon-14 contained in freshly made cloth of the same material.[5] How old is the Shroud of Turin, according to these data?

23.   Before Galileo discovered that the speed of a falling body with no air resistance is proportional to the time since it was dropped, he mistakenly conjectured that the speed was proportional to the distance it had fallen.

(a)   Assume the mistaken conjecture to be true and write an equation relating the distance fallen, $D(t)$, at time t, and its derivative.

(b)   Using your answer to part (a) and the correct initial conditions, show that D would have to be equal to 0 for all t, and therefore the conjecture must be wrong.

10.6 APPLICATIONS AND MODELING

Much of this book involves functions that represent real processes, such as how the temperature of a yam or the population of Mexico is changing with time. You may wonder where such functions come from. In some cases, we fit functions to experimental data by trial and error. In other cases, we take a more theoretical approach, leading to a differential equation whose solution is the function we want. In this section we give examples of the more theoretical approach.

How a Layer of Ice Forms

When ice forms on a lake, the water on the surface freezes first. As heat from the water travels up through the ice and is lost to the air, more ice is formed. The question we will consider is: How thick is the layer of ice as a function of time? Since the thickness of the ice increases with time, the thickness function is increasing. In addition, as the ice gets thicker, it insulates better, therefore we expect the layer of ice to form more slowly as time goes on. Hence the thickness function is increasing at a decreasing rate, so its graph is concave down.

A Differential Equation for the Thickness of the Ice

To get more detailed information about the thickness function, we have to make some assumptions. Suppose y represents the thickness of the ice as a function of time, t. Since the thicker the ice, the longer it takes the heat to get through it, we will assume that the rate at which ice is formed is inversely proportional to the thickness. In other words, we assume that for some constant k,

$$\text{Rate thickness is increasing} = \frac{k}{\text{Thickness}},$$

so

$$\frac{dy}{dt} = \frac{k}{y} \quad \text{where} \quad k > 0.$$

_____

[5] *The New York Times*, October 18, 1988

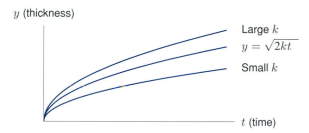

Figure 10.39: Thickness of ice as a function of time

This differential equation enables us to find a formula for the function y. Using separation of variables:

$$\int y\, dy = \int k\, dt$$

$$\frac{y^2}{2} = kt + C.$$

If we measure time so that $y = 0$ when $t = 0$, then $C = 0$. Since y must be non-negative, we have

$$y = \sqrt{2kt}\,.$$

Graphs of y against t are in Figure 10.39.

Notice that the larger y is, the more slowly y increases. In addition, this model suggests that y increases indefinitely as time passes. (Of course, the value of y cannot increase beyond the depth of the lake.)

The Net Worth of a Company

In the preceding section, we saw an example in which money in a bank account was earning interest (Example 1, page 511). Consider a company whose revenues are proportional to its net worth (like interest on a bank account) but which must also make payroll payments. The question is: under what circumstances does the company make money and under what circumstances does it go bankrupt?

Common sense says that if the payroll exceeds the rate at which revenue is earned, the company will eventually be in trouble, whereas if revenue exceeds payroll, the company should do well. In order to make this more precise, we make several assumptions. We assume that revenue is earned continuously and that payments are made continuously. (In practice, revenue and payments are not made continuously, but for a large company this is a good approximation.) We also assume that the only factors affecting net worth are revenue and payroll.

Example 1   Suppose a company's revenue is earned at a continuous annual rate of 5% of its net worth. At the same time, the company's payroll obligations amount to $200 million a year, paid out continuously.
(a)   Write a differential equation that governs the net worth of the company, W million dollars.
(b)   Solve the differential equation, assuming an initial net worth of W_0 million dollars.
(c)   Sketch the solution for $W_0 = 3000$, 4000, and 5000.

Solution   First, let's see what we can learn without writing a differential equation. For example, we can ask if there is any initial net worth W_0 which will exactly keep the net worth constant. If there's such an equilibrium, the rate at which revenue is earned must exactly balance the payments made, so

$$\text{Rate at which revenue is earned} = \text{Rate at which payroll payments are made.}$$

If net worth is a constant W_0, revenue must be earned at a constant rate of $0.05W_0$ per year, so we must have

$$0.05W_0 = 200 \quad \text{giving} \quad W_0 = 4000.$$

Therefore, if the net worth starts at \$4000 million, the revenue and payments are equal, and the net worth will remain constant. Therefore \$4000 million is an equilibrium solution.

Suppose, however, the initial net worth is above \$4000 million. Then, the revenue earned will be more than the payroll expenses, and the net worth of the company will increase, thereby increasing the revenue still further. Thus the net worth will increase faster and faster. On the other hand, if the initial net worth is below \$4000 million, the revenue will not be enough to meet the payments, and the net worth of the company will decline. This decreases the revenue; making the net worth decrease still faster. Hence the net worth will eventually go to zero, and the company will go bankrupt.

(a)   Now we set up a differential equation for the net worth, using the fact that

$$
\begin{array}{ccc}
\text{Rate at which} & = & \text{Rate revenue} \\
\text{net worth is increasing} & & \text{is earned}
\end{array}
\quad - \quad
\begin{array}{c}
\text{Rate payroll payments} \\
\text{are made}
\end{array} .
$$

In millions of dollars per year, revenue is earned at a rate of $0.05W$, and payments are made at a rate of 200 per year, so we have

$$\frac{dW}{dt} = 0.05W - 200$$

where t is in years. Notice that the initial net worth does not enter into the differential equation. The equilibrium solution, $W = 4000$, is obtained by setting $dW/dt = 0$.

(b)   We solve this equation by separation of variables. It is helpful to factor out 0.05 before separating, so that the W moves over to the left-hand side without a coefficient. This makes the manipulations less messy. So we write

$$\frac{dW}{dt} = 0.05(W - 4000).$$

Separating and integrating gives

$$\int \frac{dW}{W - 4000} = \int 0.05 \, dt,$$

so

$$\ln|W - 4000| = 0.05t + C,$$

or

$$|W - 4000| = e^{0.05t + C} = e^C e^{0.05t}.$$

This means

$$W - 4000 = Ae^{0.05t} \qquad \text{where } A = \pm e^C.$$

To find A we use the initial condition that $W = W_0$ when $t = 0$.

$$W_0 - 4000 = Ae^0 = A.$$

Substituting this value for A into $W = 4000 + Ae^{0.05t}$ gives

$$W = 4000 + (W_0 - 4000)e^{0.05t}.$$

(c)   If $W_0 = 4000$, then $W = 4000$, the equilibrium solution.

If $W_0 = 5000$, then $W = 4000 + 1000e^{0.05t}$.

If $W_0 = 3000$, then $W = 4000 - 1000e^{0.05t}$. Substituting $t \approx 27.7$ gives $W = 0$, so this solution means that the company goes bankrupt in its twenty-eighth year.

The graphs of these functions are in Figure 10.40. Notice that if the net worth starts with W_0 near, but not equal to, \$4000 million, then W moves further away. Thus, $W = 4000$ is an unstable equilibrium.

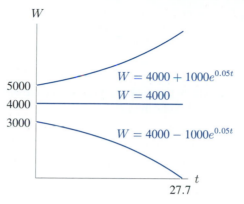

Figure 10.40: Solutions to $\frac{dW}{dt} = 0.05W - 200$

The Velocity of a Falling Body

Imagine a sky-diver jumping out of a plane. Think about how the person's velocity varies during the time before his parachute opens. When the sky-diver first jumps, his velocity is zero. Then the pull of gravity comes into play, making the velocity increase. As the sky-diver speeds up, the air resistance also increases. Since the air resistance partly balances the pull of gravity, as the sky-diver speeds up, the force causing him to accelerate decreases. Thus the velocity must be an increasing function of time, and it must be increasing at a decreasing rate. If the air resistance increases until it balances gravity, then the sky-diver's velocity will stop increasing—the velocity levels off. Thus we expect the velocity to be increasing, and to have a graph that is concave down with a horizontal asymptote.

A Differential Equation: Air Resistance Proportional to Velocity

In order to compute the velocity function, we must make some assumptions about the forces acting on the falling body and the accelerations they produce. It is reasonable to assume that the faster the body is falling, the larger the air resistance. However, to decide whether air resistance is proportional to the velocity, say, or is some other function of velocity, requires either lab experiments or a theoretical idea of how the air resistance is created. In the case of a very small object, such as a dust particle settling on a computer component during manufacturing,[5] we assume that air resistance is proportional to velocity.

Newton's Second Law of Motion, which states that

$$\text{Force} = \text{Mass} \times \text{Acceleration},$$

relates force to velocity, since acceleration is dv/dt. The force is the net force on the object, which has two components: the gravitational force, mg, which acts downward, and the air resistance, kv, which acts upward, in the opposite direction to the motion (we assume $k > 0$). (See Figure 10.41.) Thus

$$\text{Gravity} - \text{Air resistance} = \text{Mass} \times \text{Acceleration},$$

so

$$mg - kv = m\frac{dv}{dt}.$$

This differential equation can be solved by separation of variables. It is easier if we factor out $-k/m$ before separating, giving

$$\frac{dv}{dt} = -\frac{k}{m}\left(v - \frac{mg}{k}\right).$$

Separating and integrating gives

$$\int \frac{dv}{v - mg/k} = -\frac{k}{m}\int dt$$

$$\ln\left|v - \frac{mg}{k}\right| = -\frac{k}{m}t + C.$$

[5]Example suggested by Howard Stone.

Solving for v:

$$\left| v - \frac{mg}{k} \right| = e^{-kt/m+C} = e^C e^{-kt/m}$$

$$v - \frac{mg}{k} = A e^{-kt/m},$$

where A is an arbitrary constant. We find A from the initial condition that the object starts from rest, so $v = 0$ when $t = 0$. Substituting

$$0 - \frac{mg}{k} = A e^0$$

gives

$$A = -\frac{mg}{k}.$$

Thus

$$v = \frac{mg}{k} - \frac{mg}{k} e^{-kt/m} = \frac{mg}{k}(1 - e^{-kt/m}).$$

The graph of this function is in Figure 10.42. As expected, at the start the object is moving slowly, which means the air resistance is small, so it speeds up. As the object starts to move faster, the air resistance builds up until it balances the gravitational force. At this point the object has reached its *terminal velocity*, mg/k. In this model, the object never actually reaches terminal velocity (because $e^{-kt/m}$ is never actually equal to zero).

Air resistance, kv

Force due to gravity, mg

Figure 10.41: Forces acting on a falling object

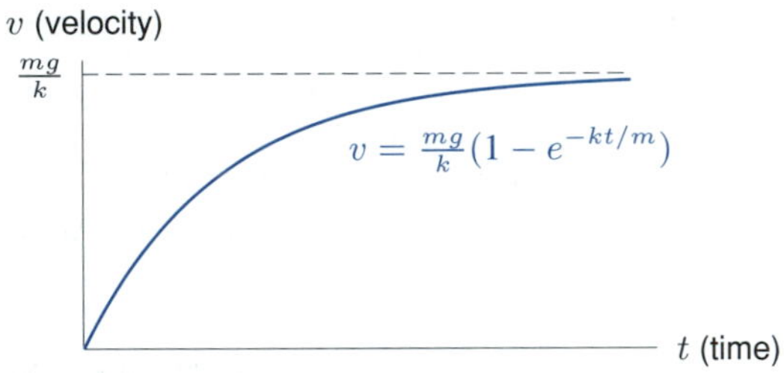

v (velocity)

$\frac{mg}{k}$

$v = \frac{mg}{k}(1 - e^{-kt/m})$

t (time)

Figure 10.42: Velocity under the assumption that air resistance $= kv$

Alternative Method of Finding the Terminal Velocity

Notice that the terminal velocity can also be obtained from the differential equation by setting $dv/dt = 0$ and solving for v:

$$m \frac{dv}{dt} = mg - kv = 0$$

so

$$v = \frac{mg}{k}.$$

Compartmental Analysis: A Reservoir

Many processes can be modeled as a container with various solutions flowing in and out—for example, drugs given intravenously or the discharge of pollutants into a lake. We will consider a city's water reservoir, fed partly by clean water from a spring and partly by run-off from the surrounding land. In New England, or many other areas with much snow in the winter, the run-off contains salt which has been put on the roads to make them safe for driving. We will consider the concentration of salt in the reservoir. If there is no salt in the reservoir initially, we expect the concentration to build up until the rate at which the salt is being brought into the reservoir by the run-off is balanced by the rate at which salt flows out. If, on the other hand, the reservoir starts with a great deal of salt in it, then initially the rate at which the salt is flowing in is less than the rate at which it is flowing out, and the quantity of salt in the lake should decrease. In either case, the salt concentration levels off at an equilibrium value.

A Differential Equation for Salt Concentration

Suppose the water reservoir holds 100 million gallons of water and supplies a city with 1 million gallons a day. The reservoir is partly refilled by a spring which provides 0.9 million gallons a day, and the rest of the water, 0.1 million gallons a day, comes from run-off from the surrounding land. The spring is clean, but the run-off contains salt with a concentration of 0.0001 pound per gallon. Assume that there was no salt in the reservoir initially and that the reservoir is well mixed (that is, that the water taken by the city residents contains the concentration of salt in the tank at that instant). We find the concentration of salt in the reservoir as a function of time, assuming the reservoir remains full.

In this type of problem, it is very important to distinguish between the total quantity and the concentration of salt, where

$$\text{Concentration} = \frac{\text{Quantity of salt}}{\text{Volume of water}}.$$

If C is the concentration of salt (in pounds/gallon) and Q is the quantity (in pounds), then since the volume of the reservoir is 100 million gallons:

$$C = \frac{Q}{100 \, \text{million}} \left(\frac{\text{lb}}{\text{gal}} \right).$$

The differential equation we use is based on the relation

$$\begin{array}{l} \text{Rate of change of} \\ \text{quantity of salt in reservoir} \end{array} = \text{Rate salt entering} - \text{Rate salt leaving},$$

which describes how the quantity is changing with time. We will find Q first, and then C.

To find the rate salt is entering, notice that all the salt is entering through the run-off. The run-off is 0.1 million gallons per day, with each gallon containing 0.0001 pound of salt. Therefore

$$\text{Rate salt entering} = \text{Concentration} \times \text{Volume per day}$$
$$= 0.0001 \left(\frac{\text{lb}}{\text{gal}} \right) \times 0.1 \left(\frac{\text{million gal}}{\text{day}} \right)$$
$$= 0.00001 \left(\frac{\text{million lb}}{\text{day}} \right) = 10 \left(\frac{\text{lb}}{\text{day}} \right).$$

Thus

$$\text{Rate salt entering} = 10 \, \text{lb/day}.$$

Salt is leaving in the water used by the city. The city uses a million gallons of water each day with a concentration of $\dfrac{Q}{100 \, \text{million}} \dfrac{\text{lb}}{\text{gal}}$. Thus

$$\text{Rate salt leaving} = \text{Concentration} \times \text{Volume per day}$$
$$= \frac{Q}{100 \, \text{million}} \left(\frac{\text{lb}}{\text{gal}} \right) \times 1 \left(\frac{\text{million gal}}{\text{day}} \right)$$
$$= \frac{Q}{100} \left(\frac{\text{lb}}{\text{day}} \right).$$

So

$$\text{Rate salt leaving} = \frac{Q}{100} \, \text{lb/day}.$$

Therefore Q must satisfy

$$\frac{dQ}{dt} = 10 - \frac{Q}{100}.$$

We factor out $-1/100$:

$$\frac{dQ}{dt} = -\frac{1}{100}(Q - 1000) = -0.01(Q - 1000),$$

and separate variables:

$$\int \frac{dQ}{Q - 1000} = -\int 0.01\, dt$$
$$\ln|Q - 1000| = -0.01t + k,$$

so

$$Q - 1000 = Ae^{-0.01t}.$$

There is no salt initially, so we substitute $Q = 0$ when $t = 0$:

$$0 - 1000 = Ae^0 \qquad \text{giving} \qquad A = -1000.$$

Thus

$$Q - 1000 = -1000e^{-0.01t}$$

so

$$Q = 1000(1 - e^{-0.01t}) \quad \text{pounds.}$$

Therefore

$$\text{Concentration} = C = \frac{Q}{100 \text{ million}} = \frac{1000}{10^8}(1 - e^{-0.01t}) = 10^{-5}(1 - e^{-0.01t}) \text{ lbs/gal.}$$

A sketch of concentration against time is in Figure 10.43.

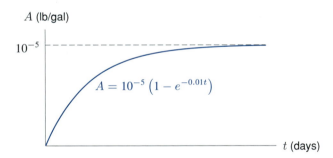

Figure 10.43: Concentration of salt in reservoir

Qualitative Information from a Differential Equation

In the last part of this section we see how information can sometimes be obtained from a differential equation without solving it. The example we take is a second-order differential equation. We first integrate to get a first-order equation, and then investigate qualitative behavior of the solution.

The Expansion of the Universe

Observations show that the universe is currently expanding, so astronomers are interested in exactly how fast it is expanding. One model[6] assumes that the universe is a giant sphere of radius $R(t)$, where t is time, and that

$$R'' = -\frac{GM_0}{R^2}$$

---

[6]From M. Rowan-Robinson, *Cosmology,* Oxford University Press (1977).

where G is the universal gravitational constant and M_0 is the mass of the universe. This is a second-order differential equation for R. Even without solving the equation, we can get information about the expansion of the universe.

The first step is to get a first-order equation from this second-order equation. In this case we multiply both sides of the equation for R'' by R', giving

$$R''R' = -\frac{GM_0}{R^2}R'.$$

Since $\frac{d}{dt}\left[(R')^2\right] = 2R''R'$ and $\frac{d}{dt}\left(R^{-1}\right) = -R^{-2}R'$, we can write the preceding equation as

$$\frac{1}{2}\frac{d}{dt}\left[(R')^2\right] = GM_0\frac{d}{dt}\left(\frac{1}{R}\right).$$

Multiplying through by 2 and integrating with respect to t gives

$$(R')^2 = \frac{2GM_0}{R} + C.$$

We use this relation between R' and R to determine the behavior of R over time. We look only at the case in which $C > 0$. The cases where $C < 0$ and $C = 0$ are investigated in Problems 19 and 20 on page 530. Since

$$R' = \pm\sqrt{\frac{2GM_0}{R} + C}$$

and since the universe is currently expanding, we know that R' must be positive, so R' must be given by the positive root:

$$R' = \sqrt{\frac{2GM_0}{R} + C}.$$

As R gets larger, R' remains positive, so the universe continues to expand. However, R', the rate of expansion, decreases to $\sqrt{C}$ as $R \to \infty$. This model predicts that the universe will expand forever, but that the rate of the expansion is slowing down to $\sqrt{C}$.

Problems for Section 10.6

1. Dead leaves accumulate on the ground in a forest at a rate of 3 grams per square centimeter per year. At the same time, these leaves decompose at a continuous rate of 75% a year. Write a differential equation for the total quantity of dead leaves (per square centimeter) at time t. Sketch a solution showing that the quantity of dead leaves tends towards an equilibrium level. What is that equilibrium level?

2. According to a simple physiological model, an athletic adult needs 20 calories per day per pound of body weight to maintain his weight. If he consumes more or fewer calories than those required to maintain his weight, his weight will change at a rate proportional to the difference between the number of calories consumed and the number needed to maintain his current weight; the constant of proportionality is $1/3500$ pounds per calorie. Suppose that a particular person has a constant caloric intake of I calories per day. Let $W(t)$ be the person's weight in pounds at time t (measured in days).

   (a) What differential equation has solution $W(t)$?
   (b) Solve this differential equation.
   (c) Draw a graph of $W(t)$ if the person starts out weighing 160 pounds and consumes 3000 calories a day. Label your axes and any intercepts and asymptotes clearly.

3.   When a gas expands without gain or loss of heat, the rate of change of pressure with respect to volume is proportional to pressure divided by volume. Find a law connecting pressure and volume in this case.

4.   In 1692, Johann Bernoulli was teaching the Marquis de l'Hopital calculus in Paris. Solve the following problem, which is similar to the one that they did. What is the equation of the curve which has subtangent (distance BC in Figure 10.44) equal to twice its abscissa (distance OC)?

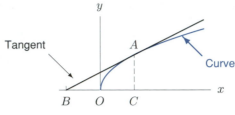

Figure 10.44

5.   Water leaks out of a barrel at a rate proportional to the square root of the depth of the water at that time. If the water level starts at 36 inches and drops to 35 inches in 1 hour, how long will it take for all of the water to leak out of the barrel?

6.   In Brazil, interest rates increased steadily in the second half of the 1970s. As 1975 began, the interest rate was 50%, and it increased 25% a year after that (to 75% in 1976, to 100% in 1977 and so on). Assuming both that the interest rate climbed continuously and that interest accrued continuously:

   (a)   Write a differential equation for the amount of money, M, in a bank account as a function of time, t, measured from 1975. Assume the only deposit was 100,000 cruzieros on January 1, 1975.
   (b)   Solve this differential equation.
   (c)   How much money was in the account on January 1, 1980?

7.   (Continuation of Problem 6)

   (a)   Suppose now Brazilian banks had offered a constant interest rate between 1975 and 1980. What would that interest rate have had to be to produce the same growth in the bank balance between 1975 and 1980?
   (b)   Sketch a graph from 1975 into the early 1980's showing the balance in the account under both schemes, the rising and the constant interest rates.

8.   Consider the sky-diver whose velocity is described by the differential equation

$$m\frac{dv}{dt} = mg - kv, \quad v(0) = 0.$$

   Find and solve a differential equation for a, the acceleration of the sky-diver.

9.   Suppose a chemical reaction involves one molecule of a substance A combining with one molecule of substance B to form one molecule of substance C, written $A + B \to C$. The Law of Mass Action states that the rate at which C is formed is proportional to the product of the quantities of A and B present. Assume a and b are the initial quantities of A and B, and x is the quantity of C present at time t.

   (a)   Write a differential equation for x.
   (b)   Solve the equation with $x(0) = 0$.

10.   If the initial quantities, a and b, in Problem 9 are the same, write and solve a differential equation for x, with $x(0) = 0$.

11.   A patient is given the drug theophylline intravenously at a rate of 43.2 mg/hour to relieve acute asthma. You can imagine the drug as entering a compartment of volume 35,000 ml. (This is the volume of the part of the body through which the drug circulates.) The rate at which the drug leaves the patient is proportional to the quantity there, with proportionality constant 0.082. Assume the patient's body contains none of the drug initially.

   (a)   Describe in words how you would expect the concentration of theophylline in the patient to vary with time and sketch an approximate graph.
   (b)   Write a differential equation satisfied by the concentration of the drug, $c(t)$.
   (c)   Solve the differential equation and graph the solution. What happens to the concentration in the long run?

12. A drug is administered intravenously at a constant rate of r mg/hour and is excreted at a rate proportional to the quantity present, with constant of proportionality α.

    (a) Solve a differential equation for quantity, Q, in milligrams, of the drug in the body at time t hours. Your answer will contain r and α. Sketch a graph of Q against t. What is Q_∞, the limiting long-run value of Q?

    (b) What effect does doubling r have on Q_∞? What effect does doubling r have on the time to reach half the limiting value, $\frac{1}{2}Q_\infty$?

    (c) What effect does doubling α have on Q_∞? On the time to reach $\frac{1}{2}Q_\infty$?

13. As you know, when a course ends, students start to forget the material they have learned. One model (called the Ebbinghaus model) assumes that the rate at which a student forgets material is proportional to the difference between the material he or she currently remembers and some positive constant, a.

    (a) Let $y = f(t)$ be the fraction of the original material remembered t weeks after the course has ended. Set up a differential equation for y. Your equation will contain two constants; the constant a is less than y for all t.

    (b) Solve the differential equation.

    (c) Describe the practical meaning (in terms of the amount remembered) of the constants in the solution $y = f(t)$.

14. A certain commodity is currently being sold at a price of $\$p$ per unit. Over a period of time market forces will make this price tend toward the equilibrium price, which we call $\$p_0$, at which supply exactly balances demand. The rate at which the price changes is described by the Evans Price Adjustment model, which says that the rate of change in the actual market price ($\$p$) is proportional to the difference between the actual market price and the equilibrium price.

    (a) Write a differential equation for p as a function of t.

    (b) Solve for p.

    (c) Sketch solutions for various different initial prices, both above and below the equilibrium price.

    (d) What happens to p as $t \to \infty$?

15. When people smoke, carbon monoxide is released into the air. Suppose, in a room of volume 60 m³, air containing 5% carbon monoxide is introduced at a rate of 0.002 m³/min. (This means that 5% of the volume of the incoming air is carbon monoxide.) Assume that the carbon monoxide mixes immediately with the rest of the air, and that the mixture leaves the room at the same rate as it enters.

    (a) Write a differential equation for $c(t)$, the concentration of carbon monoxide at time t, in minutes.

    (b) Solve the differential equation, assuming there was no carbon monoxide in the room initially.

    (c) What happens to the value of $c(t)$ in the long run?

16. (Continuation of Problem 15.) Medical texts[7] warn that exposure to air containing 0.1% carbon monoxide for some time can lead to a coma. How long does it take for the concentration of carbon monoxide in the room in Problem 15 to reach this level?

17. An aquarium pool has volume 2×10^6 liters. The pool initially contains pure fresh water. At $t = 0$ minutes, water containing 10 grams/liter of salt is poured into the pool at a rate of 60 liters/minute. The salt water instantly and totally mixes with the fresh water, and the excess mixture is drained out of the bottom of the pool at the same rate (60 liters/minute). Let $S(t)$ = the mass of salt in the pool at time t.

    (a) Write a differential equation for the amount of salt in the pool.

    (b) Solve the differential equation to find $S(t)$.

    (c) What happens to $S(t)$ as $t \to \infty$?

18. Imagine an object of mass m thrown vertically upward from the surface of the earth with initial velocity v_0. In this problem, we calculate the value of v_0, called the *escape velocity*, with which the object can escape the pull of the gravity and never return to earth. Suppose v is the velocity of the object (measured upward) at time t.

    Since the object is moving far from the surface of the earth, we must take into account the variation of gravity with altitude. If the acceleration due to gravity at sea level is g, the gravitational force, F, on

---

[7]See, for example, R.G. Petersdorf et al., *Harrison's Principles of Internal Medicine* (New York: McGraw Hill, 1983).

the object of mass m at an altitude h above the surface of the earth is given by

$$F = \frac{mgR^2}{(R+h)^2},$$

where R is the radius of the earth.

(a) Use Newton's Law of Motion to show that

$$\frac{dv}{dt} = -\frac{gR^2}{(R+h)^2}.$$

(b) Rewrite this equation with h instead of t as the independent variable using the chain rule $\frac{dv}{dt} = \frac{dv}{dh} \cdot \frac{dh}{dt}$. Hence show that

$$v\frac{dv}{dh} = -\frac{gR^2}{(R+h)^2}.$$

(c) Solve the differential equation in part (b).

(d) Find the escape velocity, the value of v_0 such that v is never zero.

Problems 19–21 refer to the model of the expansion of the universe given on page 526 by the equations

$$R'' = -\frac{GM_0}{R^2} \quad \text{and} \quad (R')^2 = \frac{2GM_0}{R} + C,$$

where $R(t)$ is the radius of the universe (assumed spherical), t is time, G is the universal gravitational constant, and M_0 is the mass of the universe. (In the text we assumed that $C > 0$.)

19. In this problem we look at the case where $C < 0$. Writing $C = -K$, where $K > 0$, we have

$$R' = \pm\sqrt{\frac{2GM_0}{R} - K}.$$

Since the universe is currently expanding, at this time $R' > 0$, so R' is currently given by the positive root. Show that R increases to some value $R_{\max}$, and then decreases again. In addition, show that when this happens and R again approaches zero, then R' has a large negative value; in other words, a "big crunch" happens.

20. In this problem we look at the case where $C = 0$. Then we know that

$$R' = \sqrt{\frac{2GM_0}{R}}.$$

Solve this differential equation, assuming $R = 0$ when $t = 0$. Give R as a function of t. The resulting formula for R is called the "flat universe model." What does this model predict about the expansion of the universe?

21. (a) Einstein, who formulated the model of the universe described before Problem 19, wanted the universe to be stable—neither expanding nor shrinking. Why don't these differential equations allow for a stable universe?

(b) One estimate for the age of the universe is $R(t_0)/R'(t_0)$ where t_0 is the current time. (This number is called the *Hubble constant.*) Why is this a reasonable estimate for the age of the universe? Is it an overestimate or an underestimate?

10.7 MODELS OF POPULATION GROWTH

Population projections have been important to political philosophers since at least the late eighteenth century. As concern for scarce resources has grown, so has interest in accurate population projections. In this section we will look at two differential equations which are used to model both human and animal population growth. These differential equations have applications in economics and medicine, such as modeling the spread of an innovation or the growth of a tumor.

Relative versus Absolute Growth Rates

When describing population growth, we often use percentages rather than absolute numbers. For example, we say the population of the world is now about 5.8 billion people and growing at a continuous rate of about 1.7% a year, meaning that the relative growth rate is 1.7%:

$$\frac{dP/dt}{P} = \frac{1}{P}\frac{dP}{dt} = 0.017.$$

The quantity dP/dt is called the *absolute growth rate* and measures the growth rate in, say, people per year. The quantity $(dP/dt)/P$ is called the *relative growth rate* and represents the absolute growth rate as a fraction of the whole population. Its units are, say, % per year. When talking about populations we often use the relative growth rate.

The US Population: 1790--1860

Every ten years the population of the United States is recorded by a census. The first such census was in 1790. Table 10.5 contains the census data from 1790 to 1940.

TABLE 10.5 *US Population in millions, 1790–1940*

Year	Population	Year	Population	Year	Population
1790	3.9	1850	23.1	1910	92.0
1800	5.3	1860	31.4	1920	105.7
1810	7.2	1870	38.6	1930	122.8
1820	9.6	1880	50.2	1940	131.7
1830	12.9	1890	62.9		
1840	17.1	1900	76.0		

You might note that the population is given only to the nearest 0.1 million (or 100,000), although the population is reported by the US Census Bureau down to the last digit (for example, 131,669,275 in 1940). We have rounded off because census figures are notoriously inaccurate. For example, in the census of 1990, New York City claimed the census had missed a million people in that city alone. Thus, giving more digits in the population does not necessarily give more accuracy.

Let us concentrate first on the relative growth rate, $(dP/dt)/P$, of the US population from 1790 to 1860. If we want to estimate $(dP/dt)/P$ in 1830, we take the rate of change in the population and divide it by the population itself:[8]

$$\frac{1}{P}\frac{dP}{dt} \approx \frac{1}{\text{Population in 1830}} \cdot \frac{\text{Population in 1840} - \text{Population in 1830}}{10\text{ years}}$$

$$= \frac{1}{12.9} \cdot \frac{17.1 - 12.9}{10} = 0.0326 = 3.26\%.$$

Similar calculations for 1790, 1800, ..., 1850 give the percentages in Table 10.6:

TABLE 10.6 *Rough estimates of yearly growth rate of US population*

Year	1790	1800	1810	1820	1830	1840	1850
Relative growth rate	3.59%	3.58%	3.33%	3.44%	3.26%	3.57%	3.53%

These percentages are pretty close. The relative growth rate, while not precisely constant, is nearly so. In fact, political and economic events such as war or recession affect the population, so we don't expect the growth rate to be exactly constant.

The simplest model for population growth is to assume that the relative growth rate *is* constant, in other words

$$\frac{1}{P}\frac{dP}{dt} = k,$$

[8]In Problem 12 at the end of this section we look at an alternative way of estimating the relative growth rate in 1830, using data from 1820 as well as from 1840.

where k is the continuous growth rate. This is equivalent to assuming that the population grows exponentially:

$$P = P_0 e^{kt}.$$

What should we take for the value of k? One possibility would be the average of the percentages we just calculated, namely 3.47%. However, there is a serious objection to using this percentage as an estimate for k. Remember that k is a continuous growth rate, but the populations are given at 10-year intervals. A 10-year population growth of 34.7% doesn't come from a continuous yearly rate of 3.47%. (This would be ignoring the effects of compounding.) If the population increases by a factor of 34.7% in 10 years, then $P(10)/P_0 = 1.347$, where $P(10)$ is the population when $t = 10$. Assuming that $P = P_0 e^{kt}$, we need k to satisfy

$$\frac{P(10)}{P_0} = e^{k \cdot 10} = 1.347.$$

Now we can find the continuous growth rate, k:

$$k = \frac{\ln(1.347)}{10} = 0.0298.$$

Let's compare predicted and actual values if we model the US population by the differential equation

$$\frac{dP}{dt} = 0.0298P.$$

We start with initial population $P_0 = 3.9$ in 1790. Notice that this says we will consider 1790 as time $t = 0$, so 1800 is $t = 10$, and 1810 is $t = 20$, etc. The solution to the differential equation is

$$P = 3.9e^{0.0298t}.$$

If we put $t = 0, 10, 20, \ldots, 70$ into this function we get the populations predicted by our model for the years 1790, 1800, $\ldots$, 1860. Table 10.7 contains the comparison to the actual populations.

TABLE 10.7   *Predicted versus actual US population 1790–1860 (exponential model)*

Year	Actual	Predicted	Year	Actual	Predicted
1790	3.9	3.9	1830	12.9	12.8
1800	5.3	5.3	1840	17.1	17.3
1810	7.2	7.1	1850	23.1	23.3
1820	9.6	9.5	1860	31.4	31.4

The agreement is remarkable. Of course, since we used the data from the entire 70 year period to estimate k, we should expect good agreement throughout that period. What is surprising is that if we had used only the populations in 1790 and 1800 to estimate k, the predictions are still quite good. Let's find a new value of k using only the 1790 and 1800 data and compare predictions. The 10 year growth from 1790 to 1800 is 35.9% so $k = \ln(1.359)/10 = 0.0307$. We would then predict the population in 1860 to be

$$3.9e^{0.0307(70)} = 33.4,$$

which is within about 6% of the actual population of 31.4. It is remarkable that a person in 1800 could accurately predict what the population of the US would be 60 years later, especially considering all the wars, recessions, epidemics, additions of new territory, and immigration that took place from 1800 to 1860.

Predictions From The Exponential Model

The grim predictions of the exponential model are reflected in the ideas of Thomas Malthus, an early nineteenth-century clergyman and political philosopher, who believed that, if unchecked, a population would grow exponentially, whereas the food supply would grow linearly, and therefore the population would eventually outstrip the food supply.

Interestingly enough, an exponential model has fit the growth of world population and the population of many regions remarkably well for decades, even centuries. However, the model must break down at some point because it predicts that the population will continue to grow without bound as time goes on—and this cannot be true forever. Eventually the effects of crowding, emigration, disease, war, and lack of food will have to curb growth. In searching for an improvement, then, we should look for a model whose solution is approximately an exponential function for small values of the population, but which levels off later.

How to Estimate dP/dt from Data

If, as is often the case, all we know about a population, P, is its values at certain points in time, we have to approximate dP/dt by $\Delta P/\Delta t$. However there are several different ways this approximation can be made. As in the example above, we can say

$$\frac{dP}{dt} \text{ at } 1830 \approx \frac{\text{Population in 1840} - \text{Population in 1830}}{10}.$$

However we could equally well have said

$$\frac{dP}{dt} \text{ at } 1830 \approx \frac{\text{Population in 1830} - \text{Population in 1820}}{10}.$$

Both of these are called *one-sided estimates* because they involve using the population to one side of 1830 but not the other. The first one involving 1840 is the forward, or right-hand, estimate; the second one involving 1820 is the backward, or left-hand estimate. In general, both are equally good or bad estimates for dP/dt. A more accurate approximation can be obtained by averaging the two one-sided estimates, giving

$$\frac{dP}{dt} \text{ at } 1830 \approx \frac{1}{2}\left(\frac{\text{Population in 1840} - \text{Population in 1830}}{10} + \frac{\text{Population in 1830} - \text{Population in 1820}}{10}\right).$$

When we add the two fractions the population in 1830 cancels and we get the two-sided estimate, so called because it involves data on both sides of 1830:

$$\frac{dP}{dt} \text{ at } 1830 \approx \frac{1}{2}\left(\frac{\text{Population 1840} - \text{Population 1820}}{10}\right) = \left(\frac{\text{Population 1840} - \text{Population 1820}}{20}\right).$$

In the exponential model above we used a one-sided estimate. This turned out to give good results after adjusting from 10-year to continuous growth rates. If we had been unable to get accurate predictions using one-sided estimates, we might have tried two-sided estimates instead. In the logistic model below, we use two-sided estimates, as they turn out to give noticeably better predictions.

The US Population: 1790--1940

The exponential model for the US population works well for reasonable periods of time, but exponential growth cannot go on forever. For example, if we tried to predict the population of the US in 1990 using the exponential model we developed above with $k = 0.0298$, we would get

$$\text{US Population in 1990} = 3.9e^{(0.0298)(200)} = 1,512 \text{ million},$$

which is far from the actual figure of around 250 million.

The problem is that growth rate in the US population between 1790 and 1860 did not stay constant in later decades. The 10-year percentage growths[9] from 1860 to 1930 are listed in Table 10.8:

TABLE 10.8 *Estimated 10-year growth rate of US population, 1860–1930*

Year	1860	1870	1880	1890	1900	1910	1920	1930
Relative growth rate	22.9%	30.1%	25.3%	20.8%	21.1%	14.9%	16.2%	7.2%

[9]Calculated using one-sided forward estimates, so that $\frac{1}{P}\frac{dP}{dt}$ at $1860 \approx \frac{\text{Population in 1870} - \text{Population in 1860}}{10(\text{Population in 1860})}$.

These figures are nothing like those during the period 1790 to 1860, where they hovered around 34%. The dramatic drop to 22.9% for the decade 1860–1870 is explainable by the Civil War (but don't ascribe the entire drop to deaths of the war—see Problem 15, page 541). The growth rate goes back up during the decade 1870–1880, but by 1890–1900 it has dropped below the rate during the Civil War. There is a slight increase in the rate during the immigrations of 1900 to 1910, another drop during World War I, a small bounce back, and finally the rate plummets during the recession of the 1930s. Notice that the effect of the recession is more dramatic than the effect of wars, which suggests that it is a decrease in birth rate rather than an increase in death rate that is a major factor in declines in the relative population growth rate.

Our exponential model gives accurate predictions up to 1860. But for the years following 1860, the exponential model is inadequate. We look for a new model that will take into account the effects of overcrowding. Because of the effects of crowding, we expect the relative growth rate to decrease as the population increases. Thus we look at how $(dP/dt)/P$ changes as P changes. This time we use a two-sided estimate for dP/dt. For example:

$$\frac{dP}{dt} \text{ at } 1830 \approx \frac{\text{Population in 1840} - \text{Population 1820}}{20}.$$

Table 10.9 contains estimates for $(dP/dt)/P$ computed this way for some years between 1790 and 1940. Comparing the last two columns suggests that the values in the last column may be an approximately linear function of P. To get a better picture of how $(dP/dt)/P$ varies with P, we plot $(dP/dt)/P$ versus P and see whether the points lie on a line. Figure 10.45 shows the scatterplot of the points together with the line that best fits the points. The equation for the line, which fits quite well, is

$$\frac{1}{P}\frac{dP}{dt} = 0.0318 - 0.000170P.$$

TABLE 10.9   *Some estimates for* $\dfrac{1}{P}\dfrac{dP}{dt}$.

Year	P	$\left(P(t+10) - P(t-10)\right)/(20P)$
1860	31.4	0.0245
1890	62.9	0.0205
1910	92.0	0.0161
1930	122.8	0.0106

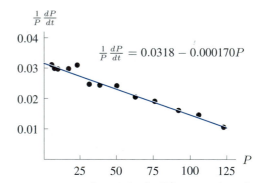

Figure 10.45: Scatterplot for US census data of $\frac{1}{P}\frac{dP}{dt}$ versus P.

Therefore, in our new model P satisfies the differential equation

$$\frac{dP}{dt} = 0.0318P - 0.000170P^2.$$

This is known as a *logistic equation*. Its slope field is shown in Figure 10.46, together with the solution with $P(0) = 3.9$ superimposed. (Notice that $t = 0$ in 1790.)

The most striking difference in this model compared with the exponential model is that it predicts that the US population will level off somewhere below 200 million. The population will continue to grow until $dP/dt = 0$, which occurs when

$$0 = 0.0318P - 0.000170P^2$$

so

$$P = 0 \quad \text{or} \quad P \approx 187 \text{ million}.$$

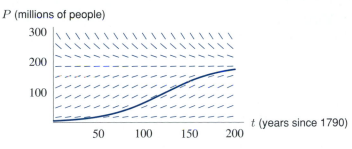

Figure 10.46: Solution to $\frac{dP}{dt} = 0.0318P - 0.000170P^2$ with $P(0) = 3.9$

Looking at the shape of the solution curve in Figure 10.46, we see that initially the population grows faster and faster and then slows down as the limiting value of 187 is approached; the fastest growth rate appears to be about half-way to the limiting value.

Later on in this section we derive the formula for the solution to the logistic equation. For now, you can check by substitution that the function

$$P = \frac{187}{1 + 47e^{-0.0318t}}$$

is a solution to our logistic equation modeling the US population. (The numbers 187 and the 47 are not exact values, but have been rounded.) The values predicted by this equation for P agree very well with the actual populations up to 1940. See Table 10.10. During the period from 1700 to 1940, the largest deviation is about 3% in 1840 and 1870 (the Civil War accounts for the second one). All other errors are less than 2%.

Of course, the final test is how well our model, based on data from 1790 to 1940, predicts the population in the "future," 1950 to 1990. Table 10.10 contains the predicted and actual data.

TABLE 10.10   *Predicted versus actual US population in millions, 1790–1980 (logistic model)*

Year	Actual	Predicted	Year	Actual	Predicted	Year	Actual	Predicted
1790	3.9	3.9	1860	31.4	30.8	1930	122.8	120.8
1800	5.3	5.3	1870	38.6	39.9	1940	131.7	133.7
1810	7.2	7.2	1880	50.2	50.7	1950	150.7	145.0
1820	9.6	9.8	1890	62.9	63.3	1960	179.3	154.4
1830	12.9	13.2	1900	76.0	77.2	1970	203.3	162.1
1840	17.1	17.7	1910	92.0	91.9	1980	226.5	168.2
1850	23.1	23.5	1920	105.7	106.7	1990	248.7	172.9

The fit between predicted and actual population values is clearly not good from 1950 on. Despite World War II, which undoubtedly depressed population growth between 1942 and 1945, in the last half of the 1940s the US population surged, wiping out in five years a deficit caused by 15 years of depression and war. The 1950s saw a population growth of 28 million, leaving our logistic model in the dust. This surge in population is referred to as the baby boom. All one can say is that based on 150 years of data, what happened in the US in the 20 years after World War II was completely without precedent. The baby boom could well end up being one of the most important sociological events of the twentieth century in the United States, and its consequences will be felt for many years to come.

Once again we have reached a point where our model is no longer useful. This should not lead you to believe that a reasonable mathematical model cannot be found; rather it should serve to point out that no model is perfect and that when one model fails, we seek a better one. Just as we abandoned the exponential model in favor of the logistic model for the US population, we could look further. (See Problems 9 and 10 on page 540.)

The Logistic Model

The logistic model we used to model the US population from 1790 to 1940 assumed that the relative growth rate of the population was a linearly decreasing function of P:

$$\frac{1}{P}\frac{dP}{dt} = k - aP.$$

(We took $k = 0.0318$ and $a = 0.000170$). For small P, we have approximately

$$\frac{1}{P}\frac{dP}{dt} \approx k.$$

The solution to this equation is an exponential function. This is why an exponential model fit the US population well during the years 1790-1860 when the population was relatively small. In the logistic model, as P increases, the relative growth rate decreases to zero; it reaches zero when P is given by

$$k - aP = 0.$$

Solving for P, we get

$$P = \frac{k}{a}.$$

This is the limiting value of the population, which we call L:

$$L = \frac{k}{a}.$$

The value L is called the *carrying capacity* of the environment, and represents the largest population the environment can support. Writing $a = k/L$, the logistic equation becomes

$$\frac{1}{P}\frac{dP}{dt} = k - \frac{k}{L}P$$
$$\frac{dP}{dt} = kP - \frac{k}{L}P^2$$

or

$$\frac{dP}{dt} = kP\left(1 - \frac{P}{L}\right).$$

This is the general logistic differential equation, first proposed as a model for population growth by the Belgian mathematician P. F. Verhulst in the 1830s.

Qualitative Solution to the Logistic Equation

Figure 10.47 shows the slope field and characteristic *sigmoid*, or S-shaped, solution curve for the logistic model. Notice that for each fixed value of P, that is, along each horizontal line, the slopes are all the same because dP/dt depends only on P and not on t. The slopes are small near $P = 0$ and near $P = L$; they are steepest around $P = L/2$. For $P > L$, the slopes are negative, meaning that if the population is above the carrying capacity, the population will decrease.

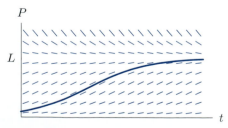

Figure 10.47: Slope field for $\frac{dP}{dt} = kP(1 - \frac{P}{L})$

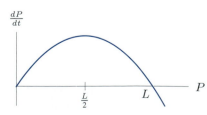

Figure 10.48: $\frac{dP}{dt} = kP(1 - \frac{P}{L})$

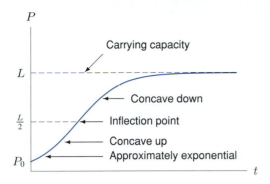

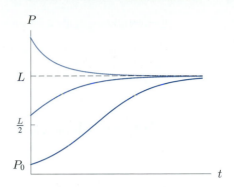

Figure 10.49: Logistic growth with inflection point

Figure 10.50: Solutions to the logistic equation

We can locate precisely the inflection point where the slopes are greatest using the graph of dP/dt against P in Figure 10.48. The graph is a parabola because dP/dt is a quadratic function of P. The horizontal intercepts are at $P = 0$ and $P = L$, so the maximum, where the slope is greatest, is at $P = L/2$. The graph in Figure 10.48 also tells us that for $0 < P < L/2$, the slope dP/dt is positive and increasing, so the graph of P against t is concave up. (See Figure 10.49.) For $L/2 < P < L$, the slope dP/dt is positive and decreasing, so the graph of P against t is concave down. For $P > L$, the slope dP/dt is negative, so the graph of P against t is decreasing.

If $P = 0$ or $P = L$, there is an equilibrium solution (not a very interesting one if $P = 0$). Figure 10.50 shows that $P = 0$ is an unstable equilibrium because solutions which start near 0 move away. However, $P = L$ is a stable equilibrium.

The Analytic Solution to the Logistic Equation

We have already obtained a lot of information about logistic growth without finding a formula for the solution. However, the equation can be solved analytically by separating variables:

$$\frac{dP}{dt} = kP\left(1 - \frac{P}{L}\right) = kP\left(\frac{L - P}{L}\right)$$

giving

$$\int \frac{dP}{P(L - P)} = \int \frac{k}{L}\, dt.$$

We can integrate the left side using the integral tables (Formula 26), or by rewriting

$$\frac{1}{P(L - P)} = \frac{1}{L}\left(\frac{1}{P} + \frac{1}{L - P}\right).$$

Thus, we have

$$\int \frac{1}{L}\left(\frac{1}{P} + \frac{1}{L - P}\right) dP = \int \frac{k}{L}\, dt.$$

Canceling the constant L, we get

$$\int \left(\frac{1}{P} + \frac{1}{L - P}\right) dP = \int k\, dt$$

which can be integrated to give

$$\ln|P| - \ln|L - P| = kt + C.$$

Multiplying through by (-1) and using the fact that $\ln M - \ln N = \ln(M/N)$, we have

$$\ln\left|\frac{L - P}{P}\right| = -kt - C.$$

Exponentiating both sides gives

$$\left| \frac{L - P}{P} \right| = e^{-kt-C} = e^{-C}e^{-kt},$$

so

$$\frac{L - P}{P} = Ae^{-kt} \quad \text{where} \quad A = \pm e^{-C}.$$

We find A by substituting $P = P_0$ when $t = 0$, which gives

$$\frac{L - P_0}{P_0} = Ae^0 = A.$$

Thus

$$\frac{L - P}{P} = Ae^{-kt} \quad \text{where} \quad A = \frac{L - P_0}{P_0}.$$

Since $(L - P)/P = (L/P) - 1$, we have

$$\frac{L}{P} = 1 + Ae^{-kt}$$

giving the formula for the logistic curve:

$$P = \frac{L}{1 + Ae^{-kt}} \quad \text{where} \quad A = \frac{L - P_0}{P_0}.$$

Problems for Section 10.7

1. Assuming that Switzerland's population is growing exponentially at a continuous rate of 0.2% a year and that its 1988 population was 6.6 million, write an expression for the population as a function of time in years. (Let $t = 0$ in 1988.)

2. Consider the logistic model

$$\frac{dP}{dt} = 3P - 3P^2.$$

   (a) On the slope field in Figure 10.51, sketch three solution curves showing different types of behavior.
   (b) Is there a stable value of the population? If so, what is it?
   (c) Describe the meaning of the shape of the solution curves for the population: Where is P increasing? Decreasing? What happens in the long run? Are there any inflection points? Where? What do they mean for the population?

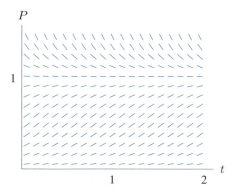

Figure 10.51: Slope field for $dP/dt = 3P - 3P^2$.

   (d) Sketch a graph of dP/dt against P. Where is dP/dt positive? Negative? Zero? Maximum? How do your observations about dP/dt explain the shapes of your solution curves?

3. The total number of people infected with a virus often grows like a logistic curve. Suppose that 10 people originally have the virus, and that in the early stages of the virus (with time, t, measured in weeks), the number of people infected is increasing exponentially with $k = 1.78$. It is estimated that, in the long run, approximately 5000 people become infected.

   (a) Use this information to find a logistic function to model this situation.
   (b) Sketch a graph of your answer to part (a).
   (c) Use your graph to estimate the length of time until the rate at which people are becoming infected starts to decrease. What is the vertical coordinate at this point?

4. Table 10.11 gives the percentage, P, of households with a VCR, as a function of year.

   (a) Explain why a logistic model is a reasonable one to use for this data.
   (b) Use the data to estimate the point of inflection of P. What limiting value L does this point of inflection predict? Does this limiting value appear to be accurate given the percentages for 1990 and 1991?
   (c) The best logistic equation for this data turns out to be the following. What limiting value does this model predict?
$$P = \frac{75}{1 + 316.75e^{-0.699t}}.$$

   TABLE 10.11 *Percentage of households with a VCR*

Year	1978	1979	1980	1981	1982	1983	1984
$P(\%)$	0.3	0.5	1.1	1.8	3.1	5.5	10.6
Year	1985	1986	1987	1988	1989	1990	1991
$P(\%)$	20.8	36.0	48.7	58.0	64.6	71.9	71.9

5. The growth of a certain animal population is governed by the equation
$$\frac{1000}{P}\frac{dP}{dt} = 100 - P,$$
   where $P(t)$ is the number of individuals in the colony at time t. The initial population is known to be 200 individuals. Sketch a graph of $P(t)$. Will there ever be more than 200 individuals in the colony? Will there ever be fewer than 100 individuals? Explain.

6. It is of considerable interest to policy makers to model the spread of information through a population. For example, various agricultural ministries use models to help them understand the spread of technical innovations or new seed types through their countries. Two models, based on how the information is spread, are given below. Assume the population is of a constant size M.

   (a) If the information is spread by mass media (TV, radio, newspapers), the rate at which information is spread is believed to be proportional to the number of people not having the information at that time. Write a differential equation for the number of people having the information by time t. Sketch a solution assuming that no one (except the mass media) has the information initially.
   (b) If the information is spread by word of mouth, the rate of spread of information is believed to be proportional to the product of the number of people who know and the number who don't. Write a differential equation for the number of people having the information by time t. Sketch the solution for the cases in which
      (i) no one
      (ii) 5% of the population
      (iii) 75% of the population
      knows initially. In each case, when is the information spreading fastest?

7. The population of a species of elk on Reading Island in Canada has been monitored for some years. When the population was 600, the relative birth rate was found to be 35% and the relative death rate was 15%. As the population grew to 800, the corresponding figures were 30% and 20%. The island is isolated so there is no hunting or migration.

   (a) Write a differential equation to model the population as a function of time. Assume that relative growth rate is a linear function of population.

(b)   Find the equilibrium size of the population. Today there are 900 elk on Reading Island. How do you expect the population to change in the future?

(c)   Oil has been discovered on a neighboring island and the oil companies want to move 450 elk of the same species to Reading Island. What effect would this move have on the elk population on Reading Island in the future?

(d)   Assuming the elk are moved to Reading Island, sketch the population on Reading Island as a function of time. Start before the elk are transferred and continue for some time into the future. Comment on the significance of your results.

8.   Many organ pipes in old European churches are made of tin. In cold climates such pipes can be affected with *tin pest*, when the tin becomes brittle and crumbles into a grey powder. This transformation can appear to take place very suddenly because the presence of the grey powder encourages the reaction to proceed. At the start, when there is little grey powder, the reaction proceeds slowly. Similarly, toward the end, when there is little metallic tin left, the reaction is also slow. In between, however, when there is plenty of both metallic tin and powder, the reaction can be alarmingly fast.

Suppose that the rate of the reaction is proportional to the product of the amount of tin left and the quantity of grey powder, p, present at time t. Assume also that when metallic tin is converted to grey powder, its mass does not change.

(a)   Write a differential equation for p. Let the total quantity of metallic tin present originally be B.

(b)   Sketch a graph of the solution $p = f(t)$ if there is a small quantity of powder initially. How much metallic tin has crumbled when it is crumbling fastest?

(c)   Suppose there is no grey powder initially. (For example, suppose the tin is completely new.) What does this model predict will happen? How do you reconcile this with the fact that many organ pipes do get tin pest?

9.   (a)   In the text we fitted a logistic model to the US population from 1790–1940. In this problem, we try to fit a logistic equation to the US population all the way from 1790 to 1990. No logistic equation fits the data exactly over this entire period, but we can use the method of page 534 to find an equation that does reasonably well throughout.

To fit a logistic equation to the data in Table 10.12, we estimate the relative growth rate, $(dP/dt)/P$, and plot it against P. To approximate $(dP/dt)/P$, calculate $(\Delta P/\Delta t)/P$ from the data for seven fairly spread-out points. Draw a reasonable line (by eye) through your points, and thus estimate k and a for the equation $(dP/dt)/P = k - aP$.

TABLE 10.12   *US population in millions, 1790–1990*

Year	Population	Year	Population	Year	Population
1790	3.9	1860	31.4	1930	122.8
1800	5.3	1870	38.6	1940	131.7
1810	7.2	1880	50.2	1950	150.7
1820	9.6	1890	62.9	1960	179.0
1830	12.9	1900	76.0	1970	205.0
1840	17.1	1910	92.0	1980	226.5
1850	23.1	1920	105.7	1990	248.7

(b)   What does this model predict about the US population in the long run?

10.   (a)   In Problem 9 we saw that a logistic model cannot be made to fit the US population very closely, because the points on the graph of $(dP/dt)/P$ against P are not exactly on a line. In this problem we try another model. This time we assume that $(dP/dt)/P$ is a linear function of t, so you plot $(dP/dt)/P$ against t (using the same approximate values you calculated in Problem 9). Put a line through the points by eye, and estimate a and b to fit the equation

$$\frac{1}{P}\frac{dP}{dt} = a - bt.$$

(b)   When, if ever, does this model predict that the US population will be at its maximum?

(c)   Solve the differential equation and sketch its solution.

11. An alternative method of finding the analytic solution to the logistic equation

$$\frac{dP}{dt} = kP\left(1 - \frac{P}{L}\right)$$

uses the substitution $P = 1/u$.

    (a)   Show that

$$\frac{dP}{dt} = -\frac{1}{u^2}\frac{du}{dt}$$

    (b)   Rewrite the logistic equation in terms of u and t, and solve for u in terms of t.

    (c)   Using your answer to part (b), find P as a function of t.

12. On page 531, we used one-sided estimates for dP/dt to fit an exponential model to the US population from 1790–1860. In this problem we use two-sided estimates for the years 1800–1850. Suppose P is the US population in millions.

    (a)   Estimate dP/dt by the symmetric difference quotient $\left(P(t + 10) - P(t - 10)\right)/20$. Then compute $(dP/dt)/P$ for each of these years and average them to estimate k for the exponential model $dP/dt = kP$. Compare your value of k with the estimate $k \approx 3.47\%$ obtained on page 532 by using $\left(P(t + 10) - P(t)\right)/10$ for dP/dt.

    (b)   Using your answer to part (a), compute k as the continuous rate of change that leads to the observed 10-year percentage change. Then compare your k with the value $k \approx 2.98\%$ obtained on page 532.

13. This section suggested two ways of estimating dP/dt: the one-sided $\left(P(t + h) - P(t)\right)/h$, and the two-sided $\left(P(t + h) - P(t - h)\right)/(2h)$. Let $f(x) = x^3$. Consider the approximations $f'(2) \approx \left(f(2 + h) - f(2)\right)/h$ and $f'(2) \approx \left(f(2 + h) - f(2 - h)\right)/2h$ for $h = 0.1, 0.01, 0.001$. Which is the better approximation? Is there a pattern to the errors in the approximation as you decrease h? If so, describe it.

14. Show that if $f(x) = x^2$, then $(f(x + h) - f(x - h))/(2h) = f'(x)$ for any value of h. This shows that if $f(x) = x^2$, the two-sided estimate for the derivative of $f(x)$ is exactly equal to the derivative.

15. Estimate the US population in 1870 from the 1860 population of 31.4 million, assuming that the population increased at the same percentage rate during the 1860s as it did in the decades previous to 1860 (about 34.7% each decade). Compare your estimate with the actual 1870 population of 38.6 million. Find some estimate of the number of people who died in the Civil War. Does the number of deaths in the Civil War explain the shortfall in the actual population in 1870? What else might influence the shortfall?

16. Another way to estimate the limiting population L for the logistic differential equation is to note that the derivative dP/dt is largest when $P = L/2$. Compute dP/dt for the US census data for 1790–1940 using the two-sided difference quotient $\left(P(t + 10) - P(t - 10)\right)/20$. Estimate L by doubling the population when dP/dt is largest. Compare this estimate with the estimate $L = 187$ given in the text.

Any population, P, for which we can ignore immigration, satisfies

$$\frac{dP}{dt} = \text{Birth rate} - \text{Death rate}.$$

For organisms which need a partner for reproduction but rely on a chance encounter for meeting a mate, the birth rate is proportional to the square of the population. Thus, the population of such a type of organism satisfies a differential equation of the form

$$\frac{dP}{dt} = aP^2 - bP \quad \text{with } a,\, b > 0.$$

Problems 17–19 investigate the solutions to such an equation.

17. Consider the equation

$$\frac{dP}{dt} = 0.02P^2 - 0.08P.$$

    (a)   Sketch the slope field for this differential equation for $0 \leq t \leq 50$, $0 \leq P \leq 8$.

    (b)   Use your slope field to sketch the general shape of the solutions to the differential equation satisfying the following initial conditions:

        (i) $P(0) = 1$ (ii) $P(0) = 3$ (iii) $P(0) = 4$ (iv) $P(0) = 5$

    (c)   Are there any equilibrium values of the population? If so, are they stable?

18. Consider the equation

$$\frac{dP}{dt} = P^2 - 6P.$$

(a) Sketch a graph of dP/dt against P for positive P.

(b) Use the graph you drew in part (a) to sketch the approximate shape of the solution curve with $P(0) = 5$. To do this, consider the following question. For $0 < P < 6$, is dP/dt positive or negative? What does this tell you about the graph of P against t? As you move along the solution curve with $P(0) = 5$, how does the value of dP/dt change? What does this tell you about the concavity of the graph of P against t?

(c) Use the graph you drew in part (a) to sketch the solution curve with $P(0) = 8$.

(d) Describe the qualitative differences in the behavior of populations with initial value less than 6 and initial value more than 6. Why do you think $P = 6$ is called the *threshold population*?

19. Consider a population satisfying

$$\frac{dP}{dt} = aP^2 - bP \quad \text{with constants } a, \, b > 0.$$

(a) Sketch a graph of dP/dt against P.

(b) Use this graph to sketch the shape of solution curves with various initial values. Use your graph from part (a) to decide where dP/dt is positive or negative, and where it is increasing or decreasing. What does this tell you about the graph of P against t?

(c) Why is $P = b/a$ called the threshold population? What happens if $P(0) = b/a$? What happens in the long-run if $P(0) > b/a$? What if $P(0) < b/a$?

10.8 SECOND-ORDER DIFFERENTIAL EQUATIONS: OSCILLATIONS

A Second-Order Differential Equation

When a body moves freely under gravity, we know that

$$\frac{d^2s}{dt^2} = -g,$$

where s is the height of the body above ground at time t and g is the acceleration due to gravity. To solve this equation, we first integrate to get the velocity, $v = ds/dt$:

$$\frac{ds}{dt} = -gt + v_0,$$

where v_0 is the initial velocity. Then we integrate again, giving

$$s = -\frac{1}{2}gt^2 + v_0 t + s_0,$$

where s_0 is the initial height.

The differential equation $d^2s/dt^2 = -g$ is called *second order* because the equation contains a second derivative but no higher derivatives. The general solution to a second-order differential equation will be a family of functions with two parameters, here v_0 and s_0. Finding values for the two constants corresponds to picking a particular function out of this family.

A Mass on a Spring

Not every second-order differential equation can be solved simply by integrating twice. Consider a mass m attached to the end of a spring hanging from the ceiling. We assume that the mass of the spring itself is negligible in comparison with the mass m. (See Figure 10.52.)

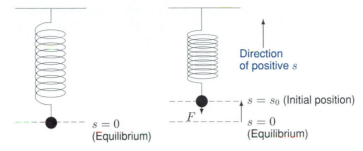

Figure 10.52: Spring and mass in equilibrium position and after upward displacement

When the system is left undisturbed, no net force acts on the mass. The force of gravity is balanced by the force the spring exerts on the mass, and the spring is in the *equilibrium position.* If we pull down on the mass, we feel a force pulling upwards. If instead, we push upward on the mass, the opposite happens: a force pushes the mass down.[10]

What happens if we push the mass upward and then release it? We'd expect the mass to oscillate up and down around the equilibrium position.

Springs and Hooke's Law

In order to figure out how the mass moves, we need to know the exact relationship between its displacement, s, from the equilibrium position and the net force, F, exerted on the mass. (See Figure 10.52.) We would expect that the further the mass is from the equilibrium position, and the more the spring is stretched or compressed, the larger the force. In fact, provided that the displacement is not large enough to deform the spring permanently, experiments show that the net force, F, is approximately proportional to the displacement, s:

$$F = -ks,$$

where k is the *spring constant* ($k > 0$) and the negative sign means that the net force is in the opposite direction to the displacement. The value of k depends on the physical properties of the particular spring. This relationship is known as *Hooke's Law.* Suppose we push the mass upward some distance and then release it. After we let go the net force causes the mass to accelerate toward the equilibrium position. By Newton's Second Law of Motion we have

$$\text{Force} = \text{Mass} \times \text{Acceleration}.$$

Since acceleration is d^2s/dt^2 and force is $F = -ks$ by Hooke's law, we have

$$-ks = m\frac{d^2s}{dt^2},$$

which is equivalent to the following differential equation.

Equation for Oscillations of a Mass on a Spring

$$\frac{d^2s}{dt^2} = -\frac{k}{m}s$$

---

[10]Pulling down on the mass stretches the spring, increasing the tension, so the combination of gravity and spring force is upward. Pushing up the spring decreases tension in the spring, so the combination of gravity and spring force is downward.

Thus, the motion of the mass is described by a second-order differential equation. Since we expect the mass to oscillate, we guess that the solution to this equation involves trigonometric functions.

Solving the Differential Equation by Guess-and-Check

Guessing is the tried-and-true method for solving differential equations. It may surprise you to learn that there is no systematic method for solving most differential equations analytically, so guesswork is often extremely important. Before looking for a general solution to the equation

$$\frac{d^2 s}{dt^2} = -\frac{k}{m} s,$$

we consider the special case $k/m = 1$.

Example 1   Find the general solution to the equation

$$\frac{d^2 s}{dt^2} = -s.$$

Solution   We want to find functions whose second derivative is the negative of the original function. We are already familiar with two functions that have this property: $s(t) = \cos t$ and $s(t) = \sin t$. We check that they are solutions by substituting:

$$\frac{d^2}{dt^2}(\cos t) = \frac{d}{dt}(-\sin t) = -\cos t,$$

and

$$\frac{d^2}{dt^2}(\sin t) = \frac{d}{dt}(\cos t) = -\sin t.$$

The remarkable thing is that starting from these two particular solutions, we can build up *all* the solutions to our equation. Here's how: If C is a constant, then $C \sin t$ and $C \cos t$ are also solutions to the differential equation. (Why?) Also, $\sin t + \cos t$ is a solution to the differential equation. (Why?) In fact, given two constants C_1 and C_2, the function

$$s(t) = C_1 \cos t + C_2 \sin t$$

satisfies the differential equation, since

$$\frac{d^2}{dt^2}(C_1 \cos t + C_2 \sin t) = \frac{d}{dt}(-C_1 \sin t + C_2 \cos t)$$
$$= -C_1 \cos t - C_2 \sin t$$
$$= -(C_1 \cos t + C_2 \sin t).$$

It can be shown (though we will not do it) that $s(t) = C_1 \cos t + C_2 \sin t$ is the most general form of the solution. As expected, it contains two constants, C_1 and C_2.

If the differential equation represents a physical problem, then C_1 and C_2 are often determined by certain conditions of that physical problem. For example if the mass is pushed up to an initial displacement of s_0 and then released, then C_1 and C_2 can be computed, as shown in the next example.

Example 2   Find the solution to

$$\frac{d^2 s}{dt^2} = -s$$

if the mass is displaced by a distance of s_0 and then released.

Solution   The position of the mass is given by the equation

$$s(t) = C_1 \cos t + C_2 \sin t.$$

We also know that the initial position is s_0; thus,

$$s(0) = C_1 \cos 0 + C_2 \sin 0 = C_1 \cdot 1 + C_2 \cdot 0 = s_0,$$

so $C_1 = s_0$, the initial displacement. What is C_2? To find it, we use the fact that at $t = 0$, when the mass has just been released, its velocity is 0. Velocity is the derivative of the displacement, so

$$\left.\frac{ds}{dt}\right|_{t=0} = (-C_1 \sin t + C_2 \cos t)\Big|_{t=0} = -C_1 \cdot 0 + C_2 \cdot 1 = 0,$$

so $C_2 = 0$. Therefore the solution is $s = s_0 \cos t$.

Solution to the General Spring Equation

Having found the general solution to the equation $\dfrac{d^2 s}{dt^2} = -s$, let us return to the equation

$$\frac{d^2 s}{dt^2} = -\frac{k}{m} s.$$

To solve this we let $\omega = \sqrt{\frac{k}{m}}$ (why will be clear in a moment); then $\dfrac{d^2 s}{dt^2} = -\omega^2 s$ or $\dfrac{d^2 s}{dt^2} + \omega^2 s = 0$. Unfortunately, this equation no longer has $\sin t$ and $\cos t$ as solutions. For example,

$$\frac{d^2}{dt^2}(\sin t) = -\sin t \neq -\omega^2 \sin t.$$

However, we can adapt our guess as follows: we know that the functions $\sin t$ and $\cos t$ are *almost* solutions; we just have to deal with the extra ω^2 factor somehow. Ideally, we would like to be able to obtain a factor of ω^2 in the process of differentiating the sine or cosine function twice. This would be accomplished if, each time we differentiated, we obtained a factor of ω. The chain rule suggests an answer: try the function $\sin \omega t$. Checking this:

$$\frac{d^2}{dt^2}(\sin \omega t) = \frac{d}{dt}(\omega \cos \omega t) = -\omega^2 \sin \omega t$$

we see that $\sin \omega t$ is a solution, and you can check that $\cos \omega t$ is a solution, too.

> The general solution to the equation
>
> $$\frac{d^2 s}{dt^2} + \omega^2 s = 0$$
>
> is of the form
>
> $$s(t) = C_1 \cos \omega t + C_2 \sin \omega t,$$
>
> where C_1 and C_2 are arbitrary constants. (We assume $\omega > 0$.) The period of this oscillation is
>
> $$T = \frac{2\pi}{\omega}.$$
>
> Such oscillations are called **simple harmonic motion**.

The solution to our original equation, $\dfrac{d^2 s}{dt^2} + \dfrac{k}{m} s = 0$, is thus $s = C_1 \cos \sqrt{\frac{k}{m}} t + C_2 \sin \sqrt{\frac{k}{m}} t$.

Initial-Value and Boundary-Value Problems

A problem in which the initial position and the initial velocity are used to determine the particular solution is called an *initial-value problem*. (See Example 2.) Alternatively, we may be given the position at two known times. Such a problem is known as a *boundary-value problem*.

Example 3   Find a solution to the differential equation satisfying each set of conditions below

$$\frac{d^2 s}{dt^2} + 4s = 0.$$

(a)   The boundary conditions $s(0) = 0$, $s(\pi/4) = 20$.
(b)   The initial conditions $s(0) = 1$, $s'(0) = -6$.

Solution   Since $\omega^2 = 4$, $\omega = 2$, the general solution to the differential equation is

$$s(t) = C_1 \cos 2t + C_2 \sin 2t.$$

(a)   Substituting the boundary condition $s(0) = 0$ into the general solution gives

$$s(0) = C_1 \cos(2 \cdot 0) + C_2 \sin(2 \cdot 0) = C_1 \cdot 1 + C_2 \cdot 0 = C_1 = 0.$$

Thus $s(t)$ must have the form $s(t) = C_2 \sin 2t$. The second condition yields the value of C_2:

$$s\left(\frac{\pi}{4}\right) = C_2 \sin\left(2 \cdot \frac{\pi}{4}\right) = C_2 = 20.$$

Therefore, the solution satisfying the boundary conditions is

$$s(t) = 20 \sin 2t.$$

(b)   For the initial-value problem, we start from the same general solution: $s(t) = C_1 \cos 2t + C_2 \sin 2t$. Substituting 0 for t once again, we find

$$s(0) = C_1 \cos(2 \cdot 0) + C_2 \sin(2 \cdot 0) = C_1 = 1.$$

Differentiating $s(t) = \cos 2t + C_2 \sin 2t$ gives

$$s'(t) = -2 \sin 2t + 2C_2 \cos 2t,$$

and applying the second initial condition gives us C_2:

$$s'(0) = -2 \sin(2 \cdot 0) + 2C_2 \cos(2 \cdot 0) = 2C_2 = -6,$$

so $C_2 = -3$ and our solution is

$$s(t) = \cos 2t - 3 \sin 2t.$$

What do the Graphs of Our Solutions Look Like?

Since the general solution of the equation $d^2 s/dt^2 + \omega^2 s = 0$ is of the form

$$s(t) = C_1 \cos \omega t + C_2 \sin \omega t$$

it would be useful to know what the graph of such a sum of sines and cosines looks like. We start with the example $s(t) = \cos t + \sin t$, which is graphed in Figure 10.53.

Interestingly, the graph in Figure 10.53 looks precisely like another sine function, and in fact it is one. If we measure the graph carefully, we find that its period is 2π, the same as $\sin t$ and $\cos t$, but the amplitude is approximately 1.414 (In fact, it's $\sqrt{2}$.), and the graph is shifted along the t-axis (in fact by $\pi/4$). If we plot $C_1 \cos t + C_2 \sin t$ for any C_1 and C_2, the resulting graph is always a sine function with period 2π, though the amplitude and shift can vary. For example, the graph of $s = 6\cos t - 8\sin t$ is in Figure 10.54; the graph has period 2π.

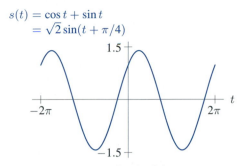

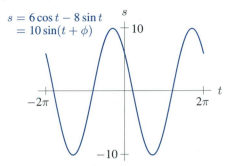

Figure 10.53: Graph of the sum:
$s(t) = \cos t + \sin t = \sqrt{2}\sin(t + \pi/4)$

Figure 10.54: Graph of a sum of sine and cosine. Amplitude $A = 10 = \sqrt{6^2 + 8^2}$

These graphs suggest that we can write the sum of a sine and a cosine of the same argument as one single sine function.[11] It turns out that this can always be done using the following relations. (See Problem 17.)

If $C_1\cos\omega t + C_2\sin\omega t = A\sin(\omega t + \phi)$, the *amplitude*, A, is given by

$$A = \sqrt{C_1^2 + C_2^2}.$$

The angle ϕ is called the *phase shift* and satisfies

$$\tan\phi = \frac{C_1}{C_2}.$$

We choose ϕ in $(-\pi, \pi]$ such that if $C_1 > 0$, ϕ is positive, and if $C_1 < 0$, ϕ is negative.

The reason that it is often useful to write the solution to the differential equation as a single sine function $A\sin(\omega t + \phi)$, as opposed to the sum of a sine and a cosine, is that the amplitude A and the phase shift ϕ are easier to recognize in this form.

Warning: Phase Shift Is Not the Same as Horizontal Translation

Let's look at $s = \sin(3t + \pi/2)$. If we rewrite this as $s = \sin(3(t + \pi/6))$, we can see from Figure 10.55 that the graph of this function is the graph of $s = \sin 3t$ shifted to the left a distance of $\pi/6$. (Remember that replacing x by $x - 2$ shifts a graph by 2 to the right.) But $\pi/6$ is *not* the phase shift; the phase shift[12] is $\pi/2$. From the point of view of a scientist, the important question is often not the distance the curve has shifted, but the relation between the distance shifted and the period.

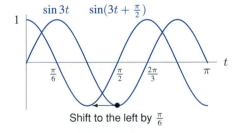

Shift to the left by $\frac{\pi}{6}$

Figure 10.55: Phase shift is $\frac{\pi}{2}$; horizontal translation is $\frac{\pi}{6}$

---

[11]The sum of a sine and a cosine can also be written as a single cosine function.
[12]This definition of phase shift is the one usually used in the sciences.

Problems for Section 10.8

1. Check by differentiation that $y = 2\cos t + 3\sin t$ is a solution to $y'' + y = 0$.

2. Check that $y = A\cos t + B\sin t$ is a solution to $y'' + y = 0$ for any constants A and B.

3. What values of α and A make $y = A\cos\alpha t$ a solution to $y'' + 5y = 0$ such that $y'(1) = 3$?

The functions in Problems 4–6 describe the motion of a mass on a spring satisfying the differential equation $y'' = -9y$, where y is the displacement of the mass from the equilibrium position at time t, with upwards as positive. In each case, describe in words how the motion starts when $t = 0$. For example, is the mass at the highest point, the lowest point, or in the middle? Is it moving up or down or is it at rest?

4. $y = 2\cos 3t$ 5. $y = -0.5\sin 3t$ 6. $y = -\cos 3t$

7. The following differential equations represent oscillating springs.

(i)	$s'' + 4s = 0$	$s(0) = 5,$	$s'(0) = 0$
(ii)	$4s'' + s = 0$	$s(0) = 10,$	$s'(0) = 0$
(iii)	$s'' + 6s = 0$	$s(0) = 4,$	$s'(0) = 0$
(iv)	$6s'' + s = 0$	$s(0) = 20,$	$s'(0) = 0$

    Which differential equation represents:

    (a) The spring oscillating most quickly (with the shortest period)?
    (b) The spring oscillating with the largest amplitude?
    (c) The spring oscillating most slowly (with the longest period)?
    (d) The spring with largest maximum velocity?

8. (a) Find the general solution of the differential equation

    $$y'' + 9y = 0$$

    (b) For each of the following initial conditions find a particular solution.
        (i) $y(0) = 0, y'(0) = 1$ (ii) $y(0) = 1, y'(0) = 0$
        (iii) $y(0) = 1, y(1) = 0$ (iv) $y(0) = 0, y(1) = 1$
    (c) Sketch a graph of the solutions found in (b).

9. Each graph in Figure 10.56 represents a solution to one of the differential equations:
    (a) $x'' + x = 0$,   (b) $x'' + 4x = 0$,   (c) $x'' + 16x = 0$.

    Assuming the t-scales on the four graphs are the same, which graph represents a solution to which equation? Find an equation for each graph.

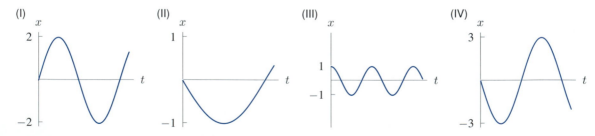

Figure 10.56: Graphs of solutions to $x'' + kx = 0$, for various values of k

10. A pendulum of length l makes an angle of x (radians) with the vertical (see Figure 10.57). When x is small, it can be shown that, approximately:

$$\frac{d^2x}{dt^2} = -\frac{g}{l}x,$$

where g is the acceleration due to gravity.

Figure 10.57

(a) Solve this equation assuming that $x(0) = 0$ and $x'(0) = v_0$.

(b) Solve this equation assuming that the pendulum is let go from the position where $x = x_0$. ("Let go" means that the velocity of the pendulum is zero when $x = x_0$. Measure t from the moment when the pendulum is let go.)

11. Look at the pendulum motion in Problem 10. What effect does it have on x as a function of time if:

(a) x_0 is increased? 　　　　(b) l is increased?

12. (a) Using a calculator or computer, graph $s = 4\cos t + 3\sin t$ for $-2\pi \leq t \leq 2\pi$.

(b) If $4\cos t + 3\sin t = A\sin(t + \phi)$, use your graph to estimate the values of A and ϕ.

(c) Calculate A and ϕ analytically.

13. Find the amplitude of $3\sin 2t + 7\cos 2t$.

14. Find the amplitude of the oscillation $y = 3\sin(2t) + 4\cos(2t)$.

15. Write $y(t) = 5\sin(2t) + 12\cos(2t)$ in the form $y(t) = A\sin(\omega t + \psi)$.

16. Write $7\sin \omega t + 24\cos \omega t$ in the form $A\sin(\omega t + \phi)$.

17. (a) Expand $A\sin(\omega t + \phi)$ using the trigonometric identity $\sin(x + y) = \sin x \cos y + \cos x \sin y$.

(b) Assume $A > 0$. If $A\sin(\omega t + \phi) = C_1 \cos \omega t + C_2 \sin \omega t$, show that we must have

$$A = \sqrt{C_1^2 + C_2^2} \quad \text{and} \quad \tan \phi = C_1/C_2.$$

The potential energy, U, of a molecule in various configurations can be expressed by

$$U = \frac{V_0}{2}\left(1 + \cos(n\phi - \phi_0)\right)$$

where ϕ is the (variable) angle which describes the configuration and V_0, n, ϕ_0 are constants, with $n = 2$ or 3. For each of the graphs[13] of U against ϕ in Problems 18- 19:

(a) Give reasonable values for V_0, n, and ϕ_0.

(b) Find a differential equation (satisfied by U) of the form

$$\frac{d^2U}{d\phi^2} + \omega^2(U - K) = 0 \qquad U(0) = A \text{ and } U'(0) = B,$$

where ω^2, K, A, and B are constants to be determined.

18.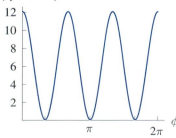

Figure 10.58

19.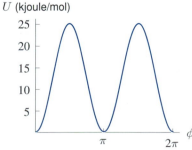

Figure 10.59

[13] Adapted from Ignacio Tinoco Jr., Kenneth Sauer, James C. Wang, *Physical Chemistry*. (New Jersey: Prentice Hall, 1978).

Problems 20–22 show how a second order differential equation can be used to describe an electric circuit. A charged capacitor connected to an inductor (as shown in Figure 10.60) will cause a current to flow through the inductor until the capacitor is fully discharged. The current in the inductor will, in turn, charge up the capacitor until the capacitor is fully charged again. We will let $Q(t)$ be the amount of charge on the capacitor at time t. We can then find the current, the rate at which charge is moving in the circuit, by looking at the rate of change of Q. Thus we have the relation between the current I and the charge Q:

$$I = \frac{dQ}{dt}.$$

The current can be positive or negative; its sign tells us which way the current is flowing. Q can also be positive or negative; its sign tells us whether the charge on the capacitor is positive or negative.

If we assume the circuit resistance is zero, then it can be shown that the charge Q and the current I in the circuit satisfy the differential equation

$$L\frac{dI}{dt} + \frac{Q}{C} = 0.$$

So

$$L\frac{d^2Q}{dt^2} + \frac{Q}{C} = 0$$

where C, the capacitance, and L, the inductance, are constants depending on the particular capacitor and inductor. The unit of charge is the coulomb, the unit of capacitance the farad, the unit of inductance the henry, the unit of current is the ampere, and time is measured in seconds. This equation is similar to the differential equations we have seen for springs based on Hooke's law.

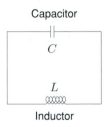

Capacitor

C

L

Inductor

Figure 10.60

20.   If $L = 36$ henry and $C = 9$ farad, find a formula for $Q(t)$ if

  (a) $Q(0) = 0$ $I(0) = 2$
  (b) $Q(0) = 6$ $I(0) = 0$

21.   Suppose we wanted to set up our circuit so that $Q(0) = 0, Q'(0) = I(0) = 4$, and the maximum possible charge is $2\sqrt{2}$ coulombs. What should the capacitance of our capacitor be if we have an inductor with inductance 10 henry?

22.   What happens to the charge and current as t goes to infinity? What does it mean that the charge and current are sometimes positive and sometimes negative?

10.9 LINEAR SECOND-ORDER DIFFERENTIAL EQUATIONS

A Spring with Friction: Damped Oscillations

The differential equation $d^2s/dt^2 = -(k/m)s$, which we used to describe the motion of a spring, disregards friction. But there is friction in every real system. For a mass on a spring, the frictional force from air resistance increases with the velocity of the mass. The frictional force is often approximately proportional to velocity, and so we introduce a *damping term* of the form $a(ds/dt)$, where a is a constant called the *damping coefficient* and ds/dt is the velocity of the mass.

Remember that without damping, the differential equation was obtained from

Force $=$ Mass $\times$ Acceleration,

or

$$F = m\frac{d^2s}{dt^2}.$$

With damping, the force $-ks$ is replaced by $-ks - a(ds/dt)$, where a is positive and the $a(ds/dt)$ term is subtracted because the frictional force is in the direction opposite to the motion. The new differential equation is therefore

$$-ks - a\frac{ds}{dt} = m\frac{d^2s}{dt^2}$$

which is equivalent to the following differential equation:

Equation for Damped Oscillations of a Spring

$$\frac{d^2s}{dt^2} + \frac{a}{m}\frac{ds}{dt} + \frac{k}{m}s = 0$$

We expect the solution to this equation to die away with time, as friction brings the motion to a stop.

The General Solution to a Linear Differential Equation

The equation for damped oscillations is an example of a *linear second-order differential equation with constant coefficients.* This section explains an analytic method of solving any such equation,

$$\frac{d^2y}{dt^2} + b\frac{dy}{dt} + cy = 0,$$

where b and c are constants. As we have seen for the spring equation, if $f_1(t)$ and $f_2(t)$ satisfy the differential equation, then the *principle of superposition* says that, for any constants C_1 and C_2, the function

$$y(t) = C_1 f_1(t) + C_2 f_2(t)$$

is also a solution. It can be shown that the general solution is of this form, provided $f_1(t)$ is not a multiple of $f_2(t)$.

Finding Solutions: The Characteristic Equation

We will now see how to use complex numbers to solve the differential equation

$$\frac{d^2y}{dt^2} + b\frac{dy}{dt} + cy = 0.$$

The method is a form of guess-and-check. This time we ask what kind of function might satisfy a differential equation which says that the second derivative d^2y/dt^2 is a sum of multiples of dy/dt and y. One possibility is the exponential function, so we try to find a solution of the form:

$$y = Ce^{rt},$$

where r may be a complex number.[14] In order to find r, we substitute into the differential equation:

$$\frac{d^2y}{dt^2} + b\frac{dy}{dt} + cy = r^2Ce^{rt} + b \cdot rCe^{rt} + c \cdot Ce^{rt} = Ce^{rt}(r^2 + br + c) = 0.$$

We can divide this equation through by $y = Ce^{rt}$ so long as $C \neq 0$, because the exponential function is never zero. If $C = 0$, then $y = 0$, which is not a very interesting solution (though it is a solution). So we assume $C \neq 0$. Then $y = Ce^{rt}$ will be a solution to the differential equation if

$$r^2 + br + c = 0.$$

---

[14]See Appendix D on complex numbers.

This quadratic equation is called the *characteristic equation* of the differential equation. Its solutions are

$$r = -\frac{1}{2}b \pm \frac{1}{2}\sqrt{b^2 - 4c}.$$

There are three different types of solutions to the differential equation, depending on whether the solutions to the characteristic equation are real and distinct, complex, or repeated. The sign of $b^2 - 4c$ determines the type of solutions.

The Case with $b^2 - 4c > 0$

There are two real solutions r_1 and r_2 to the characteristic equation, and so the following two functions satisfy the differential equation:

$$C_1 e^{r_1 t} \quad \text{and} \quad C_2 e^{r_2 t}.$$

The sum of these two solutions is the general solution to the differential equation:

If $b^2 - 4c > 0$, the general solution to

$$\frac{d^2 y}{dt^2} + b\frac{dy}{dt} + cy = 0$$

is

$$y(t) = C_1 e^{r_1 t} + C_2 e^{r_2 t}$$

where r_1 and r_2 are the solutions to the characteristic equation.
If $r_1, r_2 < 0$, the motion is called **overdamped**.

A physical system satisfying a differential equation of this type is said to be overdamped because it occurs when there is a lot of friction in the system. For example, a spring moving in a thick fluid such as oil or molasses is overdamped: it will not oscillate.

Example 1   Suppose a spring is placed in oil, where it satisfies the differential equation

$$\frac{d^2 s}{dt^2} + 3\frac{ds}{dt} + 2s = 0.$$

Solve this equation with the initial conditions $s = -0.5$ and $\frac{ds}{dt} = 3$ when $t = 0$.

Solution   The characteristic equation is

$$r^2 + 3r + 2 = 0,$$

with solutions $r = -1$ and $r = -2$, and so the general form of the solution to the differential equation is

$$s(t) = C_1 e^{-t} + C_2 e^{-2t}.$$

We use the initial conditions to find C_1 and C_2. At $t = 0$, we have

$$s = C_1 e^{-0} + C_2 e^{-2(0)} = C_1 + C_2 = -0.5.$$

Furthermore, since $ds/dt = -C_1 e^{-t} - 2C_2 e^{-2t}$, we have

$$\left.\frac{ds}{dt}\right|_{t=0} = -C_1 e^{-0} - 2C_2 e^{-2(0)} = -C_1 - 2C_2 = 3.$$

Solving these equations simultaneously, we find $C_1 = 2$ and $C_2 = -2.5$, so that the solution is

$$s(t) = 2e^{-t} - 2.5e^{-2t}.$$

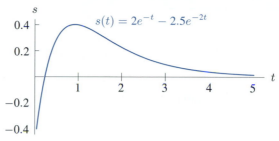

Figure 10.61: Solution to overdamped equation
$$\frac{d^2s}{dt^2} + 3\frac{ds}{dt} + 2s = 0$$

The graph of this function is in Figure 10.61. The mass is so slowed by the oil that it passes through the equilibrium point only once (when $t \approx 1/4$) and for all practical purposes, it comes to rest after a short time. The motion has been "damped out" by the oil.

The Case with $b^2 - 4c = 0$

In this case, the characteristic equation has only one solution, $r = -b/2$. By substitution, we can check that both $y = e^{-bt/2}$ and $y = te^{-bt/2}$ are solutions.

If $b^2 - 4c = 0$,

$$\frac{d^2y}{dt^2} + b\frac{dy}{dt} + cy = 0$$

has general solution

$$y(t) = (C_1t + C_2)e^{-bt/2}.$$

If $b > 0$, the system is said to be **critically damped**.

The Case with $b^2 - 4c < 0$

Example 2   An object of mass $m = 10$ kg is attached to a spring with spring constant $k = 20$ kg/sec^2, and the object experiences a frictional force proportional to the velocity, with constant of proportionality $a = 20$ kg/sec. At time $t = 0$, the object is released from rest 2 meters above the equilibrium position. Write the differential equation that describes the motion.

Solution   The differential equation that describes the motion can be obtained from the following general expression for the damped motion of a spring:

$$\underbrace{m\frac{d^2s}{dt^2}}_{\text{Mass} \times \text{Acceleration}} = \underbrace{F}_{\text{Net force}} = - \underbrace{ks}_{\text{Spring force}} - \underbrace{a\frac{ds}{dt}}_{\text{Frictional force}}.$$

Substituting $m = 10$ kg, $k = 20$ kg/sec^2, and $a = 20$ kg/sec, we obtain the differential equation for the motion:

$$\frac{d^2s}{dt^2} + 2\frac{ds}{dt} + 2s = 0.$$

At $t = 0$, the object is at rest 2 meters above equilibrium, so the initial conditions are $s(0) = 2$ and $s'(0) = 0$, where s is in meters and t in seconds.

Notice that this is the same differential equation as in Example 1 except that the coefficient of ds/dt has decreased from 3 to 2, which means that the frictional force has been reduced. This time, the roots of the characteristic equation have imaginary parts which lead to oscillations.

Example 3   Solve the differential equation

$$\frac{d^2s}{dt^2} + 2\frac{ds}{dt} + 2s = 0,$$

subject to $s(0) = 2$, $s'(0) = 0$.

Solution   The characteristic equation is

$$r^2 + 2r + 2 = 0 \quad \text{giving} \quad r = -1 \pm i.$$

The solution to the differential equation is

$$s(t) = A_1 e^{(-1+i)t} + A_2 e^{(-1-i)t},$$

where A_1 and A_2 are arbitrary complex numbers. The initial condition $s(0) = 2$ gives

$$2 = A_1 e^{(-1+i)\cdot 0} + A_2 e^{(-1-i)\cdot 0} = A_1 + A_2.$$

Also,

$$s'(t) = A_1(-1+i)e^{(-1+i)t} + A_2(-1-i)e^{(-1-i)t},$$

so $s'(0) = 0$ gives

$$0 = A_1(-1+i) + A_2(-1-i).$$

Solving the simultaneous equations for A_1 and A_2 gives (after some algebra)

$$A_1 = 1 - i \quad \text{and} \quad A_2 = 1 + i.$$

The solution is therefore

$$s(t) = (1-i)e^{(-1+i)t} + (1+i)e^{(-1-i)t} = (1-i)e^{-t}e^{it} + (1+i)e^{-t}e^{-it}.$$

Using Euler's formula[15], $e^{it} = \cos t + i \sin t$ and $e^{-it} = \cos t - i \sin t$, we get:

$$s(t) = (1-i)e^{-t}(\cos t + i \sin t) + (1+i)e^{-t}(\cos t - i \sin t).$$

Multiplying out and simplifying, all the complex terms drop out, giving

$$s(t) = e^{-t}\cos t + ie^{-t}\sin t - ie^{-t}\cos t + e^{-t}\sin t$$
$$+ e^{-t}\cos t - ie^{-t}\sin t + ie^{-t}\cos t + e^{-t}\sin t$$
$$= 2e^{-t}\cos t + 2e^{-t}\sin t.$$

The $\cos t$ and $\sin t$ terms tell us that the solution oscillates. However the oscillations are damped extremely quickly by the factor e^{-t}. (See Figure 10.62.) Alternatively, you can see the behavior of the solution by rewriting it in the form

$$s(t) = 2e^{-t}(\cos t + \sin t) = 2\sqrt{2}e^{-t}\sin\left(t + \frac{\pi}{4}\right).$$

Physically the most interesting feature of the solution is that the period of the oscillations does not change as the amplitude decreases. This is why a spring-driven clock can keep accurate time even as it is running down.

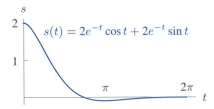

Figure 10.62: Solution to underdamped equation $\frac{d^2s}{dt^2} + 2\frac{ds}{dt} + 2s = 0$

In Example 3, notice that although the coefficients A_1 and A_2 are complex, the solution, $s(t)$, is real. (We expect this, since $s(t)$ represents a real displacement.) In general, provided the coefficients

---

[15]See Appendix D on complex numbers

b and c in the original differential equation and the initial values are real, the solution will always be real too. The complex coefficients A_1 and A_2 will always be complex conjugates.

If $b^2 - 4c < 0$, to solve

$$\frac{d^2y}{dt^2} + b\frac{dy}{dt} + cy = 0,$$

- Find the solutions $r = \alpha \pm i\beta$ to the characteristic equation $r^2 + br + c = 0$.
- The general solution to the differential equation is, for some real C_1 and C_2,

$$y = C_1 e^{\alpha t} \cos \beta t + C_2 e^{\alpha t} \sin \beta t.$$

If $\alpha < 0$, such oscillations are called **underdamped**.

Example 4   Find the general solution of the equations   (a) $y'' = 9y$ (b) $y'' = -9y$.

Solution   (a)   The characteristic equation is $r^2 - 9 = 0$, so $r = \pm 3$. Thus the general solution is

$$y = C_1 e^{3t} + C_2 e^{-3t}.$$

(b)   The characteristic equation is $r^2 + 9 = 0$, so $r = 0 \pm 3i$. The general solution is

$$y = C_1 e^{0t} \cos 3t + C_2 e^{0t} \sin 3t = C_1 \cos 3t + C_2 \sin 3t.$$

Notice that we have seen the solution to this equation in Section 10.8; it's the equation of undamped simple harmonic motion.

Example 5   Solve the initial value problem

$$y'' + 4y' + 13y = 0, \quad y(0) = 0, \, y'(0) = 30.$$

Solution   We solve the characteristic equation

$$r^2 + 4r + 13 = 0, \quad \text{getting} \quad r = -2 \pm 3i.$$

The general solution to the differential equation is

$$y = C_1 e^{-2t} \cos 3t + C_2 e^{-2t} \sin 3t.$$

Substituting $t = 0$ gives

$$y(0) = C_1 \cdot 1 \cdot 1 + C_2 \cdot 1 \cdot 0 = 0, \quad \text{so} \quad C_1 = 0.$$

Differentiating $y(t) = C_2 e^{-2t} \sin 3t$ gives

$$y'(t) = C_2(-2e^{-2t} \sin 3t + 3e^{-2t} \cos 3t).$$

Substituting $t = 0$ gives

$$y'(0) = C_2(-2 \cdot 1 \cdot 0 + 3 \cdot 1 \cdot 1) = 3C_2 = 30 \quad \text{so} \quad C_2 = 10.$$

The solution is therefore

$$y(t) = 10e^{-2t} \sin 3t.$$

Summary of Solutions to $y'' + by' + cy = 0$

If $b^2 - 4c > 0$,   then $y = C_1 e^{r_1 t} + C_2 e^{r_2 t}$
If $b^2 - 4c = 0$,   then $y = (C_1 t + C_2)e^{-bt/2}$
If $b^2 - 4c < 0$,   then $y = C_1 e^{\alpha t} \cos \beta t + C_2 e^{\alpha t} \sin \beta t$

Problems for Section 10.9

For Problems 1–12, find the general solution to the given differential equation.

1. $y'' + 4y' + 3y = 0$

2. $y'' + 4y' + 4y = 0$

3. $y'' + 4y' + 5y = 0$

4. $s'' - 7s = 0$

5. $s'' + 7s = 0$

6. $y'' - 3y' + 2y = 0$

7. $4z'' + 8z' + 3z = 0$

8. $\dfrac{d^2x}{dt^2} + 4\dfrac{dx}{dt} + 8x = 0$

9. $\dfrac{d^2p}{dt^2} + \dfrac{dp}{dt} + p = 0$

10. $z'' + 2z = 0$

11. $z'' + 2z' = 0$

12. $P'' + 2P' + P = 0$

For Problems 13–16, solve the initial value problem.

13. $y'' + 6y' + 5y = 0, \quad y(0) = 1, \quad y'(0) = 0$

14. $y'' + 6y' + 5y = 0, \quad y(0) = 5, \quad y'(0) = 5$

15. $y'' + 6y' + 10y = 0, \quad y(0) = 0, \quad y'(0) = 2$

16. $y'' + 6y' + 10y = 0, \quad y(0) = 0, \quad y'(0) = 0$

For Problems 17–18, solve the boundary value problem.

17. $p'' + 2p' + 2p = 0, \quad p(0) = 0, \quad p(\pi/2) = 20$

18. $p'' + 4p' + 5p = 0, \quad p(0) = 1, \quad p(\pi/2) = 5$

19. Match the graphs of solutions in Figure 10.63 with the differential equations below.

(a) $x'' + 4x = 0$
(b) $x'' - 4x = 0$
(c) $x'' - 0.2x' + 1.01x = 0$
(d) $x'' + 0.2x' + 1.01x = 0$

(I) 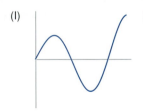 (II) 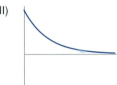 (III) 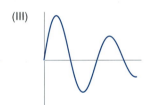 (IV)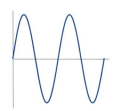

Figure 10.63

20. Match the differential equations to the solution graphs (I)-(IV). Use each graph only once.
(a) $y'' + 5y' + 6y = 0$
(b) $y'' + y' - 6y = 0$
(c) $y'' + 4y' + 9y = 0$
(d) $y'' = -9y$

(I) 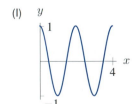 (II) 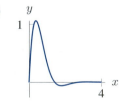 (III) 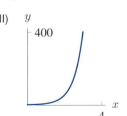 (IV)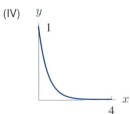

21. If $y = e^{2t}$ is a solution to the differential equation

$$\frac{d^2y}{dt^2} - 5\frac{dy}{dt} + ky = 0,$$

find the value of the constant k and the general solution to this equation.

22. Assuming $b, c > 0$, explain how you know that the solutions of an underdamped differential equation must go to 0 as $t \to \infty$.

Each of the differential equations below represents the position of a 1 gram mass oscillating on the end of a damped spring. For Problems 23–27 below, pick the differential equation representing the system which answers the question.

(i) $s'' + s' + 4s = 0$

(iii) $s'' + 3s' + 3s = 0$

(ii) $s'' + 2s' + 5s = 0$

(iv) $s'' + 0.5s' + 2s = 0$

23.   Which spring has the largest coefficient of damping?

24.   Which spring exerts the smallest restoring force for a given displacement?

25.   In which system does the mass experience the frictional force of smallest magnitude for a given velocity?

26.   Which oscillation has the longest period?

27.   Which spring is the stiffest? [Hint: You need to determine what it means for a spring to be stiff. Think of an industrial strength spring and a slinky.]

For each of the differential equations in Problems 28–30, find the values of c that make the general solution:
(a)   overdamped,   (b)   underdamped,   (c)   critically damped.

28. $s'' + 4s' + cs = 0$ 29. $s'' + 2\sqrt{2}s' + cs = 0$ 30. $s'' + 6s' + cs = 0$

31.   Find a solution to the following equation which satisfies $z(0) = 3$ and does not tend to infinity as $t \to \infty$:

$$\frac{d^2z}{dt^2} + \frac{dz}{dt} - 2z = 0.$$

32.   Consider an overdamped differential equation with $b, c > 0$.

(a)   Show that both roots of the characteristic equation are negative.

(b)   Show that any solution to the differential equation goes to 0 as $t \to \infty$.

Recall the discussion of electric circuits in Section 10.8 on page 550. Just as a spring can have a damping force which affects its motion, so can a circuit. Problems 33–36 illustrate this situation. The damping force is caused by a resistor shown in Figure 10.64.

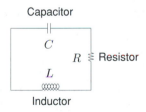

Capacitor

C

R ⩵ Resistor

L

Inductor

Figure 10.64

Just as the damping force for a mass on a spring is proportional to the velocity of the mass, the damping force in the circuit is proportional to the current. We get an extra term $RI = R(dQ/dt)$ in the differential equation. The resistance, R, is a constant that depends on the resistor, and is measured in ohms. The equation for the charge Q on a capacitor in a circuit with inductance L, capacitance C, and resistance R satisfies the differential equation

$$L\frac{d^2Q}{dt^2} + R\frac{dQ}{dt} + \frac{1}{C}Q = 0.$$

33.   If $L = 1$ henry, $R = 2$ ohms, and $C = 4$ farads, find a formula for the charge when
(a) $Q(0) = 0, Q'(0) = 2$.   (b) $Q(0) = 2, Q'(0) = 0$.

34.   If $L = 1$ henry, $R = 1$ ohm, and $C = 4$ farads, find a formula for the charge when
(a) $Q(0) = 0, Q'(0) = 2$.   (b) $Q(0) = 2, Q'(0) = 0$.
(c)   How did reducing the resistance affect the charge? Compare with your solution to Problem 33.

35.   If $L = 8$ henry, $R = 2$ ohm, and $C = 4$ farads, find a formula for the charge when
(a) $Q(0) = 0, Q'(0) = 2$.   (b) $Q(0) = 2, Q'(0) = 0$.
(c)   How did increasing the inductance affect the charge? Compare with your solution to Problem 33.

36.   Given any positive values for R, L and C, what happens to the charge as t goes to infinity?

37. Could the graph in Figure 10.65 show the position of a mass oscillating at the end of an overdamped spring? Why or why not?

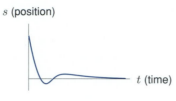

s (position)

t (time)

Figure 10.65

38. Consider the system of differential equations

$$\frac{dx}{dt} = -y \qquad\qquad \frac{dy}{dt} = -x.$$

(a) Convert this system to a second order differential equation in y by differentiating the second equation with respect to t and substituting for x from the first equation.

(b) Solve the equation you obtained for y as a function of t; hence find x as a function of t.

CHAPTER SUMMARY

- **Differential equations terminology**
  Order, initial conditions, families of solutions, stable/unstable equilibrium solutions
- **Solving first-order differential equations**
  Slope fields (graphical), Euler's method (numerical), separation of variables (analytical)
- **Modeling with differential equations**
  Growth and decay, Newton's law of heating and cooling, compartment models
- **Second-order differential equations**
  Initial and boundary conditions, oscillations (with and without damping), characteristic equation, solutions using complex numbers

REVIEW PROBLEMS FOR CHAPTER TEN

For the differential equations in Problems 1–16, find a solution which passes through the given point.

1. $\dfrac{dy}{dt} = y(10 - y)$, $y(0) = 1$

2. $\dfrac{dP}{dt} = 0.03P + 400$, $P(0) = 0$

3. $\dfrac{dy}{dx} = \dfrac{y(3 - x)}{x(0.5y - 4)}$, $y(1) = 5$

4. $\dfrac{dy}{dx} = e^{x-y}$, $y(0) = 1$

5. $\dfrac{df}{dx} = \sqrt{xf(x)}$, $f(1) = 1$

6. $1 + y^2 - \dfrac{dy}{dx} = 0$, $y(0) = 0$

7. $2\sin x - y^2\dfrac{dy}{dx} = 0$, $y(0) = 3$

8. $\dfrac{dk}{dt} = (1 + \ln t)k$, $k(1) = 1$

9. $\dfrac{dy}{dx} + xy^2 = 0$, $y(1) = 1$

10. $\dfrac{dy}{dx} = e^{x+y}$, $y = 0$ where $x = 1$

11. $\dfrac{dz}{dt} = z(z-1), \quad z(0) = 10$

12. $\dfrac{dy}{dx} = \dfrac{y(100-x)}{x(20-y)}, \quad (1,20)$

13. $\dfrac{dy}{dt} = 2^y \sin^3 t, \quad y(0) = 0$

14. $e^{-\cos\theta}\dfrac{dz}{d\theta} = \sqrt{1-z^2}\sin\theta, \quad z(0) = \tfrac{1}{2}$

15. $(1+t^2)y\dfrac{dy}{dt} = 1 - y, \quad y(1) = 0$

16. $\dfrac{dy}{dx} = \dfrac{0.2y(18+0.1x)}{x(100+0.5y)}, \quad (10,10)$

Solve the differential equations in Problems 17–20. Assume $x \geq 0, y \geq 0$.

17. $\dfrac{dQ}{dt} + t^2Q^2 + Q^2 - 4t^2 - 4 = 0$

18. $\dfrac{dy}{dt} = \dfrac{y \ln y}{t^2}$

19. $\dfrac{x}{y}\dfrac{dx}{dy} = e^{(x/a)^2} \ln y \quad (a \text{ is constant})$

20. $(y\sqrt{x^3+1})\dfrac{dy}{dx} + x^2y^2 + x^2 = 0$

21. Consider the initial value problem

$$y' = 5 - y, \quad y(0) = 1.$$

  (a) Use Euler's method with five steps to estimate $y(1)$.
  (b) Sketch the slope field for this differential equation in the first quadrant, and use it to decide if your estimate is an over- or underestimate.
  (c) Find the exact solution to the differential equation and hence find $y(1)$ exactly.
  (d) Without doing the calculation, roughly what would you expect the approximation for $y(1)$ to be with ten steps?

22. Solve the differential equation

$$\frac{dy}{dx} = \frac{1}{(\cos x)(\cos y)}$$

by Euler's method, for the solution passing through $(0,0)$. Find y when $x = \tfrac{1}{2}$.

  (a) Find the solution from Euler's method when $\Delta x = \tfrac{1}{2}$ (1 step),
  (b) $\Delta x = \tfrac{1}{4}$ (2 steps),
  (c) $\Delta x = \tfrac{1}{8}$ (4 steps).
  (d) Solve the differential equation by separation of variables for y passing through $(0,0)$ and find $y(\tfrac{1}{2})$. Compare with your results in parts (a)–(c).

Find a general solution to the differential equations in Problems 23–26.

23. $y'' + 6y' + 8y = 0$

24. $9z'' + z = 0$

25. $9z'' - z = 0$

26. $x'' + 2x' + 10x = 0$

For each of the differential equations in Problems 27–28, find the values of b that make the general solution:
(a) overdamped,   (b) underdamped,   (c) critically damped.

27. $s'' + bs' + 5s = 0$

28. $s'' + bs' - 16s = 0$

29. Suppose that the rate at which a drug leaves the bloodstream and passes into the urine is proportional to the quantity of the drug in the blood at that time. If an initial dose of Q_0 is injected directly into the blood, only 20% is left in the blood after 3 hours.

  (a) Write and solve a differential equation for the quantity, Q, of the drug in the blood at time, t, measured in hours.
  (b) How much of this drug is in a patient's body after 6 hours if the patient is given 100 mg initially?

30. Water leaks from a vertical cylindrical tank through a small hole in its base at a rate proportional to the square root of the volume of water remaining. If the tank initially contains 200 liters and 20 liters leak out during the first day, when will the tank be half empty? How much water will there be after 4 days?

31. (a) A cup of coffee is made with boiling water and stands in a room where the temperature is 20° C. If $T(t)$ is the temperature of the coffee at time t, explain what the differential equation

$$\frac{dT}{dt} = -k(T - 20)$$

says in everyday terms. What is the sign of k?

  (b) Solve this differential equation. If the coffee cools to 90° C in 2 minutes, how long will it take to cool to 60° C degrees?

32. A roast is taken from the refrigerator (where the temperature is 40°F) and put in a 350°F oven. One hour later, the meat thermometer shows a temperature of 90°F. If the roast is done when its temperature reaches 140°F, what is the total time the roast should be in the oven? Newton's Law of Cooling says that the rate of change of temperature is proportional to the difference between the temperature of the object and the temperature of the surrounding air.

33. A bank account earns 10% annual interest, compounded continuously. Money is deposited in a continuous cash flow at a rate of $1200 per year into the account.

  (a) Write a differential equation that describes the rate at which the balance $B = f(t)$ is changing.
  (b) Solve the differential equation given an initial balance $B_0 = 0$.
  (c) Find the balance after 5 years.

34. (Continuation of Problem 33.) Now suppose the money is deposited once a month (instead of continuously) but still at a rate of $1200 per year.

  (a) Write down the sum that gives the balance after 5 years, assuming the first deposit is made one month from today, and today is $t = 0$.
  (b) The sum you wrote in part (a) is a Riemann sum approximation to the integral

$$\int_0^5 1200e^{0.1t}\,dt.$$

Determine whether it is a left sum or right sum, and determine what Δt and N are. Then use your calculator to evaluate the sum.

  (c) Compare your answer in part (b) to your answer to Problem 33(c).

35. The spread of a non-fatal disease through a population of fixed size M can be modeled as follows. The rate that healthy people are infected, in people per day, is proportional to the product of the numbers of healthy and infected people. The constant of proportionality is $0.01/M$. The rate of recovery, in people per day, is 0.009 times the number of people infected. Construct a differential equation which models the spread of the disease. Assuming that initially only a small number of people are infected, plot a graph of the number of infected people against time. What fraction of the population is infected in the long run?

36. A model for the population, $P(t)$, of carp in a landlocked lake is given by the differential equation

$$\frac{dP}{dt} = 0.25P(1 - 0.0004P).$$

  (a) What is the long-term equilibrium population of carp in the lake?
  (b) Ten years ago a census was taken and there were found to be 1000 carp in the lake. Estimate the current size of the population.
  (c) It is planned to join the lake to a nearby river so that the fish will be able to leave the lake. It is estimated that there will be a net loss of 10% of the carp each year but that the patterns of birth and death are not expected to change. Revise the differential equation to take this into account. Use the revised differential equation to predict the future development of the carp population.

37. Juliet is in love with Romeo, who happens (in our version of this story) to be a fickle lover. The more Juliet loves him, the more he begins to dislike her. When she hates him, his feelings for her warm up. On the other hand, her love for him grows when he loves her and withers when he hates her. A model for their ill-fated romance is

$$\frac{dj}{dt} = Ar, \qquad \frac{dr}{dt} = -Bj,$$

where A and B are positive constants, $r(t)$ represents Romeo's love for Juliet at time t, and $j(t)$ represents Juliet's love for Romeo at time t. (Negative love is hate.)

(a) The constant on the right-hand side of Juliet's equation (the one including dj/dt) has a positive sign, whereas the constant in Romeo's equation is negative. Explain why these signs follow from the story.

(b) Derive a second-order differential equation for $r(t)$ and solve it. (Your equation should involve r and its derivatives, but not j and its derivatives.)

(c) Express $r(t)$ and $j(t)$ as functions of t, given $r(0) = 1$ and $j(0) = 0$. Your answer will contain A and B.

(d) As you may have discovered, the outcome of the relationship is a never-ending cycle of love and hate. Find what fraction of the time they both love one another.

38. The behavior of an electron is described by Schrödinger's equation. Under certain conditions, this equation becomes

$$\frac{-h^2}{8\pi^2 m} \frac{d^2\Psi}{dx^2} = E\Psi,$$

where Ψ, called a *wave function*, describes the electron's behavior as a function of x, the distance from a fixed point; the constants h, m, E are positive. In addition, Ψ satisfies the boundary conditions

$$\Psi(0) = \Psi(l) = 0$$

where l is another positive constant.

(a) Solutions to Schrödinger's equation are of the form $\Psi = C_1 \cos(\omega x) + C_2 \sin(\omega x)$ for some ω. Show that if Ψ satisfies the boundary conditions, then Ψ must be of the form

$$\Psi = C_2 \sin\left(\frac{n\pi x}{l}\right) \quad \text{for some positive integer } n.$$

(b) Using part (a), find E in terms of other constants.

(c) The constant E represents the energy level of the electron. Assuming that only positive integral values of n are possible, explain why not all values of E are possible. (This is described by saying that energy is quantized.) In addition, show that the possible energy levels are one of the following multiples of the lowest energy level: $1, 4, 9, 16 \ldots$.

PROJECTS

1. Pareto's Law

In analyzing a society, sociologists are often interested in how incomes are distributed through the society. Pareto's law asserts that each society has a constant k such that the average income of all people wealthier than you is k times your income ($k > 1$). If $p(x)$ is the number of people in the society with an income of x or above, we define $\Delta p = p(x + \Delta x) - p(x)$, for small Δx.

(a) Explain why the number of people with incomes between x and $x + \Delta x$ is represented by $-\Delta p$. Then show that the total amount of money earned by people with incomes between x and $x + \Delta x$ is approximated by $-x\Delta p$.

(b) Use Pareto's law to show that the total amount of money earned by people with incomes of x or above is $kxp(x)$. Then show that the total amount of money earned by people with incomes between x and $x + \Delta x$ is approximated by $-kp\Delta x - kx\Delta p$.

(c) Using your answers to parts (a) and (b), show that p satisfies the differential equation

$$(1 - k)xp' = kp.$$

(d) Solve this differential equation for $p(x)$.

(e) Sketch a graph of $p(x)$ for various values of k, some large and some near 1. How does the value of k alter the shape of the graph?

2. **Vibrations in a Molecule**

    Suppose the force, F, between two atoms a distance r apart in a molecule is given by

    $$F(r) = b\left(\frac{a^2}{r^3} - \frac{a}{r^2}\right)$$

    where a and b are positive constants.

    (a) Find the equilibrium distance, r, at which $F = 0$.
    (b) Expand F in a series about the equilibrium position. Give the first two nonzero terms.
    (c) Suppose one atom is fixed. Let x be the displacement of the other atom from the equilibrium position. Give the first two nonzero terms of the series for F in terms of x.
    (d) Using Newton's Second Law, show that if x is small, the second atom oscillates about the equilibrium position. Find the period of the oscillation.

3. **Harvesting and Logistic Growth**

    In this project, we look at the effects of *harvesting* a population which is growing logistically. Harvesting could be, for example, fishing or logging. An important question is what level of harvesting leads to a *sustainable yield*. In other words, how much can be harvested without having the population depleted in the long run?

    (a) When there is no fishing, suppose a population of fish is governed by the differential equation

    $$\frac{dN}{dt} = 2N - 0.01N^2,$$

    where N is the number of fish at time t in years. Sketch a graph of dN/dt against N. Mark on your graph the equilibrium values of N. Now sketch N against t for various different initial values.

    (b) Suppose now that fish are removed by fishermen at a continuous rate of 75 fish/year. Let P be the number of fish at time t with harvesting. Explain why P satisfies the differential equation

    $$\frac{dP}{dt} = 2P - 0.01P^2 - 75.$$

    (c) Sketch dP/dt against P. Find and label the intercepts.
    (d) Sketch the slope field for the differential equation for P.
    (e) Use the graph you drew in part (c) to sketch graphs of P against t, with the initial values:

        (i) $P(0) = 40$ (ii) $P(0) = 60$ (iii) $P(0) = 160$ (iv) $P(0) = 140$

    (f) Use the slope field in part (d) to sketch graphs of P against t, with the initial values:

        (i) $P(0) = 40$ (ii) $P(0) = 50$ (iii) $P(0) = 60$
        (iv) $P(0) = 150$ (v) $P(0) = 170$

    (g) Using the graphs you drew, decide what the equilibrium values of the populations are and whether or not they are stable.
    (h) We now look at the effect of different levels of fishing on a fish population. If fishing takes place at a continuous rate of H fish/year, the fish population P satisfies the differential equation

    $$\frac{dP}{dt} = 2P - 0.01P^2 - H.$$

        (i)   For each of the values $H = 75, 100, 200$, plot a graph of dP/dt against P.
        (ii)   For which of the three values of H that you considered in part (i) is there an initial condition such that the fish population does not die out eventually?
        (iii)   Looking at your answer to part (ii), decide for what values of H there is an initial value for P such that the population does not die out eventually.
        (iv)   Recommend a policy to ensure long-term survival of the fish population.

CHAPTER ELEVEN

FUNCTIONS OF SEVERAL VARIABLES

Many quantities depend on more than one variable: the amount of food grown depends on the amount of rain and the amount of fertilizer used; the rate of a chemical reaction depends on the temperature and the pressure of the environment in which it proceeds; the strength of the gravitational attraction between two bodies depends on their masses and their distance apart; and the rate of fallout from a volcanic explosion depends on the distance from the volcano and the time since the explosion. Each example involves a function of two or more variables. In this chapter, we will see many different ways of looking at functions of several variables.

11.1 FUNCTIONS OF TWO VARIABLES

Function Notation

Suppose you are planning to take out a five-year loan to buy a car and you need to calculate what your monthly payment will be; this depends on both the amount of money you borrow and the interest rate. These quantities can vary separately: the loan amount can change while the interest rate remains the same, or the interest rate can change while the loan amount remains the same. To calculate your monthly payment you need to know both. If the monthly payment is m, the loan amount is L, and the interest rate is r%, then we express the fact that m is a function of L and r by writing:

$$m = f(L, r).$$

This is just like the function notation of one-variable calculus. The variable m is called the dependent variable, and the variables L and r are called the independent variables. The letter f stands for the *function* or rule that gives the value of m corresponding to given values of L and r.

A function of two variables can be represented graphically, numerically by a table of values, or algebraically by a formula. In this section we will give examples of each of these three ways of viewing a function.

Graphical Example: A Weather Map

Figure 11.1 shows a weather map from a newspaper. What information does it convey? It displays the predicted high temperature, T, in degrees Fahrenheit (°F), throughout the US on that day. The curves on the map, called *isotherms*, separate the country into zones, according to whether T is in the 60s, 70s, 80s, 90s, or 100s. (*Iso* means same and *therm* means heat.) Notice that the isotherm separating the 80s and 90s zones connects all the points where the temperature is exactly 90°F.

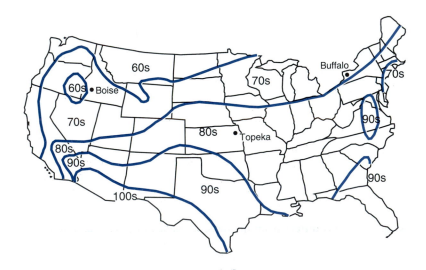

Figure 11.1: Weather map showing predicted high temperatures, T, on a summer day

---

Example 1   Estimate the predicted value of T in Boise, Idaho; Topeka, Kansas; and Buffalo, New York.

Solution   Boise and Buffalo are in the 70s region, and Topeka is in the 80s region. Thus, the predicted temperature in Boise and Buffalo is between 70 and 80 while the predicted temperature in Topeka is between 80 and 90.

In fact, we can say more. Although both Boise and Buffalo are in the 70s, Boise is quite close to the $T = 70$ isotherm, whereas Buffalo is quite close to the $T = 80$ isotherm. So we estimate that the temperature will be in the low 70s in Boise and in the high 70s in Buffalo. Topeka is about halfway between the $T = 80$ isotherm and the $T = 90$ isotherm. Thus, we guess that the temperature in Topeka will be in the mid 80s. In fact, the actual high temperatures for that day were 71°F for Boise, 79°F for Buffalo, and 86°F for Topeka.

The predicted high temperature, T, illustrated by the weather map is a function of (that is, depends on) two variables, often longitude and latitude, or miles east-west and miles north-south of a fixed point, say, Topeka. The weather map in Figure 11.1 is called a *contour map* or *contour diagram* of that function. Section 11.3 shows another way of visualizing functions of two variables using surfaces; Section 11.4 looks at contour maps in detail.

Numerical Example: Beef Consumption

Suppose you are a beef producer and you want to know how much beef people will buy. This depends on how much money people have and on the price of beef. The consumption of beef, C (in pounds per week per household) is a function of household income, I (in thousands of dollars per year), and the price of beef, p (in dollars per pound). In function notation, we write:

$$C = f(I, p).$$

Table 11.1 contains values of this function. Values of p are shown across the top, values of I are down the left side, and corresponding values of $f(I, p)$ are given in the table.[1] For example, to find the value of $f(40, 3.50)$, we look in the row corresponding to $I = 40$ under $p = 3.50$, where we find the number 4.05. Thus,

$$f(40, 3.50) = 4.05.$$

This means that, on average, if a household's income is $40,000 a year and the price of beef is $3.50/lb, the family will buy 4.05 lbs of beef per week.

TABLE 11.1   *Quanity of beef bought (pounds/household/week)*

		Price of beef, ($/lb)			
		3.00	3.50	4.00	4.50
Household	20	2.65	2.59	2.51	2.43
income	40	4.14	4.05	3.94	3.88
per year,	60	5.11	5.00	4.97	4.84
I	80	5.35	5.29	5.19	5.07
(1000)	100	5.79	5.77	5.60	5.53

Notice how this differs from the table of values of a one-variable function, where one row or one column is enough to list the values of the function. Here many rows and columns are needed because the function has a value for every *pair* of values of the independent variables.

Algebraic Examples: Formulas

In both the weather map and beef consumption examples, there is no formula for the underlying function. That is usually the case for functions representing real-life data. On the other hand, for many idealized models in physics, engineering, or economics, there are exact formulas.

---

[1] Adapted from Richard G. Lipsey, *An Introduction to Positive Economics*, 3rd ed., (London: Weidenfeld and Nicolson, 1971).

Example 2   Give a formula for the function $M = f(B, t)$ where M is the amount of money in a bank account t years after an initial investment of B dollars, if interest is accrued at a rate of 5% per year compounded (a)   annually   (b)   continuously.

Solution   (a)   Annual compounding means that M increases by a factor of 1.05 every year, so

$$M = f(B, t) = B(1.05)^t.$$

(b)   Continuous compounding means that M grows according to the exponential function e^{kt}, with $k = 0.05$, so

$$M = f(B, t) = Be^{0.05t}.$$

Example 3   A cylinder with closed ends has a radius r and a height h. If its volume is V and its surface area is A, find formulas for the functions $V = f(r, h)$ and $A = g(r, h)$.

Solution   Since the area of the circular base is πr^2, we have

$$V = f(r, h) = \text{Area of base} \cdot \text{Height} = \pi r^2 h.$$

The surface area of the side is the circumference of the bottom, $2\pi r$, times the height h, giving $2\pi rh$. Thus,

$$A = g(r, h) = 2 \cdot \text{Area of base} + \text{Area of side} = 2\pi r^2 + 2\pi rh.$$

Strategy to Investigate Functions of Two Variables: Vary One Variable at a Time

We can learn a great deal about functions of two or more variables by letting one variable vary at a time while holding the others fixed, thus obtaining a function of one variable.

The Wave

Suppose you are in a stadium where the audience is doing the wave. This is a ritual in which members of the audience stand up and sit down in such a way as to create a wave that moves around the stadium. Normally a single wave travels all the way around the stadium, but we assume there is a continuous sequence of waves. What sort of function will describe the motion of the audience? To keep things simple, we consider just one row of spectators. We consider the function which describes the motion of each individual in the row. This is a function of two variables: x (the seat number) and t (the time in seconds). For each value of x and t, we write $h(x, t)$ for the height (in feet) above the ground of the head of the spectator in seat number x at time t seconds. Suppose we are told that

$$h(x, t) = 5 + \cos(0.5x - t).$$

Example 4   (a)   Explain the significance of $h(x, 5)$ in terms of the wave. Find the period of $h(x, 5)$. What does this period represent?

(b)   Explain the significance of $h(2, t)$ in terms of the wave. Find the period of $h(2, t)$. What does this period represent?

Solution   (a)   Fixing $t = 5$ means we are taking a particular moment in time; letting x vary means we are looking along the whole row at that instant. Thus, the function $h(x, 5) = 5 + \cos(0.5x - 5)$ gives the heights along the row at the instant $t = 5$. Figure 11.2 shows the graph of $h(x, 5)$ which is a snapshot of the row at $t = 5$. The heights form a wave of period 4π, or about 12.6 seats. This period tells us that the length of the wave is about 13 seats.

(b)   Fixing $x = 2$ means we are concentrating on the spectator in seat number 2; letting t vary means we are watching the motion of that spectator as time passes. Thus, the function $h(2, t)$ describes the motion of the spectator in seat 2 as a function of time. Figure 11.3 shows the graph of $h(2, t) = 5 + \cos(1 - t)$. Notice that the value of h varies between 4 feet and 6 feet as the spectator sits down and stands up. The period is 2π, or about 6.3 seconds. This period represents the time it takes for the spectator to stand up and sit down once.

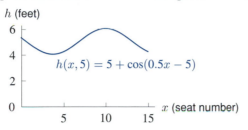

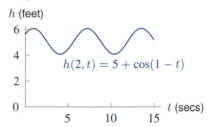

Figure 11.2: The function $h(x, 5)$ shows the shape of the wave at time $t = 5$

Figure 11.3: The function $h(2, t)$ shows the motion of the spectator in seat number 2

In general, the one-variable function $h(a, t)$ gives the motion of the spectator in seat a. Figure 11.3 shows the motion of the person in seat number 2. If we pick someone in another seat, we get a similar function, except that the graph may be shifted to the right or to the left.

Example 5   Show that the graph of $h(7, t)$ has the same shape as the graph of $h(2, t)$.

Solution   The motion of the person in seat 7 is described by

$$h(7, t) = 5 + \cos(0.5(7) - t) = 5 + \cos(3.5 - t).$$

Since $h(2, t) = 5 + \cos(1 - t)$, we can rewrite $h(7, t)$ as

$$h(7, t) = 5 + \cos(1 + 2.5 - t) = 5 + \cos(1 - (t - 2.5)) = h(2, t - 2.5).$$

We see that $h(7, t)$ is the same function as $h(2, t)$, except that the t has been replaced by $t - 2.5$. Thus, the graph of $h(7, t)$ is the graph of $h(2, t)$ shifted 2.5 seconds to the right, which means 2.5 seconds later. This means the spectator in seat 7 stands up 2.5 seconds later than the person in seat 2. (See Figure 11.4.) This lag is what makes the wave travel around the stadium. If all the spectators stood up and sat down at the same time, there would be no wave.

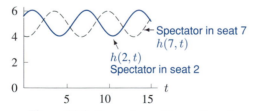

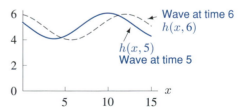

Figure 11.4: Comparison of the motion of spectators in seats 2 and 7

Figure 11.5: The shape of the wave at $t = 5$ and $t = 6$

Example 6   Use the result of Example 5 to find the speed of the wave.

Solution   Since the spectator in seat 7 does the same thing as the spectator in seat 2 but 2.5 seconds later, the wave has moved 5 seats in 2.5 seconds. Thus, the speed is $5/2.5 = 2$ seats per second.

Example 7   Use the functions $h(x, 5)$ and $h(x, 6)$ to show that the speed of the wave is 2 seats per second.

Solution   Figure 11.5 shows graphs of $h(x, 5) = 5 + \cos(0.5x - 5)$ and $h(x, 6) = 5 + \cos(0.5x - 6)$, that is, snapshots of the wave at $t = 5$ and $t = 6$. The shape of the wave at $t = 6$ is the same as the shape at $t = 5$, only shifted to the right by about two seats. Thus, the wave is moving at a rate of about 2 seats per second. To confirm that the speed is exactly 2 seats per second, we must use algebra. When $t = 5$, the equation of the wave is

$$h(x, 5) = 5 + \cos{(0.5x - 5)},$$

which has a peak where

$$0.5x - 5 = 0,$$

so, at

$$x = 10,$$

that is, in the 10th seat. One second later, at $t = 6$, the equation of the wave is

$$h = 5 + \cos(0.5x - 6),$$

which has a peak where

$$0.5x - 6 = 0,$$

so, at

$$x = 12,$$

that is, in the 12th seat. Thus, the wave moved 2 seats in one second.

The Beef Data

For a function given by a table of values, such as the beef consumption data, we allow one variable to vary at a time by looking at one row or one column. For example, to fix the income at 40, we look at the row $I = 40$. This row gives values of the function $f(40, p)$. Since I is fixed, we now have a function of one variable that shows how much beef is bought at various prices by people who earn $40,000 a year. Table 11.2 shows that $f(40, p)$ decreases as p increases. The other rows tell the same story; for each income, I, the consumption of beef goes down as the price, p, increases.

TABLE 11.2   *Beef consumption by households making \$40,000*

p	3.00	3.50	4.00	4.50
$f(40, p)$	4.14	4.05	3.94	3.88

Weather Map

What happens on the weather map in Figure 11.1 when we allow only one variable at a time to vary? Suppose x represents miles west-east of Topeka and y represents miles north-south, and suppose we move along the west-east line through Topeka. We keep y fixed at 0 and let x vary. Along this line, the high temperature T goes from the 60s along the west coast, to the 70s in Nevada and Utah, to the 80s in Topeka, to the 90s just before the east coast, then returns to the 80s. A possible graph is shown in Figure 11.6. Other graphs are possible because we don't know for sure how the temperature varies between contours.

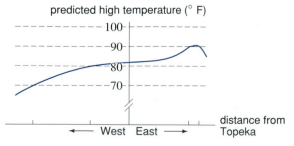

Figure 11.6: Predicted high temperature on an east-west line through Topeka

Problems for Section 11.1

Problems 1–3 refer to the weather map in Figure 11.1 on page 564.

1. Give the range of daily high temperatures for
    (a) Pennsylvania   (b) North Dakota   (c) California.

2. Sketch the graph of the predicted high temperature T on a line north-south through Topeka.

3. Sketch the graphs of the predicted high temperature on a north-south line and an east-west line through Boise.

For Problems 4–8 refer to Table 11.1 on page 565, where p is the price of beef and I is annual household income.

4. Make a table showing the amount of money, M, that the average household spends on beef (in dollars per household per week) as a function of the price of beef and household income.

5. Give tables for beef consumption as a function of p, with I fixed at $I = 20$ and $I = 100$. Give tables for beef consumption as a function of I, with p fixed at $p = 3.00$ and $p = 4.00$. Comment on what you see in the tables.

6. How does beef consumption vary as a function of household income if the price of beef is held constant?

7. Make a table of the proportion, P, of household income spent on beef per week as a function of price and income. (Note that P is the fraction of income spent on beef.)

8. Express P, the proportion of household income spent on beef per week, in terms of the original function $f(I, p)$ which gave consumption as a function of p and I.

9. Sketch the graph of the bank account function f in Example 2(a) on page 566, holding B fixed at three different values and letting only t vary. Then sketch the graph of f, holding t fixed at three different values and letting only B vary. Explain what you see.

10. You are planning a long driving trip and your principal cost will be gasoline.

     (a) Make a table showing how the daily fuel cost varies as a function of the price of gasoline (in dollars per gallon) and the number of gallons you buy each day.
     (b) If your car goes 30 miles on each gallon of gasoline, make a table showing how your daily fuel cost varies as a function of your daily travel distance and the price of gas.

11. Consider the acceleration due to gravity, g, at a height h above the surface of a planet of mass m.

     (a) If m is held constant, is g an increasing or decreasing function of h? Why?
     (b) If h is held constant, is g an increasing or decreasing function of m? Why?

12. The *temperature adjusted for wind-chill* is a temperature which tells you how cold it feels, as a result of the combination of wind and temperature. Table 11.3 shows the temperature adjusted for wind-chill as a function of wind speed and temperature.

     TABLE 11.3   *Temperature adjusted for wind-chill (°F)*

		Temperature (°F)							
		35	30	25	20	15	10	5	0
	5	33	27	21	16	12	7	0	−5
Wind	10	22	16	10	3	−3	−9	−15	−22
Speed	15	16	9	2	−5	−11	−18	−25	−31
(mph)	20	12	4	−3	−10	−17	−24	−31	−39
	25	8	1	−7	−15	−22	−29	−36	−44

     (a) If the temperature is 0°F and the wind speed is 15 mph, how cold does it feel?
     (b) If the temperature is 35°F, what wind speed makes it feel like 22°F?
     (c) If the temperature is 25°F, what wind speed makes it feel like 20°F?
     (d) If the wind is blowing at 15 mph, what temperature feels like 0°F?

13. Using Table 11.3, make tables of the temperature adjusted for wind-chill as a function of wind speed for temperatures of 20°F and 0°F.

14. Using Table 11.3, make tables of the temperature adjusted for wind-chill as a function of temperature for wind speeds of 5 mph and 20 mph.

15. Suppose the function for the stadium wave on page 566 was given by $h(x, t) = 5 + \cos(x - 2t)$. How does this wave compare with the original wave? What is the speed of this wave (in seats per second)?

16. Suppose the stadium wave on page 566 was moving in the opposite direction, right-to-left instead of left-to-right. Give a possible formula for h.

Problems 17–20 concern a vibrating guitar string. Suppose you pluck a guitar string and watch it vibrate. Snapshots of the guitar string at millisecond intervals are shown in Figure 11.7.

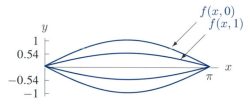

Figure 11.7: A vibrating guitar string:
$f(x, t) = \cos t \sin x$ for four t values.

Think of the guitar string stretched tight along the x-axis from $x = 0$ to $x = \pi$. Each point on the string has an x-value, $0 \le x \le \pi$. As the string vibrates, each point on the string moves back and forth on either side of the x-axis. Let $y = f(x, t)$ be the displacement at time t of the point on the string located x units from the left end. Then a possible formula for $y = f(x, t)$ is

$$y = f(x, t) = \cos t \sin x, \quad 0 \le x \le \pi, \quad t \text{ in milliseconds.}$$

17. (a) Sketch graphs of y versus x for fixed t values, $t = 0, \pi/4, \pi/2, 3\pi/4, \pi$.
     (b) Use your graphs to explain why this function could represent a vibrating guitar string.

18. Explain what the functions $f(x, 0)$ and $f(x, 1)$ represent in terms of the vibrating string.

19. Explain what the functions $f(0, t)$ and $f(1, t)$ represent in terms of the vibrating string.

20. Describe the motion of the guitar strings whose displacements are given by the following:
     (a) $y = g(x, t) = \cos 2t \sin x$ (b) $y = h(x, t) = \cos t \sin 2x$

11.2 A TOUR OF THREE-DIMENSIONAL SPACE

Cartesian Coordinates in Three-Space

Imagine three coordinate axes meeting at the *origin*: a vertical axis, and two horizontal axes at right angles to each other. (See Figure 11.8.) Think of the xy-plane as being horizontal, while the z-axis extends vertically above and below the plane. The labels x, y, and z show which part of each axis is positive; the other side is negative. We generally use *right-handed axes* in which looking down the positive z-axis gives the usual view of the xy-plane. We specify a point in 3-space by giving its coordinates (x, y, z) with respect to these axes. Think of the coordinates as instructions telling you how to get to the point; start at the origin, go x units along the x-axis, then y units in the direction parallel to the y-axis and finally z units in the direction parallel to the z-axis. The coordinates can be positive, zero or negative; a zero coordinate means "don't move in this direction," and a negative coordinate means "go in the negative direction parallel to this axis." For example, the origin has coordinates $(0, 0, 0)$, since we get there from the origin by doing nothing at all.

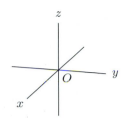

Figure 11.8: Coordinate axes in three-dimensional space

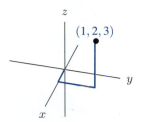

Figure 11.9: The point $(1, 2, 3)$ in 3-space

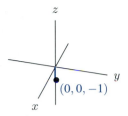

Figure 11.10: The point $(0, 0, -1)$ in 3-space

Example 1   Describe the position of the points with coordinates $(1, 2, 3)$ and $(0, 0, -1)$.

Solution   We get to the point $(1, 2, 3)$ by starting at the origin, going 1 unit along the x-axis, 2 units in the direction parallel to the y-axis, and 3 units up in the direction parallel to the z-axis. (See Figure 11.9.)

To get to $(0, 0, -1)$, we don't move at all in the x and y directions, but move 1 unit in the negative z direction. So the point is on the negative z-axis. (See Figure 11.10.) You can check that the position of the point is independent of the order of the x, y, and z displacements.

Example 2   You start at the origin, go along the y-axis a distance of 2 units in the positive direction, and then move vertically upward a distance of 1 unit. What are the coordinates of your final position?

Solution   You started at the point $(0, 0, 0)$. When you went along the y-axis, your y-coordinate increased to 2. Moving vertically increased your z-coordinate to 1; your x-coordinate didn't change because you did not move in the x direction. So your final coordinates are $(0, 2, 1)$. (See Figure 11.11.)

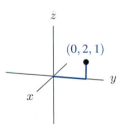

Figure 11.11: The point $(0, 2, 1)$ is reached by moving 2 along the y-axis and 1 upward

It is often helpful to picture a three dimensional coordinate system in terms of a room. The origin is a corner at floor level where two walls meet the floor. The z-axis is the vertical intersection of the two walls; the x- and y-axes are the intersections of each wall with the floor. Points with negative coordinates lie behind a wall in the next room or below the floor.

Graphing Equations in Three-Dimensional Space

We can graph equations involving the variables x, y, and z in three-dimensional space.

Example 3   What do the graphs of the equations $z = 0$, $z = 3$, and $z = -1$ look like?

Solution   Graphing an equation means drawing the set of all points in space whose coordinates satisfy the equation. So to graph $z = 0$ we need to visualize the set of points whose z-coordinate is zero. If the z-coordinate is 0, then we must be at the same vertical level as the origin, that is, we are in the

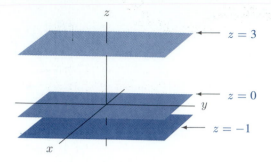

Figure 11.12: The planes $z = -1$, $z = 0$, and $z = 3$

horizontal plane containing the origin. So the graph of $z = 0$ is the middle plane in Figure 11.12. The graph of $z = 3$ is a plane parallel to the graph of $z = 0$, but three units above it. The graph of $z = -1$ is a plane parallel to the graph of $z = 0$, but one unit below it.

The plane $z = 0$ contains the x- and y-coordinate axes, and hence is called the xy-coordinate plane, or xy-plane for short. There are two other coordinate planes. The yz-plane contains both the y- and the z-axes, and the xz-plane contains the x- and z-axes. (See Figure 11.13.)

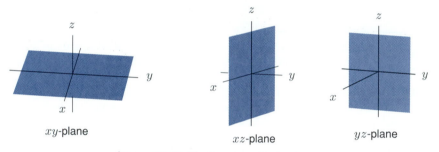

xy-plane xz-plane yz-plane

Figure 11.13: The three coordinate planes

Example 4   Which of the points $A = (1, -1, 0)$, $B = (0, 3, 4)$, $C = (2, 2, 1)$, and $D = (0, -4, 0)$ lies closest to the xz-plane? Which point lies on the y-axis?

Solution   The magnitude of the y-coordinate gives the distance to the xz-plane. The point A lies closest to that plane, because it has the smallest y-coordinate in magnitude. To get to a point on the y-axis, we move along the y-axis, but we don't move at all in the x or z directions. Thus, a point on the y-axis has both its x- and z-coordinates equal to zero. The only point of the four that satisfies this is D. (See Figure 11.14.)

In general, if a point has one of its coordinates equal to zero, it lies in one of the coordinate planes. If a point has two of its coordinates equal to zero, it lies on one of the coordinate axes.

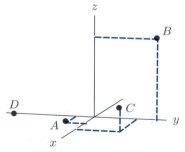

Figure 11.14: Which point lies closest to the xz-plane? Which point lies on the y-axis?

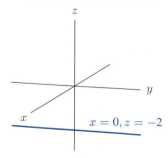

Figure 11.15: The line $x = 0$, $z = -2$

Example 5 You are 2 units below the xy-plane and in the yz-plane. What are your coordinates?

Solution Since you are 2 units below the xy-plane, your z-coordinate is -2. Since you are in the yz-plane, your x-coordinate is 0; your y-coordinate can be anything. Thus, you are at the point $(0, y, -2)$. The set of all such points forms a line parallel to the y-axis, 2 units below the xy-plane, and in the yz-plane. (See Figure 11.15.)

Example 6 You are standing at the point $(4, 5, 2)$, looking at the point $(0.5, 0, 3)$. Are you looking up or down?

Solution The point you are standing at has z-coordinate 2, whereas the point you are looking at has z-coordinate 3; hence you are looking up.

Example 7 Imagine that the yz-plane in Figure 11.15 is a page of this book. Describe the region behind the page.

Solution The positive part of the x-axis pokes out of the page; moving in the positive x direction brings you out in front of the page. The region behind the page corresponds to negative values of x, and so it is the set of all points in three-dimensional space satisfying the inequality $x < 0$.

Distance Between Two Points

In 2-space, the formula for the distance between two points (x, y) and (a, b) is given by

$$\text{Distance} = \sqrt{(x - a)^2 + (y - b)^2}.$$

The distance between two points (x, y, z) and (a, b, c) in 3-space is represented by PG in Figure 11.16. The side PE is parallel to the x-axis, EF is parallel to the y-axis, and FG is parallel to the z-axis.

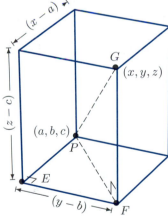

Figure 11.16: The diagonal PG gives the distance between the points (x, y, z) and (a, b, c)

Using Pythagoras' theorem twice gives

$$(PG)^2 = (PF)^2 + (FG)^2 = (PE)^2 + (EF)^2 + (FG)^2 = (x - a)^2 + (y - b)^2 + (z - c)^2.$$

Thus, a formula for the distance between the points (x, y, z) and (a, b, c) in 3-space is

$$\boxed{\text{Distance} = \sqrt{(x - a)^2 + (y - b)^2 + (z - c)^2}.}$$

Example 8   Find the distance between $(1, 2, 1)$ and $(-3, 1, 2)$.

Solution   The formula gives

$$\text{Distance} = \sqrt{(-3 - 1)^2 + (1 - 2)^2 + (2 - 1)^2} = \sqrt{18} = 4.24.$$

Example 9   Find an expression for the distance from the origin to the point (x, y, z).

Solution   The origin has coordinates $(0, 0, 0)$, so the distance from the origin to (x, y, z) is given by

$$\text{Distance} = \sqrt{(x - 0)^2 + (y - 0)^2 + (z - 0)^2} = \sqrt{x^2 + y^2 + z^2}.$$

Example 10   Find the equation for a sphere of radius 1 with center at the origin.

Solution   The sphere consists of all points (x, y, z) whose distance from the origin is 1, that is, which satisfy the equation

$$\sqrt{x^2 + y^2 + z^2} = 1.$$

This is an equation for the sphere. If we square both sides we get the equation in the form

$$x^2 + y^2 + z^2 = 1.$$

Note that this equation represents the *surface* of the sphere. The solid ball enclosed by the sphere is represented by the inequality $x^2 + y^2 + z^2 \leq 1$.

Problems for Section 11.2

1. Which of the points $A = (1.3, -2.7, 0)$, $B = (0.9, 0, 3.2)$, $C = (2.5, 0.1, -0.3)$ is closest to the yz-plane? Which one lies on the xz-plane? Which one is farthest from the xy-plane?

2. Which of the points $A = (23, 92, 48)$, $B = (-60, 0, 0)$, $C = (60, 1, -92)$ is closest to the yz-plane? Which one lies on the xz-plane? Which one is farthest from the xy-plane?

3. You are at the point $(-1, -3, -3)$, standing upright and facing the yz-plane. You walk 2 units forward, turn left, and walk for another 2 units. What is your final position? From the point of view of an observer looking at the coordinate system in Figure 11.8 on page 571, are you in front of or behind the yz-plane? Are you to the left or to the right of the xz-plane? Are you above or below the xy-plane?

4. You are at the point $(3, 1, 1)$ facing the yz-plane. Assume you are standing upright. You walk 2 units forward, turn left, and walk for another 2 units. What is your final position? From the point of view of an observer looking at the coordinate system in Figure 11.8 on page 571, are you in front of or behind the yz-plane? Are you to the left of or to the right of the xz-plane? Are you above or below the xy-plane?

Sketch graphs of the equations in Problems 5–7 in 3-space.

5. $x = -3$ 6. $y = 1$ 7. $z = 2$ and $y = 4$

8. Find a formula for the shortest distance between a point (a, b, c) and the y-axis.

9. Describe the set of points whose distance from the x-axis is 2.

10. Describe the set of points whose distance from the x-axis equals the distance from the yz-plane.

11. Which of the points $P = (1, 2, 1)$ and $Q = (2, 0, 0)$ is closest to the origin?

12. Which two of the three points $P_1 = (1, 2, 3)$, $P_2 = (3, 2, 1)$ and $P_3 = (1, 1, 0)$ are closest to each other?

13. A cube is located such that its top four corners have the coordinates $(-1, -2, 2), (-1, 3, 2), (4, -2, 2)$ and $(4, 3, 2)$. Give the coordinates of the center of the cube.

14. A rectangular solid lies with its length parallel to the y-axis, and its top and bottom faces parallel to the plane $z = 0$. If the center of the object is at $(1, 1, -2)$ and it has a length of 13, a height of 5 and a width of 6, give the coordinates of all eight corners and draw the figure labeling the eight corners.

15. Which of the points $P_1 = (-3, 2, 15)$, $P_2 = (0, -10, 0)$, $P_3 = (-6, 5, 3)$ and $P_4 = (-4, 2, 7)$ is closest to $P = (6, 0, 4)$?

16. On a set of x, y, and z axes oriented as in Figure 11.8 on page 571, draw a straight line through the origin, lying in the xz-plane and such that if you move along the line with your x-coordinate increasing, your z-coordinate is decreasing.

17. On a set of x, y and z axes oriented as in Figure 11.8 on page 571, draw a straight line through the origin, lying in the yz-plane and such that if you move along the line with your y-coordinate increasing, your z-coordinate is increasing.

18. Find the equation of the sphere of radius 5 centered at the origin.

19. Find the equation of the sphere of radius 5 centered at $(1, 2, 3)$.

20. Given the sphere
$$(x - 1)^2 + (y + 3)^2 + (z - 2)^2 = 4,$$

(a) Find the equations of the circles (if any) where the sphere intersects each coordinate plane.

(b) Find the points (if any) where the sphere intersects each coordinate axis.

11.3 GRAPHS OF FUNCTIONS OF TWO VARIABLES

How do You Visualize a Function of Two Variables?

The weather map on page 564 is one way of visualizing a function of two variables. In this section we see how to visualize a function of two variables in another way, using a surface in 3-space.

The Graph of a Function and How to Plot One

For a function of one variable, $y = f(x)$, the graph of f is the set of all points (x, y) in 2-space such that $y = f(x)$. In general, these points lie on a curve in the plane. When a computer or calculator graphs f, it approximates by plotting points in the xy-plane and joining consecutive points by line segments. The more points, the better the approximation.

Now consider a function of two variables.

> The **graph** of a function of two variables, f, is the set of all points (x, y, z) such that $z = f(x, y)$. In general, the graph of a function of two variables is a surface in 3-space.

Plotting the Graph of the Function $f(x, y) = x^2 + y^2$

To sketch the graph of f we connect points as for a function of one variable. We first make a table of values of f, such as in Table 11.4.

TABLE 11.4 *Table of values of $f(x, y) = x^2 + y^2$*

		y						
		-3	-2	-1	0	1	2	3
	-3	18	13	10	9	10	13	18
	-2	13	8	5	4	5	8	13
	-1	10	5	2	1	2	5	10
x	0	9	4	1	0	1	4	9
	1	10	5	2	1	2	5	10
	2	13	8	5	4	5	8	13
	3	18	13	10	9	10	13	18

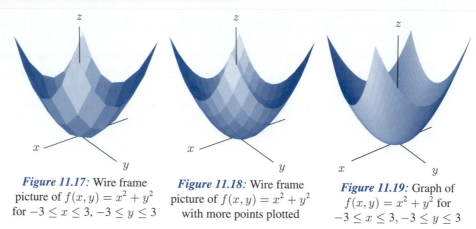

Figure 11.17: Wire frame picture of $f(x, y) = x^2 + y^2$ for $-3 \leq x \leq 3, -3 \leq y \leq 3$

Figure 11.18: Wire frame picture of $f(x, y) = x^2 + y^2$ with more points plotted

Figure 11.19: Graph of $f(x, y) = x^2 + y^2$ for $-3 \leq x \leq 3, -3 \leq y \leq 3$

Now we plot points. For example, we plot $(1, 2, 5)$ because $f(1, 2) = 5$ and we plot $(0, 2, 4)$ because $f(0, 2) = 4$. Then, we connect the points corresponding to the rows and columns in the table. The result is called a *wire-frame* picture of the graph. Filling in between the wires gives a surface. That is the way a computer drew the graphs in Figure 11.17 and 11.18. The more points that are plotted, the more the picture looks like the surface in Figure 11.19.

You should check to see if the sketches make sense. Notice that the graph goes through the origin since $(x, y, z) = (0, 0, 0)$ satisfies $z = x^2 + y^2$. Observe that if x is held fixed and y is allowed to vary, the graph dips down and then goes back up, just like the entries in the rows of Table 11.4. Similarly, if y is held fixed and x is allowed to vary, the graph dips down and then goes back up, just like the columns of Table 11.4.

New Graphs from Old

We can use the graph of a function to visualize the graphs of related functions.

Example 1   Let $f(x, y) = x^2 + y^2$. Describe in words the graphs of the following functions:
(a) $g(x, y) = x^2 + y^2 + 3$,   (b) $h(x, y) = 5 - x^2 - y^2$,   (c) $k(x, y) = x^2 + (y - 1)^2$.

Solution   We know from Figure 11.19 that the graph of f is a bowl with its vertex at the origin. From this we can work out what the graphs of g, h, and k will look like.
(a)   The function $g(x, y) = x^2 + y^2 + 3 = f(x, y) + 3$, so the graph of g is the graph of f, but raised by 3 units. See Figure 11.20.
(b)   Since $-x^2 - y^2$ is the negative of $x^2 + y^2$, the graph of $-x^2 - y^2$ is an upside down bowl. Thus, the graph of $h(x, y) = 5 - x^2 - y^2 = 5 - f(x, y)$ looks like an upside down bowl with vertex at $(0, 0, 5)$, as in Figure 11.21.
(c)   The graph of $k(x, y) = x^2 + (y - 1)^2 = f(x, y - 1)$ is a bowl with vertex at $x = 0$, $y = 1$, since that is where $k(x, y) = 0$, as in Figure 11.22.

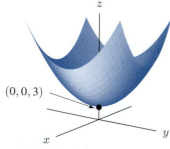

Figure 11.20: Graph of $g(x, y) = x^2 + y^2 + 3$

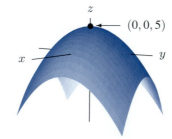

Figure 11.21: Graph of $h(x, y) = 5 - x^2 - y^2$

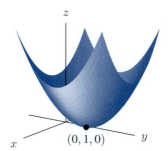

Figure 11.22: Graph of $k(x, y) = x^2 + (y - 1)^2$

Example 2   Describe the graph of $G(x, y) = e^{-(x^2+y^2)}$. What symmetry does it have?

Solution   Since the exponential function is always positive, the graph lies entirely above the xy-plane. From the graph of $x^2 + y^2$ we see that $x^2 + y^2$ is zero at the origin and gets larger as we move farther from the origin in any direction. Thus, $e^{-(x^2+y^2)}$ is 1 at the origin, and gets smaller as we move away from the origin in any direction. It can't go below the xy-plane; instead it flattens out, getting closer and closer to the plane. We say the surface is *asymptotic* to the xy-plane. (See Figure 11.23.)

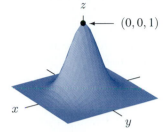

Figure 11.23: Graph of $G(x, y) = e^{-(x^2+y^2)}$

Now consider a point (x, y) on the circle $x^2 + y^2 = r^2$. Since

$$G(x, y) = e^{-(x^2+y^2)} = e^{-r^2},$$

the value of the function G is the same at all points on this circle. Thus, we say the graph of G has *circular symmetry*.

Cross-Sections and the Graph of a Function

We have seen that a good way to analyze a function of two variables is to let one variable vary at a time while the other is kept fixed.

> For a function $f(x, y)$, the function we get by holding x fixed and letting y vary is called a **cross-section** of f with x fixed. The graph of the cross-section of $f(x, y)$ with $x = c$ is the curve, or cross-section, we get by intersecting the graph of f with the plane $x = c$. We define a cross-section of f with y fixed similarly.

For example, the cross-section of $f(x, y) = x^2 + y^2$ with $x = 2$ is $f(2, y) = 4 + y^2$. The graph of this cross-section is the curve we get by intersecting the graph of f with the plane perpendicular to the x-axis at $x = 2$. (See Figure 11.24.)

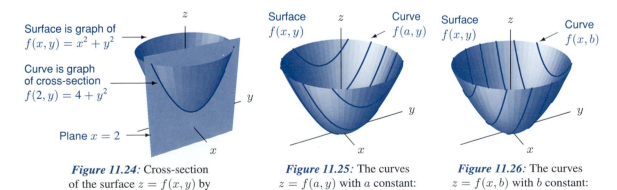

Figure 11.24: Cross-section of the surface $z = f(x, y)$ by the plane $x = 2$

Figure 11.25: The curves $z = f(a, y)$ with a constant: cross-sections with x fixed

Figure 11.26: The curves $z = f(x, b)$ with b constant: cross-sections with y fixed

Figure 11.25 shows graphs of other cross-sections of f with x fixed; Figure 11.26 shows graphs of cross-sections with y fixed.

Example 3   Describe the cross-sections of the function $g(x, y) = x^2 - y^2$ with y fixed and then with x fixed. Use these cross-sections to describe the shape of the graph of g.

Solution   The cross-sections with y fixed at $y = b$ are given by

$$z = g(x, b) = x^2 - b^2.$$

Thus, each cross-section with y fixed gives a parabola opening upwards, with minimum $z = -b^2$. The cross-sections with x fixed are of the form

$$z = g(a, y) = a^2 - y^2,$$

which are parabolas opening downwards with a maximum of $z = a^2$. (See Figures 11.27 and 11.28.) The graph of g is shown in Figure 11.29. Notice the upward opening parabolas in the x-direction and the downward opening parabolas in the y-direction. We say that the surface is *saddle-shaped*.

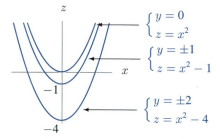

Figure 11.27: Cross-sections of $g(x, y) = x^2 - y^2$ with y fixed

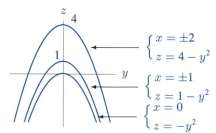

Figure 11.28: Cross-sections of $g(x, y) = x^2 - y^2$ with x fixed

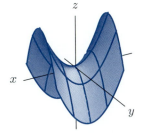

Figure 11.29: Graph of $g(x, y) = x^2 - y^2$ showing cross sections

Linear Functions

Linear functions are central to single variable calculus; they are equally important in multivariable calculus. You may be able to guess the shape of the graph of a linear function of two variables. (It's a plane.) Let's look at an example.

Example 4   Describe the graph of $f(x, y) = 1 + x - y$.

Solution   The plane $x = a$ is vertical and parallel to the yz-plane. Thus, the cross-section with $x = a$ is the line $z = 1 + a - y$ which slopes downward in the y-direction. Similarly, the plane $y = b$ is parallel to the xz-plane. Thus, the cross-section with $y = b$ is the line $z = 1 + x - b$ which slopes upward in the x-direction. Since all the cross-sections are lines, you might expect the graph to be a flat plane, sloping down in the y-direction and up in the x-direction. This is indeed the case. (See Figure 11.30.)

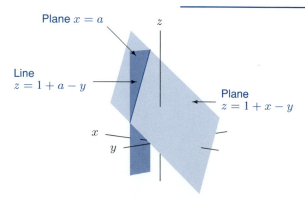

Figure 11.30: Graph of the plane $z = 1 + x - y$ showing cross-section with $x = a$

When One Variable is Missing: Cylinders

Suppose we graph an equation like $z = x^2$ which has one variable missing. What does the surface look like? Since y is missing from the equation, the cross-sections with y fixed are all the same parabola, $z = x^2$. Letting y vary up and down the y-axis, this parabola sweeps out the trough-shaped surface shown in Figure 11.31. The cross-sections with x fixed are horizontal lines, obtained by cutting the surface by a plane perpendicular to the x-axis.

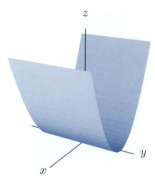

Figure 11.31: A parabolic cylinder $z = x^2$

Figure 11.32: Circular cylinder $x^2 + y^2 = 1$

This surface is called a *parabolic cylinder*, because it is formed from a parabola in the same way that an ordinary cylinder is formed from a circle; it has a parabolic cross-section instead of a circular one.

Example 5   Graph the equation $x^2 + y^2 = 1$ in 3-space.

Solution   Although the equation $x^2 + y^2 = 1$ does not represent a function, the surface representing it can be graphed by the method used for $z = x^2$. The graph of $x^2 + y^2 = 1$ in the xy-plane is a circle. Since z does not appear in the equation, the intersection of the surface with any horizontal plane will be the same circle $x^2 + y^2 = 1$. Thus, the surface is the cylinder shown in Figure 11.32.

Problems for Section 11.3

1.   The surface in Figure 11.33 is the graph of the function $z = f(x, y)$ for positive x and y.

   (a)   Suppose y is fixed and positive. Does z increase or decrease as x increases? Sketch a graph of z against x.

   (b)   Suppose x is fixed and positive. Does z increase or decrease as y increases? Sketch a graph of z against y.

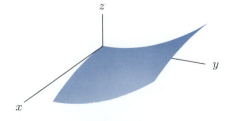

Figure 11.33

2.   Match the following descriptions of a company's success with the graphs in Figure 11.34.

   (a)   Our success is measured in dollars, plain and simple. More hard work won't hurt, but it also won't help.
   (b)   No matter how much money or hard work we put into the company, we just couldn't make a go of it.
   (c)   Although we aren't always totally successful, it seems that the amount of money invested doesn't matter. As long as we put hard work into the company our success will increase.
   (d)   The company's success is based on both hard work and investment.

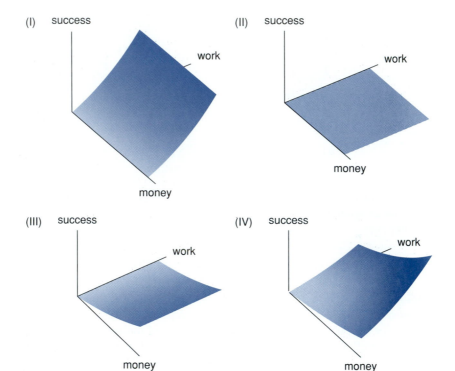

Figure 11.34

3.   For each of the following functions, decide whether it could be a bowl, a plate, or neither. Consider a plate to be any fairly flat surface and a bowl to be anything that could hold water, assuming the positive z-axis is up.
   (a) $z = x^2 + y^2$ (b) $z = 1 - x^2 - y^2$ (c) $x + y + z = 1$
   (d) $z = -\sqrt{5 - x^2 - y^2}$ (e) $z = 3$

4.   For each function in Problem 3 sketch cross-sections:

   (i)   With x fixed at $x = 0$ and $x = 1$.
   (ii)   With y fixed at $y = 0$ and $y = 1$.

5.   Match the following functions with their graphs in Figure 11.35.

   (a) $z = \dfrac{1}{x^2 + y^2}$ (b) $z = -e^{-x^2 - y^2}$ (c) $z = x + 2y + 3$

   (d) $z = -y^2$ (e) $z = x^3 - \sin y$.

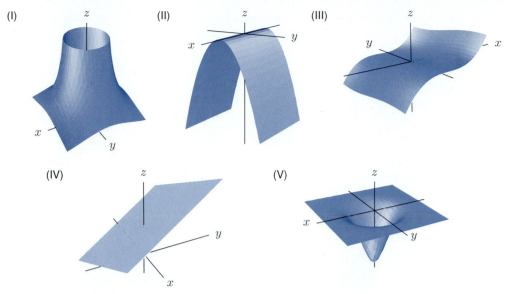

Figure 11.35

6. You like pizza and you like cola. Which of the graphs in Figure 11.36 represents your happiness as a function of how many pizzas and how much cola you have if

    (a)   There is no such thing as too many pizzas and too much cola?
    (b)   There is such a thing as too many pizzas or too much cola?
    (c)   There is such a thing as too much cola but no such thing as too many pizzas?

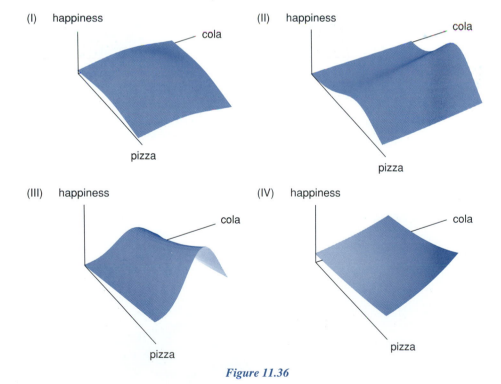

Figure 11.36

7. For each of the graphs I-IV in Problem 6 draw:

    (a)   Two cross-sections with pizza fixed   (b)   Two cross-sections with cola fixed.

8. Figure 11.37 contains graphs of the parabolas $z = f(x, b)$ for $b = -2, -1, 0, 1, 2$. Which of the graphs of $z = f(x, y)$ in Figure 11.38 best fits this information?

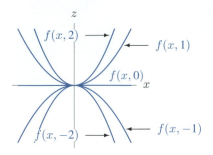

Figure 11.37

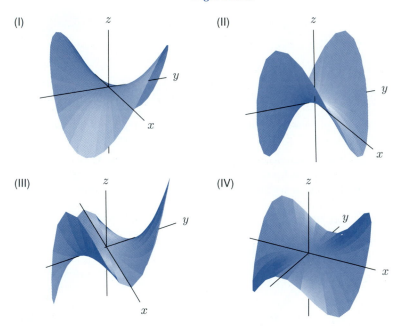

Figure 11.38

9. Imagine a single wave traveling along a canal. Suppose x is the distance along the canal from the middle, t is the time, and z is the height of the water above the equilibrium level. The graph of z as a function of x and t is shown in Figure 11.39.

    (a) Draw the profile of the wave for $t = -1, 0, 1, 2$. (Show the x-axis to the right and the z-axis vertically.)
    (b) Is the wave traveling in the direction of increasing or decreasing x?
    (c) Sketch a surface representing a wave traveling in the opposite direction.

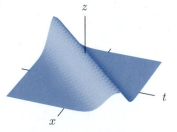

Figure 11.39

10. Describe in words the cross-sections with t fixed and the cross-sections with x fixed of the vibrating guitar string function

$$f(x, t) = \cos t \sin x, \quad 0 \le x \le \pi, \quad 0 \le t \le 2\pi$$

on page 570. Explain the relation of these cross-sections to the graph of f.

11. Use a computer or calculator to draw the graph of the vibrating guitar string function:

$$g(x, t) = \cos t \sin 2x, \quad 0 \le x \le \pi, \quad 0 \le t \le 2\pi.$$

Relate the shape of the graph to the cross-sections with t fixed and those with x fixed.

12. Use a computer or a calculator to draw the graph of the traveling wave function:

$$h(x, t) = 3 + \cos(x - 0.5t), \quad 0 \le x \le 2\pi, \quad 0 \le t \le 2\pi.$$

Relate the shape of the graph to the cross-sections with t fixed and those with x fixed.

13. Consider the function f given by $f(x, y) = y^3 + xy$. Draw graphs of cross-sections with:
    (a) x fixed at $x = -1$, $x = 0$, and $x = 1$.   (b) y fixed at $y = -1$, $y = 0$, and $y = 1$.

14. A swinging pendulum consists of a mass at the end of a string. At one moment the string makes an angle x with the vertical and the mass has speed y. At that time, the energy, E, of the pendulum is given by the expression[2]

$$E = 1 - \cos x + \frac{y^2}{2}.$$

   (a) Consider the surface representing the energy. Sketch a cross-section of the surface:
       (i) Perpendicular to the x-axis at $x = c$.
       (ii) Perpendicular to the y-axis at $y = c$.
   (b) For each of the graphs in Figures 11.40 and 11.41 use your answer to part (a) to decide which is the x-axis and which is the y-axis and to put reasonable units on each one.

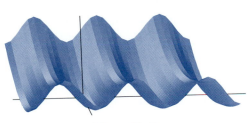

Figure 11.40

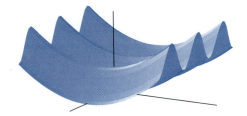

Figure 11.41

11.4 CONTOUR DIAGRAMS

The surface which represents a function of two variables often gives a good idea of the function's general behavior—for example, whether it is increasing or decreasing as one of the variables increases. However it is difficult to read numerical values off a surface and it can be hard to see all of the function's behavior from a surface. Thus, functions of two variables are often represented by contour diagrams like the weather map on page 564. Contour diagrams have the additional advantage that they can be extended to functions of three variables.

---

[2] Adapted from *Calculus in Context*, by James Callahan, Kenneth Hoffman, (New York: W.H. Freeman, 1995)

Topographical Maps

One of the most common examples of a contour diagram is a topographical map like that shown in Figure 11.42. It gives the elevation in the region and is a good way of getting an overall picture of the terrain: where the mountains are, where the flat areas are. Such topographical maps are frequently colored green at the lower elevations and brown, red, or white at the higher elevations.

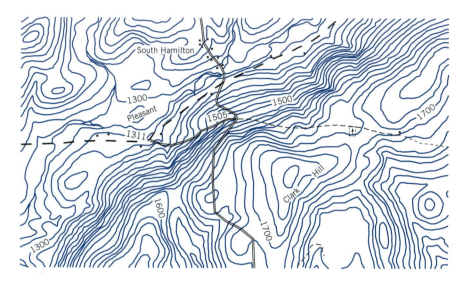

Figure 11.42: A topographical map showing the region around South Hamilton, NY

The curves on a topographical map that separate lower elevations from higher elevations are called *contour lines* because they outline the contour or shape of the land.[3] Because every point along the same contour has the same elevation, contour lines are also called *level curves* or *level sets*. The more closely spaced the contours, the steeper the terrain; the more widely spaced the contours, the flatter the terrain. (Provided, of course, that the elevation between contours varies by a constant amount.) Certain features have distinctive characteristics. A mountain peak is typically surrounded by contour lines like those in Figure 11.43. A pass in a range of mountains may have contours that look like Figure 11.44. A long valley has parallel contour lines indicating the rising elevations on both sides of the valley (see Figure 11.45); a long ridge of mountains has the same type of contour lines, only the elevations decrease on both sides of the ridge. Notice that the elevation numbers on the contour lines are as important as the curves themselves.

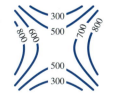

Figure 11.43: Mountain peak

Figure 11.44: Pass between two mountains

Figure 11.45: Long valley

Figure 11.46: Impossible contour lines

There are some things contour lines cannot do. Two contours corresponding to different elevations cannot cross each other as shown in Figure 11.46. If they did, the point of intersection of the two curves would have two different elevations, which is impossible (assuming the terrain has no overhangs). We will often follow the convention of drawing contours for equally spaced values of z.

---

[3]In fact they are usually not straight lines, but curves. They may also be in disconnected pieces.

Corn Production

Contour maps can also be useful to display information about a function of two variables without reference to a surface. Consider how to represent the effect of different weather conditions on US corn production. What would happen if the average temperature were to increase (due to global warming, for example)? What would happen if the rainfall were to decrease (due to a drought)? One way of estimating the effect of these climatic changes is to use Figure 11.47. This is a contour diagram giving the corn production $f(R, T)$ in the United States as a function of the total rainfall, R, in inches and average temperature, T, in degrees Fahrenheit, during the growing season.[4] Suppose at the present time, $R = 15$ inches and $T = 76°$F. Production is measured as a percentage of the present production; thus, the contour through $R = 15, T = 76$, has value 100, that is, $f(15, 76) = 100$.

Example 1   Use Figure 11.47 to estimate $f(18, 78)$ and $f(12, 76)$ and explain the answer in terms of corn production.

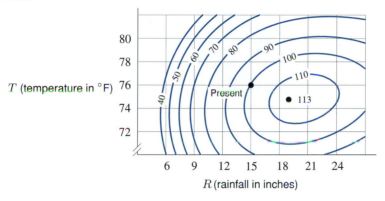

Figure 11.47: Corn production as a function of rainfall and temperature

Solution   The point with R-coordinate 18 and T-coordinate 78 is on the contour $C = 100$, so $f(18, 78) = 100$. This means that if the annual rainfall were 18 inches and the temperature were 78°F, the country would produce about the same amount of corn as at present, although it would be wetter and warmer than it is now.

The point with R-coordinate 12 and T-coordinate 76 is about halfway between the $C = 80$ and the $C = 90$ contours, so $f(12, 76) \approx 85$. This means that if the rainfall fell to 12 inches and the temperature stayed at 76°, then corn production would drop to about 85% of what it is now.

Example 2   Describe in words the cross-sections with T and R constant through the point representing present conditions. Give a common sense explanation of your answer.

Solution   To see what happens to corn production if the temperature stays fixed at 76°F but the rainfall changes, look along the horizontal line $T = 76$. Starting from the present and moving left along the line $T = 76$, the values on the contours decrease. In other words, if there is a drought, corn production decreases. Conversely, as rainfall increases, that is, as we move from the present to the right along the line $T = 76$, corn production increases, reaching a maximum of more than 110% when $R = 21$, and then decreases (too much rainfall floods the fields).

If, instead, rainfall remains at the present value and temperature increases, we move up the vertical line $R = 15$. Under these circumstances corn production decreases; a 2° increase causes a 10% drop in production. This makes sense since hotter temperatures lead to greater evaporation and hence drier conditions, even with rainfall constant at 15 inches. Similarly, a decrease in temperature leads to a very slight increase in production, reaching a maximum of around 102% when $T = 74$, followed by a decrease (the corn won't grow if it is too cold).

[4] Adapted from S. Beaty and R. Healy, "The Future of American Agriculture," *Scientific American* 248, No.2, February 1983.

Contour Diagrams and Graphs

Contour diagrams and graphs are two different ways of representing a function of two variables. How do we go from one to the other? In the case of the topographical map, the contour diagram was created by joining all the points at the same height on the surface and dropping the curve into the xy-plane.

How do we go the other way? Suppose we wanted to plot the surface representing the corn production function $C = f(R, T)$ given by the contour diagram in Figure 11.47. Along each contour the function has a constant value; if we take each contour and lift it above the plane to a height equal to this value, we get the surface in Figure 11.48.

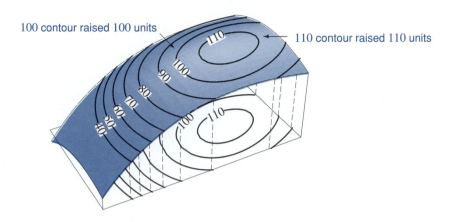

Figure 11.48: Getting the graph of the corn yield function from the contour diagram

Notice that the raised contours are the curves we get by slicing the surface horizontally. In general, we have the following result:

> Contour lines, or level curves, are obtained from a surface by slicing it with horizontal planes.

Finding Contours Algebraically

Algebraic equations for the contours of a function f are easy to find if we have a formula for $f(x, y)$. Suppose the surface has equation

$$z = f(x, y).$$

A contour is obtained by slicing the surface with a horizontal plane with equation $z = c$. Thus, the equation for the contour at height c is given by:

$$f(x, y) = c.$$

Example 3   Find equations for the contours of $f(x, y) = x^2 + y^2$ and draw a contour diagram for f. Relate the contour diagram to the graph of f.

Solution   The contour at height c is given by

$$f(x, y) = x^2 + y^2 = c.$$

This is a contour only for $c \geq 0$, For $c > 0$ it is a circle of radius $\sqrt{c}$. For $c = 0$, it is a single point (the origin). Thus, the contours at an elevation of $c = 1, 2, 3, 4, \ldots$ are all circles centered at the origin of radius $1, \sqrt{2}, \sqrt{3}, 2, \ldots$. The contour diagram is shown in Figure 11.49. The bowl–shaped graph of f is shown in Figure 11.50. Notice that the graph of f gets steeper as we move further away from the origin. This is reflected in the fact that the contours become more closely packed as we move further from the origin; for example, the contours for $c = 6$ and $c = 8$ are closer together than the contours for $c = 2$ and $c = 4$.

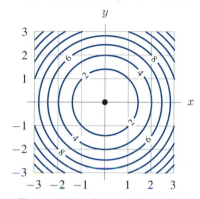

Figure 11.49: Contour diagram for $f(x, y) = x^2 + y^2$ (even values of c only)

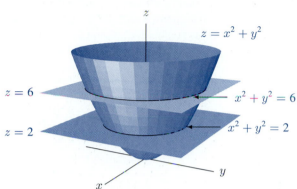

Figure 11.50: The graph of $f(x, y) = x^2 + y^2$

Example 4   Draw a contour diagram for $f(x, y) = \sqrt{x^2 + y^2}$ and relate it to the graph of f.

Solution   The contour at level c is given by

$$f(x, y) = \sqrt{x^2 + y^2} = c.$$

For $c > 0$ this is a circle, just as in the previous example, but here the radius is c instead of $\sqrt{c}$. For $c = 0$, it is the origin. Thus, if the level c increases by 1, the radius of the contour increases by 1. This means the contours are equally spaced concentric circles (see Figure 11.51) which do not become more closely packed further from the origin. Thus, the graph of f has the same constant slope as we move away from the origin (see Figure 11.52), making it a cone rather than a bowl.

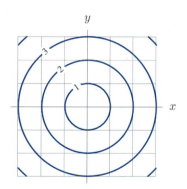

Figure 11.51: A contour diagram for $f(x, y) = \sqrt{x^2 + y^2}$

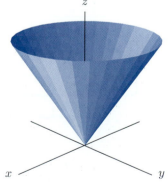

Figure 11.52: The graph of $f(x, y) = \sqrt{x^2 + y^2}$

In both of the previous examples the level curves are concentric circles because the surfaces have circular symmetry. Any function of two variables which depends only on the quantity $(x^2 + y^2)$ has such symmetry: for example, $G(x, y) = e^{-(x^2 + y^2)}$ or $H(x, y) = \sin(\sqrt{x^2 + y^2})$.

Example 5   Draw a contour diagram for $f(x, y) = 2x + 3y + 1$.

Solution   The contour at level c has equation $2x + 3y + 1 = c$. Rewriting this as $y = -(2/3)x + (c - 1)/3$, we see that the contours are parallel lines with slope $-2/3$. The y-intercept for the contour at level c is $(c - 1)/3$; each time c increases by 3, the y-intercept moves up by 1. The contour diagram is shown in Figure 11.53.

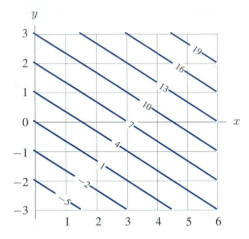

Figure 11.53: A contour diagram for $f(x, y) = 2x + 3y + 1$

Contour Diagrams and Tables

Sometimes we can get an idea of what the contour diagram of a function looks like from its table.

Example 6   Relate the values of $f(x, y) = x^2 - y^2$ in Table 11.5 to its contour diagram in Figure 11.54.

TABLE 11.5   *Table of values of $f(x, y) = x^2 - y^2$*

	3	0	−5	−8	−9	−8	−5	0
	2	5	0	−3	−4	−3	0	5
	1	8	3	0	−1	0	3	8
y	0	9	4	1	0	1	4	9
	−1	8	3	0	−1	0	3	8
	−2	5	0	−3	−4	−3	0	5
	−3	0	−5	−8	−9	−8	−5	0
		−3	−2	−1	0	1	2	3
					x			

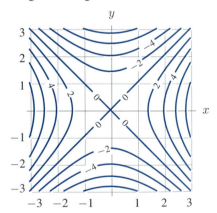

Figure 11.54: Contour map of $f(x, y) = x^2 - y^2$

Solution   One striking feature of the values in Table 11.5 is the zeros along the diagonals. This occurs because $x^2 - y^2 = 0$ along the lines $y = x$ and $y = -x$. So the $z = 0$ contour consists of these two lines. In the triangular region of the table that lies to the right of both diagonals, the entries are positive. To the left of both diagonals, the entries are also positive. Thus, in the contour diagram, the positive contours will lie in the triangular regions to the right and left of the lines $y = x$ and $y = -x$. Further, the table shows that the numbers on the left are the same as the numbers on the right; thus, each contour will have two pieces, one on the left and one on the right. See Figure 11.54. As we move away from the origin along the x-axis, we cross contours corresponding to successively larger

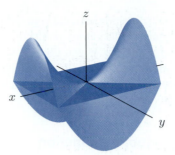

Figure 11.55: Graph of $f(x, y) = x^2 - y^2$ showing plane $z = 0$

values. On the saddle-shaped graph of $f(x, y) = x^2 - y^2$ shown in Figure 11.55, this corresponds to climbing out of the saddle along one of the ridges. Similarly, the negative contours occur in pairs in the top and bottom triangular regions; the values get more and more negative as we go out along the y-axis. This corresponds to descending from the saddle along the valleys that are submerged below the xy-plane in Figure 11.55. Notice that we could also get the contour diagram by graphing the family of hyperbolas $x^2 - y^2 = 0, \pm 2, \pm 4, \dots$.

Using Contour Diagrams: The Cobb-Douglas Production Function

Suppose you are running a small printing business, and decide to expand because you have more orders than you can handle. How should you expand? Should you start a night-shift and hire more workers? Should you buy more expensive but faster computers which will enable the current staff to keep up with the work? Or should you do some combination of the two?

Obviously, the way such a decision is made in practice involves many other considerations—such as whether you could get a suitably trained night shift, or whether there are any faster computers available. Nevertheless, you might model the quantity, P, of work produced by your business as a function of two variables: your total number, N, of workers, and the total value, V, of your equipment.

How would you expect such a production function to behave? In general, having more equipment and more workers enables you to produce more. However, increasing equipment without increasing the number of workers will increase production a bit, but not beyond a point. (If equipment is already lying idle, having more of it won't help.) Similarly, increasing the number of workers without increasing equipment will increase production, but not past the point where the equipment is fully utilized, as any new workers would have no equipment available to them.

Example 7   Explain why the contour diagram in Figure 11.56 does not model the behavior expected of the production function, whereas the contour diagram in Figure 11.57 does.

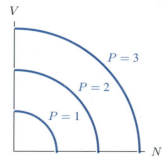

Figure 11.56: Incorrect contours for printing production

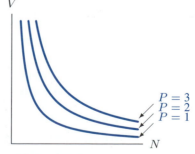

Figure 11.57: Correct contours for printing production

Solution   Look at the contour diagram in Figure 11.56. Fixing V at a particular value and letting N increase means moving to the right on the contour diagram. As you do so, you cross contours with larger and larger P values, meaning that production increases indefinitely. On the other hand, in Figure 11.57, as you move in the same direction you find yourself moving nearly parallel to the contours, crossing them less and less frequently. Therefore, production increases more and more slowly as N increases while V is held fixed. Similarly, if you hold N fixed and let V increase, the contour diagram in Figure 11.56 shows production increasing at a steady rate, whereas Figure 11.57 shows production increasing, but at a decreasing rate. Thus, Figure 11.57 fits the expected behavior of the production function best.

Formula for a Production Function

Production functions with the qualitative behavior we want are often approximated by formulas of the form

$$P = f(N, V) = cN^{\alpha}V^{\beta}$$

where P is the total quantity produced and c, α, and β are positive constants with $0 < \alpha < 1$ and $0 < \beta < 1$.

Example 8   Show that the contours of the function $P = cN^{\alpha}V^{\beta}$ have approximately the shape of the contours in Figure 11.57.

Solution   The contours are the curves where P is equal to a constant value, say P_0, that is, where

$$cN^{\alpha}V^{\beta} = P_0.$$

Solving for V we get

$$V = \left(\frac{P_0}{c}\right)^{1/\beta} N^{-\alpha/\beta}.$$

Thus, V is a power function of N with a negative exponent, and hence its graph has the shape shown in Figure 11.57.

The Cobb-Douglas Production Model

In 1928, Cobb and Douglas used a similar function to model the production of the entire US economy in the first quarter of this century. Using government estimates of P, the total yearly production between 1899 and 1922, of K, the total capital investment over the same period, and of L, the total labor force, they found that P was well approximated by the *Cobb-Douglas production function*

$$P = 1.01L^{0.75}K^{0.25}.$$

This function turned out to model the US economy surprisingly well, both for the period on which it was based, and for some time afterwards.

Problems for Section 11.4

For the functions in Problems 1–9, sketch a contour diagram with at least four labeled contours. Describe in words the contours and how they are spaced.

1. $f(x, y) = x + y$
2. $f(x, y) = xy$
3. $f(x, y) = x^2 + y^2$
4. $f(x, y) = 3x + 3y$
5. $f(x, y) = -x^2 - y^2 + 1$
6. $f(x, y) = x^2 + 2y^2$
7. $f(x, y) = \sqrt{x^2 + 2y^2}$
8. $f(x, y) = y - x^2$
9. $f(x, y) = \cos\sqrt{x^2 + y^2}$

10. Figure 11.58 is a contour diagram of the monthly payment on a 5-year car loan as a function of the interest rate and the amount you borrow. Suppose that the interest rate is 13% and that you decide to borrow $6000.

   (a) What is your monthly payment?

(b)   If interest rates drop to 11%, how much more can you borrow without increasing your monthly payment?

(c)   Make a table of how much you can borrow, without increasing your monthly payment, as a function of the interest rate.

loan amount ($)

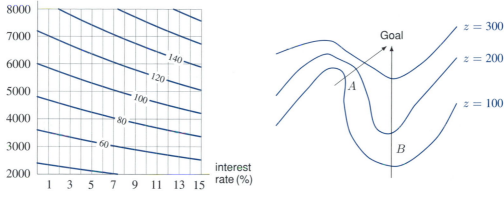

Figure 11.58   *Figure 11.59*

11.   Figure 11.59 shows a contour map of a hill with two paths, A and B.

(a)   On which path, A or B, will you have to climb more steeply?

(b)   On which path, A or B, will you probably have a better view of the surrounding countryside? (Assuming trees do not block your view.)

(c)   Alongside which path is there more likely to be a stream?

12.   Each of the contour diagrams in Figure 11.60 shows population density in a certain region. Choose the contour diagram that best corresponds to each of the following situations. Many different matchings are possible. Pick any reasonable one and justify your choice.

(a)   The center of the diagram is a city.

(b)   The center of the diagram is a lake.

(c)   The center of the diagram is a power plant.

(I)   (II)   (III)

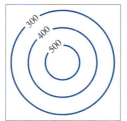 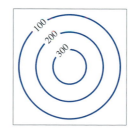 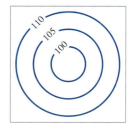

Figure 11.60

For each of the surfaces in Problems 13–15, sketch a possible contour diagram, marked with reasonable z-values. (Note: There are many possible answers.)

13.   14.   15.

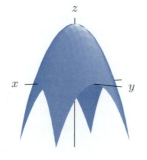 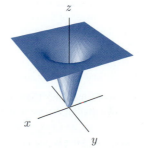 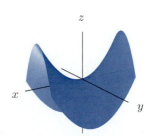

16.   Figure 11.61 shows the density of the fox population P (in foxes per square kilometer) for southern England. Draw two different cross-sections along a north-south line and two different cross-sections along an east-west line of the population density P.

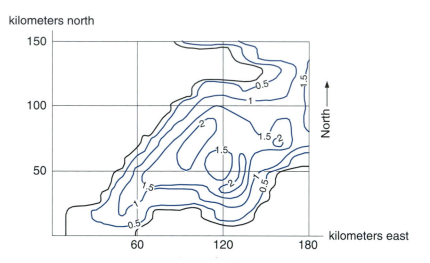

Figure 11.61: Population density of foxes in southwestern England

17.   Use a computer or a calculator to sketch a contour diagram for the vibrating string function

$$f(x, t) = \cos t \sin 2x, \qquad 0 \le x \le \pi, \ 0 \le t \le \pi.$$

Use $c = -2/3, -1/3, 0, 1/3, 2/3$. (You will not be able to do this algebraically.)

18.   On page 566 we introduced the traveling wave function

$$h(x, t) = 5 + \cos(0.5x - t).$$

Draw a contour diagram using $c = 4, 4.5, 5, 5.5, 6$ for this function. Explain how your contour diagram relates to the cross-sections of h discussed on page 566. Where are the contours most closely spaced? Most widely spaced?

19.   Draw contour diagrams for each of the pizza-cola-happiness graphs given in Problem 6 on page 581.

20.   A manufacturer sells two goods, one at a price of $3000 a unit and the other at a price of $12,000 a unit. Suppose a quantity q_1 of the first good and q_2 of the second good are sold at a total cost of $4000 to the manufacturer.

(a)   Express the manufacturer's profit, π, as a function of q_1 and q_2.
(b)   Sketch curves of constant profit in the $q_1 q_2$-plane for $\pi = 10,000$, $\pi = 20,000$, and $\pi = 30,000$ and the break-even curve $\pi = 0$.

21.   The cornea is the front surface of the eye. Corneal specialists use a TMS, or Topographical Modeling System, to produce a "map" of the curvature of the eye's surface. A computer analyzes light reflected off the eye and draws level curves joining points of constant curvature. The regions between these curves are colored different colors.

The first two pictures in Figure 11.62 are cross-sections of eyes with constant curvature, the smaller being about 38 units and the larger about 50 units. For contrast, the third eye has varying curvature.

(a)   Describe in words how the TMS map of an eye of constant curvature will look.
(b)   Draw the TMS map of an eye with the cross-section in Figure 11.63. Assume the eye is circular when viewed from the front, and the cross-section is the same in every direction. Put reasonable numeric labels on your level curves.

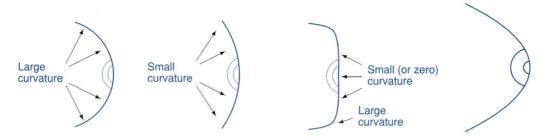

Figure 11.62: Pictures of eyes with different curvature   *Figure 11.63*

22.   Match Tables 11.6–11.9 with the contour diagrams (I) - (IV) in Figure 11.64.

TABLE 11.6

$y\backslash x$	-1	0	1
-1	2	1	2
0	1	0	1
1	2	1	2

TABLE 11.7

$y\backslash x$	-1	0	1
-1	0	1	0
0	1	2	1
1	0	1	0

TABLE 11.8

$y\backslash x$	-1	0	1
-1	2	0	2
0	2	0	2
1	2	0	2

TABLE 11.9

$y\backslash x$	-1	0	1
-1	2	2	2
0	0	0	0
1	2	2	2

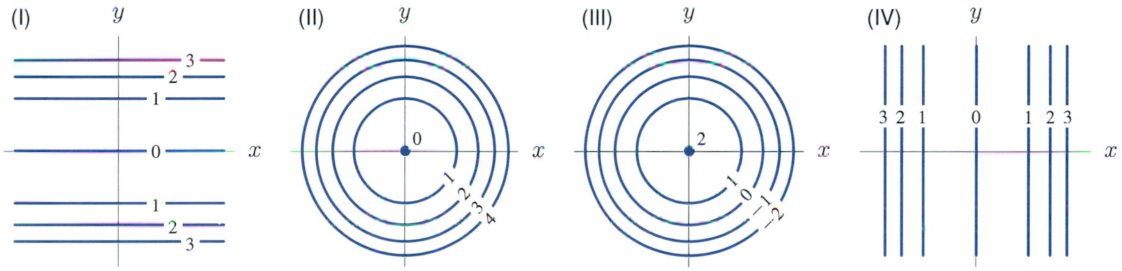

Figure 11.64

23.   Match the surfaces (a)–(e) in Figure 11.65 with the contour diagrams (I)–(V) in Figure 11.66.

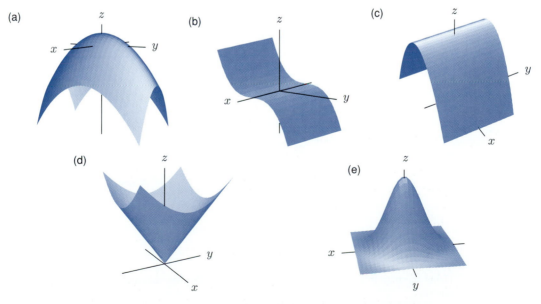

Figure 11.65

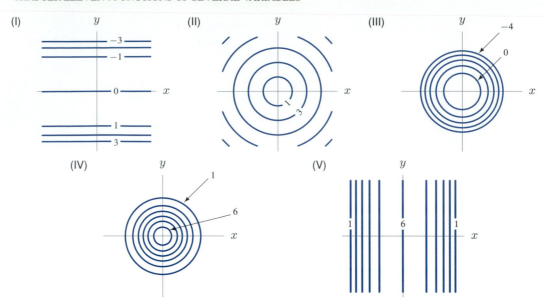

Figure 11.66

24. The map in Figure 11.67 is from the undergraduate senior thesis of Professor Robert Cook, Director of Harvard's Arnold Arboretum. It shows level curves of the function giving the species density of breeding birds at each point in the US, Canada, and Mexico.

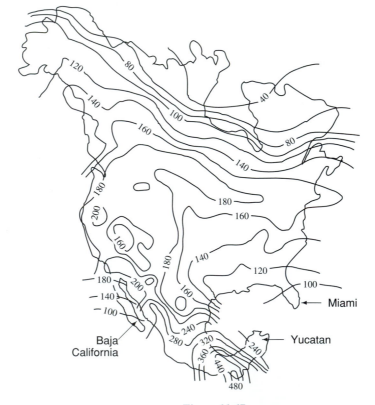

Figure 11.67

Using the map in Figure 11.67, decide if the the following statements are true or false. Explain your answers.

(a) Moving from south to north across Canada, the species density increases.

(b)   The species density in the area around Miami is over 100.

(c)   In general, peninsulas (for example, Florida, Baja California, the Yucatan) have lower species densities than the areas around them.

(d)   The greatest rate of change in species density with distance is in Mexico. If you think this is true, mark the point and direction which give the maximum rate of change and explain why you picked the point and direction you did.

25.   The temperature T (in $^\circ$C) at any point in the region $-10 \le x \le 10, -10 \le y \le 10$ is given by the function

$$T(x, y) = 100 - x^2 - y^2.$$

(a)   Sketch isothermal curves (curves of constant temperature) for $T = 100^\circ$C, $T = 75^\circ$C, $T = 50^\circ$C, $T = 25^\circ$C, and $T = 0^\circ$C.

(b)   Suppose a heat-seeking bug is put down at any point on the xy-plane. In which direction should it move to increase its temperature fastest? How is that direction related to the level curve through that point?

26.

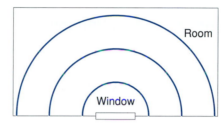

Figure 11.68

Figure 11.68 shows the level curves of the temperature H in a room near a recently opened window. Label the three level curves with reasonable values of H if the house is in the following locations.

(a)   Minnesota in winter (where winters are harsh).

(b)   San Francisco in winter (where winters are mild).

(c)   Houston in summer (where summers are hot).

(d)   Oregon in summer (where summers are mild).

27.   Match each Cobb-Douglas production function with the correct graph in Figure 11.69 and with the correct statements.

(a) $F(L, K) = L^{0.25} K^{0.25}$ (D)   Tripling each input triples output.

(b) $F(L, K) = L^{0.5} K^{0.5}$ (E)   Quadrupling each input doubles output.

(c) $F(L, K) = L^{0.75} K^{0.75}$ (G)   Doubling each input almost triples output.

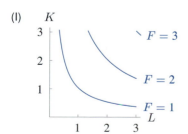 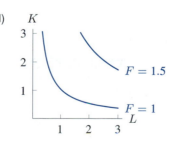 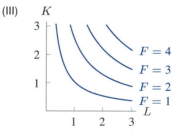

Figure 11.69

28.   Consider the Cobb-Douglas production function $P = f(L, K) = 1.01L^{0.75} K^{0.25}$. What is the effect on production of doubling both labor and capital?

29. A general Cobb-Douglas production function has the form

$$P = cL^{\alpha}K^{\beta}.$$

An important economic question concerns what happens to production if labor and capital are both scaled up? For example, does production double if both labor and capital are doubled?

Economists talk about

- *increasing returns to scale* if doubling L and K more than doubles P,
- *constant returns to scale* if doubling L and K exactly doubles P,
- *decreasing returns to scale* if doubling L and K less than doubles P.

What conditions on α and β lead to increasing, constant, or decreasing returns to scale?

11.5 LINEAR FUNCTIONS

What is a Linear Function of Two Variables?

Linear functions played a central role in one-variable calculus because many one-variable functions are locally linear. (That is, their graphs look like a line when we zoom in.) In two-variable calculus, *linear functions* are those whose graph is a plane. In Chapter 13, we will see that most of the two-variable functions that we work with have graphs which look like planes when we zoom in.

What Makes a Plane Flat?

What is it about a function $f(x, y)$ that makes its graph $z = f(x, y)$ a plane? Linear functions of *one* variable have straight line graphs because they have constant slope. No matter where we are on the graph, the slope is the same: a given increase in the x-coordinate always gives the same increase in the y-coordinate. In other words, the ratio $\Delta y/\Delta x$ is constant. On a plane, the situation is a bit more complicated. If we walk around on a tilted plane, the slope is not always the same: it depends on the direction in which we walk. However, it is still true that at every point on the plane, the slope is the same as long as we choose the same direction. If we walk in the direction parallel to the x-axis, we always find ourselves walking up or down with the same slope; the same thing is true if we walk in the direction parallel to the y-axis. In other words, the slope ratios $\Delta z/\Delta x$ (with y fixed) and $\Delta z/\Delta y$ (with x fixed) are both constant.

Example 1 We are on a plane that cuts the z-axis at $z = 5$, has slope 2 in the x direction and slope -1 in the y direction. What is the equation of the plane?

Solution Finding the equation of the plane means constructing a formula for the z-coordinate of the point on the plane directly above the point (x, y) in the xy-plane. To get to that point start from the point above the origin, where $z = 5$. Then walk x units in the x direction. Since the slope in the x direction is 2, our height increases by $2x$. Then walk y units in the y direction; since the slope in the y direction is -1, our height decreases by y units. Since our height has changed by $2x - y$ units, our z-coordinate is $5 + 2x - y$. Thus, the equation for the plane is

$$z = 5 + 2x - y.$$

For any linear function, if we know its value at a point (x_0, y_0), its slope in the x direction, and its slope in the y direction, then we can write the equation of the function. This is just like the equation of a line in the one-variable case, except that there are two slopes instead of one.

If a plane has slope m in the x direction, slope n in the y direction, and passes through the point (x_0, y_0, z_0), then its equation is

$$z = z_0 + m(x - x_0) + n(y - y_0).$$

This plane is the graph of the linear function

$$f(x, y) = z_0 + m(x - x_0) + n(y - y_0).$$

If we write $c = z_0 - mx_0 - ny_0$, then we can write $f(x, y)$ in the equivalent form

$$f(x, y) = c + mx + ny.$$

Just as in 2-dimensional space a line is determined by 2 points, so in 3-dimensional space a plane is determined by 3 points, provided they do not lie on a line.

Example 2 Find the equation of the plane that passes through the three points $(1, 0, 1)$, $(1, -1, 3)$, and $(3, 0, -1)$.

Solution The first two points have the same x-coordinate, so we use them to find the slope of the plane in the y-direction. As the y-coordinate changes from 0 to -1, the z-coordinate changes from 1 to 3, so the slope in the y-direction is $n = \Delta z/\Delta y = (3 - 1)/(-1 - 0) = -2$. The first and third points have the same y-coordinate, so we use them to find the slope in the x-direction; it is $m = \Delta z/\Delta x = (-1 - 1)/(3 - 1) = -1$. Because the plane passes through $(1, 0, 1)$, its equation is

$$z = 1 - (x - 1) - 2(y - 0) \quad \text{or} \quad z = 2 - x - 2y.$$

You should check that this equation is also satisfied by the points $(1, -1, 3)$ and $(3, 0, -1)$.

The previous example was made easier by the fact that two of the points had the same x-coordinate and two of them had the same y-coordinate. An alternative method, which works for any three points, is to substitute the x, y, and z-values of each of the three points into the equation $z = c + mx + ny$. The resulting three equations in c, m, n can then be solved simultaneously. See Problem 2 on page 599.

Linear Functions from a Numerical Point of View

To avoid flying planes with too many empty seats, airlines sell some tickets at full price and some at a discount. Table 11.10 shows an airline's revenue in dollars from tickets sold on a particular route, as a function of the number of full-price tickets sold, f, and the number of discount tickets sold, d.

TABLE 11.10 *Revenue from ticket sales (dollars)*

		Full-price tickets (f)			
		100	200	300	400
Discount tickets (d)	200	39,700	63,600	87,500	111,400
	400	55,500	79,400	103,300	127,200
	600	71,300	95,200	119,100	143,000
	800	87,100	111,000	134,900	158,800
	1000	102,900	126,800	150,700	174,600

Looking down any column, we see that the revenue jumps by \$15,800 for each extra 200 discount tickets. Thus, each column is a linear function of the number of discount tickets sold. In addition, every column has the same slope, which is $15{,}800/200 = 79$ dollars/ticket. This is the price of a discount ticket. Similarly, each row is a linear function and all the rows have the same slope, 239, which is the price of a full-fare ticket. Thus, R is a linear function of f and d, given by:

$$R = 239f + 79d.$$

We have the following general result:

A **linear function** can be recognized from its table by the following features:
- Each row and each column is linear.

- All the rows have the same slope.

- All the columns have the same slope (although the slope of the rows and the slope of the columns are generally different).

Example 3   The table contains some values of a linear function. Fill in the blank and give a formula for the function.

$x \backslash y$	1.5	2.0
2	0.5	1.5
3	−0.5	?

Solution   In the first column the function decreases by 1 (from 0.5 to −0.5) as x goes from 2 to 3. Since the function is linear, it must decrease by the same amount in the second column. So the missing entry must be $1.5 - 1 = 0.5$. The slope of the function in the x-direction is -1. The slope in the y-direction is 2, since in each row the function increases by 1 when y increases by 0.5. From the table we get $f(2, 1.5) = 0.5$. Therefore, the formula is

$$f(x, y) = 0.5 - (x - 2) + 2(y - 1.5) = -0.5 - x + 2y.$$

What Does the Contour Diagram of a Linear Function Look Like?

Consider the function for airline revenue given in Table 11.10. Its formula is

$$R = 239f + 79d,$$

where f is the number of full-fares and d is the number of discount fares sold. Figure 11.70 gives the contour diagram for this function.

Notice that the contours are parallel straight lines. What is the practical significance of the slope? Consider the contour $R = 100{,}000$; that means we are looking at combinations of ticket sales that yield \$100,000 in revenue. If we move down and to the right on the contour, the f-coordinate increases and the d-coordinate decreases, so we sell more full-fares and fewer discount fares. This makes sense, because to receive a fixed revenue of \$100,000, we must sell more full-fares if we sell fewer discount fares. The exact trade-off depends on the slope of the contour; the diagram shows that each contour has a slope of about -3. This means that for a fixed revenue, we must sell three discount fares to replace one full-fare. This can also be seen by comparing prices. Each full fare brings in \$239; to earn the same amount in discount fares we need to sell $239/79 \approx 3.03 \approx 3$ fares. Since the price ratio is independent of how many of each type of fare we sell, this slope remains constant over the whole contour map; thus, the contours are all parallel straight lines.

You should also observe that the contours are evenly spaced. This means that no matter which contour we are on, a fixed increase in one of the variables will cause the same increase in the value of the function. In terms of revenue, this says that no matter how many fares we have sold, an extra fare, whether full or discount, will bring the same revenue as it did before.

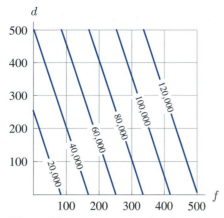

Figure 11.70: Revenue as a function of the number of full and discount fares sold

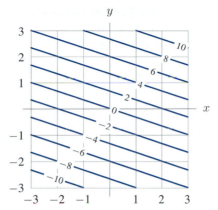

Figure 11.71: Contour map of linear function $f(x, y)$

Example 4   Find the equation of the linear function whose contour diagram is in Figure 11.71.

Solution   Suppose we start at the origin on the $z = 0$ contour. Moving 2 units in the y direction takes us to the $z = 6$ contour; so the slope in the y direction is $\Delta z / \Delta y = 6/2 = 3$. Similarly, a move of 2 in the x-direction from the origin takes us to the $z = 2$ contour, so the slope in the x direction is $\Delta z / \Delta x = 2/2 = 1$. Since $f(0,0) = 0$, we have $f(x,y) = x + 3y$.

Problems for Section 11.5

1. Suppose that z is a linear function of x and y with slope 2 in the x direction and slope 3 in the y direction.

    (a)   A change of 0.5 in x and -0.2 in y produces what change in z?
    (b)   If $z = 2$ when $x = 5$ and $y = 7$, what is the value of z when $x = 4.9$ and $y = 7.2$?

2. Find the equation of the linear function $z = c + mx + ny$ whose graph contains the points $(0, 0, 0)$, $(0, 2, -1)$, and $(-3, 0, -4)$.

3. Find the linear function whose graph is the plane through the points $(4, 0, 0)$, $(0, 3, 0)$ and $(0, 0, 2)$.

4. Find an equation for the plane containing the line in the xy-plane where $y = 1$, and the line in the xz-plane where $z = 2$.

5. Find the equation of the linear function $z = c + mx + ny$ whose graph intersects the xz-plane in the line $z = 3x + 4$ and intersects the yz-plane in the line $z = y + 4$.

6. Find the linear function $z = c + mx + ny$ whose graph intersects the xy-plane in the line $y = 3x + 4$ and contains the point $(0, 0, 5)$.

7. Is the function represented in Table 11.11 linear? Give reasons for your answer.

TABLE 11.11

$u \backslash v$	1.1	1.2	1.3	1.4
3.2	11.06	12.06	13.07	14.07
3.4	11.75	12.82	13.89	14.95
3.6	12.44	13.57	13.89	14.95
3.8	13.13	14.33	15.52	16.71
4.0	13.82	15.08	16.34	17.59

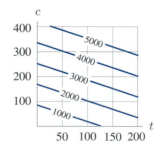

Figure 11.72

8. A discount music store sells compact discs at one price and cassette tapes at another price. Figure 11.72 shows the revenue (in dollars) of the music store as a function of the number, t, of tapes and the number, c, of compact discs that it sells. What is the price of tapes? What is the price of compact discs?

For Problems 9–10, find equations for the linear functions with the given tables.

9.

$x\backslash y$	-1	0	1	2
0	1.5	1	0.5	0
1	3.5	3	2.5	2
2	5.5	5	4.5	4
3	7.5	7	6.5	6

10.

$x\backslash y$	10	20	30	40
100	3	6	9	12
200	2	5	8	11
300	1	4	7	10
400	0	3	6	9

Problems 11–12 each contain a partial table of values for a linear function. Fill in the blanks.

11.

$x\backslash y$	0.0	1.0
0.0		1.0
2.0	3.0	5.0

12.

$x\backslash y$	-1.0	0.0	1.0
2.0	4.0		
3.0		3.0	5.0

For Problems 13–14, find equations for the linear functions with the given contour diagrams.

13.

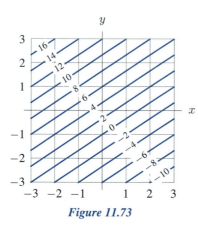

Figure 11.73

14.

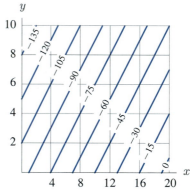

Figure 11.74

It is difficult to sketch a plane that is the graph of a linear function by hand. One method that works if the x, y, and z-intercepts of the graph are positive is to plot the intercepts and join them by a triangle as shown in Figure 11.75; this shows the part of the graph in the octant where $x \geq 0$, $y \geq 0$, $z \geq 0$. If the intercepts are not all positive, the same method works if the x, y, and z-axes are drawn from a different perspective. Use this method to sketch the graphs of the linear functions in Problems 15–17.

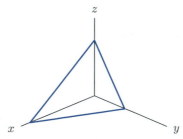

Figure 11.75: The graph of a linear function with positive x, y, and z-intercepts

15. $z = 6 - 2x - 3y$ 16. $z = 4 + x - 2y$ 17. $z = 2 - 2x + y$

18. A manufacturer makes two products out of two raw materials. Let q_1, q_2 be the quantities sold of the two products, p_1, p_2 their prices, and m_1, m_2 the quantities purchased of the two raw materials. Which of the following functions do you expect to be linear, and why? In each case, assume that all variables except the ones mentioned are held fixed.

   (a) Expenditure on raw materials as a function of m_1 and m_2.
   (b) Revenue as a function of q_1 and q_2.
   (c) Revenue as a function of p_1 and q_1.

19. A college admissions office uses the following linear equation to predict the grade point average of an incoming student:

$$z = 0.003x + 0.8y - 4,$$

where z is the predicted college GPA on a scale of 0 to 4.3, and x is the sum of the student's SAT Math and SAT Verbal on a scale of 400 to 1600, and y is the student's high school GPA on a scale of 0 to 4.3. The college admits students whose predicted GPA is at least 2.3.

   (a) Will a student with SATs of 1050 and high school GPA of 3.0 be admitted?
   (b) Will every student with SATs of 1600 be admitted?
   (c) Will every student with a high school GPA of 4.3 be admitted?
   (d) Draw a contour diagram for the predicted GPA z with $400 \leq x \leq 1600$ and $0 \leq y \leq 4.3$. Shade the points corresponding to students who will be admitted.
   (e) Which is more important, an extra 100 points on the SAT or an extra 0.5 of high school GPA?

20. Let f be the linear function $f(x, y) = c + mx + ny$, where c, m, n are constants and $n \neq 0$.

   (a) Show that all the contours of f are lines of slope $-m/n$.
   (b) Show that, for all x and y

$$f(x + n, y - m) = f(x, y)$$

   (c) Explain the relation between parts (a) and (b).

11.6 FUNCTIONS OF MORE THAN TWO VARIABLES

In applications of calculus, functions of any number of variables can arise. The density of matter in the universe is a function of three variables, since it takes three numbers to specify a point in space. If we want to study the evolution of this density over time, we need to add the fourth variable, time. We need to be able to apply calculus to functions of arbitrarily many variables.

The main problem with functions of more than two variables is that it is hard to visualize them. The graph of a function of one variable is a curve in 2-space, the graph of a function of two variables is a surface in 3-space, and so the graph of a function of three variables would be a solid in 4-space. We can't easily visualize 4-space, or any higher dimensional space, and so we won't use the graphs of functions of three or more variables.

On the other hand, the idea of cross-sectioning a function by keeping one variable fixed allows us to represent a function of three variables. It is also possible to give contour diagrams for these functions, only now the contours are surfaces in 3-space.

Representing a Function of Three Variables Using Contour Diagrams

One way to analyze a function of three variables, $f(x, y, z)$, is to look at cross-sections with one variable fixed. Suppose we keep z fixed and view f as a function of the remaining two variables, x and y. For each fixed value, $z = c$, we represent the two-variable function $f(x, y, c)$ by a contour diagram. As c varies, we get a whole collection of contour diagrams.

Example 1   A pond is 30 meters deep in the middle and 200 meters across. The pond is infested with algae. Suppose the density of algae at a point z meters below, x meters east, and y meters north of the center of the surface of the pond is approximated by the formula

$$f(x, y, z) = \frac{1}{10} \left(50 + \sqrt{x^2 + y^2} \right) (30 - z),$$

where density is measured in gm/m^3. Draw contour diagrams for f at the surface of the pond and at a depth of 10 m. Describe in words how the density of algae varies with depth and distance from the center of the pond.

Solution   At the surface of the pond, $z = 0$, so the formula for the $z = 0$ cross-section is

$$f(x, y, 0) = \frac{1}{10}\left(50 + \sqrt{x^2 + y^2}\right)(30 - 0) = 150 + 3\sqrt{x^2 + y^2}.$$

Notice that we now have a function of two variables. The contours of this function are circles, since $f(x, y, 0)$ is constant when $\sqrt{x^2 + y^2}$ is constant. The contour diagram is shown in Figure 11.76. At a depth of $z = 10$, we have

$$f(x, y, 10) = \frac{1}{10}\left(50 + \sqrt{x^2 + y^2}\right)(30 - 10) = 100 + 2\sqrt{x^2 + y^2}.$$

The contour diagram for the $z = 10$ cross-section is shown in Figure 11.77. Comparing the two contour diagrams we see that there is more algae near the surface than deeper down in the pond and that there is more algae near the shoreline than near the center of the pond.

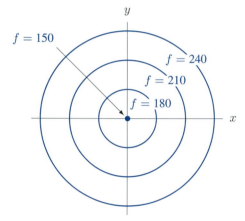

Figure 11.76: Density of algae in the pond at the surface: Contour diagram for $z = 0$

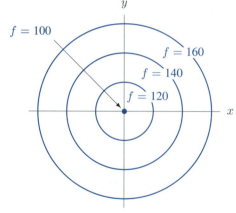

Figure 11.77: Density of algae in the pond 10 meters down: Contour diagram for $z = 10$

For weather forecasting, it is important to know the barometric pressure and temperature both at the surface of the earth and also aloft in the atmosphere above the earth. On TV, you may hear about a "surface" low pressure system or an "upper level" low pressure system (upper level lows influence the jet stream). The surface and upper level maps studied by meteorologists are contour diagrams at different altitudes of the pressure or temperature function. These maps are similar to the contour diagrams of algae density of the previous example.

Representing a Function of Three Variables Using Tables

In the previous example, we viewed the cross-sections of a three-variable function $f(x, y, z)$ as contour diagrams of a two-variable function. We can also view the cross-sections as tables of values.

Example 2   In Section 11.1, the consumption of beef as a function of the price, p, of beef and household income, I, was given in a table of values. In fact, a household's beef consumption depends on many other things as well, such as the prices of competing products. Let us view the consumption, C, of beef as a function of the price q of a competing meat, say chicken, as well as the price of beef and household income. That is, $C = f(I, p, q)$. This function can be given by a collection of tables, one for each different value of the price, q, of chicken. Tables 11.12 and 11.13 show two different cross-sections of the consumption function f with q held fixed. Explain how the two tables are related.

TABLE 11.12   *Beef consumption when price of chicken* $q = 1.50$

		p	
	3.00	3.50	4.00
20	2.65	2.59	2.51
I 60	5.11	5.00	4.97
100	5.80	5.77	5.60

TABLE 11.13   *Beef consumption when price of chicken* $q = 2.00$

		p	
	3.00	3.50	4.00
20	2.75	2.75	2.71
I 60	5.21	5.12	5.11
100	5.80	5.77	5.60

Solution   Comparing tables shows that for households with large incomes (say $I = 100$), changes in the price of chicken have little effect on the amount of beef consumed (because price is not a factor in their buying). However for small incomes (say $I = 20$), an increase in the price of chicken (from $q = 1.50$ to $q = 2.00$) causes families to spend more on beef.

Representing a Function of Three Variables Using a Family of Level Surfaces

A function of two variables, $f(x, y)$, can be represented by a *family* of level curves of the form $f(x, y) = c$ for various values of the constant, c.

> A function of three variables, $f(x, y, z)$, can be represented by a *family* of surfaces of the form $f(x, y, z) = c$, each one of which is called a *level surface*.

Example 3   Suppose the temperature, in °C, at a point (x, y, z) is given by $T = f(x, y, z) = x^2 + y^2 + z^2$. What do the level surfaces of the function f look like and what do they mean in terms of temperature?

Solution   The level surface corresponding to $T = 100$ is the set of all points where the temperature is $100°$C. That is, where $f(x, y, z) = 100$, so

$$x^2 + y^2 + z^2 = 100.$$

This is the equation of a sphere of radius 10, with center at the origin. Similarly, the level surface corresponding to $T = 200$ is the sphere with radius $\sqrt{200}$. The other level surfaces will be concentric spheres. The temperature is constant on each sphere. We may view the temperature distribution as a set of nested spheres, like concentric layers of an onion, each one labeled with a different temperature, starting from low temperatures in the middle and getting hotter as we go out from the center. (See Figure 11.78.) The level surfaces become more closely spaced as we move farther from the origin because the temperature increases more rapidly the farther we get from the origin.

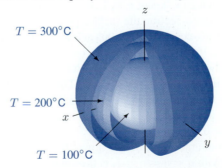

Figure 11.78: Level surfaces of $T = f(x, y, z) = x^2 + y^2 + z^2$, each one having a constant temperature

In general, the level surfaces of a function are nested together in some way, so it is often difficult to draw them. We generally use cut-outs to show the inner surfaces, as we did in Figure 11.78.

Example 4   What do the level surfaces of $f(x, y, z) = x^2 + y^2$ and $g(x, y, z) = z - y$ look like?

Solution   The level surface of f corresponding to the constant c is the surface consisting of all points satisfying the equation

$$x^2 + y^2 = c.$$

Since there is no z-coordinate in the equation, z can take any value whatsoever. For $c > 0$, this is a circular cylinder of radius $\sqrt{c}$ around the z-axis. The level surfaces are concentric cylinders; the narrow ones near the z-axis are the ones where f has small values, and the wider ones are where it has larger values. See Figure 11.79. The level surface of g corresponding to the constant c is the surface

$$z - y = c.$$

This time there is no x variable, so this surface is the one you get by taking each point on the straight line $z - y = c$ in the yz-plane and letting x roam back and forth. You get a plane which cuts the yz-plane diagonally; the x-axis is parallel to this plane. See Figure 11.80.

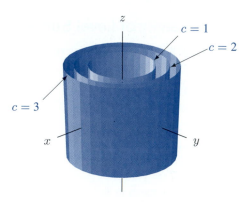

Figure 11.79: Level surfaces of $f(x, y, z) = x^2 + y^2$

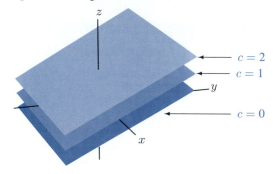

Figure 11.80: Level surfaces of $g(x, y, z) = z - y$

Example 5   What do the level surfaces of $f(x, y, z) = x^2 + y^2 - z^2$ look like?

Solution   From Section 11.4, you may recall that the two-variable quadratic function $g(x, y) = x^2 - y^2$ has a saddle-shaped graph and three types of contours. The contour equation $x^2 - y^2 = c$ gives a hyperbola opening right-left when $c > 0$, a hyperbola opening up-down when $c < 0$, and a pair of intersecting lines when $c = 0$. Similarly, the three-variable quadratic function $f(x, y, z) = x^2 + y^2 - z^2$ has three types of level surfaces depending on the value of c in the equation $x^2 + y^2 - z^2 = c$.

   Suppose that $c > 0$, say $c = 1$. Rewrite the equation as $x^2 + y^2 = z^2 + 1$ and think of what happens as we cross-section the surface perpendicular to the z-axis by holding z fixed. The result is a circle, $x^2 + y^2 = $ constant, of radius at least 1 (since the constant $z^2 + 1 \geq 1$). The circles get larger as z gets larger. If we took the $x = 0$ cross-section instead we would get the hyperbola $y^2 - z^2 = 1$. The result is shown in Figure 11.84, with $a = b = c = 1$.

   Suppose instead $c < 0$, say $c = -1$. Then the horizontal cross-sections of $x^2 + y^2 = z^2 - 1$ are again circles except that the radii shrink to 0 at $z = \pm 1$ and between $z = -1$ and $z = 1$ there are no cross-sections at all. The result is shown in Figure 11.85 with $a = b = c = 1$.

   When $c = 0$, we get the equation $x^2 + y^2 = z^2$. Again the horizontal cross-sections are circles, this time with the radius shrinking down to exactly 0 when $z = 0$. The resulting surface, shown in Figure 11.86 with $a = b = c = 1$, is the cone $z = \sqrt{x^2 + y^2}$ studied in Section 11.4, together with the lower cone $z = -\sqrt{x^2 + y^2}$.

A Catalog of Surfaces

For later reference, here is a small catalog of the surfaces we have encountered.

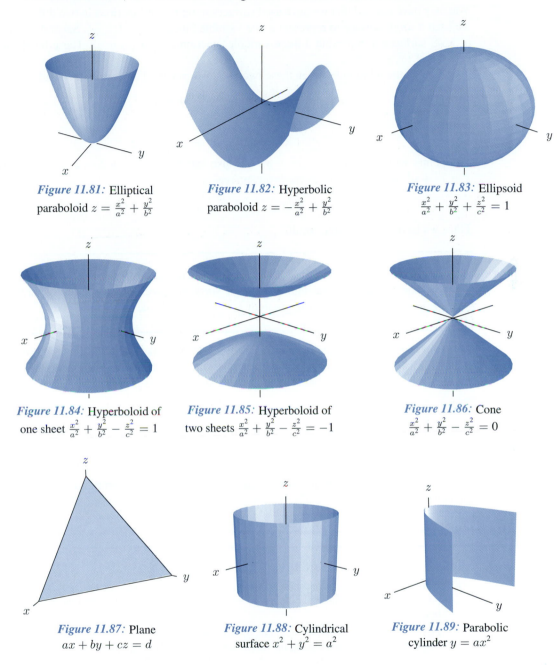

Figure 11.81: Elliptical paraboloid $z = \frac{x^2}{a^2} + \frac{y^2}{b^2}$

Figure 11.82: Hyperbolic paraboloid $z = -\frac{x^2}{a^2} + \frac{y^2}{b^2}$

Figure 11.83: Ellipsoid $\frac{x^2}{a^2} + \frac{y^2}{b^2} + \frac{z^2}{c^2} = 1$

Figure 11.84: Hyperboloid of one sheet $\frac{x^2}{a^2} + \frac{y^2}{b^2} - \frac{z^2}{c^2} = 1$

Figure 11.85: Hyperboloid of two sheets $\frac{x^2}{a^2} + \frac{y^2}{b^2} - \frac{z^2}{c^2} = -1$

Figure 11.86: Cone $\frac{x^2}{a^2} + \frac{y^2}{b^2} - \frac{z^2}{c^2} = 0$

Figure 11.87: Plane $ax + by + cz = d$

Figure 11.88: Cylindrical surface $x^2 + y^2 = a^2$

Figure 11.89: Parabolic cylinder $y = ax^2$

(These are viewed as equations in three variables $x, y,$ and z)

How Surfaces Can Represent Functions of Two Variables and Functions of Three Variables

You may have noticed that we have used surfaces to represent functions in two different ways. First, we used a *single* surface to represent a two-variable function $z = f(x, y)$. Second, we used a *family* of level surfaces to represent a three-variable function $w = F(x, y, z)$. These level surfaces have equation $F(x, y, z) = c$.

What is the relation between these two uses of surfaces? For example, consider the equation

$$z = x^2 + y^2 + 3.$$

Define

$$G(x, y, z) = x^2 + y^2 + 3 - z$$

The points which satisfy $z = x^2 + y^2 + 3$ also satisfy $x^2 + y^2 + 3 - z = 0$. Thus the surface $z = x^2 + y^2 + 3$ is the same as the level surface

$$G(x, y, z) = x^2 + y^2 + 3 - z = 0.$$

Thus, we have the following result:

> A single surface representing a two-variable function $z = f(x, y)$ can always be thought of as one member of the family of level surfaces representing a three-variable function $G(x, y, z) = f(x, y) - z$. The graph of $z = f(x, y)$ is the level surface $G = 0$.

Conversely, a single member of a family of level surfaces can be regarded as the graph of a function of the form $z = f(x, y)$ if it is possible to solve for z. For example, if $F(x, y, z) = x^2 + y^2 + z^2$, then one member of the family of level surfaces is the sphere

$$x^2 + y^2 + z^2 = 1.$$

This equation defines z implicitly as a function of x and y. Solving it gives two functions

$$z = \sqrt{1 - x^2 - y^2} \quad \text{and} \quad z = -\sqrt{1 - x^2 - y^2}.$$

The graph of the first function is the top half of the sphere and the graph of the second function is the bottom half.

Problems for Section 11.6

1. Hot water is entering a rectangular swimming pool at the surface of the pool in one corner. Sketch possible contour diagrams for the temperature of the pool at the surface and the temperature one meter below the surface.

2. Figure 11.90 shows contour diagrams of temperature in degrees Celsius in a room at three different times. Describe the heat flow in the room. What could be causing this?

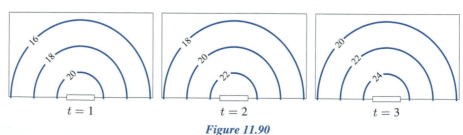

$t = 1$ $t = 2$ $t = 3$

Figure 11.90

3. Match the following functions with the level surfaces in Figure 11.91.

    (a) $f(x, y, z) = y^2 + z^2$ (b) $h(x, y, z) = x^2 + z^2.$

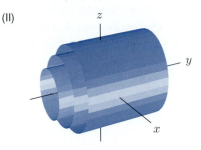

Figure 11.91

4. Describe in words the level surfaces of $f(x, y, z) = \sin(x + y + z)$.

5. Draw contour diagrams for three different cross-sections with t fixed, of the function

$$f(x, y, t) = \cos t \cos \sqrt{x^2 + y^2}, \quad 0 \leq \sqrt{x^2 + y^2} \leq \pi/2.$$

6. The height (in meters) of the water above the bottom of a pond at time t is given by the function $h(x, y, t) = 20 + \sin(x + y - t)$, where x and y are measured horizontally with the positive y-axis north and the positive x-axis east, and where t is in seconds. By considering contour diagrams for different t values, describe the motion of the water surface in the pond.

7. Find the linear function $f(x, y, z) = ax + by + cz + d$ that has the contour diagrams for the cross-sections with $z = 3$ and $z = 4$ shown in Figures 11.92 and 11.93.

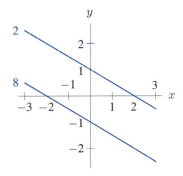
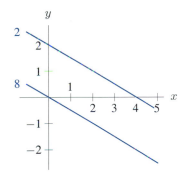

Figure 11.92: Cross-section with $z = 3$ *Figure 11.93:* Cross-section with $z = 4$

In Problems 8–9, give the equation of the linear function $f(x, y, z) = ax + by + cz + d$ that has the given values for the cross-sections with $z = 1$ and $z = 4$.

8.

TABLE 11.14
Cross-section with $z = 1$

y \ x	3	5
0	4	8
1		

TABLE 11.15
Cross-section with $z = 4$

y \ x	3	5
0		14
1		9

9.

TABLE 11.16
Cross-section with $z = 1$

y \ x	3	5
0	4	
1		8

TABLE 11.17
Cross-section with $z = 4$

y \ x	3	5
0		
1	14	9

10. In Problems 8–9, suppose the two tables had only three values filled in. Could you determine f? Suppose the two tables had five values filled in. Could you always determine f?

11. Describe in words the level surfaces of $g(x, y, z) = e^{-(x^2+y^2+z^2)}$.

12. What do the level surfaces of $f(x, y, z) = x^2 - y^2 + z^2$ look like? [Hint: Use cross-sections with y constant instead of cross-sections with z constant.]

Use the catalog of surfaces to identify the surfaces in Problems 13–19.

13. $-x^2 - y^2 + z^2 = 1$

14. $-x^2 + y^2 - z^2 = 0$

15. $x^2 + y^2 - z = 0$

16. $x^2 + z^2 = 1$

17. $x^2 + y^2/4 + z^2 = 1$

18. $x + y = 1$

19. $(x - 1)^2 + y^2 + z^2 = 1$

20. Describe the surface $x^2 + y^2 = (2 + \sin z)^2$. In general, if $f(z) \geq 0$ for all z, describe the surface $x^2 + y^2 = (f(z))^2$.

CHAPTER SUMMARY

- **3-Space**
    Cartesian coordinates, x-, y- and z-axes, xy-, xz- and yz-planes, distance formula.

- **Functions of Two Variables**
    Represented by: Tables, graphs, formulas, cross-sections (one variable fixed), contours (function value fixed); cylinders (one variable missing).

- **Linear Functions**
    Recognizing linear functions from tables, graphs, contour diagrams, formulas. Converting from one representation to another.

- **Functions of Three Variables**
    Sketching level surfaces (function value fixed) in 3-space; graph of $z = f(x, y)$ is same as level surface $F(x, y, z) = f(x, y) - z = 0$.

REVIEW PROBLEMS FOR CHAPTER ELEVEN

1. Describe the set of points whose x coordinate is 2 and whose y coordinate is 1.

2. Find the center and radius of the sphere with equation $x^2 + 4x + y^2 - 6y + z^2 + 12z = 0$.

3. Find the equation of the plane through the points $(0, 0, 2), (0, 3, 0), (5, 0, 0)$.

4. Find a linear function whose graph is the plane that intersects the xy-plane along the line $y = 2x + 2$ and contains the point $(1, 2, 2)$.

5. Consider the function $f(r, h) = \pi r^2 h$ which gives the volume of a cylinder of radius r and height h. Sketch the cross-sections of f, first by keeping h fixed, then by keeping r fixed.

6. Consider the function $z = \cos \sqrt{x^2 + y^2}$.

    (a) Sketch the level curves of this function.
    (b) Sketch a cross-section through the surface $z = \cos \sqrt{x^2 + y^2}$ in the plane containing the x- and z-axes. Put units on your axes.
    (c) Sketch the cross-section through the surface $z = \cos \sqrt{x^2 + y^2}$ in the plane containing the z-axis and the line $y = x$ in the xy-plane.

For Problems 7–10, use a computer or calculator to sketch the graph of a function with the given shapes. Include the axes and the equation used to generate it in your sketch.

7. A cone of circular cross-section opening downward and with its vertex at the origin.

8. A bowl which opens upward and has its vertex at 5 on the z-axis.

9. A plane which has its x, y, and z intercepts all positive.

10. A parabolic cylinder opening upward from along the line $y = x$ in the xy-plane.

Decide if the statements in Problems 11–15 must be true, might be true, or could not be true. The function $z = f(x, y)$ is defined everywhere.

11. The level curves corresponding to $z = 1$ and $z = -1$ cross at the origin.

12. The level curve $z = 1$ consists of the circle $x^2 + y^2 = 2$ and the circle $x^2 + y^2 = 3$, but no other points.

13. The level curve $z = 1$ consists of two lines which intersect at the origin.

14. If $z = e^{-(x^2+y^2)}$, there is a level curve for every value of z.

15. If $z = e^{-(x^2+y^2)}$, there is a level curve through every point (x, y).

For each of the functions in Problems 16–19, make a contour plot in the region $-2 < x < 2$ and $-2 < y < 2$. In each case, what is the equation and the shape of the contour lines?

16. $z = \sin y$ 　　　　17. $z = 3x - 5y + 1$ 　　18. $z = 2x^2 + y^2$ 　　　19. $z = e^{-2x^2-y^2}$

20. Suppose you are in a room 30 feet long with a heater at one end. In the morning the room is $65°$F. You turn on the heater, which quickly warms up to $85°$F. Let $H(x, t)$ be the temperature x feet from the heater, t minutes after the heater is turned on. Figure 11.94 shows the contour diagram for H. How warm is it 10 feet from the heater 5 minutes after it was turned on? 10 minutes after it was turned on?

21. Using the contour diagram in Figure 11.94, sketch the graphs of the one-variable functions $H(x, 5)$ and $H(x, 20)$. Interpret the two graphs in practical terms, and explain the difference between them.

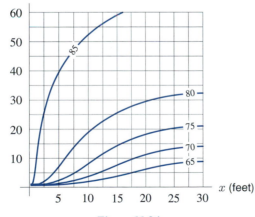

Figure 11.94

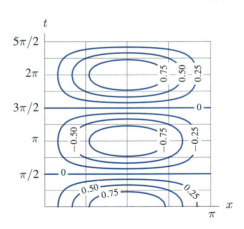

Figure 11.95

22. Figure 11.95 shows the contour diagram for the vibrating string function from page 570:

$$f(x, t) = \cos t \sin x, \quad 0 \le x \le \pi.$$

Using the diagram, describe in words the cross-sections of f with t fixed and the cross-sections of f with x fixed. Explain what you see in terms of the behavior of the string.

Find equations for the linear functions with the contour diagrams in Problems 23–24.

23.

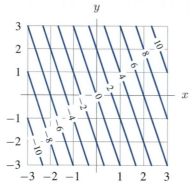

Figure 11.96: Contour map of $g(x, y)$

24.

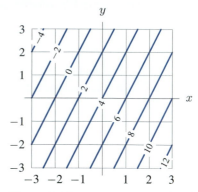

Figure 11.97: Contour map of $h(x, y)$

PROJECTS

1. **A Heater in a Room**

   Figure 11.98 shows the contours of the temperature along one wall of a heated room through one winter day, with time indicated as on a 24-hour clock. The room has a heater located at the left-most corner of the wall and one window in the wall. The heater is controlled by a thermostat about 2 feet from the window.

   (a) Where is the window?
   (b) When is the window open?
   (c) When is the heat on?
   (d) Draw graphs of the temperature along the wall of the room at 6 am, at 11 am, at 3 pm (15 hours) and at 5 pm (17 hours).
   (e) Draw a graph of the temperature as a function of time at the heater, at the window and midway between them.
   (f) The temperature at the window at 5 pm (17 hours) is less than at 11 am. Why do you think this might be?
   (g) To what temperature do you think the thermostat is set? How do you know?
   (h) Where is the thermostat?

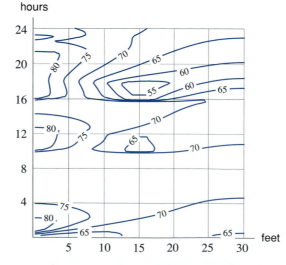

Figure 11.98

2. **Light in a Wave-guide**

    Figure 11.99 shows the contours of light intensity as a function of location and time in a microscopic wave-guide.

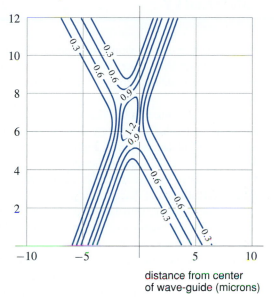

time (nanoseconds)

distance from center
of wave-guide (microns)

Figure 11.99

(a) Draw graphs showing intensity as a function of location at times 0, 2, 4, 6, 8, and 10 nanoseconds.

(b) If you could create an animation showing how the graph of intensity as a function of location varies as time passes, what would it look like?

(c) Draw a graph of intensity as a function of time at locations −5, 0, and 5 microns from center of wave-guide.

(d) Describe what the light beams are doing in this wave-guide.

FOCUS ON THEORY

LIMITS AND CONTINUITY

The sheer vertical face of Half Dome, in Yosemite National Park in California, was caused by glacial activity during the Ice Age. (See Figure 11.100.) The height of the terrain rises abruptly by nearly 1000 feet as we scale the rock from the west, whereas it is possible to make a gradual climb to the top from the east.

If we consider the function h giving the height of the terrain above sea level in terms of longitude and latitude, then h has a *discontinuity* along the path at the base of the cliff of Half Dome. Looking at the contour map of the region in Figure 11.101, we see that in most places a small change in position results in a small change in height, except near the cliff. There, no matter how small a step we take, we get a large change in height. (You can see how crowded the contours get near the cliff; some end abruptly along the discontinuity.)

This geological feature illustrates the ideas of continuity and discontinuity. Roughly speaking, a function is said to be *continuous* at a point if its values at places near the point are close to the value at the point. If this is not the case, the function is said to be *discontinuous*.

The property of continuity is one that, practically speaking, we usually assume of the functions we are studying. Informally, we expect (except under special circumstances) that values of a function do not change drastically when making small changes to the input variables. Whenever we model a one-variable function by an unbroken curve, we are making this assumption. Even when functions come to us as tables of data, we usually make the assumption that the missing function values between data points are close to the measured ones.

In this section we study limits and continuity a bit more formally in the context of functions of several variables. For simplicity we study these concepts for functions of two variables, but our discussion can be adapted to functions of three or more variables.

One can show that sums, products, and compositions of continuous functions are continuous, while the quotient of two continuous functions is continuous everywhere the denominator function is nonzero. Thus, each of the functions

$$\cos(x^2 y), \qquad \ln(x^2 + y^2), \qquad \frac{e^{x+y}}{x + y}, \qquad \ln(\sin(x^2 + y^2))$$

is continuous at all points (x, y) where it is defined. As for functions of one variable, the graph of a continuous function over an unbroken domain is unbroken—that is, the surface has no holes or rips in it.

Figure 11.100: Half Dome in Yosemite National Park

Figure 11.101: A contour map of Half Dome

Example 1   From Figures 11.102–11.105, which of the following functions appear to be continuous at $(0,0)$?

(a) $f(x,y) = \begin{cases} \dfrac{x^2 y}{x^2 + y^2}, & (x,y) \neq (0,0), \\ 0, & (x,y) = (0,0). \end{cases}$

(b) $g(x,y) = \begin{cases} \dfrac{x^2}{x^2 + y^2}, & (x,y) \neq (0,0), \\ 0, & (x,y) = (0,0). \end{cases}$

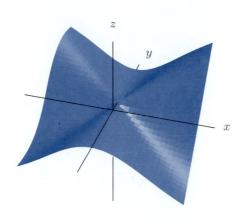

Figure 11.102: Graph of $z = x^2 y/(x^2 + y^2)$

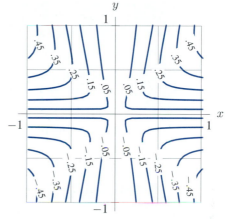

Figure 11.103: Contour diagram of $z = x^2 y/(x^2 + y^2)$

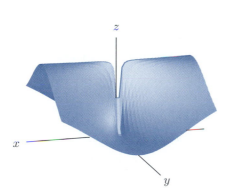

Figure 11.104: Graph of $z = x^2/(x^2 + y^2)$

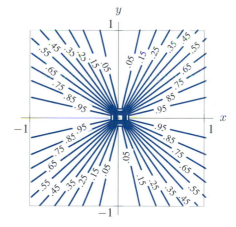

Figure 11.105: Contour diagram of $z = x^2/(x^2 + y^2)$

Solution   (a)   The graph and contour diagram of f in Figures 11.102 and 11.103 suggest that f is close to 0 when (x,y) is close to $(0,0)$. That is, the figures suggest that f is continuous at the point $(0,0)$; the graph appears to have no rips or holes there.

However, the figures cannot tell us for sure whether f is continuous. To be certain we must investigate the limit analytically, as is done in Example 2(a) on page 614.

(b)   The graph of g and its contours near $(0,0)$ in Figure 11.104 and 11.105 suggest that g behaves differently from f: The contours of g seem to "crash" at the origin and the graph rises rapidly from 0 to 1 near $(0,0)$. Small changes in (x,y) near $(0,0)$ can yield large changes in g, so we expect that g is not continuous at the point $(0,0)$. Again, a more precise analysis is given in Example 2(b) on page 614.

The previous example suggests that continuity *at* a point depends on a function's behavior *near* the point. To study behavior near a point more formally we need to define the limit of a function of two variables. Suppose that $f(x, y)$ is a function defined on a set in 2-space, not necessarily containing the point (a, b), but containing points (x, y) arbitrarily close to (a, b); suppose that L is a number.

The function f has a **limit** L at the point (a, b), written

$$\lim_{(x,y) \to (a,b)} f(x, y) = L,$$

if the difference $|f(x, y) - L|$ is as small as we wish whenever the distance from the point (x, y) to the point (a, b) is sufficiently small, but not zero.

We define continuity for functions of two variables in the same way as for functions of one variable:

A function f is **continuous at the point** (a, b) if

$$\lim_{(x,y) \to (a,b)} f(x, y) = f(a, b).$$

A function is **continuous** if it is continuous at each point of its domain.

Thus, if f is continuous at the point (a, b), then f must be defined at (a, b) and the limit, $\lim_{(x,y) \to (a,b)} f(x, y)$, must exist and be equal to the value $f(a, b)$. If a function is defined at a point (a, b) but is not continuous there, then we say that f is *discontinuous* at (a, b).

We now apply the definition of continuity to the functions in Example 1, showing that f is continuous at $(0, 0)$ and that g is discontinuous at $(0, 0)$.

Example 2   Let f and g be the functions defined everywhere on 2-space except at the origin as follows

(a) $f(x, y) = \dfrac{x^2 y}{x^2 + y^2}$ (b) $g(x, y) = \dfrac{x^2}{x^2 + y^2}$

Use the definition of the limit to show that $\displaystyle\lim_{(x,y) \to (0,0)} f(x, y) = 0$ and that $\displaystyle\lim_{(x,y) \to (0,0)} g(x, y)$ does not exist.

Solution   (a)   The graph and contour diagram of f both suggest that $\displaystyle\lim_{(x,y) \to (0,0)} f(x, y) = 0$. To use the definition of the limit, we must estimate $|f(x, y) - L|$ with $L = 0$:

$$|f(x, y) - L| = \left| \frac{x^2 y}{x^2 + y^2} - 0 \right| = \left| \frac{x^2}{x^2 + y^2} \right| |y| \le |y| \le \sqrt{x^2 + y^2},$$

Now $\sqrt{x^2 + y^2}$ is the distance from (x, y) to $(0, 0)$. Thus, to make $|f(x, y) - 0| < 0.001$, for example, we need only require (x, y) be within 0.001 of $(0, 0)$. More generally, for any positive number u, no matter how small, we are sure that $|f(x, y) - 0| < u$ whenever (x, y) is no farther than u from $(0, 0)$. This is what we mean by saying that the difference $|f(x, y) - 0|$ can be made as small as we wish by choosing the distance to be sufficiently small. Thus, we conclude that

$$\lim_{(x,y) \to (0,0)} \frac{x^2 y}{x^2 + y^2} = 0.$$

Notice that the function f has a limit at the point $(0, 0)$ even though f was not defined at $(0, 0)$. To make f continuous at $(0, 0)$ we must define its value there to be 0, as we did in Example 1.

(b)   Although the formula defining the function g looks similar to that of f, we saw in Example 1 that g's behavior near the origin is quite different. If we consider points $(x, 0)$ lying along the x-axis near $(0, 0)$, then the values $g(x, 0)$ are equal to 1, while if we consider points $(0, y)$ lying along the y-axis near $(0, 0)$, then the values $g(0, y)$ are equal to 0. Thus, within any disk (no matter how small) centered at the origin, there are points where $g = 0$ and points where $g = 1$. Therefore the limit $\lim_{(x,y)\to(0,0)} g(x, y)$ does not exist.

While the notions of limit and continuity look formally the same for one- and two-variable functions, they are somewhat more subtle in the multivariable case. The reason for this is that on the line (1-space), we can approach a point from just two directions (left or right) but in 2-space there are an infinite number of ways to approach a given point.

Problems on Limits and Continuity

1.   Show that the function f does not have a limit at $(0, 0)$ by examining the limits of f as $(x, y) \to (0, 0)$ along the curve $y = kx^2$ for different values of k. The function is given by

$$f(x, y) = \frac{x^2}{x^2 + y}, \qquad x^2 + y \neq 0.$$

2.   Show that the function f does not have a limit at $(0, 0)$ by examining the limits of f as $(x, y) \to (0, 0)$ along the line $y = x$ and along the parabola $y = x^2$. The function is given by

$$f(x, y) = \frac{x^2 y}{x^4 + y^2}, \qquad (x, y) \neq (0, 0).$$

3.   Consider the following function:

$$f(x, y) = \begin{cases} \dfrac{xy(x^2 - y^2)}{x^2 + y^2}, & (x, y) \neq (0, 0), \\ 0, & (x, y) = (0, 0). \end{cases}$$

   (a)   Use a computer to draw the graph and the contour diagram of f.
   (b)   Do your answers to part (a) suggest that f is continuous at $(0, 0)$? Explain your answer.

4.   Consider the function f, whose graph and contour diagram are in Figures 11.106 and 11.107, and which is given by

$$f(x, y) = \begin{cases} \dfrac{xy}{x^2 + y^2}, & (x, y) \neq (0, 0), \\ 0, & (x, y) = (0, 0). \end{cases}$$

   (a)   Show that $f(0, y)$ and $f(x, 0)$ are each continuous functions of one variable.
   (b)   Show that rays emanating from the origin are contained in contours of f.
   (c)   Is f continuous at $(0, 0)$?

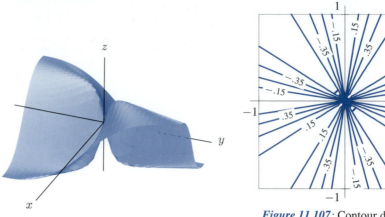

Figure 11.106: Graph of $z = xy/(x^2 + y^2)$

Figure 11.107: Contour diagram of $z = xy/(x^2 + y^2)$

For Problems 5–9 compute the limits of the functions $f(x, y)$ as $(x, y) \to (0, 0)$. You may assume that polynomials, exponentials, logarithmic, and trigonometric functions are continuous.

5. $f(x, y) = x^2 + y^2$

6. $f(x, y) = e^{-x-y}$

7. $f(x, y) = \dfrac{x}{x^2 + 1}$

8. $f(x, y) = \dfrac{x + y}{(\sin y) + 2}$

9. $f(x, y) = \dfrac{\sin(x^2 + y^2)}{x^2 + y^2}$ [Hint: You may assume that $\lim_{t \to 0}(\sin t)/t = 1$.]

For the functions in Problems 10–12, show that $\lim_{(x,y) \to (0,0)} f(x, y)$ does not exist.

10. $f(x, y) = \dfrac{x + y}{x - y}$, $x \neq y$

11. $f(x, y) = \dfrac{x^2 - y^2}{x^2 + y^2}$

12. $f(x, y) = \dfrac{xy}{|xy|}$, $x \neq 0$ and $y \neq 0$

13.   Show that the contours of the function g defined in Example 1(b) on page 613 are rays emanating from the origin. Find the slope of the contour $g(x, y) = c$.

14.   Explain why the following function is not continuous along the line $y = 0$.

$$f(x, y) = \begin{cases} 1 - x, & y \geq 0, \\ -2, & y < 0, \end{cases}$$

In Problems 15–16, determine whether there is a value for c making the function continuous everywhere. If so, find it. If not, explain why not.

15. $f(x, y) = \begin{cases} c + y, & x \leq 3, \\ 5 - y, & x > 3. \end{cases}$

16. $f(x, y) = \begin{cases} c + y, & x \leq 3, \\ 5 - x, & x > 3. \end{cases}$

CHAPTER TWELVE

A FUNDAMENTAL TOOL: VECTORS

In one-variable calculus we represented quantities such as velocity by numbers. However, to specify the velocity of a moving object in space, we need to say how fast it is moving and in what direction it is moving. In this chapter *vectors* are used to represent quantities that have direction as well as magnitude.

12.1 DISPLACEMENT VECTORS

Suppose you are a pilot planning a flight from Dallas to Pittsburgh. There are two things you must know: the distance to be traveled (so you have enough fuel to make it) and in what direction to go (so you don't miss Pittsburgh). Both these quantities together specify the displacement or *displacement vector* between the two cities.

> The **displacement vector** from one point to another is an arrow with its tail at the first point and its tip at the second. The **magnitude** (or length) of the displacement vector is the distance between the points, and is represented by the length of the arrow. The **direction** of the displacement vector is the direction of the arrow.

Figure 12.1 shows the displacement vectors from Dallas to Pittsburgh, from Albuquerque to Oshkosh, and from Los Angeles to Buffalo, SD. These displacement vectors have the same length and the same direction. We say that the displacement vectors between the corresponding cities are the same, even though they do not coincide. In other words

> Displacement vectors which point in the same direction and have the same magnitude are considered to be the same, even if they do not coincide.

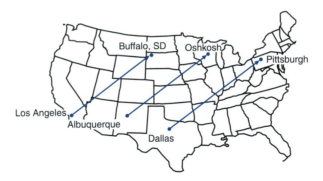

Figure 12.1: Displacement vectors between cities

Notation and Terminology

The displacement vector is our first example of a vector. Vectors have both magnitude and direction; in comparison, a quantity specified only by a number, but no direction, is called a *scalar*.[1] For instance, the time taken by the flight from Dallas to Pittsburgh is a scalar quantity. Displacement is a vector since it requires both distance and direction to specify it.

In this book, vectors are written with an arrow over them, $\vec{v}$, to distinguish them from scalars. Other books use a bold **v** to denote a vector. We use the notation $\overrightarrow{PQ}$ to denote the displacement vector from a point P to a point Q. The magnitude, or length, of a vector $\vec{v}$ is written $\|\vec{v}\|$.

Addition and Subtraction of Displacement Vectors

Suppose NASA commands a robot on Mars to move 75 meters in one direction and then 50 meters in another direction. (See Figure 12.2.) Where does the robot end up? Suppose the displacements are represented by the vectors $\vec{v}$ and $\vec{w}$, respectively. Then the sum $\vec{v} + \vec{w}$ gives the final position.

---

[1]So named by W. R. Hamilton because they are merely numbers on the *scale* from $-\infty$ to ∞.

The **sum**, $\vec{v} + \vec{w}$, of two vectors $\vec{v}$ and $\vec{w}$ is the combined displacement resulting from first applying $\vec{v}$ and then $\vec{w}$. (See Figure 12.3.) The sum $\vec{w} + \vec{v}$ gives the same displacement.

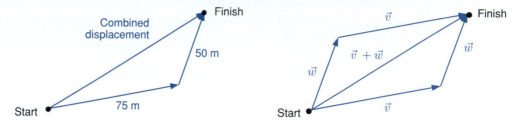

Figure 12.2: Sum of displacements of robots on Mars   *Figure 12.3:* The sum $\vec{v} + \vec{w} = \vec{w} + \vec{v}$

Suppose two different robots start from the same location. One moves along a displacement vector $\vec{v}$ and the second along a displacement vector $\vec{w}$. What is the displacement vector, $\vec{x}$, from the first robot to the second? (See Figure 12.4.) Since $\vec{v} + \vec{x} = \vec{w}$, we define $\vec{x}$ to be the difference $\vec{x} = \vec{w} - \vec{v}$. In other words, $\vec{w} - \vec{v}$ gets you from $\vec{v}$ to $\vec{w}$.

The **difference**, $\vec{w} - \vec{v}$, is the displacement vector which when added to $\vec{v}$ gives $\vec{w}$. That is, $\vec{w} = \vec{v} + (\vec{w} - \vec{v})$. (See Figure 12.4.)

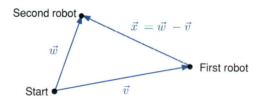

Figure 12.4: The difference $\vec{w} - \vec{v}$

If the robot ends up where it started, then its total displacement vector is the *zero vector*, $\vec{0}$. The zero vector has no direction.

The **zero vector**, $\vec{0}$, is a displacement vector with zero length.

Scalar Multiplication of Displacement Vectors

If $\vec{v}$ represents a displacement vector, the vector $2\vec{v}$ represents a displacement of twice the magnitude in the same direction as $\vec{v}$. Similarly, $-2\vec{v}$ represents a displacement of twice the magnitude in the opposite direction. (See Figure 12.5.)

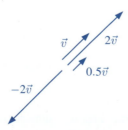

Figure 12.5: Scalar multiples of the vector $\vec{v}$

> If λ is a scalar and $\vec{v}$ is a displacement vector, the **scalar multiple of** $\vec{v}$ **by** λ, written $\lambda\vec{v}$, is the displacement vector with the following properties:
>
> - The displacement vector $\lambda\vec{v}$ is parallel to $\vec{v}$, pointing in the same direction if $\lambda > 0$, and in the opposite direction if $\lambda < 0$.
> - The magnitude of $\lambda\vec{v}$ is $|\lambda|$ times the magnitude of $\vec{v}$, that is, $\|\lambda\vec{v}\| = |\lambda|\,\|\vec{v}\|$.

Note that $|\lambda|$ represents the absolute value of the scalar λ while $\|\lambda\vec{v}\|$ represents the magnitude of the vector $\lambda\vec{v}$.

Example 1   Explain why $\vec{w} - \vec{v} = \vec{w} + (-1)\vec{v}$.

Solution   The vector $(-1)\vec{v}$ has the same magnitude as $\vec{v}$, but points in the opposite direction. Figure 12.6 shows that the combined displacement $\vec{w} + (-1)\vec{v}$ is the same as the displacement $\vec{w} - \vec{v}$.

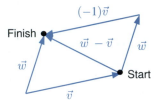

Figure 12.6: Explanation for why $\vec{w} - \vec{v} = \vec{w} + (-1)\vec{v}$

Parallel Vectors

Two vectors $\vec{v}$ and $\vec{w}$ are *parallel* if one is a scalar multiple of the other, that is, if $\vec{w} = \lambda\vec{v}$.

Components of Displacement Vectors: The Vectors $\vec{i}$, $\vec{j}$, and $\vec{k}$

Suppose that you live in a city with equally spaced streets running east-west and north-south and that you want to tell someone how to get from one place to another. You'd be likely to tell them how many blocks east-west and how many blocks north-south to go. For example, to get from P to Q in Figure 12.7, we go 4 blocks east and 1 block south. If $\vec{i}$ and $\vec{j}$ are as shown in Figure 12.7, then the displacement vector from P to Q is $4\vec{i} - \vec{j}$.

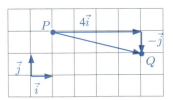

Figure 12.7: The displacement vector from P to Q is $4\vec{i} - \vec{j}$

We extend the same idea to 3-dimensions. First we choose a Cartesian system of coordinate axes. The three vectors of length 1 shown in Figure 12.8 are the vector $\vec{i}$, which points along the positive x-axis, the vector $\vec{j}$, along the positive y-axis, and the vector $\vec{k}$, along the positive z-axis.

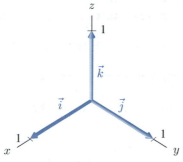

Figure 12.8: The vectors $\vec{i}$, $\vec{j}$ and $\vec{k}$ in 3-space

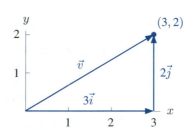

Figure 12.9: We resolve $\vec{v}$ into components by writing $\vec{v} = 3\vec{i} + 2\vec{j}$

Writing Displacement Vectors Using $\vec{i}$, $\vec{j}$, $\vec{k}$

Any displacement in 3-space or the plane can be expressed as a combination of displacements in the coordinate directions. For example, Figure 12.9 shows that the displacement vector $\vec{v}$ from the origin to the point $(3, 2)$ can be written as a sum of displacement vectors along the x and y-axes:

$$\vec{v} = 3\vec{i} + 2\vec{j}.$$

This is called *resolving $\vec{v}$ into components*. In general:

> We **resolve** $\vec{v}$ into components by writing $\vec{v}$ in the form
>
> $$\vec{v} = v_1\vec{i} + v_2\vec{j} + v_3\vec{k}.$$
>
> We call $v_1\vec{i}$, $v_2\vec{j}$, and $v_3\vec{k}$ the **components** of $\vec{v}$.

An Alternative Notation for Vectors

Many people write a vector in 3-dimensions as a string of three numbers, that is, as

$$\vec{v} = (v_1, v_2, v_3) \quad \text{instead of} \quad \vec{v} = v_1\vec{i} + v_2\vec{j} + v_3\vec{k}.$$

Since the first notation can be confused with a point and the second cannot, we usually use the second form.

Example 2   Resolve the displacement vector, $\vec{v}$, from the point $P_1 = (2, 4, 10)$ to the point $P_2 = (3, 7, 6)$ into components.

Solution   To get from P_1 to P_2, we move 1 unit in the positive x-direction, 3 units in the positive y-direction, and 4 units in the negative z-direction. Hence $\vec{v} = \vec{i} + 3\vec{j} - 4\vec{k}$.

Example 3   Decide whether the vector $\vec{v} = 2\vec{i} + 3\vec{j} + 5\vec{k}$ is parallel to each of the following vectors:

$$\vec{w} = 4\vec{i} + 6\vec{j} + 10\vec{k}, \quad \vec{a} = -\vec{i} - 1.5\vec{j} - 2.5\vec{k}, \quad \vec{b} = 4\vec{i} + 6\vec{j} + 9\vec{k}$$

Solution   Since $\vec{w} = 2\vec{v}$ and $\vec{a} = -0.5\vec{v}$, the vectors $\vec{v}$, $\vec{w}$, and $\vec{a}$ are parallel. However, $\vec{b}$ is not a multiple of $\vec{v}$ (since, for example, $4/2 \neq 9/5$), so $\vec{v}$ and $\vec{b}$ are not parallel.

In general, Figure 12.10 shows us how to express the displacement vector between two points in components:

Components of Displacement Vectors

The displacement vector from the point $P_1 = (x_1, y_1, z_1)$ to the point $P_2 = (x_2, y_2, z_2)$ is given in components by

$$\overrightarrow{P_1P_2} = (x_2 - x_1)\vec{i} + (y_2 - y_1)\vec{j} + (z_2 - z_1)\vec{k} .$$

Position Vectors: Displacement of a Point from the Origin

A displacement vector whose tail is at the origin is called a *position vector*. Thus, any point (x_0, y_0, z_0) in space has associated with it the position vector $\vec{r}_0 = x_0\vec{i} + y_0\vec{j} + z_0\vec{k}$. (See Figure 12.11.) In general, a position vector gives the displacement of a point from the origin.

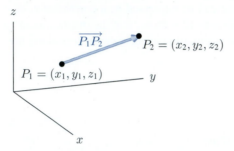

Figure 12.10: The displacement vector
$\overrightarrow{P_1P_2} = (x_2 - x_1)\vec{i} + (y_2 - y_1)\vec{j} + (z_2 - z_1)\vec{k}$

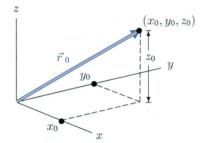

Figure 12.11: The position vector
$\vec{r}_0 = x_0\vec{i} + y_0\vec{j} + z_0\vec{k}$

The Components of the Zero Vector

The zero displacement vector has magnitude equal to zero and is written $\vec{0}$. So $\vec{0} = 0\vec{i} + 0\vec{j} + 0\vec{k}$.

The Magnitude of a Vector in Components

For a vector, $\vec{v} = v_1\vec{i} + v_2\vec{j}$, the Pythagorean theorem is used to find its magnitude, $\|\vec{v}\|$. (See Figure 12.12.)

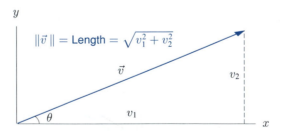

Figure 12.12: Magnitude, $\|\vec{v}\|$, of a 2-dimensional vector, $\vec{v}$

In 3-dimensions, for a vector $\vec{v} = v_1\vec{i} + v_2\vec{j} + v_3\vec{k}$, we have

---

Magnitude of $\vec{v}$ $= \|\vec{v}\| =$ Length of the arrow $= \sqrt{v_1^2 + v_2^2 + v_3^2}.$

For instance, if $\vec{v} = 3\vec{i} - 4\vec{j} + 5\vec{k}$, then $\|\vec{v}\| = \sqrt{3^2 + (-4)^2 + 5^2} = \sqrt{50}.$

Addition and Scalar Multiplication of Vectors in Components

Suppose the vectors $\vec{v}$ and $\vec{w}$ are given in components:

$$\vec{v} = v_1\vec{i} + v_2\vec{j} + v_3\vec{k} \quad \text{and} \quad \vec{w} = w_1\vec{i} + w_2\vec{j} + w_3\vec{k}.$$

Then

$$\vec{v} + \vec{w} = (v_1 + w_1)\vec{i} + (v_2 + w_2)\vec{j} + (v_3 + w_3)\vec{k},$$

and

$$\lambda\vec{v} = \lambda v_1\vec{i} + \lambda v_2\vec{j} + \lambda v_3\vec{k}.$$

Figures 12.13 and 12.14 illustrate these properties in two dimensions. Finally, $\vec{v} - \vec{w} = \vec{v} + (-1)\vec{w}$, so we can write $\vec{v} - \vec{w} = (v_1 - w_1)\vec{i} + (v_2 - w_2)\vec{j} + (v_3 - w_3)\vec{k}$.

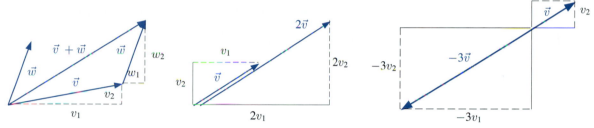

Figure 12.13: Sum $\vec{v} + \vec{w}$ in components

Figure 12.14: Scalar multiples of vectors showing $\vec{v}$, $2\vec{v}$, and $-3\vec{v}$

How to Resolve a Vector into Components

You may wonder how we find the components of a 2-dimensional vector, given its length and direction. Suppose the vector $\vec{v}$ has length v and makes an angle of θ with the x-axis, measured counterclockwise, as in Figure 12.15. If $\vec{v} = v_1\vec{i} + v_2\vec{j}$, Figure 12.15 shows that

$$v_1 = v\cos\theta \quad \text{and} \quad v_2 = v\sin\theta.$$

Thus, we resolve $\vec{v}$ into components by writing

$$\vec{v} = (v\cos\theta)\vec{i} + (v\sin\theta)\vec{j}.$$

Vectors in 3-space are resolved using direction cosines; see Problem 29 on page 651.

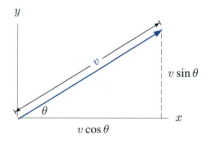

Figure 12.15: Resolving a vector: $\vec{v} = (v\cos\theta)\vec{i} + (v\sin\theta)\vec{j}$

Example 4   Resolve $\vec{v}$ into components if $v = 2$ and $\theta = \pi/6$.

Solution   We have $\vec{v} = 2\cos(\pi/6)\vec{i} + 2\sin(\pi/6)\vec{j} = 2(\sqrt{3}/2)\vec{i} + 2(1/2)\vec{j} = \sqrt{3}\vec{i} + \vec{j}$.

Unit Vectors

A *unit vector* is a vector whose magnitude is 1. The vectors $\vec{i}$, $\vec{j}$, and $\vec{k}$ are unit vectors in the directions of the coordinate axes. It is often helpful to find a unit vector in the same direction as a given vector $\vec{v}$. Suppose that $\|\vec{v}\| = 10$; a unit vector in the same direction as $\vec{v}$ is $\vec{v}/10$. In general, a unit vector in the direction of any nonzero vector $\vec{v}$ is

$$\vec{u} = \frac{\vec{v}}{\|\vec{v}\|}.$$

Example 5   Find a unit vector, $\vec{u}$, in the direction of the vector $\vec{v} = \vec{i} + 3\vec{j}$.

Solution   If $\vec{v} = \vec{i} + 3\vec{j}$, then $\|\vec{v}\| = \sqrt{1^2 + 3^2} = \sqrt{10}$. Thus, a unit vector in the same direction is given by

$$\vec{u} = \frac{\vec{v}}{\sqrt{10}} = \frac{1}{\sqrt{10}}(\vec{i} + 3\vec{j}) = \frac{1}{\sqrt{10}}\vec{i} + \frac{3}{\sqrt{10}}\vec{j} \approx 0.32\vec{i} + 0.95\vec{j}.$$

Example 6   Find a unit vector at the point (x, y, z) that points radially outward away from the origin.

Solution   The vector from the origin to (x, y, z) is the position vector

$$\vec{r} = x\vec{i} + y\vec{j} + z\vec{k}.$$

Thus, if we put its tail at (x, y, z) it will point away from the origin. Its magnitude is

$$\|\vec{r}\| = \sqrt{x^2 + y^2 + z^2},$$

so a unit vector pointing in the same direction is

$$\frac{\vec{r}}{\|\vec{r}\|} = \frac{x\vec{i} + y\vec{j} + z\vec{k}}{\sqrt{x^2 + y^2 + z^2}} = \frac{x}{\sqrt{x^2 + y^2 + z^2}}\vec{i} + \frac{y}{\sqrt{x^2 + y^2 + z^2}}\vec{j} + \frac{z}{\sqrt{x^2 + y^2 + z^2}}\vec{k}.$$

Problems for Section 12.1

1.   The vectors $\vec{w}$ and $\vec{u}$ are in Figure 12.16. Match the vectors $\vec{p}, \vec{q}, \vec{r}, \vec{s}, \vec{t}$ with five of the following vectors: $\vec{u} + \vec{w}$, $\vec{u} - \vec{w}$, $\vec{w} - \vec{u}$, $2\vec{w} - \vec{u}$, $\vec{u} - 2\vec{w}$, $2\vec{w}$, $-2\vec{w}$, $2\vec{u}$, $-2\vec{u}$, $-\vec{w}$, $-\vec{u}$.

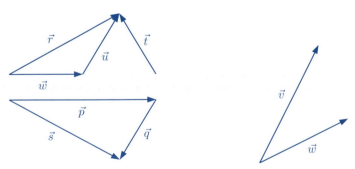

Figure 12.16   **Figure 12.17**

2.   Given the displacement vectors $\vec{v}$ and $\vec{w}$ in Figure 12.17, draw the following vectors:
(a) $\vec{v} + \vec{w}$ (b) $\vec{v} - \vec{w}$ (c) $2\vec{v}$ (d) $2\vec{v} + \vec{w}$ (e) $\vec{v} - 2\vec{w}$.

For Problems 3–8, perform the indicated operations on the following vectors:

$$\vec{a} = 2\vec{j} + \vec{k}, \quad \vec{b} = -3\vec{i} + 5\vec{j} + 4\vec{k}, \quad \vec{c} = \vec{i} + 6\vec{j},$$

$$\vec{x} = -2\vec{i} + 9\vec{j}, \quad \vec{y} = 4\vec{i} - 7\vec{j}, \quad \vec{z} = \vec{i} - 3\vec{j} - \vec{k}.$$

3. $\|\vec{z}\|$

4. $\vec{a} + \vec{z}$

5. $5\vec{b}$

6. $2\vec{c} + \vec{x}$

7. $\|\vec{y}\|$

8. $2\vec{a} + 7\vec{b} - 5\vec{z}$

For Problems 9–12, perform the indicated computation.

9. $(4\vec{i} + 2\vec{j}) - (3\vec{i} - \vec{j})$

10. $(\vec{i} + 2\vec{j}) + (-3)(2\vec{i} + \vec{j})$

11. $-4(\vec{i} - 2\vec{j}) - 0.5(\vec{i} - \vec{k})$

12. $2(0.45\vec{i} - 0.9\vec{j} - 0.01\vec{k}) - 0.5(1.2\vec{i} - 0.1\vec{k})$

Find the length of the vectors in Problems 13–16.

13. $\vec{v} = \vec{i} - \vec{j} + 3\vec{k}$

14. $\vec{v} = \vec{i} - \vec{j} + 2\vec{k}$

15. $\vec{v} = 1.2\vec{i} - 3.6\vec{j} + 4.1\vec{k}$

16. $\vec{v} = 7.2\vec{i} - 1.5\vec{j} + 2.1\vec{k}$

A cat is sitting on the ground at the point $(1, 4, 0)$ watching a squirrel at the top of a tree. The tree is one unit high and its base is at the point $(2, 4, 0)$. Find the displacement vectors in Problems 17–20.

17. From the origin to the cat.

18. From the bottom of the tree to the squirrel.

19. From the bottom of the tree to the cat.

20. From the cat to the squirrel.

21. On the graph of Figure 12.18, draw the vector $\vec{v} = 4\vec{i} + \vec{j}$ twice, once with its tail at the origin and once with its tail at the point $(3, 2)$.

Figure 12.18

Figure 12.19: Scale: 1 grid length = 0.25 inches

Resolve the vectors in Problems 22–26 into components.

22. The vector shown in Figure 12.19.

23. A vector starting at the point $P = (1, 2)$ and ending at the point $Q = (4, 6)$.

24. A vector starting at the point $Q = (4, 6)$ and ending at the point $P = (1, 2)$.

25.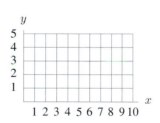

26.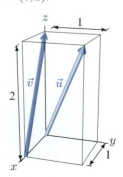

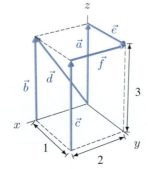

27. Find the length of the vectors $\vec{u}$ and $\vec{v}$ in Problem 26.

28. Find a vector that points in the same direction as $\vec{i} - \vec{j} + 2\vec{k}$, but has length 2.

29. (a) Find a unit vector from the point $P = (1, 2)$ and toward the point $Q = (4, 6)$.
    (b) Find a vector of length 10 pointing in the same direction.

30. Which of the following vectors are parallel?

$$\vec{u} = 2\vec{i} + 4\vec{j} - 2\vec{k}, \quad \vec{v} = \vec{i} - \vec{j} + 3\vec{k}, \quad \vec{w} = -\vec{i} - 2\vec{j} + \vec{k},$$
$$\vec{p} = \vec{i} + \vec{j} + \vec{k}, \quad \vec{q} = 4\vec{i} - 4\vec{j} + 12\vec{k}, \quad \vec{r} = \vec{i} - \vec{j} + \vec{k}.$$

31. Figure 12.20 shows a molecule with four atoms at O, A, B and C. Verify that every atom in the molecule is 2 units away from every other atom.

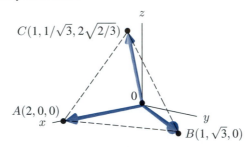

Figure 12.20

For Problems 32–33, consider the map in Figure 11.1 on page 564.

32. If you leave Topeka along the following vectors, does the temperature increase or decrease?
    (a) $\vec{u} = 3\vec{i} + 2\vec{j}$ (b) $\vec{v} = -\vec{i} - \vec{j}$ (c) $\vec{w} = -5\vec{i} - 5\vec{j}$.

33. Starting in Buffalo, sketch a vector pointing in the direction in which the temperature is increasing most rapidly. Starting in Boise, sketch a vector pointing in the direction in which the temperature is decreasing most rapidly.

34. A truck is traveling due north at 30 km/hr approaching a crossroad. On a perpendicular road a police car is traveling west toward the intersection at 40 km/hr. Suppose that both vehicles will reach the crossroad in exactly one hour. Find the vector currently representing the displacement of the truck with respect to the police car.

35. Show that the medians of a triangle intersect at a point $1/3$ of the way along each median from the side it bisects.

36. Show that the lines joining the centroid (the intersection point of the medians) of a face of the tetrahedron and the opposite vertex meet at a point $\frac{1}{4}$ of the way from each centroid to its opposite vertex.

12.2 VECTORS IN GENERAL

Besides displacement, there are many quantities that have both magnitude and direction and are added and multiplied by scalars in the same way as displacements. Any such quantity is called a *vector* and is represented by an arrow in the same manner we represent displacements. The length of the arrow is the *magnitude* of the vector, and the direction of the arrow is the direction of the vector.

Velocity Versus Speed

The speed of a moving body tells us how fast it is moving, say 80 km/hr. The speed is just a number; it is therefore a scalar. The velocity, on the other hand, tells us both how fast the body is moving and the direction of motion; it is a vector. For instance, if a car is heading northeast at 80 km/hr, then its velocity is a vector of length 80 pointing northeast.

> The **velocity vector** of a moving object is a vector whose magnitude is the speed of the object and whose direction is the direction of its motion.

Example 1   A car is traveling north at a speed of 100 km/hr, while a plane above is flying horizontally south-west at a speed of 500 km/hr. Draw the velocity vectors of the car and the plane.

Solution   Figure 12.21 shows the velocity vectors. The plane's velocity vector is five times as long as the car's, because its speed is five times as great.

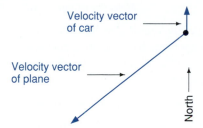

Figure 12.21: Velocity vector of the car is 100 km/hr north and of the plane is 500 km/hr south-west

The next example illustrates that the velocity vectors for two motions add to give the velocity vector for the combined motion, just as displacements do.

Example 2   A riverboat is moving with velocity $\vec{v}$ and a speed of 8 km/hr relative to the water. In addition, the river has a current $\vec{c}$ and a speed of 1 km/hr. (See Figure 12.22.) What is the physical significance of the vector $\vec{v} + \vec{c}$?

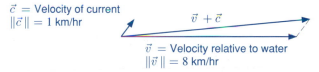

Figure 12.22: Boat's velocity relative to the river bed is the sum, $\vec{v} + \vec{c}$

Solution   The vector $\vec{v}$ shows how the boat is moving relative to the water, while $\vec{c}$ shows how the water is moving relative to the riverbed. During an hour, imagine that the boat first moves 8 km relative to the water, which remains still; this displacement is represented by $\vec{v}$. Then imagine the water moving 1 km while the boat remains stationary relative to the water; this displacement is represented by $\vec{c}$. The combined displacement is represented by $\vec{v} + \vec{c}$. Thus, the vector $\vec{v} + \vec{c}$ is the velocity of the boat relative to the riverbed.

Note that the effective speed of the boat is not necessarily 9 km/hr unless the boat is moving in the direction of the current. Although we add the velocity vectors, we do not necessarily add their lengths.

Scalar multiplication also makes sense for velocity vectors. For example, if $\vec{v}$ is a velocity vector, then $-2\vec{v}$ represents a velocity of twice the magnitude in the opposite direction.

Example 3   A ball is moving with velocity $\vec{v}$ when it hits a wall at a right angle and bounces straight back, with its speed reduced by 20%. Express its new velocity in terms of the old one.

Solution   The new velocity is $-0.8\vec{v}$, where the negative sign expresses the fact that the new velocity is in the direction opposite to the old.

We can represent velocity vectors in components in the same way we did on page 623.

Example 4   Represent the velocity vectors of the car and the plane in Example 1 using components. Take north to be the positive y-axis, east to be the positive x-axis, and upward to be the positive z-axis.

Solution   The car is traveling north at 100 km/hr, so the y-component of its velocity is $100\vec{j}$ and the x-component is $0\vec{i}$. Since it is traveling horizontally, the z-component is $0\vec{k}$. So we have

$$\text{Velocity of car} = 0\vec{i} + 100\vec{j} + 0\vec{k} = 100\vec{j}.$$

The plane's velocity vector also has $\vec{k}$ component equal to zero. Since it is traveling southwest, its $\vec{i}$ and $\vec{j}$ components have negative coefficients (north and east are positive). Since the plane is traveling at 500 km/hr, in one hour it is displaced $500/\sqrt{2} \approx 354$ km to the west and 354 km to the south. (See Figure 12.23.) Thus,

$$\text{Velocity of plane} = -(500\cos 45°)\vec{i} - (500\sin 45°)\vec{j} \approx -354\vec{i} - 354\vec{j}.$$

Of course, if the car were climbing a hill or if the plane were descending for a landing, then the $\vec{k}$ component would not be zero.

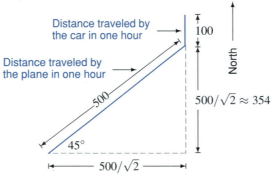

Figure 12.23: Distance traveled by the plane and car in one hour

Acceleration

Another example of a vector quantity is acceleration. Acceleration, like velocity, is specified by both a magnitude and a direction — for example, the acceleration due to gravity is 9.81 m/sec^2 vertically downward.

Force

Force is another example of a vector quantity. Suppose you push on an open door. The result depends both on how hard you push and in what direction. Thus, to specify a force we must give its magnitude (or strength) and the direction in which it is acting. For example, the gravitational force exerted on an object by the earth is a vector pointing from the object toward the center of the earth; its magnitude is the strength of the gravitational force.

Example 5   The earth travels around the sun in an ellipse. The gravitational force on the earth and the velocity of the earth are governed by the following laws:
Newton's Law of Gravitation: The magnitude of the gravitational attraction, F, between two masses m_1 and m_2 at a distance r apart is given by $F = Gm_1m_2/r^2$, where G is a constant. The force vector lies along the line between the masses.
Kepler's Second Law: The line joining a planet to the sun sweeps out equal areas in equal times.
(a)   Sketch vectors representing the gravitational force of the sun on the earth at two different positions in the earth's orbit.
(b)   Sketch the velocity vector of the earth at two points in its orbit.

Solution (a) Figure 12.24 shows the earth orbiting the sun. Note that the gravitational force vector always points toward the sun and is larger when the earth is closer to the sun because of the r^2 term in the denominator. (In fact, the real orbit looks much more like a circle than we have shown here.)

(b) The velocity vector points in the direction of motion of the earth. Thus, the velocity vector is tangent to the ellipse. See Figure 12.25. Furthermore, the velocity vector is longer at points of the orbit where the planet is moving quickly, because the magnitude of the velocity vector is the speed. Kepler's Second Law enables us to determine when the earth is moving quickly and when it is moving slowly. Over a fixed period of time, say one month, the line joining the earth to the sun sweeps out a sector having a certain area. Figure 12.25 shows two sectors swept out in two different one-month time-intervals. Kepler's law says that the areas of the two sectors are the same. Thus, the earth must move farther in a month when it is close to the sun than when it is far from the sun. Therefore, the earth moves faster when it is closer to the sun and slower when it is farther away.

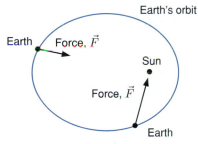

Figure 12.24: Gravitational force, $\vec{F}$, exerted by the sun on the earth: Greater magnitude closer to sun

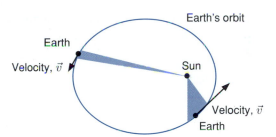

Figure 12.25: The velocity vector, $\vec{v}$, of the earth: Greater magnitude closer to the sun

Properties of Addition and Scalar Multiplication

In general, vectors add, subtract, and are multiplied by scalars in the same way as displacement vectors. Thus, for any vectors $\vec{u}, \vec{v}$, and $\vec{w}$ and any scalars α and β, we have the following properties:

Commutativity   **Associativity**
1. $\vec{v} + \vec{w} = \vec{w} + \vec{v}$ 2. $(\vec{u} + \vec{v}) + \vec{w} = \vec{u} + (\vec{v} + \vec{w})$
                                  3. $\alpha(\beta\vec{v}) = (\alpha\beta)\vec{v}$

Distributivity   **Identity**
4. $(\alpha + \beta)\vec{v} = \alpha\vec{v} + \beta\vec{v}$ 6. $1 \cdot \vec{v} = \vec{v}$ 8. $\vec{v} + \vec{0} = \vec{v}$
5. $\alpha(\vec{v} + \vec{w}) = \alpha\vec{v} + \alpha\vec{w}$ 7. $0 \cdot \vec{v} = \vec{0}$ 9. $\vec{w} + (-1) \cdot \vec{v} = \vec{w} - \vec{v}$

Problems 16–23 at the end of this section ask for a justification of these results in terms of displacement vectors.

Using Components

Example 6   A plane, heading due east at an airspeed of 600 km/hr, experiences a wind of 50 km/hr blowing toward the northeast. Find the plane's direction and ground speed.

Solution   We choose a coordinate system with the x-axis pointing east and the y-axis pointing north. See Figure 12.26.

The airspeed tells us the speed of the plane relative to still air. Thus, the plane is moving due east with velocity $\vec{v} = 600\vec{i}$ relative to still air. In addition, the air is moving with a velocity $\vec{w}$. Writing $\vec{w}$ in components, we have

$$\vec{w} = (50\cos45°)\vec{i} + (50\sin45°)\vec{j} = 35.4\vec{i} + 35.4\vec{j}.$$

The vector $\vec{v} + \vec{w}$ represents the displacement of the plane in one hour relative to the ground. Therefore, $\vec{v} + \vec{w}$ is the velocity of the plane relative to the ground. In components, we have

$$\vec{v} + \vec{w} = 600\vec{i} + \left(35.4\vec{i} + 35.4\vec{j}\right) = 635.4\vec{i} + 35.4\vec{j}.$$

The direction of the plane's motion relative to the ground is given by the angle θ in Figure 12.26, where

$$\tan\theta = \frac{35.4}{635.4}$$

so

$$\theta = \arctan\left(\frac{35.4}{635.4}\right) = 3.2°.$$

The ground speed is the speed of the plane relative to the ground, so

$$\text{Groundspeed} = \sqrt{635.4^2 + 35.4^2} = 636.4 \text{ km/hr.}$$

Thus, the speed of the plane relative to the ground has been increased slightly by the wind. (This is as we would expect, as the wind has a component in the direction in which the plane is traveling.) The angle θ shows how far the plane is blown off course by the wind.

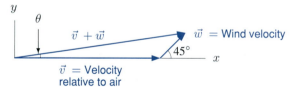

Figure 12.26: Plane's velocity relative to the ground is the sum $\vec{v} + \vec{w}$

Vectors in n Dimensions

Using the alternative notation $\vec{v} = (v_1, v_2, v_3)$ for a vector in 3-space, we can define a vector in n dimensions as a string of n numbers. Thus, a vector in n dimensions can be written as

$$\vec{c} = (c_1, c_2, \ldots, c_n).$$

Addition and scalar multiplication are defined by the formulas

$$\vec{v} + \vec{w} = (v_1, v_2, \ldots, v_n) + (w_1, w_2, \ldots, w_n) = (v_1 + w_1, v_2 + w_2, \ldots, v_n + w_n)$$

and

$$\lambda\vec{v} = \lambda(v_1, v_2, \ldots, v_n) = (\lambda v_1, \lambda v_2, \ldots, \lambda v_n).$$

Why Do We Want Vectors in n Dimensions?

Vectors in two and three dimensions can be used to model displacement, velocities, or forces. But what about vectors in n dimensions? There is another interpretation of 3-dimensional vectors (or 3-vectors) which is useful: they can be thought of as listing 3 different quantities — for example, the displacements parallel to the x, y, and z axes. Similarly, the n-vector

$$\vec{c} = (c_1, c_2, \ldots, c_n)$$

can be thought of as a way of keeping n different quantities organized. For example, a *population* vector $\vec{N}$ shows the number of children and adults in a population:

$$\vec{N} = (\text{Number of children, Number of adults}),$$

or, if we are interested in a more detailed breakdown of ages, we might give the number in each ten-year age bracket in the population (up to age 110) in the form

$$\vec{N} = (N_1, N_2, N_3, N_4, \ldots, N_{10}, N_{11}),$$

where N_1 is the population aged 0–9, and N_2 is the population aged 10–19, and so on.

A *consumption* vector ,

$$\vec{q} = (q_1, q_2, \ldots, q_n)$$

shows the quantities q_1, q_2, ..., q_n consumed of each of n different goods. A *price* vector

$$\vec{p} = (p_1, p_2, \ldots, p_n)$$

contains the prices of n different items.

In 1907, Hermann Minkowski used vectors with four components when he introduced *space-time coordinates*, whereby each event is assigned a vector position $\vec{v}$ with four coordinates, three for its position in space and one for time:

$$\vec{v} = (x, y, z, t).$$

Example 7   Suppose the vector $\vec{I}$ represents the number of copies, in thousands, made by each of four copy centers in the month of December and $\vec{J}$ represents the number of copies made at the same four copy centers during the previous eleven months (the "year-to-date"). If $\vec{I} = (25, 211, 818, 642)$, and $\vec{J} = (331, 3227, 1377, 2570)$, compute $\vec{I} + \vec{J}$. What does this sum represent?

Solution   The sum is

$$\vec{I} + \vec{J} = (25 + 331, 211 + 3227, 818 + 1377, 642 + 2570) = (356, 3438, 2195, 3212).$$

Each term in $\vec{I} + \vec{J}$ represents the sum of the number of copies made in December plus those in the previous eleven months, that is, the total number of copies made during the entire year at that particular copy center.

Example 8   The price vector $\vec{p} = (p_1, p_2, p_3)$ represents the prices in dollars of three goods. Write a vector which gives the prices of the same goods in cents.

Solution   The prices in cents are $100p_1$, $100p_2$, and $100p_3$ respectively, so the new price vector is

$$(100p_1, 100p_2, 100p_3) = 100\vec{p}.$$

Problems for Section 12.2

In Problems 1–4, say whether the given quantity is a vector or a scalar.

1. The distance from Seattle to St. Louis.

2. The population of the US.

3. The magnetic field at a point on the earth's surface.

4. The temperature at a point on the earth's surface.

5. A car is traveling at a speed of 50 km/hr. Assume the positive y-axis is north and the positive x-axis is east. Resolve the car's velocity vector (in 2-space) into components if the car is traveling in each of the following directions:
    (a) East   (b) South   (c) Southeast   (d) Northwest.

6. Which is traveling faster, a car whose velocity vector is $21\vec{i} + 35\vec{j}$, or a car whose velocity vector is $40\vec{i}$, assuming that the units are the same for both directions?

7. A moving object has velocity vector $50\vec{i} + 20\vec{j}$ in meters per second. Express the velocity in kilometers per hour.

8. A car drives clockwise around the track in Figure 12.27, slowing down at the curves and speeding up along the straight portions. Sketch velocity vectors at the points P, Q, and R.

9. A racing car drives clockwise around the track shown in Figure 12.27 at a constant speed. At what point on the track does the car have the longest acceleration vector, and in roughly what direction is it pointing? (Recall that acceleration is the rate of change of velocity.)

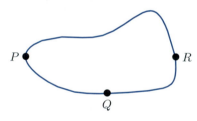

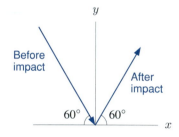

Figure 12.27   *Figure 12.28*

10. A particle moving with speed v hits a barrier at an angle of $60°$ and bounces off at an angle of $60°$ in the opposite direction with speed reduced by 20 percent, as shown in Figure 12.28. Find the velocity vector of the object after impact.

11. There are five students in a class. Their scores on the midterm (out of 100) are given by the vector $\vec{v} = (73, 80, 91, 65, 84)$. Their scores on the final (out of 100) are given by $\vec{w} = (82, 79, 88, 70, 92)$. If the final counts twice as much as the midterm, find a vector giving the total scores (as a percentage) of the students.

12. Shortly after taking off, a plane is climbing northwest through still air at an airspeed of 200 km/hr, and rising at a rate of 300 m/min. Resolve into components its velocity vector in a coordinate system in which the x-axis points east, the y-axis points north, and the z-axis points up.

13. An airplane is heading northeast at an airspeed of 700 km/hr, but there is a wind blowing from the west at 60 km/hr. In what direction does the plane end up flying? What is its speed relative to the ground?

14. An airplane is flying at an airspeed of 600 km/hr in a cross-wind that is blowing from the northeast at a speed of 50 km/hr. In what direction should the plane head to end up going due east?

15. A plane is heading due east and climbing at the rate of 80 km/hr. If its airspeed is 480 km/hr and there is a wind blowing 100 km/hr to the northeast, what is the ground speed of the plane?

Use the geometric definition of addition and scalar multiplication to explain each of the properties in Problems 16–23.

16. $\vec{w} + \vec{v} = \vec{v} + \vec{w}$

17. $(\alpha + \beta)\vec{v} = \alpha\vec{v} + \beta\vec{v}$

18. $\alpha(\vec{v} + \vec{w}) = \alpha\vec{v} + \alpha\vec{w}$

19. $(\vec{u} + \vec{v}) + \vec{w} = \vec{u} + (\vec{v} + \vec{w})$

20. $\alpha(\beta\vec{v}) = (\alpha\beta)\vec{v}$

21. $\vec{v} + \vec{0} = \vec{v}$

22. $1\vec{v} = \vec{v}$

23. $\vec{v} + (-1)\vec{w} = \vec{v} - \vec{w}$

12.3 THE DOT PRODUCT

We have seen how to add vectors; can we multiply two vectors together? In the next two sections we will see two different ways of doing so: the *scalar product* (or *dot product*) which produces a scalar, and the *vector product* (or *cross product*), which produces a vector.

Definition of the Dot Product

The dot product links geometry and algebra. We already know how to calculate the length of a vector from its components; the dot product gives us a way of computing the angle between two vectors. For any two vectors $\vec{v} = v_1\vec{i} + v_2\vec{j} + v_3\vec{k}$ and $\vec{w} = w_1\vec{i} + w_2\vec{j} + w_3\vec{k}$, shown in Figure 12.29, we define a scalar as follows:

> The following two definitions of the **dot product**, or **scalar product**, $\vec{v} \cdot \vec{w}$, are equivalent:
> - **Geometric definition**
> $\vec{v} \cdot \vec{w} = \|\vec{v}\|\|\vec{w}\| \cos\theta$ where θ is the angle between $\vec{v}$ and $\vec{w}$ and $0 \leq \theta \leq \pi$.
> - **Algebraic definition**
> $\vec{v} \cdot \vec{w} = v_1 w_1 + v_2 w_2 + v_3 w_3$.
>
> Notice that the dot product of two vectors is a *number*.

Why don't we give just one definition of $\vec{v} \cdot \vec{w}$? The reason is that both definitions are equally important; the geometric definition gives us a picture of what the dot product means and the algebraic definition gives us a way of calculating it.

How do we know the two definitions are equivalent — that is, they really do define the same thing? First, we observe that the two definitions give the same result in a particular example. Then we show why they are equivalent in general.

Figure 12.29: The vectors $\vec{v}$ and $\vec{w}$

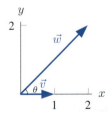

Figure 12.30: Calculating the dot product of the vectors $v = \vec{i}$ and $\vec{w} = 2\vec{i} + 2\vec{j}$ geometrically and algebraically gives the same result

Example 1 Suppose $\vec{v} = \vec{i}$ and $\vec{w} = 2\vec{i} + 2\vec{j}$. Compute $\vec{v} \cdot \vec{w}$ both geometrically and algebraically.

Solution To use the geometric definition, see Figure 12.30. The angle between the vectors is $\pi/4$, or $45°$, and the lengths of the vectors are given by

$$\|\vec{v}\| = 1 \quad \text{and} \quad \|\vec{w}\| = 2\sqrt{2}.$$

Thus,

$$\vec{v} \cdot \vec{w} = \|\vec{v}\|\|\vec{w}\| \cos\theta = 1 \cdot 2\sqrt{2} \cos\left(\frac{\pi}{4}\right) = 2.$$

Using the algebraic definition, we get the same result:

$$\vec{v} \cdot \vec{w} = 1 \cdot 2 + 0 \cdot 2 = 2.$$

Why the Two Definitions of the Dot Product Give the Same Result

In the previous example, the two definitions give the same value for the dot product. To show that the geometric and algebraic definitions of the dot product always give the same result, we must show that, for any vectors $\vec{v} = v_1\vec{i} + v_2\vec{j} + v_3\vec{k}$ and $\vec{w} = w_1\vec{i} + w_2\vec{j} + w_3\vec{k}$ with an angle θ between them:

$$\|\vec{v}\|\|\vec{w}\| \cos\theta = v_1 w_1 + v_2 w_2 + v_3 w_3.$$

One method follows; a method which does not use trigonometry is given in Problem 32 on page 641.

Using the Law of Cosines. Suppose that $0 < \theta < \pi$, so that the vectors $\vec{v}$ and $\vec{w}$ form a triangle. (See Figure 12.31.) By the Law of Cosines, we have

$$\|\vec{v} - \vec{w}\|^2 = \|\vec{v}\|^2 + \|\vec{w}\|^2 - 2\|\vec{v}\|\|\vec{w}\| \cos\theta.$$

This result is also true for $\theta = 0$ and $\theta = \pi$. We calculate the lengths using components:

$$\|\vec{v}\|^2 = v_1^2 + v_2^2 + v_3^2$$
$$\|\vec{w}\|^2 = w_1^2 + w_2^2 + w_3^2$$
$$\|\vec{v} - \vec{w}\|^2 = (v_1 - w_1)^2 + (v_2 - w_2)^2 + (v_3 - w_3)^2$$
$$= v_1^2 - 2v_1 w_1 + w_1^2 + v_2^2 - 2v_2 w_2 + w_2^2 + v_3^2 - 2v_3 w_3 + w_3^2.$$

Substituting into the Law of Cosines and canceling, we see that

$$-2v_1 w_1 - 2v_2 w_2 - 2v_3 w_3 = -2\|\vec{v}\|\|\vec{w}\| \cos\theta.$$

Therefore we have the result we wanted, namely that:

$$v_1 w_1 + v_2 w_2 + v_3 w_3 = \|\vec{v}\|\|\vec{w}\| \cos\theta.$$

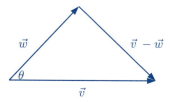

Figure 12.31: Triangle used in the justification of $\|\vec{v}\|\|\vec{w}\| \cos\theta = v_1 w_1 + v_2 w_2 + v_3 w_3$

Properties of the Dot Product

The following properties of the dot product can be justified using the algebraic definition; see Problem 25 on page 640. For a geometric interpretation of Property 3, see Problem 29.

Properties of the Dot Product. For any vectors $\vec{u}$, $\vec{v}$, and $\vec{w}$ and any scalar λ,

1. $\vec{v} \cdot \vec{w} = \vec{w} \cdot \vec{v}$
2. $\vec{v} \cdot (\lambda\vec{w}) = \lambda(\vec{v} \cdot \vec{w}) = (\lambda\vec{v}) \cdot \vec{w}$
3. $(\vec{v} + \vec{w}) \cdot \vec{u} = \vec{v} \cdot \vec{u} + \vec{w} \cdot \vec{u}$

Perpendicularity, Magnitude, and Dot Products

Two vectors are perpendicular if the angle between them is $\pi/2$ or $90°$. Since $\cos(\pi/2) = 0$, if $\vec{v}$ and $\vec{w}$ are perpendicular, then $\vec{v} \cdot \vec{w} = 0$. Conversely, provided that $\vec{v} \cdot \vec{w} = 0$, then $\cos\theta = 0$, so $\theta = \pi/2$ and the vectors are perpendicular. Thus, we have the following result:

> Two nonzero vectors $\vec{v}$ and $\vec{w}$ are **perpendicular**, or **orthogonal**, if and only if
>
> $$\vec{v} \cdot \vec{w} = 0.$$

For example: $\vec{i} \cdot \vec{j} = 0, \vec{j} \cdot \vec{k} = 0, \vec{i} \cdot \vec{k} = 0$.

If we take the dot product of a vector with itself, then $\theta = 0$ and $\cos\theta = 1$. For any vector $\vec{v}$:

> Magnitude and dot product are related as follows:
>
> $$\vec{v} \cdot \vec{v} = \|\vec{v}\|^2.$$

For example: $\vec{i} \cdot \vec{i} = 1, \vec{j} \cdot \vec{j} = 1, \vec{k} \cdot \vec{k} = 1$.

Using the Dot Product

Depending on the situation, one definition of the dot product may be more convenient to use than the other. In Example 2 which follows, the geometric definition is the only one which can be used because we are not given components. In Example 3, the algebraic definition is used.

Example 2   Suppose the vector $\vec{b}$ is fixed and has length 2; the vector $\vec{a}$ is free to rotate and has length 3. What are the maximum and minimum values of the dot product $\vec{a} \cdot \vec{b}$ as the vector $\vec{a}$ rotates through all possible positions? What positions of $\vec{a}$ and $\vec{b}$ lead to these values?

Solution   The geometric definition gives $\vec{a} \cdot \vec{b} = \|\vec{a}\|\|\vec{b}\|\cos\theta = 3 \cdot 2\cos\theta = 6\cos\theta$. Thus, the maximum value of $\vec{a} \cdot \vec{b}$ is 6, and it occurs when $\cos\theta = 1$ so $\theta = 0$, that is, when $\vec{a}$ and $\vec{b}$ point in the same direction. The minimum value of $\vec{a} \cdot \vec{b}$ is -6, and it occurs when $\cos\theta = -1$ so $\theta = \pi$, that is, when $\vec{a}$ and $\vec{b}$ point in opposite directions. (See Figure 12.32.)

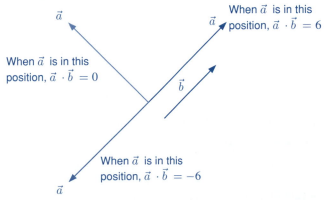

Figure 12.32: Maximum and minimum values of $\vec{a} \cdot \vec{b}$ obtained from a fixed vector $\vec{b}$ of length 2 and rotating vector $\vec{a}$ of length 3

Example 3   Which pairs from the following list of 3-dimensional vectors are perpendicular to one another?

$$\vec{u} = \vec{i} + \sqrt{3}\,\vec{k}, \quad \vec{v} = \vec{i} + \sqrt{3}\,\vec{j}, \quad \vec{w} = \sqrt{3}\,\vec{i} + \vec{j} - \vec{k}.$$

Solution   The geometric definition tells us that two vectors are perpendicular if and only if their dot product is zero. Since the vectors are given in components, we calculate dot products using the algebraic definition:

$$\vec{v} \cdot \vec{u} = (\vec{i} + \sqrt{3}\,\vec{j} + 0\vec{k}) \cdot (\vec{i} + 0\vec{j} + \sqrt{3}\,\vec{k}) = 1 \cdot 1 + \sqrt{3} \cdot 0 + 0 \cdot \sqrt{3} = 1,$$
$$\vec{v} \cdot \vec{w} = (\vec{i} + \sqrt{3}\,\vec{j} + 0\vec{k}) \cdot (\sqrt{3}\,\vec{i} + \vec{j} - \vec{k}) = 1 \cdot \sqrt{3} + \sqrt{3} \cdot 1 + 0(-1) = 2\sqrt{3},$$
$$\vec{w} \cdot \vec{u} = (\sqrt{3}\,\vec{i} + \vec{j} - \vec{k}) \cdot (\vec{i} + 0\vec{j} + \sqrt{3}\,\vec{k}) = \sqrt{3} \cdot 1 + 1 \cdot 0 + (-1) \cdot \sqrt{3} = 0.$$

So the only two vectors which are perpendicular are $\vec{w}$ and $\vec{u}$.

Normal Vectors and the Equation of a Plane

In Section 11.5 we wrote the equation of a plane given its x-slope, y-slope and z-intercept. Now we write the equation of a plane using a vector and a point lying on the plane. A *normal vector* to a plane is a vector that is perpendicular to the plane, that is, it is perpendicular to every displacement vector between any two points in the plane. Let $\vec{n} = a\vec{i} + b\vec{j} + c\vec{k}$ be a normal vector to the plane, let $P_0 = (x_0, y_0, z_0)$ be a fixed point in the plane, and let $P = (x, y, z)$ be any other point in the plane. Then $\overrightarrow{P_0P} = (x - x_0)\vec{i} + (y - y_0)\vec{j} + (z - z_0)\vec{k}$ is a vector whose head and tail both lie in the plane. (See Figure 12.33.) Thus, the vectors $\vec{n}$ and $\overrightarrow{P_0P}$ are perpendicular, so $\vec{n} \cdot \overrightarrow{P_0P} = 0$. The algebraic definition of the dot product gives $\vec{n} \cdot \overrightarrow{P_0P} = a(x - x_0) + b(y - y_0) + c(z - z_0)$, so we obtain the following result:

The **equation of the plane** with normal vector $\vec{n} = a\vec{i} + b\vec{j} + c\vec{k}$ and containing the point $P_0 = (x_0, y_0, z_0)$ is

$$a(x - x_0) + b(y - y_0) + c(z - z_0) = 0.$$

Letting $d = ax_0 + by_0 + cz_0$ (a constant), we can write the equation of the plane in the form

$$ax + by + cz = d.$$

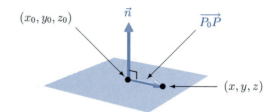

Figure 12.33: Plane with normal $\vec{n}$ and containing a fixed point (x_0, y_0, z_0)

Example 4   Find the equation of the plane perpendicular to $-\vec{i} + 3\vec{j} + 2\vec{k}$ and passing through the point $(1, 0, 4)$.

Solution   The equation of the plane is

$$-(x - 1) + 3(y - 0) + 2(z - 4) = 0,$$

which simplifies to

$$-x + 3y + 2z = 7.$$

Example 5   Find a normal vector to the plane with equation   (a) $x - y + 2z = 5$ (b) $z = 0.5x + 1.2y$.

Solution   (a)   Since the coefficients of $\vec{i}$, $\vec{j}$, and $\vec{k}$ in a normal vector are the coefficients of x, y, z in the equation of the plane, a normal vector is $\vec{n} = \vec{i} - \vec{j} + 2\vec{k}$.

(b)   Before we can find a normal vector, we rewrite the equation of the plane in the form

$$0.5x + 1.2y - z = 0.$$

Thus, a normal vector is $\vec{n} = 0.5\vec{i} + 1.2\vec{j} - \vec{k}$.

The Dot Product in n Dimensions

The algebraic definition of the dot product can be extended to vectors in higher dimensions.

> If $\vec{u} = (u_1, \ldots, u_n)$ and $\vec{v} = (v_1, \ldots, v_n)$ then the dot product of $\vec{u}$ and $\vec{v}$ is the **scalar**
>
> $$\vec{u} \cdot \vec{v} = u_1 v_1 + \ldots + u_n v_n.$$

Example 6   A video store sells videos, tapes, CDs, and computer games. We define the quantity vector $\vec{q} = (q_1, q_2, q_3, q_4)$, where q_1, q_2, q_3, q_4 denote the quantities sold of each of the items, and the price vector $\vec{p} = (p_1, p_2, p_3, p_4)$, where p_1, p_2, p_3, p_4 denote the price per unit of each item. What does the dot product $\vec{p} \cdot \vec{q}$ represent?

Solution   The dot product is $\vec{p} \cdot \vec{q} = p_1 q_1 + p_2 q_2 + p_3 q_3 + p_4 q_4$. The quantity $p_1 q_1$ represents the revenue received by the store for the videos, $p_2 q_2$ represents the revenue for the tapes, and so on. The dot product represents the total revenue received by the store for the sale of these four items.

Resolving a Vector into Components: Projections

In Section 12.1, we resolved a vector into components parallel to the axes. Now we see how to resolve a vector, $\vec{v}$, into components, called $\vec{v}_{\text{parallel}}$ and $\vec{v}_{\text{perp}}$, which are parallel and perpendicular, respectively, to a given nonzero vector, $\vec{u}$. (See Figure 12.34.)

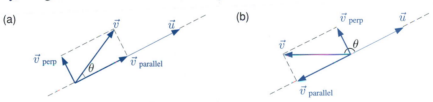

Figure 12.34: Resolving $\vec{v}$ into components parallel and perpendicular to $\vec{u}$
(a) $0 < \theta < \pi/2$ (b) $\pi/2 < \theta < \pi$

The projection of $\vec{v}$ on $\vec{u}$, written $\vec{v}_{\text{parallel}}$, measures (in some sense) how much the vector $\vec{v}$ is aligned with the vector $\vec{u}$. The length of $\vec{v}_{\text{parallel}}$ is the length of the shadow cast by $\vec{v}$ on a line in the direction of $\vec{u}$.

To compute $\vec{v}_{\text{parallel}}$, we assume $\vec{u}$ is a unit vector. (If not, create one by dividing by its length.) Then Figure 12.34(a) shows that, if $0 \le \theta \le \pi/2$:

$$\|\vec{v}_{\text{parallel}}\| = \|\vec{v}\| \cos\theta = \vec{v} \cdot \vec{u} \qquad (\text{since } \|\vec{u}\| = 1).$$

Now $\vec{v}_{\text{parallel}}$ is a scalar multiple of $\vec{u}$, and since $\vec{u}$ is a unit vector,

$$\vec{v}_{\text{parallel}} = (\|\vec{v}\| \cos\theta)\vec{u} = (\vec{v} \cdot \vec{u})\vec{u}.$$

A similar argument shows that if $\pi/2 < \theta \le \pi$, as in Figure 12.34(b), this formula for $\vec{v}_{\text{parallel}}$ still holds. The vector $\vec{v}_{\text{perp}}$ is specified by

$$\vec{v}_{\text{perp}} = \vec{v} - \vec{v}_{\text{parallel}}.$$

Thus, we have the following results:

Projection of $\vec{v}$ on the Line in the Direction of the Unit Vector $\vec{u}$

If $\vec{v}_{\text{parallel}}$ and $\vec{v}_{\text{perp}}$ are components of $\vec{v}$ which are parallel and perpendicular, respectively, to $\vec{u}$, then

$$\text{Projection of } \vec{v} \text{ on to } \vec{u} = \vec{v}_{\text{parallel}} = (\vec{v} \cdot \vec{u})\vec{u} \qquad \text{provided } \|\vec{u}\| = 1$$

and $\qquad \vec{v} = \vec{v}_{\text{parallel}} + \vec{v}_{\text{perp}} \qquad$ so $\qquad \vec{v}_{\text{perp}} = \vec{v} - \vec{v}_{\text{parallel}}.$

Example 7   Figure 12.35 shows the force the wind exerts on the sail of a sailboat. Find the component of the force in the direction in which the sailboat is traveling.

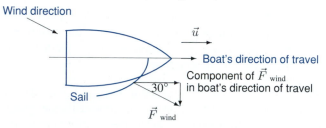

Figure 12.35: Wind moving a sailboat

Solution   Let $\vec{u}$ be a unit vector in the direction of travel. The force of the wind on the sail makes an angle of $30°$ with $\vec{u}$. Thus, the component of this force in the direction of $\vec{u}$ is

$$\vec{F}_{\text{parallel}} = (\vec{F} \cdot \vec{u})\vec{u} = \|\vec{F}\|(\cos 30°)\vec{u} = 0.87\|\vec{F}\|\vec{u}.$$

Thus, the boat is being pushed forward with about 87% of the total force due to the wind. (In fact, the interaction of wind and sail is much more complex than this model suggests.)

A Physical Interpretation of the Dot Product: Work

In physics, the word "work" has a slightly different meaning from its everyday meaning. In physics, when a force of magnitude F acts on an object through a distance d, we say the *work*, W, done by the force is

$$W = Fd,$$

provided the force and the displacement are in the same direction. For example, if a 1 kg body falls 10 meters under the force of gravity, which is 9.8 newtons, then the work done by gravity is

$$W = (9.8 \text{ newtons}) \cdot (10 \text{ meters}) = 98 \text{ joules}.$$

What if the force and the displacement are not in the same direction? Suppose a force $\vec{F}$ acts on an object as it moves along a displacement vector $\vec{d}$. Let θ be the angle between $\vec{F}$ and $\vec{d}$. First, we assume $0 \le \theta \le \pi/2$. Figure 12.36 shows how we can resolve $\vec{F}$ into components that are parallel and perpendicular to $\vec{d}$:

$$\vec{F} = \vec{F}_{\text{parallel}} + \vec{F}_{\text{perp}},$$

Then the work done by $\vec{F}$ is defined to be

$$W = \|\vec{F}_{\text{parallel}}\| \, \|\vec{d}\|.$$

We see from Figure 12.36 that $\vec{F}_{\text{parallel}}$ has magnitude $\|\vec{F}\| \cos\theta$. So the work is given by the dot product:

$$W = (\|\vec{F}\| \cos\theta)\|\vec{d}\| = \|\vec{F}\|\|\vec{d}\| \cos\theta = \vec{F} \cdot \vec{d}.$$

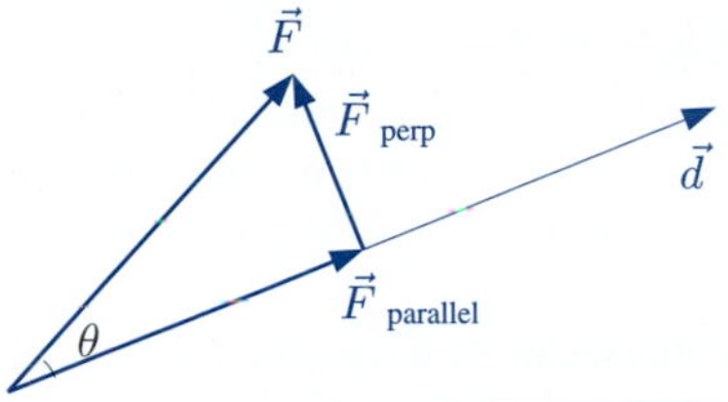

Figure 12.36: Resolving the force $\vec{F}$ into two forces, one parallel to $\vec{d}$, one perpendicular to $\vec{d}$

The formula $W = \vec{F} \cdot \vec{d}$ holds when $\pi/2 < \theta \leq \pi$ also. In that case, the work done by the force is negative and the object is moving against the force. Thus, we have the following definition:

The **work**, W, done by a force $\vec{F}$ acting on an object through a displacement $\vec{d}$ is given by

$$W = \vec{F} \cdot \vec{d}.$$

Notice that if the vectors $\vec{F}$ and $\vec{d}$ are parallel and in the same direction, with magnitudes F and d, then $\cos \theta = \cos 0 = 1$, so $W = \|\vec{F}\|\|\vec{d}\| = Fd$, which is the original definition. When the vectors are perpendicular, $\cos \theta = \cos \frac{\pi}{2} = 0$, so $W = 0$ and no work is done in the technical definition of the word. For example, if you carry a heavy box across the room at the same horizontal height, no work is done by gravity because the force of gravity is vertical but the motion is horizontal.

Problems for Section 12.3

For Problems 1–6, perform the following operations on the given 3-dimensional vectors.

$$\vec{a} = 2\vec{j} + \vec{k} \qquad \vec{b} = -3\vec{i} + 5\vec{j} + 4\vec{k} \qquad \vec{c} = \vec{i} + 6\vec{j} \qquad \vec{y} = 4\vec{i} - 7\vec{j} \qquad \vec{z} = \vec{i} - 3\vec{j} - \vec{k}$$

1. $\vec{c} \cdot \vec{y}$
2. $\vec{a} \cdot \vec{z}$
3. $\vec{a} \cdot \vec{b}$
4. $(\vec{a} \cdot \vec{b})\vec{a}$
5. $(\vec{a} \cdot \vec{y})(\vec{c} \cdot \vec{z})$
6. $((\vec{c} \cdot \vec{c})\vec{a}) \cdot \vec{a}$

7. Compute the angle between the vectors $\vec{i} + \vec{j} + \vec{k}$ and $\vec{i} - \vec{j} - \vec{k}$.
8. Which pairs of the vectors $\sqrt{3}\vec{i} + \vec{j}, 3\vec{i} + \sqrt{3}\vec{j}, \vec{i} - \sqrt{3}\vec{j}$ are parallel and which are perpendicular?
9. For what values of t are $\vec{u} = t\vec{i} - \vec{j} + \vec{k}$ and $\vec{v} = t\vec{i} + t\vec{j} - 2\vec{k}$ perpendicular? Are there values of t for which $\vec{u}$ and $\vec{v}$ are parallel?

In Problems 10–12, find a normal vector to the given plane.

10. $2x + y - z = 5$
11. $z = 3x + 4y - 7$
12. $2(x - z) = 3(x + y)$

In Problems 13–17, find an equation of a plane that satisfies the given conditions.

13. Perpendicular to the vector $-\vec{i} + 2\vec{j} + \vec{k}$ and passing through the point $(1, 0, 2)$.
14. Perpendicular to the vector $5\vec{i} + \vec{j} - 2\vec{k}$ and passing through the point $(0, 1, -1)$.
15. Perpendicular to the vector $2\vec{i} - 3\vec{j} + 7\vec{k}$ and passing through the point $(1, -1, 2)$.
16. Parallel to the plane $2x + 4y - 3z = 1$ and through the point $(1, 0, -1)$.
17. Through the point $(-2, 3, 2)$ and parallel to $3x + y + z = 4$.
18. Let S be the triangle with vertices $A = (2, 2, 2), B = (4, 2, 1)$, and $C = (2, 3, 1)$.

    (a) Find the length of the shortest side of S.
    (b) Find the cosine of the angle BAC at vertex A.

19. Write $\vec{a} = 3\vec{i} + 2\vec{j} - 6\vec{k}$ as the sum of two vectors, one parallel to $\vec{d} = 2\vec{i} - 4\vec{j} + \vec{k}$ and the other perpendicular to $\vec{d}$.
20. Find the points where the plane $z = 5x - 4y + 3$ intersects each of the coordinate axes. Then find the lengths of the sides and the angles of the triangle formed by these points.

21. Find the angle between the planes $5(x-1)+3(y+2)+2z=0$ and $x+3(y-1)+2(z+4)=0$.

22. A basketball gymnasium is 25 meters high, 80 meters wide and 200 meters long. For a half time stunt, the cheerleaders want to run two strings, one from each of the two corners above one basket to the diagonally opposite corners of the gym floor. What is the cosine of the angle made by the strings as they cross?

23. A consumption vector of three goods is defined by $\vec{x} = (x_1, x_2, x_3)$, where x_1, x_2 and x_3 are the quantities consumed of the three goods. Consider a budget constraint represented by the equation $\vec{p} \cdot \vec{x} = k$, where $\vec{p}$ is the price vector of the three goods and k is a constant. Show that the difference between two consumption vectors corresponding to points satisfying the same budget constraint is perpendicular to the price vector $\vec{p}$.

24. A 100-meter dash is run on a track in the direction of the vector $\vec{v} = 2\vec{i} + 6\vec{j}$. The wind velocity $\vec{w}$ is $5\vec{i} + \vec{j}$ km/hr. The rules say that a legal wind speed measured in the direction of the dash must not exceed 5 km/hr. Will the race results be disqualified due to an illegal wind? Justify your answer.

25. Show why each of the properties of the dot product in the box on page 634 follows from the algebraic definition of the dot product:

$$\vec{v} \cdot \vec{w} = v_1 w_1 + v_2 w_2 + v_3 w_3$$

26. What does Property 2 of the dot product in the box on page 634 say geometrically?

27. Show that the vectors $(\vec{b} \cdot \vec{c})\vec{a} - (\vec{a} \cdot \vec{c})\vec{b}$ and $\vec{c}$ are perpendicular.

28. Show that if $\vec{u}$ and $\vec{v}$ are two vectors such that

$$\vec{u} \cdot \vec{w} = \vec{v} \cdot \vec{w}$$

for every vector $\vec{w}$, then

$$\vec{u} = \vec{v}.$$

29. Figure 12.37 shows that, given three vectors $\vec{u}$, $\vec{v}$, and $\vec{w}$, the sum of the components of $\vec{v}$ and $\vec{w}$ in the direction of $\vec{u}$ is the component of $\vec{v} + \vec{w}$ in the direction of $\vec{u}$. (Although the figure is drawn in two dimensions, this result is also true in three dimensions.) Use this figure to explain why the geometric definition of the dot product satisfies $(\vec{v} + \vec{w}) \cdot \vec{u} = \vec{v} \cdot \vec{u} + \vec{w} \cdot \vec{u}$.

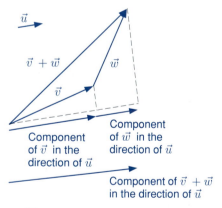

Figure 12.37: The component of $\vec{v} + \vec{w}$ in the direction of $\vec{u}$ is the sum of the components of $\vec{v}$ and $\vec{w}$ in that direction

30. (a) Using the geometric definition of the dot product, show that

$$\vec{u} \cdot (-\vec{v}) = -(\vec{u} \cdot \vec{v}).$$

[Hint: What happens to the angle when you multiply $\vec{v}$ by -1?]

(b) Using the geometric definition of the dot product, show that for any negative scalar λ

$$\vec{u} \cdot (\lambda\vec{v}) = \lambda(\vec{u} \cdot \vec{v})$$
$$(\lambda\vec{u}) \cdot \vec{v} = \lambda(\vec{u} \cdot \vec{v}).$$

31. The Law of Cosines for a triangle with side lengths a, b, and c, and with angle C opposite side c, says

$$c^2 = a^2 + b^2 - 2ab\cos C.$$

On page 634, we used the Law of Cosines to show that the two definitions of the dot product are equivalent. In this problem, you will use the geometric definition of the dot product and its properties in the box on page 634 to prove the Law of Cosines. [Hint: Let $\vec{u}$ and $\vec{v}$ be the displacement vectors from C to the other two vertices, and express c^2 in terms of $\vec{u}$ and $\vec{v}$.]

32. Use the following steps and the results of Problems 29–30 to show (without trigonometry) that the geometric and algebraic definitions of the dot product are equivalent.

    Follow these steps. Let $\vec{u} = u_1\vec{i} + u_2\vec{j} + u_3\vec{k}$ and $\vec{v} = v_1\vec{i} + v_2\vec{j} + v_3\vec{k}$ be any vectors. Write $(\vec{u} \cdot \vec{v})_{\text{geom}}$ for the result of the dot product computed geometrically. Substitute $\vec{u} = u_1\vec{i} + u_2\vec{j} + u_3\vec{k}$ and use Problems 29–30 to expand $(\vec{u} \cdot \vec{v})_{\text{geom}}$. Next substitute for $\vec{v}$ and expand. Finally, calculate geometrically the dot products $\vec{i} \cdot \vec{i}$, $\vec{i} \cdot \vec{j}$, etc.

33. Suppose that $\vec{v}$ and $\vec{w}$ are any two vectors. Consider the following function of t:

$$q(t) = (\vec{v} + t\vec{w}) \cdot (\vec{v} + t\vec{w})$$

(a) Explain why $q(t) \geq 0$ for all real t.
(b) Expand $q(t)$ as a quadratic polynomial in t using the properties on page 634.
(c) Using the discriminant of the quadratic, show that,

$$|\vec{v} \cdot \vec{w}| \leq \|\vec{v}\|\|\vec{w}\|.$$

12.4 THE CROSS PRODUCT

In the previous section we combined two vectors to get a number, the dot product. In this section we see another way of combining two vectors, this time to get a vector, the *cross product*. Any two vectors in 3-space form a parallelogram. We define the cross product using this parallelogram.

The Area of a Parallelogram

Consider the parallelogram formed by the vectors $\vec{v}$ and $\vec{w}$ with an angle of θ between them. Then Figure 12.38 shows

$$\text{Area of parallelogram} = \text{Base} \cdot \text{Height} = \|\vec{v}\|\|\vec{w}\|\sin\theta.$$

How would we compute the area of the parallelogram if we were given $\vec{v}$ and $\vec{w}$ in components, $\vec{v} = v_1\vec{i} + v_2\vec{j} + v_3\vec{k}$ and $\vec{w} = w_1\vec{i} + w_2\vec{j} + w_3\vec{k}$? Project 1 on page 652 shows that if $\vec{v}$ and $\vec{w}$ are in the xy-plane, so $v_3 = w_3 = 0$, then

$$\text{Area of parallelogram} = |v_1 w_2 - v_2 w_1|.$$

What if $\vec{v}$ and $\vec{w}$ do not lie in the xy-plane? The cross product will enable us to compute the area of the parallelogram formed by any two vectors.

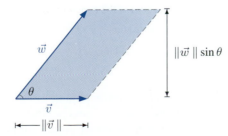

Figure 12.38: Parallelogram formed by $\vec{v}$ and $\vec{w}$ has
Area $= \|\vec{v}\|\|\vec{w}\|\sin\theta$

Definition of the Cross Product

We define the cross product of the vectors $\vec{v}$ and $\vec{w}$, written $\vec{v} \times \vec{w}$, to be a vector perpendicular to both $\vec{v}$ and $\vec{w}$. The magnitude of this vector is the area of the parallelogram formed by the two vectors. The direction of $\vec{v} \times \vec{w}$ is given by the normal vector, $\vec{n}$, to the plane defined by $\vec{v}$ and $\vec{w}$. If we require that $\vec{n}$ be a unit vector, there are two choices for $\vec{n}$, pointing out of the plane in opposite directions. We pick one by the following rule (see Figure 12.39):

> **The right-hand rule:** Place $\vec{v}$ and $\vec{w}$ so that their tails coincide and curl the fingers of your right hand through the smaller of the two angles from $\vec{v}$ to $\vec{w}$; your thumb points in the direction of the normal vector, $\vec{n}$.

Like the dot product, there are two equivalent definitions of the cross product:

> The following two definitions of the **cross product** or **vector product** $\vec{v} \times \vec{w}$ are equivalent:
> * **Geometric definition**
>   If $\vec{v}$ and $\vec{w}$ are not parallel, then
>
> $$\vec{v} \times \vec{w} = \left(\begin{array}{c}\text{Area of parallelogram} \\ \text{with edges } \vec{v} \text{ and } \vec{w}\end{array}\right)\vec{n} = (\|\vec{v}\|\|\vec{w}\|\sin\theta)\vec{n},$$
>
>   where $0 \le \theta \le \pi$ is the angle between $\vec{v}$ and $\vec{w}$ and $\vec{n}$ is the unit vector perpendicular to $\vec{v}$ and $\vec{w}$ pointing in the direction given by the right-hand rule. If $\vec{v}$ and $\vec{w}$ are parallel, then $\vec{v} \times \vec{w} = \vec{0}$.
> * **Algebraic definition**
>
> $$\vec{v} \times \vec{w} = (v_2w_3 - v_3w_2)\vec{i} + (v_3w_1 - v_1w_3)\vec{j} + (v_1w_2 - v_2w_1)\vec{k}$$
>
>   where $\vec{v} = v_1\vec{i} + v_2\vec{j} + v_3\vec{k}$ and $\vec{w} = w_1\vec{i} + w_2\vec{j} + w_3\vec{k}$.

Notice that the magnitude of the $\vec{k}$ component is the area of a 2-dimensional parallelogram and the other components have a similar form. Problems 25 and 26 at the end of this section show that the geometric and algebraic definitions of the cross product give the same result.

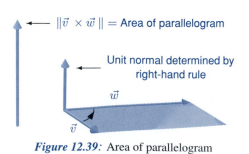

Figure 12.39: Area of parallelogram $= \|\vec{v} \times \vec{w}\|$

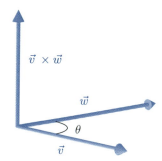

Figure 12.40: The cross product $\vec{v} \times \vec{w}$

Unlike the dot product, the cross product is only defined for three-dimensional vectors. The geometric definition shows us that the cross product is *rotation invariant*. Imagine the two vectors $\vec{v}$ and $\vec{w}$ as two metal rods welded together. Attach a third rod whose direction and length correspond to $\vec{v} \times \vec{w}$. (See Figure 12.40.) Then, no matter how we turn this set of rods, the third will still be the cross product of the first two.

The algebraic definition is more easily remembered by writing it as a 3×3 determinant. (See Appendix C.)

$$\vec{v} \times \vec{w} = \begin{vmatrix} \vec{i} & \vec{j} & \vec{k} \\ v_1 & v_2 & v_3 \\ w_1 & w_2 & w_3 \end{vmatrix} = (v_2 w_3 - v_3 w_2)\vec{i} + (v_3 w_1 - v_1 w_3)\vec{j} + (v_1 w_2 - v_2 w_1)\vec{k}.$$

Example 1   Find $\vec{i} \times \vec{j}$ and $\vec{j} \times \vec{i}$.

Solution   The vectors $\vec{i}$ and $\vec{j}$ both have magnitude 1 and the angle between them is $\pi/2$. By the right-hand rule, the vector $\vec{i} \times \vec{j}$ is in the direction of $\vec{k}$, so $\vec{n} = \vec{k}$ and we have

$$\vec{i} \times \vec{j} = \left(\|\vec{i}\| \|\vec{j}\| \sin \frac{\pi}{2} \right) \vec{k} = \vec{k}.$$

Similarly, the right-hand rule says that the direction of $\vec{j} \times \vec{i}$ is $-\vec{k}$, so

$$\vec{j} \times \vec{i} = (\|\vec{j}\| \|\vec{i}\| \sin \frac{\pi}{2})(-\vec{k}) = -\vec{k}.$$

Similar calculations show that $\vec{j} \times \vec{k} = \vec{i}$ and $\vec{k} \times \vec{i} = \vec{j}$.

Example 2   For any vector $\vec{v}$, find $\vec{v} \times \vec{v}$.

Solution   Since $\vec{v}$ is parallel to itself, $\vec{v} \times \vec{v} = \vec{0}$.

Example 3   Find the cross product of $\vec{v} = 2\vec{i} + \vec{j} - 2\vec{k}$ and $\vec{w} = 3\vec{i} + \vec{k}$ and check that the cross product is perpendicular to both $\vec{v}$ and $\vec{w}$.

Solution   Writing $\vec{v} \times \vec{w}$ as a determinant and expanding it into three two-by-two determinants, we have

$$\vec{v} \times \vec{w} = \begin{vmatrix} \vec{i} & \vec{j} & \vec{k} \\ 2 & 1 & -2 \\ 3 & 0 & 1 \end{vmatrix} = \vec{i} \begin{vmatrix} 1 & -2 \\ 0 & 1 \end{vmatrix} - \vec{j} \begin{vmatrix} 2 & -2 \\ 3 & 1 \end{vmatrix} + \vec{k} \begin{vmatrix} 2 & 1 \\ 3 & 0 \end{vmatrix}$$

$$= \vec{i} \left(1(1) - 0(-2) \right) - \vec{j} \left(2(1) - 3(-2) \right) + \vec{k} \left(2(0) - 3(1) \right)$$

$$= \vec{i} - 8\vec{j} - 3\vec{k}.$$

To check that $\vec{v} \times \vec{w}$ is perpendicular to $\vec{v}$, we compute the dot product:

$$\vec{v} \cdot (\vec{v} \times \vec{w}) = (2\vec{i} + \vec{j} - 2\vec{k}) \cdot (\vec{i} - 8\vec{j} - 3\vec{k}) = 2 - 8 + 6 = 0.$$

Similarly,

$$\vec{w} \cdot (\vec{v} \times \vec{w}) = (3\vec{i} + 0\vec{j} + \vec{k}) \cdot (\vec{i} - 8\vec{j} - 3\vec{k}) = 3 + 0 - 3 = 0.$$

Thus, $\vec{v} \times \vec{w}$ is perpendicular to both $\vec{v}$ and $\vec{w}$.

Properties of the Cross Product

The right-hand rule tells us that $\vec{v} \times \vec{w}$ and $\vec{w} \times \vec{v}$ point in opposite directions. The magnitudes of $\vec{v} \times \vec{w}$ and $\vec{w} \times \vec{v}$ are the same, so $\vec{w} \times \vec{v} = -(\vec{v} \times \vec{w})$. (See Figure 12.41.)

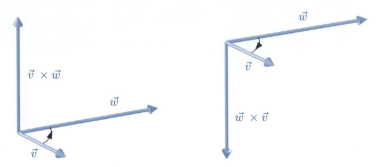

Figure 12.41: Diagram showing $\vec{v} \times \vec{w} = -(\vec{w} \times \vec{v})$

This explains the first of the following properties. The other two are derived in Problems 17, 21, and 25 at the end of this section.

Properties of the Cross Product

For vectors $\vec{u}, \vec{v}, \vec{w}$ and scalar λ
1. $\vec{w} \times \vec{v} = -(\vec{v} \times \vec{w})$
2. $(\lambda \vec{v}) \times \vec{w} = \lambda(\vec{v} \times \vec{w}) = \vec{v} \times (\lambda \vec{w})$
3. $\vec{u} \times (\vec{v} + \vec{w}) = \vec{u} \times \vec{v} + \vec{u} \times \vec{w}$.

The Equivalence of the Two Definitions of the Cross Product

Problems 22 and 25 on page 648 use geometric arguments to show that the cross product distributes over addition. Problem 26 then shows how the formula in the algebraic definition of the cross product can be derived from the geometric definition.

The Equation of a Plane Through Three Points

The equation of a plane is determined by a point $P_0 = (x_0, y_0, z_0)$ on the plane, and a normal vector, $\vec{n} = a\vec{i} + b\vec{j} + c\vec{k}$:

$$a(x - x_0) + b(y - y_0) + c(z - z_0) = 0.$$

However, a plane can also be determined by three points on it (provided they do not lie on a line). In that case we can find an equation of the plane by first determining two vectors in the plane and then finding a normal vector using the cross product, as in the following example.

Example 4   Find an equation of the plane containing the points $P = (1, 3, 0)$, $Q = (3, 4, -3)$, and $R = (3, 6, 2)$.

Solution   Since the points P and Q are in the plane, the displacement vector between them, $\overrightarrow{PQ}$, is in the plane, where

$$\overrightarrow{PQ} = (3 - 1)\vec{i} + (4 - 3)\vec{j} + (-3 - 0)\vec{k} = 2\vec{i} + \vec{j} - 3\vec{k}.$$

The displacement vector $\overrightarrow{PR}$ is also in the plane, where

$$\overrightarrow{PR} = (3 - 1)\vec{i} + (6 - 3)\vec{j} + (2 - 0)\vec{k} = 2\vec{i} + 3\vec{j} + 2\vec{k}.$$

Thus, a normal vector, $\vec{n}$, to the plane is given by

$$\vec{n} = \overrightarrow{PQ} \times \overrightarrow{PR} = \begin{vmatrix} \vec{i} & \vec{j} & \vec{k} \\ 2 & 1 & -3 \\ 2 & 3 & 2 \end{vmatrix} = 11\vec{i} - 10\vec{j} + 4\vec{k}.$$

Since the point $(1, 3, 0)$ is on the plane, the equation of the plane is

$$11(x - 1) - 10(y - 3) + 4(z - 0) = 0,$$

which simplifies to

$$11x - 10y + 4z = -19.$$

You should check that P, Q, and R satisfy the equation of the plane.

Areas and Volumes Using the Cross Product and Determinants

We can use the cross product to calculate the area of the parallelogram with sides $\vec{v}$ and $\vec{w}$. We say that $\vec{v} \times \vec{w}$ is the *area vector* of the parallelogram. The geometric definition of the cross product tells us that $\vec{v} \times \vec{w}$ is normal to the parallelogram and gives us the following result:

> **Area of a parallelogram** with edges $\vec{v} = v_1\vec{i} + v_2\vec{j} + v_3\vec{k}$ and $\vec{w} = w_1\vec{i} + w_2\vec{j} + w_3\vec{k}$ is given by
>
> $$\text{Area} = \|\vec{v} \times \vec{w}\|, \qquad \text{where} \quad \vec{v} \times \vec{w} = \begin{vmatrix} \vec{i} & \vec{j} & \vec{k} \\ v_1 & v_2 & v_3 \\ w_1 & w_2 & w_3 \end{vmatrix}.$$

Example 5   Find the area of the parallelogram with edges $\vec{v} = 2\vec{i} + \vec{j} - 3\vec{k}$ and $\vec{w} = \vec{i} + 3\vec{j} + 2\vec{k}$.

Solution   We calculate the cross product:

$$\vec{v} \times \vec{w} = \begin{vmatrix} \vec{i} & \vec{j} & \vec{k} \\ 2 & 1 & -3 \\ 1 & 3 & 2 \end{vmatrix} = (2 + 9)\vec{i} - (4 + 3)\vec{j} + (6 - 1)\vec{k} = 11\vec{i} - 7\vec{j} + 5\vec{k}.$$

The area of the parallelogram with edges $\vec{v}$ and $\vec{w}$ is the magnitude of the vector $\vec{v} \times \vec{w}$:

$$\text{Area} = \|\vec{v} \times \vec{w}\| = \sqrt{11^2 + (-7)^2 + 5^2} = \sqrt{195}.$$

Volume of a Parallelepiped

Consider the parallelepiped with sides formed by $\vec{a}$, $\vec{b}$, and $\vec{c}$. (See Figure 12.42.) Since the base is formed by the vectors $\vec{b}$ and $\vec{c}$, we have

$$\text{Area of base of parallelepiped} = \|\vec{b} \times \vec{c}\|.$$

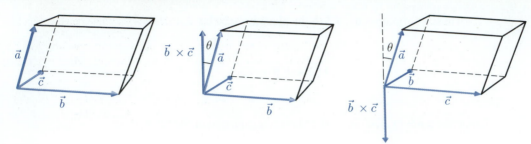

Figure 12.42: Volume of a Parallelepiped

Figure 12.43: The vectors $\vec{a}$, $\vec{b}$, $\vec{c}$ are called a right-handed set

Figure 12.44: The vectors $\vec{a}$, $\vec{b}$, $\vec{c}$ are called a left-handed set

The vectors $\vec{a}$, $\vec{b}$, and $\vec{c}$ can be arranged either as in Figure 12.43 or as in Figure 12.44. In either case,

$$\text{Height of parallelepiped} = \|\vec{a}\| \cos \theta,$$

where θ is the angle between $\vec{a}$ and $\vec{b} \times \vec{c}$. In Figure 12.43 the angle θ is less than $\pi/2$, so the product, $(\vec{b} \times \vec{c}) \cdot \vec{a}$, called the *triple product*, is positive. Thus, in this case

$$\text{Volume of parallelepiped} = \text{Base} \cdot \text{Height} = \|\vec{b} \times \vec{c}\| \cdot \|\vec{a}\| \cos \theta = (\vec{b} \times \vec{c}) \cdot \vec{a}.$$

In Figure 12.44, the angle, $\pi - \theta$, between $\vec{a}$ and $\vec{b} \times \vec{c}$ is more than $\pi/2$, so the product $(\vec{b} \times \vec{c}) \cdot \vec{a}$ is negative. Thus, in this case we have

$$\text{Volume} = \text{Base} \cdot \text{Height} = \|\vec{b} \times \vec{c}\| \cdot \|\vec{a}\| \cos \theta = -\|\vec{b} \times \vec{c}\| \cdot \|\vec{a}\| \cos(\pi - \theta)$$
$$= -(\vec{b} \times \vec{c}) \cdot \vec{a} = \left|(\vec{b} \times \vec{c}) \cdot \vec{a}\right|.$$

Therefore, in both cases the volume is given by $\left|(\vec{b} \times \vec{c}) \cdot \vec{a}\right|$. Using determinants, we can write

---

Volume of a parallelepiped with edges $\vec{a}$, $\vec{b}$, $\vec{c}$ is given by

$$\text{Volume} = \left|(\vec{b} \times \vec{c}) \cdot \vec{a}\right| = \text{Absolute value of the determinant} \begin{vmatrix} a_1 & a_2 & a_3 \\ b_1 & b_2 & b_3 \\ c_1 & c_2 & c_3 \end{vmatrix}.$$

Problems for Section 12.4

1. Find $\vec{k} \times \vec{j}$.

2. Does $\vec{i} \times \vec{i} = \vec{i} \cdot \vec{i}$? Explain your answer.

In Problems 3–6, find $\vec{a} \times \vec{b}$.

3. $\vec{a} = \vec{i} + \vec{k}$ and $\vec{b} = \vec{i} + \vec{j}$.

4. $\vec{a} = -\vec{i}$ and $\vec{b} = \vec{j} + \vec{k}$.

5. $\vec{a} = \vec{i} + \vec{j} + \vec{k}$ and $\vec{b} = \vec{i} + \vec{j} + -\vec{k}$.

6. $\vec{a} = 2\vec{i} - 3\vec{j} + \vec{k}$ and $\vec{b} = \vec{i} + 2\vec{j} - \vec{k}$.

7. Given $\vec{a} = 3\vec{i} + \vec{j} - \vec{k}$ and $\vec{b} = \vec{i} - 4\vec{j} + 2\vec{k}$, find $\vec{a} \times \vec{b}$ and check that $\vec{a} \times \vec{b}$ is perpendicular to both $\vec{a}$ and $\vec{b}$.

8. If $\vec{v} \times \vec{w} = 2\vec{i} - 3\vec{j} + 5\vec{k}$, and $\vec{v} \cdot \vec{w} = 3$, find $\tan \theta$ where θ is the angle between $\vec{v}$ and $\vec{w}$.

9. Suppose $\vec{a}$ is a fixed vector of length 3 in the direction of the positive x-axis and the vector $\vec{b}$ of length 2 is free to rotate in the xy-plane. What are the maximum and minimum values of the magnitude of $\vec{a} \times \vec{b}$? In what direction is $\vec{a} \times \vec{b}$ as $\vec{b}$ rotates?

10. You are using a jet pilot dogfight simulator. Your monitor tells you that two missiles have honed in on your plane along the directions $3\vec{i} + 5\vec{j} + 2\vec{k}$ and $\vec{i} - 3\vec{j} - 2\vec{k}$. In what direction should you turn to have the maximum chance of avoiding both missiles?

Find an equation for the plane through the points in Problems 11–12.

11. $(1, 0, 0), (0, 1, 0), (0, 0, 1)$.

12. $(3, 4, 2), (-2, 1, 0), (0, 2, 1)$.

13. Given the points $P = (0, 1, 0)$, $Q = (-1, 1, 2)$, $R = (2, 1, -1)$, find

 (a) The area of the triangle PQR.
 (b) The equation for a plane that contains P, Q, and R.

14. Find a vector parallel to the intersection of the planes $2x - 3y + 5z = 2$ and $4x + y - 3z = 7$.

15. Find the equation of the plane through the origin which is perpendicular to the line of intersection of the planes in Problem 14.

16. Find the equation of the plane through the point $(4, 5, 6)$ which is perpendicular to the line of intersection of the planes in Problem 14.

17. Use the algebraic definition for the cross product to check that

$$\vec{a} \times (\vec{b} + \vec{c}) = (\vec{a} \times \vec{b}) + (\vec{a} \times \vec{c}).$$

18. In this problem, we arrive at the algebraic definition for the cross product by a different route. Let $\vec{a} = a_1\vec{i} + a_2\vec{j} + a_3\vec{k}$ and $\vec{b} = b_1\vec{i} + b_2\vec{j} + b_3\vec{k}$. We seek a vector $\vec{v} = x\vec{i} + y\vec{j} + z\vec{k}$ which is perpendicular to both $\vec{a}$ and $\vec{b}$. Use this requirement to construct two equations for x, y, and z. Eliminate x and solve for y in terms of z. Then eliminate y and solve for x in terms of z. Since z can be any value whatsoever (the direction of $\vec{v}$ is unaffected), select the value for z which eliminates the denominator in the equation you obtained. How does the resulting expression for $\vec{v}$ compare to the formula we derived on page 643?

19. Suppose $\vec{a}$ and $\vec{b}$ are vectors in the xy-plane, such that $\vec{a} = a_1\vec{i} + a_2\vec{j}$ and $\vec{b} = b_1\vec{i} + b_2\vec{j}$ with $0 < a_2 < a_1$ and $0 < b_1 < b_2$.

 (a) Sketch $\vec{a}$ and $\vec{b}$ and the vector $\vec{c} = -a_2\vec{i} + a_1\vec{j}$. Shade the parallelogram formed by $\vec{a}$ and $\vec{b}$.
 (b) What is the relation between $\vec{a}$ and $\vec{c}$? [Hint: Find $\vec{c} \cdot \vec{a}$ and $\vec{c} \cdot \vec{c}$.]
 (c) Find $\vec{c} \cdot \vec{b}$.
 (d) Explain why $\vec{c} \cdot \vec{b}$ gives the area of the parallelogram formed by $\vec{a}$ and $\vec{b}$.
 (e) Verify that in this case $\vec{a} \times \vec{b} = (a_1b_2 - a_2b_1)\vec{k}$.

20. If $\vec{a} + \vec{b} + \vec{c} = \vec{0}$, show that

$$\vec{a} \times \vec{b} = \vec{b} \times \vec{c} = \vec{c} \times \vec{a}.$$

Geometrically, what does this imply about $\vec{a}, \vec{b}$, and $\vec{c}$?

21. If $\vec{v}$ and $\vec{w}$ are nonzero vectors, use the geometric definition of the cross product to explain why

$$(\lambda\vec{v}) \times \vec{w} = \lambda(\vec{v} \times \vec{w}) = \vec{v} \times (\lambda\vec{w}).$$

Consider the cases $\lambda > 0$, and $\lambda = 0$, and $\lambda < 0$ separately.

22. Use a parallelepiped to show that $\vec{a} \cdot (\vec{b} \times \vec{c}) = (\vec{a} \times \vec{b}) \cdot \vec{c}$ for any vectors $\vec{a}$, $\vec{b}$, and $\vec{c}$.

23. Show that $\|\vec{a} \times \vec{b}\|^2 = \|\vec{a}\|^2\|\vec{b}\|^2 - (\vec{a} \cdot \vec{b})^2$.

24. For vectors $\vec{a}$ and $\vec{b}$, let $\vec{c} = \vec{a} \times (\vec{b} \times \vec{a})$.

    (a) Show that $\vec{c}$ lies in the plane containing $\vec{a}$ and $\vec{b}$.
    (b) Show that $\vec{a} \cdot \vec{c} = 0$ and $\vec{b} \cdot \vec{c} = \|\vec{a}\|^2\|\vec{b}\|^2 - (\vec{a} \cdot \vec{b})^2$. [Hint: Use Problems 22 and 23.]
    (c) Show that
    $$\vec{a} \times (\vec{b} \times \vec{a}) = \|\vec{a}\|^2\vec{b} - (\vec{a} \cdot \vec{b})\vec{a}.$$

25. Use the result of Problem 22 to show that the cross product distributes over addition. First, use distributivity for the dot product to show that for any vector $\vec{d}$,
    $$[(\vec{a} + \vec{b}) \times \vec{c}] \cdot \vec{d} = [(\vec{a} \times \vec{c}) + (\vec{b} \times \vec{c})] \cdot \vec{d}.$$

    Next, show that for any vector $\vec{d}$,
    $$[((\vec{a} + \vec{b}) \times \vec{c}) - (\vec{a} \times \vec{c}) - (\vec{b} \times \vec{c})] \cdot \vec{d} = 0.$$

    Finally, explain why you can conclude that
    $$(\vec{a} + \vec{b}) \times \vec{c} = (\vec{a} \times \vec{c}) + (\vec{b} \times \vec{c}).$$

26. Use the fact that $\vec{i} \times \vec{i} = \vec{0}$, $\vec{i} \times \vec{j} = \vec{k}$, $\vec{i} \times \vec{k} = -\vec{j}$, and so on, together with the properties of the cross product on page 644 to derive the algebraic definition for the cross product.

27. Consider the tetrahedron determined by three vectors $\vec{a}$, $\vec{b}$, $\vec{c}$ in Figure 12.45. The *area vector* of a face is a vector perpendicular to the face, pointing outward, whose magnitude is the area of the face. Show that the sum of the four outward pointing area vectors of the faces equals the zero vector.

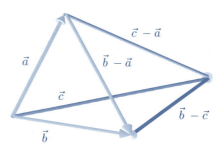

Figure 12.45

CHAPTER SUMMARY

- **Vectors**
    Geometric definition of vector addition, subtraction and scalar multiplication, resolving into $\vec{i}$, $\vec{j}$, and $\vec{k}$ components, magnitude of a vector, algebraic properties of addition and scalar multiplication.

- **Dot Product**
    Geometric and algebraic definition, algebraic properties, using dot products to find magnitudes and determine perpendicularity, the equation of a plane with given normal vector passing through a given point, projection of a vector in a direction given by a unit vector.

- **Cross Product**
    Geometric and algebraic definition, algebraic properties, cross product and volume.

REVIEW PROBLEMS FOR CHAPTER TWELVE

1. Given the displacement vectors $\vec{u}$ and $\vec{v}$ in Figure 12.46, draw the following vectors:
   (a) $(\vec{u} + \vec{v}) + \vec{u}$
   (b) $\vec{v} + (\vec{v} + \vec{u})$
   (c) $(\vec{u} + \vec{u}) + \vec{u}$.

Figure 12.46

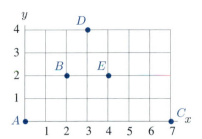

Figure 12.47

2. Figure 12.47 shows five points A, B, C, D, and E.

   (a) Read off the coordinates of the five points and thus resolve into components the following two vectors: $\vec{u} = (2.5)\overrightarrow{AB} + (-0.8)\overrightarrow{CD}$, $\vec{v} = (2.5)\overrightarrow{BA} - (-0.8)\overrightarrow{CD}$

   (b) What is the relation between $\vec{u}$ and $\vec{v}$? Why was this to be expected?

3. Find the components of a vector $\vec{p}$ which has the same direction as $\overrightarrow{EA}$ in Figure 12.47 and whose length equals two units.

4. For each of the four statements below, answer the following questions: Does the statement make sense? If yes, is it true for all possible choices of $\vec{a}$ and $\vec{b}$? If no, why not? Use complete sentences for your answers.

   (a) $\vec{a} + \vec{b} = \vec{b} + \vec{a}$
   (b) $\vec{a} + \|\vec{b}\| = \|\vec{a} + \vec{b}\|$
   (c) $\|\vec{b} + \vec{a}\| = \|\vec{a} + \vec{b}\|$
   (d) $\|\vec{a} + \vec{b}\| = \|\vec{a}\| + \|\vec{b}\|$.

5. Two adjacent sides of a regular hexagon are given as the vectors $\vec{u}$ and $\vec{v}$ in Figure 12.48. Label the remaining sides in terms of $\vec{u}$ and $\vec{v}$.

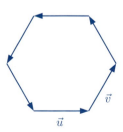

Figure 12.48

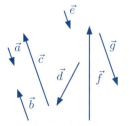

Figure 12.49

6. Figure 12.49 shows seven vectors $\vec{a}$, $\vec{b}$, $\vec{c}$, $\vec{d}$, $\vec{e}$, $\vec{f}$ and $\vec{g}$.

   (a) Are any of these two vectors equal? Write down all equal pairs.
   (b) Can you find a scalar x so that $\vec{a} = x\vec{g}$? If yes, find such an x; if no, explain why not.
   (c) Same question as in part (b), but for the equation $\vec{b} = x\vec{d}$.
   (d) Can you solve the equation $\vec{f} = u\vec{c} + v\vec{d}$ for the scalars u and v? If yes, find them; if no, explain why not.

7. For what values of t are the following pairs of vectors parallel?

(a) $2\vec{i} + (t^2 + \frac{2}{3}t + 1)\vec{j} + t\vec{k}$, $6\vec{i} + 8\vec{j} + 3\vec{k}$

(b) $t\vec{i} + \vec{j} + (t-1)\vec{k}$, $2\vec{i} - 4\vec{j} + \vec{k}$

(c) $2t\vec{i} + t\vec{j} + t\vec{k}$, $6\vec{i} + 3\vec{j} + 3\vec{k}$.

Use the geometric definition to compute the cross products in Problems 8–9.

8. $2\vec{i} \times (\vec{i} + \vec{j})$

9. $(\vec{i} + \vec{j}) \times (\vec{i} - \vec{j})$

Compute the cross products for Problems 10–11, using the algebraic definition.

10. $\left((\vec{i} + \vec{j}) \times \vec{i}\right) \times \vec{j}$

11. $(\vec{i} + \vec{j}) \times (\vec{i} \times \vec{j})$

12. True or false? $\vec{a} \times \vec{b} = -(\vec{b} \times \vec{a})$ for all $\vec{a}$ and $\vec{b}$. Explain your answer.

13. Find the area of the triangle with vertices $P = (-2, 2, 0)$, $Q = (1, 3, -1)$, and $R = (-4, 2, 1)$ using the cross product.

14. Find the equation of the plane through the origin which is parallel to $z = 4x - 3y + 8$.

15. Find a vector normal to the plane $4(x - 1) + 6(z + 3) = 12$.

16. Find the equation of the plane through the points $(0, 0, 2), (0, 3, 0), (5, 0, 0)$.

17. Consider the plane $5x - y + 7z = 21$.

(a) Find a point on the x-axis on this plane.

(b) Find two other points on the plane.

(c) Find a vector perpendicular to the plane.

(d) Find a vector parallel to the plane.

18. Given the points $P = (1, 2, 3)$, $Q = (3, 5, 7)$, and $R = (2, 5, 3)$, find:

(a) A unit vector perpendicular to a plane containing P, Q, R.

(b) The angle between PQ and PR.

(c) The area of the triangle PQR.

(d) The distance from R to the line through P and Q.

19. Find all vectors $\vec{v}$ in the plane such that $\|\vec{v}\| = 1$ and $\|\vec{v} + \vec{i}\| = 1$.

20. Find all vectors $\vec{w}$ in 3-space such that $\|\vec{w}\| = 1$ and $\|\vec{w} + \vec{i}\| = 1$. Describe this set geometrically.

21. Is the collection of populations of each of the 50 states a vector or scalar quantity?

22. The price vector of beans, rice, and tofu is $(0.30, 0.20, 0.50)$ in dollars per pound. Express it in dollars per ounce.

23. An object is attached by an inelastic string to a fixed point and rotates 30 times per minute in a horizontal plane. Show that the speed of the object is constant but the velocity is not. What does this imply about the acceleration?

24. An object is moving counterclockwise at a constant speed around the circle $x^2 + y^2 = 1$, where x and y are measured in meters. It completes one revolution every minute.

(a) What is its speed?

(b) What is its velocity vector 30 seconds after it passes the point $(1, 0)$? Does your answer change if the object is moving clockwise? Explain.

25. In the game of laser tag, you shoot a harmless laser gun and try to hit a target worn at the waist by other players. Suppose you are standing at the origin of a three dimensional coordinate system and that the xy-plane is the floor. Suppose that waist-high is 3 feet above floor level and that eye level is 5 feet above the floor. Three of your friends are your opponents. One is standing so that his target is 30 feet along the x-axis, the other lying down so that his target is at the point $x = 20$, $y = 15$, and the third lying in ambush so that his target is at a point 8 feet above the point $x = 12$, $y = 30$.

(a) If you aim with your gun at eye level, find the vector from your gun to each of the three targets.

(b) If you shoot from waist height, with your gun one foot to the right of the center of your body as you face along the x-axis, find the vector from your gun to each of the three targets.

26. An airport is at the point $(200, 10, 0)$ and an approaching plane is at the point $(550, 60, 4)$. Assume that the xy-plane is horizontal, with the x-axis pointing eastward and the y-axis pointing northward. Also assume that the z-axis is upward and that all distances are measured in kilometers. The plane flies due west at a constant altitude at a speed of 500 km/hr for half an hour. It then descends at 200 km/hr, heading straight for the airport.

   (a) Find the velocity vector of the plane while it is flying at constant altitude.
   (b) Find the coordinates of the point at which the plane starts to descend.
   (c) Find a vector representing the velocity of the plane when it is descending.

27. A large ship is being towed by two tugs. The larger tug exerts a force which is 25% greater than the smaller tug and at an angle of 30 degrees north of east. Which direction must the smaller tug pull to ensure that the ship travels due east?

28. A man wishes to row the shortest possible distance from north to south across a river which is flowing at 4 km/hr from the east. He can row at 5 km/hr.

   (a) In which direction should he steer?
   (b) If there is a wind of 10 km/hr from the southwest, in which direction should he steer to try and go directly across the river? What happens?

29. (a) A vector $\vec{v}$ of magnitude v makes an angle α with the positive x-axis, angle β with the positive y-axis, and angle γ with the positive z-axis. Show that

$$\vec{v} = v \cos \alpha \vec{i} + v \cos \beta \vec{j} + v \cos \gamma \vec{k}.$$

   (b) $\cos \alpha$, $\cos \beta$, and $\cos \gamma$ are called *direction cosines*. Show that

$$\cos^2 \alpha + \cos^2 \beta + \cos^2 \gamma = 1.$$

30. Find the vector $\vec{v}$ with all of the following properties:

   (i) Magnitude 10   (ii) Angle of $45°$ with positive x-axis
   (iii) Angle of $75°$ with positive y-axis   (iv) Positive $\vec{k}$-component.

31. Using vectors, show that the perpendicular bisectors of a triangle intersect at a point.

32. Find the distance from the point $P = (2, -1, 3)$ to the plane $2x + 4y - z = -1$.

33. Find an equation of the plane passing through the three points $(1, 1, 1)$, $(1, 4, 5)$, $(-3, -2, 0)$. Find the distance from the origin to the plane.

34. Two lines in space are skew if they are not parallel and do not intersect. Determine the minimum distance between two such lines.

PROJECTS

1. **Cross Product of Vectors in the Plane**
Let $\vec{a} = a_1\vec{i} + a_2\vec{j}$ and $\vec{b} = b_1\vec{i} + b_2\vec{j}$ be two nonparallel vectors in 2-space, as in Figure 12.50.

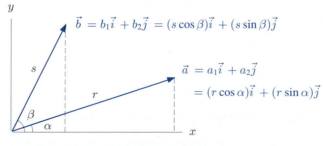

Figure 12.50

(a) Use the identity $\sin(\beta - \alpha) = (\sin \beta \cos \alpha - \cos \beta \sin \alpha)$ to verify the formula for the area of the parallelogram formed by $\vec{a}$ and $\vec{b}$:

$$\text{Area of parallelogram} = |a_1 b_2 - a_2 b_1|.$$

(b) Show that $a_1 b_2 - a_2 b_1$ is positive when the rotation from $\vec{a}$ to $\vec{b}$ is counterclockwise, and negative when it is clockwise.

(c) Use parts (a) and (b) to show that the geometric and algebraic definitions of $\vec{a} \times \vec{b}$ give the same result.

2. **The Dot Product in Genetics**[2]

Recall that in 2 or 3 dimensions, if θ is the angle between $\vec{v}$ and $\vec{w}$, the dot product is given by

$$\vec{v} \cdot \vec{w} = \|\vec{v}\|\|\vec{w}\| \cos \theta.$$

We use this relationship to define the angle between two vectors in n-dimensions. If $\vec{v}, \vec{w}$ are n-vectors, then the dot product, $\vec{v} \cdot \vec{w} = v_1 w_1 + v_2 w_2 + \cdots + v_n w_n$, is used to define[3] the angle θ by

$$\cos \theta = \frac{\vec{v} \cdot \vec{w}}{\|\vec{v}\|\|\vec{w}\|}, \qquad \text{provided } \|\vec{v}\|, \|\vec{w}\| \neq 0.$$

We now use this idea of angle to measure how close two populations are to one another genetically. Table 12.1 shows the relative frequencies of four alleles (variants of a gene) in four populations.

TABLE 12.1

Allele	Eskimo	Bantu	English	Korean
A_1	0.29	0.10	0.20	0.22
A_2	0.00	0.08	0.06	0.00
B	0.03	0.12	0.06	0.20
O	0.67	0.69	0.66	0.57

Let $\vec{a}_1$ be the 4-vector showing the relative frequencies in the Eskimo population;

$\vec{a}_2$ be the 4-vector showing the relative frequencies in the Bantu population;

$\vec{a}_3$ be the 4-vector showing the relative frequencies in the English population;

$\vec{a}_4$ be the 4-vector showing the relative frequencies in the Korean population.

The genetic distance between two populations is defined as the angle between the corresponding vectors.

(a) Using this definition, is the English population closer genetically to the Bantus or to the Koreans? Explain.

(b) Is the English population closer to a half Eskimo, half Bantu population than to the Bantu population alone?

(c) Among all possible populations that are a mix of Eskimo and Bantu, find the mix that is closest to the English population.

---

[2] Adapted from Cavalli-Sforza and Edwards, "Models and Estimation Procedures," Am J. Hum. Genet., Vol. 19 (1967), pp. 223-57.

[3] The result of Problem 33 on page 641 shows that the quantity on the right-hand side of this equation is between -1 and 1, so this definition makes sense.

CHAPTER THIRTEEN

DIFFERENTIATING FUNCTIONS OF MANY VARIABLES

For a function of one variable, $y = f(x)$, the derivative $dy/dx = f'(x)$ gives the rate of change of y with respect to x. For a function of two variables, $z = f(x, y)$, there is no such thing as *the* rate of change, since x and y can each vary while the other is held fixed or both can vary at once. However, we can consider the rate of change with respect to each one of the independent variables. This chapter introduces these *partial derivatives* and several ways they can be used to get a complete picture of the way the function varies.

13.1 THE PARTIAL DERIVATIVE

The derivative of a one-variable function measures its rate of change. In this section we see how a two variable function has two rates of change: one as x changes (with y held constant) and one as y changes (with x held constant).

Rate of Change of Temperature in a Metal Rod: a One-Variable Problem

Imagine an unevenly heated metal rod lying along the x-axis, with its left end at the origin and x measured in meters. (See Figure 13.1.) Let $u(x)$ be the temperature (in °C) of the rod at the point x. Table 13.1 gives values of $u(x)$. We see that the temperature increases as we move along the rod, reaching its maximum at $x = 4$, after which it starts to decrease.

x (m)

0   1   2   3   4   5

Figure 13.1: Unevenly heated metal rod

TABLE 13.1   *Temperature $u(x)$ of the rod*

x (m)	0	1	2	3	4	5
$u(x)$ (°C)	125	128	135	160	175	160

Example 1   Estimate the derivative $u'(2)$ using Table 13.1 and explain what the answer means in terms of temperature.

Solution   The derivative $u'(2)$ is defined as a limit of difference quotients:

$$u'(2) = \lim_{h \to 0} \frac{u(2+h) - u(2)}{h}.$$

Choosing $h = 1$ so that we can use the data in Table 13.1, we get

$$u'(2) \approx \frac{u(2+1) - u(2)}{1} = \frac{160 - 135}{1} = 25.$$

This means that the temperature increases at a rate of approximately 25°C per meter as we go from left to right, past $x = 2$.

Rate of Change of Temperature in a Metal Plate

Imagine an unevenly heated thin rectangular metal plate lying in the xy-plane with its lower left corner at the origin and x and y measured in meters. The temperature (in °C) at the point (x, y) is $T(x, y)$. See Figure 13.2 and Table 13.2. How does T vary near the point $(2, 1)$? We consider the horizontal line $y = 1$ containing the point $(2, 1)$. The temperature along this line is the cross-section, $T(x, 1)$, of the function $T(x, y)$ with $y = 1$. Suppose we write $u(x) = T(x, 1)$.

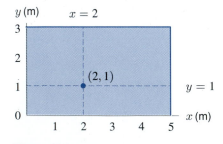

Figure 13.2: Unevenly heated metal plate

TABLE 13.2   *Temperature (°C) of a metal plate*

	3	85	90	**110**	135	155	180
	2	100	110	**120**	145	190	170
y (m)	1	**125**	**128**	**135**	**160**	**175**	**160**
	0	120	135	**155**	160	160	150
		0	1	2	3	4	5

x (m)

What is the meaning of the derivative $u'(2)$? It is the rate of change of temperature T *in the x-direction* at the point $(2, 1)$, keeping y fixed. Denote this rate of change by $T_x(2, 1)$, so that

$$T_x(2, 1) = u'(2) = \lim_{h \to 0} \frac{u(2+h) - u(2)}{h} = \lim_{h \to 0} \frac{T(2+h, 1) - T(2, 1)}{h}.$$

We call $T_x(2, 1)$ the *partial derivative of T with respect to x at the point* $(2, 1)$. Taking $h = 1$, we can read values of T from the row with $y = 1$ in Table 13.2, giving

$$T_x(2, 1) \approx \frac{T(3, 1) - T(2, 1)}{1} = \frac{160 - 135}{1} = 25^\circ\text{C/m}.$$

The fact that $T_x(2, 1)$ is positive means that the temperature of the plate is increasing as we move past the point $(2, 1)$ in the direction of increasing x (that is, horizontally from left to right in Figure 13.2).

Example 2   Estimate the rate of change of T in the y-direction at the point $(2, 1)$.

Solution   The temperature along the line $x = 2$ is the cross-section of T with $x = 2$, that is, the function $v(y) = T(2, y)$. If we denote the rate of change of T in the y-direction at $(2, 1)$ by $T_y(2, 1)$, then

$$T_y(2, 1) = v'(1) = \lim_{h \to 0} \frac{v(1 + h) - v(1)}{h} = \lim_{h \to 0} \frac{T(2, 1 + h) - T(2, 1)}{h}.$$

We call $T_y(2, 1)$ the *partial derivative of T with respect to y at the point* $(2, 1)$. Taking $h = 1$ so that we can use the column with $x = 2$ in Table 13.2, we get

$$T_y(2, 1) \approx \frac{T(2, 1 + 1) - T(2, 1)}{1} = \frac{120 - 135}{1} = -15^\circ\text{C/m}.$$

The fact that $T_y(2, 1)$ is negative means that the temperature decreases as y increases.

Definition of the Partial Derivative

We study the influence of x and y separately on the value of the function $f(x, y)$ by holding one fixed and letting the other vary. This leads to the following definitions.

Partial Derivatives of f With Respect to x and y

For all points at which the limits exist, we define the **partial derivatives at the point** $(\mathbf{a}, \mathbf{b})$ by

$$f_x(a, b) = \quad \begin{array}{c} \text{Rate of change of } f \text{ with respect to } x \\ \text{at the point } (a, b) \end{array} \quad = \lim_{h \to 0} \frac{f(a + h, b) - f(a, b)}{h},$$

$$f_y(a, b) = \quad \begin{array}{c} \text{Rate of change of } f \text{ with respect to } y \\ \text{at the point } (a, b) \end{array} \quad = \lim_{h \to 0} \frac{f(a, b + h) - f(a, b)}{h}.$$

If we let a and b vary, we have the **partial derivative functions** $f_x(x, y)$ and $f_y(x, y)$.

Just as with ordinary derivatives, there is an alternative notation:

Alternative Notation for Partial Derivatives

If $z = f(x, y)$, we can write

$$f_x(x, y) = \frac{\partial z}{\partial x} \quad \text{and} \quad f_y(x, y) = \frac{\partial z}{\partial y},$$

$$f_x(a, b) = \left.\frac{\partial z}{\partial x}\right|_{(a,b)} \quad \text{and} \quad f_y(a, b) = \left.\frac{\partial z}{\partial y}\right|_{(a,b)}.$$

We use the symbol ∂ to distinguish partial derivatives from ordinary derivatives. In cases where the independent variables have names different from x and y, we adjust the notation accordingly. For example, the partial derivatives of $f(u, v)$ are denoted by f_u and f_v.

Visualizing Partial Derivatives on a Graph

The ordinary derivative of a one-variable function is the slope of its graph. How do we visualize the partial derivative $f_x(a, b)$? The graph of the one-variable function $f(x, b)$ is the curve where the vertical plane $y = b$ cuts the graph of $f(x, y)$. (See Figure 13.3.) Thus, $f_x(a, b)$ is the slope of the tangent line to this curve at $x = a$.

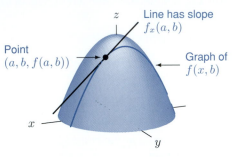

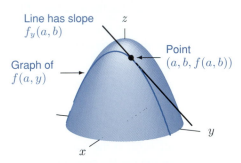

Figure 13.3: The curve $z = f(x, b)$ on the graph of f has slope $f_x(a, b)$ at $x = a$

Figure 13.4: The curve $z = f(a, y)$ on the graph of f has slope $f_y(a, b)$ at $y = b$

Similarly, the graph of the function $f(a, y)$ is the curve where the vertical plane $x = a$ cuts the graph of f, and the partial derivative $f_y(a, b)$ is the slope of this curve at $y = b$. (See Figure 13.4.)

Example 3   At each point labeled on the graph of the surface $z = f(x, y)$ in Figure 13.5, say whether each partial derivative is positive or negative.

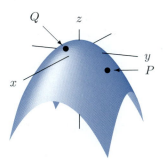

Figure 13.5: Decide the signs of f_x and f_y at P and Q

Solution   The positive x-axis points out of the page. Imagine heading off in this direction from the point marked P; we descend steeply. So the partial derivative with respect to x is negative at P, with quite a large absolute value. The same is true for the partial derivative with respect to y at P, since there is also a steep descent in the positive y-direction.

At the point marked Q, heading in the positive x-direction results in a gentle descent, whereas heading in the positive y-direction results in a gentle ascent. Thus, the partial derivative f_x at Q is negative but small (that is, near zero), and the partial derivative f_y is positive but small.

Estimating Partial Derivatives From a Contour Diagram

The graph of a function $f(x, y)$ often makes clear the sign of the partial derivatives. However, numerical estimates of these derivatives are more easily made from a contour diagram than a surface graph. If we move parallel to one of the axes on a contour diagram, the partial derivative is the rate of change of the value of the function on the contours. For example, if the values on the contours are increasing as we move in the positive direction, then the partial derivative must be positive.

Example 4 Figure 13.6 shows the contour diagram for the temperature $H(x, t)$ (in °C) in a room as a function of distance x (in meters) from a heater and time t (in minutes) after the heater has been turned on. What are the signs of $H_x(10, 20)$ and $H_t(10, 20)$? Estimate these partial derivatives and explain the answers in practical terms.

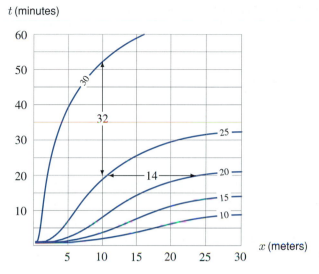

Figure 13.6: Temperature in a heated room: Heater at $x = 0$ is turned on at $t = 0$

Solution The point $(10, 20)$ is nearly on the $H = 25$ contour. As x increases, we move towards the $H = 20$ contour, so H is decreasing and $H_x(10, 20)$ is negative. This makes sense because the $H = 30$ contour is to the left: As we move further from the heater, the temperature drops. On the other hand, as t increases, we move towards the $H = 30$ contour, so H is increasing; as t decreases H decreases. Thus, $H_t(10, 20)$ is positive. This says that as time passes, the room warms up.

To estimate the partial derivatives, use a difference quotient. Looking at the contour diagram, we see there is a point on the $H = 20$ contour about 14 units to the right of the point $(10, 20)$. Hence, H decreases by 5 when x increases by 14, so we find

$$\text{Rate of change of } H \text{ with respect to } x = H_x(10, 20) \approx \frac{-5}{14} \approx -0.36°\text{C/meter}.$$

This means that near the point 10 m from the heater, after 20 minutes the temperature drops about 0.36, or one third, of a degree, for each meter we move away from the heater.

To estimate $H_t(10, 20)$, we notice that the $H = 30$ contour is about 32 units directly above the point $(10, 20)$. So H increases by 5 when t increases by 32. Hence,

$$\text{Rate of change of } H \text{ with respect to } t = H_t(10, 20) \approx \frac{5}{32} = 0.16°\text{C/meter}.$$

This means that after 20 minutes the temperature is going up about 0.16, or 1/6, of a degree each minute at the point 10 m from the heater.

Using Units to Interpret Partial Derivatives

The meaning of a partial derivative can often be explained using units.

Example 5 Suppose that your weight w in pounds is a function $f(c, n)$ of the number c of calories you consume daily and the number n of minutes you exercise daily. Using the units for w, c and n, interpret in everyday terms the statements

$$\frac{\partial w}{\partial c}(2000, 15) = 0.02 \quad \text{and} \quad \frac{\partial w}{\partial n}(2000, 15) = -0.025.$$

Solution   The units of $\partial w/\partial c$ are pounds per calorie. The statement

$$\frac{\partial w}{\partial c}(2000, 15) = 0.02$$

means that if you are presently consuming 2000 calories daily and exercising 15 minutes daily, you will weigh 0.02 pounds more for each extra calorie you consume daily, or about 2 pounds for each extra 100 calories per day. The units of $\partial w/\partial n$ are pounds per minute. The statement

$$\frac{\partial w}{\partial n}(2000, 15) = -0.025$$

means that for the same calorie consumption and number of minutes of exercise, you will weigh 0.025 pounds less for each extra minute you exercise daily, or about 1 pound less for each extra 40 minutes per day. So if you eat an extra 100 calories each day and exercise about 80 minutes more each day, your weight should remain roughly steady.

Problems for Section 13.1

1. Using difference quotients, estimate $f_x(3, 2)$ and $f_y(3, 2)$ for the function given by

    $$f(x, y) = \frac{x^2}{y + 1}.$$

    [Recall: A difference quotient is an expression of the form $(f(a + h, b) - f(a, b))/h$.]

2. Use difference quotients with $\Delta x = 0.1$ and $\Delta y = 0.1$ to estimate $f_x(1, 3)$ and $f_y(1, 3)$ where

    $$f(x, y) = e^{-x}\sin y.$$

    Then give better estimates by using $\Delta x = 0.01$ and $\Delta y = 0.01$.

3. The monthly mortgage payment in dollars, P, for a house is a function of three variables

    $$P = f(A, r, N),$$

    where A is the amount borrowed in dollars, r is the interest rate, and N is the number of years before the mortgage is paid off.

    (a) $f(92000, 14, 30) = 1090.08$. What does this tell you, in financial terms?
    (b) $\dfrac{\partial P}{\partial r}(92000, 14, 30) = 72.82$. What is the financial significance of the number 72.82?
    (c) Would you expect $\partial P/\partial A$ to be positive or negative? Why?
    (d) Would you expect $\partial P/\partial N$ to be positive or negative? Why?

4. Suppose you borrow $\$A$ at an interest rate of $r\%$ (per month) and pay it off over t months by making monthly payments of $\$P$, as determined by the function $P = g(A, r, t)$. In financial terms, what do the following statements tell you?

    (a) $g(8000, 1, 24) = 376.59$ (b) $\dfrac{\partial g}{\partial A}(8000, 1, 24) = 0.047$

    (c) $\dfrac{\partial g}{\partial r}(8000, 1, 24) = 44.83$

5. Suppose that x is the average price of a new car and that y is the average price of a gallon of gasoline. Then q_1, the number of new cars bought in a year, depends on both x and y, so $q_1 = f(x, y)$. Similarly, if q_2 is the quantity of gas bought in a year, then $q_2 = g(x, y)$.

    (a) What do you expect the signs of $\partial q_1/\partial x$ and $\partial q_2/\partial y$ to be? Explain.
    (b) What do you expect the signs of $\partial q_1/\partial y$ and $\partial q_2/\partial x$ to be? Explain.

6. A drug is injected into a patient's blood vessel. The function $c = f(x, t)$ represents the concentration of the drug at a distance x mm in the direction of the blood flow measured from the point of injection and at time t seconds since the injection. What are the units of the following partial derivatives? What are their practical interpretations? What do you expect their signs to be?
    (a) $\partial c/\partial x$ (b) $\partial c/\partial t$

7. Suppose P is your monthly car payment in dollars and $P = f(P_0, t, r)$, where $\$P_0$ is the amount you borrowed, t is the number of months it takes to pay off the loan, and $r\%$ is the interest rate. What are the units, the financial meanings, and the signs of $\partial P/\partial t$ and $\partial P/\partial r$?

8. The surface $z = f(x, y)$ is shown in Figure 13.7. The points A and B are in the xy-plane.
    (a) What is the sign of $f_x(A)$?   (b) What is the sign of $f_y(A)$?

    (c) Suppose P is a point in the xy-plane which moves along a straight line from A to B. How does the sign of $f_x(P)$ change? How does the sign of $f_y(P)$ change?

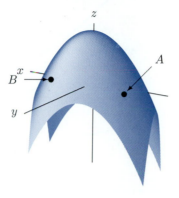

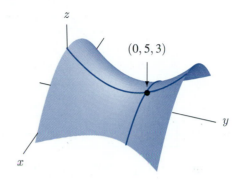

Figure 13.7   **Figure 13.8**

9. Consider the saddle-shaped surface $z = f(x, y)$ graphed in Figure 13.8.
    (a) What is the sign of $f_x(0, 5)$?   (b) What is the sign of $f_y(0, 5)$?

For Problems 10–12, refer to Table 11.3 on page 569 giving the temperature adjusted for wind-chill, C, in °F, as a function $f(w, T)$ of the wind speed, w, in mph, and the temperature, T, in °F. The temperature adjusted for wind-chill tells you how cold it feels, as a result of the combination of wind and temperature.

10. Estimate $f_w(10, 25)$. What does your answer mean in practical terms?

11. Estimate $f_T(5, 20)$. What does your answer mean in practical terms?

12. From Table 11.3 on page 569 you can see that when the temperature is 20°F, the temperature adjusted for wind-chill drops by an average of about 2.6°F with every 1 mph increase in wind speed from 5 mph to 10 mph. Which partial derivative is this telling you about?

13. Figure 13.9 shows a contour diagram for the monthly payment P as a function of the interest rate, $r\%$, and the amount, L, of a 5-year loan. Estimate $\partial P/\partial r$ and $\partial P/\partial L$ at the following points. In each case, give the units and the everyday meaning of your answer.

    (a) $r = 8, L = 4000$ (b) $r = 8, L = 6000$ (c) $r = 13, L = 7000$

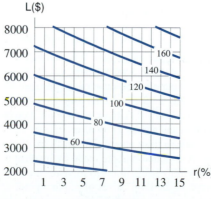

Figure 13.9

14.   Figure 13.10 gives a contour diagram for the number n of foxes per square kilometer in southwestern England. Estimate $\partial n/\partial x$ and $\partial n/\partial y$ at the points marked A, B, and C, where x is kilometers east-west and y is kilometers north-south.

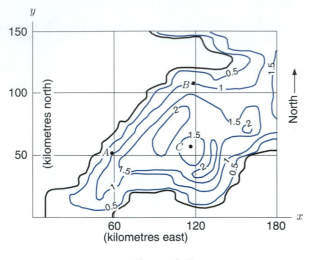

Figure 13.10   **Figure 13.11**

An airport can be cleared of fog by heating the air. The amount of heat required to do the job depends on the air temperature and the wetness of the fog. Problems 15–17 involve Figure 13.11 which shows the heat $H(T, w)$ required (in calories per cubic meter of fog) as a function of the temperature T (in degrees Celsius) and the water content w (in grams per cubic meter of fog). Note that Figure 13.11 is not a contour diagram, but shows cross-sections of H with w fixed at 0.1, 0.2, 0.3, and 0.4.

15.   Use Figure 13.11 to find an approximate value for $H_T(10, 0.1)$. Interpret the partial derivative in practical terms.

16.   Make a table of values for $H(T, w)$ from Figure 13.11, and use it to estimate $H_T(T, w)$ for $T = 10$, 20, and 30 and $w = 0.1, 0.2$, and 0.3.

17.   Repeat Problem 16 for $H_w(T, w)$ at $T = 10, 20$, and 30 and $w = 0.1, 0.2$, and 0.3. What is the practical meaning of these partial derivatives?

18.   The cardiac output, represented by c, is the volume of blood flowing through a person's heart, per unit time. The systemic vascular resistance (SVR), represented by s, is the resistance to blood flowing through veins and arteries. Let p be a person's blood pressure. Then p is a function of c and s, so $p = f(c, s)$.

(a)   What does $\partial p/\partial c$ represent?

Suppose now that $p = kcs$, where k is a constant.

(b)   Sketch the level curves of p. What do they represent? Label your axes.
(c)   For a person with a weak heart, it is desirable to have the heart pumping against less resistance, while maintaining the same blood pressure. Such a person may be given the drug Nitroglycerine to decrease the SVR and the drug Dopamine to increase the cardiac output. Represent this on a graph showing level curves. Put a point A on the graph representing the person's state before drugs are given and a point B for after.
(d)   Right after a heart attack, a patient's cardiac output drops, thereby causing the blood pressure to drop. A common mistake made by medical residents is to get the patient's blood pressure back to normal by using drugs to increase the SVR, rather than by increasing the cardiac output. On a graph of the level curves of p, put a point D representing the patient before the heart attack, a point E representing the patient right after the heart attack, and a third point F representing the patient after the resident has given the drugs to increase the SVR.

13.2 COMPUTING PARTIAL DERIVATIVES ALGEBRAICALLY

Since the partial derivative $f_x(x, y)$ is the ordinary derivative of the function $f(x, y)$ with y held constant and $f_y(x, y)$ is the ordinary derivative of $f(x, y)$ with x held constant, we can use all the differentiation formulas from one-variable calculus to find partial derivatives.

Example 1   Let $f(x, y) = \dfrac{x^2}{y + 1}$. Find $f_x(3, 2)$ algebraically.

Solution   We use the fact that $f_x(3, 2)$ equals the derivative of $f(x, 2)$ at $x = 3$. Since

$$f(x, 2) = \frac{x^2}{2 + 1} = \frac{x^2}{3},$$

differentiating with respect to x, we have

$$f_x(x, 2) = \frac{\partial}{\partial x}\left(\frac{x^2}{3}\right) = \frac{2x}{3}, \qquad \text{and so} \qquad f_x(3, 2) = 2.$$

Example 2   Compute the partial derivatives with respect to x and with respect to y for the following functions.
(a) $f(x, y) = y^2 e^{3x}$ (b) $z = (3xy + 2x)^5$ (c) $g(x, y) = e^{x+3y} \sin(xy)$

Solution   (a) This is the product of a function of x (namely e^{3x}) and a function of y (namely y^2). When we differentiate with respect to x, we think of the function of y as a constant, and vice versa. Thus,

$$f_x(x, y) = y^2 \frac{\partial}{\partial x}\left(e^{3x}\right) = 3y^2 e^{3x},$$

$$f_y(x, y) = e^{3x} \frac{\partial}{\partial y}(y^2) = 2y e^{3x}.$$

(b) Here we use the chain rule:

$$\frac{\partial z}{\partial x} = 5(3xy + 2x)^4 \frac{\partial}{\partial x}(3xy + 2x) = 5(3xy + 2x)^4 (3y + 2),$$

$$\frac{\partial z}{\partial y} = 5(3xy + 2x)^4 \frac{\partial}{\partial y}(3xy + 2x) = 5(3xy + 2x)^4 3x = 15x(3xy + 2x)^4.$$

(c) Since each function in the product is a function of both x and y, we need to use the product rule for each partial derivative:

$$g_x(x, y) = \left(\frac{\partial}{\partial x}(e^{x+3y})\right)\sin(xy) + e^{x+3y}\frac{\partial}{\partial x}(\sin(xy)) = e^{x+3y}\sin(xy) + e^{x+3y}y\cos(xy),$$

$$g_y(x, y) = \left(\frac{\partial}{\partial y}(e^{x+3y})\right)\sin(xy) + e^{x+3y}\frac{\partial}{\partial y}(\sin(xy)) = 3e^{x+3y}\sin(xy) + e^{x+3y}x\cos(xy).$$

For functions of three or more variables, we find partial derivatives by the same method: Differentiate with respect to one variable, regarding the other variables as constants. For a function $f(x, y, z)$, the partial derivative $f_x(a, b, c)$ gives the rate of change of f with respect to x along the line $y = b, z = c$.

Example 3   Find all the partial derivatives of $f(x, y, z) = \dfrac{x^2 y^3}{z}$.

Solution   To find $f_x(x, y, z)$, for example, we consider y and z as fixed, giving

$$f_x(x, y, z) = \frac{2xy^3}{z}, \quad \text{and} \quad f_y(x, y, z) = \frac{3x^2 y^2}{z}, \quad \text{and} \quad f_z(x, y, z) = -\frac{x^2 y^3}{z^2}.$$

Interpretation of Partial Derivatives

Example 4   A vibrating guitar string, originally at rest along the x-axis, is shown in Figure 13.12. Let x be the distance in meters from the left end of the string. At time t seconds the point x has been displaced $y = f(x, t)$ meters vertically from its rest position, where

$$y = f(x, t) = 0.003 \sin(\pi x) \sin(2765t).$$

Evaluate $f_x(0.3, 1)$ and $f_t(0.3, 1)$ and explain what each means in practical terms.

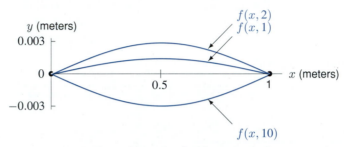

Figure 13.12: The position of a vibrating guitar string at several different times: Graph of $f(x, t)$ for $t = 1, 2, 10$.

Solution   Differentiating $f(x, t) = 0.003 \sin(\pi x) \sin(2765t)$ with respect to x, we have

$$f_x(x, t) = 0.003\pi \cos(\pi x) \sin(2765t).$$

In particular, substituting $x = 0.3$ and $t = 1$ gives

$$f_x(0.3, 1) = 0.003\pi \cos(\pi(0.3)) \sin(2765) \approx 0.002.$$

To see what $f_x(0.3, 1)$ means, think about the function $f(x, 1)$. The graph of $f(x, 1)$ in Figure 13.13 is a snapshot of the string at the time $t = 1$. Thus, the derivative $f_x(0.3, 1)$ is the slope of the string at the point $x = 0.3$ at the instant when $t = 1$.

Similarly, taking the derivative of $f(x, t) = 0.003 \sin(\pi x) \sin(2765t)$ with respect to t, we get

$$f_t(x, t) = (0.003)(2765) \sin(\pi x) \cos(2765t) = 8.3 \sin(\pi x) \cos(2765t).$$

Since $f(x, t)$ is in meters and t is in seconds, the derivative $f_t(0.3, 1)$ is in m/sec. Thus, substituting $x = 0.3$ and $t = 1$,

$$f_t(0.3, 1) = 8.3 \sin(\pi(0.3)) \cos(2765(1)) \approx 6 \text{ m/sec}.$$

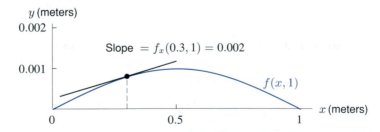

Figure 13.13: Graph of $f(x, 1)$: Snapshot of the shape of the string at $t = 1$ sec

To see what $f_t(0.3, 1)$ means, think about the function $f(0.3, t)$. The graph of $f(0.3, t)$ is a position versus time graph that tracks the up and down movement of the point on the string where $x = 0.3$.

(See Figure 13.14.) The derivative $f_t(0.3, 1) = 6$ m/sec is the velocity of that point on the string at time $t = 1$. The fact that $f_t(0.3, 1)$ is positive indicates that the point is moving upward when $t = 1$.

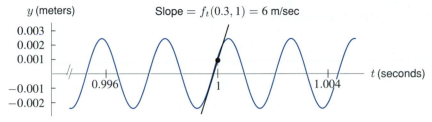

Figure 13.14: Graph of $f(0.3, t)$: Position versus time graph of the point $x = 0.3$ m from the end of the guitar string

Problems for Section 13.2

1. Let $f(x, y) = \dfrac{x^2}{y + 1}$. Find $f_y(3, 2)$ algebraically.

2. Let $f(u, v) = u(u^2 + v^2)^{3/2}$.

   (a) Use a difference quotient to approximate $f_u(1, 3)$ with $h = 0.001$.
   (b) Now evaluate $f_u(1, 3)$ exactly. Was the approximation in part (a) reasonable?

Find the indicated partial derivatives for Problems 3–34. Assume the variables are restricted to a domain on which the function is defined.

3. z_x if $z = x^2 y + 2x^5 y$

4. z_x if $z = \sin(5x^3 y - 3xy^2)$

5. g_x if $g(x, y) = \ln(ye^{xy})$

6. F_m if $F = mg$

7. $\dfrac{\partial}{\partial x}(a\sqrt{x})$

8. $\dfrac{\partial}{\partial x}(xe^{\sqrt{xy}})$

9. $\dfrac{\partial}{\partial y}(3x^5 y^7 - 32x^4 y^3 + 5xy)$

10. z_y if $z = \dfrac{3x^2 y^7 - y^2}{15xy - 8}$

11. $\dfrac{\partial A}{\partial h}$ if $A = \frac{1}{2}(a + b)h$

12. $\dfrac{\partial}{\partial m}\left(\frac{1}{2}mv^2\right)$

13. $\dfrac{\partial}{\partial B}\left(\dfrac{1}{u_0}B^2\right)$

14. $\dfrac{\partial}{\partial r}\left(\dfrac{2\pi r}{v}\right)$

15. F_v if $F = \dfrac{mv^2}{r}$

16. $\dfrac{\partial}{\partial v_0}(v_0 + at)$

17. $\dfrac{\partial F}{\partial m_2}$ if $F = \dfrac{Gm_1 m_2}{r^2}$

18. a_v if $a = \dfrac{v^2}{r}$

19. $\dfrac{\partial}{\partial T}\left(\dfrac{2\pi r}{T}\right)$

20. $\dfrac{\partial}{\partial t}\left(v_0 t + \frac{1}{2}at^2\right)$

21. u_E if $u = \dfrac{1}{2}\epsilon_0 E^2 + \dfrac{1}{2\mu_0}B^2$

22. $\dfrac{\partial f_0}{\partial L}$ if $f_0 = \dfrac{1}{2\pi\sqrt{LC}}$

23. $\dfrac{\partial y}{\partial t}$ if $y = \sin(ct - 5x)$

24. $\dfrac{\partial}{\partial M}\left(\dfrac{2\pi r^{3/2}}{\sqrt{GM}}\right)$

25. z_x if $z = \dfrac{1}{2x^2 ay} + \dfrac{3x^5 abc}{y}$

26. $\dfrac{\partial \alpha}{\partial \beta}$ if $\alpha = \dfrac{e^{x\beta - 3}}{2y\beta + 5}$

27. $\dfrac{\partial}{\partial \lambda}\left(\dfrac{x^2 y\lambda - 3\lambda^5}{\sqrt{\lambda^2 - 3\lambda + 5}}\right)$

28. $\dfrac{\partial m}{\partial v}$ if $m = \dfrac{m_0}{\sqrt{1 - v^2/c^2}}$

29. $\dfrac{\partial}{\partial w}\left(\sqrt{2\pi xyw - 13x^7 y^3 v}\right)$

30. $\dfrac{\partial}{\partial w}\left(\dfrac{x^2 yw - xy^3 w^7}{w - 1}\right)^{-7/2}$

31. z_x and z_y for $z = x^7 + 2^y + x^y$

32. $z_x(2,3)$ if $z = (\cos x) + y$

33. $\left.\dfrac{\partial z}{\partial y}\right|_{(1,0.5)}$ if $z = e^{x+2y}\sin y$

34. $\left.\dfrac{\partial f}{\partial x}\right|_{(\pi/3,1)}$ if $f(x,y) = x\ln(y\cos x)$

35. Consider the function $f(x,y) = x^2 + y^2$.

    (a) Estimate $f_x(2,1)$ and $f_y(2,1)$ using the contour diagram for f in Figure 13.15.

    (b) Estimate $f_x(2,1)$ and $f_y(2,1)$ from a table of values for f with $x = 1.9, 2, 2.1$ and $y = 0.9, 1, 1.1$.

    (c) Compare your estimates in parts (a) and (b) with the exact values of $f_x(2,1)$ and $f_y(2,1)$ found algebraically.

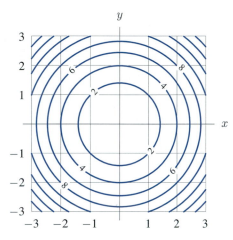

Figure 13.15

36. Show that the Cobb-Douglas function

$$Q = bK^\alpha L^{1-\alpha} \quad \text{where} \quad 0 < \alpha < 1$$

satisfies the equation

$$K\dfrac{\partial Q}{\partial K} + L\dfrac{\partial Q}{\partial L} = Q.$$

37. Money in a bank account earns interest at a continuous rate, r. The amount of money, $\$B$, in the account depends on the amount deposited, $\$P$, and the time, t, it has been in the bank according to the formula

$$B = Pe^{rt}.$$

Find $\partial B/\partial t$ and $\partial B/\partial P$ and interpret each in financial terms.

38. A one-meter long bar is heated unevenly, with temperature in $°C$ at a distance x meters from one end at time t given by

$$H(x,t) = 100e^{-0.1t}\sin(\pi x) \quad 0 \le x \le 1.$$

    (a) Sketch a graph of H against x for $t = 0$ and $t = 1$.

    (b) Calculate $H_x(0.2, t)$ and $H_x(0.8, t)$. What is the practical interpretation (in terms of temperature) of these two partial derivatives? Explain why each one has the sign it does.

    (c) Calculate $H_t(x, t)$. What is its sign? What is its interpretation in terms of temperature?

39. Let $h(x,t) = 5 + \cos(0.5x - t)$ be the function describing the stadium wave on page 566. The value of $h(x,t)$ gives the height of the head of the spectator in seat x at time t seconds. Evaluate $h_x(2,5)$ and $h_t(2,5)$ and interpret each in terms of the wave.

40. Is there a function f which has the following partial derivatives? If so what is it? Are there any others?

$$f_x(x,y) = 4x^3 y^2 - 3y^4,$$
$$f_y(x,y) = 2x^4 y - 12xy^3.$$

13.3 LOCAL LINEARITY AND THE DIFFERENTIAL

In Sections 13.1 and 13.2 we studied a function of two variables by allowing one variable at a time to change. We now let both variables change at once to develop a linear approximation for functions of two variables.

Zooming In to See Local Linearity

For a function of one variable, local linearity means that as we zoom in on the graph, it looks like a straight line. As we zoom in on the graph of a two-variable function, the graph usually looks like a plane, which is the graph of a linear function of two variables. (See Figure 13.16.)

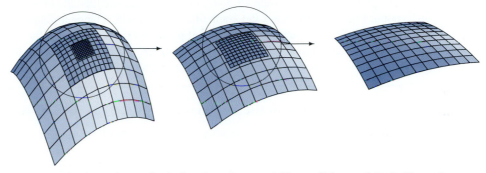

Figure 13.16: Zooming in on the graph of a function of two variables until the graph looks like a plane

Similarly, Figure 13.17 shows three successive views of the contours near a point. As we zoom in, the contours look more like equally spaced parallel lines, which are the contours of a linear function. (As we zoom in, we have to add more contours.)

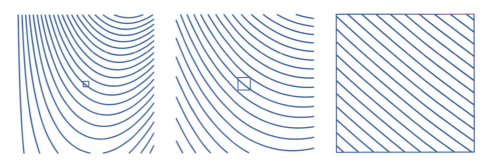

Figure 13.17: Zooming in on a contour diagram until the lines look parallel and equally spaced

This effect can also be seen numerically by zooming in with tables of values. Table 13.3 shows three tables of values for $f(x, y) = x^2 + y^3$ near $x = 2$, $y = 1$, each one a closer view than the previous one. Notice how each table looks more like the table of a linear function.

TABLE 13.3 *Zooming in on values of $f(x, y) = x^2 + y^3$ near $(2, 1)$ until the table looks linear*

		y		
		0	1	2
	1	1	2	9
x	2	4	5	12
	3	9	10	17

		y		
		0.9	1.0	1.1
	1.9	4.34	4.61	4.94
x	2.0	4.73	5.00	5.33
	2.1	5.14	5.41	5.74

		y		
		0.99	1.00	1.01
	1.99	4.93	4.96	4.99
x	2.00	4.97	5.00	5.03
	2.01	5.01	5.04	5.07

Zooming in Algebraically: Differentiability

Seeing a plane when we zoom in at a point tells us (provided the plane is not vertical) that $f(x, y)$ is closely approximated near that point by a linear function, $L(x, y)$:

$$f(x, y) \approx L(x, y).$$

The graph of the function $z = L(x, y)$ is the tangent plane at that point. Provided the approximation is sufficiently good, we say that $f(x, y)$ is *differentiable* at the point. The Focus on Theory section on page 709 explains how to tell if the approximation is sufficiently good. The functions we encounter will be differentiable at most points in their domain.

The Tangent Plane

The plane that we see when we zoom in on a surface is called the *tangent plane* to the surface at the point. Figure 13.18 shows the graph of a function with a tangent plane.

What is the equation of the tangent plane? At the point (a, b), the x-slope of the graph of f is the partial derivative $f_x(a, b)$ and the y-slope is $f_y(a, b)$. Thus, using the equation for a plane on page 597 of Chapter 11, we have the following result:

Tangent Plane to the Surface $z = f(x, y)$ at the Point (a, b)

Assuming f is differentiable at (a, b), the equation of the tangent plane is

$$z = f(a, b) + f_x(a, b)(x - a) + f_y(a, b)(y - b).$$

Here we are thinking of a and b as fixed, so $f(a, b)$, and $f_x(a, b)$, and $f_y(a, b)$ are constants. Thus, the right side of the equation is a linear function of x and y.

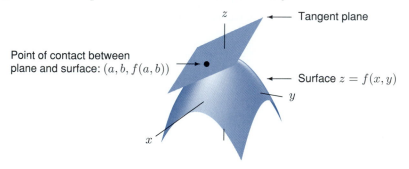

Figure 13.18: The tangent plane to the surface $z = f(x, y)$ at the point (a, b)

Example 1 Find the equation for the tangent plane to the surface $z = x^2 + y^2$ at the point $(3, 4)$.

Solution We have $f_x(x, y) = 2x$, so $f_x(3, 4) = 6$, and $f_y(x, y) = 2y$, so $f_y(3, 4) = 8$. Also, $f(3, 4) = 3^2 + 4^2 = 25$. Thus, the equation for the tangent plane at $(3, 4)$ is

$$z = 25 + 6(x - 3) + 8(y - 4) = -25 + 6x + 8y.$$

Local Linearization

Since the tangent plane lies close to the surface near the point at which they meet, z-values on the tangent plane are close to values of $f(x, y)$ for points near (a, b). Thus, replacing z by $f(x, y)$ in the equation of the tangent plane, we get the following approximation:

Tangent Plane Approximation to $f(x, y)$ for (x, y) Near the Point (a, b)

Provided f is differentiable at (a, b), we can approximate $f(x, y)$:

$$f(x, y) \approx f(a, b) + f_x(a, b)(x - a) + f_y(a, b)(y - b).$$

We are thinking of a and b as fixed, so the expression on the right side is linear in x and y. The right side of this approximation is called the **local linearization** of f near $x = a$, $y = b$.

Figure 13.19 shows the tangent plane approximation graphically.

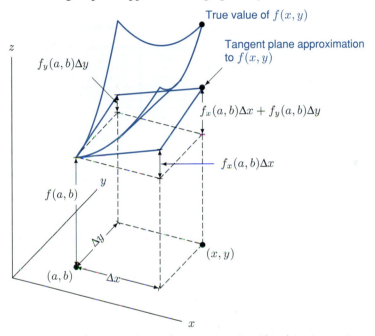

Figure 13.19: Local linearization: Approximating $f(x, y)$ by the z-value from the tangent plane

Example 2 Find the local linearization of $f(x, y) = x^2 + y^2$ at the point $(3, 4)$. Estimate $f(2.9, 4.2)$ and $f(2, 2)$ using the linearization and compare your answers to the true values.

Solution Let $z = f(x, y) = x^2 + y^2$. In Example 1 on page 666, we found the equation of the tangent plane at $(3, 4)$ to be

$$z = 25 + 6(x - 3) + 8(y - 4).$$

Therefore, for (x, y) near $(3, 4)$, we have the local linearization

$$f(x, y) \approx 25 + 6(x - 3) + 8(y - 4).$$

Substituting $x = 2.9, y = 4.2$ gives

$$f(2.9, 4.2) \approx 25 + 6(-0.1) + 8(0.2) = 26.$$

This compares favorably with the true value $f(2.9, 4.2) = (2.9)^2 + (4.2)^2 = 26.05$.

However, the local linearization does not give a good approximation at points far away from $(3, 4)$. For example, if $x = 2, y = 2$, the local linearization gives

$$f(2, 2) \approx 25 + 6(-1) + 8(-2) = 3,$$

whereas the true value of the function is $f(2, 2) = 2^2 + 2^2 = 8$.

Example 3   Designing safe boilers depends on knowing how steam behaves under changes in temperature and pressure. Steam tables, such as Table 13.4, are published giving values of the function $V = f(T, P)$ where V is the volume (in ft^3) of one pound of steam at a temperature T (in °F) and pressure P (in lb/in^2).

(a) Give a linear function approximating $V = f(T, P)$ for T near 500°F and P near 24 lb/in^2.

(b) Estimate the volume of a pound of steam at a temperature of 505°F and a pressure of 24.3 lb/in^2.

TABLE 13.4   *Volume (in cubic feet) of one pound of steam at various temperatures and pressures*

	Pressure P (lb/in^2)			
	20	22	24	26
480	27.85	25.31	23.19	21.39
500	28.46	25.86	23.69	21.86
520	29.06	26.41	24.20	22.33
540	29.66	26.95	24.70	22.79

Temperature T (°F)

Solution   (a) We want the local linearization around the point $T = 500$, $P = 24$, which is

$$f(T, P) \approx f(500, 24) + f_T(500, 24)(T - 500) + f_P(500, 24)(P - 24).$$

We read the value $f(500, 24) = 23.69$ from the table.

   Next we approximate $f_T(500, 24)$ by a difference quotient. From the $P = 24$ column, we compute the average rate of change between $T = 500$ and $T = 520$:

$$f_T(500, 24) \approx \frac{f(520, 24) - f(500, 24)}{520 - 500} = \frac{24.20 - 23.69}{20} = 0.0255.$$

Note that $f_T(500, 24)$ is positive, because steam expands when heated.

   Next we approximate $f_P(500, 24)$ by looking at the $T = 500$ row and computing the average rate of change between $P = 24$ and $P = 26$:

$$f_P(500, 24) \approx \frac{f(500, 26) - f(500, 24)}{26 - 24} = \frac{21.86 - 23.69}{2} = -0.915.$$

Note that $f_P(500, 24)$ is negative, because increasing the pressure on steam decreases its volume. Using these approximations for the partial derivatives, we obtain the local linearization:

$$V = f(T, P) \approx 23.69 + 0.0255(T - 500) - 0.915(P - 24) \text{ ft}^3 \quad \begin{array}{l} \text{for } T \text{ near } 500 \text{ °F} \\ \text{and } P \text{ near } 24 \text{ lb/in}^2. \end{array}$$

(b) We are interested in the volume at $T = 505$°F and $P = 24.3$ lb/in^2. Since these values are close to $T = 500$°F and $P = 24$ lb/in^2, we use the linear relation obtained in part (a).

$$V \approx 23.69 + 0.0255(505 - 500) - 0.915(24.3 - 24) = 23.54 \text{ ft}^3.$$

Local Linearity With Three or More Variables

Local linear approximations for functions of three or more variables follow the same pattern as for functions of two variables. The local linearization of $f(x, y, z)$ at (a, b, c) is given by

$$f(x, y, z) \approx f(a, b, c) + f_x(a, b, c)(x - a) + f_y(a, b, c)(y - b) + f_z(a, b, c)(z - c).$$

The Differential

We are often interested in the change in the value of the function as we move from the point (a, b) to a nearby point (x, y). Then we use the notation

$$\Delta f = f(x, y) - f(a, b) \quad \text{and} \quad \Delta x = x - a \quad \text{and} \quad \Delta y = y - b$$

to rewrite the tangent plane approximation

$$f(x, y) \approx f(a, b) + f_x(a, b)(x - a) + f_y(a, b)(y - b)$$

in the form

$$\Delta f \approx f_x(a, b)\Delta x + f_y(a, b)\Delta y.$$

For fixed a and b, the right side of this is a linear function of Δx and Δy that can be used to estimate Δf. We call this linear function the *differential*. To define the differential in general, we introduce new variables dx and dy to represent changes in x and y.

> ### The Differential of a Function $z = f(x, y)$
>
> The **differential**, df (or dz), at a point (a, b) is the linear function of dx and dy given by the formula
>
> $$df = f_x(a, b)\, dx + f_y(a, b)\, dy.$$
>
> The differential at a general point is often written $df = f_x\, dx + f_y\, dy$.

Note that the differential, df, is a function of four variables a, b, and dx, dy.

Example 4 Compute the differentials of the following functions.
  (a) $f(x, y) = x^2 e^{5y}$ (b) $z = x \sin(xy)$ (c) $f(x, y) = x \cos(2x)$

Solution (a) Since $f_x(x, y) = 2xe^{5y}$ and $f_y(x, y) = 5x^2 e^{5y}$, we have

$$df = 2xe^{5y}\, dx + 5x^2 e^{5y}\, dy.$$

(b) Since $\partial z/\partial x = \sin(xy) + xy\cos(xy)$ and $\partial z/\partial y = x^2 \cos(xy)$, we have

$$dz = (\sin(xy) + xy\cos(xy))\, dx + x^2 \cos(xy)\, dy.$$

(c) Since $f_x(x, y) = \cos(2x) - 2x\sin(2x)$ and $f_y(x, y) = 0$, we have

$$df = (\cos(2x) - 2x\sin(2x))\, dx + 0\, dy = (\cos(2x) - 2x\sin(2x))\, dx.$$

Example 5 The density ρ (in g/cm^3) of carbon dioxide gas CO_2 depends upon its temperature T (in °C) and pressure P (in atmospheres). The ideal gas model for CO_2 gives what is called the state equation

$$\rho = \frac{0.5363P}{T + 273.15}.$$

Compute the differential $d\rho$. Explain the signs of the coefficients of dT and dP.

Solution The differential for $\rho = f(T, P)$ is

$$d\rho = f_T(T, P)\, dT + f_P(T, P)dP = \frac{-0.5363P}{(T + 273.15)^2}\, dT + \frac{0.5363}{T + 273.15}\, dP.$$

The coefficient of dT is negative because increasing the temperature expands the gas (if the pressure is kept constant) and therefore decreases its density. The coefficient of dP is positive because increasing the pressure compresses the gas (if the temperature is kept constant) and therefore increases its density.

Where Does the Notation for the Differential Come From?

We write the differential as a linear function of the new variables dx and dy. You may wonder why we chose these names for our variables. The reason is historical: The people who invented calculus thought of dx and dy as "infinitesimal" changes in x and y. The equation

$$df = f_x dx + f_y dy$$

was regarded as an infinitesimal version of the local linear approximation

$$\Delta f \approx f_x \Delta x + f_y \Delta y.$$

In spite of the problems with defining exactly what "infinitesimal" means, some mathematicians, scientists, and engineers think of the differential in terms of infinitesimals.

Figure 13.20 illustrates a way of thinking about differentials that combines the definition with this informal point of view. It shows the graph of f along with a view of the graph around the point $(a, b, f(a, b))$ under a microscope. Since f is locally linear at the point, the magnified view looks like the tangent plane. Under the microscope, we use a magnified coordinate system with its origin at the point $(a, b, f(a, b))$ and with coordinates dx, dy, and dz along the three axes. The graph of the differential df is the tangent plane, which has equation $df = f_x(a, b)\, dx + f_y(a, b)\, dy$ in the magnified coordinates.

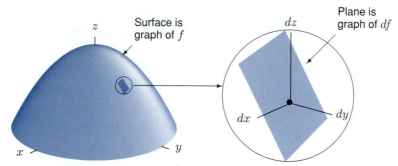

Figure 13.20: The graph of f, with a view through a microscope showing the tangent plane in the magnified coordinate system

Problems for Section 13.3

For the functions in Problems 1–3, find the equation of the tangent plane at the given point.

1. $z = e^y + x + x^2 + 6$ at the point $(1, 0, 9)$.   2. $z = ye^{x/y}$ at the point $(1, 1, e)$.

3. $z = \frac{1}{2}(x^2 + 4y^2)$ at the point $(2, 1, 4)$.

4. A student was asked to find the equation of the tangent plane to the surface $z = x^3 - y^2$ at the point $(x, y) = (2, 3)$. The student's answer was

$$z = 3x^2(x - 2) - 2y(y - 3) - 1.$$

   (a) At a glance, how do you know this is wrong?
   (b) What mistake did the student make?
   (c) Answer the question correctly.

5. Find the local linearization of the function $f(x, y) = x^2y$ at the point $(3, 1)$.

6. (a) Verify the local linearity of $f(x, y) = e^{-x} \sin y$ near $x = 1$, $y = 2$ by making a table of values of f for $x = 0.9$, 1.0, 1.1 and $y = 1.9$, 2.0, 2.1. Express values of f with 4 digits after the decimal point. Then make a table of values for $x = 0.99$, 1.00, 1.01 and $y = 1.99$, 2.00. 2.01, again showing 4 digits after the decimal point. Do both tables look nearly linear? Does the second table look more linear than the first?

    (b) Give the local linearization of $f(x, y) = e^{-x} \sin y$ at $(1, 2)$, first using your tables, and second using the fact that $f_x(x, y) = -e^{-x} \sin y$ and $f_y(x, y) = e^{-x} \cos y$.

7. Give the local linearization for the monthly car-loan payment function at each of the points investigated in Problem 13 on page 659.

8. In Example 3 on page 668 we found a linear approximation for $V = f(T, P)$ near $(500, 24)$. Now find a linear approximation near $(480, 20)$.

9. In Example 3 on page 668 we found a linear approximation for $V = f(T, p)$ near $(500, 24)$.

    (a) Test the accuracy of this approximation by comparing its predicted value with the four neighboring values in the table. What do you notice? Which predicted values are accurate? Which are not? Explain your answer.

    (b) Suggest a linear approximation for $f(T, p)$ near $(500, 24)$ that does not have the property you noticed in part (a). [Hint: Estimate the partial derivatives in a different way.]

10. Figure 13.21 shows a transistor whose state at any moment is determined by the three currents i_b, i_c, and i_e and the two voltages v_b and v_c. These five quantities can be determined from measurements of i_b and v_c alone, because there are functions f and g (called the *characteristics* of the transistor) such that $i_c = f(i_b, v_c)$ and $v_b = g(i_b, v_c)$ and $i_e = -i_b - i_c$. The units are microamps (μA) for i_b, volts (V) for v_c, and milliamps (mA) for i_c.

    Use Figure 13.22 to find a linear approximation for f that is valid when the transistor is operating with i_b near $-300\,\mu A$ and v_c near $-8V$.

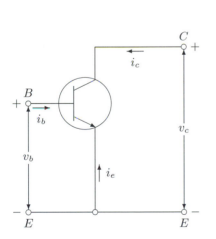

Figure 13.21

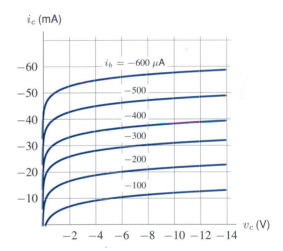

Figure 13.22: Graphs of f as a function of v_c with i_b fixed

Find the differentials of the functions in Problems 11–14.

11. $f(x, y) = \sin(xy)$

12. $z = e^{-x} \cos y$

13. $g(u, v) = u^2 + uv$

14. $h(x, t) = e^{-3t} \sin(x + 5t)$

Find differentials of the functions in Problems 15–18 at the given point.

15. $f(x, y) = xe^{-y}$ at $(1, 0)$

16. $g(x, t) = x^2 \sin(2t)$ at $(2, \pi/4)$

17. $P(L, K) = 1.01L^{0.25}K^{0.75}$ at $(100, 1)$

18. $F(m, r) = Gm/r^2$ at $(100, 10)$

19. Find the differential of $f(x, y) = \sqrt{x^2 + y^3}$ at the point $(1, 2)$. Use it to estimate $f(1.04, 1.98)$.

20. An unevenly heated plate has temperature $T(x, y)$ in °C at the point (x, y). If $T(2, 1) = 135$, and $T_x(2, 1) = 16$, and $T_y(2, 1) = -15$, estimate the temperature at the point $(2.04, 0.97)$.

21. One mole of ammonia gas is contained in a vessel which is capable of changing its volume (a compartment sealed by a piston, for example). The total energy U (in joules) of the ammonia, is a function of the volume V (in m^3) of the container, and the temperature T (in K) of the gas. The differential dU is given by

$$dU = 840\, dV + 27.32\, dT.$$

(a) How does the energy change if the volume is held constant and the temperature is increased slightly?

(b) How does the energy change if the temperature is held constant and the volume is increased slightly?

(c) Find the approximate change in energy if the gas is compressed by 100 cm^3 and heated by 2 K.

22. The coefficient, β, of thermal expansion of a liquid relates the change in the volume V (in m^3) of a fixed quantity of a liquid to an increase in its temperature T (in °C):

$$dV = \beta V\, dT.$$

(a) Let ρ be the density (in kg/m^3) of water as a function of temperature. Write an expression for $d\rho$ in terms of ρ and dT.

(b) The graph in Figure 13.23 shows density of water as a function of temperature. Use it to estimate β when $T = 20°$C and when $T = 80°$C.

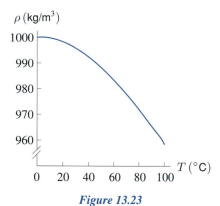

Figure 13.23

23. A pendulum of period T and length l was used to determine g from the formulas

$$g = \frac{4\pi^2 l}{T^2} \quad \text{and} \quad l = s + \frac{k^2}{s}, \quad k < s.$$

If the measurements of k and s are accurate to within 1%, find the maximum percentage error in l. If the measurement of T is accurate to 0.5%, find the maximum percentage error in the computed value of g.

24. The period, T, of oscillation in seconds of a pendulum clock is given by $T = 2\pi\sqrt{l/g}$, where g is the acceleration due to gravity. The length of the pendulum, l, depends on the temperature, t, according to the formula $l = l_0(1 + \alpha t)$ where l_0 is the length of the pendulum at temperature t_0 and α is a constant which characterizes the clock. The clock is set to the correct period at the temperature t_0. How many seconds a day does the clock lose or gain when the temperature is $t_0 + \Delta t$? Show that this loss or gain is independent of l_0.

25. (a) Write a formula for the number π using only the perimeter L and the area A of a circle.

(b) Suppose that L and A are determined experimentally. Show that if the relative, or percent, errors in the measured values of L and A are λ and μ, respectively, then the resulting relative, or percent, error in π is $2\lambda - \mu$.

13.4 GRADIENTS AND DIRECTIONAL DERIVATIVES IN THE PLANE

The Rate of Change in an Arbitrary Direction: The Directional Derivative

The partial derivatives of a function f tell us the rate of change of f in the directions parallel to the coordinate axes. In this section we see how to compute the rate of change of f in an arbitrary direction.

Example 1 Figure 13.24 shows the temperature, in °C, at the point (x, y). Estimate the average rate of change of temperature as we walk from point A to point B.

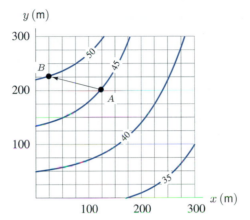

Figure 13.24: Estimating rate of change on a temperature map

Solution At the point A we are on the $H = 45°C$ contour. At B we are on the $H = 50°C$ contour. The displacement vector from A to B has x component approximately $-100\vec{i}$ and y component approximately $25\vec{j}$, so its length is $\sqrt{(-100)^2 + 25^2} \approx 103$. Thus the temperature rises by 5°C as we move 103 meters, so the average rate of change of the temperature in that direction is about $5/103 \approx 0.05°C/m$.

Suppose we want to compute the rate of change of a function $f(x, y)$ at the point $P = (a, b)$ in the direction of the unit vector $\vec{u} = u_1\vec{i} + u_2\vec{j}$. For $h > 0$, consider the point $Q = (a + hu_1, b + hu_2)$ whose displacement from P is $h\vec{u}$. (See Figure 13.25.) Since $\|\vec{u}\| = 1$, the distance from P to Q is h. Thus,

$$\begin{array}{l}\text{Average rate of change} \\ \text{in } f \text{ from } P \text{ to } Q\end{array} = \frac{\text{Change in } f}{\text{Distance from } P \text{ to } Q} = \frac{f(a + hu_1, b + hu_2) - f(a, b)}{h}.$$

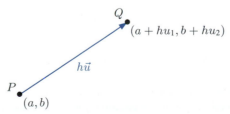

Figure 13.25: Displacement of $h\vec{u}$ from the point (a, b)

Taking the limit as $h \to 0$ gives the instantaneous rate of change and the following definition:

Directional Derivative of f at (a, b) in the Direction of a Unit Vector $\vec{u}$

If $\vec{u} = u_1\vec{i} + u_2\vec{j}$ is a unit vector, we define the directional derivative, $f_{\vec{u}}$, by

$$f_{\vec{u}}(a, b) = \begin{matrix} \text{Rate of change} \\ \text{of } f \text{ in direction} \\ \text{of } \vec{u} \text{ at } (a, b) \end{matrix} = \lim_{h \to 0} \frac{f(a + hu_1, b + hu_2) - f(a, b)}{h},$$

provided the limit exists.

Notice that if $\vec{u} = \vec{i}$, so $u_1 = 1, u_2 = 0$, then the directional derivative is f_x, since

$$f_{\vec{i}}(a, b) = \lim_{h \to 0} \frac{f(a + h, b) - f(a, b)}{h} = f_x(a, b).$$

Similarly, if $\vec{u} = \vec{j}$ then the directional derivative $f_{\vec{j}} = f_y$.

What If We Do Not Have a Unit Vector?

We defined $f_{\vec{u}}$ for $\vec{u}$ a unit vector. If $\vec{v}$ is not a unit vector, $\vec{v} \neq \vec{0}$, we construct a unit vector $\vec{u} = \vec{v}/\|\vec{v}\|$ in the same direction as $\vec{v}$ and define the rate of change of f in the direction of $\vec{v}$ as $f_{\vec{u}}$.

Example 2   For each of the functions f, g, and h in Figure 13.26, decide whether the directional derivative at the indicated point is positive, negative, or zero, in the direction of the vector $\vec{v} = \vec{i} + 2\vec{j}$, and in the direction of the vector $\vec{w} = 2\vec{i} + \vec{j}$.

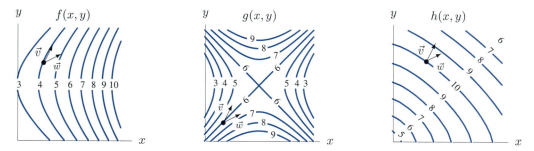

Figure 13.26: Contour diagrams of three functions with direction vectors $\vec{v} = \vec{i} + 2\vec{j}$ and $\vec{w} = 2\vec{i} + \vec{j}$ marked on each

Solution   On the contour diagram for f, the vector $\vec{v} = \vec{i} + 2\vec{j}$ appears to be tangent to the contour. Thus, in this direction, the value of the function is not changing, so the directional derivative in the direction of $\vec{v}$ is zero. The vector $\vec{w} = 2\vec{i} + \vec{j}$ points from the contour marked 4 toward the contour marked 5. Thus, the values of the function are increasing and the directional derivative in the direction of $\vec{w}$ is positive.

On the contour diagram for g, the vector $\vec{v} = \vec{i} + 2\vec{j}$ points from the contour marked 6 toward the contour marked 5, so the function is decreasing in that direction. Thus, the rate of change is negative. On the other hand, the vector $\vec{w} = 2\vec{i} + \vec{j}$ points from the contour marked 6 toward the contour marked 7, and hence the directional derivative in the direction of $\vec{w}$ is positive.

Finally, on the contour diagram for h, both vectors point from the $h = 10$ contour to the $h = 9$ contour, so both directional derivatives are negative.

Example 3 Calculate the directional derivative of $f(x, y) = x^2 + y^2$ at $(1, 0)$ in the direction of the vector $\vec{i} + \vec{j}$.

Solution First we have to find the unit vector in the same direction as the vector $\vec{i} + \vec{j}$. Since this vector has magnitude $\sqrt{2}$, the unit vector is

$$\vec{u} = \frac{1}{\sqrt{2}}(\vec{i} + \vec{j}) = \frac{1}{\sqrt{2}}\vec{i} + \frac{1}{\sqrt{2}}\vec{j}.$$

Thus,

$$f_{\vec{u}}(1, 0) = \lim_{h \to 0} \frac{f(1 + h/\sqrt{2}, h/\sqrt{2}) - f(1, 0)}{h} = \lim_{h \to 0} \frac{(1 + h/\sqrt{2})^2 + (h/\sqrt{2})^2 - 1}{h}$$

$$= \lim_{h \to 0} \frac{\sqrt{2}h + h^2}{h} = \lim_{h \to 0} (\sqrt{2} + h) = \sqrt{2}.$$

Computing Directional Derivatives From Partial Derivatives

If f is differentiable, we will now see how to use local linearity to find a formula for the directional derivative which does not involve a limit. If $\vec{u}$ is a unit vector, the definition of $f_{\vec{u}}$ says

$$f_{\vec{u}}(a, b) = \lim_{h \to 0} \frac{f(a + hu_1, b + hu_2) - f(a, b)}{h} = \lim_{h \to 0} \frac{\Delta f}{h},$$

where $\Delta f = f(a + hu_1, b + hu_2) - f(a, b)$ is the change in f. We write Δx for the change in x, so $\Delta x = (a + hu_1) - a = hu_1$; similarly $\Delta y = hu_2$. Using local linearity, we have

$$\Delta f \approx f_x(a, b)\Delta x + f_y(a, b)\Delta y = f_x(a, b)hu_1 + f_y(a, b)hu_2.$$

Thus, dividing by h gives

$$\frac{\Delta f}{h} \approx \frac{f_x(a, b)hu_1 + f_y(a, b)hu_2}{h} = f_x(a, b)u_1 + f_y(a, b)u_2.$$

This approximation becomes exact as $h \to 0$, so we have the following formula:

$$f_{\vec{u}}(a, b) = f_x(a, b)u_1 + f_y(a, b)u_2.$$

Example 4 Use the preceding formula to compute the directional derivative in Example 3. Check that we get the same answer as before.

Solution We calculate $f_{\vec{u}}(1, 0)$, where $f(x, y) = x^2 + y^2$ and $\vec{u} = \frac{1}{\sqrt{2}}\vec{i} + \frac{1}{\sqrt{2}}\vec{j}$.

The partial derivatives are $f_x(x, y) = 2x$ and $f_y(x, y) = 2y$. So, as before

$$f_{\vec{u}}(1, 0) = f_x(1, 0)u_1 + f_y(1, 0)u_2 = (2)\left(\frac{1}{\sqrt{2}}\right) + (0)\left(\frac{1}{\sqrt{2}}\right) = \sqrt{2}.$$

The Gradient Vector

Notice that the expression for $f_{\vec{u}}(a, b)$ can be written as a dot product of $\vec{u}$ and a new vector:

$$f_{\vec{u}}(a, b) = f_x(a, b)u_1 + f_y(a, b)u_2 = (f_x(a, b)\vec{i} + f_y(a, b)\vec{j}) \cdot (u_1\vec{i} + u_2\vec{j}).$$

The new vector, $f_x(a, b)\vec{i} + f_y(a, b)\vec{j}$, turns out to be important. Thus, we make the following definition:

The Gradient Vector of a differentiable function at the point (a, b) is

$$\text{grad } f(a, b) = f_x(a, b)\vec{i} + f_y(a, b)\vec{j}$$

The formula for the directional derivative can be written in terms of the gradient as follows:

The Directional Derivative and the Gradient

If f is differentiable at (a, b), then

$$f_{\vec{u}}\,(a, b) = f_x(a, b)u_1 + f_y(a, b)u_2 = \operatorname{grad} f(a, b) \cdot \vec{u}\,,$$

where $\vec{u} = u_1\vec{i} + u_2\vec{j}$ is a unit vector.

Example 5   Find the gradient vector of $f(x, y) = x + e^y$ at the point $(1, 1)$.

Solution   Using the definition we have

$$\operatorname{grad} f = f_x\vec{i} + f_y\vec{j} = \vec{i} + e^y\vec{j}\,,$$

so at the point $(1, 1)$

$$\operatorname{grad} f(1, 1) = \vec{i} + e\vec{j}\,.$$

Alternative Notation for the Gradient

You can think of $\dfrac{\partial f}{\partial x}\vec{i} + \dfrac{\partial f}{\partial y}\vec{j}$ as the result of applying the vector operator (pronounced "del")

$$\nabla = \frac{\partial}{\partial x}\vec{i} + \frac{\partial}{\partial y}\vec{j}$$

to the function f. Thus, we get the alternative notation

$$\operatorname{grad} f = \nabla f.$$

What Does the Gradient Tell Us?

The fact that $f_{\vec{u}} = \operatorname{grad} f \cdot \vec{u}$ enables us to see what the gradient vector represents. Suppose θ is the angle between the vectors $\operatorname{grad} f$ and $\vec{u}$. At the point (a, b), we have

$$f_{\vec{u}} = \operatorname{grad} f \cdot \vec{u} = \| \operatorname{grad} f\| \underbrace{\|\vec{u}\,\|}_{1} \cos\theta = \| \operatorname{grad} f\| \cos\theta.$$

Imagine that $\operatorname{grad} f$ is fixed and that $\vec{u}$ can rotate. (See Figure 13.27.) The maximum value of $f_{\vec{u}}$ occurs when $\cos\theta = 1$, so $\theta = 0$ and $\vec{u}$ is pointing in the direction of $\operatorname{grad} f$. Then

$$\text{Maximum } f_{\vec{u}} = \| \operatorname{grad} f\| \cos 0 = \| \operatorname{grad} f\|.$$

The minimum value of $f_{\vec{u}}$ occurs when $\cos\theta = -1$, so $\theta = \pi$ and $\vec{u}$ is pointing in the direction opposite to $\operatorname{grad} f$. Then

$$\text{Minimum } f_{\vec{u}} = \| \operatorname{grad} f\| \cos\pi = -\| \operatorname{grad} f\|.$$

When $\theta = \pi/2$ or $3\pi/2$, so $\cos\theta = 0$, the directional derivative is zero.

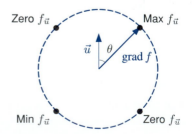

Figure 13.27: Values of the directional derivative at different angles to the gradient

Properties of The Gradient Vector

We have seen that the gradient vector points in the direction of the greatest rate of change at a point and the magnitude of the gradient vector is that rate of change.

Figure 13.28 shows that the gradient vector at any point is perpendicular to the contour through that point. Assuming f is differentiable at the point (a, b), local linearity tells us that the contours of f around the point (a, b) appear straight, parallel, and equally spaced. The greatest rate of change is obtained by moving in the direction that takes us to the next contour in the shortest possible distance, which is the direction perpendicular to the contour. Thus, we have the following:

Geometric Properties of the Gradient Vector

If f is a differentiable function at the point (a, b) and grad $f(a, b) \neq \vec{0}$, then:
- The direction of grad $f(a, b)$ is
  - Perpendicular to the contour of f through (a, b)
  - In the direction of increasing f
- The magnitude of the gradient vector, $\| \text{grad } f \|$, is
  - The maximum rate of change of f at that point
  - Large when the contours are close together and small when they are far apart.

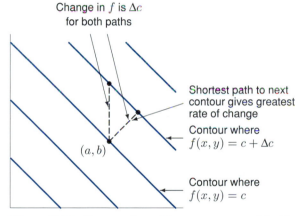

Change in f is Δc for both paths

Shortest path to next contour gives greatest rate of change

Contour where $f(x, y) = c + \Delta c$

(a, b)

Contour where $f(x, y) = c$

Figure 13.28: Close-up view of the contours around (a, b), showing the gradient is perpendicular to the contours

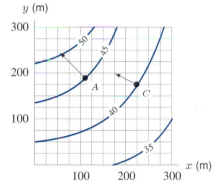

Figure 13.29: A temperature map showing directions and relative magnitudes of two gradient vectors

Examples of Directional Derivatives and Gradient Vectors

Example 6   Explain why the gradient vectors at points A and C in Figure 13.29 have the direction and the relative magnitudes they do.

Solution   The gradient vector points in the direction of greatest increase of the function. This means that in Figure 13.29, the gradient points directly towards warmer temperatures. The magnitude of the gradient vector measures the rate of change. The gradient vector at A is longer than the gradient vector at C because the contours are closer together at A, so the rate of change is larger.

Example 2 on page 674 shows how the contour diagram can tell us the sign of the directional derivative. In the next example we compute the directional derivative in three directions, two close to that of the gradient vector and one that is not.

Example 7　Use the gradient to find the directional derivative of $f(x, y) = x + e^y$ at the point $(1, 1)$ in the direction of the vectors $\vec{i} - \vec{j}, \vec{i} + 2\vec{j}, \vec{i} + 3\vec{j}$.

Solution　In Example 5 we found

$$\operatorname{grad} f(1, 1) = \vec{i} + e\vec{j}.$$

A unit vector in the direction of $\vec{i} - \vec{j}$ is $\vec{s} = (\vec{i} - \vec{j})/\sqrt{2}$, so

$$f_{\vec{s}}(1, 1) = \operatorname{grad} f(1, 1) \cdot \vec{s} = (\vec{i} + e\vec{j}) \cdot \left(\frac{\vec{i} - \vec{j}}{\sqrt{2}} \right) = \frac{1 - e}{\sqrt{2}} \approx -1.215.$$

A unit vector in the direction of $\vec{i} + 2\vec{j}$ is $\vec{v} = (\vec{i} + 2\vec{j})/\sqrt{5}$, so

$$f_{\vec{v}}(1, 1) = \operatorname{grad} f(1, 1) \cdot \vec{v} = (\vec{i} + e\vec{j}) \cdot \left(\frac{\vec{i} + 2\vec{j}}{\sqrt{5}} \right) = \frac{1 + 2e}{\sqrt{5}} \approx 2.879.$$

A unit vector in the direction of $\vec{i} + 3\vec{j}$ is $\vec{w} = (\vec{i} + 3\vec{j})/\sqrt{10}$, so

$$f_{\vec{w}}(1, 1) = \operatorname{grad} f(1, 1) \cdot \vec{w} = (\vec{i} + e\vec{j}) \cdot \left(\frac{\vec{i} + 3\vec{j}}{\sqrt{10}} \right) = \frac{1 + 3e}{\sqrt{10}} \approx 2.895.$$

Now look back at the answers and compare with the value of $\| \operatorname{grad} f \| = \sqrt{1 + e^2} \approx 2.896$. One answer is not close to this value; the other two, $f_{\vec{v}} = 2.879$ and $f_{\vec{w}} = 2.895$, are close but slightly smaller than $\| \operatorname{grad} f \|$. Since $\| \operatorname{grad} f \|$ is the maximum rate of change of f at the point, we would expect for *any* unit vector $\vec{u}$:

$$f_{\vec{u}}(1, 1) \le \| \operatorname{grad} f \|.$$

with equality when $\vec{u}$ is in the direction of $\operatorname{grad} f$. Since $e \approx 2.718$, the vectors $\vec{i} + 2\vec{j}$ and $\vec{i} + 3\vec{j}$ both point roughly, but not exactly, in the direction of the gradient vector $\operatorname{grad} f(1, 1) = \vec{i} + e\vec{j}$. Thus, the values of $f_{\vec{v}}$ and $f_{\vec{w}}$ are both close to the value of $\| \operatorname{grad} f \|$. The direction of the vector $\vec{i} - \vec{j}$ is not close to the direction of $\operatorname{grad} f$ and the value of $f_{\vec{s}}$ is not close to the value of $\| \operatorname{grad} f \|$.

Problems for Section 13.4

1. Suppose $f(x, y) = x + \ln y$. Using difference quotients as in Example 1 on page 673, estimate
   (a) The rate of change of f as you leave the point $(1, 4)$ going towards the point $(3, 5)$.
   (b) The rate of change of f as you arrive at the point $(3, 5)$.

2. Using the limit of a difference quotient, compute the rate of change of $f(x, y) = 2x^2 + y^2$ at the point $(2, 1)$ as you move in the direction of the vector $\vec{u} = (\vec{i} + \vec{j})/\sqrt{2}$.

For Problems 3–8 use Figure 13.30, showing level curves of $f(x, y)$, to estimate the directional derivatives.

3. $f_{\vec{i}}(3, 1)$

4. $f_{\vec{j}}(3, 1)$

5. $f_{\vec{u}}(3, 1)$ where $\vec{u} = (\vec{i} - \vec{j})/\sqrt{2}$

6. $f_{\vec{u}}(3, 1)$ where $\vec{u} = (-\vec{i} + \vec{j})/\sqrt{2}$

7. For what part of the rectangular region shown in Figure 13.30 is $f_{\vec{i}}$ positive?

8. For what part of the rectangular region shown in Figure 13.30 is $f_{\vec{j}}$ negative?

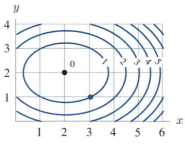

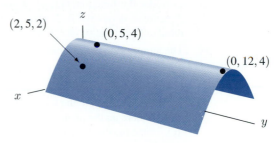

Figure 13.30   **Figure 13.31**

9. The surface $z = g(x, y)$ is shown in Figure 13.31. What is the sign of each of the following directional derivatives?
    (a) $g_{\vec{u}}\,(2, 5)$ where $\vec{u} = (\vec{i} - \vec{j}\,)/\sqrt{2}$.   (b) $g_{\vec{u}}\,(2, 5)$ where $\vec{u} = (\vec{i} + \vec{j}\,)/\sqrt{2}$.

10. If $f(x, y) = x^2 + \ln y$, find (a) grad f (b) grad f at $(4, 1)$.

In Problems 11–22 find the gradient of the given function z. Assume the variables are restricted to a domain on which the function is defined.

11. $z = \sin(x/y)$ 12. $z = xe^y$

13. $z = (x + y)e^y$ 14. $z = \tan^{-1}(x/y)$

15. $z = \sin(x^2 + y^2)$ 16. $z = xe^y/(x + y)$

17. $f(m, n) = m^2 + n^2$ 18. $f(x, y) = \frac{3}{2}x^5 - \frac{4}{7}y^6$

19. $f(s, t) = \dfrac{1}{\sqrt{s}}(t^2 - 2t + 4)$ 20. $f(\alpha, \beta) = \dfrac{2\alpha + 3\beta}{2\alpha - 3\beta}$

21. $f(\alpha, \beta) = \sqrt{5\alpha^2 + \beta}$ 22. $f(x, y) = \sin(xy) + \cos(xy)$

In Problems 23–26 compute the gradient at the specified point.

23. $f(m, n) = 5m^2 + 3n^4$, at $(5, 2)$ 24. $f(x, y) = x^2y + 7xy^3$, at $(1, 2)$

25. $f(x, y) = \sqrt{\tan x + y}$, at $(0, 1)$ 26. $f(x, y) = \sin(x^2) + \cos y$, at $(\frac{\sqrt{\pi}}{2}, 0)$

27. Let $f(x, y) = (x + y)/(1 + x^2)$. Find the directional derivative at $P = (1, -2)$ in the direction of the vectors
    (a) $\vec{v} = 3\vec{i} - 2\vec{j}$,   (b) $\vec{v} = -\vec{i} + 4\vec{j}$.
    (c) What is the direction of greatest increase at P ?

28. A student was asked to find the directional derivative of $f(x, y) = x^2e^y$ at the point $(1, 0)$ in the direction of $\vec{v} = 4\vec{i} + 3\vec{j}$. The student's answer was

$$f_{\vec{u}}\,(1, 0) = \nabla f(1, 0) \cdot \vec{u} = \frac{8}{5}\vec{i} + \frac{3}{5}\vec{j}.$$

    (a) At a glance, how do you know this is wrong?
    (b) What is the correct answer?

29. Find the directional derivative of $f(x, y) = e^x \tan(y) + 2x^2y$ at the point $(0, \pi/4)$ in the following directions (a) $\vec{i} - \vec{j}$ (b) $\vec{i} + \sqrt{3}\vec{j}$.

30. Find the directional derivative of $z = x^2y$ at the point $(1, 2)$ in the direction making an angle of $5\pi/4$ with the x-axis. In which direction is the directional derivative the largest?

31. Find the rate of change of $f(x, y) = x^2 + y^2$ at the point $(1, 2)$ in the direction of the vector $\vec{u} = 0.6\vec{i} + 0.8\vec{j}$.

32. The directional derivative of $z = f(x, y)$ at the point $(2, 1)$ in the direction toward the point $(1, 3)$ is $-2/\sqrt{5}$, and the directional derivative in the direction toward the point $(5, 5)$ is 1. Compute $\partial z/\partial x$ and $\partial z/\partial y$ at the point $(2, 1)$.

33.   Consider the function $f(x, y)$. If you start at the point $(4, 5)$ and move towards the point $(5, 6)$, the directional derivative is 2. Starting at the point $(4, 5)$ and moving towards the point $(6, 6)$ gives a directional derivative of 3. Find ∇f at the point $(4, 5)$.

34.   The temperature at any point in the plane is given by the function

$$T(x, y) = \frac{100}{x^2 + y^2 + 1}.$$

(a)   What shape are the level curves of T?

(b)   Where on the plane is it hottest? What is the temperature at that point?

(c)   Find the direction of the greatest increase in temperature at the point $(3, 2)$. What is the magnitude of that greatest increase?

(d)   Find the direction of the greatest decrease in temperature at the point $(3, 2)$.

(e)   Find a direction at the point $(3, 2)$ in which the temperature does not increase or decrease.

35.   A differentiable function $f(x, y)$ has the property that $f_x(4, 1) = 2$ and $f_y(4, 1) = -1$. Find the equation of the tangent line to the level curve of f through the point $(4, 1)$.

36.   Figure 13.32 represents the level curves $f(x, y) = c$; the values of f on each curve are marked. In each of the following parts, decide whether the given quantity is positive, negative or zero. Explain your answer.

(a)   The value of $\nabla f \cdot \vec{i}$ at P.   (b)   The value of $\nabla f \cdot \vec{j}$ at P.

(c) $\partial f / \partial x$ at Q.   (d) $\partial f / \partial y$ at Q.

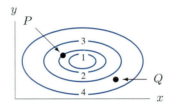

Figure 13.32

37.   In Figure 13.32, which is larger: $\|\nabla f\|$ at P or $\|\nabla f\|$ at Q? Explain how you know.

38.   The sketch in Figure 13.33 shows the level curves of a function $z = f(x, y)$. At the points $(1, 1)$ and $(1, 4)$ on the sketch, draw a vector representing grad f. Explain how you decided the approximate direction and length of each vector.

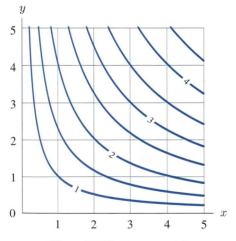

Figure 13.33: Contours of f

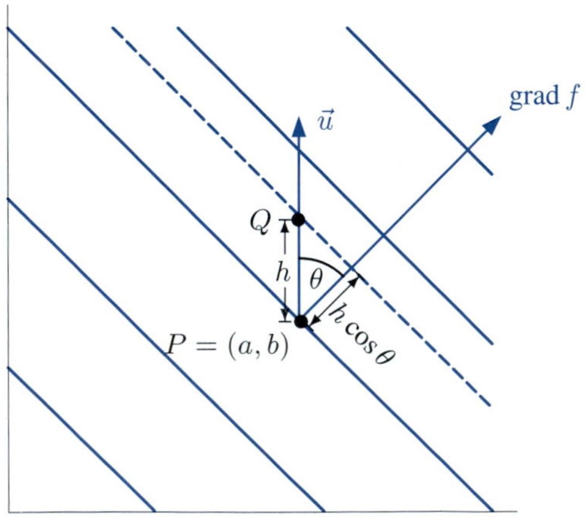

Figure 13.34

39.   In this problem we see another way of explaining the formula $f_{\vec{u}}(a, b) = \text{grad } f(a, b) \cdot \vec{u}$. Imagine zooming in on a function $f(x, y)$ at a point (a, b). By local linearity, the contours around (a, b) look like the contours of a linear function. See Figure 13.34. Suppose you want to find the directional derivative $f_{\vec{u}}(a, b)$ in the direction of a unit vector $\vec{u}$. If you move from P to Q, a small distance h in the direction

of $\vec{u}$, then the directional derivative is approximated by the difference quotient

$$\frac{\text{Change in } f \text{ between } P \text{ and } Q}{h}.$$

    (a)   Use the gradient to show that

$$\text{Change in } f \approx \| \operatorname{grad} f \| (h \cos \theta).$$

    (b)   Use part (a) to justify the formula $f_{\vec{u}}(a,b) = \operatorname{grad} f(a,b) \cdot \vec{u}$.

40.   Figure 13.36 is a graph of the directional derivative, $f_{\vec{u}}$, at the point (a,b) versus θ, the angle shown in Figure 13.35.

    (a)   Which points on the graph in Figure 13.36 correspond to the greatest rate of increase of f? The greatest rate of decrease?

    (b)   Mark points on the circle in Figure 13.35 corresponding to the points P, Q, R, S.

    (c)   What is the amplitude of the function graphed in Figure 13.36? What is its formula?

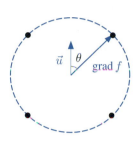

Figure 13.35

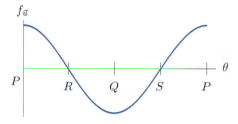

Figure 13.36

13.5 GRADIENTS AND DIRECTIONAL DERIVATIVES IN SPACE

Directional Derivatives of Functions of Three Variables

We calculate directional derivatives of a function of three variables in the same way as for a function of two variables. If the function f is differentiable at the point (a,b,c) and if $\vec{u} = u_1 \vec{i} + u_2 \vec{j} + u_3 \vec{k}$ is a unit vector, then the rate of change of $f(x,y,z)$ in the direction of $\vec{u}$ at the point (a,b,c) is

$$f_{\vec{u}}(a,b,c) = f_x(a,b,c)u_1 + f_y(a,b,c)u_2 + f_z(a,b,c)u_3.$$

This can be justified using local linearity in the same way as for functions of two variables.

Example 1   Find the directional derivative of $f(x,y,z) = xy + z$ at the point $(-1,0,1)$ in the direction of the vector $\vec{v} = 2\vec{i} + \vec{k}$.

Solution   The magnitude of $\vec{v}$ is $\|\vec{v}\| = \sqrt{2^2 + 1} = \sqrt{5}$, so a unit vector in the same direction as $\vec{v}$ is

$$\vec{u} = \frac{\vec{v}}{\|\vec{v}\|} = \frac{2}{\sqrt{5}}\vec{i} + 0\vec{j} + \frac{1}{\sqrt{5}}\vec{k}.$$

The partial derivatives of f are

$$f_x(x,y,z) = y \quad \text{and} \quad f_y(x,y,z) = x \quad \text{and} \quad f_z(x,y,z) = 1.$$

Thus,

$$f_{\vec{u}}(-1,0,1) = f_x(-1,0,1)u_1 + f_y(-1,0,1)u_2 + f_z(-1,0,1)u_3$$

$$= (0)\left(\frac{2}{\sqrt{5}}\right) + (-1)(0) + (1)\left(\frac{1}{\sqrt{5}}\right) = \frac{1}{\sqrt{5}}.$$

The Gradient Vector of a Function of Three Variables

The gradient of a function of three variables is defined in the same way as for two variables:

$$\text{grad } f(a, b, c) = f_x(a, b, c)\vec{i} + f_y(a, b, c)\vec{j} + f_z(a, b, c)\vec{k}.$$

Geometrically, the gradient is the vector pointing in the direction of greatest increase of f, whose magnitude is the rate of increase in that direction.

Just as the gradient vector of a function of two variables is perpendicular to the level curves, so the gradient of a function of three variables is perpendicular to the level surfaces. The reason is the same: The function has the same value all over a level surface, so to change the value as quickly as possible we need to move directly away from the level surface, that is, perpendicular to the surface.

Example 2 Let $f(x, y, z) = x^2 + y^2$ and $g(x, y, z) = -x^2 - y^2 - z^2$. What can we say about the direction of the following vectors?

(a) grad $f(0, 1, 1)$ (b) grad $f(1, 0, 1)$ (c) grad $g(0, 1, 1)$ (d) grad $g(1, 0, 1)$.

Solution The cylinder $x^2 + y^2 = 1$ in Figure 13.37 is a level surface of f and contains both the points $(0, 1, 1)$ and $(1, 0, 1)$. Since the value of f does not change at all in the z-direction, all the gradient vectors are horizontal. They are perpendicular to the cylinder and point outward because the value of f increases as we move out.

Similarly, the points $(0, 1, 1)$ and $(1, 0, 1)$ also lie on the same level surface of g, namely $g(x, y, z) = -x^2 - y^2 - z^2 = -2$, which is the sphere $x^2 + y^2 + z^2 = 2$. Part of this level surface is shown in Figure 13.38. This time the gradient vectors point inward, since the negative signs mean that the function increases (from large negative values to small negative values) as we move inward.

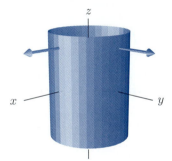

Figure 13.37: The level surface $f(x, y, z) = x^2 + y^2 = 1$ with two gradient vectors

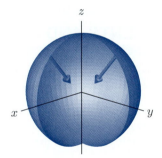

Figure 13.38: The level surface $g(x, y, z) = -x^2 - y^2 - z^2 = -2$ with two gradient vectors

---

Example 3 Consider the functions $f(x, y) = 4 - x^2 - 2y^2$ and $g(x, y) = 4 - x^2$. Calculate a vector perpendicular to each of the following:

(a) The level curve of f at the point $(1, 1)$ (b) The surface $z = f(x, y)$ at the point $(1, 1, 1)$
(c) The level curve of g at the point $(1, 1)$ (d) The surface $z = g(x, y)$ at the point $(1, 1, 3)$

Solution (a) The vector we want is a 2-vector in the plane. Since grad $f = -2x\vec{i} - 4y\vec{j}$, we have

$$\text{grad } f(1, 1) = -2\vec{i} - 4\vec{j}.$$

Any nonzero multiple of this vector is perpendicular to the level curve at the point $(1, 1)$.

(b) In this case we want a 3-vector in space. To find it we rewrite $z = 4 - x^2 - 2y^2$ as the level surface of the function F, where

$$F(x, y, z) = 4 - x^2 - 2y^2 - z = 0.$$

Then

$$\text{grad } F = -2x\vec{i} - 4y\vec{j} - \vec{k},$$

so

$$\text{grad } F(1,1,1) = -2\vec{i} - 4\vec{j} - \vec{k},$$

and grad $F(1,1,1)$ is perpendicular to the surface $z = 4 - x^2 - 2y^2$ at the point $(1,1,1)$. Notice that $-2\vec{i} - 4\vec{j} - \vec{k}$ is not the only possible answer: any multiple of this vector will do.

(c)   We are looking for a 2-vector. Since grad $g = -2x\vec{i} + 0\vec{j}$, we have

$$\text{grad } g(1,1) = -2\vec{i}.$$

Any multiple of this vector is perpendicular to the level curve also.

(d)   We are looking for a 3-vector. We rewrite $z = 4 - x^2$ as the level surface of the function G, where

$$G(x,y,z) = 4 - x^2 - z = 0.$$

Then

$$\text{grad } G = -2x\vec{i} - \vec{k}$$

So

$$\text{grad } G(1,1,3) = -2\vec{i} - \vec{k},$$

and any multiple of grad $G(1,1,3)$ is perpendicular to the surface $z = 4 - x^2$ at this point.

Example 4   (a)   A hiker on the surface $f(x,y) = 4 - x^2 - 2y^2$ at the point $(1,-1,1)$ starts to climb along the path of steepest ascent. What is the relation between the vector grad $f(1,-1)$ and a vector tangent to the path at the point $(1,-1,1)$ and pointing uphill?

(b)   Consider the surface $g(x,y) = 4 - x^2$. What is the relation between grad $g(-1,-1)$ and a vector tangent to the path of steepest ascent at $(-1,-1,3)$?

(c)   At the point $(1,-1,1)$ on the surface $f(x,y) = 4 - x^2 - 2y^2$, calculate a vector perpendicular to the surface and a vector, $\vec{T}$, tangent to the curve of steepest ascent.

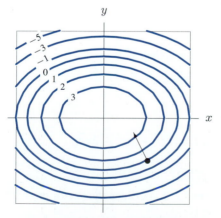

Figure 13.39: Contour diagram for $z = f(x,y) = 4 - x^2 - 2y^2$ showing direction of grad $f(1,-1)$

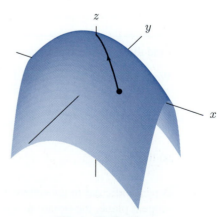

Figure 13.40: Graph of $f(x,y) = 4 - x^2 - 2y^2$ showing path of steepest ascent from the point $(1,-1,1)$

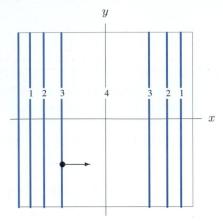

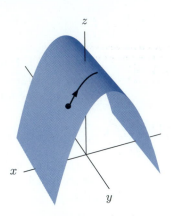

Figure 13.41: Contour diagram for $z = g(x, y) = 4 - x^2$ showing direction of grad $g(-1, -1)$

Figure 13.42: Graph of $g(x, y) = 4 - x^2$ showing path of steepest ascent from the point $(-1, -1, 3)$

Solution   (a)   The hiker at the point $(1, -1, 1)$ lies directly above the point $(1, -1)$ in the xy-plane. The vector grad $f(1, -1)$ lies in 2-space, pointing like a compass in the direction in which f increases most rapidly. Therefore, grad $f(1, -1)$ lies directly under a vector tangent to the hiker's path at $(1, -1, 1)$ and pointing uphill. (See Figures 13.39 and 13.40.)

(b)   The point $(-1, -1, 3)$ lies above the point $(-1, -1)$. The vector grad $g(-1, -1)$ points in the direction in which g increases most rapidly and lies directly under the path of steepest ascent. (See Figures 13.41 and 13.42.)

(c)   The surface is represented by $F(x, y, z) = 4 - x^2 - 2y^2 = 0$. Since grad $F = -2x\vec{i} - 4y\vec{j} - \vec{k}$, the normal, $\vec{N}$, to the surface is given by

$$\vec{N} = \text{grad } F(1, -1, 1) = -2(1)\vec{i} - 4(-1)\vec{j} - \vec{k} = -2\vec{i} + 4\vec{j} - \vec{k}.$$

We take the $\vec{i}$ and $\vec{j}$ components of $\vec{T}$ to be the vector grad $f(1, -1) = -2\vec{i} + 4\vec{j}$. Thus we have that, for some $a > 0$,

$$\vec{T} = -2\vec{i} + 4\vec{j} + a\vec{k}$$

We want $\vec{N} \cdot \vec{T} = 0$, so

$$\vec{N} \cdot \vec{T} = (-2\vec{i} + 4\vec{j} - \vec{k}) \cdot (-2\vec{i} + 4\vec{j} + a\vec{k}) = 4 + 16 - a = 0$$

So $a = 20$ and hence

$$\vec{T} = -2\vec{i} + 4\vec{j} + 20\vec{k}.$$

Example 5   Find the equation of the tangent plane to the sphere $x^2 + y^2 + z^2 = 14$ at the point $(1, 2, 3)$.

Solution   We write the sphere as a level surface as follows:

$$f(x, y, z) = x^2 + y^2 + z^2 = 14.$$

We have

$$\text{grad } f = 2x\vec{i} + 2y\vec{j} + 2z\vec{k},$$

so the vector

$$\text{grad } f(1, 2, 3) = 2\vec{i} + 4\vec{j} + 6\vec{k}$$

is perpendicular to the sphere at the point $(1, 2, 3)$. Since the vector grad $f(1, 2, 3)$ is normal to the tangent plane, the equation of the plane is

$$2x + 4y + 6z = 2(1) + 4(2) + 6(3) = 28 \quad \text{or} \quad x + 2y + 3z = 14.$$

Caution: Units and the Geometric Interpretation of the Gradient

When we interpreted the gradient of a function geometrically (page 677), we tacitly assumed that the scales along the x and y axes were the same. If they are not, the gradient vector may not look perpendicular to the contours. Consider the function $f(x, y) = x^2 + y$ with gradient vector given by grad $f = 2x\vec{i} + \vec{j}$. Figure 13.43 shows the gradient vector at $(1, 1)$ using the same scales in the x and y directions. As expected, the gradient vector is perpendicular to the contour line. Figure 13.44 shows contours of the same function with unequal scales on the two axes. Notice that the gradient vector no longer appears perpendicular to the contour lines. Thus we see that the geometric interpretation of the gradient vector requires that the same scale be used on the two axes.

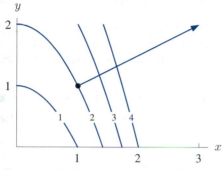

Figure 13.43: The gradient vector with x and y scales equal   *Figure 13.44:* The gradient vector with x and y scales unequal

Problems for Section 13.5

1. If $f(x, y, z) = x^2 + 3xy + 2z$, find the directional derivative at the point $(2, 0, -1)$ in the direction of $2\vec{i} + \vec{j} - 2\vec{k}$.

2. Find the directional derivative of $f(x, y, z) = 3x^2y^2 + 2yz$ at the point $(-1, 0, 4)$ in the following directions   (a) $\vec{i} - \vec{k}$ (b) $-\vec{i} + 3\vec{j} + 3\vec{k}$.

3. Find the equation of the tangent plane to $z = \sqrt{17 - x^2 - y^2}$ at the point $(3, 2, 2)$.

4. Find the equation of the tangent plane to $z = 8/(xy)$ at the point $(1, 2, 4)$.

5. Find an equation of the tangent plane and of a normal vector to the surface $x = y^3z^7$ at the point $(1, -1, -1)$.

6. Let $f(x, y) = \cos x \sin y$ and let S be the surface $z = f(x, y)$.

    (a) Find a normal vector to the surface S at the point $(0, \pi/2, 1)$.
    (b) What is the equation of the tangent plane to the surface S at the point $(0, \pi/2, 1)$?

7. Consider the function $f(x, y) = (e^x - x) \cos y$. Suppose S is the surface $z = f(x, y)$.

    (a) Find a vector which is perpendicular to the level curve of f through the point $(2, 3)$ in the direction in which f decreases most rapidly.
    (b) Suppose $\vec{v} = 5\vec{i} + 4\vec{j} + a\vec{k}$ is a vector in 3-space which is tangent to the surface S at the point P lying on the surface above $(2, 3)$. What is a?

8. (a) Sketch the surface $z = f(x, y) = y^2$ in three dimensions.
    (b) Sketch the level curves of f in the xy plane.
    (c) If you are standing on the surface $z = y^2$ at the point $(2, 3, 9)$, in which direction should you move to climb the fastest? (Give your answer as a 2-vector.)

9. Suppose $z = y - \sin x$.

    (a) Sketch the contours for $z = -1, 0, 1, 2$.

    (b) A bug starts on the surface at the point $(\pi/2, 1, 0)$ and walks on the surface in the direction of the y-axis. Is the bug walking in a valley or on top of a ridge? Explain.

    (c) On the contour $z = 0$ in your sketch for part (a), draw the gradients of z at $x = 0$, $x = \pi/2$, and $x = \pi$.

10. Suppose that $F(x, y, z) = x^2 + y^4 + x^2 z^2$ gives the concentration of salt in a fluid at the point (x, y, z), and that you are at the point $(-1, 1, 1)$.

    (a) In which direction should you move if you want the concentration to increase the fastest?

    (b) Suppose you start to move in the direction you found in part (a) at a speed of 4 units/sec. How fast is the concentration changing? Explain your answer.

11. Consider S to be the surface represented by the equation $F = 0$, where

$$F(x, y, z) = x^2 - \left(\frac{y}{z^2}\right).$$

    (a) Find all points on S where a normal vector is parallel to the xy-plane.

    (b) Find the tangent plane to S at the points $(0, 0, 1)$ and $(1, 1, 1)$.

    (c) Find the unit vectors $\vec{u}_1$ and $\vec{u}_2$ pointing in the direction of maximum increase of F at the points $(0, 0, 1)$ and $(1, 1, 1)$ respectively.

12. At what point on the surface $z = 1 + x^2 + y^2$ is its tangent plane parallel to the following planes?

    (a) $z = 5$ (b) $z = 5 + 6x - 10y$.

13. A differentiable function $f(x, y)$ has the property that $f(4, 1) = 3$ and $f_x(4, 1) = 2$ and $f_y(4, 1) = -1$. Find the equation of the tangent plane at the point on the surface $z = f(x, y)$ where $x = 4$, $y = 1$.

14. A differentiable function $f(x, y)$ has the property that $f(1, 3) = 7$ and grad $f(1, 3) = 2\vec{i} - 5\vec{j}$.

    (a) Find the equation of the tangent line to the level curve of f through the point $(1, 3)$.

    (b) Find the equation of the tangent plane to the surface $z = f(x, y)$ at the point $(1, 3, 7)$.

15. Two surfaces are said to be *tangential* at a point P if they have the same tangent plane at P. Show that the surfaces $z = \sqrt{2x^2 + 2y^2 - 25}$ and $z = \frac{1}{5}(x^2 + y^2)$ are tangential at the point $(4, 3, 5)$.

16. Two surfaces are said to be *orthogonal* to each other at a point P if the normals to their tangent planes are perpendicular at P. Show that the surfaces $z = \frac{1}{2}(x^2 + y^2 - 1)$ and $z = \frac{1}{2}(1 - x^2 - y^2)$ are orthogonal at all points of intersection.

17. Let f and g be functions on 3-space. Suppose f is differentiable and that

$$\text{grad } f(x, y, z) = (x\vec{i} + y\vec{j} + z\vec{k})g(x, y, z).$$

Explain why f must be constant on any sphere centered at the origin.

18. Let $\vec{r}$ be the position vector of the point (x, y, z). If $\vec{\mu} = \mu_1\vec{i} + \mu_2\vec{j} + \mu_3\vec{k}$ is a constant vector, show that

$$\text{grad}(\vec{\mu} \cdot \vec{r}) = \vec{\mu}.$$

19. Let $\vec{r}$ be the position vector of the point (x, y, z). Show that, if a is a constant,

$$\text{grad}(\|\vec{r}\|^a) = a\|\vec{r}\|^{a-2}\vec{r}, \qquad \vec{r} \neq \vec{0}.$$

20. Suppose the earth has mass M and is located at the origin in 3-space, while the moon has mass m. Newton's Law of Gravitation states that if the moon is located at the point (x, y, z) then the attractive force exerted by the earth on the moon is given by the vector

$$\vec{F} = -GMm\frac{\vec{r}}{\|\vec{r}\|^3},$$

where $\vec{r} = x\vec{i} + y\vec{j} + z\vec{k}$. Show that $\vec{F} = \text{grad } \varphi$, where φ is the function given by

$$\varphi(x, y, z) = \frac{GMm}{\|\vec{r}\|}.$$

13.6 THE CHAIN RULE

Composition of Functions of Many Variables and Rates of Change

The chain rule enables us to differentiate *composite functions*. If we have a function of two variables $z = f(x, y)$ and we substitute $x = g(t), y = h(t)$ into $z = f(x, y)$, then we have a composite function in which z is a function of t:

$$z = f(g(t), h(t)).$$

If, on the other hand, we substitute $x = g(u, v), y = h(u, v)$, then we have a different composite function in which z is a function of u and v:

$$z = f(g(u, v), h(u, v)).$$

The next example shows how to calculate the rate of change of a composite function.

Example 1   Corn production, C, depends on annual rainfall, R, and average temperature, T, so $C = f(R, T)$. Global warming predicts that both rainfall and temperature depend on time. Suppose that according to a particular model of global warming, rainfall is decreasing at 0.2 cm per year and temperature is increasing at $0.1°C$ per year. Use the fact that at current levels of production, $f_R = 3.3$ and $f_T = -5$ to estimate the current rate of change, dC/dt.

Solution   By local linearity, we know that changes ΔR and ΔT generate a change, ΔC, in C given approximately by

$$\Delta C \approx f_R \Delta R + f_T \Delta T = 3.3 \Delta R - 5 \Delta T.$$

We want to know how ΔC depends on the time increment, Δt. The model of global warming tells us that

$$\frac{dR}{dt} = -0.2 \quad \text{and} \quad \frac{dT}{dt} = 0.1.$$

Thus, a time increment, Δt, generates changes of ΔR and ΔT given by

$$\Delta R \approx -0.2 \Delta t \quad \text{and} \quad \Delta T \approx 0.1 \Delta t.$$

Substituting for ΔR and ΔT in the expression for ΔC gives us

$$\Delta C \approx 3.3(-0.2 \Delta t) - 5(0.1 \Delta t) = -1.16 \Delta t.$$

Thus,

$$\frac{\Delta C}{\Delta t} \approx -1.16 \quad \text{and, therefore,} \quad \frac{dC}{dt} \approx -1.16$$

Thus, a change Δt causes changes ΔR and ΔT, which in turn cause a change ΔC. The relationship between ΔC and Δt, which gives the value of dC/dt, is an example of the *chain rule*. The argument in Example 1 can be used to generate more general statements of the chain rule.

The Chain Rule for $z = f(x, y), x = g(t), y = h(t)$

Since $z = f(g(t), h(t))$ is a function of t, we can consider the derivative dz/dt. The chain rule shows how dz/dt is related to the derivatives of f, g, and h. Since dz/dt represents the rate of change of z with t, we look at how a small change, Δt, in t is propagated to z.

We substitute the local linearizations

$$\Delta x \approx \frac{dx}{dt} \Delta t \quad \text{and} \quad \Delta y \approx \frac{dy}{dt} \Delta t$$

into the local linearization

$$\Delta z \approx \frac{\partial z}{\partial x} \Delta x + \frac{\partial z}{\partial y} \Delta y,$$

yielding

$$\Delta z \approx \frac{\partial z}{\partial x}\frac{dx}{dt} \Delta t + \frac{\partial z}{\partial y}\frac{dy}{dt} \Delta t$$

$$= \left(\frac{\partial z}{\partial x}\frac{dx}{dt} + \frac{\partial z}{\partial y}\frac{dy}{dt} \right) \Delta t.$$

Thus,

$$\frac{\Delta z}{\Delta t} \approx \frac{\partial z}{\partial x}\frac{dx}{dt} + \frac{\partial z}{\partial y}\frac{dy}{dt}.$$

Taking the limit as $\Delta t \to 0$, we get the following result.

If f, g, and h are differentiable and if $z = f(x, y)$, and $x = g(t)$, and $y = h(t)$, then

$$\frac{dz}{dt} = \frac{\partial z}{\partial x}\frac{dx}{dt} + \frac{\partial z}{\partial y}\frac{dy}{dt}.$$

Visualizing the Chain Rule with a Tree Diagram

The "tree diagram" in Figure 13.45 provides a way of remembering the chain rule. It shows the chain of dependence: z depends on x and y, which in turn depend on t. Each line in the diagram is labeled with a derivative relating the variables at its ends.

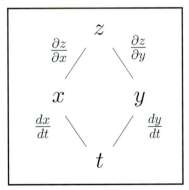

Figure 13.45: Tree diagram for $z = f(x, y)$, $x = g(t)$, $y = h(t)$

The diagram keeps track of how a change in t propagates through the chain of composed functions. There are two paths from t to z, one through x and one through y. For each path, we multiply together the derivatives along the path. Then, to calculate dz/dt, we add up the contributions from the two paths.

Example 2 Suppose that $z = f(x, y) = x \sin y$, where $x = t^2$ and $y = 2t + 1$. Let $z = g(t)$. Compute $g'(t)$ by two different methods.

Solution Since $z = g(t) = f(t^2, 2t + 1) = t^2 \sin(2t + 1)$, it is possible to compute $g'(t)$ directly by one-variable methods:

$$g'(t) = t^2 \frac{d}{dt}(\sin(2t + 1)) + \left(\frac{d}{dt}(t^2) \right) \sin(2t + 1) = 2t^2 \cos(2t + 1) + 2t \sin(2t + 1).$$

The chain rule provides an alternative route to the same answer. We have

$$\frac{dz}{dt} = \frac{\partial z}{\partial x}\frac{dx}{dt} + \frac{\partial z}{\partial y}\frac{dy}{dt} = (\sin y)(2t) + (x \cos y)(2) = 2t \sin(2t + 1) + 2t^2 \cos(2t + 1).$$

The Chain Rule in General

> To find the rate of change of one variable with respect to another in a chain of composed differentiable functions:
> - Draw a tree diagram expressing the relationship between the variables, and label each link in the diagram with the derivative relating the variables at its ends.
> - For each path between the two variables, multiply together the derivatives from each step along the path.
> - Add the contributions from each path.

A tree diagram keeps track of all the ways in which a change in one variable can cause a change in another; the diagram generates all the terms we would get from the appropriate substitutions into the local linearizations.

Example 3   Suppose that $z = f(x, y)$, with $x = g(u, v)$ and $y = h(u, v)$. Find formulas for $\partial z / \partial u$ and $\partial z / \partial v$.

Solution   Figure 13.46 shows the tree diagram for these variables. Adding the contributions for the two paths from z to u, we get

$$\frac{\partial z}{\partial u} = \frac{\partial z}{\partial x}\frac{\partial x}{\partial u} + \frac{\partial z}{\partial y}\frac{\partial y}{\partial u}.$$

Similarly, looking at the paths from z to v, we get

$$\frac{\partial z}{\partial v} = \frac{\partial z}{\partial x}\frac{\partial x}{\partial v} + \frac{\partial z}{\partial y}\frac{\partial y}{\partial v}.$$

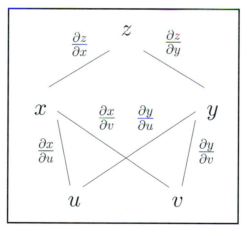

Figure 13.46: Tree diagram for $z = f(x, y)$, $x = g(u, v)$, $y = h(u, v)$

Example 4   Let $w = x^2 e^y$, $x = 4u$, and $y = 3u^2 - 2v$. Compute $\partial w / \partial u$ and $\partial w / \partial v$ using the chain rule.

Solution   Using the result of the previous example, we have

$$\frac{\partial w}{\partial u} = \frac{\partial w}{\partial x}\frac{\partial x}{\partial u} + \frac{\partial w}{\partial y}\frac{\partial y}{\partial u} = 2xe^y(4) + x^2 e^y(6u) = (8x + 6x^2 u)e^y$$

$$= (32u + 96u^3)e^{3u^2 - 2v}.$$

Similarly,

$$\frac{\partial w}{\partial v} = \frac{\partial w}{\partial x}\frac{\partial x}{\partial v} + \frac{\partial w}{\partial y}\frac{\partial y}{\partial v} = 2xe^y(0) + x^2 e^y(-2) = -2x^2 e^y$$

$$= -32u^2 e^{3u^2 - 2v}.$$

Example 5   A quantity z can be expressed either as a function of x and y, so that $z = f(x, y)$, or as a function of u and v, so that $z = g(u, v)$. The two coordinate systems are related by

$$x = u + v, \quad y = u - v.$$

(a)   Use the chain rule to express $\partial z / \partial u$ and $\partial z / \partial v$ in terms of $\partial z / \partial x$ and $\partial z / \partial y$.
(b)   Solve the equations in part (a) for $\partial z / \partial x$ and $\partial z / \partial y$.
(c)   Show that the expressions we get in part (b) are the same as we get by expressing u and v in terms of x and y and using the chain rule.

Solution   (a)   We have $\partial x / \partial u = 1$ and $\partial x / \partial v = 1$, and also $\partial y / \partial u = 1$ and $\partial y / \partial v = -1$. Thus,

$$\frac{\partial z}{\partial u} = \frac{\partial z}{\partial x}(1) + \frac{\partial z}{\partial y}(1) = \frac{\partial z}{\partial x} + \frac{\partial z}{\partial y}$$

and

$$\frac{\partial z}{\partial v} = \frac{\partial z}{\partial x}(1) + \frac{\partial z}{\partial y}(-1) = \frac{\partial z}{\partial x} - \frac{\partial z}{\partial y}.$$

(b)   Adding together the equations for $\partial z / \partial u$ and $\partial z / \partial v$ we get

$$\frac{\partial z}{\partial u} + \frac{\partial z}{\partial v} = 2\frac{\partial z}{\partial x}, \quad \text{so} \quad \frac{\partial z}{\partial x} = \frac{1}{2}\frac{\partial z}{\partial u} + \frac{1}{2}\frac{\partial z}{\partial v}.$$

Similarly, subtracting the equations for $\partial z / \partial u$ and $\partial z / \partial v$ yields

$$\frac{\partial z}{\partial y} = \frac{1}{2}\frac{\partial z}{\partial u} - \frac{1}{2}\frac{\partial z}{\partial v}.$$

(c)   Alternatively, we can solve the equations

$$x = u + v, \quad y = u - v$$

for u and v, which yields

$$u = \frac{1}{2}x + \frac{1}{2}y, \quad v = \frac{1}{2}x - \frac{1}{2}y.$$

Now we can think of z as a function of u and v, and u and v as functions of x and y, and apply the chain rule again. This gives us

$$\frac{\partial z}{\partial x} = \frac{\partial z}{\partial u}\frac{\partial u}{\partial x} + \frac{\partial z}{\partial v}\frac{\partial v}{\partial x} = \frac{1}{2}\frac{\partial z}{\partial u} + \frac{1}{2}\frac{\partial z}{\partial v}$$

and

$$\frac{\partial z}{\partial y} = \frac{\partial z}{\partial u}\frac{\partial u}{\partial y} + \frac{\partial z}{\partial v}\frac{\partial v}{\partial y} = \frac{1}{2}\frac{\partial z}{\partial u} - \frac{1}{2}\frac{\partial z}{\partial v}.$$

These are the same expressions we got in part (b).

An Example from Physical Chemistry

A chemist investigating the properties of a gas such as carbon dioxide may want to know how the internal energy U of a given quantity of the gas depends on its temperature, T, pressure, P, and volume, V. The three quantities T, P, and V are not independent, however. For instance, according to the ideal gas law, they satisfy the equation

$$PV = kT$$

where k is a constant which depends only upon the quantity of the gas. The internal energy can then be thought of as a function of any two of the three quantities T, P, and V:

$$U = U_1(T, P) = U_2(T, V) = U_3(P, V).$$

The chemist writes, for example, $\left(\frac{\partial U}{\partial T}\right)_P$ to indicate the partial derivative of U with respect to T *holding P constant*, signifying that for this computation U is viewed as a function of T and P. Thus, we interpret $\left(\frac{\partial U}{\partial T}\right)_P$ as

$$\left(\frac{\partial U}{\partial T}\right)_P = \frac{\partial U_1(T, P)}{\partial T}.$$

If U is to be viewed as a function of T and V, the chemist writes $\left(\frac{\partial U}{\partial T}\right)_V$ for the partial derivative of U with respect to T holding V constant: thus, $\left(\frac{\partial U}{\partial T}\right)_V = \frac{\partial U_2(T,V)}{\partial T}$.

Each of the functions U_1, U_2, U_3 gives rise to one of the following formulas for the differential dU:

$$dU = \left(\frac{\partial U}{\partial T}\right)_P dT + \left(\frac{\partial U}{\partial P}\right)_T dP \qquad \text{corresponds to } U_1$$

$$dU = \left(\frac{\partial U}{\partial T}\right)_V dT + \left(\frac{\partial U}{\partial V}\right)_T dV \qquad \text{corresponds to } U_2,$$

$$dU = \left(\frac{\partial U}{\partial P}\right)_V dP + \left(\frac{\partial U}{\partial V}\right)_P dV \qquad \text{corresponds to } U_3.$$

All the six partial derivatives appearing in formulas for dU have physical meaning, but they are not all equally easy to measure experimentally. A relationship among the partial derivatives, usually derived from the chain rule, may make it possible to evaluate one of the partials in terms of others that are more easily measured.

Example 6   Express $\left(\frac{\partial U}{\partial T}\right)_P$ in terms of $\left(\frac{\partial U}{\partial T}\right)_V$ and $\left(\frac{\partial U}{\partial V}\right)_T$ and $\left(\frac{\partial V}{\partial T}\right)_P$

Solution   Since we are interested in the derivatives $\left(\frac{\partial U}{\partial T}\right)_V$ and $\left(\frac{\partial U}{\partial V}\right)_T$, we think of U as a function of T and V and use the formula

$$dU = \left(\frac{\partial U}{\partial T}\right)_V dT + \left(\frac{\partial U}{\partial V}\right)_T dV \qquad \text{corresponding to } U_2.$$

We want to find a formula for $\left(\frac{\partial U}{\partial T}\right)_P$, which means thinking of U as a function of T and P. Thus, we want to substitute for dV. Since V is a function of T and P, we have

$$dV = \left(\frac{\partial V}{\partial T}\right)_P dT + \left(\frac{\partial V}{\partial P}\right)_T dP.$$

Substituting for dV into the formula for dU corresponding to U_2 gives

$$dU = \left(\frac{\partial U}{\partial T}\right)_V dT + \left(\frac{\partial U}{\partial V}\right)_T \left(\left(\frac{\partial V}{\partial T}\right)_P dT + \left(\frac{\partial V}{\partial P}\right)_T dP\right).$$

Collecting the terms containing dT and the terms containing dP gives

$$dU = \left(\left(\frac{\partial U}{\partial T}\right)_V + \left(\frac{\partial U}{\partial V}\right)_T \left(\frac{\partial V}{\partial T}\right)_P\right) dT + \left(\frac{\partial U}{\partial V}\right)_T \left(\frac{\partial V}{\partial P}\right)_T dP.$$

But we also have the formula

$$dU = \left(\frac{\partial U}{\partial T}\right)_P dT + \left(\frac{\partial U}{\partial P}\right)_T dP \qquad \text{corresponding to } U_1.$$

We now have two formulas for dU in terms of dT and dP. The coefficients of dT must be identical, so we conclude

$$\left(\frac{\partial U}{\partial T}\right)_P = \left(\frac{\partial U}{\partial T}\right)_V + \left(\frac{\partial U}{\partial V}\right)_T \left(\frac{\partial V}{\partial T}\right)_P.$$

Example 6 expresses $\left(\frac{\partial U}{\partial T}\right)_P$ in terms of three other partial derivatives. Two of them, namely $\left(\frac{\partial U}{\partial T}\right)_V$, the constant-volume heat capacity, and $\left(\frac{\partial V}{\partial T}\right)_P$, the expansion coefficient, can be easily measured experimentally. The third, the internal pressure, $\left(\frac{\partial U}{\partial V}\right)_T$, cannot be measured directly but can be related to $\left(\frac{\partial P}{\partial T}\right)_V$, which is measurable. Thus, $\left(\frac{\partial U}{\partial T}\right)_P$ can be determined indirectly using this identity.

Problems for Section 13.6

For Problems 1–6, find dz/dt using the chain rule. Assume the variables are restricted to domains on which the functions are defined.

1. $z = xy^2$, $x = e^{-t}$, $y = \sin t$
2. $z = x \sin y + y \sin x$, $x = t^2$, $y = \ln t$
3. $z = \ln(x^2 + y^2)$, $x = 1/t$, $y = \sqrt{t}$
4. $z = \sin(x/y)$, $x = 2t$, $y = 1 - t^2$
5. $z = xe^y$, $x = 2t$, $y = 1 - t^2$
6. $z = (x + y)e^y$, $x = 2t$, $y = 1 - t^2$

For Problems 7–14, find $\partial z/\partial u$ and $\partial z/\partial v$. Assume the variables are restricted to domains on which the functions are defined.

7. $z = xe^{-y} + ye^{-x}$, $x = u \sin v$, $y = v \cos u$
8. $z = \cos(x^2 + y^2)$, $x = u \cos v$, $y = u \sin v$
9. $z = xe^y$, $x = \ln u$, $y = v$
10. $z = (x + y)e^y$, $x = \ln u$, $y = v$
11. $z = xe^y$, $x = u^2 + v^2$, $y = u^2 - v^2$
12. $z = (x + y)e^y$, $x = u^2 + v^2$, $y = u^2 - v^2$
13. $z = \sin(x/y)$, $x = \ln u$, $y = v$
14. $z = \tan^{-1}(x/y)$, $x = u^2 + v^2$, $y = u^2 - v^2$

15. Suppose $w = f(x, y, z)$ and that x, y, z are functions of u and v. Use a tree diagram to write down the chain rule formula for $\partial w/\partial u$ and $\partial w/\partial v$.

16. Suppose $w = f(x, y, z)$ and that x, y, z are all functions of t. Use a tree diagram to write down the chain rule for dw/dt.

17. Corn production, C, is a function of rainfall, R, and temperature, T. Figures 13.47 and 13.48 show how rainfall and temperature are predicted to vary with time because of global warming. Suppose we know that $\Delta C \approx 3.3\Delta R - 5\Delta T$. Use this to estimate the change in corn production between the year 2020 and the year 2021. Hence, estimate dC/dt when $t = 2020$.

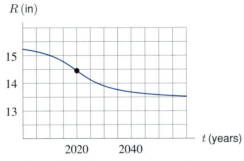

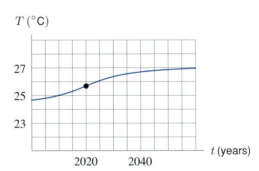

Figure 13.47: Global warming predictions: Rainfall as a function of time

Figure 13.48: Global warming predictions: Temperature as a function of time

18. The voltage, V, (in volts) across a circuit is given by Ohm's law: $V = IR$, where I is the current (in amps) flowing through the circuit and R is the resistance (in ohms). If we place two circuits, with resistance R_1 and R_2, in parallel, then their combined resistance, R, is given by

$$\frac{1}{R} = \frac{1}{R_1} + \frac{1}{R_2}.$$

Suppose the current is 2 amps and increasing at 10^{-2} amp/sec and R_1 is 3 ohms and increasing at 0.5 ohm/sec, while R_2 is 5 ohms and decreasing at 0.1 ohm/sec. Calculate the rate at which the voltage is changing.

19. A function $f(x, y)$ is *homogeneous of degree p* if $f(tx, ty) = t^p f(x, y)$ for all t. Show that any differentiable, homogeneous function of degree p satisfies Euler's Theorem:

$$x\, f_x(x, y) + y\, f_y(x, y) = p\, f(x, y).$$

[Hint: Define $g(t) = f(tx, ty)$ and compute $g'(1)$.]

Problems 20–22 are continuations of the physical chemistry example on page 691.

20. Write $\left(\frac{\partial U}{\partial P}\right)_V$ as a partial derivative of one of the functions U_1, U_2, or U_3.

21. Express $\left(\frac{\partial U}{\partial P}\right)_T$ in terms of $\left(\frac{\partial U}{\partial V}\right)_T$ and $\left(\frac{\partial V}{\partial P}\right)_T$.

22. Use the result of Problem 21 to show that

$$\left(\frac{\partial U}{\partial V}\right)_P = \frac{\left(\frac{\partial U}{\partial T}\right)_V}{\left(\frac{\partial V}{\partial T}\right)_P} + \left(\frac{\partial U}{\partial V}\right)_T.$$

For Problems 23–24, suppose the quantity z can be expressed either as a function of Cartesian coordinates (x, y) or as a function of polar coordinates (r, θ), so that $z = f(x, y) = g(r, \theta)$. [Recall that $x = r\cos\theta, y = r\sin\theta$ and $r = \sqrt{x^2 + y^2}, \theta = \arctan(y/x)$]

23. (a) Use the chain rule to find $\partial z/\partial r$ and $\partial z/\partial \theta$ in terms of $\partial z/\partial x$ and $\partial z/\partial y$.
    (b) Solve the equations you have just written down for $\partial z/\partial x$ and $\partial z/\partial y$ in terms of $\partial z/\partial r$ and $\partial z/\partial \theta$.
    (c) Show that the expressions you get in part (b) are the same as you would get by using the chain rule to find $\partial z/\partial x$ and $\partial z/\partial y$ in terms of $\partial z/\partial r$ and $\partial z/\partial \theta$.

24. Show that

$$\left(\frac{\partial z}{\partial x}\right)^2 + \left(\frac{\partial z}{\partial y}\right)^2 = \left(\frac{\partial z}{\partial r}\right)^2 + \frac{1}{r^2}\left(\frac{\partial z}{\partial \theta}\right)^2.$$

25. Let $F(x, y, z)$ be a function and define a function $z = f(x, y)$ implicitly by letting $F(x, y, f(x, y)) = 0$. Use the chain rule to show that

$$\frac{\partial z}{\partial x} = -\frac{\partial F/\partial x}{\partial F/\partial z} \quad \text{and} \quad \frac{\partial z}{\partial y} = -\frac{\partial F/\partial y}{\partial F/\partial z}.$$

13.7 SECOND-ORDER PARTIAL DERIVATIVES

Since the partial derivatives of a function are themselves functions, we can differentiate them, giving *second-order partial derivatives*. A function $z = f(x, y)$ has two first-order partial derivatives, f_x and f_y, and four second-order partial derivatives.

The Second-Order Partial Derivatives of $z = f(x, y)$

$$\frac{\partial^2 z}{\partial x^2} = f_{xx} = (f_x)_x, \qquad \frac{\partial^2 z}{\partial x \partial y} = f_{yx} = (f_y)_x,$$

$$\frac{\partial^2 z}{\partial y \partial x} = f_{xy} = (f_x)_y, \qquad \frac{\partial^2 z}{\partial y^2} = f_{yy} = (f_y)_y.$$

It is usual to omit the parentheses, writing f_{xy} instead of $(f_x)_y$ and $\frac{\partial^2 z}{\partial y\, \partial x}$ instead of $\frac{\partial}{\partial y}\left(\frac{\partial z}{\partial x}\right)$.

Example 1   Compute the four second-order partial derivatives of $f(x, y) = xy^2 + 3x^2 e^y$.

Solution   From $f_x(x, y) = y^2 + 6xe^y$ we get

$$f_{xx}(x, y) = \frac{\partial}{\partial x}(y^2 + 6xe^y) = 6e^y \quad \text{and} \quad f_{xy}(x, y) = \frac{\partial}{\partial y}(y^2 + 6xe^y) = 2y + 6xe^y.$$

From $f_y(x, y) = 2xy + 3x^2 e^y$ we get

$$f_{yx}(x, y) = \frac{\partial}{\partial x}(2xy + 3x^2 e^y) = 2y + 6xe^y \quad \text{and} \quad f_{yy}(x, y) = \frac{\partial}{\partial y}(2xy + 3x^2 e^y) = 2x + 3x^2 e^y.$$

Observe that $f_{xy} = f_{yx}$ in this example.

Example 2   Use the values of the function $f(x, y)$ in Table 13.5 to estimate $f_{xy}(1, 2)$ and $f_{yx}(1, 2)$.

TABLE 13.5   *Values of $f(x, y)$*

$y\backslash x$	0.9	1.0	1.1
1.8	4.72	5.83	7.06
2.0	6.48	8.00	9.60
2.2	8.62	10.65	12.88

Solution   Since $f_{xy} = (f_x)_y$, we first estimate f_x

$$f_x(1, 2) \approx \frac{f(1.1, 2) - f(1, 2)}{0.1} = \frac{9.60 - 8.00}{0.1} = 16.0,$$

$$f_x(1, 2.2) \approx \frac{f(1.1, 2.2) - f(1, 2.2)}{0.1} = \frac{12.88 - 10.65}{0.1} = 22.3.$$

Thus,

$$f_{xy}(1, 2) \approx \frac{f_x(1, 2.2) - f_x(1, 2)}{0.2} = \frac{22.3 - 16.0}{0.2} = 31.5.$$

Similarly,

$$f_{yx}(1, 2) \approx \frac{f_y(1.1, 2) - f_y(1, 2)}{0.1} \approx \frac{1}{0.1}\left(\frac{f(1.1, 2.2) - f(1.1, 2)}{0.2} - \frac{f(1, 2.2) - f(1, 2)}{0.2}\right)$$

$$= \frac{1}{0.1}\left(\frac{12.88 - 9.60}{0.2} - \frac{10.65 - 8.00}{0.2}\right) = 31.5.$$

Observe that in this example also, $f_{xy} = f_{yx}$.

What Do the Second-Order Partial Derivatives Tell Us?

Example 3   Let us return to the guitar string of Example 13.2, page 662. The string is 1 meter long and at time t seconds, the point x meters from one end is displaced $f(x, t)$ meters from its rest position, where

$$f(x, t) = 0.003 \sin(\pi x) \sin(2765t).$$

Compute the four second-order partial derivatives of f at the point $(x, t) = (0.3, 1)$ and describe the meaning of their signs in practical terms.

Solution   First we compute $f_x(x, t) = 0.003\pi \cos(\pi x) \sin(2765t)$, from which we get

$$f_{xx}(x, t) = \frac{\partial}{\partial x}(f_x(x, t)) = -0.003\pi^2 \sin(\pi x) \sin(2765t), \quad \text{so} \quad f_{xx}(0.3, 1) \approx -0.01;$$

and

$$f_{xt}(x, t) = \frac{\partial}{\partial t}(f_x(x, t)) = (0.003)(2765)\pi \cos(\pi x) \cos(2765t), \quad \text{so} \quad f_{xt}(0.3, 1) \approx 14.$$

On page 662 we saw that $f_x(x, t)$ gives the slope of the string at any point and time. Therefore, $f_{xx}(x, t)$ measures the concavity of the string. The fact that $f_{xx}(0.3, 1) < 0$ means the string is concave down at the point $x = 0.3$ when $t = 1$. (See Figure 13.49.)

On the other hand, $f_{xt}(x, t)$ is the rate of change of the slope of the string with respect to time. Thus $f_{xt}(0.3, 1) > 0$ means that at time $t = 1$ the slope at the point $x = 0.3$ is increasing. (See Figure 13.50.)

The slope at B is less than the slope at A

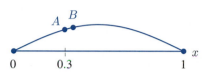

The slope at B is greater than the slope at A

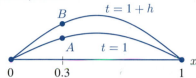

Figure 13.49: Interpretation of $f_{xx}(0.3, 1) < 0$: The concavity of the string at $t = 1$

Figure 13.50: Interpretation of $f_{xt}(0.3, 1) > 0$: The slope of one point on the string at two different times

Now we compute $f_t(x, t) = (0.003)(2765) \sin(\pi x) \cos(2765t)$, from which we get

$$f_{tx}(x, t) = \frac{\partial}{\partial x}(f_t(x, t)) = (0.003)(2765)\pi \cos(\pi x) \cos(2765t), \quad \text{so} \quad f_{tx}(0.3, 1) \approx 14$$

and

$$f_{tt}(x, t) = \frac{\partial}{\partial t}(f_t(x, t)) = -(0.003)(2765)^2 \sin(\pi x) \sin(2765t), \quad \text{so} \quad f_{tt}(0.3, 1) \approx -7200.$$

On page 662 we saw that $f_t(x, t)$ gives the velocity of the string at any point and time. Therefore, $f_{tx}(x, t)$ and $f_{tt}(x, t)$ will both be rates of change of velocity. That $f_{tx}(0.3, 1) > 0$ means that at time $t = 1$ the velocities of points just to the right of $x = 0.3$ are greater than the velocity at $x = 0.3$. (See Figure 13.51.) That $f_{tt}(0.3, 1) < 0$ means that the velocity of the point $x = 0.3$ is decreasing at time $t = 1$. Thus, $f_{tt}(0.3, 1) = -7200$ m/sec^2 is the acceleration of this point. (See Figure 13.52.)

The velocity at B is greater than the velocity at A

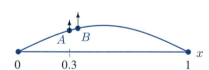

The velocity at B is less than the velocity at A

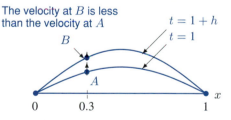

Figure 13.51: Interpretation of $f_{tx}(0.3, 1) > 0$: The velocity of different points on the string at $t = 1$

Figure 13.52: Interpretation of $f_{tt}(0.3, 1) < 0$: Negative acceleration. The velocity of one point on the string at two different times

The Mixed Partial Derivatives Are Equal

It is not an accident that the estimates for $f_{xy}(1, 2)$ and $f_{yx}(1, 2)$ are equal in Example 2, because the same values of the function are used to calculate each one. The fact that $f_{xy} = f_{yx}$ in Examples 1 and 2 corroborates the following general result.

> If f_{xy} and f_{yx} are continuous at (a, b), then
>
> $$f_{xy}(a, b) = f_{yx}(a, b).$$

Most of the functions we will encounter not only have f_{xy} and f_{yx} continuous, but all their higher order partial derivatives (such as f_{xxy} or f_{xyyy}) will exist and be continuous. We call such functions *smooth*.

Problems for Section 13.7 ━━

In Problems 1–8, calculate all four second-order partial derivatives and show that $f_{xy} = f_{yx}$. Assume the variables are restricted to a domain on which the function is defined.

1. $f(x, y) = (x + y)^2$ 2. $f(x, y) = (x + y)^3$

3. $f(x, y) = xe^y$ 4. $f(x, y) = (x + y)e^y$

5. $f(x, y) = \sin(x^2 + y^2)$ 6. $f(x, y) = \sqrt{x^2 + y^2}$

7. $f(x, y) = \sin(x/y)$ 8. $f(x, y) = \tan^{-1}(x + y)$

9. If $z = f(x) + yg(x)$, what can you say about z_{yy}? Explain your answer.

10. If $z_{xy} = 4y$, what can you say about the value of (a) z_{yx}?   (b) z_{xyx}?   (c) z_{xyy}?

In Problems 11–19, use the level curves of the function $z = f(x, y)$ to decide the sign (positive, negative, or zero) of each of the following partial derivatives at the point P. Assume the x- and y-axes are in the usual positions.

(a) $f_x(P)$ (b) $f_y(P)$ (c) $f_{xx}(P)$ (d) $f_{yy}(P)$ (e) $f_{xy}(P)$

11.

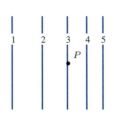

12.

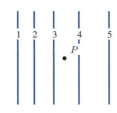

13.

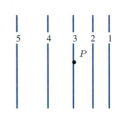

14.

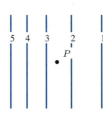

15.

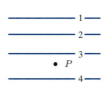

16.

17.

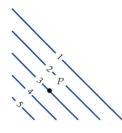

18.

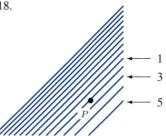

19.

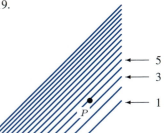

20. Give an explanation of why you might expect $f_{xy}(a, b) = f_{yx}(a, b)$ using the following steps.
   (a) Write the definition of $f_x(a, b)$.
   (b) Write a definition of $f_{xy}(a, b)$ as $(f_x)_y$.
   (c) Substitute the expression for f_x into the definition of f_{xy}.
   (d) Write an expression for f_{yx} similar to the one for f_{xy} you obtained in part (c).
   (e) Compare your answers to parts (c) and (d). What do you have to assume to conclude that f_{xy} and f_{yx} are equal?

13.8 TAYLOR APPROXIMATIONS

Just as a function of one variable can usually be better approximated by a quadratic function than by a linear function, so can a function of several variables. In Section 13.3, we saw how to approximate $f(x, y)$ by a linear function (its local linearization). In this section, we see how to improve this approximation of $f(x, y)$ using a quadratic function.

Linear and Quadratic Approximations Near (0,0)

For a function of one variable, local linearity tells us that the best *linear* approximation is the degree 1 Taylor polynomial

$$f(x) \approx f(a) + f'(a)(x - a) \quad \text{for } x \text{ near } a.$$

A better approximation to $f(x)$ is given by the degree 2 Taylor polynomial:

$$f(x) \approx f(a) + f'(a)(x - a) + \frac{f''(a)}{2}(x - a)^2 \quad \text{for } x \text{ near } a.$$

For a function of two variables the local linearization for (x, y) near (a, b) is

$$f(x, y) \approx L(x, y) = f(a, b) + f_x(a, b)(x - a) + f_y(a, b)(y - b).$$

In the case $(a, b) = (0, 0)$, we have:

> **Taylor Polynomial of Degree 1 Approximating $f(x, y)$ for (x, y) near (0,0)**
> If f has continuous first-order partial derivatives, then
>
> $$f(x, y) \approx L(x, y) = f(0, 0) + f_x(0, 0)x + f_y(0, 0)y.$$

We get a better approximation to f by using a quadratic polynomial. We choose a quadratic polynomial $Q(x, y)$, with the same partial derivatives as the original function f. You can check that the following polynomial has this property.

> **Taylor Polynomial of Degree 2 Approximating $f(x, y)$ for (x, y) near (0,0)** If f has continuous second-order partial derivatives, then
>
> $$f(x, y) \approx Q(x, y)$$
> $$= f(0, 0) + f_x(0, 0)x + f_y(0, 0)y + \frac{f_{xx}(0, 0)}{2}x^2 + f_{xy}(0, 0)xy + \frac{f_{yy}(0, 0)}{2}y^2.$$

Example 1 Let $f(x, y) = \cos(2x + y) + 3\sin(x + y)$

(a) Compute the linear and quadratic Taylor polynomials, L and Q, approximating f near $(0, 0)$.

(b) Explain why the contour plots of L and Q for $-1 \leq x \leq 1, -1 \leq y \leq 1$ look the way they do.

Solution (a) We have $f(0, 0) = 1$. The derivatives we need are as follows:

$$
\begin{aligned}
f_x(x, y) &= -2\sin(2x + y) + 3\cos(x + y) &&\text{so} && f_x(0, 0) = 3, \\
f_y(x, y) &= -\sin(2x + y) + 3\cos(x + y) &&\text{so} && f_y(0, 0) = 3, \\
f_{xx}(x, y) &= -4\cos(2x + y) - 3\sin(x + y) &&\text{so} && f_{xx}(0, 0) = -4, \\
f_{xy}(x, y) &= -2\cos(2x + y) - 3\sin(x + y) &&\text{so} && f_{xy}(0, 0) = -2, \\
f_{yy}(x, y) &= -\cos(2x + y) - 3\sin(x + y) &&\text{so} && f_{yy}(0, 0) = -1.
\end{aligned}
$$

Thus, the linear approximation, $L(x, y)$, to $f(x, y)$ at $(0, 0)$ is given by

$$f(x, y) \approx L(x, y) = f(0, 0) + f_x(0, 0)x + f_y(0, 0)y = 1 + 3x + 3y.$$

The quadratic approximation, $Q(x, y)$, to $f(x, y)$ near $(0, 0)$ is given by

$$f(x, y) \approx Q(x, y)$$

$$= f(0, 0) + f_x(0, 0)x + f_y(0, 0)y + \frac{f_{xx}(0, 0)}{2}x^2 + f_{xy}(0, 0)xy + \frac{f_{yy}(0, 0)}{2}y^2$$

$$= 1 + 3x + 3y - 2x^2 - 2xy - \frac{1}{2}y^2.$$

Notice that the linear terms in $Q(x, y)$ are the same as the linear terms in $L(x, y)$. The quadratic terms in $Q(x, y)$ can be thought of as "correction terms" to the linear approximation.

(b)   The contour plots of $f(x, y)$, $L(x, y)$, and $Q(x, y)$ are in Figures 13.53–13.55.

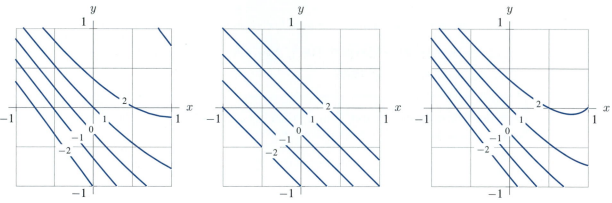

Figure 13.53: Original function, $f(x, y)$ *Figure 13.54:* Linear approximation, $L(x, y)$ *Figure 13.55:* Quadratic approximation, $Q(x, y)$

Notice that the contour plot of Q is more similar to the contour plot of f than is the contour plot of L. Since L is linear, the contour plot of L consists of parallel, equally spaced lines.

An alternative, and much quicker, way to find the Taylor polynomial in the previous example is to use the single-variable approximations. For example, since

$$\cos u = 1 - \frac{u^2}{2!} + \frac{u^4}{4!} + \cdots \quad \text{and} \quad \sin v = v - \frac{v^3}{3!} + \cdots,$$

we can substitute $u = 2x + y$ and $v = x + y$ and expand. We discard terms beyond the second (since we want the quadratic polynomial) getting

$$\cos(2x + y) = 1 - \frac{(2x + y)^2}{2!} + \frac{(2x + y)^4}{4!} + \cdots \approx 1 - \frac{1}{2}(4x^2 + 4xy + y^2) = 1 - 2x^2 - 2xy - \frac{1}{2}y^2$$

and

$$\sin(x + y) = (x + y) - \frac{(x + y)^3}{3!} + \cdots \approx x + y.$$

Combining these results, we get

$$\cos(2x + y) + 3\sin(x + y) \approx 1 - 2x^2 - 2xy - \frac{1}{2}y^2 + 3(x + y) = 1 + 3x + 3y - 2x^2 - 2xy - \frac{1}{2}y^2.$$

Linear and Quadratic Approximations Near (a, b)

The local linearization for a function $f(x, y)$ at a point (a, b) is

> **Taylor Polynomial of Degree 1 Approximating $f(x, y)$ for (x, y) near (a, b)**
> If f has continuous first-order partial derivatives, then
>
> $$f(x, y) \approx L(x, y) = f(a, b) + f_x(a, b)(x - a) + f_y(a, b)(y - b).$$

This suggests that a quadratic polynomial approximation $Q(x, y)$ for $f(x, y)$ near a point (a, b) should be written in terms of $(x - a)$ and $(y - b)$ instead of x and y. If we require that $Q(a, b) = f(a, b)$ and that the first- and second-order partial derivatives of Q and f at (a, b) be equal, then we get the following polynomial:

> **Taylor Polynomial of Degree 2 Approximating $f(x, y)$ for (x, y) near (a, b)**
> If f has continuous second-order partial derivatives, then
>
> $$f(x, y) \approx Q(x, y)$$
> $$= f(a, b) + f_x(a, b)(x - a) + f_y(a, b)(y - b)$$
> $$+ \frac{f_{xx}(a, b)}{2}(x - a)^2 + f_{xy}(a, b)(x - a)(y - b) + \frac{f_{yy}(a, b)}{2}(y - b)^2.$$

These coefficients are derived in exactly the same way as for $(a, b) = (0, 0)$.

Example 2 Find the Taylor polynomial of degree 2 at the point $(1, 2)$ for the function $f(x, y) = \dfrac{1}{xy}$.

Solution Table 13.6 contains the partial derivatives and their values at the point $(1, 2)$.

TABLE 13.6 *Partial derivatives of $f(x, y) = 1/(xy)$*

Derivative	Formula	Value at $(1, 2)$	Derivative	Formula	Value at $(1, 2)$
$f(x, y)$	$1/(xy)$	$1/2$	$f_{xx}(x, y)$	$2/(x^3 y)$	1
$f_x(x, y)$	$-1/(x^2 y)$	$-1/2$	$f_{xy}(x, y)$	$1/(x^2 y^2)$	$1/4$
$f_y(x, y)$	$-1/(xy^2)$	$-1/4$	$f_{yy}(x, y)$	$2/(xy^3)$	$1/4$

So, the quadratic Taylor polynomial for f near $(1, 2)$ is

$$\frac{1}{xy} \approx Q(x, y)$$

$$= \frac{1}{2} - \frac{1}{2}(x - 1) - \frac{1}{4}(y - 2) + \frac{1}{2}(1)(x - 1)^2 + \frac{1}{4}(x - 1)(y - 2) + \left(\frac{1}{2}\right)\left(\frac{1}{4}\right)(y - 2)^2$$

$$= \frac{1}{2} - \frac{x - 1}{2} - \frac{y - 2}{4} + \frac{(x - 1)^2}{2} + \frac{(x - 1)(y - 2)}{4} + \frac{(y - 2)^2}{8}.$$

The Error in Linear and Quadratic Approximations

Let's return to the function $f(x, y) = \cos(2x + y) + \sin(x + y)$ and its linear and quadratic approximations, $L(x, y)$ and $Q(x, y)$. The contour plots in Example 1 are evidence that Q is a better approximation to f than L. Now we'll look at exactly how much better.

We begin by considering approximations about the point $(0, 0)$. We define the *error* in the linear approximation as the difference

$$E_L = f(x, y) - L(x, y).$$

The error in the quadratic approximation is defined similarly as

$$E_Q = f(x, y) - Q(x, y).$$

Table 13.7 shows how the magnitudes of these errors, $|E_L|$ and $|E_Q|$, depend on the distance, $d(x, y) = \sqrt{x^2 + y^2}$, of the point (x, y) from $(0, 0)$. The values in Table 13.7 suggest that, in this example,

$$E_L \text{ is proportional to } d^2 \quad \text{and} \quad E_Q \text{ is proportional to } d^3.$$

In general, the errors E_L and E_Q can be shown to be proportional to d^2 and d^3, respectively.

TABLE 13.7   *Magnitude of the error in the linear and quadratic approximations to*
$f(x, y) = \cos(2x + y) + \sin(x + y)$

Point, (x, y)	Distance, d	Error, $|E_L|$	Error, $|E_Q|$
$x = y = 0$	0	0	0
$x = y = 10^{-1}$	$1.4 \cdot 10^{-1}$	$5 \cdot 10^{-2}$	$4 \cdot 10^{-3}$
$x = y = 10^{-2}$	$1.4 \cdot 10^{-2}$	$5 \cdot 10^{-4}$	$4 \cdot 10^{-6}$
$x = y = 10^{-3}$	$1.4 \cdot 10^{-3}$	$5 \cdot 10^{-6}$	$4 \cdot 10^{-9}$
$x = y = 10^{-4}$	$1.4 \cdot 10^{-4}$	$5 \cdot 10^{-8}$	$4 \cdot 10^{-12}$

To use these errors in practice, we need bounds on their magnitudes. If the distance between (x, y) and (a, b) is represented by $d(x, y) = \sqrt{(x - a)^2 + (y - b)^2}$, it can be shown that the following results hold:

Error Bound for Linear Approximation

Suppose $f(x, y)$ is a function with continuous second-order partial derivatives such that for $d(x, y) \leq d_0$,

$$|f_{xx}|, \ |f_{xy}|, \ |f_{yy}| \leq M_L.$$

Suppose

$$
\begin{aligned}
f(x, y) &= L(x, y) + E_L(x, y) \\
&= f(a, b) + f_x(a, b)(x - a) + f_y(a, b)(y - b) + E_L(x, y).
\end{aligned}
$$

Then we have

$$|E_L(x, y)| \leq 2M_L d(x, y)^2 \quad \text{for} \quad d(x, y) \leq d_0.$$

Note that the upper bound for the error term $E_L(x, y)$ has a form reminiscent of the second-order term in the Taylor formula for $f(x, y)$.

Error Bound for Quadratic Approximation

Suppose $f(x, y)$ is a function with continuous third-order partial derivatives such that for $d(x, y) \leq d_0$,

$$|f_{xxx}|, \ |f_{xxy}|, \ |f_{xyy}|, \ |f_{yyy}| \leq M_Q.$$

Suppose

$$
\begin{aligned}
f(x, y) &= Q(x, y) + E_Q(x, y) \\
&= f(a, b) + f_x(a, b)(x - a) + f_y(a, b)(y - b) \\
&\quad + \frac{f_{xx}(a, b)}{2}(x - a)^2 + f_{xy}(a, b)(x - a)(y - b) + \frac{f_{yy}(a, b)}{2}(y - b)^2 + E_Q(x, y).
\end{aligned}
$$

Then we have

$$|E_Q(x, y)| \leq \frac{4}{3} M_Q d(x, y)^3 \quad \text{for} \quad d(x, y) \leq d_0.$$

Problem 20 shows how these error estimates and the coefficients (2 and 4/3) are obtained. The important thing to notice is the fact that, for small d, the magnitude of E_L is much smaller than d and the magnitude of E_Q is much smaller than d^2. In other words we have the following result:

As $d(x, y) \to 0$:

$$\frac{E_L(x, y)}{d(x, y)} \to 0 \quad \text{and} \quad \frac{E_Q(x, y)}{(d(x, y))^2} \to 0.$$

This means that near the point (a, b), we can view the original function and the approximation as indistinguishable and behaving the same way.

Example 3   Suppose that the Taylor polynomial of degree 2 for f at $(0, 0)$ is $Q(x, y) = 5x^2 + 3y^2$. Suppose we are also told that

$$|f_{xxx}|, |f_{xxy}|, |f_{xyy}|, |f_{yyy}| \leq 9.$$

Notice that $Q(x, y) > 0$ for all (x, y) except $(0, 0)$. Show that, except at $(0, 0)$, we have

$$f(x, y) > 0 \quad \text{for all } (x, y) \text{ such that } \sqrt{x^2 + y^2} = d < 0.25.$$

Solution   By the error bound for the Taylor polynomial of degree 2, we have

$$|E_Q(x, y)| = |f(x, y) - Q(x, y)| \leq \frac{4}{3}(9)d^3 = 12d^3$$

which can be written as

$$-12d^3 \leq f(x, y) - Q(x, y) \leq 12d^3.$$

Therefore we know that

$$Q(x, y) - 12d^3 \leq f(x, y).$$

Since $Q(x, y) = 5x^2 + 3y^2$, we have

$$3x^2 + 5y^2 - 12d^3 \leq f(x, y).$$

Since $3x^2 + 5y^2 \geq 3x^2 + 3y^2 = 3d^2$, we have

$$3d^2 - 12d^3 \leq f(x, y).$$

Now d^3 approaches 0 faster than d^2, so when d is small, we have

$$0 \leq 3d^2 - 12d^3 \leq f(x,y).$$

In fact, writing $3d^2 - 12d^3 = 3d^2(1 - 4d)$ shows that $d < 1/4$ ensures that $f(x,y) > 0$, except at $(0,0)$ where $f = 0$. Thus, f has the same sign as Q for points near $(0,0)$.

Problems for Section 13.8

Find the quadratic Taylor polynomials about $(0,0)$ for the functions in Problems 1–3.

1. $e^{-2x^2 - y^2}$
2. $\sin 2x + \cos y$
3. $\ln(1 + x^2 - y)$

For each of the functions in Problems 4–11, find the linear and quadratic approximations valid near $(1,1)$. Compare the values of the approximations at $(1.1, 1.1)$ with the exact value of the function.

4. $z = xe^y$
5. $z = (x+y)e^y$
6. $z = \sin(x^2 + y^2)$

7. $z = \sqrt{x^2 + y^2}$
8. $z = \arctan(x + y)$
9. $z = \dfrac{xe^y}{x+y}$

10. $z = \sin(x/y)$
11. $z = \arctan(x/y)$

12. Let $f(x,y) = \sqrt{x + 2y + 1}$.

    (a) Compute the local linearization of f at $(0,0)$.
    (b) Compute the Taylor polynomial of degree 2 for f at $(0,0)$.
    (c) Compare the values of the linear and quadratic approximations in part (a) and part (b) with the true values for $f(x,y)$ at the points $(0.1, 0.1)$, $(-0.1, 0.1)$, $(0.1, -0.1)$, $(-0.1, -0.1)$. Which approximation gives the closest values?

13. Using a computer and your answer to Problem 12, draw the six contour diagrams of $f(x,y) = \sqrt{x + 2y + 1}$ and its linear and quadratic approximations, $L(x,y)$ and $Q(x,y)$, in the two windows $[-0.6, 0.6] \times [-0.6, 0.6]$ and $[-2, 2] \times [-2, 2]$. Explain the shape of the contours, their spacing, and the relationship between the contours of f, L, and Q.

14. Figure 13.56 shows the level curves of a function $f(x,y)$ around a maximum or minimum, M. One of the points P and Q has coordinates (x_1, y_1) and the other has coordinates (x_2, y_2). Suppose $b > 0$ and $c > 0$. Consider the two linear approximations to f given by

$$f(x,y) \approx a + b(x - x_1) + c(y - y_1)$$
$$f(x,y) \approx k + m(x - x_2) + n(y - y_2).$$

    (a) What is the relationship between the values of a and k?
    (b) What are the coordinates of P?
    (c) Is M a maximum or a minimum?
    (d) What can you say about the sign of the constants m and n?

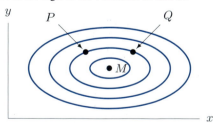

Figure 13.56

15. Consider the function $f(x,y) = (\sin x)(\sin y)$.

    (a) Find the Taylor polynomials of degree 2 for f about the points $(0,0)$ and $(\pi/2, \pi/2)$.
    (b) Use the Taylor polynomials to sketch the contours of f close to each of the points $(0,0)$ and $(\pi/2, \pi/2)$.

For Problems 16–19:

(a) Find the local linearization, $L(x, y)$, to the function $f(x, y)$ at the origin. Estimate the error $E_L(x, y) = f(x, y) - L(x, y)$ if $|x| \leq 0.1$ and $|y| \leq 0.1$.

(b) Find the degree 2 Taylor polynomial, $Q(x, y)$, for the function $f(x, y)$ at the origin. Estimate the error $E_Q(x, y) = f(x, y) - Q(x, y)$ if $|x| \leq 0.1$ and $|y| \leq 0.1$.

(c) Use a calculator to compute exactly $f(0.1, 0.1)$ and the errors $E_L(0.1, 0.1)$ and $E_Q(0.1, 0.1)$. How do these values compare with the errors predicted in parts (a) and (b)?

16. $f(x, y) = (\cos x)(\cos y)$

17. $f(x, y) = (e^x - x) \cos y$

18. $f(x, y) = e^{x+y}$

19. $f(x, y) = (x^2 + y^2)e^{x+y}$

20. It is known that if the derivatives of a one-variable function, $g(t)$, satisfy

$$|g^{(n+1)}(t)| \leq K \quad \text{for } |t| \leq d_0,$$

then the error, E_n, in the nth Taylor approximation, $P_n(x)$, is bounded as follows:

$$|E_n| = |g(t) - P_n(t)| \leq \frac{K}{(n+1)!}|t|^{n+1} \quad \text{for } |t| \leq d_0.$$

In this problem, we use this result for $g(t)$ to get the error bounds for the linear and quadratic Taylor approximations to $f(x, y)$. For a particular function $f(x, y)$, let $x = ht$ and $y = kt$ for fixed h and k, and define $g(t)$ as follows:

$$g(t) = f(ht, kt) \quad \text{for } 0 \leq t \leq 1.$$

(a) Calculate $g'(t)$, $g''(t)$, and $g'''(t)$ using the chain rule.

(b) Show that $L(ht, kt) = P_1(t)$ and that $Q(ht, kt) = P_2(t)$, where L is the linear approximation to f at $(0, 0)$ and Q is the Taylor polynomial of degree 2 for f at $(0, 0)$.

(c) What is the relation between $E_L = f(x, y) - L(x, y)$ and E_1? What is the relation between $E_Q = f(x, y) - Q(x, y)$ and E_2?

(d) Assuming that the second and third-order partial derivatives of f are bounded for $d(x, y) \leq d_0$, show that $|E_L|$ and $|E_Q|$ are bounded as on page 700.

CHAPTER SUMMARY

- **Partial Derivatives**
  Definition as a difference quotient, visualizing on a graph, estimating from a contour diagram, computing from a formula, interpreting using units, alternative notation.

- **Local Linearity**
  Zooming on a surface, contour diagram or table to see local linearity, the idea of tangent plane, formula for a tangent plane in terms of partials, the differential.

- **Directional Derivatives**
  Definition as a difference quotient, interpretation as a rate of change, computation using partial derivatives.

- **Gradient Vector**
  Definition in terms of partial derivatives, geometric properties of gradient's length and direction, relation to directional derivative, relation to contours and level surfaces.

- **Chain Rule**

- **Second and Higher Order Partial Derivatives**
  Interpretations, mixed partials are equal.

- **Taylor Approximations**
  Linear and quadratic polynomial approximations to functions near a point, behavior of errors.

REVIEW PROBLEMS FOR CHAPTER THIRTEEN

For Problems 1–4, find the indicated partial derivatives. Assume the variables are restricted to a domain on which the function is defined.

1. $\dfrac{\partial z}{\partial x}$ and $\dfrac{\partial z}{\partial y}$ if $z = (x^2 + x - y)^7$

2. $\dfrac{\partial F}{\partial L}$ if $F(L, K) = 3\sqrt{LK}$

3. $\dfrac{\partial f}{\partial p}$ and $\dfrac{\partial f}{\partial q}$ if $f(p, q) = e^{p/q}$

4. $\dfrac{\partial f}{\partial x}$ if $f(x, y) = e^{xy}(\ln y)$

Find both partial derivatives for the functions in Problems 5–8. Assume the variables are restricted to a domain on which the function is defined.

5. $z = x^4 - x^7 y^3 + 5xy^2$

6. $z = \tan(\theta)/r$

7. $w = s\ln(s + t)$

8. $w = \arctan(ue^{-v})$

9. If $f(x, y) = x^2 y$ and $\vec{v} = 4\vec{i} - 3\vec{j}$, find the directional derivative at the point $(2, 6)$ in the direction of $\vec{v}$.

Assume that $f(x, y)$ is a differentiable function. Are the statements in Problems 10–16 true or false? Explain your answer.

10. $f_{\vec{u}}(x_0, y_0)$ is a scalar.

11. $f_{\vec{u}}(a, b) = \|\nabla f(a, b)\|$

12. If $\vec{u}$ is tangent to the level curve of f at some point, then grad $f \cdot \vec{u} = 0$ there.

13. Suppose that f is differentiable at (a, b). Then there is always a direction in which the rate of change of f at (a, b) is 0.

14. There is a function with a point in its domain where $\|\operatorname{grad} f\| = 0$ and where there is a nonzero directional derivative.

15. There is a function with $\|\operatorname{grad} f\| = 4$ and $f_{\vec{i}} = 5$ at some point.

16. There is a function with $\|\operatorname{grad} f\| = 4$ and $f_{\vec{j}} = -3$ at some point.

17. Let $f(w, z) = w^2 z + 3z^2$.

    (a)   Use difference quotients with $h = 0.01$ to approximate $f_w(2, 2)$ and $f_z(2, 2)$.

    (b)   Now evaluate $f_w(2, 2)$ and $f_z(2, 2)$ exactly.

18. Figure 13.57 shows a contour diagram for the temperature T (in $^\circ$C) along a wall in a heated room as a function of distance x along the wall and time t in minutes. Estimate $\partial T/\partial x$ and $\partial T/\partial t$ at the given points. Give units for your answers and say what the answers mean.

    (a) $x = 15, t = 20$ (b) $x = 5, t = 12$

t (minutes)

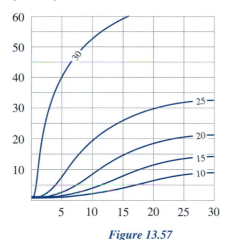

x (meters)

Figure 13.57

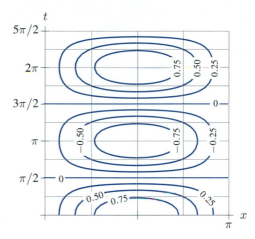

Figure 13.58

19. Figure 13.58 shows a contour diagram for a vibrating string function, $f(x, t)$.

   (a) Is $f_t(\pi/2, \pi/2)$ positive or negative? How about $f_t(\pi/2, \pi)$? What does the sign of $f_t(\pi/2, b)$ tell you about the motion of the point on the string at $x = \pi/2$ when $t = b$?

   (b) Find all t for which f_t is positive, for $0 \leq t \leq 5\pi/2$.

   (c) Find all x and t such that f_x is positive.

20. The quantity Q (in pounds) of beef that a certain community buys during a week is a function $Q = f(b, c)$ of the prices of beef, b, and chicken, c, during the week. Do you expect $\partial Q/\partial b$ to be positive or negative? What about $\partial Q/\partial c$?

21. Suppose the cost of producing one unit of a certain product is given by

$$c = a + bx + ky,$$

where x is the amount of labor used (in man hours) and y is the amount of raw material used (by weight) and a and b and k are constants. What does $\partial c/\partial x = b$ mean? What is the practical interpretation of b?

22. The acceleration g due to gravity, at a distance r from the center of a planet of mass m, is given by

$$g = \frac{Gm}{r^2},$$

where G is the universal gravitational constant.

   (a) Find $\partial g/\partial m$ and $\partial g/\partial r$.

   (b) Interpret each of the partial derivatives you found in part (a) as the slope of a graph in the plane and sketch the graph.

23. Suppose that the function $P = f(K, L)$ expresses the production of a firm as a function of the capital invested, K, and its labor costs, L.

   (a) Suppose $f(K, L) = 60K^{1/3}L^{2/3}$. Find the relationship between K and L if the marginal productivity of capital (that is, the rate of change of production with capital) equals the marginal productivity of labor cost (that is, the rate of change of production with labor cost). Simplify your answer.

   (b) Now suppose $f(K, L) = cK^aL^b$, with a, b, c positive constants. Answer the same question as in part (a).

24. In analyzing a factory and deciding whether or not to hire more workers, it is useful to know under what circumstances productivity increases. Suppose $P = f(x_1, x_2, x_3)$ is the total quantity produced as a function of x_1, the number of workers, and any other variables x_2, x_3. We define the average productivity of a worker as P/x_1. Show that the average productivity increases as x_1 increases when marginal production, $\partial P/\partial x_1$, is greater than the average productivity, P/x_1.

25. For the Cobb-Douglas production function $P = 40L^{0.25}K^{0.75}$, find the differential dP when $L = 2$ and $K = 16$.

26. The area of a triangle can be calculated from the formula $S = \frac{1}{2}ab \sin C$. Show that if an error of $10'$ (or $\pi/1080$ radians) is made in measuring C then the error in S is approximately $\pi S/(1080 \tan C)$. [Note: $10'$ means 10 minutes, where 1 minute$= 1/60$ degree.]

27. The gas equation for one mole of oxygen relates its pressure, P (in atmospheres), its temperature, T (in K), and its volume, V (in cubic decimeters, dm^3):

$$T = 16.574\frac{1}{V} - 0.52754\frac{1}{V^2} + 0.3879P + 12.187VP.$$

    (a) Find the temperature T and differential dT if the volume is 25 dm^3 and the pressure is 1 atmosphere.

    (b) Use your answer to part (a) to estimate how much the volume would have to change if the pressure increased by 0.1 atmosphere and the temperature remained constant.

28. Find the rate of change of $f(x, y) = xe^y$ at the point $(1, 1)$ in the direction of $\vec{i} + 2\vec{j}$.

29. Find the directional derivative of $z = x^2 - y^2$ at the point $(3, -1)$ in the direction making an angle $\theta = \pi/4$ with the x-axis. In which direction is the directional derivative the largest?

30. Figure 13.59 shows the level curves of a function $f(x, y)$. Give the approximate value of $f_{\vec{u}}(3, 1)$ with $\vec{u} = (-2\vec{i} + \vec{j})/\sqrt{5}$. Explain your answer.

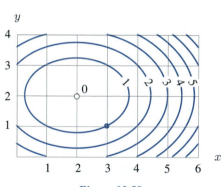

Figure 13.59

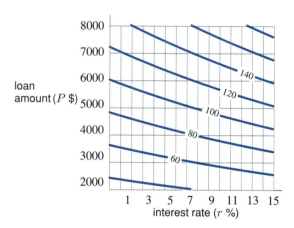

Figure 13.60

31. Figure 13.60 shows the monthly payment, m, on a 5-year car loan if you borrow P dollars at r percent interest. Find a formula for a linear function which approximates m. What is the practical significance of the constants in your formula?

32. Find the point(s) on $x^2 + y^2 + z^2 = 8$ where the tangent plane is parallel to the plane $x - y + 3z = 0$.

33. Suppose the temperature at a point (x, y) is given by the function $T(x, y) = 100 - x^2 - y^2$. In which direction should a heat-seeking bug move from the point (x, y) to increase its temperature fastest?

34. Suppose that the values of the function $f(x, y)$ near the point $x = 2$, $y = 3$ are given in Table 13.8. Estimate the following.

    (a) $\left.\dfrac{\partial f}{\partial x}\right|_{(2,3)}$ and $\left.\dfrac{\partial f}{\partial y}\right|_{(2,3)}$.

    (b) The rate of change of f at $(2, 3)$ in the direction of the vector $\vec{i} + 3\vec{j}$.

    (c) The maximum possible rate of change of f as you move away from the point $(2, 3)$. In which direction should you move to obtain this rate of change?

    (d) Write an equation for the level curve through the point $(2, 3)$.

    (e) Find a vector tangent to the level curve of f through the point $(2, 3)$.

    (f) Find the differential of f at the point $(2, 3)$. If $dx = 0.03$ and $dy = 0.04$, find df. What does df represent in this case?

TABLE 13.8

		x	
		2.00	2.01
y	3.00	7.56	7.42
	3.02	7.61	7.47

35. The function $g(x, y)$ is differentiable and has the property that $g(1, 3) = 4$ and $g_x(1, 3) = -1$ and $g_y(1, 3) = 2$.

    (a)   Find the equation of the level curve of g through the point $(1, 3)$.

    (b)   Find the coordinates of the point on the surface $z = g(x, y)$ above the point $(1, 3)$.

    (c)   Find the equation of the tangent plane to the surface $z = g(x, y)$ at the point you found in part (b).

36. Suppose $w = f(x, y, z) = 3xy + yz$ and that x, y, z are functions of u and v such that

$$x = \ln u + \cos v, \quad y = 1 + u \sin v, \quad z = uv.$$

    (a)   Find $\partial w/\partial u$ and $\partial w/\partial v$ at $(u, v) = (1, \pi)$.

    (b)   Suppose now that u and v are also functions of t such that

$$u = 1 + \sin(\pi t), \quad v = \pi t^2.$$

Use your answer to part (a) to find dw/dt at $t = 1$.

37. A circular city has radius r km and an average population density of ρ people/km^2. In 1997 the population was 3 million, the radius was 25 km and growing at 0.1 km/year. If the density was increasing at 200 people/km^2/year, find the rate at which the total population of the city was growing.

38. Find the quadratic Taylor polynomial about $(0, 0)$ for $f(x, y) = \cos(x + 2y) \sin(x - y)$.

39. Suppose $f(x, y) = e^{(x-1)^2 + (y-3)^2}$.

    (a)   Find the first-order Taylor polynomial about $(0, 0)$.

    (b)   Find the second-order (quadratic) Taylor polynomial about the point $(1, 3)$.

    (c)   Find a 2-vector perpendicular to the level curve through $(0, 0)$.

    (d)   Find a 3-vector perpendicular to the surface $z = f(x, y)$ at the point $(0, 0)$.

40. Each diagram (I) – (IV) in Figure 13.61 represents the level curves of a function $f(x, y)$. For each function f, consider the point above P on the surface $z = f(x, y)$ and choose from the lists which follow:

    (a)   A vector which could be the normal to the surface at that point;

    (b)   An equation which could be the equation of the tangent plane to the surface at that point.

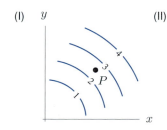 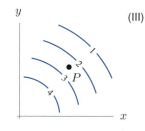 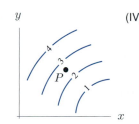 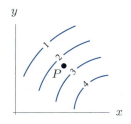

Figure 13.61

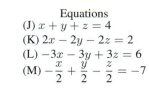

Vectors

(E) $2\vec{i} + 2\vec{j} - 2\vec{k}$

(F) $2\vec{i} + 2\vec{j} + 2\vec{k}$

(G) $2\vec{i} - 2\vec{j} + 2\vec{k}$

(H) $-2\vec{i} + 2\vec{j} + 2\vec{k}$

Equations

(J) $x + y + z = 4$

(K) $2x - 2y - 2z = 2$

(L) $-3x - 3y + 3z = 6$

(M) $-\dfrac{x}{2} + \dfrac{y}{2} - \dfrac{z}{2} = -7$

PROJECTS

1. **Heat Equation**

   The function $T(x, y, z, t)$ is a solution to the *heat equation*

   $$T_t = K(T_{xx} + T_{yy} + T_{zz}),$$

   and gives the temperature at the point (x, y, z) in 3-space and time t. The constant K is the *thermal conductivity* of the medium through which the heat is flowing.

   (a) Show that the function

   $$T(x, y, z, t) = \frac{1}{(4\pi Kt)^{3/2}} e^{-(x^2+y^2+z^2)/4Kt}$$

   is a solution to the heat equation for all (x, y, z) in 3-space and $t > 0$.

   (b) For each fixed time t, what are the level surfaces of the function $T(x, y, z, t)$ in 3-space?

   (c) Regard t as fixed and compute $\operatorname{grad} T(x, y, z, t)$. What does $\operatorname{grad} T(x, y, z, t)$ tell us about the direction and magnitude of the heat flow?

2. **Matching Birthdays**

   Consider a class of m students and a year with n days. Let $q(m, n)$ denote the probability, expressed as a number between 0 and 1, that at least two students have the same birthday. Surprisingly, $q(23, 365) \approx 0.5073$. (This means that there is slightly better than an even chance that at least two students in a class of 23 have the same birthday.) A general formula for q is complicated, but it can be shown that

   $$\frac{\partial q}{\partial m} \approx +\frac{m}{n}(1 - q) \qquad \text{and} \qquad \frac{\partial q}{\partial n} \approx -\frac{m^2}{2n^2}(1 - q).$$

   (These approximations hold when n is a good deal larger than m, and m is a good deal larger than 1.)

   (a) Explain why the $+$ and $-$ signs in the approximations for $\partial q / \partial m$ and $\partial q / \partial n$ are to be expected.

   (b) Suppose there are 21 students in a class. What is the approximate probability that at least two students in the class have the same birthday? (Assume that a year always has 365 days.)

   (c) Suppose there is a class of 24 students and you know that no one was born in the first week of the year. (This has the effect of making $n = 358$.) What is the approximate value of q for this class?

   (d) If you want to bet that a certain class of 23 students has at least two matching birthdays, would you prefer to have two more students added to the class or to be told that no one in the class was born in December?

   (e) (Optional) Find the actual formula for q. [Hint: It's easier to find $1 - q$. There are $n \cdot n \cdot n \cdots n = n^m$ different choices for the students' birthdays. How many such choices have no matching birthdays?]

FOCUS ON THEORY

DIFFERENTIABILITY

Notes on Differentiability

In Section 13.3 we gave an informal introduction to the concept of differentiability. We called a function $f(x, y)$ *differentiable* at a point (a, b) if it is well-approximated by a linear function near (a, b). This section focuses on the precise meaning of the phrase "well-approximated." By looking at examples, we shall see that local linearity requires the existence of partial derivatives, but they do not tell the whole story. In particular, existence of partial derivatives at a point is not sufficient to guarantee local linearity at that point.

We begin by discussing the relation between continuity and differentiability. As an illustration, take a sheet of paper, crumple it into a ball and smooth it out again. Wherever there is a crease it would be difficult to approximate the surface by a plane—these are points of nondifferentiability of the function giving the height of the paper above the floor. Yet the sheet of paper models a graph which is continuous—there are no breaks. As in the case of one-variable calculus, continuity does not imply differentiability. But differentiability does *require* continuity: there cannot be linear approximations to a surface at points where there are abrupt changes in height.

Starting from the definition of differentiability for single-variable functions, we develop a definition of differentiability for two-variable functions.

Differentiability For Functions Of One Variable

We recall that a function $g(x)$ is *differentiable* at the point a if the limit

$$g'(a) = \lim_{h \to 0} \frac{g(a + h) - g(a)}{h}$$

exists. Geometrically, the definition means that the graph of $y = g(x)$ can be "well-approximated" by the line $y = L(x) = g(a) + g'(a)(x - a)$. How well does this line have to approximate the function $g(x)$ near the point a before we can say that g is differentiable at a? To answer this question, suppose g is differentiable at a and let $E(x)$ be the error between the function $g(x)$ and the line $L(x)$, so that

$$E(x) = g(x) - L(x)$$
$$= g(x) - g(a) - g'(a)(x - a).$$

This means that at the point $x = a + h$ near a, the error $E(x)$ is given by

$$E(a + h) = g(a + h) - g(a) - g'(a)h.$$

Suppose we consider the *relative error* $E(a + h)/h$. We have

$$\frac{E(a + h)}{h} = \frac{g(a + h) - g(a)}{h} - g'(a).$$

Thus, in the limit as $h \to 0$, we have

$$\lim_{h \to 0} \frac{E(a + h)}{h} = \lim_{h \to 0} \frac{g(a + h) - g(a)}{h} - g'(a).$$

By the definition of the derivative, the right-hand side of the last equation is 0.

Therefore we see that if f is differentiable, the relative error tends to 0 as h tends to 0:

$$\lim_{h \to 0} \frac{E(a+h)}{h} = 0.$$

We will take "well approximated" to mean that this limit is zero. We use this idea to give a new definition of differentiability which can be generalized to functions of several variables

A function $g(x)$ is **differentiable at the point** a if there is a linear function $L(x) = g(a) + m(x - a)$ such that if the *error*, $E(x)$, is defined by

$$g(x) = L(x) + E(x)$$

and if $h = x - a$ then the *relative error* $E(a+h)/h$ satisfies

$$\lim_{h \to 0} \frac{E(a+h)}{h} = 0.$$

The function $L(x)$ is called the *local linearization* of $g(x)$ near a. The function g is *differentiable* if it is differentiable at each point of its domain.

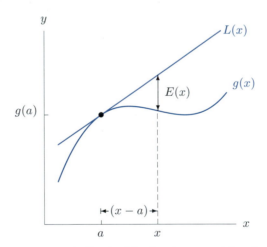

Figure 13.62: Graph of the function $y = g(x)$ and its local linearization $y = L(x)$ near the point a

This new definition tells us that the ratio $E(x)/(x - a)$ in Figure 13.62, the error divided by the distance from the point a, tends to 0 as $x \to a$. In addition, it can be shown that we must have $m = g'(a)$.

Differentiability For Functions Of Two Variables

Based on our new definition of differentiability, we define differentiability of a function of two variables at a point in terms of the error and the distance from the point. If the point is (a, b) and the nearby point is $(a + h, b + k)$, the distance is $\sqrt{h^2 + k^2}$. (See Figure 13.63.)

A function $f(x, y)$ is **differentiable at the point** (a, b) if there is a linear function $L(x, y) = f(a, b) + m(x - a) + n(y - b)$ such that if the *error* $E(x, y)$ is defined by

$$f(x, y) = L(x, y) + E(x, y),$$

and if $h = x - a, k = y - b$, then the *relative error* $E(a + h, b + k)/\sqrt{h^2 + k^2}$ satisfies

$$\lim_{\substack{h \to 0 \\ k \to 0}} \frac{E(a + h, b + k)}{\sqrt{h^2 + k^2}} = 0.$$

The function f is **differentiable** if it is differentiable at each point of its domain. The function $L(x, y)$ is called the *local linearization* of $f(x, y)$ near (a, b).

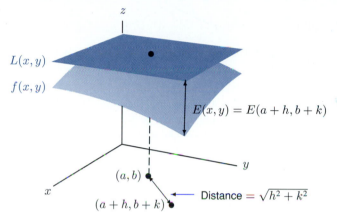

Figure 13.63: Graph of function $z = f(x, y)$ and its local linearization $z = L(x, y)$ near the point (a, b)

Partial Derivatives and Differentiability

In the next example, we show that this definition of differentiability is consistent with our previous notion — that is, that $m = f_x$ and $n = f_y$ and that the graph of $L(x, y)$ is the tangent plane.

Example 1   Show that if f is a differentiable function with local linearization $L(x, y) = f(a, b) + m(x - a) + n(y - b)$, then $m = f_x(a, b)$ and $n = f_y(a, b)$.

Solution   Since f is differentiable, we know that the relative error in $L(x, y)$ tends to 0 as we get close to (a, b). Suppose $h > 0$ and $k = 0$. Then we know that

$$0 = \lim_{h \to 0} \frac{E(a + h, b)}{\sqrt{h^2 + k^2}} = \lim_{h \to 0} \frac{E(a + h, b)}{h} = \lim_{h \to 0} \frac{f(a + h, b) - L(a + h, b)}{h}$$

$$= \lim_{h \to 0} \frac{f(a + h, b) - f(a, b) - mh}{h}$$

$$= \lim_{h \to 0} \left(\frac{f(a + h, b) - f(a, b)}{h} \right) - m = f_x(a, b) - m.$$

A similar result holds if $h < 0$, so we have $m = f_x(a, b)$. The result $n = f_y(a, b)$ is found in a similar manner.

The previous example shows that if a function is differentiable at a point, it has partial derivatives there. Therefore, if any of the partial derivatives fail to exist, then the function cannot be differentiable. This is what happens in the following example of a cone.

Example 2   Consider the function $f(x, y) = \sqrt{x^2 + y^2}$. Is f differentiable at the origin?

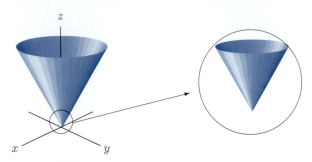

Figure 13.64: The function $f(x, y) = \sqrt{x^2 + y^2}$ is not locally linear at $(0, 0)$: Zooming in around $(0, 0)$ does not make the graph look like a plane

Solution   If we zoom in on the graph of the function $f(x, y) = \sqrt{x^2 + y^2}$ at the origin, as shown in Figure 13.64, the sharp point remains; the graph never flattens out to look like a plane. Near its vertex, the graph does not look like it is well approximated (in any reasonable sense) by any plane.

Judging from the graph of f, we would not expect f to be differentiable at $(0, 0)$. Let us check this by trying to compute the partial derivatives of f at $(0, 0)$:

$$f_x(0, 0) = \lim_{h \to 0} \frac{f(h, 0) - f(0, 0)}{h} = \lim_{h \to 0} \frac{\sqrt{h^2 + 0} - 0}{h} = \lim_{h \to 0} \frac{|h|}{h}.$$

Since $|h|/h = \pm 1$, depending on whether h approaches 0 from the left or right, this limit does not exist and so neither does the partial derivative $f_x(0, 0)$. Thus, f cannot be differentiable at the origin. If it were, both of the partial derivatives, $f_x(0, 0)$ and $f_y(0, 0)$, would exist.

Alternatively, we could show directly that there is no linear approximation near $(0, 0)$ that satisfies the small relative error criterion for differentiability. Any plane passing through the point $(0, 0, 0)$ has the form $L(x, y) = mx + ny$ for some constants m and n. If $E(x, y) = f(x, y) - L(x, y)$, then

$$E(x, y) = \sqrt{x^2 + y^2} - mx - ny.$$

Then for f to be differentiable at the origin, we would need to show that

$$\lim_{\substack{h \to 0 \\ k \to 0}} \frac{\sqrt{h^2 + k^2} - mh - nk}{\sqrt{h^2 + k^2}} = 0.$$

Taking $k = 0$ gives

$$\lim_{h \to 0} \frac{|h| - mh}{|h|} = 1 - m \lim_{h \to 0} \frac{h}{|h|}.$$

This limit exists only if $m = 0$ for the same reason as before. But then the value of the limit is 1 and not 0 as required. Thus, we again conclude f is not differentiable.

In Example 2 the partial derivatives f_x and f_y did not exist at the origin and this was sufficient to establish nondifferentiability there. We might expect that if both partial derivatives do exist, then f *is* differentiable. But the next example shows that this not necessarily true: the existence of both partial derivatives at a point is *not* sufficient to guarantee differentiability.

Example 3   Consider the function $f(x,y) = x^{1/3}y^{1/3}$. Show that the partial derivatives $f_x(0,0)$ and $f_y(0,0)$ exist, but that f is not differentiable at $(0,0)$.

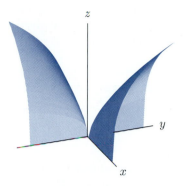

Figure 13.65: Graph of $z = x^{1/3}y^{1/3}$ for $z \geq 0$

Solution   See Figure 13.65 for the part of the graph of $z = x^{1/3}y^{1/3}$ when $z \geq 0$. We have $f(0,0) = 0$ and we compute the partial derivatives using the definition:

$$f_x(0,0) = \lim_{h \to 0} \frac{f(h,0) - f(0,0)}{h} = \lim_{h \to 0} \frac{0 - 0}{h} = 0,$$

and similarly

$$f_y(0,0) = 0.$$

So, if there did exist a linear approximation near the origin, it would have to be $L(x,y) = 0$. But we can show that this choice of $L(x,y)$ doesn't result in the small relative error that is required for differentiability. In fact, since $E(x,y) = f(x,y) - L(x,y) = f(x,y)$, we need to look at the limit

$$\lim_{\substack{h \to 0 \\ k \to 0}} \frac{h^{1/3}k^{1/3}}{\sqrt{h^2 + k^2}}.$$

If this limit exists, we get the same value no matter how h and k approach 0. Suppose we take $k = h > 0$. Then the limit becomes

$$\lim_{h \to 0} \frac{h^{1/3}h^{1/3}}{\sqrt{h^2 + h^2}} = \lim_{h \to 0} \frac{h^{2/3}}{h\sqrt{2}} = \lim_{h \to 0} \frac{1}{h^{1/3}\sqrt{2}}.$$

But this limit does not exist, since small values for h will make the fraction arbitrarily large. So the only possible candidate for a linear approximation at the origin does not have a sufficiently small relative error. Thus, this function is *not* differentiable at the origin, even though the partial derivatives $f_x(0,0)$ and $f_y(0,0)$ exist. Figure 13.65 confirms that near the origin the graph of $z = f(x,y)$ is not well approximated by any plane.

In summary,

- If a function is differentiable at a point, then both partial derivatives exist there.
- Having both partial derivatives at a point does not guarantee that a function is differentiable there.

Continuity and Differentiability

We know that differentiable functions of one variable are continuous. Similarly, it can be shown that if a function of two variables is differentiable at a point, then the function is continuous there.

In Example 3 the function f was continuous at the point where it was not differentiable. Example 4 shows that even if the partial derivatives of a function exist at a point, the function is not necessarily continuous at that point if it is not differentiable there.

Example 4   Suppose that f is the function of two variables defined by

$$f(x,y) = \begin{cases} \dfrac{xy}{x^2 + y^2}, & (x,y) \neq (0,0), \\ 0, & (x,y) = (0,0). \end{cases}$$

Problem 4 on page 615 showed that $f(x,y)$ is not continuous at the origin. Show that the partial derivatives $f_x(0,0)$ and $f_y(0,0)$ exist. Could f be differentiable at $(0,0)$?

Solution   From the definition of the partial derivative we see that

$$f_x(0,0) = \lim_{h \to 0} \frac{f(h,0) - f(0,0)}{h} = \lim_{h \to 0} \left(\frac{1}{h} \cdot \frac{0}{h^2 + 0^2} \right) = \lim_{h \to 0} \frac{0}{h} = 0,$$

and similarly

$$f_y(0,0) = 0.$$

So, the partial derivatives $f_x(0,0)$ and $f_y(0,0)$ exist. However, f cannot be differentiable at the origin since it is not continuous there.

In summary,

> - If a function is differentiable at a point, then it is continuous there.
> - Having both partial derivatives at a point does not guarantee that a function is continuous there.

How Do We Know If a Function Is Differentiable?

Can we use partial derivatives to tell us if a function is differentiable? As we see from Examples 3 and 4, it is not enough that the partial derivatives exist. However, the following condition *does* guarantee differentiability:

> If the partial derivatives, f_x and f_y, of a function f exist and are continuous on a small disk centered at the point (a, b), then f is differentiable at (a, b).

We will not prove this fact, although it provides a criterion for differentiability which is often simpler to use than the definition. It turns out that the requirement of continuous partial derivatives is more stringent than that of differentiability, so there exist differentiable functions which do not have continuous partial derivatives. However, most functions we encounter will have continuous partial derivatives. The class of functions with continuous partial derivatives is given the name C^1.

Example 5 Show that the function $f(x, y) = \ln(x^2 + y^2)$ is differentiable everywhere in its domain.

Solution The domain of f is all of 2-space except for the origin. We shall show that f has continuous partial derivatives everywhere in its domain (that is, the function f is in C^1). The partial derivatives are

$$f_x = \frac{2x}{x^2 + y^2} \quad \text{and} \quad f_y = \frac{2y}{x^2 + y^2}.$$

Since each of f_x and f_y is the quotient of continuous functions, the partial derivatives are continuous everywhere except the origin (where the denominators are zero). Thus, f is differentiable everywhere in its domain.

Most functions built up from elementary functions have continuous partial derivatives, except perhaps at a few obvious points. Thus, in practice, we can often identify functions as being C^1 without explicitly computing the partial derivatives.

Problems on Differentiability

For the functions f in Problems 1–4 answer the following questions. Justify your answers.

(a) Use a computer to draw a contour diagram for f.
(b) Is f differentiable at all points $(x, y) \neq (0, 0)$?
(c) Do the partial derivatives f_x and f_y exist and are they continuous at all points $(x, y) \neq (0, 0)$?
(d) Is f differentiable at $(0, 0)$?
(e) Do the partial derivatives f_x and f_y exist and are they continuous at $(0, 0)$?

1. $f(x, y) = \begin{cases} \dfrac{x}{y} + \dfrac{y}{x}, & x \neq 0 \text{ and } y \neq 0, \\[2mm] 0, & x = 0 \text{ or } y = 0. \end{cases}$

2. $f(x, y) = \begin{cases} \dfrac{2xy}{(x^2 + y^2)^2}, & (x, y) \neq (0, 0), \\[2mm] 0, & (x, y) = (0, 0). \end{cases}$

3. $f(x, y) = \begin{cases} \dfrac{xy}{\sqrt{x^2 + y^2}}, & (x, y) \neq (0, 0), \\[2mm] 0, & (x, y) = (0, 0). \end{cases}$

4. $f(x, y) = \begin{cases} \dfrac{x^2 y}{x^4 + y^2}, & (x, y) \neq (0, 0), \\[2mm] 0, & (x, y) = (0, 0). \end{cases}$

5. Consider the function

$$f(x, y) = \begin{cases} \dfrac{xy^2}{x^2 + y^2}, & (x, y) \neq (0, 0), \\[2mm] 0, & (x, y) = (0, 0). \end{cases}$$

(a) Use a computer to draw the contour diagram for f.
(b) Is f differentiable for $(x, y) \neq (0, 0)$?
(c) Show that $f_x(0, 0)$ and $f_y(0, 0)$ exist.
(d) Is f differentiable at $(0, 0)$?
(e) Suppose $x(t) = at$ and $y(t) = bt$, where a and b are constants, not both zero. If $g(t) = f(x(t), y(t))$, show that

$$g'(0) = \frac{ab^2}{a^2 + b^2}.$$

(f) Show that

$$f_x(0, 0)x'(0) + f_y(0, 0)y'(0) = 0.$$

Does the chain rule hold for the composite function $g(t)$ at $t = 0$? Explain.
(g) Show that the directional derivative $f_{\vec{u}}(0, 0)$ exists for each unit vector $\vec{u}$. Does this imply that f is differentiable at $(0, 0)$?

6. Consider the function

$$f(x, y) = \begin{cases} \dfrac{xy^2}{x^2 + y^4}, & (x, y) \neq (0, 0), \\ 0, & (x, y) = (0, 0). \end{cases}$$

(a) Use a computer to draw the contour diagram for f.
(b) Show that the directional derivative $f_{\vec{u}}(0, 0)$ exists for each unit vector $\vec{u}$.
(c) Is f continuous at $(0, 0)$? Is f differentiable at $(0, 0)$? Explain.

7. Consider the function $f(x, y) = \sqrt{|xy|}$.

(a) Use a computer to draw the contour diagram for f. Does the contour diagram look like that of a plane when we zoom in on the origin?
(b) Use a computer to draw the graph of f. Does the graph look like a plane when we zoom in on the origin?
(c) Is f differentiable for $(x, y) \neq (0, 0)$?
(d) Show that $f_x(0, 0)$ and $f_y(0, 0)$ exist.
(e) Is f differentiable at $(0, 0)$? [Hint: Consider the directional derivative $f_{\vec{u}}(0, 0)$ for $\vec{u} = (\vec{i} + \vec{j})/\sqrt{2}$.]

8. Suppose a function f is differentiable at the point (a, b). Show that f is continuous at (a, b).

9. Suppose $f(x, y)$ is a function such that $f_x(0, 0) = 0$ and $f_y(0, 0) = 0$, and $f_{\vec{u}}(0, 0) = 3$ for $\vec{u} = (\vec{i} + \vec{j})/\sqrt{2}$.

(a) Is f differentiable at $(0, 0)$? Explain.
(b) Give an example of a function f defined on 2-space which satisfies these conditions. [Hint: The function f does not have to be defined by a single formula valid over all of 2-space.]

10. Consider the following function:

$$f(x, y) = \begin{cases} \dfrac{xy(x^2 - y^2)}{x^2 + y^2}, & (x, y) \neq (0, 0), \\ 0, & (x, y) = (0, 0). \end{cases}$$

The graph of f is shown in Figure 13.66, and the contour diagram of f is shown in Figure 13.67.

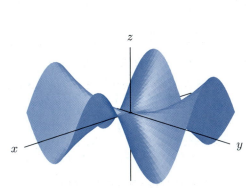

Figure 13.66: Graph of $\dfrac{xy(x^2 - y^2)}{x^2 + y^2}$

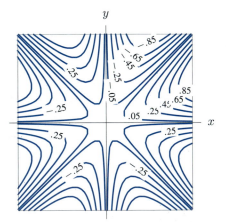

Figure 13.67: Contour diagram of $\dfrac{xy(x^2 - y^2)}{x^2 + y^2}$

(a) Find $f_x(x, y)$ and $f_y(x, y)$ for $(x, y) \neq (0, 0)$.
(b) Show that $f_x(0, 0) = 0$ and $f_y(0, 0) = 0$.
(c) Are the functions f_x and f_y continuous at $(0, 0)$?
(d) Is f differentiable at $(0, 0)$?

CHAPTER FOURTEEN

OPTIMIZATION: LOCAL AND GLOBAL EXTREMA

In one-variable calculus we saw how to find the maximum and minimum values of a function of one variable. In practice, there are often several variables in an optimization problem. For example, you may have $10,000 to invest in new equipment and advertising for your business. What combination of equipment and advertising will yield the greatest profit? Or, what combination of drugs will lower a patient's temperature the most? In this chapter we consider optimization problems, where the variables are completely free to vary (unconstrained optimization) and where there is a constraint on the variables (for example, a budget constraint).

14.1 LOCAL EXTREMA

Functions of several variables, like functions of one variable, can have *local* and *global* extrema. (That is, local and global maxima and minima.) A function has a local extremum at a point where it takes on the largest or smallest values in a small region around the point. Global extrema are the largest or smallest values anywhere on the domain under consideration. (See Figures 14.1 and 14.2.)

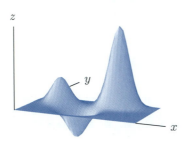

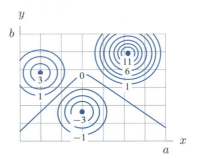

Figure 14.1: Local and global extrema for a function of two variables on $0 \leq x \leq a$, $0 \leq y \leq b$

Figure 14.2: Contour map of the function in Figure 14.1

More precisely, considering only points at which f is defined, we say:

- f has a **local maximum** at the point P_0 if $f(P_0) \geq f(P)$ for all points P near P_0.
- f has a **local minimum** at the point P_0 if $f(P_0) \leq f(P)$ for all points P near P_0.

How Do We Detect a Local Maximum or Minimum?

Recall that if the gradient vector of a function is defined and nonzero, then it points in a direction in which the function increases. Suppose that a function f has a local maximum at a point P_0 which is not on the boundary of the domain. If the vector grad $f(P_0)$ were defined and nonzero, then we could increase f by moving in the direction of grad $f(P_0)$. Since f has a local maximum at P_0, there is no direction in which f is increasing. Thus, if grad $f(P_0)$ is defined, we must have

$$\text{grad } f(P_0) = \vec{0}.$$

Similarly, suppose f has a local minimum at the point P_0. If grad $f(P_0)$ were defined and nonzero, then we could decrease f by moving in the direction opposite to grad $f(P_0)$, and so we must again have grad $f(P_0) = \vec{0}$ Therefore, we make the following definition:

Points where the gradient is either $\vec{0}$ or undefined are called **critical points** of the function. If a function has a local maximum or minimum at a point P_0, not on the boundary of its domain, then P_0 is a critical point.

For a function of two variables, we can also see that the gradient vector must be zero or undefined at a local maximum by looking at its contour diagram and a plot of its gradient vectors. (See

Figures 14.3 and 14.4.) Around the maximum the vectors are all pointing inward, perpendicularly to the contours. At the maximum the gradient vector must be zero or undefined. A similar argument shows that the gradient must be zero at a local minimum.

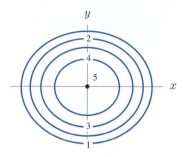

Figure 14.3: Contour diagram around a local maximum of a function

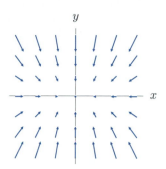

Figure 14.4: Gradients pointing toward the local maximum of the function in Figure 14.3

Finding Critical Points

To find critical points we set grad $f = f_x \vec{i} + f_y \vec{j} + f_z \vec{k} = \vec{0}$, which means setting all the partial derivatives of f equal to zero. We must also look for the points where one or more of the partial derivatives is undefined.

Example 1   Find and analyze the critical points of $f(x, y) = x^2 - 2x + y^2 - 4y + 5$.

Solution   To find the critical points, we set both partial derivatives equal to zero:

$$f_x = 2x - 2 = 0$$
$$f_y = 2y - 4 = 0.$$

Solving these equations gives $x = 1$, $y = 2$. Hence, f has only one critical point, namely $(1, 2)$. To see the behavior of f near $(1, 2)$, look at the values of the function in Table 14.1.

TABLE 14.1   *Values of $f(x, y)$ near the point $(1, 2)$*

		x				
		0.8	0.9	1.0	1.1	1.2
	1.8	0.08	0.05	0.04	0.05	0.08
	1.9	0.05	0.02	0.01	0.02	0.05
y	2.0	0.04	0.01	0.00	0.01	0.04
	2.1	0.05	0.02	0.01	0.02	0.05
	2.2	0.08	0.05	0.04	0.05	0.08

The table suggests that the function has a local minimum value of 0 at $(1, 2)$. We can verify this by completing the square:

$$f(x, y) = x^2 - 2x + y^2 - 4y + 5 = (x - 1)^2 + (y - 2)^2.$$

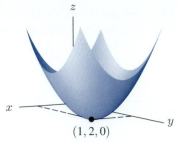

$(1, 2, 0)$

Figure 14.5: The graph of $f(x, y) = x^2 - 2x + y^2 - 4y + 5$
with a local minimum at the point $(1, 2)$

Figure 14.5 shows that the graph of f is a parabolic bowl with vertex at the point $(1, 2, 0)$. It is the same shape as the graph of $z = x^2 + y^2$ (shown in Figure 11.18 on page 576), except that the vertex has been shifted to $(1, 2)$. So the point $(1, 2)$ is a local minimum of f (as well as a global minimum).

Example 2   Find and analyze any critical points of $f(x, y) = -\sqrt{x^2 + y^2}$.

Solution   We look for points where grad $f = \vec{0}$ or is undefined. The partial derivatives are given by

$$\frac{\partial f}{\partial x} = -\frac{x}{\sqrt{x^2 + y^2}},$$

$$\frac{\partial f}{\partial y} = -\frac{y}{\sqrt{x^2 + y^2}}.$$

These are never both zero; but they are both undefined at $x = 0$, $y = 0$. Thus, $(0, 0)$ is a critical point and a possible extreme point. The graph of f (see Figure 14.6) is a cone, with vertex at $(0, 0)$. So f has a local and global maximum at $(0, 0)$.

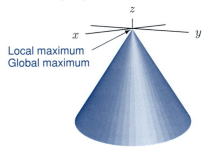

Local maximum
Global maximum

Figure 14.6: Graph of $f(x, y) = -\sqrt{x^2 + y^2}$

Example 3   Find the local extrema of the function $f(x, y) = 8y^3 + 12x^2 - 24xy$.

Solution   We begin by looking for critical points:

$$f_x = 24x - 24y,$$
$$f_y = 24y^2 - 24x.$$

Setting these expressions equal to zero gives the system of equations

$$x = y, \qquad x = y^2,$$

which has two solutions, $(0, 0)$ and $(1, 1)$. Are these maxima, minima or neither? Let's look at the contours near the points: Figure 14.7 shows the contour diagram of this function. Notice that

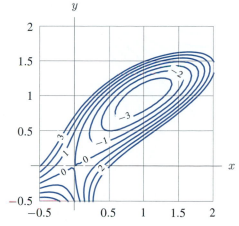

Figure 14.7: Contour diagram of $f(x, y) = 8y^3 + 12x^2 - 24xy$ showing critical points at $(0, 0)$ and $(1, 1)$

$f(1, 1) = -4$ and that there is no other -4 contour. The contours near $P = (1, 1)$ appear oval in shape and show that f is increasing in value no matter in which direction you move away from P. This suggests that f has a local minimum at the point $(1, 1)$.

The level curves near $Q = (0, 0)$ show a very different behavior. While $f(0, 0) = 0$, we see that f takes on both positive and negative values at nearby points. Thus, the point $(0, 0)$ is a critical point which is neither a local maximum nor a local minimum.

Saddle Points

The previous example shows that critical points can occur at local maxima or minima, or at points which are neither — the value of the function is larger in some directions and smaller in others. We make the following definition:

A function, f, has a **saddle point** at P_0 if P_0 is a critical point of f and within any distance of P_0, no matter how small, there are points, P_1 and P_2, with

$$f(P_1) > f(P_0) \quad \text{and} \quad f(P_2) < f(P_0).$$

Thus, we see from Figure 14.7 that the function $f(x, y) = 8y^3 + 12x^2 - 24xy$ in Example 3 has a saddle point at the origin.

For another example, look at the graph of $g(x, y) = x^2 - y^2$ in Figure 14.8. The origin is a critical point and $g(0, 0) = 0$. Since there are positive values on the x-axis and negative values on the y-axis, the origin is a saddle point. Notice that the graph of g looks like a saddle there.

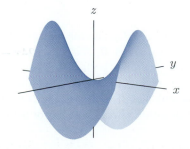

Figure 14.8: Graph of $g(x, y) = x^2 - y^2$, showing a saddle point at the origin

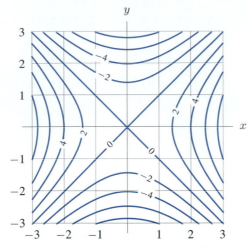

Figure 14.9: Contours of $g(x, y) = x^2 - y^2$, showing a saddle point at the origin

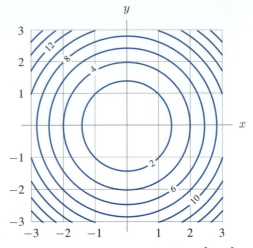

Figure 14.10: Contours of $h(x, y) = x^2 + y^2$, showing a local minimum at the origin

Figure 14.9 shows the level curves of g near the saddle point $(0, 0)$. They are hyperbolas showing both positive and negative values of g near $(0, 0)$. Contrast this with the appearance of level curves near a local maximum or minimum. For example, Figure 14.10 shows $h(x, y) = x^2 + y^2$ near $(0, 0)$.

Is a Critical Point a Local Maximum, Local Minimum, or Saddle Point?

We can see whether a critical point of a function, f, is a maximum, minimum, or saddle point by looking at the contour diagram. There is also a simple analytic method for making the distinction if the critical point is one at which the partial derivatives of f are zero. Near most critical points, a function has the same behavior as its quadratic Taylor approximation about that point, so we must first understand quadratic functions.

Quadratic Functions of the form $f(x, y) = ax^2 + bxy + cy^2$

We begin by looking at what can happen at critical points of quadratic functions of the form $f(x, y) = ax^2 + bxy + cy^2$, where a, b and c are constants.

Example 4 Find and analyze the local extrema of the function $f(x, y) = x^2 + xy + y^2$.

Solution To find critical points, we set

$$f_x = 2x + y = 0,$$
$$f_y = x + 2y = 0.$$

The only critical point is $(0, 0)$, and the value of the function there is $f(0, 0) = 0$. If f is always positive or zero near $(0, 0)$, then $(0, 0)$ is a local minimum; if f is always negative or zero near $(0, 0)$, it is a local maximum; if f takes both positive and negative values it is a saddle point. The graph in Figure 14.11 suggests that $(0, 0)$ is a local minimum.

How can we be sure that $(0, 0)$ is a local minimum? The algebraic way to determine if a quadratic function is always negative, always positive, or neither, is to complete the square. Writing

$$f(x, y) = x^2 + xy + y^2 = \left(x + \frac{1}{2}y\right)^2 + \frac{3}{4}y^2,$$

shows that $f(x, y)$ is a sum of two squares, so it must always be greater than or equal to zero. Thus, the critical point is both a local and a global minimum.

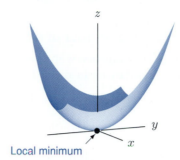

Local minimum

Figure 14.11: Graph of
$f(x, y) = x^2 + xy + y^2 = (x + \frac{1}{2}y)^2 + \frac{3}{4}y^2$ showing local
minimum at the origin

The Shape of the Graph of $f(x, y) = ax^2 + bxy + cy^2$

In general, a function of the form $f(x, y) = ax^2 + bxy + cy^2$ has one critical point at $(0, 0)$. To analyze its graph, we complete the square. Assuming $a \neq 0$, we write

$$ax^2 + bxy + cy^2 = a\left[x^2 + \frac{b}{a}xy + \frac{c}{a}y^2\right]$$

$$= a\left[\left(x + \frac{b}{2a}y\right)^2 + \left(\frac{c}{a} - \frac{b^2}{4a^2}\right)y^2\right]$$

$$= a\left[\left(x + \frac{b}{2a}y\right)^2 + \left(\frac{4ac - b^2}{4a^2}\right)y^2\right].$$

The shape of the graph of f depends on whether the coefficient of y^2 is positive, negative, or zero. The sign of $D = 4ac - b^2$, called the *discriminant*, determines the sign of the coefficient of y^2.

- If $D > 0$, then the expression inside the brackets is positive or zero, so the function has a local maximum or a local minimum.

  - If $a > 0$, the function has a local minimum, since the graph is a right-side-up paraboloid, like $z = x^2 + y^2$. (See Figure 14.12.)
  - If $a < 0$, the function has a local maximum, since the graph is an upside-down paraboloid, like $z = -x^2 - y^2$. (See Figure 14.13.)

- If $D < 0$, then the function goes up in some directions and goes down in others, like $z = x^2 - y^2$. Hence it has a saddle point. (See Figure 14.14.)

- If $D = 0$, then the quadratic function is $a(x + by/2a)^2$, whose graph is a parabolic cylinder. (See Figure 14.15.)

Figure 14.12: Concave up: $D > 0$ and $a > 0$

Figure 14.13: Concave down: $D > 0$ and $a < 0$

Figure 14.14: Saddle-shaped: $D < 0$

Figure 14.15: Parabolic cylinder: $D = 0$

More generally, the graph of $g(x, y) = a(x - x_0)^2 + b(x - x_0)(y - y_0) + c(y - y_0)^2$ has exactly the same shape as the graph of $f(x, y) = ax^2 + bxy + cy^2$, except that the critical point is at (x_0, y_0) rather than $(0, 0)$. Therefore, the discriminant test[1] gives the same results for the behavior of g near (x_0, y_0).

Classifying the Critical Points of a Function

Now, suppose that f is any function with $f(0, 0) = 0$ and grad $f(0, 0) = \vec{0}$. Recall from page 699 that f can be approximated by its quadratic Taylor polynomial near $(0, 0)$:

$$f(x, y) \approx f(0, 0) + f_x(0, 0)x + f_y(0, 0)y$$

$$+ \frac{1}{2}f_{xx}(0, 0)x^2 + f_{xy}(0, 0)xy + \frac{1}{2}f_{yy}(0, 0)y^2.$$

Since $f(0, 0) = 0$ and $f_x(0, 0) = f_y(0, 0) = 0$, the quadratic polynomial simplifies to

$$f(x, y) \approx \frac{1}{2}f_{xx}(0, 0)x^2 + f_{xy}(0, 0)xy + \frac{1}{2}f_{yy}(0, 0)y^2.$$

The discriminant is

$$D = 4ac - b^2 = 4\left(\frac{1}{2}f_{xx}(0, 0)\right)\left(\frac{1}{2}f_{yy}(0, 0)\right) - \left(f_{xy}(0, 0)\right)^2,$$

which simplifies to

$$D = f_{xx}(0, 0)f_{yy}(0, 0) - (f_{xy}(0, 0))^2.$$

There is a similar formula for D if $f(0, 0) \neq 0$ or if the critical point is at (x_0, y_0). Thus, we get the following test:

Second Derivative Test for Functions of Two Variables

Suppose (x_0, y_0) is a point where grad $f(x_0, y_0) = \vec{0}$. Let

$$D = f_{xx}(x_0, y_0)f_{yy}(x_0, y_0) - (f_{xy}(x_0, y_0))^2.$$

- If $D > 0$ and $f_{xx}(x_0, y_0) > 0$, then f has a local minimum at (x_0, y_0).
- If $D > 0$ and $f_{xx}(x_0, y_0) < 0$, then f has a local maximum at (x_0, y_0).
- If $D < 0$, then f has a saddle point at (x_0, y_0).
- If $D = 0$, anything can happen: f can have a local maximum, or a local minimum or a saddle point at (x_0, y_0).

Example 5   Find the local maxima, minima, and saddle points of the function

$$f(x, y) = \frac{x^2}{2} + 3y^3 + 9y^2 - 3xy + 9y - 9x.$$

Solution   The partial derivatives of f are $f_x = x - 3y - 9$ and $f_y = 9y^2 + 18y - 3x + 9$. The equations $f_x = 0$ and $f_y = 0$ give

$$9y^2 + 18y + 9 - 3x = 0,$$

$$x - 3y - 9 = 0.$$

Eliminating x gives

$$9y^2 + 9y - 18 = 0,$$

which has solutions $y = -2$ and $y = 1$. We find the corresponding values of x, so the critical points of f are $(3, -2)$ and $(12, 1)$. The discriminant is

$$D(x, y) = f_{xx}f_{yy} - f_{xy}^2 = (1)(18y + 18) - (-3)^2 = 18y + 9.$$

---

[1]We assumed that $a \neq 0$. If $a = 0$ and $c \neq 0$, the same argument works. If both $a = 0$ and $c = 0$, then $f(x, y) = bxy$, which is a saddle.

Since $D(3, -2) = -36 + 9 < 0$, we know that $(3, -2)$ is a saddle point of f. Since $D(12, 1) = 18 + 9 > 0$ and $f_{xx}(12, 1) = 1 > 0$, we know that $(12, 1)$ is a local minimum of f.

The second derivative test does not give any information in the case $D = 0$. However, as the following example illustrates, we can still classify the critical points by looking at the graph of the function.

Example 6   Classify the critical point $(0, 0)$ of the functions $f(x, y) = x^4 + y^4$, and $g(x, y) = -x^4 - y^4$, and $h(x, y) = x^4 - y^4$.

Solution   Each of these functions has a critical point at $(0, 0)$. However, all the second partial derivatives are 0 there, so each function has $D = 0$. Near the origin, the graphs of f, g and h look like the surfaces in Figures 14.12–14.14, respectively, and so we see that f has a minimum at $(0, 0)$, and g has a maximum at $(0, 0)$, and h has a saddle point at $(0, 0)$.

We can get the same results algebraically. Since $f(0, 0) = 0$ and $f(x, y) > 0$ elsewhere, f must have a minimum at the origin. Since $g(0, 0) = 0$ and $g(x, y) < 0$ elsewhere, g has a maximum at the origin. Lastly, h has a saddle point at the origin since $h(0, 0) = 0$ and $h(x, y) > 0$ on the x-axis and $h(x, y) < 0$ on the y-axis.

Problems for Section 14.1

1. Consider the points marked A, B, C in the contour plot in Figure 14.16. Which of these appear to be critical points? Classify those that are critical points.

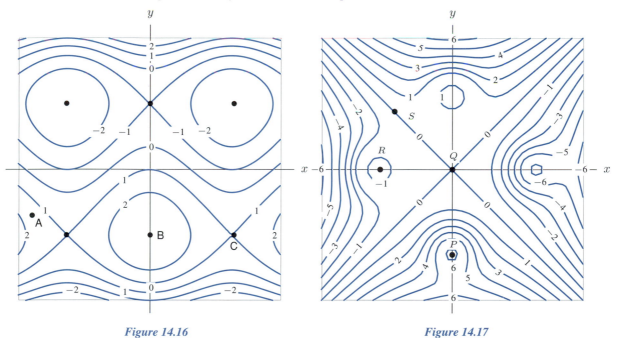

Figure 14.16   Figure 14.17

For Problems 2–4, use Figure 14.17, which shows level curves of some function $f(x, y)$.

2. Decide whether you think each point is a local maximum, local minimum, saddle point, or none of these.

   (a) P (b) Q (c) R (d) S

3. Sketch the direction of ∇f at several points around each of P, Q, and R.

4. Put arrows showing the direction of ∇f at the points where $\|\nabla f\|$ is largest.

For Problems 5–11, find the local maxima, local minima, and saddle points of the given function.

5. $f(x, y) = x^3 - 3x + y^3 - 3y$ 6. $f(x, y) = x^3 + e^{-y^2}$

7. $f(x, y) = (x + y)(xy + 1)$ 8. $f(x, y) = 8xy - \frac{1}{4}(x + y)^4$

9. $E(x, y) = 1 - \cos x + y^2/2$ 10. $f(x, y) = \sin x \sin y$

11. $P(x, y) = 400 - 3x^2 - 4x + 2xy - 5y^2 + 48y$

12. Suppose $f(x, y) = A - (x^2 + Bx + y^2 + Cy)$. What values of A, B, and C give $f(x, y)$ a local maximum value of 15 at the point $(-2, 1)$?

Each function in Problems 13–15 has a critical point at $(0, 0)$. What sort of critical point is it?

13. $f(x, y) = x^6 + y^6$ 14. $g(x, y) = x^4 + y^3$ 15. $h(x, y) = \cos x \cos y$

16. (a) Find all critical points of
$$f(x, y) = e^x(1 - \cos y).$$
   (b) Are these critical points local maxima, local minima, or saddle points?

17. Suppose $f_x = f_y = 0$ at $(1, 3)$ and $f_{xx} > 0$, $f_{yy} > 0$, $f_{xy} = 0$.
   (a) What can you conclude about the behavior of the function near the point $(1, 3)$?
   (b) Sketch a possible contour diagram.

18. Suppose that for some function $f(x, y)$ at the point (a, b), we have $f_x = f_y = 0$, $f_{xx} > 0$, $f_{yy} = 0$, $f_{xy} > 0$.
   (a) What can you conclude about the shape of the graph of f near the point (a, b)?
   (b) Sketch a possible contour diagram.

19. The behavior of a function can be complicated near a critical point where $D = 0$. Suppose that
$$f(x, y) = x^3 - 3xy^2.$$
Show that there is one critical point at $(0, 0)$ and that $D = 0$ there. Then show that the contour for $f(x, y) = 0$ consists of three lines intersecting at the origin and that these lines divide the plane into six regions around the origin where f alternates from positive to negative. Sketch a contour map for f near $(0, 0)$. The graph of this function is called a *monkey saddle*.

20. On a computer, draw contour diagrams for the family of functions
$$f(x, y) = k(x^2 + y^2) - 2xy$$
for $k = -2, -1, 0, 1, 2$. Use these figures to classify the critical point at $(0, 0)$ for each value of k. Explain your observations using the discriminant, D.

14.2 GLOBAL EXTREMA: UNCONSTRAINED OPTIMIZATION

Suppose we want to find the highest and the lowest points in some region of the country. First of all, it makes a difference what the region is, the whole United States, an individual state, or a county. Let's suppose the region is Colorado (a contour map is shown in Figure 14.18). The highest point is the top of a mountain peak (point A on the map, Mt. Elbert, 14,431 feet high). What about the lowest point? Colorado does not have large pits without drainage, like Death Valley in California. A drop

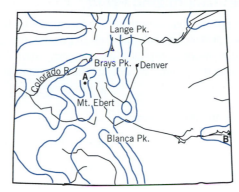

Figure 14.18: The highest and lowest points in the state of Colorado

of rain falling at any point in Colorado will flow eventually out of the state to either the Pacific or the Atlantic Ocean. If there is no local minimum inside the state, where is the lowest point? It must be on the state boundary at a point where a river is flowing out of the state (point B where the Arkansas River leaves the state, 3,400 feet high). The highest point in Colorado is a global maximum for the elevation function in Colorado and the lowest point is the global minimum.

In general, if we are given a function f defined on a region R, we say:

- f has a **global maximum on** R at the point P_0 if $f(P_0) \geq f(P)$ for all points P in R.
- f has a **global minimum on** R at the point P_0 if $f(P_0) \leq f(P)$ for all points P in R.

The process of finding a global maximum or minimum for a function f on a region R is called *optimization*. If the region R is the entire xy-plane, we speak of *unconstrained optimization*; if the region R is not the entire xy-plane, that is, if x or y is restricted in some way, then we speak of *constrained optimization*. If the region R is not stated explicitly, it is understood to be the whole xy-plane.

How Do We Find Global Maxima and Minima?

As the Colorado example illustrates, a global extremum can occur either at a critical point inside the region or at a point on the boundary of the region. This is analogous to single-variable calculus, where a function achieves its global extrema on an interval either at a critical point inside the interval or at an endpoint of the interval. Optimization for functions of more than one variable, however, is more difficult because regions in 2-space can have very complicated boundaries.

For an Unconstrained Optimization Problem

- Find the critical points.
- Investigate whether the critical points give global maxima or minima.

Not all functions have a global maximum or minimum: it depends on the function and the region. For the remainder of this section we consider applications in which global extrema are expected from practical considerations. In general, the fact that a function has a single local maximum or minimum does not guarantee that the point is the global maximum or minimum. (See Problem 22.) An exception is if the function is quadratic, in which case the local maximum or minimum is the global maximum or minimum. See Example 1 on page 719.

Economic Example: Maximizing Profit

In planning production, a company is concerned with how much of a particular item to manufacture and the price at which to sell the item. In general, the higher the price, the less that can be sold. To determine how much to produce, the company often chooses the combination of price and quantity that maximizes the profit. To calculate the maximum we use the fact that

$$\text{Profit} = \text{Revenue} - \text{Cost},$$

and, provided the price is constant,

$$\text{Revenue} = \text{Price} \times \text{Quantity} = pq.$$

In addition, we need to know how the cost and price depend on quantity.

Example 1   A company manufactures two items which are sold in two separate markets. The quantities, q_1 and q_2, demanded by consumers, and the prices, p_1 and p_2 (in dollars), of each item are related by

$$p_1 = 600 - 0.3q_1 \quad \text{and} \quad p_2 = 500 - 0.2q_2.$$

Thus, if the price for either item increases, the demand for it decreases. The company's total production cost is given by

$$C = 16 + 1.2q_1 + 1.5q_2 + 0.2q_1q_2.$$

If the company wants to maximize its total profits, how much of each product should it produce? What will be the maximum profit? [2]

Solution   The total revenue, R, is the sum of the revenues, p_1q_1 and p_2q_2, from each market. Substituting for p_1 and p_2, we get

$$
\begin{aligned}
R &= p_1q_1 + p_2q_2 \\
&= (600 - 0.3q_1)q_1 + (500 - 0.2q_2)q_2 \\
&= 600q_1 - 0.3q_1^2 + 500q_2 - 0.2q_2^2.
\end{aligned}
$$

Thus, the total profit P is given by

$$
\begin{aligned}
P &= R - C \\
&= 600q_1 - 0.3q_1^2 + 500q_2 - 0.2q_2^2 - (16 + 1.2q_1 + 1.5q_2 + 0.2q_1q_2) \\
&= -16 + 598.8q_1 - 0.3q_1^2 + 498.5q_2 - 0.2q_2^2 - 0.2q_1q_2.
\end{aligned}
$$

To maximize P, we compute partial derivatives and set them equal to 0:

$$\frac{\partial P}{\partial q_1} = 598.8 - 0.6q_1 - 0.2q_2 = 0,$$

$$\frac{\partial P}{\partial q_2} = 498.5 - 0.4q_2 - 0.2q_1 = 0.$$

Since grad P is defined everywhere, the only critical points of P are those where grad $P = \vec{0}$. Thus, solving for q_1, q_2, we find that

$$q_1 = 699.1 \quad \text{and} \quad q_2 = 896.7.$$

The corresponding prices are

$$p_1 = 390.27 \quad \text{and} \quad p_2 = 320.66.$$

To see whether or not we have found a maximum, we compute second partial derivatives:

$$\frac{\partial^2 P}{\partial q_1^2} = -0.6, \quad \frac{\partial^2 P}{\partial q_2^2} = -0.4, \quad \frac{\partial^2 P}{\partial q_1 \partial q_2} = -0.2,$$

so,

$$D = \frac{\partial^2 P}{\partial q_1^2}\frac{\partial^2 P}{\partial q_2^2} - \left(\frac{\partial^2 P}{\partial q_1 \partial q_2}\right)^2 = (-0.6)(-0.4) - (-0.2)^2 = 0.2.$$

[2] Adapted from M. Rosser, *Basic Mathematics for Economists*, p. 316 (New York: Routledge, 1993).

Therefore we have found a local maximum. The graph of P is an upside-down paraboloid and so $(699.1, 896.7)$ is in fact a global maximum. The company should produce 699.1 units of the first item priced at \$390.27 per unit, and 896.7 units of the second item priced at \$320.66 per unit. The maximum profit $P(699.1, 896.7) \approx \$433,000$.

Fitting a Line to Data

An important application of optimization is to the problem of fitting the "best" line to some data. Suppose the data is plotted in the plane. We measure the distance from a line to the data points by adding the squares of the vertical distances from each point to the line. The smaller this sum of squares is, the better the line fits the data. The line with the minimum sum of square distances is called the *least squares line*, or the *regression line*. If the data is nearly linear, the least squares line will be a good fit; otherwise it may not be. (See Figure 14.19.)

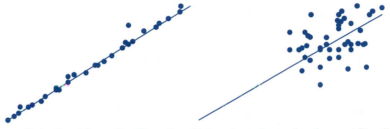

Data almost linear: line fits well   Data not very linear: line does not fit well

Figure 14.19: Fitting lines to data points

Example 2 Find a least squares line for the following data points: $(1, 1)$, $(2, 1)$, and $(3, 3)$.

Solution Suppose the line has equation $y = b + mx$. If we find b and m then we have found the line. So, for this problem, b and m are the two variables. We want to minimize the function $f(b, m)$ that gives the sum of the three squared vertical distances from the points to the line in Figure 14.20.

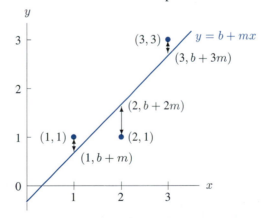

Figure 14.20: The least squares line minimizes the sum of the squares of these vertical distances

The vertical distance from the point $(1, 1)$ to the line is the difference in the y-coordinates $1 - (b + m)$; similarly for the other points. Thus, the sum of squares is

$$f(b, m) = (1 - (b + m))^2 + (1 - (b + 2m))^2 + (3 - (b + 3m))^2.$$

To minimize f we look for critical points. First we differentiate f with respect to b:

$$\frac{\partial f}{\partial b} = -2(1 - (b + m)) - 2(1 - (b + 2m)) - 2(3 - (b + 3m))$$

$$= -2 + 2b + 2m - 2 + 2b + 4m - 6 + 2b + 6m$$

$$= -10 + 6b + 12m.$$

Now we differentiate with respect to m:

$$\frac{\partial f}{\partial m} = 2(1 - (b + m))(-1) + 2(1 - (b + 2m))(-2) + 2(3 - (b + 3m))(-3)$$
$$= -2 + 2b + 2m - 4 + 4b + 8m - 18 + 6b + 18m$$
$$= -24 + 12b + 28m.$$

The equations $\dfrac{\partial f}{\partial b} = 0$ and $\dfrac{\partial f}{\partial m} = 0$ give a system of two linear equations in two unknowns:

$$-10 + 6b + 12m = 0,$$
$$-24 + 12b + 28m = 0.$$

The solution to this pair of equations is the critical point $b = -1/3$ and $m = 1$. Since

$$D = f_{bb}f_{mm} - (f_{mb})^2 = (6)(28) - 12^2 = 24 \quad \text{and} \quad f_{bb} = 6 > 0,$$

we have found a local minimum. The graph of $f(b, m)$ is a parabolic bowl, so the local minimum is the global minimum of f. Thus, the least squares line is

$$y = x - \frac{1}{3}.$$

As a check, notice that the line $y = x$ passes through the points $(1, 1)$ and $(3, 3)$. It is reasonable that introducing the point $(2, 1)$ moves the y-intercept down from 0 to $-1/3$.

The general formulas for the slope and y-intercept of a least squares line are in Project 2 at the end of this chapter. Many calculators have these formulas built in, so that when you enter the data, out come the values of b and m. At the same time, you get the *correlation coefficient*, which measures how close the data points actually come to fitting the least squares line.

Gradient Search for Finding Local Extrema

So far we have searched for values that maximize or minimize a function $f(x, y)$ by first finding the critical points of f. Finding the critical points amounts to solving the equation grad $f = \vec{0}$, which is really a pair of simultaneous equations for x and y:

$$\frac{\partial f}{\partial x}(x_0, y_0) = 0 \quad \text{and} \quad \frac{\partial f}{\partial y}(x_0, y_0) = 0.$$

However, solving such equations can be very difficult. In practice, most optimization problems are solved by numerical methods such as the *gradient search*. The gradient search method can be explained by analogy with a mountain climber who wishes to maximize his elevation by getting to the top of the highest mountain. All he has to do is keep going up and eventually he will get to the top of some mountain. If he starts near the highest mountain, that is probably the mountain he will conquer. If not, he may go up a lower mountain, ending up at a local rather than a global maximum.

The gradient search method is illustrated in the next example. It is a minimization problem, so imagine a hiker seeking the bottom of the lowest valley by always going down.

Example 3   Twenty cubic meters of gravel are to be delivered to a landfill by a trucker. She plans to purchase an open-top box in which to transport the gravel in numerous trips. The cost to her is the cost of the box plus $2 per trip. The box must have height 0.5 m, but she can choose the length and width. The cost of the box will be $20/m² for the ends and $10/m² for the bottom and sides. Notice the tradeoff she faces: A smaller box is cheaper to buy but requires more trips. What size box should she buy to minimize her over-all costs? [3]

[3] Adapted from Claude McMillan, Jr., *Mathematical Programming*, 2nd ed., p. 156-157 (New York: Wiley, 1978).

Solution   We first get an algebraic expression for the trucker's cost. Let the length of the box be x meters and the width be y meters and let the height be 0.5 m (See Figure 14.21.)

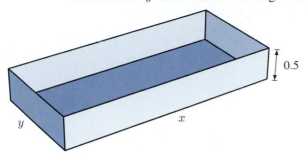

TABLE 14.2   *Trucker's itemized cost*

$20/(0.5xy)$ at \$2/trip	$80/(xy)$
2 ends at \$20/m² × 0.5y m²	$20y$
2 sides at \$10/m² × 0.5x m²	$10x$
1 bottom at \$10/m² × xy m²	$10xy$
Total cost	$f(x, y)$

Figure 14.21: The box for transporting gravel

The volume of the box is $0.5xy$ m³, so delivery of 20 m³ of gravel will require $20/(0.5xy)$ trips. The trucker's cost is itemized in Table 14.2. The problem is to choose x and y to minimize

$$\text{Total cost} = f(x, y) = \frac{80}{xy} + 20y + 10x + 10xy.$$

We pick a starting point (x_0, y_0) that may not minimize f but which we hope is not too far from the minimum point. In this example we start with $(x_0, y_0) = (5, 5)$, which is definitely not a critical point of f because

$$\text{grad } f(5, 5) = 59.4\vec{i} + 69.4\vec{j} \neq \vec{0}.$$

We plan to move from (x_0, y_0) to a new point (x_1, y_1) in such a way that f decreases, that is $f(x_1, y_1) < f(x_0, y_0)$. Moving in the direction of grad $f(x_0, y_0)$ increases f as rapidly as possible, so we move in the opposite direction, namely $-\text{grad } f(x_0, y_0)$. We continue to move in this direction until the f values begin to increase again. If we move parallel to $-\text{grad } f(x_0, y_0)$, our displacement from the original point is of the form $-t\,\text{grad } f(x_0, y_0)$, where t is a scalar to be determined. Since grad $f(x_0, y_0) = 59.4\vec{i} + 69.4\vec{j}$, the coordinates of our final point are

$$(x_0 - 59.4t, y_0 - 69.4t).$$

We want to find the minimum value of the function f as t increases. Figure 14.22 gives the graph of

$$f\left((x_0, y_0) - t\,\text{grad } f(x_0, y_0)\right) = f(5 - 59.4t, 5 - 69.4t)$$

for positive t. Zooming in shows that the local minimum is at $t \approx 0.0554$. So we move to the point given by

$$(x_1, y_1) = (5 - (59.4)(0.0554), 5 - (69.4)(0.0554)) \approx (1.71, 1.16).$$

Notice that the cost at the initial point is $f(5, 5) = 403.2$ and the cost at the new point is

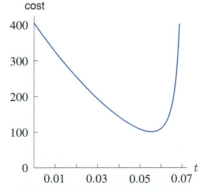

Figure 14.22: First step: Graph of $f\left((x_0, y_0) - t\,\text{grad } f(x_0, y_0)\right)$ showing local minimum at $t \approx 0.0554$

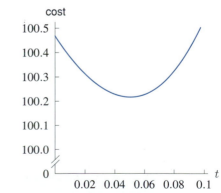

Figure 14.23: Second step: Graph of $f((x_1, y_1) - t\,\text{grad } f(x_1, y_1))$ showing local minimum at $t \approx 0.050$

$f(1.71, 1.16) = 100.47$, so we have decreased the cost considerably.

We can further decrease the cost by moving away from $(1.71, 1.16)$ in the direction opposite to grad $f(1.71, 1.16) = -1.99\vec{i} + 2.33\vec{j}$. Figure 14.23 gives the graph of

$$f\left((x_1, y_1) - t\operatorname{grad} f(x_1, y_1)\right) = f(1.71 + 1.99t, 1.16 - 2.33t)$$

which achieves a local minimum at $t \approx 0.050$. So we take

$$(x_2, y_2) = (1.71 + (1.99)(0.050), 1.16 - (2.33)(0.050)) \approx (1.81, 1.04).$$

Notice that $f(1.81, 1.04) = 100.22$ whereas $f(1.71, 1.16) = 100.47$. The move from (x_1, y_1) to (x_2, y_2) decreased the cost f by a rather small amount, only \$0.25. Thus, although we have not actually achieved a minimum, we may feel that for practical purposes we are probably close enough. The trucker will round off, buying a box of dimensions about 1.8m × 1m × 0.5m.

How Do We Know Whether a Function Has a Global Maximum or Minimum?

Under what circumstances does a function of two variables have a global maximum or minimum? The next example shows that a function may have both a global maximum and a global minimum on a region, or just one, or neither.

Example 4   Investigate the global maxima and minima of the following functions:

(a) $h(x, y) = 1 + x^2 + y^2$ on the disk $x^2 + y^2 \le 1$.
(b) $f(x, y) = x^2 - 2x + y^2 - 4y + 5$ on the xy-plane.
(c) $g(x, y) = x^2 - y^2$ on the xy-plane.

Solution   (a)   The graph of $h(x, y) = 1 + x^2 + y^2$ is a bowl shaped paraboloid with a global minimum of 1 at $(0, 0)$, and a global maximum of 2 on the edge of the region, $x^2 + y^2 = 1$.
(b)   The graph of f in Figure 14.5 on page 720 shows that f has a global minimum at the point $(1, 2)$ and no global maximum (because the value of f increases without bound as $x \to \infty$, $y \to \infty$).
(c)   The graph of g in Figure 14.8 on page 721 shows that g has no global maximum because $g(x, y) \to \infty$ as $x \to \infty$ if y is constant. Similarly, g has no global minimum because $g(x, y) \to -\infty$ as $y \to \infty$ if x is constant.

There are, however, conditions that guarantee that a function has a global maximum and minimum. For $h(x)$, a function of one variable, the function must be continuous on a closed interval $a \le x \le b$. If h is continuous on a non-closed interval, such as $a \le x < b$ or $a < x < b$, or on an interval which is not bounded, such as $a < x < \infty$, then h need not have a maximum or minimum value. What is the situation for functions of two variables? As it turns out, a similar result is true for continuous functions defined on regions which are closed and bounded, analogous to the closed and bounded interval $a \le x \le b$. In everyday language we say

- A **closed** region is one which contains its boundary;
- A **bounded** region is one which does not stretch to infinity in any direction.

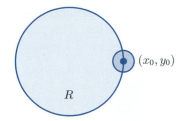

Figure 14.24: Boundary point (x_0, y_0) of R *Figure 14.25:* Interior point (x_0, y_0) of R

More precise definitions are as follows. Suppose R is a region in 2-space. A point (x_0, y_0) is a *boundary point* of R if, for every $r > 0$, the disk $(x - x_0)^2 + (y - y_0)^2 < r^2$ with center (x_0, y_0) and radius r contains both points which are in R and points which are not in R. See Figure 14.24. A point (x_0, y_0) can be a boundary point of the region R without actually belonging to R. A point (x_0, y_0) in R is an *interior point* if it is not a boundary point; thus, for small enough $r > 0$, the disk of radius r centered at (x_0, y_0) lies entirely in the region R. See Figure 14.25. The collection of all the boundary points is the *boundary of R* and the collection of all the interior points is the *interior of R*. The region R is *closed* if it contains its boundary, while it is *open* if every point in R is an interior point.

A region R in 2-space is *bounded* if the distance between every point (x, y) in R and the origin is less than or equal to some constant number K. Closed and bounded regions in 3-space are defined in the same way.

Example 5   (a)   The square $-1 \leq x \leq 1$, $-1 \leq y \leq 1$ is closed and bounded.
(b)   The first quadrant $x \geq 0$, $y \geq 0$ is closed but is not bounded.
(c)   The disk $x^2 + y^2 < 1$ is open and bounded, but is not closed.
(d)   The half-plane $y > 0$ is open, but is neither closed nor bounded.

The reason that closed and bounded regions are useful is the following result[4]:

If f is a continuous function on a closed and bounded region R, then f has a global maximum at some point (x_0, y_0) in R and a global minimum at some point (x_1, y_1) in R.

The result is also true for functions of three or more variables.

If f is not continuous or the region R is not closed and bounded, there is no guarantee that f will achieve a global maximum or global minimum on R. In Example 4, the function g is continuous but does not achieve a global maximum or minimum in 2-space, a region which is closed but not bounded. The following example illustrates what can go wrong when the region is bounded but not closed.

Example 6   Does the following function have a global maximum or minimum on the region R given by $0 < x^2 + y^2 \leq 1$?

$$f(x, y) = \frac{1}{x^2 + y^2}$$

---

[4]For a proof, see W. Rudin, *Principles of Mathematical Analysis*, 2nd ed., p. 89, (New York: McGraw-Hill, 1976)

Solution   The region R is bounded, but it is not closed since it does not contain the boundary point $(0,0)$. We see from the graph of $z = f(x, y)$ in Figure 14.26 that f has a global minimum on the circle $x^2 + y^2 = 1$. However, $f(x, y) \to \infty$ as $(x, y) \to (0, 0)$, so f has no global maximum.

Figure 14.26: Graph showing $f(x, y) = \frac{1}{x^2 + y^2}$ has no global maximum on $0 < x^2 + y^2 \leq 1$

Problems for Section 14.2

1. By looking at the weather map in Figure 11.1 on page 564, find the maximum and minimum daily high temperatures in the states of Mississippi, Alabama, Pennsylvania, New York, California, Arizona, and Massachusetts.

2. Does the function $f(x, y) = -2x^2 - 7y^2$ have global maxima and minima? Explain.

3. Does the function $f(x, y) = x^2/2 + 3y^3 + 9y^2 - 3x + 9y - 9$ have global maxima and minima? Explain.

4. Which of the following functions have both a global maximum and global minimum? Which have neither?
    $$f(x, y) = 5 + x^2 - 2y^2, \quad g(x, y) = x^2 y^2, \quad h(x, y) = x^3 + y^3.$$

In Problems 5–7, find the global maximum and minimum of the given function over the square $-1 \leq x \leq 1$, $-1 \leq y \leq 1$, and say whether it occurs on the boundary of the square. (Hint: Consider the graph of the function.)

5. $z = x^2 + y^2$ 6. $z = x^2 - y^2$ 7. $z = -x^2 - y^2$

8. The quantity of a product demanded by consumers is a function of its price. The quantity of one product demanded may also depend on the price of other products. For example, the demand for tea is affected by the price of coffee; the demand for cars is affected by the price of gas. Suppose the quantities demanded, q_1 and q_2, of two products depend on the prices, p_1 and p_2, as follows
    $$q_1 = 150 - 2p_1 - p_2$$
    $$q_2 = 200 - p_1 - 3p_2.$$
    (a) What does the fact that the coefficients of p_1 and p_2 are negative tell you? Give an example of two products that might be related this way.
    (b) Suppose one manufacturer sells both of these products. How should the manufacturer set prices to earn the maximum possible revenue? What is that maximum possible revenue?

9. A company operates two plants which manufacture the same item and whose total cost functions are
    $$C_1 = 8.5 + 0.03q_1^2 \quad \text{and} \quad C_2 = 5.2 + 0.04q_2^2,$$
    where q_1 and q_2 are the quantities produced by each plant. The total quantity demanded, $q = q_1 + q_2$, is related to the price, p, by
    $$p = 60 - 0.04q.$$
    How much should each plant produce in order to maximize the company's profit? [5]

---

[5] Adapted from M. Rosser, *Basic Mathematics for Economists*, p. 318 (New York: Routledge, 1993).

10. Assume that two products are manufactured in quantities q_1 and q_2 and sold at prices of p_1 and p_2 respectively, and that the cost of producing them is given by

$$C = 2q_1^2 + 2q_2^2 + 10.$$

(a) Find the maximum profit that can be made, assuming the prices are fixed.

(b) Find the rate of change of that maximum profit as p_1 increases.

11. A missile has a remote guidance device which is sensitive to both temperature and humidity. If t is the temperature in $°C$ and h is percent humidity, the range over which the missile can be controlled is given by:

$$\text{Range in km} = 27{,}800 - 5t^2 - 6ht - 3h^2 + 400t + 300h,$$

What are the optimal atmospheric conditions for controlling the missile?

12. Some items are sold at different prices to different groups of people. For example, there are sometimes discounts for senior citizens or for children. The reason is that these groups may be more sensitive to price, so a discount will have greater impact on their purchasing decisions. The seller faces an optimization problem: How large a discount to offer in order to maximize profits?

A theater can sell q_c child tickets and q_a adult tickets at prices p_c and p_a, according to the following demand functions:

$$q_c = rp_c^{-4} \quad \text{and} \quad q_a = sp_a^{-2},$$

and has operating costs proportional to the total number of tickets sold. What should be the relative price of children's and adults' tickets?

13. Compute the regression line for the points $(-1, 2)$, $(0, -1)$, $(1, 1)$ using least squares.

When data is not linear, it can sometimes be transformed in such a way that it looks more linear. For example, suppose we expect that data points (x, y) lie approximately on an exponential curve, say

$$y = Ce^{ax},$$

where a and C are constants. Taking the natural log of both sides, we find that $\ln y$ is a linear function of x.

$$\ln y = ax + \ln C.$$

To find a and C, we use least squares for $\ln y$ against x. Use this method in Problems 14–15.

14. The population of the United States was about 180 million in 1960, grew to 206 million in 1970, and 226 million in 1980.

(a) Based on this data and assuming that the population was growing at an exponential rate, use the method of least squares to estimate the population in 1990.

(b) According to the national census, the 1990 population was 249 million. What does this say about the assumption of exponential growth?

(c) Predict the population in the year 2010.

15. The data in Table 14.3 shows the cost of a first class stamp in the US over the last 70 years.

TABLE 14.3 *Cost of a first class stamp*

Year	1920	1932	1958	1963	1968	1971	1974
Postage	0.02	0.03	0.04	0.05	0.06	0.08	0.10
Year	1975	1978	1981	1985	1988	1991	1995
Postage	0.13	0.15	0.20	0.22	0.25	0.29	0.32

(a) Find the line of best fit through the data. Using this line, predict the cost of a postage stamp in the year 2010.

(b) Plot the data. Does it look linear?

(c) Plot the year against the natural logarithm of the price: Does this look linear? If it is linear, what does that tell you about the price of a stamp as a function of time? Find the line of best fit through this data, and use your answer to again predict the cost of a postage stamp in the year 2010.

16. Design a rectangular milk carton box of width w, length l, and height h which holds 512 cm^3 of milk. The sides of the box cost 1 cent/cm^2 and the top and bottom cost 2 cent/cm^2. Find the dimensions of the box that minimize the total cost of materials used.

17. An international airline has a regulation that each passenger can carry a suitcase having the sum of its width, length and height less than or equal to 135 cm. Find the dimensions of the suitcase of maximum volume that a passenger may carry under this regulation.

18. A company manufactures a product which requires capital and labor to produce. The quantity, Q, of the product manufactured is given by the Cobb-Douglas production function

$$Q = AK^a L^b,$$

where K is the quantity of capital and L is the quantity of labor used and A, a, and b are positive constants with $0 < a < 1$ and $0 < b < 1$. Suppose one unit of capital costs $\$k$ and one unit of labor costs $\$\ell$. The price of the product is fixed at $\$p$ per unit.

   (a) If $a + b < 1$, how much capital and labor should the company use to maximize its profit?
   (b) Is there a maximum profit in the case $a + b = 1$? What about $a + b \geq 1$? Explain.

   [Note: See page 589 for a discussion of the Cobb-Douglas production function. The three cases considered above, namely $a + b < 1$, $a + b = 1$, and $a + b > 1$ are, respectively, the cases of *decreasing returns to scale*, *constant returns to scale*, and *increasing returns to scale*.]

19. Show analytically that the function $f(x, y)$ in Example 3 has a local minimum at $(2, 1)$.

20. We wish to find the minimum value of

$$f(x, y) = (x + 1)^4 + (y - 1)^4 + \frac{1}{x^2 y^2 + 1}.$$

   (a) Use a computer to plot the contour diagram for f.
   (b) Minimize f using the gradient search method.

21. The government wants to build a pipe that will pump water up from a dam to a reservoir, as in Figure 14.27. The cost, C, (in millions of dollars) will depend on the diameter, d, of the pipe (in meters) and the number, n, of pumping stations, according to the following formula[6]:

$$C = 0.15n + 3\left(\frac{4d}{5}\right)^{-4.87} + \left(\frac{4d}{5}\right)^{1.8} + 3\left(\frac{4d}{5}\right)^{1.8} n^{-1}.$$

Using the gradient search method, find the optimal number of pumping stations and pipe diameter.

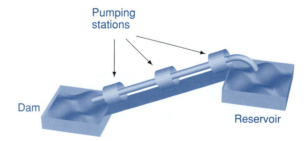

Figure 14.27

22. Consider the function given by $f(x, y) = x^2(y + 1)^3 + y^2$. Show that f has only one critical point, namely $(0, 0)$, and that point is a local minimum but not a global minimum. Contrast this with the case of a function with a single local minimum in one-variable calculus.

---

[6]From Douglass J. Wilde, *Globally Optimal Design*, (New York: John Wiley & Sons, 1978).

14.3 CONSTRAINED OPTIMIZATION: LAGRANGE MULTIPLIERS

Many, perhaps most, real optimization problems are constrained by external circumstances. For example, a city wanting to build a public transportation system has only a limited number of tax dollars it can spend on the project. In this section, we will see how to find an optimum value under such constraints.

Graphical Approach: Maximizing Production Subject to a Budget Constraint

Suppose we want to maximize the production of a firm under a budget constraint. Suppose production, f, is a function of two variables, x and y, which are quantities of two raw materials, and that

$$f(x, y) = x^{2/3} y^{1/3}.$$

If x and y are purchased at prices of p_1 and p_2 thousands of dollars per unit, what is the maximum production f that can be obtained with a budget of c thousand dollars?

To maximize f without regard to the budget, we simply increase x and y. However, the budget constraint prevents us from increasing x and y beyond a certain point. Exactly how does the budget constrain us? With prices of p_1 and p_2, the amount spent on x is $p_1 x$ and the amount spent on y is $p_2 y$, so we must have

$$g(x, y) = p_1 x + p_2 y \le c,$$

where $g(x, y)$ is the total cost of the raw materials x and y and c is the budget in thousands of dollars.

Let's look at the case when $p_1 = p_2 = 1$ and $c = 3.78$. Then

$$x + y \le 3.78.$$

Figure 14.28 shows some contours of f and the budget constraint represented by the line $x + y = 3.78$. Any point on or below the line represents a pair of values of x and y that we can afford. A point on the line completely exhausts the budget, a point below the line represents values of x and y which can be bought without using up the budget. Any point above the line represents a pair of values that we cannot afford. To maximize f, we find the point which lies on the level curve with the largest possible value of f *and* which lies within the budget. The point must lie on the budget constraint because we should spend all the available money. Unless we are at the point where the budget constraint is tangent to the contour $f = 2$, we can increase f by moving along the line representing the budget constraint in Figure 14.28. For example, if we are on the line to the left of the point of tangency, moving right will increase f; if we are on the line to the right of the point of tangency, moving left will increase f. Thus, the maximum value of f on the budget constraint occurs at the point where the budget constraint is tangent to the contour $f = 2$.

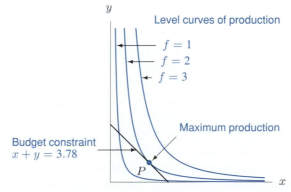

Figure 14.28: Optimal point, P, where budget constraint is tangent to a level of production function

Analytical Solution: Lagrange Multipliers

We know that maximum production is achieved at the point where the budget constraint is tangent to a level curve of the production function. The method of Lagrange multipliers uses this fact in algebraic form. Figure 14.29 shows that at the optimum point, P, the gradient of f and the normal to the budget line $g(x, y) = 3.78$ are parallel. Thus, at P, grad f and grad g are parallel, so for some scalar λ, called the *Lagrange multiplier*

$$\text{grad } f = \lambda \text{ grad } g$$

Since grad $f = \left(\frac{2}{3}x^{-1/3}y^{1/3}\right)\vec{i} + \left(\frac{1}{3}x^{2/3}y^{-2/3}\right)\vec{j}$ and grad $g = \vec{i} + \vec{j}$, we have, by equating components,

$$\frac{2}{3}x^{-1/3}y^{1/3} = \lambda \quad \text{and} \quad \frac{1}{3}x^{2/3}y^{-2/3} = \lambda.$$

Eliminating λ gives

$$\frac{2}{3}x^{-1/3}y^{1/3} = \frac{1}{3}x^{2/3}y^{-2/3}, \quad \text{which leads to} \quad 2y = x.$$

Since we must also satisfy the constraint $x + y = 3.78$, we have $x = 2.52$ and $y = 1.26$. For these values,

$$f(2.52, 1.26) = (2.52)^{2/3}(1.26)^{1/3} \approx 2.$$

Thus, as before, we see that the maximum value of f is approximately 2; we also learn that this maximum occurs at $x = 2.52$ and $y = 1.26$.

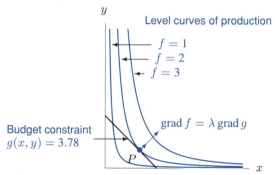

Figure 14.29: At the point, P, of maximum production, the vectors grad f and grad g are parallel

Lagrange Multipliers in General

Suppose we want to optimize an *objective function* $f(x, y)$ subject to a *constraint* $g(x, y) = c$. We consider only those points which satisfy the constraint and look for extrema among them. We make the following definition.

> Suppose P_0 is a point satisfying the constraint $g(x, y) = c$.
> - f has a **local maximum** at P_0 **subject to the constraint** if $f(P_0) \geq f(P)$ for all points P near P_0 satisfying the constraint.
> - f has a **global maximum** at P_0 **subject to the constraint** if $f(P_0) \geq f(P)$ for all points P satisfying the constraint.
> Local and global minima are defined similarly.

As we saw in the production example, constrained extrema occur at points where the contours of f are tangent to the contours of g; they can also occur at endpoints of the constraint. We sometimes find these extrema by substituting from the constraint into the objective function. However, the method of Lagrange multipliers works when substitution is not possible.

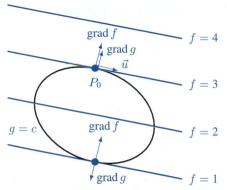

Figure 14.30: Maximum and minimum values
of $f(x, y)$ on $g(x, y) = c$ are at points where
grad f is parallel to grad g

At any point, grad f points in the direction in which f increases most rapidly. Suppose $\vec{u}$ is a unit vector tangent to the constraint. If grad $f \cdot \vec{u} > 0$, then the directional derivative, $f_{\vec{u}}$, is positive and moving in the direction of $\vec{u}$ increases f. If grad $f \cdot \vec{u} < 0$, then $f_{\vec{u}}$ is negative and moving in the direction of $-\vec{u}$ increases f. Thus, at a point P_0 where f has a constrained local maximum, we must have grad $f \cdot \vec{u} = 0$. Therefore, at P_0, both grad f and grad g are perpendicular to $\vec{u}$ so grad f and grad g are parallel. (See Figure 14.30.) Thus, provided grad $g \neq \vec{0}$ at P_0, we can use the following method:

To optimize f subject to the constraint $g = c$, solve the equations

$$\text{grad } f = \lambda \text{ grad } g \quad \text{and} \quad g = c$$

where λ is called the **Lagrange multiplier**.

If f and g are functions of two variables, the Lagrange method gives us three equations for three unknowns, x, y, λ:

$$f_x = \lambda g_x, \quad f_y = \lambda g_y, \quad g(x, y) = c.$$

If f and g are functions of three variables, the Lagrange method gives us four equations for four unknowns, x, y, z, λ:

$$f_x = \lambda g_x, \quad f_y = \lambda g_y, \quad f_z = \lambda g_z, \quad g(x, y, z) = c$$

Example 1   Find the maximum and minimum values of $x + y$ on the circle $x^2 + y^2 = 4$.

Solution   The objective function is

$$f(x, y) = x + y,$$

and the constraint is

$$g(x, y) = x^2 + y^2 = 4.$$

Since grad $f = f_x \vec{i} + f_y \vec{j} = \vec{i} + \vec{j}$ and grad $g = g_x \vec{i} + g_y \vec{j} = 2x\vec{i} + 2y\vec{j}$, then grad $f = \lambda$ grad g gives

$$1 = 2\lambda x,$$
$$1 = 2\lambda y,$$

so

$$x = y.$$

We also know that

$$x^2 + y^2 = 4,$$

giving $x = y = \sqrt{2}$ or $x = y = -\sqrt{2}$.

Since $f(x, y) = x + y$, the maximum value of f is $f(\sqrt{2}, \sqrt{2}) = 2\sqrt{2}$, and occurs when $x = y = \sqrt{2}$; the minimum value is $f(-\sqrt{2}, -\sqrt{2}) = -2\sqrt{2}$, and occurs when $x = y = -\sqrt{2}$. (See Figure 14.31.)

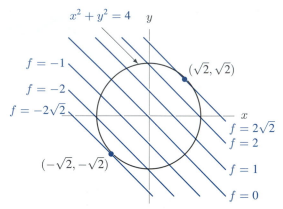

Figure 14.31: Maximum and minimum values of $f(x, y) = x + y$ on the circle $x^2 + y^2 = 4$ are at points where contours of f are tangent to the circle

How to Distinguish Maxima from Minima

There is a second derivative test[7] for classifying the critical points of constrained optimization problems, but it is more complicated than the test in Section 14.1. However, as you see from the examples, a graph of the constraint and some level curves can usually make it clear which points are maxima, which points are minima, and which are neither.

Optimization with Inequality Constraints

The production problem that we looked at first was to maximize production $f(x, y)$ subject to a budget constraint

$$g(x, y) = p_1 x + p_2 y \leq c.$$

This budget constraint is an inequality constraint, which restricts (x, y) to a region of the plane rather than to a curve in the plane. In principle, we should first check to see whether or not $f(x, y)$ has any critical points in the interior defined by

$$p_1 x + p_2 y < c.$$

However, in the case of a budget constraint, we can see that the maximum of f must occur when the budget is exhausted, and so we look for the maximum value of f on the boundary line:

$$p_1 x + p_2 y = c.$$

Strategy for Optimizing $f(x, y)$ Subject to the Constraint $g(x, y) \leq c$

- Find all points in the interior $g(x, y) < c$ where grad f is zero or undefined.
- Use Lagrange multipliers to find the local extrema of f on the boundary $g(x, y) = c$.
- Evaluate f at the points found in the previous two steps and compare the values.

From Section 14.2 we know that if f is continuous on a closed and bounded region, R, then f is guaranteed to attain its global maximum and minimum values on R.

---

[7]See J. E. Marsden and A. J. Tromba, *Vector Calculus*, 2nd ed., pp. 224–230 (San Francisco: W.H. Freeman, 1981).

Example 2   Find the maximum and minimum values of $f(x, y) = (x - 1)^2 + (y - 2)^2$ subject to the constraint $x^2 + y^2 \leq 45$.

Solution   First, we look for all critical points for f in the interior of the region. Setting

$$f_x = 2(x - 1) = 0$$
$$f_y = 2(y - 2) = 0$$

we find f has exactly one critical point at $x = 1$, $y = 2$. Since $1^2 + 2^2 < 45$, that critical point is in the interior of the region.

Next, we find the local extrema of f on the boundary curve $x^2 + y^2 = 45$. To do this, we use Lagrange multipliers with constraint $g(x, y) = x^2 + y^2 = 45$. Setting grad $f = \lambda$ grad g, we get

$$2(x - 1) = \lambda \cdot 2x,$$
$$2(y - 2) = \lambda \cdot 2y.$$

If $\lambda = 0$, then $x = 1$, $y = 2$, the interior critical point. Thus, on the boundary we have

$$\frac{x}{x - 1} = \frac{y}{y - 2}$$

and so

$$y = 2x.$$

Combining this with the constraint $x^2 + y^2 = 45$, we get

$$5x^2 = 45$$

so

$$x = \pm 3.$$

Since $y = 2x$, we have possible local extrema at $x = 3$, $y = 6$ and $x = -3$, $y = -6$.

We conclude that the only candidates for the maximum and minimum values of f in the region occur at $(1, 2)$, $(3, 6)$, and $(-3, -6)$. Evaluating f at these three points we find

$$f(1, 2) = 0, \qquad f(3, 6) = 20, \qquad f(-3, -6) = 80.$$

Therefore, the minimum value of f is 0 at $(1, 2)$ and the maximum value is 80 at $(-3, -6)$.

Problems for Section 14.3

In Problems 1–17, use Lagrange multipliers to find the maximum and minimum values of $f(x, y)$ subject to the given constraints.

1. $f(x, y) = x + y, \quad x^2 + y^2 = 1$

2. $f(x, y) = 3x - 2y, \quad x^2 + 2y^2 = 44$

3. $f(x, y) = x^2 + y, \quad x^2 - y^2 = 1$

4. $f(x, y) = xy, \quad 4x^2 + y^2 = 8$

5. $f(x, y) = x^2 + y^2, \quad x^4 + y^4 = 2$

6. $f(x, y) = x^2 - xy + y^2, \quad x^2 - y^2 = 1$

7. $f(x, y, z) = x + 3y + 5z, \quad x^2 + y^2 + z^2 = 1$

8. $f(x, y, z) = 2x + y + 4z, \quad x^2 + y + z^2 = 16$

9. $f(x, y, z) = x^2 - y^2 - 2z, \quad x^2 + y^2 = z$

10. $f(x, y, z) = x^2 - 2y + 2z^2, \quad x^2 + y^2 + z^2 = 1$

11. $f(x, y, z) = x + y + z$, subject to $x^2 + y^2 + z^2 = 1$ and $x - y = 1$

12. $f(x, y) = x^2 + 2y^2, \quad x^2 + y^2 \leq 4$

13. $f(x, y) = xy, \quad x^2 + 2y^2 \leq 1$

14. $f(x, y) = x^2 - y^2, \quad x^2 \geq y$

15. $f(x, y) = x + 3y, \quad x^2 + y^2 \leq 2$

16. $f(x, y) = x^3 + y, \quad x + y \geq 1$

17. $f(x, y) = x^3 - y^2, \quad x^2 + y^2 \leq 1$

18. A company manufactures a product using inputs x, y, and z according to the production function

$$Q(x, y, z) = 20x^{1/2}y^{1/4}z^{2/5}.$$

The prices per unit are \$20 for x, and \$10 for y, and \$5 for z. What quantity of each input should the company use in order to manufacture 1,200 products at minimum cost?[8]

19. Consider a firm which manufactures a commodity at two different factories. The total cost of manufacturing depends on the quantities, q_1 and q_2, supplied by each factory, and is expressed by the *joint cost function*, $C = f(q_1, q_2)$. Suppose the joint cost function is approximated by

$$f(q_1, q_2) = 2q_1^2 + q_1q_2 + q_2^2 + 500$$

and that the company's objective is to produce 200 units, at the same time minimizing production costs. How many units should be supplied by each factory?

20. An industry manufactures a product from two raw materials. The quantity produced, Q, can be given by the Cobb-Douglas function:

$$Q = cx^ay^b,$$

where x and y are quantities of each of the two raw materials used and a, b, and c are positive constants. Suppose the first raw material costs \$$P_1$ per unit and the second costs \$$P_2$ per unit. Find the maximum production possible if no more than \$$K$ can be spent on raw materials.

21. Each person tries to balance his or her time between leisure and work. The tradeoff is that as you work less your income falls. Therefore each person has *indifference curves* which connect the number of hours of leisure, l, and income, s. If, for example, you are indifferent between 0 hours of leisure and an income of \$1125 a week on the one hand, and 10 hours of leisure and an income of \$750 a week on the other hand, then the points $l = 0$, $s = 1125$, and $l = 10$, $s = 750$ both lie on the same indifference curve. Table 14.4 gives information on three indifference curves, I, II, and III.

TABLE 14.4

Weekly income			Weekly leisure hours		
I	II	III	I	II	III
1125	1250	1375	0	20	40
750	875	1000	10	30	50
500	625	750	20	40	60
375	500	625	30	50	70
250	375	500	50	70	90

(a) Sketch the three indifference curves on graph paper.
(b) Suppose you have 100 hours a week available for work and leisure combined, and that you earn \$10/hour. Write an equation in terms of l and s which represents this constraint.
(c) On the same graph paper, sketch a graph of this constraint.
(d) Estimate from the graph what combination of leisure hours and income you would choose under these circumstances. Give the corresponding number of hours per week you would work. Explain how you made this estimate.

22. Figure 14.32 shows ∇f for a function $f(x, y)$ and two curves $g(x, y) = 1$ and $g(x, y) = 2$. Notice that $g = 1$ is the inside curve and $g = 2$ is the outside curve. Mark the following points on a copy of the figure.

(a) The point(s) A where f has a local maximum.
(b) The point(s) B where f has a saddle point.
(c) The point C where f has a maximum on $g = 1$.
(d) The point D where f has a minimum on $g = 1$.

---

[8] Adapted from M. Rosser, *Basic Mathematics for Economists*, p.363 (New York: Routledge, 1993).

(e)   If you used Lagrange multipliers to find C, what would the sign of λ be? Why?

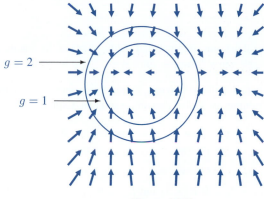

Figure 14.32

23.   Design a closed cylindrical container which holds 100 cm^3 and has the minimal possible surface area. What should its dimensions be?

24.   A mountain climber at the summit of a mountain wants to descend to a lower altitude as fast as possible. Suppose the altitude of the mountain is given approximately by

$$h(x,y) = 3000 - \frac{1}{10000}(5x^2 + 4xy + 2y^2) \quad \text{meters,}$$

where x, y are horizontal coordinates on the earth (in meters), with the mountain summit located above the origin. In thirty minutes, the climber can reach any point (x, y) on a circle of radius 1000 m. In which direction should she travel in order to descend as far as possible?

25.   Minimize

$$f(x, y, z) = \sqrt{(x-a)^2 + (y-b)^2 + (z-c)^2},$$

subject to the constraint $Ax + By + Cz + D = 0$. What is the geometric meaning of your solution?

26.   Let $f(x, y)$ be a linear function, so that $f(x, y) = ax + by + c$ where a, b and c are constants, and let R be a region in the xy-plane.

(a)   If R is any disk, show that the maximum and minimum values of f on R occur on the boundary of the disk.

(b)   If R is any rectangle, show that the maximum and minimum values of f on R occur at the corners of the rectangle. They may occur at other points of the rectangle as well.

(c)   Explain, with the aid of a graph of the plane $z = f(x, y)$, why you expect the answers you obtained in parts (a) and (b).

CHAPTER SUMMARY

- **Unconstrained Optimization**

  Definitions of local and global extrema, critical points, saddle points, finding critical points algebraically, behavior of contours near critical points, second derivative test to classify critical points, existence of global extrema.

  Method of Least Squares, gradient search.

- **Constrained Optimization**

  Geometric interpretation of Lagrange Multiplier method, solving Lagrange Multiplier problems algebraically, inequality constraints.

REVIEW PROBLEMS FOR CHAPTER FOURTEEN

1. Find the local maxima, minima, and saddle points of the function
$$f(x, y) = \sin x + \sin y + \sin(x + y), \quad 0 < x < \pi, \quad 0 < y < \pi.$$

For Problems 2–4, find the local maxima, local minima, and saddle points of the functions given. Decide if the local maxima or minima are global maxima or minima. Explain.

2. $f(x, y) = x^2 + y^3 - 3xy$

3. $f(x, y) = xy + \ln x + y^2 - 10 \quad (x > 0)$

4. $f(x, y) = x + y + \dfrac{1}{x} + \dfrac{4}{y}$

5. Suppose $f_x = f_y = 0$ at $(1, 3)$ and $f_{xx} < 0$, $f_{yy} < 0$, $f_{xy} = 0$. Draw a possible contour diagram.

6. Find the least squares line for the data points $(0, 4)$, $(1, 3)$, $(2, 1)$.

7. Find the minimum and maximum of the function $z = 4x^2 - xy + 4y^2$ over the closed disk $x^2 + y^2 \leq 2$.

8. What are the maximum and minimum values of $f(x, y) = -3x^2 - 2y^2 + 20xy$ on the line $x + y = 100$?

9. A company sells two products which are partial substitutes for each other, such as coffee and tea. If the price of one product rises, then the demand for the other product rises. The quantities demanded, q_1 and q_2, are given as a function of the prices, p_1 and p_2, by

$$q_1 = 517 - 3.5p_1 + 0.8p_2 \quad \text{and} \quad q_2 = 770 - 4.4p_2 + 1.4p_1.$$

What prices should the company charge in order to maximize the total sales revenue? [9]

10. A biological rule of thumb states that as the area A of an island increases tenfold, the number of animal species, N, living on it doubles. Table 14.5 shows the area (in square km) of several islands in the West Indies and the number of species living on each one. Assume that N is a power function of A. Using the biological rule of thumb, find

    (a) N as a function of A.
    (b) $\ln N$ as a function of $\ln A$.

    (c) Using the data given, tabulate $\ln N$ against $\ln A$ and find the line of best fit. Does your answer agree with the biological rule of thumb?

TABLE 14.5 *Number of species on various islands*

Island	Area (sq km)	Number
Redonda	3	5
Saba	20	9
Montserrat	192	15
Puerto Rico	8858	75
Jamaica	10854	70
Hispaniola (Haiti & Dominican Rep.)	75571	130
Cuba	113715	125

11. Suppose that the quantity, Q, manufactured of a certain product depends on the quantity of labor, L, and of capital, K, used according to the function

$$Q = 900L^{1/2}K^{2/3}.$$

Suppose that labor costs \$100 per unit and that capital costs \$200 per unit. What combination of labor and capital should be used to produce 36,000 units of the goods at minimum cost? What is that minimum cost?

---

[9] Adapted from M. Rosser, *Basic Mathematics for Economists*, p. 318 (New York: Routledge, 1993).

12. An international organization must decide how to spend the $2000 they have been allotted for famine relief in a remote area. They expect to divide the money between buying rice at $5/sack and beans at $10/sack. The number, P, of people who would be fed if they buy x sacks of rice and y sacks of beans is given by

$$P = x + 2y + \frac{x^2 y^2}{2 \cdot 10^8}.$$

What is the maximum number of people that can be fed, and how should the organization allocate its money?

13. The quantity, Q, of a product manufactured by a company is given by

$$Q = aK^{0.6}L^{0.4},$$

where a is a positive constant, K is the quantity of capital and L is the quantity of labor used. Capital costs are $20 per unit, labor costs are $10 per unit, and the company wants costs for capital and labor combined to be no higher than $150. Suppose you are asked to consult for the company, and learn that 5 units each of capital and labor are being used.

  (a) What do you advise? Should the plant use more or less labor? More or less capital? If so, by how much?
  (b) Write a one sentence summary that could be used to sell your advice to the board of directors.

14. The Cobb-Douglas equation models the total quantity, q, of a commodity produced as a function of the number of workers, W, and the amount of capital invested, K, by the production function

$$q = cW^{1-a}K^a$$

where a and c are positive constants. Assume labor costs are $\$p_1$ per worker, capital costs are $\$p_2$ per unit, and there is a fixed budget of $\$b$. Show that when W and K are at their optimal levels, the ratio of marginal productivity of labor to marginal productivity of capital equals the ratio of the cost of one unit of labor to one unit of capital.

15. An irrigation canal has a trapezoidal cross-section of area 50 m^2, as in Figure 14.33. The average flow rate in the canal is inversely proportional to the wetted perimeter, p, of the canal, that is, to the perimeter of the trapezoid in Figure 14.33, excluding the top. Thus, to maximize the flow rate we must minimize p. Find the depth d, base width w, and angle θ that give the maximum flow rate.[10]

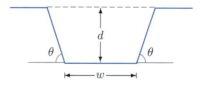

Figure 14.33

16. Find the minimum distance from the point $(1, 2, 10)$ to the paraboloid given by the equation $z = x^2 + y^2$. Give a geometric justification for your answer.

17. The energy required to compress a gas from pressure p_1 to pressure p_{N+1} in N stages is proportional to

$$E = \left(\frac{p_2}{p_1}\right)^2 + \left(\frac{p_3}{p_2}\right)^2 + \cdots + \left(\frac{p_{N+1}}{p_N}\right)^2 - N.$$

Show how to choose the intermediate pressures $p_2, \ldots, p_N$ so as to minimize the energy requirement. [11]

[10] Adapted from Robert M. Stark and Robert L. Nichols, *Mathematical Foundations of Design: Civil Engineering Systems*, (New York: McGraw-Hill, 1972).

[11] Adapted from Rutherford, Aris, *Discrete Dynamic Programming*, p. 35 (New York: Blaisdell, 1964).

18. A family wants to move to a house at a point that is better situated with respect to the children's school and both the parents' places of work. See Figure 14.34.

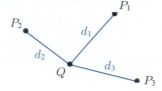

Figure 14.34

Currently they live at Q, the school is at P_1, the mother's work is at P_2, and the father's work is at P_3. They want to minimize $d = d_1 + d_2 + d_3$, where

$$d_1 = \text{distance to school},$$
$$d_2 = \text{distance to mother's work},$$
$$d_3 = \text{distance to father's work}.$$

(a) Show that $\operatorname{grad} d_i$ is a unit vector pointing directly away from P_i, for $i = 1, 2, 3$. [Hint: Draw contours of d_i, paying careful attention to the spacing of the contours.]

(b) Use your answer to part (a) to draw $\operatorname{grad} d$ at the point Q in Figure 14.34. In what direction should the family move to decrease d?

(c) Find as best you can the point on the diagram where $\operatorname{grad} d = 0$. What geometric condition characterizes this location?

19. Consider the function $f(x, y) = 2x^2 - 3xy + 8y^2 + x - y$.

(a) Compute the critical points of f and classify them.

(b) By completing the square, plot the contour diagram of f and show that the local extremum found in part (a) is a global one.

(c) Starting at the point $(1, 1)$, minimize f using the gradient search method and compare your answer after two iterations with your answer from part (a).

PROJECTS

1. **Hockey and Entropy**

Thirty teams compete for the Stanley Cup in the National Hockey League (after expansion in 2000). At the beginning of the season an experienced fan estimates that the probability that team i will win is some number p_i, where $0 \leq p_i \leq 1$ and

$$\sum_{i=1}^{30} p_i = 1.$$

Exactly one team will actually win, so the probabilities have to add to 1. If one of the teams, say team i, is certain to win then p_i is equal to 1 and all the other p_j are zero. Another extreme case occurs if all the teams are equally likely to win, so all the p_i are equal to $1/30$, and the outcome of the hockey season is completely unpredictable. Thus, the *uncertainty* in the outcome of the hockey season depends on the probabilities $p_1, \ldots, p_{30}$. In this problem we measure this uncertainty quantitatively using the following function:

$$S(p_1, \ldots, p_{30}) = -\sum_{i=1}^{30} p_i \frac{\ln p_i}{\ln 2}.$$

Note that as $p_i \leq 1$, we have $-\ln p_i \geq 0$ and hence $S \geq 0$.

(a) Show that $\lim_{p \to 0} p \ln p = 0$. (This means that S is a continuous function of the p_i, where $0 \le p_i \le 1$ and $1 \le i \le 30$, if we set $p \ln p|_{p=0}$ equal to zero. Since S is then a continuous function on a closed and bounded region, it attains a maximum and a minimum value on this region.)

(b) Find the maximum value of $S(p_1, \ldots, p_{30})$ subject to the constraint $p_1 + \cdots + p_{30} = 1$. What are the values of p_i in this case? What does your answer mean in terms of the uncertainty in the outcome of the hockey season?

(c) Find the minimum value of $S(p_1, \ldots, p_{30})$, subject to the constraint $p_1 + \cdots + p_{30} = 1$. What are the values of p_i in this case? What does your answer mean in terms of the uncertainty in the outcome of the hockey season?

[Note: The function S is an example of an *entropy* function; the concept of entropy is used in information theory, statistical mechanics, and thermodynamics when measuring the uncertainty in an experiment (the hockey season in this problem) or physical system.]

2. **Fitting a Line to Data Using Least Squares**
    In this problem you will derive the general formulas for the slope and y-intercept of a least squares line. Assume that you have n data points $(x_1, y_1), (x_2, y_2), \cdots, (x_n, y_n)$. Let the equation of the least squares line be $y = b + mx$.

(a) For each data point (x_i, y_i), show that the corresponding point directly above or below it on the least squares line has y-coordinate $b + mx_i$.

(b) For each data point (x_i, y_i), show that the square of the vertical distance from it to the point found in part (a) is $(y_i - (b + mx_i))^2$.

(c) Form the function $f(b, m)$ which is the sum of all of the n squared distances found in part (b). That is,

$$f(b, m) = \sum_{i=1}^{n} (y_i - (b + mx_i))^2.$$

Show that the partial derivatives $\dfrac{\partial f}{\partial b}$ and $\dfrac{\partial f}{\partial m}$ are given by

$$\frac{\partial f}{\partial b} = -2 \sum_{i=1}^{n} (y_i - (b + mx_i))$$

and

$$\frac{\partial f}{\partial m} = -2 \sum_{i=1}^{n} (y_i - (b + mx_i)) \cdot x_i.$$

(d) Show that the critical point equations $\dfrac{\partial f}{\partial b} = 0$ and $\dfrac{\partial f}{\partial m} = 0$ lead to a pair of simultaneous linear equations in b and m:

$$nb + \left(\sum x_i \right) m = \sum y_i$$

$$\left(\sum x_i \right) b + \left(\sum x_i^2 \right) m = \sum x_i y_i$$

(e) Solve the equations in part (d) for b and m, getting

$$b = \left(\sum_{i=1}^{n} x_i^2 \sum_{i=1}^{n} y_i - \sum_{i=1}^{n} x_i \sum_{i=1}^{n} x_i y_i \right) \Big/ \left(n \sum_{i=1}^{n} x_i^2 - \left(\sum_{i=1}^{n} x_i \right)^2 \right)$$

$$m = \left(n \sum_{i=1}^{n} x_i y_i - \sum_{i=1}^{n} x_i \sum_{i=1}^{n} y_i \right) \Big/ \left(n \sum_{i=1}^{n} x_i^2 - \left(\sum_{i=1}^{n} x_i \right)^2 \right)$$

(f) Apply the formulas of part (e) to the data points $(1, 1), (2, 1), (3, 3)$ to verify that you get the same result as in Example 2 on page 729.

FOCUS ON MODELING

THE LAGRANGIAN AND ITS INTERPRETATION

In the uses of Lagrange multipliers so far, we never found (or needed) the value of λ. However, λ does have a practical interpretation. We will first consider this interpretation in the production example on page 737 of Section 14.3. Then we will give a general interpretation.

The Meaning of λ in the Production Example

The production function to be maximized was

$$f(x, y) = x^{2/3} y^{1/3}$$

subject to the constraint

$$g(x, y) = x + y = 3.78.$$

We solved the equations

$$\frac{2}{3} x^{-1/3} y^{1/3} = \lambda,$$

$$\frac{1}{3} x^{2/3} y^{-2/3} = \lambda,$$

$$x + y = 3.78,$$

to get $x = 2.52, y = 1.26$ and $f(2.52, 1.26) \approx 2$. Continuing to find λ gives us

$$\lambda \approx 0.53.$$

Suppose now we did another, apparently unrelated calculation. Suppose our budget is increased slightly, from 3.78 to 4.78, giving a new budget constraint of $x + y = 4.78$. Then the corresponding solution is at $x = 3.19$ and $y = 1.59$ and the new maximum value (instead of $f = 2$) is

$$f = (3.19)^{2/3}(1.59)^{1/3} \approx 2.53.$$

Notice that the amount by which f has increased is 0.53, the value of λ. Thus, in this example, the value of λ represents the extra production achieved by increasing the budget by one — in other words, the extra "bang" you get for an extra "buck" of budget. In fact, this is true in general:

- The value of λ is approximately the increase in the optimum value of f when the budget is increased by 1 unit.

More precisely:

- The value of λ represents the rate of change of the optimum value of f as the budget increases.

The Meaning of λ in General

To interpret λ, we look at how the optimum value of the objective function f changes as the value c of the constraint function g is varied. The optimum point (x_0, y_0) will, in general, depend on the constraint value c. So, provided x_0 and y_0 are differentiable functions of c, we can use the chain rule to differentiate the optimum value $f(x_0(c), y_0(c))$ with respect to c:

$$\frac{df}{dc} = \frac{\partial f}{\partial x} \frac{dx_0}{dc} + \frac{\partial f}{\partial y} \frac{dy_0}{dc}.$$

At the optimum point (x_0, y_0), we have $f_x = \lambda g_x$ and $f_y = \lambda g_y$, and therefore

$$\frac{df}{dc} = \lambda \left(\frac{\partial g}{\partial x} \frac{dx_0}{dc} + \frac{\partial g}{\partial y} \frac{dy_0}{dc} \right) = \lambda \frac{dg}{dc}.$$

But, as $g(x_0(c), y_0(c)) = c$, we see that $dg/dc = 1$, and so $df/dc = \lambda$. Thus, we have the following interpretation of the Lagrange multiplier λ:

The value of λ is the rate of change of the optimum value of f as c increases (where $g(x, y) = c$). If the optimum value of f is written as $f(x_0(c), y_0(c))$, then we have

$$\frac{d}{dc} f(x_0(c), y_0(c)) = \lambda.$$

Example 1   Suppose the quantity of goods produced according to the function $f(x, y) = x^{2/3} y^{1/3}$ is maximized subject to the budget constraint $x + y \leq 3.78$. Suppose the budget is increased to allow for a small increase in production. What price must the product sell for if it is to be worth the increased budget for its production?

Solution   We know that $\lambda = 0.53$, which tells us that $df/dc = 0.53$. Therefore increasing the budget by $1 increases production by about 0.53 unit. In order to make the increase in budget profitable, the extra goods produced must sell for more than \$1. Thus, if p is the price of each unit of the good, then $0.53p$ is the revenue from the extra 0.53 units sold. Thus, we need $0.53p \geq 1$ so $p \geq 1/0.53 = \$1.89$.

The Lagrangian Function

Constrained optimization problems are frequently solved using a *Lagrangian function*, $\mathcal{L}$, particularly in economics. For example, to optimize the function $f(x, y)$ subject to the constraint $g(x, y) = c$, we use the Lagrangian function

$$\mathcal{L}(x, y, \lambda) = f(x, y) - \lambda(g(x, y) - c).$$

To see why the function $\mathcal{L}$ is useful, compute the partial derivatives of $\mathcal{L}$:

$$\frac{\partial \mathcal{L}}{\partial x} = \frac{\partial f}{\partial x} - \lambda \frac{\partial g}{\partial x},$$
$$\frac{\partial \mathcal{L}}{\partial y} = \frac{\partial f}{\partial y} - \lambda \frac{\partial g}{\partial y},$$
$$\frac{\partial \mathcal{L}}{\partial \lambda} = -(g(x, y) - c).$$

Notice that if (x_0, y_0) is a critical point of $f(x, y)$ subject to the constraint $g(x, y) = c$ and λ_0 is the corresponding Lagrange multiplier, then at the point (x_0, y_0, λ_0) we have

$$\frac{\partial \mathcal{L}}{\partial x} = 0 \quad \text{and} \quad \frac{\partial \mathcal{L}}{\partial y} = 0 \quad \text{and} \quad \frac{\partial \mathcal{L}}{\partial \lambda} = 0.$$

In other words, (x_0, y_0, λ_0) is a critical point for the unconstrained problem of optimization of the Lagrangian, $\mathcal{L}(x, y, \lambda)$. Thus, the Lagrangian enables us to convert a constrained optimization problem to an unconstrained problem.

Example 2   A company has a production function with three inputs x, y, and z given by

$$f(x, y, z) = 50x^{2/5} y^{1/5} z^{1/5}.$$

The total budget is \$24,000 and the company can buy x, y, and z at \$80, \$12, and \$10 per unit, respectively. What combination of inputs will maximize production? [12]

---

[12] Adapted from M. Rosser, *Basic Mathematics for Economists*, p. 363 (New York: Routledge, 1993).

Solution   We need to maximize the objective function

$$f(x, y, z) = 50x^{2/5}y^{1/5}z^{1/5},$$

subject to the constraint

$$g(x, y, z) = 80x + 12y + 10z = 24,000.$$

Therefore, the Lagrangian function is

$$\mathcal{L}(x, y, z) = 50x^{2/5}y^{1/5}z^{1/5} - \lambda(80x + 12y + 10z - 24,000),$$

and so we look for solutions to the system of equations we get from grad $\mathcal{L} = 0$:

$$\frac{\partial \mathcal{L}}{\partial x} = 20x^{-3/5}y^{1/5}z^{1/5} - 80\lambda = 0,$$

$$\frac{\partial \mathcal{L}}{\partial y} = 10x^{2/5}y^{-4/5}z^{1/5} - 12\lambda = 0,$$

$$\frac{\partial \mathcal{L}}{\partial z} = 10x^{2/5}y^{1/5}z^{-4/5} - 10\lambda = 0,$$

$$\frac{\partial \mathcal{L}}{\partial \lambda} = -(80x + 12y + 10z - 24,000) = 0.$$

We simplify this system to give

$$\lambda = \frac{1}{4}x^{-3/5}y^{1/5}z^{1/5},$$

$$\lambda = \frac{5}{6}x^{2/5}y^{-4/5}z^{1/5},$$

$$\lambda = x^{2/5}y^{1/5}z^{-4/5},$$

$$80x + 12y + 10z = 24,000.$$

Eliminating z from the first two equations gives $x = 0.3y$. Eliminating x from the second and third equations gives $z = 1.2y$. Substituting for x and z into $80x + 12y + 10z = 24,000$ gives

$$80(0.3y) + 12y + 10(1.2y) = 24,000,$$

so $y = 500$. Hence we get $x = 150$ and $z = 600$, and the corresponding value of f is $f(150, 500, 600) = 4,622$ units.

The graph of the constraint is a plane in 3-space. Since the inputs x, y, z must be nonnegative, the constraint is a triangle in the first quadrant, with edges on the coordinate planes. On the boundary of the triangle, one (or more) of the variables x, y, z is zero and so the function f is zero. Thus, $x = 150$, $y = 500$, $z = 600$ is a maximum.

Interpretation of the Lagrangian Function

It may seem surprising that the Lagrangian function transforms a constrained optimization problem into an unconstrained problem. An understanding of the Lagrangian function suggests why this works. Suppose $f(x, y)$ is the cost of running a factory that uses quantities x of labor and y of raw material; suppose $g(x, y)$ is the quantity produced by the factory. The factory's owners wish to produce a fixed amount, c, at the minimum cost.

View $\mathcal{L}(x, y, \lambda)$ as a modification of $f(x, y)$, the cost function to be minimized. In particular, $\mathcal{L}(x, y, \lambda)$ is equal to $f(x, y)$ plus a *penalty term* $-\lambda(g(x, y) - c)$ that depends on the degree to which the constraint is violated. So we can think of $\mathcal{L}(x, y, \lambda)$ as a modified cost.

If the constraint $g(x, y) = c$ is satisfied, the penalty term is 0 and $\mathcal{L}(x, y, \lambda)$ is exactly $f(x, y)$. But if $g(x, y)$ is less than c (a cheater's way to minimize the cost by underproducing), then the penalty term is positive. If $g(x, y)$ is greater than c (overproducing while trying to minimize cost), then the penalty term is negative; it is a bonus term. The coefficient λ in the penalty term represents the penalty cost of each unit of underproduction. In other words, λ gives the strength of the penalty.

We know that the minimum cost for a fixed production level occurs where

$$\text{grad } f = \lambda \text{ grad } g \quad \text{and} \quad g(x, y) = c.$$

Now we must explain why the minimum cost occurs at a critical point of $\mathcal{L}(x, y, \lambda)$. This may be seen either by taking partial derivatives of $\mathcal{L}$ or by the following argument. If x and y change by Δx and Δy, then, by local linearity,

$$\Delta \mathcal{L} \approx (f_x \Delta x + f_y \Delta y) - \lambda (g_x \Delta x + g_y \Delta y)$$
$$= (\text{grad } f - \lambda \text{ grad } g) \cdot (\Delta x \vec{i} + \Delta y \vec{j}).$$

If (x, y, λ) is a critical point for $\mathcal{L}$, then, to the first order, we must have $\Delta \mathcal{L} = 0$ for all Δx and Δy. So we must have

$$\text{grad } f = \lambda \text{ grad } g.$$

The reason the unconstrained minimum of $\mathcal{L}(x, y, \lambda)$ occurs when the constraint $g(x, y) = c$ is satisfied lies in the form of the penalty term, $-\lambda(g(x, y) - c)$. If the unconstrained minimum of $\mathcal{L}$ were to occur at a point where the constraint were not satisfied, then the factor $(g(x, y) - c)$ in the penalty term would be nonzero. If this factor were nonzero, say positive, then $\mathcal{L}$ could be easily lowered by reducing λ. In other words, an unconstrained minimum of $\mathcal{L}(x, y, \lambda)$ cannot occur if the constraint $g(x, y) = c$ is not satisfied.

Problems on Economics and Lagrange Multipliers

1. A company manufactures x units of one item and y units of another. The total cost in dollars, C, of producing these two items is approximated by the function

$$C = 5x^2 + 2xy + 3y^2 + 800.$$

   (a) If the production quota for the total number of items (both types combined) is 39, find the minimum production cost.
   (b) Estimate the additional production cost or savings if the production quota is raised to 40 or lowered to 38.

2. Suppose the quantity, q, of a product manufactured depends on the number of workers, W, and the amount of capital invested, K, and is represented by the Cobb-Douglas function

$$q = 6W^{3/4} K^{1/4}.$$

   In addition, labor costs are \$10 per worker and capital costs are \$20 per unit, and the budget is \$3000.

   (a) What are the optimum number of workers and the optimum number of units of capital?
   (b) Show that at the optimum values of W and K, the ratio of the marginal productivity of labor $(\partial q / \partial W)$ to the marginal productivity of capital $(\partial q / \partial K)$ is the same as the ratio of the cost of a unit of labor to the cost of a unit of capital.
   (c) Recompute the optimum values of W and K when the budget is increased by one dollar. Check that increasing the budget by \$1 allows the production of λ extra units of the good, where λ is the Lagrange multiplier.

3. The director of a neighborhood health clinic has an annual budget of $600,000. He wants to allocate his budget so as to maximize the number of patient visits, V, which is given as a function of the number of doctors, D, and the number of nurses, N, by

$$V = 1000D^{0.6}N^{0.3}.$$

Doctors receive a salary of $40,000, while nurses get $10,000.

(a) Set up the director's constrained optimization problem.
(b) Describe, in words, the conditions which must be satisfied by $\partial V/\partial D$ and $\partial V/\partial N$ for V to have an optimum value.
(c) Solve the problem formulated in part (a).
(d) Find the value of the Lagrange multiplier and interpret its meaning in this problem.
(e) At the optimum point, what is the marginal cost of a patient visit (that is, the cost of an additional visit)? Will that marginal cost rise or fall with the number of visits? Why?

4. (a) In Problem 2, does the value of λ change if the budget changes from $3000 to $4000?
(b) In Problem 3, does the value of λ change if the budget changes from $600,000 to $700,000?
(c) What condition must a Cobb-Douglas production function

$$Q = cK^aL^b$$

satisfy to ensure that the marginal increase of production (that is, the rate of increase of production with budget) is not affected by the size of the budget?

CHAPTER FIFTEEN

INTEGRATING FUNCTIONS OF MANY VARIABLES

A definite integral is the limit of a sum. We use a definite integral to calculate the total population of a region given the population density as a function of position. If the population density is a function of one variable only (for example, the distance from the center of a city), we have an ordinary one-variable definite integral. If the density depends on more than one variable, we need a multivariable integral to calculate the total population. In this chapter we develop double and triple integrals in Cartesian and polar coordinates.

15.1 THE DEFINITE INTEGRAL OF A FUNCTION OF TWO VARIABLES

In this section we see how to estimate total population from population density in the plane. This leads to the definition of the definite integral of a function of two variables.

Population Density of Foxes in England

The fox population in parts of England is important to public health officials concerned about the disease rabies, which is spread by animals. The contour diagram in Figure 15.1 shows the population density $D = f(x, y)$ of foxes in the southwestern corner of England, where x and y are in kilometers from the southwest corner of the map and D is in foxes per square kilometer.[1] The bold contour is the coastline (approximately), and may be thought of as the $D = 0$ contour; clearly the density is zero outside it.

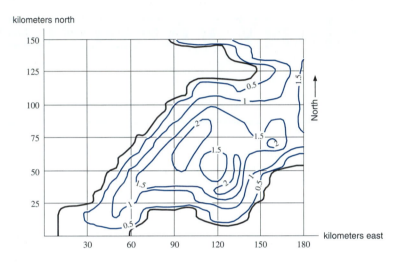

Figure 15.1: Population density of foxes in southwestern England

Example 1   Estimate the total fox population in the region represented by the map in Figure 15.1.

Solution   We want to find upper and lower bounds for the population. We subdivide the map into the 36 rectangles shown in Figure 15.1 and estimate the population in each rectangle. We find an upper bound for the population density in each rectangle, multiply it by the area of the rectangle to get an upper bound for the population in that rectangle, then add these upper bounds to get an upper bound for the total population. The bottom left rectangle includes a small region between the 0.5 and 1 contours, so we estimate the density of foxes in this rectangle to be at most 0.6 foxes per square kilometer. The next rectangle northwards is all sea and so there are no foxes. However, the rectangle to the east includes a region between the contours labeled 1 and 1.5, so we estimate the maximum density to be 1.3 foxes per square kilometer. Continuing in this way we obtain the upper bounds in Table 15.1. Similarly, we obtain the lower bounds shown in Table 15.2.

Each rectangle has an area $30 \times 25 = 750 \text{ km}^2$, so we obtain

Lower estimate $= (0.1 + 0.1 + 0.5 + 1.2 + 1.2 + 1 + 1 + 0.5 + 0.5) \cdot 750 = 4575$ foxes.

Similarly, we obtain an upper bound of

Upper estimate $= 41.6 \cdot 750 = 31{,}200$ foxes

The average of our two estimates is about 18,000, so we take that as our estimate. Notice that there is a wide discrepancy between the upper and lower estimates; by taking finer subdivisions we could make the upper and lower estimates closer.

---

[1] Adapted from J. D. Murray, *Mathematical Biology*, Springer-Verlag, 1989

TABLE 15.1   *Upper estimates of the fox population density*

0	0	0	0.8	1.5	1.5
0	0	0.5	1.5	1.5	1.5
0	0	1.5	2.5	1.9	2.3
0	1.2	2.2	2.5	2.5	2.5
0	1.3	2	2.2	2.5	0.5
0.6	1.3	1	1	1.3	0

TABLE 15.2   *Lower estimates of the fox population density*

0	0	0	0	0	0.1
0	0	0	0	0	0.1
0	0	0	0.5	1.2	1.2
0	0	0	1	1	0
0	0	0.5	0.5	0	0
0	0	0	0	0	0

Definition of the Definite Integral

The sums used to approximate the fox population are similar to the Riemann sums used to define the definite integral of a function of one variable. We now define the definite integral for a function f of two variables on a rectangular region. Given a continuous function $f(x, y)$ defined on a region $a \leq x \leq b$ and $c \leq y \leq d$, we subdivide each of the intervals $a \leq x \leq b$ and $c \leq y \leq d$ into n and m equal subintervals respectively, giving nm subrectangles. (See Figure 15.2.)

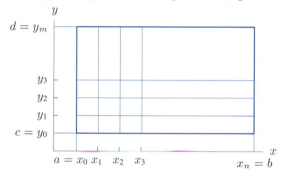

Figure 15.2: Subdivision of a rectangle into nm subrectangles

The area of each subrectangle is ΔA, where $\Delta A = \Delta x \, \Delta y$ and $\Delta x = (b - a)/n$ is the width of each subdivision along the x-axis, and $\Delta y = (d - c)/m$ is the width of each subdivision along the y-axis. To compute the Riemann sum, we multiply the area of each subrectangle by the value of the function at a point in the rectangle and add all the resulting numbers. Choosing the point which gives the maximum value, M_{ij}, of the function on each rectangle, we get the *upper sum*, $\sum_{i,j} M_{ij} \Delta x \Delta y$.

The *lower sum*, $\sum_{i,j} L_{ij} \Delta x \Delta y$, is obtained by taking the minimum value on each rectangle. If (x_i, y_i) is any point in the ij-th subrectangle, any other Riemann sum satisfies

$$\sum_{i,j} L_{ij} \Delta x \Delta y \leq \sum_{i,j} f(x_i, y_j) \, \Delta x \, \Delta y \leq \sum_{i,j} M_{ij} \Delta x \Delta y$$

We define the definite integral by taking the limit as the numbers of subdivisions, n and m, tend to infinity. By comparing upper and lower sums, as we did for the fox population, it can be shown that the limit exists when the function, f, is continuous. We get the same limit by letting Δx and Δy tend to 0. Thus, we have the following definition:

> Suppose the function f is continuous on R, the rectangle $a \leq x \leq b$, $c \leq y \leq d$. We define the **definite integral** of f over R
>
> $$\int_R f \, dA = \lim_{\Delta x, \Delta y \to 0} \sum_{i,j} f(x_i, y_j) \Delta x \Delta y.$$
>
> Such an integral is called a **double integral**.

Sometimes we think of dA as being the area of an infinitesimal rectangle of length dx and height dy, so that $dA = dx\, dy$. Then we use the notation[2]

$$\int_R f\, dA = \int_R f(x, y)\, dx\, dy.$$

The Riemann sum used in the definition, with equal-sized rectangular subdivisions, is just one type of Riemann sum. For a general Riemann sum, the subdivisions do not all have to be the same size.

Example 2   Let R be the rectangle $0 \le x \le 1$ and $0 \le y \le 1$. Use Riemann sums to estimate $\displaystyle\int_R e^{-(x^2+y^2)}\, dA$.

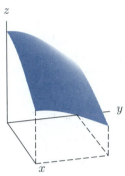

Figure 15.3: Graph of $e^{-(x^2+y^2)}$ above the rectangle R

Solution   We divide R into 16 subrectangles by dividing each edge into 4 parts. Figure 15.3 shows that $f(x, y) = e^{-(x^2+y^2)}$ decreases as we move away from the origin. Thus, to get an upper sum we evaluate f on each subrectangle at the corner nearest the origin. For example, in the rectangle $0 \le x \le 0.25, 0 \le y \le 0.25$, we evaluate f at $(0,0)$.

Using Table 15.3 we find that

$$\begin{aligned}
\text{Upper sum} = [(&\ 1 + 0.9394 + 0.7788 + 0.5698) \\
&+ (0.9394 + 0.8825 + 0.7316 + 0.5353) \\
&+ (0.7788 + 0.7316 + 0.6065 + 0.4437) \\
&+ (0.5698 + 0.5353 + 0.4437 + 0.3247)](0.0625) = 0.68.
\end{aligned}$$

To get a lower sum, we must evaluate f at the opposite corner of each rectangle because the surface slopes down in both the x and y directions. This yields a lower sum of 0.44. Thus,

$$0.44 \le \int_R e^{-(x^2+y^2)}\, dA \le 0.68.$$

To get a better approximation, we compute lower and upper estimates with more subdivisions. The results for several cases with equal numbers of subdivisions in the x and y-directions are shown in Table 15.4.

TABLE 15.3   *Values of $f(x, y) = e^{-(x^2+y^2)}$ on the rectangle R*

		0.0	0.25	0.50	0.75	1.00
	0.0	1	0.9394	0.7788	0.5698	0.3679
	0.25	0.9394	0.8825	0.7316	0.5353	0.3456
x	0.50	0.7788	0.7316	0.6065	0.4437	0.2865
	0.75	0.5698	0.5353	0.4437	0.3247	0.2096
	1.00	0.3679	0.3456	0.2865	0.2096	0.1353

The y values (0.0, 0.25, 0.50, 0.75, 1.00) are the column headers.

---

[2]Another common notation for the double integral is $\int \int_R f\, dA$.

TABLE 15.4   *Riemann sum approximations to* $\int_R e^{-(x^2+y^2)}\, dA$

	Number of subdivisions in x and y directions			
	8	16	32	64
Upper	0.6168	0.5873	0.5725	0.5651
Lower	0.4989	0.5283	0.5430	0.5504

The true value of the double integral, $0.5577\ldots$, is trapped between the lower and upper sums. Notice that the lower sum increases and the upper sum decreases as the number of subdivisions grows. (Why is this?) However, even with 64 subdivisions, leading to $64^2 = 4096$ terms in the Riemann sum, the lower and upper sums agree with the true value of the integral only in the first decimal place. Hence, we should look for better ways to approximate the integral.

The Region R

In our definition of the definite integral $\int_R f(x,y)\, dA$, the region R is a rectangle. However, the definite integral can be defined for regions of other shapes, including triangles, circles, and regions bounded by the graphs of piecewise continuous functions.

To approximate the definite integral over a region, R, which is not rectangular, we use a grid of rectangles approximating the region. We obtain this grid by enclosing R in a large rectangle and subdividing that rectangle; we consider just the subrectangles which are inside R.

As before, we pick a point (x_i, y_j) in each subrectangle and form a Riemann sum

$$\sum_{i,j} f(x_i, y_j)\Delta x \Delta y.$$

This time, however, the sum is over only those subrectangles within R. For example, in the case of the fox population we can use the rectangles which are entirely on land. As the subdivisions become finer, the grid resembles the region R more closely. For a function, f, which is continuous on R, we define the definite integral as follows:

$$\int_R f\, dA = \lim_{\Delta x, \Delta y \to 0} \sum_{i,j} f(x_i, y_j)\Delta x \Delta y$$

where the Riemann sum is taken over the subrectangles inside R.

You may wonder why we can leave out the rectangles which cover the edge of R—if we included them might we get a different value for the integral? The answer is that for any region that we are likely to meet, the area of the subrectangles covering the edge tends to 0 as the grid becomes finer. Therefore, omitting these rectangles does not affect the limit.

Interpretations of the Double Integral

Interpretation as an Area

Suppose $f(x, y) = 1$ for all points (x, y) in the region R. Then each term in the Riemann sum is of the form $1 \cdot \Delta A = \Delta A$ and the double integral gives the area of the region R.

$$\text{Area}(R) = \int_R 1\, dA = \int_R dA$$

Interpretation as a Volume

Just as the definite integral of a positive one-variable function can be interpreted as an area under the graph of the function, so the definite integral of a positive two-variable function can be interpreted as a volume under its graph. In the one-variable case we visualize the Riemann sums as the total area of rectangles above the subdivisions. In the two-variable case we get solid bars instead of rectangles. As the number of subdivisions grows, the tops of the bars approximate the surface better, and the volume of the bars gets closer to the volume under the surface and above the region R. (See Figure 15.4.)

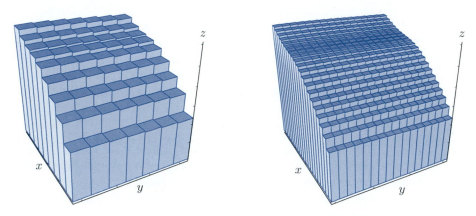

Figure 15.4: Approximating volume under a graph with finer and finer Riemann sums

---

Example 3   Find the volume under the graph of $f(x, y) = 2 - x^2 - y^2$ lying above the rectangle $-1 \leq x \leq 1$ and $-1 \leq y \leq 1$. (See Figure 15.5.)

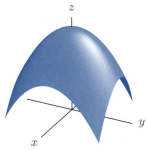

Figure 15.5: Graph of $f(x, y) = 2 - x^2 - y^2$ above $-1 \leq x \leq 1$ and $-1 \leq y \leq 1$

Solution   If R is the rectangle $-1 \leq x \leq 1$, $-1 \leq y \leq 1$, the volume we want is given by

$$\text{Volume} = \int_R (2 - x^2 - y^2) \, dA$$

Table 15.5 contains values of Riemann sums, S_n, calculated by subdividing the rectangle into n^2 subrectangles and evaluating f at the point with minimum x and y-values. The table suggests the value of the integral is about 5.3.

TABLE 15.5   *Riemann sums for $\int_R (2 - x^2 - y^2) \, dA$*

n	5	10	20	40
S_n	5.12	5.28	5.32	5.33

Interpretation of the Definite Integral When f is a Density Function

A function of two variables can represent a density per unit area, for example, the fox population density (in foxes per unit area), or the mass density of a thin metal plate. Then the integral $\int_R f \, dA$ represents the total population or total mass in the region R.

Interpretation of the Definite Integral as an Average Value

As in the one-variable case, the definite integral can be used to compute the average value of a function:

$$\text{Average value of } f \text{ on the region } R = \frac{1}{\text{Area of } R} \int_R f \, dA$$

We can rewrite this as

$$\text{Average value } \times \text{ Area of } R = \int_R f \, dA.$$

Thus, if we interpret the integral as the volume under the graph of f, then we can think of the average value of f as the height of the box with the same volume that is on the same base. (See Figure 15.6.)

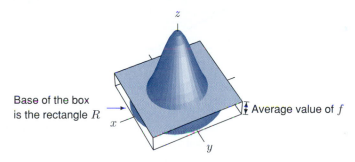

Base of the box is the rectangle R

Average value of f

Figure 15.6: Volume and average value

One way to think of this is to imagine that the volume under the graph is made out of wax; if the wax melted and leveled out within walls erected on the perimeter of R, then it would end up box-shaped with height equal to the average value of f.

Problems for Section 15.1

1. A function $f(x, y)$ has values in Table 15.6. Let R be the rectangle $1 \le x \le 1.2, 2 \le y \le 2.4$. Find the Riemann sums which are reasonable over- and under-estimates for $\int_R f(x, y) \, dA$ with $\Delta x = 0.1$ and $\Delta y = 0.2$.

TABLE 15.6

		x		
		1.0	1.1	1.2
y	2.0	5	7	10
	2.2	4	6	8
	2.4	3	5	4

2. A solid is formed above the rectangle R with $0 \le x \le 2$, $0 \le y \le 4$ by the graph of $f(x, y) = 2 + xy$. Using Riemann sums with four subdivisions, find upper and lower bounds for the volume of this solid.

3. Let R be the rectangle with vertices $(0,0)$, $(4,0)$, $(4,4)$, and $(0,4)$ and let $f(x,y) = \sqrt{xy}$.

    (a) Find reasonable upper and lower bounds for $\int_R f\, dA$ without subdividing R.

    (b) Estimate $\int_R f\, dA$ by partitioning R into four subrectangles and evaluating f at its maximum and minimum values on each subrectangle.

4. Figure 15.7 shows the distribution of temperature, in °C, in a 5 meter by 5 meter heated room. Using Riemann sums, estimate the average temperature in the room.

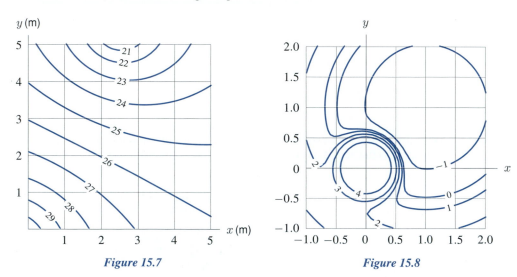

Figure 15.7   Figure 15.8

5. Figure 15.8 shows the contour diagram of a function $z = f(x,y)$. Let R be the square $-0.5 \le x \le 1$, $-0.5 \le y \le 1$. Is the integral $\int_R f\, dA$ positive or negative? Explain your reasoning.

6. A biologist studying insect populations measures the population density of flies and mosquitos at various points in a rectangular study region. The graphs of the two population densities for the region are shown in Figures 15.9 and 15.10. Assuming that the units along the corresponding axes are the same in the two graphs, are there more flies or more mosquitos in the region?

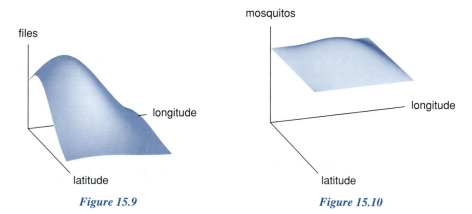

Figure 15.9   Figure 15.10

In Problems 7–8, use a computer or calculator program that finds two-dimensional Riemann sums to estimate the given integral.

7. If R is the rectangle $1 \le x \le 2$, $1 \le y \le 3$, estimate $\int_R (x^2 + y^2)\, dA$.

8. If R is the rectangle $-\pi \le x \le 0$, $0 \le y \le \pi/2$, estimate $\int_R \sin(xy)\, dA$.

9.   The hull of a certain boat has width $w(x, y)$ feet at a point x feet from the front and y feet below the water line. A table of values of w follows. Set up a definite integral that gives the volume of the hull below the waterline, and then estimate the value of the integral.

TABLE 15.7

		Front of boat → Back of boat					
		0	10	20	30	40	50	60
	0	2	8	13	16	17	16	10
Depth	2	1	4	8	10	11	10	8
below	4	0	3	4	6	7	6	4
waterline	6	0	1	2	3	4	3	2
(in feet)	8	0	0	1	1	1	1	1

10.   Let $f(x, y)$ be a function of x and y which is independent of y, that is, $f(x, y) = g(x)$ for some one-variable function g.

  (a)   What does the graph of f look like?
  (b)   Let R be the rectangle $a \leq x \leq b, c \leq y \leq d$. By interpreting the integral as a volume, and using your answer to part (a), express $\int_R f \, dA$ in terms of a one-variable integral.

11.   Figure 15.11 shows contours for the annual frequency of tornados per 10,000 square miles in the U.S.[3] Each grid square is 100 miles on a side. Use the map to estimate the total number of tornados per year in (a)   Texas   (b)   Florida   (c)   Arizona.

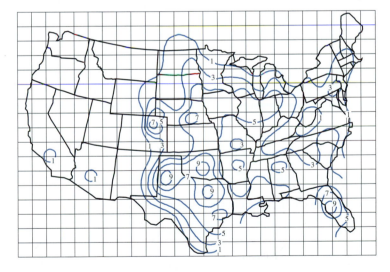

Figure 15.11

12.   Figure 15.12 on page 762 shows contours of annual rainfall (in centimeters) in Oregon.[4] Use it to estimate how much rain falls in Oregon in one year. Each grid square is 100 kilometers on a side.

---

[3]From *Modern Physical Geography*, Alan H. Strahler and Arthur H. Strahler, Fourth Edition, John Wiley & Sons, New York, 1992, p. 128
[4]From *Physical Geography of the Global Environment*, H. J. de Blij and Peter O. Muller, John Wiley & Sons, New York, 1993, p. 133

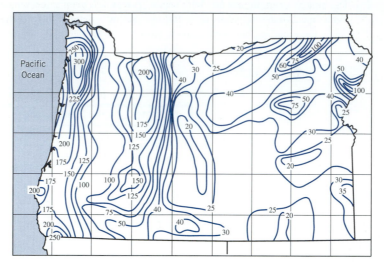

Figure 15.12

13. Let D be the region inside the unit circle centered at the origin, let R be the right half of D and let B be the bottom half of D. Decide (without calculating the value of any of the integrals) whether each integral is positive, negative, or zero.

(a) $\int_D dA$ (b) $\int_B dA$ (c) $\int_R 5x\, dA$ (d) $\int_B 5x\, dA$

(e) $\int_D 5x\, dA$ (f) $\int_D (y^3 + y^5)\, dA$ (g) $\int_B (y^3 + y^5)\, dA$ (h) $\int_R (y^3 + y^5)\, dA$

(i) $\int_B (y - y^3)\, dA$ (j) $\int_D (y - y^3)\, dA$ (k) $\int_D \sin y\, dA$ (l) $\int_D \cos y\, dA$

(m) $\int_D e^x\, dA$ (n) $\int_D xe^x\, dA$ (o) $\int_D xy^2\, dA$ (p) $\int_B x\cos y\, dA$

14. For any numbers a and b, assume that $|a + b| \leq |a| + |b|$. Use this to explain why

$$\left| \int_R f\, dA \right| \leq \int_R |f|\, dA.$$

15.2 ITERATED INTEGRALS

In Section 15.1 we approximated double integrals using Riemann sums. In this section we see how to compute double integrals exactly using ordinary one-variable integrals.

The Fox Population Again: Expressing a Double Integral as an Iterated Integral

To estimate the fox population, we computed a sum of the form

$$\text{Total population} \approx \sum_{i,j} f(x_i, y_j)\Delta x\, \Delta y,$$

where $1 \leq i \leq n$ and $1 \leq j \leq m$ and the values $f(x_i, y_j)$ can be arranged as in Table 15.8.

TABLE 15.8 *Upper bounds for fox population densities for $n = m = 6$*

0	0	0	0.8	1.5	1.5
0	0	0.5	1.5	1.5	1.5
0	0	1.5	2.5	1.9	2.3
0	1.2	2.2	2.5	2.5	2.5
0	1.3	2	2.2	2.5	0.5
0.6	1.3	1	1	1.3	0

For any values of n and m, there are two ways to compute this sum: one is to add across the rows first, the other is to add down the columns first. If we add rows first, we can write the sum in the form

$$\text{Total population} \approx \sum_{j=1}^{m} \left(\sum_{i=1}^{n} f(x_i, y_j) \Delta x \right) \Delta y.$$

The inner sum, $\sum_{i=1}^{n} f(x_i, y_j) \Delta x$, approximates the integral $\int_{0}^{180} f(x, y_j) \, dx$. Thus, we have

$$\text{Total population} \approx \sum_{j=1}^{m} \left(\int_{0}^{180} f(x, y_j) \, dx \right) \Delta y.$$

The outer sum represents a Riemann sum approximating another integral, this time with integrand $\int_{0}^{180} f(x, y) \, dx$, which is a function of y. Thus, we can write the total population in terms of nested one-variable integrals:

$$\text{Total population} = \int_{0}^{150} \left(\int_{0}^{180} f(x, y) \, dx \right) dy.$$

Since the total population is represented by $\int_{R} f \, dA$, we have discovered a way of expressing double integrals:

If R is the rectangle $a \leq x \leq b$, $c \leq y \leq d$ and if $f(x, y)$ is a continuous function of two variables, then

$$\int_{R} f \, dA = \int_{c}^{d} \left(\int_{a}^{b} f(x, y) \, dx \right) dy.$$

The expression $\int_{c}^{d} \left(\int_{a}^{b} f(x, y) \, dx \right) dy$, or simply $\int_{c}^{d} \int_{a}^{b} f(x, y) \, dx \, dy$, is called an **iterated integral.**

The inside integral is performed with respect to x, holding y constant, and then the result is integrated with respect to y.

The Parallel Between Repeated Summation and Repeated Integration

It is helpful to notice how the summation and integral notation parallel each other. Viewing a Riemann sum as a sum of sums,

$$\sum_{i,j} f(x_i, y_j) \, \Delta x \, \Delta y = \sum_{j} \left(\sum_{i} f(x_i, y_j) \Delta x \right) \Delta y$$

leads us to see a double integral as an integral of integrals:

$$\int_{R} f(x, y) \, dA = \int_{c}^{d} \left(\int_{a}^{b} f(x, y) \, dx \right) dy,$$

where R is the rectangle $a \leq x \leq b$ and $c \leq y \leq d$.

Example 1   A building is 8 feet wide and 16 feet long. It has a flat roof that is 12 feet high at one corner, and 10 feet high at each of the adjacent corners. What is the volume of the building?

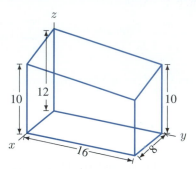

Figure 15.13: A slant-roofed building

Solution   If we put the high corner on the z-axis, the long side along the y-axis, and the short side along the x-axis, as in Figure 15.13, then the roof is a plane with z-intercept 12, and x slope $(-2)/8 = -1/4$, and y slope $(-2)/16 = -1/8$. Hence, the equation of the roof is

$$z = 12 - \tfrac{1}{4}x - \tfrac{1}{8}y.$$

To calculate the volume, we will integrate over the rectangle $0 \leq x \leq 8, 0 \leq y \leq 16$. Setting up an iterated integral, we get

$$\text{Volume} = \int_0^{16} \int_0^8 (12 - \tfrac{1}{4}x - \tfrac{1}{8}y)\, dx\, dy.$$

The inside integral is

$$\int_0^8 (12 - \tfrac{1}{4}x - \tfrac{1}{8}y)\, dx = \left(12x - \tfrac{1}{8}x^2 - \tfrac{1}{8}xy\right)\bigg|_0^8 = 88 - y.$$

Then the outside integral is

$$\int_0^{16} (88 - y)\, dy = \left(88y - \tfrac{1}{2}y^2\right)\bigg|_0^{16} = 1280.$$

So the volume of the building is 1280 cubic feet.

The Order of Integration

In computing the fox population, we could have chosen to add columns (fixed x) first, instead of the rows. This leads to an iterated integral where x is constant in the inner integral instead of y. Thus,

$$\int_R f(x,y)\, dA = \int_a^b \left(\int_c^d f(x,y)\, dy \right) dx$$

where R is the rectangle $a \leq x \leq b$ and $c \leq y \leq d$.

For any function we are likely to meet, it doesn't matter in which order we integrate over a rectangular region R; we get the same value for the double integral either way.

$$\int_R f\, dA = \int_c^d \left(\int_a^b f(x,y)\, dx \right) dy = \int_a^b \left(\int_c^d f(x,y)\, dy \right) dx$$

Example 2   Compute the volume of Example 1 as an iterated integral with y fixed in the inner integral.

Solution   Rewriting the integral, we have

$$\text{Volume} = \int_0^8 \left(\int_0^{16} (12 - \tfrac{1}{4}x - \tfrac{1}{8}y)\, dy \right) dx = \int_0^8 \left((12y - \tfrac{1}{4}xy - \tfrac{1}{16}y^2) \Big|_0^{16} \right) dx$$

$$= \int_0^8 (176 - 4x)\, dx = (176x - 2x^2)\Big|_0^8 = 1280.$$

Iterated Integrals Over Non-Rectangular Regions

Example 3   The density at the point (x, y) of a right triangular metal plate, as shown in Figure 15.14, is $\delta(x, y)$. Express its mass as an iterated integral.

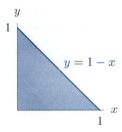

Figure 15.14: A triangular metal plate with density $\delta(x, y)$ at the point (x, y)

Solution   Using a grid, divide the region into small rectangles of side Δx and Δy. Thus, the mass of one piece is given by

$$\text{Mass of rectangle} \approx \text{Density} \cdot \text{Area} \approx \delta(x, y)\Delta x \Delta y.$$

Summing over all rectangles gives a Riemann sum which approximates a double integral:

$$\text{Mass} = \int_R \delta(x, y)\, dA,$$

where R is the triangle. The sloping edge of the triangle is the line $y = 1 - x$. We want to compute this integral using an iterated integral. Think about how an iterated integral over a rectangle, such as

$$\int_a^b \int_c^d f(x, y)\, dy\, dx,$$

works. This integral is over the rectangle $a \leq x \leq b$, $c \leq y \leq d$. The inside integral with respect to y is along vertical strips from $y = c$ to $y = d$. There is one such strip for each x value between $x = a$ and $x = b$. Thus, the value of the inside integral depends on the value of x. After computing the inside integral with respect to y, we compute the outside integral with respect to x, which means adding the contributions from the individual vertical strips that make up the rectangle. (See Figure 15.15).

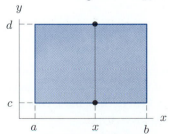

Figure 15.15: Integrating over a rectangle using vertical strips

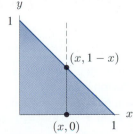

Figure 15.16: Integrating over a triangle using vertical strips

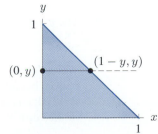

Figure 15.17: Integrating over a triangle using horizontal strips

For the triangular region in Figure 15.14, the idea is the same. The only difference is that the individual vertical strips no longer all go from $y = c$ to $y = d$. The vertical strip that enters the triangle at the point $(x, 0)$ leaves it at the point $(x, 1 - x)$, because the top edge of the triangle is the line $y = 1 - x$. See Figure 15.16. Thus, on this vertical strip, y goes from 0 to $1 - x$. Hence, the inside integral is

$$\int_0^{1-x} \delta(x, y)\, dy.$$

Finally, since there is one of these integrals for each x value between 0 and 1, the outside integral goes from 0 to 1. Thus, the iterated integral we want is

$$\text{Mass} = \int_0^1 \int_0^{1-x} \delta(x, y)\, dy\, dx.$$

We could have chosen to integrate in the opposite order, keeping y fixed in the inner integral instead of x. The limits are formed by looking at horizontal strips instead of vertical ones, and expressing the x-values at the end points in terms of y. A typical horizontal strip goes from $x = 0$ to $x = 1 - y$, and since y-values overall range between 0 and 1, the iterated integral is

$$\text{Mass} = \int_0^1 \int_0^{1-y} \delta(x, y)\, dx\, dy.$$

Limits on Iterated Integrals

- The limits on the outer integral must be constants.
- If the inner integral is with respect to x, its limits should be constants or expressions in terms of y, and vice versa.

Example 4   Find the mass M of a metal plate R bounded by $y = x$ and $y = x^2$, with density given by $\delta(x, y) = 1 + xy$ kg/meter^2. (See Figure 15.18.)

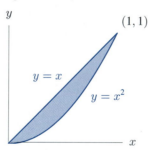

Figure 15.18: A metal plate
with density $\delta(x, y)$

Solution   The mass is given by

$$M = \int_R \delta(x, y)\, dA.$$

We integrate along vertical strips first; this means we do the y integral first, which goes from the bottom boundary $y = x^2$ to the top boundary $y = x$. The left edge of the region is at $x = 0$ and the right edge is at the intersection point of $y = x$ and $y = x^2$, which is $(1, 1)$. Thus, the x-coordinate of the vertical strips can vary from $x = 0$ to $x = 1$, and so the mass is given by

$$M = \int_0^1 \int_{x^2}^{x} \delta(x, y)\, dy\, dx = \int_0^1 \int_{x^2}^{x} (1 + xy)\, dy\, dx.$$

Calculating the inner integral first gives

$$
M = \int_0^1 \left(\int_{x^2}^x (1 + xy)\, dy \right) dx = \int_0^1 \left((y + x\frac{y^2}{2}) \Big|_{y=x^2}^{y=x} \right) dx
$$

$$
= \int_0^1 (x - x^2 + \frac{x^3}{2} - \frac{x^5}{2})\, dx = \left(\frac{x^2}{2} - \frac{x^3}{3} + \frac{x^4}{8} - \frac{x^6}{12} \right) \Big|_0^1
$$

$$
= \frac{5}{24} \text{ kg.}
$$

Example 5   A city is in the form of a semicircular region of radius 3 km bordering on the ocean. Find the average distance from any point in the city to the ocean.

Solution   Think of the ocean as everything below the x-axis in the xy-plane and think of the city as the upper half of the circular disk of radius 3 bounded by $x^2 + y^2 = 9$. (See Figure 15.19).

The distance from any point (x, y) in the city to the ocean is the vertical distance to the x-axis, namely y. Thus, we want to compute

$$
\text{Average distance} = \frac{1}{\text{Area}(R)} \int_R y\, dA,
$$

where R is the region between the upper half of the circle $x^2 + y^2 = 9$ and the x-axis. The area of R is $\pi 3^2/2 = 9\pi/2$. To compute the integral, let's take the inner integral with respect to y. Then a typical vertical strip goes from the x-axis, namely $y = 0$, to the semicircle. The upper limit must be expressed in terms of x so we solve $x^2 + y^2 = 9$ to get $y = \sqrt{9 - x^2}$. Since x varies from -3 to 3 throughout the region, the integral is:

$$
\int_R y\, dA = \int_{-3}^3 \left(\int_0^{\sqrt{9-x^2}} y\, dy \right) dx = \int_{-3}^3 \left(\frac{y^2}{2} \Big|_0^{\sqrt{9-x^2}} \right) dx
$$

$$
= \int_{-3}^3 \frac{1}{2}(9 - x^2)\, dx = \frac{1}{2}(9x - \frac{x^3}{3}) \Big|_{-3}^3 = \frac{1}{2}(18 - (-18)) = 18.
$$

Therefore, the average distance is $18/(9\pi/2) = 4/\pi$ km.

What if we choose the inner integral to be with respect to x? Then we get the limits by looking at horizontal strips, not vertical, and we solve $x^2 + y^2 = 9$ for x in terms of y. We get $x = -\sqrt{9 - y^2}$ at the left end of the strip and $x = \sqrt{9 - y^2}$ at the right. Now y varies from 0 to 3 so the integral becomes:

$$
\int_R y\, dA = \int_0^3 \left(\int_{-\sqrt{9-y^2}}^{\sqrt{9-y^2}} y\, dx \right) dy = \int_0^3 \left(yx \Big|_{x=-\sqrt{9-y^2}}^{x=\sqrt{9-y^2}} \right) dy = \int_0^3 2y\sqrt{9 - y^2}\, dy
$$

$$
= -\frac{2}{3}(9 - y^2)^{3/2} \Big|_0^3 = -\frac{2}{3}(0 - 27) = 18.
$$

We get the same result as before. Thus, the average distance to the ocean is $(2/(9\pi))18 = 4/\pi$ km.

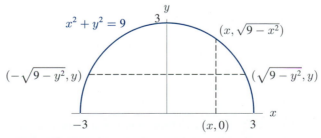

Figure 15.19: The city by the ocean showing a typical vertical strip and a typical horizontal strip

In the examples so far, the region was given and the problem was to determine the limits for an iterated integral. Sometimes only the limits are known and we want to determine the region.

Example 6   Sketch the region of integration for the iterated integral $\int_0^6 \int_{x/3}^2 x\sqrt{y^3 + 1}\, dy\, dx$.

Solution   The inner integral is with respect to y, so we imagine vertical strips crossing the region of integration. The bottom of each strip is $y = x/3$, a line through the origin, and the top is $y = 2$, a horizontal line. Since the limits of the outer integral are 0 and 6, the whole region is contained between the vertical lines $x = 0$ and $x = 6$. Notice that the lines $y = 2$ and $y = x/3$ meet where $x = 6$. The region is shown in Figure 15.20 .

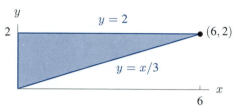

Figure 15.20: The region of integration for
Example 6

Reversing the Order of Integration

It can sometimes be helpful to reverse the order of integration in an iterated integral. Surprisingly enough, an integral which is difficult or impossible with the limits in one order can be quite straightforward in the other. The next example is such a case.

Example 7   Evaluate $\int_0^6 \int_{x/3}^2 x\sqrt{y^3 + 1}\, dy\, dx$ using the region sketched in Figure 15.20.

Solution   Since $\sqrt{y^3 + 1}$ has no elementary antiderivative, we cannot calculate the inner integral symbolically. We try reversing the order of integration. From Figure 15.20, we see that horizontal strips go from $x = 0$ to $x = 3y$. For the whole region, y varies from 0 to 2. Thus, when we change the order of integration we get

$$\int_0^6 \int_{x/3}^2 x\sqrt{y^3 + 1}\, dy\, dx = \int_0^2 \int_0^{3y} x\sqrt{y^3 + 1}\, dx\, dy.$$

Now we can at least do the inner integral because we know the antiderivative of x. What about the outer integral?

$$\int_0^2 \int_0^{3y} x\sqrt{y^3 + 1}\, dx\, dy = \int_0^2 \left(\frac{x^2}{2}\sqrt{y^3 + 1} \right) \Big|_{x=0}^{x=3y} dy = \int_0^2 \frac{9y^2}{2}(y^3 + 1)^{1/2}\, dy$$

$$= (y^3 + 1)^{3/2} \Big|_0^2 = 27 - 1 = 26.$$

Thus, reversing the order of integration made the integral in the previous problem much easier. Notice that to reverse the order it is essential first to sketch the region over which the integration is being performed.

Problems for Section 15.2

For Problems 1–4, evaluate the given integral.

1. $\int_R \sqrt{x+y} \, dA$, where R is the rectangle $0 \le x \le 1, 0 \le y \le 2$.

2. Calculate the integral in Problem 1 using the other order of integration.

3. $\int_R (5x^2 + 1) \sin 3y \, dA$, where R is the rectangle $-1 \le x \le 1, 0 \le y \le \pi/3$.

4. $\int_R (2x + 3y)^2 \, dA$, where R is the triangle with vertices at $(-1, 0)$, $(0, 1)$, and $(1, 0)$.

For each of the regions R in Problems 5–8, write $\int_R f \, dA$ as an iterated integral.

5.
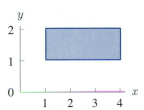

6.
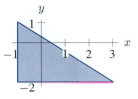

7.
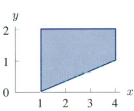

8.
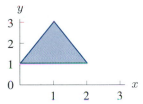

For Problems 9–13, sketch the region of integration and evaluate the integral.

9. $\int_1^3 \int_0^4 e^{x+y} \, dy \, dx$

10. $\int_0^2 \int_0^x e^{x^2} \, dy \, dx$

11. $\int_1^5 \int_x^{2x} \sin(x) \, dy \, dx$

12. $\int_1^4 \int_{\sqrt{y}}^y x^2 y^3 \, dx \, dy$

13. $\int_{-2}^0 \int_{-\sqrt{9-x^2}}^0 2xy \, dy \, dx$

14. Consider the integral $\int_0^4 \int_0^{-(y-4)/2} g(x, y) \, dx \, dy$.

   (a) Sketch the region over which the integration is being performed.
   (b) Write the integral with the order of the integration reversed.

Evaluate the integral in Problems 15–17 by reversing the order of integration.

15. $\int_0^1 \int_y^1 e^{x^2} \, dx \, dy$

16. $\int_0^3 \int_{y^2}^9 y \sin(x^2) \, dx \, dy$

17. $\int_0^1 \int_{\sqrt{y}}^1 \sqrt{2 + x^3} \, dx \, dy$

18. Reverse the order of integration: $\int_{-4}^0 \int_0^{2x+8} f(x, y) \, dy \, dx + \int_0^4 \int_0^{-2x+8} f(x, y) \, dy \, dx$.

For Problems 19–21 set up, but do not evaluate, an iterated integral for the volume of the solid.

19. Under the graph of $f(x, y) = 25 - x^2 - y^2$ and above the xy-plane.

20. Below the graph of $f(x, y) = 25 - x^2 - y^2$ and above the plane $z = 16$.

21. The three-sided pyramid whose base is on the xy-plane and whose three sides are the vertical planes $y = 0$ and $y - x = 4$, and the slanted plane $2x + y + z = 4$.

For Problems 22–24, find the volume of the given region.

22.   Under the graph of $f(x, y) = xy$ and above the square $0 \le x \le 2, 0 \le y \le 2$ in the xy-plane.

23.   The solid between the planes $z = 3x + 2y + 1$ and $z = x + y$, and above the triangle with vertices $(1, 0, 0)$, $(2, 2, 0)$, and $(0, 1, 0)$ in the xy-plane. See Figure 15.21.

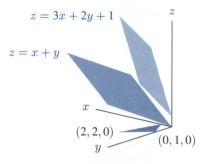

Figure 15.21

24.   The region R bounded by the graph of $ax + by + cz = 1$ and the coordinate planes. Assume a, b, and c are positive.

25.   Find the average distance to the x-axis for points in the region bounded by the x-axis and the graph of $y = x - x^2$.

26.   Prove that for a right triangle the average distance from any point in the triangle to one of the legs is one-third the length of the other leg. (The legs of a right triangle are the two sides that are not the hypotenuse.)

27.   Evaluate $\displaystyle\int_0^1 \int_y^1 \sin(x^2)\, dx\, dy$

28.   Evaluate $\displaystyle\int_0^1 \int_{e^y}^e \frac{x}{\ln x}\, dx\, dy$

15.3 TRIPLE INTEGRALS

A continuous function of three variables can be integrated over a solid region W in 3-space in the same way as a function of two variables is integrated over a flat region in 2-space. Again, we start with a Riemann sum. First we subdivide W into smaller regions, then we multiply the volume of each region by a value of the function in that region, and then we add the results. For example, if W is the box $a \le x \le b$, $c \le y \le d$, $p \le z \le q$, then we subdivide each side into l, m, and n pieces, thereby chopping W into lmn smaller boxes, as shown in Figure 15.22.

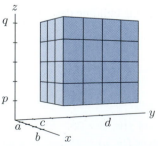

Figure 15.22: Subdividing a three-dimensional box

The volume of each smaller box is

$$\Delta V = \Delta x \Delta y \Delta z,$$

where $\Delta x = (b-a)/l$, and $\Delta y = (d-c)/m$, and $\Delta z = (q-p)/n$. Using this subdivision, we pick a point (x_i, y_j, z_k) in the ijk-th small box and construct a Riemann sum

$$\sum_{i,j,k} f(x_i, y_j, z_k) \Delta V.$$

If f is continuous, as Δx, Δy, and Δz approach 0, this Riemann sum approaches the definite integral, $\int_W f \, dV$, called a *triple integral*, which is defined as

$$\int_W f \, dV = \lim_{l,m,n \to \infty} \sum_{i,j,k} f(x_i, y_j, z_k) \Delta V.$$

As in the case of a double integral, we can evaluate this integral as an iterated integral:

Triple integral as an iterated integral

$$\int_W f \, dV = \int_p^q \left(\int_c^d \left(\int_a^b f(x, y, z) \, dx \right) dy \right) dz,$$

where y and z are treated as constants in the innermost (dx) integral, and z is treated as a constant in the middle (dy) integral. The integration can be performed in any order.

Example 1   A cube C has sides of length 4 cm and is made of a material of variable density. If one corner is at the origin and the adjacent corners are on the positive x, y, and z axes, then the density (in gm/cm^3) at the point (x, y, z) is $\delta(x, y, z) = 1 + xyz$ gm/cm^3. Find the mass of the cube.

Solution   Consider a small piece ΔV of the cube, small enough so that the density remains close to constant over the piece. Then

$$\text{Mass of small piece} = \text{Density} \cdot \text{Volume} \approx \delta(x, y, z) \, \Delta V.$$

To get the total mass, we add the masses of the small pieces and take the limit as $\Delta V \to 0$. Thus, the mass is the triple integral

$$M = \int_C \delta \, dV = \int_0^4 \int_0^4 \int_0^4 (1 + xyz) \, dx \, dy \, dz = \int_0^4 \int_0^4 \left[x + \frac{1}{2}x^2 yz \right]_{x=0}^{x=4} dy \, dz$$

$$= \int_0^4 \int_0^4 (4 + 8yz) \, dy \, dz = \int_0^4 \left[4y + 4y^2 z \right]_{y=0}^{y=4} dz = \int_0^4 (16 + 64z) \, dz = 576 \, \text{gm}.$$

Example 2   Express the volume of the building described in Example 1 on page 764 as a triple integral.

Solution   The building is given by $0 \leq x \leq 8$, $0 \leq y \leq 16$, and $0 \leq z \leq 12 - x/4 - y/8$. (See Figure 15.23 on page 772.) To find its volume, divide it into small cubes of volume $\Delta V = \Delta x \, \Delta y \, \Delta z$ and add. First, make a vertical stack of cubes above the point $(x, y, 0)$. This stack goes from $z = 0$ to $z = 12 - x/4 - y/8$, so

$$\text{Volume of vertical stack} \approx \sum_z \Delta V = \sum_z \Delta x \, \Delta y \, \Delta z = \left(\sum_z \Delta z \right) \Delta x \, \Delta y.$$

Next, line up these stacks parallel to the y-axis to form a slice from $y = 0$ to $y = 16$. So

$$\text{Volume of slice} \approx \left(\sum_y \sum_z \Delta z \, \Delta y \right) \Delta x.$$

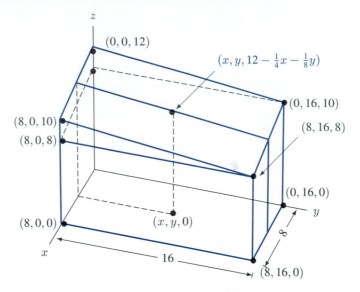

Figure 15.23: Volume of building as triple integral

Finally, line up the slices along the x-axis from $x = 0$ to $x = 8$ and add up their volumes, to get

$$\text{Volume of building} \approx \sum_x \sum_y \sum_z \Delta z \, \Delta y \, \Delta x.$$

Thus, in the limit,

$$\text{Volume of building} = \int_0^8 \int_0^{16} \int_0^{12-x/4-y/8} 1 \, dz \, dy \, dx.$$

Example 3   Set up an iterated integral to compute the mass of the solid cone bounded by $z = \sqrt{x^2 + y^2}$ and $z = 3$, if the density is given by $\delta(x, y, z) = z$.

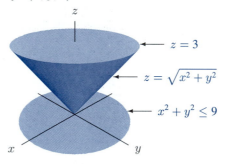

Figure 15.24

Solution   We break the cone in Figure 15.24 into small cubes of volume $\Delta V = \Delta x \, \Delta y \, \Delta z$, on which the density is approximately constant, and approximate the mass of each cube by $\delta(x, y, z) \, \Delta x \, \Delta y \, \Delta z$. Stacking the cubes vertically above the point $(x, y, 0)$, starting on the cone at height $z = \sqrt{x^2 + y^2}$ and going up to $z = 3$, tells us that the inner integral is

$$\int_{\sqrt{x^2+y^2}}^3 \delta(x, y, z) \, dz = \int_{\sqrt{x^2+y^2}}^3 z \, dz.$$

There is a stack for every point in the xy-plane in the shadow of the cone. The cone $z = \sqrt{x^2 + y^2}$ intersects the horizontal plane $z = 3$ in the circle $x^2 + y^2 = 9$, so there is a stack for all (x, y) in the region $x^2 + y^2 \le 9$. Lining up the stacks parallel to the y-axis gives a slice from $y = -\sqrt{9 - x^2}$ to $y = \sqrt{9 - x^2}$, for each fixed value of x. Thus, the limits on the middle integral are

$$\int_{-\sqrt{9-x^2}}^{\sqrt{9-x^2}} \int_{\sqrt{x^2+y^2}}^{3} z \, dz \, dy.$$

Finally, there is a slice for each x between -3 and 3, so the integral we want is

$$\text{Mass} = \int_{-3}^{3} \int_{-\sqrt{9-x^2}}^{\sqrt{9-x^2}} \int_{\sqrt{x^2+y^2}}^{3} z \, dz \, dy \, dx.$$

Notice that setting up the limits on the two outer integrals was just like setting up the limits for a double integral over the region $x^2 + y^2 \leq 9$.

As the previous example illustrates, for a region W contained between two surfaces, the innermost limits correspond to these surfaces. The middle and outer limits ensure that we integrate over the "shadow" of W in the xy-plane.

Limits on Triple Integrals

- The limits for the outer integral are constants.
- The limits for the middle integral can involve only one variable (that in the outer integral).
- The limits for the inner integral can involve two variables (those on the two outer integrals).

Problems for Section 15.3

In Problems 1–4, find the triple integrals of the given functions over the given regions.

1. $f(x, y, z) = x^2 + 5y^2 - z$, W is the rectangular box $0 \leq x \leq 2, -1 \leq y \leq 1, 2 \leq z \leq 3$.
2. $f(x, y, z) = \sin x \cos(y + z)$, W is the cube $0 \leq x \leq \pi, 0 \leq y \leq \pi, 0 \leq z \leq \pi$.
3. $h(x, y, z) = ax + by + cz$, W is the rectangular box $0 \leq x \leq 1, 0 \leq y \leq 1, 0 \leq z \leq 2$.
4. $f(x, y, z) = e^{-x-y-z}$, W is the rectangular box with corners at $(0, 0, 0), (a, 0, 0), (0, b, 0)$, and $(0, 0, c)$.

For Problems 5–11 describe or sketch the region of integration for the triple integrals. If the limits do not make sense, say why.

5. $\displaystyle\int_{0}^{6} \int_{0}^{3-x/2} \int_{0}^{6-x-2y} f(x, y, z) \, dz \, dy \, dx$ 6. $\displaystyle\int_{0}^{1} \int_{0}^{x} \int_{0}^{x} f(x, y, z) \, dz \, dy \, dx$

7. $\displaystyle\int_{0}^{1} \int_{0}^{z} \int_{0}^{x} f(x, y, z) \, dz \, dy \, dx$ 8. $\displaystyle\int_{0}^{3} \int_{-\sqrt{9-y^2}}^{0} \int_{\sqrt{x^2+y^2}}^{3} f(x, y, z) \, dz \, dx \, dy$

9. $\displaystyle\int_{1}^{3} \int_{1}^{x+y} \int_{0}^{y} f(x, y, z) \, dz \, dx \, dy$ 10. $\displaystyle\int_{0}^{1} \int_{0}^{2-x} \int_{0}^{3} f(x, y, z) \, dz \, dy \, dx$

11. $\displaystyle\int_{-1}^{1} \int_{0}^{\sqrt{1-x^2}} \int_{0}^{\sqrt{2-x^2-y^2}} f(x, y, z) \, dz \, dy \, dx$

12. Find the volume of the pyramid with base in the plane $z = -6$ and sides formed by the three planes $y = 0$ and $y - x = 4$ and $2x + y + z = 4$.

13. Find the mass of the solid bounded by the xy-plane, yz-plane, xz-plane, and the plane $(x/3) + (y/2) + (z/6) = 1$, if the density of the solid is given by $\delta(x, y, z) = x + y$.

14. Find the average value of the sum of the squares of three numbers x, y, z, where each number is between 0 and 2.

15. Set up, but do not evaluate, an iterated integral for the volume of the solid formed by the intersections of the cylinders $x^2 + z^2 = 1$ and $y^2 + z^2 = 1$.

The motion of a solid object can be analyzed by thinking of the mass as concentrated at a single point, the *center of mass*. If the object has density $\rho(x, y, z)$ at the point (x, y, z) and occupies a region W, then the coordinates $(\bar{x}, \bar{y}, \bar{z})$ of the center of mass are given by

$$\bar{x} = \frac{1}{m} \int_W x\rho \, dV \qquad \bar{y} = \frac{1}{m} \int_W y\rho \, dV \qquad \bar{z} = \frac{1}{m} \int_W z\rho \, dV$$

where $m = \int_W \rho \, dV$ is the total mass of the body. Use these definitions for Problems 16–17.

16. A solid is bounded below by the square $z = 0, 0 \le x \le 1, 0 \le y \le 1$ and above by the surface $z = x + y + 1$. Find the total mass and the coordinates of the center of mass if the density is 1 gm/cm^3 and x, y, z are measured in centimeters.

17. Find the center of mass of the tetrahedron that is bounded by the x, y, and z planes and the plane $x + 2y + 3z = 1$. Assume the density is 1 gm/cm^3 and x, y, z are in centimeters.

The *moment of inertia* of a solid body about an axis in 3-space gives the angular acceleration about this axis for a given torque (force twisting the body). The moments of inertia about the coordinate axes of a body of constant density and mass m occupying a region W of volume V are defined to be

$$I_x = \frac{m}{V} \int_W (y^2 + z^2) \, dV \qquad I_y = \frac{m}{V} \int_W (x^2 + z^2) \, dV \qquad I_z = \frac{m}{V} \int_W (x^2 + y^2) \, dV$$

Use these definitions for Problems 18–20.

18. Find the moment of inertia about the z-axis of the rectangular solid of mass m given by $0 \le x \le 1$, $0 \le y \le 2, 0 \le z \le 3$.

19. Find the moment of inertia about the x-axis of the rectangular solid $-a \le x \le a$, $-b \le y \le b$ and $-c \le z \le c$ of mass m.

20. Let a, b, and c denote the moments of inertia of a homogeneous solid object about the x, y and z-axes respectively. Explain why $a + b > c$.

15.4 DOUBLE INTEGRALS IN POLAR COORDINATES

Integration in Polar Coordinates

We started this chapter by putting a rectangular grid on the fox population density map, to estimate the total population using a Riemann sum. However, sometimes a polar grid is more appropriate. A review of polar coordinates is in Appendix A.

Example 1   A biologist studying insect populations around a circular lake divides the area into polar sectors as in Figure 15.25. The population density in each sector is shown in millions per square km. Estimate the total insect population around the lake.

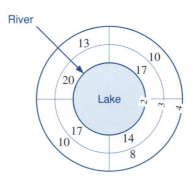

Figure 15.25: An insect infested lake showing the insect population density by sector

Solution   To get our estimate we will multiply the population density in each sector by the area of that sector. Unlike the rectangles in a rectangular grid, the sectors in this grid do not all have the same area. The inner sectors have area

$$\frac{1}{4}(\pi3^2 - \pi2^2) = \frac{5\pi}{4} \approx 3.93 \text{ km}^2,$$

and the outer sectors have area

$$\frac{1}{4}(\pi4^2 - \pi3^2) = \frac{7\pi}{4} \approx 5.50 \text{ km}^2,$$

so we estimate

$$\text{Population} \approx (20)(3.93) + (17)(3.93) + (14)(3.93) + (17)(3.93) +$$
$$(13)(5.50) + (10)(5.50) + (8)(5.50) + (10)(5.50)$$
$$\approx 493 \text{ million insects.}$$

What is dA in Polar Coordinates?

The previous example used a polar grid rather than a rectangular grid. A rectangular grid is constructed from vertical and horizontal lines corresponding to $x = k$ (a constant) and $y = l$ (another constant), respectively. In polar coordinates, setting $r = k$ gives a circle of radius k centered at the origin and setting $\theta = l$ gives a ray emanating from the origin (at angle l with the x-axis). A polar grid is built out of these circles and rays. Figure 15.26 shows a subdivision of the polar region $a \leq r \leq b$, $\alpha \leq \theta \leq \beta$, using n subdivisions each way. This bent rectangle is the sort of region that is naturally represented in polar coordinates.

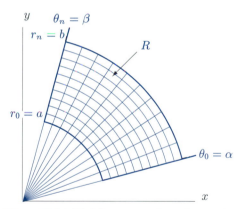

Figure 15.26: Dividing up a region using a polar grid

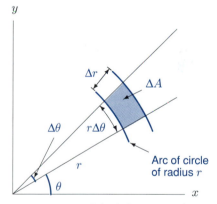

Figure 15.27: Calculating area ΔA in polar coordinates

In general, dividing R as shown in Figure 15.26, gives a Riemann sum:

$$\sum_{i,j} f(r_i, \theta_j) \Delta A.$$

However, calculating the area ΔA is more complicated in polar coordinates than in Cartesian coordinates. Figure 15.27 shows ΔA. If Δr and $\Delta \theta$ are small, the shaded region is approximately a rectangle with sides $r \Delta \theta$ and Δr, so

$$\Delta A \approx r \Delta \theta \Delta r.$$

Thus, the Riemann sum is approximately

$$\sum_{i,j} f(r_i, \theta_j) r_i \Delta \theta \Delta r.$$

If we take the limit as Δr and $\Delta\theta$ approach 0, we obtain

$$\int_R f\, dA = \int_\alpha^\beta \int_a^b f(r,\theta)\, r\, dr\, d\theta.$$

When computing integrals in polar coordinates, put $dA = r\, dr\, d\theta$ or $dA = r\, d\theta\, dr$.

Example 2 Compute the integral of $f(x,y) = 1/(x^2+y^2)^{3/2}$ over the region R shown in Figure 15.28.

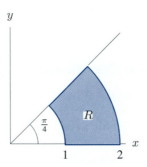

Figure 15.28: Integrate f
over the polar region

Solution The region R is described by the inequalities $1 \le r \le 2$, $0 \le \theta \le \pi/4$. In polar coordinates, $r = \sqrt{x^2+y^2}$, so we can write f as

$$f(x,y) = \frac{1}{(r^2)^{3/2}} = \frac{1}{r^3}.$$

Then

$$\int_R f\, dA = \int_0^{\pi/4} \int_1^2 \frac{1}{r^3}\, r\, dr\, d\theta = \int_0^{\pi/4}\left(\int_1^2 r^{-2}\, dr\right) d\theta$$

$$= \int_0^{\pi/4}\left[-\frac{1}{r}\right]_{r=1}^{r=2} d\theta = \int_0^{\pi/4}\frac{1}{2}\, d\theta = \frac{\pi}{8}.$$

Example 3 For each region in Figure 15.29, decide whether to integrate using polar or Cartesian coordinates. On the basis of its shape, write an iterated integral of an arbitrary function $f(x,y)$ over the region.

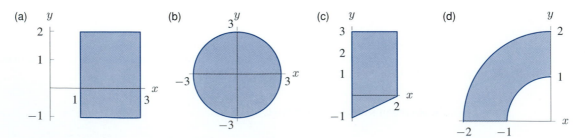

Figure 15.29

Solution (a) Since this is a rectangular region, Cartesian coordinates are likely to be a better choice. The rectangle is described by the inequalities $1 \le x \le 3$ and $-1 \le y \le 2$, so the integral is

$$\int_{-1}^{2} \int_{1}^{3} f(x, y) \, dx \, dy.$$

(b) A circle is best described in polar coordinates. The radius is 3, so r goes from 0 to 3, and to describe the whole circle, θ goes from 0 to 2π. The integral is

$$\int_{0}^{2\pi} \int_{0}^{3} f(r \cos\theta, r \sin\theta) \, r \, dr \, d\theta.$$

(c) The bottom boundary of this trapezoid is the line $y = (x/2) - 1$ and the top is the line $y = 3$, so we use Cartesian coordinates. If we integrate with respect to y first, the lower limit of the integral is $(x/2) - 1$ and the upper limit is 3. The x limits are $x = 0$ to $x = 2$. So the integral is

$$\int_{0}^{2} \int_{(x/2)-1}^{3} f(x, y) \, dy \, dx.$$

(d) This is another polar region: it is a piece of a ring in which r goes from 1 to 2. Since it is in the second quadrant, θ goes from $\pi/2$ to π. The integral is

$$\int_{\pi/2}^{\pi} \int_{1}^{2} f(r \cos\theta, r \sin\theta) \, r \, dr \, d\theta.$$

Problems for Section 15.4

For each of the regions R in Problems 1–4, write $\int_{R} f \, dA$ as an iterated integral in polar coordinates.

1.

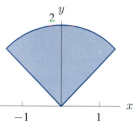

2.

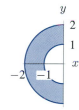

3.

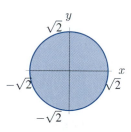

4.

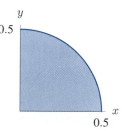

Sketch the region over which the integrals in Problems 5–11 are computed.

5. $\displaystyle\int_{0}^{2\pi} \int_{1}^{2} f(r, \theta) \, r \, dr \, d\theta.$

6. $\displaystyle\int_{\pi/2}^{\pi} \int_{0}^{1} f(r, \theta) \, r \, dr \, d\theta.$

7. $\displaystyle\int_{\pi/6}^{\pi/3} \int_{0}^{1} f(r, \theta) \, r \, dr \, d\theta.$

8. $\displaystyle\int_{3}^{4} \int_{3\pi/4}^{3\pi/2} f(r, \theta) \, r \, d\theta \, dr.$

9. $\displaystyle\int_{0}^{\pi/4} \int_{0}^{1/\cos\theta} f(r, \theta) \, r \, dr \, d\theta.$

10. $\displaystyle\int_{\pi/4}^{\pi/2} \int_{0}^{2/\sin\theta} f(r, \theta) \, r \, dr \, d\theta.$

11. $\displaystyle\int_{0}^{4} \int_{-\pi/2}^{\pi/2} f(r, \theta) \, r \, d\theta \, dr.$

12. Evaluate $\displaystyle\int_R \sin(x^2 + y^2)\, dA$, where R is the disk of radius 2 centered at the origin.

13. Evaluate $\displaystyle\int_R (x^2 - y^2)\, dA$, where R is the first quadrant region between the circles of radius 1 and radius 2.

14. Consider the integral $\displaystyle\int_0^3 \int_{x/3}^1 f(x, y)\, dy\, dx$.

     (a) Sketch the region R over which the integration is being performed.
     (b) Rewrite the integral with the order of integration reversed.
     (c) Rewrite the integral in polar coordinates.

Convert the integrals in Problems 15–17 to polar coordinates and evaluate.

15. $\displaystyle\int_{-1}^0 \int_{-\sqrt{1-x^2}}^{\sqrt{1-x^2}} x\, dy\, dx$ 16. $\displaystyle\int_0^{\sqrt{2}} \int_y^{\sqrt{4-y^2}} xy\, dx\, dy$ 17. $\displaystyle\int_0^{\sqrt{6}} \int_{-x}^x dy\, dx$

18. Find the volume of the region between the graph of $f(x, y) = 25 - x^2 - y^2$ and the xy plane.

19. An ice cream cone can be modeled by the region bounded by the hemisphere $z = \sqrt{8 - x^2 - y^2}$ and the cone $z = \sqrt{x^2 + y^2}$. Find its volume.

20. A disk of radius 5 cm has density 10 gm/cm^2 at its center, density 0 at its edge, and its density is a linear function of the distance from the center. Find the mass of the disk.

21. A city by the ocean surrounds a bay as shown in Figure 15.30. The population density of the city (in thousands of people per square km) is given by the function $\delta(r, \theta)$, where r and θ are polar coordinates with respect to the x and y-axes shown, and the distances indicated on the y-axis are in km.

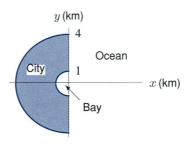

Figure 15.30

     (a) Set up an iterated integral in polar coordinates giving the total population of the city.
     (b) The population density decreases the farther you live from the shoreline of the bay; it also decreases the farther you live from the ocean. Which of the following functions best describes this situation?

     (i) $\delta(r, \theta) = (4 - r)(2 + \cos\theta)$ (ii) $\delta(r, \theta) = (4 - r)(2 + \sin\theta)$
     (iii) $\delta(r, \theta) = (r + 4)(2 + \cos\theta)$

     (c) Estimate the population using your answers to parts (a) and (b).

22. A watch spring lies flat on the table. It is made of a coiled steel strip standing a height of 0.2 inches above the table. The inner edge is the spiral $r = 0.25 + 0.04\theta$, where $0 \le \theta \le 4\pi$ (so the spiral makes two complete turns). The outer edge is given by $r = 0.26 + 0.04\theta$. Find the volume of the spring.

15.5 INTEGRALS IN CYLINDRICAL AND SPHERICAL COORDINATES

Some double integrals are easier to evaluate in polar, rather than Cartesian, coordinates. Similarly, some triple integrals are easier in non-Cartesian coordinates.

Cylindrical Coordinates

The cylindrical coordinates of a point (x, y, z) in 3-space are obtained by representing the x and y coordinates in polar coordinates and letting the z-coordinate be the z-coordinate of the Cartesian coordinate system. (See Figure 15.31.)

Relation between Cartesian and Cylindrical Coordinates

Each point in 3-space is represented using $0 \leq r < \infty, 0 \leq \theta \leq 2\pi, -\infty < z < \infty$.

$$x = r \cos \theta,$$
$$y = r \sin \theta,$$
$$z = z.$$

As with polar coordinates in the plane, note that $x^2 + y^2 = r^2$.

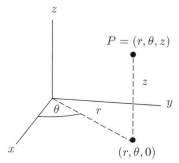

Figure 15.31: Cylindrical coordinates: (r, θ, z)

A useful way to visualize cylindrical coordinates is to sketch the surfaces obtained by setting one of the coordinates equal to a constant. See Figures 15.32–15.34.

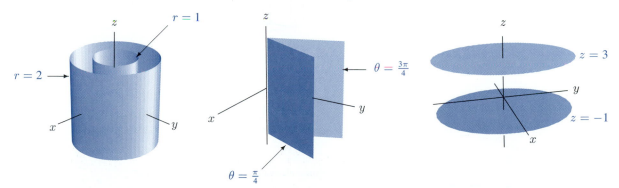

Figure 15.32: The surfaces $r = 1$ and $r = 2$

Figure 15.33: The surfaces $\theta = \pi/4$ and $\theta = 3\pi/4$

Figure 15.34: The surfaces $z = -1$ and $z = 3$

Setting $r = c$ (where c is constant) gives a cylinder around the z-axis whose radius is c. Setting $\theta = c$ gives a half-plane perpendicular to the xy plane, with one edge along the z-axis, making an angle c with the x-axis. Setting $z = c$ gives a horizontal plane $|c|$ units from the xy-plane. We call these *fundamental surfaces*.

The regions that can most easily be described in cylindrical coordinates are those regions whose boundaries are such fundamental surfaces. (For example, vertical cylinders, or wedge-shaped parts of vertical cylinders.)

Example 1   Describe in cylindrical coordinates a wedge of cheese cut from a cylinder 4 cm high and 6 cm in radius; this wedge subtends an angle of $\pi/6$ at the center. (See Figure 15.35.)

Solution   The wedge is described by the inequalities $0 \leq r \leq 6$, and $0 \leq z \leq 4$, and $0 \leq \theta \leq \pi/6$.

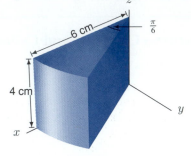

Figure 15.35: A wedge of cheese

Integration in Cylindrical Coordinates

To integrate in polar coordinates, we had to express the area element dA in terms of polar coordinates: $dA = r\, dr\, d\theta$. To evaluate a triple integral $\int_W f\, dV$ in cylindrical coordinates, we need to express the volume element dV in cylindrical coordinates.

Consider the volume element ΔV, shown in Figure 15.36. It is a wedge bounded by fundamental surfaces. The area of the base is $\Delta A \approx r\Delta r\Delta\theta$. Since the height is Δz, the volume element is given approximately by $\Delta V \approx r\, \Delta r\, \Delta\theta\, \Delta z$.

> When computing integrals in cylindrical coordinates, put $dV = r\, dr\, d\theta\, dz$. Other orders of integration are also possible.

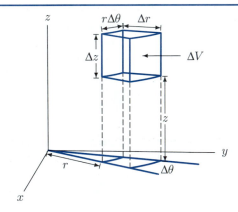

Figure 15.36: Volume element in cylindrical coordinates

Example 2   Find the mass of the wedge of cheese in Example 1, if its density is 1.2 grams/cm^3.

Solution   If the wedge is W, its mass is

$$\int_W 1.2\, dV.$$

In cylindrical coordinates this integral is

$$\int_0^4 \int_0^{\pi/6} \int_0^6 1.2\, r\, dr\, d\theta\, dz = \int_0^4 \int_0^{\pi/6} 0.6r^2 \Big|_0^6 d\theta\, dz = 21.6 \int_0^4 \int_0^{\pi/6} d\theta\, dz$$

$$= 21.6 (\frac{\pi}{6})4 \approx 45.24 \text{ grams.}$$

Example 3   A water tank in the shape of a hemisphere has radius a; its base is its plane face. Find the volume, V, of water in the tank as a function of h, the depth of the water.

Solution   In Cartesian coordinates a sphere of radius a has the equation $x^2 + y^2 + z^2 = a^2$. (See Figure 15.37.) In cylindrical coordinates, $r^2 = x^2 + y^2$, so this becomes

$$r^2 + z^2 = a^2.$$

Thus, if we want to describe the amount of water in the tank in cylindrical coordinates, we let r go from 0 to $\sqrt{a^2 - z^2}$, we let θ go from 0 to 2π, and we let z go from 0 to h, giving

$$\begin{aligned}
\text{Volume of water} &= \int_W dV = \int_0^{2\pi} \int_0^h \int_0^{\sqrt{a^2-z^2}} r\, dr\, dz\, d\theta = \int_0^{2\pi} \int_0^h \left.\frac{r^2}{2}\right|_{r=0}^{r=\sqrt{a^2-z^2}} dz\, d\theta \\
&= \int_0^{2\pi} \int_0^h \frac{1}{2}(a^2 - z^2)\, dz\, d\theta = \int_0^{2\pi} \left.\frac{1}{2}\left(a^2 z - \frac{z^3}{3}\right)\right|_{z=0}^{z=h} d\theta \\
&= \int_0^{2\pi} \frac{1}{2}\left(a^2 h - \frac{h^3}{3}\right) d\theta = \pi\left(a^2 h - \frac{h^3}{3}\right).
\end{aligned}$$

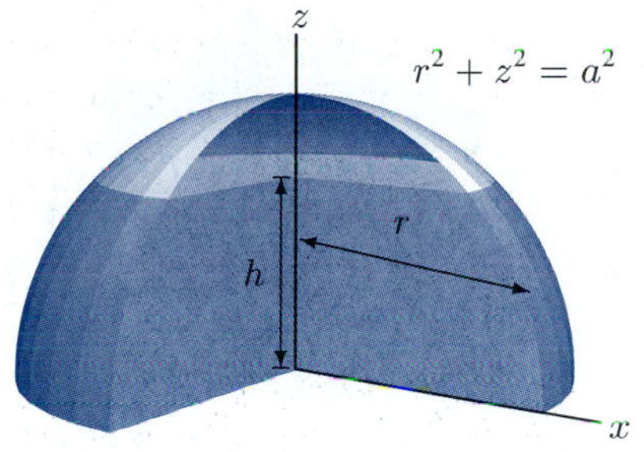

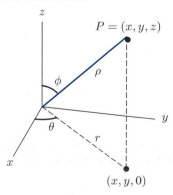

Figure 15.37: Hemispherical water tank with radius a and water of depth h

Spherical Coordinates

In Figure 15.38, the point P has coordinates (x, y, z) in the Cartesian coordinate system. We define spherical coordinates ρ, ϕ, and θ for P as follows: $\rho = \sqrt{x^2 + y^2 + z^2}$ is the distance of P from the origin; ϕ is the angle between the positive z-axis and the line through the origin and the point P; and θ is the same as in cylindrical coordinates.

Figure 15.38: Spherical coordinates

In cylindrical coordinates,

$$x = r\cos\theta, \quad \text{and} \quad y = r\sin\theta, \quad \text{and} \quad z = z.$$

From Figure 15.38 we have $z = \rho\cos\phi$ and $r = \rho\sin\phi$, giving the following relationship:

Relation between Cartesian and Spherical Coordinates

Each point in 3-space is represented using $0 \leq \rho < \infty, 0 \leq \phi \leq \pi$, and $0 \leq \theta \leq 2\pi$.

$$x = \rho \sin \phi \cos \theta$$
$$y = \rho \sin \phi \sin \theta$$
$$z = \rho \cos \phi.$$

Also, $\rho^2 = x^2 + y^2 + z^2$.

This system of coordinates is useful when there is spherical symmetry with respect to the origin, either in the region of integration or in the integrand. The fundamental surfaces in spherical coordinates are $\rho = k$ (a constant), which is a sphere of radius k centered at the origin, $\theta = k$ (a constant), which is the half-plane with its edge along the z-axis, and $\phi = k$ (a constant), which is a cone if $k \neq \pi/2$ and the xy-plane if $k = \pi/2$. (See Figures 15.39-15.41.)

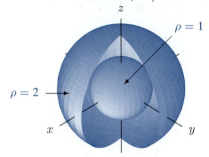

Figure 15.39: The surfaces $\rho = 1$ and $\rho = 2$

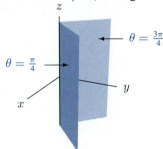

Figure 15.40: The surfaces $\theta = \pi/4$ and $\theta = 3\pi/4$

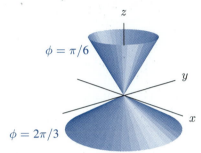

Figure 15.41: The surfaces $\phi = \pi/6$ and $\phi = 2\pi/3$

Integration in Spherical Coordinates

To use spherical coordinates in triple integrals we need to express the volume element, dV, in spherical coordinates. From Figure 15.42, we see that the volume element can be approximated by a box with curved edges. One edge has length $\Delta \rho$. The edge parallel to the xy-plane is an arc of a circle made from rotating the cylindrical radius r ($= \rho \sin \phi$) through an angle $\Delta \theta$, and so has length $\rho \sin \phi \, \Delta \theta$. The remaining side comes from rotating the radius ρ through an angle $\Delta \phi$, and so has length $\rho \, \Delta \phi$. Therefore, $\Delta V \approx \Delta \rho (\rho \, \Delta \phi)(\rho \sin \phi \, \Delta \theta) = \rho^2 \sin \phi \, \Delta \rho \, \Delta \phi \, \Delta \theta$.

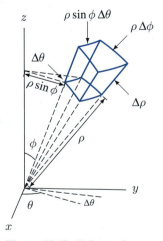

Figure 15.42: Volume element in spherical coordinates

Thus,

> When computing integrals in spherical coordinates, put $dV = \rho^2 \sin\phi \, d\rho \, d\phi \, d\theta$. Other orders of integration are also possible.

Example 4   Use spherical coordinates to derive the formula for the volume of a solid sphere of radius a.

Solution   In spherical coordinates, a sphere of radius a is described by the inequalities $0 \le \rho \le a, 0 \le \theta \le 2\pi$, and $0 \le \phi \le \pi$. Note that θ goes all the way around the circle, whereas ϕ only goes from 0 to π. We find the volume by integrating the constant density function 1 over the sphere:

$$\text{Volume} = \int_R 1 \, dV = \int_0^{2\pi} \int_0^{\pi} \int_0^a \rho^2 \sin\phi \, d\rho \, d\phi \, d\theta = \int_0^{2\pi} \int_0^{\pi} \frac{1}{3} a^3 \sin\phi \, d\phi \, d\theta$$

$$= \frac{1}{3} a^3 \int_0^{2\pi} -\cos\phi \Big|_0^{\pi} d\theta = \frac{2}{3} a^3 \int_0^{2\pi} d\theta = \frac{4\pi a^3}{3}.$$

Example 5   Find the magnitude of the gravitational force exerted by a solid hemisphere of radius a and constant density δ on a unit mass located at the center of the base of the hemisphere.

Solution   Assume the base of the hemisphere rests on the xy-plane with center at the origin. (See Figure 15.43.) Newton's law of gravitation says that the force between two masses m_1 and m_2 at a distance r apart is $F = Gm_1m_2/r^2$. In this example, symmetry shows that the net component of the force on the particle at the origin due to the hemisphere is in the z direction only. Any force in the x or y direction from some part of the hemisphere will be canceled by the force from another part of the hemisphere directly opposite the first. To compute the net z component of the gravitational force, we imagine a small piece of the hemisphere with volume ΔV, located at spherical coordinates (ρ, θ, ϕ). This piece has mass $\delta \Delta V$, and exerts a force of magnitude F on the unit mass at the origin. The z-component of this force is given by its projection onto the z-axis, which can be seen from the figure to be $F \cos\phi$. The distance from the mass $\delta \Delta V$ to the unit mass at the origin is the spherical coordinate ρ. Therefore the z-component of the force due to the small piece ΔV is

$$\begin{array}{c} z\text{-component} \\ \text{of force} \end{array} = \frac{G(\delta \Delta V)(1)}{\rho^2} \cos\phi.$$

Adding the contributions of the small pieces, we get a vertical force with magnitude

$$F = \int_0^{2\pi} \int_0^{\pi/2} \int_0^a \left(\frac{G\delta}{\rho^2}\right)(\cos\phi)\rho^2 \sin\phi \, d\rho \, d\phi \, d\theta = \int_0^{2\pi} \int_0^{\pi/2} G\delta(\cos\phi \sin\phi)\rho \Big|_{\rho=0}^{\rho=a} d\phi \, d\theta$$

$$= \int_0^{2\pi} \int_0^{\pi/2} G\delta a \cos\phi \sin\phi \, d\phi \, d\theta = \int_0^{2\pi} G\delta a \left(-\frac{(\cos\phi)^2}{2}\right) \Big|_{\phi=0}^{\phi=\pi/2} d\theta$$

$$= \int_0^{2\pi} G\delta a \left(\frac{1}{2}\right) d\theta = G\delta a\pi.$$

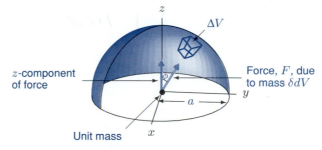

Figure 15.43: Gravitational force of hemisphere on mass at origin

Problems for Section 15.5

In Problems 1–2, evaluate the triple integrals in cylindrical coordinates.

1. $f(x, y, z) = x^2 + y^2 + z^2$, W is the region $0 \leq r \leq 4$, $\pi/4 \leq \theta \leq 3\pi/4$, $-1 \leq z \leq 1$.

2. $f(x, y, z) = \sin(x^2 + y^2)$, W is the solid cylinder with height 4 and with base of radius 1 centered on the z axis at $z = -1$.

In Problems 3–4, evaluate the triple integrals in spherical coordinates.

3. $f(x, y, z) = 1/(x^2 + y^2 + z^2)^{1/2}$ over the bottom half of the sphere of radius 5 centered at the origin.

4. $f(\rho, \theta, \phi) = \sin \phi$, over the region $0 \leq \theta \leq 2\pi$, $0 \leq \phi \leq \pi/4$, $1 \leq \rho \leq 2$.

For Problems 5–9, choose a set of coordinate axes, and then set up the three-variable integral in an appropriate coordinate system for integrating a density function δ over the given region.

5.

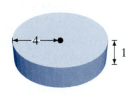

6.

7. A piece of a sphere; angle at the center is $\pi/3$.

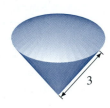

8.

9.

10. Sketch the region R over which the integration is being performed:

$$\int_0^{\pi/2} \int_{\pi/2}^{\pi} \int_0^1 f(\rho, \phi, \theta) \rho^2 \sin \phi \, d\rho \, d\phi \, d\theta.$$

Evaluate the integrals in Problems 11–12.

11. $\displaystyle \int_0^1 \int_{-\sqrt{1-x^2}}^{\sqrt{1-x^2}} \int_{-\sqrt{1-x^2-z^2}}^{\sqrt{1-x^2-z^2}} \frac{1}{(x^2 + y^2 + z^2)^{1/2}} \, dy \, dz \, dx$

12. $\displaystyle \int_0^1 \int_{-1}^1 \int_{-\sqrt{1-x^2}}^{\sqrt{1-x^2}} \frac{1}{(x^2 + y^2)^{1/2}} \, dy \, dx \, dz$

Without performing the integration, decide whether each of the integrals in Problems 13–14 is positive, negative, or zero. Give reasons for your decision.

13. W_1 is the unit ball, $x^2 + y^2 + z^2 \leq 1$.

    (a) $\int_{W_1} \sin \phi \, dV$ (b) $\int_{W_1} \cos \phi \, dV$

14. W_2 is the top half of the unit ball, $0 \leq z \leq \sqrt{1 - x^2 - y^2}$.

    (a) $\int_{W_2} (z^2 - z) \, dV$ (b) $\int_{W_2} (-xz) \, dV$

15. Write a triple integral representing the volume of a slice of the cylindrical cake of height 2 and radius 5 between the planes $\theta = \pi/6$ and $\theta = \pi/3$. Evaluate this integral.

16. Find the mass M of the solid region W given in spherical coordinates by

$$W = \{(\rho, \phi, \theta) : 0 \leq \rho \leq 3, 0 \leq \theta < 2\pi, 0 \leq \phi \leq \pi/4\}$$

if the density, $\delta(P)$, at any point P is given by the distance of P from the origin.

17. A particular spherical cloud of gas of radius 3 km is more dense at the center than towards the edge. The density, D, of the gas at a distance ρ km from the center is given by $D(\rho) = 3 - \rho$. Write an integral representing the total mass of the cloud of gas, and evaluate it.

18. Find the volume that remains after a cylindrical hole of radius R is bored through a sphere of radius a, where $0 < R < a$, passing through the center of the sphere along the pole.

19. Use appropriate coordinates to find the average distance to the origin for points in the ice cream cone region bounded by the hemisphere $z = \sqrt{8 - x^2 - y^2}$ and the cone $z = \sqrt{x^2 + y^2}$. [Hint: The volume of this region is computed in Problem 19 on page 778.]

20. Compute the force of gravity exerted by a solid cylinder of radius R, height H, and constant density δ on a unit mass at the center of the base of the cylinder.

For Problems 21–24, use the definition of center of mass given on page 774.

21. Let C be a solid cone with both height and radius 1 and contained between the surfaces $z = \sqrt{x^2 + y^2}$ and $z = 1$. If C has constant mass density of 1 gm/cm^3, find the z-coordinate of C's center of mass.

22. Suppose that the density of the cone C in Problem 21 is given by $\rho(z) = z^2$ gm/cm^3. Find

    (a) The mass of C.   (b) The z-coordinate of C's center of mass.

23. For $a > 0$, consider the family of solids bounded below by the paraboloid $z = a(x^2 + y^2)$ and above by the plane $z = 1$. If the solids all have constant mass density 1 gm/cm^3, show that the z-coordinate of the center of mass is 2/3, and hence is independent of the parameter a.

24. Find the location of the center of mass of a hemisphere of radius a and density b gm/cm^3.

For Problems 25–26, use the definition of moment of inertia given on page 774.

25. The moment of inertia of a solid homogeneous ball B of mass 1 and radius a centered at the origin is the same about any of the coordinate axes (due to the symmetry of the ball). It is easier to evaluate the sum of the three integrals involved in computing the moment of inertia about each of the axes than to compute them individually. Find the sum of the moments of inertia about the x, y and z-axes and thus find the individual moments of inertia.

26. Find the moment of inertia about the z axis of the solid "fat ice cream cone" given in spherical coordinates by $0 \leq \rho \leq a, 0 \leq \phi \leq \frac{\pi}{3}$ and $0 \leq \theta \leq 2\pi$. Assume that the solid is homogeneous with mass m.

15.6 APPLICATIONS OF INTEGRATION TO PROBABILITY

To represent how a quantity such as height or weight is distributed throughout a population, we use a density function. To study two or more quantities at the same time and see how they are related, we use a multivariable density function.

Density Functions

Distribution of Weight and Height in Expectant Mothers

Table 15.9 shows the distribution of weight and height in a survey of expectant mothers. The histogram in Figure 15.44 is constructed in such a way that the volume of each bar represents the percentage in the corresponding weight and height range. For example, the bar representing the mothers who weighed 60–70 kg and were 160–165 cm tall has base of area 10 kg · 5 cm = 50 kg cm. The volume of this bar is 12%, so its height is 12%/50 kg cm = 0.24%/ kg cm. Notice that the units on the vertical axis are percent/ kg cm. Thus, volumes under the histogram are in units of %. The total volume is 100% = 1.

TABLE 15.9 *Distribution of weight and height in a survey of expectant mothers*

	45-50 kg	50-60 kg	60-70 kg	70-80 kg	80-105 kg	Totals by height
150-155 cm	2	4	4	2	1	13
155-160 cm	0	12	8	2	1	23
160-165 cm	1	7	12	4	3	27
165-170 cm	0	8	12	6	2	28
170-180 cm	0	1	3	4	1	9
Totals by weight	3	32	39	18	8	100

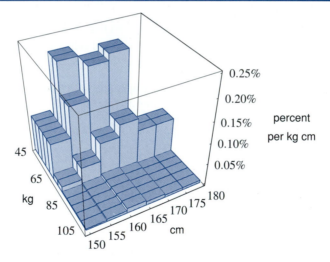

Figure 15.44: Histogram representing the data in Table 15.9.

Example 1 Find the percentage of mothers in the survey with height between 170 and 180 cm.

Solution We add the percentages across the row corresponding to the 170–180 cm height range; this is equivalent to adding the volumes of the corresponding rectangular solids in the histogram.

$$\text{Percentage of mothers} = 0 + 1 + 3 + 4 + 1 = 9\%.$$

Smoothing the Histogram

If we have smaller weight and height groups (and a larger sample), we can draw a smoother histogram and get finer estimates. In the limit, we replace the histogram with a smooth surface, in such a way that the volume under the surface above a rectangle is the percentage of mothers in that rectangle. We define a *density function*, $p(w, h)$, to be the function whose graph is the smooth surface. It has the property that

$$\begin{array}{c} \text{Fraction of sample with} \\ \text{weight between } a \text{ and } b \text{ and} \\ \text{height between } c \text{ and } d \end{array} = \begin{array}{c} \text{Volume under graph of } p \\ \text{over the rectangle} \\ a \le w \le b, c \le h \le d \end{array} = \int_a^b \int_c^d p(w, h) \, dh \, dw.$$

Joint Density Functions

We generalize this idea to represent any two characteristics, x and y, distributed throughout a population.

> A function $p(x, y)$ is called a **joint density function** for x and y if
>
> $$\begin{array}{c} \text{Fraction of population with} \\ x \text{ between } a \text{ and } b \text{ and} \\ y \text{ between } c \text{ and } d \end{array} = \begin{array}{c} \text{Volume under graph of } p \\ \text{above the rectangle} \\ a \le x \le b, c \le y \le d \end{array} = \int_a^b \int_c^d p(x, y) \, dy \, dx$$
>
> where
>
> $$\int_{-\infty}^{\infty} \int_{-\infty}^{\infty} p(x, y) \, dy \, dx = 1 \quad \text{and} \quad p(x, y) \ge 0 \text{ for all } x \text{ and } y.$$

A joint density function need not be continuous, as in Example 2 which follows. In addition, as in Example 4, the integrals involved may be improper and must be computed by methods similar to those used for improper one-variable integrals.

Example 2   Let $p(x, y)$ be defined on the square $0 \le x \le 1, 0 \le y \le 1$ by $p(x, y) = x + y$; let $p(x, y) = 0$ if (x, y) is outside this square. Verify that p is a joint density function. In terms of the distribution of x and y in the population, what does it mean that $p(x, y) = 0$ outside the square?

Solution   First, we have $p(x, y) \ge 0$ for all x and y. To verify that p is a joint density function, we check that the total volume under the graph is 1:

$$\int_{-\infty}^{\infty} \int_{-\infty}^{\infty} p(x, y) \, dy \, dx = \int_0^1 \int_0^1 (x + y) \, dy \, dx$$

$$= \int_0^1 \left(xy + \frac{y^2}{2} \right) \Big|_0^1 dx = \int_0^1 \left(x + \frac{1}{2} \right) dx = \left(\frac{x^2}{2} + \frac{x}{2} \right) \Big|_0^1 = 1$$

The fact that $p(x, y) = 0$ outside the square means that the variables x and y never take values outside the interval $[0, 1]$, that is, the value of x and y for any individual in the population is always between 0 and 1.

Example 3   Suppose two variables x and y are distributed in a population according to the density function of Example 2. Find the fraction of the population with $x \le 1/2$, the fraction with $y \le 1/2$, and the fraction with both $x \le 1/2$ and $y \le 1/2$.

Solution   The fraction with $x \le 1/2$ is the volume under the graph to the left of the line $x = 1/2$:

$$\int_0^{1/2} \int_0^1 (x + y) \, dy \, dx = \int_0^{1/2} \left(xy + \frac{y^2}{2} \right) \Big|_0^1 dx = \int_0^{1/2} \left(x + \frac{1}{2} \right) dx$$

$$= \left(\frac{x^2}{2} + \frac{x}{2} \right) \Big|_0^{1/2} = \frac{1}{8} + \frac{1}{4} = \frac{3}{8}.$$

Since the function is symmetric in x and y, the fraction with $y \le 1/2$ is also $3/8$.

Finally, the fraction with both $x \le 1/2$ and $y \le 1/2$ is

$$\int_0^{1/2} \int_0^{1/2} (x + y) \, dy \, dx = \int_0^{1/2} \left(xy + \frac{y^2}{2} \right) \Big|_0^{1/2} dx = \int_0^{1/2} \left(\frac{1}{2}x + \frac{1}{8} \right) dx$$

$$= \left(\frac{1}{4}x^2 + \frac{1}{8}x \right) \Big|_0^{1/2} = \frac{1}{16} + \frac{1}{16} = \frac{1}{8}$$

Recall that a one-variable density function $p(x)$ is a function such that $p(x) \ge 0$ for all x, and $\int_{-\infty}^{\infty} p(x) \, dx = 1$. (See Chapter 8.)

Example 4   Let p_1 and p_2 be one-variable density functions for x and y, respectively. Verify that $p(x, y) = p_1(x)p_2(y)$ is a joint density function.

Solution   Since both p_1 and p_2 are density functions, they are nonnegative everywhere. Thus, their product $p_1(x)p_2(x) = p(x, y)$ is nonnegative everywhere. Now we must check that the volume under the graph of p is 1. Since $\int_{-\infty}^{\infty} p_2(y) \, dy = 1$ and $\int_{-\infty}^{\infty} p_1(x) \, dx = 1$, we have

$$\int_{-\infty}^{\infty} \int_{-\infty}^{\infty} p(x, y) \, dy \, dx = \int_{-\infty}^{\infty} \int_{-\infty}^{\infty} p_1(x)p_2(y) \, dy \, dx = \int_{-\infty}^{\infty} p_1(x) \left(\int_{-\infty}^{\infty} p_2(y) \, dy \right) dx$$

$$= \int_{-\infty}^{\infty} p_1(x)(1) \, dx = \int_{-\infty}^{\infty} p_1(x) \, dx = 1.$$

Density Functions and Probability

What is the probability that an expectant mother weighs 60–70 kg and is 155–160 cm tall? Table 15.9 shows that 8% of mothers fall in this group, so the probability that a randomly chosen mother falls in this group is 0.08.

$$\begin{array}{ccc} \text{Probability that a mother} & \text{Volume under graph of } p & \\ \text{has weight between } a \text{ and } b & = & \text{above the rectangle} & = \int_a^b \int_c^d p(w, h) \, dh \, dw. \\ \text{and height between } c \text{ and } d & & a \le w \le b, c \le h \le d \end{array}$$

For a joint density function $p(x, y)$, the probability that x falls in an interval of width Δx around x_0 and y falls in an interval of width Δy around y_0 is approximately $p(x_0, y_0)\Delta x \Delta y$. Thus, p is often called a *probability density function*.

Example 5   A machine in a factory is set to produce components 10 cm long and 5 cm in diameter. In fact, there is a slight variation from one component to the next. A component is usable if its length and diameter deviate from the correct values by less than 0.1 cm. If the length is x cm and the diameter is y cm, the probability density function for the variation in x and y is

$$p(x, y) = \frac{50\sqrt{2}}{\pi} e^{-100(x-10)^2} e^{-50(y-5)^2}.$$

What is the probability that a component will be usable? (See Figure 15.45.)

Figure 15.45: The density function $p(x, y) = \frac{50\sqrt{2}}{\pi} e^{-100(x-10)^2} e^{-50(y-5)^2}$.

Solution We know that

Probability that x and y satisfy
$$x_0 - \Delta x \le x \le x_0 + \Delta x$$
$$y_0 - \Delta y \le y \le y_0 + \Delta y$$
$$= \frac{50\sqrt{2}}{\pi} \int_{y_0-\Delta y}^{y_0+\Delta y} \int_{x_0-\Delta x}^{x_0+\Delta x} e^{-100(x-10)^2} e^{-50(y-5)^2} \, dx \, dy.$$

Thus,

$$\begin{array}{c}\text{Probability that}\\ \text{component is usable}\end{array} = \frac{50\sqrt{2}}{\pi} \int_{4.9}^{5.1} \int_{9.9}^{10.1} e^{-100(x-10)^2} e^{-50(y-5)^2} \, dx \, dy.$$

The double integral must be evaluated numerically. This yields

$$\begin{array}{c}\text{Probability that}\\ \text{component is usable}\end{array} = \frac{50\sqrt{2}}{\pi} (0.02556) \approx 0.57530.$$

Thus, there is a 57.5% chance that the component will be usable.

Problems for Section 15.6

1. Let x and y have joint density function

$$p(x, y) = \begin{cases} \frac{2}{3}(x + 2y) & \text{for } 0 \le x \le 1, 0 \le y \le 1, \\ 0 & \text{otherwise.} \end{cases}$$

Find the probability that (a) $x > 1/3.$ (b) $x < (1/3) + y.$

2. Table 15.10 gives some values of the joint density function for two variables x and y. We assume x can take the values 1, 2, 3 and 4 and y can take the values 1, 2 and 3.
    (a) Explain why this table defines a joint density function.
    (b) What is the probability that $x = 2$?
    (c) Find the probability that $y \le 2$.
    (d) Find the probability that $x \le 3$ and $y \le 2$.

TABLE 15.10

	y = 1	2	3
x = 1	0.3	0.2	0.1
2	0.2	0.1	0
3	0.1	0	0
4	0	0	0

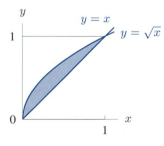

Figure 15.46

3. Assume that the joint density function for x, y is given by

$$f(x, y) = \begin{cases} kxy & \text{for } 0 \le x \le y \le 1, \\ 0 & \text{otherwise.} \end{cases}$$

    (a) Determine the value of k.
    (b) Find the probability that (x, y) lies in the shaded region in Figure 15.46.

4. A joint density function is given by

$$f(x, y) = \begin{cases} kx^2 & \text{for } 0 \leq x \leq 2 \text{ and } 0 \leq y \leq 1, \\ 0 & \text{otherwise.} \end{cases}$$

 (a) Find the value of the constant k.
 (b) Find the probability that a point (x, y) satisfies $x + y \leq 2$.
 (c) Find the probability that a point (x, y) satisfies $x \leq 1$ and $y \leq 1/2$.

5. A health insurance company wants to know what proportion of its policies are going to cost them a lot of money because the insured people are over 65 and sick. In order to compute this proportion, the company defines a *disability index*, x, with $0 \leq x \leq 1$, where $x = 0$ represents perfect health and $x = 1$ represents total disability. In addition, the company uses a density function, $f(x, y)$, defined in such a way that the quantity

$$f(x, y) \, \Delta x \, \Delta y$$

 approximates the fraction of the population with disability index between x and $x + \Delta x$, and aged between y and $y + \Delta y$. The company knows from experience that a policy no longer covers its costs if the insured person is over 65 and has a disability index exceeding 0.8. Write an expression for the fraction of the company's policies held by people meeting these criteria.

6. Assume that a point is chosen at random from the region S in the xy-plane containing all points (x, y) such that $-1 \leq x \leq 1, -2 \leq y \leq 2$ and $x - y \geq 0$ (at random means that the density function is constant on S).
 (a) Determine the joint density function for x and y.
 (b) If T is a subset of S with area α, then find the probability that a point (x, y) is in T.

7. Suppose Figure 15.47 represents a baseball field, with the bases at $(1, 0)$, $(1, 1)$, $(0, 1)$, and home plate at $(0, 0)$. The outer bound of the outfield is a piece of a circle about the origin with radius 4. Whenever a ball is hit by a batter we can record the spot on the field (i.e., in the plane) where the ball is caught.

 Let $p(r, \theta)$ be a function in the plane that gives the density of the distribution of such spots. Write an expression that represents the probability that a hit will be caught in

 (a) The right field (region R).   (b) The center field (region C).

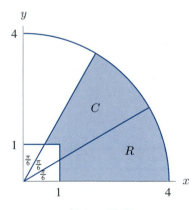

Figure 15.47

CHAPTER SUMMARY

- **Double Integral**
  Definition as a limit of Riemann sum, interpretation as volume under graph, as area, or as average value, estimating from contour diagrams, evaluating using iterated integrals, setting up in polar coordinates.

- **Triple Integral**

  Definition as a limit of Riemann sum, interpretation as volume of solid, as total mass, or as average value, evaluating using iterated integrals, setting up in cylindrical or spherical coordinates.

- **Probability**

  Joint density functions, using integrals to calculate probability .

REVIEW PROBLEMS FOR CHAPTER FIFTEEN

1. Figure 15.48 shows contours of average annual rainfall (in inches) in South America.[5] Each grid square is 500 miles on a side. Estimate the total volume of rain that falls on the considered area in a year.

2. Figure 15.49 gives isotherms for low winter temperature in Washington, DC.[6] The grid squares are 1 mile on a side. Find the average low temperature over the whole city (the city is the shaded region).

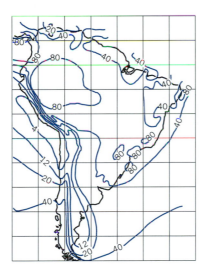

Figure 15.48   *Figure 15.49*

Sketch the regions over which the integrals in Problems 3–6 are being performed.

3. $\displaystyle\int_{1}^{4}\int_{-\sqrt{y}}^{\sqrt{y}} f(x,y)\,dx\,dy$

4. $\displaystyle\int_{0}^{1}\int_{0}^{\sin^{-1} y} f(x,y)\,dx\,dy$

5. $\displaystyle\int_{-1}^{1}\int_{-\sqrt{1-x^2}}^{\sqrt{1-x^2}} f(x,y)\,dy\,dx$

6. $\displaystyle\int_{0}^{2}\int_{-\sqrt{4-y^2}}^{0} f(x,y)\,dx\,dy$

Calculate exactly the integrals in Problems 7–8. (Your answer may contain e, π, $\sqrt{2}$, and so on.)

7. $\displaystyle\int_{0}^{1}\int_{0}^{z}\int_{0}^{2} (y+z)^7\,dx\,dy\,dz$

8. $\displaystyle\int_{0}^{1}\int_{3}^{4} \left(\sin\left(2-y\right)\right)\cos\left(3x-7\right)\,dx\,dy$

---

[5] Adapted from Alan H. Strahler and Arthur H. Strahler, *Modern Physical Geography*, Fourth Edition, p. 144 (New York: John Wiley & Sons, 1992).

[6] Adapted from H. J. de Blij and Peter O. Muller, *Physical Geography of the Global Environment*, p. 220 (New York: John Wiley & Sons, 1993).

Calculate exactly the integrals in Problems 9–12. (Your answer may contain e, π, $\sqrt{2}$, and so on.)

9. $\displaystyle\int_0^{10}\int_0^{0.1} xe^{xy}\,dy\,dx$

10. $\displaystyle\int_0^1\int_0^y (\sin^3 x)(\cos x)(\cos y)\,dx\,dy$

11. $\displaystyle\int_3^4\int_0^1 x^2 y\cos(xy)\,dy\,dx$

12. $\displaystyle\int_0^1\int_{-\sqrt{1-x^2}}^{\sqrt{1-x^2}} e^{-(x^2+y^2)}\,dy\,dx$

13. Write $\int_R f(x,y)\,dA$ as an iterated integral if R is the region in Figure 15.50.

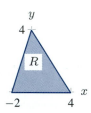

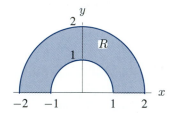

Figure 15.50   **Figure 15.51**

14. Evaluate $\int_R \sqrt{x^2+y^2}\,dA$ where R is the region in Figure 15.51.

15. Set up $\int_R f\,dV$ as an iterated integral in all six possible orders of integration, where R is the hemisphere bounded by the upper half of $x^2+y^2+z^2=1$ and the xy-plane.

Evaluate the integrals in Problems 16–18 by changing them to cylindrical or spherical coordinates as appropriate.

16. $\displaystyle\int_{-\sqrt{3}}^{\sqrt{3}}\int_{-\sqrt{3-x^2}}^{\sqrt{3-x^2}}\int_1^{4-x^2-y^2}\frac{1}{z^2}\,dz\,dy\,dx$

17. $\displaystyle\int_0^3\int_{-\sqrt{9-z^2}}^{\sqrt{9-z^2}}\int_{-\sqrt{9-y^2-z^2}}^{\sqrt{9-y^2-z^2}} x^2\,dx\,dy\,dz$

18. $\displaystyle\int_0^1\int_0^{\sqrt{1-x^2}}\int_0^{\sqrt{x^2+y^2}} (z+\sqrt{x^2+y^2})\,dz\,dy\,dx$

19. If $W=\{(x,y,z):1\le x^2+y^2\le 4, 0\le z\le 4\}$ evaluate the integral $\displaystyle\int_W \frac{z}{(x^2+y^2)^{3/2}}\,dV$.

20. Write an integral representing the mass of a sphere of radius 3 if the density of the sphere at any point is twice the distance of that point from the center of the sphere.

21. A forest next to a road has the shape in Figure 15.52. The population density of rabbits is proportional to the distance from the road. It is 0 at the road, and 10 rabbits per square mile at the opposite edge of the forest. Find the total rabbit population in the forest.

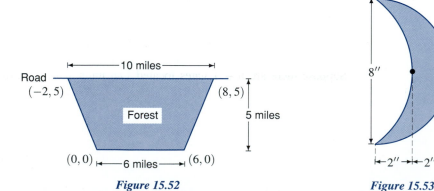

Figure 15.52   **Figure 15.53**

22. Find the area of the half-moon shape with circular arcs as edges and the dimensions shown in Figure 15.53.

For Problems 23–24, use the definition of moment of inertia on Page 774.

23. Consider a rectangular brick with length 5, width 3, and height 1, and of uniform density 1. Compute the moment of inertia about each of the three axes passing through the center of the brick, perpendicular to one of the sides.

24. Compute the moment of inertia of a ball of radius R about an axis passing through its center. Assume that the ball has a constant density of 1.

25. A particle of mass m is placed at the center of one base of a circular cylindrical shell of inner radius r_1, outer radius r_2, height h, and constant density δ. Find the force of gravitational attraction exerted by the cylinder on the particle.

PROJECTS

1. **A Connection Between e and π**

   In this problem you will derive one of the remarkable formulas of mathematics, namely that

   $$\int_{-\infty}^{\infty} e^{-x^2} dx = \sqrt{\pi}.$$

   (a) Change the following double integral into polar coordinates and evaluate it:

   $$\int_{-\infty}^{\infty} \int_{-\infty}^{\infty} e^{-(x^2+y^2)} dx dy.$$

   (b) Explain why

   $$\int_{-\infty}^{\infty} \int_{-\infty}^{\infty} e^{-(x^2+y^2)} dx dy = \left(\int_{-\infty}^{\infty} e^{-x^2} dx \right)^2.$$

   (c) Explain why the answers to parts (a) and (b) give the formula we want.

2. **Average Distance Walked to an Airport Gate**

   At airports, departure gates are often lined up in a terminal like points along a line. If you arrive at one gate and proceed to another gate for a connecting flight, what proportion of the length of the terminal will you have to walk, on average?

   (a) One way to model this situation is to randomly choose two numbers, $0 \le x \le 1$ and $0 \le y \le 1$, and calculate the average value of $|x - y|$. Use a double integral to show that, on average, you have to walk $1/3$ the length of the terminal.

   (b) The terminal gates are not actually located continuously from 0 to 1, as we assumed in part (a). There are only a finite number of gates and they are likely to be equally spaced. Suppose there are $n + 1$ gates located $1/n$ units apart from one end of the terminal ($x_0 = 0$) to the other ($x_n = 1$). Assuming that all pairs (i, j) of arrival and departure gates are equally likely, show that

   $$\text{Average distance between gates} = \frac{1}{(n+1)^2} \cdot \sum_{i=0}^{n} \sum_{j=0}^{n} \left| \frac{i}{n} - \frac{j}{n} \right|.$$

   Identify this sum as approximately (but not exactly) a Riemann sum with n subdivisions for the integrand used in part (a). Compute this sum for $n = 5$ and $n = 10$ and compare to the answer of $1/3$ obtained in part (a).

FOCUS ON THEORY

CHANGE OF VARIABLES IN A MULTIPLE INTEGRAL

In the previous sections, we used polar, cylindrical, and spherical coordinates to simplify iterated integrals. In this section, we discuss more general changes of variable. In the process, we will see where the extra factor of r comes from when we change from Cartesian to polar coordinates and the factor $\rho^2 \sin\phi$ when we change from Cartesian to spherical coordinates.

Polar Change of Variables Revisited

Consider the integral $\int_R (x+y)\,dA$ where R is the region in the first quadrant bounded by the circle $x^2 + y^2 = 16$ and the x and y-axes. Writing the integral in Cartesian and polar coordinates we have

$$\int_R (x+y)\,dA = \int_0^4 \int_0^{\sqrt{16-x^2}} (x+y)\,dy\,dx = \int_0^{\pi/2} \int_0^4 (r\cos\theta + r\sin\theta)r\,dr\,d\theta.$$

This is an integral over the rectangle in the $r\theta$-space given by $0 \le r \le 4,\ 0 \le \theta \le \pi/2$. The conversion from polar to Cartesian coordinates changes this rectangle into a quarter-disk. Figure 15.54 shows how a typical rectangle (shaded) in the $r\theta$-plane with sides of length Δr and $\Delta\theta$ corresponds to a curved rectangle in the xy-plane with sides of length Δr and $r\Delta\theta$. The extra r is needed because the correspondence between r, θ and x, y not only curves the lines $r = 1, 2, 3 \ldots$ into circles, it also stretches those lines around larger and larger circles.

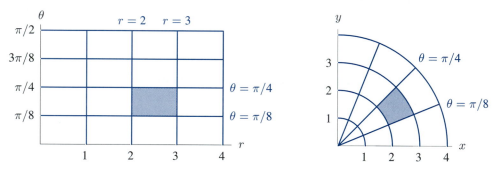

Figure 15.54: A grid in the $r\theta$-plane and the corresponding curved grid in the xy-plane

General Change of Variables

We now consider a general change of variable, where x, y coordinates are related to s, t coordinates by the differentiable functions

$$x = x(s,t) \quad y = y(s,t).$$

Just as a rectangular region in the $r\theta$-plane corresponds to a circular region in the xy-plane, a rectangular region, T, in the st-plane corresponds to a curved region, R, in the xy-plane. We assume that the change of coordinates is one-to-one, that is, that each point R corresponds to one point in T.

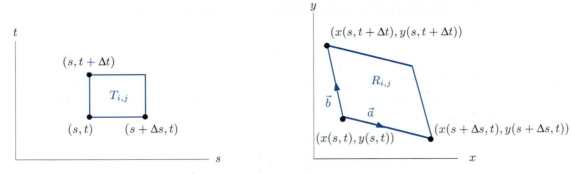

Figure 15.55: A small rectangle $T_{i,j}$ in the st-plane and the corresponding region $R_{i,j}$ of the xy-plane

We divide T into small rectangles $T_{i,j}$ with sides of length Δs and Δt. (See Figure 15.55.) The corresponding piece $R_{i,j}$ of the xy-plane is a quadrilateral with curved sides. If we choose Δs and Δt very small, then by local linearity, $R_{i,j}$ is approximately a parallelogram.

Recall from Chapter 12 that the area of the parallelogram with sides $\vec{a}$ and $\vec{b}$ is $\|\vec{a} \times \vec{b}\|$. Thus, we need to find the sides of $R_{i,j}$ as vectors. The side of $R_{i,j}$ corresponding to the bottom side of $T_{i,j}$ has endpoints $(x(s,t), y(s,t))$ and $(x(s+\Delta s, t), y(s+\Delta s, t))$, so in vector form that side is

$$\vec{a} = (x(s+\Delta s, t) - x(s,t))\vec{i} + (y(s+\Delta s, t) - y(s,t))\vec{j} + 0\vec{k} \approx \left(\frac{\partial x}{\partial s}\Delta s\right)\vec{i} + \left(\frac{\partial y}{\partial s}\Delta s\right)\vec{j} + 0\vec{k}.$$

Similarly, the side of $R_{i,j}$ corresponding to the left edge of $T_{i,j}$ is given by

$$\vec{b} \approx \left(\frac{\partial x}{\partial t}\Delta t\right)\vec{i} + \left(\frac{\partial y}{\partial t}\Delta t\right)\vec{j} + 0\vec{k}.$$

Computing the cross product, we get

$$\text{Area } R_{i,j} \approx \|\vec{a} \times \vec{b}\| \approx \left|\left(\frac{\partial x}{\partial s}\Delta s\right)\left(\frac{\partial y}{\partial t}\Delta t\right) - \left(\frac{\partial x}{\partial t}\Delta t\right)\left(\frac{\partial y}{\partial s}\Delta s\right)\right|$$

$$= \left|\frac{\partial x}{\partial s} \cdot \frac{\partial y}{\partial t} - \frac{\partial x}{\partial t} \cdot \frac{\partial y}{\partial s}\right|\Delta s \Delta t.$$

Using determinant notation, we define the *Jacobian*, $\dfrac{\partial(x,y)}{\partial(s,t)}$, as follows

$$\frac{\partial(x,y)}{\partial(s,t)} = \frac{\partial x}{\partial s} \cdot \frac{\partial y}{\partial t} - \frac{\partial x}{\partial t} \cdot \frac{\partial y}{\partial s} = \begin{vmatrix} \dfrac{\partial x}{\partial s} & \dfrac{\partial y}{\partial s} \\ \dfrac{\partial x}{\partial t} & \dfrac{\partial y}{\partial t} \end{vmatrix}.$$

Thus, we can write

$$\text{Area } R_{i,j} \approx \left|\frac{\partial(x,y)}{\partial(s,t)}\right|\Delta s \, \Delta t.$$

To compute $\int_R f(x,y)\, dA$, where f is a continuous function, we look at the Riemann sum obtained by dividing the region R into the small curved regions $R_{i,j}$, giving

$$\int_R f(x,y)\, dA \approx \sum_{i,j} f(x_i, y_j) \cdot \text{Area of } R_{i,j} \approx \sum_{i,j} f(x_i, y_j)\left|\frac{\partial(x,y)}{\partial(s,t)}\right|\Delta s \, \Delta t.$$

Each point (x_i, y_j) corresponds to a point (s_i, t_j), so the sum can be written in terms of s and t:

$$\sum_{i,j} f(x(s_i, t_j), y(s_i, t_j))\left|\frac{\partial(x,y)}{\partial(s,t)}\right|\Delta s \, \Delta t.$$

This is a Riemann sum in terms of s and t, so as Δs and Δt approach 0, we get

$$\int_R f(x,y)\, dA = \int_T f(x(s,t), y(s,t)) \left| \frac{\partial(x,y)}{\partial(s,t)} \right| ds\, dt.$$

To convert an integral from x, y to s, t coordinates we make three changes:
1. Substitute for x and y in the integrand in terms of s and t.
2. Change the xy region R into an st region T.
3. Introduce the absolute value of the Jacobian, $\left| \dfrac{\partial(x,y)}{\partial(s,t)} \right|$, representing the change in the area element.

Example 1   Verify that the Jacobian $\dfrac{\partial(x,y)}{\partial(r,\theta)} = r$ for polar coordinates $x = r\cos\theta$, $y = r\sin\theta$.

Solution $\dfrac{\partial(x,y)}{\partial(r,\theta)} = \begin{vmatrix} \frac{\partial x}{\partial r} & \frac{\partial y}{\partial r} \\ \frac{\partial x}{\partial \theta} & \frac{\partial y}{\partial \theta} \end{vmatrix} = \begin{vmatrix} \cos\theta & \sin\theta \\ -r\sin\theta & r\cos\theta \end{vmatrix} = r\cos^2\theta + r\sin^2\theta = r.$

Example 2   Find the area of the ellipse $\dfrac{x^2}{a^2} + \dfrac{y^2}{b^2} = 1$.

Solution   Let $x = as$, $y = bt$. Then the ellipse $x^2/a^2 + y^2/b^2 = 1$ in the xy-plane corresponds to the circle $s^2 + t^2 = 1$ in the st-plane. The Jacobian is $\begin{vmatrix} a & 0 \\ 0 & b \end{vmatrix} = ab$. Thus, if we let R be the ellipse in the xy-plane and T the unit circle in the st-plane, we get

$$\text{Area of } xy\text{-ellipse} = \int_R 1\, dA = \int_T 1ab\, ds\, dt = ab\int_T ds\, dt = ab \cdot \text{Area of } st\text{-circle} = \pi ab.$$

Change of Variables in Triple Integrals

For triple integrals, there is a similar formula. Suppose the differentiable functions

$$x = x(s,t,u), \quad y = y(s,t,u), \quad z = z(s,t,u)$$

define a change of variables from a region S in stu-space to a region W in xyz-space. Then, the Jacobian of this change of variables is given by the determinant

$$\frac{\partial(x,y,z)}{\partial(s,t,u)} = \begin{vmatrix} \frac{\partial x}{\partial s} & \frac{\partial y}{\partial s} & \frac{\partial z}{\partial s} \\ \frac{\partial x}{\partial t} & \frac{\partial y}{\partial t} & \frac{\partial z}{\partial t} \\ \frac{\partial x}{\partial u} & \frac{\partial y}{\partial u} & \frac{\partial z}{\partial u} \end{vmatrix}.$$

Just as the Jacobian in two dimensions gives us the change in the area element, the Jacobian in three dimensions represents the change in the volume element. Thus, we have

$$\int_W f(x,y,z)\, dx\, dy\, dz = \int_S f(x(s,t,u), y(s,t,u), z(s,t,u)) \left| \frac{\partial(x,y,z)}{\partial(s,t,u)} \right| ds\, dt\, du.$$

Problem 3 at the end of this section asks you to verify that the Jacobian for the change of variables for spherical coordinates is $\rho^2 \sin\phi$. The next example generalizes Example 2 to ellipsoids.

Example 3   Find the volume of the ellipsoid $\dfrac{x^2}{a^2} + \dfrac{y^2}{b^2} + \dfrac{z^2}{c^2} = 1$.

Solution   Let $x = as$, $y = bt$, $z = cu$. The Jacobian is computed to be abc. The xyz-ellipsoid corresponds to the stu-sphere $s^2 + t^2 + u^2 = 1$. Thus, as in Example 2,

$$\text{Volume of } xyz\text{-ellipsoid} = abc \cdot \text{Volume of } stu\text{-sphere} = abc\frac{4}{3}\pi = \frac{4}{3}\pi abc.$$

Problems on Change of Variables in a Multiple Integral

1.   Find the region R in the xy-plane corresponding to the region $T = \{(s,t) \mid 0 \le s \le 3,\, 0 \le t \le 2\}$ under the change of variables $x = 2s - 3t$, $y = s - 2t$. Check that

$$\int_R dx\, dy = \int_T \left| \frac{\partial(x,y)}{\partial(s,t)} \right| ds\, dt.$$

2.   Find the region R in the xy-plane corresponding to the region $T = \{(s,t) \mid 0 \le s \le 2,\, s \le t \le 2\}$ under the change of variables $x = s^2$, $y = t$. Check that

$$\int_R dx\, dy = \int_T \left| \frac{\partial(x,y)}{\partial(s,t)} \right| ds\, dt.$$

3.   Compute the Jacobian for the change of variables into spherical coordinates:

$$x = \rho \sin\phi \cos\theta, \quad y = \rho \sin\phi \sin\theta, \quad z = \rho \cos\phi.$$

4.   For the change of variables $x = 3s - 4t$, $y = 5s + 2t$, show that

$$\frac{\partial(x,y)}{\partial(s,t)} \cdot \frac{\partial(s,t)}{\partial(x,y)} = 1$$

5.   Use the change of variables $x = 2s + t$, $y = s - t$ to compute the integral $\int_R (x + y)\, dA$, where R is the parallelogram formed by $(0,0)$, $(3,-3)$, $(5,-2)$, and $(2,1)$.

6.   Use the change of variables $x = \frac{1}{2}s$, $y = \frac{1}{3}t$ to compute the integral $\int_R (x^2 + y^2)\, dA$, where R is the region bounded by the curve $4x^2 + 9y^2 = 36$.

7.   Use the change of variables $s = xy$, $t = xy^2$ to compute $\int_R xy^2\, dA$, where R is the region bounded by $xy = 1$, $xy = 4$, $xy^2 = 1$, $xy^2 = 4$.

8.   Evaluate the integral $\displaystyle\int_R \cos\left(\frac{x - y}{x + y}\right) dx\, dy$ where R is the triangle bounded by $x + y = 1$, $x = 0$, and $y = 0$.

9.   Find the area of the metal frames with one or four cutouts shown in Figure 15.56. Start with Cartesian coordinates x, y aligned along one side. Consider slanted coordinates $u = x - y$, $v = y$ in which the frame is "straightened". [Hint: First describe the shape of the cut-out in the uv-plane; second, calculate its area in the uv-plane; third, using Jacobians, calculate its area in the xy-plane.]

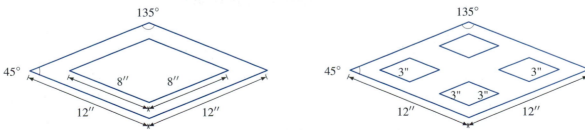

Figure 15.56

10.   A river follows the path $y = f(x)$ where x, y are in kilometers. Near the sea, it widens into a lagoon, then narrows again at its mouth. See Figure 15.57. At the point (x, y), the depth, $d(x, y)$, of the lagoon is given by

$$d(x, y) = 40 - 160(y - f(x))^2 - 40x^2 \text{ meters.}$$

The lagoon itself is described by $d(x, y) \geq 0$. What is the volume of the lagoon in cubic meters? [Hint: Use new coordinates $u = x/2$, $v = y - f(x)$ and Jacobians.]

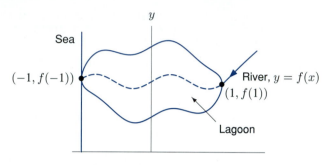

Figure 15.57

CHAPTER SIXTEEN

PARAMETERIZED CURVES

In single-variable calculus, we study the motion of a particle along a line. For example, we represent the motion of an object thrown straight up into the air by a single function $h(t)$, the height of the object above the ground at time t.

To study the motion of a particle in space, we must express all the coordinates of the particle in terms of t, giving $x(t)$, $y(t)$, and $z(t)$ if the motion is in 3-space. This is called the *parametric representation* of the path of motion, which is a curve. The parametric representation enables us to find the velocity and acceleration of the particle.

16.1 PARAMETERIZED CURVES

How Do We Represent Motion?

To represent the motion of a particle in the xy-plane we use two equations, one for the x-coordinate of the particle, $x = f(t)$, and another for the y-coordinate, $y = g(t)$. Thus at time t the particle is at the point $(f(t), g(t))$. The equation for x describes the right-left motion; the equation for y describes the up-down motion. The two equations for x and y are called *parametric equations* with *parameter t*.

Example 1   Describe the motion of the particle whose coordinates at time t are $x = \cos t, y = \sin t$.

Solution   Since $(\cos t)^2 + (\sin t)^2 = 1$, we have $x^2 + y^2 = 1$. That is, at any time t the particle is at a point (x, y) somewhere on the unit circle $x^2 + y^2 = 1$. We plot points at different times to see how the particle moves on the circle. (See Figure 16.1 and Table 16.1.) The particle moves at a uniform speed, completing one full trip counterclockwise around the circle every 2π units of time. Notice how the x-coordinate goes repeatedly back and forth from -1 to 1 while the y-coordinate goes repeatedly up and down from -1 to 1. The two motions combine to trace out a circle.

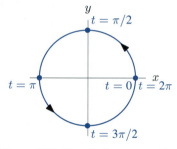

Figure 16.1: The circle parameterized by $x = \cos t, y = \sin t$

TABLE 16.1   *Points on the circle with* $x = \cos t, y = \sin t$

t	x	y
0	1	0
$\pi/2$	0	1
π	-1	0
$3\pi/2$	0	-1
2π	1	0

Example 2   Figure 16.2 shows the graphs of two functions, $f(t)$ and $g(t)$. Describe the motion of the particle whose coordinates at time t are $x = f(t),\quad y = g(t)$.

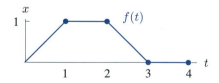 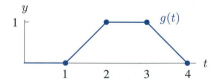

Figure 16.2: Graphs of $x = f(t)$ and $y = g(t)$ used to trace out the path $(f(t), g(t))$ in Figure 16.3

Solution   Between times $t = 0$ and $t = 1$, the x-coordinate goes from 0 to 1, while the y-coordinate stays fixed at 0. So the particle moves along the x-axis from $(0, 0)$ to $(1, 0)$. Then, between times $t = 1$ and $t = 2$, the x-coordinate stays fixed at $x = 1$, while the y-coordinate goes from 0 to 1. Thus, the particle moves along the vertical line from $(1, 0)$ to $(1, 1)$. Similarly, between times $t = 2$ and $t = 3$, it moves horizontally back to $(0, 1)$, and between times $t = 3$ and $t = 4$ it moves down the y-axis to $(0, 0)$. Thus, it traces out the square in Figure 16.3.

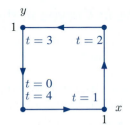

Figure 16.3: The square parameterized by $(f(t), g(t))$

Different Motions Along the Same Path

Example 3   Describe the motion of the particle whose x and y coordinates at time t are given by the equations

$$x = \cos 3t, \quad y = \sin 3t.$$

Solution   Since $(\cos 3t)^2 + (\sin 3t)^2 = 1$, we have $x^2 + y^2 = 1$, giving motion around the unit circle. But if we plot points at different times, we see that in this case the particle is moving three times as fast as in Example 1 on page 800. (See Figure 16.4 and Table 16.2.)

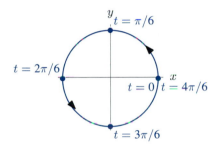

Figure 16.4: The circle parameterized by $x = \cos 3t$, $y = \sin 3t$

TABLE 16.2   *Points on circle with $x = \cos 3t$, $y = \sin 3t$*

t	x	y
0	1	0
$\pi/6$	0	1
$2\pi/6$	-1	0
$3\pi/6$	0	-1
$4\pi/6$	1	0

Example 3 is obtained from Example 1 by replacing t by $3t$; this is called a *change in parameter*. If we make a change in parameter, the particle traces out the same curve (or a part of it) but at a different speed or in a different direction. Section 16.2 shows how to compute the speed of a moving particle.

Example 4   Describe the motion of the particle whose x and y coordinates at time t are

$$x = \cos(e^{-t^2}), \quad y = \sin(e^{-t^2}).$$

Solution   As in Examples 1 and 3, we have $x^2 + y^2 = 1$ so the motion lies on the unit circle. As time t goes from $-\infty$ (way back in the past) to 0 (the present) to ∞ (way off in the future), e^{-t^2} goes from near 0 to 1 back to near 0. So $(x, y) = (\cos(e^{-t^2}), \sin(e^{-t^2}))$ goes from near $(1, 0)$ to $(\cos 1, \sin 1)$ and back to near $(1, 0)$. The particle does not actually reach the point $(1, 0)$. (See Figure 16.5 and Table 16.3.)

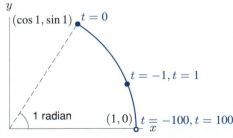

Figure 16.5: The circle parameterized by $x = \cos(e^{-t^2})$, $y = \sin(e^{-t^2})$

TABLE 16.3   *Points on circle with $x = \cos(e^{-t^2})$, $y = \sin(e^{-t^2})$*

t	x	y
-100	~ 1	~ 0
-1	0.93	0.36
0	0.54	0.84
1	0.93	0.36
100	~ 1	~ 0

Parametric Representations of Curves in the Plane

Sometimes we are more interested in the curve traced out by the particle than we are in the motion itself. In that case we will call the parametric equations a *parameterization* of the curve. As we can see by comparing Examples 1 and 3, two different parameterizations can describe the same curve in 2-space. Though the parameter, which we usually denote by t, may not have physical meaning it is still helpful to think of it as time.

Example 5   Give a parameterization of the semicircle of radius 1 shown in Figure 16.6.

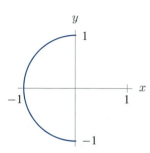

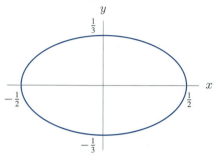

Figure 16.6: Find a parameterization of this semicircle

Figure 16.7: Find a parameterization of the ellipse $4x^2 + 9y^2 = 1$

Solution   We can use the equations $x = \cos t$ and $y = \sin t$ for counterclockwise motion in a circle, from Example 1 on page 800. The particle passes $(0, 1)$ at $t = \pi/2$, moves counterclockwise around the circle, and reaches $(0, -1)$ at $t = 3\pi/2$. So a parameterization is

$$x = \cos t, \ y = \sin t, \quad \frac{\pi}{2} \leq t \leq \frac{3\pi}{2}.$$

Example 6   Give a parameterization of the ellipse $4x^2 + 9y^2 = 1$ shown in Figure 16.7.

Solution   Since $(2x)^2 + (3y)^2 = 1$, we adapt the parameterization of the circle in Example 1. Replacing x by $2x$ and y by $3y$ gives the equations $2x = \cos t, 3y = \sin t$. A parameterization of the ellipse is thus

$$x = \tfrac{1}{2} \cos t, \qquad y = \tfrac{1}{3} \sin t, \qquad 0 \leq t \leq 2\pi.$$

We usually require that the parameterization of a curve go from one end of the curve to the other without retracing any portion of the curve. This is different from parameterizing the motion of a particle, where, for example, a particle may move around the same circle many times.

Parameterizing the Graph of a Function

The graph of any function $y = f(x)$ can be parameterized by letting the parameter t be x:

$$x = t, \quad y = f(t).$$

Example 7   Give parametric equations for the curve $y = x^3 - x$. In which direction does this parameterization trace out the curve?

Solution   Let $x = t$, $y = t^3 - t$. Thus, $y = t^3 - t = x^3 - x$. Since $x = t$, as time increases the x-coordinate moves from left to right, so the particle traces out the curve $y = x^3 - x$ from left to right.

Curves Given Parametrically

Some complicated curves can be graphed more easily using parametric equations; the next example shows such a curve.

Example 8   Assume t is time in seconds. Sketch the curve traced out by the particle whose motion is given by

$$x = \cos 3t, \quad y = \sin 5t.$$

Solution   The x-coordinate oscillates back and forth between 1 and -1, completing 3 oscillations every 2π seconds. The y-coordinate oscillates up and down between 1 and -1, completing 5 oscillations every 2π seconds. Since both the x- and y-coordinates return to their original values every 2π seconds, the curve is retraced every 2π seconds. The result is a pattern called a Lissajous figure. (See Figure 16.8.) Problems 35–38 concern Lissajous figures $x = \cos at$, $y = \sin bt$ for other values of a and b.

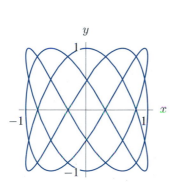

Figure 16.8: A Lissajous figure:
$x = \cos 3t, y = \sin 5t$

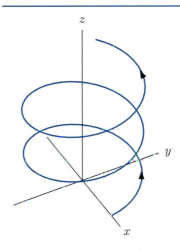

Figure 16.9: The helix $x = \cos t, y = \sin t, z = t$

Parametric Equations in Three Dimensions

To describe a motion in 3-dimensional space parametrically, we need a third equation giving z in terms of t.

Example 9   Describe in words the motion given parametrically by

$$x = \cos t, \quad y = \sin t, \quad z = t.$$

Solution   The particle's x- and y-coordinates are the same as in Example 1, which gives circular motion in the xy-plane, while the z-coordinate increases steadily. Thus, the particle traces out a rising spiral, like a coiled spring. (See Figure 16.9.) This curve is called a *helix*.

Example 10   Find parametric equations for the line through the point $(1, 5, 7)$ and parallel to the vector $2\vec{i} + 3\vec{j} + 4\vec{k}$.

Solution   Let's imagine a particle at the point $(1, 5, 7)$ at time $t = 0$ and moving through a displacement of $2\vec{i} + 3\vec{j} + 4\vec{k}$ for each unit of time, t. When $t = 0$, $x = 1$ and x increases by 2 units for every unit of time. Thus, at time t, the x-coordinate of the particle is given by

$$x = 1 + 2t.$$

Similarly, the y-coordinate starts at $y = 5$ and increases at a rate of 3 units for every unit of time. The z-coordinate starts at $y = 7$ and increases by 4 units for every unit of time. Thus, the parametric equations of the line are

$$x = 1 + 2t, \quad y = 5 + 3t, \quad z = 7 + 4t.$$

We can generalize the previous example as follows:

Parametric Equations of a Line through the point (x_0, y_0, z_0) and parallel to the vector $a\vec{i} + b\vec{j} + c\vec{k}$ are

$$x = x_0 + at, \quad y = y_0 + bt, \quad z = z_0 + ct.$$

Notice that the parameterization of a line given above expresses the coordinates x, y, and z as linear functions of the parameter t.

Example 11   (a)   Describe in words the curve given by these parametric equations:

$$x = 3 + t, \quad y = 2t, \quad z = 1 - t.$$

(b)   Find parametric equations for the line through the points $(1, 2, -1)$ and $(3, 3, 4)$.

Solution   (a)   The curve is a line through the point $(3, 0, 1)$ and parallel to the vector $\vec{i} + 2\vec{j} - \vec{k}$.

(b)   The line is parallel to the displacement vector between the points $P = (1, 2, -1)$ and $Q = (3, 3, 4)$.

$$\overrightarrow{PQ} = (3 - 1)\vec{i} + (3 - 2)\vec{j} + (4 - (-1))\vec{k} = 2\vec{i} + \vec{j} + 5\vec{k}.$$

Thus, the parametric equations are

$$x = 1 + 2t, \quad y = 2 + t, \quad z = -1 + 5t.$$

Note that the equations $x = 3 + 2t, y = 3 + t, z = 4 + 5t$ represent the same line.

Where Does a Curve Pierce a Surface?

Parametric equations for a curve enable us to find where the curve intersects a given surface.

Example 12   Find the points at which the line $x = t, y = 2t, z = 1 + t$ pierces the sphere of radius 10 centered at the origin.

Solution   The equation for the sphere of radius 10 and center at the origin is

$$x^2 + y^2 + z^2 = 100.$$

To find the intersection points of the line and the sphere, substitute the parametric equations of the line into the equation of the sphere, giving

$$t^2 + 4t^2 + (1 + t)^2 = 100,$$

so

$$6t^2 + 2t - 99 = 0,$$

which has the two solutions at approximately $t = -4.23$ and $t = 3.90$. Using the parametric equation for the line, $(x, y, z) = (t, 2t, 1 + t)$, we see that the line cuts the sphere at the two points

$$(x, y, z) = (-4.23, 2(-4.23), 1 + (-4.23)) = (-4.23, -8.46, -3.23),$$

and

$$(x, y, z) = (3.90, 2(3.90), 1 + 3.90) = (3.90, 7.80, 4.90).$$

Problems for Section 16.1

For Problems 1–4, describe the motion of a particle whose position at time t is $x = f(t)$, $y = g(t)$, where the graphs of f and g are as shown.

1.

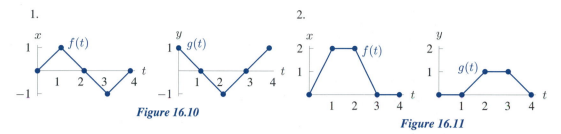

Figure 16.10

2.

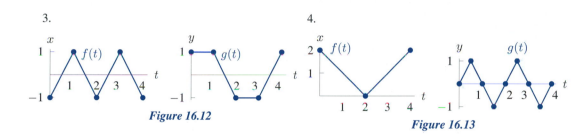

Figure 16.11

3.

Figure 16.12

4.

Figure 16.13

Problems 5–10 give parameterizations of the unit circle or a part of it. In each case, describe in words how the circle is traced out, including when and where the particle is moving clockwise and when and where the particle is moving counterclockwise.

5. $x = \cos t, \quad y = -\sin t$

6. $x = \sin t, \quad y = \cos t$

7. $x = \cos(t^2), \quad y = \sin(t^2)$

8. $x = \cos(t^3 - t), \quad y = \sin(t^3 - t)$

9. $x = \cos(\ln t), \quad y = \sin(\ln t)$

10. $x = \cos(\cos t), \quad y = \sin(\cos t)$

11. Describe the similarities and differences among the motions in the plane given by the following three pairs of parametric equations:
    (a) $x = t, \quad y = t^2$ (b) $x = t^2, \quad y = t^4$ (c) $x = t^3, \quad y = t^6$.

Write a parameterization for each of the curves in the xy-plane in Problems 12–18.

12. A circle of radius 3 centered at the origin and traced out clockwise.

13. A vertical line through the point $(-2, -3)$.

14. A circle of radius 5 centered at the point $(2, 1)$ and traced out counterclockwise.

15. A circle of radius 2 centered at the origin traced clockwise starting from $(-2, 0)$ when $t = 0$.

16. The line through the points $(2, -1)$ and $(1, 3)$.

17. An ellipse centered at the origin and crossing the x-axis at ± 5 and the y-axis at ± 7.

18. An ellipse centered at the origin, crossing the x-axis at ± 3 and the y-axis at ± 7. Start at the point $(-3, 0)$ and trace out the ellipse counterclockwise.

19. As t varies, the following parametric equations trace out a line in the plane

$$x = 2 + 3t, \quad y = 4 + 7t.$$

    (a) What part of the line is obtained by restricting t to nonnegative numbers?
    (b) What part of the line is obtained if t is restricted to $-1 \leq t \leq 0$?
    (c) How should t be restricted to give the part of the line to the left of the y-axis?

20. Suppose $a, b, c, d, m, n, p, q > 0$. Match each pair of parametric equations below with one of the lines l_1, l_2, l_3, l_4 in Figure 16.14.

I. $\begin{cases} x = a + ct, \\ y = -b + dt. \end{cases}$ II. $\begin{cases} x = m + pt, \\ y = n - qt. \end{cases}$

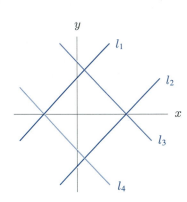

Figure 16.14

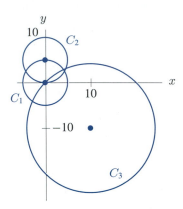

Figure 16.15

21. What can you say about the values of a, b and k if the equations

$$x = a + k \cos t, \quad y = b + k \sin t, \quad 0 \le t \le 2\pi,$$

trace out each of the circles in Figure 16.15? (a) C_1 (b) C_2 (c) C_3

22. Describe in words the curve represented by the parametric equations

$$x = 3 + t^3, \quad y = 5 - t^3, \quad z = 7 + 2t^3.$$

Write a parameterization in 3-space for each of the curves in Problems 23–24.

23. The circle of radius 2 in the xz-plane, centered at the origin.

24. The circle of radius 3 centered at the point $(0, 0, 2)$ parallel to the xy-plane.

For Problems 25–29, find parametric equations for the given line.

25. The line through the points $(2, 3, -1)$ and $(5, 2, 0)$.

26. The line pointing in the direction of the vector $3\vec{i} - 3\vec{j} + \vec{k}$ and through the point $(1, 2, 3)$.

27. The line parallel to the z-axis passing through the point $(1, 0, 0)$.

28. The line of intersection of the planes $x - y + z = 3$ and $2x + y - z = 5$.

29. The line perpendicular to the surface $z = x^2 + y^2$ at the point $(1, 2, 5)$.

30. Do the lines in Problems 25 and 26 intersect?

31. Is the point $(-3, -4, 2)$ visible from the point $(4, 5, 0)$ if there is an opaque ball of radius 1 centered at the origin?

32. Show that the equations

$$x = 3 + t, \quad y = 2t, \quad z = 1 - t$$

satisfy the equations $x + y + 3z = 6$ and $x - y - z = 2$. What does this tell you about the curve parameterized by these equations?

33. Two particles are traveling through space. At time t the first particle is at the point $(-1 + t, 4 - t, -1 + 2t)$ and the second particle is at $(-7 + 2t, -6 + 2t, -1 + t)$.
    (a) Describe the two paths.
    (b) Do the two particles collide? If so, when and where?
    (c) Do the paths of the two particles cross? If so, where?

34. Imagine a light shining on the helix of Example 9 on page 803 from far down each of the axes. Sketch the shadow cast by the helix on each of the coordinate planes: xy, xz, and yz.

Graph the Lissajous figures in Problems 35–38 using a calculator or computer.

35. $x = \cos 2t$, $\quad y = \sin 5t$

36. $x = \cos 3t$, $\quad y = \sin 7t$

37. $x = \cos 2t$, $\quad y = \sin 4t$

38. $x = \cos 2t$, $\quad y = \sin \sqrt{3}t$

39. Motion along a straight line is given by a single equation, say, $x = t^3 - t$ where x is distance along the line. It is difficult to see the motion from a plot; it just traces out the x-line, as in Figure 16.16. To visualize the motion, we introduce a y-coordinate and let it slowly increase, giving Figure 16.17. Try the following on a calculator or computer. Let $y = t$. Now plot the parametric equations $x = t^3 - t$, $y = t$ for, say, $-3 \le t \le 3$. What does the plot in Figure 16.17 tell you about the particle's motion?

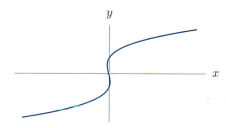

Figure 16.16   *Figure 16.17*

For Problems 40–42, plot the motion along the x-line by the method of Problem 39. What does the plot tell you about the particle's motion?

40. $x = \cos t$, $\quad -10 \le t \le 10$

41. $x = t^4 - 2t^2 + 3t - 7$, $\quad -3 \le t \le 2$

42. $x = t \ln t$, $\quad 0.01 \le t \le 10$

16.2 MOTION, VELOCITY, AND ACCELERATION

In this section, we write parametric equations using position vectors. This enables us to calculate the velocity and acceleration of a particle moving in 2-space or 3-space.

Using Position Vectors to Write Parameterized Curves as Vector Functions

Recall that a point in the plane with coordinates (x, y) can be represented by the position vector $\vec{r} = x\vec{i} + y\vec{j}$ shown in Figure 16.18. Similarly, in 3-space we write $\vec{r} = x\vec{i} + y\vec{j} + z\vec{k}$. (See Figure 16.19.)

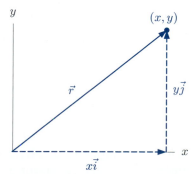

Figure 16.18: Position vector $\vec{r}$ for the point (x, y)

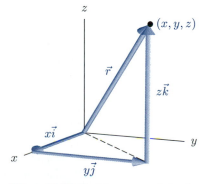

Figure 16.19: Position vector $\vec{r}$ for the point (x, y, z)

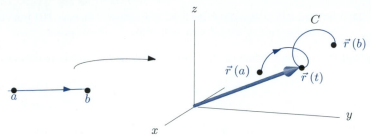

Figure 16.20: The parameterization sends the interval, $a \le t \le b$, to the curve, C, in 3-space

We can write the parametric equations $x = f(t)$, $y = g(t)$, $z = h(t)$ as a single vector equation

$$\vec{r}(t) = f(t)\vec{i} + g(t)\vec{j} + h(t)\vec{k}$$

called *parameterization*. As the parameter t varies, the point with position vector $\vec{r}(t)$ traces out a curve in 3-space. For example, the circular motion

$$x = \cos t, y = \sin t \quad \text{can be written as} \quad \vec{r} = (\cos t)\vec{i} + (\sin t)\vec{j}$$

and the helix

$$x = \cos t, y = \sin t, z = t \quad \text{can be written as} \quad \vec{r} = (\cos t)\vec{i} + (\sin t)\vec{j} + t\vec{k}\,.$$

See Figure 16.20.

Example 1   Give a parameterization for the circle of radius $\frac{1}{2}$ centered at the point $(-1, 2)$.

Solution   The circle of radius 1 centered at the origin is parameterized by the vector-valued function

$$\vec{r_1}(t) = \cos t\vec{i} + \sin t\vec{j}, \quad 0 \le t \le 2\pi.$$

The point $(-1, 2)$ has the position vector $\vec{r}_0 = -\vec{i} + 2\vec{j}$. The position vector, $\vec{r}(t)$, of a point on the circle of radius $\frac{1}{2}$ centered at $(-1, 2)$ is found by adding $\frac{1}{2}\vec{r_1}$ to $\vec{r}_0$. (See Figures 16.21 and 16.22.) Thus,

$$\vec{r}(t) = \vec{r}_0 + \tfrac{1}{2}\vec{r}_1(t) = -\vec{i} + 2\vec{j} + \tfrac{1}{2}(\cos t\vec{i} + \sin t\vec{j}) = (-1 + \tfrac{1}{2}\cos t)\vec{i} + (2 + \tfrac{1}{2}\sin t)\vec{j},$$

or, equivalently,

$$x = -1 + \tfrac{1}{2}\cos t, \quad y = 2 + \tfrac{1}{2}\sin t, \quad 0 \le t \le 2\pi.$$

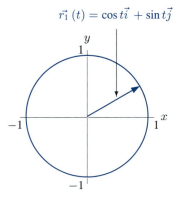

Figure 16.21: The circle $x^2 + y^2 = 1$
parameterized by $\vec{r_1}(t) = \cos t\vec{i} + \sin t\vec{j}$

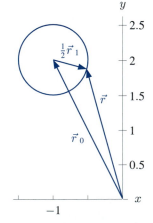

Figure 16.22: The circle of radius $\frac{1}{2}$ and center
$(-1, 2)$ parameterized by $\vec{r}(t) = \vec{r}_0 + \tfrac{1}{2}\vec{r}_1(t)$

Parametric Equation of a Line

Consider a straight line in the direction of a vector $\vec{v}$ passing through the point (x_0, y_0, z_0) with position vector $\vec{r}_0$. We start at $\vec{r}_0$ and move up and down the line, adding different multiples of $\vec{v}$ to $\vec{r}_0$. (See Figure 16.23.)

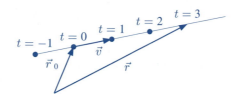

Figure 16.23: The line $\vec{r}(t) = \vec{r}_0 + t\vec{v}$

In this way, every point on the line can be written as $\vec{r}_0 + t\vec{v}$, which yields the following:

Parametric Equation of a Line

The line through the point with position vector $\vec{r}_0 = x_0\vec{i} + y_0\vec{j} + z_0\vec{k}$ in the direction of the vector $\vec{v} = a\vec{i} + b\vec{j} + c\vec{k}$ has parametric equation

$$\vec{r}(t) = \vec{r}_0 + t\vec{v}.$$

Example 2 Find the parametric equation for

(a) The line passing through the points $(2, -1, 3)$ and $(-1, 5, 4)$.

(b) The line segment from $(2, -1, 3)$ to $(-1, 5, 4)$.

Solution (a) The line passes through $(2, -1, 3)$ and is parallel to the displacement vector $\vec{v} = -3\vec{i} + 6\vec{j} + \vec{k}$ from $(2, -1, 3)$ to $(-1, 5, 4)$. Thus the parametric equation is

$$\vec{r}(t) = 2\vec{i} - \vec{j} + 3\vec{k} + t(-3\vec{i} + 6\vec{j} + \vec{k}).$$

(b) In the parameterization in part (a), $t = 0$ corresponds to the point $(2, -1, 3)$ and $t = 1$ corresponds to the point $(-1, 5, 4)$. So the parameterization of the segment is

$$\vec{r}(t) = 2\vec{i} - \vec{j} + 3\vec{k} + t(-3\vec{i} + 6\vec{j} + \vec{k}), \qquad 0 \le t \le 1.$$

The Velocity Vector

The velocity of a moving particle can be represented by a vector with the following properties:

The **velocity vector** of a moving object is a vector $\vec{v}$ such that:
- The magnitude of $\vec{v}$ is the speed of the object.
- The direction of $\vec{v}$ is the direction of motion.

Thus the speed of the object is $\|\vec{v}\|$ and the velocity vector is tangent to the object's path.

Example 3 A child is sitting on a ferris wheel of diameter 10 meters, making one revolution every 2 minutes. Find the speed of the child and draw velocity vectors at two different times.

Solution The child moves at a constant speed around a circle of radius 5 meters, completing one revolution every 2 minutes. One revolution around a circle of radius 5 is a distance of 10π, so the child's speed is $10\pi/2 = 5\pi \approx 15.7$ m/min. Hence, the magnitude of the velocity vector is 15.7 m/min. The direction of motion is tangent to the circle, and hence perpendicular to the radius at that point.

Figure 16.24 shows the direction of the vector at two different times.

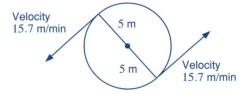

Figure 16.24: Velocity vectors of a child on a ferris wheel (Note that vectors
would be in opposite direction if viewed from the other side.)

Computing the Velocity

We find the velocity, as in one-variable calculus, by taking a limit. If the position vector of the
particle is $\vec{r}(t)$ at time t, then the displacement vector between its positions at times t and $t + \Delta t$ is
$\Delta \vec{r} = \vec{r}(t + \Delta t) - \vec{r}(t)$. (See Figure 16.25.) Over this interval,

$$\text{Average velocity} = \frac{\Delta \vec{r}}{\Delta t}.$$

In the limit as Δt goes to zero we have the instantaneous velocity at time t:

> The **velocity vector**, $\vec{v}(t)$, of a moving object with position vector $\vec{r}(t)$ at time t is
>
> $$\vec{v}(t) = \lim_{\Delta t \to 0} \frac{\Delta \vec{r}}{\Delta t} = \lim_{\Delta t \to 0} \frac{\vec{r}(t + \Delta t) - \vec{r}(t)}{\Delta t},$$
>
> whenever the limit exists. We use the notation $\vec{v} = \dfrac{d\vec{r}}{dt} = \vec{r}\,'(t)$.

Notice that the direction of the velocity vector $\vec{r}\,'(t)$ in Figure 16.25 is approximated by the
direction of the vector $\Delta \vec{r}$ and that the approximation gets better as $\Delta t \to 0$.

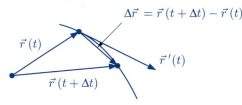

Figure 16.25: The change, $\Delta \vec{r}$, in the position vector for a particle moving on a
curve and the velocity vector $\vec{v} = \vec{r}\,'(t)$

The Components of the Velocity Vector

If we represent a curve parametrically by $x = f(t), y = g(t), z = h(t)$, then we can write its
position vector as: $\vec{r}(t) = f(t)\vec{i} + g(t)\vec{j} + h(t)\vec{k}$. Now we can compute the velocity vector:

$$
\begin{aligned}
\vec{v}(t) &= \lim_{\Delta t \to 0} \frac{\vec{r}(t + \Delta t) - \vec{r}(t)}{\Delta t} \\
&= \lim_{\Delta t \to 0} \frac{(f(t + \Delta t)\vec{i} + g(t + \Delta t)\vec{j} + h(t + \Delta t)\vec{k}) - (f(t)\vec{i} + g(t)\vec{j} + h(t)\vec{k})}{\Delta t} \\
&= \lim_{\Delta t \to 0} \left(\frac{f(t + \Delta t) - f(t)}{\Delta t}\vec{i} + \frac{g(t + \Delta t) - g(t)}{\Delta t}\vec{j} + \frac{h(t + \Delta t) - h(t)}{\Delta t}\vec{k} \right) \\
&= f'(t)\vec{i} + g'(t)\vec{j} + h'(t)\vec{k} \\
&= \frac{dx}{dt}\vec{i} + \frac{dy}{dt}\vec{j} + \frac{dz}{dt}\vec{k}.
\end{aligned}
$$

Thus we have the following result:

> The **components of the velocity vector** of a particle moving in space with position vector $\vec{r}(t) = f(t)\vec{i} + g(t)\vec{j} + h(t)\vec{k}$ at time t are given by
>
> $$\vec{v}(t) = f'(t)\vec{i} + g'(t)\vec{j} + h'(t)\vec{k} = \frac{dx}{dt}\vec{i} + \frac{dy}{dt}\vec{j} + \frac{dz}{dt}\vec{k}.$$

Example 4   Find the components of the velocity vector for the child on the ferris wheel in Example 3 using a coordinate system which has its origin at the center of the ferris wheel and which makes the rotation counterclockwise.

Solution   The ferris wheel has radius 5 meters and completes 1 revolution counterclockwise every 2 minutes. The motion is parameterized by an equation of the form

$$\vec{r}(t) = 5\cos(\omega t)\vec{i} + 5\sin(\omega t)\vec{j},$$

where ω is chosen to make the period 2 minutes. Since the period of $\cos(\omega t)$ and $\sin(\omega t)$ is $2\pi/\omega$, we must have

$$\frac{2\pi}{\omega} = 2, \quad \text{so} \quad \omega = \pi.$$

Thus, the motion is described by the equation

$$\vec{r}(t) = 5\cos(\pi t)\vec{i} + 5\sin(\pi t)\vec{j},$$

where t is in minutes. The velocity is given by

$$\vec{v} = \frac{dx}{dt}\vec{i} + \frac{dy}{dt}\vec{j} = -5\pi\sin(\pi t)\vec{i} + 5\pi\cos(\pi t)\vec{j}.$$

To check, we calculate the magnitude of $\vec{v}$,

$$\|\vec{v}\| = \sqrt{(-5\pi)^2\sin^2(\pi t) + (5\pi)^2\cos^2(\pi t)} = 5\pi\sqrt{\sin^2(\pi t) + \cos^2(\pi t)} = 5\pi \approx 15.7,$$

which agrees with the speed we calculated in Example 3. To see that the direction is correct, we must show that the vector $\vec{v}$ at any time t is perpendicular to the position vector of the particle at time t. To do this, we compute the dot product of $\vec{v}$ and $\vec{r}$:

$$\vec{v} \cdot \vec{r} = (-5\pi\sin(\pi t)\vec{i} + 5\pi\cos(\pi t)\vec{j}) \cdot (5\cos(\pi t)\vec{i} + 5\sin(\pi t)\vec{j})$$
$$= -25\pi\sin(\pi t)\cos(\pi t) + 25\pi\cos(\pi t)\sin(\pi t) = 0.$$

So the velocity vector, $\vec{v}$, is perpendicular to $\vec{r}$ and hence tangent to the circle. (See Figure 16.26.)

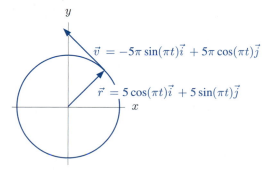

Figure 16.26: Velocity and radius vector of motion around a circle

Velocity Vectors and Tangent Lines

Since the velocity vector is tangent to the path of motion, it can be used to find parametric equations for the tangent line, if there is one.

Example 5   Find the tangent line at the point $(1, 1, 2)$ to the curve defined by the parametric equation

$$\vec{r}(t) = t^2\vec{i} + t^3\vec{j} + 2t\vec{k}.$$

Solution   At time $t = 1$ the particle is at the point $(1, 1, 2)$ with position vector $\vec{r}_0 = \vec{i} + \vec{j} + 2\vec{k}$. The velocity vector at time t is $\vec{r}'(t) = 2t\vec{i} + 3t^2\vec{j} + 2\vec{k}$, so at time $t = 1$ the velocity is $\vec{v} = \vec{r}'(1) = 2\vec{i} + 3\vec{j} + 2\vec{k}$. The tangent line passes through $(1, 1, 2)$ in the direction of $\vec{v}$, so it has the parametric equation

$$\vec{r}(t) = \vec{r}_0 + t\vec{v} = (\vec{i} + \vec{j} + 2\vec{k}) + t(2\vec{i} + 3\vec{j} + 2\vec{k}).$$

The Acceleration Vector

Just as the velocity of a particle moving in 2-space or 3-space is a vector quantity, so is the rate of change of the velocity of the particle, namely its acceleration. Figure 16.27 shows a particle at time t with velocity vector $\vec{v}(t)$ and then a little later at time $t + \Delta t$. The vector $\Delta\vec{v} = \vec{v}(t + \Delta t) - \vec{v}(t)$ is the change in velocity and points approximately in the direction of the acceleration. So,

$$\text{Average acceleration} = \frac{\Delta\vec{v}}{\Delta t}.$$

In the limit as $\Delta t \to 0$, we have the instantaneous acceleration at time t:

The **acceleration vector** of an object moving with velocity $\vec{v}(t)$ at time t is

$$\vec{a}(t) = \lim_{\Delta t \to 0} \frac{\Delta\vec{v}}{\Delta t} = \lim_{\Delta t \to 0} \frac{\vec{v}(t + \Delta t) - \vec{v}(t)}{\Delta t},$$

if the limit exists. We use the notation $\vec{a} = \dfrac{d\vec{v}}{dt} = \dfrac{d^2\vec{r}}{dt^2} = \vec{r}''(t).$

Figure 16.27: Computing the difference between two velocity vectors

Components of the Acceleration Vector

If we represent a curve in space parametrically by $x = f(t), y = g(t), z = h(t)$, we can express the acceleration in components. The velocity vector $\vec{v}(t)$ is given by

$$\vec{v}(t) = f'(t)\vec{i} + g'(t)\vec{j} + h'(t)\vec{k}.$$

From the definition of the acceleration vector, we have

$$\vec{a}(t) = \lim_{\Delta t \to 0} \frac{\vec{v}(t + \Delta t) - \vec{v}(t)}{\Delta t} = \frac{d\vec{v}}{dt}.$$

Using the same method to compute $d\vec{v}/dt$ as we used to compute $d\vec{r}/dt$ on page 811, we obtain

> The **components of the acceleration vector**, $\vec{a}(t)$, at time t of a particle moving in space with position vector $\vec{r}(t) = f(t)\vec{i} + g(t)\vec{j} + h(t)\vec{k}$ at time t are given by
>
> $$\vec{a}(t) = f''(t)\vec{i} + g''(t)\vec{j} + h''(t)\vec{k} = \frac{d^2x}{dt^2}\vec{i} + \frac{d^2y}{dt^2}\vec{j} + \frac{d^2z}{dt^2}\vec{k}.$$

Motion In a Circle and Along a Line

Example 6   Find the acceleration vector for the child on the ferris wheel in Examples 3 and 4.

Solution   The child's position vector is given by $\vec{r}(t) = 5\cos(\pi t)\vec{i} + 5\sin(\pi t)\vec{j}$. In Example 4 we saw that the velocity vector is

$$\vec{v}(t) = \frac{dx}{dt}\vec{i} + \frac{dy}{dt}\vec{j} = -5\pi\sin(\pi t)\vec{i} + 5\pi\cos(\pi t)\vec{j}.$$

Thus, the acceleration vector is

$$\vec{a}(t) = \frac{d^2x}{dt^2}\vec{i} + \frac{d^2y}{dt^2}\vec{j} = -(5\pi)\cdot\pi\cos(\pi t)\vec{i} - (5\pi)\cdot\pi\sin(\pi t)\vec{j}$$

$$= -5\pi^2\cos(\pi t)\vec{i} - 5\pi^2\sin(\pi t)\vec{j}.$$

Notice that $\vec{a}(t) = -\pi^2\vec{r}(t)$. Thus, the acceleration vector is a multiple of $\vec{r}(t)$ and points toward the origin.

The motion of the child on the ferris wheel is an example of uniform circular motion, whose properties follow. (See Problem 26.)

> ## Uniform Circular Motion
>
> A particle whose motion is described by
>
> $$\vec{r}(t) = R\cos(\omega t)\vec{i} + R\sin(\omega t)\vec{j}$$
>
> - Moves in a circle of radius R with period $2\pi/\omega$.
> - Velocity, $\vec{v}$, is tangent to the circle and speed is constant $\|\vec{v}\| = \omega R$.
> - Acceleration, $\vec{a}$, points toward the center of the circle with $\|\vec{a}\| = \|\vec{v}\|^2/R$.

In uniform circular motion, the acceleration vector reflects the fact that the velocity vector does not change in magnitude, only in direction. We now look at straight line motion where the velocity vector always has the same direction but the magnitude changes. We expect that the acceleration vector will point in the same direction as the velocity vector if the speed is increasing and in the opposite direction to the velocity vector if the speed is decreasing.

Example 7   Consider the motion given by the vector equation

$$\vec{r}(t) = 2\vec{i} + 6\vec{j} + (t^3 + t)(4\vec{i} + 3\vec{j} + \vec{k}).$$

Show that this is straight line motion in the direction of the vector $4\vec{i} + 3\vec{j} + \vec{k}$ and relate the acceleration vector to the velocity vector.

Solution   The velocity vector is

$$\vec{v} = (3t^2 + 1)(4\vec{i} + 3\vec{j} + \vec{k}).$$

Since $(3t^2 + 1)$ is a positive scalar, the velocity vector $\vec{v}$ always points in the direction of the vector $4\vec{i} + 3\vec{j} + \vec{k}$. In addition,

$$\text{Speed} = \|\vec{v}\| = (3t^2 + 1)\sqrt{4^2 + 3^2 + 1^2} = \sqrt{26}(3t^2 + 1).$$

Notice that the speed is decreasing until $t = 0$, then starts increasing. The acceleration vector is

$$\vec{a} = 6t(4\vec{i} + 3\vec{j} + \vec{k}).$$

For $t > 0$, the acceleration vector points in the same direction as $4\vec{i} + 3\vec{j} + \vec{k}$, which is the same direction as $\vec{v}$. This makes sense because the object is speeding up. For $t < 0$, the acceleration vector $6t(4\vec{i} + 3\vec{j} + \vec{k})$ points in the opposite direction to $\vec{v}$ because the object is slowing down.

The Length of a Curve

The speed of a particle is the magnitude of its velocity vector:

$$\text{Speed} = \|\vec{v}\| = \sqrt{\left(\frac{dx}{dt}\right)^2 + \left(\frac{dy}{dt}\right)^2 + \left(\frac{dz}{dt}\right)^2}.$$

As in one dimension, we can find the distance traveled by a particle along a curve by integrating its speed. Thus,

$$\text{Distance traveled} = \int_a^b \|\vec{v}(t)\| \, dt.$$

If the particle never stops or reverses its direction as it moves along the curve, the distance it travels will be the same as the length of the curve. This suggests the following formula, which is justified in Problem 32:

> If the curve C is given parametrically for $a \leq t \leq b$ by smooth functions and if the velocity vector $\vec{v}$ is not $\vec{0}$ for $a < t < b$, then
>
> $$\text{Length of } C = \int_a^b \|\vec{v}\| \, dt.$$

Example 8   Find the circumference of the ellipse given by the parametric equations

$$x = 2\cos t, \quad y = \sin t, \quad 0 \leq t \leq 2\pi.$$

Solution   The circumference of this curve is given by an integral which must be calculated numerically:

$$\text{Circumference} = \int_0^{2\pi} \sqrt{\left(\frac{dx}{dt}\right)^2 + \left(\frac{dy}{dt}\right)^2} \, dt = \int_0^{2\pi} \sqrt{(-2\sin t)^2 + (\cos t)^2} \, dt$$

$$= \int_0^{2\pi} \sqrt{4\sin^2 t + \cos^2 t} \, dt = 9.69.$$

Since the ellipse is inscribed in a circle of radius 2 and circumscribes a circle of radius 1, we would expect the length of the ellipse to be between $2\pi(2) \approx 12.57$ and $2\pi(1) \approx 6.28$, so the value of 9.69 is reasonable.

Problems for Section 16.2

1. (a) Explain how you know that the following two equations
$\vec{r} = (2+t)\vec{i} + (4+3t)\vec{j}$, $\vec{r} = (1-2t)\vec{i} + (1-6t)\vec{j}$ parameterize the same line.
   (b) What are the slope and y intercept of this line?

2. The equation $\vec{r} = 10\vec{k} + t(\vec{i} + 2\vec{j} + 3\vec{k})$ parameterizes a line.

   (a) Suppose we restrict ourselves to $t < 0$. What part of the line do we get?
   (b) Suppose we restrict ourselves to $0 \le t \le 1$. What part of the line do we get?

3. (a) Explain why the line of intersection of two planes must be parallel to the cross product of a normal vector to the first plane and a normal vector to the second.
   (b) Find a vector parallel to the line of intersection of the two planes $x+2y-3z = 7$ and $3x-y+z = 0$.
   (c) Find parametric equations for the line in part (b).

4. Using time increments of 0.01, give a table of values near $t = 1$ for the position vector of the circular motion
$$\vec{r}(t) = (\cos t)\vec{i} + (\sin t)\vec{j}.$$
Use the table to approximate the velocity vector, $\vec{v}$, at time $t = 1$. Show that $\vec{v}$ is perpendicular to the radius from the origin at $t = 1$.

5. (a) Sketch the parameterized curve $x = t\cos t$, $y = t\sin t$ for $0 \le t \le 4\pi$.
   (b) Use difference quotients to approximate the velocity vectors $\vec{v}(t)$ for $t = 2, 4$, and 6.
   (c) Compute the velocity vectors $\vec{v}(t)$ for $t = 2, 4$, and 6, exactly and sketch them on the graph of the curve.

For Problems 6–9, find the velocity vector $\vec{v}(t)$ for the given motion of a particle. Also find the speed $\|\vec{v}(t)\|$ and any times when the particle comes to a stop.

6. $x = t^2$, $y = t^3$

7. $x = \cos(t^2)$, $y = \sin(t^2)$

8. $x = \cos 2t$, $y = \sin t$

9. $x = t^2 - 2t$, $y = t^3 - 3t$, $z = 3t^4 - 4t^3$

10. Find parametric equations for the tangent line at $t = 2$ for Problem 6.

For Problems 11–14, find the velocity and acceleration vectors for the given motions.

11. $x = 3\cos t$, $y = 4\sin t$

12. $x = t$, $y = t^3 - t$

13. $x = 2 + 3t$, $y = 4 + t$, $z = 1 - t$

14. $x = 3\cos(t^2)$, $y = 3\sin(t^2)$, $z = t^2$

Find the length of the curves in Problems 15–17.

15. $x = 3 + 5t$, $y = 1 + 4t$, $z = 3 - t$ for $1 \le t \le 2$. Explain your answer.

16. $x = \cos(e^t)$, $y = \sin(e^t)$ for $0 \le t \le 1$. Explain why your answer is reasonable.

17. $x = \cos 3t$, $y = \sin 5t$ for $0 \le t \le 2\pi$.

18. A particle that passes through the point $P = (5, 4, -2)$ at time $t = 4$ is moving with constant velocity $\vec{v} = 2\vec{i} - 3\vec{j} + \vec{k}$. Find the parametric equations for its motion.

19. A particle that passes through the point $P = (5, 4, 3)$ at time $t = 7$ is moving with constant velocity $\vec{v} = 3\vec{i} + \vec{j} + 2\vec{k}$. Find the equations for its position at time t.

20. An object moving with constant velocity in 3-space (with coordinates in meters) passes through the point $(1, 1, 1)$, and then passes through the point $(2, -1, 3)$ five seconds later. What is its velocity vector? What is its acceleration vector?

21. Table 16.4 gives x and y coordinates of a particle in the plane at time t. Assuming the path is smooth, estimate the following quantities:

    (a) The velocity vector and speed at time $t = 2$.
    (b) Any times when the particle is moving parallel to the y-axis.
    (c) Any times when the particle has come to a stop.

    TABLE 16.4

t	0	0.5	1.0	1.5	2.0	2.5	3.0	3.5	4.0
x	1	4	6	7	6	3	2	3	5
y	3	2	3	5	8	10	11	10	9

22. Consider the motion of the particle given by the parametric equations
$$x = t^3 - 3t, \quad y = t^2 - 2t,$$
where the y-axis is vertical and the x-axis is horizontal.

    (a) Does the particle ever come to a stop? If so, when and where?
    (b) Is the particle ever moving straight up or down? If so, when and where?
    (c) Is the particle ever moving straight horizontally right or left? If so, when and where?

23. Suppose $\vec{r}(t) = \cos t\,\vec{i} + \sin t\,\vec{j} + 2t\,\vec{k}$ represents the position of a particle on a helix, where z is the height of the particle above the ground.

    (a) Is the particle ever moving downwards? When?
    (b) When does the particle reach a point 10 units above the ground?
    (c) What is the velocity of the particle when it is 10 units above the ground?
    (d) Suppose the particle leaves the helix and moves along the tangent line to the spiral at this point. Find parametric equations for this tangent line.

24. Mr. Skywalker is traveling along the curve given by
$$\vec{r}(t) = -2e^{3t}\vec{i} + 5\cos t\,\vec{j} - 3\sin(2t)\vec{k}.$$
If the power thrusters are turned off, his ship flies off on a tangent line to $\vec{r}(t)$. He is almost out of power when he notices that a station on Xardon is open at the point with coordinates $(1.5, 5, 3.5)$. Quickly calculating his position, he turns off the thrusters at $t = 0$. Does he make it to the Xardon station? Explain.

25. Determine the position vector $\vec{r}(t)$ for a rocket which is launched from the origin at time $t = 0$ seconds, reaches its highest point of $(x, y, z) = (1000, 3000, 10{,}000)$, where x, y, z are measured in meters, and after the launch is subject only to the acceleration due to gravity, 9.8 m/sec^2.

26. The motion of a particle is given by $\vec{r}(t) = R\cos(\omega t)\vec{i} + R\sin(\omega t)\vec{j}$, with $R > 0, \omega > 0$.

    (a) Show that the particle moves on a circle and find the radius, direction, and period.
    (b) Determine the velocity vector of the particle and its direction and speed.
    (c) What are the direction and magnitude of the acceleration vector of the particle?

27. A stone is swung around on a string at a constant speed with period 2π seconds in a horizontal circle centered at the point $(0, 0, 8)$. When $t = 0$, the stone is at the point $(0, 5, 8)$; it travels clockwise when viewed from above. When the stone is at the point $(5, 0, 8)$, the string breaks and it moves under gravity.

    (a) Parameterize the stone's circular trajectory.
    (b) Find the velocity and acceleration of the stone at the moment before the string breaks.
    (c) Write, but do not solve, the differential equations (with initial conditions) satisfied by the coordinates x, y, z giving the position of the stone after it has left the circle.

28. Emily is standing on the outer edge of a merry-go-round, 10 meters from the center. The merry-go-round completes one full revolution every 20 seconds. As Emily passes over a point P on the ground, she drops a ball from 3 meters above the ground.

    (a) How fast is Emily going?
    (b) How far from P does the ball hit the ground? (The acceleration due to gravity is 9.8 m/sec^2.)
    (c) How far from Emily does the ball hit the ground?

29. A lighthouse L is located on an island in the middle of a lake as shown in Figure 16.28. Consider the motion of the point where the light beam from L hits the shore of the lake.

(a) Suppose the beam rotates counterclockwise about L at a constant angular velocity. At which of $A, B, C, D,$ or E is the speed of the point greatest? At which point is it smallest?

(b) Repeat part (a), supposing the beam rotates counterclockwise so that it sweeps out equal areas of the lake in equal times.

(c) What happens if you place the lighthouse at different points in the lake? Can the speed of the point on the shore ever be infinite for part (a)?

(d) Suppose now that the lake is rectangular instead. What happens to the velocity vector at the corners? For part (b) show that the speed is constant along each side (possibly a different constant on each side).

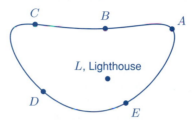

Figure 16.28: The lighthouse in the lake

30. Suppose $F(x, y) = 1/(x^2 + y^2 + 1)$ gives the temperature at the point (x, y) in the plane. A ladybug moves along a parabola according to the parametric equations

$$x = t, \quad y = t^2.$$

Find the rate of change in the temperature of the ladybug at time t.

31. This problem generalizes the result of Problem 30. Suppose $F(x, y)$ gives the temperature at any point (x, y) in the plane and that a ladybug moves in the plane with position vector at time t given by $\vec{r}(t) = x(t)\vec{i} + y(t)\vec{j}$ and velocity vector $\vec{r}'(t)$. Use the chain rule to show that

Rate of change in the temperature of the bug at time $t \ = \ \operatorname{grad} F(\vec{r}(t)) \cdot \vec{r}'(t)$.

32. In this problem we justify the formula for the length of a curve given on page 814. Suppose the curve C is given by smooth parametric equations $x = x(t)$, $y = y(t)$, $z = z(t)$ for $a \le t \le b$. By dividing the parameter interval $a \le t \le b$ at points $t_1, \ldots, t_{n-1}$ into small segments of length $\Delta t = t_{i+1} - t_i$, we get a corresponding division of the curve C into small pieces. See Figure 16.29, where the points $P_i = (x(t_i), y(t_i), z(t_i))$ on the curve C correspond to parameter values $t = t_i$. Let C_i be the portion of the curve C between P_i and P_{i+1}.

(a) Use local linearity to show that

$$\text{Length of } C_i \approx \sqrt{x'(t_i)^2 + y'(t_i)^2 + z'(t_i)^2}\, \Delta t.$$

(b) Use part (a) and a Riemann sum to explain why

$$\text{Length of } C = \int_a^b \sqrt{x'(t)^2 + y'(t)^2 + z'(t)^2}\, dt.$$

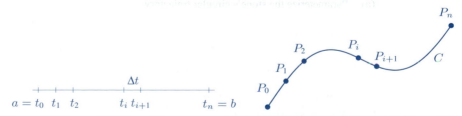

Figure 16.29: A subdivision of the parameter interval and the corresponding subdivision of the curve C

- **Parameterized Curves**

    Parameterizations representing motion in 2- and 3-space, parameterizing the graph of a function, parametric equation of a line.

- **Velocity and Acceleration Vectors**

    Computing velocity and acceleration, uniform circular motion, the length of a parametric curve.

REVIEW PROBLEMS FOR CHAPTER SIXTEEN

Write a parameterization for each of the curves in Problems 1–10.

1. The horizontal line through the point $(0, 5)$.

2. The circle of radius 2 centered at the origin starting at the point $(0, 2)$ when $t = 0$.

3. The circle of radius 4 centered at the point $(4, 4)$ starting on the x-axis when $t = 0$.

4. The circle of radius 1 in the xy-plane centered at the origin, traversed counterclockwise when viewed from above.

5. The line through the points $(2, -1, 4)$ and $(1, 2, 5)$.

6. The line through the point $(1, 3, 2)$ perpendicular to the xz-plane.

7. The line through the point $(1, 1, 1)$ perpendicular to the plane $2x - 3y + 5z = 4$.

8. The circle of radius 2 parallel to the xy-plane, centered at the point $(0, 0, 1)$, and traversed counterclockwise when viewed from below.

9. The circle of radius 3 parallel to the xz-plane, centered at the point $(0, 5, 0)$, and traversed counterclockwise when viewed from $(0, 10, 0)$.

10. The circle of radius 2 centered at $(0, 1, 0)$ lying in the plane $x + z = 0$.

11. On a graphing calculator or a computer, plot $x = 2t/(t^2 + 1)$, $y = (t^2 - 1)/(t^2 + 1)$, first for $-50 \le t \le 50$ then for $-5 \le t \le 5$. Explain what you see. Is the curve really a circle?

12. Let $f(x, y) = \dfrac{x^2 - y^2}{x^2 + y^2}$.

    (a) In which direction should you move from the point $(1, 1)$ to obtain the maximum rate of increase of f?

    (b) Find a direction in which the directional derivative at the point $(1, 1)$ is equal to zero.

    (c) Suppose you move along the curve $x = e^{2t}$, $y = 2t^3 + 6t + 1$. What is df/dt at $t = 0$?

13. Find the parametric equation for the line of intersection of the planes $z = 4 + 2x + 5y$ and $z = 3 + x + 3y$.

14. Find parametric equations of the line passing through the points $(1, 2, 3)$, $(3, 5, 7)$ and calculate the shortest distance from the line to the origin.

15. Are the lines $x = 3 + 2t$, $y = 5 - t$, $z = 7 + 3t$ and $x = 3 + t$, $y = 5 + 2t$, $z = 7 + 2t$ parallel?

16. Plot the Lissajous figure given by $x = \cos 2t$, $y = \sin t$ using a graphing calculator or computer. Explain why it looks like part of a parabola. [Hint: Use a double angle identity from trigonometry.]

17. A particle travels along a line, with position at time t given by

$$\vec{r}(t) = (2 + 5t)\vec{i} + (3 + t)\vec{j} + 2t\vec{k}.$$

    (a) Where is the particle when $t = 0$?

    (b) At what time does the particle reach the point $(12, 5, 4)$?

    (c) Does the particle ever reach the point $(12, 4, 4)$? Why or why not?

18.   An ant, starting at the origin, moves at 2 units/sec along the x-axis to the point $(1, 0)$. The ant then moves counterclockwise along the unit circle to $(0, 1)$ at a speed of $3\pi/2$ units/sec, then straight down to the origin at a speed of 2 units/sec along the y-axis.

   (a)   Express the ant's coordinates as a function of time, t, in secs.
   (b)   Express the reverse path as a function of time.

19.   Consider the parametric equations below for $0 \le t \le \pi$.

   (I) $\vec{r} = \cos(2t)\vec{i} + \sin(2t)\vec{j}$ (II) $\vec{r} = 2\cos t\vec{i} + 2\sin t\vec{j}$
   (III) $\vec{r} = \cos(t/2)\vec{i} + \sin(t/2)\vec{j}$ (IV) $\vec{r} = 2\cos t\vec{i} - 2\sin t\vec{j}$

   (a)   Match the equations above with four of the curves C_1, C_2, C_3, C_4, C_5 and C_6 in Figure 16.30. (Each curve is part of a circle.)
   (b)   Give parametric equations for the curves which have not been matched, again assuming $0 \le t \le \pi$.

20.   A wheel of radius 1 meter rests on the x-axis with its center on the y-axis. There is a spot on the rim at the point $(1, 1)$. See Figure 16.31. At time $t = 0$ the wheel starts rolling on the x-axis in the direction shown at a rate of 1 radian per second.

   (a)   Find parametric equations describing the motion of the center of the wheel.
   (b)   Find parametric equations describing the motion of the spot on the rim. Plot its path.

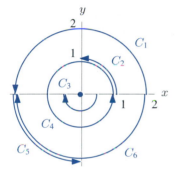

Figure 16.30

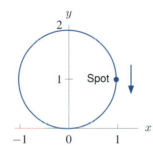

Figure 16.31

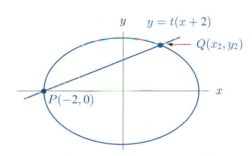

Figure 16.32

21.   (a)   The ellipse $2x^2 + 3y^2 = 8$ intersects the line of slope t through the point $P = (-2, 0)$ in two points, one of which is P. Compute the coordinate of the other point Q. See Figure 16.32.
   (b)   Give a parameterization of the ellipse $2x^2 + 3y^2 = 8$ by rational functions.

22.   A cheerleader has a 0.4 m long baton with a light on one end. She throws the baton in such a way that its center moves along a parabola, and the baton rotates counterclockwise around the center with a constant angular velocity. The baton is initially horizontal and 1.5 m above the ground; its initial velocity is 8 m/sec horizontally and 10 m/sec vertically, and its angular velocity is 2 revolutions per second. Find parametric equations describing the following motions:

   (a)   The center of the baton relative to the ground.
   (b)   The end of the baton relative to its center.
   (c)   The path traced out by the end of the baton relative to the ground.
   (d)   Sketch a graph of the motion of the end of the baton.

PROJECTS

1. **Shooting a Basketball**

   A basketball player shoots the ball from 6 feet above the ground towards a basket that is 10 feet above the ground and 15 feet away horizontally.

   (a) Suppose she shoots the ball at an angle of A degrees above the horizontal $(0 < A < \pi/2)$ with an initial speed V. Give the x- and y-coordinates of the position of the basketball at time t. Assume the x-coordinate of the basket is 0 and that the x-coordinate of the shooter is -15. [Hint: There is an acceleration of -32 ft/sec^2 in the y-direction; there is no acceleration in the x-direction. Ignore air resistance.]

   (b) Using the parametric equations you obtained in part (a), experiment with different values for V and A, plotting the path of the ball on a graphing calculator or computer to see how close the ball comes to the basket. (The tick marks on the y-axis can be used to locate the basket.) Find some values of V and A for which the shot goes in.

   (c) Find the angle A that minimizes the velocity needed for the ball to reach the basket. (This is a lengthy computation. First find an equation in V and A that holds if the path of the ball passes through the point 15 feet from the shooter and 10 feet above the ground. Then minimize V.)

2. **The Orbit of a Planet and its Moon**

   A hypothetical moon orbits a planet which in turn orbits a star. Suppose that the orbits are circular and that the moon orbits the planet 12 times in the time it takes for the planet to orbit the star once. In this problem we will investigate whether the moon could come to a stop at some instant. (See Figure 16.33.)

   (a) Suppose the radius of the moon's orbit around the planet is 1 unit and the radius of the planet's orbit around the star is R units. Explain why the motion of the moon relative to the star can be described by the parametric equations

   $$x = R\cos t + \cos(12t), \quad y = R\sin t + \sin(12t).$$

   (b) Find a value for R and t such that the moon stops relative to the star at time t.

   (c) On a graphing calculator, plot the path of the moon for the value of R you obtained in part (b). Experiment with other values for R.

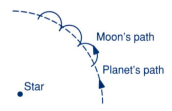

Moon's path

Planet's path

Star

Figure 16.33

CHAPTER SEVENTEEN

VECTOR FIELDS

Some physical quantities (such as temperature) are best represented by scalars; others (such as velocity) are best represented by vectors. We have looked at functions of many variables whose values are scalars, for example, temperature as a function of position on a weather map. Such functions are called scalar-valued functions.

Some weather maps indicate wind velocity at various points by arrows. Wind velocity is an example of a vector-valued function, since its value at any point is the vector indicating the direction and strength of the wind. Such functions are also called *vector fields*. We have already seen one important example of a vector field, namely the gradient of a scalar-valued function. In this chapter we will look at other examples, such as velocity vector fields describing a fluid flow. We will also look at the path followed by a particle moving with the flow, which is called a *flow line* of the vector field.

17.1 VECTOR FIELDS

Introduction to Vector Fields

A *vector field* is a function that assigns a vector to each point in the plane or in 3-space. One example of a vector field is the gradient of a function $f(x, y)$; at each point (x, y) the vector grad $f(x, y)$ points in the direction of maximum rate of increase of f. In this section we will look at other vector fields representing velocities and forces.

Velocity Vector Fields

Figure 17.1 shows the flow of a part of the Gulf stream, a current in the Atlantic Ocean.[1] It is an example of a *velocity vector field*: each vector shows the velocity of the current at that point. The current is fastest where the velocity vectors are longest in the middle of the stream. Beside the stream are eddies where the water flows round and round in circles.

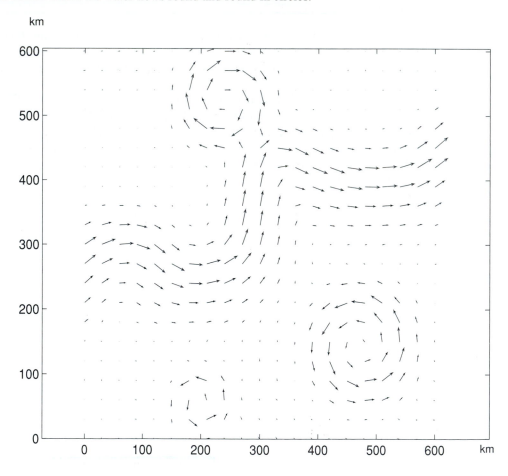

Figure 17.1: The velocity vector field of the Gulf stream

Force Fields

Another physical quantity represented by a vector is force. When we experience a force, sometimes it results from direct contact with the object that supplies the force (for example, a push). Many forces, however, can be felt at all points in space. For example, the earth exerts a gravitational pull on all other masses. Such forces can be represented by vector fields.

[1] Based on data supplied by Avijit Gangopadhyay of the Jet Propulsion Laboratory

Figure 17.2: The gravitational field of the earth

Figure 17.2 shows the gravitational force exerted by the earth on a mass of one kilogram at different points in space. This is a sketch of the vector field in 3-space. You can see that the vectors all point towards the earth (which is not shown in the diagram) and that the vectors further from the earth are smaller in magnitude.

Definition of a Vector Field

Now that you have seen some examples of vector fields we give a more formal definition.

> A **vector field** in 2-space is a function $\vec{F}(x, y)$ whose value at a point (x, y) is a 2-dimensional vector. Similarly, a vector field in 3-space is a function $\vec{F}(x, y, z)$ whose values are 3-dimensional vectors.

Notice the arrow over the function, $\vec{F}$, indicating that its value is a vector, not a scalar. We often represent the point (x, y) or (x, y, z) by its position vector $\vec{r}$ and write the vector field as $\vec{F}(\vec{r})$.

Visualizing a Vector Field Given by a Formula

Since a vector field is a function that assigns a vector to each point, a vector field can often be given by a formula.

Example 1   Sketch the vector field in 2-space given by $\vec{F}(x, y) = -y\vec{i} + x\vec{j}$.

Solution   Table 17.1 shows the value of the vector field at a few points. Notice that each value is a vector. To plot the vector field, we plot $\vec{F}(x, y)$ with its tail at (x, y). (See Figure 17.3.)

TABLE 17.1   *A few values of* $\vec{F}(x, y) = -y\vec{i} + x\vec{j}$

		x		
		-1	0	1
x	-1	$\vec{i} - \vec{j}$	$-\vec{j}$	$-\vec{i} - \vec{j}$
	0	$\vec{i}$	$\vec{0}$	$-\vec{i}$
	1	$\vec{i} + \vec{j}$	$\vec{j}$	$-\vec{i} + \vec{j}$

Now we look at the formula to get a better sketch. The magnitude of the vector at (x, y) is $\|\vec{F}(x, y)\| = \|-y\vec{i} + x\vec{j}\| = \sqrt{x^2 + y^2}$, which is the distance from (x, y) to the origin. Therefore, all the vectors at a fixed distance from the origin (that is, on a circle centered at the origin) have the same magnitude. The magnitude gets larger as we move further from the origin. What about the direction?

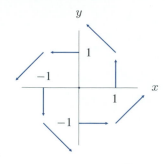

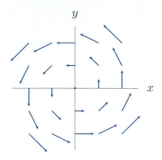

Figure 17.3: The value $\vec{F}(x,y)$
is placed at the point (x,y).

Figure 17.4: The vector field
$\vec{F}(x,y) = -y\vec{i} + x\vec{j}$, vectors
scaled smaller to fit in diagram.

Figure 17.3 suggests that at each point (x,y) the vector $\vec{F}(x,y)$ is perpendicular to the position vector $\vec{r} = x\vec{i} + y\vec{j}$. We verify this using the dot product: $\vec{r} \cdot \vec{F}(x,y) = (x\vec{i} + y\vec{j}) \cdot (-y\vec{i} + x\vec{j}) = 0$. This means that the vectors in this vector field are tangent to circles centered at the origin and get longer as we go out. In Figure 17.4, the vectors have been scaled so that they do not obscure each other.

Example 2   Draw pictures of the vector fields in 2-space given by (a) $\vec{F}(x,y) = x\vec{j}$ (b) $\vec{G}(x,y) = x\vec{i}$.

Solution   (a)   The vector $x\vec{j}$ is parallel to the y-direction, pointing up when x is positive and down when x is negative. Also, the larger $|x|$ is, the longer the vector. The vectors in the field are constant along vertical lines since the vector field does not depend on y. (See Figure 17.5.)

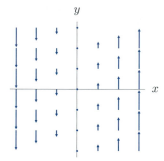

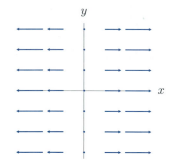

Figure 17.5: The vector field
$\vec{F}(x,y) = x\vec{j}$

Figure 17.6: The vector field
$\vec{F}(x,y) = x\vec{i}$

(b)   This is similar to the previous example, except that the vector $x\vec{i}$ is parallel to the x-direction, pointing to the right when x is positive and to the left when x is negative. Again, the larger $|x|$ is the longer the vector, and the vectors are constant along vertical lines, since the vector field does not depend on y. (See Figure 17.6.)

Example 3   Describe the vector field in 3-space given by $\vec{F}(\vec{r}) = \vec{r}$, where $\vec{r} = x\vec{i} + y\vec{j} + z\vec{k}$.

Solution   The notation $\vec{F}(\vec{r}) = \vec{r}$ means that the value of $\vec{F}$ at the point (x,y,z) with position vector $\vec{r}$ is the vector $\vec{r}$ with its tail at (x,y,z). Thus, the vector field points outward. See Figure 17.7. Note that the lengths of the vectors have been scaled down so as to fit into the diagram. This vector field can also be written as $\vec{F}(x,y,z) = x\vec{i} + y\vec{j} + z\vec{k}$. You can see the notation using $\vec{r}$ is more concise.

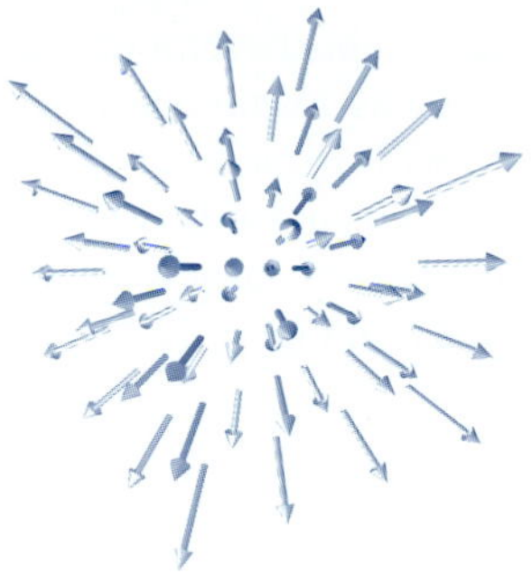

Figure 17.7: The vector field
$\vec{F}(\vec{r}) = \vec{r}$

Figure 17.8: Force exerted
on mass m by mass M

Finding a Formula for a Vector Field

Example 4   Newton's Law of Gravitation states that the magnitude of the gravitational force exerted by an object of mass M on an object of mass m is proportional to M and m and inversely proportional to the square of the distance between them. The direction of the force is from m to M along the line connecting them. (See Figure 17.8.) Find a formula for the vector field $\vec{F}(\vec{r})$ that represents the gravitational force, assuming M is located at the origin and m is located at the point with position vector $\vec{r}$.

Solution   Since the mass m is located at $\vec{r}$, Newton's law says that the magnitude of the force is given by

$$\|\vec{F}(\vec{r})\| = \frac{GMm}{\|\vec{r}\|^2},$$

where G is called the universal gravitational constant. A unit vector in the direction of the force is $-\vec{r}/\|\vec{r}\|$, where the negative sign indicates that the direction of force is towards the origin (gravity is attractive). By taking the product of the magnitude of the force and a unit vector in the direction of the force we obtain an expression for the force vector field:

$$\vec{F}(\vec{r}) = \frac{GMm}{\|\vec{r}\|^2}\left(-\frac{\vec{r}}{\|\vec{r}\|}\right) = \frac{-GMm\vec{r}}{\|\vec{r}\|^3}.$$

We have already seen a picture of this vector field in Figure 17.2.

Gradient Vector Fields

The gradient of a scalar function f is a function that assigns a vector to each point, and is therefore a vector field. It is called the *gradient field* of f. Many vector fields in physics are gradient fields.

Example 5   Sketch the gradient field of the functions in Figures 17.9–17.11.

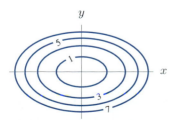

Figure 17.9: The contour map of
$f(x,y) = x^2 + 2y^2$

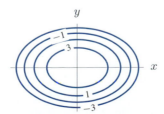

Figure 17.10: The contour map
of $g(x,y) = 5 - x^2 - 2y^2$

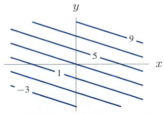

Figure 17.11: The contour map
of $h(x,y) = x + 2y + 3$

Solution   See Figures 17.12–17.14. For a function $f(x, y)$, the gradient vector of f at a point is perpendicular to the contours in the direction of increasing f and its magnitude is the rate of change in that direction. The rate of change is large when the contours are close together and small when they are far apart. Notice that in Figure 17.12 the vectors all point outward, away from the local minimum of f, and in Figure 17.13 the vectors of grad g all point inward, toward the local maximum of g. Since h is a linear function, its gradient is constant, so grad h in Figure 17.14 is a constant vector field.

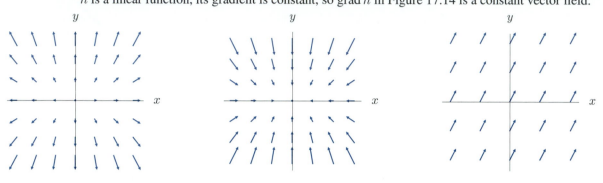

Figure 17.12: grad f *Figure 17.13:* grad g *Figure 17.14:* grad h

Problems for Section 17.1

1. Each vector field in Figures (I)–(IV) represents the force on a particle at different points in space as a result of another particle at the origin. Match up the vector fields with the descriptions below.

    (a)   A repulsive force whose magnitude decreases as distance increases, such as between electric charges of the same sign.

    (b)   A repulsive force whose magnitude increases as distance increases.

    (c)   An attractive force whose magnitude decreases as distance increases, such as gravity.

    (d)   An attractive force whose magnitude increases as distance increases.

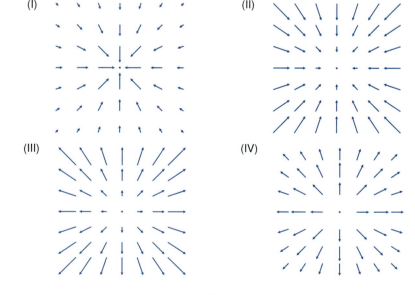

Sketch the vector fields in Problems 2–7.

2. $\vec{F}(x, y) = 2\vec{i} + 3\vec{j}$ 3. $\vec{F}(x, y) = y\vec{i}$ 4. $\vec{F}(x, y) = 2x\vec{i} + x\vec{j}$

5. $\vec{F}(\vec{r}) = 2\vec{r}$ 6. $\vec{F}(\vec{r}) = \dfrac{\vec{r}}{\|\vec{r}\|}$

7. $\vec{F}(x, y) = (x + y)\vec{i} + (x - y)\vec{j}$

For Problems 8–13, find formulas for the vector fields. (There are many possible answers.)

8.

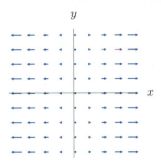

9.

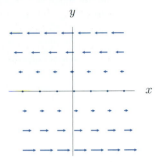

10.

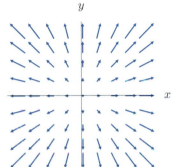

11.

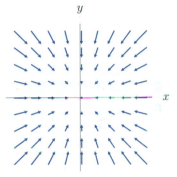

12.

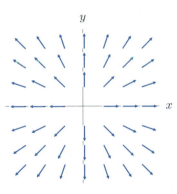

13.

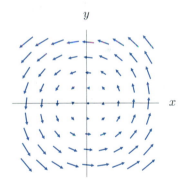

14. Figures 17.15 and 17.16 show the gradient of the functions $z = f(x, y)$ and $z = g(x, y)$.

(a) For each function, draw a rough sketch of the level curves, showing possible z-values.

(b) The xz-plane cuts each of the surfaces $z = f(x, y)$ and $z = g(x, y)$ in a curve. Sketch each of these curves, making clear how they are similar and how they are different from one another.

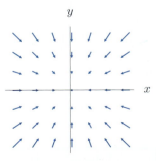

Figure 17.15: Gradient of
$z = f(x, y)$

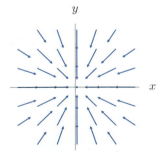

Figure 17.16: Gradient of
$z = g(x, y)$

In Problems 15–17, use a computer to print out vector fields with the given properties. Show on your printout the formula used to generate it. (There are many possible answers.)

15. All vectors are parallel to the x-axis; all vectors on a vertical line have the same magnitude.

16. All vectors point towards the origin and have constant length.

17. All vectors are of unit length and perpendicular to the position vector at that point.

18. Imagine a wide, steadily flowing river in the middle of which there is a fountain that spouts water horizontally in all directions.

   (a) Suppose that the river flows in the $\vec{i}$-direction in the xy-plane and that the fountain is at the origin. Explain why the expression

   $$\vec{v} = A\vec{i} + K(x^2 + y^2)^{-1}(x\vec{i} + y\vec{j}), \quad A > 0, K > 0$$

   could represent the velocity field for the combined flow of the river and the fountain.
   (b) What is the significance of the constants A and K?
   (c) Using a computer, sketch the vector field $\vec{v}$ for $K = 1$ and $A = 1$ and $A = 2$, and for $A = 0.2, K = 2$.

17.2   THE FLOW OF A VECTOR FIELD

When an iceberg is spotted in the North Atlantic, it is important to be able to predict where the iceberg is likely to be a day or a week later. To do this, one needs to know the velocity vector field of the ocean currents, that is, how fast and in what direction the water is moving at each point.

In this section we use differential equations to find the path of an object in a fluid flow. This path is called a *flow line*. Figure 17.17 shows several flow lines for the Gulf stream velocity vector field in Figure 17.1 on page 822. The arrows on each flow line indicate the direction of flow along it.

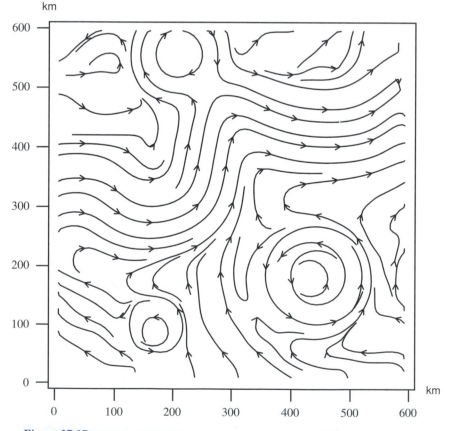

Figure 17.17: Flow lines for objects in the Gulf stream with different starting points

How Do We Find a Flow Line?

Suppose that $\vec{F}$ is the velocity vector field of water on the surface of a creek and imagine a seed being carried along by the current. We want to know the position vector $\vec{r}\,(t)$ of the seed at time t. We know

$$\text{Velocity of seed at time } t = \text{Velocity of current at seed's position at time } t$$

that is,

$$\vec{r}\,'(t) = \vec{F}\,(\vec{r}\,(t)).$$

For an arbitrary vector field we make the following definition:

A **flow line** of a vector field $\vec{v} = \vec{F}\,(\vec{r}\,)$ is a path $\vec{r}\,(t)$ whose velocity vector equals $\vec{v}$. Thus

$$\vec{r}\,'(t) = \vec{v} = \vec{F}\,(\vec{r}\,(t)).$$

The **flow** of a vector field is the family of all of its flow lines.

A flow line is also called an *integral curve* or a *streamline*. We define flow lines for any vector field, as it turns out to be useful to study the flow of fields (for example, electric and magnetic) that are not velocity fields.

Resolving $\vec{F}$ and $\vec{r}$ into components, $\vec{F} = F_1\vec{i} + F_2\vec{j}$ and $\vec{r}\,(t) = x(t)\vec{i} + y(t)\vec{j}$, the definition of a flow line tells us that $x(t)$ and $y(t)$ satisfy the system of differential equations

$$x'(t) = F_1(x(t), y(t)) \quad \text{and} \quad y'(t) = F_2(x(t), y(t)).$$

Solving these differential equations gives a parameterization of the flow line.

Example 1 Find the flow line of the constant velocity field $\vec{v} = 3\vec{i} + 4\vec{j}$ cm/sec that passes through the point $(1, 2)$ at time $t = 0$.

Solution Let $\vec{r}\,(t) = x(t)\vec{i} + y(t)\vec{j}$ be the position in cm of a particle at time t, where t is in seconds. We have

$$x'(t) = 3 \quad \text{and} \quad y'(t) = 4.$$

Thus,

$$x(t) = 3t + x_0 \quad \text{and} \quad y(t) = 4t + y_0.$$

Since the path passes the point $(1, 2)$ at $t = 0$, we have $x_0 = 1$ and $y_0 = 2$ and so

$$x(t) = 3t + 1 \quad \text{and} \quad y(t) = 4t + 2.$$

Thus, the path is the line given parametrically by

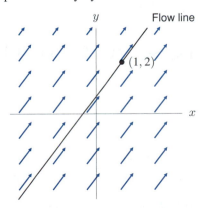

Figure 17.18: Vector field $\vec{F} = 3\vec{i} + 4\vec{j}$ with the flow line through $(1, 2)$

$$\vec{r}(t) = (3t + 1)\vec{i} + (4t + 2)\vec{j}.$$

(See Figure 17.18.) To find an explicit equation for the path, eliminate t between these expressions to get

$$\frac{x - 1}{3} = \frac{y - 2}{4} \qquad \text{or} \qquad y = \frac{4}{3}x + \frac{2}{3}.$$

Example 2 The velocity of a flow at the point (x, y) is $\vec{F}(x, y) = \vec{i} + x\vec{j}$. Find the path of motion of an object in the flow that is at the point $(-2, 2)$ at time $t = 0$.

Solution Figure 17.19 shows a sketch of this field. Since $\vec{r}'(t) = \vec{F}(\vec{r}(t))$, we are looking for the flow line that satisfies the system of differential equations

$$x'(t) = 1, \quad y'(t) = x(t).$$

Solving for $x(t)$ first, we get $x(t) = t + x_0$, where x_0 is a constant of integration. Thus, $y'(t) = t + x_0$, so $y(t) = \frac{1}{2}t^2 + x_0 t + y_0$, where y_0 is also a constant of integration. Since $x(0) = x_0 = -2$ and $y(0) = y_0 = 2$, the path of motion is given by

$$x(t) = t - 2, \quad y(t) = \tfrac{1}{2}t^2 - 2t + 2,$$

or, equivalently,

$$\vec{r}(t) = (t - 2)\vec{i} + (\tfrac{1}{2}t^2 - 2t + 2)\vec{j}.$$

The graph of this flow line in Figure 17.20 looks like a parabola. We check this by seeing that an explicit equation for the path is $y = \frac{1}{2}x^2$.

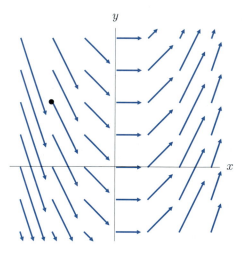

Figure 17.19: The velocity field $\vec{v} = \vec{i} + x\vec{j}$

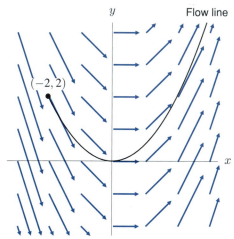

Figure 17.20: A flow line of the velocity field $\vec{v} = \vec{i} + x\vec{j}$

Example 3 Determine the flow of the vector field $\vec{v} = -y\vec{i} + x\vec{j}$.

Solution Figure 17.21 suggests that the flow consists of concentric counterclockwise circles, centered at the origin. The system of differential equations for the flow is

$$x'(t) = -y(t) \qquad y'(t) = x(t).$$

The equations $(x(t), y(t)) = (a \cos t, a \sin t)$ parameterize a family of counterclockwise circles of radius a, centered at the origin. We check that this family satisfies the system of differential equations:

$$x'(t) = -a \sin t = -y(t) \quad \text{and} \quad y'(t) = a \cos t = x(t).$$

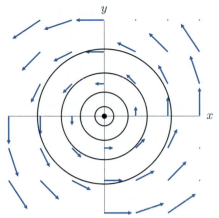

Figure 17.21: The flow of the vector field
$$\vec{v} = -y\vec{i} + x\vec{j}$$

Approximating Flow Lines Numerically

Often it is not possible to find formulas for the flow lines of a vector field. However, we can approximate them numerically by Euler's method for solving differential equations. Since the flow lines $\vec{r}(t) = x(t)\vec{i} + y(t)\vec{j}$ of a vector field $\vec{v} = \vec{F}(x,y)$ satisfy the differential equation $\vec{r}'(t) = \vec{F}(\vec{r}(t))$, we have

$$\vec{r}(t + \Delta t) \approx \vec{r}(t) + (\Delta t)\vec{r}'(t)$$
$$= \vec{r}(t) + (\Delta t)\vec{F}(\vec{r}(t)) \quad \text{for } \Delta t \text{ near } 0.$$

To approximate a flow line, we start at a point $\vec{r}_0 = \vec{r}(0)$ and estimate the position $\vec{r}_1$ of a particle at time Δt later:

$$\vec{r}_1 = \vec{r}(\Delta t) \approx \vec{r}(0) + (\Delta t)\vec{F}(\vec{r}(0))$$
$$= \vec{r}_0 + (\Delta t)\vec{F}(\vec{r}_0).$$

We then repeat the same procedure starting at $\vec{r}_1$, and so on. The general formula for getting from one point to the next is

$$\vec{r}_{n+1} = \vec{r}_n + (\Delta t)\vec{F}(\vec{r}_n).$$

The points with position vectors $\vec{r}_0, \vec{r}_1, \ldots$ trace out the path, as shown in the next example.

Example 4   Use Euler's method to approximate the flow line through $(1, 2)$ for the vector field $\vec{v} = y^2\vec{i} + 2x^2\vec{j}$.

Solution   The flow is determined by the differential equations $\vec{r}'(t) = \vec{v}$, or equivalently

$$x'(t) = y^2, \qquad y'(t) = 2x^2.$$

We use Euler's method with $\Delta t = 0.02$, giving

$$\vec{r}_{n+1} = \vec{r}_n + 0.02\,\vec{v}(x_n, y_n)$$
$$= x_n\vec{i} + y_n\vec{j} + 0.02(y_n^2\vec{i} + 2x_n^2\vec{j}),$$

or equivalently,

$$x_{n+1} = x_n + 0.02y_n^2, \qquad y_{n+1} = y_n + 0.02 \cdot 2x_n^2.$$

When $t = 0$, we have $(x_0, y_0) = (1, 2)$. Then

$$x_1 = x_0 + 0.02 \cdot y_0^2 = 1 + 0.02 \cdot 2^2 = 1.08,$$
$$y_1 = y_0 + 0.02 \cdot 2x_0^2 = 2 + 0.02 \cdot 2 \cdot 1^2 = 2.04.$$

So after one step $x(0.02) \approx 1.08$ and $y(0.02) \approx 2.04$. Similarly, $x(0.04) = x(2\Delta t) \approx 1.16$, $y(0.04) = y(2\Delta t) \approx 2.08$ and so on. Further values along the flow line are given in Table 17.2 and plotted in Figure 17.22.

TABLE 17.2   *The approximated flow line for the vector field $\vec{v} = y^2\vec{i} + 2x^2\vec{j}$ starting at the point $(1, 2)$*

t	0	0.02	0.04	0.06	0.08	0.1	0.12	0.14	0.16	0.18
x	1	1.08	1.16	1.25	1.34	1.44	1.54	1.65	1.77	1.90
y	2	2.04	2.08	2.14	2.20	2.28	2.36	2.45	2.56	2.69

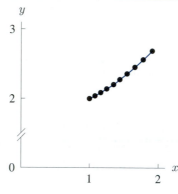

Figure 17.22: Euler's method solution to $x' = y^2$, $y' = 2x^2$

Problems for Section 17.2

For Problems 1–3, sketch the vector field and its flow.

1. $\vec{v} = 3\vec{i}$
2. $\vec{v} = 2\vec{j}$
3. $\vec{v} = 3\vec{i} - 2\vec{j}$

For Problems 4–7, sketch the vector field and the flow. Then find the system of differential equations associated with the vector field and verify that the flow satisfies the system.

4. $\vec{v} = y\vec{i} + x\vec{j}$; $x(t) = a(e^t + e^{-t})$, $y(t) = a(e^t - e^{-t})$.
5. $\vec{v} = y\vec{i} - x\vec{j}$; $x(t) = a\sin t$, $y(t) = a\cos t$.
6. $\vec{v} = x\vec{i} + y\vec{j}$; $x(t) = ae^t$, $y(t) = be^t$.
7. $\vec{v} = x\vec{i} - y\vec{j}$; $x(t) = ae^t$, $y(t) = be^{-t}$.
8. Use a computer or calculator with Euler's method to approximate the flow line through $(1, 2)$ for the vector field $\vec{v} = y^2\vec{i} + 2x^2\vec{j}$ using 5 steps with a time interval $\Delta t = 0.1$.
9. Match the following vector fields with their flow lines. Put arrows on the flow lines indicating the direction of flow.

   (a) $y\vec{i} + x\vec{j}$
   (b) $-y\vec{i} + x\vec{j}$
   (c) $x\vec{i} + y\vec{j}$
   (d) $-y\vec{i} + (x + y/10)\vec{j}$
   (e) $-y\vec{i} + (x - y/10)\vec{j}$
   (f) $(x - y)\vec{i} + (x - y)\vec{j}$

(I)

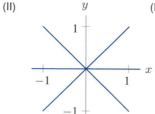

(II)

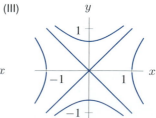

(III)

(IV)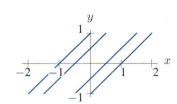

(V)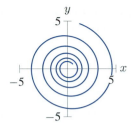

(VI)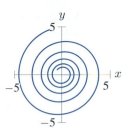

10. We have a fixed set of axes and a solid metal ball whose center is at the origin. The ball rotates once every 24 hours around the z-axis. The direction of rotation is counterclockwise when viewed from above. Consider the point (x, y, z) in our coordinate system lying within the ball. Let $\vec{v}(x, y, z)$ be the velocity vector of the particle of metal at this point. Assume x, y, z are in meters and time is in hours.

   (a) Find a formula for the vector field $\vec{v}$. Give units for your answer.
   (b) Describe in words the flow lines of $\vec{v}$.

CHAPTER SUMMARY

- **Vector Fields**
  Definition of vector field, visualizing fields, gradient fields.
- **Flow Lines**
  Parametric equations of flow lines, approximating flow lines numerically.

REVIEW PROBLEMS FOR CHAPTER SEVENTEEN

1. (a) What is meant by a vector field?
   (b) Suppose $\vec{a} = a_1\vec{i} + a_2\vec{j} + a_3\vec{k}$ is a constant vector. Which of the following are vector fields? Explain.

   (i) $\vec{r} + \vec{a}$
   (iii) $x^2\vec{i} + y^2\vec{j} + z^2\vec{k}$

   (ii) $\vec{r} \cdot \vec{a}$
   (iv) $x^2 + y^2 + z^2$

Sketch the vector fields in Problems 2–4.

2. $\vec{F} = \left(\dfrac{y}{\sqrt{x^2 + y^2}}\right)\vec{i} - \left(\dfrac{x}{\sqrt{x^2 + y^2}}\right)\vec{j}$

3. $\vec{F} = \left(\dfrac{y}{x^2 + y^2}\right)\vec{i} - \left(\dfrac{x}{x^2 + y^2}\right)\vec{j}$

4. $\vec{F} = y\vec{i} - x\vec{j}$

5. If $\vec{F} = \vec{r}/\|\vec{r}\|^3$, find the following quantities in terms of x, y, z, or t.

   (a) $\|\vec{F}\|$
   (b) $\vec{F} \cdot \vec{r}$
   (c) A unit vector parallel to $\vec{F}$ and pointing in the same direction.
   (d) A unit vector parallel to $\vec{F}$ and pointing in the opposite direction.
   (e) $\vec{F}$ if $\vec{r} = \cos t\vec{i} + \sin t\vec{j} + \vec{k}$
   (f) $\vec{F} \cdot \vec{r}$ if $\vec{r} = \cos t\vec{i} + \sin t\vec{j} + \vec{k}$

For Problems 6–9, find the region of the Gulf stream velocity field in figure 17.23 represented by the given table of velocity vectors (in cm/sec).

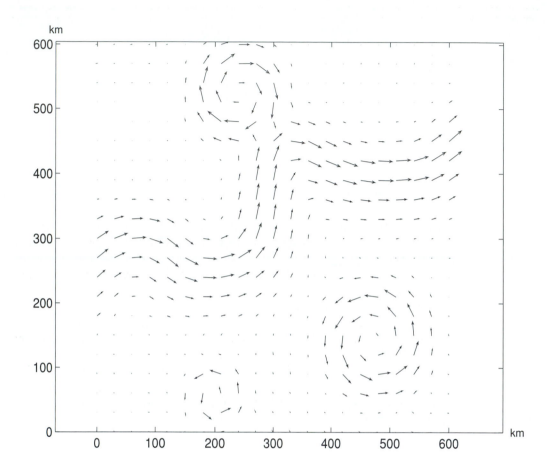

Figure 17.23: The velocity field of the Gulf stream

6.

$35\vec{i} + 131\vec{j}$	$48\vec{i} + 92\vec{j}$	$47\vec{i} + \vec{j}$
$-32\vec{i} + 132\vec{j}$	$-44\vec{i} + 92\vec{j}$	$-42\vec{i} + \vec{j}$
$-51\vec{i} + 73\vec{j}$	$-119\vec{i} + 84\vec{j}$	$-128\vec{i} + 6\vec{j}$

7.

$10\vec{i} - 3\vec{j}$	$11\vec{i} + 16\vec{j}$	$20\vec{i} + 75\vec{j}$
$53\vec{i} - 7\vec{j}$	$58\vec{i} + 23\vec{j}$	$64\vec{i} + 80\vec{j}$
$119\vec{i} - 8\vec{j}$	$121\vec{i} + 31\vec{j}$	$114\vec{i} + 66\vec{j}$

8.

$97\vec{i} - 41\vec{j}$	$72\vec{i} - 24\vec{j}$	$54\vec{i} - 10\vec{j}$
$134\vec{i} - 49\vec{j}$	$131\vec{i} - 44\vec{j}$	$129\vec{i} - 18\vec{j}$
$103\vec{i} - 36\vec{j}$	$122\vec{i} - 30\vec{j}$	$131\vec{i} - 17\vec{j}$

9.

$-95\vec{i} - 60\vec{j}$	$18\vec{i} - 48\vec{j}$	$82\vec{i} - 22\vec{j}$
$-29\vec{i} + 48\vec{j}$	$76\vec{i} + 63\vec{j}$	$128\vec{i} - 16\vec{j}$
$26\vec{i} + 105\vec{j}$	$49\vec{i} + 119\vec{j}$	$88\vec{i} + 13\vec{j}$

10. Each of the following vector fields represents an ocean current. Sketch the vector field, and sketch the path of an iceberg in this current. Determine the location of an iceberg at time $t = 7$ if it is at the point $(1, 3)$ at time $t = 0$.

(a) The current everywhere is $\vec{i}$. (b) The current at (x, y) is $2x\vec{i} + y\vec{j}$.

(c) The current at (x, y) is $-y\vec{i} + x\vec{j}$.

CHAPTER EIGHTEEN

LINE INTEGRALS

When a constant force, $\vec{F}$, acts on an object as it moves through a displacement, $\vec{d}$, the work done by the force is the dot product, $\vec{F} \cdot \vec{d}$. What if the object moves along a curved path through a variable force field? In this chapter we define a *line integral* that calculates the work in this situation. We also define the *circulation* which is used to measure the strength of eddies in a fluid flow.

Line integrals provide us with an analogue of the Fundamental Theorem of Calculus, which tells us how to recover a function from its derivative. The line integral analog of the Fundamental Theorem shows how a line integral can be used to recover a multivariable function from its gradient field.

In contrast with the one-variable situation, not all vector fields are gradient fields. The line integral can be used to distinguish those that are, using the concept of a *path-independent* (or *conservative*) vector field. We study path-independent fields, which are central in physics, and end with Green's Theorem.

18.1   THE IDEA OF A LINE INTEGRAL

Imagine that you are rowing in a river with a noticeable current. At times you may be working against the current and at other times you may be moving with it. At the end you have a sense of whether, overall, you were helped or hindered by the current. The line integral, defined in this section, measures the extent to which a curve in a vector field is, overall, going with the vector field or against it.

Orientation of a Curve

A curve can be traced out in two directions, as shown in Figure 18.1. We need to choose one direction before we can define a line integral.

> A curve is said to be **oriented** if we have chosen a direction of travel on it.

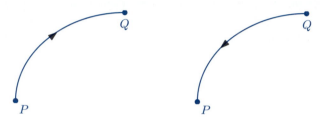

Figure 18.1: A curve with two different orientations represented by arrowheads

Definition of the Line Integral

Consider a vector field $\vec{F}$ and an oriented curve C. We begin by dividing C into n small, almost straight pieces along which $\vec{F}$ is approximately constant. Each piece can be represented by a displacement vector $\Delta \vec{r}_i = \vec{r}_{i+1} - \vec{r}_i$ and the value of $\vec{F}$ at each point of this small piece of C is approximately $\vec{F}(\vec{r}_i)$. See Figures 18.2 and 18.3.

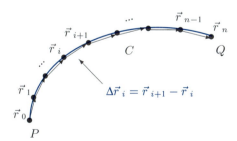

Figure 18.2: The curve C, oriented from P to Q, approximated by straight line segments represented by displacement vectors
$$\Delta \vec{r}_i = \vec{r}_{i+1} - \vec{r}_i$$

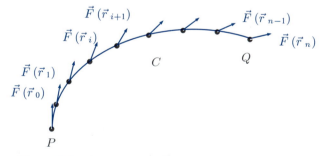

Figure 18.3: The vector field $\vec{F}$ evaluated at the points with position vector $\vec{r}_i$ on the curve C oriented from P to Q

For each point with position vector $\vec{r}_i$ on C, we form the dot product $\vec{F}(\vec{r}_i) \cdot \Delta \vec{r}_i$. Summing over all such pieces, we get a Riemann sum:

$$\sum_{i=0}^{n-1} \vec{F}(\vec{r}_i) \cdot \Delta \vec{r}_i.$$

We define the line integral, written $\int_C \vec{F} \cdot d\vec{r}$, by taking the limit as $\|\Delta \vec{r}_i\| \to 0$. Provided the limit exists, we say

> The **line integral** of a vector field $\vec{F}$ along an oriented curve C is
> $$\int_C \vec{F} \cdot d\vec{r} = \lim_{\|\Delta\vec{r}_i\| \to 0} \sum_{i=0}^{n-1} \vec{F}(\vec{r}_i) \cdot \Delta\vec{r}_i.$$

How Does the Limit Defining a Line Integral Work?

The limit in the definition of a line integral exists if $\vec{F}$ is continuous on the curve C and if C is made by joining end to end a finite number of smooth curves, that is, curves which can be parameterized by smooth functions. We can use the parameterization to subdivide a smooth curve, by subdividing the parameter interval in the same way as for ordinary one-variable integrals. The parameterization must go from one end of the curve to the other, in the forward direction, without retracing any portion of the curve. Under these conditions, the line integral is independent of the way in which the subdivisions are made. All the curves we consider in this book are *piecewise smooth* in this sense. Section 18.2 shows how to use a parameterization to compute a line integral.

Example 1   Find the line integral of the constant vector field $\vec{F} = \vec{i} + 2\vec{j}$ along the path from $(1, 1)$ to $(10, 10)$ shown in Figure 18.4.

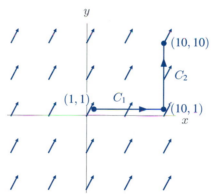

Figure 18.4: The constant vector field $\vec{F} = \vec{i} + 2\vec{j}$ and the path from $(1, 1)$ to $(10, 10)$

Solution   Let C_1 be the horizontal segment of the path going from $(1, 1)$ to $(10, 1)$. When we break this path into pieces, each piece $\Delta\vec{r}$ is horizontal, so $\Delta\vec{r} = \Delta x\vec{i}$ and $\vec{F} \cdot \Delta\vec{r} = (\vec{i} + 2\vec{j}) \cdot \Delta x\vec{i} = \Delta x$. Hence,

$$\int_{C_1} \vec{F} \cdot d\vec{r} = \int_{x=1}^{x=10} dx = 9.$$

Similarly, along the vertical segment C_2, we have $\Delta\vec{r} = \Delta y\vec{j}$ and $\vec{F} \cdot \Delta\vec{r} = (\vec{i} + 2\vec{j}) \cdot \Delta y\vec{j} = 2\Delta y$, so

$$\int_{C_2} \vec{F} \cdot d\vec{r} = \int_{y=1}^{y=10} 2\, dy = 18.$$

Thus,

$$\int_C \vec{F} \cdot d\vec{r} = \int_{C_1} \vec{F} \cdot d\vec{r} + \int_{C_2} \vec{F} \cdot d\vec{r} = 9 + 18 = 27.$$

What Does the Line Integral Tell Us?

Remember that for any two vectors $\vec{u}$ and $\vec{v}$, the dot product $\vec{u} \cdot \vec{v}$ is positive if $\vec{u}$ and $\vec{v}$ point roughly in the same direction (that is, if the angle between them is less than $\pi/2$). The dot product is zero if $\vec{u}$ is perpendicular to $\vec{v}$ and is negative if they point roughly in opposite directions (that is, if the angle between them is greater than $\pi/2$).

The line integral of $\vec{F}$ adds up the dot products of $\vec{F}$ and $\Delta\vec{r}$ along the path. If $\|\vec{F}\|$ is constant, the line integral gives a positive number if $\vec{F}$ is mostly pointing in the same direction as $\Delta\vec{r}$, a negative number if $\vec{F}$ is mostly pointing in the opposite direction. The line integral is zero if $\vec{F}$ is perpendicular to the path at all points or if the positive and negative contributions cancel out. In general, the line integral of a vector field $\vec{F}$ along a curve C measures the extent to which C is going with $\vec{F}$ or against it.

Example 2   The vector field $\vec{F}$ and the oriented curves C_1, C_2, C_3, C_4 are shown in Figure 18.5. The curves C_1 and C_3 are the same length. Which of the line integrals $\int_{C_i} \vec{F} \cdot d\vec{r}$, for $i = 1, 2, 3, 4$, are positive? Which are negative? Arrange these line integrals in ascending order.

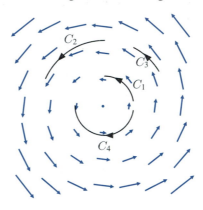

Figure 18.5: Vector field and paths C_1, C_2, C_3, C_4

Solution   The vector field $\vec{F}$ and the line segments $\Delta\vec{r}$ are approximately parallel and in the same direction for the curves C_1, C_2, and C_3. So the contributions of each term $\vec{F} \cdot \Delta\vec{r}$ are positive for these curves. Thus, $\int_{C_1} \vec{F} \cdot d\vec{r}$, $\int_{C_2} \vec{F} \cdot d\vec{r}$, and $\int_{C_3} \vec{F} \cdot d\vec{r}$ are each positive. For the curve C_4, the vector field and the line segments are in opposite directions, so each term $\vec{F} \cdot \Delta\vec{r}$ is negative, and therefore the integral $\int_{C_4} \vec{F} \cdot d\vec{r}$ is negative.

Since the magnitude of the vector field is smaller along C_1 than along C_3, and these two curves are the same length, we have

$$\int_{C_1} \vec{F} \cdot d\vec{r} < \int_{C_3} \vec{F} \cdot d\vec{r}.$$

In addition, the magnitude of the vector field is the same along C_2 and C_3, but the curve C_2 is longer than the curve C_3. Thus,

$$\int_{C_3} \vec{F} \cdot d\vec{r} < \int_{C_2} \vec{F} \cdot d\vec{r}.$$

Putting these results together with the fact that $\int_{C_4} \vec{F} \cdot d\vec{r}$ is negative, we have

$$\int_{C_4} \vec{F} \cdot d\vec{r} < \int_{C_1} \vec{F} \cdot d\vec{r} < \int_{C_3} \vec{F} \cdot d\vec{r} < \int_{C_2} \vec{F} \cdot d\vec{r}.$$

Interpretations of the Line Integral

Work

Recall from Section 12.3 that if a constant force $\vec{F}$ acts on an object while it moves along a straight line through a displacement $\vec{d}$, the work done by the force on the object is

$$\text{Work done} = \vec{F} \cdot \vec{d}.$$

Now suppose we want to find the work done by gravity on an object moving far above the surface of the earth. Since the force of gravity varies with distance from the earth and the path may not be straight, we can't use the formula $\vec{F} \cdot \vec{d}$. We approximate the path by line segments which are small enough that the force is approximately constant on each one. Suppose the force at a point with position vector $\vec{r}$ is $\vec{F}(\vec{r})$, as in Figures 18.2 and 18.3. Then

$$\begin{array}{c} \text{Work done by force } \vec{F}(\vec{r}_i) \\ \text{over small displacement } \Delta\vec{r}_i \end{array} \quad \approx \quad \vec{F}(\vec{r}_i) \cdot \Delta\vec{r}_i,$$

and so,

$$\begin{array}{c} \text{Total work done by force} \\ \text{along oriented curve } C \end{array} \quad \approx \quad \sum_i \vec{F}(\vec{r}_i) \cdot \Delta\vec{r}_i.$$

Taking the limit as $\|\Delta\vec{r}_i\| \to 0$, we get

$$\begin{array}{c} \text{Work done by force } \vec{F}(\vec{r}) \\ \text{along curve } C \end{array} = \lim_{\|\Delta\vec{r}_i\| \to 0} \sum_i \vec{F}(\vec{r}_i) \cdot \Delta\vec{r}_i = \int_C \vec{F} \cdot d\vec{r}.$$

Example 3 A mass lying on a flat table is attached to a spring whose other end is fastened to the wall. (See Figure 18.6.) The spring is extended 20 cm beyond its rest position and released. If the axes are as shown in Figure 18.6, when the spring is extended by a distance of x, the force exerted by the spring on the mass is given by

$$\vec{F}(x) = -kx\vec{i},$$

where k is a positive constant that depends on the strength of the spring.

Suppose the mass moves back to the rest position. How much work is done by the force exerted by the spring?

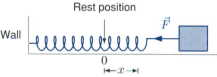

Figure 18.6: Force on mass due to an extended spring

Figure 18.7: Dividing up the interval $0 \leq x \leq 20$ in order to calculate the work done

Solution The path from $x = 20$ to $x = 0$ is divided as shown in Figure 18.7, with a typical segment represented by

$$\Delta\vec{r} = \Delta x\vec{i}.$$

Since we are moving from $x = 20$ to $x = 0$, the quantity Δx will be negative. The work done by the force as the mass moves through this segment is approximated by

$$\text{Work done} \approx \vec{F} \cdot \Delta\vec{r} = (-kx\vec{i}) \cdot (\Delta x\vec{i}) = -kx\,\Delta x.$$

Thus, we have

$$\text{Total work done} \approx \sum -kx\,\Delta x.$$

In the limit, as $\|\Delta x\| \to 0$, this sum becomes an ordinary definite integral. Since the path starts at $x = 20$, this is the lower limit of integration; $x = 0$ is the upper limit. Thus, we get

$$\text{Total work done} = \int_{x=20}^{x=0} -kx\,dx = -\frac{kx^2}{2}\Big|_{20}^{0} = \frac{k(20)^2}{2} = 200k.$$

Note that the work done is positive, since the force acts in the direction of motion.

Example 3 shows how a line integral over a path parallel to the x-axis reduces to a one-variable integral. Section 18.2 shows how to convert *any* line integral into a one-variable integral.

Example 4   A particle with position vector $\vec{r}$ is subject to a force, $\vec{F}$, due to gravity. What is the *sign* of the work done by $\vec{F}$ as the particle moves along the path C_1, a radial line through the center of the earth, starting 8000 km from the center and ending 10,000 km from the center? (See Figure 18.8.)

Solution   We divide the path into small radial segments, $\Delta\vec{r}$, pointing away from the center of the earth and parallel to the gravitational force. The vectors $\vec{F}$ and $\Delta\vec{r}$ point in opposite directions, so each term $\vec{F} \cdot \Delta\vec{r}$ is negative. Adding all these negative quantities and taking the limit results in a negative value for the total work. Thus, the work done by gravity is negative. The negative sign indicates that we would have to do work *against* gravity to move the particle along the path C_1.

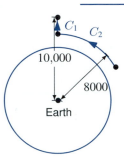

Figure 18.8: The earth

Example 5   Find the sign of the work done by gravity along the curve C_1 in Example 4, but with the opposite orientation.

Solution   Tracing a curve in the opposite direction changes the sign of the line integral because all the segments $\Delta\vec{r}$ change direction, and so every term $\vec{F} \cdot \Delta\vec{r}$ changes sign. Thus, the result will be the negative of the answer found in Example 4. Therefore, the work done by gravity as a particle moves along C_1 toward the center of the earth is positive.

Example 6   Find the work done by gravity as a particle moves along C_2, an arc of a circle 8000 km long at a distance of 8000 km from the center of the earth. (See Figure 18.8.)

Solution   Since C_2 is everywhere perpendicular to the gravitational force, $\vec{F} \cdot \Delta\vec{r} = 0$ for all $\Delta\vec{r}$ along C_2. Thus,

$$\text{Work done} = \int_{C_2} \vec{F} \cdot d\vec{r} = 0,$$

so the work done is zero. This is why satellites can remain in orbit without expending any fuel, once they have attained the correct altitude and velocity.

Circulation

The velocity vector field for the Gulf stream on page 822 shows distinct eddies or regions where the water circulates. We can measure this circulation using a *closed curve*, that is, one that starts and ends at the same point.

> If C is an oriented closed curve, the line integral of a vector field $\vec{F}$ around C is called the **circulation** of $\vec{F}$ around C.

Circulation is a measure of the net tendency of the vector field to point around the curve C. To emphasize that C is closed, the circulation is sometimes denoted $\oint_C \vec{F} \cdot d\vec{r}$, with a small circle on the integral sign.

Example 7   Describe the rotation of the vector fields in Figures 18.9 and 18.10. Find the sign of the circulation of the vector fields around the indicated paths.

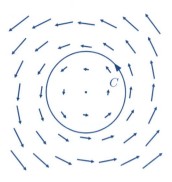

Figure 18.9: A circulating flow

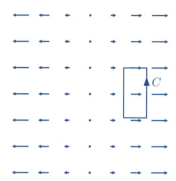

Figure 18.10: A flow with zero circulation

Solution   Consider the vector field in Figure 18.9. If you think of this as representing the velocity of water flowing in a pond, you see that the water is circulating. The line integral around C, measuring the circulation around C, is positive, because the vectors of the field are all pointing in the direction of the path. By way of contrast, look at the vector field in Figure 18.10. Here the line integral around C is zero because the vertical portions of the path are perpendicular to the field and the contributions from the two horizontal portions cancel out. This means that there is no net tendency for the water to circulate around C.

It turns out that the vector field in Figure 18.10 has the property that its circulation around *any* closed path is zero. Water moving according to this vector field has no tendency to circulate around any point and a leaf dropped into the water will not spin. We'll look at such special fields again later when we introduce the notion of the *curl* of a vector field.

Properties of Line Integrals

Line integrals share some basic properties with ordinary one-variable integrals:

For a scalar constant λ, vector fields $\vec{F}$ and $\vec{G}$, and oriented curves C, C_1, and C_2

1. $\displaystyle\int_C \lambda\vec{F} \cdot d\vec{r} = \lambda \int_C \vec{F} \cdot d\vec{r}.$ 2. $\displaystyle\int_C (\vec{F} + \vec{G}) \cdot d\vec{r} = \int_C \vec{F} \cdot d\vec{r} + \int_C \vec{G} \cdot d\vec{r}.$

3. $\displaystyle\int_{-C} \vec{F} \cdot d\vec{r} = -\int_C \vec{F} \cdot d\vec{r}.$ 4. $\displaystyle\int_{C_1+C_2} \vec{F} \cdot d\vec{r} = \int_{C_1} \vec{F} \cdot d\vec{r} + \int_{C_2} \vec{F} \cdot d\vec{r}.$

Properties 3 and 4 are concerned with the curve C over which the line integral is taken. If C is an oriented curve, then $-C$ is the same curve traversed in the opposite direction, that is, with the opposite orientation. (See Figure 18.11 on page 842.) Property 3 holds because if we integrate along $-C$, the vectors $\Delta\vec{r}$ point in the opposite direction and the dot products $\vec{F} \cdot \Delta\vec{r}$ are the negatives of what they were along C.

If C_1 and C_2 are oriented curves with C_1 ending where C_2 begins, we construct a new oriented curve, called $C_1 + C_2$, by joining them together. (See Figure 18.12.) Property 4 is the analogue for line integrals of the property for definite integrals which says that

$$\int_a^b f(x)\,dx = \int_a^c f(x)\,dx + \int_c^b f(x)\,dx.$$

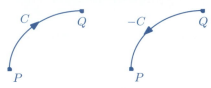
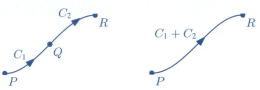

Figure 18.11: A curve, C, and its opposite, $-C$

Figure 18.12: Joining two curves, C_1, and C_2, to make a new one, $C_1 + C_2$

Problems for Section 18.1

In Problems 1–4, say whether you expect the line integral of the pictured vector field over the given curve to be positive, negative, or zero.

1.

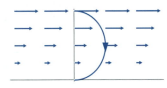

2.

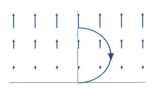

3.

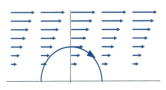

4.

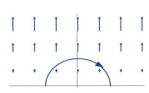

5.   Consider the vector field $\vec{F}$ shown in Figure 18.13, together with the paths C_1, C_2, and C_3. Arrange the line integrals $\int_{C_1} \vec{F} \cdot d\vec{r}$, $\int_{C_2} \vec{F} \cdot d\vec{r}$ and $\int_{C_3} \vec{F} \cdot d\vec{r}$ in ascending order.

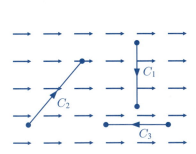

Figure 18.13

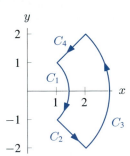

Figure 18.14

For Problems 6–10, say whether you expect the given vector field to have positive, negative, or zero circulation around the closed curve $C = C_1 + C_2 + C_3 + C_4$ in Figure 18.14. The segments C_1 and C_3 are circular arcs centered at the origin; C_2 and C_4 are radial line segments. You may find it helpful to sketch the vector field.

6. $\vec{F}(x,y) = x\vec{i} + y\vec{j}$

7. $\vec{F}(x,y) = -y\vec{i} + x\vec{j}$

8. $\vec{F}(x,y) = y\vec{i} - x\vec{j}$

9. $\vec{F}(x,y) = x^2\vec{i}$

10. $\vec{F}(x,y) = -\dfrac{y}{x^2+y^2}\vec{i} + \dfrac{x}{x^2+y^2}\vec{j}$

In Problems 11–16, calculate the line integral along the line between the given points.

11. $\vec{F} = x\vec{j}$,   from $(1,0)$ to $(3,0)$

12. $\vec{F} = x\vec{j}$,   from $(2,0)$ to $(2,5)$

13. $\vec{F} = x\vec{i}$,   from $(2,0)$ to $(6,0)$

14. $\vec{F} = x\vec{i} + y\vec{j}$,   from $(2,0)$ to $(6,0)$

15. $\vec{F} = \vec{r}$,   from $(2,2)$ to $(6,6)$

16. $\vec{F} = 3\vec{i} + 4\vec{j}$,   from $(0,6)$ to $(0,13)$

17. Draw an oriented curve C and a vector field $\vec{F}$ along C that is not always perpendicular to C, but for which $\int_C \vec{F} \cdot d\vec{r} = 0$.

18. Given the force field $\vec{F}(x,y) = y\vec{i} + x^2\vec{j}$ and the right-angle curve, C, from the points $(0,-1)$ to $(4,-1)$ to $(4,3)$ shown in Figure 18.15:

    (a) Evaluate $\vec{F}$ at the points $(0,-1)$, $(1,-1)$, $(2,-1)$, $(3,-1)$, $(4,-1)$, $(4,0)$, $(4,1)$, $(4,2)$, $(4,3)$.

    (b) Make a sketch showing the force field along C.

    (c) Estimate the work done by the indicated force field on an object traversing the curve C.

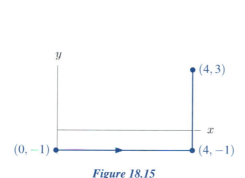

Figure 18.15   **Figure 18.16**

19. If $\vec{F}$ is the constant force field $\vec{j}$, consider the work done by the field on particles traveling on paths C_1, C_2, and C_3 of Figure 18.16. On which of these paths will zero work be done? Explain.

For Problems 20–23, use a computer to calculate the following integrals.

(a) The line integral of $\vec{F}$ around several closed curves. What do you get?

(b) The line integral of $\vec{F}$ along three curves, each of which starts at the origin and ends at the point $(\frac{1}{2}, \frac{1}{2})$. What do you notice?

20. $\vec{F} = x\vec{i} + y\vec{j}$ 21. $\vec{F} = -y\vec{i} + x\vec{j}$ 22. $\vec{F} = \vec{i} + y\vec{j}$ 23. $\vec{F} = \vec{i} + x\vec{j}$

24. As a result of your answers to Problems 20–23, you should have noticed that the following statement is true: Whenever the line integral of a vector field around any closed curve is zero, the line integral along a curve with fixed endpoints has a constant value (that is, the line integral is independent of the path the curve takes between the endpoints). Explain why this is so.

25. As a result of your answers to Problems 20–23, you should have noticed that the converse to the statement in Problem 24 is also true: Whenever the line integral of a vector field depends only on endpoints and not on paths, the circulation is always zero. Explain why this is so.

In Problems 26–27, use the fact that the force of gravity on a particle of mass m at the point with position vector $\vec{r}$ is given by

$$\vec{F} = -\frac{GMm\vec{r}}{r^3}$$

where $r = \|\vec{r}\|$, and G is the gravitational constant, and M is the mass of the earth.

26. Calculate the work done by the force of gravity on a particle of mass m as it moves from 8000 km to 10,000 km from the center of the earth.

27. Calculate the work done by the force of gravity on a particle of mass m as it moves from 8000 km from the center of the earth to infinitely far away.

28.   The fact that an electric current gives rise to a magnetic field is the basis for some electric motors. Ampère's Law relates the magnetic field $\vec{B}$ to a steady current I. It says

$$\int_C \vec{B} \cdot d\vec{r} = kI$$

where I is the current[1] flowing through a closed curve C and k is a constant. Figure 18.17 shows a rod carrying a current and the magnetic field induced around the rod. If the rod is very long and thin, experiments show that the magnetic field $\vec{B}$ is tangent to every circle that is perpendicular to the rod and has center on the axis of the rod (like C in Figure 18.17). The magnitude of $\vec{B}$ is constant along every such circle. Use Ampère's Law to show that around a circle of radius r, the magnetic field due to a current I has magnitude given by

$$\|\vec{B}\| = \frac{kI}{2\pi r}.$$

(In other words, the strength of the field is inversely proportional to the radial distance from the rod.)

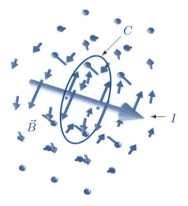

Figure 18.17

18.2 COMPUTING LINE INTEGRALS OVER PARAMETERIZED CURVES

The goal of this section is to show how to use a parameterization of a curve to convert a line integral into an ordinary one-variable integral.

Using a Parameterization to Evaluate a Line Integral

Recall the definition of the line integral,

$$\int_C \vec{F} \cdot d\vec{r} = \lim_{\|\Delta \vec{r}_i\| \to 0} \sum \vec{F}(\vec{r}_i) \cdot \Delta \vec{r}_i,$$

where the $\vec{r}_i$ are the position vectors of points subdividing the curve into short pieces. Now suppose we have a smooth parameterization, $\vec{r}(t)$, of C for $a \leq t \leq b$, so that $\vec{r}(a)$ is the position vector of the beginning of the curve and $\vec{r}(b)$ is the position vector of the end. Then we can divide C into n pieces by dividing the interval $a \leq t \leq b$ into n pieces, each of size $\Delta t = (b-a)/n$. See Figures 18.18 and 18.19.

At each point $\vec{r}_i = \vec{r}(t_i)$ we want to compute

$$\vec{F}(\vec{r}_i) \cdot \Delta \vec{r}_i.$$

---

[1]More precisely, I is the net current through any surface that has C as its boundary.

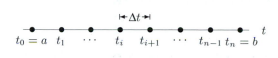

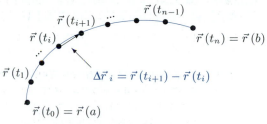

Figure 18.19: Corresponding subdivision of the parameterized path C

Figure 18.18: Subdivision of the interval $a \leq t \leq b$

Since $t_{i+1} = t_i + \Delta t$, the displacement vectors $\Delta \vec{r_i}$ are given by

$$
\begin{aligned}
\Delta \vec{r_i} &= \vec{r}(t_{i+1}) - \vec{r}(t_i) \\
&= \vec{r}(t_i + \Delta t) - \vec{r}(t_i) \\
&= \frac{\vec{r}(t_i + \Delta t) - \vec{r}(t_i)}{\Delta t} \cdot \Delta t \\
&\approx \vec{r}'(t_i)\Delta t,
\end{aligned}
$$

where we use the facts that Δt is small and that $\vec{r}(t)$ is differentiable to obtain the last approximation. Therefore,

$$
\int_C \vec{F} \cdot d\vec{r} \approx \sum \vec{F}(\vec{r}_i) \cdot \Delta \vec{r}_i \approx \sum \vec{F}(\vec{r}(t_i)) \cdot \vec{r}'(t_i)\,\Delta t.
$$

Notice that $\vec{F}(\vec{r}(t_i)) \cdot \vec{r}'(t_i)$ is the value at t_i of a one-variable function of t, so this last sum is really a one-variable Riemann sum. In the limit as $\Delta t \to 0$, we get a definite integral:

$$
\lim_{\Delta t \to 0} \sum \vec{F}(\vec{r}(t_i)) \cdot \vec{r}'(t_i)\,\Delta t = \int_a^b \vec{F}(\vec{r}(t)) \cdot \vec{r}'(t)\,dt.
$$

Thus, we have the following result:

> If $\vec{r}(t)$, for $a \leq t \leq b$, is a smooth parameterization of an oriented curve C and $\vec{F}$ is a vector field which is continuous on C, then
>
> $$
> \int_C \vec{F} \cdot d\vec{r} = \int_a^b \vec{F}(\vec{r}(t)) \cdot \vec{r}'(t)\,dt.
> $$
>
> In words: To compute the line integral of $\vec{F}$ over C, take the dot product of $\vec{F}$ evaluated on C with the velocity vector, $\vec{r}'(t)$, of the parameterization of C, then integrate along the curve.

Even though we assumed that C is smooth, we can use the same formula to compute line integrals over curves which are only *piecewise smooth*, such as the boundary of a rectangle: If C is piecewise smooth, we apply the formula to each one of the smooth pieces and add the results.

Example 1 Compute $\int_C \vec{F} \cdot d\vec{r}$ where $\vec{F} = (x+y)\vec{i} + y\vec{j}$ and C is the quarter unit circle, oriented counterclockwise as shown in Figure 18.20.

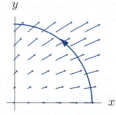

Figure 18.20: The vector field $\vec{F} = (x+y)\vec{i} + y\vec{j}$ and the quarter circle C

Solution   Since all of the vectors in $\vec{F}$ along C point generally in a direction opposite to the orientation of C, we expect our answer to be negative. The first step is to parameterize C by

$$\vec{r}(t) = x(t)\vec{i} + y(t)\vec{j} = \cos t\vec{i} + \sin t\vec{j}, \quad 0 \le t \le \frac{\pi}{2}.$$

Substituting the parameterization into $\vec{F}$ we get $\vec{F}(x(t), y(t)) = (\cos t + \sin t)\vec{i} + \sin t\vec{j}$. The vector $\vec{r}'(t) = x'(t)\vec{i} + y'(t)\vec{j} = -\sin t\vec{i} + \cos t\vec{j}$. Then

$$\int_C \vec{F} \cdot d\vec{r} = \int_0^{\pi/2} ((\cos t + \sin t)\vec{i} + \sin t\vec{j}) \cdot (-\sin t\vec{i} + \cos t\vec{j})dt$$

$$= \int_0^{\pi/2} (-\cos t \sin t - \sin^2 t + \sin t \cos t)dt$$

$$= \int_0^{\pi/2} -\sin^2 t \, dt = -\frac{\pi}{4} \approx -0.7854.$$

So the answer is negative, as expected.

Example 2   Consider the vector field $\vec{F} = x\vec{i} + y\vec{j}$.

(a)   Suppose C_1 is the line segment joining $(1,0)$ to $(0,2)$ and C_2 is a part of a parabola with its vertex at $(0,2)$, joining the same points in the same order. (See Figure 18.21.) Verify that

$$\int_{C_1} \vec{F} \cdot d\vec{r} = \int_{C_2} \vec{F} \cdot d\vec{r}.$$

(b)   If C is the triangle shown in Figure 18.22, show that $\int_C \vec{F} \cdot d\vec{r} = 0$.

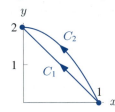

Figure 18.21

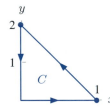

Figure 18.22

Solution   (a)   We parameterize C_1 by $\vec{r}(t) = (1-t)\vec{i} + 2t\vec{j}$ with $0 \le t \le 1$. Then $\vec{r}'(t) = -\vec{i} + 2\vec{j}$, so

$$\int_{C_1} \vec{F} \cdot d\vec{r} = \int_0^1 \vec{F}(1-t, 2t) \cdot (-\vec{i} + 2\vec{j}) \, dt = \int_0^1 ((1-t)\vec{i} + 2t\vec{j}) \cdot (-\vec{i} + 2\vec{j}) \, dt$$

$$= \int_0^1 (5t - 1) \, dt = \frac{3}{2}.$$

To parameterize C_2, we use the fact that it is part of a parabola with vertex at $(0,2)$, so its equation is of the form $y = -kx^2 + 2$ for some k. Since the parabola crosses the x-axis at $(1,0)$, we find that $k = 2$ and $y = -2x^2 + 2$. Therefore, we use the parameterization $\vec{r}(t) = t\vec{i} + (-2t^2 + 2)\vec{j}$ with $0 \le t \le 1$, which has $\vec{r}' = \vec{i} - 4t\vec{j}$. This traces out C_2 in reverse, since $t = 0$ gives $(0,2)$, and $t = 1$ gives $(1,0)$. Thus, we make $t = 0$ the upper limit of integration and $t = 1$ the lower limit:

$$\int_{C_2} \vec{F} \cdot d\vec{r} = \int_1^0 \vec{F}(t, -2t^2 + 2) \cdot (\vec{i} - 4t\vec{j}) \, dt = -\int_0^1 (t\vec{i} + (-2t^2 + 2)\vec{j}) \cdot (\vec{i} - 4t\vec{j}) \, dt$$

$$= -\int_0^1 (8t^3 - 7t) \, dt = \frac{3}{2}.$$

So the line integrals along C_1 and C_2 have the same value.

(b)   We break $\int_C \vec{F} \cdot d\vec{r}$ into three pieces, one of which we have already computed (namely, the piece connecting $(1,0)$ to $(0,2)$, where the line integral has value $3/2$). The piece running from

$(0, 2)$ to $(0, 0)$ can be parameterized by $\vec{r}(t) = (2-t)\vec{j}$ with $0 \le t \le 2$. The piece running from $(0, 0)$ to $(1, 0)$ can be parameterized by $\vec{r}(t) = t\vec{i}$ with $0 \le t \le 1$. Then

$$\int_C \vec{F} \cdot d\vec{r} = \frac{3}{2} + \int_0^2 \vec{F}(0, 2-t) \cdot (-\vec{j})\, dt + \int_0^1 \vec{F}(t, 0) \cdot \vec{i}\, dt$$

$$= \frac{3}{2} + \int_0^2 (2-t)\vec{j} \cdot (-\vec{j})\, dt + \int_0^1 t\vec{i} \cdot \vec{i}\, dt$$

$$= \frac{3}{2} + \int_0^2 (t-2)\, dt + \int_0^1 t\, dt = \frac{3}{2} + (-2) + \frac{1}{2} = 0.$$

Example 3   Let C be the closed curve consisting of the upper half-circle of radius 1 and the line forming its diameter along the x-axis, oriented counterclockwise. (See Figure 18.23.) Find $\int_C \vec{F} \cdot d\vec{r}$ where $\vec{F}(x, y) = -y\vec{i} + x\vec{j}$.

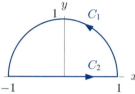

Figure 18.23: The curve $C = C_1 + C_2$ for Example 3

Solution   We write $C = C_1 + C_2$ where C_1 is the half-circle and C_2 is the line, and compute $\int_{C_1} \vec{F} \cdot d\vec{r}$ and $\int_{C_2} \vec{F} \cdot d\vec{r}$ separately. We parameterize C_1 by $\vec{r}(t) = \cos t\vec{i} + \sin t\vec{j}$, with $0 \le t \le \pi$. Then

$$\int_{C_1} \vec{F} \cdot d\vec{r} = \int_0^\pi (-\sin t\vec{i} + \cos t\vec{j}) \cdot (-\sin t\vec{i} + \cos t\vec{j})\, dt$$

$$= \int_0^\pi (\sin^2 t + \cos^2 t)\, dt = \int_0^\pi 1\, dt = \pi.$$

For C_2, we have $\int_{C_2} \vec{F} \cdot d\vec{r} = 0$, since the vector field $\vec{F}$ has no $\vec{i}$ component along the x-axis (where $y = 0$) and is therefore perpendicular to C_2 at all points.

Finally, we can write

$$\int_C \vec{F} \cdot d\vec{r} = \int_{C_1} \vec{F} \cdot d\vec{r} + \int_{C_2} \vec{F} \cdot d\vec{r} = \pi + 0 = \pi.$$

It is no accident that the result for $\int_{C_1} \vec{F} \cdot d\vec{r}$ is the same as the length of the curve C_1. See Problems 14–15 on page 850.

The next example illustrates the computation of a line integral over a path in 3-space.

Example 4   A particle travels along the helix C given by $\vec{r}(t) = \cos t\vec{i} + \sin t\vec{j} + 2t\vec{k}$ and is subject to a force $\vec{F} = x\vec{i} + z\vec{j} - xy\vec{k}$. Find the total work done on the particle by the force for $0 \le t \le 3\pi$.

Solution   The work done is given by a line integral, which we evaluate using the given parameterization:

$$\text{Work done} = \int_C \vec{F} \cdot d\vec{r} = \int_0^{3\pi} \vec{F}(\vec{r}(t)) \cdot \vec{r}\,'(t)\, dt$$

$$= \int_0^{3\pi} (\cos t\vec{i} + 2t\vec{j} - \cos t \sin t\vec{k}) \cdot (-\sin t\vec{i} + \cos t\vec{j} + 2\vec{k})\, dt$$

$$= \int_0^{3\pi} (-\cos t \sin t + 2t \cos t - 2\cos t \sin t)\, dt$$

$$= \int_0^{3\pi} (-3\cos t \sin t + 2t \cos t)\, dt = -4.$$

The Notation $\int_C P\,dx + Q\,dy + R\,dz$

There is an alternative notation for line integrals that is quite common. Given functions $P(x, y, z)$ and $Q(x, y, z)$ and $R(x, y, z)$, and an oriented curve C, we consider the vector field $\vec{F} = P\vec{i} + Q\vec{j} + R\vec{k}$. We can write

$$\int_C \vec{F} \cdot d\vec{r} = \int_C P(x, y, z)dx + Q(x, y, z)dy + R(x, y, z)dz,$$

The relation between the two notations can be remembered using $d\vec{r} = dx\vec{i} + dy\vec{j} + dz\vec{k}$.

Example 5   Evaluate $\int_C xy\,dx - y^2\,dy$ where C is the line segment from $(0, 0)$ to $(2, 6)$.

Solution   We parameterize C by $\vec{r}(t) = x(t)\vec{i} + y(t)\vec{j} = t\vec{i} + 3t\vec{j}$, for $0 \le t \le 2$. Thus,

$$\int_C xy\,dx - y^2\,dy = \int_C (xy\vec{i} - y^2\vec{j}) \cdot d\vec{r} = \int_0^2 (3t^2\vec{i} - 9t^2\vec{j}) \cdot (\vec{i} + 3\vec{j})\,dt = \int_0^2 (-24t^2)\,dt = -64.$$

Independence of Parameterization

Since there are many different ways of parameterizing a given oriented curve, you may be wondering what happens to the value of a given line integral if you choose another parameterization. The answer is that the choice of parameterization makes no difference. Since we initially defined the line integral without reference to any particular parameterization, this is exactly as we would expect.

Example 6   Consider the oriented path which is a straight line segment L running from $(0, 0)$ to $(1, 1)$. Calculate the line integral of the vector field $\vec{F} = (3x - y)\vec{i} + x\vec{j}$ along L using each of the parameterizations

(a) $A(t) = (t, t), \quad 0 \le t \le 1,$ (b) $D(t) = (e^t - 1, e^t - 1), \quad 0 \le t \le \ln 2.$

Solution   The line L has equation $y = x$. Both $A(t)$ and $D(t)$ give a parameterization of L: each has both coordinates equal and each begins at (0,0) and ends at (1,1). Now let's calculate the line integral of the vector field $\vec{F} = (3x - y)\vec{i} + x\vec{j}$ using each parameterization.
(a)   Using $A(t)$, we get

$$\int_L \vec{F} \cdot d\vec{r} = \int_0^1 ((3t - t)\vec{i} + t\vec{j}) \cdot (\vec{i} + \vec{j})\,dt = \int_0^1 3t\,dt = \frac{3t^2}{2}\bigg|_0^1 = \frac{3}{2}.$$

(b)   Using $D(t)$, we get

$$\int_L \vec{F} \cdot d\vec{r} = \int_0^{\ln 2} \left((3(e^t - 1) - (e^t - 1))\,\vec{i} + (e^t - 1)\vec{j} \right) \cdot (e^t\vec{i} + e^t\vec{j})\,dt$$

$$= \int_0^{\ln 2} 3(e^{2t} - e^t)\,dt = 3\left(\frac{e^{2t}}{2} - e^t\right)\bigg|_0^{\ln 2} = \frac{3}{2}.$$

The fact that both answers are the same illustrates that the value of a line integral is independent of the parameterization of the path. Problems 17–19 at the end of this section give another way of seeing this.

Problems for Section 18.2

In Problems 1–10, compute the line integral of the given vector field along the given path.

1. $\vec{F}(x, y) = \ln y \vec{i} + \ln x \vec{j}$ and C is the curve given parametrically by $(2t, t^3)$, for $2 \le t \le 4$.

2. $\vec{F} = x\vec{i} + y\vec{j}$ and C is the line from the origin to the point $(3, 3)$.

3. $\vec{F}(x, y) = x^2\vec{i} + y^2\vec{j}$ and C is the line from the point $(1, 2)$ to the point $(3, 4)$.

4. $\vec{F} = 2y\vec{i} - (\sin y)\vec{j}$ counterclockwise around the unit circle C starting at the point $(1, 0)$.

5. $\vec{F}(x, y) = e^x\vec{i} + e^y\vec{j}$ and C is the part of the ellipse $x^2 + 4y^2 = 4$ joining the point $(0, 1)$ to the point $(2, 0)$ in the clockwise direction.

6. $\vec{F}(x, y) = xy\vec{i} + (x - y)\vec{j}$ and C is the triangle joining the points $(1, 0)$, $(0, 1)$ and $(-1, 0)$ in the clockwise direction.

7. $\vec{F} = x\vec{i} + 2zy\vec{j} + x\vec{k}$ and C is given by $\vec{r} = t\vec{i} + t^2\vec{j} + t^3\vec{k}$ for $1 \le t \le 2$.

8. $\vec{F} = x^3\vec{i} + y^2\vec{j} + z\vec{k}$ and C is the line from the origin to the point $(2, 3, 4)$.

9. $\vec{F} = -y\vec{i} + x\vec{j} + 5\vec{k}$ and C is the helix $x = \cos t, y = \sin t, z = t$, for $0 \le t \le 4\pi$.

10. $\vec{F} = e^y\vec{i} + \ln(x^2 + 1)\vec{j} + \vec{k}$ and C is the circle of radius 2 in the yz-plane centered at the origin and traversed as shown in Figure 18.24.

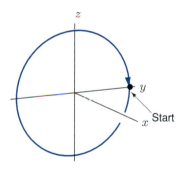

Figure 18.24

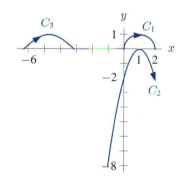

Figure 18.25

11. Find parameterizations for the oriented curves shown in Figure 18.25. The curve C_1 is a semicircle of radius 1, centered at the point $(1, 0)$. The curve C_2 is a portion of a parabola, with vertex at the point $(1, 0)$ and y-intercept -2. The curve C_3 is one arc of the sine curve.

12. Suppose C is the line segment from the point $(0, 0)$ to the point $(4, 12)$ and $\vec{F} = xy\vec{i} + x\vec{j}$.

    (a) Is $\int_C \vec{F} \cdot d\vec{r}$ greater than, less than, or equal to zero? Give a geometric explanation.

    (b) A parameterization of C is $(x(t), y(t)) = (t, 3t)$ for $0 \le t \le 4$. Use this to compute $\int_C \vec{F} \cdot d\vec{r}$.

    (c) Suppose a particle leaves the point $(0, 0)$, moves along the line towards the point $(4, 12)$, stops before reaching it and backs up, stops again and reverses direction, then completes its journey to the endpoint. All travel takes place along the line segment joining the point $(0, 0)$ to the point $(4, 12)$. If we call this path C', explain why $\int_{C'} \vec{F} \cdot d\vec{r} = \int_C \vec{F} \cdot d\vec{r}$.

    (d) A parameterization for a path like C' is given by

    $$(x(t), y(t)) = \left(\frac{1}{3}(t^3 - 6t^2 + 11t), (t^3 - 6t^2 + 11t)\right), \qquad 0 \le t \le 4.$$

    Check that this begins at the point $(0, 0)$ and ends at the point $(4, 12)$. Check also that all points of C' lie on the line segment connecting the point $(0, 0)$ to the point $(4, 12)$. What are the values of t at which the particle changes direction?

    (e) Find $\int_{C'} \vec{F} \cdot d\vec{r}$ using the parameterization in part (d). Do you get the same answer as in part (b)?

13. In Example 6 on page 848 we integrated $\vec{F} = (3x - y)\vec{i} + x\vec{j}$ over two parameterizations of the line from $(0,0)$ to $(1,1)$, getting $3/2$ each time. Now compute the line integral along two different paths with the same endpoints, and show that the answers are different.

    (a) The path (t, t^2), with $0 \le t \le 1$ (b) The path (t^2, t), with $0 \le t \le 1$

14. Consider the vector field $\vec{F} = -y\vec{i} + x\vec{j}$. Let C be the unit circle oriented counterclockwise.

    (a) Show that $\vec{F}$ has a constant magnitude of 1 on the circle C.
    (b) Show that $\vec{F}$ is always tangent to the circle C.
    (c) Show that $\int_C \vec{F} \cdot d\vec{r} = $ Length of C.

15. Suppose that along a curve C, a vector field $\vec{F}$ is always tangent to C in the direction of orientation and has constant magnitude $\|\vec{F}\| = m$. Use the definition of the line integral to explain why

$$\int_C \vec{F} \cdot d\vec{r} = m \cdot \text{Length of } C.$$

16. Consider the oriented path which is a straight line segment L running from $(0,0)$ to $(1,1)$. Calculate the line integral of the vector field $\vec{F} = (3x - y)\vec{i} + x\vec{j}$ along L using each of the parameterizations

    (a) $B(t) = (2t, 2t), \quad 0 \le t \le 1/2,$ (b) $C(t) = \left(\dfrac{t^2 - 1}{3}, \dfrac{t^2 - 1}{3} \right), \quad 1 \le t \le 2,$

In Example 6 on page 848 two parameterizations, $A(t)$, and $D(t)$, are used to convert a line integral into a definite integral. In Problem 16, two other parameterizations, $B(t)$ and $C(t)$, are used on the same line integral. In Problems 17–19 show that two definite integrals corresponding to two of the given parameterizations are equal by finding a substitution which converts one integral to the other. This gives us another way of seeing why changing the parameterization of the curve does not change the value of the line integral.

17. $A(t)$ and $B(t)$ 18. $A(t)$ and $C(t)$ 19. $A(t)$ and $D(t)$

20. A spiral staircase in a building is in the shape of a helix of radius 5 meters. Between two floors of the building, the stairs make one full revolution and climb by 4 meters. A person carries a bag of groceries up two floors. The combined mass of the person and the groceries is 70 kg and the gravitational force is $70g$ downward, where g is the acceleration due to gravity. Calculate the work done by the person against gravity.

18.3 GRADIENT FIELDS AND PATH-INDEPENDENT FIELDS

For a function, f, of one variable, the Fundamental Theorem of Calculus tells us that the definite integral of a rate of change, f', gives the total change in f:

$$\int_a^b f'(t)\, dt = f(b) - f(a).$$

What about functions of two or more variables? The quantity that describes the rate of change is the gradient vector field. If we know the gradient of a function f, can we compute the total change in f between two points? The answer is yes, using a line integral.

Finding the Total Change in f from grad f: The Fundamental Theorem

To find the change in f between two points P and Q, we choose a smooth path C from P to Q, then divide the path into many small pieces. See Figure 18.26.

First we estimate the change in f as we move through a displacement $\Delta \vec{r}_i$ from $\vec{r}_i$ to $\vec{r}_{i+1}$. Suppose $\vec{u}$ is a unit vector in the direction of $\Delta \vec{r}_i$. Then the change in f is given by

$$f(\vec{r}_{i+1}) - f(\vec{r}_i) \approx \text{Rate of change of } f \times \text{Distance moved in direction of } \vec{u}$$
$$= f_{\vec{u}}(\vec{r}_i)\|\Delta \vec{r}_i\|$$
$$= \text{grad } f \cdot \vec{u} \, \|\Delta \vec{r}_i\|$$
$$= \text{grad } f \cdot \Delta \vec{r}_i. \qquad \text{since } \Delta \vec{r}_i = \|\Delta \vec{r}_i\|\vec{u}$$

Therefore, summing over all pieces of the path, the total change in f is given by

$$\text{Total change} = f(Q) - f(P) \approx \sum_{i=0}^{n-1} \text{grad } f(\vec{r}_i) \cdot \Delta \vec{r}_i.$$

In the limit as $\|\Delta \vec{r}_i\|$ approaches zero, we obtain the following result:

The Fundamental Theorem of Calculus for Line Integrals

Suppose C is a piecewise smooth oriented path with starting point P and endpoint Q. If f is a function whose gradient is continuous on the path C, then

$$\int_C \text{grad } f \cdot d\vec{r} = f(Q) - f(P).$$

Notice that there are many different paths from P to Q. (See Figure 18.27.) However, the value of the line integral $\int_C \text{grad } f \cdot d\vec{r}$ depends only on the endpoints of C; it does not depend on where C goes in between.[2]

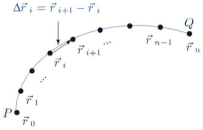

Figure 18.26: Subdivision of the path from P to Q. We estimate the change in f along $\Delta \vec{r}_i$

Figure 18.27: There are many different paths from P to Q: all give the same value of $\int_C \text{grad } f \cdot d\vec{r}$

Example 1   Suppose that grad f is everywhere perpendicular to the curve joining P and Q shown in Figure 18.28.
(a)   Explain why you expect the path joining P and Q to be a contour.
(b)   Using a line integral, show that $f(P) = f(Q)$.

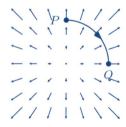

Figure 18.28: The gradient vector field of the function f

Solution   (a)   The gradient of f is everywhere perpendicular to the path from P to Q, as you expect along a contour.
(b)   Consider the path from P to Q shown in Figure 18.28 and evaluate the line integral

$$\int_C \text{grad } f \cdot d\vec{r} = f(Q) - f(P).$$

Since grad f is everywhere perpendicular to the path, the line integral is 0. Thus, $f(Q) = f(P)$.

---

[2]Problem 13 on page 868 shows how the Fundamental Theorem for Line Integrals can be derived from the one-variable Fundamental Theorem of Calculus.

Example 2   Consider the vector field $\vec{F} = x\vec{i} + y\vec{j}$. In Example 2 on page 846 we calculated $\int_{C_1} \vec{F} \cdot d\vec{r}$ and $\int_{C_2} \vec{F} \cdot d\vec{r}$ over the oriented curves shown in Figure 18.29 and found they were the same. Find a scalar function f with grad $f = \vec{F}$. Hence, find an easy way to calculate the line integrals, and explain how we could have expected them to be the same.

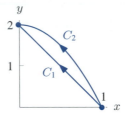

Figure 18.29: Find the line integral of $\vec{F} = x\vec{i} + y\vec{j}$ over the curves C_1 and C_2

Solution   One possibility for f is

$$f(x, y) = \frac{x^2}{2} + \frac{y^2}{2}.$$

You can check that grad $f = x\vec{i} + y\vec{j}$. Now we can use the Fundamental Theorem to compute the line integral. Since $\vec{F} = $ grad f we have

$$\int_{C_1} \vec{F} \cdot d\vec{r} = \int_{C_1} \text{grad } f \cdot d\vec{r} = f(0, 2) - f(1, 0) = \frac{3}{2}.$$

Notice that the calculation looks exactly the same for C_2. Since the value of the integral depends only on the value of f at the endpoints, it is the same no matter what path we choose.

Path-Independent, or Conservative, Vector Fields

In the previous example, the line integral was independent of the path taken between the two (fixed) endpoints. We give vector fields whose line integrals have this property a special name.

A vector field $\vec{F}$ is said to be **path-independent**, or **conservative**, if for any two points P and Q, the line integral $\int_C \vec{F} \cdot d\vec{r}$ has the same value along any path C from P to Q lying in the domain of $\vec{F}$.

If, on the other hand, the line integral $\int_C \vec{F} \cdot d\vec{r}$ does depend on the path C joining P to Q, then $\vec{F}$ is said to be a *path-dependent* vector field.

Now suppose that $\vec{F}$ is any gradient field, so $\vec{F} = $ grad f. If C is a path from P to Q, the Fundamental Theorem for Line Integrals tells us that

$$\int_C \vec{F} \cdot d\vec{r} = f(Q) - f(P).$$

Since the right-hand side of this equation does not depend on the path, but only on the endpoints of the path, the vector field $\vec{F}$ is path-independent. Thus, we have the following important result:

If $\vec{F}$ is a gradient vector field, then $\vec{F}$ is path-independent.

Why Do We Care about Path-Independent, or Conservative, Vector Fields?

Many of the fundamental vector fields of nature are path-independent — for example the gravitational field and the electric field of particles at rest. The fact that the gravitational field is path-independent means that the work done by gravity when an object moves depends only on the starting and ending points and not on the path taken. For example the work done by gravity (computed by the line integral) on a bicycle being carried to a sixth floor apartment is the same whether it is carried up the stairs in a zig-zag path or taken straight up in an elevator.

When a vector field is path-independent we can define the *potential energy* of a body. When the body moves to another position, the potential energy changes by an amount equal to the work done by the vector field, which depends only on the starting and ending positions. If the work done had not been path-independent, the potential energy would depend both on the body's current position *and* on how it got there, making it impossible to define a useful potential energy.

Project 1 on page 870 explains why path-independent force vector fields are also called *conservative* vector fields: When a particle moves under the influence of a conservative vector field, the *total energy* of the particle is *conserved*. It turns out that the force field is obtained from the gradient of the potential energy function.

Path-Independent Fields and Gradient Fields

We have seen that every gradient field is path-independent. What about the converse? That is, given a path-independent vector field $\vec{F}$, can we find a function f such that $\vec{F} = \operatorname{grad} f$? The answer is yes.

How to Construct f from $\overrightarrow{F}$

First, notice that there are many different choices for f, since we can add a constant to f without changing $\operatorname{grad} f$. If we pick a fixed starting point P, then by adding or subtracting a constant to f we can ensure that $f(P) = 0$. For any other point Q, we define $f(Q)$ by the formula

$$f(Q) = \int_C \vec{F} \cdot d\vec{r}, \quad \text{where } C \text{ is any path from } P \text{ to } Q.$$

Since $\vec{F}$ is path-independent, it doesn't matter which path we choose from P to Q. On the other hand, if $\vec{F}$ is not path-independent, then different choices might give different values for $f(Q)$, so f would not be a function (a function has to have a single value at each point).

We still have to show that the gradient of the function f really is $\vec{F}$; we do this on page 855. However, by constructing a function f in this manner, we have the following result:

> If $\vec{F}$ is a path-independent vector field, then $\vec{F} = \operatorname{grad} f$ for some f.

Combining the two results, we have

> A vector field $\vec{F}$ is path-independent if and only if $\vec{F}$ is a gradient vector field.

The function f is sufficiently important that it is given a special name:

> If a vector field $\vec{F}$ is of the form $\vec{F} = \operatorname{grad} f$ for some scalar function f, then f is called a **potential function** for the vector field $\vec{F}$.

Warning

Physicists use the convention that a function ϕ is a potential function for a vector field $\vec{F}$ if $\vec{F} = -\operatorname{grad} \phi$. See Problem 20 on page 858.

Example 3   Show that the vector field $\vec{F}(x, y) = y \cos x \vec{i} + \sin x \vec{j}$ is path-independent.

Solution   If we can find a potential function f, then $\vec{F}$ must be path-independent. We want grad $f = \vec{F}$. Since

$$\frac{\partial f}{\partial x} = y \cos x,$$

f must be of the form

$$f(x, y) = y \sin x + g(y) \qquad \text{where } g(y) \text{ is a function of } y \text{ only.}$$

In addition, since grad $f = \vec{F}$, we must have

$$\frac{\partial f}{\partial y} = \sin x,$$

and differentiating $f(x, y) = y \sin x + g(y)$ gives

$$\frac{\partial f}{\partial y} = \sin x + g'(y).$$

Thus, we must have $g'(y) = 0$, so $g(y) = C$ where C is some constant. Thus,

$$f(x, y) = y \sin x + C$$

is a potential function for $\vec{F}$. Therefore $\vec{F}$ is path-independent.

Example 4   The gravitational force field, $\vec{F}$, of an object of mass M, is given by

$$\vec{F} = -\frac{GM}{r^3} \vec{r}.$$

Show that $\vec{F}$ is a gradient field by finding a potential function for $\vec{F}$.

Solution   All the force vectors point in toward the origin. If $\vec{F} = \text{grad } f$, the force vectors must be perpendicular to the level surfaces of f, so the level surfaces of f must be spheres. Also, if grad $f = \vec{F}$, then $\| \text{grad } f \| = \| \vec{F} \| = GM/r^2$ is the rate of change of f in the direction toward the origin. Now, differentiating with respect to r gives the rate of change in a radially outward direction. Thus, if $w = f(x, y, z)$ we have

$$\frac{dw}{dr} = -\frac{GM}{r^2} = GM\left(-\frac{1}{r^2}\right) = GM \frac{d}{dr}\left(\frac{1}{r}\right).$$

So let's try

$$w = \frac{GM}{r} \quad \text{or} \quad f(x, y, z) = \frac{GM}{\sqrt{x^2 + y^2 + z^2}}.$$

We calculate

$$f_x = \frac{\partial}{\partial x} \frac{GM}{\sqrt{x^2 + y^2 + z^2}} = \frac{-GMx}{(x^2 + y^2 + z^2)^{3/2}},$$

$$f_y = \frac{\partial}{\partial y} \frac{GM}{\sqrt{x^2 + y^2 + z^2}} = \frac{-GMy}{(x^2 + y^2 + z^2)^{3/2}},$$

$$f_z = \frac{\partial}{\partial z} \frac{GM}{\sqrt{x^2 + y^2 + z^2}} = \frac{-GMz}{(x^2 + y^2 + z^2)^{3/2}}.$$

So

$$\text{grad } f = f_x \vec{i} + f_y \vec{j} + f_z \vec{k} = \frac{-GM}{(x^2 + y^2 + z^2)^{3/2}}(x\vec{i} + y\vec{j} + z\vec{k}) = \frac{-GM}{r^3} \vec{r} = \vec{F}.$$

Our computations show that $\vec{F}$ is a gradient field and that $f = GM/r$ is a potential function for $\vec{F}$.

Why Path-Independent Vector Fields are Gradient Fields: Showing grad $f = \overrightarrow{F}$

Suppose $\vec{F}$ is a path-independent vector field. On page 853 we defined the function f, which we hope will satisfy grad $f = \vec{F}$, as follows:

$$f(x_0, y_0) = \int_C \vec{F} \cdot d\vec{r},$$

where C is a path from a fixed starting point P to a point $Q = (x_0, y_0)$. This integral has the same value for any path from P to Q because $\vec{F}$ is path-independent. Now we show why grad $f = \vec{F}$. We consider vector fields in 2-space; the argument in 3-space is essentially the same.

First, we write the line integral in terms of the components $\vec{F}(x, y) = F_1(x, y)\vec{i} + F_2(x, y)\vec{j}$ and the components $d\vec{r} = dx\vec{i} + dy\vec{j}$:

$$f(x_0, y_0) = \int_C F_1(x, y)dx + F_2(x, y)dy.$$

We want to compute the partial derivatives of f, that is, the rate of change of f at (x_0, y_0) parallel to the axes. To do this easily, we choose a path which reaches the point (x_0, y_0) on a horizontal or vertical line segment. Let C' be a path from P which stops short of Q at a fixed point (a, b) and let L_x and L_y be the paths shown in Figure 18.30. Then we can split the line integral into three pieces. Since $d\vec{r} = \vec{j}\,dy$ on L_y and $d\vec{r} = \vec{i}\,dx$ on L_x, we have:

$$f(x_0, y_0) = \int_{C'} \vec{F}\cdot d\vec{r} + \int_{L_y} \vec{F}\cdot d\vec{r} + \int_{L_x} \vec{F}\cdot d\vec{r} = \int_{C'} \vec{F}\cdot d\vec{r} + \int_b^{y_0} F_2(a, y)dy + \int_a^{x_0} F_1(x, y_0)dx.$$

The first two integrals do not involve x_0. Thinking of x_0 as a variable and differentiating with respect to it gives

$$f_{x_0}(x_0, y_0) = \frac{\partial}{\partial x_0} \int_{C'} \vec{F} \cdot d\vec{r} + \frac{\partial}{\partial x_0} \int_b^{y_0} F_2(a, y)dy + \frac{\partial}{\partial x_0} \int_a^{x_0} F_1(x, y_0)dx$$

$$= 0 + 0 + F_1(x_0, y_0) = F_1(x_0, y_0),$$

and thus

$$f_x(x, y) = F_1(x, y).$$

A similar calculation for y using the path from P to Q shown in Figure 18.31 gives

$$f_{y_0}(x_0, y_0) = F_2(x_0, y_0).$$

Therefore, as we claimed

$$\text{grad } f = f_x\vec{i} + f_y\vec{j} = F_1\vec{i} + F_2\vec{j} = \vec{F}.$$

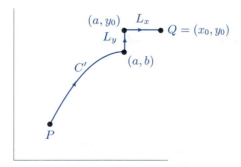

Figure 18.30: The path $C' + L_y + L_x$ is used to show $f_x = F_1$

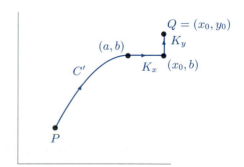

Figure 18.31: The path $C' + K_x + K_y$ is used to show $f_y = F_2$

Summary

We have studied two apparently different types of vector field: path-independent vector fields and gradient vector fields. It turns out that these are the same. Here is a summary of the definitions and properties of these vector fields:

- **Path-independent vector fields** have the property that for any two points P and Q, the line integral along a path from P to Q is the same no matter what path we choose.

- **Gradient vector fields** are of the form grad f for some scalar function f, called the potential function of the vector field.

- **Gradient vector fields are path-independent** by the Fundamental Theorem of Line Integrals,

$$\int_C \text{grad } f \cdot d\vec{r} = f(Q) - f(P).$$

- **Path-independent vector fields are gradient fields** because we can use a line integral and path independence to construct a potential function.

Problems for Section 18.3

1. The vector field $\vec{F}(x, y) = x\vec{i} + y\vec{j}$ is path-independent. Compute geometrically the line integrals over the three paths A, B, and C shown in Figure 18.32 from $(1, 0)$ to $(0, 1)$ and verify that they are equal. Here A is a portion of a circle, B is a line, and C consists of two line segments meeting at a right angle.

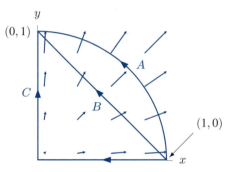

Figure 18.32

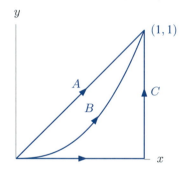

Figure 18.33

2. The vector field $\vec{F}(x, y) = x\vec{i} + y\vec{j}$ is path-independent. Compute algebraically the line integrals over the three paths A, B, and C shown in Figure 18.33 from $(0, 0)$ to $(1, 1)$ and verify that they are equal. Here A is a line segment, B is part of the graph of $f(x) = x^2$, and C consists of two line segments meeting at a right angle.

For Problems 3–6, decide whether or not the given vector fields could be gradient vector fields. Give a justification for your answer.

3. $\vec{F}(x, y) = x\vec{i}$

4. $\vec{G}(x, y) = (x^2 - y^2)\vec{i} - 2xy\vec{j}$

5. $\vec{F}(x, y, z) = \dfrac{-z}{\sqrt{x^2 + z^2}}\vec{i} + \dfrac{y}{\sqrt{x^2 + z^2}}\vec{j} + \dfrac{x}{\sqrt{x^2 + z^2}}\vec{k}$

6. $\vec{F}(\vec{r}) = \vec{r}/\|\vec{r}\|^3$,   where $\vec{r} = x\vec{i} + y\vec{j} + z\vec{k}$

For the vector fields in Problems 7–10, find the line integral along the curve C from the origin along the x-axis to the point $(3, 0)$ and then counterclockwise around the circumference of the circle $x^2 + y^2 = 9$ to the point $(3/\sqrt{2}, 3/\sqrt{2})$.

7. $\vec{F} = x\vec{i} + y\vec{j}$

8. $\vec{H} = -y\vec{i} + x\vec{j}$

9. $\vec{F} = y(x + 1)^{-1}\vec{i} + \ln(x + 1)\vec{j}$

10. $\vec{G} = (ye^{xy} + \cos(x + y))\vec{i} + (xe^{xy} + \cos(x + y))\vec{j}$

11. Suppose that grad $f = 2xe^{x^2} \sin y \vec{i} + e^{x^2} \cos y \vec{j}$. Find the change in f between $(0,0)$ and $(1, \pi/2)$:
    (a) By computing a line integral (b) By computing f.

12. The line integral of $\vec{F} = (x+y)\vec{i} + x\vec{j}$ along each of the following paths is $3/2$:

    (i) The path (t, t^2), with $0 \le t \le 1$
    (ii) The path (t^2, t), with $0 \le t \le 1$
    (iii) The path (t, t^n), with $n > 0$ and $0 \le t \le 1$

    Verify this

    (a) Using the given parameterization to compute the line integral.
    (b) Using the Fundamental Theorem of Calculus for Line Integrals.

13. Consider the vector field $\vec{F}(x, y) = x\vec{j}$ shown in Figure 18.34.

    (a) Find paths C_1, C_2, and C_3 from P to Q such that

    $$\int_{C_1} \vec{F} \cdot d\vec{r} = 0, \qquad \int_{C_2} \vec{F} \cdot d\vec{r} > 0, \qquad \text{and} \qquad \int_{C_3} \vec{F} \cdot d\vec{r} < 0.$$

    (b) Is $\vec{F}$ a gradient field?

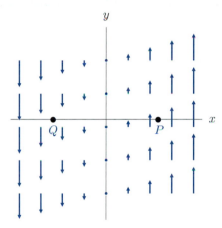

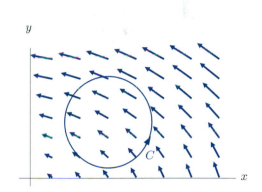

Figure 18.34   **Figure 18.35**

14. Consider the vector field $\vec{F}$ graphed in Figure 18.35.

    (a) Is the line integral $\int_C \vec{F} \cdot d\vec{r}$ positive, negative, or zero?
    (b) From your answer to part (a), can you determine whether or not $\vec{F} = \text{grad } f$ for some function f?
    (c) Which of the following formulas best fits this vector field?

    $$\vec{F_1} = \frac{x}{x^2 + y^2}\vec{i} + \frac{y}{x^2 + y^2}\vec{j}, \quad \vec{F_2} = -y\vec{i} + x\vec{j}, \quad \vec{F_3} = \frac{-y}{(x^2 + y^2)^2}\vec{i} + \frac{x}{(x^2 + y^2)^2}\vec{j}.$$

In Problems 15–18, each of the statements is *false*. Explain why or give a counterexample.

15. If $\int_C \vec{F} \cdot d\vec{r} = 0$ for one particular closed path C, then $\vec{F}$ is path-independent.

16. $\int_C \vec{F} \cdot d\vec{r}$ is the total change in $\vec{F}$ along C.

17. If the vector fields $\vec{F}$ and $\vec{G}$ have $\int_C \vec{F} \cdot d\vec{r} = \int_C \vec{G} \cdot d\vec{r}$ for a particular path C, then $\vec{F} = \vec{G}$.

18. If the total change of a function f along a curve C is zero, then C must be a contour of f.

19. Suppose a particle subject to a force $\vec{F}(x, y) = y\vec{i} - x\vec{j}$ moves clockwise along the arc of the unit circle, centered at the origin, that begins at $(-1, 0)$ and ends at $(0, 1)$.

    (a) Find the work done by $\vec{F}$. Explain the sign of your answer.
    (b) Is $\vec{F}$ path-independent? Explain.

20. Let $\vec{F}$ be a path-independent vector field. In physics, the potential function ϕ is usually required to satisfy the equation $\vec{F} = -\nabla\phi$. This problem illustrates the significance of the negative sign.[3]

    (a) Let the xy-plane represent part of the earth's surface with the z-axis pointing away from the earth. (We assume the scale is small enough so that a flat plane is a good approximation to the earth's surface.) Let $\vec{r} = x\vec{i} + y\vec{j} + z\vec{k}$, with $z \geq 0$, and x, y, z in meters, be the position vector of a rock of unit mass. The gravitational potential energy function for the rock is $\phi(x, y, z) = gz$, where $g \approx 9.8$ m/sec^2. Describe in words the level surfaces of ϕ. Does the potential energy increase or decrease with height above the earth?

    (b) What is the relation between the gravitational vector, $\vec{F}$, and the vector $\nabla\phi$? Explain the significance of the negative sign in the equation $\vec{F} = -\nabla\phi$.

18.4 PATH-DEPENDENT VECTOR FIELDS AND GREEN'S THEOREM

Suppose we are given a vector field but are not told whether it is path-independent. How can we tell if it has a potential function, that is, if it is a gradient field?

How to Tell if a Vector Field is Path-Dependent Using Line Integrals

Example 1   Is the vector field, $\vec{F}$, shown in Figure 18.36 path-independent? At any point $\vec{F}$ has magnitude equal to the distance from the origin and direction perpendicular to the line joining the point to the origin.

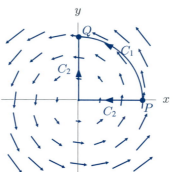

Figure 18.36: Is this vector field path-independent?

Solution   We choose $P = (1, 0)$ and $Q = (0, 1)$ and two paths between them: C_1, a quarter circle of radius 1, and C_2, formed by parts of the x- and y-axes. (See Figure 18.36.) Along C_1, the line integral $\int_{C_1} \vec{F} \cdot d\vec{r} > 0$, since $\vec{F}$ points in the direction of the curve. Along C_2, however, we have $\int_{C_2} \vec{F} \cdot d\vec{r} = 0$, since $\vec{F}$ is perpendicular to C_2 everywhere. Thus, $\vec{F}$ is not path-independent.

Path-Dependent Fields and Circulation

Notice that the vector field in the previous example has nonzero circulation around the origin. What can we say about the circulation of a general path-independent vector field around a closed curve, C? Suppose C is a *simple* closed curve, that is, a curve which does not cross itself. If P and Q are any two points on the path, then we can think of C (oriented as shown in Figure 18.37) as made up of the path C_1 followed by $-C_2$. Since $\vec{F}$ is path-independent, we know that

$$\int_{C_1} \vec{F} \cdot d\vec{r} = \int_{C_2} \vec{F} \cdot d\vec{r}.$$

[3] Adapted from V.I. Arnold, *Mathematical Methods of Classical Mechanics*, 2nd Edition, Graduate Texts in Mathematics, Springer

Thus, we see that the circulation around C is zero:

$$\int_C \vec{F} \cdot d\vec{r} = \int_{C_1} \vec{F} \cdot d\vec{r} + \int_{-C_2} \vec{F} \cdot d\vec{r} = \int_{C_1} \vec{F} \cdot d\vec{r} - \int_{C_2} \vec{F} \cdot d\vec{r} = 0.$$

If the curve C does cross itself, we break it into simple closed curves as shown in Figure 18.38 and apply the same argument to each one.

Now suppose we know that the line integral around any closed curve is zero. For any two points, P and Q, with two paths, C_1 and C_2, between them, create a closed curve, C, as in Figure 18.37. Since the circulation around this closed curve, C, is zero, the line integrals along the two paths, C_1 and C_2, are equal. Thus, $\vec{F}$ is path-independent.

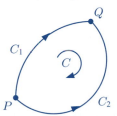

Figure 18.37: A simple closed curve C broken into two pieces, C_1 and C_2

Figure 18.38: A curve C which crosses itself can be broken into simple closed curves

Thus, we have the following result:

> A vector field is path-independent if and only if $\int_C \vec{F} \cdot d\vec{r} = 0$ for every closed curve C.

Hence, to see if a field is *path-dependent*, we look for a closed path with nonzero circulation. For instance, the vector field in Example 1 has nonzero circulation around a circle around the origin, showing it is path-dependent.

How to Tell If a Vector Field is Path-Dependent Algebraically: The Curl

Example 2   Does the vector field $\vec{F} = 2xy\vec{i} + xy\vec{j}$ have a potential function? If so, find it.

Solution   Let's suppose $\vec{F}$ does have a potential function, f, so $\vec{F} = \text{grad } f$. This means that

$$\frac{\partial f}{\partial x} = 2xy \quad \text{and} \quad \frac{\partial f}{\partial y} = xy.$$

Integrating the expression for $\partial f/\partial x$ shows that we must have

$$f(x, y) = x^2 y + C(y) \qquad \text{where } C(y) \text{ is a function of } y.$$

Differentiating this expression for $f(x, y)$ with respect to y and using the fact that $\partial f/\partial y = xy$, we get

$$\frac{\partial f}{\partial y} = x^2 + C'(y) = xy.$$

Thus, we must have

$$C'(y) = xy - x^2.$$

But this expression for $C'(y)$ is impossible because $C'(y)$ is a function of y alone. This argument shows that there is no potential function for the vector field $\vec{F}$.

Is there an easier way to see that a vector field has no potential function, other than by trying to find the potential function and failing? The answer is yes. First we look at a 2-dimensional vector field $\vec{F} = F_1\vec{i} + F_2\vec{j}$. If $\vec{F}$ is a gradient field, then there is a potential function f such that

$$\vec{F} = F_1\vec{i} + F_2\vec{j} = \frac{\partial f}{\partial x}\vec{i} + \frac{\partial f}{\partial y}\vec{j}.$$

Thus,

$$F_1 = \frac{\partial f}{\partial x} \quad \text{and} \quad F_2 = \frac{\partial f}{\partial y}.$$

Let us assume that f has continuous second partial derivatives. Then, by the equality of mixed partial derivatives

$$\frac{\partial F_1}{\partial y} = \frac{\partial^2 f}{\partial y \partial x} = \frac{\partial^2 f}{\partial x \partial y} = \frac{\partial F_2}{\partial x}$$

Thus we have the following result:

If $\vec{F}(x,y) = F_1\vec{i} + F_2\vec{j}$ is a gradient vector field with continuous partial derivatives, then

$$\frac{\partial F_2}{\partial x} - \frac{\partial F_1}{\partial y} = 0.$$

We call $\dfrac{\partial F_2}{\partial x} - \dfrac{\partial F_1}{\partial y}$ the 2-dimensional or scalar **curl** of the vector field $\vec{F}$.

Notice that we now know that if $\vec{F}$ is a gradient field, then its curl is 0. We do not (yet) know whether the converse is true. (That is: If the curl is 0, does $\vec{F}$ have to be a gradient field?) However, the curl already enables us to show that a vector field is *not* a gradient field.

Example 3   Show that $\vec{F} = 2xy\vec{i} + xy\vec{j}$ cannot be a gradient vector field.

Solution   We have $F_1 = 2xy$ and $F_2 = xy$. Since $\partial F_1/\partial y = 2x$ and $\partial F_2/\partial x = y$, in this case

$$\partial F_2/\partial x - \partial F_1/\partial y \neq 0$$

so $\vec{F}$ cannot be a gradient field.

We now have two ways of seeing that a vector field $\vec{F}$ in the plane is path-dependent. We can evaluate $\int_C \vec{F} \cdot d\vec{r}$ for some closed curve and find it is not zero, or we can show that $\partial F_2/\partial x - \partial F_1/\partial y \neq 0$. It's natural to think that

$$\int_C \vec{F} \cdot d\vec{r} \quad \text{and} \quad \frac{\partial F_2}{\partial x} - \frac{\partial F_1}{\partial y}$$

might be related. The relation is called Green's Theorem.

Green's Theorem

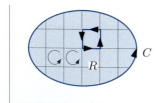

Figure 18.39: Region R bounded by a closed curve C and split into many small regions, ΔR

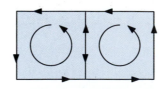

Figure 18.40: Two adjacent small closed curves

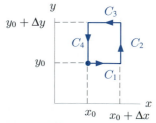

Figure 18.41: A small closed curve ΔC broken into C_1, C_2, C_3, C_4

We obtain the statement of Green's Theorem by splitting a region up into small pieces and looking at the relation between the line integral and the curl on each one.

In 2-space, a simple closed curve C separates the plane into points inside and points outside C. We consider $\int_C \vec{F} \cdot d\vec{r}$, where C is oriented as shown in Figure 18.39. We divide the region R inside C into small pieces, each bounded by a closed curve with the orientation shown. Figure 18.40 shows that if we add the circulation around all of these small closed curves, each common edge of a pair of adjacent curves is counted twice, once in each direction. Hence the integrals along these edges cancel. Thus, the line integrals along all the edges inside the region R cancel, giving

$$\text{Circulation of } \vec{F} \text{ around } C = \sum_{\Delta C} \text{Circulation of } \vec{F} \text{ around small curve, } \Delta C$$

Now we estimate the line integral around one of these small closed curves, ΔC. We break ΔC into C_1, C_2, C_3, and C_4, as shown in Figure 18.41. Then we calculate the line integrals along C_1 and C_3, where $\Delta \vec{r}$ is parallel to the x-axis, so

$$\Delta \vec{r} = \Delta x \vec{i}.$$

Thus, on C_1 and C_3,

$$\vec{F} \cdot \Delta \vec{r} = (F_1 \vec{i} + F_2 \vec{j}) \cdot \Delta x \vec{i} = F_1 \Delta x.$$

However, the function F_1 is evaluated at (x, y_0) on C_1 and at $(x, y_0 + \Delta y)$ on C_3, so

$$\int_{C_1} \vec{F} \cdot d\vec{r} + \int_{C_3} \vec{F} \cdot d\vec{r} = = \int_{C_1} F_1(x, y_0)\, dx + \int_{C_3} F_1(x, y_0 + \Delta y)\, dx$$

$$= \int_{x=x_0}^{x_0+\Delta x} F_1(x, y_0)\, dx + \int_{x_0+\Delta x}^{x=x_0} F_1(x, y_0 + \Delta y)\, dx$$

$$= \int_{x=x_0}^{x_0+\Delta x} F_1(x, y_0)\, dx - \int_{x=x_0}^{x_0+\Delta x} F_1(x, y_0 + \Delta y)\, dx$$

$$= \int_{x=x_0}^{x_0+\Delta x} (F_1(x, y_0) - F_1(x, y_0 + \Delta y))\, dx.$$

Since F_1 is differentiable and Δy is small,

$$F_1(x, y_0) - F_1(x, y_0 + \Delta y) = -(F_1(x, y_0 + \Delta y) - F_1(x, y_0)) \approx -\frac{\partial F_1}{\partial y}(x, y_0) \Delta y,$$

so we have

$$\int_{x=x_0}^{x_0+\Delta x} (F_1(x, y_0) - F_1(x, y_0 + \Delta y))\, dx \approx -\left(\int_{x=x_0}^{x_0+\Delta x} \frac{\partial F_1}{\partial y}\, dx \right) \Delta y.$$

Assuming $\dfrac{\partial F_1}{\partial y}(x, y_0)$ is approximately constant over the interval $x_0 \le x \le x_0 + \Delta x$, we get

$$-\left(\int_{x=x_0}^{x_0+\Delta x} \frac{\partial F_1}{\partial y}\, dx \right) \Delta y \approx -\frac{\partial F_1}{\partial y}(x_0, y_0) \left(\int_{x=x_0}^{x_0+\Delta x} dx \right) \Delta y = -\frac{\partial F_1}{\partial y}(x_0, y_0) \Delta x\, \Delta y.$$

By a similar argument on C_2 and C_4, we have

$$\int_{C_2} \vec{F} \cdot d\vec{r} + \int_{C_4} \vec{F} \cdot d\vec{r} \approx \frac{\partial F_2}{\partial x} \Delta x\, \Delta y.$$

Combining these results for C_1, C_2, C_3, and C_4, we get

$$\int_{\Delta C} \vec{F} \cdot d\vec{r} = \int_{C_1} \vec{F} \cdot d\vec{r} + \int_{C_2} \vec{F} \cdot d\vec{r} + \int_{C_3} \vec{F} \cdot d\vec{r} + \int_{C_4} \vec{F} \cdot d\vec{r} \approx \frac{\partial F_2}{\partial x} \Delta x \, \Delta y - \frac{\partial F_1}{\partial y} \Delta x \, \Delta y.$$

Summing over all small regions ΔC gives

$$\int_C \vec{F} \cdot d\vec{r} \approx \sum_{\Delta C} \int_{\Delta C} \vec{F} \cdot d\vec{r} \approx \sum_{\Delta R} \left(\frac{\partial F_2}{\partial x} - \frac{\partial F_1}{\partial y} \right) \Delta x \, \Delta y.$$

The last sum is a Riemann sum approximating a double integral; taking the limit as $\Delta x, \Delta y$ tend to zero, we get

---

Green's Theorem

Suppose C is a simple closed curve surrounding a region R in the plane and oriented so that the region is on the left as we move around the curve. Suppose $\vec{F} = F_1 \vec{i} + F_2 \vec{j}$ is a smooth vector field defined at every point of the region R and boundary C. Then

$$\int_C \vec{F} \cdot d\vec{r} = \int_R \left(\frac{\partial F_2}{\partial x} - \frac{\partial F_1}{\partial y} \right) dx \, dy.$$

The Focus on Theory section on page 871 contains a proof of Green's Theorem using the change of variables formula for double integrals.

The Curl Test for Vector Fields in the Plane

We already know that if $\vec{F} = F_1 \vec{i} + F_2 \vec{j}$ is a gradient field with continuous partial derivatives, then

$$\frac{\partial F_2}{\partial x} - \frac{\partial F_1}{\partial y} = 0.$$

Now we show that the converse is true if the domain of $\vec{F}$ has no holes in it. This means that we assume that

$$\frac{\partial F_2}{\partial x} - \frac{\partial F_1}{\partial y} = 0$$

and show that $\vec{F}$ is path-independent. If C is any oriented closed curve in the domain of $\vec{F}$ and R is the region inside C, then

$$\int_R \left(\frac{\partial F_2}{\partial x} - \frac{\partial F_1}{\partial y} \right) dx \, dy = 0$$

since the integrand is identically 0. Therefore, by Green's Theorem

$$\int_C \vec{F} \cdot d\vec{r} = \int_R \left(\frac{\partial F_2}{\partial x} - \frac{\partial F_1}{\partial y} \right) dx dy = 0.$$

Thus, $\vec{F}$ is path-independent and therefore a gradient field. This argument is valid for every closed curve, C, provided the region R is entirely in the domain of $\vec{F}$. Thus we have the following result:

The Curl Test for Vector Fields in 2-Space

Suppose $\vec{F} = F_1 \vec{i} + F_2 \vec{j}$ is a vector field with continuous partial derivatives, such that
- The domain of $\vec{F}$ has the property that every closed curve in it encircles a region that lies entirely within the domain. In particular, the domain of $\vec{F}$ has no holes.
- $\dfrac{\partial F_2}{\partial x} - \dfrac{\partial F_1}{\partial y} = 0.$

Then $\vec{F}$ is path-independent, so $\vec{F}$ is a gradient field and has a potential function.

Why Are Holes in the Domain of the Vector Field Important?

The reason for assuming that the domain of the vector field $\vec{F}$ has no holes is to ensure that the region R inside C is actually contained in the domain of $\vec{F}$. Otherwise, we cannot apply Green's Theorem. The next two examples show that if $\partial F_2/\partial x - \partial F_1/\partial y = 0$ but the domain of $\vec{F}$ contains a hole, then $\vec{F}$ can either be path-independent or path-dependent.

Example 4   Let $\vec{F}$ be the vector field given by $\vec{F}(x,y) = \dfrac{-y\vec{i} + x\vec{j}}{x^2 + y^2}$.

(a)   Calculate $\dfrac{\partial F_2}{\partial x} - \dfrac{\partial F_1}{\partial y}$. Does the curl test imply that $\vec{F}$ is path-independent?

(b)   Calculate $\displaystyle\int_C \vec{F}\cdot d\vec{r}$, where C is the unit circle centered at the origin and oriented counter-clockwise. Is $\vec{F}$ a path-independent vector field?

(c)   Explain why the answers to parts (a) and (b) do not contradict Green's Theorem.

Solution   (a)   Taking partial derivatives, we have

$$\frac{\partial F_2}{\partial x} = \frac{\partial}{\partial x}\left(\frac{x}{x^2+y^2}\right) = \frac{1}{x^2+y^2} - \frac{x\cdot 2x}{(x^2+y^2)^2} = \frac{y^2-x^2}{(x^2+y^2)^2}.$$

Similarly,

$$\frac{\partial F_1}{\partial y} = \frac{\partial}{\partial y}\left(\frac{-y}{x^2+y^2}\right) = \frac{-1}{x^2+y^2} + \frac{y\cdot 2y}{(x^2+y^2)^2} = \frac{y^2-x^2}{(x^2+y^2)^2}.$$

Thus,

$$\frac{\partial F_2}{\partial x} - \frac{\partial F_1}{\partial y} = 0.$$

Since $\vec{F}$ is undefined at the origin, the domain of $\vec{F}$ contains a hole. Therefore, the curl test does not apply.

(b)   On the unit circle, $\vec{F}$ is tangent to the circle and $||\vec{F}|| = 1$. Thus,

$$\int_C \vec{F}\cdot d\vec{r} = ||\vec{F}||\cdot \text{Length of curve} = 1\cdot 2\pi = 2\pi.$$

Since the line integral around the closed curve C is nonzero, $\vec{F}$ is not path-independent.

(c)   The domain of $\vec{F}$ is the "punctured plane," as shown in Figure 18.42. Since $\vec{F}$ is not defined at the origin, which is inside C, Green's Theorem does not apply. In this case

$$2\pi = \int_C \vec{F}\cdot d\vec{r} \neq \int_R \left(\frac{\partial F_2}{\partial x} - \frac{\partial F_1}{\partial y}\right)dx\,dy = 0.$$

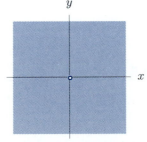

Figure 18.42: The domain of $\vec{F}(x,y) = \frac{-y\vec{i}+x\vec{j}}{x^2+y^2}$ is the plane minus the origin

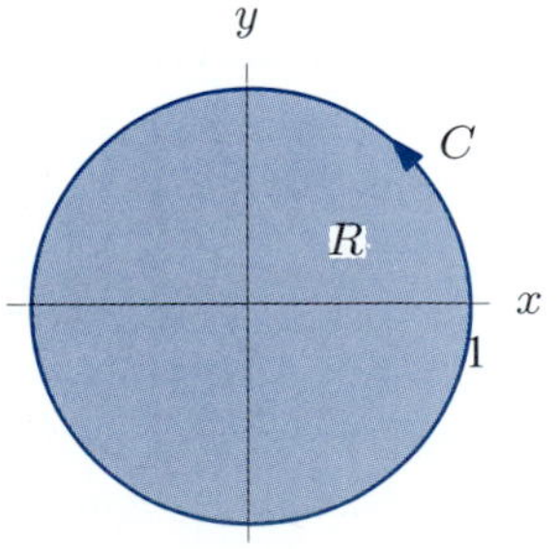

Figure 18.43: The region R is *not* contained in the domain of $\vec{F}(x,y) = \frac{-y\vec{i}+x\vec{j}}{x^2+y^2}$

Although the vector field $\vec{F}$ in the last example was not defined at the origin, this by itself does not prevent the vector field from being path-independent as we see in the following example.

Example 5   Consider the vector field $\vec{F}$ given by $\vec{F}(x,y) = \dfrac{x\vec{i} + y\vec{j}}{x^2 + y^2}$.

(a)   Calculate $\dfrac{\partial F_2}{\partial x} - \dfrac{\partial F_1}{\partial y}$. Does the curl test imply that $\vec{F}$ is path-independent?

(b)   Explain how we know that $\displaystyle\int_C \vec{F} \cdot d\vec{r} = 0$, where C is the unit circle centered at the origin and oriented counterclockwise. Does this imply that $\vec{F}$ is path-independent?

(c)   Check that $f(x,y) = \frac{1}{2}\ln(x^2 + y^2)$ is a potential function for $\vec{F}$. Does this imply that $\vec{F}$ is path-independent?

Solution   (a)   Taking partial derivatives, we have

$$\frac{\partial F_2}{\partial x} = \frac{\partial}{\partial x}\left(\frac{y}{x^2 + y^2}\right) = \frac{-2xy}{(x^2 + y^2)^2}, \qquad \text{and} \qquad \frac{\partial F_1}{\partial y} = \frac{\partial}{\partial y}\left(\frac{x}{x^2 + y^2}\right) = \frac{-2xy}{(x^2 + y^2)^2}.$$

Therefore,

$$\frac{\partial F_2}{\partial x} - \frac{\partial F_1}{\partial y} = 0.$$

This does *not* imply that $\vec{F}$ is path-independent: The domain of $\vec{F}$ contains a hole since $\vec{F}$ is undefined at the origin. Thus, the curl test does not apply.

(b)   Since $\vec{F}(x,y) = x\vec{i} + y\vec{j} = \vec{r}$ on the unit circle C, the field $\vec{F}$ is everywhere perpendicular to C. Thus

$$\int_C \vec{F} \cdot d\vec{r} = 0.$$

The fact that $\int_C \vec{F} \cdot d\vec{r} = 0$ when C is the unit circle does *not* imply that $\vec{F}$ is path-independent. To be sure that $\vec{F}$ is path-independent, we would have to show that $\int_C \vec{F} \cdot d\vec{r} = 0$ for *every* closed curve C in the domain of $\vec{F}$, not just the unit circle.

(c)   To check that grad $f = \vec{F}$, we differentiate f:

$$f_x = \frac{1}{2}\frac{\partial}{\partial x}\ln(x^2 + y^2) = \frac{1}{2}\frac{2x}{x^2 + y^2} = \frac{x}{x^2 + y^2},$$

and

$$f_y = \frac{1}{2}\frac{\partial}{\partial y}\ln(x^2 + y^2) = \frac{1}{2}\frac{2y}{x^2 + y^2} = \frac{y}{x^2 + y^2},$$

so that

$$\text{grad } f = \frac{x\vec{i} + y\vec{j}}{x^2 + y^2} = \vec{F}.$$

Thus, $\vec{F}$ is a gradient field and therefore is path-independent — even though $\vec{F}$ is undefined at the origin.

The Curl Test for Vector Fields in 3-Space

The curl test is a convenient way of deciding whether a 2-dimensional vector field is path-independent. Fortunately, there is an analogous test for 3-dimensional vector fields, although we cannot justify it until Chapter 20.

If $\vec{F}(x,y,z) = F_1\vec{i} + F_2\vec{i} + F_3\vec{k}$ is a vector field on 3-space we define a new vector field, curl $\vec{F}$, on 3-space by

$$\text{curl } \vec{F} = \left(\frac{\partial F_3}{\partial y} - \frac{\partial F_2}{\partial z}\right)\vec{i} + \left(\frac{\partial F_1}{\partial z} - \frac{\partial F_3}{\partial x}\right)\vec{j} + \left(\frac{\partial F_2}{\partial x} - \frac{\partial F_1}{\partial y}\right)\vec{k}.$$

The vector field curl $\vec{F}$ can be used to determine whether the vector field $\vec{F}$ is path-independent.

> ## The Curl Test for Vector Fields in 3-Space
>
> Suppose $\vec{F}$ is a vector field on 3-space with continuous partial derivatives, such that
> - The domain of $\vec{F}$ has the property that every closed curve in it can be contracted to a point in a smooth way, staying at all times within the domain.
> - curl $\vec{F} = \vec{0}$.
>
> Then $\vec{F}$ is path-independent, so $\vec{F}$ is a gradient field and has a potential function.

For the 2-dimensional curl test, the domain of $\vec{F}$ must have no holes. This meant that if $\vec{F}$ was defined on a closed curve C, then it was also defined at all points inside C. One way to test for holes is to try to "lasso" them with a closed curve. If every closed curve in the domain can be pulled to a point without hitting a hole, that is, without straying outside the domain, then the domain has no holes. In 3-space, we need the same condition to be satisfied; we must be able to pull every closed curve to a point, like a lasso, without straying outside the domain.

Example 6   Decide if the following vector fields are path-independent and whether or not the curl test applies.

(a) $\vec{F} = \dfrac{x\vec{i} + y\vec{j} + z\vec{k}}{(x^2 + y^2 + z^2)^{3/2}}$
(b) $\vec{G} = \dfrac{-y\vec{i} + x\vec{j}}{x^2 + y^2}$

Solution   (a)   Suppose $f = -(x^2 + y^2 + z^2)^{-1/2}$. Then $f_x = x(x^2 + y^2 + z^2)^{-3/2}$ and similarly grad $f = \vec{F}$. Thus, $\vec{F}$ is a gradient field and therefore path-independent. Calculations show curl $\vec{F} = \vec{0}$. The domain of $\vec{F}$ is all of 3-space minus the origin, and any closed curve in the domain can be pulled to a point without leaving the domain. Thus, the curl test applies.

(b)   Let C be the circle $x^2 + y^2 = 1, z = 0$ traversed counterclockwise when viewed from the positive z-axis. The vector field is everywhere tangent to this curve and of magnitude 1, so

$$\int_C \vec{G} \cdot d\vec{r} = \|\vec{G}\| \cdot \text{Length of curve} = 1 \cdot 2\pi = 2\pi.$$

Since the line integral around this closed curve is nonzero, $\vec{G}$ is path-dependent. Computations show curl $\vec{G} = \vec{0}$. However, the domain of $\vec{G}$ is all of 3-space minus the z-axis, and it does not satisfy the curl test domain criterion. For example, the circle, C, is lassoed around the z-axis, and cannot be pulled to a point without hitting the z-axis. Thus the curl test does not apply.

Problems for Section 18.4

1. Example 1 on page 858 showed that the vector field in Figure 18.44 could not be a gradient field by showing that it is not path-independent. Here is another way to see the same thing. Suppose that the vector field were the gradient of a function f. Draw and label a diagram showing what the contours of f would have to look like, and explain why it would not be possible for f to have a single value at any given point.

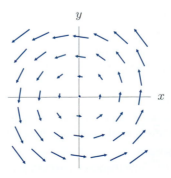

Figure 18.44

2. Repeat Problem 1 for the vector field in Problem 13 on page 857.

3. Find f if grad $f = 2xy\vec{i} + (x^2 + 8y^3)\vec{j}$

4. Find f if grad $f = (yze^{xyz} + z^2\cos(xz^2))\vec{i} + xze^{xyz}\vec{j} + (xye^{xyz} + 2xz\cos(xz^2))\vec{k}$

For Problems 5–6, decide whether the given vector field is the gradient of a function f. If so, find such an f. If not, explain why not.

5. $y\vec{i} + y\vec{j}$ 6. $(x^2 + y^2)\vec{i} + 2xy\vec{j}$

Do the vector fields in Problems 7–10 have potential functions, given that $\vec{F} = $ grad f? If so, compute them.

7. $\vec{F} = (2xy^3 + y)\vec{i} + (3x^2y^2 + x)\vec{j}$

8. $\vec{F} = \dfrac{\vec{i}}{x} + \dfrac{\vec{j}}{y} + \dfrac{\vec{k}}{xy}$ 9. $\vec{F} = \dfrac{\vec{i}}{x} + \dfrac{\vec{j}}{y} + \dfrac{\vec{k}}{z}$

10. $\vec{F} = 2x\cos(x^2 + z^2)\vec{i} + \sin(x^2 + z^2)\vec{j} + 2z\cos(x^2 + z^2)\vec{k}$

11. Consider the vector field $\vec{F} = y\vec{i}$.

    (a) Sketch $\vec{F}$ and hence decide the sign of the circulation of $\vec{F}$ around the unit circle centered at the origin and traversed counterclockwise.

    (b) Use Green's Theorem to compute the circulation in part (a) exactly.

12. Suppose $\vec{F} = x\vec{j}$. Show that the line integral of $\vec{F}$ around a closed curve in the xy-plane, oriented as in Green's Theorem, measures the area of the region enclosed by the curve.

Use the result of Problem 12 to calculate the area of the region within the parameterized curves in Problems 13–15. In each case, sketch the curve.

13. The ellipse $x^2/a^2 + y^2/b^2 = 1$ parameterized by $x = a\cos t$, $y = b\sin t$, for $0 \le t \le 2\pi$.

14. The hypocycloid $x^{2/3} + y^{2/3} = a^{2/3}$ parameterized by $x = a\cos^3 t$, $y = a\sin^3 t$, $0 \le t \le 2\pi$.

15. The folium of Descartes, $x^3 + y^3 = 3xy$, parameterized by $x = \dfrac{3t}{1 + t^3}$, $y = \dfrac{3t^2}{1 + t^3}$, for $0 \le t < \infty$.

16. Suppose that R is the region inside the top half of the unit circle, C, centered at the origin and that we want to compute the double integral

$$\int_R (2x - 2y)e^{x^2 + y^2}\,dA$$

    (a) Explain why converting this integral to an iterated integral in Cartesian coordinates does not help us compute it.

    (b) Since iterated integration fails, we use a numerical method. Show how Green's Theorem can be used to convert the integral to a one-variable integral. Then evaluate the one-variable integral, using ordinary numerical methods.

CHAPTER SUMMARY

- **Line Integrals**
  Oriented curves, definition as a limit of a Riemann sum, work interpretation, circulation, algebraic properties, computing line integrals over parameterized curves, independence of parameterization.

- **Gradient Fields**
  Fundamental Theorem for line integrals, path-independent (conservative) fields and their relation to gradient fields, potential functions.

- **Green's Theorem**
  Statement of the theorem, curl test for path-independence.

REVIEW PROBLEMS FOR CHAPTER EIGHTEEN

For Problems 1–2, consider the vector field $\vec{F}$ shown. Say whether you expect the line integral $\int_C \vec{F} \cdot d\vec{r}$ to be positive, negative, or zero along

(a) A (b) C_1, C_2, C_3, C_4 (c) C, the closed curve consisting of all the C's together.

1.
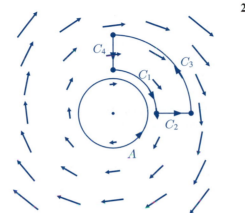

2.
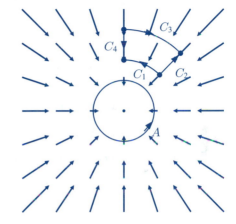

For Problems 3–5, compute $\int_C \vec{F} \cdot d\vec{r}$ for the given $\vec{F}$ and C.

3. $\vec{F} = (x^2 - y)\vec{i} + (y^2 + x)\vec{j}$ and C is the parabola $y = x^2 + 1$ traversed from $(0, 1)$ to $(1, 2)$.

4. $\vec{F} = (3x - 2y)\vec{i} + (y + 2z)\vec{j} - x^2\vec{k}$ and C is the path consisting of the line joining the point $(0, 0, 0)$ to $(1, 1, 1)$.

5. $\vec{F} = (2x - y + 4)\vec{i} + (5y + 3x - 6)\vec{j}$ and C is the triangle with vertices $(0, 0)$, $(3, 0)$, $(3, 2)$ traversed counterclockwise.

Are the statements in Problems 6–9 true or false? Explain why or give a counterexample.

6. $\int_C \vec{F} \cdot d\vec{r}$ is a vector.

7. $\int_C \vec{F} \cdot d\vec{r} = \vec{F}(Q) - \vec{F}(P)$ when P and Q are the endpoints of C.

8. The fact that the line integral of a vector field $\vec{F}$ is zero around the unit circle $x^2 + y^2 = 1$ means that $\vec{F}$ must be a gradient vector field.

9. Suppose C_1 is the unit square joining the points $(0, 0)$, $(1, 0)$, $(1, 1)$, $(0, 1)$ oriented clockwise and C_2 is the same square but traversed twice in the opposite direction. If $\int_{C_1} \vec{F} \cdot d\vec{r} = 3$, then $\int_{C_2} \vec{F} \cdot d\vec{r} = -6$.

10. Let $\vec{F} = x\vec{i} + y\vec{j}$, and let C_1 be the line joining the point $(1, 0)$ to the point $(0, 2)$ and let C_2 be the line joining the point $(0, 2)$ to the point $(-1, 0)$. Is $\int_{C_1} \vec{F} \cdot d\vec{r} = -\int_{C_2} \vec{F} \cdot d\vec{r}$? Explain.

11. What is the value of $\int_C \vec{F} \cdot d\vec{r}$ if C is an oriented curve that runs from the point $(2, -6)$ to the point $(4, 4)$ and if $\vec{F} = 6\vec{i} - 7\vec{j}$?

12. Suppose P and Q both lie on the same contour of f. What can you say about the total change in f from P to Q? Explain your answer in terms of $\int_C \operatorname{grad} f \cdot d\vec{r}$ where C is a portion of the contour that goes from P to Q.

13. In this problem, we see how the Fundamental Theorem for Line Integrals can be derived from the Fundamental Theorem for ordinary definite integrals. Suppose that $(x(t), y(t))$, for $a \le t \le b$, is a parameterization of C, with endpoints $P = (x(a), y(a))$ and $Q = (x(b), y(b))$. The values of f along C are given by the single variable function $h(t) = f(x(t), y(t))$.

    (a) Use the chain rule to show that

    $$h'(t) = f_x(x(t), y(t))x'(t) + f_y(x(t), y(t))y'(t).$$

    (b) Use the Fundamental Theorem of Calculus applied to $h(t)$ to show

    $$\int_C \operatorname{grad} f \cdot d\vec{r} = f(Q) - f(P).$$

14. Let $\vec{F}(x, y)$ be the path-independent vector field in Figure 18.45. The vector field $\vec{F}$ associates with each point a unit vector pointing radially outward. The curves $C_1, C_2, \ldots, C_7$ have the directions shown. Consider the line integrals $\int_{C_i} \vec{F} \cdot d\vec{r}$, $i = 1, \ldots, 7$. Without computing any integrals

    (a) List all the line integrals which you expect to be zero.
    (b) List all the line integrals which you expect to be negative.
    (c) Arrange the positive line integrals in what you believe to be ascending order.

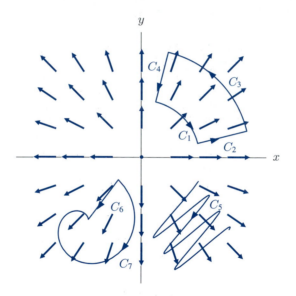

Figure 18.45

15. A *free vortex* circulating about the origin in the xy-plane (or about the z-axis in 3-space) has vector field $\vec{v} = K(x^2 + y^2)^{-1}(-y\vec{i} + x\vec{j})$ where K is a constant. The Rankine model of a tornado hypothesizes an inner core that rotates at constant angular velocity, surrounded by a free vortex. Suppose that the inner core has radius 100 meters and that $\|\vec{v}\| = 3 \cdot 10^5$ meters/hr at a distance of 100 meters from the center.

    (a) Assuming that the tornado rotates counterclockwise (viewed from above the xy-plane) and that $\vec{v}$ is continuous, determine ω and K such that

    $$\vec{v} = \begin{cases} \omega(-y\vec{i} + x\vec{j}) & \text{if } \sqrt{x^2 + y^2} < 100 \\ K(x^2 + y^2)^{-1}(-y\vec{i} + x\vec{j}) & \text{if } \sqrt{x^2 + y^2} \ge 100. \end{cases}$$

    (b) Sketch the vector field $\vec{v}$.
    (c) Compute the circulation of $\vec{v}$ about the circle of radius r centered at the origin, traversed counterclockwise.

16. Figure 18.46 shows the tangential velocity as a function of radius for the tornado that hit Dallas on April 2, 1957. Use it and Problem 15 to estimate K and ω for the Rankine model of this tornado.[4]

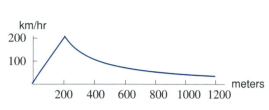

Figure 18.46

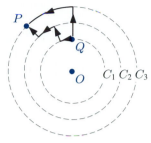

Figure 18.47

17. A *central vector field* is a vector field whose direction is always toward (or away from) a fixed point O (the center) and whose magnitude at a point P is a function only of the distance from P to O. In two dimensions this means that the vector field has constant magnitude on circles centered at O. The gravitational and electrical fields of spherically symmetric sources are both central fields.

   (a) Sketch an example of a central vector field.
   (b) Suppose that the central field $\vec{F}$ is a gradient field, that is, $\vec{F} = \text{grad } f$. What must be the shape of the contours of f? Sketch some contours for this case.
   (c) Is every gradient field a central vector field? Explain.
   (d) In Figure 18.47, two paths are shown between the points Q and P. Assuming that the three circles $C_1, C_2,$ and C_3 are centered at O, explain why the work done by a central vector field $\vec{F}$ is the same for either path.
   (e) It is in fact true that every central vector field is a gradient field. Use an argument suggested by Figure 18.47 to explain why any central vector field must be path-independent.

18. Consider the vector field

$$\vec{F} = (-y^3 + y\sin(xy))\vec{i} + (4x(1-y^2) + x\sin(xy))\vec{j}$$

   defined on the disk D of radius 5 centered at the origin in the plane. Consider the line integral $\int_C \vec{F} \cdot d\vec{r}$, where C is some closed curve contained in D. For which C is the value of this integral the largest? [Hint: Assume C is a closed curve, made up of smooth pieces and never crossing itself, and oriented counterclockwise.]

PROJECTS

1. **Conservation of Energy**
   (a) A particle moves with position vector $\vec{r}(t) = x(t)\vec{i} + y(t)\vec{j} + z(t)\vec{k}$. Let $\vec{v}(t)$ and $\vec{a}(t)$ be its velocity and acceleration vectors. Show that

   $$\frac{1}{2}\frac{d}{dt}\|\vec{v}(t)\|^2 = \vec{a}(t) \cdot \vec{v}(t).$$

   (b) We now derive the principle of Conservation of Energy. The kinetic energy of a particle of mass m moving with speed v is $(1/2)mv^2$. Suppose the particle has potential energy

---

[4]Adapted from *Encyclopedia Britannica, Macropedia*, Vol. 16, page 477, "Climate and the Weather", Tornados and Waterspouts, 1991.

$f(\vec{r})$ at the position $\vec{r}$ due to a force field $\vec{F} = -\nabla f$. If the particle moves with position vector $\vec{r}(t)$ and velocity $\vec{v}(t)$, then the Conservation of Energy principle says that

$$\text{Total energy} = \text{Kinetic energy} + \text{Potential energy} = \frac{1}{2}m\|\vec{v}(t)\|^2 + f(\vec{r}(t)) = \text{Constant}.$$

Let P and Q be two points in space and let C be a path from P to Q parameterized by $\vec{r}(t)$ for $t_0 \leq t \leq t_1$, where $\vec{r}(t_0) = P$ and $\vec{r}(t_1) = Q$.

(i)   Using part (a) and Newton's law $\vec{F} = m\vec{a}$, show

$$\begin{array}{c} \text{Work done by } \vec{F} \\ \text{as particle moves along } C \end{array} = \text{Kinetic energy at } Q - \text{Kinetic energy at } P.$$

(ii)   Use the Fundamental Theorem of Calculus for Line Integrals to show that

$$\begin{array}{c} \text{Work done by } \vec{F} \\ \text{as particle moves along } C \end{array} = \text{Potential energy at } P - \text{Potential energy at } Q.$$

(iii)   Use parts (a) and (b) to show that the total energy at P is the same as at Q.

This problem explains why force vector fields which are *path-independent* are usually called *conservative* (force) vector fields.

2.   **Ampère's Law**

Ampère's Law, introduced in Problem 28 on page 844 , relates the net current, I, flowing through a surface to the magnetic field around the boundary, C, of the surface. The law says that

$$\int_C \vec{B} \cdot d\vec{r} = kI, \quad \text{for some constant } k.$$

The orientation of C is given by the right hand rule; if the fingers of your right hand lie against the surface and curl in the direction of the oriented curve C, then your thumb points in the direction of positive current.

(a)   We consider an infinitely long cylindrical wire having a radius r_0, where $r_0 > 0$. Suppose the wire is centered on the z-axis and carries a constant current, I, uniformly distributed across a cross-section of the wire. The magnitude of $\vec{B}$ is constant along any circle which is centered on and perpendicular to the z-axis. The direction of $\vec{B}$ is tangent to such circles. Show that the magnitude of $\vec{B}$, at a distance r from the z-axis, is given by

$$\|\vec{B}\| = \begin{cases} \dfrac{aI}{2\pi r} & \text{for } r \geq r_0 \\ \dfrac{aIr}{2\pi r_0^2} & \text{for } r < r_0. \end{cases}$$

(b)   A *torus* is a doughnut-shaped surface obtained by rotating around the z-axis the circle $(x - \beta)^2 + z^2 = \alpha^2$ of radius α and center $(\beta, 0, 0)$, where $0 < \alpha < \beta$. A *toroidal solenoid* is constructed by wrapping a thin wire a large number, say N, times around the torus. Experiments show that if a constant current I flows though the wire, then the magnitude of the magnetic field is constant on circles inside the torus which are centered on and perpendicular to the z-axis. The direction of $\vec{B}$ is tangent to such circles. Explain why, at all points inside the torus,

$$\|\vec{B}\| = \frac{aNI}{2\pi r},$$

and $\|\vec{B}\| = 0$ otherwise. [Hint: Apply Ampère's Law to a suitably chosen surface S with boundary curve C. What is the net current passing through S?]

FOCUS ON THEORY

PROOF OF GREEN'S THEOREM

In this section we will give a proof of Green's Theorem based on the change of variables formula for double integrals. Assume the vector field $\vec{F}$ is given in components by

$$\vec{F}(x,y) = F_1(x,y)\vec{i} + F_2(x,y)\vec{j}.$$

Proof for Rectangles

We prove Green's Theorem first when R is a rectangular region, as shown in Figure 18.48. The line integral in Green's theorem can be written as

$$\int_C \vec{F} \cdot d\vec{r} = \int_{C_1} \vec{F} \cdot d\vec{r} + \int_{C_2} \vec{F} \cdot d\vec{r} + \int_{C_3} \vec{F} \cdot d\vec{r} + \int_{C_4} \vec{F} \cdot d\vec{r}$$

$$= \int_a^b F_1(x,c)\,dx + \int_c^d F_2(b,y)\,dy - \int_a^b F_1(x,d)\,dx - \int_c^d F_2(a,y)\,dy$$

$$= \int_c^d (F_2(b,y) - F_2(a,y))\,dy + \int_a^b (-F_1(x,d) + F_1(x,c))\,dx.$$

On the other hand, the double integral in Green's theorem can be written as an iterated integral. We evaluate the inside integral using the Fundamental Theorem of Calculus.

$$\int_R \left(\frac{\partial F_2}{\partial x} - \frac{\partial F_1}{\partial y} \right) dx\,dy = \int_R \frac{\partial F_2}{\partial x}\,dx\,dy + \int_R -\frac{\partial F_1}{\partial y}\,dx\,dy$$

$$= \int_c^d \int_a^b \frac{\partial F_2}{\partial x}\,dx\,dy + \int_a^b \int_c^d -\frac{\partial F_1}{\partial y}\,dy\,dx$$

$$= \int_c^d (F_2(b,y) - F_2(a,y))\,dy + \int_a^b (-F_1(x,d) + F_1(x,c))\,dx.$$

Since the line integral and the double integral are equal, we have proved Green's theorem for rectangles.

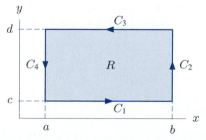

Figure 18.48: A rectangular region R with boundary C broken into C_1, C_2, C_3, and C_4

Proof for Regions Parameterized by Rectangles

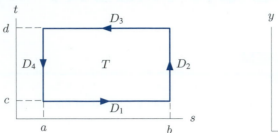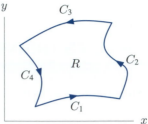

Figure 18.49: A curved region R in the xy-plane corresponding to a rectangular region T in the st-plane

Now we prove Green's Theorem for a region R which can be transformed into a rectangular region. Suppose we have a smooth change of coordinates

$$x = x(s,t), \qquad y = y(s,t).$$

Consider a curved region R in the xy-plane corresponding to a rectangular region T in the st-plane, as in Figure 18.49. We suppose that the change of coordinates is one-to-one on the interior of T.

We prove Green's theorem for R using Green's theorem for T and the change of variables formula for double integrals given on page 796. First we express the line integral around C

$$\int_C \vec{F} \cdot d\vec{r},$$

as a line integral in the st-plane around the rectangle $D = D_1 + D_2 + D_3 + D_4$. In vector notation, the change of coordinates is

$$\vec{r} = \vec{r}(s,t) = x(s,t)\vec{i} + y(s,t)\vec{j}$$

and so

$$\vec{F} \cdot d\vec{r} = \vec{F}(\vec{r}(s,t)) \cdot \frac{\partial \vec{r}}{\partial s}\, ds + \vec{F}(\vec{r}(s,t)) \cdot \frac{\partial \vec{r}}{\partial t}\, dt.$$

We define a vector field $\vec{G}$ on the st-plane with components

$$G_1 = \vec{F} \cdot \frac{\partial \vec{r}}{\partial s} \quad \text{and} \quad G_2 = \vec{F} \cdot \frac{\partial \vec{r}}{\partial t}.$$

Then, if $\vec{u}$ is the position vector of a point in the st-plane, we have $\vec{F} \cdot d\vec{r} = G_1\, ds + G_2\, dt = \vec{G} \cdot d\vec{u}$. Problem 5 at the end of this section asks you to show that the formula for line integrals along parameterized paths leads to the following result:

$$\int_C \vec{F} \cdot d\vec{r} = \int_D \vec{G} \cdot d\vec{u}.$$

In addition, using the product rule and chain rule we can show that

$$\frac{\partial G_2}{\partial s} - \frac{\partial G_1}{\partial t} = \left(\frac{\partial F_2}{\partial x} - \frac{\partial F_1}{\partial y} \right) \begin{vmatrix} \frac{\partial x}{\partial s} & \frac{\partial y}{\partial s} \\ \frac{\partial x}{\partial t} & \frac{\partial y}{\partial t} \end{vmatrix}.$$

(See Problem 6 at the end of this section.) Hence, by the change of variables formula for double integrals on page 796,

$$\int_R \left(\frac{\partial F_2}{\partial x} - \frac{\partial F_1}{\partial y} \right) dx\, dy = \int_T \left(\frac{\partial F_2}{\partial x} - \frac{\partial F_1}{\partial y} \right) \begin{vmatrix} \frac{\partial x}{\partial s} & \frac{\partial y}{\partial s} \\ \frac{\partial x}{\partial t} & \frac{\partial y}{\partial t} \end{vmatrix} ds\, dt = \int_T \left(\frac{\partial G_2}{\partial s} - \frac{\partial G_1}{\partial t} \right) ds\, dt.$$

Thus we have shown that

$$\int_C \vec{F} \cdot d\vec{r} = \int_D \vec{G} \cdot d\vec{u}$$

and that

$$\int_R \left(\frac{\partial F_2}{\partial x} - \frac{\partial F_1}{\partial y} \right) dx\,dy = \int_T \left(\frac{\partial G_2}{\partial s} - \frac{\partial G_1}{\partial t} \right) ds\,dt.$$

The integrals on the right are equal, by Green's Theorem for rectangles; hence the integrals on the left are equal as well, which is Green's Theorem for the region R.

Pasting Regions Together

Lastly we show that Green's Theorem holds for a region formed by pasting together regions which can be transformed into rectangles. Figure 18.50 shows two regions R_1 and R_2 that fit together to form a region R. We break the boundary of R into C_1, the part shared with R_1, and C_2, the part shared with R_2. We let C be the part of the the boundary of R_1 which it shares with R_2. So

$$\text{Boundary of } R = C_1 + C_2, \quad \text{Boundary of } R_1 = C_1 + C, \quad \text{Boundary of } R_2 = C_2 + (-C).$$

Note that when the curve C is considered as part of the boundary of R_2, it receives the opposite orientation from the one it receives as the boundary of R_1. Thus

$$\int_{\text{Boundary of } R_1} \vec{F} \cdot d\vec{r} + \int_{\text{Boundary of } R_2} \vec{F} \cdot d\vec{r} = \int_{C_1+C} \vec{F} \cdot d\vec{r} + \int_{C_2+(-C)} \vec{F} \cdot d\vec{r}$$

$$= \int_{C_1} \vec{F} \cdot d\vec{r} + \int_C \vec{F} \cdot d\vec{r} + \int_{C_2} \vec{F} \cdot d\vec{r} - \int_C \vec{F} \cdot d\vec{r}$$

$$= \int_{C_1} \vec{F} \cdot d\vec{r} + \int_{C_2} \vec{F} \cdot d\vec{r}$$

$$= \int_{\text{Boundary of } R} \vec{F} \cdot d\vec{r}.$$

So, applying Green's Theorem for R_1 and R_2, we get

$$\int_R \left(\frac{\partial F_2}{\partial x} - \frac{\partial F_1}{\partial y} \right) dx\,dy = \int_{R_1} \left(\frac{\partial F_2}{\partial x} - \frac{\partial F_1}{\partial y} \right) dx\,dy + \int_{R_2} \left(\frac{\partial F_2}{\partial x} - \frac{\partial F_1}{\partial y} \right) dx\,dy$$

$$= \int_{\text{Boundary of } R_1} \vec{F} \cdot d\vec{r} + \int_{\text{Boundary of } R_2} \vec{F} \cdot d\vec{r}$$

$$= \int_{\text{Boundary of } R} \vec{F} \cdot d\vec{r},$$

which is Green's Theorem for R. Thus, we have proved Green's Theorem for any region formed by pasting together regions that are smoothly parameterized by rectangles.

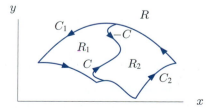

Figure 18.50: Two regions R_1 and R_2 pasted together to form a region R

Example 1   Let R be the annulus (ring) centered at the origin with inner radius 1 and outer radius 2. Using polar coordinates, show that the proof of Green's Theorem applies to R. See Figure 18.51.

Solution   In polar coordinates, $x = r \cos t$ and $y = r \sin t$, the annulus corresponds to the rectangle in the rt-plane $1 \le r \le 2$, $0 \le t \le 2\pi$. The sides $t = 0$ and $t = 2\pi$ are pasted together in the xy-plane along the x-axis; the other two sides become the inner and outer circles of the annulus. Thus R is formed by pasting the ends of a rectangle together.

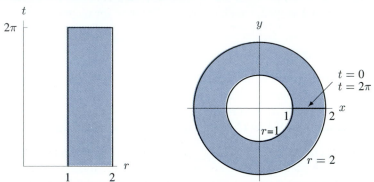

Figure 18.51: The annulus R in the xy-plane and the corresponding rectangle $1 \le r \le 2$, $0 \le t \le 2\pi$ in the rt-plane

Problems on Proof of Green's Theorem

1. Let R be the annulus centered at $(-1, 2)$ with inner radius 2 and outer radius 3. Show that R can be parameterized by a rectangle.

2. Let R be the region under the first arc of the graph of the sine function. Show that R can be parameterized by a rectangle.

3. Let $f(x)$ and $g(x)$ be two smooth functions, and suppose that $f(x) \le g(x)$ for $a \le x \le b$. Let R be the region $f(x) \le y \le g(x)$, $a \le x \le b$.

    (a) Sketch an example of such a region.
    (b) For a constant x_0, parameterize the vertical line segment in R where $x = x_0$. Choose your parameterization so that the parameter starts at 0 and ends at 1.
    (c) By putting together the parameterizations in part (b) for different values of x_0, show that R can be parameterized by a rectangle.

4. Let $f(y)$ and $g(y)$ be two smooth functions, and suppose that $f(y) \le g(y)$ for $c \le y \le d$. Let R be the region $f(y) \le x \le g(y)$, $c \le y \le d$.

    (a) Sketch an example of such a region.
    (b) For a constant y_0, parameterize the horizontal line segment in R where $y = y_0$. Choose your parameterization so that the parameter starts at 0 and ends at 1.
    (c) By putting together the parameterizations in part (b) for different values of y_0, show that R can be parameterized by a rectangle.

5. Use the formula for calculating line integrals by parameterization to prove the statement on page 872:

$$\int_C \vec{F} \cdot d\vec{r} = \int_D \vec{G} \cdot d\vec{u}\,.$$

6. Use the product rule and the chain rule to prove the formula on page 872:

$$\frac{\partial G_2}{\partial s} - \frac{\partial G_1}{\partial t} = \left(\frac{\partial F_2}{\partial x} - \frac{\partial F_1}{\partial y} \right) \begin{vmatrix} \frac{\partial x}{\partial s} & \frac{\partial y}{\partial s} \\ \frac{\partial x}{\partial t} & \frac{\partial y}{\partial t} \end{vmatrix}\,.$$

CHAPTER NINETEEN

FLUX INTEGRALS

In the previous chapter we saw how to integrate vector fields along curves. In this chapter we shall define a new sort of integral, the flux integral, which goes over a surface rather than along a curve. If we view a vector field as representing the velocity of a fluid flow, the flux integral tells us about the rate at which fluid is flowing through the surface. In addition, the flux integral appears in the theory of electricity and magnetism.

19.1   THE IDEA OF A FLUX INTEGRAL

Flow Through a Surface

Imagine water flowing through a fishing net stretched across a stream. Suppose we want to measure the flow rate of water through the net, that is, the volume of fluid that passes through the surface per unit time. (See Figure 19.1.) This flow rate is called the *flux* of the fluid through the surface. We can also compute the flux of vector fields, such as electric and magnetic fields, where no flow is actually taking place.

Figure 19.1: Flux measures rate of flow
through a surface

Orientation of a Surface

Before computing the flux of a vector field through a surface, we need to decide which direction of flow through the surface is the positive direction; this is described as choosing an orientation. [1]

> At each point on a smooth surface there are two unit normals, one in each direction. **Choosing an orientation** means picking one of these normals at every point of the surface in a continuous way. The normal vector in the direction of the orientation is denoted by $\vec{n}$. For a closed surface, we usually choose the outward orientation.

We say the flux through a piece of surface is positive if the flow is in the direction of the orientation and negative if it is in the opposite direction. (See Figure 19.2.)

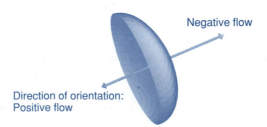

Figure 19.2: An oriented surface showing
directions of positive and negative flow

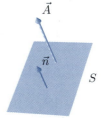

Figure 19.3: Area vector $\vec{A} = \vec{n}\,A$ of flat surface
with area A and orientation $\vec{n}$

The Area Vector

The flux through a flat surface depends both on the area of the surface and its orientation. Thus, it is useful to represent its area by a vector as shown in Figure 19.3.

---

[1] Although we will not study them, there are a few surfaces for which this cannot be done. See page 881.

The **area vector** of a flat, oriented surface is a vector $\vec{A}$ such that
- The magnitude of $\vec{A}$ is the area of the surface.
- The direction of $\vec{A}$ is the direction of the orientation vector $\vec{n}$.

The Flux of a Constant Vector Field Through a Flat Surface

Suppose the velocity vector field, $\vec{v}$, of a fluid is constant and $\vec{A}$ is the area vector of a flat surface. The flux through this surface is the volume of fluid that flows through in one unit of time. The volume of the skewed box in Figure 19.4 has cross-sectional area $\|\vec{A}\|$ and height $\|\vec{v}\| \cos\theta$, so its volume is $\left(\|\vec{v}\| \cos\theta\right) \|\vec{A}\| = \vec{v} \cdot \vec{A}$. Thus we have the following result:

If $\vec{v}$ is constant and $\vec{A}$ is the area vector of a flat surface, then

$$\text{Flux through surface} = \vec{v} \cdot \vec{A}.$$

Figure 19.4: Flux of $\vec{v}$ through a surface with area vector $\vec{A}$ is the volume of this skewed box

---

Example 1   Water is flowing down a cylindrical pipe 2 cm in radius with a velocity of 3 cm/sec. Find the flux of the velocity vector field through the ellipse-shaped region shown in Figure 19.5. The normal to the ellipse makes an angle of θ with the direction of flow and the area of the ellipse is $4\pi/(\cos\theta)$ cm^2.

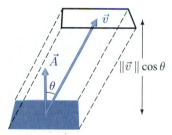

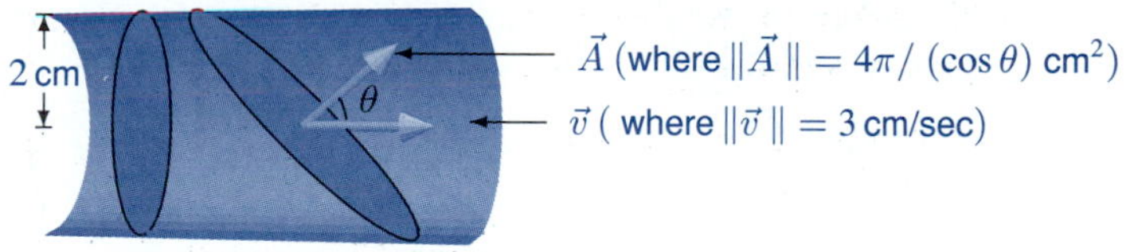

Figure 19.5: Flux through ellipse-shaped region across a cylindrical pipe

Solution   There are two ways to approach this problem. One is to use the formula we just derived which gives

$$\text{Flux through ellipse} = \vec{v} \cdot \vec{A} = \|\vec{v}\|\|\vec{A}\| \cos\theta = 3(\text{Area of ellipse}) \cos\theta$$

$$= 3\left(\frac{4\pi}{\cos\theta}\right) \cos\theta = 12\pi \text{ cm}^3/\text{sec}.$$

The second way is to notice that the flux through the ellipse is equal to the flux through the circle perpendicular to the pipe in Figure 19.5. Since the flux is the rate at which water is flowing down the pipe, we have

$$\text{Flux through circle} = \frac{\text{Velocity}}{\text{of water}} \times \frac{\text{Area of}}{\text{circle}} = \left(3\,\frac{\text{cm}}{\text{sec}}\right)(\pi 2^2 \text{ cm}^2) = 12\pi \text{ cm}^3/\text{sec}.$$

When the vector field is not constant or the surface is not flat, we divide the surface up into small, almost flat pieces such that the vector field is approximately constant on each one, as follows.

The Flux Integral

To calculate the flux of a vector field $\vec{F}$ which is not necessarily constant through a curved, oriented surface S, we divide the surface into a patchwork of small, approximately flat pieces (like a wire-frame representation of the surface) as shown in Figure 19.6. Suppose a particular patch has area ΔA. We pick an orientation vector $\vec{n}$ at a point on the patch and define the area vector of the patch, $\Delta \vec{A}$, as

$$\Delta \vec{A} = \vec{n}\,\Delta A.$$

(See Figure 19.6.) If the patches are small enough, we can assume that $\vec{F}$ is approximately constant on each piece. Then we know that

$$\text{Flux through patch} \approx \vec{F} \cdot \Delta \vec{A}$$

and so

$$\text{Flux through whole surface} \approx \sum \vec{F} \cdot \Delta \vec{A},$$

where the sum adds the fluxes through all the small pieces. As each patch becomes smaller and $\|\Delta \vec{A}\| \to 0$, the approximation gets better and we get

$$\text{Flux through } S = \lim_{\|\Delta \vec{A}\| \to 0} \sum \vec{F} \cdot \Delta \vec{A}.$$

Thus, we make the following definition:

> The **flux integral** of the vector field $\vec{F}$ through the oriented surface S is
>
> $$\int_S \vec{F} \cdot d\vec{A} = \lim_{\|\Delta \vec{A}\| \to 0} \sum \vec{F} \cdot \Delta \vec{A}.$$

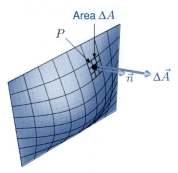

Figure 19.6: Surface S divided into small, almost flat pieces, showing a typical orientation vector $\vec{n}$ and area vector $\Delta \vec{A}$

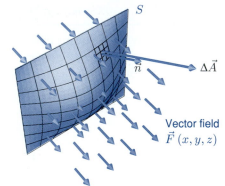

Figure 19.7: Flux of a vector field through a curved surface S

In computing a flux integral, we have to divide the surface up in a reasonable way, or the limit might not exist. In practice this problem seldom arises; however, one way to avoid it is to define flux integrals by the method used to compute them.

Flux and Fluid Flow

If $\vec{v}$ is the velocity vector field of a fluid, we have

$$\boxed{\begin{array}{ccccc} \text{Rate fluid flows} & = & \text{Flux of } \vec{v} & = & \displaystyle\int_S \vec{v} \cdot d\vec{A} \\ \text{through surface } S & & \text{through } S & & \end{array}}$$

The rate of fluid flow is measured in units of volume per unit time.

Example 2   Find the flux of the vector field $\vec{B}(x, y, z)$ shown in Figure 19.8 through the square S of side 2 shown in Figure 19.9, oriented in the $\vec{j}$ direction, where

$$\vec{B}(x, y, z) = \frac{-y\vec{i} + x\vec{j}}{x^2 + y^2}.$$

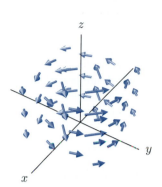

Figure 19.8: The vector field $\vec{B}(x, y, z) = \frac{-y\vec{i} + x\vec{j}}{x^2 + y^2}$

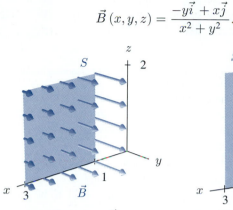

Figure 19.9: Flux of $\vec{B}$ through the square S of side 2 in xy-plane and oriented in $\vec{j}$ direction

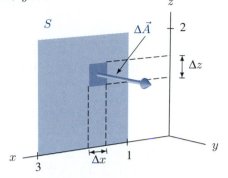

Figure 19.10: A small patch of surface with area $\|\Delta\vec{A}\| = \Delta x \Delta z$

Solution   Consider a small rectangular patch with area vector $\Delta\vec{A}$ in S, with sides Δx and Δz so that $\|\Delta\vec{A}\| = \Delta x \Delta z$. Since $\Delta\vec{A}$ points in the $\vec{j}$ direction we have $\Delta\vec{A} = \vec{j}\Delta x\Delta z$. (See Figure 19.10.)

At the point $(x, 0, z)$ in S, substituting $y = 0$ into $\vec{B}$ gives $\vec{B}(x, 0, z) = (1/x)\vec{j}$. Thus, we have

$$\text{Flux through small patch} \approx \vec{B}\cdot\Delta\vec{A} = \left(\frac{1}{x}\vec{j}\right)\cdot(\vec{j}\,\Delta x\Delta z) = \frac{1}{x}\Delta x\,\Delta z.$$

Therefore,

$$\text{Flux through surface} = \int_S \vec{B}\cdot d\vec{A} = \lim_{\|\Delta\vec{A}\|\to 0}\sum\vec{B}\cdot\Delta\vec{A} = \lim_{\substack{\Delta x\to 0 \\ \Delta z\to 0}}\sum\frac{1}{x}\Delta x\,\Delta z.$$

This last expression is a Riemann sum for the double integral $\int_R \frac{1}{x}\,dA$, where R is the square $1 \le x \le 3$, $0 \le z \le 2$. Thus,

$$\text{Flux through surface} = \int_S \vec{B}\cdot d\vec{A} = \int_R \frac{1}{x}\,dA = \int_0^2\int_1^3 \frac{1}{x}\,dx\,dz = 2\ln 3.$$

The result is positive since the vector field is passing through the surface in the positive direction.

Example 3   Each of the vector fields in Figure 19.11 consists entirely of vectors parallel to the xy-plane, and is constant in the z direction (that is, the vector field looks the same in any plane parallel to the xy-plane). For each one, say whether you expect the flux through a closed surface surrounding the origin to be positive, negative, or zero. In part (a) the surface is a closed cube with faces parallel to the axes; in parts (b) and (c) the surface is a closed cylinder. In each case we choose the outward orientation. (See Figure 19.12 on page 880.)

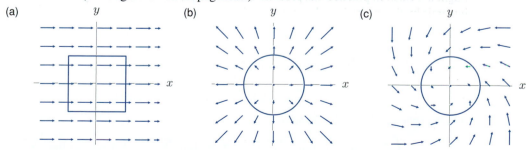

Figure 19.11: Flux of a vector field through the closed surfaces whose cross-sections are shown in the xy-plane

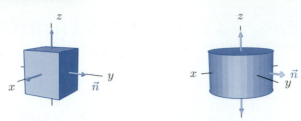

Figure 19.12: The closed cube and closed cylinder, both oriented outward

Solution   (a)   Since the vector field appears to be parallel to the faces of the cube which are perpendicular to the y- and z-axes, we expect the flux through these faces to be zero. The fluxes through the two faces perpendicular to the x-axis appear to be equal in magnitude and opposite in sign, so we expect the net flux to be zero.

(b)   Since the top and bottom of the cylinder are parallel to the flow, the flux through them is zero. On the round surface of the cylinder, $\vec{v}$ and $\Delta \vec{A}$ appear to be everywhere parallel and in the same direction, so we expect each term $\vec{v} \cdot \Delta \vec{A}$ to be positive, and therefore the flux integral $\int_S \vec{v} \cdot d\vec{A}$ to be positive.

(c)   As in part (b), the flux through the top and bottom of the cylinder is zero. In this case $\vec{v}$ and $\Delta \vec{A}$ are not parallel on the round surface of the cylinder, but since the fluid appears to be flowing inwards as well as swirling, we expect each term $\vec{v} \cdot \Delta \vec{A}$ to be negative, and therefore the flux integral to be negative.

Calculating Flux Integrals Using $d\vec{A} = \vec{n}\,dA$

For a small patch of surface ΔS with normal $\vec{n}$ and area ΔA, the area vector is $\Delta \vec{A} = \vec{n}\,\Delta A$. The next example shows how we can use this relationship to compute a flux integral.

Example 4   An electric charge q is placed at the origin in 3-space. The resulting electric field $\vec{E}\,(\vec{r}\,)$ at the point with position vector $\vec{r}$ is given by

$$\vec{E}\,(\vec{r}\,) = q\frac{\vec{r}}{\|\vec{r}\,\|^3}, \qquad \vec{r} \neq \vec{0}\,.$$

Find the flux of $\vec{E}$ out of the sphere of radius R centered at the origin. (See Figure 19.13.)

Figure 19.13: Flux of $\vec{E} = q\vec{r}\,/\|\vec{r}\,\|^3$ through the surface of a sphere of radius R centered at the origin

Solution   This vector field points radially outward from the origin in the same direction as $\vec{n}$. Thus, since $\vec{n}$ is a unit vector,

$$\vec{E} \cdot \Delta \vec{A} = \vec{E} \cdot \vec{n}\,\Delta A = \|\vec{E}\,\|\Delta A.$$

On the sphere, $\|\vec{E}\,\| = q/R^2$, so

$$\int_S \vec{E} \cdot d\vec{A} = \lim_{\|\Delta \vec{A}\,\| \to 0} \sum \vec{E} \cdot \Delta \vec{A} = \lim_{\Delta A \to 0} \sum \frac{q}{R^2}\,\Delta A = \frac{q}{R^2}\lim_{\Delta A \to 0} \sum \Delta A.$$

The last sum approximates the surface area of the sphere. In the limit as the subdivisions get finer we have

$$\lim_{\Delta A \to 0} \sum \Delta A = \text{Surface area of sphere.}$$

Thus, the flux is given by

$$\int_S \vec{E} \cdot d\vec{A} = \frac{q}{R^2} \lim_{\Delta A \to 0} \sum \Delta A = \frac{q}{R^2} \cdot \text{Surface area of sphere} = \frac{q}{R^2}(4\pi R^2) = 4\pi q.$$

This result is known as Gauss's law.

Instead of using Riemann sums, we often write $d\vec{A} = \vec{n}\, dA$, as in the next example.

Example 5   Suppose S is the surface of the cube bounded by the six planes $x = \pm 1$, $y = \pm 1$, and $z = \pm 1$. Compute the flux of the electric field $\vec{E}$ of the previous example outward through S.

Solution   It is enough to compute the flux of $\vec{E}$ through a single face, say the top face S_1 defined by $z = 1$, where $-1 \leq x \leq 1$ and $-1 \leq y \leq 1$. By symmetry, the flux of $\vec{E}$ through the other five faces of S must be the same.

On the top face, S_1, we have $d\vec{A} = \vec{k}\, dx\, dy$ and

$$\vec{E}(x, y, 1) = q\frac{x\vec{i} + y\vec{j} + \vec{k}}{(x^2 + y^2 + 1)^{3/2}}.$$

The corresponding flux integral is given by

$$\int_{S_1} \vec{E} \cdot d\vec{A} = q \int_{-1}^{1}\int_{-1}^{1} \frac{x\vec{i} + y\vec{j} + \vec{k}}{(x^2 + y^2 + 1)^{3/2}} \cdot \vec{k}\, dx\, dy = q\int_{-1}^{1}\int_{-1}^{1} \frac{1}{(x^2 + y^2 + 1)^{3/2}}\, dx\, dy.$$

Computing this integral numerically shows that

$$\text{Flux through top face} = \int_{S_1} \vec{E} \cdot d\vec{A} \approx 2.0944q.$$

Thus,

$$\text{Total flux of } \vec{E} \text{ out of cube} = \int_S \vec{E} \cdot d\vec{A} \approx 6(2.0944q) = 12.5664q.$$

Example 4 on page 880 showed that the flux of $\vec{E}$ through a sphere of radius R centered at the origin is $4\pi q$. Since $4\pi \approx 12.5664$, Example 5 suggests that

$$\text{Total flux of } \vec{E} \text{ out of cube} = 4\pi q.$$

By computing the flux integral in Example 5 exactly, it is possible to verify that the flux of $\vec{E}$ through the cube and the sphere are exactly equal. When we encounter the Divergence Theorem in Chapter 20 we will see why this is so.

Notes on Orientation

Two difficulties can occur in choosing an orientation. The first is that if the surface is not smooth, it may not have a normal vector at every point. For example, a cube does not have a normal vector along its edges. When we have a surface, such as a cube, which is made of a finite number of smooth pieces, we choose an orientation for each piece separately. The best way to do this is usually clear. For example, on the cube we choose the outward orientation on each face. (See Figure 19.14.)

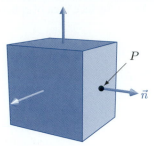

Figure 19.14: The orientation vector field $\vec{n}$ on the cube surface determined by the choice of unit normal vector at the point P

Figure 19.15: The Möbius strip is an example of a non-orientable surface

The second difficulty is that there are some surfaces which cannot be oriented at all, such as the *Möbius strip* in Figure 19.15.

Problems for Section 19.1

1. Let $\vec{F}(x, y, z) = z\vec{i}$. For each of the surfaces in (a)–(e), say whether the flux of $\vec{F}$ through the surface is positive, negative, or zero. In each case, the orientation of the surface is indicated by the given normal vector.

(a)  (b)  (c)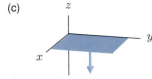

(d) 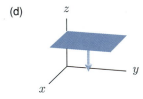   (e)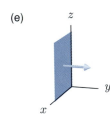

2. Repeat Problem 1 with $\vec{F}(x, y, z) = -z\vec{i} + x\vec{k}$.

3. Repeat Problem 1 with the vector field $\vec{F}(\vec{r}) = \vec{r}$.

4. Arrange the following flux integrals,

$$\int_{S_i} \vec{F} \cdot d\vec{A},$$

with $i = 1, 2, 3, 4$, in ascending order if $\vec{F} = -\vec{i} - \vec{j} + \vec{k}$ and S_i are the following surfaces:

-   S_1 is a horizontal square of side 1 with one corner at $(0, 0, 2)$, above the first quadrant of the xy-plane, oriented upward.
-   S_2 is a horizontal square of side 1 with one corner at $(0, 0, 3)$, above the third quadrant of the xy-plane, oriented upward.
-   S_3 is a square of side $\sqrt{2}$ in the xz-plane with one corner at the origin, one edge along the positive x-axis, one along the negative z-axis, oriented in the negative y- direction.
-   S_4 is a square of side $\sqrt{2}$ with one corner at the origin, one edge along the positive y-axis, one corner at $(1, 0, 1)$, oriented upward.

5. Compute the flux of the vector field $\vec{v} = 2\vec{i} + 3\vec{j} + 5\vec{k}$ through each of the rectangular regions in (a)–(d), assuming each is oriented as shown.

(a)
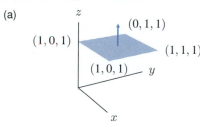

(b)
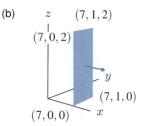

(c)
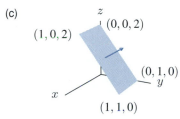

(d)
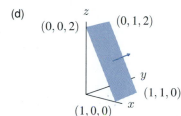

6. Figure 19.16 shows a cross-section of the earth's magnetic field. Say whether the magnetic flux through a horizontal plate, oriented skyward, is positive, negative, or zero if the plate is
    (a) At the north pole.   (b) At the south pole.   (c) On the equator.
    [Note: You may assume that the earth's magnetic and geographic poles coincide.]

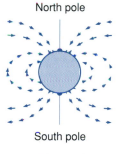

Figure 19.16   **Figure 19.17**   **Figure 19.18**

7. (a) What do you think will be the electric flux through the cylindrical surface that is placed as shown in the constant electric field in Figure 19.17? Why?
    (b) What if the cylinder is placed upright, as shown in Figure 19.18? Explain.

For Problems 8–14, compute the flux integral of the given vector field through the given surface S.

8. $\vec{F} = 2\vec{i}$ and S is a disk of radius 2 on the plane $x + y + z = 2$, oriented upward.

9. $\vec{F} = -y\vec{i} + x\vec{j}$ and S is the square plate in the yz-plane with corners at $(0, 1, 1)$, $(0, -1, 1)$, $(0, 1, -1)$, and $(0, -1, -1)$, oriented in the positive x-direction.

10. $\vec{F} = -y\vec{i} + x\vec{j}$ and S is the disk in the xy-plane with radius 2, oriented upwards and centered at the origin.

11. $\vec{F} = \vec{r}$ and S is the disk of radius 2 parallel to the xy-plane oriented upwards and centered at $(0, 0, 2)$.

12. $\vec{F} = (2 - x)\vec{i}$ and S is the cube whose vertices include the points $(0, 0, 0)$, $(3, 0, 0)$, $(0, 3, 0)$, $(0, 0, 3)$, and oriented outward.

13. $\vec{F} = (x^2 + y^2)\vec{i} + xy\vec{j}$ and S is the square in the xy-plane with corners at $(1, 1, 0)$, $(-1, 1, 0)$, $(1, -1, 0)$, $(-1, -1, 0)$, and oriented upward.

14. $\vec{F} = \vec{r}/r^2$ and S is the sphere of radius R centered at the origin, oriented outwards.

15. Explain why if $\vec{F}$ has constant magnitude on S and is everywhere normal to S and in the direction of orientation, then

$$\int_S \vec{F} \cdot d\vec{A} = \|\vec{F}\| \cdot \text{Area of } S.$$

16.   Let S be the cube with side length 2, faces parallel to the coordinate planes, and centered at the origin.

(a)   Calculate the total flux of the constant vector field $\vec{v} = -\vec{i} + 2\vec{j} + \vec{k}$ out of S by computing the flux through each face separately.

(b)   Calculate the flux out of S for any constant vector field $\vec{v} = a\vec{i} + b\vec{j} + c\vec{k}$.

(c)   Do your answers in parts (a) and (b) make sense? Explain.

17.   Let S be the tetrahedron with vertices at the origin and at $(1, 0, 0)$, $(0, 1, 0)$ and $(0, 0, 1)$.

(a)   Calculate the total flux of the constant vector field $\vec{v} = -\vec{i} + 2\vec{j} + \vec{k}$ out of S by computing the flux through each face separately.

(b)   Calculate the flux out of S in part (a) for any constant vector field $\vec{v}$.

(c)   Do your answers in parts (a) and (b) make sense? Explain.

18.   Suppose the z-axis carries a constant electric charge density of λ units of charge per unit length, with $\lambda > 0$, and that $\vec{E}$ is the resulting electric field.

(a)   Sketch the electric field, $\vec{E}$, in the xy-plane, given that

$$\vec{E}(x, y, z) = 2\lambda \frac{x\vec{i} + y\vec{j}}{x^2 + y^2}.$$

(b)   Compute the flux of $\vec{E}$ outward through the cylinder $x^2 + y^2 = R^2$, for $0 \le z \le h$.

19.   Let $P(x, y, z)$ be the pressure at the point (x, y, z) in a fluid. Let $\vec{F}(x, y, z) = P(x, y, z)\vec{k}$. Let S be the surface of a body submerged in the fluid. If S is oriented inward, show that $\int_S \vec{F} \cdot d\vec{A}$ is the buoyant force on the body, that is, the force upwards on the body due to the pressure of the fluid surrounding it. [Hint: $\vec{F} \cdot d\vec{A} = P(x, y, z)\vec{k} \cdot d\vec{A} = (P(x, y, z) \, d\vec{A}) \cdot \vec{k}$.]

20.   Consider the function $\rho(x, y, z)$ which gives the electrical charge density at points in space. The vector field $\vec{J}(x, y, z)$ gives the electric current density at any point in space and is defined so that the current through a small area $d\vec{A}$ is given by

$$\text{Current through small area} \approx \vec{J} \cdot d\vec{A}.$$

Suppose S is a closed surface enclosing a volume W.

(a)   What do the following integrals represent, in terms of electricity?

(i) $\displaystyle \int_W \rho \, dV$ (ii) $\displaystyle \int_S \vec{J} \cdot d\vec{A}$

(b)   Using the fact that an electric current through a surface is the rate at which electric charge passes through the surface per unit time, explain why

$$\int_S \vec{J} \cdot d\vec{A} = -\frac{\partial}{\partial t}\left(\int_W \rho \, dV\right).$$

21.   A fluid is flowing along a cylindrical pipe of radius a in the $\vec{i}$ direction. The velocity of the fluid at a radial distance r from the center of the pipe is $\vec{v} = u(1 - r^2/a^2)\vec{i}$.

(a)   What is the significance of the constant u?

(b)   What is the velocity of the fluid at the wall of the pipe?

(c)   Find the flux through a circular cross-section of the pipe.

22.   Suppose a region of 3-space has a temperature which varies from point to point. Let $T(x, y, z)$ be the temperature at a point (x, y, z). Newton's law of cooling says that $\operatorname{grad} T$ is proportional to the heat flow vector field, $\vec{F}$, where $\vec{F}$ points in the direction in which heat is flowing and has magnitude equal to the rate of flow of heat.

(a)   Suppose $\vec{F} = k \operatorname{grad} T$ for some constant k. What is the sign of k?

(b)   Explain why this form of Newton's law of cooling makes sense.

(c)   Let W be a region of space bounded by the surface S. Explain why

$$\begin{array}{l} \text{Rate of heat} \\ \text{loss from } W \end{array} = k \int_S (\operatorname{grad} T) \cdot d\vec{A}.$$

19.2 FLUX INTEGRALS FOR GRAPHS, CYLINDERS, AND SPHERES

In Section 19.1 we computed flux integrals in certain simple cases. In this section we see how to compute flux through surfaces that are graphs of functions, through cylinders, and through spheres.

Flux of a Vector Field through the Graph of $z = f(x, y)$

Suppose S is the graph of the differentiable function $z = f(x, y)$, oriented upward, and that $\vec{F}$ is a smooth vector field. In Section 19.1 we subdivided the surface into small pieces with area vector $\Delta \vec{A}$ and defined the flux of $\vec{F}$ through S as follows:

$$\int_S \vec{F} \cdot d\vec{A} = \lim_{\|\Delta\vec{A}\| \to 0} \sum \vec{F} \cdot \Delta\vec{A}.$$

How do we divide S into small pieces? One way is to use the cross-sections of f with x or y constant and take the patches in a wire frame representation of the surface. So we must calculate the area vector of one of these patches, which is approximately a parallelogram.

The Area Vector of a Parallelogram-Shaped Patch

According to the geometric definition of the cross product on page 642, the vector $\vec{v} \times \vec{w}$ has magnitude equal to the area of the parallelogram formed by $\vec{v}$ and $\vec{w}$ and direction perpendicular to this parallelogram and determined by the right-hand rule. Thus, we have

$$\boxed{\text{Area vector of parallelogram} = \vec{A} = \vec{v} \times \vec{w}.}$$

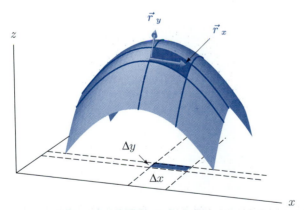

Figure 19.19: Surface showing parameter rectangle and tangent vectors $\vec{r}_x$ and $\vec{r}_y$

Figure 19.20: Parallelogram-shaped patch in the tangent plane to the surface

Consider the patch of surface above the rectangular region with sides Δx and Δy in the xy-plane shown in Figure 19.19. We approximate the area vector, $\Delta\vec{A}$, of this patch by the area vector of the corresponding patch on the tangent plane to the surface. See Figure 19.20 This patch is the parallelogram determined by the vectors $\vec{v}_x$ and $\vec{v}_y$, so its area vector is given by

$$\Delta\vec{A} \approx \vec{v}_x \times \vec{v}_y.$$

To find $\vec{v}_x$ and $\vec{v}_y$, notice that a point on the surface has position vector $\vec{r} = x\vec{i} + y\vec{j} + f(x,y)\vec{k}$. Thus, a cross-section of S with y constant has tangent vector

$$\vec{r}_x = \frac{\partial \vec{r}}{\partial x} = \vec{i} + f_x\vec{k},$$

and a cross-section with x constant has tangent vector

$$\vec{r}_y = \frac{\partial \vec{r}}{\partial y} = \vec{j} + f_y\vec{k}.$$

The vectors $\vec{r}_x$ and $\vec{v}_x$ are parallel because they are both tangent to the surface and in the xz-plane. Since the x-component of $\vec{r}_x$ is $\vec{i}$ and the x-component of $\vec{v}_x$ is $(\Delta x)\vec{i}$, we have $\vec{v}_x = (\Delta x)\vec{r}_x$. Similarly, we have $\vec{v}_y = (\Delta y)\vec{r}_y$. So the upward pointing area vector of the parallelogram is

$$\Delta\vec{A} \approx \vec{v}_x \times \vec{v}_y = (\vec{r}_x \times \vec{r}_y)\,\Delta x\,\Delta y = \left(-f_x\vec{i} - f_y\vec{j} + \vec{k}\right)\Delta x\,\Delta y.$$

This is our approximation for the area vector $\Delta\vec{A}$ on the surface. Replacing $\Delta\vec{A}$, Δx, and Δy by $d\vec{A}$, dx and dy, we write

$$d\vec{A} = \left(-f_x\vec{i} - f_y\vec{j} + \vec{k}\right)dx\,dy.$$

The Flux of $\vec{F}$ through a Surface given by a Graph of $z = f(x, y)$

Suppose the surface S is the part of the graph of $z = f(x,y)$ above a region R in the xy-plane, and suppose S is oriented upward. The flux of $\vec{F}$ through S is

$$\int_S \vec{F} \cdot d\vec{A} = \int_R \vec{F}(x,y,f(x,y)) \cdot \left(-f_x\vec{i} - f_y\vec{j} + \vec{k}\right)dx\,dy.$$

Example 1 Compute $\int_S \vec{F} \cdot d\vec{A}$ where $\vec{F}(x,y,z) = z\vec{k}$ and S is the rectangular plate with corners $(0,0,0)$, $(1,0,0)$, $(0,1,3)$, $(1,1,3)$, oriented upwards.

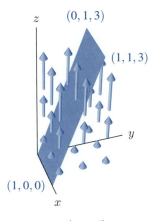

Figure 19.21: The vector field $\vec{F} = z\vec{k}$ on the rectangular surface S

Solution We find the equation for the plane S in the form $z = f(x,y)$. Since f is linear, with x-slope equal to 0 and y-slope equal to 3, and $f(0,0) = 0$, we have

$$z = f(x,y) = 0 + 0x + 3y = 3y.$$

Thus, we have

$$d\vec{A} = (-f_x\vec{i} - f_y\vec{j} + \vec{k})\,dx\,dy = (0\vec{i} - 3\vec{j} + \vec{k})\,dx\,dy = (-3\vec{j} + \vec{k})\,dx\,dy.$$

The flux integral is therefore

$$\int_S \vec{F} \cdot d\vec{A} = \int_0^1 \int_0^1 3y\vec{k} \cdot (-3\vec{j} + \vec{k}) \, dx \, dy = \int_0^1 \int_0^1 3y \, dx \, dy = 1.5.$$

Flux of a Vector Field through a Cylindrical Surface

Consider the cylinder of radius R centered on the z-axis illustrated in Figure 19.22 and oriented away from the z-axis. The small patch of surface, or parameter rectangle, in Figure 19.23 has surface area given by

$$\Delta A \approx R \, \Delta\theta \, \Delta z.$$

The outward unit normal $\vec{n}$ points in the direction of $x\vec{i} + y\vec{j}$, so

$$\vec{n} = \frac{x\vec{i} + y\vec{j}}{\|x\vec{i} + y\vec{j}\|} = \frac{R\cos\theta\vec{i} + R\sin\theta\vec{j}}{R} = \cos\theta\vec{i} + \sin\theta\vec{j}.$$

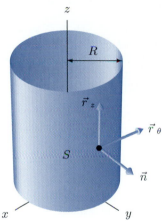

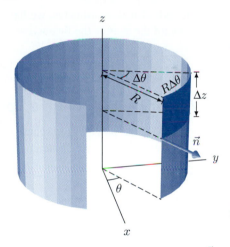

Figure 19.22: Outward oriented cylinder

Figure 19.23: Small patch with area $\Delta\vec{A}$ on surface of a cylinder

Therefore, the area vector of the parameter rectangle is approximated by

$$\Delta\vec{A} = \vec{n} \, \Delta A \approx \left(\cos\theta\vec{i} + \sin\theta\vec{j}\right) R \, \Delta z \, \Delta\theta.$$

Replacing $\Delta\vec{A}$, Δz, and $\Delta\theta$ by $d\vec{A}$, dz, and $d\theta$, we write

$$d\vec{A} = \left(\cos\theta\vec{i} + \sin\theta\vec{j}\right) R \, dz \, d\theta.$$

This gives the following result:

The Flux of a Vector Field through a Cylinder

The flux of $\vec{F}$ through the cylindrical surface S, of radius R and oriented away from the z-axis, is given by

$$\int_S \vec{F} \cdot d\vec{A} = \int_T \vec{F}(R, \theta, z) \cdot \left(\cos\theta\vec{i} + \sin\theta\vec{j}\right) R \, dz \, d\theta,$$

where T is the θz-region corresponding to S.

Example 2   Compute $\int_S \vec{F} \cdot d\vec{A}$ where $\vec{F}(x, y, z) = y\vec{j}$ and S is the part of the cylinder of radius 2 centered on the z-axis with $x \geq 0, y \geq 0$, and $0 \leq z \leq 3$. The surface is oriented towards the z-axis.

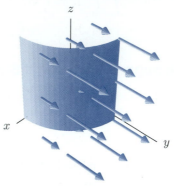

Figure 19.24: The vector field $\vec{F} = y\vec{j}$ on the surface S

Solution   In cylindrical coordinates, we have $R = 2$ and $\vec{F} = y\vec{j} = 2\sin\theta\vec{j}$. Since the orientation of S is toward the z-axis, the flux across S is given by

$$\int_S \vec{F} \cdot d\vec{A} = -\int_T 2\sin\theta\vec{j} \cdot (\cos\theta\vec{i} + \sin\theta\vec{j})2\, dz\, d\theta = -4\int_0^{\pi/2}\int_0^3 \sin^2\theta\, dz\, d\theta = -3\pi.$$

Flux of a Vector Field through a Spherical Surface

Consider the piece of the sphere of radius R centered at the origin, oriented outward illustrated in Figure 19.25. The small parameter rectangle in Figure 19.25 has surface area given by

$$\Delta A \approx R^2 \sin\phi\, \Delta\phi\, \Delta\theta.$$

The outward unit normal $\vec{n}$ points in the direction of $\vec{r} = x\vec{i} + y\vec{j} + z\vec{k}$, so

$$\vec{n} = \frac{\vec{r}}{\|\vec{r}\|} = \sin\phi\cos\theta\vec{i} + \sin\phi\sin\theta\vec{j} + \cos\phi\vec{k}.$$

Therefore, the area vector of the parameter rectangle is approximated by

$$\Delta\vec{A} \approx \vec{n}\,\Delta A = \frac{\vec{r}}{\|\vec{r}\|}\Delta A = \left(\sin\phi\cos\theta\vec{i} + \sin\phi\sin\theta\vec{j} + \cos\phi\vec{k}\right)R^2\sin\phi\,\Delta\phi\,\Delta\theta.$$

Replacing $\Delta\vec{A}$, $\Delta\phi$, and $\Delta\theta$ by $d\vec{A}$, $d\phi$, and $d\theta$, we write

$$d\vec{A} = \frac{\vec{r}}{\|\vec{r}\|}dA = \left(\sin\phi\cos\theta\vec{i} + \sin\phi\sin\theta\vec{j} + \cos\phi\vec{k}\right)R^2\sin\phi\,d\phi\,d\theta.$$

Thus, we obtain the following result:

The Flux of a Vector Field through a Sphere

The flux of $\vec{F}$ through the spherical surface S, with radius R and oriented away from the origin, is given by

$$\int_S \vec{F} \cdot d\vec{A} = \int_S \vec{F} \cdot \frac{\vec{r}}{\|\vec{r}\|}dA$$

$$= \int_T \vec{F}(R, \theta, \phi) \cdot \left(\sin\phi\cos\theta\vec{i} + \sin\phi\sin\theta\vec{j} + \cos\phi\vec{k}\right)R^2\sin\phi\,d\phi\,d\theta,$$

where T is the $\theta\phi$-region corresponding to S.

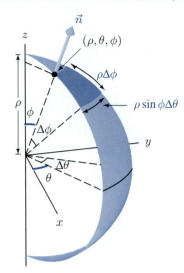

Figure 19.25: Small patch with area $\Delta \vec{A}$ on surface of a sphere

Example 3 Find the flux of $\vec{F} = z\vec{k}$ through S, the upper hemisphere of radius 2 centered at the origin, oriented outward.

Solution The hemisphere S is parameterized by spherical coordinates θ and ϕ, with $0 \le \theta \le 2\pi$ and $0 \le \phi \le \pi/2$. Since $R = 2$ and $\vec{F} = z\vec{k} = 2\cos\phi\vec{k}$, the flux is

$$\int_S \vec{F} \cdot d\vec{A} = \int_S 2\cos\phi\vec{k} \cdot (\sin\phi\cos\theta\vec{i} + \sin\phi\sin\theta\vec{j} + \cos\phi\vec{k})4\sin\phi \, d\phi \, d\theta$$

$$= \int_0^{2\pi} \int_0^{\pi/2} 8\sin\phi\cos^2\phi \, d\phi \, d\theta = 2\pi \left(8 \left(\frac{-\cos^3\phi}{3} \right) \Big|_{\phi=0}^{\pi/2} \right) = \frac{16\pi}{3}.$$

Example 4 The magnetic field $\vec{B}$ due to an *ideal magnetic dipole*, $\vec{\mu}$, located at the origin is defined to be

$$\vec{B}(\vec{r}) = -\frac{\vec{\mu}}{\|\vec{r}\|^3} + \frac{3(\vec{\mu} \cdot \vec{r})\vec{r}}{\|\vec{r}\|^5}.$$

Figure 19.26 shows a sketch of $\vec{B}$ in the plane $z = 0$ for the dipole $\vec{\mu} = \vec{i}$. Notice that $\vec{B}$ is similar to the magnetic field of a bar magnet with its north pole at the tip of the vector $\vec{i}$ and its south pole at the tail of the vector $\vec{i}$.

Compute the flux of $\vec{B}$ outward through the sphere S with center at the origin and radius R.

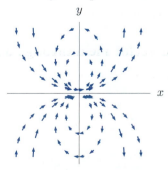

Figure 19.26: The magnetic field of a dipole, $\vec{i}$, at the origin:

$$\vec{B} = \frac{-\vec{i}}{\|\vec{r}\|^3} + \frac{3(\vec{i} \cdot \vec{r})\vec{r}}{\|\vec{r}\|^5}$$

Solution   Since $\vec{i} \cdot \vec{r} = x$ and $\|\vec{r}\| = R$ on the sphere of radius R, we have

$$\int_S \vec{B} \cdot d\vec{A} = \int_S \left(-\frac{\vec{i}}{\|\vec{r}\|^3} + \frac{3(\vec{i} \cdot \vec{r})\vec{r}}{\|\vec{r}\|^5} \right) \cdot \frac{\vec{r}}{\|\vec{r}\|} \, dA = \int_S \left(-\frac{\vec{i} \cdot \vec{r}}{\|\vec{r}\|^4} + \frac{3(\vec{i} \cdot \vec{r})\|\vec{r}\|^2}{\|\vec{r}\|^6} \right) dA$$

$$= \int_S \frac{2\vec{i} \cdot \vec{r}}{\|\vec{r}\|^4} \, dA = \int_S \frac{2x}{\|\vec{r}\|^4} \, dA = \frac{2}{R^4} \int_S x \, dA,$$

But the sphere S is centered at the origin. Thus, the contribution to the integral from each positive x- value is canceled by the contribution from the corresponding negative x-value; so $\int_S x \, dA = 0$. Therefore,

$$\int_S \vec{B} \cdot d\vec{A} = \frac{2}{R^4} \int_S x \, dA = 0.$$

Problems for Section 19.2

In Problems 1–12 compute the flux of the vector field, $\vec{F}$, through the surface, S.

1. $\vec{F} = (x - y)\vec{i} + z\vec{j} + 3x\vec{k}$ and S is the part of the plane $z = x + y$ above the rectangle $0 \le x \le 2$, $0 \le y \le 3$, oriented upward.

2. $\vec{F} = \vec{r}$ and S is the part of the plane $x + y + z = 1$ above the rectangle $0 \le x \le 2$, $0 \le y \le 3$, oriented downward.

3. $\vec{F} = \vec{r}$ and S is the part of the surface $z = x^2 + y^2$ above the disk $x^2 + y^2 \le 1$, oriented downward.

4. $\vec{F}(x, y, z) = 2x\vec{j} + y\vec{k}$ and S is the part of the surface $z = -y + 1$ above the square $0 \le x \le 1$, $0 \le y \le 1$, oriented upward.

5. $\vec{F} = 3x\vec{i} + y\vec{j} + z\vec{k}$ and S is the part of the surface $z = -2x - 4y + 1$ above the triangle R in the xy-plane with vertices $(0, 0)$, $(0, 2)$, $(1, 0)$, oriented upward.

6. $\vec{F} = x\vec{i} + y\vec{j}$ and S is the part of the surface $z = 25 - (x^2 + y^2)$ above the disk of radius 5 centered at the origin, oriented upward.

7. $\vec{F} = \cos(x^2 + y^2)\vec{k}$ and S is as in Problem 6.

8. $\vec{F} = -y\vec{j} + z\vec{k}$ and S is the part of the surface $z = y^2 + 5$ over the rectangle $-2 \le x \le 1$, $0 \le y \le 1$, oriented upward.

9. $\vec{F}(x, y, z) = -xz\vec{i} - yz\vec{j} + z^2\vec{k}$ and S is the cone $z = \sqrt{x^2 + y^2}$ for $0 \le z \le 6$, oriented upward.

10. $\vec{F} = y\vec{i} + \vec{j} - xz\vec{k}$ and S is the surface $y = x^2 + z^2$, with $x^2 + z^2 \le 1$, oriented in the positive y-direction.

11. $\vec{F} = xz\vec{i} + y\vec{k}$ and S is the hemisphere $x^2 + y^2 + z^2 = 9$, $z \ge 0$, oriented upward.

12. $\vec{F} = x^2\vec{i} + y^2\vec{j} + z^2\vec{k}$ and S is the oriented triangular surface shown in Figure 19.27.

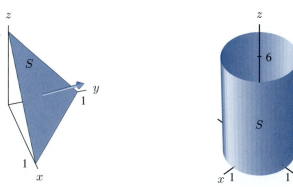

Figure 19.27   **Figure 19.28**

In Problems 13–14, compute the flux of the vector field, $\vec{F}$, through the cylindrical surface shown in Figure 19.28, oriented away from the z-axis.

13. $\vec{F} = x\vec{i} + y\vec{j}$ 14. $\vec{F} = xz\vec{i} + yz\vec{j} + z^3\vec{k}$

In Problems 15–16 compute the flux of the vector field, $\vec{F}$, through the given spherical surface, S.

15. $\vec{F} = z^2\vec{k}$ and S is the upper hemisphere of the sphere $x^2 + y^2 + z^2 = 25$, oriented away from the origin.

16. $\vec{F} = x\vec{i} + y\vec{j} + z\vec{k}$ and S is the surface of the sphere $x^2 + y^2 + z^2 = a^2$, oriented outward.

17. Calculate the flux of

$$\vec{F} = (xze^{yz})\vec{i} + xz\vec{j} + (5 + x^2 + y^2)\vec{k}$$

through the disk $x^2 + y^2 \leq 1$ in the xy-plane, oriented upward.

18. Calculate the flux of

$$\vec{H} = (e^{xy} + 3z + 5)\vec{i} + (e^{xy} + 5z + 3)\vec{j} + (3z + e^{xy})\vec{k}$$

through the square of side 2 with one vertex at the origin, one edge along the positive y-axis, one edge in the xz-plane with $x > 0$, $z > 0$, and the normal $\vec{n} = \vec{i} - \vec{k}$.

CHAPTER SUMMARY

- **Flux Integrals**
  Oriented surfaces, definition of flux through a surface, definition of flux integral as a limit of a Riemann sum

- **Calculating flux integrals** over surfaces:
  Graphs: $d\vec{A} = (-f_x\vec{i} - f_y\vec{j} + \vec{k})\,dx\,dy$
  Cylinders: $d\vec{A} = (\cos\theta\vec{i} + \sin\theta\vec{j})R\,dz\,d\theta$
  Spheres: $d\vec{A} = (\sin\phi\cos\theta\vec{i} + \sin\phi\sin\theta\vec{j} + \cos\phi\vec{k})R^2\,d\phi\,d\theta$

REVIEW PROBLEMS FOR CHAPTER NINETEEN

For Problems 1–2, let $\vec{F}(\vec{r}) = \vec{r}$ and let S be a square plate perpendicular to the z-axis and centered on the z-axis. Sketch as a function of time the flux of $\vec{F}$ through S as S moves in the given manner.

1. S moves from far up the positive z-axis to far down the negative z-axis. Assume S is oriented upward.

2. S rotates about an axis parallel to the x-axis, through the center of S. Assume that S is far up the z-axis so that $\vec{r}$ is approximately constant on S as S rotates, and that S is initially oriented upward.

3. Repeat Problems 1–2 with $\vec{F}(\vec{r}) = \vec{r}/r^3$, where $r = \|\vec{r}\|$.

For Problems 4–7 find the flux of the constant vector field $\vec{v} = \vec{i} - \vec{j} + 3\vec{k}$ through the given surfaces.

4. A disk of radius 2 in the xy-plane oriented upward.

5. A triangular plate of area 4 in the yz-plane oriented in the positive x-direction.

6. A square plate of area 4 in the yz-plane oriented in the positive x-direction.

7. The triangular plate with vertices $(1, 0, 0)$, $(0, 1, 0)$, $(0, 0, 1)$, oriented away from the origin.

8. Suppose water is flowing down a cylindrical pipe of radius 2 cm and that the speed is $(3 - (3/4)r^2)$ cm/sec at a distance r cm from the center of the pipe. Find the flux through the circular cross-section of the pipe, oriented so that the flow is positive.

In Problems 9–16 compute the flux of the given vector field, $\vec{F}$, through the given surface, S.

9. $\vec{F} = x\vec{i} + y\vec{j} + (z^2 + 3)\vec{k}$ and S is the rectangle $z = 4$, $0 \leq x \leq 2$, $0 \leq y \leq 3$, oriented in the positive z-direction.

10. $\vec{F} = z\vec{i} + y\vec{j} + 2x\vec{k}$ and S is the rectangle $z = 4$, $0 \leq x \leq 2$, $0 \leq y \leq 3$, oriented in the positive z-direction.

11. $\vec{F} = (x + \cos z)\vec{i} + y\vec{j} + 2x\vec{k}$ and S is the rectangle $x = 2$, $0 \leq y \leq 3$, $0 \leq z \leq 4$, oriented in the positive x-direction.

12. $\vec{F} = x^2\vec{i} + (x + e^y)\vec{j} - \vec{k}$, and S is the rectangle $y = -1$, $0 \leq x \leq 2$, $0 \leq z \leq 4$, oriented in the negative y-direction.

13. $\vec{F} = (5 + xy)\vec{i} + z\vec{j} + yz\vec{k}$ and S is the 2×2 square plate in the yz-plane centered at the origin, oriented in the positive x-direction.

14. $\vec{F} = x\vec{i} + y\vec{j}$ and S is the surface of a closed cylinder of radius 2 and height 3 centered on the z-axis with its base in the xy-plane.

15. $\vec{F} = -y\vec{i} + x\vec{j} + z\vec{k}$ and S is the surface of a closed cylinder of radius 1 centered on the z-axis with base in the plane $z = -1$ and top in the plane $z = 1$.

16. $\vec{F} = x^2\vec{i} + y^2\vec{j} + z\vec{k}$ and S is the cone $z = \sqrt{x^2 + y^2}$, oriented upward with $x^2 + y^2 \leq 1$, $x \geq 0$, $y \geq 0$.

17. Suppose that $\vec{E}$ is a *uniform* electric field on 3-space, so $\vec{E}(x, y, z) = a\vec{i} + b\vec{j} + c\vec{k}$, for all points (x, y, z), where a, b, c are constants. Show, with the aid of symmetry, that the flux of $\vec{E}$ through each of the following closed surfaces S is zero:

    (a) S is the cube bounded by the planes $x = \pm 1$, $y = \pm 1$, and $z = \pm 1$
    (b) S is the sphere $x^2 + y^2 + z^2 = 1$
    (c) S is the cylinder bounded by $x^2 + y^2 = 1$, $z = 0$, and $z = 2$

18. According to Coulomb's Law the electrostatic field $\vec{E}$, at the point with position vector $\vec{r}$ in 3-space, due to a charge q at the origin is given by

$$\vec{E}(\vec{r}) = q\frac{\vec{r}}{\|\vec{r}\|^3}.$$

    Let S_a be the outward oriented sphere in 3-space of radius $a > 0$ and center at the origin. Show that the flux of the resulting electric field $\vec{E}$ through the surface S_a is equal to $4\pi q$, for any radius a. This is Gauss's Law for a single point charge.

19. An infinitely long straight wire lying along the z-axis carries an electric current I flowing in the $\vec{k}$ direction. Ampère's Law in magnetostatics says that the current gives rise to a magnetic field $\vec{B}$ given by

$$\vec{B}(x, y, z) = \frac{I}{2\pi}\frac{-y\vec{i} + x\vec{j}}{x^2 + y^2}.$$

    (a) Sketch the field $\vec{B}$ in the xy-plane.
    (b) Suppose S_1 is a disk with center at $(0, 0, h)$, radius a, and parallel to the xy-plane, oriented in the $\vec{k}$ direction. What is the flux of $\vec{B}$ through S_1? Is your answer reasonable?
    (c) Suppose S_2 is the rectangle given by $x = 0$, $a \leq y \leq b$, $0 \leq z \leq h$, and oriented in the $-\vec{i}$ direction. What is the flux of $\vec{B}$ through S_2? Does your answer seem reasonable?

20. An *ideal electric dipole* in electrostatics is characterized by its position in 3-space and its dipole moment vector $\vec{p}$. The electric field $\vec{D}$, at the point with position vector $\vec{r}$, of an ideal electric dipole located at the origin with dipole moment $\vec{p}$ is given by

$$\vec{D}(\vec{r}) = 3\frac{(\vec{r} \cdot \vec{p})\vec{r}}{\|\vec{r}\|^5} - \frac{\vec{p}}{\|\vec{r}\|^3}.$$

    Assume $\vec{p} = p\vec{k}$, so the dipole points in the $\vec{k}$ direction and has magnitude p.

    (a) What is the flux of $\vec{D}$ through a sphere S with center at the origin and radius $a > 0$?

(b)   The field $\vec{D}$ is a useful approximation to the electric field $\vec{E}$ produced by two "equal and opposite" charges, q at $\vec{r}_2$ and $-q$ at $\vec{r}_1$, where the distance $\|\vec{r}_2 - \vec{r}_1\|$ is small. The dipole moment of this configuration of charges is defined to be $q(\vec{r}_2 - \vec{r}_1)$. Gauss's Law in electrostatics says that the flux of $\vec{E}$ through S is equal to 4π times the total charge enclosed by S. What is the flux of $\vec{E}$ through S if the charges at $\vec{r}_1$ and $\vec{r}_2$ are enclosed by S? How does this compare with your answer for the flux of $\vec{D}$ through S if $\vec{p} = q(\vec{r}_2 - \vec{r}_1)$?

PROJECTS

1. **Gauss' Law Applied to a Charged Wire and a Charged Sheet**

    Gauss' Law states that the flux of an electric field through a closed surface, S, is proportional to the quantity of charge, q, enclosed within S. That is,

    $$\int_S \vec{E} \cdot d\vec{A} = kq.$$

    In this project we use Gauss' Law to calculate the electric field of a uniformly charged wire in part (a) and a flat sheet of charge in part (b).

    (a)   Consider the electric field due to an infinitely long, straight, uniformly charged wire. (There is no current running through the wire—all charges are fixed.) Assuming that the wire is infinitely long means that we can assume that the electric field is perpendicular to any cylinder that has the wire as an axis and that the magnitude of the field is constant on any such cylinder. Denote by E_r the magnitude of the electric field due to the wire on a cylinder of radius r. (See Figure 19.29.)

    Imagine a closed surface S made up of two cylinders, one of radius a and one of larger radius b, both coaxial with the wire, and the two washers that cap the ends. (See Figure 19.30.) The outward orientation of S means that a normal on the outer cylinder points away from the wire and a normal on the inner cylinder points towards the wire.

    (i)   Explain why the flux of $\vec{E}$, the electric field, through the washers is 0.

    (ii)   Explain why Gauss' Law implies that the flux through the inner cylinder is the same as the flux through the outer cylinder. [Hint: The charge on the wire is not inside the surface S].

    (iii) Use part (ii) to show that $E_b/E_a = a/b$.

    (iv)   Explain why part (iii) shows that the strength of the field due to an infinitely long uniformly charged wire is proportional to $1/r$.

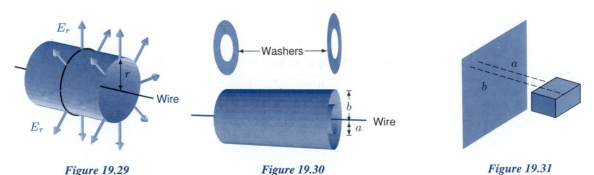

Figure 19.29   Figure 19.30   Figure 19.31

    (b)   Now consider an infinite flat sheet uniformly covered with charge. As in part (a), symmetry shows that the electric field $\vec{E}$ is perpendicular to the sheet and has the same magnitude at all points that are the same distance from the sheet. Use Gauss' Law to explain why, on any one side of the sheet, the electric field is the same at all points in space off the sheet. [Hint: Consider the flux through the box with sides parallel to the sheet shown in Figure 19.31.]

2. **Flux Across a Cylinder Due to a Point Charge: Obtaining Gauss' Law from Coulomb's Law**

    An electric charge q is placed at the origin in 3-space. The induced electric field $\vec{E}\,(\vec{r}\,)$ at the point with position vector $\vec{r}$ is given by Coulomb's Law, which says

    $$\vec{E}\,(\vec{r}\,) = q\frac{\vec{r}}{\|\vec{r}\,\|^3}, \qquad \vec{r} \neq \vec{0}.$$

    In this project, Gauss' Law is obtained for a cylinder enclosing a point charge by direct calculation from Coulomb's Law.

    (a) Let S be the open cylinder of height $2H$ and radius R given by $x^2 + y^2 = R^2$, $-H \leq z \leq H$, oriented outward.

        (i) Show that the flux of $\vec{E}$, the electric field, through S is given by

            $$\int_S \vec{E} \cdot d\vec{A} = 4\pi q\frac{H}{\sqrt{H^2 + R^2}}.$$

        (ii) What are the limits of the flux $\int_S \vec{E} \cdot d\vec{A}$ if

            • $H \to 0$ or $H \to \infty$ when R is fixed?
            • $R \to 0$ or $R \to \infty$ when H is fixed?

    (b) Let T be the outward oriented, closed cylinder of height $2H$ and radius R whose curved side is given by $x^2 + y^2 = R^2$, $-H \leq z \leq H$, whose top is given by $z = H$, $x^2 + y^2 \leq R^2$, and bottom by $z = -H$, $x^2 + y^2 \leq R^2$. Use part (a) to show that the flux of the electric field, $\vec{E}$, through T is given by

        $$\int_T \vec{E} \cdot d\vec{A} = 4\pi q.$$

        Notice that this is Gauss' Law. In particular, the flux is independent of both the height, H, and radius, R, of the cylinder.

CHAPTER TWENTY

CALCULUS OF VECTOR FIELDS

We have seen two ways of integrating vector fields in three dimensions: along curves and over surfaces. Now we will look at two ways of differentiating them. If we view the vector field as the velocity field of a fluid flow, then one method of differentiation (the divergence) tells us about the net strength of outflow from a point, and the other method (the curl) tells us about the strength of rotation around a point. Each method fits together with one of the ways of integrating to form a vector analogue of the Fundamental Theorem of Calculus: the Divergence Theorem relating the divergence to flux, and Stokes' theorem relating the curl to circulation around a closed path.

20.1 THE DIVERGENCE OF A VECTOR FIELD

Imagine that the vector fields in Figures 20.1 and 20.2 are velocity vector fields describing the flow of a fluid.[1] Figure 20.1 suggests outflow from the origin; for example, it could represent the expanding cloud of matter in the big bang theory of the origin of the universe. We say that the origin is a *source*. Figure 20.2 suggests flow into the origin; in this case we say that the origin is a *sink*.

In this section we will use the flux out of a closed surface surrounding a point to measure the outflow per unit volume there, also called the *divergence*, or *flux density*.

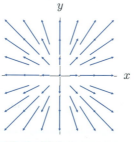

Figure 20.1: Vector field showing a source

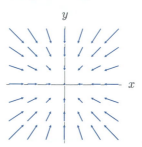

Figure 20.2: Vector field showing a sink

Definition of Divergence

To measure the outflow per unit volume of a vector field at a point, we calculate the flux out of a small sphere centered at the point, divide by the volume enclosed by the sphere, then take the limit of this flux to volume ratio as the sphere contracts around the point.

Geometric Definition of Divergence

The **divergence**, or **flux density**, of a smooth vector field $\vec{F}$, written **div** $\vec{F}$, is a scalar-valued function defined by

$$\operatorname{div}\vec{F}(x, y, z) = \lim_{\text{Volume}\to 0} \frac{\int_S \vec{F} \cdot d\vec{A}}{\text{Volume of } S}.$$

Here S is a sphere centered at (x, y, z), oriented outwards, that contracts down to (x, y, z) in the limit.

The limit can be computed using other shapes as well, such as the cubes in Example 2.

Cartesian Coordinate Definition of Divergence

If $\vec{F} = F_1\vec{i} + F_2\vec{j} + F_3\vec{k}$, then

$$\operatorname{div}\vec{F} = \frac{\partial F_1}{\partial x} + \frac{\partial F_2}{\partial y} + \frac{\partial F_3}{\partial z}.$$

Example 1   Calculate the divergence of $\vec{F}(\vec{r}) = \vec{r}$ at the origin

(a)   Using the geometric definition.

(b)   Using the Cartesian coordinate definition.

[1] Although not all vector fields represent physically realistic fluid flows, it is useful to think of them this way.

Solution   (a)   Using the method of Example 4 on page 880, you can calculate the flux of $\vec{F}$ out of the sphere of radius a, centered at the origin; it is $4\pi a^3$. So we have

$$\operatorname{div}\vec{F}(0,0,0) = \lim_{a \to 0}\frac{\text{Flux}}{\text{Volume}} = \lim_{a \to 0}\frac{4\pi a^3}{\frac{4}{3}\pi a^3} = \lim_{a \to 0}3 = 3.$$

(b)   In coordinates, $\vec{F}(x,y,z) = x\vec{i} + y\vec{j} + z\vec{k}$, so

$$\operatorname{div}\vec{F} = \frac{\partial}{\partial x}(x) + \frac{\partial}{\partial y}(y) + \frac{\partial}{\partial z}(z) = 1 + 1 + 1 = 3.$$

The next example shows that the divergence can be negative if there is net inflow to a point.

Example 2   (a)   Using the geometric definition, find the divergence of $\vec{v} = -x\vec{i}$ at: (i) $(0,0,0)$ (ii) $(2,2,0)$.
(b)   Confirm that the coordinate definition gives the same results.

Solution   (a)   (i)   The vector field $\vec{v} = -x\vec{i}$ is parallel to the x axis, as shown in Figure 20.3. To compute the flux density at $(0,0,0)$, we use a cube S_1, centered at the origin with edges parallel to the axes, of length $2c$. Then the flux through the faces perpendicular to the y- and z-axes is zero (because the vector field is parallel to these faces). On the faces perpendicular to the x-axis, the vector field and the outward normal are parallel but point in opposite directions. On the face at $x = c$, we have

$$\vec{v} \cdot \Delta\vec{A} = -c\,\|\Delta\vec{A}\|.$$

On the face at $x = -c$, the dot product is still negative, and

$$\vec{v} \cdot \Delta\vec{A} = -c\,\|\Delta\vec{A}\|.$$

Therefore, the flux through the cube is given by

$$\int_{S_1}\vec{v} \cdot d\vec{A} = \int_{\text{Face }x=-c}\vec{v} \cdot d\vec{A} + \int_{\text{Face }x=c}\vec{v} \cdot d\vec{A}$$
$$= -c \cdot \text{Area of one face} + (-c) \cdot \text{Area of other face} = -2c(2c)^2 = -8c^3.$$

Thus,

$$\operatorname{div}\vec{v}(0,0,0) = \lim_{\text{Volume} \to 0}\frac{\int_S\vec{v} \cdot d\vec{A}}{\text{Volume of cube}} = \lim_{c \to 0}\left(\frac{-8c^3}{(2c)^3}\right) = -1.$$

Since the vector field points inward towards the yz-plane, it makes sense that the divergence is negative at the origin.

(ii)   Take S_2 to be a cube as before, but centered this time at the point $(2,2,0)$. See Figure 20.3. As before, the flux through the faces perpendicular to the y- and z-axes is zero. On the face at $x = 2 + c$,
$$\vec{v} \cdot \Delta\vec{A} = -(2 + c)\,\|\Delta\vec{A}\|.$$

On the face at $x = 2 - c$ with outward normal, the dot product is positive, and

$$\vec{v} \cdot \Delta\vec{A} = (2 - c)\,\|\Delta\vec{A}\|.$$

Therefore, the flux through the cube is given by

$$\int_{S_2}\vec{v} \cdot d\vec{A} = \int_{\text{Face }x=2-c}\vec{v} \cdot d\vec{A} + \int_{\text{Face }x=2+c}\vec{v} \cdot d\vec{A}$$
$$= (2 - c) \cdot \text{Area of one face} - (2 + c) \cdot \text{Area of other face} = -2c(2c)^2 = -8c^3.$$

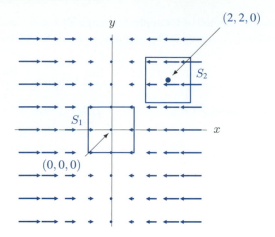

Figure 20.3: Vector field $\vec{v} = -x\vec{i}$

Then, as before,

$$\text{div}\,\vec{v}\,(2, 2, 0) = \lim_{\text{Volume}\to 0} \frac{\int_S \vec{v}\cdot d\vec{A}}{\text{Volume of cube}} = \lim_{c\to 0}\left(\frac{-8c^3}{(2c)^3}\right) = -1.$$

Although the vector field is flowing away from the point $(2, 2, 0)$ on the left, this outflow is smaller in magnitude than the inflow on the right, so the the net outflow is negative.

(b)   Since $\vec{v} = -x\vec{i} + 0\vec{j} + 0\vec{k}$, the formula gives

$$\text{div}\,\vec{v} = \frac{\partial}{\partial x}(-x) + \frac{\partial}{\partial y}(0) + \frac{\partial}{\partial z}(0) = -1 + 0 + 0 = -1.$$

Why Do the Two Definitions of Divergence Give the Same Result?

The geometric definition defines $\text{div}\,\vec{F}$ as the flux density of $\vec{F}$. To see why the coordinate definition is also the flux density, imagine computing the flux out of a small box-shaped surface S at (x_0, y_0, z_0), with sides of length Δx, Δy, and Δz parallel to the axes. On S_1 (the back face of the box shown in Figure 20.4, where $x = x_0$), the outward normal is in the negative x-direction, so $d\vec{A} = -dy\,dz\,\vec{i}$. Assuming $\vec{F}$ is approximately constant on S_1, we have

$$\int_{S_1} \vec{F}\cdot d\vec{A} = \int_{S_1} \vec{F}\cdot(-\vec{i})\,dy\,dz \approx -F_1(x_0, y_0, z_0)\int_{S_1} dy\,dz$$
$$= -F_1(x_0, y_0, z_0)\cdot\text{Area of }S_1 = -F_1(x_0, y_0, z_0)\,\Delta y\,\Delta z.$$

On S_2, the face where $x = x_0 + \Delta x$, the outward normal points in the positive x-direction, so $d\vec{A} = dy\,dz\,\vec{i}$. Therefore,

$$\int_{S_2} \vec{F}\cdot d\vec{A} = \int_{S_2} \vec{F}\cdot\vec{i}\,dy\,dz \approx F_1(x_0 + \Delta x, y_0, z_0)\int_{S_2} dy\,dz$$
$$= F_1(x_0 + \Delta x, y_0, z_0)\cdot\text{Area of }S_2 = F_1(x_0 + \Delta x, y_0, z_0)\,\Delta y\,\Delta z.$$

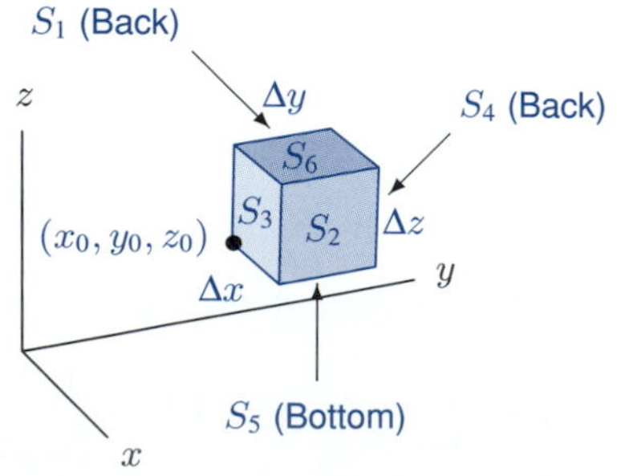

Figure 20.4: Box used to find div $\vec{F}$ at (x_0, y_0, z_0)

Thus

$$\int_{S_1} \vec{F} \cdot d\vec{A} + \int_{S_2} \vec{F} \cdot d\vec{A} \approx F_1(x_0 + \Delta x, y_0, z_0)\Delta y \Delta z - F_1(x_0, y_0, z_0)\Delta y \Delta z$$

$$= \frac{F_1(x_0 + \Delta x, y_0, z_0) - F_1(x_0, y_0, z_0)}{\Delta x}\Delta x \Delta y \Delta z$$

$$\approx \frac{\partial F_1}{\partial x}\Delta x \Delta y \Delta z.$$

By an analogous argument, the contribution to the flux from S_3 and S_4 (the surfaces perpendicular to the y-axis) is approximately

$$\frac{\partial F_2}{\partial y}\Delta x \, \Delta y \, \Delta z,$$

and the contribution to the flux from S_5 and S_6 is approximately

$$\frac{\partial F_3}{\partial z}\Delta x \, \Delta y \, \Delta z.$$

Thus, adding these contributions, we have

$$\text{Total flux through } S \approx \frac{\partial F_1}{\partial x}\Delta x \, \Delta y \, \Delta z + \frac{\partial F_2}{\partial y}\Delta x \, \Delta y \, \Delta z + \frac{\partial F_3}{\partial z}\Delta x \, \Delta y \, \Delta z.$$

Since the volume of the box is $\Delta x \, \Delta y \, \Delta z$, the flux density is

$$\frac{\text{Total flux through } S}{\text{Volume of box}} \approx \frac{\frac{\partial F_1}{\partial x}\Delta x \Delta y \Delta z + \frac{\partial F_2}{\partial y}\Delta x \Delta y \Delta z + \frac{\partial F_3}{\partial z}\Delta x \Delta y \Delta z}{\Delta x \Delta y \Delta z}$$

$$= \frac{\partial F_1}{\partial x} + \frac{\partial F_2}{\partial y} + \frac{\partial F_3}{\partial z}.$$

Divergence Free Vector Fields

A vector field $\vec{F}$ is said to be *divergence free* or *solenoidal* if $\text{div}\vec{F} = 0$ everywhere that $\vec{F}$ is defined.

Example 3   Figure 20.5 shows, for three values of the constant p, the vector field

$$\vec{E} = \frac{\vec{r}}{\|\vec{r}\|^p} \qquad \vec{r} \neq \vec{0}.$$

(a)   Find a formula for div $\vec{E}$.
(b)   Is there a value of p for which $\vec{E}$ is divergence free? If so, find it.

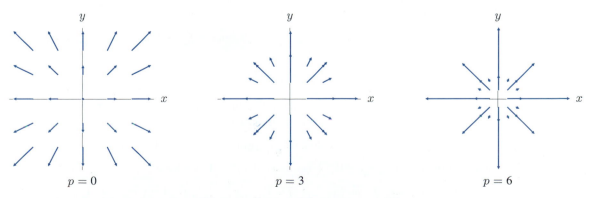

Figure 20.5: The vector field $\vec{E}(\vec{r}) = \vec{r}/\|\vec{r}\|^p$ for $p = 0$, 3, and 6

Solution　(a)　The components of $\vec{E}$ are

$$\vec{E} = \frac{x}{(x^2 + y^2 + z^2)^{p/2}}\vec{i} + \frac{y}{(x^2 + y^2 + z^2)^{p/2}}\vec{j} + \frac{z}{(x^2 + y^2 + z^2)^{p/2}}\vec{k}.$$

We compute the partial derivatives

$$\frac{\partial}{\partial x}\left(\frac{x}{(x^2 + y^2 + z^2)^{p/2}}\right) = \frac{1}{(x^2 + y^2 + z^2)^{p/2}} - \frac{px^2}{(x^2 + y^2 + z^2)^{(p/2)+1}}$$

$$\frac{\partial}{\partial y}\left(\frac{y}{(x^2 + y^2 + z^2)^{p/2}}\right) = \frac{1}{(x^2 + y^2 + z^2)^{p/2}} - \frac{py^2}{(x^2 + y^2 + z^2)^{(p/2)+1}}$$

$$\frac{\partial}{\partial z}\left(\frac{z}{(x^2 + y^2 + z^2)^{p/2}}\right) = \frac{1}{(x^2 + y^2 + z^2)^{p/2}} - \frac{pz^2}{(x^2 + y^2 + z^2)^{(p/2)+1}}.$$

So

$$\text{div}\,\vec{E} = \frac{3}{(x^2 + y^2 + z^2)^{p/2}} - \frac{p(x^2 + y^2 + z^2)}{(x^2 + y^2 + z^2)^{(p/2)+1}}$$

$$= \frac{3 - p}{(x^2 + y^2 + z^2)^{p/2}} = \frac{3 - p}{\|\vec{r}\|^p}.$$

(b)　The divergence is zero when $p = 3$, so $\vec{F}(\vec{r}) = \vec{r}/\|\vec{r}\|^3$ is a divergence free vector field.

Magnetic Fields

An important class of divergence free vector fields is the magnetic fields. One of Maxwell's Laws of Electromagnetism is that the magnetic field $\vec{B}$ satisfies

$$\text{div}\,\vec{B} = 0.$$

Example 4　An infinitesimal current loop, similar to that shown in Figure 20.6, is called a *magnetic dipole*. Its magnitude is described by a constant vector $\vec{\mu}$, called the dipole moment. The magnetic field due to a magnetic dipole with moment $\vec{\mu}$ is

$$\vec{B} = -\frac{\vec{\mu}}{\|\vec{r}\|^3} + \frac{3(\vec{\mu} \cdot \vec{r})\vec{r}}{\|\vec{r}\|^5}, \qquad \vec{r} \neq \vec{0}.$$

Show that $\text{div}\,\vec{B} = 0$.

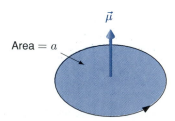

$\vec{\mu}$

Area $= a$

Figure 20.6: A current loop

Solution　To show that $\text{div}\,\vec{B} = 0$ we can use the following version of the product rule for the divergence: if g is a scalar function and $\vec{F}$ is a vector field, then

$$\text{div}(g\vec{F}) = (\text{grad}\,g) \cdot \vec{F} + g\,\text{div}\,\vec{F}.$$

(See Problem 12 on page 901.) Thus, since $\text{div}\,\vec{\mu} = 0$, we have

$$\text{div}\left(\frac{\vec{\mu}}{\|\vec{r}\|^3}\right) = \text{div}\left(\frac{1}{\|\vec{r}\|^3}\vec{\mu}\right) = \text{grad}\left(\frac{1}{\|\vec{r}\|^3}\right) \cdot \vec{\mu} + \frac{1}{\|\vec{r}\|^3} \cdot 0$$

and

$$\operatorname{div}\left(\frac{(\vec{\mu}\cdot\vec{r})\vec{r}}{\|\vec{r}\|^5}\right) = \operatorname{grad}(\vec{\mu}\cdot\vec{r})\cdot\frac{\vec{r}}{\|\vec{r}\|^5} + (\vec{\mu}\cdot\vec{r})\operatorname{div}\left(\frac{\vec{r}}{\|\vec{r}\|^5}\right).$$

From Problems 18 and 19 on page 686 and Example 3 on page 899 we have

$$\operatorname{grad}\left(\frac{1}{\|\vec{r}\|^3}\right) = \frac{-3\vec{r}}{\|\vec{r}\|^5}, \qquad \operatorname{grad}(\vec{\mu}\cdot\vec{r}) = \vec{\mu}, \qquad \operatorname{div}\left(\frac{\vec{r}}{\|\vec{r}\|^5}\right) = \frac{-2}{\|\vec{r}\|^5}.$$

Putting these results together gives

$$\begin{aligned}
\operatorname{div}\vec{B} &= -\operatorname{grad}\left(\frac{1}{\|\vec{r}\|^3}\right)\cdot\vec{\mu} + 3\operatorname{grad}(\vec{\mu}\cdot\vec{r})\cdot\frac{\vec{r}}{\|\vec{r}\|^5} + 3(\vec{\mu}\cdot\vec{r})\operatorname{div}\left(\frac{\vec{r}}{\|\vec{r}\|^5}\right) \\
&= \frac{3\vec{r}\cdot\vec{\mu}}{\|\vec{r}\|^5} + \frac{3\vec{\mu}\cdot\vec{r}}{\|\vec{r}\|^5} - \frac{6\vec{\mu}\cdot\vec{r}}{\|\vec{r}\|^5} \\
&= 0.
\end{aligned}$$

Alternative Notation for Divergence

Using $\nabla = \dfrac{\partial}{\partial x}\vec{i} + \dfrac{\partial}{\partial y}\vec{j} + \dfrac{\partial}{\partial z}\vec{k}$, we can write

$$\operatorname{div}\vec{F} = \nabla\cdot\vec{F} = \left(\frac{\partial}{\partial x}\vec{i} + \frac{\partial}{\partial y}\vec{j} + \frac{\partial}{\partial z}\vec{k}\right)\cdot(F_1\vec{i} + F_2\vec{j} + F_3\vec{k}) = \frac{\partial F_1}{\partial x} + \frac{\partial F_2}{\partial y} + \frac{\partial F_3}{\partial z}.$$

Problems for Section 20.1

1. Draw two vector fields that have positive divergence everywhere.
2. Draw two vector fields that have negative divergence everywhere.
3. Draw two vector fields that have zero divergence everywhere.

In Problems 4–10, find the divergence of the given vector field. (Note: $\vec{r} = x\vec{i} + y\vec{j} + z\vec{k}$.)

4. $\vec{F}(x,y) = -x\vec{i} + y\vec{j}$

5. $\vec{F}(x,y) = -y\vec{i} + x\vec{j}$

6. $\vec{F}(x,y) = (x^2 - y^2)\vec{i} + 2xy\vec{j}$

7. $\vec{F}(\vec{r}) = \vec{a}\times\vec{r}$

8. $\vec{F}(x,y) = \dfrac{-y\vec{i} + x\vec{j}}{x^2 + y^2}$

9. $\vec{F}(\vec{r}) = \dfrac{\vec{r} - \vec{r}_0}{\|\vec{r} - \vec{r}_0\|}, \quad \vec{r}\neq\vec{r}_0$

10. $\vec{F}(x,y,z) = (-x+y)\vec{i} + (y+z)\vec{j} + (-z+x)\vec{k}$

11. Show that if $\vec{a}$ is a constant vector and $f(x,y,z)$ is a function, then $\operatorname{div}(f\vec{a}) = (\operatorname{grad}f)\cdot\vec{a}$.

12. Show that if $g(x,y,z)$ is a scalar valued function and $\vec{F}(x,y,z)$ is a vector field, then

$$\operatorname{div}(g\vec{F}) = (\operatorname{grad}g)\cdot\vec{F} + g\operatorname{div}\vec{F}.$$

In Problems 13–15, use Problem 12 with $\vec{r} = x\vec{i} + y\vec{j} + z\vec{k}$ to find the divergence of the given vector field.

13. $\vec{F}(\vec{r}) = \dfrac{1}{\|\vec{r}\|^p}\vec{a}\times\vec{r}$

14. $\vec{B} = \dfrac{1}{x^a}\vec{r}$

15. $\vec{G}(\vec{r}) = (\vec{b}\cdot\vec{r})\vec{a}\times\vec{r}$

16. Which of the following two vector fields has the greater divergence at the origin? Assume the scales are the same on each.

(a)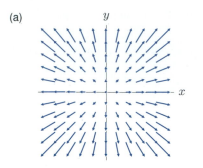

(b)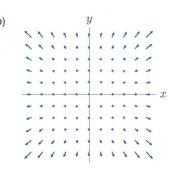

17. For each of the following vector fields, say whether the divergence is positive, zero, or negative at the indicated point.

(a)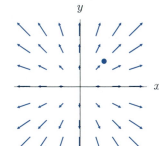

(b)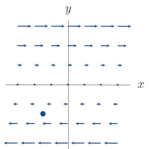

(c)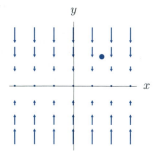

18. Let $\vec{F}(x, y, z) = z\vec{k}$.

   (a) Calculate div $\vec{F}$.
   (b) Sketch $\vec{F}$. Does it appear to be diverging? Does this agree with your answer to part (a)?

19. Let $\vec{F}(\vec{r}) = \vec{r}/\|\vec{r}\|^3$ (in 3-space), $\vec{r} \neq \vec{0}$.

   (a) Calculate div $\vec{F}$.
   (b) Sketch $\vec{F}$. Does it appear to be diverging? Does this agree with your answer to part (a)?

For Problems 20–22,

(a) Find the flux of the given vector field through a cube in the first octant with edge length c, one corner at the origin and edges along the axes.

(b) Use your answer to part (a) to find div $\vec{F}$ at the origin using the geometric definition.

(c) Compute div $\vec{F}$ at the origin using partial derivatives.

20. $\vec{F} = x\vec{i}$ 21. $\vec{F} = 2\vec{i} + y\vec{j} + 3\vec{k}$ 22. $\vec{F} = x\vec{i} + y\vec{j}$

23. (a) Find the flux of the vector field $\vec{F} = 2x\vec{i} - 3y\vec{j} + 5z\vec{k}$ through a box with four of its corners at the points $(a, b, c), (a + w, b, c), (a, b + w, c), (a, b, c + w)$ and edge length w. See Figure 20.7.
   (b) Use the geometric definition and part (a) to find div $\vec{F}$ at the point (a, b, c).
   (c) Find div $\vec{F}$ using partial derivatives.

24. Suppose $\vec{F} = (3x+2)\vec{i} + 4x\vec{j} + (5x+1)\vec{k}$. Use the method of Problem 23 to find div $\vec{F}$ at the point (a, b, c) by two different methods.

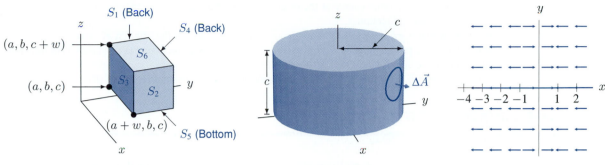

Figure 20.7   Figure 20.8   Figure 20.9

25. (a) Find the flux of the vector field $\vec{F} = x\vec{i} + y\vec{j}$ through the surface of the closed cylinder of radius c and height c, centered on the z-axis with base in the xy-plane. (See Figure 20.8.)
    (b) Use your answer to part (a) to find div $\vec{F}$ at the origin using the geometric definition.
    (c) Compute div $\vec{F}$ at the origin using partial derivatives.

Problems 26–27 involve electric fields. Electric charge produces a vector field $\vec{E}$, called the electric field, which represents the force on a unit positive charge placed at the point. Two positive or two negative charges repel one another, whereas two charges of opposite sign attract one another. The divergence of $\vec{E}$ is proportional to the density of the electric charge (that is, the charge per unit volume), with a positive constant of proportionality.

26. Suppose a certain distribution of electric charge produces the electric field shown in Figure 20.9. Where are the charges that produced this electric field concentrated? Which concentrations are positive and which are negative?

27. The electric field at the point $\vec{r}$ as a result of a point charge at the origin is $\vec{E}(\vec{r}) = k\vec{r}/\|\vec{r}\|^3$.

    (a) Calculate div $\vec{E}$ for $\vec{r} \neq \vec{0}$.
    (b) Calculate the limit suggested by the geometric definition of div $\vec{E}$ at the point $(0, 0, 0)$.
    (c) Explain what your answers mean in terms of charge density.

28. The divergence of a magnetic vector field $\vec{B}$ must be zero everywhere. Which of the following vector fields cannot be a magnetic vector field?

    (a) $\vec{B}(x, y, z) = -y\vec{i} + x\vec{j} + (x+y)\vec{k}$
    (b) $\vec{B}(x, y, z) = -z\vec{i} + y\vec{j} + x\vec{k}$
    (c) $\vec{B}(x, y, z) = (x^2 - y^2 - x)\vec{i} + (y - 2xy)\vec{j}$

29. If $f(x, y, z)$ and $g(x, y, z)$ are functions with continuous second partial derivatives, show that

$$\text{div}(\text{grad}\, f \times \text{grad}\, g) = 0.$$

30. In Problem 22 on page 884 it was shown that the rate of heat loss from a volume V in a region of non-uniform temperature equals $k \int_S (\text{grad}\, T) \cdot d\vec{A}$, where k is a constant, S is the surface bounding V, and $T(x, y, z)$ is the temperature at the point (x, y, z) in space. By taking the limit as V contracts to a point, show that

$$\frac{\partial T}{\partial t} = B\, \text{div}\, \text{grad}\, T$$

at that point, where B is a constant with respect to x, y, z, but may depend on time, t.

31. A vector field in the plane is a *point source* at the origin if its direction is away from the origin at every point, its magnitude depends only on the distance from the origin, and its divergence is zero away from the origin.

   (a) Explain why a point source at the origin must be of the form $\vec{v} = \left[f(x^2 + y^2) \right] (x\vec{i} + y\vec{j})$ for some positive function f.
   (b) Show that $\vec{v} = K(x^2 + y^2)^{-1}(x\vec{i} + y\vec{j})$ is a point source at the origin if $K > 0$.
   (c) Determine the magnitude $\|\vec{v}\|$ of the source in part (b) as a function of the distance from its center.
   (d) Sketch the vector field $\vec{v} = (x^2 + y^2)^{-1}(x\vec{i} + y\vec{j})$.
   (e) Show that $\phi = \frac{K}{2} \log(x^2 + y^2)$ is a potential function for the source in part (b).

20.2 THE DIVERGENCE THEOREM

The Divergence Theorem is a multivariable analogue of the Fundamental Theorem of Calculus; it says that the integral of the flux density over a solid region equals the flux integral through the boundary of the region.

The Boundary of a Solid Region

The boundary of a solid region may be thought of as the skin between the interior of the region and the space around it. For example, the boundary of a solid ball is a spherical surface, the boundary of a solid cube is its six faces, and the boundary of a solid cylinder is a tube sealed at both ends by disks. (See Figure 20.10). A surface which is the boundary of a solid region is called a *closed surface*.

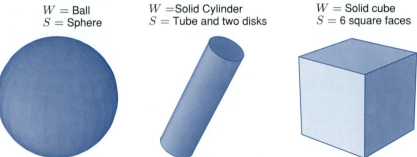

$W =$ Ball
$S =$ Sphere

$W =$ Solid Cylinder
$S =$ Tube and two disks

$W =$ Solid cube
$S =$ 6 square faces

Figure 20.10: Several solid regions and their boundaries

Calculating the Flux from the Flux Density

Consider a solid region W in 3-space whose boundary is the closed surface S. There are two ways to find the total flux of a vector field $\vec{F}$ out of W. One is to calculate the flux of $\vec{F}$ through S:

$$\text{Flux out of } W = \int_S \vec{F} \cdot d\vec{A}.$$

Another way is to use div $\vec{F}$, which gives the flux density at any point in W. We subdivide W into small boxes, as shown in Figure 20.11. Then, for a small box of volume ΔV,

$$\text{Flux out of box} \approx \text{Flux density} \cdot \text{Volume} = \text{div } \vec{F} \, \Delta V.$$

What happens when we add the fluxes out of all the boxes? Consider two adjacent boxes, as shown in Figure 20.12. The flux through the shared wall is counted twice, once out of the box on each side. When we add the fluxes, these two contributions cancel, so we get the flux out of the solid region formed by joining the two boxes. Continuing in this way, we find that

$$\text{Flux out of } W = \sum \text{Flux out of small boxes} \approx \sum \text{div } \vec{F} \, \Delta V.$$

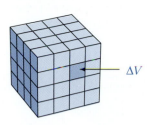

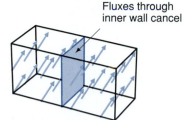

Fluxes through
inner wall cancel

Figure 20.11: Subdivision
of region into small boxes

Figure 20.12: Adding the flux out of
adjacent boxes

We have approximated the flux by a Riemann sum. As the subdivision gets finer, the sum approaches an integral, so

$$\text{Flux out of } W = \int_W \text{div } \vec{F} \, dV.$$

We have calculated the flux in two ways, as a flux integral and as a volume integral. Therefore, these two integrals must be equal. This result holds even if W is not a rectangular solid, as shown in Figure 20.11. Thus, we have the following result.[2]

The Divergence Theorem

If W is a solid region whose boundary S is a piecewise smooth surface, and if $\vec{F}$ is a smooth vector field which is defined everywhere in W and on S, then

$$\int_S \vec{F} \cdot d\vec{A} = \int_W \text{div } \vec{F} \, dV,$$

where S is given the outward orientation.

Example 1   Use the Divergence Theorem to calculate the flux of the vector field $\vec{F}(\vec{r}) = \vec{r}$ through the sphere of radius a centered at the origin.

Solution   In Example 4 on page 880 we computed the flux integral directly:

$$\int_S \vec{r} \cdot d\vec{A} = 4\pi a^3.$$

Now we use div $\vec{F} = 3$ and the Divergence Theorem:

$$\int_S \vec{r} \cdot d\vec{A} = \int_W \text{div } \vec{F} \, dV = \int_W 3 \, dV = 3\left(\frac{4}{3}\pi a^3\right) = 4\pi a^3.$$

Example 2   Use the Divergence Theorem to calculate the flux of the vector field

$$\vec{F}(x, y, z) = (x^2 + y^2)\vec{i} + (y^2 + z^2)\vec{j} + (x^2 + z^2)\vec{k}$$

through the cube in Figure 20.13.

---

[2]A proof of the Divergence Theorem using the coordinate definition of the divergence can be found in *Multivariable Calculus*, by William McCallum et al. (New York: Wiley, 1997).

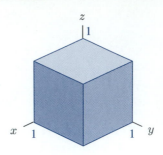

Figure 20.13

Solution   The divergence of $\vec{F}$ is div $\vec{F} = 2x + 2y + 2z$. Since div $\vec{F}$ is positive everywhere in the first quadrant, the flux through S is positive. By the Divergence Theorem,

$$\int_S \vec{F} \cdot d\vec{A} = \int_0^1 \int_0^1 \int_0^1 2(x + y + z) \, dx \, dy \, dz = \int_0^1 \int_0^1 x^2 + 2x(y + z) \Big|_0^1 \, dy \, dz$$

$$= \int_0^1 \int_0^1 1 + 2(y + z) \, dy \, dz = \int_0^1 y + y^2 + 2yz \Big|_0^1 \, dz$$

$$= \int_0^1 (2 + 2z) \, dz = 2z + z^2 \Big|_0^1 = 3.$$

The Divergence Theorem and Divergence Free Vector Fields

An important application of the Divergence Theorem is the study of divergence free vector fields.

Example 3   In Example 3 on page 899 we saw that the following vector field is divergence free:

$$\vec{F}(\vec{r}) = \frac{\vec{r}}{\|\vec{r}\|^3}, \qquad \vec{r} \neq \vec{0}.$$

Calculate $\int_S \vec{F} \cdot d\vec{A}$, using the Divergence Theorem if possible, for the following surfaces:

(a) S_1 is the sphere of radius a centered at the origin.
(b) S_2 is the sphere of radius a centered at the point $(2a, 0, 0)$.

Solution   (a)   We cannot use the Divergence Theorem directly because $\vec{F}$ is not defined everywhere inside the sphere (it is not defined at the origin). Since $\vec{F}$ points outward everywhere on S_1, the flux out of S_1 is positive. On S_1,

$$\vec{F} \cdot d\vec{A} = \|\vec{F}\| dA = \frac{a}{a^3} dA,$$

so

$$\int_{S_1} \vec{F} \cdot d\vec{A} = \frac{1}{a^2} \int_{S_1} dA = \frac{1}{a^2} (\text{Area of } S_1) = \frac{1}{a^2} 4\pi a^2 = 4\pi.$$

Notice that the flux is not zero, although div $\vec{F}$ is zero everywhere it is defined.

(b)   Suppose W is the solid region enclosed by S_2. Since div $\vec{F} = 0$ everywhere in W, we can use the Divergence Theorem in this case, giving

$$\int_{S_2} \vec{F} \cdot d\vec{A} = \int_W \text{div} \, \vec{F} \, dV = \int_W 0 \, dV = 0.$$

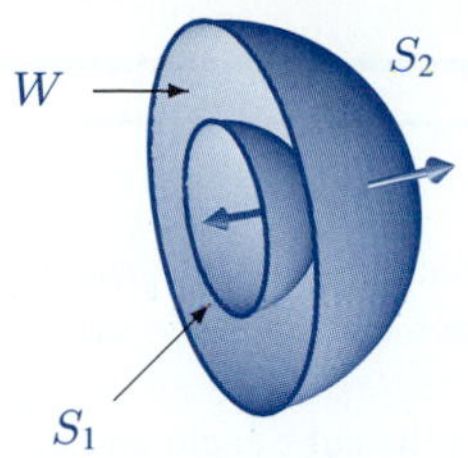

Figure 20.14: Cut-away view of the region W between two spheres, showing orientation vectors

The Divergence Theorem applies to any solid region W and its boundary S, even in cases where the boundary consists of two or more surfaces. For example, if W is the solid region between the sphere S_1 of radius 1 and the sphere S_2 of radius 2, both centered at the same point, then the boundary of W consists of both S_1 and S_2. The Divergence Theorem requires the outward orientation, which on S_2 points away from the center and on S_1 points towards the center. (See Figure 20.14.)

Example 4   Let S_1 be the sphere of radius 1 centered at the origin and let S_2 be the ellipsoid $x^2 + y^2 + 4z^2 = 16$, both oriented outward. For

$$\vec{F}(\vec{r}) = \frac{\vec{r}}{\|\vec{r}\|^3}, \qquad \vec{r} \neq \vec{0},$$

show that

$$\int_{S_1} \vec{F} \cdot d\vec{A} = \int_{S_2} \vec{F} \cdot d\vec{A}.$$

Solution   The ellipsoid contains the sphere; let W be the solid region between them. Since W does not contain the origin, div $\vec{F}$ is defined and equal to zero everywhere in W. Thus, if S is the boundary of W, then

$$\int_{S} \vec{F} \cdot d\vec{A} = \int_{W} \text{div} \, \vec{F} \, dV = 0.$$

But S consists of S_2 oriented outwards and S_1 oriented inwards, so

$$0 = \int_{S} \vec{F} \cdot d\vec{A} = \int_{S_2} \vec{F} \cdot d\vec{A} - \int_{S_1} \vec{F} \cdot d\vec{A},$$

and thus

$$\int_{S_2} \vec{F} \cdot d\vec{A} = \int_{S_1} \vec{F} \cdot d\vec{A}.$$

In Example 3 we showed that $\int_{S_1} \vec{F} \cdot d\vec{A} = 4\pi$, so $\int_{S_2} \vec{F} \cdot d\vec{A} = 4\pi$ also. Note that it would have been more difficult to compute the integral over the ellipsoid directly.

Electric Fields

The electric field produced by a positive point charge q placed at the origin is

$$\vec{E} = q \frac{\vec{r}}{\|\vec{r}\|^3}.$$

Using Example 3 we see that the flux of the electric field through any sphere centered at the origin is $4\pi q$. In fact, using the idea of Example 4, we can show that the flux of $\vec{E}$ through any simple closed surface containing the origin is $4\pi q$. See Problems 18 and 19 on page 923. This is a special case of Gauss's Law, which states that the flux of an electric field through any closed surface is proportional to the total charge enclosed by the surface. Carl Friedrich Gauss (1777-1855) also discovered the Divergence Theorem, which is sometimes called Gauss's Theorem.

Problems for Section 20.2

For Problems 1–3, compute the flux integral $\int_S \vec{F} \cdot d\vec{A}$ in two ways, if possible, directly and using the Divergence Theorem. In each case, S is closed and oriented outwards.

1. $\vec{F}(\vec{r}) = \vec{r}$ and S is the cube enclosing the volume $0 \le x \le 2, 0 \le y \le 2$, and $0 \le z \le 2$.

2. $\vec{F}(x, y, z) = y\vec{j}$ and S is a vertical cylinder of height 2, with its base a circle of radius 1 on the xy-plane, centered at the origin. S includes the disks that close it off at top and bottom.

3. $\vec{F}(x, y, z) = -z\vec{i} + x\vec{k}$ and S is a square pyramid with height 3 and base on the xy-plane of side length 1.

4. Suppose V_1 and V_2 are the rectangular solids in the first quadrant shown in Figure 20.15. Both have sides of length 1 parallel to the axes; V_1 has one corner at the origin, while V_2 has the corresponding corner at the point $(1, 0, 0)$. Suppose S_1 and S_2 are the six-faced surfaces of V_1 and V_2, respectively. Suppose V is the box-shaped volume consisting of V_1 and V_2 together, and having outside surface S. Are the following true or false? Give reasons.

    (a) If $\vec{F}$ is a constant vector field, $\int_S \vec{F} \cdot d\vec{A} = 0$

    (b) If S_1, S_2 and S are all oriented outward and $\vec{F}$ is any vector field

$$\int_S \vec{F} \cdot d\vec{A} = \int_{S_1} \vec{F} \cdot d\vec{A} + \int_{S_2} \vec{F} \cdot d\vec{A}.$$

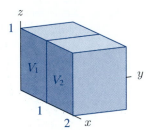

Figure 20.15

5. Calculate $\int_{S_2} \vec{F} \cdot d\vec{A}$ where $\vec{F} = x^2\vec{i} + 2y^2\vec{j} + 3z^2\vec{k}$ and S_2 is as in Problem 4. Do this directly and using the Divergence Theorem.

6. Use the Divergence Theorem to evaluate the flux integral $\int_S (x^2\vec{i} + (y - 2xy)\vec{j} + 10z\vec{k}) \cdot d\vec{A}$, where S is the sphere of radius 5 centered at the origin, oriented outward.

7. Use the Divergence Theorem to calculate the flux of the vector field $\vec{F}(x, y, z) = -z\vec{i} + x\vec{k}$ through the sphere of radius a centered at the origin. Give a geometric explanation for your answer.

8. Suppose $\vec{F}$ is a vector field with div $\vec{F} = 10$. Find the flux of $\vec{F}$ out of a cylinder of height a and radius a, centered on the z-axis and with base in the xy-plane.

9. Consider the vector field $\vec{F} = \vec{r}/\|\vec{r}\|^3$.

    (a) Calculate div $\vec{F}$ for $\vec{r} \ne \vec{0}$.
    (b) Find the flux of $\vec{F}$ out of a box of side a centered at the origin and with edges parallel to the axes.

10. Suppose $\vec{G}$ is a vector field with the property that $\vec{G} = \vec{r}$ for $2 \le \|\vec{r}\| \le 7$ and that the flux of $\vec{G}$ through the sphere of radius 3 centered at the origin is 8π. Find the flux of $\vec{G}$ through the sphere of radius 5 centered at the origin.

11. Suppose that a vector field $\vec{F}$ satisfies div$\vec{F} = 0$ everywhere. Show that $\int_S \vec{F} \cdot d\vec{A} = 0$ for every closed surface S.

12. The gravitational field, $\vec{F}$, of a planet of mass m at the origin is given by

$$\vec{F} = -Gm\frac{\vec{r}}{\|\vec{r}\|^3}.$$

Use the Divergence Theorem to show that the flux of the gravitational field through the sphere of radius a is independent of a. [Hint: Consider the region bounded by two concentric spheres.]

13. At the point with position vector $\vec{r}$, the electric field $\vec{E}$ of an ideal electric dipole with moment $\vec{p}$ located at the origin is given by

$$\vec{E}(\vec{r}) = 3\frac{(\vec{r} \cdot \vec{p})\vec{r}}{\|\vec{r}\|^5} - \frac{\vec{p}}{\|\vec{r}\|^3}.$$

   (a) What is div $\vec{E}$?
   (b) Suppose S is an outward oriented, closed smooth surface surrounding the origin. Compute the flux of $\vec{E}$ through S. Can you use the Divergence Theorem directly to compute the flux? Explain why or why not. [Hint: First compute the flux of the dipole field $\vec{E}$ through an outward-oriented sphere S_a with center at the origin and radius a. Then apply the Divergence Theorem to the region W lying between S and a small sphere S_a.]

14. If a surface S is submerged in an incompressible fluid, a force $\vec{F}$ is exerted on one side of the surface by the pressure in the fluid. If we choose a coordinate system where the z-axis is vertical, with the positive direction upward and the fluid level at $z = 0$, then the component of force in the direction of a unit vector $\vec{u}$ is given by the following:

$$\vec{F} \cdot \vec{u} = -\int_S z\rho g\vec{u} \cdot d\vec{A},$$

where ρ is the density of the fluid (mass/volume), g is the acceleration due to gravity, and the surface is oriented away from the side on which the force is exerted. In this problem we will consider a totally submerged closed surface enclosing a volume V. We are interested in the force of the liquid on the external surface, so S is oriented inward.

   (a) Use the Divergence Theorem to show that the force in the $\vec{i}$ and $\vec{j}$ directions is zero.
   (b) Use the Divergence Theorem to show that the force in the $\vec{k}$ direction is $\rho g V$, the weight of the volume of fluid with the same volume as V. This is *Archimedes' Principle*.

20.3   THE CURL OF A VECTOR FIELD

The divergence of a vector field is a sort of scalar derivative which measures its outflow per unit volume. Now we introduce a vector derivative, the curl, which measures the circulation of a vector field. Imagine holding the paddle-wheel in Figure 20.16 in the flow shown by Figure 20.17. The speed at which the paddle-wheel spins (its angular velocity) measures the strength of circulation. Note that the angular velocity depends on the direction in which the stick is pointing. The paddle-wheel spins one way if the stick is pointing up and the opposite way if it is pointing down. If the stick is pointing horizontally it doesn't spin at all, because the velocity field strikes opposite vanes of the paddle with equal force.

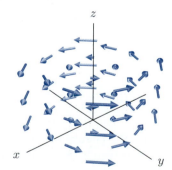

Figure 20.16: A device for measuring circulation

Figure 20.17: A vector field with circulation about the z-axis

Circulation Density

We measure the strength of the circulation using a closed curve. Suppose C is a circle with center (x, y, z) in the plane perpendicular to $\vec{n}$, traversed in the direction determined from $\vec{n}$ by the right-hand rule. (See Figures 20.18 and 20.19.)

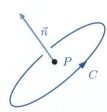

Figure 20.18: Direction of C relates to direction of $\vec{n}$ by the right-hand rule

Figure 20.19: When the thumb points in the direction of $\vec{n}$, the fingers curl in the forward direction around C

We make the following definition:

> The **circulation density** of a smooth vector field $\vec{F}$ at (x, y, z) around the direction of the unit vector $\vec{n}$ is defined to be
>
> $$\text{circ}_{\vec{n}} \, \vec{F} \, (x, y, z) = \lim_{\text{Area} \to 0} \frac{\text{Circulation around } C}{\text{Area inside } C} = \lim_{\text{Area} \to 0} \frac{\displaystyle\int_C \vec{F} \cdot d\vec{r}}{\text{Area inside } C},$$
>
> provided the limit exists.

The circulation density determines the angular velocity[3] of the paddle-wheel in Figure 20.16 provided you could make one sufficiently small and light and insert it without disturbing the flow.

Example 1   Consider the vector field $\vec{F}$ in Figure 20.17. Suppose that $\vec{F}$ is parallel to the xy-plane and that at a distance r from the z-axis it has magnitude $2r$. Calculate $\text{circ}_{\vec{n}} \, \vec{F}$ at the origin for
(a) $\vec{n} = \vec{k}$ (b) $\vec{n} = -\vec{k}$ (c) $\vec{n} = \vec{i}$.

Solution   (a) Take a circle C of radius a in the xy-plane, centered at the origin, traversed in a direction determined from $\vec{k}$ by the right hand rule. Then, since $\vec{F}$ is tangent to C everywhere and points in the forward direction around C, we have

$$\text{Circulation around } C = \int_C \vec{F} \cdot d\vec{r} = \|\vec{F}\| \cdot \text{Circumference of } C = 2a(2\pi a) = 4\pi a^2.$$

Thus, the circulation density is

$$\text{circ}_{\vec{k}} \, \vec{F} = \lim_{a \to 0} \frac{\text{Circulation around } C}{\text{Area inside } C} = \lim_{a \to 0} \frac{4\pi a^2}{\pi a^2} = 4.$$

(b) If $\vec{n} = -\vec{k}$ the circle is traversed in the opposite direction, so the line integral changes sign. Thus

$$\text{circ}_{-\vec{k}} \, \vec{F} = -4.$$

(c) The circulation around $\vec{i}$ is calculated using circles in the yz-plane. Since $\vec{F}$ is everywhere perpendicular to such a circle C,

$$\int_C \vec{F} \cdot d\vec{r} = 0.$$

Thus, we have

$$\text{circ}_{\vec{i}} \, \vec{F} = \lim_{a \to 0} \frac{\int_C \vec{F} \cdot d\vec{r}}{\pi a^2} = \lim_{a \to 0} \frac{0}{\pi a^2} = 0.$$

---

[3]In fact it is twice the angular velocity. See Example 3 on page 912.

Definition of the Curl

Example 1 shows that the circulation density of a vector field can be positive, negative, or zero, depending on the direction. We assume that there is one direction in which the circulation density is greatest. Now we define a single vector quantity that incorporates all these different circulation densities.

Geometric Definition of Curl

The curl of a smooth vector field $\vec{F}$, written curl $\vec{F}$, is the vector field with the following properties

- The direction of curl $\vec{F}(x, y, z)$ is the direction $\vec{n}$ for which $\mathrm{circ}_{\vec{n}}(x, y, z)$ is the greatest.

- The magnitude of curl $\vec{F}(x, y, z)$ is the circulation density of $\vec{F}$ around that direction.

If the circulation density is zero around every direction, then we define the curl to be $\vec{0}$.

Cartesian Coordinate Definition of Curl

If $\vec{F} = F_1\vec{i} + F_2\vec{j} + F_3\vec{k}$, then

$$\mathrm{curl}\,\vec{F} = \left(\frac{\partial F_3}{\partial y} - \frac{\partial F_2}{\partial z}\right)\vec{i} + \left(\frac{\partial F_1}{\partial z} - \frac{\partial F_3}{\partial x}\right)\vec{j} + \left(\frac{\partial F_2}{\partial x} - \frac{\partial F_1}{\partial y}\right)\vec{k}.$$

Example 2 For each field in Figure 20.20, use the sketch and the geometric definition to decide whether the curl at the origin points up, down, or is the zero vector. Then check your answer using the coordinate definition of curl. Note that the vector fields have no z-components and are independent of z.

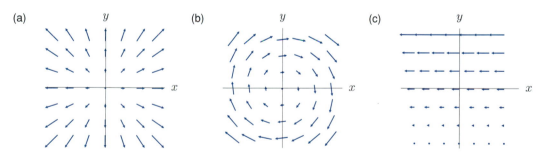

Figure 20.20: Sketches in the xy-plane of (a) $\vec{F} = x\vec{i} + y\vec{j}$ (b) $\vec{F} = y\vec{i} - x\vec{j}$ (c) $\vec{F} = -(y+1)\vec{i}$

Solution (a) This vector field shows no rotation, and the circulation around any closed curve appears to be zero, so we suspect that the curl is the zero vector. The coordinate definition of curl gives

$$\mathrm{curl}\,\vec{F} = \left(\frac{\partial(0)}{\partial y} - \frac{\partial y}{\partial z}\right)\vec{i} + \left(\frac{\partial x}{\partial z} - \frac{\partial(0)}{\partial x}\right)\vec{j} + \left(\frac{\partial y}{\partial x} - \frac{\partial x}{\partial y}\right)\vec{k} = \vec{0}.$$

(b) This vector field is rotating around the z-axis. By the right-hand rule, the circulation density around $\vec{k}$ is negative, so we expect the z-component of the curl points down. The coordinate definition gives

$$\mathrm{curl}\,\vec{F} = \left(\frac{\partial(0)}{\partial y} - \frac{\partial(-x)}{\partial z}\right)\vec{i} + \left(\frac{\partial y}{\partial z} - \frac{\partial(0)}{\partial x}\right)\vec{j} + \left(\frac{\partial(-x)}{\partial x} - \frac{\partial y}{\partial y}\right)\vec{k} = -2\vec{k}.$$

(c)   At first glance, you might expect this vector field to have zero curl, as all the vectors are parallel to the x-axis. However, if you find the circulation around the curve C in Figure 20.21, the sides contribute nothing (they are perpendicular to the vector field), the bottom contributes a negative quantity (the curve is in the opposite direction to the vector field), and the top contributes a larger positive quantity (the curve is in the same direction as the vector field and the magnitude of the vector field is larger at the top than at the bottom). Thus, the circulation around C is positive and hence we expect the curl to be nonzero and point up. The coordinate definition gives

$$\text{curl } \vec{F} = \left(\frac{\partial(0)}{\partial y} - \frac{\partial(0)}{\partial z} \right) \vec{i} + \left(\frac{\partial(-(y+1))}{\partial z} - \frac{\partial(0)}{\partial x} \right) \vec{j} + \left(\frac{\partial(0)}{\partial x} - \frac{\partial(-(y+1))}{\partial y} \right) \vec{k} = \vec{k}.$$

Another way to see that the curl is nonzero in this case is to imagine the vector field representing the velocity of moving water. A boat sitting in the water tends to rotate, as the water moves faster on one side than the other.

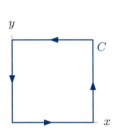

Figure 20.21

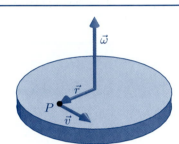

Figure 20.22: Rotating flywheel

Example 3   A flywheel is rotating with angular velocity $\vec{\omega}$ and the velocity of a point P with position vector $\vec{r}$ is given by $\vec{v} = \vec{\omega} \times \vec{r}$. (See Figure 20.22.) Calculate curl $\vec{v}$.

Solution   If $\vec{\omega} = \omega_1 \vec{i} + \omega_2 \vec{j} + \omega_3 \vec{k}$, we have

$$\vec{v} = \vec{\omega} \times \vec{r} = \begin{vmatrix} \vec{i} & \vec{j} & \vec{k} \\ \omega_1 & \omega_2 & \omega_3 \\ x & y & z \end{vmatrix} = (\omega_2 z - \omega_3 y)\vec{i} + (\omega_3 x - \omega_1 z)\vec{j} + (\omega_1 y - \omega_2 x)\vec{k}.$$

Thus,

$$\text{curl } \vec{v} = \begin{vmatrix} \vec{i} & \vec{j} & \vec{k} \\ \frac{\partial}{\partial x} & \frac{\partial}{\partial y} & \frac{\partial}{\partial z} \\ \omega_2 z - \omega_3 y & \omega_3 x - \omega_1 z & \omega_1 y - \omega_2 x \end{vmatrix}$$

$$= \left(\frac{\partial}{\partial y}(\omega_1 y - \omega_2 x) - \frac{\partial}{\partial z}(\omega_3 x - \omega_1 z) \right)\vec{i} + \left(\frac{\partial}{\partial z}(\omega_2 z - \omega_3 y) - \frac{\partial}{\partial x}(\omega_1 y - \omega_2 x) \right)\vec{j}$$

$$+ \left(\frac{\partial}{\partial x}(\omega_3 x - \omega_1 z) - \frac{\partial}{\partial y}(\omega_2 z - \omega_3 y) \right)\vec{k}$$

$$= 2\omega_1 \vec{i} + 2\omega_2 \vec{j} + 2\omega_3 \vec{k} = 2\vec{\omega}.$$

Thus, as we would expect, curl $\vec{v}$ is parallel to the axis of rotation of the flywheel (namely, the direction of $\vec{\omega}$) and the magnitude of curl $\vec{v}$ is larger the faster the flywheel is rotating (that is, the larger the magnitude of $\vec{\omega}$).

Why Do the Two Definitions of Curl Give the Same Result?

Using Green's Theorem in Cartesian coordinates, we can show that for curl $\vec{F}$ defined in Cartesian coordinates

$$\text{curl } \vec{F} \cdot \vec{n} = \text{circ}_{\vec{n}} \ \vec{F}.$$

This shows that curl $\vec{F}$ defined in Cartesian coordinates satisfies the geometric definition, since the left hand side takes its maximum value when $\vec{n}$ points in the same direction as curl $\vec{F}$, and in that case its value is $\| \text{curl } \vec{F} \|$.

The following example justifies this formula in a specific case. Problem 24 on page 916 shows how to prove curl $\vec{F} \cdot \vec{n} = \text{circ}_{\vec{n}} \ \vec{F}$ in general.

Example 4   Use the definition of curl in Cartesian coordinates and Green's Theorem to show that

$$\left(\text{curl } \vec{F} \right) \cdot \vec{k} = \text{circ}_{\vec{k}} \ \vec{F}.$$

Solution   Using the definition of curl in Cartesian coordinates, the left hand side of the formula is

$$\left(\text{curl } \vec{F} \right) \cdot \vec{k} = \frac{\partial F_2}{\partial x} - \frac{\partial F_1}{\partial y}.$$

Now let's look at the right hand side. The circulation density around $\vec{k}$ is calculated using circles perpendicular to $\vec{k}$; hence the $\vec{k}$-component of $\vec{F}$ does not contribute to it, that is, the circulation density of $\vec{F}$ around $\vec{k}$ is the same as the circulation density of $F_1 \vec{i} + F_2 \vec{j}$ around $\vec{k}$. But in any plane perpendicular to $\vec{k}$, z is constant, so in that plane F_1 and F_2 are functions of x and y alone. Thus $F_1 \vec{i} + F_2 \vec{j}$ can be thought of as a two-dimensional vector field on the horizontal plane through the point (x, y, z) where the circulation density is being calculated. Let C be a circle in this plane, with radius a and centered at (x, y, z), and let R be the region enclosed by C. Green's Theorem says that

$$\int_C (F_1 \vec{i} + F_2 \vec{j}) \cdot d\vec{r} = \int_R \left(\frac{\partial F_2}{\partial x} - \frac{\partial F_1}{\partial y} \right) dA.$$

When the circle is small, $\partial F_2 / \partial x - \partial F_1 / \partial y$ is approximately constant on R, so

$$\int_R \left(\frac{\partial F_2}{\partial x} - \frac{\partial F_1}{\partial y} \right) dA \approx \left(\frac{\partial F_2}{\partial x} - \frac{\partial F_1}{\partial y} \right) \cdot \text{Area of } R = \left(\frac{\partial F_2}{\partial x} - \frac{\partial F_1}{\partial y} \right) \pi a^2.$$

Thus, taking a limit as the radius of the circle goes to zero, we have

$$\text{circ}_{\vec{k}} \ \vec{F}(x, y, z) = \lim_{a \to 0} \frac{\int_C (F_1 \vec{i} + F_2 \vec{j}) \cdot d\vec{r}}{\pi a^2} = \lim_{a \to 0} \frac{\int_R \left(\frac{\partial F_2}{\partial x} - \frac{\partial F_1}{\partial y} \right) dA}{\pi a^2} = \frac{\partial F_2}{\partial x} - \frac{\partial F_1}{\partial y}.$$

Curl Free Vector Fields

A vector field is said to be *curl free* or *irrotational* if curl $\vec{F} = \vec{0}$ everywhere that $\vec{F}$ is defined.

Example 5   Figure 20.23 shows the vector field $\vec{B}$ for three values of the constant p, where $\vec{B}$ is defined on 3-space by

$$\vec{B} = \frac{-y\vec{i} + x\vec{j}}{(x^2 + y^2)^{p/2}}.$$

(a)   Find a formula for curl $\vec{B}$.
(b)   Is there a value of p for which $\vec{B}$ is curl free? If so, find it.

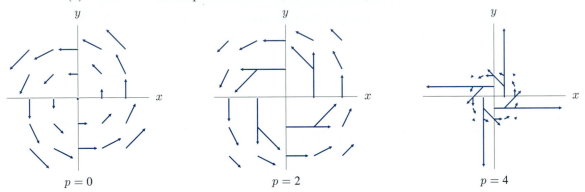

Figure 20.23: The vector field $\vec{B}\,(\vec{r}) = (-y\vec{i} + x\vec{j}\,)/(x^2 + y^2)^{p/2}$ for $p = 0, 2,$ and 4

Solution   (a)   We can use the following version of the product rule for curl. If ϕ is a scalar function and $\vec{F}$ is a vector field, then

$$\text{curl}(\phi\vec{F}\,) = \phi\,\text{curl}\,\vec{F}\, + (\text{grad}\,\phi) \times \vec{F}\, .$$

(See Problem 17 on page 915.) We write $\vec{B}\, = \phi\vec{F}\, = \dfrac{1}{(x^2 + y^2)^{p/2}}(-y\vec{i} + x\vec{j}\,)$. Then

$$\text{curl}\,\vec{F}\, = \text{curl}(-y\vec{i} + x\vec{j}\,) = 2\vec{k}$$

$$\text{grad}\,\phi = \text{grad}\left(\frac{1}{(x^2 + y^2)^{p/2}}\right) = \frac{-p}{(x^2 + y^2)^{(p/2)+1}}(x\vec{i} + y\vec{j}\,).$$

Thus, we have

$$\text{curl}\,\vec{B}\, = \frac{1}{(x^2 + y^2)^{p/2}}\text{curl}(-y\vec{i} + x\vec{j}\,) + \text{grad}\left(\frac{1}{(x^2 + y^2)^{p/2}}\right) \times (-y\vec{i} + x\vec{j}\,)$$

$$= \frac{1}{(x^2 + y^2)^{p/2}}2\vec{k}\, + \frac{-p}{(x^2 + y^2)^{(p/2)+1}}(x\vec{i} + y\vec{j}\,) \times (-y\vec{i} + x\vec{j}\,)$$

$$= \frac{1}{(x^2 + y^2)^{p/2}}2\vec{k}\, + \frac{-p}{(x^2 + y^2)^{(p/2)+1}}(x^2 + y^2)\vec{k}$$

$$= \frac{2 - p}{(x^2 + y^2)^{p/2}}\vec{k}\, .$$

(b)   The curl is zero when $p = 2$. Thus, when $p = 2$ the vector field is curl free:

$$\vec{B}\, = \frac{-y\vec{i} + x\vec{j}}{x^2 + y^2}.$$

Alternative Notation for Curl

Using $\nabla = \dfrac{\partial}{\partial x}\vec{i}\, + \dfrac{\partial}{\partial y}\vec{j}\, + \dfrac{\partial}{\partial z}\vec{k}$, we can write

$$\text{curl}\,\vec{F}\, = \nabla \times \vec{F}\, = \begin{vmatrix} \vec{i} & \vec{j} & \vec{k} \\ \dfrac{\partial}{\partial x} & \dfrac{\partial}{\partial y} & \dfrac{\partial}{\partial z} \\ F_1 & F_2 & F_3 \end{vmatrix}.$$

Problems for Section 20.3

Compute the curl of the vector fields in Problems 1–7.

1. $\vec{F} = (x^2 - y^2)\vec{i} + 2xy\vec{j}$

2. $\vec{F}(\vec{r}) = \vec{r}/\|\vec{r}\|$

3. $\vec{F} = x^2\vec{i} + y^3\vec{j} + z^4\vec{k}$

4. $\vec{F} = e^x\vec{i} + \cos y\vec{j} + e^{z^2}\vec{k}$

5. $\vec{F} = 2yz\vec{i} + 3xz\vec{j} + 7xy\vec{k}$

6. $\vec{F} = (-x + y)\vec{i} + (y + z)\vec{j} + (-z + x)\vec{k}$

7. $\vec{F} = (x + yz)\vec{i} + (y^2 + xzy)\vec{j} + (zx^3y^2 + x^7y^6)\vec{k}$

8. Use the geometric definition to find the curl of the vector field $\vec{F}(\vec{r}) = \vec{r}$. Check your answer using the coordinate definition.

9. Using your answers to Problems 3–4, make a conjecture about the value of curl $\vec{F}$ when the vector field $\vec{F}$ has a certain form. (What form?) Show why your conjecture is true.

10. Let $\vec{F}$ be the vector field in Figure 20.17 on page 909. It is rotating counterclockwise around the z-axis when viewed from above. Suppose that at a distance r from the z-axis $\vec{F}$ has magnitude $2r$.

    (a) Find a formula for $\vec{F}$.
    (b) Find curl $\vec{F}$ using the coordinate definition and relate your answer to circulation density.

11. Decide whether each of the following vector fields has a nonzero curl at the origin. In each case, the vector field is shown in the xy-plane; assume it has no z-component and is independent of z.

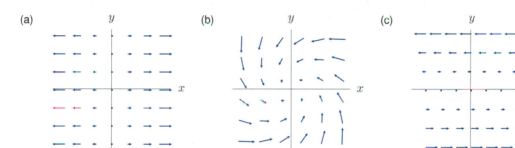

12. A large fire becomes a fire-storm when the nearby air acquires a circulatory motion. The associated updraft has the effect of bringing more air to the fire, causing it to burn faster. Records show that a fire-storm developed during the Chicago Fire of 1871 and during the Second World War bombing of Hamburg, Germany, but there was no fire-storm during the Great Fire of London in 1666. Explain how a fire-storm could be identified using the curl of a vector field.

13. Show that curl $(\vec{F} + \vec{C}) = $ curl $\vec{F}$ for a constant vector field $\vec{C}$.

14. For any constant vector field $\vec{c}$, and any vector field, $\vec{F}$, show that div$(\vec{F} \times \vec{c}) = \vec{c} \cdot$ curl $\vec{F}$.

15. In Chapter 18 we saw how the Fundamental Theorem of Calculus for Line Integrals implies $\int_C$ grad $f \cdot d\vec{r} = 0$ for any smooth closed path C and any smooth function f.

    (a) Use the geometric definition of curl to deduce that curl grad $f = \vec{0}$.
    (b) Verify that curl grad $f = \vec{0}$ using the coordinate definition.

16. If $\vec{F}$ is any vector field whose components have continuous second partial derivatives, show that div curl $\vec{F} = 0$.

17. Show that curl $(\phi\vec{F}) = \phi$ curl $\vec{F} + ($grad $\phi) \times \vec{F}$ for a scalar function ϕ and a vector field $\vec{F}$.

18. A vortex that rotates at constant angular velocity ω about the z-axis has velocity vector field $\vec{v} = \omega(-y\vec{i} + x\vec{j})$.

    (a) Sketch the vector field with $\omega = 1$ and the vector field with $\omega = -1$.
    (b) Determine the speed $\|\vec{v}\|$ of the vortex as a function of the distance from its center.
    (c) Compute div $\vec{v}$ and curl $\vec{v}$.
    (d) Compute the circulation of $\vec{v}$ counterclockwise about the circle of radius R in the xy-plane, centered at the origin.

Problems 19–21 concern the vector fields in Figure 20.24. In each case, assume that the cross-section is the same in all other planes parallel to the given cross-section.

19.   Three of the vector fields have zero curl at each point shown. Which are they? How do you know?

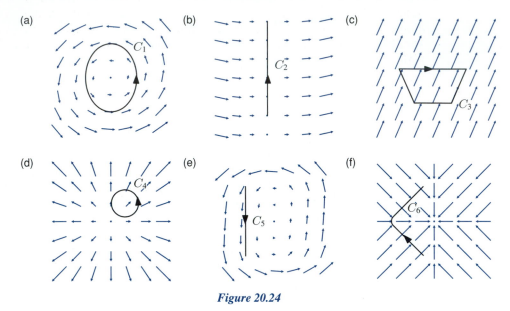

Figure 20.24

20.   Three of the vector fields have zero divergence at each point shown. Which are they? How do you know?

21.   Four of the line integrals $\int_{C_i} \vec{F} \cdot d\vec{r}$ are zero. Which are they? How do you know?

22.   Find a vector field $\vec{F}$ such that curl $\vec{F} = 2\vec{i} - 3\vec{j} + 4\vec{k}$. [Hint: Try $\vec{F} = \vec{v} \times \vec{r}$ for some vector $\vec{v}$.]

23.   Figure 20.25 gives a sketch of a velocity vector field $\vec{F} = y\vec{i} + x\vec{j}$ in the xy-plane.

    (a)   What is the direction of rotation of a thin twig placed at the origin along the x-axis?

    (b)   What is the direction of rotation of a thin twig placed at the origin along the y-axis?

    (c)   Compute curl $\vec{F}$.

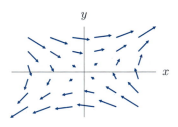

Figure 20.25

24.   Let $\vec{F}$ be a smooth vector field and let $\vec{u}$ and $\vec{v}$ be constant vectors. Using the definition of curl $\vec{F}$ in Cartesian coordinates, show that

$$\text{grad}(\vec{F} \cdot \vec{v}) \cdot \vec{u} - \text{grad}(\vec{F} \cdot \vec{u}) \cdot \vec{v} = (\text{curl } \vec{F}) \cdot \vec{u} \times \vec{v}.$$

20.4 STOKES' THEOREM

The Divergence Theorem says that the integral of the flux density over a solid region is equal to the flux through the surface bounding the region. Similarly, Stokes' Theorem says that the integral of the circulation density over a surface is equal to the circulation around the boundary of the surface.

The Boundary of a Surface

The *boundary* of a surface S is the curve running around the edge of S (like the hem around the edge of a piece of cloth). An orientation of S determines an orientation for its boundary, C, as follows. Pick a positive normal vector $\vec{n}$ on S, near C, and use the right hand rule to determine a direction of travel around $\vec{n}$. This in turn determines a direction of travel around the boundary C. See Figure 20.26. Another way of describing the orientation on C is that someone walking along C in the forward direction, body upright in the direction of the positive normal on S, would have the surface on their left. Notice that the boundary can consist of two or more curves, as the surface on the right in Figure 20.26 shows.

Figure 20.26: Two oriented surfaces and their boundaries

Calculating the Circulation from the Circulation Density

Consider a closed, oriented curve C in 3-space. We can find the circulation of a vector field $\vec{F}$ around C by calculating the line integral:

$$\begin{array}{c} \text{Circulation} \\ \text{around } C \end{array} = \int_C \vec{F} \cdot d\vec{r}.$$

If C is the boundary of an oriented surface S, there is another way to calculate the circulation using curl $\vec{F}$. We subdivide S into pieces as shown on the surface on the left in Figure 20.26. If $\vec{n}$ is a positive unit normal vector to a piece of surface with area ΔA, then $\Delta\vec{A} = \vec{n}\,\Delta A$. In addition, $\text{circ}_{\vec{n}}\,\vec{F}$ is the circulation density of $\vec{F}$ around $\vec{n}$, so

$$\begin{array}{c} \text{Circulation of } \vec{F} \text{ around} \\ \text{boundary of the piece} \end{array} \approx \left(\text{circ}_{\vec{n}}\,\vec{F}\right)\Delta A = ((\text{curl }\vec{F})\cdot\vec{n})\Delta A = (\text{curl}\vec{F})\cdot\Delta\vec{A}.$$

Next we add up the circulations around all the small pieces. The line integral along the common edge of a pair of adjacent pieces appears with opposite sign in each piece, so it cancels out. (See Figure 20.27.) When we add up all the pieces the internal edges cancel and we are left with the circulation around C, the boundary of the entire surface. Thus,

$$\begin{array}{c} \text{Circulation} \\ \text{around } C \end{array} = \sum \begin{array}{c} \text{Circulation around} \\ \text{boundary of pieces} \end{array} \approx \sum \text{curl }\vec{F} \cdot \Delta\vec{A}.$$

Taking the limit as $\Delta A \to 0$, we get

$$\begin{array}{c} \text{Circulation} \\ \text{around } C \end{array} = \int_S \text{curl }\vec{F} \cdot d\vec{A}.$$

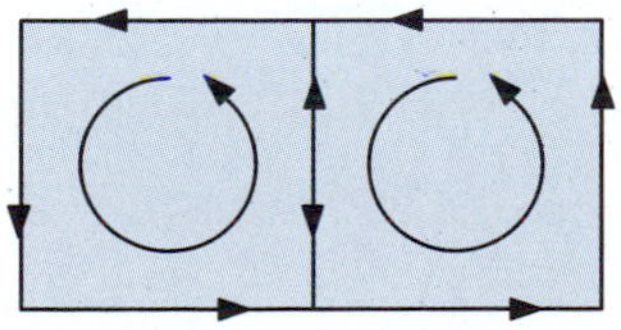

Figure 20.27: Two adjacent pieces of the surface

We have expressed the circulation as a line integral around C and as a flux integral over S; thus, the two integrals must be equal. Hence we have[4]

Stokes' Theorem

If S is a smooth oriented surface with piecewise smooth, oriented boundary C, and if $\vec{F}$ is a smooth vector field which is defined on S and C, then

$$\int_C \vec{F} \cdot d\vec{r} = \int_S \text{curl}\,\vec{F} \cdot d\vec{A}.$$

The orientation of C is determined from the orientation of S according to the right hand rule.

Example 1   Let $\vec{F}(x, y, z) = -2y\vec{i} + 2x\vec{j}$. Use Stokes' Theorem to find $\int_C \vec{F} \cdot d\vec{r}$, where C is a circle

(a)   Parallel to the yz-plane, of radius a, centered at a point on the x-axis, with either orientation.

(b)   Parallel to the xy-plane, of radius a, centered at a point on the z-axis, oriented counterclockwise as viewed from a point on the z-axis above the circle.

Solution   We have $\text{curl}\,\vec{F} = 4\vec{k}$. Figure 20.28 shows sketches of $\vec{F}$ and $\text{curl}\,\vec{F}$.

(a)   Let S be the disk enclosed by C. Since S lies in a vertical plane and $\text{curl}\,\vec{F}$ points vertically everywhere, the flux of $\text{curl}\,\vec{F}$ through S is zero. Hence, by Stokes' Theorem,

$$\int_C \vec{F} \cdot d\vec{r} = \int_S \text{curl}\,\vec{F} \cdot d\vec{A} = 0.$$

It makes sense that the line integral is zero. If C is parallel to the yz-plane (even if it is not lying in the plane), the symmetry of the vector field means that the line integral of $\vec{F}$ over the top half of the circle cancels the line integral over the bottom half.

(b)   Let S be the horizontal disk enclosed by C. Since $\text{curl}\,\vec{F}$ is a constant vector field pointing in the direction of $\vec{k}$, we have, by Stokes' Theorem,

$$\int_C \vec{F} \cdot d\vec{r} = \int_S \text{curl}\,\vec{F} \cdot d\vec{A} = \|\,\text{curl}\,\vec{F}\,\| \cdot \text{Area of } S = 4\pi a^2.$$

Since $\vec{F}$ is circling around the z-axis in the same direction as C, we expect the line integral to be positive. In fact, in Example 1 on page 910, we computed this line integral directly.

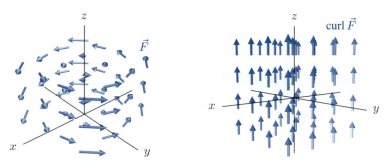

Figure 20.28: The vector fields $\vec{F}$ and $\text{curl}\,\vec{F}$

---

[4]A proof of Stokes' Theorem using the coordinate definition of curl can be found in *Multivariable Calculus*, by William McCallum et al. (New York: Wiley, 1997).

Curl Free Vector Fields

Stokes' Theorem applies to any oriented surface S and its boundary C, even in cases where the boundary consists of two or more curves. This is useful in studying curl free vector fields.

Example 2   A current I flows along the z-axis in the $\vec{k}$ direction. The induced magnetic field $\vec{B}(x, y, z)$ is

$$\vec{B}(x, y, z) = \frac{2I}{c}\left(\frac{-y\vec{i} + x\vec{j}}{x^2 + y^2}\right),$$

where c is the speed of light. In Example 5 on page 914 we showed that curl $\vec{B} = \vec{0}$.

(a) Compute the circulation of $\vec{B}$ around the circle C_1 in the xy-plane of radius a, centered at the origin, and oriented counterclockwise when viewed from above.

(b) Use part (a) and Stokes' Theorem to compute $\int_{C_2} \vec{B} \cdot d\vec{r}$, where C_2 is the ellipse $x^2 + 9y^2 = 9$ in the plane $z = 2$, oriented counterclockwise when viewed from above.

Solution   (a) On the circle C_1, we have $\|\vec{B}\| = 2I/(ca)$. Since $\vec{B}$ is tangent to C_1 everywhere and points in the forward direction around C_1,

$$\int_{C_1} \vec{B} \cdot d\vec{r} = \int_{C_1} \|\vec{B}\|\, dr = \frac{2I}{ca} \cdot \text{Length of } C_1 = \frac{2I}{ca} \cdot 2\pi a = \frac{4\pi I}{c}.$$

(b) Let S be the conical surface extending from C_1 to C_2 in Figure 20.29 on page 919. The boundary of this surface has two pieces, $-C_2$ and C_1. The orientation of C_1 leads to the outward normal on S, which forces us to choose the clockwise orientation on C_2. By Stokes' Theorem,

$$\int_S \text{curl}\,\vec{B} \cdot d\vec{A} = \int_{-C_2} \vec{B} \cdot d\vec{r} + \int_{C_1} \vec{B} \cdot d\vec{r} = -\int_{C_2} \vec{B} \cdot d\vec{r} + \int_{C_1} \vec{B} \cdot d\vec{r}.$$

Since curl $\vec{B} = \vec{0}$, we have $\int_S \text{curl}\,\vec{B} \cdot d\vec{A} = 0$, so the two line integrals must be equal:

$$\int_{C_2} \vec{B} \cdot d\vec{r} = \int_{C_1} \vec{B} \cdot d\vec{r} = \frac{4\pi I}{c}.$$

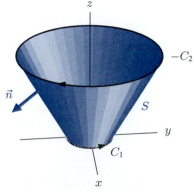

Figure 20.29: Surface joining C_1 to C_2, oriented to satisfy the conditions of Stoke's Theorem

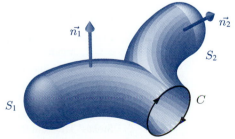

Figure 20.30: The flux of a curl is the same through the two surfaces S_1 and S_2 if they determine the same orientation on the boundary, C

Curl Fields

A vector field $\vec{F}$ is called a *curl field* if $\vec{F} = \operatorname{curl}\vec{G}$ for some vector field $\vec{G}$. Recall that if $\vec{F} = \operatorname{grad} f$, then f is called a potential function. By analogy, if a vector field $\vec{F} = \operatorname{curl}\vec{G}$, then $\vec{G}$ is called a *vector potential* for $\vec{F}$. The following example shows that the flux of a curl field through a surface depends only on the boundary of the surface. This is analogous to the fact that the line integral of a gradient field depends only on the endpoints of the path.

Example 3   Suppose $\vec{F} = \operatorname{curl}\vec{G}$. Suppose that S_1 and S_2 are two oriented surfaces with the same boundary C. Show that, if S_1 and S_2 determine the same orientation on C (as in Figure 20.30), then

$$\int_{S_1} \vec{F} \cdot d\vec{A} = \int_{S_2} \vec{F} \cdot d\vec{A}.$$

If S_1 and S_2 determine opposite orientations on C, then

$$\int_{S_1} \vec{F} \cdot d\vec{A} = -\int_{S_2} \vec{F} \cdot d\vec{A}.$$

Solution   Since $\vec{F} = \operatorname{curl}\vec{G}$, by Stokes' Theorem we have

$$\int_{S_1} \vec{F} \cdot d\vec{A} = \int_{S_1} \operatorname{curl}\vec{G} \cdot d\vec{A} = \int_C \vec{G} \cdot d\vec{r}$$

and

$$\int_{S_2} \vec{F} \cdot d\vec{A} = \int_{S_2} \operatorname{curl}\vec{G} \cdot d\vec{A} = \int_C \vec{G} \cdot d\vec{r}.$$

In each case the line integral on the right must be computed using the orientation determined by the surface. Thus, the two flux integrals of $\vec{F}$ are the same if the orientations are the same and they are opposite if the orientations are opposite.

Problems for Section 20.4

1.   Can you use Stokes' Theorem to compute the line integral $\int_C (2x\vec{i} + 2y\vec{j} + 2z\vec{k}) \cdot d\vec{r}$ where C is the straight line from the point $(1, 2, 3)$ to the point $(4, 5, 6)$? Why or why not?

In Problems 2–5 compute the given line integral using Stokes' Theorem.

2. $\int_C \vec{F} \cdot d\vec{r}$, where $\vec{F} = (z - 2y)\vec{i} + (3x - 4y)\vec{j} + (z + 3y)\vec{k}$ and C is the circle $x^2 + y^2 = 4$, $z = 1$, oriented counterclockwise when viewed from above.

3. $\int_C \vec{F} \cdot d\vec{r}$ where $\vec{F} = (2x - y)\vec{i} + (x + 4y)\vec{j}$ and C is a circle of radius 10, centered at the origin

   (a)   In the xy-plane, oriented clockwise as viewed from the positive z-axis.
   (b)   In the yz-plane, oriented clockwise as viewed from the positive x-axis.

4. $\int_C \vec{F} \cdot d\vec{r}$, with $\vec{F} = \vec{r}/\|\vec{r}\|^3$ where C is the path consisting of line segments from $(1, 0, 1)$ to $(1, 0, 0)$ to $(0, 0, 1)$ back to $(1, 0, 1)$.

5.   Find the circulation of the vector field $\vec{F} = xz\vec{i} + (x + yz)\vec{j} + x^2\vec{k}$ around the circle $x^2 + y^2 = 1$, $z = 2$, oriented counterclockwise when viewed from above.

6.   Compute the line integral $\int_C ((yz^2 - y)\vec{i} + (xz^2 + x)\vec{j} + 2xyz\vec{k}) \cdot d\vec{r}$ where C is the circle of radius 3 in the xy-plane, centered at the origin, oriented counterclockwise as viewed from the positive z-axis. Do it two ways: (a) Directly   (b) Using Stokes' Theorem

7.   Let S be the surface given by $z = 1 - x^2$ for $0 \le x \le 1$ and $-2 \le y \le 2$, oriented upward. Verify Stokes' Theorem for $\vec{F} = xy\vec{i} + yz\vec{j} + xz\vec{k}$. Sketch the surface S and the curve C that bounds S.

8.   Verify Stokes' Theorem for $\vec{F} = y\vec{i} + z\vec{j} + x\vec{k}$ and S, the paraboloid $z = 1 - (x^2 + y^2)$, $z \ge 0$ oriented upward. [Hint: Use polar coordinates.]

9.   Suppose that C is a closed curve in the xy-plane, oriented counterclockwise when viewed from above. Show that $\frac{1}{2}\int_C (-y\vec{i} + x\vec{j}) \cdot d\vec{r}$ equals the area of the region R in the xy-plane enclosed by C.

10. The vector fields $\vec{F}$ and $\vec{G}$ are sketched in Figures 20.31 and 20.32. Each vector field has no z-component and is independent of z. Assume all the axes have the same scales.

   (a) What can you say about div $\vec{F}$ and div $\vec{G}$ at the origin?
   (b) What can you say about curl $\vec{F}$ and curl $\vec{G}$ at the origin?
   (c) Is there a closed surface around the origin such that $\vec{F}$ has a nonzero flux through it?
   (d) Repeat part (c) for $\vec{G}$.
   (e) Is there a closed curve around the origin such that $\vec{F}$ has a nonzero circulation around it?
   (f) Repeat part (e) for $\vec{G}$.

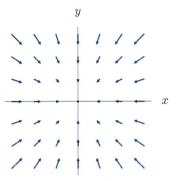

Figure 20.31: Cross-section of $\vec{F}$ **Figure 20.32**: Cross-section of $\vec{G}$

11. Use Stokes' Theorem to show that $\int_S$ curl $\vec{F} \cdot d\vec{A} = 0$ for any surface S which is the boundary surface of a solid region W. Use the geometric definition of the divergence to deduce that div curl $\vec{F} = 0$.

12. Show that Green's Theorem is a special case of Stokes' Theorem.

13. A vector field $\vec{F}$ is defined everywhere except on the z-axis and curl $\vec{F} = \vec{0}$ everywhere where $\vec{F}$ is defined. What can you say about $\int_C \vec{F} \cdot d\vec{r}$ if C is a circle of radius 1 in the xy-plane, and if the center of C is at   (a) the origin,   (b) the point $(2, 0)$?

14. Evaluate $\int_C (-z\vec{i} + y\vec{j} + x\vec{k}) \cdot d\vec{r}$, where C is a circle of radius 2 around the y-axis with orientation indicated in Figure 20.33.

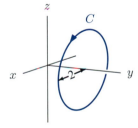

Figure 20.33

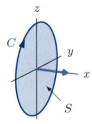

Figure 20.34

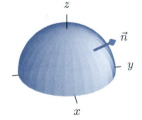

Figure 20.35

15. Let $\vec{F} = -z\vec{j} + y\vec{k}$, let C be the circle of radius a in the yz-plane oriented clockwise as viewed from the positive x-axis, and let S be the disk in the yz-plane enclosed by C, oriented in the positive x-direction. See Figure 20.34.

   (a) Evaluate directly $\int_C \vec{F} \cdot d\vec{r}$.
   (b) Evaluate directly $\int_S$ curl $\vec{F} \cdot d\vec{A}$.
   (c) The answers in parts (a) and (b) are not equal. Explain why this does not contradict Stokes' Theorem.

16. Let $\vec{F} = (8yz - z)\vec{j} + (3 - 4z^2)\vec{k}$.

   (a) Show that $\vec{G} = 4yz^2\vec{i} + 3x\vec{j} + xz\vec{k}$ is a vector potential for $\vec{F}$.
   (b) Evaluate $\int_S \vec{F} \cdot d\vec{A}$ where S is the hemisphere of radius 5 shown in Figure 20.35, oriented upwards. [Hint: Use Example 3 on page 920 to simplify the calculation.]

17.   Let $\vec{F} = -y\vec{i} + x\vec{j} + \cos(xy)z\vec{k}$ and let S be the surface of the lower unit hemisphere $x^2 + y^2 + z^2 = 1$, $z \leq 0$, oriented with outward pointing normal. Find $\int_S \text{curl}\,\vec{F} \cdot d\vec{A}$.

CHAPTER SUMMARY

- **Divergence**
  Geometric and coordinate definition of divergence, calculating divergence, interpretation in terms of outflow per unit volume.
- **Curl**
  Geometric and coordinate definition of curl, calculating curl, interpretation in terms of circulation per unit area.
- **The Divergence Theorem**
  Statement of the theorem, divergence free fields, harmonic functions.
- **Stokes' Theorem**
  Statement of the theorem, curl free and curl fields.

REVIEW PROBLEMS FOR CHAPTER TWENTY

1.   Use the geometric definition of divergence to find div $\vec{v}$ at the origin, where $\vec{v} = -2\vec{r}$. Check that you get the same result using the definition in Cartesian coordinates.

2.   Can you evaluate the flux integral in Problem 13 on page 892 by application of the Divergence Theorem? Why or why not?

3.   If V is a volume surrounded by a closed surface S, show that $\frac{1}{3} \int_S \vec{r} \cdot d\vec{A} = V$.

4.   Use Problem 3 to compute the volume of a sphere of radius R given that its surface area is $4\pi R^2$.

5.   Use Problem 3 to compute the volume of a cone of base radius b and height h.
      [Hint: Stand the cone with its point downward and its axis along the positive z-axis.]

6.   Compute the flux integral $\int_S (x^3\vec{i} + 2y\vec{j} + 3\vec{k}) \cdot d\vec{A}$, where S is the $2 \times 2 \times 2$ rectangular surface centered at the origin, oriented outward. Do this in two ways:

      (a)   Directly   (b)   By means of the Divergence Theorem

7.   Suppose $\vec{F} = \vec{r}/\|\vec{r}\|^3$. Find $\int_S \vec{F} \cdot d\vec{A}$ where S is the ellipsoid $x^2 + 2y^2 + 3z^2 = 6$. Give reasons for your computation.

8.   A central vector field is one of the form $\vec{F} = f(r)\vec{r}$ where f is any function of $r = \|\vec{r}\|$. Show that any central vector field is irrotational.

Are the statements in Problems 9–16 true or false? Assume $\vec{F}$ and $\vec{G}$ are smooth vector fields in 3-space. Explain your answer.

9.   curl $\vec{F}$ is a vector field.

10. $\text{grad}(fg) = (\text{grad}\,f) \cdot (\text{grad}\,g)$

11. $\text{div}(\vec{F} + \vec{G}) = \text{div}\,\vec{F} + \text{div}\,\vec{G}$

12. $\text{grad}(\vec{F} \cdot \vec{G}) = \vec{F}(\text{div}\,\vec{G}) + (\text{div}\,\vec{F})\vec{G}$

13. $\text{curl}(f\vec{G}) = (\text{grad}\,f) \times \vec{G} + f(\text{curl}\,\vec{G})$

14.   div $\vec{F}$ is a scalar whose value can vary from point to point.

15.   If $\int_S \vec{F} \cdot d\vec{A} = 12$ and S is a flat disk of area 4π, then div $\vec{F} = 3/\pi$.

16.   If $\vec{F}$ is as shown in Figure 20.36, then curl $\vec{F} \cdot \vec{j} > 0$.

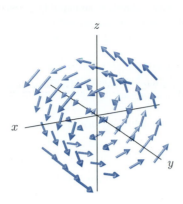

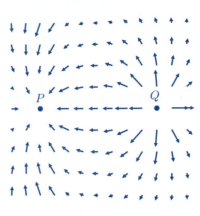

Figure 20.36   *Figure 20.37*

17. Figure 20.37 shows the part of a vector field $\vec{E}$ that lies in the xy-plane. Assume that the vector field is independent of z, so that any horizontal cross section looks the same. What can you say about div $\vec{E}$ at the points marked P and Q, assuming it is defined?

18. According to Coulomb's Law, the electrostatic field $\vec{E}$ at the point $\vec{r}$ due to a charge q at the origin is given by

$$\vec{E}\,(\vec{r}\,) = q\frac{\vec{r}}{\|\vec{r}\,\|^3}.$$

(a) Compute div $\vec{E}$.

(b) Let S_a be the sphere of radius a centered at the origin and oriented outwards. Show that the flux of $\vec{E}$ through S_a is $4\pi q$.

(c) Could you have used the Divergence Theorem in part (b)? Explain why or why not.

(d) Let S be an arbitrary, closed, outward-oriented surface surrounding the origin. Show that the flux of $\vec{E}$ through S is again $4\pi q$. [Hint: Apply the Divergence Theorem to the solid region lying between a small sphere S_a and the surface S.]

19. According to Coulomb's Law, the electric field $\vec{E}$ at the point $\vec{r}$ due to a charge q at the point $\vec{r}_0$ is given by

$$\vec{E}\,(\vec{r}\,) = q\frac{(\vec{r}-\vec{r}_0)}{\|\vec{r}-\vec{r}_0\|^3}.$$

Suppose S is a closed, outward-oriented surface and that $\vec{r}_0$ does not lie on S. Use Problem 18 to show that

$$\int_S \vec{E}\cdot d\vec{A} = \begin{cases} 4\pi q & \text{if } q \text{ lies inside } S, \\ 0 & \text{if } q \text{ lies outside } S. \end{cases}$$

20. Due to roadwork ahead, the traffic on a highway slows linearly from 55 miles/hour to 15 miles/hour over a 2000 foot stretch of road, then crawls along at 15 miles/hour for 5000 feet, then speeds back up linearly to 55 miles/hour in the next 1000 feet, after which it moves steadily at 80 km/hr.

(a) Sketch a velocity vector field for the traffic flow.

(b) Write a formula for the velocity vector field $\vec{v}$ (miles/hour) as a function of the distance x feet from the initial point of slowdown. (Take the direction of motion to be $\vec{i}$ and consider the various sections of the road separately.)

(c) Compute div $\vec{v}$ at $x = 1000, 5000, 7500, 10{,}000$. Be sure to include the proper units.

21. The velocity field $\vec{v}$ in Problem 20 does not give a complete description of the traffic flow, for it takes no account of the spacing between vehicles. Let ρ be the density (cars/mile) of highway, where we assume that ρ depends only on x.

(a) Using your highway experience, arrange in ascending order: $\rho(0), \rho(1000), \rho(5000)$.

(b) What are the units and interpretation of the vector field $\rho\vec{v}$?

(c) Would you expect $\rho\vec{v}$ to be constant? Why? What does this mean for div$(\rho\vec{v})$?

(d) Determine $\rho(x)$ if $\rho(0) = 75$ cars/mile and $\rho\vec{v}$ is constant.

(e) If the highway has two lanes, find the approximate number of feet between cars at $x = 0, 1000$, and 5000.

22. (a) A river flows across the xy-plane in the positive x-direction and around a circular rock of radius 1 centered at the origin. The velocity of the river can be modeled using the potential function $\phi = x + (x/(x^2 + y^2))$. Compute the velocity vector field, $\vec{v} = \operatorname{grad} \phi$.
     (b) Show that div $\vec{v} = 0$.
     (c) Show that the flow of $\vec{v}$ is tangent to the circle $x^2 + y^2 = 1$. This means that no water crosses the circle. The water on the outside must therefore all flow around the circle.
     (d) Use a computer to sketch the vector field $\vec{v}$ in the region outside the unit circle.

Problems 23–24 use the fact that the electric field, $\vec{E}$, is related to the charge density, $\rho(x, y, z)$, in units of charge/volume, by the equation

$$\operatorname{div} \vec{E} = 4\pi\rho.$$

In addition, there is an electric potential, ϕ, whose gradient gives the electric field:

$$\vec{E} = -\operatorname{grad} \phi.$$

23. Calculate and describe in words the electric field and the charge distribution corresponding to the potential function defined as follows:

$$\phi = \begin{cases} x^2 + y^2 + z^2 & \text{for } x^2 + y^2 + z^2 \le \frac{b^2}{4} \\ \dfrac{b^2}{4} - \dfrac{b^3}{4(x^2 + y^2 + z^2)^{1/2}} & \text{for } \frac{b^2}{4} \le x^2 + y^2 + z^2 \end{cases}$$

24. A vector field which could possibly represent an electric field is given by

$$\vec{E} = 10xy\vec{i} + (5x^2 - 5y^2)\vec{j}.$$

     (a) Calculate the line integral of $\vec{E}$ from the origin to the point (a, b) along the path which runs straight from the origin to the point $(a, 0)$ and then straight from $(a, 0)$ to the point (a, b).
     (b) Calculate the line integral of $\vec{E}$ between the same points as in part (a) but via the point $(0, b)$.
     (c) Why do your answers to part (a) and (b) suggest that $\vec{E}$ could indeed be an electric field?
     (d) Find the electric potential, ϕ, and calculate grad ϕ to confirm that $\vec{E} = -\operatorname{grad} \phi$.

25. A vector field is a *point source* at the origin in 3-space if its direction is away from the origin at every point, its magnitude depends only on the distance from the origin, and its divergence is zero except at the origin. (Such a vector field might be used to model the photon flow out of a star or the neutrino flow out of a supernova.)

     (a) Show that $\vec{v} = K(x^2 + y^2 + z^2)^{-3/2}(x\vec{i} + y\vec{j} + z\vec{k})$ is a point source at the origin if $K > 0$.
     (b) Determine the magnitude $\|\vec{v}\|$ of the source in part (a) as a function of the distance from its center.
     (c) Compute the flux of $\vec{v}$ through a sphere of radius r centered at the origin.
     (d) Compute the flux of $\vec{v}$ through a closed surface that does not contain the origin.

26. The speed of a naturally occurring vortex (tornado, waterspout, whirlpool) is a decreasing function of the distance from its center, so the constant angular velocity model of Problem 18 on page 915 is inappropriate. A *free vortex* circulating about the z-axis has vector field $\vec{v} = K(x^2 + y^2)^{-1}(-y\vec{i} + x\vec{j})$ where K is a constant.

     (a) Sketch the vector field with $K = 1$ and the vector field with $K = -1$.
     (b) Determine the speed $\|\vec{v}\|$ of the vortex as a function of the distance from its center.
     (c) Compute div $\vec{v}$.
     (d) Show that curl $\vec{v} = \vec{0}$.
     (e) Compute the circulation of $\vec{v}$ counterclockwise about the circle of radius R at the origin.
     (f) The computations in parts (d) and (e) show that $\vec{v}$ has curl $\vec{0}$, but has nonzero circulation around the closed curve in part (e). Explain why this does not contradict Stokes' Theorem.

PROJECTS

1. **Water Draining From a Bathtub**

   Suppose water in a bathtub has velocity vector field near the drain given, for x, y, z in cm, by

   $$\vec{F} = -\frac{y + xz}{(z^2 + 1)^2}\vec{i} - \frac{yz - x}{(z^2 + 1)^2}\vec{j} - \frac{1}{z^2 + 1}\vec{k} \qquad \text{cm/sec.}$$

   (a) Rewriting $\vec{F}$ as follows, describe in words how the water is moving:

   $$\vec{F} = \frac{1}{(z^2 + 1)^2}\left(-y\vec{i} + x\vec{j}\right) + \frac{-z}{(z^2 + 1)^2}\left(x\vec{i} + y\vec{j}\right) - \frac{1}{z^2 + 1}\vec{k}.$$

   (b) The drain in the bathtub is a disk in the xy-plane with center at the origin and radius 1 cm. Find the rate at which the water is leaving the bathtub. (That is, find the rate at which water is flowing through the disk.) Give units for your answer.

   (c) Find the divergence of $\vec{F}$.

   (d) Find the flux of the water through the hemisphere of radius 1, centered at the origin, lying below the xy-plane and oriented downward.

   (e) Find $\int_C \vec{G} \cdot d\vec{r}$ where C is the edge of the drain, oriented clockwise when viewed from above, and where

   $$\vec{G} = \frac{1}{2}\left(\frac{y}{z^2 + 1}\vec{i} - \frac{x}{z^2 + 1}\vec{j} - \frac{x^2 + y^2}{(z^2 + 1)^2}\vec{k}\right).$$

   (f) Calculate curl $\vec{G}$.

   (g) Explain why your answers to parts (d) and (e) are equal.

2. **Heat Inside the Earth**

   Heat is generated inside the earth by radioactive decay. Assume it is generated uniformly throughout the earth at a rate of 30 watts per cubic kilometer. (A watt is a rate of heat production.) The heat then flows to the earth's surface where it is lost to space. Let $\vec{F}(x, y, z)$ denote the rate of flow of heat measured in watts per square kilometer. By definition, the flux of $\vec{F}$ across a surface is the quantity of heat flowing through the surface per unit of time.

   (a) What is the value of div $\vec{F}$? Include units.

   (b) Assume the heat flows outward symmetrically. Verify that $\vec{F} = \alpha\vec{r}$, where $\vec{r} = x\vec{i} + y\vec{j} + z\vec{k}$ and α is a suitable constant, satisfies the given conditions. Find α.

   (c) Let $T(x, y, z)$ denote the temperature inside the earth. Heat flows according to the equation $\vec{F} = -k\,\text{grad}\,T$, where k is a constant. Explain why this makes sense physically.

   (d) If T is in °C, then $k = 30{,}000$ watts/km°C. Assuming the earth is a sphere with radius 6400 km and surface temperature 20°C, what is the temperature at the center?

FOCUS ON THEORY

THE THREE FUNDAMENTAL THEOREMS

We have now seen three multivariable versions of the Fundamental Theorem of Calculus. In this section we will examine some consequences of these theorems.

Fundamental Theorem of Calculus for Line Integrals

$$\int_C \operatorname{grad} f \cdot d\vec{r} = f(Q) - f(P).$$

Stokes' Theorem

$$\int_S \operatorname{curl} \vec{F} \cdot d\vec{A} = \int_C \vec{F} \cdot d\vec{r}.$$

Divergence Theorem

$$\int_W \operatorname{div} \vec{F}\, dV = \int_S \vec{F} \cdot d\vec{A}.$$

Notice that, in each case, the region of integration on the right is the boundary of the region on the left (except that for the first theorem we simply evaluate f at the boundary points); the integrand on the left is a sort of derivative of the integrand on the right; see Figure 20.38.

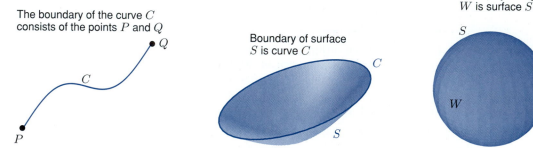

Figure 20.38: Regions and their boundaries for the three fundamental theorems

The Gradient and the Curl

Suppose that $\vec{F}$ is a smooth gradient field, so $\vec{F} = \operatorname{grad} f$ for some function f. Using the Fundamental Theorem for Line Integrals, we saw in Chapter 18 that

$$\int_C \vec{F} \cdot d\vec{r} = 0$$

for any closed curve C. Thus, for any unit vector $\vec{n}$

$$\operatorname{circ}_{\vec{n}} \vec{F} = \lim_{\text{Area}\to 0} \frac{\displaystyle\int_C \vec{F} \cdot d\vec{r}}{\text{Area of } C} = \lim_{\text{Area}\to 0} \frac{0}{\text{Area}} = 0,$$

where the limit is taken over circles C in a plane perpendicular to $\vec{n}$, and oriented by the right hand rule. Thus the circulation density of $\vec{F}$ is zero in every direction, so $\operatorname{curl} \vec{F} = \vec{0}$, that is,

$$\boxed{\operatorname{curl} \operatorname{grad} f = \vec{0}.}$$

(This formula can also be verified using the coordinate definition of curl. See Problem 15 on page 915.)

Is the converse true? Is any vector field whose curl is zero a gradient field? Suppose that curl $\vec{F} = \vec{0}$ and let us consider the line integral $\int_C \vec{F} \cdot d\vec{A}$ for a closed curve C contained in the domain of $\vec{F}$. If C is the boundary curve of an oriented surface S that lies wholly in the domain of curl $\vec{F}$, then Stokes' Theorem asserts that

$$\int_C \vec{F} \cdot d\vec{r} = \int_S \text{curl}\,\vec{F} \cdot d\vec{A} = \int_S \vec{0} \cdot d\vec{A} = 0.$$

If we knew that $\int_C \vec{F} \cdot d\vec{r} = 0$ for every closed curve C, then $\vec{F}$ would be path-independent, and hence a gradient field. Thus we need to know whether every closed curve in the domain of $\vec{F}$ is the boundary of an oriented surface contained in the domain. It can be quite difficult to determine if a given curve is the boundary of a surface (suppose, for example, that the curve is knotted in a complicated way). However, if the curve can be contracted smoothly to a point, remaining all the time in the domain of $\vec{F}$, then it is the boundary of a surface, namely, the surface it sweeps through as it contracts. Thus, we have proved the test for a gradient field that we stated in Chapter 18.

The Curl Test for Vector Fields in 3-Space

Suppose $\vec{F}$ is a smooth vector field on 3-space such that
- The domain of $\vec{F}$ has the property that every closed curve in it can be contracted to a point in a smooth way, staying at all times within the domain.
- curl $\vec{F} = \vec{0}$.

Then $\vec{F}$ is path-independent, and thus is a gradient field.

Example 6 on page 865 shows how the curl test is applied.

The Curl and The Divergence

In this section we will use the second two fundamental theorems to get a test for a vector field to be a curl field, that is, a field of the form $\vec{F} = \text{curl}\,\vec{G}$ for some $\vec{G}$.

Example 1   Suppose that $\vec{F}$ is a smooth curl field. Use Stokes' Theorem to show that for any closed surface, S, contained in the domain of $\vec{F}$

$$\int_S \vec{F} \cdot d\vec{A} = 0.$$

Solution   Suppose $\vec{F} = \text{curl}\,\vec{G}$. Draw a closed curve C on the surface S, thus dividing S into two surfaces S_1 and S_2 as shown in Figure 20.39. Pick the orientation for C corresponding to S_1; then the orientation of C corresponding to S_2 is the opposite. Thus, using Stokes' Theorem,

$$\int_{S_1} \vec{F} \cdot d\vec{A} = \int_{S_1} \text{curl}\,\vec{G} \cdot d\vec{A} = \int_C \vec{G} \cdot d\vec{r} = -\int_{S_2} \text{curl}\,\vec{G} \cdot d\vec{A} = -\int_{S_2} \vec{F} \cdot d\vec{A}.$$

Thus, for any closed surface S, we have

$$\int_S \vec{F} \cdot d\vec{A} = \int_{S_1} \vec{F} \cdot d\vec{A} + \int_{S_2} \vec{F} \cdot d\vec{A} = 0.$$

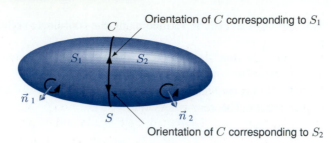

Figure 20.39: The closed surface S divided into two surfaces S_1 and S_2

Thus, if $\vec{F} = \text{curl } \vec{G}$, we use the result of Example 1 to see that

$$
\text{div } \vec{F} = \lim_{\text{Volume} \to 0} \frac{\displaystyle\int_S \vec{F} \cdot d\vec{A}}{\text{Volume enclosed by } S} = \lim_{\text{Volume} \to 0} \frac{0}{\text{Volume}} = 0,
$$

where the limit is taken over spheres S contracting down to a point. So we conclude that:

$$
\boxed{\text{div curl } \vec{G} = 0.}
$$

(This formula can also be verified using coordinates. See Problem 16 on page 915.)

Is every vector field whose divergence is zero a curl field? It turns out that we have the following analogue of the curl test, though we will not prove it.

The Divergence Test for Vector Fields in 3-Space

Suppose $\vec{F}$ is a smooth vector field on 3-space such that
- The domain of $\vec{F}$ has the property that every closed surface in it is the boundary of a solid region completely contained in the domain.
- $\text{div } \vec{F} = 0$.

Then $\vec{F}$ is a curl field.

Example 2   Consider the vector fields $\vec{E} = q \dfrac{\vec{r}}{\|\vec{r}\|^3}$ and $\vec{B} = \dfrac{2I}{c} \left(\dfrac{-y\vec{i} + x\vec{j}}{x^2 + y^2} \right)$.

(a) Calculate $\text{div } \vec{E}$ and $\text{div } \vec{B}$.
(b) Do $\vec{E}$ and $\vec{B}$ satisfy the divergence test?
(c) Is either $\vec{E}$ or $\vec{B}$ a curl field?

Solution   (a) Example 3 on page 899 shows that $\text{div } \vec{E} = 0$. The following calculation shows $\text{div } \vec{B} = 0$ also:

$$
\text{div } \vec{B} = \frac{2I}{c} \left(\frac{\partial}{\partial x} \left(\frac{-y}{x^2 + y^2} \right) + \frac{\partial}{\partial y} \left(\frac{x}{x^2 + y^2} \right) + \frac{\partial}{\partial z} (0) \right)
$$
$$
= \frac{2I}{c} \left(\frac{2xy}{(x^2 + y^2)^2} + \frac{-2yx}{(x^2 + y^2)^2} \right) = 0.
$$

(b) The domain of $\vec{E}$ is 3-space minus the origin, so a region is contained in the domain if it misses the origin. Thus the surface of a sphere centered at the origin is contained in the domain of E, but the solid ball inside is not. Hence $\vec{E}$ does not satisfy the divergence test.

The domain of $\vec{B}$ is 3-space minus the z-axis, so a region is contained in the domain if it avoids the z-axis. If S is a surface bounding a solid region W, then the z-axis cannot pierce W without piercing S as well. Hence, if S avoids the z-axis, so does W. Thus $\vec{B}$ satisfies the divergence test.

(c) In Example 3 on page 906 we computed the flux of $\vec{r}/\|\vec{r}\|^3$ through a sphere centered at the origin, and found it was 4π, so the flux of $\vec{E}$ through this sphere is $4\pi q$. Thus, $\vec{E}$ cannot be a curl field, because by Example 1, the flux of a curl field through a closed surface is zero.

On the other hand, $\vec{B}$ satisfies the divergence test, so it must be a curl field. In fact, Problem 5 below shows that

$$\vec{B} = \text{curl}\left(\frac{-I}{c}\ln(x^2 + y^2)\vec{k}\right).$$

Problems on the Three Fundamental Theorems

Which of the vector fields in Problems 1–2 is a gradient field?

1. $\vec{F} = yz\vec{i} + (xz + z^2)\vec{j} + (xy + 2yz)\vec{k}$

2. $\vec{G} = -y\vec{i} + x\vec{j}$

3. Let $\vec{B} = b\vec{k}$, for some constant b. Show that the following are all possible vector potentials for $\vec{B}$:
   (a) $\vec{A} = -by\vec{i}$ (b) $\vec{A} = bx\vec{j}$ (c) $\vec{A} = \frac{1}{2}\vec{B} \times \vec{r}$.

4. Find a vector potential for the constant vector field $\vec{B}$ whose value at every point is $\vec{b}$.

5. Show that $\vec{A} = \dfrac{-I}{c}\ln(x^2 + y^2)\vec{k}$ is a vector potential for $\vec{B} = \dfrac{2I}{c}\left(\dfrac{-y\vec{i} + x\vec{j}}{x^2 + y^2}\right)$.

6. Is there a vector field $\vec{G}$ such that $\text{curl}\,\vec{G} = y\vec{i} + x\vec{j}$? How do you know?

For each vector field in Problems 7–8, determine whether a vector potential exists. If so, find one.

7. $\vec{F} = 2x\vec{i} + (3y - z^2)\vec{j} + (x - 5z)\vec{k}$

8. $\vec{G} = x^2\vec{i} + y^2\vec{j} + z^2\vec{k}$

9. An electric charge q at the origin produces an electric field $\vec{E} = q\vec{r}/\|\vec{r}\|^3$.
   (a) Does $\text{curl}\,\vec{E} = \vec{0}$?
   (b) Does $\vec{E}$ satisfy the curl test?
   (c) Is $\vec{E}$ a gradient field?

10. Suppose c is the speed of light. A thin wire along the z-axis carrying a current I produces a magnetic field

$$\vec{B} = \frac{2I}{c}\left(\frac{-y\vec{i} + x\vec{j}}{x^2 + y^2}\right),$$

   (a) Does $\text{curl}\,\vec{B} = \vec{0}$?
   (b) Does $\vec{B}$ satisfy the curl test?
   (c) Is $\vec{B}$ a gradient field?

11. For constant p, consider the vector field $\vec{E} = \dfrac{\vec{r}}{\|\vec{r}\|^p}$.
   (a) Find $\text{curl}\,\vec{E}$.
   (b) Find the domain of $\vec{E}$.
   (c) For which values of p does $\vec{E}$ satisfy the curl test? For those values of p, find a potential function for $\vec{E}$.

12. The magnetic field, $\vec{B}$, due to a magnetic dipole with moment $\vec{\mu}$ satisfies div $\vec{B} = 0$ and is given by

$$\vec{B} = -\frac{\vec{\mu}}{\|\vec{r}\|^3} + \frac{3(\vec{\mu} \cdot \vec{r})\vec{r}}{\|\vec{r}\|^5}, \qquad \vec{r} \neq \vec{0}.$$

(a) Does $\vec{B}$ satisfy the divergence test?

(b) Show that a vector potential for $\vec{B}$ is given by $\vec{A} = \dfrac{\vec{\mu} \times \vec{r}}{\|\vec{r}\|^3}$.

[Hint: Use Problem 17 on page 915. The identities in Example 3 on page 912, Problem 19 on page 686, and Problem 24 on page 648 might also be useful.]

(c) Does your answer to part (a) contradict your answer to part (b)? Explain.

13. Suppose that $\vec{A}$ is a vector potential for $\vec{B}$.

(a) Show that $\vec{A} + \operatorname{grad} \psi$ is also a vector potential for $\vec{B}$, for any function ψ with continuous second-order partial derivatives. (The vector potentials $\vec{A}$ and $\vec{A} + \operatorname{grad} \psi$ are called *gauge equivalent* and the transformation, for any ψ, from $\vec{A}$ to $\vec{A} + \operatorname{grad} \psi$ is called a *gauge transformation*.)

(b) What is the divergence of $\vec{A} + \operatorname{grad} \psi$? How should ψ be chosen such that $\vec{A} + \operatorname{grad} \psi$ has zero divergence? (If div $\vec{A} = 0$, the magnetic vector potential $\vec{A}$ is said to be in *Coulomb gauge*.)

The condition in the 3-dimensional curl test about contracting curves can be stated more precisely as follows. A curve C is said to be *smoothly contractible* to a point P if there is a family of parameterized closed curves $C_s, 0 \leq s \leq 1$, with parameterizations

$$\vec{r} = \vec{r}_s(t), \qquad a \leq t \leq b,$$

such that C_0 is the original curve C and C_1 is the point P. (Thus, as s moves from 0 to 1, the curve C_s shrinks from C to P; imagine an animated picture, which at time s shows C_s.) We require that $\vec{r}_s(t)$ be smooth as a function of the two variables s and t. Notice that, since the curves C_s are closed, we must have $\vec{r}_s(a) = \vec{r}_s(b)$ for each s. Then the condition in the curl test is that every closed curve in the domain of $\vec{F}$ be smoothly contractible to a point in such a way that C_s is in the domain of $\vec{F}$ for every s. Problems 14–15 use these ideas.

14. Let C be the circle of radius 1 in the xy-plane, centered at the origin. Write down a family of curves C_s that smoothly contracts C to the origin.

15. Show that any smoothly parameterized curve C in 3-space can be smoothly contracted to any point P. [Hint: Contract along straight lines joining the curve to P.]

APPENDICES

A POLAR COORDINATES

A point P with Cartesian coordinates (x, y) can be referred to by its *polar coordinates, r* and θ. The number r is the distance between P and the origin, and θ is the angle between the positive x-axis and the line joining P to the origin (with the convention that counterclockwise is positive). Figure A.1 shows the connection between Cartesian and polar coordinates:

Relation between Cartesian and Polar Coordinates

$$x = r \cos \theta \qquad r = \sqrt{x^2 + y^2}$$
$$y = r \sin \theta \qquad \tan \theta = \frac{y}{x}$$

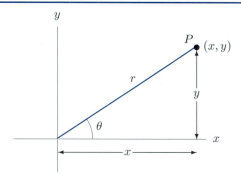

Figure A.1: Cartesian and polar coordinates

Example 1   Convert the point $(x, y) = (-3, 4)$ to polar coordinates.

Solution   The formula for r gives
$$r = \sqrt{x^2 + y^2} = \sqrt{(-3)^2 + 4^2} = 5.$$
The formula involving the angle θ is
$$\tan \theta = \frac{y}{x} = -\frac{4}{3}.$$
This means $\theta = \arctan(-4/3) \approx -0.927$ or $\theta = \arctan(-4/3) + \pi \approx 2.21$. Since the point is in the second quadrant, we choose $\theta \approx 2.21$.

Problems for Section A

For Problems 1–7, give Cartesian coordinates for the points with the following polar coordinates (r, θ). The angles are measured in radians.

1. $(1, 0)$
2. $(0, 1)$
3. $(2, \pi)$
4. $(\sqrt{2}, 5\pi/4)$
5. $(5, -\pi/6)$
6. $(3, \pi/2)$
7. $(1, 1)$

For Problems 8–15, give polar coordinates for the points with the following Cartesian coordinates. Choose $0 \le \theta < 2\pi$.

8. $(1, 0)$
9. $(0, 2)$
10. $(1, 1)$
11. $(-1, 1)$
12. $(-3, -3)$
13. $(0.2, -0.2)$
14. $(3, 4)$
15. $(-3, 1)$

In Problems 16–21, graph the functions $r = g(\theta)$ in the xy-plane using a calculator. Explain how the equations are related to the shapes of your graphs.

16. $r = 1$

17. $\theta = \dfrac{\pi}{3}$

18. $r = \dfrac{\theta}{10}$

19. $r = \dfrac{2}{\cos \theta}$

20. $r = 2 \cos \theta$

21. $r = \dfrac{\sin \theta}{\cos^2 \theta}$

22.   Every point in the plane can be represented by some pair of polar coordinates, but are the polar coordinates (r, θ) uniquely determined by the Cartesian coordinates (x, y)? In other words, for each pair of Cartesian coordinates, is there one and only one pair of polar coordinates for that point? Why or why not?

B COMPLEX NUMBERS

The quadratic equation

$$x^2 - 2x + 2 = 0$$

is not satisfied by any real number x. If you try applying the quadratic formula, you get

$$x = \frac{2 \pm \sqrt{4 - 8}}{2} = 1 \pm \frac{\sqrt{-4}}{2}.$$

Apparently, you need to take a square root of -4. But -4 doesn't have a square root, at least, not one which is a real number. Let's give it a square root.

We define the imaginary number i to be a number such that

$$i^2 = -1.$$

Using this i, we see that $(2i)^2 = -4$, so

$$x = 1 \pm \frac{\sqrt{-4}}{2} = 1 \pm \frac{2i}{2} = 1 \pm i.$$

This solves our quadratic equation. The numbers $1 + i$ and $1 - i$ are examples of complex numbers.

A **complex number** is defined as any number that can be written in the form

$$z = a + bi,$$

where a and b are real numbers and $i = \sqrt{-1}$.
The *real part* of z is the number a; the *imaginary part* is the number bi.

Calling the number i imaginary makes it sound as if i doesn't exist in the same way that real numbers exist. In some cases, it is useful to make such a distinction between real and imaginary numbers. For example, if we measure mass or position, we want our answers to be real numbers. But the imaginary numbers are just as legitimate mathematically as the real numbers are.

As an analogy, consider the distinction between positive and negative numbers. Originally, people thought of numbers only as tools to count with; their concept of "five" or "ten" was not far removed from "five arrows" or "ten stones." They were unaware that negative numbers existed at all. When negative numbers were introduced, they were viewed only as a device for solving equations like $x + 2 = 1$. They were considered "non-numbers," or, in Latin, "negative numbers." Thus, even though people started to use negative numbers, they did not view them as existing in the same way that positive numbers did. An early mathematician might have reasoned: "The number 5 exists because I can have 5 coins in my hand. But how can I have -5 coins in my hand?" Today we have an answer: "I have -5 coins" means I owe somebody 5 coins. We have realized that negative numbers are just as real as positive ones, and that in some cases, negative numbers can have physical meaning, even though they are useless for measuring length or keeping baseball scores. As we will see, complex numbers can have physical meaning as well. For example, complex numbers are used in studying wave motion in electric circuits.

Algebra of Complex Numbers

Numbers such as 0, 1, $\frac{1}{2}$, π, and $\sqrt{2}$ are called *purely real* because they contain no imaginary components. Numbers such as i, $2i$, and $\sqrt{2}i$ are called *purely imaginary* because they contain only the number i multiplied by a nonzero real coefficient.

Two complex numbers are called *conjugates* if their real parts are equal and if their imaginary parts are opposites. The complex conjugate of the complex number $z = a + bi$ is denoted $\bar{z}$ (pronounced "z bar"), so we have

$$\bar{z} = a - bi.$$

(Note that z is real if and only if $z = \bar{z}$.) Complex conjugates have the following remarkable property: if $f(x)$ is any polynomial with real coefficients ($x^3 + 1$, say) and $f(z) = 0$, then $f(\bar{z}) = 0$. This means that if z is the solution to a polynomial equation with real coefficients, then so is $\bar{z}$.

- Adding two complex numbers is done by adding real and imaginary parts separately:

$$(a + bi) + (c + di) = (a + c) + (b + d)i.$$

- Subtracting is similar:

$$(a + bi) - (c + di) = (a - c) + (b - d)i.$$

- Multiplication works just like for polynomials, using $i^2 = -1$:

$$(a + bi)(c + di) = a(c + di) + bi(c + di)$$
$$= ac + adi + bci + bdi^2$$
$$= ac + adi + bci - bd = (ac - bd) + (ad + bc)i.$$

- Powers of i: We know that $i^2 = -1$; then $i^3 = i \cdot i^2 = -i$, and $i^4 = (i^2)^2 = (-1)^2 = 1$. Then $i^5 = i \cdot i^4 = i$, and so on. Thus we have

$$(bi)^n = b^n i^n = \begin{cases} b^n i & \text{for } n = 1, 5, 9, 13, \ldots \\ -b^n & \text{for } n = 2, 6, 10, 14, \ldots \\ -b^n i & \text{for } n = 3, 7, 11, 15, \ldots \\ b^n & \text{for } n = 4, 8, 12, 16, \ldots \end{cases}$$

- The product of a number and its conjugate is always real and nonnegative:

$$z \cdot \bar{z} = (a + bi)(a - bi) = a^2 - abi + abi - b^2 i^2 = a^2 + b^2.$$

- Dividing is done by multiplying the denominator by its conjugate, thereby making the denominator real:

$$\frac{a + bi}{c + di} = \frac{a + bi}{c + di} \cdot \frac{c - di}{c - di} = \frac{ac - adi + bci - bdi^2}{c^2 + d^2} = \frac{ac + bd}{c^2 + d^2} + \frac{bc - ad}{c^2 + d^2}i.$$

Example 1   Compute $(2 + 7i)(4 - 6i) - i$.

Solution $(2 + 7i)(4 - 6i) - i = 8 + 28i - 12i - 42i^2 - i = 8 + 15i + 42 = 50 + 15i.$

Example 2   Compute $\dfrac{2 + 7i}{4 - 6i}$.

Solution

$$\frac{2 + 7i}{4 - 6i} = \frac{2 + 7i}{4 - 6i} \cdot \frac{4 + 6i}{4 + 6i} = \frac{8 + 12i + 28i + 42i^2}{4^2 + 6^2} = \frac{-34 + 40i}{52} = \frac{-17}{26} + \frac{10}{13}i.$$

The Complex Plane and Polar Coordinates

It is often useful to picture a complex number $z = x + iy$ in the plane, with x along the horizontal axis and y along the vertical. The xy-plane is then called the *complex plane*. Figure B.2 shows the complex numbers $-2i$, $1 + i$, and $-2 + 3i$.

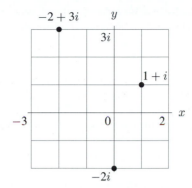

Figure B.2: Points in the complex plane

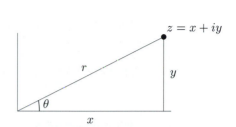

Figure B.3: The point $z = x + iy$ in the complex plane, showing polar coordinates

The triangle in Figure B.3 shows that a complex number can be written using polar coordinates as follows:

$$z = x + iy = r \cos \theta + ir \sin \theta.$$

Example 3   Express $z = -2i$ and $z = -2 + 3i$ using polar coordinates. (See Figure B.2.)

Solution   For $z = -2i$, the distance of z from the origin is 2, so $r = 2$. Also, one value for θ is $\theta = 3\pi/2$. Using polar coordinates, $-2i = 2\cos(3\pi/2) + i\,2(\sin 3\pi/2)$.
For $z = -2 + 3i$, we have $x = -2$, $y = 3$. So $r = \sqrt{(-2)^2 + 3^2} \approx 3.61$, and one solution of $\tan \theta = 3/(-2)$ is $\theta \approx 2.16$. So $-2 + 3i \approx 3.61 \cos(2.16) + i\,3.61 \sin(2.16)$.

Example 4   Consider the point with polar coordinates $r = 5$ and $\theta = 3\pi/4$. What complex number does this point represent?

Solution   Since $x = r \cos \theta$ and $y = r \sin \theta$ we see that $x = 5 \cos 3\pi/4 = -5/\sqrt{2}$, and $y = 5 \sin 3\pi/4 = 5/\sqrt{2}$, so $z = -5/\sqrt{2} + i\,5/\sqrt{2}$.

Euler's Formula

Consider the complex number z lying on the unit circle in Figure B.4. Writing z in polar coordinates, and using the fact that $r = 1$, we have

$$z = f(\theta) = \cos \theta + i \sin \theta.$$

It turns out that there is a particularly beautiful and compact way of rewriting $f(\theta)$ using complex exponentials. Let's take the derivative of f, treating i like any other constant but using the fact that $i^2 = -1$:

$$f'(\theta) = -\sin \theta + i \cos \theta = i \cos \theta + i^2 \sin \theta.$$

Factoring out an i gives

$$f'(\theta) = i(\cos \theta + i \sin \theta) = i \cdot f(\theta).$$

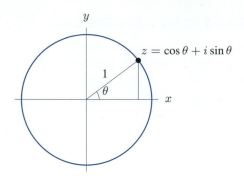

Figure B.4: Complex number represented by a point on the unit circle

As you know from Chapter 10, page 510, the only real-valued function whose derivative is proportional to the function itself is the exponential function. In other words, we know that if

$$g'(x) = k \cdot g(x), \quad \text{then} \quad g(x) = Ce^{kx}$$

for some constant C. If we assume that a similar result holds for complex-valued functions, then we have

$$f'(\theta) = i \cdot f(\theta), \quad \text{so} \quad f(\theta) = Ce^{i\theta}$$

for some constant C. To find C we substitute $\theta = 0$. Now $f(0) = Ce^{i \cdot 0} = C$, and since $f(0) = \cos 0 + i \sin 0 = 1$, we must have $C = 1$. Therefore $f(\theta) = e^{i\theta}$. Thus we have *Euler's formula:*

$$\boxed{e^{i\theta} = \cos \theta + i \sin \theta}$$

This elegant and surprising relationship was discovered by the Swiss mathematician Leonhard Euler in the eighteenth century, and it is particularly useful in solving second-order differential equations. Another derivation of Euler's formula (using Taylor series) is given in Problem 16 on page 487. It allows us to write the complex number represented by the point with polar coordinates (r, θ) in the following form:

$$\boxed{z = r(\cos \theta + i \sin \theta) = re^{i\theta}.}$$

Similarly, since $\cos(-\theta) = \cos \theta$ and $\sin(-\theta) = -\sin \theta$, we have

$$re^{-i\theta} = r\left(\cos(-\theta) + i \sin(-\theta)\right) = r(\cos \theta - i \sin \theta).$$

Example 5   Evaluate $e^{i\pi}$.

Solution   Using Euler's formula, $e^{i\pi} = \cos \pi + i \sin \pi = -1$.

Example 6   Express the complex number represented by the point $r = 8$, $\theta = 3\pi/4$ in Cartesian form and polar form, $z = re^{i\theta}$.

Solution   Using Cartesian coordinates, the complex number is

$$z = 8\left(\cos\left(\frac{3\pi}{4}\right) + i \sin\left(\frac{3\pi}{4}\right)\right) = \frac{-8}{\sqrt{2}} + i\frac{8}{\sqrt{2}}.$$

Using polar coordinates, we have

$$z = 8e^{i\,3\pi/4}.$$

Among its many benefits, the polar form of complex numbers makes finding powers and roots of complex numbers much easier. Using the polar form, $z = re^{i\theta}$, for a complex number, we may find any power of z as follows:

$$z^p = (re^{i\theta})^p = r^p e^{ip\theta}.$$

To find roots, we let p be a fraction, as in the following example.

Example 7   Find a cube root of the complex number represented by the point with polar coordinates $(8, 3\pi/4)$.

Solution   In Example 6, we saw that this complex number could be written as $z = 8e^{i3\pi/4}$. So,

$$\sqrt[3]{z} = \left(8e^{i\,3\pi/4}\right)^{1/3} = 8^{1/3}e^{i(3\pi/4)\cdot(1/3)} = 2e^{\pi i/4} = 2\left(\cos(\pi/4) + i\sin(\pi/4)\right)$$

$$= 2\left(1/\sqrt{2} + i/\sqrt{2}\right) = \sqrt{2}(1 + i).$$

Problems for Section B

For Problems 1–8, express the given complex number in polar form, $z = re^{i\theta}$.

1. $2i$
2. -5
3. $1 + i$
4. $-3 - 4i$
5. 0
6. $-i$
7. $-1 + 3i$
8. $5 - 12i$

For Problems 9–18, perform the indicated calculations. Give your answer in Cartesian form, $z = x + iy$.

9. $(2 + 3i) + (-5 - 7i)$
10. $(2 + 3i)(5 + 7i)$
11. $(2 + 3i)^2$
12. $(1 + i)^2 + (1 + i)$
13. $(0.5 - i)(1 - i/4)$
14. $(2i)^3 - (2i)^2 + 2i - 1$
15. $(e^{i\pi/3})^2$
16. $\sqrt{e^{i\pi/3}}$
17. $(5e^{i7\pi/6})^3$
18. $\sqrt[4]{10e^{i\pi/2}}$

By writing the complex numbers in polar form, $z = re^{i\theta}$, find a value for the quantities in Problems 19–28. Give your answer in Cartesian form, $z = x + iy$.

19. $\sqrt{i}$
20. $\sqrt{-i}$
21. $\sqrt[3]{i}$
22. $\sqrt{7i}$
23. $(1 + i)^{100}$
24. $(1 + i)^{2/3}$
25. $(-4 + 4i)^{2/3}$
26. $(\sqrt{3} + i)^{1/2}$
27. $(\sqrt{3} + i)^{-1/2}$
28. $(\sqrt{5} + 2i)^{\sqrt{2}}$

Solve the simultaneous equations in Problems 29–30 for A_1 and A_2.

29. $A_1 + A_2 = 2$
     $(1 - i)A_1 + (1 + i)A_2 = 3$

30. $A_1 + A_2 = 2$
     $(i - 1)A_1 + (1 + i)A_2 = 0$

31.   Let $z_1 = -3 - i\sqrt{3}$ and $z_2 = -1 + i\sqrt{3}$.

     (a)   Find $z_1 z_2$ and z_1/z_2. Give your answer in Cartesian form, $z = x + iy$.
     (b)   Put z_1 and z_2 into polar form, $z = re^{i\theta}$. Find $z_1 z_2$ and z_1/z_2 using the polar form, and verify that you get the same answer as in part (a).

32.   If the roots of the equation $x^2 + 2bx + c = 0$ are the complex numbers $p \pm iq$, find expressions for p and q in terms of b and c.

Are the statements in Problems 33–38 true or false? Explain your answer.

33. Every nonnegative real number has a real square root.

34. For any complex number z, the product $z \cdot \bar{z}$ is a real number.

35. The square of any complex number is a real number.

36. If f is a polynomial, and $f(z) = i$, then $f(\bar{z}) = i$.

37. Every nonzero complex number z can be written in the form $z = e^w$, where w is another complex number.

38. If $z = x + iy$, where x and y are positive, then $z^2 = a + ib$ has a and b positive.

For Problems 39–43, use Euler's formula to derive the following relationships. (Note that if a, b, c, d are real numbers, $a + bi = c + di$ means that $a = c$ and $b = d$.)

39. $\sin^2 \theta + \cos^2 \theta = 1$ 40. $\sin 2\theta = 2 \sin \theta \cos \theta$ 41. $\cos 2\theta = \cos^2 \theta - \sin^2 \theta$

42. $\dfrac{d}{d\theta} \sin \theta = \cos \theta$ 43. $\dfrac{d^2}{d\theta^2} \cos \theta = -\cos \theta$

C DETERMINANTS

We introduce the determinant of an array of numbers. Each 2 by 2 array of numbers has another number associated with it, called its determinant, which is given by

$$\begin{vmatrix} a_1 & a_2 \\ b_1 & b_2 \end{vmatrix} = a_1 b_2 - a_2 b_1.$$

For example

$$\begin{vmatrix} 2 & 5 \\ -4 & -6 \end{vmatrix} = 2(-6) - 5(-4) = 8.$$

Each 3 by 3 array of numbers also has a number associated with it, also called a determinant, which is defined in terms of 2 by 2 determinants as follows:

$$\begin{vmatrix} a_1 & a_2 & a_3 \\ b_1 & b_2 & b_3 \\ c_1 & c_2 & c_3 \end{vmatrix} = a_1 \begin{vmatrix} b_2 & b_3 \\ c_2 & c_3 \end{vmatrix} - a_2 \begin{vmatrix} b_1 & b_3 \\ c_1 & c_3 \end{vmatrix} + a_3 \begin{vmatrix} b_1 & b_2 \\ c_1 & c_2 \end{vmatrix}.$$

Notice that the determinant of the 2 by 2 array multiplied by a_i is the determinant of the array found by removing the row and column containing a_i. Also, note the minus sign in the second term. An example is given by

$$\begin{vmatrix} 2 & 1 & -3 \\ 0 & 3 & -1 \\ 4 & 0 & 5 \end{vmatrix} = 2 \begin{vmatrix} 3 & -1 \\ 0 & 5 \end{vmatrix} - 1 \begin{vmatrix} 0 & -1 \\ 4 & 5 \end{vmatrix} + (-3) \begin{vmatrix} 0 & 3 \\ 4 & 0 \end{vmatrix} = 2(15 + 0) - 1(0 - (-4)) + (-3)(0 - 12) = 62.$$

Suppose the vectors $\vec{a}$ and $\vec{b}$ have components $\vec{a} = a_1 \vec{i} + a_2 \vec{j} + a_3 \vec{k}$ and $\vec{b} = b_1 \vec{i} + b_2 \vec{j} + b_3 \vec{k}$. Recall that the cross product $\vec{a} \times \vec{b}$ is given by the expression

$$\vec{a} \times \vec{b} = (a_2 b_3 - a_3 b_2)\vec{i} + (a_3 b_1 - a_1 b_3)\vec{j} + (a_1 b_2 - a_2 b_1)\vec{k}.$$

Notice that if we expand the following determinant, we get the cross product:

$$\begin{vmatrix} \vec{i} & \vec{j} & \vec{k} \\ a_1 & a_2 & a_3 \\ b_1 & b_2 & b_3 \end{vmatrix} = \vec{i}(a_2 b_3 - a_3 b_2) - \vec{j}(a_1 b_3 - a_3 b_1) + \vec{k}(a_1 b_2 - a_2 b_1) = \vec{a} \times \vec{b}.$$

Determinants give a useful way of computing cross products.

D PROJECTS

1. AN ORBITING SATELLITE   **Chapter 3**

Overview: Consider a NASA satellite which orbits the earth every 90 minutes. During an orbit, the satellite's electric power comes either from solar array wings, when these are illuminated by the sun, or from batteries. The batteries discharge whenever the satellite uses more electricity than the solar array can provide or whenever the satellite is in the shadow of the earth (where the solar array cannot be used). If the batteries are overused, however, they can be damaged.

The Problem:[1] You are to determine whether the batteries could be damaged in either of the following operations. You are told that the battery capacity is 50 ampere-hours. If the total battery discharge does not exceed 40% of battery capacity, the batteries will not be damaged.

(a) Operation 1 is performed by the satellite while orbiting the earth. At the beginning of a given 90-minute orbit, the satellite performs a 15-minute maneuver which requires more current than the solar array can deliver, causing the batteries to discharge. The maneuver causes a sinusoidally varying battery discharge of period 30 minutes with a maximum discharge of ten amperes at 7.5 minutes. For the next 45 minutes the solar array meets the total satellite current demand, and the batteries do not discharge. During the last 30 minutes, the satellite is in the shadow of the earth and the batteries supply the total current demand of 30 amperes.

  (i) The battery current in amperes is a function of time. Plot the function, showing the current in amperes as a function of time for the 90-minute orbit. Write a formula (or formulas) for the battery current function.

  (ii) Calculate the total battery discharge (in units of ampere-hours) for the 90-minute orbit for Operation 1.

  (iii) What is your recommendation regarding the advisability of Operation 1?

(b) Operation 2 is simulated at NASA's laboratory in Houston. The following graph was produced by the laboratory simulation of the current demands on the battery during the 90-minute orbit required for Operation 2.

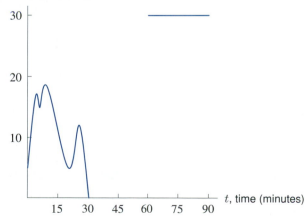

Figure D.5: Battery discharge simulation graph for Operation 2

  (i) Calculate the total battery discharge (in units of ampere-hours) for the 90-minute orbit for Operation 2.

  (ii) What is your recommendation regarding the advisability of Operation 2?

---

[1] Adapted from Amy C. R. Gerson, "Electrical Engineering: Space Systems," in *She Does Math! Real Life Problems from Women on the Job*, ed. Marla Parker, p. 61 (Washington, DC: Mathematical Association of America, 1995).

2. **THE EARTH'S PATH**   **Chapter 3**

Overview: The orbits of the earth and Halley's comet are both ellipses. In this project we identify some parameters of these orbits.

The Problem:[2] The equation of an ellipse is

$$\frac{x^2}{a^2} + \frac{y^2}{b^2} = 1,$$

where a and b are positive constants, $a \geq b$. The foci of this ellipse are located at $(c, 0)$ and $(-c, 0)$, where $c = \sqrt{a^2 - b^2}$. The eccentricity[3] of the ellipse is defined as $e = c/a$. The length of the ellipse's perimeter is given by

$$\int_0^{\pi/2} 4a(1 - e^2 \sin^2 \theta)^{1/2} d\theta.$$

The path of the earth is an ellipse, with the sun at one of its foci. The closest the earth comes to the sun is 91.5 million miles and the furthest is 94.5 million miles.

The path of Halley's comet is also an ellipse, with the sun at one of its foci. The semi-major axis, a, of the ellipse is 1674 million miles and its eccentricity is 0.97. It takes about 76 years for Halley's comet to orbit the sun. In 1705 Halley (1656-1742) predicted correctly that the comet would be seen in 1758.

(a) What is the eccentricity of a circle?
(b) What does the magnitude of e say about the shape of an ellipse?
(c) For the earth and sun configuration, what are the values of a, b, c, and e?
(d) How far does the earth travel in one orbit of the sun?
(e) What is the average speed of the earth around the sun in miles per second?
(f) If the earth's path were a circle but it still traveled the same distance as in part (c), what would the radius have to be?
(g) How close does Halley's comet come to the sun? How far does it travel in one orbit? What is the average speed around the sun? Compare this to part (e).
(h) In what year did Halley see the comet? Did he know whether his prediction came true?

3. **GAS TANKS**   **Chapter 3**

Overview: Gasoline is stored at a gas station in an underground tank which is a circular cylinder lying on its side. You may have seen a gas station attendant dropping a long measuring stick into the tank to measure the amount of gasoline remaining. Suppose the remaining fuel comes to a height of h inches and the diameter of the tank is d inches. It might be tempting to guess that the tank is h/d full. The purpose of this project is to see how far off such a guess is.

The Problem: Assume a cylindrical gas tank has diameter d and length l, as in Figure D.6.

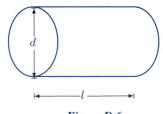

Figure D.6

(a) Suppose another gas tank has the shape of a w by d by l rectangular box. What is w if this tank holds the same amount as the cylindrical tank? Does h/d give an exact measure of how full the rectangular tank is?

---

[2]This project is adapted from M.E. Moody and K. Shannon, *Microcomputer Exercises for Calculus*, (St. Paul: West, 1988).

[3]Although it is traditionally represented by the letter e, the eccentricity has nothing to do with the number 2.71828 . . . .

(b) When computing the accuracy of the stick method for determining what fraction of the tank is full, why can we ignore the length of the tank?

(c) Draw cross-sections of both tanks on one set of axes, with the x-axis measuring height and with both tanks centered at the origin.

(d) Let $E(h)$ be the error, as a function of h, when using h/d to measure how full the cylindrical tank is. Use the picture you drew in part (c) to write $E(h)$ as an integral.

(e) For what value of h is the error maximal?

4. **THE GRAPHS OF $y = b^x$ AND $y = x^b$**   **Chapter 4**

Overview: This project investigates how the number of intersections of the curves $y = b^x$ and $y = x^b$ depends on the value of the parameter b. Throughout the project, we restrict our attention to $x, y > 0$ and $b > 1$.

The Problem:[4]

(a) Where do the graphs of $y = 2^x$ and $y = x^2$ cross?

(b) Where do the graphs of $y = 3^x$ and $y = x^3$ cross?

(c) Where do the graphs of $y = 4^x$ and $y = x^4$ cross?

We now consider the graphs of $y = b^x$ and $y = x^b$ for any $b > 1$.

(d) There is one obvious crossing point for the graphs of $y = b^x$ and $y = x^b$. Find this point.

(e) Using parts (a)-(c), decide if there is a second crossing point for the graphs of $y = b^x$ and $y = x^b$. If so, is it to the left or right of the point you found in part (d)?

(f) Is there a value of b for which the graphs of $y = b^x$ and $y = x^b$ touch only once? Guess an answer and confirm it using a calculator or computer.

(g) For the value of b you chose in part (f), show analytically that the graphs of $y = b^x$ and $y = x^b$ touch.

(h) (Optional; Uses Chapter 5.) Show that for any $b > 1$, the graphs of $y = b^x$ and $y = x^b$ cross. Do this by considering the fact that $b^x = x^b$ whenever the function $f(x) = x \ln b - b \ln x$ is zero.

5. **THE ROAD AND THE RIVER**   **Chapter 5**

Overview: The equation of a road, a river, and a dam are given; we will find the location of a spillway and a road.

The Problem: A particular river follows the quadratic curve $f(x) = x^2$, while a nearby road follows the cubic curve

$$g(x) = ax^3 + bx^2 + cx + d.$$

There are three bridges on the road and they cross the river at the points $(-1, 1)$, $(0, 0)$ and $(1, 1)$. (The units are kilometers.) The bridge at $(1, 1)$ is at right angles to the river. A dam is located at $(1, 2)$. The dam is not on this river. There is a straight spillway from the dam to the nearest point on the river.

(a) What are the values of a, b, c, and d?

(b) What are the distances between the bridges, as the crow flies?

(c) At what angles (in radians and degrees) do the other two bridges cross the river?

(d) Where does the spillway from the dam intersect the river?

(e) A straight access road is to be built from the existing road to the dam. Where on the original road should it start to have minimum length?

(f) A farmer owns the land between the river and the road, bounded by the first and last bridges. How many square kilometers does the farmer own?

---

[4]Based on an idea from Gil Strang.

6. **INTERSTATE TRUCKING**   **Chapter 5**

Overview: In this project we will investigate the effect of the price of diesel fuel and drivers' wages on a trucker's optimal driving speed. We will analyze the relationship between fuel cost and wages and the 55 mph speed limit.

The Problem:[5] After an extensive study, the Interstate Commerce Commission has concluded that the primary variable expenses for any over-the-road freight hauler are the wages for the driver and the cost of fuel. The study concluded that maintenance and replacement costs for vehicles, although considerable, did not vary significantly from carrier to carrier. On the other hand, the fuel costs and wages did vary. Given this result, your project is to determine how these two variables affect the cost of transporting freight. In particular, you are to determine if there is an optimal wage to fuel cost relationship for which most haulers would encourage their drivers to abide by a 55 mile per hour speed limit.

Reading the study tells you that under ideal conditions, an interstate freight hauling vehicle (18 wheeler) gets 6 miles per gallon of fuel. This mileage is affected by the speed of the vehicle and its weight. In general, the miles per gallon of fuel decreases by 0.2 for each increase of 10,000 pounds in the weight of the truck and freight over 25,000 pounds. Also, the miles per gallon of fuel decreases by 0.1 mile per gallon for each mile per hour the truck averages above 45 miles per hour.

(a)   Using this information, create an expression for the cost per mile of driving, taking into account only the drivers' wages and fuel cost.

(b)   Suppose that the national average for diesel fuel is $1.25 per gallon, that drivers on the average earn $15.00 per hour, and that the average weight of a loaded truck is 75,000 pounds. What is the optimal average speed under these conditions?

(c)   Compute the cost per mile for 55 and 60 miles per hour. Suppose you hauled produce from California to Maine. What average speed would you choose?

(d)   In solving parts (a)-(c), you found an equation relating cost, wages, fuel cost, weight of the truck, and the average speed of the truck. This equation is a mathematical model of interstate freight hauling costs. Using this model, determine the fuel cost that makes 55 miles per hour the optimum average speed for trucking companies.

(e)   Each year drivers' wages change as a result of contract renewals and inflation. The cost of fuel also fluctuates, but can be adjusted by surcharges and fuel taxes. Thus, the relationship between wages and the cost of fuel can be altered by various government agencies. Assuming that 75,000 pounds remains the national average for the weight of a truck involved in interstate freight hauling, what should be the relationship between fuel cost and wages to maintain 55 miles per hour as the optimal speed?

(f)   There have been some suggestions of changing the road use tax to lower the average weight of over-the-road freight haulers. Assuming the relationship of wages to fuel cost found in part (e) is maintained, how would lowering the average weight affect the optimal average speed?

7. **APPROXIMATING π WITH INTEGRALS**   **Chapter 7**

Overview: Since π is irrational, it is often useful to approximate it by a rational number. We derive an explicit formula for finding rational approximations to π. The inequality we derive was first discovered in about 1650 by John Wallis, though using a different method than the one we shall follow.

The Problem:

(a)   Use integration by parts to show that if n is an even positive integer, then

$$\int_0^{\pi/2} \sin^n x\, dx = \frac{n-1}{n} \cdot \frac{n-3}{n-2} \cdots \frac{3}{4} \cdot \frac{1}{2} \cdot \frac{\pi}{2}.$$

[5]Adapted from L. Carl Leinbach, *Calculus Laboratories Using Derive*, (Belmont, CA: Wadsworth, 1991). Reprinted by permission of Brooks/Cole Publishing Company, a division of International Thompson Publishing, Inc.

(b)   Use integration by parts to show that if n is an odd integer and $n > 1$, then

$$\int_0^{\pi/2} \sin^n x \, dx = \frac{n-1}{n} \cdot \frac{n-3}{n-2} \cdots \frac{4}{5} \cdot \frac{2}{3}.$$

(c)   Explain why the following inequality is true for any integer k:

$$\int_0^{\pi/2} \sin^{2k+1} x \, dx < \int_0^{\pi/2} \sin^{2k} x \, dx < \int_0^{\pi/2} \sin^{2k-1} x \, dx.$$

Use this inequality to show, for any integer $k > 1$, that

$$2 \cdot \frac{(2k)(2k)(2k-2)(2k-2) \cdots 4 \cdot 4 \cdot 2 \cdot 2}{(2k+1)(2k-1)(2k-1) \cdots 5 \cdot 5 \cdot 3 \cdot 3 \cdot 1} < \pi < 2 \cdot \frac{(2k)(2k-2)(2k-2) \cdots 4 \cdot 4 \cdot 2 \cdot 2}{(2k-1)(2k-1) \cdots \cdot 5 \cdot 5 \cdot 3 \cdot 3 \cdot 1}.$$

(d)   How large is the gap between the upper and the lower bounds for π given by these inequalities? Try different values of k and make a table showing the size of the gap against the value of k. Does the size of the gap appear proportional to some function of k? What does this tell us about the accuracy of the approximation?

8.   **SURFACE AREA OF AN UNPAINTABLE CAN OF PAINT**   **Chapter 8**

Overview: This project introduces the formula for the surface area of a volume of revolution. It then shows that it is possible to have a solid whose volume is finite but whose surface area is infinite.

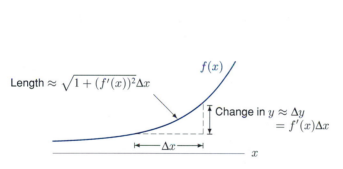

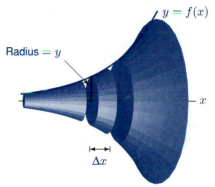

Figure D.7   *Figure D.8*

You know that the arclength of the curve $y = f(x)$ from a to b can be found using the integral

$$\text{Arc length} = \int_a^b \sqrt{1 + (f'(x))^2} \, dx.$$

Figure D.7 shows the corresponding arc length of a small piece of the curve.

Similarly, it can be shown that the surface area from a to b of the solid obtained by revolving $y = f(x)$ around the x-axis is given by

$$\text{Surface area} = 2\pi \int_a^b f(x) \sqrt{1 + (f'(x))^2} \, dx.$$

We can see this why this might be true by looking at Figure D.8.

We approximate the small piece of the surface by a slanted cylinder of radius y and "height" equal to the arclength of the curve, so that

$$\text{Surface area of edge of slice} \approx 2\pi y \sqrt{1 + (f'(x))^2} \, \Delta x.$$

Integrating this expression from $x = a$ to $x = b$ gives the surface area of the solid.

The Problem:

(a)   Calculate the surface area of a sphere of radius r.

(b)   Calculate the surface area of a cone of radius r and height h.

(c)   Rotate the curve $y = 1/x$ for $x \geq 1$ around the x-axis. Find the volume of this solid.

(d)   Show that the surface area of the solid in part (c) is infinite. [Hint: You might not be able to find an antiderivative of the integrand in the surface area formula; can you get a lower bound on the integral?]

(e)   (Optional. Requires Chapter 10.) Find a curve such that when the portion of the curve from $x = a$ to $x = b$ is rotated around the x-axis (for any a and b), the volume of the solid of revolution is equal to its surface area. You may assume $dy/dx \geq 0$.

9.   **QUININE**   **Chapter 9**

Overview: Malaria is a parasitic infection transmitted by mosquito bites, mainly in tropical areas of the world. The disease has existed since ancient times, and currently there are hundreds of millions of cases each year, with millions of deaths. Around 1630, Jesuits in Peru introduced the bark of the cinchona tree to the West as the first treatment for malaria. The drug quinine is the active ingredient in the bark, and it is still used today. However, the concentration in the body of quinine, and of most drugs, must be kept within certain parameters. If the concentration is too low, the drug is ineffective. If the concentration is too high, toxic side effects can result.

The Problem: Suppose you are a doctor who prescribes quinine for a 70 kg malaria patient. At 8 am each day, she is to receive 50 mg of the drug.[6] To be effective, the average concentration in the body must be at least 0.4 mg/kg. However, concentrations above 3.0 mg/kg can be fatal. The half-life of quinine in the body is 11.5 hours.

(a)   What is the continuous rate of decay of quinine (in units of %/min)?

(b)   How much quinine is in the patient's system just before and just after the first day's dose? After the second day's dose?

(c)   How much is in her system after a few more doses? After the nth dose? Explain what happens in the long run.

(d)   Graph the concentration of the drug versus time showing what the steady-state looks like. Find a formula for the repeating function.

(e)   What is the average (over time) of the concentration of quinine in the patient's body?

(f)   Is this treatment both effective and safe?

(g)   Suppose instead that the patient were to receive two doses of 25 mg each day, one at 8 am and one at 8 pm. Is this a safe and effective treatment?

(h)   Determine the average value of an exponentially decaying function between two points (x_0, y_0) and (x_1, y_1).

(i)   Notice that the average concentration of quinine is the same for the 50 mg once per day and 25 mg twice per day treatments. Use part (h) to explain why. Would the average concentration be the same for a 100 mg dose once every two days?

(j)   Suppose the original quinine regimen is stopped. How long after the last dose will the amount of quinine be less than 10^{-10} times the patient's body mass?

10.   **US ELECTRICITY CONSUMPTION**   **Chapter 10**

Overview: In this project we are given US electricity consumption over much of the 20th century and asked to predict consumption in the future. Several different predictions are made by fitting various differential equations to the data.

The Problem: The following data shows US electricity consumption, measured in billions of kilowatt hours per year, for much of this century.

---

[6]This is a simplified model; actual treatments involve different drugs and more complicated dosage regimens. See, for example, *The Pharmacological Basis of Therapeutics*, 9th Ed., ed. Joel G. Hardman, Alfred Goodman Gilman, and Lee E. Limbird, (New York: McGraw Hill, 1996).

Year	1912	1917	1920	1929	1936	1945
Electricity consumption (E)	12	25	39	92	109	222

Year	1955	1960	1965	1970	1980	1987
Electricity consumption (E)	547	755	1055	1531	2286	2455

(a) Decide how well each of the following differential equations model the growth of electricity consumption, E, over time, t, measured in years. Do this by using the data to approximate dE/dt. For each differential equation, make a plot to which you can fit a line and use it to estimate the parameters a, b, or c.

(i) $\dfrac{1}{E}\dfrac{dE}{dt} = c$

(ii) $\dfrac{dE}{dt} = a - bE$

(iii) $\dfrac{1}{E}\dfrac{dE}{dt} = a - bE$

(iv) $\dfrac{1}{E}\dfrac{dE}{dt} = a - bt$

(b) Use each model to predict energy consumption for the year 2000. Which prediction do you think is most reliable? Why? Discuss the long-run predictions from each model. Which differential equation do you think fits best?

11. TERMINAL VELOCITY   Chapter 10

Overview: The motion of falling objects is often modeled by the equation $dv/dt = -g$. However, since the earth is surrounded by air, any accurate description of falling objects must take account of air resistance. We frequently use the model:

$$m\frac{dv}{dt} = mg - f(v),$$

where t is time, v is the speed of the object measured downwards and m is its mass. The term mg represents the force of gravity; it is positive and acts to speed up the motion of a falling object. The value of g varies slightly from place to place, but is very close to 9.8 m/sec² everywhere on the earth. The function $f(v)$ represents the force exerted by the air. It will oppose the motion of a falling object and so we have written it with a negative sign. Common sense says $f(0) = 0$. Let us assume $f(v)$ strictly increases with v; that is, the faster the object moves, the greater the air resistance. The actual function f may be obtained either experimentally or theoretically. If the falling object is irregular, like a person, the air resistance depends on the orientation of the object as it falls. To avoid this complication, we assume that the object is a ball.

The Problem:

(a) Make up a few possible f's. The only requirements are that f be strictly increasing and that $\lim_{v \to \infty} f = \infty$. Look at the corresponding slope fields (on v vs. t axes) and sketch solutions for each. What do these solutions have in common? Explain, in terms of derivatives, the concavity of your solutions. [Hint: Take $m = 1$. Consider f proportional to v, v^2, v^3.]

(b) Under the assumptions of part (a), the differential equation has exactly one constant solution. If $f(v) = kv$, then what is the constant solution? How about for $f(v) = kv^2$? How about for a general increasing f? Explain why this velocity is called the terminal velocity.

(c) Suppose now that we have a ball which can be opened and filled with various materials to change its mass. We drop this ball from the Washington Monument (height 160 m). Let's assume that this is high enough to ensure the ball will reach terminal velocity, and that f has one of the forms $f(v) = kv$, $f(v) = kv^{3/2}$, or $f(v) = kv^2$. When the ball weighs 50 gm, it hits the ground at 24 m/sec, and when it weighs 100 gm it hits the ground at 33.5 m/sec. What is the correct formula for f?

For the rest of the problem, assume f is given by the formula you found in part (c).

(d)   Use the ball with a mass of 50 gm. Suppose the falling ball increases its speed from 12 m/sec to 13 m/sec in a short time interval. How far has the ball fallen in this time interval? (Give lower and upper bounds).

(e)   Suppose instead the speed of this ball goes from v to $v + \Delta v$. Approximately how long is this time interval? About how far does the ball fall? Using this result, write integrals for how long it takes the ball to go from a speed of v_1 to a speed of v_2 and for how far the ball travels during this interval.

(f)   Use the integral from part (e) for the time to go from speed v_1 to speed v_2 to show that the balls falling according to this model never exactly reach their terminal velocity. [Hint: It is easier if you retain the symbols for the various constants and use the expression for terminal velocity you got in part (b).]

(g)   How long did it take the 50 gm ball to fall from the top of the Washington Monument? Was it close to its terminal velocity? [Hint: Realize that you don't know the exact velocity with which the ball hit the ground.]

(h)   Assume our ball has the same dimensions as a baseball and that a baseball has a mass of 145 gm. What is its terminal velocity? How fast would it be going when it hits the ground after being dropped from the top of the Washington Monument?

12.   THE MASS BALANCE EQUATION   Chapter 13

Overview: A fluid flows in a pipe of variable width, as shown in Figure D.9. Let $A(x)$ be the cross-sectional area of the pipe (in m^2) at a position x meters along the pipe. Let $v(x, t)$ be the velocity (in meters/sec) of the fluid at time t (in seconds) and at position x, and let $\rho(x, t)$ be the density of the fluid (in kg/m^3). We assume that the velocity and density do not vary across the pipe.

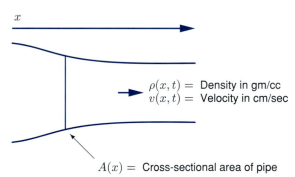

x

$\rho(x, t) =$ Density in gm/cc
$v(x, t) =$ Velocity in cm/sec

$A(x) =$ Cross-sectional area of pipe

Figure D.9: Fluid flowing in a pipe of varying width

The aim of this project is to understand the following partial differential equation, called the *mass balance equation*, relating these quantities:

$$A\frac{\partial \rho}{\partial t} + \frac{\partial}{\partial x}(\rho v A) = 0.$$

The Problem: Suppose that ρ and v are differentiable functions of x and t, and that A is a differentiable function of x.

(a)   Suppose that the density is constant and that the velocity of flow is steady (that is, independent of time). What does the differential equation tell you in this case? Does your answer make sense?

(b)   Let $x(t)$ be the position at time t of a particle being carried by the fluid. We say that the fluid is *incompressible* if, for any given particle, $\rho(x(t), t)$ is constant. Explain why this agrees with the common sense meaning of incompressible. Use the chain rule to express incompressibility in terms of v and the partial derivatives of ρ.

(c)   Now suppose that the cross-sectional area is constant and that the fluid is incompressible. What does common sense tell you about the velocity function in this case? Is it the same as what the differential equation tells you?

(d)   Imagine a small piece of the pipe of length Δx. By approximating the increase of fluid in the piece during a short time Δt, and comparing it with the inflow and outflow at both ends of the piece, explain where the differential equation comes from.

13.   **OLIGOPOLIES AND COLLUSION**   **Chapter 14**

Overview: A *monopoly* is a market in which there is only one supplier for a particular good. ("Mono" means one and "polein" means sell in Greek.) For example, many local electric companies in the US are monopolies, though deregulation is starting to change this. A market in which a small number of companies compete, so that the behavior of any one firm can significantly influence the rest, is called an *oligopoly* ("oligo" means few). In this project, we model the behavior of an oligopoly consisting of two companies, and see why they might choose to work together as a monopoly rather than compete.

    The first person to study the behavior of an oligopoly was the French economist Cournot in 1838. He imagined two owners of mineral water springs, who sold the water. The question he considered was how much water each would choose to sell, and how much revenue (which in this case is the same as profit) each would make.

The Problem:[7] Suppose the two spring owners sell q_1 and q_2 liters of water a day, respectively. We assume that they charge the same price of $\$p$ per liter. Then the owner's revenues are, respectively,

$$R_1 = pq_1, \quad \text{and} \quad R_2 = pq_2.$$

In order to see how the revenues behave as the quantities sold change, we need to know how the price depends on the quantity sold.

(a)   Explain why it is plausible to assume that $p = f(q_1 + q_2)$, where f is a decreasing function.

(b)   In particular, we will assume for the rest of the project that f is a linear function of $q_1 + q_2$, so that we can write

$$p = b - m \cdot (q_1 + q_2)$$

where b and m are positive constants. What is the meaning of b and m for the spring owners?

(c)   Write the owners' revenues as functions of q_1 and q_2. (Your expressions should not contain p).

(d)   Separate action:

        Cournot decided to assume that each owner would look at the other owner's output as a given (a constant) and maximize his own output. The first owner, therefore, regards q_2 as a constant and wants to maximize R_1 as a function of q_1.

    (i)   Write an equation that holds at this maximum.

    (ii)   Write an equation that holds when the second owner wants to maximize his output.

    (iii) Find the amounts of water sold when each owner independently maximizes his revenue.

    (iv) What is the price per liter for the values of q_1 and q_2 found in (iii)? What is each owner's revenue?

(e)   Collusion:

        Instead of looking at the other's production as fixed, as in part (d), suppose the owners get together and decide to choose q_1 and q_2 so as to maximize total revenue, $R = R_1 + R_2$.

[7]Adapted from *An Application of Calculus in Economics: Oligopolistic Competition,* Donald R. Sherbert (COMAP, Inc., Lexington, MA 1982).

(i)   Write the total revenue as a function of q_1 and q_2.

(ii) Write equations that hold when the owners maximize total revenue.

(iii) Find the total quantity sold when total revenue is maximized. Can you determine q_1 and q_2?

(iv) If the owners agree to share the total quantity sold equally, how much does each sell? At what price? What is each owner's revenue?

(f) Compare prices and revenues in the cases of collusion and separate action. Which type of action would the owners prefer? Which would the consumers prefer?

14. **KEPLER'S SECOND LAW**   **Chapter 16**

Overview: We know that the planets do not orbit in circles with the sun at the center, nor does the moon orbit in a circle with the earth at the center. For example, the moon's distance from the earth varies from 220,000 to 260,000 miles. In the last half of the 16th century the Danish astronomer Tycho Brahe (1546–1601) made measurements of the positions of the planets. Johann Kepler (1571–1630) studied this data for years and, after some false starts, arrived at three laws for planetary motion:

Kepler's Laws

I.   The orbit of each planet is an ellipse with the sun at one focus. In particular, the orbit lies in a plane containing the sun.

II. As a planet orbits around the sun, the line segment from the sun to the planet sweeps out equal areas in equal times.

III. The ratio p^2/d^3 is the same for every planet orbiting around the sun, where p is the period of the orbit (time to complete one revolution) and d is the mean distance of the orbit (average of the shortest and farthest distances from the sun).

Kepler's Laws, impressive as they are, were purely descriptive; they didn't explain the motion of the planets. Newton's great achievement was to find an underlying cause for them. In this project, you will derive Kepler's Second Law from Newton's Law of Gravity.

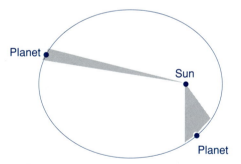

Figure D.10: The line segment joining a planet
to the sun sweeps out equal areas in equal time

The Problem: Let $\vec{r}$ be the position vector of a planet, with the origin of the coordinate system at the center of the sun.[8] Let $\vec{v}$ and $\vec{a}$ be the planet's velocity and acceleration. Define $\vec{L} = \vec{r} \times \vec{v}$. (This is a multiple of the planet's angular momentum.)

(a) Explain why $\dfrac{d\vec{L}}{dt} = \vec{r} \times \vec{a}$.

(b) Consider the planet moving from $\vec{r}$ to $\vec{r} + \Delta\vec{r}$. Explain why the area ΔA about the origin swept out by the particle is approximately $\frac{1}{2}\|\Delta\vec{r} \times \vec{r}\|$.

---

[8]We are assuming the center of the sun is the same as the center of mass of the planet/sun system. This is only approximately true.

(c)   Using part (b), explain why $\dfrac{dA}{dt} = \dfrac{1}{2}\|\vec{L}\|$.

(d)   Newton's Laws imply that the gravitational acceleration, $\vec{a}$, of the planet is directed towards the sun. Using this fact and part (a), explain why $\vec{L}$ is constant.

(e)   Use parts (c) and (d) to explain Kepler's Second Law.

(f)   Using Kepler's Second Law, determine whether a planet is moving most quickly when it is closest to, or farthest from, the sun.

15.   **ELECTRIC DIPOLES**   **Chapter 17**

Overview: Suppose $q_1, \ldots, q_n$ are electric charges at points with position vectors $\vec{r}_1, \ldots, \vec{r}_n$. Coulomb's Law states that, at the point with position vector $\vec{r}$, the resulting electric field $\vec{E}$ is given by

$$\vec{E}(\vec{r}) = \sum_{i=1}^{n} q_i \frac{(\vec{r} - \vec{r}_i)}{\|\vec{r} - \vec{r}_i\|^3}.$$

When there is a single charge q_1 at the origin, the electric field $\vec{E} = q_1\vec{r}/\|\vec{r}\|^3$ is simply a radial field whose magnitude is inversely proportional to the square of the distance from the origin. This project deals with the next most complicated configuration, that of two charges, called an *electric dipole*.

The Problem:

(a)   Suppose two charges q_1 and q_2 are located at $\vec{r}_1 = \vec{i}$ and $\vec{r}_2 = -\vec{i}$.

(i)   If $q_1 = q$ and $q_2 = -q$, use a computer to sketch the vector field $\vec{E}$ in the xy-plane produced by these two opposite charges.

(ii)   If $q_1 = q_2 = q$, sketch the vector field $\vec{E}$ in the xy-plane produced by these two like charges.

(b)   An *ideal electric dipole* can be thought of as an infinitesimal dipole; its magnitude and direction are given by its dipole moment vector $\vec{p}$. The resulting electric field $\vec{D}$ at the point with position vector $\vec{r}$ is given by

$$\vec{D}(\vec{r}) = 3\frac{(\vec{r} \cdot \vec{p})\vec{r}}{\|\vec{r}\|^5} - \frac{\vec{p}}{\|\vec{r}\|^3}.$$

Assume $\vec{p} = p\vec{i}$, so that the dipole points in the $\vec{i}$ direction and has magnitude p.

(i)   Use a computer to plot the vector field $\vec{D}$ in the xy-plane for three different values of p.

(ii)   The field $\vec{D}$ is an approximation to the electric field $\vec{E}$ produced by two opposite charges, q at $\vec{r}_2$ and $-q$ at $\vec{r}_1$, when the distance $\|\vec{r}_2 - \vec{r}_1\|$ is small. The dipole moment of this configuration of charges is defined to be $\vec{p} = q(\vec{r}_2 - \vec{r}_1)$. Suppose $\vec{r}_2 = (\ell/2)\vec{i}$ and $\vec{r}_1 = -(\ell/2)\vec{i}$, so that $\vec{p} = q\ell\vec{i}$.

A.   Plot the vector field $\vec{E}$ using the same values of $p = q\ell$ you used to plot $\vec{D}$.

B.   Where is the vector field $\vec{D}$ a good approximation to $\vec{E}$? Where is it a poor approximation?

C.   The magnitude of each term in the expression for $\vec{E}$ decays like $1/\|\vec{r}\|^2$, while the magnitude of $\vec{D}$ decays like $1/\|\vec{r}\|^3$. If the vector field $\vec{D}$ is supposed to be a good approximation to $\vec{E}$ when the distance $\|\vec{r}\|$ from the origin is large, suggest a reason for this apparent discrepancy.

16.   **MAXWELL'S EQUATIONS**   **Chapter 20**

Overview: Classical electromagnetism is concerned with electric fields $\vec{E}(x, y, z, t)$ and magnetic fields $\vec{B}(x, y, z, t)$. These fields are functions of position, (x, y, z), and time, t. *Maxwell's equations* express the relationship between these fields and a system of electric charges:

$$\text{div } \vec{E} = 4\pi\rho \qquad \text{(Gauss's Law)},$$

$$\text{div } \vec{B} = 0 \qquad \text{(No magnetic monopoles)},$$

$$\text{curl } \vec{B} - \frac{1}{c}\frac{\partial \vec{E}}{\partial t} = \frac{4\pi}{c}\vec{J} \qquad \text{(Ampère's Law)},$$

$$\text{curl } \vec{E} + \frac{1}{c}\frac{\partial \vec{B}}{\partial t} = 0 \qquad \text{(Faraday's Law)}.$$

Here c is the speed of light. The system of charges is described by the charge density, $\rho(x, y, z, t)$, in units of charge per unit volume, and the charge velocity $\vec{v}(x, y, z, t)$. The quantity $\vec{J} = \rho\vec{v}$ is called the current density and is measured in units of charge per unit area per unit time.

The Problem:

(a)   The *integral form* of *Faraday's Law of Induction* is

$$\int_C \vec{E} \cdot d\vec{r} = -\frac{1}{c}\frac{d}{dt}\int_S \vec{B} \cdot d\vec{A},$$

where S is any oriented surface with boundary curve C oriented by the right-hand rule.

   (i)   Let $\vec{v}$ and $\vec{w}$ be vectors in 3-space with the property that $\vec{v} \cdot \vec{n} = \vec{w} \cdot \vec{n}$ for *any* unit vector $\vec{n}$. Show that $\vec{v} = \vec{w}$.

   (ii)   Explain why the integral form of Faraday's Law implies the *differential form* of *Faraday's Law,* i.e. the form that we have listed as one of Maxwell's equations. You may assume $\int_S \frac{\partial \vec{B}}{\partial t} \cdot d\vec{A} = \frac{d}{dt}\int_S \vec{B} \cdot d\vec{A}$. [Hint: Circulation density.]

   (iii)   Conversely, show that the differential form implies the integral form of Faraday's Law. [Hint: Stokes' theorem.]

(b)   The *integral form* of *Ampère's Law* is

$$\int_C \vec{B} \cdot d\vec{r} = \frac{4\pi}{c}I + \frac{1}{c}\frac{d}{dt}\int_S \vec{E} \cdot d\vec{A},$$

where S is any oriented surface with boundary curve C oriented by the right-hand rule, and I is the total current flowing through S.

   (i)   Show that the integral form of Ampère's Law implies the differential form, i.e. the form in Maxwell's equations. You may assume $\int_S \frac{\partial \vec{E}}{\partial t} \cdot d\vec{A} = \frac{d}{dt}\int_S \vec{E} \cdot d\vec{A}$. [Hint: $I = \int_S \vec{J} \cdot d\vec{A}$.]

   (ii)   Show that the differential form implies the integral form of Ampère Law.

(c)   (i)   Consider the equation

$$\int_S \vec{J} \cdot d\vec{A} = -\frac{d}{dt}\left(\int_W \rho\, dV\right),$$

where W is any region of 3-space bounded by a surface S. Explain why this equation is called the Law of Charge Conservation. What is the significance of the negative sign? [Hint: See Problem 20 on page 884.]

   (ii)   Show that the Law of Charge Conservation implies the following Continuity Equation:

$$\frac{\partial \rho}{\partial t} + \text{div } \vec{J} = 0.$$

You may assume $\int_S \frac{\partial \rho}{\partial t}dV = \frac{d}{dt}\int_S \rho\, dV$.

   (iii)   Show that the continuity equation implies the law of charge conservation. [Hint: Divergence Theorem.]

   (iv)   Deduce the continuity equation directly from Maxwell's equations, specifically Gauss's and Ampère's laws.

ANSWERS TO ODD NUMBERED PROBLEMS

Section 1.1

1   (I)   None
    (II)   (b)
    (III)   (c)
    (IV)   (a)

5   $F = k/d^2$

7   $V = kr^3$

15   p_1 = maximum price any consumer
     would pay
     q_1 = quantity given away if item were
     free

17   Domain: $0 \le x \le 5$
     Range: $0 \le y \le 4$

19   Domain: all x
     Range: $y \ge 2$

21   Domain: all x
     Range: $0 < y \le 1/2$

23   All $x \ne 0, -1$
     $x = \pm 2$

Section 1.2

1   Slope:$-12/7$
    Vertical intercept: $2/7$

3   Slope: 2
    Vertical intercept: $-2/3$

5   $y = (1/2)x + 2$

7   $y = -\frac{1}{5}x + \frac{7}{5}$

9   Parallel: $y = m(x - a) + b$
    Perpendicular:
    $y = (-1/m)(x - a) + b$

11   (a)   (V)
     (b)   (VI)
     (c)   (I)
     (d)   (IV)
     (e)   (III)
     (f)   (II)

13   $W = (5/3)R + 10$

15   (a)   \$0.025/cubic foot
     (b) $c = 65 + 0.025w$
           c = cost of water
           w = cubic feet of water
     (c) $w = 2600$ cubic feet

17   (a)   Slope = 1.8
     (b) $°F = 1.8(°C) + 32$
     (c) $68°F$
     (d) $-40°$

19   $9 \cdot 10^9$ newton/meter^2

21   (a) $d = 5t$
     (b)   No, in the last 200 years the
           average speed is:
           $700/200 = 3.5$ mi/yr.
     (c)   Unlikely

23   (a) $R = k(350 - H)$
           $(k \ge 0)$

Section 1.3

1   Concave up

3   Neither

5   Exponential growth

7   Exponential growth
    $g(t) = 2^t$

9   Exponential decay
    $h(t) = 16768(1/2)^t$

15   $P = 5$

17   $Z = 3$

19   (a)   1.05
     (b)   5%

21   $y = 4(2^{-x})$

23   $y = 4(1 - 2^{-x})$

25   $f(s) = 2(1.1)^s$
     $g(s) = 3(1.05)^s$
     $h(s) = (1.03)^s$

27   (a) $Q = Q_0 \left(\frac{1}{2}\right)^{(t/1620)}$
     (b)   80.7%

29   25.5%

31   (a)   47.3%
     (b)   11054.4%

Section 1.4

1   16

3   100,000

5   1/3

7   1/4

9   $10 \cdot 2^x$

11   $0.25\sqrt{x}$

13   As $x \to \infty$, $y \to \infty$.
     As $x \to -\infty$, $y \to -\infty$.

17   $f(x) = x^3$

19   (a)   1.3 m^2
     (b)   86.8 kg
     (c) $h = 112.6s^{4/3}$

21   44.25 ft and 708 ft

23   $h(t) = ab^t$
     $g(t) = kt^3$
     $f(t) = ct^2$

25   (a) $0 \le x \le 5$,
           $0 \le y \le 9$
     (b) $0 \le x \le 5$,
           $0 \le y \le 625$
     (c) $0 \le x \le 10$,
           $0 \le x \le 50,000$

Section 1.5

3   Length of column of mercury when
    temperature is $75°F$

5   Probably not invertible

7   Not invertible

9   Not invertible

11   Not invertible

13   (a) $q = \frac{C-100}{2} = f^{-1}(C)$
     (b)   Number of articles produced at
           given cost

17   (a) -1

Section 1.6

3   $(\log 2)/(\log 17) \approx 0.24$

5   $(\log(2/5))/(\log 1.04) \approx -23.4$

7   $(\log(2/11))/(\log(7/5)) \approx -5.07$

9   $(\log a)/(\log b)$

11   $(\log Q - \log Q_0)/(n \log a)$

13   $(10/3) \log A + \log B$

15   2

17   $(A/B)^2$

19   The graph is a straight line, with slope
     1, to the right of the origin.

21   $\log x$

23   14.21 years

25   1990

27   (a) $A = 10.32e^{-0.057762t}$
     (b)   40.41 days

29   $p^{-1}(t) \approx 58.708 \log t$

Section 1.7

3   -0.347

5   0.515

7   1.01, 1.17

9   $\ln(P/P_0)/k$

11   1/2

13   2AB

15   $-1 + \ln A + \ln B$

17   $3 \ln A + \ln B$

19   $P = 2(0.61)^t$; decay

21   $P = 79(0.0821)^t$; decay

23   $P = 7(0.0432)^t$; decay

25   $P = 10e^{0.5306t}$

27   $P = 4e^{-0.6t}$

29   $f^{-1}(t) = 10 \ln \left(\frac{t}{50}\right)$

31   (a)   Decreasing
     (b) $g^{-1}(x) = \ln(\frac{2}{x} - 5)$

33 (a) $P = 10^6(e^{0.02t})$

35 (a) 47.5%
    (b) 24%

37 2010

39 7.925 hours

41 (a) 5 kg
    (b) 38.2 years

43 (a) $P = 3.6(1.034)^t$
    (b) $P = 3.6e^{0.0334t}$
    (c) Annual = 3.4%
         Continuous = 3.3%

45 It is a fake

Section 1.8

3 $2\ln(x + 3) + 1$

5 $2e^{4x+7} + 1$

7 $\ln(e^{4x+7} + 3)$

9 $4x + 7$

11 $e^7(2x + 4)^4$

13 Even

15 Neither

17 Neither

19 $f(x) = x + 1$
    $g(x) = x^3$

21 $f(x) = x^3$
    $g(x) = \ln x$

23 $2zh + h^2$

25 $4hz$

27 $-60 < x < -40$
    $-25100 < y < -24000$

31 $g(f(2)) \approx 1.1$

Section 1.9

1 Negative
   0
   Undefined

3 Positive
   Positive
   Positive

5 Positive
   Positive
   Positive

7 Positive
   Negative
   Negative

9 Negative
   Positive
   Negative

11 0.259

13 0.966

15 20.94 to 52.36 rad/sec

17 If $f(x) = \sin x$ and
    $g(x) = x^2$ then
    $\sin x^2 = f(g(x))$
    $\sin^2 x = g(f(x))$
    $\sin(\sin x) = f(f(x))$

19 (a) $h(t)$
    (b) $f(t)$
    (c) $g(t)$

21 $f(x) = 2\sin(x/4)$

23 $f(x) = 2\sin(x/4) + 2$

25 $f(x) = \sin\left(2(\pi/5)x\right)$

27 $f(x) = 3\sin(\pi x/9)$

29 (a) $\frac{1}{60}$ second
    (b) V_0 represents the amplitude of
         oscillation.

31 $f(t) = 75 - 15\cos\left(\frac{2\pi}{12}t\right)$

33 (b) Domain:
         $-1 \le x \le 1$
         Range:
         $-\frac{\pi}{2} \le x \le \frac{\pi}{2}$

35 $(0, 0)$, $(1.31, 2.29)$,
    $(-1.31, -2.29)$

37 $\theta = \pi/4$; $R = v_0^2/g$

Section 1.10

1 (a) $f(x) \to \infty$
         as $x \to \infty$;
         $f(x) \to -\infty$
         as as $x \to -\infty$
    (b) $f(x) \to -\infty$
         as as $x \to \pm\infty$
    (c) $f(x) \to 0$ as $x \to \pm\infty$
    (d) $f(x) \to 6$ as $x \to \pm\infty$

7 a, c, f and h are even;
   b, d and g are odd;
   e and i are neither

9 $x = 2$: vertical asymptote
   No horizontal asymptote
   $y \to -\infty$ as $x \to +\infty$
   $y \to +\infty$ as $x \to -\infty$
   $y \to -\infty$ as $x \to 2^+$
   $y \to +\infty$ as $x \to 2^-$

11 $x = \pm 2$: vertical asymptotes
    $y = 1$: horizontal asymptote
    $y \to +\infty$ as $x \to 2^+$
    $y \to -\infty$ as $x \to 2^-$
    $y \to -\infty$ as $x \to -2^+$
    $y \to +\infty$ as $x \to -2^-$

13 (a) 0
    (b) $t = 2v_0/g$
    (c) $t = v_0/g$
    (d) $(v_0)^2/(2g)$

15 (a) $1 = a + b + c$
    (b) $b = -2a$ and $c = 1 + a$
    (c) $c = 6$
    (d) $y = 5x^2 - 10x + 6$

17 $y = -\frac{1}{2}(x + 2)^2(x - 2)$

19 (a) $f(x) = kx(x + 3)(x - 4)$
         ($k < 0$)
    (b) Increasing:
         $-1.5 < x < 2.5$
         Decreasing:
         $x < -1.5$ and $x > 2.5$

21 (a) $f(x) =$
         $k(x + 2)(x - 2)^2(x - 5)$
         ($k < 0$)
    (b) Increasing:
         $x < -1$ and
         $2 < x < 4$
         Decreasing:
         $-1 < x < 2$ and
         $4 < x$

23 (a) III
    (b) IV
    (c) I
    (d) II

25 (a) $a(v) = \frac{1}{m}(F_E - kv^2)$
         ($k > 0$)

27 (a) $S = 2\pi r^2 + 2V/r$
    (b) $S \to \infty$ as $r \to \infty$

29 $y = x$

Section 1.11

1 No

3 Yes

5 No

7 Yes

9 Not continuous

Chapter 1 Review

1 (a) $[0, 7]$
    (b) $[-2, 5]$
    (c) 5
    (d) $(1, 7)$
    (e) Concave up
    (f) 1
    (g) No

3 (b) Concave down

7 $x = 3\left(1 - a^{-t}\right), a > 1$

9 $y = (x + 1)^3 + 1$

11 (a) $h(x) = 31 - 3x$
    (b) $g(x) = 36(1.5^x)$

13 $Q(m) = T + L + Pm$
    $T = $ fuel for take-off
    $L = $ fuel for landing
    $P = $ fuel per mile in the air
    $m = $ length of the trip (miles)

15 (a) $h^2 + 6h + 11$
    (b) 11
    (c) $h^2 + 6h$

17 (a) $f(n) + g(n) =$
         $3n^2 + n - 1$
    (b) $f(n)g(n) =$
         $3n^3 + 3n^2 - 2n - 2$
    (c) $n \ne -1$
    (d) $f(g(n)) = 3n^2 + 6n + 1$
    (e) $g(f(n)) = 3n^2 - 1$

19 Not continuous

21 Not continuous

23 $P = (5 \cdot 10^{-3})(0.9812)^t$

25 (a) $Q = 25e^{0.5423t}$

(b)　1.3 months
(c)　6.8 months

27　10 hours

29　38.87 hours

31　(a)　44%
(b)　about two years

33　Mercury: 87.8 days
Earth: 1 year
Pluto: 253 years

35　$y = e^{0.4621x}$

37　Simplest is $y = 1 - e^{-x}$

39　$y = k(x^3 + 2x^2 - x - 2)$
$(k > 0)$

41　$y = 5 \sin\left(\pi t/20\right)$

43　$z = 1 - \cos\theta$

45　(b)　It becomes a vertical asymptote at $x = -4$.

47　(a)　$f(15) \approx 48$
(b)　f is invertible because the graph of f is cut by a horizontal line at most once.
(c)　$f^{-1}(120) \approx 35$.
At a depth of 120 meters down, the rock is 35 million years old.

49　(a)　$R = k_1 - k_2 G$

51　(a)　Domain: $(0, 4000)$
Range: $(0, 10^8)$
(b)　Domain: $(0, 3000)$
Range: $(0, 10^7)$
(c)　Domain: $(0, 0.2)$
Range: $(0, 0.04)$

Theory: Binomial Theorem

1　Must be integer

Section 2.1

7　$F < B < E < 0 < D < A < C$

9　27

11　1.9 . . .

Section 2.2

1　negative

9　From smallest to largest:
$0, f'(3), f(3) - f(2), f'(2), 1$

11　(a)　$(f(b) - f(a))/(b - a)$
(b)　Slopes same
(c)　Yes

13　-1

15　12

17　$-1/4$

19　$y = 12x + 16$

21　$y = x$

23　$f'(3) = 6, y = 6x - 8$

25　(a)　$f'(0) \approx 0.01745$

27　$f'(0) \approx -1$
$f'(1) \approx 3.5$

29　(a)　$\mathrm{erf}'(0) \approx 1.12$
(b)　$\mathrm{erf}'(0) \approx 1.1283$

31　16.0 million people/year
16.4 million people/year

Section 2.3

11　$f'(x)$ positive:
$4 \le x \le 8$
$f'(x)$ negative:
$0 \le x \le 3$
$f'(x)$ greatest:
at $x \approx 8$

13　$4x$

15　$-2/x^3$

29　(a)　x_3
(b)　x_4
(c)　x_5
(d)　x_3

31　(a)　$t = 3$
(b)　$t = 9$
(c)　$t = 14$

Section 2.4

1　feet/mile; negative

3　dollars/year; negative

5　dollars/year

7　(a)　Quarts; dollars.
(b)　Quarts; dollars/quart.

11　(a)　Gal/minute
(b)　(i) 0
(ii) Negative
(iii) 0

13　mpg/mph

15　(a)　$f'(a)$ is always positive
(c)　$f'(100) = 2$: more
$f'(100) = 0.5$: less

Section 2.5

1　$f'(x) > 0$
$f''(x) > 0$

3　$f'(x) < 0$
$f''(x) = 0$

5　$f'(x) > 0$
$f''(x) < 0$

7　$0 \le t \le 1$:
acceleration = 30 ft/sec^2
$1 \le t \le 2$:
acceleration = 22 ft/sec^2

9　(b)　$\frac{dN}{dt}$ is positive.
$\frac{d^2 N}{dt^2}$ is negative.

17　(a)　B and E
(b)　A and D

Chapter 2 Review

3　(b)　0.24
(c)　0.22

5　$f'(1) = 1$
$f'(2) = 0.5$
$f'(5) = 0.2$
$f'(10) = 0.1$
$f'(x) = \frac{1}{x}$

13　$10x + 1$

17　(a)　Negative
(b)　F$^\circ$/min

19　(b)　Student C
(c)　$f'(x) = \frac{f(x+h) - f(x-h)}{2h}$

23　(a)　0.25
(b)　$y = 0.25x + 1$
(c)　$k = \frac{1}{8}$
(d)　$(-2, 0.5)$ and $(4, 2)$

25　(b)　One
(c)　It must be between
$x = 1$ and $x = 5$.
(d)　$\lim\limits_{x \to -\infty} f(x) = -\infty$
(e)　Yes
(f)　No

27　(a)　Concave down
(b)　135°
(c)　138°
(d)　$t = 48$ minutes

Theory: Limits

1　0.05

3　2

5　0.01745 . . .

7　The limit appears to be 1.

9　(a)　$x = 1/(n\pi)$,
$n = 1, 2, 3, \ldots$
(b)　$x = 2/(n\pi)$,
$n = 1, 5, 9, \ldots$
(c)　$x = 2/(n\pi)$,
$n = 3, 7, 11, \ldots$

19　0.46, 0.21, 0.09

Theory: Differentiability

1　(a)　(i)　$x = 1$
(ii)　$x = 1, 2, 3$
(b)　(i)　None
(ii)　$x = 2, 4$

3　Yes

5　Yes

7　(a)　$B = B_0$ at $r = r_0$
B_0 is max
(b)　Yes
(c)　No

9　(a)　Yes
(b)　Yes

11　(a)　$y = 2 + (x - 4)/4$
(b)　2.02485..., 2.025
(c)　True = 4, Approx = 5

13　$2x - 1$

Section 3.1

1   (a)   Lower estimate = 122 ft
        Upper estimate = 298 ft

3   (a)   Lower estimate = 5.25 mi
        Upper estimate = 5.75 mi
    (b)   Lower estimate = 11.5 mi
        Upper estimate = 14.5 mi
    (c)   Every 30 seconds

5   Left sum = 46 m
   Right sum = 118 m
   Average = 82 m

7   Area $\approx 11\frac{1}{16}$

9   Between 140 and 150 meters

11   (a)   14.73 minutes
         15.93 minutes
     (b)   0.60 minutes

Section 3.2

1   (a)   224
    (b)   96
    (c)   200
    (d)   136

3   limit $= \frac{1}{4}$
   True value is between
   0.248004 and 0.252004

5   Since $\sin(t^2)$ is *not* monotonic on
   [2, 3], we cannot be sure of the true
   value.

7   Since $\sin(1/x)$ is *not* monotonic on
   [0.2, 3], we cannot be sure of the true
   value.

9   Left-hand sum = 1.96875
   Right-hand sum = 2.71875
   The most the estimate could be off is
   0.375.

11   396

13   24.7

15   4.39

17   0.0833

19   (b)   3.08
     (c)   2.50

21   Positive

23   93.47

25   (a) $\sum_{i=1}^{n} i^4/n^5$
     (b) $(6n^4+15n^3+10n^2-1)/30n^4$
     (c)   1/5

27   $a = 2$, $b = 6$, $f(x) = x^2$; other
   answers possible

Section 3.3

1   (a)   Car 1: 1031.25 ft
        Car 2: 562.5 ft
    (b)   1.6 minutes

3   (b)   Twice
        At each intersection point,
        the distance between the two
        vehicles is at a local extremum.

5   Dollars

7   Total amount $= \int_0^{60} f(t)\, dt.$

9   15

11   (a)   8.5
     (b)   1.7

13   About $13,800

17   (a)   0.79

19   (a)   22°C
     (b)   183°C
     (c)   Smaller

21   (a)   9.9 hours
     (b)   14.4 hours
     (c)   12.0 hours

23   12 newton · meters

25   (a)   III
     (b)   I
     (c)   II and IV

Section 3.4

3   $F(0) = 0$
   $F(1) = 1$
   $F(2) = 1.5$
   $F(3) = 1$
   $F(4) = 0$
   $F(5) = -1$
   $F(6) = -1.5$

5   $f(3) - f(2)$,
   $\frac{f(4)-f(2)}{2}$,
   $f(4) - f(3)$

7   45.8°C.

9   (a)   0.375 thousand/hour
    (b)   1.75 thousands

15   9

17   $8c$

19   8

21   (a)   Positive, since $e^{x^2} > 0$

23   (a)   0
     (b)   0

25   30/7

Chapter 3 Review

1   (a)   260 ft
    (b)   Every 0.5 sec

3   60

5   0.40

7   2.00

9   $\pi - 2 \approx 1.14$

11   (iii) < (ii) < (i) < (iv)

15   (b)   Largest to smallest:
        $n = 1, n = 3$,
        $n = 4$, and $n = 2$.

17   (a) ≈ 76.8 million
     (b) $= 77.24$ million

19   (a)   18 appliances per month
     (b)   17 appliances per month
     (c)   Close, but not equal

(d)   Integral is easier to calculate

21   (a)   300 m³/sec
     (b)   250 m³/sec
     (c)   1996: 1250 m³/sec
        1957: 3500 m³/sec
     (d)   1996: 10 days
        1957: 4 months
     (e) 10^9 meter³
     (f) $2 \cdot 10^{10}$ meter³

23   (a)   At $t =17, 23, 27$ seconds
     (b)   Right: $t = 10$ seconds
        Left: $t = 40$ seconds
     (c)   Right: $t = 17$ seconds
        Left: $t = 40$ seconds
     (d) $t = 10$ to 17 seconds,
        20 to 23 seconds, and
        24 to 27 seconds
     (e)   At $t = 0$ and $t = 35$

25   (a) $\int_0^5 f(x)\, dx -$
        $\frac{1}{2}\int_{-2}^2 f(x)\, dx$
     (b) $\int_{-2}^5 f(x)\, dx -$
        $2\int_{-2}^0 f(x)\, dx$
     (c) $\frac{1}{2}(\int_{-2}^5 f(x)\, dx -$
        $\int_2^5 f(x)\, dx)$

27   0; Positive; 0; Negative

Theory: Definite Integral

5   2.852, 2.919, 2.886, $n = 30$

7   0.465,0.474,0.470, $n = 10$

9   0.0045, 0.0276, 0.016, $n = 10$

11   0.825, 0.905, 0.865, $n = 20$

Section 4.1

3   (a) -10
    (b) $\frac{d}{dx}(f[g(x)]) = f'(x)g'(x)$

5   $12x^{11}$

7   $3.2x^{2.2}$

9   $4x^{1/3}/3$

11   $-3x^{-7/4}/4$

13   $x^{-3/4}/4$

15   $6x^{1/2} - \frac{5}{2}x^{-1/2}$

17   $-12x^3 - 12x^2 - 6$

19   $6t - 6/t^{\frac{3}{2}} + 2/t^3$

21   $1 - x^{-2}$

23   $-2t^{-3} - t^{-2} + 4t^{-5}$

25   Problems 8, 10, 11, 12, 13, 15

27   Do not apply.

29   $-2/3z^3$
   (power rule and sum rule)

31   Do not apply.

33   $12/\sqrt[6]{x^5} - 18/(x^{5/3})$

35   $f'(t) = 6t^2 - 8t + 3$
   $f''(t) = 12t - 8$

37   For $x < 0$ or $2 < x < 3$

39    $r = 3\sqrt{2}$

41    $y = 2x - 1$

43    $n = 4, a = 3/32$

45    $y = 2x$ and $y = -6x$

47    $-2GMm/r^3$

49   At $x = 1$:
     $y = -x + 2$
     (tangent line approx)
     $f(2) \approx 0$
     At $x = 2$:
     $y = -0.0001x + 0.02$
     (tangent line approx)
     $f(2) \approx 0.0198$

51    $V(r) = 4\pi r^3/3$
     $\frac{dV}{dr} = 4\pi r^2 = $ surface area of a sphere

Section 4.2

1    $2e^x + 2x$

3    $(\ln 5)5^x$

5    $10x + (\ln 2)2^x$

7    $4(\ln 10)10^x - 3x^2$

9    $((\ln 3)3^x)/3 - (33x^{-3/2})/2$

11   e^{1+x}

13    $e^{\theta - 1}$

15    $(\ln 4)2^{4x}$

17    $(\ln(\ln 3))(\ln 3)^t$

19    $5 \cdot 5^t \ln 5 + 6 \cdot 6^t \ln 6$

21    $\pi^2 x^{(\pi^2 - 1)} + (\pi^2)^x \ln(\pi^2)$

23    $1/(2\sqrt{x}) + \ln 2(1/2)^x$

25   Our rules do not apply here.

27    $5e^{5x}$

29    $(\ln 2)2^z$

31    $\approx -444.3 \frac{\text{people}}{\text{year}}$

33    $22.5(1.35)^t$

35   (a) $f'(0) = -1$
     (b) $y = -x$
     (c) $y = x$

37    $g(x) = x^2/2 + x + 1$

39    $x = 1$ and $x = 2$

Section 4.3

1    $5x^4 + 10x$

3    $e^x(x + 1)$

5    $2^x/(2\sqrt{x}) + \sqrt{x}(\ln 2)2^x$

7    $4s^3 - 1$

9    $(3t^2 + 5)(t^2 - 7t + 2) + (t^3 + 5t)(2t - 7)$

11    $(1 - x)/e^x$

13    $(3.2w^{2.2} - (\ln 5)w^{3.2})/(5^w)$

15    $1/(5t + 2)^2$

17    $x^2 - 3)/x^2$

19    $\frac{(t^2+1)/2\sqrt{t} - \sqrt{t}(2t)}{(t^2+1)^2}$

21    $17e^x(1 - \ln 2)/2^x$

23    $\frac{(-4x^2 - 8x - 1)}{(2 + 3x + 4x^2)^2}$

25    $x < 2$

27    $f'(t) = -e^{-t}$

29    $f'(x) = 3e^{3x}$

31   (a) $\frac{d}{dx}\left(\frac{e^x}{x}\right) = \frac{e^x}{x} - \frac{e^x}{x^2}$
         $\frac{d}{dx}\left(\frac{e^x}{x^2}\right) = \frac{e^x}{x^2} - \frac{2e^x}{x^3}$
         $\frac{d}{dx}\left(\frac{e^x}{x^3}\right) = \frac{e^x}{x^3} - \frac{3e^x}{x^4}$
     (b) $\frac{d}{dx}\left(\frac{e^x}{x^n}\right) = \frac{e^x}{x^n} - \frac{ne^x}{x^{n+1}}$

35   (a)   19
     (b) -11

37   (a) $f(140) = 15,000$:
         If the cost \$140 per board then 15,000 skateboards are sold.
         $f'(140) = -100$:
         Every dollar of increase from \$140 will decrease the total sales by 100 boards.
     (b) $\frac{dR}{dp}|_{p=140} = 1000$
     (c)   Positive
         Increase by \$1000.

39   (c) -3776.63

43    $f'(x) = f(x)(\frac{1}{x - r_1} + \frac{1}{x - r_2} + \cdots + \frac{1}{x - r_n})$

Section 4.4

1    $99(x + 1)^{98}$

3    $200t(t^2 + 1)^{99}$

5    $50(\sqrt{t} + 1)^{99}/\sqrt{t}$

7    $(\ln 2)2^{(x+2)}$

9    $4(x^3 + e^x)^3(3x^2 + e^x)$

11    $(1 - 2z \ln 2)/(2^{z+1}\sqrt{z})$

13    $\frac{3}{2}e^{\frac{3}{2}w}$

15    $3s^2/2\sqrt{s^3 + 1}$

17    $e^{-t^2}(1 - 2t^2)$

19    $e^{-z}/2\sqrt{z} - \sqrt{z}e^{-z}$

21    $e^{5-2t}(1 - 2t)$

23    $e^{-\theta}/(1 + e^{-\theta})^2$

25    $2we^{w^2}(5w^2 + 8)$

27    $-(\ln 10)(10^{\frac{5}{2} - \frac{y}{2}})/2$

29    $2ye^{[e^{(y^2)} + y^2]}$

31    $y = 4451.66x - 3560.81$

33    $f'(x) =$
     $[(2x + 1)^9(3x - 1)^6] \times$
     $(102x + 1)$
     $f''(x) =$
     $[9(2x + 1)^8(2)(3x - 1)^6 +$
     $(2x + 1)^9(6)(3x - 1)^5(3)]$
     $\times (102x + 1) +$
     $(2x + 1)^9(3x - 1)^6(102)$

35   (a) $\pi\sqrt{2}$
     (b) $7e$
     (c) πe

37   (a) $g'(1) = 3/4$
     (b) $h'(1) = 3/2$

39    $e^{(x^6)}/6$

41   (a) $dQ/dt =$
         $-0.000121e^{-0.000121t}$

43   (a) $dB/dt =$
         $P(1 + r/100)^t \ln(1 + r/100)$
     (b) $dB/dr =$
         $Pt(1 + r/100)^{t-1}/100$

45   (a) $\frac{dm}{dv} = \frac{m_0 v}{c^2\sqrt{(1 - v^2/c^2)^3}}$
     (b)   The rate of change of mass with respect to the speed v

47    $T = RC$

Section 4.5

3    $\cos^2 \theta - \sin^2 \theta = \cos 2\theta$

5    $-4\sin(4\theta)$

7    $e^t \cos(e^t)$

9    $-\sin x e^{\cos x}$

11    $\sin x/2\sqrt{1 - \cos x}$

13    $\cos x/\cos^2(\sin x)$

15    $(\ln 2)2^{\sin x}\cos x$

17    $e^{\cos \theta} - \theta(\sin \theta)e^{\cos \theta}$

19    $2\cos(2x)\sin(3x)$
     $+3\sin(2x)\cos(3x)$

21    $e^{-2x}[\cos x - 2\sin x]$

23    $5\sin^4 \theta \cos \theta$

25    $-3e^{-3\theta}/\cos^2(e^{-3\theta})$

27    $\cos t - t\sin t + 1/\cos^2 t$

29    $\theta^2 \cos \theta$

31    $e^x/\sin x$

33   (a) $dy/dt = -\frac{4.9\pi}{6}\sin\left(\frac{\pi}{6}t\right)$
         Represents rate of change of depth of water
     (b)   Occurs when $\sin(\frac{\pi}{6}t) = 0$, or at $t = 6$ am, 12 noon, 6 pm, and 12 midnight
         $dy/dt = 0$ means that the water level is at either high tide or low tide

35   (a) $t = (\pi/2)(m/k)^{\frac{1}{2}}$;
         $t = 0$;
         $t = (3\pi/2)(m/k)^{\frac{1}{2}}$
     (b) $T = 2\pi(m/k)^{\frac{1}{2}}$
     (c) $dT/dm = \pi/\sqrt{km}$;
         Positive sign means an increase in mass causes the period to increase

37    $2/\cos^2 \theta$

Section 4.6

1    $\frac{2t}{(t^2 + 1)}$

3   2

5   $-1/z(\ln z)^2$

7   $\frac{e^{-x}}{1-e^{-x}}$

9   $\frac{e^x}{e^x+1}$

11   7

13   $-\tan(w-1)$

15   $2y/\sqrt{1-y^4}$

17   1

19   $-\sin(\ln t)/t$

21   $\arcsin w + \frac{w}{\sqrt{1-w^2}}$

23   $-1 < x < 1$

25   $\frac{d}{dx}(\log x) = \frac{1}{(\ln 10)x}$

27   (a)   10.59 million
(b) -0.02 million/year

29   (a) $f'(x) = 0$
(b) f is a constant function.

31   (a) $y = -x^2/2 + 2x - 3/2$
(b)   From graph, notice that around $x = 1$, the values of $\ln x$ and its approximation are very close.
(c)   At $x = 1.1$, $y \approx 0.095$
At $x = 2$, $y = 0.5$

33   (a) $z = \sqrt{0.25 + x^2}$
(b)   0.693 km/min
(c)   0.4 radians/min

35   (a) $k \approx 0.067$
(b) $t \approx 10.3$ hours
(c)   Formula:
$T(24) \approx 74.1°$ F,
rule of thumb: 73.6° F

Section 4.7

1   $dy/dx = -x/y$

3   1/25

5   $-y/(2x)$

7   $\frac{y^2 + x^4 y^4 - 2xy}{x^2 - 2xy - 2x^5 y^3}$

9   $dy/dx = \frac{2 - y\cos(xy)}{x\cos(xy)}$

11   $\frac{dy}{dx} = \frac{3x^2 \arctan y}{-e^{\cos y}\sin y - x^3(1+y^2)^{-1}}$

13   $y = x/2 - 3/2$

15   $y = 2$

17   $dy/dx = (m/n)x^{m/n-1}$

19   (a) $dy/dx = -9x/25y$
(b)   The slope is not defined anywhere along the line $y = 0$.

21   Tangent line is $y = 2x - 1$; circle is $(x-8)^2 + y^2 = 45$.

Section 4.8

1   $1/x \approx 2 - x$

5   (b)   This estimate is about right.
(c)   Below

7   Positive

9   Negative

11   (a)   Not defined
(b)   0
(c)   0
(d)   Not defined

13   $0.1x^7$

15   $x^{0.2}$

17   -2

19   $y = 2/3$

21   (c)   0

Chapter 4 Review

1   2/3

3   (a) $h(4) = 1$
(b) $h'(4) = 2$
(c) $h(4) = 3$
(d) $h'(4) = 3$
(e) $h'(4) = -5/16$
(f) $h'(4) = 13$

5   These functions look like the line $y = 0$:
$\sin x - \tan x$, $\frac{x^2}{x^2+1}$
$x - \sin x$, $\frac{1-\cos x}{\cos x}$
These functions look like the line $y = x$:
$\arcsin x$, $\frac{\sin x}{1+\sin x}$
$\arctan x$, $e^x - 1$
$\frac{x}{x+1}$, $\frac{x}{x^2+1}$
These functions are undefined at the origin:
$\frac{\sin x}{x} - 1$, $-x\ln x$
Defined at the origin but with a vertical tangent:
$x^{10} + \sqrt[10]{x}$

7   (a) $dg/dr = -2GM/r^3$
(b) dg/dr is rate of change of acceleration due to pull of gravity.
The further away from the earth's center, the weaker gravity's pull.
(c) -3.05×10^{-6}
(d)   It is reasonable because the magnitude of $\frac{dg}{dr}$ is so small (compared to $g = 9.8$) that for r near 6400 km, g is not varying much at all.

9   (a) $v(t) = 10e^{t/2}$
(b) $v(t) = s(t)/2$

11   (a)   Falling, 0.38 m/hr
(b)   Rising, 3.76 m/hr
(c)   Rising, 0.75 m/hr
(d)   Falling, 1.12 m/hr

13   (a) $v = -2\pi\omega y_0 \sin(2\pi\omega t)$
$a = -4\pi^2\omega^2 y_0 \cos(2\pi\omega t)$

(b)   Amplitudes: different
$(y_0,\ 2\pi\omega y_0,\ 4\pi^2\omega^2 y_0)$
Periods $= 1/\omega$

15   $1/\pi \approx 0.32\mu\text{m/day}$

17   233.2 miles/hour

19   (a)   Decreases
(b) -0.25 cm^3/min

Practice: Differentiation

1   $6t - 4$

3   $-(5x^4 + 2)/2$

5   $20x^3 - 2/x^3$

7   $2e^{2x}(x^2 - x + 1)/(x^2 + 1)^2$

9   $4x\left(x^2 + 2\right)/9$

11   $-3\cos(2 - 3x)$

13   $(z^2 - 1)/3z^2$

15   $1/(\sqrt{\sin(2z)}\sqrt{\cos^3(2z)})$

17   $ae^{ax}/(e^{ax} + b)$

19   $\cos(\tan\theta)/\cos^2\theta$

21   $\cos(\cos x + \sin x) \cdot (\cos x - \sin x)$

23   $5\sin^4\alpha\cos^4\alpha - 3\sin^6\alpha\cos^2\alpha$

25   $-2/(\sqrt{t}(3 + \sqrt{t})^2)$

27   $-(3e^{3x} + 2x)/(e^{3x} + x^2)^2$

29   $\frac{(4\theta - 2\sin(2\theta)\cos(2\theta))}{\sqrt{4\theta^2 - \sin^2(2\theta)}}$

31   $3w^2\ln(10w) + w^2$

33   π

35   $2r^3/\sqrt{r^4 + 1}$

37   $\frac{\sqrt{x + 3}(x^2 + 6x - 9)}{[2\sqrt{x^2 + 9}(x+3)^2]}$

39   $1/(1 + 2u + 2u^2)$

41   $(2t - ct^2)e^{-ct}$

43   $\ln x/(1 + \ln x)^2$

45   $8/\sin t$

47   $3\cos(3\theta)e^{\sin(3\theta)}$

49   $(\ln\pi)\pi^{(x+2)}$

51   $(\cos\theta)e^{\sin\theta}$

53   $e^{2x}\left[2x^2 + 2x + (\ln 5 + 2)5^x\right]$

55   $-8/(4 + t)^2$

57   $-8b^4 z/(a + z^2)^5$

59   $(\ln(\ln 2))(\ln 2)^z$

61   $-3\sin(\arctan 3x)/(1 + 9x^2)$

63   $6(1 + 3t)e^{(1+3t)^2}$

65   $2r(r + 1)/(2r + 1)^2$

67   $2e^t + 2te^t + 1/(2t^{3/2})$

69   $-\frac{2^w\ln 2 + e^w}{(2^w + e^w)^2}$

71   $x^2\ln x$

73 $6(3\theta - \pi)\cos[(3\theta - \pi)^2]$

75 $(-e^{-t} - 1)/(e^{-t} - t)$

77 $1/\sin^2\theta - 2\theta\cos\theta/\sin^3\theta$

79 $-1/(1 + (2 - x)^2)$

81 $e^n\cos(e^n)$

83 $\cos(\sqrt{t}e^t) - t\sin(\sqrt{t}e^t) \cdot$
    $(\sqrt{t}e^t + e^t/(2\sqrt{t}))$

85 0

87 $e^{\tan x} + xe^{\tan x}/\cos^2 x$

89 $6x/\left(9x^4 + 6x^2 + 2\right)$

91 a

93 $ke^{k\theta}$

95 $e^{-4kt}(\cos t - 4k\sin t)$

97 $-4a^2 x/(a^2 + x^2)^2$

99 $(-3a^2 s - s^3)/(a^2 + s^2)^{3/2}$

101 $(-cat^2 + 2at - bc)e^{-ct}$

103 $-2/(x^2 + 4)$

105 $ae^{au}/(a^2 + b^2)$

107 $4/(e^x + e^{-x})^2$

109 $(-4 - 6x)(6x^e - 3\pi) +$
    $(2 - 4x - 3x^2)(6ex^{e-1})$

111 0

113 $4x - 2 - 4x^{-2} + 8x^{-3}$

115 (a) $f'(2) = 20$
    (b) $f'(2) = 11/9$
    (c) $f'(2) = -4$
    (d) $f'(2) = -24$
    (e) $f'(2) = \sin 3 - 8\cos 3$
    (f) $f'(2) = 4\ln 3 - 16/3$

117 $(1 - y)/(x - 3)$

119 ax/by

121 $(8xy - 3x^2)/(3y^2 - 4x^2)$

Section 5.1

3 (a) $x \approx 2.5$ (or any $2 < x < 3$)
      $x \approx 6.5$ (or any $6 < x < 7$)
      $x \approx 9.5$ (or any $9 < x < 10$)
    (b) $x \approx 2.5$: local max;
      $x \approx 6.5$: local min;
      $x \approx 9.5$: local max

7 $x = -1, 1/2$

9 Increasing for all x, no critical point

11 Alternately incr/decr

13 Oscillating: $x < 0$
   Increasing: $x > 0$

15 Local min: $(0.37, -0.37)$

19 $a = -1/3$

21 (a) Local and global max
   (b) Local and global min
   (c) Neither
   (d) Local and global min

33 (a) Decreasing.
   (b) Local minimum at x_1.
   (c) Concave up at x_2.

Section 5.2

3 Increasing $|a|$ stretches the graph horizontally.

5 (a) $b = 20°C$, $a = 180°C$
   (b) $k = \frac{1}{90}\min^{-1}$

7 (a) $x = 0$ and $x = \pm\sqrt{-a/2}$
   (b) For any a or b, $x = 0$ is crit pt.
       Only crit pt if $a \geq 0$
       Local minimum.
   (c) If a is negative:
       $x = 0$ is local max,
       $x = \pm\sqrt{-\frac{a}{2}}$ are local min
   (d) No

11 $(1/b, 1/be)$

13 A determines y intercept; stretches or flattens bell-shaped graph.
   B affects width.

17 (a) Larger $|A|$, steeper
   (b) Shifted horizontally by B. Left for $B > 0$; right for $B < 0$.
       Vertical asymptote $x = -B$

Section 5.3

3 (a) $f(1)$ local minimum;
       $f(0)$, $f(2)$ local maxima
   (b) $f(1)$ global minimum
       $f(2)$ global maximum

5 (a) $f(\frac{2\pi}{3})$ local maximum
       $f(0)$ and $f(\pi)$ local minima
   (b) $f(\frac{2\pi}{3})$ global maximum
       $f(0)$ global minimum

7 (b) Yes, at $x = 0$
   (c) $5 > g(0) > g(2)$

9 44.1 feet

11 $r = \frac{2}{3}R$

13 $\theta = -\mu + \sqrt{\mu^2 + 1}$

15 Minimum: $x = -r_0/\sqrt{2}$
   Maximum: $x = r_0/\sqrt{2}$

17 (c) No

19 $0 \leq y$; No upper bound.

21 $0 \leq y \leq 16$

23 (a) 10
   (b) 9

25 (b) 160 mph or 320 mph; no; yes
   (c) 220 mph

Section 5.4

3 (a) Fixed costs
   (b) Decreases slowly, then increases

5 (a) $0
   (b) $96.56
   (c) Raise the price by $5

7 (b) (i) $N'(x) = 20$
       (ii) $\frac{N(x)}{x} = \frac{100}{x} + 20$

11 (b) $q = \left[Fa/(K(1 - a))\right]^a$

Section 5.5

1 $2000 - 1200/\sqrt{5}$

3 Min $v = \sqrt{2}k$; no max

5 (a) $V = Ax/4 - x^3/2$
   (c) $(A/6)^{3/2}$

7 40 feet by 80 feet

9 $h = \sqrt{50}$ meters

11 $(0.59, 0.35)$

13 When the rectangle is a square.

15 15 miles/hour

17 (a) The arithmetic mean unless $a = b$, in which case the two means are equal.
   (b) The arithmetic mean unless $a = b = c$, in which case the two means are equal.

19 (a) $E = 500e\left(\dfrac{200 - \cos\theta}{\sin\theta}\right) + 2000e$
       $(\arctan\left(\dfrac{500}{2000}\right) \leq \theta \leq \dfrac{\pi}{2})$
   (b) $\theta = \pi/3$
   (c) Independent of e, but dependent on $\overline{AB}/\overline{AL}$.

21 (a) $T = \sqrt{a^2 + (c - x)^2}/v_1$
       $+ \sqrt{b^2 + x^2}/v_2$

Section 5.6

5 $\cosh(2x) = \cosh^2 x + \sinh^2 x$

7 $2\sinh(2x)$

9 $2te^{t^2}\sinh(e^{t^2})$

11 $\tanh(1 + \theta)$

13 (b) $A = 6.325$
       (stretch factor)
       $c = 0.458$
       (horizontal shift)

15 (b) U-shaped
   (c) Incr $(A > 0)$
       or Decr $(A < 0)$
   (d) Max: $A < 0, B < 0$
       Min: $A > 0, B > 0$

17 $y \approx 715 - 100\cosh(x/100)$

Chapter 5 Review

3 (a) Increasing: $(0, \infty)$
       Decreasing: $(-\infty, 0)$
   (b) Local and global min: $f(0)$

5 (a) Increasing: $(0, 4)$
       Decreasing: $(-\infty, 0)$, $(4, \infty)$
   (b) Local max: $f(4)$
       Local min: $f(0)$

7 (a) $f'(x) = 3x(x - 2)$
       $f''(x) = 6(x - 1)$
   (b) $x = 0$
       $x = 2$
   (c) Inflection point: $x = 1$

(d) Endpoints:
$f(-1) = -4$
$f(3) = 0$
Critical Points:
$f(0) = 0$
$f(2) = -4$
Global max:
$f(0) = 0$ and $f(3) = 0$
Global min:
$f(-1) = -4$ and
$f(2) = -4$
(e) f increasing:
for $x < 0$ and $x > 2$
f decreasing:
for $0 < x < 2$
f concave up:
for $x > 1$
f concave down:
for $x < 1$.

9 (a) $f'(x) = $
$-e^{-x}\sin x + e^{-x}\cos x$
$f''(x) = -2e^{-x}\cos x$
(b) Critical points:
$x = \frac{\pi}{4}$ and $\frac{5\pi}{4}$
(c) Inflection points:
$x = \frac{\pi}{2}$ and $\frac{3\pi}{2}$
(d) Endpoints:
$f(0) = 0$
$f(2\pi) = 0$
Global max:
$f(\frac{\pi}{4}) = (e^{-\frac{\pi}{4}})(\frac{\sqrt{2}}{2})$
Global min:
$f(\frac{5\pi}{4}) = -e^{\frac{-5\pi}{4}}(\frac{\sqrt{2}}{2})$
(e) f increasing:
$0 < x < \frac{\pi}{4}$ and
$\frac{5\pi}{4} < x < 2\pi$
f decreasing:
$\frac{\pi}{4} < x < \frac{5\pi}{4}$
f concave down:
for $0 \le x < \frac{\pi}{2}$
and $\frac{3\pi}{2} < x \le 2\pi$
f concave up:
for $\frac{\pi}{2} < x < \frac{3\pi}{2}$

11 $\lim_{x \to \infty} f(x) = \infty$
$\lim_{x \to -\infty} f(x) = -\infty$
(a) $f'(x) = 6(x-2)(x-1)$
$f''(x) = 6(2x-3)$
(b) $x = 1$ and $x = 2$
(c) $x = \frac{3}{2}$
(d) Critical points:
$f(1) = 6$, $f(2) = 5$
Local max: $f(1) = 6$
Local min: $f(2) = 5$
Global max and min: none
(e) f increasing: $x < 1$ and $x > 2$
Decreasing: $1 < x < 2$
f concave up: $x > \frac{3}{2}$
f concave down: $x < \frac{3}{2}$

13 $\lim_{x \to -\infty} f(x) = -\infty$
$\lim_{x \to \infty} f(x) = 0$
(a) $f'(x) = (1-x)e^{-x}$
$f''(x) = (x-2)e^{-x}$

(b) Only critical point is at $x = 1$.
(c) Inflection point: $f(2) = \frac{2}{e^2}$
(d) Global max: $f(1) = \frac{1}{e}$
Local and global min: none
(e) f increasing: $x < 1$
f decreasing: $x > 1$
f concave up: $x > 2$
f concave down: $x < 2$

15 Local max: $f(-3) = 12$
Local min: $f(1) = -20$
Inflection pt: $x = -1$
Global max and min: none

17 Global and local min: $x = 2$
Global and local max: none
Inflection pts: none

19 Global and local max:
$f(0) = 1$
Local minimum: none
Inflection pts: $x = \pm\frac{1}{\sqrt{2}}$

25 $x = \sqrt[3]{2V}$
$y = \sqrt[3]{V/4}$

27 $-4.81 \le f(x) \le 1.82$

29 (a) $g(e)$ is a global maximum.
There is no minimum.
(b) There are exactly two solutions.
(c) $x = 5$ and $x \approx 1.75$

31 (a) x-intercept: $(a, 0)$
y-intercept: $(0, 1/(a^2+1))$
(b) Area $= a/(2(a^2+1))$
(c) $a = 1$
(d) $A = \frac{1}{4}$
(e) $a = 2$ and $a = \frac{1}{2}$

33 (a) The concavity changes at y_1 and y_3.
(b) $f(t)$ grows most quickly where the vase is skinniest and most slowly where the vase is widest.

Section 6.1

5 (a) $x = 1$, $x = 3$
(b) Local min at $x = 1$, local max at $x = 3$

7 (a) $x = 1$, $x = 4$
(b) Local max at $x = 1$, neither at $x = 4$

9 x_1 inflection pt;
x_2 local min

11 (a) $f(3) = 1$; $f(7) = 0$
(b) $x = 0, 5.5, 7$

15 Acceleration is zero at points A and C.

Section 6.2

1 $5x$

3 $x^3/3$

5 $\sin t$

7 $\ln|z|$

9 $-1/2z^2$

11 $-\cos t$

13 $t^4/4 - t^3/6 - t^2/2$

15 $5x^2/2 - 2x^{3/2}/3$

17 $-\cos 2\theta$

19 $(t+1)^3/3$

21 $\sin t + \tan t$

23 $F(x) = x^2$
(only possibility)

25 $F(x) = x^2/8$
(only possibility)

27 $F(x) = \frac{2}{3}x^{3/2}$
(only possibility)

29 $F(x) = -\cos x + 1$
(only possibility)

31 $2t^2 + 7t + C$

33 $5e^z + C$

35 $-\cos t + C$

37 $2t^{5/2}/5 - 2t^{-1/2} + C$

39 $e^{2r}/2 + C$

41 $y^3/3 - 2y - 1/y + C$

43 $\frac{1}{2}\ln|2x - 1| + C$

45 $\frac{1}{2}\sin 2x + 2\cos x + C$

47 $1 - \cos 1 \approx 0.460$

49 $16/3 \approx 5.333$

51 $-8/9 \approx -0.889$

53 $\tan\frac{\pi}{4} = 1$

55 $1 - \frac{\sqrt{3}}{4} \approx 0.567$

57 $\sqrt{2} - 1$

59 $c = 3$

61 (a) 0
(b) $2/\pi$

63 (b) 7 years
(c) $69\frac{1}{3}$ cubic yards

Section 6.3

1 $x^4/4 + 5x + C$

3 $8t^{3/2}/3 + C$

5 $2x^3 + 2x^2 - 14$

7 $-16t^2 + 100t + 50$

11 $y = 2kt^{3/2}/3$

13 (c) 200 ft
(d) 200 ft

15 (a) 6 seconds
(b) Left sum: 97.5 ft
overestimate
Right sum: 82.5 ft
underestimate
(c) 90 ft
(d) $s(t) = 30t - \frac{5}{2}t^2$; $s(6) = 90$ ft
Distance by antidiff = Average of left and right hand sums

17   (c)   height $= 400$ feet
       (d) $v(t) = -32t + 160$
             height $= 400$ feet

19   (a) $v(t) = 1.6t$
       (b) $s(t) = 0.8t^2 + s_0$

21   77 mph; 113.1 ft/sec

Section 6.4

1   500

7   (a) $\text{Si}(4) \approx 1.76$
            $\text{Si}(5) \approx 1.55$
       (b) $(\sin x)/x$ is negative on that interval

9    $(1 + x)^{200}$

11    $-\cos(t^3)$

13    $(2 \sin x^2)/x$

15   (a) $F'(x) = 1/(\ln x)$
       (b)   Increasing, concave down

17    $e^{-x}/\sqrt{\pi x}$

19    $\text{erf}(x_2/\sqrt{2}) - \text{erf}(x_1/\sqrt{2})$

Chapter 6 Review

1    $\frac{5}{2}x^2 + 7x + C$

3    $e^x + 5x + C$

5    $\tan x + C$

7    $(x + 1)^3/3 + C$

9    $\frac{1}{10}(x + 1)^{10} + C$

11    $\frac{1}{2}x^2 + x + \ln|x| + C$

13    $3 \sin x + 7 \cos x + C$

15    $2e^x - 8 \sin x + C$

17    $\ln|x| - 1/x - 1/(2x^2) + C$

19    $G(t) = 5t + \sin t + C$

21    $H(r) = 4r^{3/2}/3 + r^{1/2} + C$

23    $H(t) = t - 2\ln|t| - 1/t + C$

25    $F(z) = e^z + 3z + C$

27    $e^x + e^{1+x} + C$

29    $P(r) = \pi r^2 + C$

31    $(1 + \sin t)^{30}/30 + C$

33    $\sin(t^2)/2 + C$

35    $e^2 y + 2^y/\ln 2 + C$

37    $e^{x^2}/2 + C$

39    $\sqrt{3} - \pi/9$

41    $c = 3/4$

49   (a)   14,000 revs/min²
       (b)   180 revolutions

51   (b)   Highest pt: $t = 2.5$ sec
            Hits ground: $t = 5$ sec
       (c)   Left sum:
             136 ft (an overest.)
            Right sum:
             56 ft (an underest.)
       (d)   100 ft

53   (b) $t = 6$ hours

       (c) $t = 11$ hrs

55   Positive, zero, negative, positive, zero

Modeling: Motion

1    $t = 5; v = -160$ ft/sec

3   400 feet

7   (a)   First second: $-g/2$
            Second: $-3g/2$
            Third: $-5g/2$
            Fourth: $-7g/2$
       (b)   Galileo seems to have been correct.

Section 7.1

1   (a) $2x \cos(x^2 + 1);$
            $3x^2 \cos(x^3 + 1)$
       (b)   (i) $\frac{1}{2} \sin(x^2 + 1) + C$
               (ii) $\frac{1}{3} \sin(x^3 + 1) + C$
       (c)   (i) $-\frac{1}{2} \cos(x^2 + 1) + C$
               (ii) $-\frac{1}{3} \cos(x^3 + 1) + C$

3    $-\frac{1}{3} \cos 3x + C$

5    $4 \sin(t^3) + C$

7    $e^{\sin x} + C$

9    $\frac{1}{2} \ln(x^2 + 1) + C$

11    $\frac{1}{18}(y^2 + 5)^9 + C$

13    $\frac{1}{9}(x^2 - 4)^{9/2} + C$

15    $\frac{1}{6}(x^2 + 3)^3 + C$

17    $\frac{1}{148}(2t - 7)^{74} + C$

19    $-\frac{1}{8}(\cos\theta + 5)^8 + C$

21    $-\frac{1}{2}e^{-x^2} + C$

23    $\frac{1}{35} \sin^7 5\theta + C$

25    $\frac{1}{4} \sin^4 \alpha + C$

27    $\ln|e^t + t| + C$

29    $-\frac{1}{2} \ln|\cos 2x| + C$

31    $2e^{\sqrt{y}} + C$

33    $\ln(2 + e^x) + C$

35    $\frac{1}{5}y^5 + \frac{1}{2}y^4 + \frac{1}{3}y^3 + C$

37    $\frac{1}{6} \ln(1 + 3t^2) + C$

39    $t + 2\ln|t| - \frac{1}{t} + C$

41    $e^{x^2 + x} + C$

43   (a) $x^4 + 2x^2 + C$
       (b) $(x^2 + 1)^2 + C$
       (c)   Both correct but they differ by a constant.

45    $\frac{mg}{k}t - \frac{m^2 g}{k^2}(1 - e^{-kt/m}) + h_0$

Section 7.2

1   (a) $(\ln 2)/2$
       (b) $(\ln 2)/2$

3    $1/\pi$

5    $e(e^3 - 1)$

7   1

9   2/5

11   40

13    $\ln 3$

15   14/3

17    $49932\frac{1}{6}$

19    $\frac{1}{3}(1 - \frac{\sqrt{2}}{2})$

21    $3/2 + \ln 2$

23    $(\ln 2)/2$

25    $\pi/12$

27    $\ln 5$

29   Substitute $w = \ln x$

31   Substitute $w = x + 1$, $w = 1 + \sqrt{x}$

33    $\arctan(x + 2) + C$

35    $\frac{1}{2}(e^4 - 1)$

37   (a)   0
       (b)   2/3

39    $\frac{1}{2} \ln 3$

41   (a)   30,500 pennies
       (b)   Two 15,000-word novels

43    $-\frac{1}{k} \ln\left(\left(e^{t\sqrt{gk}} + e^{-t\sqrt{gk}}\right)/2\right) + h_0$

Section 7.3

1    $x \cdot \arctan x - \frac{1}{2} \ln(1 + x^2) + C$

3    $\frac{1}{5}t^2 e^{5t} - \frac{2}{25}te^{5t} + \frac{2}{125}e^{5t} + C$

5    $-t \cos t + \sin t + C$

7    $\frac{x^4}{4} \ln x - \frac{x^4}{16} + C$

9    $-(z + 1)e^{-z} + C$

11    $\frac{1}{3}\theta^2 \sin 3\theta + \frac{2}{9}\theta \cos 3\theta - \frac{2}{27} \sin 3\theta + C$

13    $-(\theta + 1)\cos(\theta + 1) + \sin(\theta + 1) + C$

15    $-x^{-1} \ln x - x^{-1} + C$

17    $\frac{2}{3}y(y + 3)^{3/2} - \frac{4}{15}(y + 3)^{5/2} + C$

19    $-2y(5 - y)^{1/2} - \frac{4}{3}(5 - y)^{3/2} + C$

21    $t(\ln t)^2 - 2t \ln t + 2t + C$

23    $w \arcsin w + \sqrt{1 - w^2} + C$

25    $\frac{1}{2}x^2 \arctan x^2 - \frac{1}{4} \ln(1 + x^4) + C$

27    $\frac{1}{3}x^3 \sin x^3 + \frac{1}{3} \cos x^3 + C$

29    $\cos 5 + 5 \sin 5 - \cos 3 - 3 \sin 3 \approx -3.944$

31    $\frac{9}{2} \ln 3 - 2 \approx 2.944$

33    $6 \ln 6 - 5 \approx 5.751$

35 $\frac{\pi}{2} - 1 \approx 0.571$

37 $\frac{1}{2}\theta - \frac{1}{4}\sin 2\theta + C$

39 $\frac{1}{2}e^x(\sin x - \cos x) + C$

41 $\frac{1}{2}xe^x(\sin x - \cos x) + \frac{1}{2}e^x\cos x + C$

43 Integrate by parts choosing $u = x^n$, $v' = e^x$.

45 Integrate by parts, choosing $u = x^n$, $v' = \sin ax$.

47 π

49 (a) $-a^2e^{-a} - 2ae^{-a} - 2e^{-a} + 2$
    (b) Increasing
    (c) Concave up

51 (a) $A = a/(a^2 + b^2)$,
         $B = -b/(a^2 + b^2)$
    (b) $e^{ax}(b\sin bx + a\cos bx)/(a^2 + b^2) + C$

53 (a) $C_1 = \sqrt{2}$
    (b) $C_n = \sqrt{2}$

Section 7.4

1 $\frac{1}{10}e^{(-3\theta)}(\sin\theta - 3\cos\theta) + C$

3 $-\frac{1}{5}\cos^5 w + C$

5 $\frac{1}{\sqrt{3}}\arctan\frac{y}{\sqrt{3}} + C$

7 $(\frac{1}{2}x^3 - \frac{3}{4}x^2 + \frac{3}{4}x - \frac{3}{8})e^{2x} + C$

9 $\frac{3}{16}\cos 3\theta \sin 5\theta - \frac{5}{16}\sin 3\theta \cos 5\theta + C$

11 $\frac{1}{3}e^{x^3} + C$

13 $\frac{u^6}{6}\ln 5u - \frac{1}{36}u^6 + C$

15 $-\frac{1}{2}x^2\cos x^2 + \frac{1}{2}\sin x^2 + C$

17 $-\frac{1}{2}y^2\cos 2y + \frac{1}{2}y\sin 2y + \frac{1}{4}\cos 2y + C$

19 $-\frac{1}{2\tan 2\theta} + C$

21 $\frac{1}{21}\frac{\tan 7x}{\cos^2 7x} + \frac{2}{21}\tan 7x + C$

23 $\frac{1}{3}\frac{\sin x}{\cos^3 x} - \frac{4}{3}\frac{\sin x}{\cos x} + x + C$

25 $-\frac{1}{4}(\ln|y-2| - \ln|y+2|) + C$

27 $\arctan(y+2) + C$

29 $-\frac{1}{9}(\cos^3 3\theta) + \frac{1}{15}(\cos^5 3\theta) + C$

33 $2\ln|x| + \ln|x+3| + C$

35 $\ln|x-1| - \ln|x| + C$

37 $\frac{1}{3}\ln|P/(1-P)| + C$

39 (a) 0
    (b) $V_0/\sqrt{2}$
    (c) 156 volts

41 0

Section 7.5

1 (a) LEFT(2)= 12;
         RIGHT(2)= 44
    (b) LEFT(2) underestimate;
         RIGHT(2) overestimate

5 (a) (i) LEFT(32) = 13.6961
              RIGHT(32) = 14.3437
              TRAP(32) = 14.0199
              Exact value:
              $(x\ln x - x)|_1^{10}$
              ≈ 14.02585093
         (ii) LEFT(32) = 50.3180
              RIGHT(32) = 57.0178
              TRAP(32) = 53.6679
              Exact value: $e^x|_0^4$
              ≈ 53.59815003
    (b) (i) Smallest to largest:
              LEFT(32)
              TRAP(32)
              Actual value
              RIGHT(32)
         (ii) Smallest to largest:
              LEFT(32)
              Actual value
              TRAP(32)
              RIGHT(32)

7 (a) $0.664 = $ LEFT
         $0.633 = $ TRAP
         $0.632 = $ MID
         $0.601 = $ RIGHT
    (b) Between
         0.632 and 0.633

9 (a) TRAP(4); 1027.5
    (b) Underestimate

11 MID: over; TRAP: under

13 TRAP: over; MID: under

15 (b) LEFT(5) $\approx$ 1.32350
         error $\approx$ -0.03810
         RIGHT(5) $\approx$ 1.24066
         error $\approx$ 0.04474
         TRAP(5) $\approx$ 1.28208
         error $\approx$ 0.00332
         MID(5) $\approx$ 1.28705
         error $\approx$ -0.001656

17 445 lbs. of fertilizer

23 RIGHT(10) = 5.556
    TRAP(10) = 4.356
    LEFT(20) = 3.199
    RIGHT(20) = 4.399
    TRAP(20) = 3.799

Section 7.6

1 (a) 76/3
    (b) 76/3
    (c) 0

3 $4.2365, n = 10$

5 $1.0894, n = 10$

7 $0.904524, n = 10$

9 (a) ≈ 53.598
    (b) LEFT(2)= 16.778;
         error= 36.820
         RIGHT(2)= 123.974;
         error= -70.376
         TRAP(2)= 70.376;
         error= -16.778
         MID(2)= 45.608;
         error= 7.990
         SIMP(2)= 53.864;
         error= -0.266
    (c) LEFT(4)= 31.193;
         error=22.405
         RIGHT(4)= 84.791;
         error=-31.193
         TRAP(4)= 57.992;
         error=-4.394
         MID(4)= 51.428;
         error=2.170
         SIMP(4)= 53.616;
         error=-0.018

11 (a) 3.449
    (b) 3.816
    (c) 3.980

13 (a) 9.5 years
    (b) 8.33 hours
    (c) 5 minutes

15 0.272

17 (a) ≈ 4.78
    (b) ≈ 4.7962

19 (a) Mostly increasing; mostly concave down

Section 7.7

1 $e^{-2}/2$

3 Does not converge

5 $\pi/5$

7 $2^{3/4}$

9 Does not converge

11 Does not converge

13 $\pi/8$

15 Does not converge

17 Does not converge

19 $\frac{1}{\ln 3}$

21 1/3

23 Does not converge

25 1

27 (a) Does not converge
    (b) Not differentiable

29 (a) $\Gamma(1) = 1$
         $\Gamma(2) = 1$
    (c) $\Gamma(n) = (n-1)!$

31 9×10^9 joules

33 Converges for $p > -1$
    to $pe^{p+1}/(p+1)^2$

Section 7.8

1. (a) diverges
   (b) $f(x)$: converges
   $g(x)$: impossible to tell
   $h(x)$: diverges
   $k(x)$: diverges

3. Converges

5. Converges

7. Does not converge

9. Converges

11. Converges

13. Converges

15. Does not converge

17. 0.139

19. $a = 0.399$

21. (a) $\int_3^\infty e^{-x^2}\,dx \le \frac{e^{-9}}{3}$
    (b) $\int_n^\infty e^{-x^2}\,dx \le \frac{1}{n}e^{-n^2}$

23. Converges for $p < 1$
    Diverges for $p \ge 1$

25. (a) e^t is concave up for all t

Chapter 7 Review

1. $\int_0^b h\,dx = hb$

3. $2\int_{-r}^r \sqrt{r^2 - x^2}\,dx = \pi r^2$

5. (a) (i) 0
       (ii) $\frac{2}{\pi}$
       (iii) $\frac{1}{2}$
   (b) Smallest to largest:
       Average value of $f(t)$
       Average value of $k(t)$
       Average value of $g(t)$

7. (a) $\frac{1}{2}\arcsin 2x + C$
   (b) ≈ 0.45167
   (c) SIMP[100] ≈ 0.45167

9. Substitute $w = x + 2$, $w = x^2 + 1$

11. Substitute $w = 1 - x^2$, $w = \ln x$

13. Converges
    $\int_4^\infty t^{-3/2}\,dt = 1$

15. Converges
    $\int_0^\infty w e^{-w}\,dw = 1$

17. Converges
    $\int_{-\pi/4}^{\pi/4} \tan\theta\,d\theta = 0$

19. Converges
    $\int_{10}^\infty \frac{1}{z^2 - 4}\,dz = (\ln(3/2))/4$

21. Does not converge

23. Does not converge

25. Does not converge

27. 11/3

29. Area $= 2\sqrt{2}$

31. False

33. True

35. (a) Trapezoid: overestimate.
       Midpoint: underestimate.
    (b) Trapezoid: underestimate.
       Midpoint: overestimate.
    (c) Trapezoid and midpoint give
       exact value.

37. ≈ 7.4175.

39. 45 years

Practice: Integration

1. $-\cos t + C$

3. $(1/5)e^{5z} + C$

5. $(-1/2)\cos 2\theta + C$

7. $2x^{5/2}/5 + 3x^{5/3}/3 + C$

9. $(r + 1)^4/4 + C$

11. $x^2/2 + \ln|x| - x^{-1} + C$

13. $\frac{1}{2}e^{t^2} + C$

15. $-\frac{2}{9}(2 + 3\cos x)^{3/2} + C$

17. $\frac{2}{5}(1 - x)^{5/2} - \frac{2}{3}(1 - x)^{3/2} + C$

19. $-y\cos y + \sin y + C$

21. $2x\ln x - 2x + C$

23. $(1/3)\sin^3\theta + C$

25. $\frac{1}{2}u^2 + 3u + 3\ln|u| - \frac{1}{u} + C$

27. $\tan z + C$

29. $(1/2)t^{12} - (10/11)t^{11} + C$

31. ≈ 3.8875

33. -133.8724

35. $-5/64$

37. $2e(e - 1)$

39. $(1/3)(\ln x)^3 + C$

41. $(1/3)x^3 + x^2 + \ln|x| + C$

43. $(1/2)e^{t^2+1} + C$

45. $(1/10)\sin^2(5\theta) + C$ (other forms of answer are possible)

47. $\arctan z + C$

49. $-\frac{1}{8}\cos^4 2\theta + C$

51. $(-1/4)\cos^4 z + (1/6)\cos^6 z + C$

53. $(2/3)(1 + \sin\theta)^{3/2} + C$

55. $t^3 e^t - 3t^2 e^t + 6te^t - 6e^t + C$

57. $(3z + 5)^4/12 + C$

59. $\arctan(\sin w) + C$

61. $-\cos(\ln x) + C$

63. $-\sqrt{16 - w^2} + C$

65. $2\sqrt{1 - \cos w} + C$

67. $(1/3)\ln|3u + 8| + C$

69. $\sqrt{1 + t^2}(t^2 - 2)/3 + C$

71. $(w + 5)^6/6 - (w + 5)^5 + C$

73. $r^2[(\ln r)^2 - \ln r + (1/2)]/2 + C$

75. $(u^3 \ln u)/3 - u^3/9 + C$

77. $-\cos(2x)/(4\sin^2(2x)) + \frac{1}{8}\ln\left|\frac{\cos(2x) - 1}{\cos(2x) + 1}\right| + C$

79. $-y^2\cos(cy)/c + 2y\sin(cy)/c^2 + 2\cos(cy)/c^3 + C$

81. $(1/34)\left[e^{5x}(5\cos(3x) + 3\sin(3x))\right] + C$

83. $(\sqrt{3}/4)(2x\sqrt{1 + 4x^2} + \ln|2x + \sqrt{1 + 4x^2}|) + C$

85. $(\ln|x + 1| - \ln|x + 4|)/3 + C$

87. $x^2/2 - 3x - \ln|x + 1| + 8\ln|x + 2| + C$

89. $\frac{1}{b}\left(\ln|x| - \ln|x + b/a|\right) + C$

91. $\ln|z| - \ln|z + 1| + C$

93. $(1/\ln 2)\ln|2^t + 1| + C$

95. $(1/7)x^7 + 3x^5 + 25x^3 + 125x + C$

97. $\sin^3(2\theta)/6 - \sin^5(2\theta)/10 + C$

99. $\frac{1}{2}x\sqrt{4 - x^2} + 2\arcsin(x/2) + C$

101. $-(1/2)\ln|1 + \cos^2 w| + C$

103. $x\tan x + \ln|\cos x| + C$

105. $(2/3)(\sqrt{x + 1})^3 - 2\sqrt{x + 1} + C$

107. $(1/2)\ln|e^{2y} + 1| + C$

109. $-1/(z - 5) - 5/(2(z - 5)^2) + C$

111. $e^{x^2 - x} + C$

113. $2\sin(2x) + x^3\sin(2x) + 3x^2\cos(2x)/2 + C$

Section 8.1

3. $V = (\pi r^2 h)/3$

5. $V = (4/3)\pi ab^2$

7. $V = (8/15)\pi \approx 1.68$

9. $V = (\pi(e^2 - 1)/2) \approx 10.036$

11. $V \approx 65.54$

13. $V = (\pi(e^2 - 1)/16) \approx 1.25$

15. (a) Volume ≈ 152 in^3
    (b) About 15 apples

17. 3509 cubic inches

19. π

21. (a) $4\int_0^r \sqrt{1 + (-\frac{x}{y})^2}\,dx$
    (b) $2\pi r$

23. 2.35

Section 8.2

1. (a) $\sum_{i=1}^{N}(2 + 6x_i)\Delta x$
   (b) 16 grams

3. (a) $\sum_{k=0}^{n-1} \frac{5\Delta x}{1 + x_k^4}$
   (b) 5.5

5 (a) $\int_0^5 2\pi r(0.115e^{-2r})dr$
   (b) 181 cubic meters

7 Total mass = 12; $\bar{x} = 2.06$

9 (a) Right
   (b) $2/(1+6e-e^2) \approx 0.2$

13 (a) $\pi r^2 l/2$
   (b) $2klr^3/3$

15 (a) $\sum_{i=0}^{N-1} 4\pi(r_e+h_i)^2$
      $\times 1.28e^{-0.000124h_i}\Delta h$
   (b) 6.48×10^{16}

Section 8.3

1 (a) Approximately 1.9
         seconds
   (b) Approximately 103.5
         newton-seconds
   (c) Newton-seconds
   (d) A
   (e) B

3 11,000 ft-lbs

5 1,058,591.1 ft-lb.

7 518,363 ft-lbs

9 1.489×10^{10} joules

11 $v \approx 2360$ m/sec

13 (a) Bottom: 187,200 lbs.
   (b) 15×10 side: 70,200 lbs.
   (c) 15×20 side: 140,400 lbs.

15 60 joules

17 Potential
   $= 2\pi\sigma(\sqrt{R^2+a^2}-R)$

19 $\frac{2GMmy}{a^2}\left(y^{-1} - \frac{1}{(a^2+y^2)^{1/2}}\right)$

21 $\frac{GM_1M_2}{l_1 l_2} \ln\left[\frac{(a+l_1)(a+l_2)}{a(a+l_1+l_2)}\right]$

Section 8.4

3 (a) Option 1
   (b) Option 1: $10.929 million;
         Option 2: $10.529 million

5 (a) $5820 per year
   (b) $36,787.94

7 $46,800

9 (a) 10.6 years
   (b) 624.9 million dollars

11 $85,750,000

15 (a) Less
   (b) Can't tell
   (c) Less

Chapter 8 Review

1 (b) $\sum_{i=1}^N \pi x_i \Delta x$
   (c) Volume $= \pi/2$

3 (a) $\sum_{i=1}^N \pi \frac{9x_i}{4}\Delta x$
   (b) 18π

5 (a) 39 cu.in.
   (b) 94¢

7 $\int_0^\pi \sqrt{1+\cos^2 x}\,dx$

9 16.34 kg

11 1000 ft-lb

13 1,170,000 lbs

15 (a) $\sum_{i=0}^{n-1}(2000-100t_i)$
         $\times e^{-0.1t_i}\Delta t$
   (b) $\int_0^T e^{-0.10t} \times$
         $(2000-100t)\,dt$
   (c) After 20 years
         $11,353.35

17 (a) $\sum_{i=1}^N \pi\left(\frac{3.5\cdot 10^5}{\sqrt{h+600}}\right)^2 \Delta h$
   (b) $1.05\cdot 10^{12}$ cubic feet

19 $0.43m^3$

21 (a) $\pi h^2/(2a)$
   (b) $\pi h/a$
   (c) $dh/dt = -k$
   (d) h_0/k

23 The thin spherical shell.

Modeling: Distributions

5 (a) 0.9m - 1.1m

9 (a) Cumulative distribution
         increasing

   (b) Vertical 0.2,
         horizontal 2

11 (a) 22.1%
   (b) 33.0%
   (c) 30.1%
   (d) $C(h) = 1 - e^{-0.4h}$

13 (a) 21%
   (b) 2%
   (c) 13%
   (e) 20 - 25 mins

15 (b) About 3/4

17 (a) $f(r) = 0.2\ (0 < r < 5)$
         $f(r) = 0\ (5 \le r)$
   (b) $F(r) = 0.2r$
         $(0 \le r \le 5)$
         $F(r) = 1\ (5 < r)$
   (c) $G(v) = 0.124r^{1/3}$
         $(0 \le v \le 523.6)$
         $G(v) = 1\ (523.6 < v)$
   (d) $g(v) = 0.0413v^{-2/3}$
         $(0 < v < 523.6)$
         $g(v) = 1\ (523.6 \le v)$

Modeling: Probability

3 (a) $c = 0.0176$
   (b) 9%

5 (a) $a = 0.122$
   (b) $P(x) = 1 - e^{-0.122x}$
   (c) Median = 5.68 seconds
         Mean = 8.20 seconds

9 (c) μ represents the mean of the
         distribution, while σ is the
         standard deviation.

11 (a) $-e^{-2} + 1 \approx 0.865$
   (b) $-(\ln 0.05)/2 \approx 1.5$ km

13 (a) $6/25 = 24\%$
   (b) $12,600
   (c) $\approx 8000

15 (a) $p(r) = 4r^2 e^{-2r}$
   (b) Mean: 1.5 Bohr radii
         Median: 1.33 Bohr radii
         Most likely:
            1 Bohr radius

Section 9.1

1 $P_4(x) = 1 - x + x^2 - x^3$
      $+ x^4$
   $P_6(x) = 1 - x + x^2 - x^3$
      $+ x^4 - x^5 + x^6$
   $P_8(x) = 1 - x + x^2 - x^3$
      $+ x^4 - x^5 + x^6 - x^7 + x^8$

3 $P_2(x) = 1 + \frac{1}{2}x - \frac{1}{8}x^2$
   $P_3(x) = 1 + \frac{1}{2}x - \frac{1}{8}x^2$
      $+ \frac{1}{16}x^3$
   $P_4(x) = 1 + \frac{1}{2}x - \frac{1}{8}x^2$
      $+ \frac{1}{16}x^3 - \frac{5}{128}x^4$

5 $P_3(x) = P_4(x) = x - \frac{1}{3}x^3$

7 $P_2(x) = 1 - \frac{1}{3}x - \frac{1}{9}x^2$
   $P_3(x) = 1 - \frac{1}{3}x - \frac{1}{9}x^2$
      $- \frac{5}{81}x^3$
   $P_4(x) = 1 - \frac{1}{3}x - \frac{1}{9}x^2$
      $- \frac{5}{81}x^3 - \frac{10}{243}x^4$

9 $P_2(x) = 1 - \frac{1}{2}x + \frac{3}{8}x^2$
   $P_3(x) = 1 - \frac{1}{2}x + \frac{3}{8}x^2$
      $- \frac{5}{16}x^3$
   $P_4(x) = 1 - \frac{1}{2}x + \frac{3}{8}x^2$
      $- \frac{5}{16}x^3 + \frac{35}{128}x^4$

11 (a) 0
   (b) 3
   (c) -24
   (d) 0
   (e) 3600

13 $P_4(x) = 1 - \frac{1}{2!}\left(x - \frac{\pi}{2}\right)^2$
      $+ \frac{1}{4!}\left(x - \frac{\pi}{2}\right)^4$

15 $P_4(x) = e[1 + (x-1)$
      $+ \frac{1}{2}(x-1)^2 + \frac{1}{6}(x-1)^3$
      $+ \frac{1}{24}(x-1)^4]$

17 $c < 0, b > 0, a > 0$

19 $a < 0, b > 0, c > 0$

21 $\sin x = \frac{\sqrt{2}}{2} + \frac{\sqrt{2}}{2}\left(x - \frac{\pi}{4}\right)$
$- \frac{\sqrt{2}}{4}\left(x - \frac{\pi}{4}\right)^2 - \frac{\sqrt{2}}{12}\left(x - \frac{\pi}{4}\right)^3$
$- \cdots$

23 $\sin\theta = -\frac{\sqrt{2}}{2} + \frac{\sqrt{2}}{2}\left(\theta + \frac{\pi}{4}\right)$
$+ \frac{\sqrt{2}}{4}\left(\theta + \frac{\pi}{4}\right)^2$
$- \frac{\sqrt{2}}{12}\left(\theta + \frac{\pi}{4}\right)^3 + \cdots$

25 $\frac{d}{dx}(x^2 e^{x^2})|_{x=0} = 0$
$\frac{d^6}{dx^6}(x^2 e^{x^2})|_{x=0} = \frac{6!}{2} = 360$

27 $P_2(x) = 4x^2 - 7x + 2$
$f(x) = P_2(x)$

29 (a) If $f(x)$ is a polynomial of degree n, then $P_n(x)$, the nth degree Taylor polynomial for $f(x)$ about $x = 0$, is $f(x)$ itself.

35 (a) $f(x) = x^2 - \frac{1}{3!}x^6 + \cdots$
(b) Substitute x^2 for x in the Taylor expansion of $\sin x$.

Section 9.2

1 Yes

3 No

5 $f(x) = 1 + x + x^2$
$+ x^3 + \cdots$

7 $f(x) = 1 - \frac{x}{2} + \frac{3x^2}{8}$
$- \frac{5x^3}{16} + \cdots$

9 $\frac{1}{x} = 1 - (x - 1) +$
$(x - 1)^2 - (x - 1)^3 + \cdots$

11 $\frac{1}{x} = -1 - (x + 1)$
$- (x + 1)^2 - (x + 1)^3$
$- \cdots$

13 $-1 < x < 1$

15 $-1 < x < 1$

17 1

19 1

23 32

25 Does not converge

27 e^2

29 4/3

31 $\ln(3/2)$

Section 9.3

1 $\sqrt{1 - 2x} = 1 - x - \frac{x^2}{2}$
$- \frac{x^3}{2} - \cdots$

3 $e^{-x} = 1 - x + \frac{x^2}{2!}$
$- \frac{x^3}{3!} + \cdots$

5 $\ln(1 - 2y) = -2y - 2y^2$
$- \frac{8}{3}y^3 - 4y^4 - \cdots$

7 $\frac{1}{\sqrt{1 - z^2}} = 1 + \frac{1}{2}z^2 + \frac{3}{8}z^4$
$+ \frac{5}{16}z^6 + \cdots$

9 $\frac{z}{e^{z^2}} = z - z^3 + \frac{z^5}{2!}$
$- \frac{z^7}{3!} + \cdots$

11 $e^t \cos t = 1 + t - \frac{t^3}{3}$
$- \frac{t^4}{6} + \cdots$

13 Smallest: $\cos\theta$
Largest: $1 + \sin\theta$

15 (a) I
(b) IV
(c) III
(d) II

17 $\frac{1}{2+x} = \frac{1}{2}\left(1 - \frac{x}{2} + \left(\frac{x}{2}\right)^2\right.$
$\left. - \left(\frac{x}{2}\right)^3 + \cdots\right)$

19 (a) $f(x) = 1 + (a - b)x$
$+ (b^2 - ab)x^2 + \cdots$
(b) $a = \frac{1}{2}, b = -\frac{1}{2}$

21 $(x^2/a) + x^6/(8a^5) + \cdots$

23 (b) $\phi \approx -3b$

25 (a) If $M \gg m$, then
$\mu \approx \frac{mM}{M} = m$.
(b) $\mu = m[1 - \frac{m}{M} + \left(\frac{m}{M}\right)^2$
$- \left(\frac{m}{M}\right)^3 + \cdots]$
(c) -0.0545%

27 (b) $F = mg(1 - \frac{2h}{R} + \frac{3h^2}{R^2}$
$- \frac{4h^3}{R^3} + \cdots)$
(c) 300 km

Section 9.4

1 Yes, $a = 1$, ratio $= -1/2$

3 Yes, $a = 5$, ratio $= -2$

5 No

7 Yes, $a = 1$, ratio $= -x$

9 No

11 $y^2/(1 - y), |y| < 1$

13 $1/(1 + y^2), |y| < 1$

15 (a) $7(1.02^{104} - 1)/$
$(0.02(1.02)^{100})$
(b) $7e^{0.01}$

17 $(3(2^{11} - 1))/(2^{10})$

19 32/3

21 (a) $P_n =$
$250(0.04) + 250(0.04)^2$
$+ 250(0.04)^3 + \cdots$
$+ 250(0.04)^{n-1}$
(b) $P_n = 250 \cdot 0.04(1 - (0.04)^{n-1})$
$/(1 - 0.04)$
(c) $\lim_{n \to \infty} P_n \approx 10.42$
We'd expect the difference between them to be 250 mg.

23 (a) $h_n = 10(3/4)^n$

(b) $D_1 = 10$ feet
$D_2 = h_0 + 2h_1$
$= 25$ feet
$D_3 = h_0 + 2h_1 + 2h_2$
$= 36.25$ feet
$D_4 = h_0 + 2h_1 + 2h_2$
$+ 2h_3 \approx 44.69$ feet
(c) $D_n =$
$10 + 60\left(1 - (3/4)^{n-1}\right)$

25 (a) \$1250
(b) 12.50

27 \$900 million

Section 9.5

1 Not a Fourier series

3 Fourier series

5 $F_1(x) = F_2(x) = \frac{4}{\pi}\sin x$
$F_3(x) = \frac{4}{\pi}\sin x + \frac{4}{3\pi}\sin 3x$

7 99.942% of the total energy

9 $H_n(x) =$
$\frac{\pi}{4} + \sum_{i=1}^{n} \frac{(-1)^{i+1}\sin(ix)}{i} +$
$\sum_{i=1}^{[n/2]} \frac{-2}{(2i-1)^2\pi}\cos((2i - 1)x)$,
where $[n/2]$ denotes the biggest integer smaller than or equal to $n/2$.

11 (a) $F_3(x) =$
$\frac{1}{2} + \frac{2}{\pi}\cos x - \frac{2}{3\pi}\cos 3x$.
(b) There are cosines instead of sines, but the energy spectrum remains the same

13 $F_4(x) = 1 - \frac{4}{\pi}\sin(\pi x)$
$- \frac{2}{\pi}\sin(2\pi x) - \frac{4}{3\pi}\sin(3\pi x)$
$- \frac{1}{\pi}\sin(4\pi x)$

19 (a) 15.9155%, 0.451808%
(b) $(4\sin^2\frac{k}{2})/(k^2\pi^2)$
(c) The constant term and the first five harmonics are needed.
(d) $F_5(x) = \frac{1}{2\pi} + \frac{2\sin(1/2)}{\pi}\cos x$
$+ \frac{\sin 1}{\pi}\cos 2x$
$+ \frac{2\sin(3/2)}{3\pi}\cos 3x$
$+ \frac{\sin 2}{2\pi}\cos 4x$
$+ \frac{2\sin(5/2)}{5\pi}\cos 5x$

21 (a) 31.83%, 76.91%
(b) 90.07%
(c) $F_3(x) = \frac{1}{\pi} + \frac{2\sin 1}{\pi}\cos x$
$+ \frac{\sin 2}{\pi}\cos 2x + \frac{2\sin 3}{3\pi}\cos 3x$

Chapter 9 Review

1 $e^x \approx 1 + e(x - 1) + \frac{e}{2}(x - 1)^2$

3 $\sin x \approx -\frac{1}{\sqrt{2}} + \frac{1}{\sqrt{2}}\left(x + \frac{\pi}{4}\right)$
$+ \frac{1}{2\sqrt{2}}\left(x + \frac{\pi}{4}\right)^2$

5 $\theta^2 \cos\theta^2 =$
$\theta^2 - \frac{\theta^6}{2!} + \frac{\theta^{10}}{4!} - \frac{\theta^{14}}{6!} + \cdots$

7 $\frac{1}{\sqrt{4 - x}} =$
$\frac{1}{2} + \frac{1}{8}x + \frac{3}{64}x^2 + \frac{5}{256}x^3 + \cdots$

9 $\frac{a}{a+b} = 1 - \frac{b}{a} + (\frac{b}{a})^2 - (\frac{b}{a})^3 \cdots$

11 $3/4$

13 $3e$

15 Smallest to largest:
$1 - \cos x,\ x\sqrt{1-x},$
$\ln(1+x),\ \arctan x,\ \sin x,$
$x,\ e^x - 1$

17 $1/2$

19 (a) $f(t) = t + t^2 + \frac{t^3}{2!} + \frac{t^4}{3!}$
$+ \cdots$
(b) $\int_0^x f(t)\,dt = \frac{x^2}{2} + \frac{x^3}{3}$
$+ \frac{x^4}{4\cdot 2!} + \frac{x^5}{5\cdot 3!} + \cdots$
(c) Substitute $x = 1$,
and integrate by parts.

21 (a) Set $\frac{dV}{dr} = 0$, solve for r.
Check for max or min.
(b) $V(r) = -V_0$
$+ 72V_0 r_0^{-2} \cdot (r - r_0)^2 \cdot \frac{1}{2}$
$+ \cdots$
(d) $F = 0$ when $r - r_0$

23 (a) $0.232323\ldots =$
$0.23 + 0.23(0.01)$
$+ 0.23(0.01)^2 + \cdots$
(b) $0.23/(1-0.01) = (23)/(99)$

27 (a) $g(x) \approx P_n(x) =$
$g(0) + \frac{g''(0)}{2!}x^2 + \frac{g'''(0)}{3!}x^3$
$+ \cdots + \frac{g^{(n)}(0)}{n!}x^n$
(b) If $g''(0) > 0$:
0 is a local minimum
If $g''(0) < 0$:
0 is a local maximum.

31 (b) If the amplitude of the kth
harmonic of f is A_k, then the
amplitude of the kth harmonic
of f' is kA_k.
(c) The energy of the kth harmonic
of f' is k^2 times the energy of
the kth harmonic of f.

Theory: Convergence

5 Does not converge

7 Not convergent

9 Converges

Theory: Errors

1 (a) Underestimate
(b) 1

3 (a) Overestimate:
$0 < \theta \leq 1$
Underestimate:
$-1 \leq \theta < 0$
(b) $|E_2| \leq 0.17$

5 $|E_3| \leq 16.5$

7 $|E_4| \leq 0.016$

9 For $\sin x$ and $\cos x$,
$|E_n| \leq \frac{1}{(n+1)!}$

15 $|E_0| < 0.01$ for $|x| \leq 0.1$

Section 10.1

1 (a) (III)
(b) (V)
(c) (I)
(d) (II)
(e) (IV)

7 $\lambda = -3$ or 2.

9 (A) (II), (V)
(B) (I)
(C) No solutions
(D) (IV)
(E) (III)

11 (b) $A = k$

Section 10.2

1 (b) $y = -x - 1$

3 (a) (II)
(b) (I)
(c) (V)
(d) (III)
(e) (IV)

5 Slope field (b)

7 (c) Increasing:
$-1 < y < 2$
Decreasing:
$y > 2$ or $y < -1$
Horizontal:
$y = 2$ or $y = -1$

Section 10.3

3 (a) $y(0.4) \approx 1.5282$
(b) $y(0.4) = -1.4$

5 (b) $y = \ln|t|$, so at $t = 2$, $y \approx$
0.693
(c) Bigger.

7 (a) At $x = 1$, $y \approx 1.5$
(b) At $x = 1$, $y \approx 1.75$
(c) $y = x^2 + 1$, so $y(1) = 2$
(d) Yes

9 (a) $B \approx 1050$
(b) $B \approx 1050.63$
(c) $B \approx 1050.94$

Section 10.4

1 $P = 20e^{0.02t}$

3 $m = 5e^{3t-3}$

5 $y = 10e^{-x/3}$

7 $P = 104e^t - 4$

9 $m = 3000e^{0.1t} - 2000$

11 $y = 200 - 150e^{t/2}$

13 $R(r) = 1 - 0.9e^{1-r}$

15 $z = -\ln(1 - t^2/2)$

17 $y = -2/(t^2 + 2t - 4)$

19 $w = 2/(\cos\theta^2 + 1)$

21 $z(r) = e^{r+(1/3)r^3}$

23 $R = Ae^{kt}$

25 $P = a + Ae^t$

27 $P = a + Ae^{kt}$

29 $y = 2/(1 + e^{-2t})$

31 $x = e^{At}$

33 (c) $y^2 - x^2 = 2C$

Section 10.5

1 (b) 2001

3 (a) (I)
(b) (IV)
(c) (II) and (IV)
(d) (II) and (III)

5 (a) $y = 3$ and $y = -2$
(b) $y = 3$ unstable;
$y = -2$ stable

7 (a) $y = 500$
(b) $y = Ae^{0.5t} + 500$
(d) Unstable

9 (a) $\frac{dB}{dt} = \frac{r}{100}B$
Constant $= \frac{r}{100}$
(b) $B = Ae^{\frac{r}{100}t}$

11 Michigan: 71 years
Ontario: 18 years

13 (b) $\frac{dQ}{dt} = -0.0187Q$
(c) 3 days

15 $C(t) = 30e^{-100.33t}$

17 (a) Mt. Whitney:
17.50 inches
Mt. Everest:
10.23 inches
(b) 18,661.5 feet

19 (a) $\frac{dT}{dt} = -k(T - A)$
$(A = 68°\text{F})$
(b) $T = 68 + 22.3e^{-kt}$; 3:45 am.

21 About 2150 B.C.

23 (a) $\frac{dD}{dt} = kD$

Section 10.6

1 $\frac{dD}{dt} = -0.75(D - 4)$
Equilibrium $= \ 4$ g/cm^2

3 $P = AV^k$

5 about 3 days

7 (a) 112.5%

9 (a) $dx/dt = k(a - x)(b - x)$
(b) $x = ab(e^{bkt} - e^{akt})/(be^{bkt} - ae^{akt})$

11 (b) $dc/dt = (43.2/35,000) - 0.082c$
(c) $c = 0.015(1 - e^{-0.082t})$
As $t \to \infty$,
$c \to 0.015$ mg/ml

13 (a) $\frac{dy}{dt} = -k(y - a)$
(b) $y = (1 - a)e^{-kt} + a$
(c) a: fraction remembered in the
long run;
k: rate material forgotten

15 (a) $\frac{dc}{dt} = \frac{0.0001}{60} - \frac{0.002}{60}c$
(b) $c = 0.05 - 0.05e^{-3\times 10^{-5}t}$
(c) $c \to 0.05$

17 (a) $\frac{dS}{dt} = 600 - \frac{3S}{100,000}$
(S in grams)
(b) $S(t) =$
$2 \times 10^7(1 - e^{-\frac{3}{100,000}t})$
(c) As $t \to \infty$
$S(t) \to 2 \times 10^7$

21 (b) Overestimate

Section 10.7

1    $P = (6.6 \times 10^6)e^{0.002t}$

3   (a) $P = 5000/(1 + 499e^{-1.78t})$
      (c) $t \approx 3.5$; $P \approx 2500$

5    $100 < P < 200$

9   (a) $k \approx 0.0286$
          $a \approx 0.0001$
      (b)   Increases to 286 million

11   (a)   Use chain rule.
      (b) $\frac{du}{dt} - k(u - \frac{1}{L})$;
          $u = \frac{1}{L} + Ae^{-kt}$
      (c) $P = L/(1 + LAe^{-kt})$

13   Two-sided better

15   42.3 million people; off by 3.7

17   (c) $P = 0$ (stable)
          $P = 4$ (unstable)

19   (b) $dP/dt < 0$ for $P < b/a$
          $dP/dt > 0$ for $P > b/a$
      (c) $P > b/a$: increase
          $P < b/a$: extinction

Section 10.8

3    $A \approx -1.705$
     $\alpha = \pm\sqrt{5}$

5   Middle point, moving down

7   (a)   Spring (iii)
      (b)   Spring (iv)
      (c)   Spring (iv)
      (d)   Spring (i)

9   (a) goes with (II)
     (b) goes with (I) and (IV)
     (c) goes with (III)
     (I) $x = 2\sin 2t$
     (II) $x = -\sin t$
     (III) $x = \cos 4t$
     (IV) $x = -3\sin 2t$

11   (a)   Starts higher; period same.
      (b)   Period increases.

13    $\sqrt{58}$

15    $\omega = 2$, $A = 13$, $\psi = \tan^{-1} 12/5$

17   (a) $A\sin(\omega t + \phi) = $
          $(A\sin\phi)\cos\omega t +$
          $(A\cos\phi)\sin\omega t$

19   (a) $V_0 \approx 25$, $n = 2$,
          $\phi_0 = -\pi$
      (b) $K = 25$, $\omega^2 = 4$,
          $A = B = 0$

21    $C = \frac{1}{20}$ farads

Section 10.9

1    $y(t) = C_1 e^{-t} + C_2 e^{-3t}$

3    $y(t) = $
     $C_1 e^{-2t}\cos t + C_2 e^{-2t}\sin t$

5    $s(t) = $
     $C_1\cos\sqrt{7}t + C_2\sin\sqrt{7}t$

7    $z(t) = C_1 e^{-t/2} + C_2 e^{-3t/2}$

9    $p(t) = C_1 e^{-t/2}\cos\frac{\sqrt{3}}{2}t +$
     $C_2 e^{-t/2}\sin\frac{\sqrt{3}}{2}t$

11    $y(t) = A + Be^{-2t}$

13    $y(t) = \frac{5}{4}e^{-t} - \frac{1}{4}e^{-5t}$

15    $y(t) = 2e^{-3t}\sin t$

17    $p(t) = 20e^{(\pi/2)-t}\sin t$

19   (a)   (IV)
      (b)   (II)
      (c)   (I)
      (d)   (III)

21    $k = 6$
     $y(t) = C_1 e^{2t} + C_2 e^{3t}$

23   (iii)

25   (iv)

27   (ii)

29   Overdamped: $c < 2$
     Critically damped: $c = 2$
     Underdamped: $c > 2$

31    $z(t) = 3e^{-2t}$

33   (a) $Q(t) = \frac{2}{\sqrt{3}}(e^{(-1+\frac{\sqrt{3}}{2})t}$
          $- e^{(-1-\frac{\sqrt{3}}{2})t})$
      (b) $Q(t) = $
          $-\frac{1}{\sqrt{3}}((2+\sqrt{3})e^{(-1+\frac{\sqrt{3}}{2})t} -$
          $(2-\sqrt{3})e^{(-1-\frac{\sqrt{3}}{2})t})$

35   (a) $Q(t) = 16e^{-\frac{1}{8}t}\sin\frac{t}{8}$
      (b) $Q(t) = $
          $2e^{-\frac{1}{8}t}\left(\sin\frac{t}{8} + \cos\frac{t}{8}\right)$
      (c)   Goes from overdamped to
          underdamped

37   No

Chapter 10 Review

1    $y = 10/(1 + 9e^{-10t})$

3    $\frac{1}{2}y - 4\ln|y| = 3\ln|x| - x$
     $+ \frac{7}{2} - 4\ln 5$.

5    $f(x) = (\frac{1}{3}x^{3/2} + \frac{2}{3})^2$
     (for $x \geq 0$)

7    $y = \sqrt[3]{33 - 6\cos x}$

9    $y = 2/(x^2 + 1)$

11    $z(t) = 1/(1 - 0.9e^t)$

13    $y = -\ln(\frac{\ln 2}{3}\sin^2 t\cos t$
     $2\ln 23\cos t - \frac{2\ln 2}{3} + 1)/\ln 2$

15    $-y - \ln|1 - y| = \arctan t - \frac{\pi}{4}$

17    $Q = 4/(1 - Ae^{-(4t^3/3)-4t}) - 2$
     or $Q = -2$

19    $-\frac{a^2}{2}e^{-\frac{x^2}{a^2}} = $
     $\frac{y^2}{2}\ln y - \frac{y^2}{4} + C$

21   (a) $y(1) \approx 3.689$
      (b)   Overestimate
      (c) $y = 5 - 4e^{-x}$
          $y(1) = 5 - 4e^{-1} \approx$
          3.528

     (d) ≈ 3.61

23    $y(t) = C_1 e^{-2t} + C_2 e^{-4t}$

25    $y(t) = C_1 e^{t/3} + C_2 e^{-t/3}$

27   Overdamped if:
     $b > 2\sqrt{5}$ or $b < -2\sqrt{5}$
     Critically damped if:
     $b = \pm 2\sqrt{5}$
     Underdamped if:
     $-2\sqrt{5} < b < 2\sqrt{5}$

29   (a) $\frac{dQ}{dt} = -0.5365Q$
          $Q = Q_0 e^{-0.5365t}$
      (b)   4 mg

31   (a)   Positive
      (b)   Approximately 10 minutes

33   (a) $dB/dt = 0.1B + 1200$
      (b) $B = 12000(e^{0.1t} - 1)$
      (c)   7784.66

35    $dI/dt = 0.001I(1 - 10(I/M))$;
     $M/10$

37   (b) $d^2r/dt^2 = -ABr$
          $r(t) = C_1\cos\sqrt{AB}t$
          $+ C_2\sin\sqrt{AB}t$
      (c) $r(t) = \cos\sqrt{AB}t$
          $j(t) = \sqrt{\frac{A}{B}}\sin\sqrt{AB}t$
      (d)   One quarter of the time.

Section 11.1

1   (a)   80-90°F
      (b)   60-72°F
      (c)   60-100°F

11   (a)   Decreasing
      (b)   Increasing

15   Half wave-length of original
     Speed = 2 seats/sec

Section 11.2

1   B, B, B

3    $(1, -1, -3)$; Front, left, below

9   Cylinder radius 2 along x-axis

11   Q

13    $(1.5, 0.5, -0.5)$

15    $(-4, 2, 7)$

19    $(x-1)^2 + (y-2)^2$
     $+ (z-3)^2 = 25$

Section 11.3

1   (a)   Decreases
      (b)   Increases

3   (a)   Bowl
      (b)   Neither
      (c)   Plate
      (d)   Bowl
      (e)   Plate

5   (a)   I
      (b)   V
      (c)   IV
      (d)   II
      (e)   III

9   (b)   Increasing x

Section 11.4

11   (a)   A
       (b)   B
       (c)   A

23   (a)   (III)
       (b)   (I)
       (c)   (V)
       (d)   (II)
       (e)   (IV)

27   (a) (II) (E)
       (b) (I) (D)
       (c) (III) (G)

29    $\alpha + \beta > 1$: increasing
       $\alpha + \beta = 1$: constant
       $\alpha + \beta < 1$: decreasing

Section 11.5

1    $\Delta z = 0.4$; $z = 2.4$

3    $f(x,y) = 2 - \frac{1}{2} \cdot x - \frac{2}{3} \cdot y$

5    $z = 4 + 3x + y$

7   No

9    $f(x,y) = 2x - 0.5y + 1$

11   -1.0

13    $f(x,y) = 3 - 2x + 3y$

19   (a)   no
       (b)   no
       (c)   no
       (e)   0.5 of GPA

Section 11.6

3   (a)   I
       (b)   II

7    $-(3/2)x - 3y + 3z - 4$

9    $9x - \frac{5}{2}y + \frac{1}{3}z + \frac{67}{6}$

11   Spheres

13   Hyperboloid of two sheets

15   Elliptic paraboloid

17   Ellipsoid

19   Sphere

Chapter 11 Review

1   Vertical line through $(2,1,0)$

3    $(x/5) + (y/3) + (z/2) = 1$

11   Could not be true

13   Might be true

15   True

17    $3x - 5y + 1 = c$

19    $2x^2 + y^2 = k$

23    $g(x,y) = 3x + y$

Theory: Limits

3   (b)   Yes

5   0

7   0

9   1

13    $\pm\sqrt{(1-c)/c}.$

15   No

Section 12.1

1    $\vec{p} = 2\vec{w}$
       $\vec{q} = -\vec{u}$
       $\vec{r} = \vec{u} + \vec{w}$
       $\vec{s} = 2\vec{w} - \vec{u}$
       $\vec{t} = \vec{u} - \vec{w}$

3    $\sqrt{11}$

5    $-15\vec{i} + 25\vec{j} + 20\vec{k}$

7    $\sqrt{65}$

9    $\vec{i} + 3\vec{j}$

11    $-4.5\vec{i} + 8\vec{j} + 0.5\vec{k}$

13    $\sqrt{11}$

15   5.6

17    $\vec{i} + 4\vec{j}$

19    $-\vec{i}$

23    $3\vec{i} + 4\vec{j}$

25    $\vec{a} = \vec{b} = \vec{c} = 3\vec{k}$
       $\vec{d} = 2\vec{i} + 3\vec{k}$
       $\vec{e} = \vec{j}$
       $\vec{f} = -2\vec{i}$

27    $\|\vec{u}\| = \sqrt{6}$
       $\|\vec{v}\| = \sqrt{5}$

29   (a) $(3/5)\vec{i} + (4/5)\vec{j}$
       (b) $6\vec{i} + 8\vec{j}$

Section 12.2

1   Scalar

3   Vector

5   (a) $50\vec{i}$
       (b) $-50\vec{j}$
       (c) $25\sqrt{2}\vec{i} - 25\sqrt{2}\vec{j}$
       (d) $-25\sqrt{2}\vec{i} + 25\sqrt{2}\vec{j}$

7    $180\vec{i} + 72\vec{j}$

9   P
       Towards the center

11    $(79.00, 79.33, 89.00, 68.33, 89.33)$

13    $48.3°$ east of north
       744 km/hr

15   548.6 km/hr

Section 12.3

1   -38

3   14

5   238

7   1.91 radians $(109.5°)$

9    $\vec{u} \perp \vec{v}$ for $t = 2$ or -1.
       No values of t make $\vec{u}$ parallel to $\vec{v}$.

11    $3\vec{i} + 4\vec{j} - \vec{k}$
       (multiples of)

13    $-x + 2y + z = 1$

15    $2x - 3y + 7z = 19$

17    $3x + y + z = -1$

19    $\vec{a} = -\frac{8}{21}\vec{d} + (\frac{79}{21}\vec{i} + \frac{10}{21}\vec{j} - \frac{118}{21}\vec{k})$

21    $38.7°$

Section 12.4

1    $-\vec{i}$

3    $-\vec{i} + \vec{j} + \vec{k}$

5    $-2\vec{i} + 2\vec{j}$

7    $\vec{a} \times \vec{b} = -2\vec{i} - 7\vec{j} - 13\vec{k}$
       $\vec{a} \cdot (\vec{a} \times \vec{b}) = 0$
       $\vec{b} \cdot (\vec{a} \times \vec{b}) = 0$

9   Max: 6
       Min: 0

11    $x + y + z = 1$

13   (a)   1.5
       (b) $y = 1$

15    $4x + 26y + 14z = 0$

19   (b) $\vec{c} \perp \vec{a}$ and $\|\vec{c}\| = \|\vec{a}\|$
       (c) $-a_2 b_1 + a_1 b_2$

Chapter 12 Review

3    $\vec{p} = -\frac{4\sqrt{5}}{5}\vec{i} - \frac{2\sqrt{5}}{5}\vec{j}$

5    $-\vec{u}, \vec{v}, \vec{v} - \vec{u}, \vec{u} - \vec{v}$

7   (a) $t = 1$
       (b)   no t values
       (c)   any t values

9    $-2\vec{k}$

11    $\vec{i} - \vec{j}$

13    $\sqrt{6}/2$

15    $\vec{n} = 4\vec{i} + 6\vec{k}$

17   (a) $(21/5, 0, 0)$
       (b) $(0, -21, 0)$ and $(0, 0, 3)$
           (for example)
       (c) $\vec{n} = 5\vec{i} - \vec{j} + 7\vec{k}$
           (for example)
       (d) $21\vec{j} + 3\vec{k}$
           (for example)

19    $-\vec{i}/2 + \sqrt{3}\vec{j}/2$
       $-\vec{i}/2 - \sqrt{3}\vec{j}/2$

21   Vector

25   (a) $30\vec{i} - 2\vec{k}$
           $20\vec{i} + 15\vec{j} - 5\vec{k}$
           $12\vec{i} + 30\vec{j} + 3\vec{k}$
       (b) $30\vec{i} + \vec{j}$
           $20\vec{i} + 16\vec{j} - 3\vec{k}$
           $12\vec{i} + 31\vec{j} + 5\vec{k}$

27    $38.7°$ south of east.

33    $9x - 16y + 12z = 5$
       0.23

Section 13.1

1    $f_x(3,2) \approx 2, f_y(3,2) \approx -1$

3   (c)   Positive
       (d)   Negative

5   (a)   Both negative
       (b)   Both negative

7 (a) $\partial P/\partial t$:
dollars/month
Rate of change in payments with time.
negative

   (b) $\partial P/\partial r$:
dollars/percentage point
Rate of change in payments with interest rate.
positive

9 (a) Negative
   (b) Positive

11 $f_T(5, 20) \approx 1$

13 (a) 2.5, 0.02
    (b) 3.33, 0.02
    (c) 3.33, 0.02

15 -1

Section 13.2

1 -1

3 $2xy + 10x^4y$

5 y

7 $a/(2\sqrt{x})$

9 $21x^5y^6 - 96x^4y^2 + 5x$

11 $(a + b)/2$

13 $2B/u_0$

15 $2mv/r$

17 Gm_1/r^2

19 $-2\pi r/T^2$

21 $\epsilon_0 E$

23 $c\cos(ct - 5x)$

25 $(15a^2bcx^7 - 1)/(ax^3y)$

27 $[x^2y(-3\lambda + 10) - 3\lambda^4(8\lambda^2 - 27\lambda + 50)]/2(\lambda^2 - 3\lambda + 5)^{3/2}$

29 $\pi xy/\sqrt{2\pi xyw - 13x^7y^3v}$

31 $z_x = 7x^6 + yx^{y-1}$
    $z_y = 2^y \ln 2 + x^y \ln x$

33 13.6

35 (a) 3.3, 2.5
    (b) 4.1, 2.1
    (c) 4, 2

37 (a) Pre^{rt}
    (b) e^{rt}

39 $h_x(2, 5) \approx -0.38$ ft/seat
    $h_t(2, 5) \approx 0.76$ ft/second

Section 13.3

1 $z = 6 + 3x + y$

3 $z = -4 + 2x + 4y$

5 $z = 9 + 6(x - 3) + 9(y - 1)$

7 $P(r, L) \approx 80 + 2.5(r - 8) + 0.02(L - 4000)$, $P(r, L) \approx 120 + 3.33(r - 8) + 0.02(L - 6000)$, $P(r, L) \approx 160 + 3.33(r - 13) + 0.02(L - 7000)$.

11 $df = y\cos(xy)\,dx + x\cos(xy)\,dy$

13 $dg = (2u + v)\,du + u\,dv$

15 $df = dx - dy$

17 $dP \approx 2.395\,dK + 0.008\,dL$

19 $df = \frac{1}{3}dx + 2dy$
    $f(1.04, 1.98) \approx 2.973$

21 (a) Increases
    (b) Increases
    (c) 55 joules

23 l is 1%; g is 2%

Section 13.4

1 (a) 1.01
   (b) 0.98

3 1

5 2.12

7 $x > 2$

9 (a) Negative
   (b) Negative

11 $\nabla z = \frac{1}{y}\cos\left(\frac{x}{y}\right)\vec{i} - \frac{x}{y^2}\cos\left(\frac{x}{y}\right)\vec{j}$

13 $\nabla z = e^y\vec{i} + e^y(1 + x + y)\vec{j}$

15 $\nabla z = 2x\cos(x^2 + y^2)\vec{i} + 2y\cos(x^2 + y^2)\vec{j}$

17 $2m\vec{i} + 2n\vec{j}$

19 $-\frac{(t^2 - 2t + 4)}{(2s\sqrt{s})}\vec{i} + \frac{(2t - 2)}{\sqrt{s}}\vec{j}$

21 $\left(\frac{5\alpha}{\sqrt{5\alpha^2 + \beta}}\right)\vec{i} + \left(\frac{1}{2\sqrt{5\alpha^2 + \beta}}\right)\vec{j}$

23 $50\vec{i} + 96\vec{j}$

25 $(1/2)\vec{i} + (1/2)\vec{j}$

27 (a) $2/\sqrt{13}$
    (b) $1/\sqrt{17}$
    (c) $\vec{i} + \frac{1}{2}\vec{j}$

29 (a) $-\sqrt{2}/2$
    (b) $\sqrt{3} + 1/2$

31 4.4

33 $(3\sqrt{5} - 2\sqrt{2})\vec{i} + (4\sqrt{2} - 3\sqrt{5})\vec{j}$

35 $y = 2x - 7$

37 P

Section 13.5

1 10/3

3 $2z + 3x + 2y = 17$

5 $x + 3y + 7z = -9$
   $\vec{i} + 3\vec{j} + 7\vec{k}$

7 (a) $6.33\vec{i} + 0.76\vec{j}$
   (b) -34.69

9 (b) Valley

11 (a) $x = y = 0$ and $z \neq 0$.
    (b) $y = 0$; $2x - y + 2z = 3$
    (c) $-\vec{j}$,
         $\frac{2}{3}\vec{i} - \frac{1}{3}\vec{j} + \frac{2}{3}\vec{k}$

13 $2x - y - z = 4$

Section 13.6

1 $\frac{dz}{dt} = e^{-t}\sin(t)(2\cos t - \sin t)$

3 $(t^3 - 2)/(t + t^4)$

5 $2e^{1 - t^2}(1 - 2t^2)$

7 $\frac{\partial z}{\partial u} = (e^{-v\cos u} - v\cos(u)e^{-u\sin v})\sin v - (-u\sin(v)e^{-v\cos u} + e^{-u\sin v})v\sin u$
   $\frac{\partial z}{\partial v} = (e^{-v\cos u} - v\cos(u)e^{-u\sin v})u\cos v + (-u\sin(v)e^{-v\cos u} + e^{-u\sin v})\cos u$

9 $\frac{\partial z}{\partial u} = e^v/u$
   $\frac{\partial z}{\partial v} = e^v\ln u$

11 $\frac{\partial z}{\partial u} = 2ue^{(u^2 - v^2)}(1 + u^2 + v^2)$
    $\frac{\partial z}{\partial v} = 2ve^{(u^2 - v^2)}(1 - u^2 - v^2)$

13 $\frac{\partial z}{\partial u} = \frac{1}{vu}\cos\left(\frac{\ln u}{v}\right)$
    $\frac{\partial z}{\partial v} = -\frac{\ln u}{v^2}\cos\left(\frac{\ln u}{v}\right)$

15 $\frac{\partial w}{\partial u} = \frac{\partial w}{\partial x}\frac{\partial x}{\partial u} + \frac{\partial w}{\partial y}\frac{\partial y}{\partial u} + \frac{\partial w}{\partial z}\frac{\partial z}{\partial u}$
    $\frac{\partial w}{\partial v} = \frac{\partial w}{\partial x}\frac{\partial x}{\partial v} + \frac{\partial w}{\partial y}\frac{\partial y}{\partial v} + \frac{\partial w}{\partial z}\frac{\partial z}{\partial v}$

17 -0.6

21 $\left(\frac{\partial U}{\partial P}\right)_T = \left(\frac{\partial U}{\partial V}\right)_T \left(\frac{\partial V}{\partial P}\right)_T$

23 (a) $\frac{\partial z}{\partial r} = \cos\theta\frac{\partial z}{\partial x} + \sin\theta\frac{\partial z}{\partial y}$
         $\frac{\partial z}{\partial \theta} = r(\cos\theta\frac{\partial z}{\partial y} - \sin\theta\frac{\partial z}{\partial x})$
    (b) $\frac{\partial z}{\partial y} = \sin\theta\frac{\partial z}{\partial r} + \frac{\cos\theta}{r}\frac{\partial z}{\partial \theta}$
         $\frac{\partial z}{\partial x} = \cos\theta\frac{\partial z}{\partial r} - \frac{\sin\theta}{r}\frac{\partial z}{\partial \theta}$

Section 13.7

1 $f_{xx} = 2$
   $f_{yy} = 2$
   $f_{yx} = 2$
   $f_{xy} = 2$

3 $f_{xx} = 0$
   $f_{xy} = e^y = f_{yx}$
   $f_{yy} = xe^y$

5 $f_{xx} = -(\sin{(x^2 + y^2)})4x^2$
$+ 2\cos{(x^2 + y^2)}$
$f_{xy} = -(\sin{(x^2 + y^2)})4xy$
$= f_{yx}$
$f_{yy} = -(\sin{(x^2 + y^2)})4y^2$
$+ 2\cos{(x^2 + y^2)}$

7 $f_{xx} = -(\sin{(\frac{x}{y})})(\frac{1}{y^2})$
$f_{xy} = -(\sin{(\frac{x}{y})})(\frac{-x}{y^2})(\frac{1}{y})$
$+ (\cos{(\frac{x}{y})})(\frac{-1}{y^2}) = f_{yx}$
$f_{yy} = -(\sin{(\frac{x}{y})})(\frac{-x}{y^2})^2$
$+ (\cos{(\frac{x}{y})})(\frac{2x}{y^3})$

9 $z_{yy} = 0$

11 (a) Positive
 (b) Zero
 (c) Positive
 (d) Zero
 (e) Zero

13 (a) Negative
 (b) Zero
 (c) Negative
 (d) Zero
 (e) Zero

15 (a) Zero
 (b) Negative
 (c) Zero
 (d) Negative
 (e) Zero

17 (a) Negative
 (b) Negative
 (c) Zero
 (d) Zero
 (e) Zero

19 (a) Negative
 (b) Positive
 (c) Positive
 (d) Positive
 (e) Negative

Section 13.8

1 $Q(x, y) = 1 - 2x^2 - y^2$

3 $Q(x, y) = -y + x^2 - y^2/2$

5 $L(x, y) = 2e + e(x - 1) + 3e(y - 1)$
$Q(x, y) = 2e + e(x - 1) + 3e(y - 1) + e(x - 1)(y - 1) + 2e(y - 1)^2$

7 $L(x, y) = \sqrt{2} + \frac{1}{\sqrt{2}}(x - 1) + \frac{1}{\sqrt{2}}(y - 1)$
$Q(x, y) = \sqrt{2} + \frac{1}{\sqrt{2}}(x - 1) + \frac{1}{\sqrt{2}}(y-1) + \frac{1}{4\sqrt{2}}(x-1)^2 - \frac{1}{2\sqrt{2}}(x-1)(y - 1) + \frac{1}{4\sqrt{2}}(y - 1)^2$

9 $L(x, y) = \frac{e}{2} + \frac{e}{4}(x-1) + \frac{e}{4}(y-1)$
$Q(x, y) = \frac{e}{2} + \frac{e}{4}(x - 1) + \frac{e}{4}(y-1) - \frac{e}{8}(x-1)^2 + \frac{e}{4}(x-1)(y-1) + \frac{e}{8}(y - 1)^2$

11 $L(x, y) = \frac{\pi}{4} + \frac{1}{2}(x-1) - \frac{1}{2}(y-1)$
$Q(x, y) = \frac{\pi}{4} + \frac{1}{2}(x - 1) - \frac{1}{2}(y-1) - \frac{1}{4}(x - 1)^2 + \frac{1}{4}(y - 1)^2$

15 (a) xy
$1 - \frac{1}{2}(x - \frac{\pi}{2})^2 - \frac{1}{2}(y - \frac{\pi}{2})^2$

17 (a) $L(x, y) = 1$, $|E_L(x, y)| \leq 0.047$
 (b) $Q(x, y) = 1 + (1/2)x^2 - (1/2)y^2$, $|E_Q(x, y)| \leq 0.0047$

19 (a) $L(x, y) = 0$,
$|E_L(x, y)| \leq 0.14$
 (b) $Q(x, y) = x^2 + y^2$,
$|E_Q(x, y)| \leq 0.036$

Chapter 13 Review

1 $\partial z/\partial x = \frac{14x+7}{(x^2+x-y)^{-6}}$
$\partial z/\partial y = -7(x^2 + x - y)^6$

3 $\partial f/\partial p = (1/q)e^{p/q}$
$\partial f/\partial q = -(p/q^2)e^{p/q}$

5 $\partial z/\partial x = 4x^3 - 7x^6y^3 + 5y^2$
$\partial z/\partial y = -3x^7y^2 + 10xy$

7 $\partial w/\partial s = \ln(s + t) + \frac{s}{(s+t)}$
$\partial w/\partial t = \frac{s}{(s+t)}$

9 $84/5$

11 False

13 True

15 False

17 (a) $f_w(2, 2) \approx 2.78$
$f_z(2, 2) \approx 4.01$
 (b) $f_w(2, 2) \approx 2.773$
$f_z(2, 2) = 4$

19 (a) negative, positive, up if positive, down if negative
 (b) $\pi < t < 2\pi$
 (c) $0 < x < 3\pi/2$ and $0 < t < \pi/2$ or $3\pi/2 < t < 5\pi/2$.

23 (a) $L = 2K$
 (b) $aL = bK$

25 $dP \approx 47.6\,dL + 17.8\,dK$

27 (a) $T = 305.7$
$dT = 12.16\,dV + 305.06\,dP$
 (b) ≈ 2.5 dm^3

29 $4\sqrt{2}$,
$6\vec{i} + 2\vec{j}$

33 $-2x\vec{i} - 2y\vec{j}$

35 (a) $g(x, y) = 4$
 (b) $(1, 3, 4)$
 (c) $-x + 2y - z = 1$

37 416,699 people/year

39 (a) $e^{10} - 2e^{10}x - 6e^{10}y$
 (b) $1 + (x - 1)^2 + (y - 3)^2$
 (c) $-2\vec{i} - 6\vec{j}$
 (d) $-2e^{10}\vec{i} - 6e^{10}\vec{j} - \vec{k}$

Theory: Differentiability

1 (b) No
 (c) No
 (d) No

 (e) Exist, not continuous.

3 (b) Yes
 (c) Yes
 (d) No
 (e) Exist, not continuous

5 (b) Yes
 (d) No
 (f) No

7 (c) No
 (e) No

9 (a) No

Section 14.1

1 A: no
B: yes, max
C: yes, saddle

5 Saddle pts: $(1, -1)$, $(-1, 1)$
local max $(-1, -1)$
local min $(1, 1)$.

7 $(1, -1)$ and $(-1, 1)$, both are saddle points.

9 Critical points: $(0, 0)$, $(\pm\pi, 0)$,
$(\pm 2\pi, 0)$, $(\pm 3\pi, 0)$, $\cdots$
Local minima: $(0, 0)$,
$(\pm 2\pi, 0)$, $\pm 4\pi, 0)$, $\cdots$
Saddle points: $(\pm\pi, 0)$,
$(\pm 3\pi, 0)$, $(\pm 5\pi, 0)$, $\cdots$

11 Local max: $(1, 5)$

13 Local minimum

15 Local maximum

17 (a) $(1, 3)$ is a minimum

Section 14.2

1 Mississippi:
$87 - 88$ (max), $83 - 87$ (min)
Alabama:
$88 - 89$ (max), $83 - 87$ (min)
Pennsylvania:
$89 - 90$ (max), 80 (min)
New York:
$81 - 84$ (max), $74 - 76$ (min)
California:
$100 - 101$ (max), $65 - 68$ (min)
Arizona:
$102 - 107$ (max), $85 - 87$ (min)
Massachusetts:
$81 - 84$ (max), 70 (min)

3 Neither.

5 Min $= 0$ at $(0, 0)$
(not on boundary)
Max $= 2$ at $(1, 1)$, $(1, -1)$,
$(-1, -1)$ and $(-1, 1)$
(on boundary)

7 Max $= 0$ at $(0, 0)$
(not on boundary)
Min $= -2$ at $(1, -1)$, $(-1, -1)$,
$(-1, 1)$ and $(1, 1)$
(on boundary)

9    $q_1 = 300, q_2 = 225.$

11    $h = 25\%, t = 25°C$

13    $y = 2/3 - x/2$

15   (a) $y = 0.0066(x - 1920) - 0.2135$
$0.38.

(b)   Looks linear after 1972.
(c)   Linear after 1960
$y = 0.0529(x - 1920) - 4.9329$
$0.84

17    $l = w = h = 45$ cm

21   6 pumping stations, 1.61 m diameter pipe

Section 14.3

1    $\text{Min} = -\sqrt{2}, \text{max} = \sqrt{2}$

3    $\text{Min} = \frac{3}{4}$, no max

5    $\text{Min} = \sqrt{2}, \text{max} = 2$

7    $\text{Min} = -\sqrt{35}, \text{max} = \sqrt{35}$

9   max: 0, no min

11   max: $\frac{3}{\sqrt{6}}$, min: $-\frac{3}{\sqrt{6}}$

13   Max: $\frac{\sqrt{2}}{4}$, min: $-\frac{\sqrt{2}}{4}$.

15   Max: $f(\frac{1}{\sqrt{5}}, \frac{3}{\sqrt{5}}) = 2\sqrt{5}$
Min: $f(-\frac{1}{\sqrt{5}}, -\frac{3}{\sqrt{5}}) = -2\sqrt{5}$

17   Max: 1
Min: -1.

19    $q_1 = 50$ units
$q_2 = 150$ units

21   (b) $S = 1000 - 10l$

23    $r = \sqrt[3]{\frac{50}{\pi}}$
$h = 2\sqrt[3]{\frac{50}{\pi}}$

25    $f(x, y, z) = \frac{|Aa + Bb + Cc + D|}{\sqrt{A^2 + B^2 + C^2}}$

Chapter 14 Review

1   Local maximum: $(\pi/3, \pi/3)$

3    $(\sqrt{2}, -\sqrt{2}/2)$ saddle point

7   Maxima: $(-1, 1)$ and $(1, -1)$
Minimum: $(0, 0)$

9    $p_1 = 110, p_2 = 115.$

11    $K = 20$
$L = 30$
$C = \$7,000$

13   (a)   Reduce K by 1/2 unit, increase L by 1 unit.

15    $d \approx 5.37$ m, $w \approx 6.21$ m,
$\theta = \pi/3$ radians

17    $p_i = \sqrt[N]{p_1{}^{N+1-i}(p_{N+1})^{i-1}}, i = 2, 3, \ldots, N.$

19   (a)   Local min: $(-13/55, 1/55)$

Modeling: The Lagrangian

1   (a) $C = 4349$
(b) $182

---

3   (c) $D = 10, N = 20,$
$V \approx 9,779$
(d) $\lambda = 14.67$
(e) $68, rise

Section 15.1

1   Lower sum : 0.34
Upper sum : 0.62

3   Upper sum $= 46.63$
Lower sum $= 8$
Average ≈ 27.3

5   Positive

7    $40/3$

9    $\int_R w(x, y)\, dx\, dy \approx 2700$ cubic feet, where R is the region $0 \le x \le 60$, $0 \le y \le 8$

11   (a)   About 148 tornados
(b)   About 56 tornados
(c)   About 2 tornados

13   (a)   positive
(b)   positive
(c)   positive
(d)   zero
(e)   zero
(f)   zero
(g)   negative
(h)   zero
(i)   negative
(j)   zero
(k)   zero
(l)   positive
(m)   positive
(n)   positive
(o)   zero
(p)   zero

Section 15.2

1    $\frac{4}{15}(9\sqrt{3} - 4\sqrt{2} - 1) = 2.38176$

3    $32/9$

5    $\int_1^4 \int_1^2 f\, dy\, dx$
or $\int_1^2 \int_1^4 f\, dx\, dy$

7    $\int_1^4 \int_{(x-1)/3}^2 f\, dy\, dx$

9    $(e^4 - 1)(e^2 - 1)e$

11   ≈ -2.68

13   14

15    $\frac{e-1}{2}$

17    $\frac{2}{9}(3\sqrt{3} - 2\sqrt{2})$

19    $\int_{-5}^5 \int_{-\sqrt{25-y^2}}^{\sqrt{25-y^2}} (25 - x^2 - y^2)\, dx\, dy$

21    $\int_0^4 \int_{y-4}^{(4-y)/2} (4 - 2x - y)\, dx\, dy$

23   Volume $= 6$

25    $1/10$

27    $\frac{1}{2}(1 - \cos 1) = 0.23$

Section 15.3

1   2

3    $a + b + 2c$

7   Limits do not make sense.

---

9   Limits do not make sense.

13    $\frac{15}{2}$

15    $\int_{-1}^1 \int_{-\sqrt{1-x^2}}^{\sqrt{1-x^2}} \int_{-\sqrt{1-z^2}}^{\sqrt{1-z^2}} dy\, dz\, dx$

17    $m = 1/36; (\bar{x}, \bar{y}, \bar{z}) = (1/4, 1/8, 1/12)$

19    $m(b^2 + c^2)/3$

Section 15.4

1    $\int_{\pi/4}^{3\pi/4} \int_0^2 f\, r\, dr\, d\theta$

3    $\int_0^{2\pi} \int_0^{\sqrt{2}} f\, r\, dr\, d\theta$

13   0

15    $-2/3$

17   6

19    $32\pi(\sqrt{2} - 1)/3$

21   (a) $\int_{\pi/2}^{3\pi/2} \int_1^4 \delta(r, \theta)\, r\, dr\, d\theta$
(b)   (i)
(c)   About 39,000

Section 15.5

1    $200\pi/3$

3   25π

5    $\int_0^1 \int_0^{2\pi} \int_0^4 \delta \cdot r\, dr\, d\theta\, dz$

7    $\int_0^{2\pi} \int_0^{\pi/6} \int_0^3 \delta \cdot \rho^2 \sin\phi\, d\rho\, d\phi\, d\theta$

9    $\int_0^3 \int_0^1 \int_0^5 \delta\, dz\, dy\, dx$

11   π

13   (a)   Positive
(b)   Zero

15    $25\pi/6$

17   27π

19    $3/\sqrt{2}$

21    $3/4$

25    $3I = \frac{6}{5}a^2; I = \frac{2}{5}a^2$

Section 15.6

1   (a) $20/27$
(b) $199/243$

3   (a) $k = 8$
(b) $1/3$

5    $\int_{65}^{100} \int_{0.8}^1 f(x, y)\, dx\, dy$

Chapter 15 Review

1   9200 cubic miles

7    $85/12$

9    $10(e - 2)$

11    $-4\cos 4 + 2\sin 4$
$+ 3\cos 3 - 2\sin 3 - 1$

13    $\int_0^4 \int_{\frac{y}{2}-2}^{-y+4} f(x, y)\, dx\, dy$
or $\int_{-2}^0 \int_0^{2x+4} f(x, y)\, dy\, dx$
$+ \int_0^4 \int_0^{-x+4} f(x, y)\, dy\, dx.$

17 $162\pi/5$

19 8π

21 ≈ 183

23 $I_x = 25/2$
    $I_y = 65/2$
    $I_z = 85/2$

25 $2\pi Gm(r_2 - r_1 - \sqrt{r_2^2 + h^2} + \sqrt{r_1^2 + h^2}$

Theory: Change Variables

3 $\rho^2 \sin\phi$

5 13.5

7 9

9 $40\sqrt{2}, 54\sqrt{2}$

Section 16.1

1 The particle moves on straight lines from $(0, 1)$ to $(1, 0)$ to $(0, -1)$ to $(-1, 0)$ and back to $(0, 1)$.

3 The particle moves on straight lines from $(-1, 1)$ to $(1, 1)$ to $(-1, -1)$ to $(1, -1)$ and back to $(-1, 1)$.

5 Clockwise for all t.

7 Clockwise: $t < 0$,
   Counter-clockwise: $t > 0$.

9 Counterclockwise: $t > 0$.

13 $x = -2, y = t$

15 $x = -2\cos t, y = 2\sin t$,
    $0 \le t \le 2\pi$

17 $x = 5\cos t, y = 7\sin t$,
    $0 \le t \le 2\pi$

19 (a) Right of $(2, 4)$
    (b) $(-1, -3)$ to $(2, 4)$
    (c) $t < -2/3$

21 (a) $a = b = 0, k = 5$ or -5
    (b) $a = 0, b = 5, k = 5$ or -5
    (c) $a = 10, b = -10, k = \sqrt{200}$
         or $-\sqrt{200}$

23 $x = 2\cos t, y = 0, z = 2\sin t$

25 $x = 2 + 3t,\quad y = 3 - t,\quad z = -1 + t$.

27 $x = 1,\quad y = 0,\quad z = t$.

29 $x = 1 + 2t,\quad y = 2 + 4t,\quad z = 5 - t$

31 Yes

33 (a) Straight lines
    (b) No
    (c) $(1, 2, 3)$

Section 16.2

1 (a) Both parameterize the line
         $y = 3x - 2$
   (b) Slope $= 3,0$
        y-intercept $= -2$

3 (b) $-\vec{i} - 10\vec{j} - 7\vec{k}$.

(c) $\vec{r} = (1-t)\vec{i} + (3 - 10t)\vec{j} - 7t\vec{k}$.

5 (a) Spiral
   (b) $\vec{v}(2) = -2.24\vec{i} + 0.08\vec{j}$,
        $\vec{v}(4) = 2.38\vec{i} - 3.37\vec{j}$,
        $\vec{v}(6) = 2.63\vec{i} + 5.48\vec{j}$.
   (c) $\vec{v}(2) = -2.235\vec{i} + 0.077\vec{j}$,
        $\vec{v}(4) = 2.374\vec{i} - 3.371\vec{j}$,
        $\vec{v}(6) = 2.637\vec{i} + 5.482\vec{j}$.

7 $\vec{v} = -2t\sin(t^2)\vec{i} + 2t\cos(t^2)\vec{j}$,
   Speed $= 2|t|$,
   Particle stops when $t = 0$.

9 $\vec{v} = (2t - 2)\vec{i} + (3t^2 - 3)\vec{j} + (2t^3 - 12t^2)\vec{k}$,
   Speed $= ((2t - 2)^2 + (3t^2 - 3)^2 + (12t^3 - 12t^2)^2)^{1/2}$,
   Particle stops at $t = 1$.

11 $\vec{v} = -3\sin t\vec{i} + 4\cos t\vec{j}$, $\vec{a} = -3\cos t\vec{i} - 4\sin t\vec{j}$

13 $\vec{v} = 3\vec{i} + \vec{j} - \vec{k}$, $\vec{a} = \vec{0}$

15 Length $= \sqrt{42}$

17 Length ≈ 24.6

19 $x = 5 + 3(t - 7), y = 4 + 1(t - 7)$, $z = 3 + 2(t - 7)$.

21 (a) $\vec{v}(2) \approx -4\vec{i} + 5\vec{j}$,
         Speed $\approx \sqrt{41}$
    (b) About $t = 1.5$
    (c) About $t = 3$

23 (a) No
    (b) $t = 5$
    (c) $\vec{v}(5) \approx 0.959\vec{i} + 0.284\vec{j} + 2\vec{k}$
    (d) $\vec{r} \approx 0.284\vec{i} - 0.959\vec{j} + 10\vec{k} + (t-5)(0.959\vec{i} + 0.284\vec{j} + 2\vec{k})$.

25 $\vec{r}(t) = 22.1t\vec{i} + 66.4t\vec{j} + (442.7t - 4.9t^2)\vec{k}$

27 (a) $x(t) = 5\sin t, y(t) = 5\cos t$, $z(t) = 8$
    (b) $\vec{v} = -5\vec{j}, \vec{a} = -5\vec{i}$
    (c) $x_{tt}(t) = y_{tt}(t) = 0$,
         $z_{tt}(t) = -g$,
         $x_t(0) = z_t(0) = 0$,
         $y_t(0) = -5, x_t(0) = 5$,
         $y_t(0) = 0, z_t(0) = 8$

29 (a) C, E.
    (b) E, C.
    (c) Yes, when the light beam is tangential to shoreline.
    (d) The speed is not defined at corners.

Chapter 16 Review

1 $x = t, y = 5$

3 $x = 4 + 4\sin t, y = 4 - 4\cos t$

5 $x = 2 - t, y = -1 + 3t, z = 4 + t$.

7 $x = 1 + 2t, y = 1 - 3t, z = 1 + 5t$.

9 $x = 3\cos t$
   $y = 5$
   $z = -3\sin t$

11 No, since the point $(0, 1)$ is not on the curve.

13 $\vec{r} = 3\vec{i} - 2\vec{j} + t(-2\vec{i} + \vec{j} + \vec{k})$.

15 No

17 (a) $(2, 3, 0)$
    (b) 2
    (c) No; not on line

19 (a) $(I) = C_4, (II) = C_1$,
         $(III) = C_2, (IV) = C_6$
    (b) $C_3 : 0.5\cos t\vec{i} - 0.5\sin t\vec{j}$,
         $C_5 : -2\cos(\frac{t}{2})\vec{i} - 2\sin(\frac{t}{2})\vec{j}$

21 (a) $\left(\frac{4 - 6t^2}{2 + 3t^2}, \frac{8t}{2 + 3t^2}\right)$
    (b) $x = \frac{4 - 6t^2}{2 + 3t^2}, \quad y = \frac{8t}{2 + 3t^2}$

Section 17.1

1 (a) IV
   (b) III
   (c) I
   (d) II

9 $\vec{V} = -y\vec{i}$

11 $\vec{V} = -x\vec{i} - y\vec{j} = -\vec{r}$

13 $\vec{V} = -y\vec{i} + x\vec{j}$

15 $\vec{F}(x, y) = x\vec{i}$
    (for example)

17 $\vec{F}(x, y) = \dfrac{y\vec{i} - x\vec{j}}{\sqrt{x^2 + y^2}}$
    (for example)

Section 17.2

1 $y = $ constant

3 $y = -\frac{2}{3}x + c$

9 (a) III
   (b) I
   (c) II
   (d) V
   (e) VI
   (f) IV

Chapter 17 Review

1 (b) (i)   yes
         (ii) no
         (iii) yes
         (iv) no

5 (a) $\dfrac{1}{x^2 + y^2 + z^2}$
   (b) $\dfrac{1}{\sqrt{x^2 + y^2 + z^2}}$
   (c) $\dfrac{x}{\sqrt{x^2 + y^2 + z^2}}\vec{i} + \dfrac{y}{\sqrt{x^2 + y^2 + z^2}}\vec{j} + \dfrac{z}{\sqrt{x^2 + y^y + z^2}}\vec{k}$
   (d) $\dfrac{-x}{\sqrt{x + y^2 + z^2}}\vec{i} + \dfrac{-y}{\sqrt{x^2 + y^2 + z^2}}\vec{j} + \dfrac{-z}{\sqrt{x^2 + y^y + z^2}}\vec{k}$
   (e) $\dfrac{\cos t}{2\sqrt{2}}\vec{i} + \dfrac{\sin t}{2\sqrt{2}}\vec{j} + \dfrac{1}{2\sqrt{2}}\vec{k}$
   (f) $\dfrac{1}{\sqrt{2}}$

Section 18.1

1   Positive

3   Positive

5   $\int_{C_3} \vec{F} \cdot d\vec{r} < \int_{C_1} \vec{F} \cdot d\vec{r}$
$< \int_{C_2} \vec{F} \cdot d\vec{r}$

7   Positive

9   0

11   0

13   16

15   32

19   C_1, C_2

21   (a)   Various values
    (b)   Various values

23   (a)   Various values
    (b)   Various values

27   $-GMm/8000$

Section 18.2

1   116.28

3   82/3

5   $e^2 - e$

7   85.32

9   24π

11   $C_1 : (t, \sqrt{2t - t^2}),$
    $0 \le t \le 2$
    $C_2 : (t, -2(t-1)^2),$
    $-1 \le t \le 2$
    $C_3 : (t, \sin t),$
    $-2\pi \le t \le -\pi$

13   (a)   11/6
    (b)   7/6

Section 18.3

3   Yes

5   No

7   9/2

9   $\frac{3}{\sqrt{2}} \ln(\frac{3}{\sqrt{2}} + 1)$

11   (a) e
    (b) e

13   (b)   No

19   (a) $\pi/2$
    (b)   No

Section 18.4

3   $f(x,y) = x^2 y + 2y^4 + K$
    $K = $ constant

5   No

7   Yes, $f = x^2 y^3 + xy + C$

9   Yes, $f = \ln A|xyz|$ where A is a
    positive constant.

11   (b) $-\pi$

13   πab

15   3/2

Chapter 18 Review

1   (a)   Negative
    (b) C_1: Positive
         C_2, C_4: Zero
         C_3: Negative
    (c)   Negative

3   2

5   12

7   False

9   True

11   -58

15   (a) $\omega = 3000$ rad/hr
         $K = 3 \cdot 10^7$ m²·rad/hr
    (c) $r < 100$ m, circulation is
         $2\omega\pi r^2$
         $r \ge 100$ m, circulation is
         $2K\pi$

17   (b)   Circles
    (c)   No

Theory: Proof of Green's

1   $x = -1 + r\cos\theta,$
    $y = 2 + r\sin\theta,$
    $2 \le r \le 3, 0 \le \theta \le 2\pi$

3   $x = s,$
    $y = tg(s) + (1-t)f(s),$
    $a \le s \le b, 0 \le t \le 1$

Section 19.1

1   (a)   Positive
    (b)   Negative
    (c)   Zero
    (d)   Zero
    (e)   Zero

3   (a)   Zero
    (b)   Zero
    (c)   Zero
    (d)   Negative
    (e)   Zero

5   (a)   5
    (b)   4
    (c)   11
    (d)   9

7   (a)   Zero
    (b)   Zero

9   Zero

11   8π

13   Zero

17   (a)   Zero
    (b)   Zero

21   (a)   Maximum speed
    (b)   0
    (c) $\pi u a^2 /2$

Section 19.2

1   6

3   $\pi/2$

5   7/3

7   $\pi \sin 25$

9   1296π

11   $-81\pi/4$

13   12π

15   $625\pi/2$

17   $11\pi/2$

Chapter 19 Review

5   4

7   1.5

9   114

11   $3(8 + \sin 4)$

13   20

15   2π

19   (b)   0
    (c) $Ih \ln |b/a|/2\pi$

Section 20.1

5   0

7   0

9   $2/\|\vec{r} - \vec{r}_0\|$

13   0

15   $\vec{b} \cdot (\vec{a} \times \vec{r})$

17   (a)   Positive
    (b)   Zero
    (c)   Negative

19   (a)   0

21   (a)   Flux $= c^3$
    (b)   1
    (c)   1

23   (a) $4w^3$
    (b)   4
    (c)   4

25   (a) $2\pi c^3$
    (b)   2
    (c)   2

27   (a)   0
    (b)   Undefined.

Section 20.2

1   24

3   Zero

5   $\int_{S_2} \vec{F} \cdot d\vec{A} = 8$

9   (a)   0
    (b) 4π

11   $\int_S \vec{F} \cdot d\vec{A}$
    $= \int_R \text{div } \vec{F} \, dV = 0$

13   (a)   0
    (b)   0

Section 20.3

1     $4y\vec{k}$

3     $\vec{0}$

5     $4x\vec{i} - 5y\vec{j} + z\vec{k}$

7     $(2x^3yz + 6x^7y^5 - xy)\vec{i}$
      $+ (-3x^2y^2z - 7x^6y^6 + y)\vec{j}$
      $+ (yz - z)\vec{k}$

9     $\text{curl}(F_1(x)\vec{i} + F_2(y)\vec{j} + F_3(z)\vec{k}) = 0$

11   (a)   Zero curl
       (b)   Nonzero curl
       (c)   Nonzero curl

19    $(c), (d), (f)$

21    $C_2, C_3, C_4, C_6,$

23   (a)   Counterclockwise
       (b)   Clockwise
       (c) $\vec{0}$

Section 20.4

1   No

3   (a) -200π
       (b)   0

5   π

13   (a)   Can't say anything
       (b)   0

15   (a) $-2\pi a^2$
       (b) $2\pi a^2$
       (c)   Orientations not related

17   -2π

Chapter 20 Review

1     $\text{div}\,\vec{v} = -6$

5     $\pi b^2 h/3$

7   4π

9   True

11   True

13   True

15   False

17    $\text{div}\,\vec{E}\,(P) \leq 0$, $\text{div}\,\vec{E}\,(Q) \geq 0$.

21   (a) $\rho(0) < \rho(1000) < \rho(5000)$
       (b)   cars/hour
       (d) $\rho(x) = 4125/(55 - x/50)$
            if $0 \leq x < 2000$
            $\rho(x) = 4125/15 = 275$
            if $2000 \leq x < 7000$
            $\rho(x) = \frac{4125}{(15 + (x - 7000)/25)}$
            if $7000 \leq x < 8000$
            $\rho(x) = 4125/55 = 75$
            if $x \geq 8000$
       (e)   139 ft. at $x = 0$
            89 ft. at $x = 1000$
            38 ft. at $x = 5000$

25   (b) $\|\vec{v}\| = K/r^2$
       (c)   Flux $= 4\pi K/3$
       (d)   Zero

Theory: Three Theorems

1   Yes

7   Yes
      $(-xy + 5yz)\vec{i} + (2xy + xz^2)\vec{k}$.

9   (a)   Yes
       (b)   Yes
       (c)   Yes

11   (a) $\text{curl}\,\vec{E} = \vec{0}$
       (b)   3-space minus a point if $p > 0$
            3-space if $p \leq 0$.
       (c)   Satisfies test for all p.
            $\phi(r) = r^{2-p}$ if $p \neq 2$.
            $\phi(r) = \ln r$ if $p = 2$.

15    $\vec{r} = (1 - s)\vec{r}(t) + s\vec{r}_0$
      $a \leq t \leq b$

Appendix A

1   (1,0)

3     $(-2, 0)$

5     $\left(\frac{5\sqrt{3}}{2}, -\frac{5}{2}\right)$

7     $(\cos 1, \sin 1)$

9     $(2, \pi/2)$

11    $(\sqrt{2}, 3\pi/4)$

13    $(0.28, 7\pi/4)$

15    $(3.16, 2.82)$

Appendix B

1     $2e^{\frac{i\pi}{2}}$

3     $\sqrt{2}e^{\frac{i\pi}{4}}$

5     $0e^{i\theta}$, for any θ.

7     $\sqrt{10}e^{i\arctan(-3)}$

9     $-3 - 4i$

11    $-5 + 12i$

13    $\frac{1}{4} - \frac{9i}{8}$

15    $-\frac{1}{2} + i\frac{\sqrt{3}}{2}$

17    $-125i$

19    $\frac{\sqrt{2}}{2} + i\frac{\sqrt{2}}{2}$

21    $\frac{\sqrt{3}}{2} + \frac{i}{2}$

23   -2^{50}

25    $8i\sqrt[3]{2}$

27    $\frac{1}{\sqrt{2}}\cos(\frac{-\pi}{12}) + i\frac{1}{\sqrt{2}}\sin(\frac{-\pi}{12})$

29    $A_1 = 1 - i, A_2 = 1 + i$

31   (a) $z_1 z_2 = 6 - i2\sqrt{3}$
            $\frac{z_1}{z_2} = i\sqrt{3}$
       (b)   same as (a)

33   True

35   False

37   True

INDEX